7 Day

University of Plymouth Library

Subject to status this item may be renewed
via your Voyager account

http://voyager.plymouth.ac.uk

Tel: (01752) 232323

Biology
of
MARINE
BIRDS

COVER PHOTOGRAPHS

Front Cover
(clockwise from top right)

Red-tailed Tropicbird (R.W. Schreiber)
Great Frigatebird (R.W. Schreiber)
Laysan Albatross — Adult protecting young chick (R.W. and E.A. Schreiber)

Back Cover
(top to bottom)

Shy albatrosses (H. Weimerskirch)
White tern with fish in its bill (E.A. Schreiber)
Masked Booby adult with chick (R.W. and E.A. Schreiber)
Jackass Penguin pair with young (R.W. and E.A. Schreiber)

Biology *of* MARINE BIRDS

Edited by
E. A. Schreiber
Joanna Burger

CRC
MARINE BIOLOGY
SERIES

Peter L. Lutz, Editor

CRC PRESS

Boca Raton   London   New York   Washington, D.C.

Senior Editor: John Sulzycki
Project Editor: Naomi Lynch
Marketing Manager: Carolyn Spence
Cover Designer: Shayna Murry

Library of Congress Cataloging-in-Publication Data

Biology of marine birds / edited by Elizabeth A. Schreiber and Joanna Burger.
     p. cm. — (CRC marine biology)
     Includes bibliographical references (p.).
     ISBN 0-8493-9882-7 (alk. paper)
     1. Sea birds. I. Schreiber, Elizabeth Anne. II. Burger, Joanna. III. Series.

QL673 .B53   2001
598.177—dc21

2001025898
CIP

Visit the CRC Press Web site at www.crcpress.com

Dedication

---

for
Ralph W. Schreiber, Gary A. Schenk, and Michael Gochfeld

For a lifetime of challenges, collaboration, stimulating discussions, and companionable fieldwork with the seabirds we love.

Preface

The field of seabird ornithology has developed dramatically in recent years, partially owing to the application of new technology to this diverse group of birds. For instance, the advent of satellite tracking studies has helped us learn about an aspect of seabirds that was unknown previously — their lives at sea. Because of this we now have a much better knowledge of the energy budgets of adults and what energetic constraints they experience. Another factor that has limited our ability to understand the lives of marine birds is their long life span. Without information spanning at least one generation of individuals, there are many facets of the life history of seabirds that are difficult to impossible to interpret. But, for seabirds, a generation can be 20 to 30 years, longer than most ornithological studies on any species. Over the past 20 to 30 years, however, some excellent long-term studies have been carried out. Our aim with the present work is to provide an examination and summary of the research on seabirds, and also to provide a guide to the relevant literature for those desiring further information.

This book discusses and summarizes our current knowledge of the biology of marine birds today. It provides information on the biology, ecology, physiology, evolution, behavior, environmental threats, and conservation of marine birds. It also provides information on key questions for researchers to address, and for public policy makers involved in management of coastal lands and marine reserves. We felt that marine birds needed to be examined from a wider perspective: not only that of the biologist, but also that of those who are concerned about conservation, management, and public policy. We provide the basis for understanding the biology of marine birds, as well as their role in and relationship to coastal and oceanic ecosystems.

We explore all facets of the lives of the four main orders of seabirds, examining their fossil history, taxonomy, distribution, life histories, population dynamics, foraging behavior, nesting ecology, physiology, energetics, the effects of pollution and other human activities on the birds, and needs for conservation. Each chapter presents the basics of our current knowledge about that topic and many chapters also include a guide to yet unanswered questions and suggestions of potential research paths.

Once into the project we realized that an entire book could be written about each topic we had selected as a chapter. However, we were required to limit the length of each chapter, and to include only the most important information and examples. The literature section for each chapter is extensive and provides an overview of the subject for researchers, conservationists, managers, and policy-makers.

We also faced the difficulty of defining marine birds (or seabirds; see Chapter 1). While some orders contain birds that almost entirely live in coastal and marine environments, others do not. Moreover, some birds not usually considered "marine" spend a great deal of their time in coastal environments (herons, egrets, some shorebirds), and we have included separate chapters on these groups, highlighting their marine lives.

We have included three chapters on conservation issues: Chapter 15 (Effects of Chemicals and Pollution on Seabirds), Chapter 16 (Interactions between Fisheries and Seabirds), and Chapter 17 (Seabird Conservation). As man's influence reaches the most remote parts of the world, our effect on seabirds' lives is increasing. Every aspect of seabird biology and ecology is affected. Many seabird colonies have been extirpated already and others are disappearing as human development and disturbance expand. Researchers have estimated that seabird populations today are 10% or less of what they were a thousand years ago before humans reached many islands (Steadman et al. 1984, Pregill et al. 1994). As the human population expands, invading seabird nesting habitats, we are going to see the extinction of many species unless we can learn to value this resource, learn to coexist, and have the science to effect conservation measures.

While the fossil origins of marine birds are not certain, we do know that they enjoy a worldwide distribution today, from the poles to the tropics, and from urban coasts to remote oceanic islands. And there are ornithologists studying them in all these habitats: some of us wear multiple layers and try to take measurements with heavy gloves on while conducting Arctic field work, and others wear shorts while enjoying the tropics. Our field studies often take us away from home for extended periods and we continually try to outthink birds as we devise new ways to accomplish our research goals. We frequently have to invent our own equipment — from capturing devices, to weighing scales, to temperature probes, or to adapting new cutting edge technologies to fit our needs. Studying seabirds can be a daily excursion or a seasonal expedition, and can involve hours driven by car or days flown by air. All the authors of these chapters are active field researchers working on seabirds. As you read you can try to imagine the thousands, or perhaps millions, of hours of field work the knowledge in this book represents, and, for most seabird species, we have only begun. The field is open, with many, many possibilities for new studies. As soon as we think we have found the answer to a question, some bird does it differently, so we are forced to use words like *most, sometimes, maybe, often, generally* — words you will see used often in this text. But that's part of what keeps the discovering fascinating.

In reviewing the chapters, we find that there is important research that needs to be conducted on nearly every group of seabirds, in all aspects from basic breeding biology and communication to contaminants. Effects of long-term phenomena such as El Niño–Southern Oscillation events and global warming on seabird biology and their evolution are difficult to study, making our task that much more challenging. The task for managers and policy-makers will be to translate the biology detailed in this book into action to enhance breeding populations, protect nesting and foraging seabirds, reduce adverse interactions between human activities and seabirds, and enhance the opportunities for people to study and watch seabirds, thereby ensuring their continued survival.

We feel particularly privileged to have worked with such a superb group of our colleagues in creating this book. They were all dedicated, responsive, a pleasure to work with, and, we believe, have created an exceptional volume. We know that the quality of our work herein was significantly increased by the dedicated reviewers, who, in spite of their busy schedules, took time to provide thoughtful insight and feedback: Keith Bildstein, Claus Bech, Glen Fox, Mike Gochfeld, David Goldstein, William Montevecchi, David Nettleship, Storrs Olson, Robert Ricklefs, Peter Stettenheim, Causey Whittow, and all those who must remain anonymous.

E.A. Schreiber
Joanna Burger

Chapter opening drawings by John P. Busby

Acknowledgments

Between us, we have conducted over 60 years of research on marine birds. During this time, we have had fruitful discussions with many people too numerous to acknowledge. However, several people have profoundly influenced our thinking and our research, including C. Beer, J. Coulson, J. Diamond, M. Erwin, G. Fox, R. W. Furness, J. Hickey, G. L. Hunt, J. Jehl, J. Kushlan, J. Mills, B. Nelson, D. Nettleship, I. C. T. Nisbet, R. E. Ricklefs, J. Rodgers, C. Safina, J. Saliva, S. Senner, J. Spendelow, N. Tinbergen, H. B. Tordoff, D. Warner, and G. E. Woolfenden.

There are no words to adequately thank the three people who have influenced our thinking, challenged our ideas, and conducted field work with us on a variety of seabirds throughout our careers: Michael Gochfeld, Gary Schenk, and Ralph Schreiber. They will continue to mold our thinking for years to come. Gary Schenk and Michael Gochfeld spent long hours in fruitful discussions about seabirds, organization of this book, assisting in preparation of Appendix 2, and many other preparation details. They were drafted in a hundred ways, and this volume is a tribute to their knowledge and love of seabirds, not to mention us. We thank our editor, John Sulzycki, at CRC Press for shepherding the manuscript through the process, and for understanding that deadlines are meant to encourage progress. We also thank the editor of the series, Peter Lutz, for seeing the importance of highlighting marine birds in this marine series.

Over the years, we have received support from a number of organizations, and we thank them. EAS thanks the American Association of University Women Educational Foundation, Defense Nuclear Agency, Giles W. and Elise G. Mead Foundation, National Science Foundation (OCE-8308756, OCE-8404152), the National Geographic Society, Los Angeles County Museum of Natural History, and the National Museum of Natural History, Smithsonian Institution. JB thanks the American Association of University Women Educational Foundation, National Institute of Mental Health, National Institute of Environmental Health Sciences (ESO 5022, 5955), National Science Foundation, U.S. Environmental Protection Agency, National Oceanographic and Atmospheric Adminstration, U.S. Fish & Wildlife Service, Consortium for Risk Evaluation with Stakeholder Participation (CRESP) through the Department of Energy (AI #DE-FC01-95EW55084, DE-FG 26-00NT 40938), the Endangered and Nongame Species Program of New Jersey Department of Environmental Protection, and Institute for Coastal and Marine Sciences and the Environmental and Occupational Health Sciences Institute of Rutgers University.

The Editors

E. A. Schreiber is a Research Associate in the Bird Department of the National Museum of Natural History, Smithsonian Institution. She received her Ph.D. from UCLA, studying breeding biology and energetics in Red-tailed Tropicbirds. She carries out research on tropical seabirds in many areas of the world, specializing in studies of their breeding biology, ecology, demography, and energetics. Her research activities have spanned 30 years and include long-term studies on Christmas Island (Pacific Ocean) and Johnston Atoll.

In addition to her basic research, Dr. Schreiber serves as an advisor to conservation organizations in the United States and worldwide on seabird issues, working to promote seabird conservation. She has published over 50 scientific papers and 3 books on terns, gulls, and conservation issues, and authored or co-authored 6 "Birds of North America" species accounts on Pelecaniformes and terns. She has served the professional ornithological community for many years through terms on the governing boards and councils of the Ornithological Council, the Association of Field Ornithologists, the Pacific Seabird Group, the Cooper Ornithological Society, the Waterbird Society, and the American Bird Conservancy. Prior to moving to Washington, D.C. in 1994 she was a Research Associate at the Los Angeles County Museum of Natural History in Los Angeles for 18 years. She recently co-edited with D. S. Lee the "Status and Conservation of West Indian Seabirds" (Society of Caribbean Ornithology).

Joanna Burger is a Distinguished Professor of Biology at Rutgers University where she teaches ecology and behavior and ecological risk to undergraduate and graduate students. She is a member of the Institute for Marine and Coastal Sciences, the Biodiversity Center, and the Environmental and Occupational Health Sciences Institute. She is an ecologist, behavioral biologist, and ecotoxicologist who has worked with seabirds for over 30 years in many parts of the world. She is a past-president for the Waterbird Society. She has served on the New Jersey Governor's Endangered and Nongame Species Council since 1980, on the Technical Committee for the Roseate Tern Recovery Team, on numerous National Research Council Committees, and on the Atlantic States Marine Fisheries Commission Technical Committee for Horseshoe Crabs, and presently serves on the NRC's Commission of Life Sciences. She is on the Editorial Board for *Environmental Research, Science for the Total Environment, Environmental Monitoring and Assessment,* and *Journal of Toxicology and Environmental Health.* Her research has led to over 300 papers and numerous book chapters, and she has authored or co-authored six "Birds of North America" accounts on gulls and terns. She has edited 6 volumes on avian behavior and seabirds, and co-written two books on behavior of colonial-nesting birds (Common Terns, Black Skimmers). She recently edited *The Commons Revisited: An Americas Perspective* (Island Press 2000).

The Contributors

Patricia Herron Baird
Kahiltna Research Group
Department of Biological Sciences
California State University
Long Beach, California, USA

P. Dee Boersma
Department of Zoology
University of Washington
Seattle, Washington, USA

Joël Bried
Groupe Ecologie Comportementale
Centre National de la Recherche Scientifique
Centre d'Ecologie Fonctionnelle et Evolutive
Montpellier, France

M. de L. Brooke
Department of Zoology
University of Cambridge
Cambridge, UK

Joanna Burger
Distinguished Professor of Biology
Division of Life Sciences
Environmental and Occupational Health
  Sciences Institute
Institute of Marine and Coastal Sciences
Rutgers University
Piscataway, New Jersey, USA

J. Alan Clark
Department of Zoology
University of Washington
Seattle, Washington, USA

John C. Coulson
St. Mary's Close
Durham, UK

Hugh I. Ellis
Department of Biology
University of San Diego
San Diego, California, USA

Chris Elphick
Department of Ecology and Evolutionary
  Biology
University of Connecticut
Storrs, Connecticut, USA

Peter C. Frederick
Department of Wildlife Ecology and
  Conservation
University of Florida
Gainesville, Florida, USA

Geir W. Gabrielsen
Norwegian Polar Institute
Tromsø, Norway

Michael Gochfeld
Professor of Medicine
Environmental and Community Medicine
Environmental and Occupational Health
  Sciences Institute
UMDNJ-Robert Woods Johnson Medical
  School
Piscataway, New Jersey, USA

David L. Goldstein
Department of Biological Sciences
Wright State University
Dayton, Ohio, USA

Keith C. Hamer
Department of Biological Sciences
University of Durham
Durham, UK

Nigella Hillgarth
Tracey Aviary
Department of Biology
University of Utah
Salt Lake City, Utah, USA

Pierre Jouventin
Groupe Ecologie Comportementale
Centre National de la Recherche
Scientifique
Centre d'Ecologie Fonctionnelle et
  Evolutive
Montpellier, France

William A. Montevecchi
Biophysiology Programme
Memorial University of Newfoundland
St. Johns, Newfoundland, Canada

J. Bryan Nelson
Reader Emeritus
Aberdeen University
Aberdeen, Scotland, UK

Margaret A. Rubega
Department of Ecology and Evolutionary
  Biology
University of Connecticut
Storrs, Connecticut, USA

E. A. Schreiber
Research Associate
National Museum of Natural History
Smithsonian Institution
Washington, D.C., USA

David A Shealer
Department of Biology
Loras College
Dubuque, Iowa, USA

G. Henk Visser
Zoological Laboratory and Centrum voor
  Isotopen Onderzoek
University of Groningen
Groningen, The Netherlands

Kenneth I. Warheit
Wildlife Research Division
Department of Fish and Wildlife
Olympia, Washington, USA

Nils Warnock
Point Reyes Bird Observatory
Stinson Beach, California, USA

Henri Weimerskirch
Centre d'Etudes Biologiques de Chizé
Centre National de la Recherche Scientifique
Villiers en Bois, France

G. Causey Whittow
Professor and Chairman
Department of Physiology
J. A. Burns School of Medicine
University of Hawaii
Honolulu, Hawaii, USA

Table of Contents

Courting Red-tailed Tropicbirds

1 Seabirds in the Marine Environment

E. A. Schreiber and Joanna Burger

CONTENTS

1.1 INTRODUCTION

Marine birds are equally at home on land, in the air, and in the water. While many organisms can go from land to water (amphibians, some reptiles, some insects), others generally live in only one medium during their lives. Marine birds switch from one to the other, often daily. Such flexibility requires unique physiological and morphological adaptations to the environment, a medium that has also exerted selective forces on the behavior, ecology, and demography of these birds. Amazingly, marine birds have adapted to essentially all environments on the earth, from those able to survive winters in Antarctica to those who can sit for days incubating their eggs in the tropical sun. Trying to learn about and explain this diversity may be why we find the study of them so fascinating: How does their structure and function interact with the marine environment to produce their particular life histories?

There is no one definition of marine birds or seabirds. For this book, we define marine birds as those living in and making their living from the marine environment, which includes coastal areas, islands, estuaries, wetlands, and oceanic islands (Table 1.1). But many Charadriiformes (shorebirds) and Ciconiiformes (erons, egrets, ibises) that feed near shore or along the coastlines are generally not considered to be true seabirds. Seabirds are a subset of the birds in Table 1.1, those that feed at sea, either nearshore or offshore; this excludes all the Ciconiiformes and the shorebirds from the Charadriiformes. The one common characteristic that all seabirds share is that they feed in saltwater, but, as seems to be true with any statement in biology, some do not.

In this book we have attempted to provide a thorough examination of the biology of seabirds: all the Sphenisciformes and Procellariiformes, all the Pelecaniformes except anhingas, and all the Charadriiformes except shorebirds (Figure 1.1). Because we felt the book should be useful to land managers, public policy-makers, and conservationists (who must knowledgeably manage our quickly disappearing wetlands and estuaries), we have included gulls as seabirds (although few go to sea) and also summary chapters on wading birds (Ciconiiformes) and shorebirds. These birds are particularly dependent on nearshore habitat for both feeding and nesting.

Seabirds exemplify one of the reasons for man's fascination with birds — the ability to fly and live so far from the mainland. They are among the most aerial of birds, able to spend weeks,

0-8493-9882-7/02/$0.00+$1.50
© 2002 by CRC Press LLC

TABLE 1.1
Marine Birds Include Birds in the Following Orders

Order	Types of Birds
Sphenisciformes	Penguins
Procellariiformes	Albatrosses, petrels, storm-petrels, fulmars, shearwaters
Ciconiiformes	Herons, egrets, storks, ibis, spoonbills
Pelecaniformes	Pelicans, frigatebirds, gannets, boobies, cormorants, anhingas
Charadriiformes	Shorebirds, skuas, jaegers, gulls, terns, skimmers, auks, guillemots, puffins

Note: The Ciconiiformes, anhingas, shorebirds, and skimmers are not considered to be seabirds.

(a)

(b)

FIGURE 1.1 Representatives of the four major seabird orders: (a) Sphenisciformes: King Penguins incubating their eggs on their feet; (b) Procellariiformes: a Wedge-tailed Shearwater on Midway Island; (c) Pelecaniformes: a Brown Pelican incubates its three eggs; (d) Charadriiformes: a Blue Noddy on Christmas Island. (Photos a and b by J. Burger; c and d by E. A. Schreiber.)

(c)

(d)

FIGURE 1.1 *Continued.*

months, and, in some cases, even years at sea. This habit of spending long periods at sea, out of sight of land, has also made them among the most difficult of bird species to study and understand. Much of their life is spent where we cannot observe or study them, although this is changing with advances in technology such as satellite transmitters that are light enough to be carried by a bird.

Although the open ocean seems to us to be a uniform environment, a tremendous diversity of seabirds has evolved to feed in this environment in a great variety of ways. Such diversity suggests that the marine environment is not as homogeneous as we once thought, at least to the organisms that live there. The apparent uniformity was reflected in our inability to detect and measure the heterogeneity. We now know that the seas vary on seasonal cycles as well as stochastically and spatially (see Chapters 6 and 7). We are not as at home on the ocean as seabirds and have learned to take lessons from birds. Mariners often relied on seabirds to tell them they were near land, while fishermen today still rely on feeding flocks to help locate schools of fish. Mutiny on Columbus' voyage to the New World was thwarted by seabirds: when the crew finally saw feeding flocks of seabirds, they knew they were close to land (Couper-Johnston 2000).

TABLE 1.2
Comparison of Characteristics of Seabirds and Passerines

Life History Characteristic	Seabirds	Passerines
Age of first breeding	2–9 years	1–2 years
Clutch size	1–5	4–8
Incubation period	20–69 days	12–18 days
Nestling/fledging period	30–280 days	20–35 days
Maximum life span	12–60 years	5–15 years

1.2 WHY ARE SEABIRDS DIFFERENT?

Seabirds have dramatically different life-history characteristics, or demography, from most land birds, such as members of the order Passeriformes (Table 1.2). In fact, their life history characteristics are often referred to as extreme: long life (20 to 60 years), deferred maturity (breeding age delayed to up to 10 years of age), small clutch size (in many cases one egg), and extended chick-rearing periods (often up to 6 months). Passerine birds, in comparison, have shorter lives and larger clutches of eggs, and chicks grow to fledging age much faster. Seabirds also tend to be larger than land birds, less colorful in plumage, and sexually monomorphic. Plumage colors of seabirds are mainly white, gray, black, or brown, or some combination thereof, another area that needs research.

Basically the two life styles exemplified by seabirds and passerines represent two different ways to accomplish the same end: leave enough offspring to replace yourself in the population. Red-footed Boobies (*Sula sula*) commonly live 16 years, begin reproducing (one young per year) at 3 years of age, and 35 to 40% of their young survive to reproduce (Schreiber et al. 1996). A pair thus has the potential to produce about five breeding offspring (birds), although there are generally a few failed breeding seasons owing to the occurrence of El Niño events (see Chapter 7). More coastal species, such as Black Skimmers (*Rynchops niger*), live for about the same time and are capable of raising two or three young a season, but colonies can also fail completely in some years due to heavy rains and thermal stress (Burger and Gochfeld 1991; see Chapter 7). Robins (*Turdus migratorius*), a typical passerine, commonly live 3 years, first lay at 1 year of age (lay an average of four eggs), and can raise two broods in some years; about 20% of their young survive to reproduce (Sallabanks and James 1999). So in a lifetime they can raise about five young that survive to reproduce. They also can have failed years when no young are produced, but it is less likely to occur throughout a whole region as it does in seabirds.

Why have these two very different lifestyles evolved? They may reflect conditions imposed on seabirds by living in the marine environment (Ashmole 1963, Lack 1968), and also conditions imposed on land birds by predation (Slagsvold 1982). Seabirds may not have been exposed to predation historically, although the human introduction of mammalian predators to both coastal and oceanic islands has been a major source of mortality for seabirds that did not evolve with this threat (Moors and Atkinson 1984, Burger and Gochfeld 1994).

Early hypotheses on the reasons for the life-history characteristics of seabirds have come to be called the "energy-limitation hypotheses." David Lack (1968) proposed that seabirds' unusual demography evolved owing to energetic constraints on adults' ability to supply food to chicks. Birds feeding at sea were viewed as randomly searching a vast area for patchily distributed food that then had to be caught and carried long distances back to a colony. Philip Ashmole (1963) also suggested that dense aggregations of birds in one area, such as in seabird colonies, depressed local food resources, causing density-dependent limitations on breeding and nest success (Figure 1.2). He proposed that seabirds were perhaps over-fishing the area around colonies and adults could not find enough food to raise more young or faster-growing young. Specifically then, small clutch sizes and slow growth of young were considered to be adaptations to an imposed low rate of food delivery

FIGURE 1.2 Cape Gannets (South Africa) are one of the most densely nesting seabirds. Neighbors can easily peck each other if they have a disagreement and thus much signaling of intentions (behavioral posturing) goes on to forestall any misunderstanding. Shown is Michael Gochfeld. (Photo by J. Burger.)

to chicks. Additionally, seabird chicks (particularly Procellariiformes) lay down large amounts of fat during development, which, presumably, was necessary to carry them through periods when adults could not find enough food (Lack 1968, Ashmole 1971).

These hypotheses have been the driving force behind many studies on seabirds over the past 35 years and, interestingly, they are hypotheses for which it is hard to find support. Their role in the development of seabird biology was critical. However, as with any discipline, hypotheses change as we gather more information, and the energy-limitation hypothesis proved particularly difficult to validate. Some studies do not support the hypotheses, and other studies show that they could be true. We believe that biologists will never prove one way or the other why seabirds are different from land birds. It is undoubtedly a combination of selective factors. Indeed, it may be more of a continuum than we had believed. The discussion that follows is intended to highlight some issues for future study. It is also necessary to note that marine birds may appear food limited today because of the rapidly intensifying competition with fisheries and increasing human pressure.

Potential support for the energy-limitation hypothesis comes from clutch size, colony size, and foraging area comparisons. Seabirds that feed offshore generally have smaller clutches than those that feed nearshore (Nelson 1983; see Chapter 8). Pelicans, cormorants, gulls, and skimmers feed primarily nearshore and have average clutches of two to four eggs (see Appendix 2), presumably because they feed close by, making use of highly productive nearshore and estuarine resources. Offshore-feeding seabirds, such as albatrosses, petrels, boobies, and some terns, have clutches of one. Lower clutch size in itself does not prove offshore feeders are energy limited, however.

If there were a correlation between colony size and productivity of local waters, one might expect the smallest colonies to be in tropical waters away from cold water upwelling areas such as in the Humboldt Current where food is abundant. There certainly are some very large colonies in the Humboldt and Benguela Current areas, but there are also large concentrations of breeding birds in tropical non-upwelling areas such as on Midway Island (approximately one million seabirds; U.S. Fish and Wildlife Service 1996) and on Christmas Island (an estimated 12 million seabirds; Schreiber and Schreiber 1989), both in the central Pacific.

If adults are energy limited, you might expect to see populations with high mortality rates of growing chicks when feeding conditions deteriorate at all. There is little evidence for this occurring. Nest success rates in seabird colonies on oceanic islands are frequently on the order of 75% or greater, and failed nests are often those of young, inexperienced birds (see Chapter 8). Years with

FIGURE 1.3 Multispecies assemblages of breeding seabirds often have overlapping diets, foraging zones, and foraging methods, raising the question of the significance of competition in their evolution. Least Auklets (left) and Parakeet Auklets often nest in colonies (around the Alaskan coast) with several other species. (Photo by J. Burger.)

high chick mortality occur infrequently, and are generally associated with an unusual weather occurrence such as an El Niño event, when starvation of chicks occurs because of a disappearance of, or great reduction in, the food source (see Chapter 7; Schreiber and Schreiber 1989).

If adults are limited in their ability to provide food to chicks because of an irregular or unpredictable food supply, daily feeding rates of young should be sporadic and irregular. As you might expect, with the great diversity of seabird species, there is some evidence on both sides of this prediction. Some studies of feeding rates of chicks found that chicks are fed on a more regular basis than expected by chance alone and that fat stores are not needed for periods of fasting (Taylor and Konarzewski 1989, Navarro 1992, Hamer 1994, Hamer and Hill 1994, Cook and Hamer 1997, Schreiber 1994, Reid et al. 2000). Other studies have found a degree of unpredictability in food delivery which indicates fat reserves may be useful in carrying a chick through lean times (Hamer et al. 2000). Reid et al. (2000) suggested that fat stores in albatross chicks may have evolved to carry chicks through fledging while they learn to feed themselves.

Dense aggregations of breeding seabirds trying to raise hungry young might be expected to over-fish an area, but there is little evidence for this happening, and it would be difficult to prove. With high nest success rates (in non-El Niño years) in some very huge seabird colonies, such as that on Christmas Island (Central Pacific Ocean), it appears that birds may not over-fish an area (Schreiber and Schreiber 1989). Birt et al. (1987) found some inconclusive evidence for prey depletion around a colony of Double-crested Cormorants (*Hypoleucos auritus*).

A possible indication that food supply is an energy-limiting factor would be the evolution of the reliance on separate food sources in sympatrically breeding species as a way to avoid competition for the resource (Figure 1.3). Ornithologists have reconciled the discrepancy between high reproductive success and limited food resources by claiming that seabirds are partitioning the food resource by either taking different prey species, foraging in different areas, or breeding at different times of the year. However, there is little direct support for this. Ashmole and Ashmole (1967) found a large degree of overlap in the species and sizes of fish and squid taken by eight tropical seabird species breeding on Christmas Island (central Pacific). There is also extensive overlap in the size of fish and squid taken by the Pelecaniform species nesting on Johnston Atoll (central Pacific; E. A. Schreiber unpublished). In both locations, breeding seasons of the nesting seabirds overlap extensively. Large overlap in the prey base has been found in other studies (Whittam and

Siegel-Causey 1981, Ainley 1990). Thus, diet differences may be important in some colonies, but they are far from the rule. Conversely, reliance on different types of food may have been a pre-adaptation to cohabitation, but which came first?

Studies on seabird populations of the Farallon Islands, off northern California, found that feeding-niche segregation mainly occurred during difficult times such as an El Niño event (Ainley and Boekelheide 1990). Ainley's (1990) suggestion that Farallon seabird communities appeared to be operating much like grassland shrub-steppe communities of birds with regards to food (foraging opportunistically on a highly variable, but nonlimiting resource with no evident competition) brings to mind the question: Are seabirds any more energy limited than land birds? The biological importance of differences that are detected should be examined: When differences are small, but statistically significant, was there actually selection pressure to avoid competition?

The diets of the six main seabird species breeding on Bird Island, South Georgia, show extensive overlap in krill size taken (Croxall and Prince 1980, Croxall et al. 1988, Croxall et al. 1997). However, Croxall et al. (1997, see Figure 1.2) report significant differences in the *mean* sizes of krill taken, implying dietary segregation in spite of the large degree of overlap in sizes. To seabirds, the statistical differences may not be biologically relevant, and more studies are needed to examine the significance of such differences.

Several authors have examined feeding-niche separation in species nesting and foraging in coastal habitats. The question of niche separation has been examined extensively in Common (*Sterna hirundo*) and Roseate Terns (*Sterna dougallii*) along the east coast of North America. Duffy (1986) suggested that the two species appeared to partition food on the basis of patchiness, with Common Terns being more successful over larger patches of prey than were Roseate Terns. He made the important point that it is essential to examine foraging behavior at sea, and not rely only on the traditional methods of examining diet, and identifying prey species and prey size at the colony. However, he did not measure prey availability, nor examine the foods parents brought back to their young. Safina and Burger (1985), working in the same general area, used sonar to demonstrate that terns fished in areas with high concentrations of prey fish (usually with predatory fish), but there was no correlation between number of feeding terns and prey density, as one would expect if prey were limited.

In Australia, Hulsman (1987, 1988) similarly found that the niches of several tern species varied, and that the size and type of prey in a bird's diet were a function of the bird's morphology, foraging method, foraging zones, and interactions with other birds and predatory fish. Even so, most species of terns fed solitarily (except for Black Noddy, *Anous minutus*) and fed near the colony (except for Lesser Crested Tern, *Sterna bengalensis*), and there was overlap in the sizes of prey taken (Hulsman 1988). The data suggested that the guilds are dynamic, and that terns exhibit a wide range of foraging habitats and foraging methods and take a variety of prey sizes and types (Hulsman 1988).

Tests of the energy-limitation hypotheses have also included experiments designed to determine whether adult seabirds are bringing the maximum amount of food to chicks that they can. If birds can be induced to work harder, this would prove they are not normally working at full capacity (Figure 1.4). Doubling experiments have been conducted where two chicks are put in a nest of species that normally raise only one to see if increased demand causes adults to supply more food. This also implies that adults feeding young respond to the amount of food demanded and are not just bringing the maximum amount they can. In many cases parents were able to successfully provision these enlarged broods (Harris 1970 [Swallow-tailed Gull, *Creagrus furcatus*], Nelson 1978 [Northern Gannets, *Morus bassanus*], Navarro 1991 [Cape Gannets, *Morus capensis*], Schreiber 1996 [Red-tailed Tropicbirds, *Phaethon rubricauda*]). Experiments on most Procellariiformes have failed, but the reasons why remain unknown; it may not be due to lack of ability to increase effort, but to behavioral limitations (Boersma et al. 1980, Ricklefs et al. 1987).

If the amount of food brought to the chick is somewhat regulated by the chick, mediated by food begging, as many studies have found (Nelson 1964, Henderson 1975, Navarro 1991, Anderson

FIGURE 1.4 It is hypothesized that energy limitation prevents most seabirds from raising more than one young. Brown Boobies lay two eggs but rarely raise more than one chick. However, on Johnston Atoll (Pacific Ocean), about 0.5% of nesting pairs raise two young. (Photo by E. A. Schreiber.)

and Ricklefs 1992, Schreiber 1996, Cook and Hamer 1997), then food limitation may not account for the slow growth and long fledging period of seabird chicks. Adults are simply responding to chick needs, not bringing the maximum amount of food possible. There may be physiological or genetic constraints on growth rate in chicks as found in some studies (Place et al. 1989 [Leach's Storm-petrel, *Oceanodroma leucorhoa*], Konarzewski et al. 1990 [several species of altricial and precocial birds], Ricklefs 1992 [Leach's Storm-petrel]). Or the nutritional content of food may be the limiting factor (Prince and Ricketts 1981 [Grey-headed Albatross, *Thalassarche chrysostoma*, and Black-browed Albatross, *T. melanophris*]).

Parent seabirds appear to have flexible time budgets that allow them to increase feeding effort in years of poor food availability (Drent and Daan 1980, Burger and Piatt 1990, Schreiber 1996). Spare time is notably present in many seabirds, such as boobies, gulls, terns, and alcids where both members of a pair often have time to loaf together at the nest, even during the chick-rearing period (Burger 1984, Schreiber et al. 1996, Norton and Schreiber in press). The presence of spare time in birds' lives would imply that they are not normally energy limited.

Mass loss of adult birds during breeding has often been interpreted to indicate stress or increased effort (Bleopol'skii 1956, Ricklefs 1974, Harris 1979, Gaston and Nettleship 1981). This seems to be a reasonable explanation, and there are some data in support of it (Drent and Daan 1980, Monaghan et al. 1991, Chastel et al. 1995). Yet, an alternative hypothesis proposes that loss of mass is adaptive, resulting in lower wing loading and more efficient flight that enables adults to fly farther in search of food (Blem 1976, Norberg 1981, Croll et al. 1991).

Chick growth rate might be constrained (slow in seabirds) by the inability of tissues to mature at a faster rate. There is some evidence that metabolizable energy is limited simply because the digestive tract cannot assimilate food faster (Ricklefs 1969, Konarzewski et al. 1990, Diamond and Obst 1992). In domestic fowl, the gut capacity of chicks to assimilate nutrients is closely matched to the chick's requirements, suggesting that there are constraints on growth rate (Obst and Diamond 1992). We might expect a difference in growth rate between the altricial chicks of Pelecaniformes (hatching naked and helpless) and the semiprecocial chicks of Charadriiformes (hatching with a full coat of down and able to move about; Ricklefs et al. 1998). In fact, the more mature semi-precocial chicks grow more slowly than altricial chicks, also suggesting that functional maturity of tissues might limit growth rate (Ricklefs et al. 1998). If chicks lacked physiological constraints on growth, you might also expect to see them exhibit spurts of high growth (compensatory growth) following periods of starvation, which apparently does not happen (Schew and Ricklefs 1998).

Continuing investigations of growth in seabirds, and understanding the effects of constraints on growth, are needed before we can fully understand the evolution of seabird life histories. Experimental studies across phylogenetic lines can provide one of the most fruitful avenues of investigation. We need to know if chicks can make use of extra food and alter growth rates significantly. We do not yet understand how maturation of tissues and growth are controlled. The role of nutrient reserves, in the form of fat, is not fully understood. However, as Ricklefs et al. (1998) acknowledge, "Testing an hypothesis about a growth rate-function is exceedingly difficult because several tissues may assume synmorphic relationships to a single most limiting tissue, several tissues may constrain growth simultaneously, and limiting tissues may differ between age or different developmental types."

1.3 COLONIAL LIVING

While this topic is considered in detail in Chapter 4, some mention is warranted here. Lack (1954) thought about birds living in colonies and the potential for competition for space as well as food. Seabirds must be one of the ultimate examples of colonial living! Colonies can consist of several species and millions of individuals, providing a ripe environment for investigations of topics such as competitive exclusion (see Chapter 8). There are few data on population dynamics in most seabird species. And even for those few species on which we have good data, we do not truly understand how populations are regulated or the effect of density-dependent mechanisms.

If large colonies of seabirds deplete the food resource around the colony you might see a decrease in the breeding population size or an effect in some other aspect of reproductive biology (Figure 1.5). This has been documented in a few colonies (Hunt and Butler 1980, Anderson et al. 1982, Piatt 1987, Safina et al. 1988), but not in most others (see discussion in Chapter 4). However, in many cases adults apparently have some spare time in their budget and can compensate for reductions in the food supply (Drent and Daan 1980, Burger and Piatt 1990, Schreiber 1996), implying they are able to cope with potential competition for food.

Over 95% of seabirds are colonial, with colony sizes ranging from a few pairs to many thousands. Some colonies are almost unbelievably large, numbering in the millions of pairs. Living

FIGURE 1.5 The largest Magellanic Penguin colony in the world, at Punta Tombo, Argentina, consumes many tons of fish from local waters during the nesting season. (Photo by P. D. Boersma.)

in colonies makes communication among birds a necessary part of daily life and thus colonies can be exceptionally noisy. Colonies of more densely nesting birds are often noisier and it may be that the proximity of neighbors makes communicating their intentions more important (Figure 1.2; see discussion in Chapter 10).

Understanding population dynamics of seabirds requires long-term studies of individually marked birds. Ideally a study should last at least one generation of a species, if not more, to truly understand what is driving changes in population levels, survival, and demographics. With long-lived species, such as seabirds, this can mean a researcher's entire lifetime of field work spent on one species. Studies such as John Dunnet's on Northern Fulmars (*Fulmaris glacialis*; Dunnet and Ollason 1978, Dunnet et al. 1979), John Coulson's on Black-legged Kittiwakes (*Rissa tridactyla*; Coulson 1966, 1983, 1985, Coulson and Thomas 1983, Coulson and White 1956, 1958), John Mill's on Red-billed Gulls (*Larus scopulinus*; Mills 1973, 1980, Mills et al. 1996), and the British Antarctic Surveys' long-term commitment to Antarctic studies (Croxall 1992, Croxall and Rothery 1991, 1994, Croxall et al. 1988, 1992, 1997, Prince 1985, Prince and Ricketts 1981, Prince et al. 1994) have given us tremendous insights into seabird breeding biology, ecology, physiology, and demography.

1.4 ADAPTATIONS AND LIFESTYLES OF MARINE BIRDS

Life at sea and feeding on marine organisms presents several challenges to seabirds, and it undoubtedly has played an important role in shaping their life histories and physiology. Feeding in the marine environment requires that seabirds deal with high physiological salt loads. One of the methods they use to accomplish this is through their salt glands, an extra-renal kidney located in the orbit of the eye (see Chapter 14). They also limit their ingestion of salt water, getting most of their fluids from the high water content of the food they eat. For instance, seawater contributes about 8.5% of the total water influx in Diving Petrels (*Pelecanoides* spp.; Green and Brothers 1989). Life at sea also involves other challenges, such as dealing with foraging conditions that are greatly impacted by weather (see Chapter 7), with natural and anthropogenic contaminants (see Chapter 15), and with increasing competition from fisheries worldwide (see Chapter 16).

Seabirds have diversified to live in all areas of the globe and to feed by a great variety of means (Chapter 6). Some seabird species fly vast distances to their feeding grounds (albatrosses) and their long, narrow wings make them well adapted for this. The dynamic soaring of albatrosses enables them to fly without flapping, making headway in almost any kind of weather and expending little energy to do so. Smaller birds, such as auks and puffins, flap hard and fast to stay airborne, and feed closer to shore, probably because of the high energy cost of flapping flight (Rahn and Whittow 1984: see discussion in Chapter 11). Feeding methods of seabirds are just as diverse, from piracy and cannibalism (frigatebirds, skuas) to sitting on the ocean surface plucking squid and krill (albatrosses, petrels), to plunge diving (boobies and Brown Pelicans [*Pelecanus occidentalis*]), to deep diving (penguins, see Chapter 6).

Bills, feet, and body shapes also show a myriad of adaptations to the various lifestyles of seabirds. Many of the adaptations are for swimming and diving. Most have webbed feet to aide in propulsion through the water. Frigatebirds are an exception, with greatly reduced webs, but they never enter the water. Bill adaptations for various types of feeding are diverse. They all use their bills to capture and handle food, except for pelicans (*Pelecanus* sp.) who capture fish in their large pouches. For the albatrosses and petrels, a hook on the end of the beak helps hold their food (generally squid and krill). They do not have tremendous closing strength in the bill, possibly because they do not take strong, muscular prey. Frigatebirds (*Fregata* sp.) often take large flying fish, using their hooked bills to pin the fish between the mandibles until they can flip them around and swallow them. The hooked bill of pelicans seems to be used primarily for preening, and rarely serves a purpose in feeding. Boobies, tropicbirds, cormorants, gulls, and terns that feed on fish generally catch them sideways in the bill. Some bills are serrated on the edge, with the teeth angled toward the throat so that fish cannot wriggle out of their grasp (boobies). Boobies and tropicbirds have a hinge on the upper mandible at the base which allows them to exert greater pressure at the tip, further ensuring that prey do not get away. The lower mandible of skimmers (*Rhynchops* sp.) is compressed laterally and is longer then the upper mandible. They catch fish by flying along at the water surface with the lower mandible slicing through the water, searching for prey by tactile means. The bill is snapped shut as soon as a prey item is encountered. The bill of puffins is impossible to explain in terms of a functional food-catching mechanism, and its evolution may be related to its use in courtship.

Bodies of boobies and gannets are compressed to a bullet shape, making them efficient divers. Most seabirds are black, white, or black and white, and most are basically sexually monomorphic. Given the colorful variety and wonderful sexual differences found within land birds, one wonders why seabirds are so "dull." Several polar nesting species are white, such as the Ivory Gull (*Pagophila eburnea*), providing cryptic coloration. Yet other Polar species have large amounts of black, like penguins, and even young penguins (supposedly more vulnerable to predators) are not cryptically colored. But predation is a problem for few seabirds that nest on islands or remote cliffs free of predators. White in some birds is considered conspicuous coloration, offering at-sea feeding birds an opportunity to see others who might have found food and head toward the source. White on the belly of seabirds has been considered to provide them with less conspicuous coloring to avoid being seen by the fish for which they are searching (Simmons 1972). Yet, immature birds of several species (such as Brown, *Sula leucogaster,* and Red-footed Boobies) are dark below, presumably putting these amateur fishers at a disadvantage if this theory is true. Indeed immatures are usually less efficient foragers than adults (see Chapter 6). Many aspects of seabird biology are, as yet, unexplained.

1.5 LOOKING TO THE FUTURE

The past 20 years have seen tremendous progress in our knowledge about marine birds and about their relationships with their environment, competitors, predators, and prey. Early scientists observed seabirds, but now we have multiple methodologies to examine them. New developments in technology and techniques are allowing us to examine aspects of birds' lives that were once unknowable. These include physiological studies of energetics, the connection of weather patterns to seabird ecology, DNA studies examining taxonomic relationships and populational relationships, stable isotope studies of diet and trophic level, tracking daily and annual movements at sea with satellite telemetry, and collecting dive depth and frequency data electronically.

As the chapters in this book indicate, answering questions about the biology, ecology, and conservation of marine birds is challenging, and will continue to be so for years to come. There are still many unanswered questions in need of research, particularly by those willing to make a long-term commitment to studying a single species. New improvements in technology now allow us to follow seabirds during the periods they are at sea, a new frontier in seabird research. Changing concepts of the uniformity–heterogeneity of the ocean, and of the scales (both temporal and spatial) on which the oceanic environment operates, have advanced our ability to ask the right questions (see Chapter 6). One of the threads you will find woven throughout this book is that the more we learn about seabirds, the more we find they have adapted and are adaptable to the situation at hand. For instance, the diversity of morphology in seabird families which allows them to exploit a broad range of resources and environments has resulted in differing demographic strategies worldwide (see Chapter 5). We encourage students of seabirds to keep an open mind, think broadly, and question and test what they read. We still have much to learn.

Exciting research directions that need to be taken include: comparisons of coastal- vs. oceanic-nesting species, studies of traditional seabirds in comparison with others heavily using marine environments (marine shorebirds), examinations of conspecifics nesting on oceanic vs. coastal islands, and investigations of "energy limitation" in conspecifics in large vs. small colonies.

Addressing the issue of statistical vs. biological significance to marine birds would make major contributions to the fields of ecology, evolution, and biostatistics. Consideration of the continuum from an oceanic existence to coastal, and finally to a truly land-based life-history strategy within seabirds will also advance our knowledge. While answering these questions, most seabird biologists will admit to the exhilaration of watching these fascinating birds on land or at sea, among urban waterways or amidst some of the most spectacular scenery anywhere on earth. It is an exciting time in marine bird biology.

LITERATURE CITED

AINLEY, D. G. 1990. Farallon seabirds, patterns at a community level. Pp. 349–380 in Seabirds of the Farallon Islands (D. C. Ainley and R. J. Boekelheide, Eds.). Stanford University Press, Stanford, CA.

AINLEY, D. C., AND R. J. BOEKELHEIDE. 1990. Seabirds of the Farallon Islands. Stanford University Press, Stanford, CA.

ANDERSON, D. J., AND R. E. RICKLEFS. 1992. Brood size and food provisioning I masked and blue-footed boobies (Sula spp.). Ecology 73: 1363–1374.

ANDERSON, D. W., F. GRESS, AND K. F. MAIS. 1982. Brown Pelicans: influence of food supply on reproduction. Oikos 39: 23–31.

ASHMOLE, N. P. 1963. The regulation of numbers of tropical oceanic birds. Ibis 103b: 458–473.

ASHMOLE, N. P. 1971. Seabird ecology and the marine environment. Pp. 223–286, Vol. 1 in Avian Biology (D. S. Farner and J. R. King, Eds.). Academic Press, London.

ASHMOLE, N. P., AND M. J. ASHMOLE. 1967. Comparative feeding ecology of sea birds of a tropical ocean island. Peabody Museum of Natural History Bulletin, Yale University 24: 1–131.

BIRT, V. L., T. P. BIRT, D. GOULET, D. K. CAIRNS, AND W. A. MONTEVECCHI. 1987. Ashmole's halo: direct evidence for prey depletion by a seabird. Marine Ecology Progress Series 40: 205–298.

BLEM, C. R. 1976. Patterns of lipid storage and utilization in birds. American Zoologist 16: 671–684.

BLEOPOL'SKII, L. O. 1956. Ecology of sea colony birds of the Barents Sea. Israel Program of Scientific Translations, Jerusalem.

BOERSMA, P. D., N. T. WHEELRIGHT, M. K. NERINI, AND E. S. WHEELRIGHT. 1980. The breeding biology of the Fork-tailed Storm Petrel (Oceanodroma furcata). Auk 97: 268–282.

BURGER, J. 1984. Pattern, mechanism, and adaptive significance of territoriality in Herring Gulls (Larus argentatus). American Ornithologists' Union Monograph No. 34, 92 pp.

BURGER, J., AND M. GOCHFELD. 1991. Black Skimmer: Social Dynamics of a Colonial Species. 355 pp. Columbia University Press, New York.

BURGER, J., AND M. GOCHFELD. 1994. Predation and effects of humans on island-nesting seabirds. Pp. 39–67 in Seabirds on Islands: Threats, Case Studies and Action Plans (D. N. Nettleship, J. Burger, and M. Gochfeld, Eds.). BirdLife International, Cambridge, UK.

BURGER, A. E., AND J. F. PIATT. 1990. Flexible time budgets in breeding Common Murres: buffers against variable prey abundance. Studies in Avian Biology 14: 71–83.

CHASTEL, O., H. WEIMERSKIRCH, AND P. JOUVENTIN. 1995. Body condition and seabird reproductive performance: a study of three petrel species. Ecology 76: 2240–2246.

COOK, M. I., AND K. C. HAMER. 1997. Effects of supplementary feeding on provisioning and growth rates of nestling Puffins Fratercula arctica: evidence for regulation of growth. Journal of Avian Biology 28: 56–62.

COULSON, J. C. 1966. The influence of the pair bond and age on the breeding biology of the Kittiwake Gull (Rissa tridactyla). Journal of Animal Ecology 35: 269–279.

COULSON, J. C. 1983. The changing status of the Kittiwake Rissa tridactyla in the British Isles, 1969–1979. Bird Study 30: 9–16.

COULSON, J. C. 1985. Density regulation in colonial seabird colonies. Proceedings of the XVIII International Ornithological Congress, Moscow pp. 783–791.

COULSON, J. C., AND C. S. THOMAS. 1983. Mate choice in the Kittiwake Gull. In Mate Choice (P. Bateson, Ed.). Cambridge University Press, Cambridge.

COULSON, J. C., AND E. WHITE. 1956. A study of colonies of the Kittiwake Rissa tridactyla (L.). Ibis 98: 63–79.

COULSON, J. C., AND E. WHITE. 1958. The effect of age on the breeding biology of the Kittiwake *Rissa tridactyla*. Ibis 100: 40–51.

COUPER-JOHNSTON, R. 2000. The Weather Phenomenon That Changed the World: El Niño. Hodder and Stoughton, London.

CROLL, D. A., A. J. GASTON, AND D. G. NOBLE. 1991. Adaptive mass loss in Thick-billed Murres. Condor 93: 496–502.

CROXALL, J. P. 1992. Southern Ocean environmental changes: effects on seabird, seal and whale populations. Philosophical Transactions of the Royal Society, London 338: 319–328.

CROXALL, J. P., AND P. A. PRINCE. 1980. Food, feeding ecology and ecological segregation of seabirds at South Georgia. Biological Journal of the Linnean Society 14: 103–131.

CROXALL, J. P., AND P. ROTHERY. 1991. Population regulation of seabirds: implications of their demography for conservation. Pp. 272–296 *in* Bird Population Studies: Relevance to Conservation and Management (C. M. Perrins, J.-D. Lebreton, and G. M. Hirons, Eds.). Oxford University Press, Oxford.

CROXALL, J. P., AND P. ROTHERY. 1994. Population change in Gentoo Penguins *Pygoscelis papua* at Bird Island, South Georgia: potential roles of adult survival, recruitment and deferred breeding. Pp. 26–38 *in* The Penguins (P. Dann, I. Norman, and P. Reilly, Eds.). Surrey Beatty and Sons, Chipping Norton.

CROXALL, J. P., T. S. McCANN, P. A. PRINCE, AND P. ROTHERY. 1988. Reproductive performance of seabirds and seals on South Georgia and Signey Island, South Orkney Islands, 1976–1987: implications for Southern Ocean monitoring studies. Pp. 261–285 *in* Antarctic Ocean and Resources Variability (D. Sahrahage, Ed.). Springer-Verlag, Berlin.

CROXALL, J. P., P. ROTHERY, AND A. CRISP. 1992. The effect of maternal age and experience on egg-size and hatching success in Wandering Albatross *Diomedea exulans*. Ibis 134: 219–228.

CROXALL. J. P., P. A. PRINCE, AND K. REID. 1997. Dietary segregation of krill-eating South Georgia seabirds. Journal of London Zoology 242: 531–556.

DIAMOND, J. M., AND B. S. OBST. 1992. Constraints that digestive physiology imposes on behavioral ecology and development in birds. Pp. 97–101 *in* Current Topics in Avian Biology (R. van den Elzen, K.-L. Schuchmann, and K. Schmidt-Koenig, Eds.). Verlag Deutsche Ornithologen-Gesellschaft, Stuttgart.

DRENT, R. H., AND S. DAAN. 1980. The prudent parent: energetic adjustments in avian breeding. Ardea 68: 225–252.

DUFFY, D. C. 1986. Foraging at patches: interactions between Common and Roseate Terns. Ornis Scandinavica 17: 47–52.

DUNNET, G. M., AND J. C. OLLASON. 1978. The estimation of survival rate in the fulmar, *Fulmarus glacialis*. Journal of Animal Ecology 47: 507–520.

DUNNET, G. M., J. C. OLLASON, AND A. ANDERSON. 1979. A 28 year study of breeding Fulmars *Fulmarus glacialis* in Orkney. Ibis 121: 293–300.

GASTON, A. J., AND D. N. NETTLESHIP. 1981. The Thick-billed Murres of Prince Leopold Island. Canadian Wildlife Service Monograph Series No. 6.

GREEN, B., AND N. BROTHERS. 1989. Water and sodium turnover and estimated food consumption rates in free-living fairy prisons (*Pachyptila turtur*) and common diving petrels (*Pelecanoides urinatrix*). Physiological Zoology 62: 702–705.

HAMER, K. C. 1994. Variability and stochasticity of meal size and feeding frequency in the little Shearwater *Puffinus assimilis*. Ibis 136: 271–278.

HAMER, K. C., AND J. K. HILL. 1994. The regulation of food delivery to nestling Cory's shearwaters *Calonectris diomedea*: the roles of parents and offspring. Journal of Avian Biology 25: 198–204.

HAMER, K. C., J. K. HILL, J. S. BRADLEY, AND R. D. WOOLLER. 2000. Contrasting patterns of nestling obesity and food provisioning in three species of *Puffinus shearwaters*: the role of predictability. Ibis 142: 139–158.

HARRIS, M. P. 1970. Breeding ecology of the Swallow-tailed Gull (*Creagrus furcatus*). Auk 87: 215–243.

HARRIS, M. P. 1979. Measurements and weights of British Puffins. Bird Study 26: 179–196.

HENDERSON, B. A. 1975. Role of chick's begging behavior in the regulation of parental feeding behavior of *Larus glaucescens*. Condor 77: 488–492.

HULSMAN, K. 1987. Resource partitioning among sympatric species of tern. Ardea 75: 29–32.

HULSMAN, K. 1988. The structure of seabird communities: an example from Australian waters. Pp. 59–91 *in* Seabirds and Other Marine Vertebrates: Competition, Predation, and Other Interactions (J. Burger, Ed.). Columbia University Press, New York.

HUNT, G. L., AND J. L. BUTLER. 1980. Reproductive ecology of Western Gulls and Xantus' Murrelets with respect to food resources in the southern California Bight. California Cooperative Oceanographic and Fisheries Reports 21: 62–67.

KONARZEWSKI, M., C. LILJA, J. KOZLOWSKI, AND B. LEWONCZUK. 1990. On the optimal growth of the alimentary tract in avian postembryonic development. Journal of Zoology, London 222: 89–101.

LACK, D. 1954. The Natural Regulation of Animal Numbers. Oxford University Press, Oxford.

LACK, D. 1968. Ecological Adaptations for Breeding in Birds. 409 pp. Methuen, London.

MILLS, J. A. 1973. The influence of age and pair-bond on the breeding biology of the Red-billed Gull *Larus novaehollandiae scopulinus*. Journal of Animal Ecology 42: 147.

MILLS, J. A. 1980. Red-billed Gull. Pp. 387–404 *in* Lifetime Reproduction in Birds (I. Newton, Ed.). Academic Press, New York.

MILLS, J. A., J. W. YARRALL, AND D. A. MILLS. 1996. Causes and consequences of mate fidelity in red-billed gulls. Pp. 286–304 *in* Partnerships in Birds. The Study of Monogamy (J. M. Black, Ed.). Oxford University Press, Oxford.

MONAGHAN, P., J. D. UTTLEY, AND M. D. BURNS. 1991. The influences in changes in prey availability on the breeding ecology of terns. Acta Ornithological Congress 20: 2257–2262.

MOORS, P. J., AND I. A. E. ATKINSON. 1984. Predation on seabirds by introduced animals, and factors affecting its severity. Pp. 667–690 *in* Status and Conservation of the World's Seabirds (J. P. Croxall, P. G. H. Evans, and R. W. Schreiber, Eds.). International Committee for Bird Preservation (Tech. Publ. 3), Cambridge, UK.

NAVARRO, R. A. 1991. Food addition and twinning experiments in the Cape Gannet: effects on breeding success and chick growth and behavior. Colonial Waterbirds 14: 92–102.

NAVARRO, R. A. 1992. Body composition, fat reserves, and fasting capability of cape gannet chicks. Wilson Bulletin 104: 644–655.

NELSON, J. B. 1964. Factors influencing clutch size and chick growth in the North Atlantic Gannet *Sula bassana*. Ibis 106: 63–77.

NELSON, J. B. 1978. The Gannet (T. and A. D. Poyser, Eds.). Berkhamstead, U.K.

NELSON, J. B. 1983. Contrasts in breeding strategies between some tropical and temperate marine pelecaniformes. Studies in Avian Biology 8: 95–114.

NORBERG, R. A. 1981. Temporary weight decrease in breeding birds may result in more fledged young. American Naturalist 118: 838–850.

NORTON, R., AND E. A. SCHREIBER. In press. The Brown Booby, *Sula leucogaster. In* The Birds of North America (A. Poole and F. Gill, Eds.). The Birds of North America, Inc., Philadelphia.

OBST, B. S., AND J. M. DIAMOND. 1992. Ontogenesis of intestinal nutrient transport in domestic chickens (*Gallus gallus*), and its relation to growth. Auk 109: 451–464.

PIATT, J. F. 1987. Behavioral ecology of Common Murres and Atlantic Puffin predation on capelin: implications for population biology. Ph.D. thesis, Memorial University, St. John's, Newfoundland.

PLACE, A. R., N. C. STOYAN, R. E. RICKLEFS, AND R. G. BUTLER. 1989. Physiological basis of stomach oil formation in Leach's Storm-petrel (*Oceanodroma leucorhoa*). Auk 106: 687–699.

PRINCE, P. A. 1985. Population and energetic aspects of the relationships between Black-browed and Grey-headed Albatross and the southern ocean marine environment. Pp. 473–477 *in* Antarctic Nutrient Cycles and Food Webs (W. R. Siegfried, P. R. Condy, and R. M. Laws, Eds.). Springer-Verlag, Berlin.

PRINCE, P. A., AND C. RICKETTS. 1981. Relationships between food supply and growth in albatrosses: an interspecies chick fostering experiment. Ornis Scandinavica 12: 207–210.

PRINCE, P. A., P. ROTHERY, J. P. CROXALL, AND A. G. WOOD. 1994. Population dynamics of Black-browed and Grey-headed Albatrosses *Diomedea melanophris* and *D. chrysostoma* at Bird Island, South Georgia. Ibis 136: 50–71.

RAHN, H., AND G. C. WHITTOW. 1984. Introduction. Pp. 1–32 *in* Seabird Energetics (G. C. Whittow and H. Rahn, Eds.). Plenum Press, New York.

REID, K., P. A. PRINCE, AND J. P. CROXALL. 2000. Fly or die: the role of fat stores in the growth and development of Grey-headed Albatross *Diomedea chrysostoma* chicks. Ibis 142: 188–198.

RICKLEFS, R. E. 1969. An analysis of nestling mortality in birds. Smithsonian Contributions in Zoology 9: 1–48.

RICKLEFS, R. E. 1974. Energetics of reproduction in birds. Pp. 152–292 *in* Avian Energetics (R. A. Paynter, Jr., Ed.). Nuttal Ornithological Club, No. 15.

RICKLEFS, R. E. 1992. The roles of parent and chick in determining feeding rates in Leach's Storm-petrels. Animal Behaviour 43: 895–906.

RICKLEFS, R. E., A. PLACE, AND D. J. ANDERSON. 1987. An experimental investigation of the influence of diet quality on growth in Leach's Storm Petrel. American Naturalist 130: 300–305.

RICKLEFS, R. E., J. M. STARCK, AND M. KONARZEWSKI. 1998. Internal constraints on growth in birds. Pp. 266–287 *in* Avian Growth and Development; Evolution within the Altricial-Precocial Spectrum (J. M. Starck and R. E. Ricklefs, Eds.). Oxford Ornithology Series, Oxford University Press, Oxford.

SAFINA, C., AND J. BURGER. 1985. Common Tern foraging: seasonal trends in prey fish densities and competition with Bluefish. Ecology 66: 1457–1463.

SAFINA, C., J. BURGER, M. GOCHFELD, AND R. H. WAGNER. 1988. Evidence for prey limitation of Common and Roseate Tern reproduction. Condor 90: 852–859.

SALLABANKS, R., AND F. C. JAMES. 1999. American Robin (*Turdus migratorius*). No. 462 *in* The Birds of North America (A. Poole and F. Gill, Eds.). The Birds of North America, Inc. Philadelphia.

SCHEW, W. A., AND R. E. RICKLEFS. 1998. Developmental plasticity. Pp. 288–304 *in* Avian Growth and Development: Evolution within the Altricial-Precocial Spectrum (J. M. Starck and R. E. Ricklefs, Eds.). Oxford Ornithology Series, Oxford University Press, Oxford.

SCHREIBER, E. A. 1994. El Niño–Southern Oscillation effects on chick provisioning and growth in red-tailed tropicbirds. Colonial Waterbirds 17: 105–119.

SCHREIBER, E. A. 1996. Experimental manipulation of feeding in Red-tailed Tropicbird chicks. Colonial Waterbirds 19: 45–55.

SCHREIBER, E. A., AND R. W. SCHREIBER. 1989. Insights into seabird ecology from a global natural experiment. National Geographic Research 5: 64–81.

SCHREIBER, E. A., R. W. SCHREIBER, AND G. A. SCHENK. 1996. The Red-footed Booby (*Sula sula*). No. 241. *In* The Birds of North America (A. Poole and F. Gill, Eds.). Academy of Natural Sciences, Philadelphia; American Ornithologists' Union, Washington, D.C.

SIMMONS, K. E. L. 1972. Some adaptive features of seabird plumage types. British Birds 65: 466–479.

SLAGSVOLD, T. 1982. Clutch size variation in passerine birds: the nest predation hypothesis. Oecologia 54: 159–169.

TAYLOR, J. R. E., AND M. KONARZEWSKI. 1989. On the importance of fat reserves for the little auk (*Alle alle*) chicks. Oecologia 81: 551–558.

U. S. FISH AND WILDLIFE SERVICE. 1996. Seabirds and Shorebirds of Midway Atoll National Wildlife Refuge. U. S. Fish and Wildlife Service, Midway Island.

WHITTAM, T. S., AND D. SIEGEL-CAUSEY. 1981. Species interactions and community structure in Alaskan seabird colonies. Ecology 62: 1515–1524.

Hesperonis regalis

2 The Seabird Fossil Record and the Role of Paleontology in Understanding Seabird Community Structure

Kenneth I. Warheit

CONTENTS

2.1 INTRODUCTION

Most seabird systems (e.g., species, communities, populations) are large in both temporal and spatial scale. For example, it is now firmly established that many seabird populations and communities are affected by climatic cycles, some of which operate globally and over periods extending from several years to decades (e.g., El Niño–Southern Oscillation and the North Pacific decadal oscillation; see Chapter 7). In general, seabirds are long lived with each bird experiencing a variety of climatic conditions during its lifetime. The longevity of individual seabirds and the fact that these birds live in environments that are affected by large-scale phenomena have prompted a plethora of long-term studies of seabird populations and communities (e.g., Coulson and Thomas 1985, Ainley and Boekelheide 1990, Harris 1991, Wooler et al. 1992). In fact, there is a lengthy history of long-term studies of seabird populations (e.g., Rickdale 1949, 1954, 1957, Serventy 1956) and communities (e.g., Uspenski 1958, Belopol'skii 1961).

The long-term history of seabird systems is even more remarkable when we consider the fossil record. Contrary to "common knowledge," birds have a rather extensive fossil record (Olson 1985a) that is most informative. Owing to the fact that seabirds generally live or lived in depositional environments (e.g., nearshore marine) rather than erosional environments (e.g., upland), the fossil record of seabirds represents a large percentage of the total fossil record of all birds (see Olson

1985a). Given this relatively good but clearly incomplete fossil record, it is possible to use seabird fossils as a tool not only to study the truly long-term history of seabirds, but also to help interpret the biogeographical patterns and community structure of modern-day seabird systems.

In this chapter, I summarize first the fossil history of seabirds, here defined as Sphenisciformes, Procellariiformes, Pelecaniformes (excluding Anhingidae), Laridae, and Alcidae. This summary includes a comprehensive table (Appendix 2.1) listing each fossil taxon, with its corresponding temporal, spatial, and bibliographic information. I then discuss the importance of fossils and the paleontological record in elucidating many aspects of seabird ecology and evolution. I introduce what fossils can tell us about biology, geography, and time, and provide a series of examples of how the study of seabird fossils presents essential information to our understanding of the long-term and large-scale development of seabird communities. Finally, I conclude with a discussion of the fossil history of the Alcidae. I highlight the Alcidae for several reasons. First, the fossil record of alcids is one of the best fossil records of all seabirds because of the large amount of material that has been collected and described, and the high degree of taxonomic diversity resulting from these descriptions. Second, the alcids encapsulate many of the discussions that are emphasized throughout this chapter. That is, to correctly understand the biogeographic and phylogenetic relationships of alcids requires knowledge of the alcid fossil record. Third, the fossil history of alcids is enigmatic and presents some interesting questions requiring future research.

2.2 THE FOSSIL RECORD OF SEABIRDS

I have provided a list of fossil seabird taxa in Appendix 2.1 (368 entries, including 253 taxa described to species, 28 of which are assigned or have affinities to modern species). Although this list is comprehensive, undoubtedly it is not complete, and it does not include modern seabird taxa found in Pleistocene or Holocene deposits (see Brodkorb 1963, 1967; and Tyrberg 1998 for listing of Pleistocene fossils of modern seabirds). There are at least two published revisions of a fossil taxon (penguins from New Zealand and Antarctica; Fordyce and Jones 1990, Myrcha in press) that were not included in this analysis. In Appendix 2.2, 23 additional fossil taxa are listed that are now considered synonymous with a species listed in Appendix 2.1.

It is tempting to compare the diversity among some higher taxa based on a list of species; however, these species were probably not described using the same set of procedures. For example, one author might feel justified naming a new species based on fragmentary material (e.g., Harrison 1985), while another author might be reluctant to do so or will wait until a greater number of higher quality material is in hand (Olson and Rasmussen 2001). The lack of a standard in describing new fossil species will result in some higher taxa having a greater number of described species than other taxa simply because of authors' biases rather than a product of true morphological diversity. That being said, I will still make some rudimentary comparisons among the higher taxa listed in Appendix 2.1.

Pelecaniformes is the most diverse order in this list in terms of both the number of entries (141) and described species (94). Procellariidae is the most diverse family with 68 entries and 42 described species, followed by the Alcidae (46 entries, 31 species) and Spheniscidae (45 entries, 38 species). The oldest taxon in the list is *Tytthostonyx glauconiticus*, from the late Cretaceous of New Jersey (see Figure 2.1 for time scale), tentatively placed in the Procellariiformes by Olson and Parris (1987). Following this species there are several taxa described from the Paleocene and Eocene, most of which are either archaic penguins or Pelagornithidae, an extinct group of bony-tooth pelecaniforms (see below). In fact, the Paleogene (Paleocene through Oligocene; Figure 2.1) appeared to be dominated by extinct Pelecaniformes (Pelagornithidae and Plotopteridae), Procellariidae, and large-sized penguins (Figure 2.2). Except for *Puffinus* (*P. raemdonckii*, from the early Oligocene of Belgium), modern genera of seabirds do not appear until the early Miocene or 16 to 23 million years ago (mya), and do not become taxonomically diverse until the middle Miocene (11 to 16 mya). The middle Miocene (Fauna I in Warheit 1992; see Figure 2.1) marked the onset

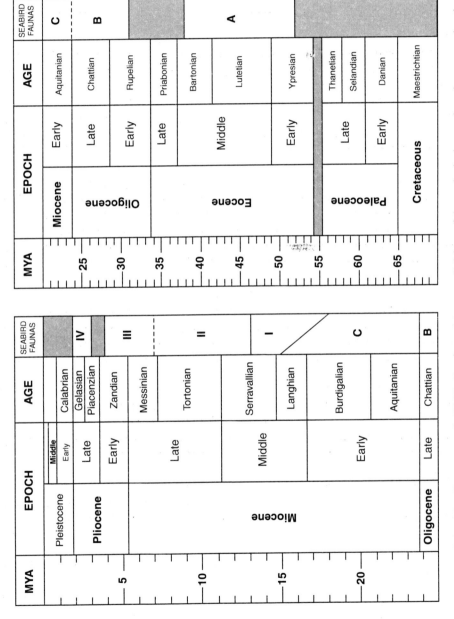

FIGURE 2.1 Cenozoic time scale based on Berggren et al. (1995). Epochs and Ages are divisions of the geologic time scale and correspond to the stratigraphic sequence of rocks and fossils. Epochs and Ages are scaled to absolute time using a combination of paleomagnetic and radioisotopic data. The seabird faunas are from Warheit (1992) and are based on the association of fossil-bearing rock formations from the North Pacific formed during a single, but broadly defined interval of time. The assemblage of seabird fossils from each of these isochronous rock formations is defined as a fauna. See Warheit (1992) for definitions of each of these North Pacific seabird faunas.

FIGURE 2.2 A reconstruction of one of the largest fossils in the Plotopteridae (Pelecaniformes). This plotopterid was larger than Emperor Penguins and had paddle-like wings similar to penguins. Its hindlimb and pelvic morphology were similar to Anhingas. It used its wings to swim underwater, an adaptation that has evolved several times in birds (Olson and Hasegawa 1979). (After Olson and Hasegawa 1979.)

of a permanent East Antarctic ice cap, a drop in sea level, and an increase in the latitudinal thermal gradient of the world's oceans (Warheit 1992). The steepening of this thermal gradient intensified the gyral circulation of surface currents, and strengthened the coastal and trade winds that promote upwelling (Barron and Bauldauf 1989). Indeed, there appears to be a temporal correlation between these climatic and oceanographic events and the taxonomic diversification of seabirds (see also Warheit 1992).

I discuss some of these issues and other aspects of the seabird fossil record in the next few sections. However, I would like to highlight here two groups of extinct seabirds: Pelagornithidae and Plotopteridae. The Pelagornithidae or pseudodontorns first appeared in the eastern North Atlantic (England) in the late Paleocene and early Eocene (49 to 61 mya) and in the eastern North Pacific and Antarctica in the middle and late Eocene, respectively. This group was truly global in distribution, occurring in fossil deposits in North and South America, Europe, Asia, Africa, New Zealand, and Antarctica, and survived some 57 to 59 million years (Appendix 2.1). The birds were also remarkable in their morphology: gigantic in size, one species was estimated to have a wingspan of almost 6 m (K. Warheit and S. Olson, unpublished data), with bony projections on their rostrum and mandible (Olson 1985a). Their mandible was also composed of a hinge-like synovial joint and lacked a bony symphysis (Zusi and Warheit 1992). Zusi and Warheit (1992) speculated that the birds captured prey on or near the surface of the water while in flight or by lunging while sitting on the water surface. Their extinction is enigmatic, but may be related to fluctuations in local or global food resources (Warheit 1992).

The Plotopteridae were pan-North Pacific in distribution and ranged in size from over 2 m in length to the size of a Brandt's Cormorant (Olson and Hasegawa 1979, Olson 1980, Olson and Hasegawa 1996; Figure 2.2). These seabirds were closely related to sulids, cormorants, and anhingas, but were flightless and possessed paddle-like wings remarkably convergent with those of penguins and flightless alcids (Olson and Hasegawa 1979, Olson 1985a). They disappeared in the early and middle Miocene from the eastern and western Pacific, respectively (Appendix 2.1). Olson and Hasegawa (1979) and Warheit and Lindberg (1988) considered the evolution and radiation of gregarious marine mammals as a possible cause for the extinction of the plotopterids, while Goedert (1988) suggested that a sharp rise in ocean temperature was a better explanation for their demise (see Warheit 1992 for discussion of both hypotheses).

2.3 THE IMPORTANCE OF SEABIRD FOSSILS

2.3.1 PALEONTOLOGY AND THE STRUCTURE OF SEABIRD COMMUNITIES

Press and Siever (1982) define paleontology as "the science of fossils of ancient life forms, and their evolution" and define a fossil as "an impression, cast, outline, track, or body part of an animal or plant that is preserved in rock after the original organic material is transformed or removed." Olson and James (1982a) extended the definition of fossil to also include subfossil bones (bones that have not become mineralized), such as those present in archeological midden sites, and I will adhere to this definition of fossil throughout this chapter. Because fossils, especially seabird fossils, occur in rocks that may also contain the fossiliferous remains of climate-sensitive microorganisms such as foraminiferans, it is possible to associate a particular climatic régime to a particular fossil community. Furthermore, since fossil-bearing rocks also can be placed geographically and dated either relatively or absolutely using a variety of methods, we can associate a fossil with a specific time and place. As such, if fossils are grouped together based on time, they can provide information on what species co-occurred during a specific period and in a specific place, and under the influence of a specific climatic régime. Therefore, fossils are not simply a collection of broken bones, but are in fact treasure troves that provide us with information about the morphology, anatomy, physiology, and behavior of individual organisms, as well as composition of past ecological communities.

Recent and historical processes contribute to the structure of seabird communities today. That is, those that can be measured in ecological time (e.g., predation, competition, dispersal) as well as factors that are measured in geological time (e.g., plate tectonics and the origin of modern oceanic currents), and perhaps random luck (see Jablonski 1986 and Gould 1989 for examples of the importance of random extinctions and historical contingencies, respectively), are responsible for the composition of the seabird communities today. I argue that in order to understand the structure of seabird communities today, we must not only study predation, competition, dispersal, etc., but we must also study fossils. Without incorporating history, an incomplete or a potentially incorrect story is built. To emphasize this point, I provide three examples of how studies of fossils and geological history have contributed essential components to our understanding of seabird communities. The first two examples (North Pacific and South African seabirds) provide information on how continental drift, sea level, and associated changes in climate and oceanography may have been responsible for profound changes in the composition of seabird communities. The final example concerns how the Polynesian colonization of oceanic islands in the Pacific Ocean resulted in extensive extinctions of both land- and seabird taxa prior to European exploration of the Pacific or written history.

2.3.1.1 North Pacific Seabird Communities

I have previously reviewed the fossil history of seabirds from the North Pacific and related this history to plate tectonics and paleooceanography (Warheit 1992). In what follows I highlight some of the findings from this study, focusing primarily on the seabird communities from central and southern California. The California Current upwelling system today is one of the primary eastern boundary systems, and, along with the Benguela and Humboldt upwelling systems of the Southern Hemisphere, currently support abundant and diverse seabird faunas. These three upwelling systems have many of the same types of seabirds. That is, each system has wing-propelled divers (e.g., alcids in the north, penguins and diving petrels in the south), foot-propelled divers (cormorants), pelicans, storm-petrels, and gulls, as well as others. Also present in both the Benguela and Humboldt systems are plunge-diving sulids, although there are no sulids, indigenous or otherwise, in the California Current today. It would be possible to develop a series of hypotheses to explain this difference; sulids are present in the Northern Hemisphere and in the North Pacific, and there are breeding sulids as close to the California Current as Baja California. However, developing such hypotheses using only ecological data collected from these communities today would be in error. Sulids existed in the California Current for the better part of nearly 16 million years and were represented by at least 11 to 13 different species (Appendix 2.1; Warheit 1992). Therefore, the question that should be asked is no longer simply "What ecological processes exist that have prevented sulids from occurring in the California Current?" but should also be "Why did sulids become extinct in the California Current, while remaining extant and thriving in other cold water upwelling systems?"

The local extinction of sulids is only one example of a dynamic seabird system. Overall, the seabird communities of the North Pacific in the past are quite different from those that exist today. There are at least 94 species of fossil seabirds in the North Pacific from at least seven distinct seabird "faunas" (Warheit 1992). Most of these species are from extant genera, but there also existed three groups of extinct and somewhat bizarre taxa: Pelagornithidae and Plotopteridae (discussed above), and the mancallids. The mancallids consisted of two, possibly three genera (*Praemancalla*, *Mancalla*, and perhaps *Alcodes*) of flightless alcids with estimated body mass ranging from 1 to 4 kg, compared with a mass of 5 kg for the Great Auk (*Pinguinus impennis*) (Livezy 1988). These were the most abundant seabirds in the California Current from at least 12 mya to the Plio-Pleistocene, especially during the late Pliocene (1.5 to 3 mya; Chandler 1990a), when there were at least three species of *Mancalla* and well over 200 specimens recovered from the San Diego Formation. The flightlessness of mancallids and the Great Auk was convergent in that these two taxa are not considered to be closely related (Storer 1945, Chandler 1990b), and the mancallids were more specialized for wing-propelled diving than the Great Auk, approaching the extreme morphology of penguins (Olson 1985a, Livezy 1988). Mancallids remained extant until the Pleistocene, but became extinct approximately 470,000 years ago (Howard 1970, Kohl 1974), perhaps as a result of competition for terrestrial space with gregarious pinnipeds (Warheit and Lindberg 1988, Warheit 1992).

In its entirety, the seabird history from the California Current upwelling system can be summarized as a transition from archaic pelecaniforms to a fauna closely resembling the system today, consisting of volant alcids, shearwaters, and storm-petrels, but a fauna that also included sulids and flightless alcids. Although competition and predation may have contributed to the various radiations and extinctions that characterized the California Current seabird faunas, the underlying physical process that governed the development of these faunas was the tectonic activities that resulted in the thermal isolation and refrigeration of Antarctica and the uplift of the Isthmus of Panama (Warheit 1992).

2.3.1.2 South African Seabird Faunas

As with the North Pacific seabird communities, there have been significant changes in the composition of the South African seabird faunas during the past several millions of years. Recent seabird faunas in both the North Pacific (in particular California and Oregon) and South African (Atlantic) coasts occur in cold-water upwelling systems. These upwelling systems are a function of continental positions and global circulation patterns, which, in turn, are products of tectonic activities. As such, these upwelling systems have had different characteristics throughout the Tertiary. According to Siesser (1980; in Olson 1983), the Benguela upwelling system off the southwest coast of South Africa did not develop until the early late Miocene. No fossil seabirds have been recovered from deposits prior to the development of this cold water system, but Olson (1983) speculated that since water temperatures were warmer than those in the Pliocene and today, cold-water taxa were either absent or present in low diversity and abundance. The appearance of the first known South African seabird fauna roughly coincided with a good depositional environment, and, more importantly, with the development of the Benguela system and the production of cold water. Olson (1983, 1985b) concluded that with the progressive development of this cold-water nutrient-rich environment, seabird taxa more typical of cold-water systems moved north from the southerly latitudes near and around Antarctica.

The early Pliocene (5 mya) deposits of South Africa have yielded a diverse seabird fauna consisting of four species of penguins possibly related to *Spheniscus*, an albatross, two species of storm-petrels (*Oceanites*), three species of prions (*Pachyptila*), at least five species of shearwaters (*Procellaria, Calonectris, Puffinus*), and at least one species each of fulmarine petrel, diving petrel (*Pelecanoides*), and booby (*Sula;* Olson 1983, 1985b,c; Table 2.1). Based on the fossil localities and their depositional environments, and the presence of juvenile individuals in the deposits, Olson (1985b,c) reasoned that this seabird fauna consisted of both breeding and nonbreeding species (see Table 2.1). Although there are similarities between this early Pliocene fauna and South African seabirds today, mostly in terms of the higher taxonomic diversity of the nonbreeding species, there are considerable differences in the diversity of the breeding taxa (Table 2.1). There are no procellariiform taxa currently breeding in South Africa today, although there were at least three species (prion, storm-petrel, diving petrel) breeding locally during the early Pliocene. Olson (1983, 1985b) concluded that, except for the cormorant species, there has been a complete change in the seabird fauna of South Africa from the early Pliocene to today and this faunal turnover was mirrored by a similar turnover in the pinniped fauna. Specifically, taxa with cold-water affinities today and present in South Africa during the early Pliocene have been eliminated from the modern breeding fauna (*Oceanites, Pachyptila, Pelecanoides*), or are present in the modern fauna, but severely reduced in diversity (*Spheniscus*). This reduction in the number of cold-water species breeding in South Africa from the Pliocene to today is enigmatic because the Benguela cold-water upwelling system has been present off South Africa since the late Miocene. Olson (1983, 1985b) reasoned that the presence of the cold-water system was not the only factor in determining the relative diversity of species, but that a combination of factors contributed to the change in seabird faunas in South Africa. In addition to changes in oceanographic conditions and possible warming of the Benguela Current, it is possible that there were substantial changes in availability of island habitats resulting from fluctuating sea levels during the late Pliocene and throughout the Pleistocene. That is, changes in the height of sea level associated with tectonic activities and polar temperatures affect the availability of breeding habitats by either creating or removing islands. Islands can be created during low sea levels through the emergence of submerged land, or during high sea levels through flooding of low lands and isolation of high lands. The opposite can be true for the destruction of suitable island habitats.

2.3.1.3 Human-Induced Extinction of Seabirds from Pacific Islands

In the previous two examples, the long-term structure of seabird communities appears to have been largely affected by geological processes, namely, those responsible for the development of particular oceanic currents and water temperature, and for changes in relative sea level. However, some of the most profound changes to seabird systems have occurred relatively recently (geologically speaking) and were the direct result of human activities. Steadman (1995) summarized information on the Holocene extinction of birds from Pacific islands resulting from activities of indigenous people from Melanesia, Micronesia, and Polynesia. He determined that approximately 8000 species or populations, mostly flightless rails, became extinct following the geographic expansion of Polynesian populations (the extinction of a local population is here referred to as extirpation; see Steadman 1995). These extinctions and extirpations dramatically reduced the diversity of birds nesting on Pacific islands prior to the arrival of Europeans (and a written history) and, as such,

TABLE 2.1
List of Fossil Seabird Species Described by Olson
(1985b,c) from Deposits in South Africa (see text)

Taxon	Number Breeding	
	Fossil[a]	Recent
Sphenisciformes	0	1
Nucleornis insolitus		
Dege hendeyi		
?Palaeospheniscus huxleyorum		
Inguza predemersus		
Diomedeidae	0	0
Diomedea sp.		
Oceanitidae	1	0
Oceanites zaloscarthmus	[b]	
Oceanites sp.		
Procellariidae	1	0
Fulmarinae sp.		
Pachyptila salax	[b]	
Pachyptila sp. B		
Pachyptila sp. C		
Procellaria sp.		
Calonectris sp.		
Puffinus sp. A		
Puffinus sp. B		
Puffinus sp. C		
Pelecanoididae	1	0
Pelecanoides cymatotrypetes	[b]	
Sulidae	0	1
Sula sp.		
Phalacrocoracidae	0	4
Phalacrocorax medium sp. A		
Phalacrocorax medium sp. B		
Phalacrocorax small sp.		

[a] The number of fossil species determined to be breeding is a minimum number and in most cases there are not enough data to determine breeding status.

[b] A fossil species is said to be breeding at a locality if remains of juveniles are found.

send a clear message that our studies of island biogeography *must not* ignore the extinct, prehistoric faunas and floras (Olson and James 1982a). In what follows, I briefly describe some of the changes that occurred to the status and distribution of seabird species throughout the Pacific as a result of the activities of these Pacific island people. This section summarizes the work of H. James, S. Olson, and D. Steadman, and I refer the reader to these original references (Olson and James 1982a,b, 1991, Steadman and Olson 1985, James 1995, Steadman 1995, and references therein). In addition, Harrison (1990) provided a popular account of the interactions between seabirds and humans on the Hawaiian Islands.

James (1995) reviewed the background of prehuman extinction rates for birds on oceanic islands. Although it is not possible to calculate annual turnover rates in species abundance and distribution, as is possible to do for islands today, the fossil record provides the means by which we can measure long-term biogeographic patterns of seabird species. After reviewing both the Pleistocene and Holocene (i.e., post-Pleistocene) fossil record of birds on Pacific islands, James (1995) and others concluded that bird diversity was relatively stable during the Pleistocene, even during periods of great climatic change, but the number of extinctions increased dramatically following human occupation. For example, on the Hawaiian island of Oahu, James (1987, in James 1995) recorded 17 species of landbirds from Pleistocene deposits. All but two of these species survived a period greater than 120,000 years, during intense global climatic change, including a complete cycle of polar glaciation and deglaciation. However, human activities may have extirpated 13 of these 17 Pleistocene birds during the past thousand years or so (James 1995). In another example, Steadman (1995) described extinction rates in the Galapagos Islands where some 500,000 bones from Holocene deposits have been unearthed; about 90% of these bones predate the arrival of humans. During a period of 4000 to 8000 years prior to human occupation, a maximum of only 3 populations were extirpated from the Galapagos; however, during the few centuries since the arrival of humans, 21 to 24 populations were extirpated (Steadman 1995).

The human-related extinction of birds from islands can be caused by any number of perturbations ranging from direct predation and habitat destruction, to the introduction of non-native predators, competitors, or pathogens (Steadman 1995). On Hawaii, where the extinction of seabird species or populations appears less severe than on the Polynesian islands to the south, Olson and James (1982a) concluded that predation by humans, or collateral predation by their pets, was most important in the extinction of populations or species of flightless and ground-nesting landbirds and burrow-nesting seabirds. However, habitat destruction in the form of clearing of lowland forests was most likely the cause of the extinction of most of the small land bird species. Steadman (1995) added that soil erosion following deforestation also might have eliminated nest sites for burrowing seabirds.

The importance of fossils in understanding modern biogeographic patterns is best demonstrated by the documentation of extinctions and extirpations of birds from these oceanic islands. Steadman (1995 and references therein) stated that the Pacific seabird biodiversity on subtropical and tropical islands is now considerably lower than that on temperate and sub-Antarctic islands, and that this difference in biodiversity has been associated by others with the fact that marine waters in the tropics are less productive. However, Steadman indicated that the difference in seabird diversity between lower and higher latitude islands becomes less when you consider the extinct or extirpated species revealed by the fossil record. For example, on Ua Huka in the Marquesas, the prehistoric diversity of seabirds included at least 7 species of shearwaters and petrels and a total of 22 species of nesting species of seabirds; today there are only four species of seabirds and no breeding shearwaters or petrels (Steadman 1995).

The reduction in biodiversity from the low-latitude Pacific islands is mostly the result of the local extirpation of a population, not the outright extinction of a species. Steadman (1995) stated that there have been few examples of seabird species extinctions throughout Oceania. In the Hawaiian Islands, Olson and James (1991) documented only one extinct species of seabird, *Pterodroma jugabilis*, although there were many examples of local extirpation of populations (Olson and James 1982b). On Henderson Island, Steadman and Olson (1985) showed that although the island still maintains a diverse seabird fauna, *Nesofregatta fuliginosa* is recorded only as a fossil and was most likely eliminated from the island and the rest of the Pitcairn Group of islands because of human activities.

Finally, and perhaps most telling of the prehistoric destruction of Oceania seabird fauna, the fossil record indicates that on Easter Island there were at least 25 species of seabirds including an albatross, fulmar, prion, several species of petrels and shearwaters, a storm-petrel, two species of tropicbirds, a frigatebird, booby, and a suite of tern species (Steadman 1995). Today, 1 of these species is extinct (unnamed Procellariidae), 12 to 15 species no longer occur in or around Easter Island, and only 1 of these 25 species (Red-tailed Tropicbird, *Phaethon rubricauda*) currently breeds on Easter Island (Steadman 1995). Steadman stated (1995, p. 1124) that "Evidently, Easter

Island lost more of its indigenous terrestrial biota than did any other island of its size in Oceania" and that this destruction occurred in a period from 1500 to 550 years ago, during human colonization. In interpreting these data, Steadman assumed that the Polynesians collected the seabirds locally on Easter Island. However, an alternative explanation is that many of these seabird taxa did not breed on Easter Island and the Polynesians captured birds at sea and brought the carcasses back to the island (S. Olson, personal communication). This would inflate the number of "breeding" seabird species on Easter Island if Steadman defined breeding as simply the presence of bones on the island.

2.3.2 THE FOSSIL RECORD OF THE ALCIDAE

The fossil record of the Alcidae is enigmatic when one attempts to reconcile the geographic distribution of certain fossil taxa with that of their modern relatives. For example, while alcid fossils are extremely abundant in western Atlantic deposits (Olson 1985a, Olson and Rasmussen 2001), the overall alcid diversity in the Atlantic was lower than that of the Pacific, and there are no pre-Pleistocene specimens of *Uria* and no fossil specimens of *Cepphus* (see Appendix 2.1). However, while there are relatively few alcid fossils from eastern Pacific deposits except those from the mancallines (see above), alcid diversity was high and there are two fossil species of *Uria* and at least one fossil species of *Cepphus*. In what follows, I briefly review the fossil history of the Alcidae in terms of when and where taxa first appeared (Appendix 2.1, Table 2.2), based on Olson (1985a), Chandler (1990a), Warheit (1992), and Olson and Rasmussen (2001). See Gaston and Jones (1998) for a general account of the fossil record of the Alcidae.

Fossils representing the earliest evolution of the Alcidae are either not described in the literature or their relationships are in question. Storrs Olson (personal communication) stated that a fossil of a "primitive auk" might be present in the London Clay material from the lower Eocene of England, which, if shown to be correct, would represent the earliest known alcid taxon. There are two published accounts of pre-Miocene alcids: *Hydrotherikornis oregonus* from the late Eocene of Oregon (Miller 1931) and *Petralca austriaca* (Mlíkovský and Kovar 1987) from the late Oligocene of Austria. It is unclear if *Hydrotherikornis* is an alcid or a procellariid (see Olson 1985a). Chandler (1990b, p. 73) considered *Hydrotherikornis* to be "a petrel very similar to *Daption*" and he provided one skeletal character to justify this relationship. Chandler (1990b) also doubted the alcid affinities of *Petralca* and placed the taxon in Aves, *Incertae Sedis*; however, he did not examine the specimen but considered the taxon's description by Mlíkovský and Kovar (1987) insufficient to justify placement in the Alcidae.

TABLE 2.2
Distribution of Alcidae and Relative Dates of First Appearance in the Fossil Record (see also Appendix 2.1)

Taxon[a]	Recent Distribution[b]		First Appearance Fossil Record		Comments
	Atlantic	Pacific	Atlantic	Pacific	
Alcini	Yes	Yes	middle Miocene	late Miocene	No *Uria* in Atlantic until Pleistocene
Cepphini	Yes	Yes	—	late Miocene	No *Cepphus* in Atlantic until Recent
Brachyramphini	No	Yes	—	late Pliocene	No *Brachyramphus* in Atlantic
Aethiini	No	Yes	early Pliocene	late Miocene	Only fossil Aethiini in Atlantic
Fraterculini	Yes	Yes	early Pliocene	late Miocene	

[a] Alcini (*Alle, Alca, Uria, Pinguinus, Miocepphus*); Cepphini (*Cepphus, Synthliboramphus*); Brachyramphini (*Brachyramphus*); Aethiini (*Ptychoramphus, Cyclorhynchus, Aethia*); Fraterculini (*Cerorhinca, Fratercula*).
[b] Pacific also includes Bering Sea.

Another 25 to 30 and 8 to 12 million years pass following *Hydrotherikornis* and *Petralca*, respectively, before the appearance of the next fossil alcids, which appear nearly simultaneously in both the western Atlantic and the eastern Pacific (Appendix 2.1, Table 2.2). However, like *Hydrotherikornis* and *Petralca*, these species were not of modern affinities and were described in extinct genera (Appendix 2.1). In the eastern Pacific, there are two alcid fossils known from middle Miocene deposits. The first of these fossils was from Baja, California, and was described as an alcid, but with indeterminate affinities. The second specimen was described in the extinct genus *Alcodes*, whose relationships within the Alcidae are uncertain (Olson 1985a, Chandler 1990b), but was tentatively considered by Howard (1968) to be closely related to the mancallids. In the Atlantic, there existed at least two species of alcids, both described in the extinct genus *Miocepphus*. *Miocepphus* was not closely related to *Cepphus*, as originally described by Wetmore (1940), but was part of the *Alca*-like radiation of Atlantic alcids (Howard 1978, Olson 1985a).

Following this initial middle Miocene radiation, alcid diversity dramatically increased in both the Atlantic and Pacific; however, the radiation within each of the ocean basins did not follow parallel paths (Table 2.2). The radiation in the Atlantic centered within the Alcinae, in particular, birds described as *Alca* (including the extinct genus *Australca*, which Olson and Rasmussen [2001] made synonymous with *Alca*). Of the nine alcid taxa from the late Miocene and early Pliocene deposits of the Atlantic, six are described as Alcini (*Alca*, *Pinguinus*, and *Alle*), while four of these six are considered *Alca* (see Appendix 2.1). The only Alcini missing from the Atlantic at this time was *Uria*. Also present in the Atlantic at this time was *Fratercula* (two species described as having affinities to the *F. arctica* and *F. cirrhata*, respectively) and an Aethiinae of indeterminate relations. During this same time, the situation in the Pacific was quite different, where at least 13 alcid species are recognized (Appendix 2.1) including *Aethia* (1 species), *Uria* (2), *Cepphus* (1), and *Cerorhinca* (2), as well as 7 species of mancallids (*Praemancalla*, *Mancalla*, and *Alcodes*). In addition to these taxa, fossils described as *Alca*, *Synthliboramphus*, and Fraterculini are present. Finally, there are late Pliocene alcid-bearing deposits in the Pacific, but not the Atlantic, and from within these deposits six additional alcid species are described, including two species of *Brachyramphus* and one species each of *Ptychoramphus*, *Synthliboramphus*, *Cerorhinca*, and *Mancalla* (see Appendix 2.1).

Olson and Rasmussen (2001) discussed the biogeographical implications of the Miocene and Pliocene Lee Creek deposits of North Carolina and highlighted two important points related to the history of the Alcidae. First, the two species of *Fratercula* (including *F. cirrhata*) and an unidentified species of Aethiinae in the early Pliocene of North Carolina require some explanation, given the fact that there is only one species of *Fratercula* and no species of Aethiinae in the Atlantic today (Table 2.2). Olson and Rasmussen (2001) considered that both taxa moved from the Pacific to the Atlantic, via the Arctic Ocean, sometime right before or during the early Pliocene. Second, given the possibility of a pre-Pleistocene movement of alcid taxa from the Pacific to the Atlantic, Olson and Rasmussen (2001) speculated that the absence of *Uria* and *Cepphus* from the Atlantic until the late Pleistocene and Recent, respectively, was a result of competition with *Alca*. Olson and Rasmussen (2001) reasoned that until appropriate "niches" became available, a product of the Pleistocene extinction of many of the *Alca* species, *Uria*, and *Cepphus* were unable to colonize the Atlantic.

For the remainder of this section I focus on this second point, and detail several important components of the alcid fossil record that contribute to our understanding of the origin of *Uria*. These components focus on the following four points associated with the fossil record: (1) the presence of *Alca* in the Pacific; (2) the presence and close association of *Uria* and *Cepphus* in the Pacific; (3) the abundance and taxonomic diversity of *Alca* in the Atlantic; and (4) the appearance of *Uria* in the Atlantic during the late Pleistocene. After I detail each of these points, I provide a hypothesis for the biogeographic history of *Uria*.

Howard (1968) described a coracoid and a humerus from late Miocene deposits in southern California as *Alca*. This material is fragmentary and Olson (1985a) was cautious in referring these specimens to a specific genus. Although Howard was reluctant to assign these fragments to a species or base a description of a new species on this material, she was definitive in her assignment of the fossils to *Alca*. If Howard's identification is correct, *Alca* is no longer restricted to the Atlantic, and this Pacific *Alca* is only slightly younger in age than the first *Alca*-like species from the Atlantic (*Miocepphus*) and older than all other species described to the genus *Alca*. Howard also described two species of murres from Tertiary deposits of California. The older of the two species was *U. brodkorbi* from the Miocene diatomite deposits of southern California and was described by Howard (1981) as a murre comparable in size to the Recent *Uria*. *Uria paleohesperis*, the second *Uria* species described by Howard (1982), was from the late Miocene San Mateo Formation of San Diego County and was younger in age and smaller than *U. brodkorbi*.

The fossil record of *Cepphus* follows closely that of *Uria*. While there are no *Cepphus* fossils from the Atlantic, Howard (1968, 1978) tentatively assigned fossil material from the Miocene of California to this genus. This material is roughly the same age as *U. brodkorbi* and suggests the origin of both taxa may be contemporaneous. In addition, *C. olsoni*, again described by Howard (1982), is from the same fossil locality as *U. paleohesperis*, further emphasizing the temporal and geographic similarity between murres and guillemots.

The most abundant alcid taxon from the Atlantic is *Alca*, in terms of both taxonomic diversity and numbers of specimens recovered. Thousands of *Alca* fossils have been recovered from the early Pliocene Lee Creek deposits of North Carolina (Olson and Rasmussen 2001), from which at least four species, including *A. torda*, are described (see Appendix 2.1). The first and only Atlantic appearance of a fossil correctly identified to *Uria* is *U. affinis*, a single humerus from the Pleistocene of Maine (12,000 years ago), which Olson (1985a) stated is likely referable to one of the extant species. It is clear from the fossil record from the western Atlantic that the Alcini underwent an extraordinary radiation, compared with that of the Pacific, and that this radiation began at essentially the same time as the Pacific radiation of the other alcid clades (Appendix 2.1).

The geographic distribution of fossil *Uria* is enigmatic given *Uria*'s relationships within the Alcini and its current distribution (north Atlantic, north Pacific, and Arctic Oceans; Gaston and Jones 1998). This fossil history has also led to several hypotheses for the evolution of *Uria* (e.g., Olson 1985a, Gaston and Jones 1998, Olson and Rasmussen 2001). These hypotheses generally concern (1) the relationships of *Uria* with the other Alcini, in particular, *Alca*; (2) the ocean of origin of the Alcini and *Uria*; (3) the historical interchange between the Atlantic and Pacific via the Arctic Ocean from the Miocene through the Pleistocene; and (4) the extinction and the loss of diversity of Alcini in the Atlantic. If *Uria* is indeed closely related to *Alca*, as both the morphological (Strauch 1985 and Chandler 1990b) and molecular (Moum 1994, Friesen et al. 1993, 1996) evidence conclusively indicate, and Howard (1968) was correct in identifying *Alca* fossils from the Pacific, the following scenario is most plausible: the Alcini evolved in the Pacific, and quickly moved into the Atlantic where it greatly diversified. In the Pacific, the diversification of Alcini was minimal and centered primarily on the genus *Uria*. *Uria* evolved in the Pacific (or the Arctic) Ocean and moved into the Atlantic sometime between the early Pliocene and the Pleistocene. Alternatively, *Uria* moved into the Atlantic at an earlier date, but remained in northerly latitudes, similar to the distribution of *U. lomvia* today, and therefore would not have occurred in the highly fossiliferous deposits of Lee Creek, North Carolina. I refer the reader to Gaston and Jones (1998) and Olson and Rasmussen (2001) for further discussion of this topic.

2.4 CONCLUSIONS

This has been a brief summary of fossil seabirds and an argument for the importance of fossils in the study of seabird ecology and evolution. Fossils are not simply a collection of bones. People who study fossils are concerned not only with naming and cataloging species. Fossils provide definite information on the history of a taxon or ecological community and, as such, are essential

FIGURE 2.3 This reconstruction of an early Eocene frigatebird (*Limnofregata azgosternon*) shows similarities to the tropicbirds which extend to its skeleton. For instance, both have coracoids of the same proportions and a four-notched sternum. (After Olson 1977.)

in our understanding of that taxon or community (Figure 2.3). I have shown that seabird communities in the California and Benguela Currents today are composed of different sets of species from those that existed in the past — related to a combination of geological (e.g., plate tectonics) and ecological (e.g., competition for space with gregarious marine mammals) processes. Therefore, the community structure of the systems today reflects these past processes and these past processes must be considered when evaluating hypotheses concerning this structure. Furthermore, past processes may also be useful in predicting changes in community structure resulting from future short- or long-term events such as habitat alteration and global climate change. Finally, it is quite apparent that we need to consider the fossil history of Pacific islands. Clearly, the seabird composition on these islands scarcely resembles that which existed prior to the expansion of Polynesian populations, and as stated by Olson, Steadman, James, and others, it would be folly to attempt to explain the relative diversity of seabirds there without considering the fossil record.

The fossil record also provides information on the presence and distribution of a particular taxon from times inaccessible to ecological study. We know from the fossil record of the Alcidae that the current distribution of alcid taxa, with *Alca* and *Alca*-like species in the Atlantic and most of the other alcid clades in the Pacific, has existed for many millions of years. Nevertheless, the presence of fossil *Alca* in the Pacific and the absence of fossil *Uria* and *Cepphus* from the Atlantic, for example, deviate from the current distributional patterns and provide important data in our understanding of the evolution of the Alcidae.

ACKNOWLEDGMENTS

I dedicate this paper to Hildegarde Howard and Storrs Olson, two giants in the field of avian paleontology whom I have had the honor and pleasure of knowing. Storrs Olson's impact on my studies of seabird paleontology is immeasurable, and without his help this paper would have been impossible. I thank Tony Gaston, Vicki Friesen, and Storrs Olson for reviewing an earlier draft of this paper, and Cheryl Niemi, Storrs Olson, Betty Anne Schreiber, and Joanna Burger for providing comments on the final draft. I thank Chris Thompson and Cheryl Niemi for making several clever suggestions in formatting Appendix 2.1. Finally, I thank Betty Anne Schreiber and Joanna Burger for inviting me to participate in this project, and for demonstrating extreme patience with my many missed deadlines.

LITERATURE CITED

AINLEY, D. G., AND R. J. BOEKELHEIDE. 1990. Seabirds of the Farallon Island. Ecology, Dynamics, and Structure of an Upwelling-System Community. Stanford University Press, Stanford, CA.

ALVAREZ, R. 1977. A Pleistocene avifauna from Jalisco, Mexico. Contributions of the Museum of Paleontology, University of Michigan 24: 205–220.

BALLMANN, P. 1976. The contribution of fossil birds to avian classification. Proceedings of the International Ornithological Congress 16: 196–200.

BARRON, J. A., AND J. G. BALDAUF. 1989. Tertiary cooling steps and paleoproductivity as reflected by diatoms and biosiliceous sediments. Pp. 341–354 in Productivity of the Ocean: Present and Past (W. H. Berger, V. S. Smetacek, and G. Wefer, Eds.). John Wiley & Sons, Chichester.

BECKER, J. J. 1987. Neogene Avian Localities of North America. Smithsonian Institution Press, Washington, D.C.

BELOPOL'SKII, L.O. 1961. The Ecology of Sea Colony Birds of the Barents Sea. Israel Program for Scientific Translations, Jerusalem.

BERGGREN, W. A., D. V. KENT, C. C. SWISHER, III, AND M. P. AUBRY. 1995. A revised Cenozoic geochronology and chronostratigraphy. Pp. 129–212 in Geochronology Time Scales and Global Stratigraphic Correlation, S.E.P.M., Special Publication No. 54 (W. A. Berggren, D. V. Kent, M. P. Aubry, and J. Hardenbol, Eds.). Society for Sedimentary Geology, Tulsa, OK.

BICKART, K. J. 1990. The birds of the Late Miocene-Early Pliocene Big Sandy Formation, Mohave County, Arizona. Ornithological Monographs 44: 1–72.

BOCHENSKI, Z. 1997. List of European fossil bird species. Acta Zoologica, Cracov 40: 292–333.

BRODKORB, P. 1955. The avifauna of the Bone Valley Formation. Florida Geological Survey Report of Investigations 14: 1–57.

BRODKORB, P. 1963. Catalogue of fossil birds: Part 1 (Archaeopterygiformes through Ardeiformes). Bulletin of the Florida State Museum, Biological Sciences 7: 179–293.

BRODKORB, P. 1967. Catalogue of fossil birds: Part 3 (Ralliformes, Ichthyornithiformes, Charadriiformes). Bulletin of the Florida State Museum, Biological Sciences 11: 99–220.

CHANDLER, R. C. 1990a. Fossil birds of the San Diego Formation, late Pliocene, Blancan, San Diego County, California. Ornithological Monographs 44: 73–161.

CHANDLER, R. C. 1990b. Phylogenetic Analysis of the Alcids. Ph.D. dissertation, University of Kansas, Lawrence.

CHENEVAL, J. 1984. Les oiseaux aquatiques (Gaviiformes a Ansériformes) du gisement Aquitanien de Saint-Gérand-le-Puy (Allier, France): Révision systematique [The water birds (Gaviiformes to Ansériformes) from the Aquitanien layer of Saint-Gérand-le-Puy (Allier, France): systematic revision]. Palaeovertebrata 14: 33–115.

CHENEVAL, J. 1993. L'avifaune Mio-Pliocène de la formation Pisco (Pérou) étude préliminaire [Preliminary study of the Mio-Pliocene avifauna of the Pisco Formation (Peru)]. Documents des Laboratoires de Géologie de la Faculte des Sciences de Lyon 125: 85–95.

CHENEVAL, J. 1995. A fossil shearwater (Aves: Procellariiformes) from the Upper Oligocene of France and the Lower Miocene of Germany. Courier Forschungsinstitut Senckenberg 181: 187–198.

CIONE, A. L., AND E. P. TONNI. 1981. Un pingüino de la formación Puerto Madryn (Miocene Tardío) de Chubut, Argentina. Commentarios acerca del origen, la paleoecología y zoogeografía de los Spheniscidae [A penguin from the Puerto Madryn formation (late Miocene) of Chubut, Argentina. Commentary about the origin, paleoecology, and zoogeography of the Spheniscidae]. An. Congr. Latino-Am. Paleont., Porto Alegre 2: 591–604.

COULSON, J. C., AND C. S. THOMAS. 1985. Changes in the biology of the Kittiwake *Rissa tridactyla*: a 31-year study of a breeding colony. Journal of Animal Ecology 54: 9–26.

DEMÉRÉ, T. A., M. A. ROEDER, R. M. CHANDLER, AND J. A. MINCH. 1984. Paleontology of the middle Miocene Los Indios Member of the Rosarito Beach Formation, northwestern Baja California, Mexico. Pp. 47–56 *in* Miocene and Cretaceous Depositional Environments, Northwestern Baja California, Mexico (J. A. Minch and J. R. Ashby, Eds.). Pacific Section, American Association of Petroleum Geologists, Los Angeles, CA.

EMSLIE, S. D. 1992. Two new late Blancan avifaunas from Florida and the extinction of wetland birds in the Plio-Pleistocene. Pp. 249–269 *in* Papers in Avian Paleontology Honoring Pierce Brodkorb. Natural History Museum of Los Angeles County, Science Series No. 36 (K. E. Campbell, Jr., Ed.). Natural History Museum of Los Angeles County, Los Angeles, CA.

EMSLIE, S. D. 1995. A catastrophic death assemblage of a new species of cormorant and other seabirds from the late Pliocene of Florida. Journal of Vertebrate Paleontology 15: 313–330.

FEDUCCIA, A., AND A. B. MCPHERSON. 1993. A petrel-like bird from the late Eocene of Louisiana: earliest record for the order Procellariiformes. Proceedings of the Biological Society of Washington 106: 749–751.

FORDYCE, R. E., AND C. M. JONES. 1990. Penguin history and new fossil material from New Zealand. Pp. 419–446 *in* Penguin Biology (L. S. Davis and J. T. Darby, Eds.). Academic Press, San Diego, CA.

FRIESEN, V. L., A. J. BAKER, AND J. F. PIATT. 1996. Phylogenetic relationships within the Alcidae (Charadriiformes: Aves) inferred from total molecular evidence. Molecular Biology and Evolution 13: 359–367.

FRIESEN, V. L., W. A. MONTEVECCHI, AND W. S. DAVIDSON. 1993. Cytochrome *b* nucleotide sequence variation among the Atlantic Alcidae. Hereditas 116: 245–252.

GASTON, A. J., AND I. L. JONES. 1998. The Auks. Oxford University Press, Oxford.

GOEDERT, J. L. 1988. A new Late Eocene species of Plotopteridae (Aves: Pelecaniformes) from northwestern Oregon. Proceedings of the California Academy of Science 45: 97–102.

GOEDERT, J. L. 1989. Giant late Eocene marine birds (Pelecaniformes: Pelagornithidae) from northwestern Oregon. Journal of Paleontology 63: 939–944.

GOULD, S. J. 1989. Wonderful Life: The Burgess Shale and the Nature of History. W.W. Norton and Company, New York.

GRIGORESCU, D., AND E. KESSLER. 1977. The middle Sarmatian avian fauna of South Dobrogea. Revue Roumaine de Géologie Géophysique et Géographie 21: 93–108.

GRIGORESCU, D., AND E. KESSLER. 1988. New contributions to the knowledge of the Sarmatian birds from South Dobrogea in the frame of the eastern Paratethyan avifauna. Revue Roumaine de Géologie Géophysique et Géographie, Géologie 32: 91–97.

HARRIS, M. P. 1991. Population changes in British Common Murres and Atlantic Puffins, 1969–88. Pp. 52–58 *in* Studies of High-Latitude Seabirds, Vol. 2: Conservation Biology of Thick-billed Murres in the Northwest Atlantic (A. J. Gaston and R. D. Elliot, Eds.). Canadian Wildlife Service, Ottawa.

HARRISON, C. J. O. 1985. A bony-toothed bird (Odontopterygiformes) from the Palaeocene of England. Tertiary Research 7: 23–25.

HARRISON, C. J. O., AND C. A. WALKER. 1976. A review of the bony-toothed birds (Odontopterygiformes): with descriptions of some new species. Tertiary Research Special Paper 2: 1–62.

HARRISON, C. J. O., AND C. A. WALKER. 1977. Birds of the British lower Eocene. Tertiary Research Special Paper 3: 1–52.

HARRISON, C. S. 1990. Seabirds of Hawaii. Natural History and Conservation. Cornell University Press, Ithaca, NY.

HOLDAWAY, R. N., AND T. H. WORTHY. 1994. A new fossil species of shearwater *Puffinus* from the late Quaternary of the South Island, New Zealand, and notes on the biogeography and evolution of the *Puffinus gavia* superspecies. Emu 94: 201–215.

HOPSON, J. A. 1964. *Pseudodontornis* and other large marine birds from the Miocene of South Carolina. Postilla 83: 1–19.

HOWARD, H. 1946. A review of the Pleistocene birds of Fossil Lake, Oregon. Carnegie Institution of Washington Publication 551: 141–195.

HOWARD, H. 1949. New avian records of the Pliocene of California. Carnegie Institution of Washington Publication 584: 177–199.

HOWARD, H. 1958. Miocene sulids of southern California. Los Angeles County Natural History Museum, Contributions in Science 25: 1–15.

HOWARD, H. 1965. A new species of cormorant from the Pliocene of Mexico. Bulletin of the Southern California Academy of Sciences 64: 50–55.

HOWARD, H. 1966. Additional avian records from the Miocene of Sharktooth Hill, California. Los Angeles County Natural History Museum, Contributions in Science 114: 1–11.

HOWARD, H. 1968. Tertiary birds from Laguna Hills, Orange County, California. Los Angeles County Natural History Museum, Contributions in Science 142: 1–21.

HOWARD, H. 1969. A new avian fossil from Kern County, California. Condor 71: 68–69.

HOWARD, H. 1970. A review of the extinct genus, *Mancalla*. Los Angeles County Natural History Museum, Contributions in Science 203: 1–12.

HOWARD, H. 1971. Pliocene avian remains from Baja California. Los Angeles County Natural History Museum, Contributions in Science 217: 1–17.

HOWARD, H. 1978. Late Miocene birds from Orange County, California. Los Angeles County Natural History Museum, Contributions in Science 290: 1–26.

HOWARD, H. 1981. A new species of murre, genus *Uria,* from the late Miocene of California (Aves: Alcidae). Bulletin of the Southern California Academy of Sciences 80: 1–12.

HOWARD, H. 1982. Fossil birds from the Tertiary marine beds at Oceanside, San Diego County, California, with descriptions of two new species of the genera *Uria* and *Cepphus* (Aves: Alcidae). Los Angeles County Natural History Museum, Contributions in Science 341: 1–15.

HOWARD, H. 1984. Additional records from the Miocene of Kern County, California with the description of a new species of fulmar (Aves: Procellariidae). Bulletin of the Southern California Academy of Sciences 83: 84–89.

HOWARD, H., AND L. G. BARNES. 1987. Middle Miocene marine birds from the foothills of the Santa Ana Mountains, Orange County, California. Los Angeles County Natural History Museum, Contributions in Science 383: 1–9.

HOWARD, H., AND S. L. WARTER. 1969. A new species of bony-toothed bird (Family Pseudodontornithidae) from the Tertiary of New Zealand. Records of the Canterbury Museum 8: 345–357.

JABLONSKI, D. 1986. Larval ecology and macroevolution in marine invertebrates. Bulletin of Marine Science 39: 565–587.

JAMES, H. F. 1987. A late Pleistocene avifauna from the island of Oahu, Hawaiian Islands. Documents des Laboratoires de Géologie de la Faculte des Sciences de Lyon 99: 221–230.

JAMES, H. F. 1995. Prehistoric extinctions and ecological changes on oceanic islands. Ecological Studies 115: 87–102.

JENKINS, R. J. F. 1974. A new giant penguin from the Eocene of Australia. Paleontology 17: 291–310.

KIMURA, M., AND K. SAKURAI. 1998. An extinct fossil bird (Plotopteridae) from the Tokoro Formation (late Oligocene) in Abashiri City, northeastern Hokkaido, Japan. Journal of Hokkaido University of Education (Section IIB) 48: 11–16.

KOHL, R. F. 1974. A new Late Pleistocene fauna from Humboldt County, California. Veliger 17: 211–219.

LAMBRECHT, K. 1930. Studien über fossile Riesenvögel. Geologica Hungarica Series Palaeontologica 7: 1–37.

LIVEZEY, B. C. 1988. Morphometrics of flightlessness in the Alcidae. Auk 105: 681–698.

MATSUOKA, H., F. SAKAKURA, AND F. OHE. 1998. A Miocene pseudodontorn (Pelecaniformes: Pelagornithidae) from the Ichishi Group of Misata, Mie Prefecture, central Japan. Paleontological Research 2: 246–252.

MCKEE, J. W. A. 1985. A pseudodontorn (Pelecaniformes: Pelagornithidae) from the middle Pliocene of Hawera, Taranaki, New Zealand). New Zealand Journal of Zoology 12: 181–184.

MICHEAUX, J., R. HUTTERER, AND N. LOPEZ-MARTINEZ. 1991. New fossil faunas from Fuerteventura, Canary Islands: evidence for Pleistocene age of endemic rodents and shrews. Comptes Rendus de l'Academie des Sciences Series 2 312: 801–806.

MILLER, A. H. 1931. An auklet from the Eocene of Oregon. University of California Publications Bulletin of the Department of Geological Sciences 20: 23–26.

MILLER, A. H. 1966. The fossil pelicans of Australia. Memoirs of the Queensland Museum 14: 181–190.

MILLER, A. H., AND C. G. SIBLEY. 1941. A Miocene gull from Nebraska. Auk 58: 563–566.

MILLER, L. H. 1929. A new cormorant from the Miocene of California. Condor 31: 167–172.

MILLER, L. H. 1951. A Miocene petrel from California. Condor 53: 78–80.

MLÍKOVSKÝ, J. 1992. The present state of knowledge of the Tertiary birds of central Europe. Pp. 433–458 *in* Papers in Avian Paleontology Honoring Pierce Brodkorb. Natural History Museum of Los Angeles County, Science Series No. 36 (K. E. Campbell, Jr., Ed.). Natural History Museum of Los Angeles County, Los Angeles, CA.

MLÍKOVSKÝ, J. 1997. A new tropicbird (Aves: Phaethontidae) from the late Miocene of Austria. Annalen des Naturhistorischen Museums in Wien 98A: 151–154.

MLÍKOVSKÝ, J., AND J. KOVAR. 1987. Eine neue Alkenart (Aves: Alcidae) aus dem Ober-Oligozän Österreichs. Annalen des Naturhistorischen Museums in Wien 88A: 131–147.

MOUM, T., S. JOHANSEN, K. E. ERIKSTAD, AND J. F. PIATT. 1994. Phylogeny and evolution of the auks (subfamily Alcinae) based on mitochondrial DNA sequences. Proceedings of the National Academy of Sciences 91: 7912–7916.

MOURER-CHAUVIRÉ, C. 1982. Les oiseaux fossiles des Phosphorites du Quercy (Éocène Supérieur a Oligocène Supérieur): implications paléobiogeographiques [The fossil birds from the Phosphorites du Quercy (Upper Eocene to Upper Oligocene): Paleobiographical implications]. Géobios Mémoire Spécial 6: 413–426.

MOURER-CHAUVIRÉ, C. 1995. Dynamics of the avifauna during the Paleogene and the early Neogene of France. Settling of the recent fauna. Acta Zoologica Cracoviensia 38: 325–342.

MYRCHA, A., A. TATUR, AND R. DEL VALLE. 1990. A new species of fossil penguin from Seymour Island, west Antarctica. Alcheringa 14: 195–205.

MYRCHA, A., P. JADWISZCZAK, C. TAMBUSSI, J. NORIEGA, A. TATUR, A. GAZDZICKI, AND R. DEL VALLE. (In press). Taxonomic revision of Antarctic Eocene penguins based on tarsometatarsus morphology. Palaeontologia Polonica.

NUNN, G. B., J. COOPER, P. JOUVENTIN, C. J. R. ROBERTSON, AND G. G. ROBERTSON. 1996. Evolutionary relationships among extant albatrosses (Procellariiformes: Diomedeidae) established from complete cytochrome-*B* gene sequences. Auk 113: 784–801.

OKAZAKI, Y. 1989. An occurrence of fossil bony-toothed bird (Odontopterygiformes) from the Ashiya Group (Oligocene), Japan. Bulletin of the Kitakyushu Museum of Natural History 9: 123–126.

OLSON, S. L. 1975. Paleornithology of St. Helena Island, South Atlantic Ocean. Smithsonian Contributions to Paleobiology 23: 1–49.

OLSON, S. L. 1977. A lower Eocene Frigatebird from the Green River Formation of Wyoming (Pelecaniformes: Fregatidae). Smithsonian Contributions to Paleobiology 35: 1–33.

OLSON, S. L. 1980. A new genus of penguin-like pelecaniform bird from the Oligocene of Washington (Pelecaniformes: Plotopteridae). Los Angeles County Natural History Museum, Contributions in Science 330: 51–57.

OLSON, S. L. 1983. Fossil seabirds and the changing marine environments in the late Tertiary of South Africa. South African Journal of Science 79: 399–402.

OLSON, S. L. 1984a. A brief synopsis of the fossil birds from the Pamunky River and other Tertiary marine deposits in Virginia. Pp. 217–223 *in* Stratigraphy and Paleontology of the Outcropping Tertiary Beds in the Pamunkey River Region, Central Virginia Coastal Plain — Guidebook for the 1984 Field Trip Atlantic Coastal Plain Geological Association (L. W. Ward and K. Krafft, Eds.). Atlantic Coastal Plain Geological Association, Norfolk, VA.

OLSON, S. L. 1984b. Evidence of a large albatross in the Miocene of Argentina (Aves: Diomedeidae). Proceedings of the Biological Society of Washington 97: 741–743.

OLSON, S. L. 1985a. The fossil record of birds. Pp. 79–252 *in* Avian Biology, 8 (D. S. Farner, J. R. King, and K. C. Parkes, Eds.). Academic Press, Orlando, FL.

OLSON, S. L. 1985b. Early Pliocene Procellariiformes (Aves) from Langebaanweg, South-Western Cape Province, South Africa. Annals of the South African Museum 95: 123–145.

OLSON, S. L. 1985c. An early Pliocene marine avifauna from Duinefontein, Cape Province, South Africa. Annals of the South African Museum 95: 147–164.

OLSON, S. L. 1985d. A new genus of tropicbird (Pelecaniformes: Phaethontidae) from the middle Miocene Calvert Formation of Maryland. Proceedings of the Biological Society of Washington 98: 851–855.

OLSON, S. L. 1986. A replacement name for the fossil penguin *Microdytes* Simpson (Aves: Spheniscidae). Journal of Paleontology 60: 785.

OLSON, S. L. 1999. A new species of pelican (Aves: Pelecanidae) from the lower Pliocene of North Carolina and Florida. Proceedings of the Biological Society of Washington 112: 503–509.

OLSON, S. L., AND Y. HASEGAWA. 1979. Fossil counterparts of giant penguins from the North Pacific. Science 206: 688–689.

OLSON, S. L., AND Y. HASEGAWA. 1985. A femur of *Plotopterum* from the early middle Miocene of Japan (Pelecaniformes: Plotopteridae). Bulletin of the Natural Science Museum, Tokyo, Series C 11: 137–140.

OLSON, S. L., AND Y. HASEGAWA. 1996. A new genus and two new species of gigantic Plotopteridae from Japan (Aves: Pelecaniformes). Journal of Vertebrate Paleontology 16: 742–751.

OLSON, S. L., AND H. F. JAMES. 1982a. Fossil birds from the Hawaiian Islands: evidence for wholesale extinction by man before western contact. Science 217: 633–635.

OLSON, S. L., AND H. F. JAMES. 1982b. Prodromus of the fossil avifauna of the Hawaiian Islands. Smithsonian Contribution to Zoology 365: 1–59.

OLSON, S. L., AND H. F. JAMES. 1991. Descriptions of thirty-two new species of birds from Hawaiian Islands: Part I. Non-passeriformes. Ornithological Monographs 45: 1–88.

OLSON, S. L., AND D. C. PARRIS. 1987. The Cretaceous birds of New Jersey. Smithsonian Contributions to Paleobiology 63: 1–22.

OLSON, S. L., AND P. C. RASMUSSEN. 2001. Miocene and Pliocene Birds from the Lee Creek Mine, North Carolina, *in* Geology and Paleontology of the Lee Creek Mine, North Carolina, III (C. E. Ray and D. J. Bohaska, Eds.). Smithsonian Contributions to Paleobiology, 90.

OLSON, S. L., AND D. W. STEADMAN. 1979. The fossil record of the Glareolidae and Haematopodidae (Aves: Charadriiformes). Proceedings of the Biological Society of Washington 91: 972–981.

ONO, K. 1983. A Miocene bird (gannet) from Chichibu Basin, central Japan. Bulletin of Saitama Museum of Natural History 1: 11–15.

ONO, K. 1989. A bony-toothed bird from the middle Miocene, Chichibu Basin, Japan. Bulletin of the Natural Science Museum, Tokyo, Series C 15: 33–38.

ONO, K., AND O. SAKAMOTO. 1991. Discovery of five Miocene birds from Chichibu Basin, central Japan. Bulletin of Saitama Museum of Natural History 9: 41–49.

PETERS, D. S., AND A. HAMEDANI. 2000. *Frigidafons babaheydariensis* n. sp., ein Sturmvogel aus dem Oligozan des Iran (Aves: Procellariidae) [*Frigidafons babaheydariensis* n. sp., a petrel from the Oligocene of Iran (Aves: Procellariidae)]. Senckenbergiana Lethaea 80: 29–37.

PRESS, F., AND R. SIEVER. 1982. Earth. W. H. Freeman and Co., San Francisco, CA.

RASMUSSEN, D. T., S. L. OLSON, AND E. L. SIMONS. 1987. Fossil birds from the Oligocene Jebel Qatrani Formation, Fayum Province, Egypt. Smithsonian Contributions to Paleobiology 62: 1–20.

RASMUSSEN, P. C. 1998. Early Miocene avifauna from the Pollack Farm site, Delaware. Pp. 149–151 *in* Geology and Paleontology of the Lower Miocene Pollack Farm Fossil Site, Delaware. Special Publication No. 21 (R. N. Benson, Ed.). Delaware Geological Survey.

RICH, P. V., AND G. F. VAN TETS. 1981. The fossil pelicans of Australasia. Records of the South Australian Museum 18: 235–264.

RICHDALE, L. E. 1949. A study of a group of penguins of known age. Biological Monographs 1: 1–88.

RICHDALE, L. E. 1954. Breeding efficiency in yellow-eyed penguins. Ibis 96: 206–224.

RICHDALE, L. E. 1957. A Population Study of Penguins. Oxford University Press, Oxford.

SCARLETT, R. J. 1972. Bone of a presumed Odontopterygian bird from the Miocene of New Zealand. New Zealand Journal of Geology and Geophysics 15: 269–274.

SERVENTY, D. L. 1956. Age at first breeding of the short-tailed shearwater, *Puffinus tenuirostris*. Ibis 98: 532–533.

SHUFELDT, R. W. 1915. Fossil birds of the Marsh collection of Yale University. Transactions of the Connecticut Academy of Arts and Sciences 19: 1–110.

SIESSER, W. G. 1980. Late Miocene origin of the Benguela upswelling (*sic*) system off northern Namibia. Science 208: 283–285.

SIMPSON, G. G. 1972. Conspectus of Patagonian fossil penguins. America Museum Novitates 2488: 1–37.

SIMPSON, G. G. 1975. Fossil penguins. Pp. 19–41 *in* The Biology of Penguins (B. Stonehouse, Ed.). Macmillan, London.

SIMPSON, G. G. 1981. Notes on some fossil penguins, including a new genus from Patagonia. Ameghiniana 18: 266–272.

STEADMAN, D. W. 1995. Prehistoric extinctions of Pacific island birds: biodiversity meets zooarchaeology. Science 267: 1123–1131.

STEADMAN, D. W., AND S. L. OLSON. 1985. Bird remains from an archaeological site on Henderson Island, South Pacific: man-caused extinctions on an "uninhabited" island. Proceedings of the National Academy of Science 82: 6191–6195.

STORER, R. W. 1945. Structural modification in the hindlimb in the Alcidae. Ibis 87: 433–456.

STRAUCH, J. G., JR. 1985. The phylogeny of the Alcidae. Auk 102: 520–539.

TYRBERG, T. 1998. Pleistocene birds of the Palearctic: a catalogue. Publications of the Nuttall Ornithological Club 27: 1–720.

USPENSKI, S. M. 1958. The Bird Bazaars of Novaya Zemlya. (Translation) Russian Game Report, Vol. 4. Queen's Printer, Ottawa.

VAN TETS, G. F., C. W. MEREDITH, P. J. FULLAGAR, AND P. M. DAVIDSON. 1988. Osteological differences between *Sula* and *Morus*, and a description of an extinct new species of *Sula* from Lord Howe and Norfolk Islands, Tasman Sea. Notornis 35: 35–57.

WALKER, C. A., G. M. WRAGG, AND C. J. O. HARRISON. 1990. A new shearwater from the Pleistocene of the Canary Islands and its bearing on the evolution of certain *Puffinus* shearwaters. Historical Biology 3: 203–224.

WARHEIT, K. I. 1990. The Phylogeny of the Sulidae (Aves: Pelecaniformes) and the Morphometry of Flight-Related Structures in Seabirds: A Study of Adaptation. Ph.D. dissertation. University of California, Berkeley.

WARHEIT, K. I. 1992. A review of the fossil seabirds from the Tertiary of the North Pacific: Plate Tectonics, Paleoceanography, and Faunal Change. Paleobiology 18: 401–424.

WARHEIT, K. I., AND D. R. LINDBERG. 1988. Interactions between seabirds and marine mammals through time: interference competition at breeding sites. Pp. 292–328 *in* Seabirds and Other Marine Vertebrates. Competition, Predation, and Other Interactions (J. Burger, Ed.). Columbia University Press, New York.

WETMORE, A. 1940. Fossil bird remains from the Tertiary deposits of the United States. Journal of Morphology 66: 25–37.

WILKINSON, H. E. 1969. Description of an Upper Miocene albatross from Beaumaris, Victoria, Australia, and a review of the fossil Diomedeidae. Memoirs of the National Museum of Victoria 29: 41–51.

WOOLLER, R. D., J. S. BRADLEY, AND J. P. CROXALL. 1992. Long-term population studies of seabirds. Trends in Ecology and Evolution 7: 111–114.

ZUSI, R. L., AND K. I. WARHEIT. 1992. On the evolution of the intramandibular joints of pseudodontorns (Aves: Odontopterygia). Pp. 351–360 *in* Papers in Avian Paleontology Honoring Pierce Brodkorb. Natural History Museum of Los Angeles County, Science Series No. 36 (K. E. Campbell, Jr., Ed.). Natural History Museum of Los Angeles County, Los Angeles, CA.

APPENDIX 2.1
List of fossil seabirds
See text and notes at bottom of table for details. [a, d, e]

Genus or higher taxon [b,c]	Species	Cretaceous latest	Palaeocene late	Eocene early	Eocene middle	Eocene late	Oligocene early	Oligocene late	Miocene early	Miocene middle	Miocene late	Pliocene early	Pliocene late	Pleistocene early	Pleistocene middle	Pleistocene late	Holocene	Geographic Region [f]	Specific Locality [f]	Comment [h]	Citation [g]
Charadriiformes																					
Haematopodidae																					
Haematopus	*sulcatus*												▨					w. Atlantic	Florida	1	Olson & Steadman 1979
Haematopus	aff. *palliatus*											■						w. Atlantic	N. Carolina		Olson & Rasmussen 2001
Haematopus	aff. *ostralegus*											■						w. Atlantic	N. Carolina		Olson & Rasmussen 2001
Stercorariidae																					
Stercorarius	sp. small									■								w. Atlantic	Maryland		Olson 1985a
Stercorarius	sp. big									■								w. Atlantic	Maryland		Olson 1985a
Catharacta	sp.											■						w. Atlantic	N. Carolina		Olson &Rasmussen 2001
Stercorarius	aff. *pomarinus*											■						w. Atlantic	N. Carolina		Olson & Rasmussen 2001
Stercorarius	aff. *parasiticus*											■						w. Atlantic	N. Carolina		Olson &Rasmussen 2001
Stercorarius	aff. *longicaudus*											■						w. Atlantic	N. Carolina		Olson &Rasmussen 2001
Stercorarius	sp.												■					w. Atlantic	Florida		Emslie 1995
Stercorarius	*shufeldti*														■			e. N. Pacific	Oregon		Howard 1946
Laridae																					
genus indeterminate	sp.				■	■												Paratethys	France	2	Mourer-Chauviré 1982
Gaviota	*lipsiensis*						▨	■										int. Europe	Germany		Bochenski 1997
Rupelornis	*definitus*							■										e. N. Atlantic	Belgium	3	Olson 1985a
Larus	*pristinus*								■									e. N. Pacific	Oregon	4	Olson 1985a
genus indeterminate	sp.								■									w. Atlantic	Delaware	5	Rasmussen 1998
Larus	*dolnicensis*								■									int. Europe	Bohemia	6	Olson 1985a

APPENDIX 2.1 (Continued)
List of fossil seabirds
See text and notes at bottom of table for details. [a,d,e]

Genus or higher taxon [b,c]	Species	Geologic age (filled = occurrence)	Geographic Region [f]	Specific Locality [f]	Comment [h]	Citation [g]
Larus	desnoyersii	early Miocene	Paratethys	France	7	Olson 1985a
Larus	elegans	early Miocene	Paratethys	France	8	Olson 1985a
Larus	totanoides	early Miocene	Paratethys	France	8	Olson 1985a
Gaviota	niobrara	late Miocene	int. N. America	Nebraska		Miller & Sibley 1941
cf. Larus	sp.	late Miocene	int. N. America	Arizona		Bickart 1990
Larus	elmorei	early Pliocene	Paratethys	Romania		Grigorescu & Kessler 1977
Larus	aff. argentatus	early Pliocene	w. Atlantic	Florida		Olson 1985a
Larus	aff. argentatus	early Pliocene	w. Atlantic	N. Carolina		Olson & Rasmussen 2001
Larus	aff. delawarensis	early Pliocene	w. Atlantic	N. Carolina	9	Olson & Rasmussen 2001
Larus	aff. atricilla	early Pliocene	w. Atlantic	N. Carolina		Olson & Rasmussen 2001
Larus	magn. ribidundus	early Pliocene	w. Atlantic	N. Carolina		Olson & Rasmussen 2001
Larus	aff. minutus	early Pliocene	w. Atlantic	N. Carolina	10	Olson & Rasmussen 2001
Larus	sp.	early Pliocene	w. Atlantic	N. Carolina		Olson & Rasmussen 2001
cf. Sterna	aff. maxima	early Pliocene	w. Atlantic	N. Carolina		Olson & Rasmussen 2001
Sterna	aff. nilotica	early Pliocene	w. Atlantic	N. Carolina		Olson & Rasmussen 2001
Larus	sp.	late Pliocene	e. N. Pacific	Calif.		Chandler 1990a
Rissa	estesi	late Pliocene	e. N. Pacific	Calif.		Chandler 1990a
Sterna	sp.	late Pliocene	e. N. Pacific	Calif.		Chandler 1990a
Larus	perpetuus	late Pliocene	w. Atlantic	N. Carolina		Emslie 1995
Larus	lacus	late Pliocene	w. Atlantic	Florida		Emslie 1995
Larus	robustus	late Pleistocene	e. N. Pacific	Oregon		Brodkorb 1967
Larus	oregonus	late Pleistocene	e. N. Pacific	Oregon		Brodkorb 1967
Pseudosterna	degener	middle–late Pleistocene	w. S. Atlantic	Argentina	11	Olson 1985a
Pseudosterna	pampeana	late Pleistocene	w. S. Atlantic	Argentina	11	Olson 1985a

Geologic age column headers (as printed across the top): Cretaceous (latest); Paleocene [e] (late); Eocene [e] (early, middle, late); Oligocene [e] (early, late); Miocene (early, middle, late); Pliocene (early, late); Pleistocene [e] (early, middle, late); Holocene [e]

APPENDIX 2.1 (Continued)

List of fossil seabirds

See text and notes at bottom of table for details. [a, d, e]

Alcidae

Genus or higher taxon [b, c]	Species	Cretaceous latest	Paleocene late	Eocene early	Eocene middle	Eocene late	Oligocene early	Oligocene late	Miocene early	Miocene middle	Miocene late	Pliocene early	Pliocene late	Pleistocene early	Pleistocene middle	Pleistocene late	Holocene late	Geographic Region [f]	Specific Locality [f]	Comment [h]	Citation [g]
Hydrotherikornis	oregonus				■													e. N. Pacific	Oregon	12	Olson 1985a
Petralca	austriaca							■										Paratethys	Austria		Mlikovsk& Kovar 1987
genus indeterminate	sp.									■								e. Pacific	Baja Calif.		Deméré et al. 1984
Alcodes	aff. A. ulnulus										■							e. N. Pacific	Calif.		Howard & Barnes 1987
Miocepphus	mcclungi										■							w. Atlantic	Maryland	13	Olson 1985a
Miocepphus	new sp.										■							w. Atlantic	Maryland		Olson 1984a
Aethia	rossmoori										■							e. N. Pacific	Calif.		Howard 1968
Alca	sp.										■							e. N. Pacific	Calif.		Howard 1968
Alcodes	ulnulus										■							e. N. Pacific	Calif.		Warheit 1992
Cepphus (?)	sp.										■							e. N. Pacific	Calif.		Warheit 1992
Cerorhinca	dubia										■							e. N. Pacific	Calif.		Warheit 1992
Fraterculini	sp.										■							e. N. Pacific	Calif.		Howard 1978
Praemancalla	lagunensis										■							e. N. Pacific	Calif.		Howard 1966
Praemancalla	wetmorei										■							e. N. Pacific	Calif.		Warheit 1992
Uria	brodkorbi										■							e. N. Pacific	Calif.		Howard 1981
Uria (?)	sp.										■							e. N. Pacific	Calif.		Howard 1978
Aethia (?)	sp.											■	■					e. N. Pacific	Calif.		Warheit 1992
Cepphus	olsoni											■	■					e. N. Pacific	Calif.		Warheit 1992
Mancalla	californicus											■	■					e. N. Pacific	Calif.		Warheit 1992
Mancalla	cf. cedrocensis											■	■					e. N. Pacific	Calif.		Warheit 1992
Praemancalla	cf. wetmorei											■	■					e. N. Pacific	Calif.		Warheit 1992
Uria	paleohesperis											■	■					e. N. Pacific	Calif.		Warheit 1992
Cerorhinca	minor											■	■					e. Pacific	Mexico		Howard 1971
Mancalla	cedrocensis											■	■					e. Pacific	Mexico		Warheit 1992
Synthliboramphus	sp.											■	■					e. Pacific	Mexico	14	Howard 1971

APPENDIX 2.1 (Continued)
List of fossil seabirds
See text and notes at bottom of table for details.[a,d,e]

Genus or higher taxon[b,c]	Species	Cretaceous latest	Paleocene late	Eocene early	Eocene middle	Eocene late	Oligocene early	Oligocene late	Miocene early	Miocene middle	Miocene late	Pliocene early	Pliocene late	Pleistocene early	Pleistocene middle	Pleistocene late	Holocene	Geographic Region[f]	Specific Locality[f]	Comment[h]	Citation[g]
Mancalla	diegensis										■	■	■				■	e. N. Pacific	Calif.		Warheit 1992
Mancalla	milleri											■	■					e. N. Pacific	Calif.		Warheit 1992
Alca	ausonia											■						Paratethys & w. Atl.	Italy, N. Carolina	1, 15	Olson & Rasmussen 2001
Aethiinae	sp.											■						w. Atlantic	N. Carolina	16	Olson & Rasmussen 2001
Alca	antiqua											■						w. Atlantic	N. Carolina	17	Olson & Rasmussen 2001
Alca	aff. torda											■						w. Atlantic	N. Carolina		Olson & Rasmussen 2001
Alca	new sp.											■						w. Atlantic	N. Carolina	18	Olson & Rasmussen 2001
Alle	aff. alle											■						w. Atlantic	N. Carolina		Olson & Rasmussen 2001
Fratercula	aff. arctica											■						w. Atlantic	N. Carolina		Olson & Rasmussen 2001
Fratercula	aff. cirrhata											■						w. Atlantic	N. Carolina		Olson & Rasmussen 2001
Pinguinus	alfrednewtoni											■						w. Atlantic	N. Carolina		Olson & Rasmussen 2001
Brachyramphus	dunkeli												■					e. N. Pacific	Calif.		Chandler 1990a
Brachyramphus	pliocenus												■					e. N. Pacific	Calif.		Warheit 1992
Cerorhinca	reai												■					e. N. Pacific	Calif.		Chandler 1990a
Cerorhinca	sp.												■					e. N. Pacific	Calif.		Chandler 1990a
Mancalla	emlongi												■					e. N. Pacific	Calif.		Warheit 1992
Ptychoramphus	tenuis												■					e. N. Pacific	Calif.		Warheit 1992
Synthliboramphus	rineyi												■					e. N. Pacific	Calif.		Chandler 1990a
genus indeterminate	sp.												■					e. N. Pacific	Calif.		Chandler 1990a
Pinguinus	impennis														■	■	■	e. N. Atlantic	Europe	19	Bochenski 1997
Uria	affinis																	w. N. Atlantic	Maine	20	Olson & Rasmussen 2001

Pelecaniformes
incertae sedis

Genus	Species	Cretaceous latest	Paleocene late	Eocene early	Eocene middle	Eocene late	Oligocene early	Oligocene late	Miocene early	Miocene middle	Miocene late	Pliocene early	Pliocene late	Pleistocene early	Pleistocene middle	Pleistocene late	Holocene	Geographic Region	Specific Locality	Comment	Citation
Eostega	lebedinskyi				■													Paratethys	Romania	21	Olson 1985a

APPENDIX 2.1 *(Continued)*
List of fossil seabirds
See text and notes at bottom of table for details. [a, d, e]

Genus or higher taxon [b, c]	Species	Cretaceous latest	Palaeocene late	Eocene early	Eocene middle	Eocene late	Oligocene early	Oligocene late	Miocene early	Miocene middle	Miocene late	Pliocene early	Pliocene late	Pleistocene early	Pleistocene middle	Pleistocene late	Holocene	Geographic Region [f]	Specific Locality [f]	Comment [h]	Citation [g]
Liptornis	*hesternus*									■								w. S. Atlantic	Argentina	22	Olson 1985a
Protopelicanus	*cavierii*												■					e. N. Atlantic	France	23	Olson 1985a
Phaethontes																					
Prophaethon	*shrubsolei*			■														e. N. Atlantic	England	24	Harrison & Walker 1976
Heliadornis	*ashbyi*									■								Atlantic	Maryland, Belgium		Olson 1985d
Heliadornis	*paratethydicus*										■							Paratethys	Austria		Mlikovsk 1997
Fregatidae																					
Limnofregata	*azygosternon*			■														int. N. America	Wyoming		Olson 1977
Pelecanidae																					
Miopelecanus	*gracilis*								■									Paratethys	France	25	Cheneval 1984
Miopelecanus	*intermedius*									■								int. Europe	Germany	26	Cheneval 1984
Pelecanus	*fraasi*									■								int. Europe	Germany		Olson 1985a
Pelecanus	*schreiberi*											■						w. Atlantic	N. Carolina		Olson 1999
Pelecanus	*odessanus*											■						Paratethys	Ukraine		Olson 1985a
Pelecanus	*cautleyi*											■						India	India		Olson 1985a
Pelecanus	*sivalensis*											■						India	India	27	Olson 1985a
Pelecanus	*halieus*												■					int. N. America	Idaho		Olson 1985a
Pelecanus	*erthrorhynchos*												▨					e. N. Pacific	Oregon	1	Becker 1987
Pelecanus	*grandiceps*															■		w. S. Pacific	Australia		Brodkorb 1963
Pelecanus	*proavus*									▨								w. S. Pacific	Australia		Brodkorb 1963
Pelecanus	*tirarensis*									▨								w. S. Pacific	Australia	28	Miller 1966
Pelecanus	*cadimurka*																▨	w. S. Pacific	Australia		Rich & Van Tetes 1981
Pelecanus	*novaezealandiae*																▨	w. S. Pacific	Australia	29	Rich & Van Tetes 1981

APPENDIX 2.1 (Continued)
List of fossil seabirds
See text and notes at bottom of table for details. [a, d, e]

Pelagornithidae

Genus or higher taxon [b, c]	Species	Cret. latest	Paleo. late	Eoc. early	Eoc. mid.	Eoc. late	Olig. early	Olig. late	Mio. early	Mio. mid.	Mio. late	Plio. early	Plio. late	Pleist. early	Pleist. mid.	Pleist. late	Holocene	Geographic Region [f]	Specific Locality [f]	Comment [h]	Citation [g]
Pseudodontornis	tenuirostris		■															e. N. Atlantic	England	30, 35	Harrison 1985
Odontopteryx	toliapica			■														e. N. Atlantic	England		Harrison & Walker 1976
Macrodontopteryx	oweni			■														e. N. Atlantic	England	31	Harrison & Walker 1976
Dasornis	londinensis			■														e. N. Atlantic	England	32, 33	Harrison & Walker 1976
Argillornis	emuinus			■														e. N. Atlantic	England	32-34	Harrison & Walker 1976
Argillornis	longipennis			■														e. N. Atlantic	England	32-34	Harrison & Walker 1976
Pseudodontornis	longidentata			■														e. N. Atlantic	England	35	Harrison & Walker 1976
Argillornis (?)	sp.				■													e. N. Pacific	Washington		Goedert 1989
genus indeterminate	sp.				■													e. N. Pacific	Washington		Goedert 1989
Gigantornis	eaglesomei				■													e. Atlantic	Nigeria		Olson 1985a
Pelagornithidae	sp.					■												Antarctic Peninsula	Seymour I.		Olson 1985a
Osteodontornis	orri										■							e. N. Pacific	Calif.		Olson 1985a, Warheit 1992
Palaeochenoides	mioceanus								■									w. Atlantic	S. Carolina		Olson 1985a
Pelagornithidae	sp. small							▨										w. Atlantic	S. Carolina	36	Warheit & Olson, unpub. data
Pelagornithidae	sp. medium								■									w. Atlantic	S. Carolina	36	Warheit & Olson, unpub. data
Pelagornithidae	sp. large								■									w. Atlantic	S. Carolina	36	Warheit & Olson, unpub. data
Tympanonesiotes	wetmorei							■										w. Atlantic	S. Carolina	37	Olson 1985a
Cyphornis	magnus								■									e. N. Pacific	British Columbia		Olson 1985a
genus indeterminate	sp.								■									w. Atlantic	Delaware		Rasmussen 1998
genus indeterminate	sp.									■								w. N. Pacific	Japan		Okazaki 1989
Osteodontornis	sp.									■								w. N. Pacific	Japan	38	Matsuoka et al. 1998
Pseudodontornis	stirtoni												▨					w. S. Pacific	New Zealand	39	Howard & Warter 1969
Pelagornithidae	sp. A									■								w. Atlantic	Maryland	40	Warheit & Olson, unpub. data
Pelagornithidae	sp. B									■								w. Atlantic	Maryland	40	Warheit & Olson, unpub. data

APPENDIX 2.1 *(Continued)*

List of fossil seabirds

See text and notes at bottom of table for details. [a, d, e]

Genus or higher taxon [b, c]	Species	Cretaceous latest	late	Paleocene early	middle	Eocene late	early	Oligocen late	early	Miocene middle	late	Pliocene early	late	Pleistocene early	middle	Holocene late	Geographic Region [f]	Specific Locality [f]	Comment [h]	Citation [g]
Pelagornis	*miocaenis*									■							e. N. Atlantic	France		Olson 1985a
Osteodontornis	sp.									■							w. N. Pacific	Japan		Ono & Sakamoto 1991
Osteodontornis	sp.									■							w. N. Pacific	Japan		Ono 1989
Pelagornithidae	sp.										■						w. N. Pacific	Japan		Okazaki 1989
Pelagornithidae	sp.										■						w. S. Pacific	New Zealand		Scarlett 1972
Pelagornis	sp. 1											■					w. Atlantic	N. Carolina	41	Olson & Rasmussen 2001
Pelagornis	sp. 2											■					w. Atlantic	N. Carolina	41	Olson & Rasmussen 2001
Pelagornis	sp.										■						e. S. Pacific	Peru		Cheneval 1993
Caspiodontornis	*kobystanicus*						▒	▒									Paratethys	Caucasus	42	Olson 1985a
Pelagornithidae	sp.											▒					w. S. Pacific	New Zealand	1	McKee 1985
Pseudodontornis	*longirostris*							▒	▒								unknown	unknown	43	Harrison & Walker 1976

Sulidae

Genus or higher taxon [b, c]	Species	Cretaceous latest	late	Paleocene early	middle	Eocene late	early	Oligocen late	early	Miocene middle	late	Pliocene early	late	Pleistocene early	middle	Holocene late	Geographic Region [f]	Specific Locality [f]	Comment [h]	Citation [g]
Sula	*ronzoni*						■										Paratethys	France		Olson 1985a
genus indeterminate	sp. 1								■								w. Atlantic	S. Carolina	44	Warheit & Olson, unpub. data
genus indeterminate	sp. 2								■								w. Atlantic	S. Carolina	44	Warheit & Olson, unpub. data
Empheresula	*arvernensis*							■									Paratethys	France	45	Olson 1985a
Sula	*universitatis*									■							w. Atlantic	Florida	46	Brodkorb 1963
Morus	*loxostylus*									■							w. Atlantic	Maryland	47	Olson & Rasmussen 2001
Morus	*vagabundus*									■							e. N. Pacific	Calif.		Warheit 1992
Morus	sp. A									■							e. N. Pacific	Calif.	48	Warheit, unpub. data
Morus	sp. B									■							e. N. Pacific	Calif.	49	Warheit, unpub. data
Morus	*avitus*									■							w. Atlantic	Maryland	50	Olson & Rasmussen 2001
Morus	*atlanticus*									■							w. Atlantic	N. Carolina	51	Olson & Rasmussen 2001
Sula	sp.									■							w. Atlantic	Maryland	52	Warheit & Olson, unpub. data
Morus	*pygmaea*									■							e. N. Atlantic	France	53	Olson 1985a

APPENDIX 2.1 *(Continued)*

List of fossil seabirds

See text and notes at bottom of table for details. [a, d, e]

Genus or higher taxon [b, c]	Species	Geographic Region [f]	Specific Locality [f]	Comment [h]	Citation [g]
Morus	*olsoni*	Paratethys	Romania		Grigorescu & Kessler 1988
Sarmatosula	*dobrogensis*	Paratethys	Romania		Grigorescu & Kessler 1977
Sula	sp.	w. N. Pacific	Japan	54	Ono 1983
Sula	sp.	w. N. Pacific	Japan	55	Ono & Sakamoto 1991
Morus	*willetti*	e. N. Pacific	Calif.	56	Warheit 1992
Sula	*pohli*	e. N. Pacific	Calif.	57	Warheit 1992
Morus	*stocktoni*	e. N. Pacific	Calif.	58	Warheit 1992
Morus	*lompocanus*	e. N. Pacific	Calif.		Warheit 1992
Morus	*magnus*	e. N. Pacific	Calif.		Warheit 1992
Morus	*media*	e. N. Pacific	Calif.	59	Warheit 1992
Morus	sp.	e. Pacific	Mexico		Howard 1971
Sula	*guano*	w. Atlantic	Florida	60	Brodkorb 1955
Sula	*phosphata*	w. Atlantic	Florida	60	Brodkorb 1955
Sula	new sp.	w. Atlantic	Florida	61	Warheit & Becker, unpub. ms
Morus	*peninsularis*	w. Atlantic	Florida		Olson & Rasmussen 2001
Morus	new sp. 1	w. Atlantic	N. Carolina		Olson & Rasmussen 2001
Morus	new sp. 2	w. Atlantic	N. Carolina		Olson & Rasmussen 2001
Sula	new sp. A	e. S. Pacific	Peru	62	Cheneval 1993
Sula	new sp. B	e. S. Pacific	Peru	62	Cheneval 1993
Sula	new sp. C	e. S. Pacific	Peru	62	Cheneval 1993
Sula	sp.	e. S. Atlantic	S. Africa		Olson 1985c
Morus	*humeralis*	e. N. Pacific	Calif.	63	Chandler 1990a
Morus	*recentior*	e. N. Pacific	Calif.	64	Chandler 1990a
Sula	*clarki*	e. N. Pacific	Calif.		Chandler 1990a
Sula	sp.	e. N. Pacific	Calif.		Chandler 1990a
Morus	*reyanus*	e. N. Pacific	Calif.		Brodkorb 1963

The stratigraphic range columns (Cretaceous latest; Paleocene late; Eocene early/middle/late; Oligocene early/late; Miocene early/middle/late; Pliocene early/late; Pleistocene early/middle/late; Holocene) are indicated by filled bars for each taxon.

APPENDIX 2.1 (Continued)
List of fossil seabirds
See text and notes at bottom of table for details. [a, d, e]

Genus or higher taxon [b, c]	Species	Geographic Region [f]	Specific Locality [f]	Comment [h]	Citation [g]
Sula	*tasmani*	w. S. Pacific	Norfolk I.	65	van Tets et al. 1988

Phalacrocoracidae

Genus or higher taxon [b, c]	Species	Geographic Region [f]	Specific Locality [f]	Comment [h]	Citation [g]
genus indeterminate	sp.	Paratethys	France		Mourer-Chauviré 1982
genus indeterminate	sp.	Paratethys	Egypt		Rasmussen et al. 1987
Phalacrocorax	*marinavis*	e. N. Pacific	Oregon		Shufeldt 1915
Phalacrocorax	*littoralis*	Paratethys, int. Eur.	France		Brodkorb 1963
Nectornis	*miocaenus*	Paratethys	France	66	Cheneval 1984
Phalacrocorax	*anatolicus*	Paratethys	Turkey		Olson 1985a
Phalacrocorax	*leptopus*	e. N. Pacific	Oregon		Brodkorb 1963
Phalacrocorax	*femoralis*	e. N. Pacific	Calif.		Miller 1929
Phalacrocorax	*wetmorei*	w. Atlantic	Florida		Olson & Rasmussen 2001
Phalacrocorax	*idahensis*	N. America	Florida, Idaho		Brodkorb 1963
Phalacrocorax	*brunhuberi*	int. Europe	Bavaria	67	Olson 1985a
Phalacrocorax	*intermedius*	e. N. Atlantic	France		Brodkorb 1963
Phalacrocorax	*ibericum*	e. Atlantic	Spain		Olson 1985a
Phalacrocorax	*lautus*	Paratethys	Moldavia		Olson 1985a
Phalacrocorax	*serdicensis*	Paratethys	Bulgaria		Bochenski 1997
Phalacrocorax	*goletensis*	e. Pacific	Mexico		Howard 1965
Phalacrocorax	sp. large	w. Atlantic	N. Carolina	68	Olson & Rasmussen 2001
Pliocarbo	*longipes*	Paratethys	Ukraine	69	Olson 1985a
Phalacrocorax	sp.	e. S. Pacific	Peru		Cheneval 1993
Phalacrocorax	sp. medium	e. S. Atlantic	S. Africa		Olson 1985c
Phala. cf. (*Microcarbo*)	sp.	e. S. Atlantic	S. Africa	70	Olson 1985c
Phalacrocorax	*macer*	int. N. America	Idaho		Brodkorb 1963
Phalacrocorax	*kennelli*	e. N. Pacific	Calif.		Warheit 1992

Geologic time columns (left to right): Cretaceous (latest), Paleocene (late), Eocene (early, middle, late), Oligocene (early, late), Miocene (early, middle, late), Pliocene (early, late), Pleistocene (early, middle, late), Holocene.

APPENDIX 2.1 (Continued)
List of fossil seabirds
See text and notes at bottom of table for details. [a, d, e]

Genus or higher taxon [b, c]	Species	Age (stratigraphic range)	Geographic Region [f]	Specific Locality [f]	Comment [h]	Citation [g]
Stictocarbo	kumeyaay	late Pliocene	e. N. Pacific	Calif.		Chandler 1990a
genus indeterminate	spp.		e. N. Pacific	Calif.	71	Chandler 1990a
Phalacrocorax	sp.		w. Atlantic	Florida		Emslie 1992
Phalacrocorax	filyawi		w. Atlantic	Florida		Emslie 1995
Phalacrocorax	rogersi	early Pleistocene	e. N. Pacific	Calif.		Brodkorb 1963
Phalacrocorax	macropus	late Pleistocene–Holocene	e. N. Pacific	Oregon		Brodkorb 1963
Phalacrocorax	pampeanus		w. S. Atlantic	Argentina		Brodkorb 1963
Phalacrocorax	gregorii		w. S. Pacific	Australia		Brodkorb 1963
Phalacrocorax	vetustus		w. S. Pacific	Australia		Brodkorb 1963
Phalacrocorax	auritus	Pleistocene–Holocene	N. America	Florida, Idaho	1	Becker 1987
Phalacrocorax	destefani		Paratethys	Italy	1	Brodkorb 1963
Phalacrocorax	mongoliensis	Pliocene	int. Asia	Mongolia	1	Olson 1985a
Phalacrocorax	religuus	Pliocene	int. Asia	Mongolia	1	Olson 1985a
Phalacrocorax	chapalensis	Pliocene–Pleistocene	e. Pacific	Mexico	72	Alvarez 1977

Plotopteridae

Genus or higher taxon [b, c]	Species	Age (stratigraphic range)	Geographic Region [f]	Specific Locality [f]	Comment [h]	Citation [g]
Phocavis	maritimus	middle Eocene	e. N. Pacific	Washington		Goedert 1988
genus indeterminate	spp.	late Oligocene	w. N. Pacific	Japan	73	Olson & Hasegawa 1996
Plotopterum	joaquinensis	late Oligocene–early Miocene	e. N. Pacific	Calif.		Howard 1969
Tonsala	hildegardae	late Oligocene	e. N. Pacific	Washington		Olson 1980
Copepteryx	hexeris	late Oligocene	w. N. Pacific	Japan		Olson & Hasegawa 1996
Copepteryx	titan	late Oligocene	w. N. Pacific	Japan	74	Olson & Hasegawa 1996
genus indeterminate	spp.	late Oligocene	w. N. Pacific	Japan	75	Olson & Hasegawa 1996
genus indeterminate	sp.	late Oligocene	w. N. Pacific	Japan	76	Kimura & Sakurai 1998
Plotopterum	sp.	middle Miocene	w. N. Pacific	Japan		Olson & Hasegawa 1985

APPENDIX 2.1 *(Continued)*
List of fossil seabirds
See text and notes at bottom of table for details. [a, d, e]

Procellariiformes

incertae sedis

Genus or higher taxon [b, c]	Species	Cretaceous latest	Paleocene late	Eocene early	Eocene middle	Eocene late	Oligocene early	Oligocene late	Miocene early	Miocene middle	Miocene late	Pliocene early	Pliocene late	Pleistocene early	Pleistocene middle	Pleistocene late	Holocene late	Geographic Region [f]	Specific Locality [f]	Comment [h]	Citation [g]
Tytthostonyx	*glauconiticus*	■																w. Atlantic	New Jersey	77	Olson & Parris 1987
Marinavis	*longirostris*			■														e. N. Atlantic	England	78	Harrison & Walker 1977

Diomedeoididae

Genus or higher taxon [b, c]	Species	Cretaceous latest	Paleocene late	Eocene early	Eocene middle	Eocene late	Oligocene early	Oligocene late	Miocene early	Miocene middle	Miocene late	Pliocene early	Pliocene late	Pleistocene early	Pleistocene middle	Pleistocene late	Holocene late	Geographic Region [f]	Specific Locality [f]	Comment [h]	Citation [g]
Diomedeoides	*minimus*						▦											int. Europe	Germany	42	Bochenski 1997

Diomedeidae

Genus or higher taxon [b, c]	Species	Cretaceous latest	Paleocene late	Eocene early	Eocene middle	Eocene late	Oligocene early	Oligocene late	Miocene early	Miocene middle	Miocene late	Pliocene early	Pliocene late	Pleistocene early	Pleistocene middle	Pleistocene late	Holocene late	Geographic Region [f]	Specific Locality [f]	Comment [h]	Citation [g]
Plotornis (?)	sp.							■										w. Atlantic	S. Carolina		Olson 1985a
Plotornis	*arvernensis*								■									Paratethys	France	79	Cheneval 1984
Diomedea	*californica*									■	■							e. N. Pacific	Calif.		Warheit 1992
Diomedea	*milleri*									■								e. N. Pacific	Calif.		Warheit 1992
Plotornis	*delfortrii*									■								e. N. Atlantic	France	80	Olson 1985a
Diomedea	*rumana*										■							Paratethys	Romania		Grigorescu & Kessler 1988
Diomedea	sp.										■							e. N. Pacific	Calif.		Warheit 1992
Diomedea	sp.											■						e. N. Pacific	Calif.		Warheit 1992
Diomedia	*thyridata*										■							w. S. Pacific	Australia		Wilkinson 1969
Diomedia	sp.										■							w. S. Atlantic	Argentina		Olson 1984b
Phoebastria	aff. *albatrus*												■					w. Atl. & e. Pacific	N. Carolina, Calif.	81	Olson & Rasmussen 2001
Phoebastria	aff. *nigripes*												■					w. Atlantic	N. Carolina		Olson & Rasmussen 2001
Phoebastria	aff. *immutabilis*												■					w. Atlantic	N. Carolina		Olson & Rasmussen 2001
Phoebastria	*rexsularum*												■					w. Atlantic	N. Carolina	82	Olson & Rasmussen 2001
Phoebastria	*anglica*												■					Atl. & e. N. Pacific	Calif, N. Carol., Engl.	83	Olson & Rasmussen 2001
Diomedea	sp.												■					e. S. Atlantic	S. Africa		Olson 1985b

APPENDIX 2.1 (Continued)
List of fossil seabirds
See text and notes at bottom of table for details. [a, d, e]

Genus or higher taxon [b,c]	Species	Geographic Region [f]	Specific Locality [f]	Comment [h]	Citation [g]
Diomedea	sp. B	e. N. Pacific	Calif.		Chandler 1990a
Procellariidae					
Neptuniavis	minor	e. N. Atlantic	England	84	Harrison & Walker 1977
Neptuniavis	miranda	e. N. Atlantic	England	84	Harrison & Walker 1977
genus indeterminate	sp.	w. Atlantic	Louisiana	85	Feduccia & McPherson 1993
Puffinus	raemdonckii	e. N. Atlantic	Belgium		Olson 1985a
Frigidafons	brodkorbi	int. Europe	Germany		Cheneval 1995
Frigidafons	babaheydariensis	Parathys	Iran	85a	Peters & Hamedani 2000
genus indeterminate	sp. 1	w. Atlantic	S. Carolina		Olson 1985a
genus indeterminate	sp. 2	w. Atlantic	S. Carolina	86	Olson 1985a
Argyrodyptes	microtarsus	w. S. Atlantic	Argentina	87	Olson 1985a
Puffinus	micraulax	w. Atlantic	Florida		Olson 1985a
Fulmarus	miocaenus	e. N. Pacific	Calif.		Howard 1984
Puffinus	inceptor	e. N. Pacific	Calif.		Warheit 1992
Puffinus	mitchelli	e. N. Pacific	Calif.		Warheit 1992
Puffinus	priscus	e. N. Pacific	Calif.		Warheit 1992
Puffinus	sp.	e. N. Pacific	Calif.		Warheit 1992
Puffinus	conradi	w. Atlantic	Maryland		Olson 1985a
Puffinus	spp.	Atlantic	Maryland & S. Africa	88	Olson 1985a
Puffinus	aquitanicus	e. N. Atlantic	France		Brodkorb 1963
Puffinus	antiquus	e. N. Atlantic	France		Brodkorb 1963
Bulweria?	sp.	w. Atlantic	N. Carolina	89	Olson & Rasmussen 2001
Puffinus(Thyellodroma)	sp.	w. Atlantic	N. Carolina	89	Olson & Rasmussen 2001
Puffinus(Ardenna)	sp.	w. Atlantic	N. Carolina	89	Olson & Rasmussen 2001
Puffinus	aff. gravis	w. Atlantic	N. Carolina	89	Olson & Rasmussen 2001

Geologic time columns (between Species and Geographic Region): Cretaceous (latest); Paleocene (late); Eocene (early, middle, late); Oligocene (early, late); Miocene (early, middle, late); Pliocene (early, late); Pleistocene (early, middle, late); Holocene.

APPENDIX 2.1 *(Continued)*
List of fossil seabirds
See text and notes at bottom of table for details. [a, d, e]

Genus or higher taxon [b,c]	Species	Geographic Region [f]	Specific Locality [f]	Comment [h]	Citation [g]
Puffinus	sp. A	w. N. Pacific	Japan		Ono & Sakamoto 1991
Puffinus	sp. B	w. N. Pacific	Japan		Ono & Sakamoto 1991
Fulmarus	*hammeri*	e. N. Pacific	Calif.		Howard 1968
Puffinus	*barnesi*	e. N. Pacific	Calif.		Warheit 1992
Puffinus	*calhouni*	e. N. Pacific	Calif.		Howard 1968
Puffinus	*diatomicus*	e. N. Pacific	Calif.		Warheit 1992
Puffinus	*felthami*	e. N. Pacific	Calif.		Howard 1949
Puffinus	*tedfordi*	e. Pacific	Mexico		Howard 1971
Calonectris	*krantzi*	w. Atlantic	N. Carolina	90	Olson & Rasmussen 2001
Puffinus	aff. *pacificoides*	w. Atlantic	N. Carolina		Olson & Rasmussen 2001
Pterodromoides	*minoricensis*	Atlantic	Medit., N. Carolina		Olson & Rasmussen 2001
Fulmarus	sp.	e. S. Pacific	Peru		Cheneval 1993
Calonectris	sp.	e. S. Atlantic	S. Africa		Olson 1985c
Fulmarinae	sp.	e. S. Atlantic	S. Africa		Olson 1985c
genus indeterminate	sp.	e. S. Atlantic	S. Africa		Olson 1985c
genus indeterminate	sp.	e. S. Atlantic	S. Africa	91	Olson 1985b
Pachyptila	*salax*	e. S. Atlantic	S. Africa		Olson 1985b,c
Pachyptila	sp. B	e. S. Atlantic	S. Africa	92	Olson 1985b,c
Pachyptila	sp. C	e. S. Atlantic	S. Africa	93	Olson 1985b,c
Procellaria	sp.	e. S. Atlantic	S. Africa		Olson 1985c
Puffinus (*Puffinus*)	sp. A	e. S. Atlantic	S. Africa		Olson 1985c
Puffinus (*Puffinus*)	sp. B	e. S. Atlantic	S. Africa		Olson 1985b,c
Puffinus (*Puffinus*)	sp. C	e. S. Atlantic	S. Africa		Olson 1985c
Calonectris	aff. *borealis*	w. Atlantic	N. Carolina	94, 95	Olson & Rasmussen 2001
Calonectris	aff. *diomedea*	w. Atlantic	N. Carolina	94, 96	Olson & Rasmussen 2001
Pachyptila	sp.	w. Atlantic	N. Carolina	94, 97	Olson & Rasmussen 2001
Procellaria	cf. *parkinsoni*	w. Atlantic	N. Carolina	94	Olson & Rasmussen 2001

Geological age columns (spanning from Cretaceous latest through Holocene: Cretaceous (latest); Paleocene (late); Eocene (early, middle, late); Oligocene (early, late); Miocene (early, middle, late); Pliocene (early, late); Pleistocene (early, middle, late); Holocene) indicate the temporal ranges of the listed taxa.

APPENDIX 2.1 (Continued)
List of fossil seabirds
See text and notes at bottom of table for details. [a,d,e]

Genus or higher taxon [b,c]	Species	Geographic Region [f]	Specific Locality [f]	Comment [h]	Citation [g]
Procellaria	cf. aequinoctialis	w. Atlantic	N. Carolina	94, 98	Olson & Rasmussen 2001
Pterodroma	magn. lessonii	w. Atlantic	N. Carolina	94	Olson & Rasmussen 2001
Puffinus	aff. tenuirostris	w. Atlantic	N. Carolina	94	Olson & Rasmussen 2001
Puffinus	cf. puffinus	w. Atlantic	N. Carolina	94	Olson & Rasmussen 2001
Puffinus	magn. lherminieri	w. Atlantic	N. Carolina	94	Olson & Rasmussen 2001
Puffinus	gilmorei	e. N. Pacific	Calif.		Chandler 1990a
Puffinus	kanakoffi	e. N. Pacific	Calif.		Warheit 1992
Puffinus	sp.	e. N. Pacific	Calif.		Chandler 1990a
genus indeterminate	sp.	e. N. Pacific	Calif.		Chandler 1990a
Puffinus	nestori	Paratethys	Ibiza		Olson & Rasmussen 2001
Puffinus	pacificoides	e. S. Atlantic	St. Helena I.	99	Olson 1975
Bulweria	bifax	e. S. Atlantic	St. Helena I.	99	Olson 1975
Pterodroma	rupinarum	e. S. Atlantic	St. Helena I.	100	Olson 1975
Puffinus	holeae	e. Atlantic	Canary I.	101	Walker et al. 1990
Pterodroma	jugabilis	c. Pacific	Hawaii		Olson & James 1991
Puffinus	olsoni	e. Atlantic	Canary I.		Olson & Rasmussen 2001
Puffinus	spelaeus	w. S. Pacific	New Zealand	102	Holdaway & Worthy 1994
Procellariidae	new sp.	c. S. Pacific	Easter I.		Steadman 1995

Pelecanoididae

Genus or higher taxon	Species	Geographic Region	Specific Locality	Comment	Citation
Pelecanoides	cymatotrypetes	e. S. Atlantic	S. Africa		Olson 1985b

Oceanitidae

Genus or higher taxon	Species	Geographic Region	Specific Locality	Comment	Citation
Primodroma	bournei	e. N. Atlantic	England	103	Harrison & Walker 1977
Oceanodroma	hubbi	e. N. Pacific	Calif.		Miller 1951
Oceanodroma	sp.	e. N. Pacific	Calif.		Howard 1978

Geological time columns (left portion of table): Cretaceous (latest); Paleocene (late); Eocene (early, middle, late); Oligocene (early, late); Miocene (early, middle, late); Pliocene (early, late); Pleistocene (early, middle, late); Holocene.

APPENDIX 2.1 (Continued)
List of fossil seabirds
See text and notes at bottom of table for details. [a, d, e]

Genus or higher taxon [b,c]	Species	Cret. latest	Paleoc. late	Eoc. early	Eoc. middle	Eoc. late	Olig. early	Olig. late	Mioc. early	Mioc. middle	Mioc. late	Plioc. early	Plioc. late	Pleist. early	Pleist. middle	Pleist. late	Holocene	Geographic Region [f]	Specific Locality [f]	Comment [h]	Citation [g]
Oceanites	zaloscarthmus											■						e. S. Atlantic	S. Africa		Olson 1985b
Oceanites	sp.											■						e. S. Atlantic	S. Africa	104	Olson 1985c
Oceanodroma	sp.												■					e. N. Pacific	Calif.		Chandler 1990a

Sphenisciformes
Spheniscidae

Genus or higher taxon [b,c]	Species	Cret. latest	Paleoc. late	Eoc. early	Eoc. middle	Eoc. late	Olig. early	Olig. late	Mioc. early	Mioc. middle	Mioc. late	Plioc. early	Plioc. late	Pleist. early	Pleist. middle	Pleist. late	Holocene	Geographic Region [f]	Specific Locality [f]	Comment [h]	Citation [g]
Palaeeudyptes	sp.			■														w. S. Pacific	Australia		Simpson 1975
Pachydyptes	simpsoni																	w. S. Pacific	Australia		Jenkins 1974
Pachydyptes	ponderosus					■												w. S. Pacific	New Zealand		Simpson 1975
Palaeeudyptes	marplesi					■												w. S. Pacific	New Zealand		Simpson 1975
Palaeeudyptes	sp.					■												w. S. Pacific	New Zealand	105	Simpson 1975
Anthropornis	nordenskjoeldii					■												Antarctic Peninsula	Seymour I.		Simpson 1975
Anthropornis	grandis					■												Antarctic Peninsula	Seymour I.		Simpson 1975
Archaeospheniscus	wimani					■												Antarctic Peninsula	Seymour I.		Simpson 1975
Delphinornis	larsenii					■												Antarctic Peninsula	Seymour I.		Simpson 1975
Palaeeudyptes	gunnari					■												Antarctic Peninsula	Seymour I.		Simpson 1975
Palaeeudyptes	klekowskii					■												Antarctic Peninsula	Seymour I.		Myrcha et al. 1990
Wimanornis	seymourensis					■												Antarctic Peninsula	Seymour I.		Simpson 1975
genus indeterminate	sp.							■										w. S. Pacific	New Zealand		Simpson 1975
?Platydyptes	marplesi						▨	▨										w. S. Pacific	New Zealand	106	Simpson 1975
Archaeospheniscus	lowei						▨	▨										w. S. Pacific	New Zealand	106	Simpson 1975
Archaeospheniscus	lopdelli						▨	▨										w. S. Pacific	New Zealand	106	Simpson 1975
Duntroonornis	parvus						▨	▨										w. S. Pacific	New Zealand	106	Simpson 1975
Palaeeudyptes	sp.						▨	▨										w. S. Pacific	New Zealand	106	Simpson 1975
Palaeeudyptes	antarcticus							■										w. S. Pacific	New Zealand		Simpson 1975

APPENDIX 2.1 *(Continued)*
List of fossil seabirds
See text and notes at bottom of table for details. [a, d, e]

Genus or higher taxon [b, c]	Species	Geographic Region [f]	Specific Locality [f]	Comment [h]	Citation [g]
Platydyptes	novaezealandiae	w. S. Pacific	New Zealand		Simpson 1975
genus indeterminate	sp.	w. S. Pacific	Australia	107	Simpson 1975
Korora	oliveri	w. S. Pacific	New Zealand		Simpson 1975
Platydyptes	amiesi	w. S. Pacific	New Zealand		Simpson 1975
?Paraptenodytes	brodkorbi	w. S. Atlantic	Argentina		Simpson 1972
Arthrodytes	grandis	w. S. Atlantic	Argentina	108	Simpson 1972
Chubutodyptes	biloculata	w. S. Atlantic	Argentina		Simpson 1972
Eretiscus	tonnii	w. S. Atlantic	Argentina	109	Olson 1986
Palaeospheniscus	patagonicus	w. S. Atlantic	Argentina	110	Simpson 1972
Palaeospheniscus	bergi	w. S. Atlantic	Argentina	111	Simpson 1972
Palaeospheniscus	gracilis	w. S. Atlantic	Argentina	112	Simpson 1972
Palaeospheniscus	wimani	w. S. Atlantic	Argentina	113	Simpson 1972
Paraptenodytes	antarcticus	w. S. Atlantic	Argentina	114	Simpson 1972
Paraptenodytes	robustus	w. S. Atlantic	Argentina	115	Simpson 1972
Anthropodytes	gilli	w. S. Pacific	Australia		Simpson 1975
genus indeterminate	sp.	w. S. Atlantic	Argentina		Cione & Tonni 1981
?Pseudaptenodytes	minor	w. S. Pacific	Australia		Simpson 1975
Pseudaptenodytes	macraei	w. S. Pacific	Australia		Simpson 1975
genus indeterminate	sp.	e. S. Pacific	Peru		Cheneval 1993
?Palaeospheniscus	huxleyorum	e. S. Atlantic	S. Africa	116	Olson 1985c
Dege	hendeyi	e. S. Atlantic	S. Africa	116	Olson 1985c
Inguz'	predemersus	e. S. Atlantic	S. Africa	117	Olson 1985c
Nucleornis	insolitus	e. S. Atlantic	S. Africa	116	Olson 1985c
Aptenodytes	ridgeni	w. S. Pacific	New Zealand		Simpson 1975
Marplesornis	novaezealandiae	w. S. Pacific	New Zealand		Simpson 1975
Pygoscelis	tyreei	w. S. Pacific	New Zealand		Simpson 1975

Stratigraphic range columns (latest Cretaceous; late Paleocene; early, middle, late Eocene; early, late Oligocene; early, middle, late Miocene; early, late Pliocene; early, middle, late Pleistocene; Holocene) are indicated by shaded time-range bars in the original table.

APPENDIX 2.1
Text and Notes

<u>Notes:</u>

[a] Except for the Pelagornithidae and Sulidae, taxa included in this table are based entirely on a review of the literature, in particular Olson (1985a), as well as Bochenski (1997), Brodkorb (1963), Chandler (1990a), Olson and Rasmussen (2001), Warheit (1992), and Simpson (1975). The Pelagornithidae and Sulidae are based on both a review of the literature and unpublished data from Warheit and Olson.

[b] I have not included modern species in this list, except for the Lee Creek fauna, as described by Olson and Rasmussen (2001), or if the modern species is described from a deposit older than Pleistocene. I have also not included a taxon if its affinities are uncertain, but it has been established that the taxon is not a seabird (e.g., *Actiornis anglicus*, see Olson, 1985:207). Furthermore, I have not attempted to sort marine and non-marine deposits; therefore, some of the taxa listed here may have been freshwater/inland species (e.g., perhaps *Phalacrocorax macer*).

[c] The generic identification for some fossils provided in this list may not reflect current taxonomy. For example, most albatross fossils were described in the genus *Diomedea*. Nunn et al. (1996) revised albatross taxonomy, based on a molecular analysis, and the albatross currently inhabiting the north Pacific, for example, are now placed in the genus *Phoebastria*. However, fossil species that would now be placed in *Phoebastria* are listed in this table by their original generic designation (e.g., *Diomedea californica*; see Olson and Rasmussen [2001]), because there has been no formal revision of these taxa.

[d] Each fossil was placed into a specific Epoch (see Figure 2.1) based on the description of the fossil locality in either the original publication or a review article (e.g., Brodkorb 1963, Olson 1985a, Warheit 1992). Fossils that were placed in more than one Epoch are those that occur across several Epochs (solid box) or those with uncertainty as to which Epoch they should be placed (hatched box).

[e] Occasionally, the age of a fossil is revised based on improved stratigraphic or radiometric analyses. I made no attempt here to review the geological literature to determine if there has been a change in the relative or absolute age of any particular fossil since it was originally described or was discussed in a review article. However, if the Age (see Figure 2.1) of a fossil was provided, I established the appropriate Epoch for that fossil based on the most recent Cenozoic geochronology (Berggren et al. 1995; Figure 2.1).

[f] I provided a general locality for each taxon to make evident that these birds occurred in geographic regions more widespread than their specific fossil locality. However, I also provided an example of the more specific locality from which the fossils were recovered. The abbreviations used here are as follows: north (n.), south (s.), east (e.), west (w.), interior (int.), central (c.), Atlantic (Atl.), England (Engl.), Europe (Eur.), Mediterranean (Medit.), California (Calif.), North Carolina (N. Carol.), and island (I.). In addition, Paratethys indicates those areas in relict Paratethys and Tethys Seas (Mediterranean, Black, and Caspian Seas).

[g] Each citation provided here is not necessarily the original reference for the species. For the most part, I have associated a single citation for each taxon listed; that citation will provide additional information for each species, beyond that which I provide in this table, or will point the reader to several additional citations, including the original reference for the species.

[h] The following are the list of comments. Each comment is based on information provided in the citation associated with that taxon, unless noted directly in the comment:

1. Age described as Pliocene.
2. Age described as Oligocene.
3. Affinities not confirmed.
4. Indeterminate affinities - probably not a gull.
5. Rasmussen identified this specimen to the Charadriiformes only, but stated that it was most likely a small species of gull.
6. Probably not *Larus* gull: Mlikovsk' (1992) considered species as Glareolidae.
7. Probably *Stercorarius*; Mourer-Chauviré (1995) listed as Laridae.
8. Ballmann (1976) considered these species to be gulls but not in *Larus*.
9. Similar to *Larus elmorei*, but slightly larger.
10. Similar to cf. *Larus elmorei* to represent extinct taxon.
11. Unlikely to represent extinct taxon.
12. May not be an alcid.
13. *Miocepphus* closely related to *Alca* and not to *Cepphus*.
14. Originally described as *?Endomychura*.
15. Originally described as *Uria*.
16. Humerus morphology similar to *Cyclorrhynchus*.
17. Also includes *Australca grandis*; *Australca* is synonymous with *Alca*.
18. Larger than *Alca antiqua*.
19. Also found at Holocene prehistoric sites in both eastern and western Atlantic (Brodkorb 1967).
20. Same size as *Uria lomvia arra*, perhaps specimen is *U. aalge* or *U. lomvia*.
21. Possibly a pelecaniform.
22. Should be placed in Aves *Incertae Sedis*.

APPENDIX 2.1
Text and Notes

23. Pelecaniform, but probably not a pelican.

24. Harrison and Walker (1977) created Prophaethontidae for this species.

25. Originally described as *Pelecanus*.

26. May be synonymous with *Miopelecanus gracilis*.

27. Tentatively *Pelecanus*.

28. Age described as from Miocene.

29. Originally described as *Pelecanus conspicillatus novaezealandiae*.

30. Warheit and Olson (unpubl. data) concluded that there are no conclusive data to differentiate this species from *Pseudodontornis longidentata*, and that the two taxa may be synonymous; Olson (1985a) suggested that *Pseudodontornis* may be synonymous with *Pelagornis*.

31. Harrison and Walker established this genus based on characters that cannot be confirmed (Warheit and Olson, unpubl. data). It is most likely that correct genus for this species is *Odontopteryx*.

32. *Dasornis* has taxonomic priority over *Argillornis*. *Dasornis londinensis* and the two species of *Argillornis* may be conspecific and, if so, the species will be referred to as *Dasornis londinensis*.

33. The specimens of *Argillornis* and *Dasornis* cannot be compared because they are described from different skeletal elements.

34. *Argillornis emuinus* and *A. longipennis* are most likely conspecific (Warheit and Olson, unpubl. data).

35. Warheit and Olson (unpubl. data) examined a cast of the specimen and made comparisons with *Macrodontopteryx* and *Dasornis*. Harrison and Walker (1976) have not clearly differentiated this specimen from either *Macrodontopteryx* or *Dasornis*, and therefore, this species may not be valid.

36. Warheit and Olson (unpubl. data) have tentatively determined that there are three species, based on relative size, present in the Chattian deposits of South Carolina; one of these species is an extremely large bird with estimated wingspan of over 18 feet. The taxonomy of pseudodontorns from this locality and age needs to be revised. The "medium-sized" bird is comparable in size to *Palaeochenoides*; the smallest bird is roughly the same size as *Tympanonesiotes*, although it is not entirely clear if that species is a pseudodontorn.

37. Olson states that the age of the specimen is not clearly known and was also uncertain that the taxon is a pseudodontorn.

38. I have not seen this reference; the taxonomic designation, age, and locality were taken from the Zoological Record.

39. The age of this specimen is uncertain; it is younger than early Miocene, but older than late Pliocene (McKee 1985).

40. There are at least two species of pseudodontorns from the middle Miocene deposits of the Chesapeake Bay area. Based on the material in hand, the smaller of the two species is somewhat intermediate in size between the small- and medium-sized birds from the Oligocene of South Carolina, and the larger of the two species is intermediate between the medium- and large-sized species from South Carolina, but closer in size to the medium-sized species.

41. There are two, possibly three species of Pseudodontorns from Lee Creek, and as with the Oligocene birds from South Carolina, these species are diagnosed by size. In an effort to simplify a very confused pseudodontorm taxonomy, Olson and Rasmussen (2001) are referring all the species from late Oligocene and Neogene deposits to the genus *Pelagornis*, which has taxonomical priority over all other pseudodontorm genera from this period.

42. Age described as middle Oligocene, but there is no middle Oligocene (see Figure 2.1).

43. This specimen was originally described in the genus *Odontopteryx*. Lambrecht (1930) established the genus *Pseudodontornis* based on this species. The type specimen for this species is lost and its age and locality are also unknown, although Brodkorb (1963) tentatively listed the species as from the Miocene. Hopson (1964) referred to this species a fragment of a lower mandible from the late Oligocene of South Carolina.

44. There are at least two species of sulids from the Oligocene of South Carolina.

45. Originally described as *Sula*.

46. Warheit and Becker (unpubl. ms) consider this species to be Sulidae, *Incertae Sedis*.

47. Approximately the size of *Morus* sp. A from middle Miocene California (Warheit and Olson, unpubl. data).

48. Approximately the size of *M. loxostyla*.

49. Approximately the size of, or slightly smaller than, *M. lompocanus*.

50. Approximately the size of, or slightly smaller than, *M. willetti* (Warheit and Olson, unpubl. data).

51. Approximately the size of *M. vagabundus* (Warheit and Olson, unpubl. data).

52. Small in size.

53. Originally described as *Sula*, then *Microsula*; *Microsula* synonymous with *Morus* (Olson and Rasmussen, in press).

54. Smaller than *S. pohli*; approximately the size of *M. willetti*, but described as *Sula*.

55. Described as *Sula (Microsula)*. Maybe conspecific with *Sula* sp. from Japan.

56. Originally described as *Sula*; moved to *Morus* by Warheit (1990).

57. Described originally by Howard (1958) from the early late Miocene (Fauna II of Warheit [1992]). There are specimens of *Sula* from the middle Miocene of California (Fauna I) referred to this species by Warheit (1992). These specimens are slightly smaller than *S. pohli* and may not be conspecific with this species.

58. Originally described as *Palaeosula*; moved to *Morus* by Warheit (1990).

APPENDIX 2.1
Text and Notes

59. Originally described as *Miosula*; moved to *Morus* by Warheit (1990).

60. *Sula guano* and *S. phosphata* are conspecific (Warheit and Becker, unpubl. ms).

61. Large species of *Sula*.

62. Cheneval stated that there were two or three species from the Pisco Formation. Based on skeletal morphometrics (see Warheit 1992), Warheit and Olson (unpubl. data) determined that there are most probably three large-sized species of *Sula*, the smallest of which is the same size or larger than the largest extant booby (*S. dactylatra*).

63. Originally described in the genus *Sula*; moved to *Morus* by Chandler (1990a).

64. Originally described in the genus *Miosula*; moved to *Morus* by Chandler (1990a).

65. Specimens found on Norfolk Island were in association with Polynesian Rat; species may have been seen on Lord Howe Island in 1788.

66. Originally described as *Phalacrocorax*; Cheneval established this new genus for this species.

67. *Phalacrocorax praecarbo* is synonymous with this species.

68. May be referable to a previously described species; perhaps *Phalacrocorax filyawi*.

69. May not be a cormorant.

70. Within size range of *Phalacrocorax* (*Microcarbo*) *coronatus*.

71. Chandler stated that there are at least two additional species of cormorant and shag present in the San Diego Formation.

72. Age described as Pliocene - Pleistocene.

73. Probably early Oligocene, perhaps late Eocene; at least six species, including a species smaller than any of the species from the late Oligocene of Japan.

74. "... a species of immense size, being the largest diving bird of any sort ever known to have existed ...." (Olson and Hasegawa 1996:750).

75. In addition to the two species of *Copepteryx*, there may be at least an additional four species including another genus in the late Oligocene deposits of Japan.

76. The taxon represented by these fossils may also be included in the material discussed by Olson and Hasegawa (1996). See note 75.

77. This taxon is based on a single specimen (humerus) and is the type for a new family (Tythostonychidae). Olson and Parris tentatively placed this and an unnamed ulna in the Procellariiformes.

78. Harrison and Walker (1977) placed this species in its own family (Marinavidae).

79. Originally described as *Puffinus*; Olson (1985) said this species was more similar to *Pterodroma*; Cheneval placed the species in *Plotornis*.

80. Olson indicated that this species may have been present in the middle Miocene of Maryland.

81. Olson and Rasmussen made synonymous *Diomedea howardae* (Chandler 1990a) and this species; also known from Pleistocene of Bermuda.

82. *Diomedea* sp. B from Chandler (1990a) may be synonymous with this species.

83. *Diomedea* sp. A from Chandler (1990a) is synonymous with this species. This species may be in the same lineage as *D. californica*.

84. Preliminary analysis by Warheit and Olson (unpubl. data) place these species closer to the Diomedeidae than the Procellariidae.

85. Feduccia and McPherson considered this specimen close in morphology to *Pterodroma*.

85a. I have not seen this reference; the age and location indicated here were taken from the title of the paper.

86. Similar in size and morphology to *Bulweria bulwerii*.

87. Originally described as a penguin.

88. There are many undescribed specimens of *Puffinus* from middle Miocene of Maryland and early Pliocene of South Africa.

89. Age is uncertain, but probably from middle Miocene.

90. Largest species in this genus.

91. Roughly same size of Fulmarinae species from early Pliocene of South Africa; based on the descriptions in Olson (1985b,c), this specimen differs from the preceding undescribed taxon.

92. The available material is indistinguishable from *Pachyptila vittata* and *P. salvini*.

93. Similar in size to the smallest extant *Pachyptila*.

94. Age is uncertain, but probably early Pliocene.

95. Olson and Rasmussen consider this taxon to be a full species rather than a subspecies.

96. Distinguished from *Calonectris borealis* by size.

97. Specimens here are indistinguishable from the medium-sized modern species of *Pachyptila*.

98. Possibly a vagrant given the current distribution of this species and its rarity in the fossil deposit.

99. Age described as Pleistocene.

100. Presumably exterminated after 1502.

101. Species was originally named *Puffinus holei*; Michaux et al. (1991) corrected the spelling of this species to *P. holeae*.

102. The extinction of this species probably resulted from the introduction of *Rattus* by the Polynesians.

103. This species is often omitted from lists or reviews (e.g., Olson 1985a) and its systematic position needs to be reviewed.

104. Smaller sized than *Oceanites zaloscarthmus*.

APPENDIX 2.1
Text and Notes

105. Not *Palaeeudyptes marplesi*.
106. Age described as early to middle Oligocene, but there is no middle Oligocene (see Figure 2.1).
107. There are at least two distinct species.
108. Includes *Paraptenodytes andrewsi*.
109. Originally described by Simpson (1981) as *Microdytes*.
110. Includes *Palaeospheniscus menzbieri, P. intermedius, P. affinis*.
111. Includes *Palaeospheniscus planus, P. rothi, Pseudospheniscus planus, P. interplanus, P. concavus, P. convexus*.
112. Includes *Palaeospheniscus nereius, P. medianus*.
113. Includes *Palaeospheniscus robustus*.
114. Includes *Isotremornis nordenskjöldi*.
115. Includes *Paraptenodytes curtus, Metancylornis curtus, Treleudytes crassa, T. crassus*.
116. *Spheniscus* or *Inguza*.
117. Perhaps *Spheniscus*.

APPENDIX 2.2
List of Seabird Species

List of seabird species that are now synonymous with a species in Appendix 2.1. The parenthetical number beside each species refers to the Comment in Appendix 2.1

Charadriiformes
Alcidae
Australca grandis (17)

Pelecaniformes
Phalacrocoracidae
Phalacrocorax praecarbo (67)

Procellariiformes
Diomedeidae
Diomedea howardae (81)
Diomedea sp. A Chandler (1990a) (83)

Sphenisciformes
Spheniscidae
Isotremornis nordenskjöldi (114)
Metancylornis curtus (115)
Palaeospheniscus affinis (110)
P. intermedius (110)
P. interruptus (110)
P. medianus (112)
P. menzbieri (110)
P. nereius (112)
P. planus (111)
P. robustus (113)
P. rothi (111)
Paraptenodytes andrewsi (108)
P. curtus (115)
Pseudospheniscus concavus (111)
P. convexus (111)
P. interplanus (111)
P. planus (111)
Treleudytes crassa (115)
T. crassus (115)

Galapagos Penguins and Flightless Cormorant

3 Seabird Systematics and Distribution: A Review of Current Knowledge

M. de L. Brooke

CONTENTS

3.1 INTRODUCTION

This review of systematics and distribution will be restricted to the groups of birds traditionally considered as seabirds. These groups are the Sphenisciformes, Procellariiformes, Pelecaniformes, and certain families among the Charadriiformes (Table 3.1). And I begin by explaining the significance of the restriction. While all species among the Sphenisciformes (penguins) and Procellariiformes (albatrosses, petrels, shearwaters, fulmars, and allies) are seabirds, this is not universally true for members of the other two orders. Among the Pelecaniformes, tropicbirds, frigatebirds, and boobies are exclusively seabirds. On the other hand, the various species of cormorant, anhinga (= darter), and pelican can be strict seabirds, or freshwater birds, or are able to thrive in both environments. But at least all members of the order are waterbirds. That is not true of the Charadriiformes, an order which comprises some 200 species of shorebirds plus five groups considered to be primarily seabirds, namely, the gulls, terns, skuas, skimmers, and auks. Of these, the auks and skuas are strict seabirds while different species of gull, tern, and skimmer are variously associated with the sea, or with freshwater, or with estuaries.

It is evident already that the distinction between seabirds and other birds is not wholly clearcut. There are, for example, species of duck, grebe, and loon that may spend a substantial fraction of the year floating on salt water — yet these species are not considered to be seabirds. On the other hand, some species traditionally considered to be seabirds spend much of their lives far from the sea. The Brown-headed Gull (*Larus brunnicephalus*), breeding on the Tibetan Plateau, springs to mind.

In this chapter, the defining characteristics of each of the four orders containing seabirds are outlined. Then the features of the seabird families are described within the orders. This provides an opportunity for considering the relationships among families, and for selectively mentioning certain within-family taxonomic issues that have engendered special debate. At this stage the geographical distributions of the families are sketched. The chapter concludes with a discussion of the broad patterns of seabird distribution. Why, for example, are penguins confined to the southern hemisphere, and how do features of seabird lifestyles influence speciation which, in turn, accounts for the difficulty of drawing species boundaries in some groups?

The broad aim of taxonomic studies is to discover the true (= evolutionary) relationships between lineages. To this end, characters indicative of a common descent from some ancestor are most useful. At a very simple level, birds are considered to be a single lineage marked out by the possession of feathers, a feature not shared with their reptilian ancestors. On the other hand, the possession of feathers, a primitive avian character, is of little use in determining the relationships between orders of birds because it is a character shared by all birds. If, in the future, some birds were to lose feathers, the presence of feathers, a primitive feature, would not allow us to deduce that those birds still feathered were closely related. The risk of relying on shared derived characters is that there may be times when it is difficult to determine whether they are shared because of common descent, and therefore indicative of relationship, or shared because of convergence, and therefore taxonomically irrelevant. The fact that the plumage of so many seabirds is some combination of black, brown, gray, or white, and lacks the vivid colors of land birds, is almost certainly the result of convergence.

By the end of the 19th century bird taxonomists, using a suite of anatomical characters including nostrils, palate, tarsus, syrinx, and certain muscles and arteries, had gained a fair understanding of the relationships between the main bird orders (van Tyne and Berger 1966). The next major advance arrived when Sibley and Ahlquist applied the technique of DNA hybridization. Because it compares the entire genome of species A with that of species B, this technique is relatively crude. Nevertheless the results, culminating in Sibley and Ahlquist's magnum opus (1990), represented a significant taxonomic advance. However, nowadays the technique has largely been superseded by other genetic techniques, especially the sequencing of the individual bases on the genes of the species of interest. Nonetheless, it is important to realize that the modern geneticist and the 19th century anatomist

TABLE 3.1
Two Classifications of Seabirds

A. Traditional Classification of Seabirds
Order Sphenisciformes
      Family Spheniscidae: Penguins (6/17)
Order Procellariiformes
      Family Diomedeidae: Albatrosses (4/21)
      Family Procellariidae: Gadfly petrels, shearwaters, fulmars, and allies (14/79)
      Family Pelecanoididae: Diving petrels (1/4)
      Family Hydrobatidae: Storm petrels (8/21)
Order Pelecaniformes
  Suborder Phaethontes
      Family Phaethontidae: Tropicbirds (1/3)
  Suborder Pelecani
      Family Pelecanidae: Pelicans (1/7)
      Family Fregatidae: Frigatebirds (1/5)
      Family Sulidae: Gannets and boobies (3/10)
      Family Phalacrocoracidae
          Subfamily Phalacrocoracinae: Cormorants (9/36)
          Subfamily Anhinginae: Anhingas or darters (1/4)
Order Charadriiformes
  Suborder Charadrii: Various shorebirds (not considered further)
  Suborder Lari
      Family Stercorariidae: Skuas and jaegers (2/7)
      Family Laridae
          Subfamily Larinae: Gulls (6/50)
          Subfamily Sterninae: Terns (7/45)
      Family Rhynchopidae: Skimmers (1/3)
  Suborder Alcae
      Family Alcidae: Auks (13/23)

B. Sibley–Ahlquist Classification of Seabirds
Order Ciconiiformes
  Suborder Charadrii
      Families various, including waders and sandgrouse
      Family Laridae
        Subfamily Larinae
          Tribe Stercorariini: Skuas and jaegers
          Tribe Rynchopini: Skimmers
          Tribe Larini: Gulls
          Tribe Sternini: Terns
  Suborder Ciconii
    Infraorder Falconides: Birds of Prey
    Infraorder Ciconiides
      Parvorder Podicipedida: Grebes
      Parvorder Phaethontida: Tropicbirds
      Parvorder Sulida:
        Superfamily Suloidea
          Family Sulidae: Boobies, gannets
          Family Anhingidae: Anhingas
        Superfamily Phalacrocoracoidea
          Family Phalacrocoracidae: Cormorants
      Parvorder Ciconiida
        Superfamilies various including herons, ibises, flamingos, storks, and New World vultures

TABLE 3.1 *(Continued)*
Two Classifications of Seabirds

> Superfamily Pelecanoidea
>> Family Pelecanidae
>>> Subfamily Balaenicipitinae: Shoebill
>>> Subfamily Pelecaninae: Pelicans
>>> Superfamily Procellariodea
>> Family Fregetidae: Frigatebirds
>> Family Spheniscidae: Penguins
>> Family Gaviiidae: Loons
>> Family Procellariidae
>>> Subfamily Procellariinae: Gadfly petrels, shearwaters, fulmars, and diving-petrels
>>> Subfamily Diomedeinae: Albatrosses
>>> Subfamily Hydrobatinae: Storm petrels

Note: (A) A "traditional" classification following Peters (1934, 1979). The number of extant genera and species is shown in brackets (genera/species) after each family or subfamily. (B) A classification that follows Sibley and Ahlquist (1990).

employ a similar rationale. Both are comparing the character states of the animals of interest, and proceeding to argue that birds with more similar character states are more closely related. The two are simply using different characters for their studies.

For various reasons, different genes evolve at different rates. Therefore studies of higher level taxonomy preferentially use more slowly evolving genes, while studies at the species level and below use rapidly evolving genes. The cytochrome *b* gene, on the mitochondrial genome, has proved especially useful for species-level studies (Meyer 1994). While there are serious problems with the idea that genes evolve at a steady clock-like rate (e.g., Nunn and Stanley 1998), the idea retains an appeal, not the least because it opens the possibility of ascribing a date to when two lineages separated. Thus if the genetic characters of lineage A and lineage B differ by X units, and Y units of difference are known to accumulate per million years of separation, then the lineages diverged X/Y million years ago. There are examples of the application of this approach both to hybridization and to sequence data later in the chapter.

In this chapter, the classification followed here at the subfamily level and upward will be a "traditional" one, espoused for example by Peters (1934, 1979) and based principally on anatomy. There are significant contrasts between the Peters classification and that suggested by Sibley and Ahlquist (1990) based on DNA hybridization data (Table 3.1). In brief, the Sibley and Ahlquist classification places **all** seabirds in a single order, the Ciconiiformes, which also includes birds of prey, shorebirds, and the long-legged waterbirds such as herons, storks, and ibises. While the validity of this general grouping is beyond the scope of this chapter, it is worth emphasizing that, in a seabird context, the principal impact of the Sibley and Ahlquist scheme is to emphasize the separateness of the various birds placed formerly in the Pelecaniformes. As will be discussed later, these birds form a heterogeneous group whose natural affinities have long been in doubt. Insofar as they relate to other nonpelecaniform seabirds, the contrasts between the two classifications outlined in Table 3.1 generally concern differences over the taxonomic level at which a group is recognized, but do not question the unity of the group. For example, the albatrosses are a family, Diomedeidae, under Peters' classification but a subfamily, Diomedeinae, under Sibley and Ahlquist's scheme. However, the Sibley and Ahlquist scheme allies the diving petrels more closely with the gadfly petrels and shearwaters than is customary in traditional classifications.

While these studies, from a decade or more in the past, provide an adequate higher level taxonomic framework for the chapter, this is not true at lower levels where the pace of taxonomic

FIGURE 3.1 Jackass Penguin pair with their chick — South Africa. (Photo by R.W. and E.A. Schreiber.)

revision is faster. In particular, molecular studies are prompting reassessment of species boundaries. I take the work of Sibley and Monroe (1990) as the starting point for the species list, but frequently deviate from it. Although space does not allow the case for each deviation to be made, at least an attempt will be made to direct the reader to a source that does make the case.

3.2 THE ORDERS OF SEABIRDS

3.2.1 ORDER SPHENISCIFORMES, FAMILY SPHENISCIDAE

Penguins are flightless and easily recognized. On land they stand upright and walk with a shuffling gait, occasionally sliding forward on their bellies. At sea, the legs, set well to the rear, serve as a rudder along with the tail. The forelimbs are modified into stiff flippers which cannot be folded and which lack flight feathers (Figure 3.1). The wing bones are flattened and more or less fused, while the scapula and coracoid are both large. Bones are not pneumatic. Many of these features are evidently adaptations for wing-propelled underwater swimming (Brooke and Birkhead 1991, Sibley and Ahlquist 1990). Penguins, densely covered with three layers of scale-like short feathers, lack the bare areas between feather tracts (apteria) found in most other birds.

While the monophyletic origin of penguins is not in question, it has proved difficult to pinpoint that origin. The earliest possible fossil penguin, from 50 to 60 million years ago (mya), is partial and undescribed. From the late Eocene (40 mya), penguin fossils are more numerous, more specialized, and already highly evolved marine divers (Fordyce and Jones 1990, Williams 1995; see Chapter 2). Thus there are no described fossils truly intermediate between the presumed flying ancestor and extinct species that are broadly similar to extant species (Simpson 1976, Williams 1995). However there are persistent pointers to an ancestry shared with the Procellariiformes.

Such pointers include not only the DNA hybridization data of Sibley and Ahlquist (1990), but also various anatomical features. Features shared by these two groups, and also by the divers (= loons in North America), are these. All have webbed feet and two sets of nestling down. There are two carotid arteries, as opposed to the one found in many birds. More technically, the nostrils are termed holorhinal which means that the posterior margin of the nasal opening is formed by a concave nasal bone. Of the four palate types into which bird palates are sometimes categorized, petrels and penguins have the type known as schizognathous (Sibley and Ahlquist 1990). However, these shared features are primitive, retained from distant ancestors, and provide suggestive but not conclusive evidence of a more recent relationship for the groups concerned (Brooke in press).

All penguins belong in a single family, the Spheniscidae, containing 6 genera and 17 species (Table 3.1; Williams 1995). Note that here and subsequently, genus and species totals refer to extant taxa only. The penguins are an exclusively southern hemisphere group, concentrated in cooler waters. Judging by the fossil record, the same has always been true in the past. The modern range extends farther north than elsewhere in southern Africa and South America because of cool currents, the Benguela and Humboldt, respectively, sweeping northward. Indeed, the Galapagos Penguin (*Spheniscus mendiculus*) is found at the Equator breeding on the archipelago swept by the Humboldt Current.

3.2.2 ORDER PROCELLARIIFORMES

All procellariiforms have tubular nostrils which are totally characteristic of this group whose monophyly has never been seriously questioned (Figure 3.2). Indeed, this feature provided the now-redundant name of the order, the Tubinares. While the nostrils of albatrosses are separated by the upper ridge of the bill, in the other petrels the left and right nostrils are merged on top of the bill in a single tube divided by a vertical septum. The prominence of the tube varies between species and its function is uncertain. It may serve in olfaction. Thanks in part to well-developed olfactory bulbs, the powers of smell of many procellariiforms are exceptionally good, at least by the standards of birds (Verheyden and Jouventin 1994). It is also possible that the tubes play some part in distributing the secretions of the densely tufted preen gland which may be responsible for the characteristic musky odor of most procellariiforms (Fisher 1952, Warham 1990).

Another unique feature of the petrels is the digestive tract. The gut of petrels does not have a crop. Instead the lower part of the esophagus is a large bag, the proventriculus. In most birds the walls of the proventriculus are smooth. Not so in petrels where the walls are thickened, glandular, and much folded. Morphological reasons for suspecting a common ancestor for penguins and procellariiforms were discussed above. This suspicion has been strengthened by Sibley and Ahlquist's work (Table 3.1B). If correct, it would suggest a southern hemisphere origin for the procellariiforms. Certainly petrels today are most diverse in the southern hemisphere (Figure 3.3). The fact that most fossil petrels have been found in northern deposits (see Chapter 2) does not necessarily argue against the southern case, since the amount of land where fossils might be unearthed is so much greater in the north.

FIGURE 3.2 Laysan Albatross feeding its chick — Midway Island, north Pacific Ocean. (Photo by J. Burger.)

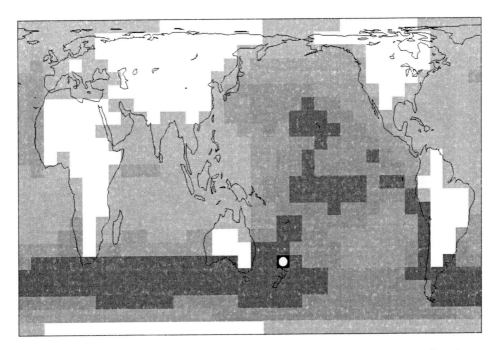

FIGURE 3.3 Map of worldwide species richness of procellariiform species, based on at-sea foraging ranges. Richness is indicated by darkness of the grid cell, and ranges from no records (white) to a maximum of 46 species (black with white circle) in the grid cell immediately north of New Zealand. (After Chown et al. 1998. With permission.)

3.2.2.1 Family Diomedeidae

Albatrosses are easily recognized by their large size and, as mentioned, by the separation of the left and right nasal tubes. An interesting feature, shared with the giant petrels (*Macronectes* spp.), is that the extended humerus can be "locked" in place by a fan of tendons that prevents the wing rising above the horizontal. Once the humerus is slightly retracted from the fully forward position, the lock no longer operates, and the wing can be raised. This shoulder lock facilitates the remarkable gliding of albatrosses (Pennycuick 1982).

The taxonomy of albatrosses is in a state of flux. Until recently there were two widely accepted genera: *Phoebetria,* containing the two sooty albatross species of the Southern Ocean, and *Diomedea,* containing all other species. However, molecular work by Nunn et al. (1996) revealed that *Phoebetria* was a sister group to the smaller Southern Ocean species, the "mollymawks," which were assigned to the genus *Thalassarche*. Meanwhile the North Pacific albatrosses were a sister group to the Southern Ocean's great albatrosses, such as the Wandering *D. exulans*. Accordingly, Nunn et al. (1996) placed these two groups, respectively, into the genera *Phoebastria* and *Diomedea* (Appendix 1). This generic revision has commanded general support among seabird biologists.

More contentious than the generic revision has been the extensive splitting advocated by Robertson and Nunn (1998), who designated 24 species in place of a former 14. While it may transpire that these splits are justified, this author's personal view is that the case for all of them is not yet made (Brooke 1999). Accordingly I (Brooke in press), along with BirdLife International (2000), adopt a slightly more conservative 21-species position; *Thalassarche* — 9 species; *Phoebetria* — 2; *Diomedea* — 6; *Phoebastria* — 4 (Appendix 1).

Today's albatrosses are largely found in higher latitudes (>20°), either in the Southern Ocean (17 species) or the North Pacific (3 species). With the exception of the Waved Albatross (*Phoebastria irrorata*) breeding on the Galapagos Islands and off Ecuador, they are absent as breeding birds from lower latitude stations. This absence has been plausibly related to the dearth, at such low latitudes, of the strong and steady winds on which albatrosses rely for gliding (Pennycuick 1982).

However, the absence of breeding albatrosses from the North Atlantic is more puzzling. Such was not the case in the past. Olson and Rasmussen (in press) report five species in Lower Pliocene marine deposits of North Carolina, dating from about 4 mya (see Chapter 2). They have also been found in Lower Pleistocene, and probably also in underlying Upper Pliocene deposits, of England. This means that albatrosses were common in the Atlantic into the late Tertiary, and disappeared during the Quaternary period (Olson 1985). Presumably Pleistocene climatic fluctuations impinged more severely in the North Atlantic than in the North Pacific. Now it may be that mere chance and the difficulty of crossing Equatorial waters are sufficient explanations of the albatrosses' failure to reestablish in the North Atlantic after the Pleistocene disappearance. The fact that individual Black-browed Albatrosses (*Thalassarche melanophrys*) have survived for over 30 years in the North Atlantic in the 19th and 20th centuries (Rogers 1996, 1998) implies that the ocean is not inimitable to the day-to-day survival of albatrosses.

3.2.2.2 Family Procellariidae

The most diverse and speciose family within the order Procellariiformes is, without question, the Procellariidae, containing 79 species (following Brooke in press). While evidently petrels, these mid-sized species (body weights 90 to 4500 g) are most conveniently defined by an absence of the features characteristic of the other three families. Within the Procellariidae there are 5 more or less distinct groups of species, namely, the fulmars and allies (7 species), the gadfly petrels (39), the prions (7), the shearwaters (21), and the larger petrels (5). Do these groupings reflect evolutionary history? Drawing principally on the cytochrome *b* data of Nunn and Stanley (1998) the answer is a qualified affirmative (Figure 3.4).

The fulmarines are generally medium to large, often scavenging species, represented by six species in the higher latitudes of the southern hemisphere and one, Northern Fulmar *Fulmarus glacialis*, in the north. The six prion species in the genus *Pachyptila* and the Blue Petrel (*Halobaena caerulea*) are united by plumage pattern, myology, and bill structure (Warham 1990). All are confined to the southern hemisphere. Also confined to the southern hemisphere are the five fairly large (700 to 1400 g) species in the genus *Procellaria*. Shearwaters include more aerial species that obtain their food at or close to the surface and those which recent research has revealed to be adept and deep divers. For instance, the mean maximum depth reached by Sooty Shearwaters

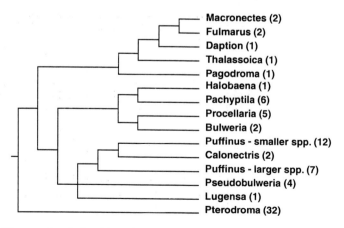

FIGURE 3.4 Possible generic relationships within the Procellariidae based on cytochrome *b* evidence from Nunn and Stanley (1998) and Bretagnolle et al. (1998). After each genus, the number of species within the genus is indicated in brackets.

FIGURE 3.5 Wedge-tailed Shearwater courting group on Johnston Atoll, Pacific Ocean. (Photo by R.W. Schreiber.)

(*Puffinus griseus*) on foraging trips was 39 m, and the greatest depth attained was 67 m (Weimerskirch and Sagar 1996). Shearwaters occur in virtually all oceans, except at the very highest latitudes (Figure 3.5). However, there is one very significant exception. No shearwaters breed in the North Pacific although huge numbers of Sooty and Short-tailed Shearwaters (*Puffinus tenuirostris*) spend the austral winter in this area, having undertaken a transequatorial migration from breeding stations mainly around Australia and New Zealand.

While Mathews and Iredale (1915) placed the two gray-plumaged shearwater species in *Calonectris*, this separation has not been supported by molecular studies. These same molecular studies (Austin 1996) have revealed an unexpectedly deep split within the genus *Puffinus* between the larger species and the smaller species (*nativitatis*, and members of the *puffinus*, *lherminieri*, and *assimilis* species complexes).

Finally the largest and most confusing procellariid group comprises the gadfly petrels, so called because of their helter-skelter flight over the waves. They are found in all oceans, but nowhere breed at high latitudes. The two *Bulweria* species, long recognized as distinct (Bourne 1975), show possible molecular, bill, and skull affinities with *Procellaria* (Imber 1985, Bretagnolle et al. 1998, Nunn and Stanley 1998). Four species in *Pseuodobulweria* have in the past been merged with *Pterodroma*. However, various authors, reviewed by Imber (1985), have recognized the case for generic differentiation, and the molecular case for a relationship with shearwaters was made by Bretagnolle et al. (1998). The Kerguelen Petrel (*Lugensa brevirostris*) is widely viewed as an "oddball" species. While Imber (1985) thought it might be allied to the fulmarine species, the molecular evidence places it closer to shearwaters (Nunn and Stanley 1998). This leaves 32 gadfly petrels in the core genus *Pterodroma*. This total (following Brooke in press) reflects some judgments about species boundaries that certainly would not be universally accepted. Why species boundaries have proved so very difficult to draw in some seabird groups like *Pterodroma*, but not in others, will be reviewed later in the chapter.

3.2.2.3 Family Pelecanoididae

The four species of diving petrel, all members of the single genus *Pelecanoides*, form a very distinct southern hemisphere group. There is no evidence that their range has ever extended into the northern hemisphere. These birds are characterized by flanges — or paraseptal processes — attached to the central septum dividing the two nostrils. The function of these processes is uncertain, but it may serve to reduce the ingress of water into the nostrils which face upward. Diving petrels are all small (100 to 130 g) and very similar in plumage, being shiny black above, and white below. Unlike the majority of petrels which often glide, the diving petrels are instantly recognizable by their rapidly whirring flight on short, stubby wings. This flight style is associated with the birds' means of underwater progression, using the half-closed wings as paddles in a manner similar to the auks of the northern hemisphere. Indeed the remarkable convergence between the smaller auks and the diving petrels has been noted for over 200 years (Latham 1785). The convergence extends to many skeletal features (Warham 1990). Interestingly, the convergence may also extend to the molt pattern. Diving petrels, like certain auks, shed the main wing and tail feathers simultaneously (Watson 1968) and become flightless. But given that the full wing area is generally not deployed during swimming underwater, this loss of feathers may be no great impediment.

Cytochrome *b* sequence data confirm that the Pelecanoididae and Procellariidae are sister taxa (Nunn and Stanley 1998). However, given the distinctiveness of the diving petrels, there is a case for retaining them as a separate family rather than merging diving petrels and procellariids into a single taxon (Table 3.1; Sibley and Ahlquist 1990).

3.2.2.4 Family Hydrobatidae

There are 21 species of storm petrel in 8 genera, with a notable concentration of species nesting off western Mexico and California. All are small seabirds, typically less than 100 g, with particularly conspicuous nostrils, often up-tilted at the ends. The 21 species are divided into two subfamilies. Recent molecular work suggests these two subfamilies represent monophyletic but separate radiations from an early petrel stock (Nunn and Stanley 1998). The subfamily Oceanitinae comprises seven southern hemisphere species split into five genera. These birds have relatively short wings with only ten secondaries, squarish tails, and long legs that extend beyond the tail. Carboneras (1992) suggested that these features are associated with the stronger winds of the southern hemisphere, and the fact that the birds feed by slow gliding. As the birds glide, they almost appear to be walking on water since their dangling feet frequently contact the surface. In contrast the 14 species of the northern subfamily Hydrobatinae are split into only three genera, of which two, *Hydrobates* and *Halocyptena,* are monotypic. The remaining 12 species belong in the genus *Oceanodroma* whose center of distribution is the Pacific Ocean. Two species breed in the North Atlantic and two visit the Indian Ocean where, however, no species breed — an unexpected gap in the distribution. Compared to the Oceanitinae, the Hydrobatinae have longer, more pointed wings with 12 or more secondary feathers and frequently their tails are forked. In the manner of swallows, they intersperse busy flying with short periods of gliding.

3.2.3 ORDER PELECANIFORMES

Taxonomic relationships within the Pelecaniformes are frankly problematical and unresolved. That in turn makes it difficult to identify with confidence the group's nearest relatives (Table 3.1). That said, features uniting the group are as follows. They are the only birds to have all four toes connected by webs, the condition known as totipalmate. A brood patch is lacking in all groups (Nelson in press). Whereas the salt gland of most seabirds lies in a cavity on top of the skull, that of the pelecaniforms is enclosed completely within the orbit (Nelson in press). All have a bare gular pouch, with the exception of the tropicbirds where the feature is inconspicuous and feathered. External nostrils are slit-like (tropicbirds), nearly closed (cormorants and anhingas), or absent (pelicans, frigatebirds, and sulids; Figure 3.6).

Even this brief account is sufficient to indicate that the relationship of the tropicbirds to other pelecaniform groups is especially uncertain. Frigatebirds also may be distantly related to the rest of the order (Nelson in press, Sibley and Ahlquist 1990). On the other hand, an ancestral relationship between sulids, cormorants, and anhingids seems likely. That said, just how closely related the

FIGURE 3.6 Courting pair of Blue-footed Boobies on the Galapagos Islands. (Photo by J. Burger.)

cormorants and anhingids, the only pelecaniform groups that might be confused in the field, are remains uncertain. Sibley and Ahlquist place the two groups in separate superfamilies (Table 3.1), and Becker (1986) has suggested that they have been separated for over 30 million years.

The general picture so far sketched uses evidence from DNA and morphology. However, the conspicuous displays of Pelecaniformes at their colonies, exhaustively documented by van Tets (1965), provide a further line of evidence. When Kennedy et al. (1996) compared a pelecaniform phylogeny based on van Tets' behavioral data with that derived from molecular and morphological data, the congruence was significantly greater than expected by chance. This suggests, perhaps counter-intuitively, that ritualized behavioral displays, such as gaping the bill during greeting, can remain stable over millions of years and thereby retain significant phylogenetic information (see Chapter 10). Further, the Kennedy et al. (1996) study reinforced the case for supposing that tropicbirds and frigatebirds are distinct from other pelecaniforms.

Siegel-Causey (1997) has discussed why the correspondence between the pelecaniform phylogenies derived from molecular, morphological, and behavioral studies may be so poor. Aside from confirming the likely sulid–cormorant–anhingid grouping, the studies are consistent only in their inconsistency. In particular Siegel-Causey wondered whether morphological characters supposed to unite the group may in fact be independently derived. There is an evident opportunity for further work.

3.2.3.1　Family Phaethontidae

There are three closely related species in the single tropicbird genus *Phaethon*. All are medium-sized, predominantly white seabirds with long (30 to 55 cm) tail streamers (Figure 3.7). While the pectoral region is well developed, allowing remarkably sustained flapping flight, the pelvic region is atrophied. Thus tropicbirds can barely stand. They shuffle on land, their bellies scraping the ground.

While Tertiary fossils showing resemblances to tropicbirds come from higher latitudes (London, England, and Maryland, USA: Olson 1985), today's species are essentially tropical. The Red-tailed Tropicbird (*Phaethon rubricauda*) occurs in waters over 22°C (Enticott and Tipling 1997). While the smallest species, the White-tailed (*P. lepturus*), has a pan-tropical distribution, the distributions of the two larger species, the Red-tailed and the Red-billed (*P. aethereus*), are nearly complementary. The former occurs across the Indo-Pacific as far east as Easter Island. The latter occurs in the

FIGURE 3.7 Red-tailed Tropicbird adult prospecting for a nest site, showing long tail streamers common to all the tropicbirds. (Photo by E.A. Schreiber.)

extreme eastern tropical Pacific, in the Caribbean and the Atlantic, and finally in the Arabian Sea where there is overlap with Red-tailed Tropicbirds.

3.2.3.2 Family Pelecanidae

The huge size and capacious throat pouch of pelicans make them easy to recognize. In fact, pelicans are among the heaviest flying birds (4 to 13 kg, depending on species; Figure 3.8; Elliott 1992; see Appendix 2). The seven species, placed in the single genus *Pelecanus*, are distributed across the world in tropical and warm temperate zones where they feed in coastal or inland waters. Like the anhingas, the status of pelicans as seabirds is open to question, and the treatment here is accordingly brief. The Brown Pelican (*Pelecanus occidentalis*) is the species most often met at sea, and is also the only species that plunge-dives in pursuit of prey.

3.2.3.3 Family Fregatidae

With long pointed wings and deeply forked tail, the frigatebirds are aerial seabirds of the tropics (Figure 3.9). Using their long hooked robust beak, they are capable of snatching prey from the sea surface, or indeed in the case of flying fish, from above the surface, without alighting on the water. In fact, their plumage is not sufficiently waterproofed with preen gland oil to allow safe swimming. The reduced webs between the toes are confined to the basal portion of the toes.

There are five decidedly similar modern species of frigatebird in a single genus *Fregata*. Two species, the Great Frigatebird (*Fregata minor*) and Lesser (*F. ariel*), have generally overlapping distributions in the Indo-Pacific. Both also breed at Trindade and Martin Vaz in the tropical south Atlantic. The Magnificent Frigatebird (*F. magnificens*) is found in the tropical Atlantic plus the eastern tropical Pacific, while two species, the Ascension (*F. aquila*) and Christmas (*F. andrewsi*), are single-island endemics.

3.2.3.4 Family Sulidae

As is true of most Pelecaniform groups, sulids are easily recognized. They are fairly large seabirds, with long, strong, tapering bills. The skull is hinged to allow more pressure to be applied to the tip of the bill, the better to grasp fish. Facial skin, bill, eyes, and feet are usually brightly colored.

FIGURE 3.8 The neck of this Brown Pelican will soon molt to brown and it will move into the nesting colony to begin courtship and pair formation. (Photo by R.W. Schreiber.)

FIGURE 3.9 A male Magnificent Frigatebird inflates its pouch and waits for a potential mate to fly over, at which time he will begin his courtship behaviors to attract her. (Photo by J. Burger.)

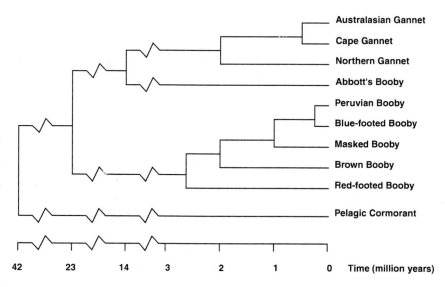

FIGURE 3.10 Approximate time frame for speciation events within the Sulidae (redrawn from Friesen and Anderson 1997). Note that Friesen and Anderson's study was completed before Pitman and Jehl (1998) recommended a split of the Nazca Booby from the Masked Booby.

The wings are long and pointed, and the tail is often diamond-shaped. The preen gland at the base of the tail opens via five apertures (Nelson 1978).

There has been sustained debate over whether the sulids should be divided into two genera, the gannets *Morus* spp. and boobies *Sula* spp. Checklists are divided on the issue. However, using cytochrome *b* evidence, Friesen and Anderson (1997) estimated the booby and gannet lineages diverged about 23 million years ago, about the time when fossils can be clearly recognized as either *Sula* or *Morus* (Nelson in press). Thus the case for the division is strong. Friesen and Anderson's study also lent support to the suggestion of Olson and Warheit (1988) that Abbott's Booby *(Papasula abbotti)* should be placed in a monospecific genus *Papasula*, allied by its long humerus with the gannets, rather than with the boobies characterized by short humeri. In fact, Friesen and Anderson estimated *Papasula* and *Morus* diverged about 14 million years ago. This study therefore proposed the time frame for sulid speciation shown in Figure 3.10. The alliance of Abbott's Booby with the gannets is also supported by behavior; they alone among the sulids have a prolonged face-to-face greeting ceremony using outspread wings (Nelson in press). Since the completion of Friesen and Anderson's study, Pitman and Jehl (1998) have recommended a split of the Nazca Booby *(Sula granti)* from the Masked Booby *(S. dactylatra)*. Subsequent cytochrome *b* analysis (Friesen et al. submitted) has confirmed the distinctiveness of the two taxa.

Gannets are plunge-diving birds of productive temperate waters of the North Atlantic and south African and Australian regions. As an adaptation to underwater wing-powered pursuit of prey, the gannets' humeri are long relative to the more distal bones of the wing. On the other hand, the boobies are essentially tropical, species occurring in all tropical oceans. Boobies catch prey on the wing or by dives that are shallower than those of gannets. Accordingly, the humeri are shorter in relation to the distal parts of the wing than in gannets (Warheit 1990).

Implicit within this brief account is the information that today no sulids breed in the temperate North Pacific, an absence which is puzzling given the Miocene and Pliocene records of both *Sula* and *Morus* species from deposits stretching from California to British Columbia (Warheit 1992). There is no evidence to support the idea that the absence represents a major contraction of range resulting from human devastation of colonies. Such contraction has occurred on massive scale in the case of Abbott's Booby which is vulnerable to hunting and habitat destruction. Formerly

FIGURE 3.11 A Flightless Cormorant in the Galapagos, the only cormorant species that cannot fly. (Photo by R.W. and E.A. Schreiber.)

distributed across the entire Indian Ocean and east into the Pacific as far as the Marquesas, the species is now confined to the Indian Ocean's Christmas Island (Steadman et al. 1988).

3.2.3.5 Subfamily Phalacrocoracinae

Cormorants are medium to large aquatic birds that obtain prey underwater by pursuit. Body, neck, head, and bill tend to be elongated (Figure 3.11). The bill is laterally flattened, hooked (c.f. anhingas), and with nostrils nearly closed (Orta 1992). Cormorants occur around most of the world's coasts, with the exception of the high Arctic. Although they breed at certain oceanic islands, such as those of the Southern Ocean and the Galapagos, they are rarely seen in pelagic waters. In addition to the wholly marine species, there are cormorants that occur in both marine and freshwater environments and species which are confined to freshwater. Thus cormorants can be met in the rivers and lakes of all continents, except at the higher northern latitudes.

While cormorants and shags are certainly the most speciose pelecaniform group, deciding just how many genera and species there are has proved exceptionally difficult. For example, Dorst and Mougin (1979) considered that there were 29 species in a single genus *Phalacrocorax*. If species are to be removed from this one genus, the most likely candidates in the past have been the Flightless Cormorant (*Compsohalieus* [= *Nannopterum*]*harrisi*) of the Galapagos and/or the five species of micro-cormorants *Microcarbo* (see Siegel-Causey 1988 for review of past studies). However, Siegel-Causey's own analysis suggested a more drastic revision of the group. He proposed 37 species in nine genera. Excluding one extinct species, his classification is followed in Appendix 1. Relying mainly on osteological characters, Siegel-Causey identified two major groups, the Phalacrocoracinae ("true" cormorants) comprising four genera of all dark littorine species and the Leucocarboninae (shags), five genera of variably plumaged, littorine, or more pelagic species. The increase in the number of species was caused because Siegel-Causey decided to split the blue-eyed shags of the Southern Ocean, often represented by different taxa on different island groups, into more species than recognized by earlier workers. The details of this re-arrangement are beyond the scope of this survey, but the general issue of how to deal with subtly different taxa on different islands, an issue also bearing on albatross and petrel taxonomy, will be considered below.

3.2.4 Order Charadriiformes

The alliance of the shorebird families with the skua/gull/tern/skimmer grouping and with the auks was originally based on a shared schizognathous palate, and further anatomical similarities in syrinx and leg tendons (Brooke and Birkhead 1991). It has been supported by Sibley and Ahlquist's (1990) DNA study (Table 3.1) which suggests that these shorebird and seabird lineages diverged at least 25 million years ago.

3.2.4.1 Family Stercorariidae

The skuas form a small, distinctive family of seven species that probably diverged from the gulls about 10 mya (Furness 1996; Figure 3.12). They combine catching their own prey (sometimes on land during the breeding season) with kleptoparasitism. All breed at moderate to high latitudes, and most migrate toward the Equator during the nonbreeding period. The three smaller well-defined species, also known as jaegers, breed in northern high latitudes and are placed in the genus *Stercorarius*. On the other hand, defining species limits in the larger *Catharacta* species has been problematical because of plumage variation within taxa (Devillers 1978). While the northern hemisphere Great Skua (*Catharacta skua*) is certainly distinct, the southern hemisphere forms are less so. Here the author recognizes the Chilean (*C. chilensis*), Brown (*C. antarctica*), and South Polar Skuas (*C. maccormicki*). While the small (<1%) mitochondrial DNA differences between these three (Cohen et al. 1997) might argue for subspecific status, Devillers (1978) has made the case for their recognition because, despite considerable overlaps in breeding range, hybridization is avoided.

Relationships among these skuas have yielded one of the most extraordinary and fascinating tales to emerge in seabird systematics in recent years. Mitochondrial DNA sequence data presented by Cohen et al. (1997) suggested that the Great Skua and the Pomarine Jaeger (*Stercorarius pomarinus*) are closely related. Albeit less convincingly, nuclear DNA data supported the close relationship between the Great Skua and the Pomarine Skua. This species pair, in turn, is most closely related to the southern hemisphere skuas and more distantly related to the other northern species, the Parasitic Jaeger (*S. parasiticus*) and Long-tailed (*S. longicaudus*). If this picture is correct, neither of the genera *Catharacta* or *Stercorarius* is monophyletic. Remarkably the feather lice found on Pomarine Skuas are also more akin to those on Great Skuas than those on Parasitic and Long-tailed Jaegers (Cohen et al. 1997).

FIGURE 3.12 A Brown Skua tends its egg and chick in the Falkland Islands. (Photo by P.D. Boersma.)

Cohen et al. (1997) suggested three evolutionary routes to this present-day picture. The first is that the skua ancestor resembled a modern Pomarine Jaeger. From this ancestor, one lineage developed into Parasitic and Long-tailed Jaegers. The other retained the Pomarine Jaeger-like species, and twice budded off *Catharacta* forms. Another idea is that the resemblance of the Pomarine Jaeger to Parasitic and Long-tailed Jaegers is a case of convergence. The third and most intriguing possibility is that interbreeding between a female Great Skua and male Parasitic or Long-tailed Jaeger introduced *Catharacta* mtDNA into the *Stercorarius* lineage, and created the hybrid that was the progenitor of today's Pomarine Jaegers. When Braun and Brumfield (1998) re-analyzed Cohen et al.'s molecular data in a maximum likelihood framework, they concluded that *Catharacta* was, after all, monophyletic. However, Andersson (1999) has supported the hybridization scenario of Cohen et al.

3.2.4.2   Subfamily Larinae

Associated with lakes, wetlands, or marine environments, gulls are fairly small (100 g) to fairly large (2 kg) birds with stout bills and webbed feet. They are long winged and, typically, some shade of gray or black above and white below. There is broad agreement that gulls and terns (Sterninae) are closely related. Gulls have a cosmopolitan distribution. They are normally absent only from deserts, high mountains, extensive tracts of forest (especially tropical rainforest), and from ice sheets. While gulls are invariably encountered on temperate coastlines, they may be absent from tropical coasts, especially from tropical oceanic islands. This absence is not because any other group of birds obviously replaces the gulls as a scavenger/predator, nor is it easily explained on the grounds that tropical coastal zones are less productive than their temperate counterparts. Therefore the explanation offered here is that gulls are relatively scarce on tropical coasts because their scavenging role is undertaken by crabs which can attain great densities on tropical shores. In the warmth of the tropics crabs are not metabolically disadvantaged, compared to homeothermic gulls, as they perhaps are in temperate regions.

For reasons that will be addressed in the discussion (Section 3.3) below, drawing species boundaries has often been problematical. However, most modern lists (e.g., Sibley and Monroe 1990, Burger and Gochfeld 1996) recognize about 50 species in 6 to 7 genera. The overwhelming majority of species are placed in the genus *Larus,* while separated into other genera are the Swallow-tailed Gull (*Creagrus furcatus*) of the Galapagos, the two Kittiwake *Rissa* species, and the high Arctic trio of Sabine's Gull (*Xema sabini*), Ivory Gull (*Pagophila eburnea*), and Ross's Gull (*Rhodostethia rosea*).

Several studies have attempted to clarify relationships between species. Dwight (1925) emphasized plumage differences, separating the large white-headed species from the smaller dark-headed species. Moynihan (1959) followed Tinbergen (1959) in arguing that behavioral patterns of gulls could reflect relationships as accurately as plumage which might be adapted to current ecology. A similar argument was adduced above in respect to sulids. However, Moynihan's work still recognized the white-headed group of gulls identified by Dwight, but split the dark-headed species into two sister groups. Using 117 skeletal and 64 integument characters, Chu (1998) constructed a gull phylogeny that indicated the dark hood was ancestral, and therefore not necessarily indicative of a relationship. This seems a reasonable conclusion given that the sister groups of the gulls (terns, skimmers, and skuas) are also characteristically dark capped. It is a conclusion supported by the recent study on the topic by Crochet et al. (2000) who used sequence data from the mitochondrial control region and cytochrome *b* gene to assess relationships among 32 gull species.

The principal conclusions of Crochet et al.'s (2000) study were as follows. Dark-headed species are not a single clade, but broadly split into two groups, one of which is allied to the large white-headed species. The several dark tropical gull species are not closely related. Their similarity in plumage is therefore interpreted as convergence, specifically the dark feathers being more resistant

FIGURE 3.13 A White Tern pair courting — Christmas Island, Pacific. (Photo by R.W. and E.A. Schreiber.)

to bleaching. The Arctic Sabine's and Ivory Gulls are sister taxa, despite their strikingly different plumages. Ross's Gull was not available for sequencing.

Noting that Sibley and Ahlquist provide a $\Delta T_{50}H$ of value of 4.5 between *Larus* and *Sterna*, and following Moum et al.'s (1994) estimate that one unit of $\Delta T_{50}H$ corresponds to 3 million years of independent evolution, Crochet et al. date the gull-tern split at 13.5 mya. If molecular evolution has proceeded at a constant rate thereafter, then the divisions within the extant gull lineages date back no farther than 6 mya. This sits thoroughly uncomfortably with possible fossil gulls from the middle Oligocene (30 mya) and more certain gulls from the Lower Miocene (Burger and Gochfeld 1996).

3.2.4.3 Subfamily Sterninae

Terns are invariably associated with water, most frequently coasts, but also freshwater wetlands and rivers or pelagic environments (Figure 3.13). They are small to medium birds with a sharp pointed bill and more or less forked tail. Many species have a black cap. Their distribution is cosmopolitan. Species breeding at higher latitudes are mostly migratory.

Most modern lists (e.g., Sibley and Monroe 1990, Gochfeld and Burger 1996) recognize about 45 species in 7 to 10 genera. Following Sibley and Monroe (7 genera, 45 species), the majority of species (32) are placed in the genus *Sterna*. This genus here includes the relatively large crested terns, sometimes split off into the genus *Thalasseus*. The four so-called marsh terns are placed in the genus *Chlidonias*, while the highly distinctive Large-billed Tern (*Phaetusa simplex*) of South American rivers and the Inca Tern (*Larosterna inca*) of the coasts of Peru and Chile belong to monospecific genera. This leaves seven species in three related genera of the noddy group, *Anous* (3), *Gygis* (2), and *Procelsterna* (2). Because some of the forms in this group (for example, the Black Noddy [*A. (tenuirostris) minutus*] and White-capped Noddy [*A. tenuirostris tenuirostris*]) have allopatric distributions, they may or may not be conspecific.

3.2.4.4 Family Rhynchopidae

The skimmers belong to a single genus *Rhynchops* where the lower jaw is markedly longer than the upper and where, uniquely among birds, the eye pupil is not round but closes to a vertical cat-like slit. The three species live on the coasts and large rivers of southeast Asia, tropical Africa, eastern Northern America, and much of Central and South America.

While most authors consider that the terns and gulls are more closely related to each other than either is to the skimmers, the possibility that the noddies are a sister group to the skimmers rather than other terns has been aired by Zusi (1996).

3.2.4.5 Family Alcidae

The auks, comprising 23 extant species and the much-lamented extinct Great Auk (*Pinguinnis impennis*), are a distinct group of diving seabirds confined to the northern hemisphere. Following Gaston and Jones (1998), the auks have a compact body, short wings (very short in the flightless Great Auk), and short tail. Because the webbed feet are set far back on the body, the auks frequently rest on their bellies when ashore. There are 11 primaries and 16 to 21 secondaries on the wings which beat hectically in flight and, slightly bent, provide most of the underwater propulsion. To increase the birds' overall density and thereby facilitate diving, the long bones and breast bone are not pneumatized. Nearly all of the features above reflect compromises imposed on birds which combine the power of flight and active underwater pursuit of prey. The bill is very variable in shape and, in some species, highly ornamented during the breeding season.

Recent studies of relationships among the auk species have principally used anatomy (Strauch 1985), protein polymorphism (Watada et al. 1987), and allozymes in combination with mtDNA data (Friesen et al. 1996). While the results of these studies were not identical, there was substantial agreement, and here Friesen et al.'s phylogeny (Figure 3.14) which identifies six lineages is presented:

1. The Dovekie (*Alle alle*) is grouped with Razorbill (*Alca torda*) and murres *Uria* spp. Had the Great Auk been included in the study, there is little doubt it would have fallen into this group.
2. The puffins are grouped with the Rhinoceros Auklet (*Cerorhinca monocerata*).
3. The planktivorous Pacific auklets form a distinct group.
4. Among the brachyramphine murrelets, the Long-billed (*Brachyramphus perdix*) of the Pacific coasts of Asia was the most divergent, strengthening the case that it should be recognized as a full species, distinct from the Marbled Murrelet (*B. marmoratus*) of the American Pacific (see also Friesen et al. 1996b).
5/6. The synthliboramphine murrelets and guillemots form the two final groups. Whether they are closely related is less certain.

Friesen et al.'s (1996a) study failed to resolve some of the relationships between and within tribes, suggesting periods of "starburst" adaptive radiation during the auks' history. This history was certainly underway 15 mya, for there are unequivocal mid-Miocene fossils. The identity of possible auk fossils from more than 10 million years prior to that is less certain (Olson 1985). Following the early radiation, many of today's auk genera evolved and are represented in the fossil record from 5 mya onward (Gaston and Jones 1998).

Auks today are most richly represented in the Pacific: 17 species confined to that ocean, 2 to the Atlantic, and 4 whose distribution spans both. While it seems likely that the auks originated in the Pacific, the subsequent history of radiation of the various groups in the two oceans is certainly complicated and discussed in some detail by Gaston and Jones (1998). But the modern paucity of Atlantic auk species appears to be a consequence of Pleistocene extinctions, rather than any failure of auk stocks to penetrate to the Atlantic. Thus Olson and Rasmussen (in press) record at least nine auk species from Lower Pliocene deposits of North Carolina (see Chapter 2). There is clearly a parallel between the scarcity of auk species in the North Atlantic today and the absence of albatrosses which likewise disappeared in the Pleistocene (see above). Furthermore, the North Atlantic supports three breeding phalacrocoracids, as compared to six in the North Pacific.

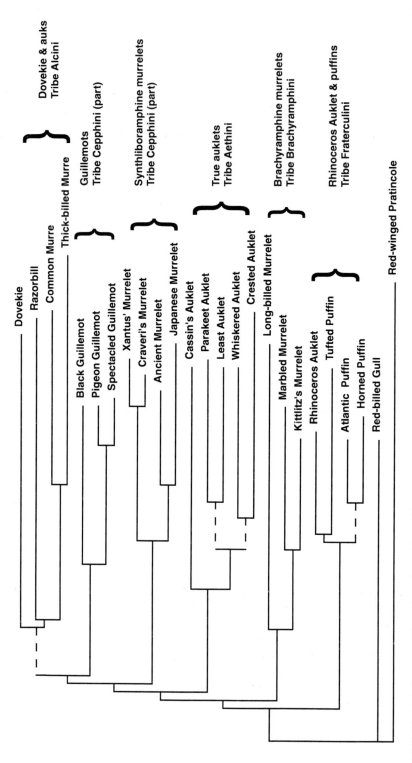

FIGURE 3.14 A phylogeny of the auks redrawn from Friesen et al. (1996a,b). Dashed lines indicate branches with <95% support.

3.3 DISCUSSION

This broad-brush account of seabird systematics and distribution has highlighted many instances of uncertainty, for example, unresolved questions surrounding relationships within the Pelecaniformes. There is not space to discuss them all. Instead two particular issues are considered here, namely, why the species-level taxonomy of some groups, but not others, has proved and continues to prove so contentious, and whether some interesting general patterns of seabird distribution can be discerned.

3.3.1 SPECIES BOUNDARIES

There are four groups of seabirds where species boundaries are difficult to define and conspicuously in a state of flux: the albatrosses, gadfly petrels, southern shags, and larger northern gulls. While this state of affairs could simply represent the fact that it is unreasonable to expect natural diversity always to slot into the constructs of biologists, I suspect the observation may be revealing something more interesting about these groups, and am quite prepared to be criticized for that suspicion.

Let us consider first the albatrosses, gadfly petrels, and southern shags. These birds characteristically nest on islands. Moreover, a significant fraction (3/21 albatrosses, c. 11/39 gadfly petrels, 7/36 cormorants/shags) breeds at just a single island or archipelago (Brooke in press, Enticott and Tipling 1997). In those species breeding at a single island, it must be the case that all individuals return to breed at the island where they themselves were hatched. In more technical terms, natal philopatry is extremely high (Brooke in press). By extension it is also likely to be high in those species breeding at only a few sites.

If, over many generations, seabirds at different stations have evolved slightly different genotypes in response to different conditions, then there might be selection against intermingling of the genotypes, against hybridization between immigrants and those faithful to the natal colony. One way of achieving that is for birds to develop isolating mechanisms, such as divergent plumage, that prevent reproduction and reinforce slight differences already evolved. This is a familiar argument with respect to the evolution of new species (Mayr 1963). However, with high philopatry, few birds will disperse to other colonies. This could reduce selection for plumage divergence. In time the upshot would be birds in widely separated colonies with similar but not identical plumage and structure. But that external similarity need not indicate recent separation of the two populations, or indeed genetic similarity. In summary, an effect of extreme philopatry could be a reduction in the tendency for populations of different colonies to diverge in external appearance. Plumage and morphology would then be a poorer guide than usual to the independent evolutionary history of the birds. Only molecular studies would reveal the extent of independent history and the possible need for redrawing of species boundaries (e.g., Robertson and Nunn 1998).

While many of the difficulties in drawing species boundaries discussed with respect to albatrosses, petrels, and shags arise because the taxa of interest are isolated on remote islands and do not interbreed, the situation is different with respect to gulls. Here the taxa do frequently interbreed and often produce viable hybrids. However, the general observation is that the hybrids are not spreading at the expense of the original taxa. This suggests that some degree of reproductive isolation does exist, and/or that there is selection against the hybrids.

These vexing taxonomic problems at the specific level most acutely affect the larger gull species of the northern temperate regions (Barth 1968, Snell 1991a,b). For example, relationships between the Herring (*Larus argentatus*), Lesser Black-backed (*L. fuscus*), Yellow-legged (*L. cachinnans*), Armenian (*L. armenicus*), and Slaty-backed Gulls (*L. schistisagus*) remain uncertain. Any profound understanding will certainly also take account of the North American cluster of Iceland (*L. glaucoides*), Thayer's (*L. [glaucoides] thayeri*), Kumlien's (*L. [glaucoides] kumlieni*), and Glaucous-winged Gulls (*L. glaucescens*). However, I suggest that the fact that the problems largely involve temperate species and do not extend at the specific level to the tropics gives a clue to the root of the problem. It is that the larid populations became fragmented during the Pleistocene glaciation. When the ice last retreated some 12,000 years ago, the populations re-established contact, allowing the possibility of interbreeding. But, as mentioned above, the evidence is that some degree of reproductive isolation has developed.

Why then are other seabirds characteristic of the northern boreal and temperate zones, for example, the Northern Fulmar and the auks, not bedeviled by similarly confusing species complexes? It would be difficult to make any case that the Fulmar, auks, and gulls differ fundamentally in philopatry. All are known from modern studies to show significant natal dispersal (Birt-Friesen et al. 1992, Dunnet et al. 1979, Harris 1984, Monaghan and Coulson 1977). Indeed significant post-Pleistocene dispersal must have been involved in the expansion of such northern species into their modern ranges.

Various factors such as generation time, population size, and metabolic rate may affect the rate of molecular evolution (Nunn and Stanley 1998 and references therein). If molecular evolution proceeded more rapidly in gulls than the other northern seabirds when their populations were fragmented by the advance of Pleistocene ice, then the gulls might have proceeded further toward reproductive isolation that would become evident when the ice retreated. While gull populations may be smaller than auk populations by roughly half an order of magnitude (Furness 1996, Nettleship 1996), it is not evident that gulls do differ sufficiently in these factors from the other seabirds to explain the matter.

The suggestion offered here is that this difference in species-level taxonomic uncertainty between the gulls and other north temperate seabirds arises because the two groups are more or less strictly associated with offshore waters. Those strictly associated (e.g., Northern Fulmar, auks) will have been pushed south along the essentially north–south axes offered by the east and west coasts of the Pacific and the Atlantic during glacial advances. They will have moved back north during interglacial periods, but, within each ocean, populations will not have been greatly fragmented. On the other hand, the more coastal large gulls will have experienced a complicated history of population fragmentation as the colonies, broadly strung along an east–west axis encompassing inter alia the North Pacific, Great Lakes, North Atlantic, Mediterranean, Black Sea, and Sea of Okhotsk, moved south and north as ice advanced and retreated.

Interestingly, the more marine northern gull species such as the kittiwakes, Ivory Gull, Ross's Gull, and Sabine's Gull present clearly defined species. So do the terns, which, while coastal inhabitants like the large gulls, differ in being long-distance migrants. Such migration may incidentally enhance population homogeneity.

3.3.2 PATTERNS OF SEABIRD DISTRIBUTION

3.3.2.1 Family Level Patterns

No seabird family is found exclusively in the Atlantic or Indian or Pacific Oceans. All seabird families except three are found in both northern and southern hemispheres. The three exceptions are the penguins, diving petrels, and auks which are largely confined to the higher latitudes of their respective hemispheres. The fact that they are also the seabirds most adapted to underwater pursuit of prey is almost certainly not a coincidence. Partly because birds adapted for underwater pursuit of prey may have sacrificed flight efficiency, thereby making the costs of travel between prey patches higher, and partly because underwater pursuit of prey is itself energetically expensive, underwater pursuit of prey is only a viable way of life when prey density is high, which is most likely where marine productivity is high. With the exception of upwelling zones, marine primary productivity is higher at higher latitudes than near the Equator (Begon et al. 1996, Robertson and Gales 1998). Thus this argument is that the penguins, diving petrels, and auks have been confined to their respective hemispheres by an inability to cross the unproductive waters of the tropics. It is tempting also to relate the lower species richness of the most speciose seabird order, the Procellariiformes, at lower latitudes to generally lower productivity there (Figure 3.3).

3.3.2.2 Contrasts between the North Pacific and North Atlantic

While the southern seabird communities either of the Antarctic or the sub-Antarctic are broadly similar wherever around Antarctica they are found, there are much more striking contrasts between the communities of the North Atlantic and North Pacific, especially between about 40 and 60°N. These contrasts include:

1. The absence of breeding shearwaters in the North Pacific.
2. The absence of albatrosses in the North Atlantic.
3. The absence of sulids in the North Pacific.
4. The far greater species (and generic) richness of auks in the North Pacific.

As has been indicated in the family accounts, points 2 to 4 appear historical accidents. The seabird family was represented in the ocean concerned until the Pliocene, and it then disappeared or dwindled during the Pleistocene. Today there are major continental barriers to seabird dispersal at the northern temperate latitudes in question and unproductive tropical waters to the south. Together these constraints have presumably impeded the restoration of the pre-Pleistocene pattern.

The situation with respect to point 1 is different. Shearwaters breed in the Hawaiian archipelago and also in Japanese waters (Streaked Shearwater [*Calonectris leucomelas*]), but none are to be found breeding in the Pacific farther north and east. As argued elsewhere (Brooke in press), this could be related to two non-exclusive factors. The first is the greater species richness of auks in the North Pacific which, like most temperate shearwaters, catch prey underwater. The second is the huge numbers of Short-tailed and Sooty Shearwaters which migrate from the Antipodes into the North Pacific during the northern summer. The second argument is given strength by the fact that North Atlantic breeding shearwaters are mostly found in the northeast Atlantic, and in puny numbers in the northwest Atlantic where transequatorial migrants (especially Greater Shearwaters, *Puffinus gravis*) are concentrated. Both the rich auk community and the huge influx of nonbreeding shearwaters to the North Pacific will reduce prey stocks, and therefore may have contributed to the absence of breeding shearwaters.

3.3.2.3 The Influence of Foraging Technique on Abundance and Distribution

It is intriguing to consider the seabird species with the largest global populations (>10 million individuals; data from del Hoyo et al. 1992, 1996). These species are Chinstrap Penguin (*Pygoscelis antarctica*), Macaroni Penguin (*Eudyptes chrysolophus*), Northern Fulmar, Short-tailed and Greater Shearwater, Antarctic Prion (*Pachyptila desolata*), Salvin's Prion (*P. salvini*), Leach's Storm Petrel (*Oceanodroma leucorhoa*), Common Diving-petrel (*Pelecanoides urinatrix*), Guanay Cormorant (*Leucocarbo bougainvilli*; before recent declines), Black-legged Kittiwake (*Rissa tridactyla*), Sooty Tern (*Sterna fuscata*), Dovekie, Common Murre (*Uria aalge*), Thick-billed Murre (*U. lomvia*), Least Auklet (*Aethia pusilla*), and Atlantic Puffin (*Fractercula arctica*).

While population numbers, of course, provide only a crude index of a species' impact on the ecosystem and may have been reduced in historical times, two points stand out. First, reflecting the higher productivity of higher latitudes, all but two (Guanay Cormorant, Sooty Tern) of the species listed are higher latitude species. Second, the majority of the species obtain their food by underwater pursuit of prey. It appears that, where prey density is high enough to render the underwater pursuit lifestyle viable, then species adopting this lifestyle can become very numerous. They are essentially harvesting prey in three dimensions while the surface feeders are restricted to two. The numerical and biomass dominance in polar or subpolar regions of seabirds feeding by underwater pursuit, using feet or wings for propulsion, is detailed in several studies (Ainley 1977). Where they breed, auks form from 28 to 97% of the breeding seabird biomass (Gaston and Jones 1998). Penguins at South Georgia form 76% of the seabird biomass (Croxall and Prince 1987).

If these arguments have any worth, then we would expect underwater pursuit specialists to be less prominent in the seabird community where productivity was lower. Precisely this argument has already been used to explain the failure of penguins, diving petrels, and auks to cross the Equator. And we might predict that, where a productivity gradient existed at a single latitude, species feeding underwater would form a greater part of the community where productivity was higher.

Among species obtaining food at the surface of the sea, those feeding offshore have potentially a greater area available in which to search for food, because of straightforward geometrical considerations, than do those feeding close to shore. They might therefore have larger populations. Diamond (1978) found support for this idea at several tropical seabird colonies. It is also notable that surface-feeding species with populations in excess of 10 million (Northern Fulmar, Antarctic and Salvin's Prions, Leach's Storm Petrel, Black-legged Kittiwake, Sooty Tern) are all offshore species.

While higher productivity may be one factor contributing to the concentration of certain seabird species or groups, especially the underwater pursuit specialists, to higher latitudes, another factor may be water temperature. As the water becomes warmer nearer the Equator, poikilothermic prey will become more mobile and more difficult to catch. This will further militate against the occurrence of underwater pursuit specialists in warmer waters.

3.3.2.4 Species Level Patterns

While no seabird family is confined by longitude to a single ocean, various species are so confined. In the northern hemisphere this is most evident in the different suite of seabirds found in the North Atlantic and North Pacific. In some cases the species of one ocean are represented by sister taxa in the other. For example, related members of the *Puffinus puffinus* complex breed in the North Atlantic and North Pacific. Similarly, the large *Larus* gulls breeding on the east and west coasts of the lower 48 states of the United States are different but closely related: the Herring and Great Black-backed Gulls (*L. marinus*) in the east, vs. the Western (*L. occidentalis*) and Glaucous-winged Gulls in the west. In other cases the replacement is by less closely related species, for example, the puffins.

Land barriers that might divide seabird species are less manifest in the southern hemisphere than in the northern. Nonetheless, there remain examples of closely related taxa occupying different oceans. Such examples can be from low latitudes (e.g., Red-tailed and Red-billed Tropicbird). However, there are comparable examples from higher southern latitudes where barriers to longitudinal dispersal appear slight. Thus the Greater and Short-tailed Shearwaters are confined, respectively, to the Atlantic and Pacific (Marchant and Higgins 1990). Since allopatric speciation caused by extrinsic barriers to gene flow seems unlikely, I have argued above that philopatry has contributed to genetic divergence in some groups (see also Friesen and Anderson 1997 for a discussion of sulids).

Species distributions are limited not only longitudinally but also latitudinally. As a result one species may replace another along a latitudinal cline, and/or at a temperature discontinuity. Thus the Grey-headed Albatross (*Thalassarche chrysostoma*) tends to have the most southerly distribution of the Southern Ocean mollymawks, and is the species most likely to be met south of the Antarctic Polar Front. Hornby's Storm Petrel (*Oceanodroma hornbyi*) is associated with the cool Humboldt upwelling off Peru and Chile (Murphy 1936). Alternatively, the replacement of one species by another may be associated with salinity differences. In the northern Indian Ocean, Jouanin's Petrel (*Bulweria fallax*) is associated with more saline waters than its congener, Bulwer's Petrel (*B. bulwerii*: Pocklington 1979).

As yet we have limited understanding of what underlies this association between seabirds and particular water bodies. Two examples of studies that indicate the sort of understanding that may emerge can be cited. At the largest possible spatial scale, the body characteristics of nine medium-sized procellariids from the Eastern Tropical Pacific were compared with those of seven species from the Southern Ocean south of 55°S by Spear and Ainley (1998). It emerged that the tropical species had longer wings and tails, bigger bills, and less fat than their polar counterparts. This was interpreted as enabling tropical species to forage economically over large expanses of ocean, catching sparse and often mobile prey. In contrast, the polar species had smaller wings to cope with stronger winds, smaller bills, to catch abundant and not very mobile prey; and larger fat deposits to weather stormy periods. Presumably this relationship between seabird morphology, prey mobility, and climate has arisen as natural selection has acted over very many thousands of years.

At the scale of a species pair with partly nonoverlapping distributions, Thick-billed Murres have a more northerly distribution than Common Murres. They also have shorter, thicker bills that are presumably more efficient for catching a diet that contains more zooplankton than the more fishy diet consumed by the relatively slender-billed Common Murre (Gaston and Jones 1998). This, in turn, raises the possibility that Thick-billed Murres tend to be more planktivorous because food chains tend to be shorter in the Arctic (Briand and Cohen 1987).

In conclusion, the large-scale patterns of seabird distribution are fairly well documented. At a smaller scale, radio-tracking, and more especially satellite-tracking, are allowing researchers to follow individual birds as they search for prey at sea. But the reasons why seabirds of one species should "choose" to forage in a different sea area to a similar, related species often remain obscure. It is such choices, made by the individual, which generate the observed species distribution. Presumably the choice is made in that individual's best interest and reflects the ability to secure prey efficiently, either at or below the sea surface. While ornithologists studying land birds have established links between morphology, habitat chosen, diet, and foraging efficiency (e.g., Partridge 1976, Winkler and Leisler 1985, Grant 1986), comparable studies on seabirds are generally less developed.

ACKNOWLEDGMENTS

I am extremely grateful to Joanna Burger and Betty Anne Schreiber for advice during the preparation of this chapter, to Steven Chown for help with map matters, and to Vicki Friesen and Bryan Nelson for supplying unpublished papers.

LITERATURE CITED

AINLEY, D. G. 1977. Feeding methods in seabirds: a comparison of polar and tropical nesting communities in the eastern Pacific Ocean. Pp. 669–685 *in* Adaptations within Antarctic Ecosystems (G. A. Llano, Ed.). Smithsonian Institution, Washington, D.C.

ANDERSSON, M. 1999. Hybridization and skua phylogeny. Proceedings of the Royal Society of London B 266: 1579–1585.

AUSTIN, J. J. 1996. Molecular phylogenetics of *Puffinus* shearwaters: preliminary evidence from mitochondrial cytochrome *b* gene sequences. Molecular Phylogenetics and Evolution 6: 77–88.

BALLANCE, L. T., R. L. PITMAN, AND S. B. REILLY. 1997. Seabird community structure along a productivity gradient: importance of competition and energetic constraint. Ecology 78: 1502–1518.

BARTH, E. K. 1968. The circumpolar systematics of *Larus argentatus* and *Larus fuscus* with special reference to Norwegian populations. Nytt Magasin for Zoologi 15 (suppl.): 1–50.

BECKER, J. J. 1986. Reidentification of "*Phalacrocorax*" *subvolans* Brodkorb as the earliest record of Anhingidae. Auk 103: 804–808.

BEGON, M., J. L. HARPER, AND C. R. TOWNSEND. 1996. Ecology — individuals, populations and communities. 3rd Edition. Blackwell Science, Oxford.

BIRDLIFE INTERNATIONAL. 2000. Threatened Birds of the World. BirdLife International, Cambridge and Lynx Edicions, Barcelona.

BIRT-FRIESEN, V. L., W. A. MONTEVECCHI, A. J. GASTON, AND W. S. DAVIDSON. 1992. Genetic structure of thick-billed murre (*Uria lomvia*) populations examined using direct sequence analysis of amplified DNA. Evolution 46: 267–272.

BOURNE, W. R. P. 1975. The lachrymal bone in the genus *Bulweria*. Ibis 117: 535.

BRAUN, M. J., AND R. T. BRUMFIELD. 1998. Enigmatic phylogeny of skuas: An alternative hypothesis. Proceedings of the Royal Society of London B 265: 995–999.

BRETAGNOLLE, V., C. ATTIÉ, AND E. PASQUET. 1998. Cytochrome-*b* evidence for validity and phylogenetic relationships of *Pseudobulweria* and *Bulweria* (Procellariidae). Auk 115: 188–195.

BRIAND, F., AND J. E. COHEN. 1987. Environmental correlates of food chain length. Science 238: 956–960.

BROOKE, M., AND T. BIRKHEAD. 1991. The Cambridge Encyclopedia of Ornithology. Cambridge University Press, Cambridge.

BROOKE, M. DE L. 1999. Review of Albatross Biology and Conservation (G. Robertson and R. Gales, Eds.). Surrey Beatty and Sons, Chipping Norton. Polar Record 35: 351–353.

BROOKE, M. DE L. In press. Albatrosses, Petrels and Allies. Oxford University Press, Oxford.

BURGER, J., AND GOCHFELD, M. 1996. Family Laridae. Pp. 572–623 *in* Handbook of the Birds of the World, Vol. 3 (J. Del Hoyo, A. Elliot, and J. Sargatal, Eds.). Lynx Edicions, Barcelona.

CARBONERAS, C. 1992. Family Hydrobatidae. Pp. 258–271 *in* Handbook of the Birds of the World, Vol. 1 (J. Del Hoyo, A. Elliot, and J. Sargatal, Eds.). Lynx Edicions, Barcelona.

CHU, P.C. 1998. A phylogeny of the gulls (Aves: Larinae) inferred from osteological and integumentary characters. Cladistics 14: 1–43.

COHEN, B. L., A. J. BAKER, K. BLECHSCHMIDT, D. L. DITTMANN, R. W. FURNESS, J. A. GERWIN, A. J. HELBIG, J. DE KORTE, H. D. MARSHALL, R. L. PALMA, H.-U. PETER, R. RAMLI, I. SIEBOLD, M. S. WILLCOX, R. H. WILSON, AND R. M. ZINK. 1997. Enigmatic phylogeny of skuas (Aves: Stercorariidae). Proceedings of the Royal Society of London B 264: 181–190.

CHOWN, S. L., K. J. GASTON, AND P. H. WILLIAMS. 1998. Global patterns in species richness of pelagic seabirds: the Procellariiformes. Ecography 21: 342–350.

CROCHET, P.-A., F. BONHOMME, AND J.-D. LEBRETON. 2000. Molecular phylogeny and plumage evolution in gulls (Larini). Journal of Evolutionary Biology 13: 47–57.

CROXALL, J. P., AND P. A. PRINCE. 1987. Seabirds as predators on marine resources, especially krill, at South Georgia. Pp. 347–368 *in* Seabirds — Feeding Ecology and Role in Marine Ecosystems (J. P. Croxall, Ed.). Cambridge University Press, Cambridge.

CROXALL, J. P., Y. NAITO, A. KATO, P. ROTHERY, AND D. R. BRIGGS. 1991. Diving patterns and performance in the Antarctic blue-eyed shag *Phalacrocorax atriceps*. Journal of Zoology (London) 225: 177–199.

DEL HOYO, J., A. ELLIOT, AND J. SARGATAL. (Eds.) 1992. Handbook of the Birds of the World, Vol. 1. Lynx Edicions, Barcelona.

DEL HOYO, J., A. ELLIOT, AND J. SARGATAL. (Eds.) 1996. Handbook of the Birds of the World, Vol. 3. Lynx Edicions, Barcelona.

DEVILLERS, P. 1978. Distribution and relationships of South American skuas. Gerfaut 68: 374–417.

DIAMOND, A. W. 1978. Feeding strategies and population size in tropical seabirds. American Naturalist 12: 215–223.

DORST, J., AND J.-L. MOUGIN. 1979. Family Phalcrocoracidae. Pp. 162–179 *in* Checklist of Birds of the World (E. Mayr and G. W. Cottrell, Eds.). Harvard University Press, Cambridge.

DUNNET, G. M., J. C. OLLASON, AND A. ANDERSON. 1979. A 28-year study of breeding Fulmars *Fulmarus glacialis* in Orkney. Ibis 121: 293–300.

DWIGHT, J. 1925. The gulls (Laridae) of the world; their plumage, moults, variations, relationships and distribution. Bulletin of the American Museum of Natural History 52: 63–401.

ELLIOTT, A. 1992. Family Pelecanidae. Pp. 290–311 *in* Handbook of the Birds of the World, Vol. 1 (J. Del Hoyo, A. Elliot, and J. Sargatal, Eds.). Lynx Edicions, Barcelona.

ENTICOTT, J., AND D. TIPLING. 1997. Photographic Handbook of the Seabirds of the World. New Holland, London.

FISHER, J. 1952. The Fulmar. Collins, London.

FORDYCE, R. E., AND C. M. JONES. 1990. Penguin history and new fossil material from New Zealand. Pp. 419–446 *in* Penguin Biology (L. S. Davis and J. T. Darby, Eds.). Academic Press, San Diego, CA.

FRIESEN, V. L., AND D. J. ANDERSON. 1997. Phylogeny and evolution of the Sulidae (Aves: Pelecaniformes): a test of alternative modes of speciation. Molecular Phylogenetics and Evolution 7: 252–260.

FRIESEN, V. L., A. J. BAKER, AND J. F. PIATT. 1996a. Phylogenetic relationships within the Alcidae (Charadriiformes: Aves) inferred from total molecular evidence. Molecular Biology and Evolution 13: 359–367.

FRIESEN, V. L., J. F. PIATT, AND A. J. BAKER. 1996b. Evidence from cytochrome *b* sequences and allozymes for a 'new' species of alcid: the long-billed murrelet (*Brachyramphus perdix*). Condor 98: 681–690.

FRIESEN, V. L., D. J. ANDERSON, T. E. STEEVES, H. JONES, AND E. A. SCHREIBER. 2000. The Nazca booby: a challenge to the classical model of speciation. Proceedings of the Royal Society of London B, submitted.

FURNESS, R. W. 1996. Family Stercorariidae. Pp. 556–571 *in* Handbook of the Birds of the World, Vol. 3 (J. Del Hoyo, A. Elliot, and J. Sargatal, Eds.). Lynx Edicions, Barcelona.

GASTON, A. J., AND I. L. JONES. 1998. The Auks. Oxford University Press, Oxford.

GOCHFELD, M., AND J. BURGER. 1996. Family Sternidae. Pp. 624–667 *in* Handbook of the Birds of the World, Vol. 3 (J. Del Hoyo, A. Elliot, and J. Sargatal, Eds.). Lynx Edicions, Barcelona.

GRANT, P. R. 1986. Ecology and evolution of Darwin's Finches. Princeton University Press, Princeton.

GRÉMILLET, D., G. ARGENTIN, B. SCHULTE, AND B. M. CULIK. 1998. Flexible foraging techniques in breeding cormorants *Phalacrocorax carbo* and Shags *Phalacrocorax aristotelis*: benthic or pelagic feeding? Ibis 140: 113–119.

HARRIS, M. P. 1984. The Puffin. T. and A. D. Poyser, Calton.

IMBER, M. J. 1985. Origins, phylogeny and taxonomy of the gadfly petrels *Pterodroma* spp. Ibis 127: 197–229.

KENNEDY, M., H. G. SPENCER, AND R. D. GRAY. 1996. Hop, step and gape: do the social displays of the Pelecaniformes reflect phylogeny? Animal Behaviour 51: 273–291.

LATHAM, J. 1785. A General Synopsis of Birds, Vol. 3. London.

MARCHANT, S., AND P. J. HIGGINS (Eds.). 1990. Handbook of Australian, New Zealand and Antarctic Birds. Vol. IA. Oxford University Press, Oxford.

MATHEWS, G. M., AND T. IREDALE. 1915. On some petrels from the north-east Pacific Ocean. Ibis (Tenth Series) 3: 572–609.

MAYR, E. 1963. Animal Species and Evolution. Belknap Press, Cambridge.

MEYER, A. 1994. Shortcomings of the cytochome *b* gene as a molecular marker. Trends in Ecology and Evolution 9: 278–280.

MONAGHAN, P., AND J. C. COULSON. 1977. Status of large gulls nesting on buildings. Bird Study 24: 89–104.

MOUM, T., S. JOHANSEN, K. E. ERIKSTAD, AND J. F. PIATT. 1994. Phylogeny and evolution of the auks (subfamily Alcinae) based on mitochondrial DNA sequences. Proceedings of the National Academy of Sciences USA 91: 7912–7916.

MOYNIHAN, M. 1959. A revision of the family Laridae (Aves). American Museum Novitates 1928: 1–42.

MURPHY, R. C. 1936. Oceanic birds of South America. American Museum of Natural History, New York.

NELSON, J. B. 1978. The Sulidae — Gannets and Boobies. Oxford University Press, Oxford.

NELSON, J. B. In press. Pelecaniformes in Bird Families Series. Oxford University Press, Oxford.

NETTLESHIP, D. N. 1996. Family Alcidae. Pp. 678–722 *in* Handbook of the Birds of the World, Vol. 3 (J. Del Hoyo, A. Elliot, and J. Sargatal, Eds.). Lynx Edicions, Barcelona.

NUNN, G. B., AND S. E. STANLEY. 1998. Body size effects and rates of cytochrome *b* evolution in tube-nosed seabirds. Molecular Biology and Evolution 15: 1360–1371.

NUNN, G. B., J. COOPER, P. JOUVENTIN, C. J. R. ROBERTSON, AND G. G. ROBERTSON. 1996. Evolutionary relationships among extant albatrosses (Procellariiformes: Diomedeidae) established from complete cytochrome-*b* gene sequences. Auk 113: 784–801.

OLSON, S. L. 1985. The fossil record of birds. Pp. 79–238 *in* Avian Biology, Vol. 8 (D. S. Farner, J. R. King, and K. C. Parkes, Eds.). Academic Press, London.

OLSON, S. L., AND P. C. RASMUSSEN. In press. Miocene and Pliocene Birds from the Lee Creek Mine, North Carolina. *In* Geology and Paleontology of the Lee Creek Mine, North Carolina III (E. R. Clayton and D. J. Bohaska, Eds.). Smithsonian Contributions to Paleobiology 90.

OLSON, S. L., AND K. I. WARHEIT. 1988. A new genus for *Sula abbotti*. Bulletin of the British Ornithologists' Club 108: 9–12.

ORTA, J. 1992. Family Phalacrocoracidae. Pp. 326–353 *in* Handbook of the Birds of the World, Vol. 1 (J. Del Hoyo, A. Elliot, and J. Sargatal, Eds.). Lynx Edicions, Barcelona.

PARTRIDGE, L. 1976. Field and laboratory observations on the foraging and feeding techniques of Blue Tits (*Parus caeruleus*) and Coal Tits (*P. ater*) in relation to their habitats. Animal Behaviour 24: 534–544.

PENNYCUICK, C. J. 1982. The flight of petrels and albatrosses (Procellariiformes), observed in South Georgia and its vicinity. Philosophical Transactions of the Royal Society of London B 300: 75–106.

PETERS, J. L. 1934. Checklist of Birds of the World. Vol. 2. Harvard University Press, Cambridge.

PETERS, J. L. 1979. Checklist of Birds of the World. Vol. 1 (2nd ed.). Harvard University Press, Cambridge.

PITMAN, R. L. AND J. R. JEHL. 1998. Geographic variation and reassessment of species limits in the "masked" boobies of the eastern Pacific Ocean. Wilson Bulletin 110: 155–170.

POCKLINGTON, R. 1979. An oceanographic interpretation of seabird distributions in the Indian Ocean. Marine Biology 51: 9–21.

ROBERTSON, C. J. R., AND G. B. NUNN. 1998. Towards a new taxonomy for albatrosses. Pp. 13–19 *in* Albatross Biology and Conservation (G. Robertson and R. Gales, Eds.). Surrey Beatty and Sons, Chipping Norton.

ROBERTSON, G., AND R. GALES (Eds.). 1998. Albatross Biology and Conservation. Surrey Beatty and Sons, Chipping Norton.

ROGERS, M. J. 1996. Report on rare birds in Great Britain in 1995. British Birds 89: 481–531.

ROGERS, M. J. 1998. Report on rare birds in Great Britain in 1997. British Birds 91: 455–517.

SIBLEY, C. G., AND J. E. AHLQUIST. 1990. Phylogeny and classification of birds — a study in molecular evolution. Yale University Press, New Haven, CT.

SIBLEY, C. G., AND B. L. MONROE. 1990. Distribution and Taxonomy of Birds of the World. Yale University Press, New Haven, CT.

SIEGEL-CAUSEY, D. 1988. Phylogeny of the Phalacrocoracidae. Condor 90: 885–905.

SIEGEL-CAUSEY, D. 1997. Phylogeny of the Pelecaniformes: molecular systematics of a primative group. Pp. 159–171 *in* Avian Molecular Evolution and Systematics (D. P. Mindell, Ed.). Academic Press, San Diego, CA.

SIMPSON, G. G. 1976. Penguins — Past and Present, Here and There. Yale University Press, New Haven, CT.

SNELL, R. R. 1991a. Interspecific allozyme differentiation among North Atlantic white-headed larid gulls. Auk 108: 319–328.

SNELL, R. R. 1991b. Variably plumaged Icelandic Herring Gulls reflect founders not hybrids. Auk 108: 329–341.

SPEAR, L. B., AND D. G. AINLEY. 1998. Morphological differences relative to ecological segregation in petrels (Family: Procellariidae) of the Southern Ocean and Tropical Pacific. Auk 115: 1017–1033.

STEADMAN, D. W., S. E. SCHUBEL, AND D. PAHLAVAN. 1988. New subspecies and new records of *Papasla abbotti* (Aves, Sulidae) from archaeological sites in the Tropical Pacific. Proceedings of the Biological Society of Washington 101: 487–495.

STRAUCH, J. G. 1985. The phylogeny of the Alcidae. Auk 102: 520–539.

TINBERGEN, N. 1959. Comparative studies of the behaviour of gulls (Laridae): a progress report. Behaviour 15: 1–70.

VAN TETS, G. F. 1965. A comparative study of some social communication patterns in the Pelecaniformes. Ornitholological Monographs 2: 1–88.

VAN TYNE, J., AND A. J. BERGER. 1966. Fundamentals of Ornithology. John Wiley & Sons, New York.

VERHEYDEN, C., AND P. JOUVENTIN. 1994. Olfactory behavior of foraging Procellariiforms. Auk 111: 285–291.

WARHAM, J. 1990. The petrels — their ecology and breeding systems. Academic Press, London.

WARHEIT, K. I. 1990. The phylogeny of the Sulidae (Aves: Pelecaniformes) and the morphometry of flight-related structures in seabirds: a study of adaptation. Ph.D. thesis, University of California, Berkeley.

WARHEIT, K. I. 1992. A review of the fossil seabirds from the Tertiary of the North Pacific: plate tectonics, paleoceanography, and faunal change. Paleobiology 18: 401–424.

WATADA, M., R. KAKIZAWA, N. KURODA, AND S. UTIDA. 1987. Genetic differentiation and phylogenetic relationships of an avian family, Alcidae (auks). Journal of the Yamashina Institute of Ornithology 19: 79–88.

WATSON, G. E. 1968. Synchronous wing and tail molt in diving petrels. Condor 70: 182–183.

WEIMERSKIRCH, H., AND P. M. SAGAR. 1996. Diving depths of Sooty Shearwaters *Puffinus griseus*. Ibis 138: 786–788.

WILLIAMS, T. D. 1995. The Penguins — Spheniscidae. Oxford University Press, Oxford.

WINKLER, H., AND B. LEISLER. 1985. Morphological aspects of habitat selection in birds. Pp. 415–434 *in* Habitat Selection in Birds (M. L. Cody, Ed.). Academic Press, New York.

ZUSI, R. L. 1996. Family Rhynchopidae. Pp. 668–677 *in* Handbook of the Birds of the World, Vol. 3 (J. Del Hoyo, A. Elliot, and J. Sargatal, Eds.). Lynx Edicions, Barcelona.

Lesser Frigatebird Males Court Females Flying Overhead

4 Colonial Breeding in Seabirds

John C. Coulson

CONTENTS

4.1 INTRODUCTION

The word "colony" has several different definitions. Its use with respect to seabirds should not be confused with the meaning when applied to human society, where, from Roman times, it has indicated a group of people under the jurisdiction of a country some distance away. In a zoological sense, "colony" describes a group of individual organisms that live close together. It also carries an implication of communication and collaboration, resulting in positive (beneficial) interactions between individuals. In ornithology, the term is usually restricted to a group of individuals at a breeding site, while "flock" is applied to birds which are gregarious at other times of the year or when away from the breeding site. Several authors have considered practical definitions of a colony (Buckley and Buckley 1979, Kushlan 1986, Kharitonov and Siegel-Causey 1988), but others have not readily accepted these definitions, mainly because they do not apply to all species or they include implications that have not been fully researched. In mathematical terms, the individuals in

FIGURE 4.1 Australian Gannet colony illustrating a dense colony in which birds nest about a beaks' distance apart resulting in the necessity for frequent communication among neighbors to signal intentions. (Photo by J.B. Nelson.)

a colony are clumped or aggregated in space, with other apparently suitable areas remaining unoccupied more frequently than would be expected by chance. One of several methods of demonstrating this is to show that, on average, the nearest neighbor to each pair or nest is significantly closer than would be expected by chance. In sessile organisms such as corals and sponges, the colony is a permanent group, persisting for the lifetime of the individuals and often for longer. The characteristic of permanence is also applicable to many seabird colonies, but in this case it is because of the breeding-site tenacity of many adult birds, the stability of the nesting area, e.g., cliffs and islands, and also because successive generations are attracted to the same colony sites.

There is a further important difference between a colony and a flock of birds. In a colony, the pairs have the same neighbors, in the same spatial positions for most of the breeding season. In many species, many of the individual neighbors are retained in successive breeding seasons. In a flock, a bird rarely retains the same neighbor for more than a few minutes, although in some species, such as geese and swans, pairs and family groups often remain together within flocks. These differences between a flock and colony are important because in the former, the change of position limits to a much greater extent the potential complexity of behavior and other interactions that could develop between individuals, whereas in a colony, interactions between the same neighboring individuals can develop and accumulate over time and present an opportunity for more complex interactions and effects to develop.

In birds, almost all species which are colonial can readily be identified as such by their spatial distribution (Figure 4.1). In most cases, there is no practical problem in recognizing which are colonial species; the clumping of nests is obvious and the individuals or pairs are in close proximity. However, in a few species which nest some distance apart from each other, actual measurement may be required to confirm that the nearest neighbor is closer than would be expected by chance. For example, the distance between pairs of some "great albatrosses" and Bonaparte's Gulls *(Larus philadelphia)* may be 100 m or more, but despite this distance, the nest sites are still clumped, with many apparently suitable nesting areas remaining unoccupied. The term "loosely colonial" (Cramp and Simmons 1977) has been applied to such situations, but this term is not very helpful because it is descriptive rather than functional and fails to take into account the manner in which the birds communicate and react with each other. The important question is "To what extent do neighboring pairs interact?" and this still needs to be investigated. Colonial breeding in seabirds has been discussed and reviewed on several occasions, particularly by Gochfeld (1980) and Wittenberger and Hunt (1985), while Brown and Brown (1996) and Orians (1961) have considered it in land birds.

It remains to be established whether the selective pressures which are and have been involved in colonial breeding in seabirds are similar in other birds, and in colonial mammals, such as seals and bats. In many animals, there are considerable disadvantages in living close to each other. This is the basis of density-dependent mortality and reduced breeding success. Colonial breeding presumably only occurs when the disadvantages are outweighed by advantages in coloniality, at least in the long term. Adverse density-dependent effects have not yet been reported in seabirds, but this may simply reflect the difficulty of studying such effects at sea and away from colonies.

This chapter includes reviews of the published literature on colonial breeding in seabirds. It leans heavily on studies over 40 years on Black-legged Kittiwakes (*Rissa tridactyla*), not primarily because it is the main study animal of the author, but because parallel studies on other seabird species are often yet to be made. It also includes personal experience and thoughts, and unpublished analyses on the topic. This author would have liked to include more between species comparisons, but as yet there are insufficient data on colonial breeding in most seabirds, and until more studies are available, in-depth between-species analyses of aspects of coloniality are not possible.

4.2 WHAT IS A SEABIRD COLONY AND WHAT ARE ITS LIMITS?

Essentially, a seabird colony is a group of breeding individuals which associate together and maintain the association to an extent that is greater (often much greater) than that expected by chance. Such a group needs a mechanism to maintain the grouping and some form of communication is necessary to achieve this. In many cases, the approximate limits of a colony are obvious because of the large distance between that group and the next. But the practical problems of recognizing whether there are one or more functional groups (= colony) involved where huge numbers of birds nest together are not easily determined. The use of "subcolony" or similar terms indicating subgroups (e.g., Buckley and Buckley 1972, Gochfeld 1979, Burger and Shisler 1980) may be useful in some respects. However, in cases where this term has been used, the interactions, if any, between the subgroups have not been investigated and the opportunity in these studies of understanding the functioning of colony structure as a whole has not been grasped.

When birds in a colony are spread out over a relatively large area, doubts may exist as to whether the group exists as a functional unit. To maintain a group or colony, some form of communication is necessary. Difficulties in achieving this exist when the birds are separated by physical features of their nesting area, e.g., around a headland sea cliff or on each side of a hill, and where all of the birds cannot see or hear each other. Such situations beg the practical question of "When does a colony become two colonies?" Fundamentally, it is to be expected that there is direct or possibly an indirect interaction between the individuals within a colony, and, in some cases, simple observations of the responses birds make to each other can resolve the extent of a colony. Often, interaction between birds in a colony is by sight and this is evident in the silent "panic flights" of some terns and gulls, where all of the individuals respond virtually simultaneously and synchronously, even without the external stimulus of a predator. There are also vocal interactions, evident by the waves of calling which repeatedly spread through colonies of some species, e.g., Common and Thick-billed Murres (*Uria aalge* and *U. lomvia*), some gulls, penguins, and albatrosses. It is very likely that calls play a particularly important part in communication in species which visit the colony at night, such as in the small petrels and other species which nest in dark areas such as caves, under boulders, and in burrows. It is conceivable that smell and even sonar responses may also be important in some nocturnal species, although this is an area which still requires more critical research and experiment.

Most diurnal species use calls and postures as a means of communicating (see Chapter 10), so that cliff-nesting seabirds which spread around a headland, or ground nesting species on both sides of a ridge, have pairs which are unlikely to be able to communicate with all others (unless the species is one which has aerial displays). Some Herring Gull (*Larus argentatus*) "colonies" extend continuously for over 2 km (Chabryzk and Coulson 1976, Burger and Shisler 1980), and Black-legged Kittiwake colonies sometimes spread around headlands and also can extend for several kilometers. In such colonies, the birds at each end of the group are most unlikely to be in direct communication. Yet the use of "subcolonies" is not really a helpful term because when used, the methods of interaction are not normally considered and the term is usually used in an arbitrary manner, for example, to achieve constancy during surveys, e.g., Northern Fulmar (*Fulmarus glacialis*) (Fisher 1952) and Black-legged Kittiwake (Coulson 1963).

4.3 FUNCTIONAL STRUCTURE OF A COLONY

Coulson and Dixon (1979) made a study of the functional nature and structure of Black-legged Kittiwake colonies, relying on the behavior of the birds to indicate the area within which the birds constituted an association. Early in the season, when birds first returned to the colony, social behavior was primarily limited to "panic" or "dread" flights from the colony. The birds on cliff nesting sites over distances up to several hundred meters would frequently, spontaneously, silently, and synchronously leave the breeding areas on the cliffs. Such responses are, apparently, synchronized by vision; the synchrony was lost when different groups of birds could not see each other. At this time of year, and at the end of breeding season, a series of such panic flights often synchronously terminated the daily occupation of the colony.

At the beginning of the breeding season, kittiwakes occupy a colony for parts of several days. They become less nervous, panic flights stop, and the distances between interacting individuals becomes restricted to much shorter distances. At this stage, the main interactions occur only between pairs and small groups of pairs, and are solely triggered by the greeting ceremony performed between members of a pair. These greeting ceremonies result in the characteristic social outbreak of "kittiwaking," calling among groups of pairs. In contrast, solitary birds on nest sites (i.e., birds whose mates are temporarily away) usually show no or only minimal reaction to a greeting ceremony performed by a neighboring pair. Detailed studies of kittiwakes based on this response show that the social reaction between pairs extended for less than 2 m (Coulson and Dixon 1979), and rarely produced responses between pairs 5 m or more apart (Figure 4.2).

This greeting reaction had a number of characteristics:

1. Each individual pair was stimulated more frequently if there were many pairs in close proximity (within 2 m), i.e., where the nest density was high.
2. The reaction by other pairs to an arrival and reuniting of the focal pair was over a very limited distance, but when considered over a period of time, the reactions linked the members of a colony together.
3. The timing of breeding in the kittiwake was closely correlated with the density of other pairs immediately around a nesting pair (Coulson and White 1960). It is presumed that this difference was the result of differences in the frequency of greeting ceremonies arising from the density of nesting birds.
4. The density of neighboring pairs tended to be lower at the edge of the colony, since other pairs were absent from the edge side, more potential nest-sites were unoccupied, and, as a result, the overall density was usually lower.
5. By inference, isolated pairs will receive little of this "social" stimulation from other pairs of kittiwakes and will be appreciably delayed in display and nest-building. As a result, breeding by isolated pairs is inhibited. (The same inhibitor also prevents relaying in pairs which have lost their eggs, resulting in relaying being restricted to early breeders, which lose their eggs soon after laying.)

These results explain a number of the characteristics of breeding in kittiwake colonies and also identify the nature of the structure within a kittiwake colony. They explain why the spread of breeding in Black-legged Kittiwake colonies is closely linked with the maximum density in the

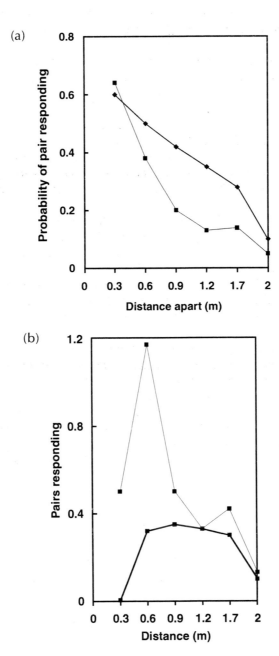

FIGURE 4.2 The responses of other pairs of Black-legged Kittiwakes to the reuniting of a pair in relation to distance between nests. (a) The probability in relation to distance of a particular pair responding to the reuniting of the focal pair. (b) The number of pairs responding at given distances from the focal pair which are reunited and engage in mutual courtship. The thick line indicates low density areas and the thin line high density areas. The data from high and low density areas are significantly different ($p < 0.01$), indicating that birds at high density make more responses, although they may actually respond to a small proportion of the times that other pairs reunite in their immediate neighborhood. (After Coulson and Dixon 1979.)

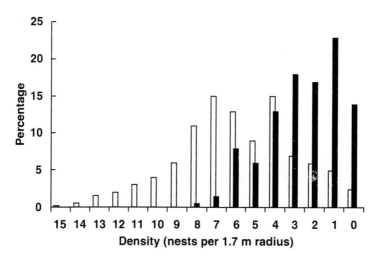

FIGURE 4.3 The variation in local density of nests in two Black-legged Kittiwake colonies. Note that both colonies have a proportion of nests at low density and the colonies differ by the maximum density reached. Breeding starts first at high density (at the left) and last at low density (at the right). Thus the spread of breeding is greatest in the colonies with the highest local densities. These effects are shown in the spread of breeding in two actual colonies in Figure 4.6. (After Coulson and White 1960.)

colony (Figure 4.3) where laying takes place first, and that laying ends synchronously in all neighboring colonies, since all have some low density areas. These observations were particularly valuable in giving an insight into the functional structure of a colony. In a kittiwake colony there is a very short distance over which successful breeding birds, even when changing mates, will normally change nesting sites (Coulson and Thomas 1983, Fairweather and Coulson 1995).

4.4 THEORIES OF THE FUNCTIONS AND ADVANTAGES OF COLONIAL BREEDING

4.4.1 SHORTAGE OF NESTING SITES

One of the earliest suggestions as to why some birds nest in colonies was that they use nesting sites which are extremely limited in space and, as a result, they crowd together. While some colonial species have precise nest-site requirements and nest in very restricted areas, many show only broad requirements found in many places. For many species, it is difficult to envisage that in the past, suitable nesting sites were sufficiently scarce to force individuals to nest very close together, unless numbers were many times greater than now.

The size and extent of some colonies are huge, but it is difficult to believe that alternative nesting areas do not exist, and this view is supported by species which are currently spreading to new areas and forming new colonies. However, some oceanic species have very few land areas on which to nest, yet have vast oceans over which to feed. This may be the reason for large numbers of Greater Shearwaters (*Puffinus gravis*) nesting on two islands in the Tristan da Cunha group and on Gough Island, where some pairs are apparently forced to lay above ground (Rowan 1952). But why this species does not breed elsewhere on other islands within its range is not known.

4.4.2 DEFENSE AGAINST PREDATION

There is much information showing the effect of large numbers intimidating or confusing the predator by the high density of potential prey. Colonies of Arctic (*Sterna paradisaea*) and Common Terns (*S. hirundo*) show the effect of large numbers of animals in reducing the effectiveness of predators (Hamilton 1971). For example, a territorial predator may exclude others of its own species, so there may be many prey in a colony but few predators to exploit them. Other effects operate through confusing or repeatedly diving at the heads of predators. Avian predators may be harried in flight by a group of terns, but the attacks are much less successful when only a few terns are involved. There is, however, a paradox about colonial breeding having evolved as an anti-predator device, since most colonies are situated in places free or relatively free from predators. Colonies on sea cliffs, in trees, on islands, and even on buildings are in situations which reduce or totally avoid predation by mammals. It seems surprising that colonial breeding should be evolved as an anti-predator device when the species involved choose to breed in localities relatively free from predators. In any case, the nature of the predator determines the likely outcome of a predation attempt. A predator wanting a single item of prey has to find the whole colony before obtaining its meal. In contrast, a predator which needs to take several prey items for a single meal is probably made to search harder and longer if the prey is evenly spread out and not colonial. Large predators, such as man, dogs, and foxes, are little deterred by diving Common Terns and species of crested terns do not even intimidate mammalian predators by swooping at them. Royal Terns (*Sterna maxima*) seem to be unable to defend their eggs from Laughing Gulls (*Larus atricilla;* Buckley and Buckley 1972) and Arctic Terns are unable to defend their eggs against individual European Starlings (*Sturnus vulgaris*) and Ruddy Turnstones *(Arenaria interpres*; J. C. Coulson unpublished). While some bird species have developed a social defense against certain predators by breeding in a colony, it seems unlikely that predation was the common factor responsible for development of colonial breeding in the first place.

Frank Fraser Darling (1938), in his book *Bird Flocks and the Breeding Cycle,* envisaged colonial breeding as an anti-predator device that produced a selective advantage as a result of colonial birds breeding more synchronously. He believed that the greater the size of the colony, the greater the synchrony, and hence the reduced impact of predators by swamping them with a temporary overabundance of prey, producing a higher breeding success (Figure 4.4). He provided quantitative evidence from studies on Herring Gull and Lesser Back-backed Gull (*Larus fuscus*) colonies, but his sample sizes were relatively small and the results did not show statistically significant differences in breeding success between large and small colonies.

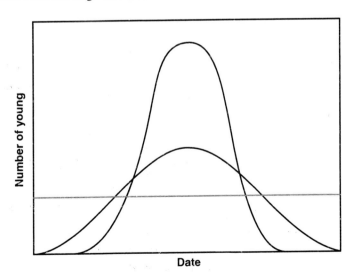

FIGURE 4.4 A diagrammatic representation of Darling's idea of synchrony of breeding reducing the impact of predation. The horizontal line represents the daily requirements of predators; through synchrony, the proportion of chicks taken by the predator is less in the more synchronized than in the less synchronized colony. This diagram should be compared with Figure 4.7 which shows actual results from two Black-legged Kittiwake colonies.

4.4.3 A Colony Is a Safe Place

Subadult birds visit their natal colony and other colonies prior to breeding as they make a decision on where to breed, although little is known about the actual decision-making process. Presumably, a safe nesting site is an important factor in such a decision and the presence of young is a good indicator of this. Individuals of many species are highly philopatric, returning to breed where they hatched, while others move into colonies elsewhere (see discussion of philopatry in Section 4.7, number 9).

4.4.4 Social Stimulation

Work by MacRoberts and MacRoberts (1972) on Herring and Lesser Black-backed Gulls found no evidence of socially induced synchrony in colonies. In contrast, Parsons (1976) showed that synchrony occurred within, but not between, 52 small areas of about 12 m × 12 m within a Herring Gull colony. He also found that breeding was earliest in central and latest in peripheral areas of the colony. The statistically significant differences were mainly caused by synchronous breeding in old birds within small areas. The later breeding in young birds, which tended to be a high proportion of those breeding at the periphery, also contributed to this effect. Parsons (1975) also demonstrated that the highest hatching success in this species was among those individuals breeding when most of their neighbors were at a similar stage of breeding (Figure 4.5), thus able to swamp predators (in this case, cannibal Herring Gulls) with excess food. Overall, very early and late laying pairs were much less successful. He went on to demonstrate this effect was indeed relative to the time of breeding of neighbors and not to actual date by field experiments which altered the timing of breeding. When substantial groups were delayed by egg removal, the time of optimal breeding was also delayed, but again was highest when synchronous with most of the immediate neighbors (Figure 4.6). The fact that the time of peak of breeding success moved shows clearly that success was not linked with a peak of food availability for the parents (and as is the case in Great Tits *Parus major;* Perrins 1970), but to a social effect within the colony.

Re-examination of the "social stimulation effect" in the Black-legged Kittiwake produced very different findings to those of Darling. Larger colonies are less synchronous, the opposite of that expected from the Darling effect (Coulson and White 1956; Figure 4.7). More detailed investigation on kittiwakes showed that local density, and not colony size or age composition, was the important

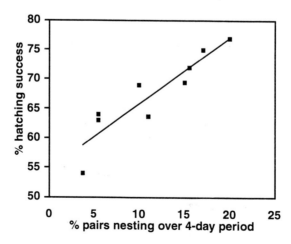

FIGURE 4.5 The hatching success in Herring Gulls in relation to the percentage of pairs nesting in each 4-day period. The correlation is $r_7 = 0.92$ and is highly significant ($p < 0.01$). The points in the bottom left-hand corner of the graph include both early and late breeding birds. Hatching success is highest in those birds which laid when many other individuals also laid. (After Parsons 1975.)

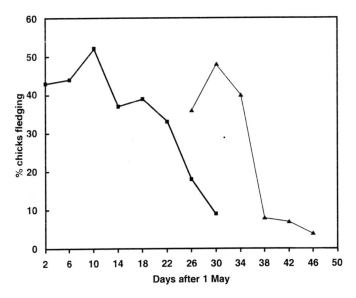

FIGURE 4.6 The fledging success in Herring Gulls laying on different dates in a control (squares and thick line) and an experimentally delayed colony (triangles and thin line) about 200 m apart. The birds in the experimental group were delayed by removing early laid eggs. In both cases, the highest success is among those breeding relatively early in each colony and so receiving protection from their immediate neighbors with eggs and chicks. Note the appreciable difference in the fledging success of the two groups which laid between 26 May and 2 June. The main cause of loss was caused by a small number of cannibal Herring Gulls. (After Parsons 1975.)

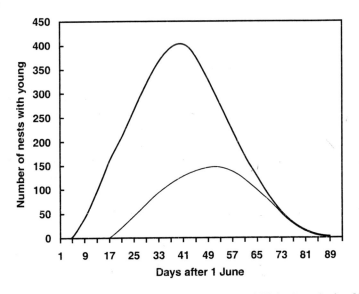

FIGURE 4.7 The timing of breeding in two neighboring Black-legged Kittiwake colonies. Note the different time of first appearance of young, but that the last chicks appeared at the same time. Thus breeding was less synchronous in the large colony, contrary to Darling's concept (after Coulson and White 1958). These colonies are those showing the distribution of nest density in Figure 4.3.

factor and that breeding was earlier in dense groups, although at all densities, late breeding occurred in young birds breeding for the first time (Coulson and White 1958). Colony size as such was not the important factor envisaged by Darling, and local density within the colony was the key to the timing of breeding.

Less-detailed investigation of other species has sometimes shown differences in timing and spread of breeding between colonies, e.g., Ashmole (1963) and Hailman (1964). While differences found have been explained by the Darling effect, these studies did not by themselves demonstrate the existence of the social stimulation effect, and these authors have been unable, for example, to exclude differences in the age composition of birds or in environmental conditions if colonies are situated some distance apart.

4.4.5 WYNNE-EDWARDS' CONCEPT OF SELF-REGULATION

Wynne-Edwards (1962, 1986) applied his views of social interactions between individuals and groups to animals as a whole and colonial breeding was seen by him as just one of the many ways animals can evaluate and respond to their own numbers. His ideas have been severely criticized (e.g., Elton 1963, Lack 1966), mainly because he invoked group selection (reduced breeding success by each pair for the benefit of the group) as the mechanism. The concept of group selection, apart from where kin (related individuals) are involved, is contrary to the outcome expected through natural selection. His ideas that animals may self-regulate their own numbers have not been effectively tested. It is extremely difficult to plan and execute such experiments while at the same time eliminate other variables. As a result, most of the attempts to examine and test this potential effect have been flawed and illustrate the advantages of applying Occam's Razor (the concept of testing the simplest hypothesis first before testing a more complex one) to research.

It is possible that some of the effects Wynne-Edwards proposed could actually be produced by natural selection. For example, there is much evidence that social interactions impose a constraint on where individuals can nest in colonies, with some individuals forced into using poorer nesting sites, or prevented from breeding until a later year. Physical sites of good quality are not in short supply, but become so when the birds have to select from within the limits of the colony. This results in an overall reduction of the reproductive output per pair. This type of effect has been reported in the Shag (*Stictocarbo aristotelis;* Potts et al. 1980) and Black-legged Kittiwake (Figure 4.8; Coulson 1971). Similar reduced reproductive output is produced by delaying the age at first breeding, e.g., Herring Gull (Chabryzk and Coulson 1976) and albatrosses (Warham 1990, Tickell 2000), induced by the high density of birds within colonies. How these situations became established by selection in the past is irrelevant; they can and do occur and have the effect of reducing reproductive output when numbers and density in the colony are high.

It must be pointed out that in most seabirds, a colony is not synonymous with a population, the latter usually being composed of several, and often many, colonies. Thus control or limitation of the size of a colony will not necessarily regulate the size of the population because surplus individuals can move and produce new colonies. Coloniality may regulate the size and growth of a colony; it is much more doubtful that it can regulate the overall numbers of the species.

4.4.6 FOOD-FINDING AND THE COLONY AS AN INFORMATION CENTER

Suggestions that colonial breeding facilitates exploitation of food sources are probably untenable, even in cases where food is highly clumped. The hypothesis that birds concentrate to breed at the place that minimizes the mean distance traveled between the nest and foraging locations (mentioned in Danchin and Wagner 1997) is perhaps not an adequate explanation since the position of food sources can change over time and so would need the optimal breeding sites to change in parallel, a situation that does not seem to occur frequently.

FIGURE 4.8 A Black-legged Kittiwake colony on a cliff. Excrement is used to cement nests to the cliff ledges. (Photo by J.B. Nelson.)

The Information Center Hypothesis proposed by Ward and Zahavi (1973) considered that within a colony (or other group) birds may obtain information about sources of food from the behavior of those individuals which have recently fed successfully. A close analogy is with the dances of beehive workers which instruct others in the hive about the direction and distance to a food source. However, worker bees are all clones of genetically identical individuals and such behavior can be readily explained by kin selection. Based on the concept of natural selection, it is argued that it is not to the advantage of an individual to pass information about food sources to unrelated individuals and, in fact, birds should actively conceal a food source from others. However, this argument may be weak since food concealing does not happen during flock feeding; perhaps the advantage in protection within a flock outweighs the disadvantages in passing on information about food availability.

The information center concept has been modified and some now propose that the bird requiring food simply identifies a successful individual by its condition and an individual short of food follows the successful bird when it leaves to feed. In this form, the bird with knowledge does not knowingly inform others but it cannot hide its well-being. Attempts to find direct evidence of information center effects in birds have been negative (Mock et al. 1988, Richner and Heeb 1995). However, circumstantial evidence was obtained by Krebs (1974), who showed that individual Great Blue Herons (*Ardea herodias*) do not leave the colony at random, but that several individuals tend to leave at the same time. The implication is that the first bird was being followed by those leaving immediately afterward. This situation is open to an alternative explanation, namely, "leaving the party syndrome." Several individuals are reaching the point when they need to feed and are stimulated to leave by seeing other individuals leaving.

4.4.7 ADJUSTMENT OF THE BREEDING SEASON

It is characteristic of the breeding areas of many seabirds that the physical and biological conditions vary from year to year. This is particularly obvious in the Arctic and Antarctic when, in some years, ice and snow melt may be early or delayed. Elsewhere, breeding is also affected by variations in weather and by periodic effects such as the well-documented El Niño–Southern Oscillation which affects water currents, ocean temperature, distribution of fish, and breeding in seabirds over large areas (see Chapter 7). Coulson (1985b) suggested that colonial breeding allows more effective adjustment of the breeding season to environmental conditions.

One of the characteristics of colonial birds is that in most species, isolated, solitary pairs are incapable of breeding. Breeding in a colony is achieved by the display and interaction of a pair *plus* necessary stimulation from other individuals. Without this stimulation, egg laying cannot occur. The amount of stimulation from other individuals is dependent upon the density of birds and when the colony is occupied in the particular year. In the Black-legged Kittiwake, the earlier that breeding starts, the larger the mean clutch size. On average, breeding is advanced by 1 day for every extra 3 days the pair can spend in the colony, that is, presence in the colony with other individuals offers a fast-track toward egg laying (Coulson 1985b). Social stimulation arising from colonial breeding is an adaptation that facilitates quicker and more effective adjustment of the time of breeding in relation to environmental conditions. In years in which adverse or extreme climatic conditions delay breeding, birds are able to respond rapidly when an improvement eventually occurs, and may still manage to breed, albeit later, but with a lower average clutch size.

4.5   RECENT HYPOTHESES

At the present time, the concepts listed below have only a theoretical basis and are not (as yet) supported by field data or experimental evidence.

4.5.1   RICHNER AND HEEB — GROUP FORAGING

Richner and Heeb (1995, 1996) propose that group foraging is the selective advantage in colonial breeding and suggest that it is easier to form a group by returning to a high-density area of conspecifics (the colony), rather than chance meetings to form a group at a suitable food source. This hypothesis suffers from the problems associated with finding suitable ways of testing the concept. Also, the concept is difficult to reconcile with the habits of long-distance feeders, such as many albatrosses, petrels, and gannets, and the length of time involved to return to their colony to form a feeding group, and then return to the feeding area again.

4.5.2   DANCHIN AND WAGNER — QUALITY SEPARATION

Danchin and Wagner (1997) raised the idea of the importance of choice of breeding sites by individuals and propose that high-quality individuals tend to group together (Figure 4.9). This undoubtedly occurs and the difference between birds at the center and edge of the colony (Coulson 1968) could be used to support this idea, but it is not clear how the overall reproductive output is enhanced by this effect. Even if it is, there is the possibility that the quality of the individuals is enhanced by the group or that the good quality birds select good quality sites, where breeding success is likely to be high anyway. Separating the importance of these effects will be difficult.

4.5.3   DANCHIN AND WAGNER — SEXUAL SELECTION

Danchin and Wagner (1997) also put forward a sexual selection hypothesis of colony formation and they raise (again) the effects of extra-pair copulation. More basic data on the extent and effectiveness of extra-pair copulations in seabirds are needed before this hypothesis can be evaluated. Emphasis is placed on sexual selection and methods of optimizing reproductive fitness. While the advocates suggest that implications and outcomes of this concept are testable, they are likely to suffer from the same problem as tests of the Wynne-Edwards idea, in that totally acceptable field tests are difficult to construct and execute.

FIGURE 4.9 A colony of Laysan Albatross on Midway Island. Generally older birds nest in the center of a colony and younger birds on the edges. (Photo by R.W. and E.A. Schreiber.)

4.5.4 DANCHIN AND WAGNER — COMMODITY SELECTION

Danchin and Wagner (1997) suggest a third possible explanation of colonial breeding, that of "commodity selection," which tends to be an integration of several of the above ideas. The reader is referred to their paper for more details of this concept. Basically, the concept suggests that animals are colonial because they evaluate suites of ecological factors to choose where to breed; several of the factors favor colonial breeding. The authors consider that "Several assumptions and predictions of the commodity selection hypothesis of coloniality can be tested using empirical, experimental and theoretical data." Perhaps this is so, but rigorous planning is needed to ensure that the tests exclude other possible interpretations. At present, the nature and structure of the tests of these hypotheses remain to be defined and disagreements already exist about how to evaluate the concept (Tella et al. 1998).

4.6 DISADVANTAGES IN COLONIAL BREEDING

There are several obvious disadvantages in colonial breeding:

1. Competition for food near the colony may be increased; however, this will be moderated as the distances individuals travel to feed increase. Many seabirds travel over 50 km to feed, the exceptions being mainly gulls and terns. In birds generally, particularly those with a short feeding range when breeding, colonial breeding may make it more difficult to exploit food resources in their environment which are evenly or randomly dispersed. Thus, colonial breeding might be expected to be linked only with a clumped food supply and/or the ability to range long distances to feed.
2. Close proximity of birds aids the transfer of microbes and parasites.
3. The re-use of sites year after year may encourage the buildup of large numbers of parasites, such as ticks and mites, which may also act as the vectors for viruses and other microbes.
4. The potential for intraspecific aggression is much greater in colonial species, since many potential recipients are close. One individual can threaten or even attack many individuals in a few minutes.

5. The nature of density-dependent effects is much modified, particularly at the breeding sites where densities are particularly high.

6. The potential adverse effects of predators, particularly mammals, on colonial seabirds can be high. This is well established in the many cases where cats and rats have been introduced to seabird islands. Yet the "swamping" effect may protect all but the edge nesting birds.

7. Møller and Birkhead (1993) have suggested that cuckoldry is a disadvantage of colonial breeding. While birds are close together in a colony the opportunity for cuckoldry is high, but it still has to be established that cuckoldry is actually higher in colonial birds.

Colonial breeding does not appear to be advantageous to many bird species, since most species breed as solitary pairs and defend large territories. Colonial vs. solitary breeding may be related to the nature of the food supply or the distribution, abundance, and method of feeding of predators, or the presence of parasites.

4.7 CHARACTERISTICS OF COLONIAL SEABIRDS AND SEABIRD COLONIES

There are a number of characteristics of colonial seabirds, most of which occur in the majority of colonial species and some also occur in noncolonial breeding birds. Sixteen characteristics are first listed below and then considered in more detail.

1. Over 96% of seabird species are colonial.
2. Maximum colony size is related to feeding range.
3. Colonial breeding evolved several times in the evolutionary history of seabirds.
4. Few species which have colonial breeding also breed as solitary pairs.
5. Colonial seabirds are present at all latitudes.
6. Coloniality is found throughout the size range of seabirds.
7. Coloniality is found in nocturnal and diurnal species.
8. Coloniality increases social interaction.
9. Coloniality exerts a constraint on recruitment.
10. Coloniality makes formation of new colonies and spread into new geographical areas more difficult.
11. Nesting areas of colonial species are often characterized by the absence of mammalian predators and ineffective predation by avian predators.
12. Most seabirds are monogamous.
13. Seabirds rarely rear more than one brood in a year.
14. Better quality birds are often partially segregated within a colony from poorer quality individuals.
15. Individual colonies are not discrete populations, and movement of individuals between colonies occurs to varying degrees.
16. Very few colonial seabirds form dense groups outside of the breeding season and away from the colony.

1. Over 96% of all seabird species are colonial: Wittenberger and Hunt (1985) report that over 95% of all seabird species are colonial and in an independent assessment; this author obtained a very similar figure of over 96%. The proportion is similar across distantly related groups and is much higher than in any other taxa of birds, being most nearly approached by the herons and egrets (Ciconiiformes). This behavior clearly represents an important component of seabird breeding biology. The difference in the extent of colonial breeding between the seabirds and other bird species within the Charadriiformes is also worthy of note. The gulls, terns, and auks are all predominantly colonial, whereas almost all shore birds ("waders" in Europe) are solitary nesters, typically holding large, defended territories in the breeding season; some have developed complex breeding systems, such as leks and polygamy, which are absent in seabirds. It is not immediately clear why seabirds should have a much higher proportion of colonial species than any other group of birds. They exploit a resource, the sea, which does not supply nesting sites, and places to breed are restricted to islands and the coastline adjacent to the seas and oceans. Many coastal areas do not provide suitable nesting sites, e.g., lack of cliffs and islands or low-lying areas free from mammalian predators. At the present time, a shortage of nesting sites does not seem to exist in most areas, although in some places it does occur. If colonial breeding has evolved to allow more birds to nest in given areas, then this selection pressure is not operating on most colonial seabird species at the present time.

Because of the spatial difference in feeding and breeding sites in seabirds, many seabirds travel much farther to collect food for their young than other birds. This probably explains why breeding in seabirds typically requires a contribution and cooperation from both members of the pair; reaching the feeding grounds, collecting food, and feeding the young may take proportionately more time than in other birds. Whether a species which engages in long-distance feeding trips nests in a colony or solitarily and is spread out over a long length of coast, does not appreciably change the distance traveled to reach the feeding grounds; and so in this situation colonial breeding does not involve an extra energy cost. In species with short breeding ranges, the energy cost is likely to be greater unless an appreciable part of the feeding area is not utilized.

2. Maximum colony size is related to feeding range: There may be a relationship between maximum colony size and maximum feeding range during the breeding season (Table 4.1; Coulson 1985a). Species which have large feeding ranges, such as Northern Gannets and most shearwaters, tend to have colonies many times larger than those of species with smaller feeding ranges. Sooty Terns (*Sterna fuscata*), which nest in vast colonies, travel longer distances to obtain food (Ashmole and Ashmole 1967, Feare 1976) than most other terns (Pearson 1968). This relationship probably also applies to the situation in the Antarctic and to petrels and albatrosses, but data on feeding

TABLE 4.1

The Relationship between the Maximum Recorded Size of Colonies in 15 European Seabird Species and the Normal Feeding Range from the Colony during the Breeding Season

Species	Maximum Colony Size	Normal Feeding Range (km)
Atlantic Puffin	100,000	250
Manx Shearwater	100,000	450
Northern Gannet	50,000	450
Northern Fulmar	45,000	450
Common Murre	40,000	100
Black-legged Kittiwake	30,000	75
Herring Gull	18,000	60
Lesser Black-backed Gull	15,000	60
Arctic Tern	3,000	15
Sandwich Tern	2,000	18
Common Tern	1,500	15
European Shag	1,000	14
Great Cormorant	500	13
Black Guillemot	150	10
Little Tern	110	8

ranges are only now becoming available. Since penguins have to swim to their feeding areas, it might be expected that the distances traveled in relation to maximum colony size, but not necessarily the time taken, would be anomalous.

Furness and Birkhead (1984) have presented evidence that colony size in kittiwakes is related to the proximity and size of the next nearest colony. While their data for Shetland are persuasive, data this author has examined from other areas do not always support this conclusion.

3. Colonial breeding evolved several times during the evolutionary history of seabirds: Each major order of birds contains some species which breed in colonies and others which breed as isolated pairs. This leads to the conclusion that colonial and solitary breeding could both have evolved many times during the evolution of the birds. This logic led Siegel-Causey and Kharitonov (1990) to suggest that colonial breeding evolved independently at least 20 times in birds. There is no reason why the forces which selected for and produced colonial breeding (for instance, location of food or nesting areas) were the same in all instances. Further, once colonial breeding developed and birds were nesting close together, secondary developments and advantages could have been selected, particularly social and group effects. As a hypothetical example, birds which had grouped to gain a defensive advantage may then develop group or social stimulation which adjusted the breeding season more precisely to peaks in food supply. The difficulty is to separate the primary from secondary effects, particularly since the primary effect could now have become lost or unimportant, with only the secondary effects remaining obvious and important in some species.

4. Few seabird species which show colonial breeding also breed as solitary pairs: Almost all seabird species breed either in colonies or as solitary pairs. There are only a few exceptions where solitary and colonial nesting occur in the same species. In Europe, Common Terns will nest as isolated pairs, usually along inland rivers, but most are intensely colonial (Cramp 1985, J.C. Coulson unpublished) and similar behavior is reported in North America (Burger and Gochfeld 1991). Great Black-backed Gulls (*Larus marinus*) often nest as isolated pairs, but others breed in single-species colonies and yet others nest scattered through colonies of other large gull species. In North America, Herring Gulls breed as isolated pairs or in colonies (Wynne-Edwards 1962). In Europe, this author's experience is that only adult Herring Gulls with previous breeding experience nest as isolated pairs. Both Masked and Brown Boobies (*Sula dactylatra, S. leucogaster*) sometimes nest as solitary pairs (Anderson 1993, Norton and Schreiber in preparation). In a detailed study of the European Shags, birds breeding for the first time (and still retaining traces of immature plumage, i.e., 2 and 3 year olds) were not found to nest in isolation, although older adults with previous breeding experience sometimes nest out of sight of other pairs.

Most young seabirds do not even try to breed as isolated pairs. Obtaining a place in the group or colony appears to be necessary in many species before a female will pair with a male and breed. To achieve and maintain coloniality, many species have evolved means by which breeding by isolated pairs is inhibited: relative time of breeding is positively linked with, and probably induced by, the density of neighboring birds. It may be that the time of laying of solitary pairs is delayed too long that it is overtaken by the seasonal inhibition of laying, which was found to stop kittiwakes laying or relaying (Coulson 1985b). In other colonial birds, e.g., in birds of prey and some passerines, the occurrence of isolated breeding is more common than in seabirds, and an investigation of this effect could be rewarding.

5. Colonial species are represented at all latitudes: There is no geographical change in the proportions of colonial seabird species in equatorial, temperate, Arctic, or Antarctic areas. Clearly, global position is not a factor which contributed to the evolution of colonial breeding nor its persistence.

6. Coloniality is found throughout the size range of seabirds: There is no evidence that the size of individuals affects whether seabird species are colonial or not. Colonial breeding may be less frequent in those species that are predatory on mammals or other seabirds (e.g., some skuas and some large gulls). Their distribution is probably influenced by the fewer individuals which can exploit a high trophic level food source in a given area. However, there are too few examples in this category to examine this effect in quantitative terms. In contrast, virtually all seabirds that feed on fish or squid are colonial.

7. Coloniality is found in nocturnal and diurnal species: Most colonial species are diurnally active, but colonial breeding occurs in seabird species which are totally or partially active at night, e.g., Swallow-tailed Gull (*Creagrus furcatus*; Hailman 1964), storm petrels, many shearwaters, and, at times, Grey Gull (*Larus modestus;* Howell et al. 1974). Many nocturnal species, such as the small petrels, are also diurnal feeders, but visit the colony at night, presumably in an attempt to avoid predators.

8. Coloniality increases social interaction: Because of the close proximity of individuals in colonies, aggressive interaction between individuals of different pairs is frequent, although typically only a small area around the nest is defended. In a colony, one bird can interact with several individuals within a few moments. In contrast, noncolonial species, where the normal distance between neighboring pairs is much larger, interactions are fewer. In a colonial situation, an individual female can visit several males in as many minutes; this proximity of other individuals increases the potential for extra-pair copulation, although the frequency of this is unknown.

9. Coloniality exerts a constraint on recruitment: Colonial breeding in some instances places a constraint on where (and when) individuals can breed. In many species, most new recruits can only breed at sites within or near the edge of the colony, because older birds move into the colony area first and occupy the central sites. Physically suitable, unoccupied nesting areas beyond the edge of the colony are apparently socially unacceptable and are eventually occupied only if the colony is expanding. This aspect of socially induced limits to a colony size appears not to be entertained by those who claim that there is no limitation or shortage of nest sites, e.g., Wittenberger and Hunt (1985), Brown and Brown (1996), and Siegel-Causey and Kharitonov (1990), because they only consider the physical properties of colony and nest sites; they ignore the socially induced limits.

This competition can increase the age at first breeding, enhance the number of nonbreeding individuals, and produce differences in the age structure between (and within) colonies where differences in densities of nesting birds occur. For some species, age at first breeding is greater in large, dense colonies (Chabrzyk and Coulson 1976) and when the number of potential recruits is high (Porter and Coulson 1987), presumably a result of the greater competition for access to sites within the colony. The recruiting male has first to obtain a socially acceptable nesting site; as in many seabird species, pair formation takes place only at the potential nest site. In these species males without a site are usually ignored by prospecting females. There are many studies which give information on the time of return to the colony of birds of different ages. In virtually all (gulls, terns, albatrosses, shearwaters), young birds usually return to the colony later in the year, by which time older birds have occupied most of the central sites. In many cases, young birds are restricted to sites where the previous site holder has died since the last breeding season, to poorer quality subcentral sites or to sites at the edge of the colony. Some are unsuccessful in obtaining a site and these produce a pool of nonbreeding birds characteristic of most colonies, the size of which has been suggested to be an indicator of the health of the colony and the population (Porter and Coulson 1987). These social pressures also may exert a limit to the growth of a colony as the ratio of edge-to-central area declines in larger colonies (Coulson 1985a in preparation). As a result, some seabird colonies show a progressively lower proportionate increase in size as they become large (Coulson 1983, Coulson and Raven 1997), although the population as a whole may be still increasing exponentially.

10. Coloniality makes formation of new colonies and spread into new geographical areas more difficult: Many seabirds visit breeding colonies for a year (or even longer) before they breed for the first time. This visit is often made during the period in which established breeders have eggs and young. Time is spent selecting a colony, a potential nesting site, and then obtaining a mate. In deciding where to breed, one good indicator may be to choose an established colony, particularly one where young are being successfully reared. This is believed to be the case in the Herring Gull, Lesser Black-backed Gull, and Black-legged Kittiwake (Duncan 1978, Coulson et al. 1982, Boulinier et al. 1996). Individuals of many seabird species show a high degree of philopatry, returning to breed where they were born, e.g., most albatrosses and Red-tailed Tropicbird (*Phaethon rubricauda*;

Schreiber and Schreiber 1993), but in others an appreciable proportion move and are recruited into other colonies, e.g., Sandwich Tern (*Sterna sandvicensis;* Langham 1974), Herring Gull (Duncan and Monaghan 1977, Vercruijsse 1999), Northern Fulmar (Dunnet and Ollason 1978, Dunnet et al. 1979), and Black Guillemot (*Cepphus grylle*; Frederiksen and Petersen 2000). Currently, little is known about hole-nesting petrels, since locating and then recapturing marked birds is much more difficult (Warham 1990). In the case of the Herring Gull, young birds apparently always visit their natal colony first, even if they reject it and then nest elsewhere (Vercruijsse 1999).

New breeding areas can be initially pioneered by a single pair in many bird species, but following the initial suggestion by Darling (1938), there is increasing evidence of the need for a threshold number of pairs for new seabird breeding colonies to form. In Black-legged Kittiwakes, this minimum number is probably at least 20 birds, and most of them visit the potential nesting site for the first time late in the breeding season and then return in the following years. Such sites are frequently first used as resting sites, often by young birds with some individuals returning in successive years. It may be 2 to 5 years after the first roosting in a new site before any birds lay eggs. Evidence of threshold numbers exists for Common Murres, Northern Gannets (*Morus bassanus*; Nelson 1978), boobies of several species, Atlantic Puffins, Sandwich and Royal Terns, and several species of penguins. It is obvious that this requirement for many individuals to be present in the same place at the same time before breeding can take place puts an additional constraint on, but does not prevent, many colonial species spreading into new geographical areas. Formation of a new colony usually occurs relatively infrequently. For example, in Britain, the numbers of breeding Black-legged Kittiwakes increased by about 50% from 1890 to 1920, but not a single new colony was formed; all the recruitment and population increase took place within the existing colonies (Coulson 1963).

11. Colonial nesting areas are often characterized by the absence of mammalian predators and few avian predators: Although the first suggestion of the function of colonies was as an anti-predator device, Lack (1954) was first to point out the illogical nature of this. Most colonial birds nest at sites relatively safe from predators, many on precipitous cliffs, on islands, in marshes, or in trees. The sites used are typically inaccessible to mammalian predators and some of the sites used make access by predatory bird species difficult (but not impossible). When mammalian predators reach islands or other areas where seabirds nest, the effects on seabirds are catastrophic, with major declines in numbers or cessation of breeding altogether. The cases of introduced cats or rats depleting seabird colonies are legion. Rats even have adverse effects on birds as large as gannets, taking eggs and killing chicks. American Mink (*Mustela vison*), introduced into the UK, has had adverse effects on terns and Common Gulls (*Larus canus*) (Craik 1995, 1997), while the spread and increase of the European Fox (*Vulpes vulpes;* Tapper 1992) on the east coast of Scotland and elsewhere in Great Britain have caused reduction and disappearance of Herring Gull and Eider Duck colonies on the mainland coast (J. C. Coulson unpublished). Burger and Gochfeld (1991) record several mammal species preying on terns and skimmer eggs, young, and adults in the northeast U.S. These included Grey Squirrels *(Sciurus carolinensis)*, Racoons *(Procyon lotor)*, Red Fox *(Vulpes fulva)*, dogs, and feral cats. In general, these predators were few and were probably limited by the isolated, tidal, and open nature of the sites. As a result, predation did not normally result in complete failure of breeding in the colonies, but such sites must be vulnerable to any increase in the abundance of the predators. Movement of the colony position seems to be frequent in the Little Tern *(Sterna albifrons)* in Europe as a strategy to defeat predators which learn the location of specific colonies, resulting in increasing predation after the first few breeding years.

Seabird colonies are less frequently located on sites that prevent avian predators from reaching them, although Cullen (1957) interpreted a typical kittiwake cliff nesting site to avoid bird predators. While Ravens (*Corvus corvus*), Great Skuas (*Catharacta skua*), Glaucous Gulls (*Larus hyperboreus*), and Great Black-backed Gulls prey on cliff-nesting colonial seabirds, it is often over a very limited area within the whole colony and on sites which are more difficult for the colonial birds to defend, e.g., where the ledges are broader or the cliff less than sheer and so sites are inferior to others. Late breeding in a colony can be a disadvantage as the colony population declines later in the season. Total breeding failure in a colony due to avian predation is uncommon, but not unknown in colonial seabirds. In general, avian predators are larger than their prey and with size they are less maneuverable on cliffs and tree tops and are not able to penetrate burrows. Often, avian predators are present in relatively small numbers and because of this their impact is not usually large.

12. Most seabirds are monogamous: Colonial breeding engenders competition for nesting sites and the need to defend the site. Defense by a pair allows time for one of the pair to feed while the site is defended by the other. The need for intensive nest site defense (which may extend for months before egg laying in some seabirds, e.g., Northern Fulmars, gannets, kittiwakes, Herring Gulls, shag, but a few weeks in terns, albatrosses, and some gulls) is probably a key factor which has maintained monogamy in seabirds. Within the Charadriiformes, those species which are seabirds appear to be, without exception, monogamous. In contrast, some of the shorebirds have developed complex mating and breeding systems, including polygamy, lekking, laying separate clutches which are cared for by the male and female, respectively, and the care of young by one parent only. The difference probably stems from the fact that the shorebirds have food available in their spaced-out nesting areas with larger territories. Most pairs of seabirds share responsibility for nest defense and feeding young. This is, presumably, the effective way to breed when the nesting areas are markedly distant from the feeding areas, although some auks (*Synthliboramphus* sp., *Endomychura* sp.) have overcome part of this restriction by taking the young to sea and to feeding areas at an early age. Possibly linked with monogamy and nest defense is the fact that few seabirds show a high degree of sexual dimorphism. In most seabirds, the male is slightly larger than the female (the reverse is true in skuas, frigatebirds, and boobies), but the difference is rarely more than 20% of mass (the exception being the Black Skimmer, *Rhynchops niger*).

13. Seabirds rarely rear more than one brood a year: Many seabirds, particularly those belonging to the Procellariiformes, have much longer incubation periods and fledging periods than do land birds of comparable size. At one extreme of size, adults of Wilson's Storm Petrel (*Oceanites oceanus*) weighs about 40 g and takes about 95 days from laying to fledging (Roberts 1940), while the European Starling, with a mass of about 80 g, takes only 33 days. Size is generally directly related to the duration of the incubation and fledging periods, although there are a few exceptions. As a result, seabirds typically rear only one brood per season and a few albatross species and frigatebirds breed biennially (Warham 1990, Tickell 2000, Diamond 1972).

FIGURE 4.10 Colony of Adelie Penguins. More experienced individuals were found to nest toward the center of the colony where there is more severe competition for nest sites than at the edge. (Photo by H. Weimerskirch.)

The successful rearing of two broods in a breeding season is only possible when the breeding period is of short-enough duration. Isolated and exceptional cases have been reported in European Shags (Cadiou 1994, Wanless and Harris 1997), but it is obvious that this could not happen where this species nests in Arctic or sub-Arctic waters. Hays (1984) and Wiggins et al. (1984) both reported Common Terns rearing a brood and then laying a further clutch of eggs. Unfortunately, the outcome from the late clutches was not known. In some tropical breeding terns, breeding seasons reoccur at about 6- to 9-month intervals and, potentially, some adults may rear two broods within a year, although in these cases two separate sets of birds are generally involved.

14. Better quality birds are often partially segregated within a colony from poorer quality individuals: Black-legged Kittiwakes nesting in the center of the colony live longer and produce more young each year than those breeding at the edge so that the lifetime reproductive output of central birds, particularly males, is up to three times greater than those breeding at the edge (Coulson 1968, 1971). This is not an age effect, with older birds nesting centrally and the young individuals at the periphery, nor do kittiwakes move from the edge into central areas. It appears to reflect a difference in the long-term quality of the individuals which manage to recruit into the center, where there is much more severe competition for sites than at the edge. Since the original observations, similar effects have been found in the Ring-billed Gull (*Larus delawarensis*; Dexheimer and Southern 1974), Herring Gull (Parsons 1975), Adelie Penguin (*Pygoscelis adeliae*; Ainley et al. 1983; Figure 4.10), Great Cormorant (*Phalacrocrax carbo*; Andrew and Day 1999), and several other species. How extensive this effect is in other species needs to be investigated, particularly since it has considerable implications in managing endangered species.

It is evident that there is much scope for further research on the relationship between position of pairs within a colony and breeding success. The original kittiwake study was made on a colony where there was no predation and the role of position requires further study in situations where some predation takes place.

15. Individual seabird colonies are not discrete populations; movement of individuals between colonies occurs: For some time, the belief was held that a colony is a discrete, self-sustaining population, with young returning to the same colony when they mature (e.g., Wynne-Edwards 1962). The increasing amount of information arising from banding and preliminary data from DNA analyses indicate that this is rarely true. The extent of philopatry varies between species but is rarely complete. Philopatry is used here to indicate the return to the place of birth and not in the incorrect usage of returning to the place where the individual previously bred; the term "natal philopatry" is tautology and "breeding philopatry" is incorrect, although frequently used synonymously with "breeding site fidelity." The decision to move and breed elsewhere is usually made before breeding for the first time and in some species may even exceed half of the surviving individuals (Chabrzyk and Coulson 1976).

In contrast, once they become breeding adults, most seabirds are usually highly site- and colony-faithful. Although movement of adults does occur, it is often linked with disturbance and/or poor breeding success. Because of high adult survival rates, most adult seabirds survive from one breeding season to the next and return to the colony in which they last bred. This behavior is the main reason why many colonies tend to stabilize in size between years. Occasionally, there is a mass movement of birds in a colony to another site, e.g., Black-legged Kittiwakes (Danchin and Monnat 1992, Fairweather and Coulson 1995). In general, the movement of adults from one colony to another colony is infrequent.

There is considerable variation in the extent to which young seabirds return and breed in their natal colony. In most seabirds an appreciable proportion of the surviving young do so. However, in the Northern Fulmar extremely few young return to the study colony (Dunnet et al. 1979). Few cases exist where all individuals are philopatric and in some of these, this may be an apparent rather than a real effect because much additional effort is needed to search other colonies to find marked individuals which emigrated. A recent study by Vercruijsse (1999) on Herring Gulls has shown that almost every chick which survived to breed visited its natal colony (usually first when 3 years old), although about 40% eventually bred in other colonies. In Black-legged Kittiwakes, all young birds visited the colony in which they would breed at least 1 year before they did so (Porter 1990). However, not all surviving young which were reared there returned. Most young returned within 10 km of their natal colony. About 15% moved between 500 and 1000 km away, and virtually none were found between 10 and 500 km from the natal colony (Coulson and Neve de Mevergnies 1992). Those which moved considerable distances did not visit their natal colony before breeding elsewhere.

Typically, philopatry in seabirds is more pronounced in male offspring, in line with generalizations for other birds (Greenwood 1980, Swingland and Greenwood 1984). Many young seabirds may make active decisions about where to breed, and not just return to where they were born. The manner in which they make the decision is not known, but it is not unlikely that they visit many colonies in the years before breeding for the first time. Identification of how they select a colony in which to breed is of considerable conservation and management importance.

16. Very few colonial seabirds form dense permanent groups outside of the breeding season and away from the colonies: Auks, petrels, albatrosses, and penguins occur individually or only in small groups when away from the colonies or when not breeding. Terns, cormorants, skimmers, frigatebirds, pelicans, and gulls often roost by day or by night in flocks. These roosting groups may occur for protection and mixed species flocks are common. In contrast, feeding at these times is usually solitary or in small groups unless large concentrations of food occur, such as at large fish shoals, at sewer outfalls, or at land-fill sites. Some gulls, e.g., Herring Gulls, Common Gulls (*Larus canus*), and Black-headed Gulls (*L. ridibundus*), feed in loose flocks on pastures, but less so at marine sites. Migration in terns and gulls takes place singly or in small flocks, involving far fewer birds than would be found in a colony. This behavior is in contrast to the situation in many shore birds where large flocks are characteristic of the nonbreeding season and migration, but not of breeding birds.

4.8 SYNCHRONY

The discerning reader may have noted that synchrony of breeding has not been listed as a characteristic of colonial breeding. For a long time, synchronous breeding has been regarded as an attribute and a consequence of colonial breeding. It is fundamental to Fraser Darling's (1938) ideas on colonial breeding. Gochfeld (1980) wrote at length on this topic in his review of colonial breeding, but did not investigate the extent to which it occurred in noncolonial species. Synchrony of breeding occurs to some extent in most bird species, inasmuch as they do not breed continuously throughout the year. Most birds have a breeding season even if this restricted period occurs at different times from year to year. Interest should therefore be focused not on whether synchrony occurs, but upon the degree of synchrony. However, it is evident that a high degree of breeding synchrony is not characteristic of many seabirds (e.g., European Shags may lay over many weeks), nor is synchrony higher is seabirds than in solitary breeders. It is therefore unlikely to have been the primary selective force in the evolution and spread of colonial breeding. In fact, breeding synchrony is no greater in colonial than in birds which breed as solitary pairs. Nevertheless, synchrony probably plays an important part in the reproductive success in some colonial species, particularly in certain years. Its influence on the evolution of colonial breeding should be reevaluated. A new appraisal of synchrony is made below.

4.8.1 COMPARISON BETWEEN COLONIAL AND NONCOLONIAL SPECIES

In different colonial species, the onset of laying by individuals can be spread over a few days, or at the other extreme, weeks or even months. However, in a preliminary analysis of the spread of breeding in birds which typically rear only one brood in a season, there is considerable variation between species of colonial and noncolonial birds.

High degrees of breeding synchrony are not consistently found in colonial species of birds. This variation is well illustrated in the Procellariiformes. Blue Petrels (*Halobaena caerulea*) have a spread of laying of only 9 days (based on a sample of 23 pairs; Jouventin et al. 1985) and the Sooty Albatross (*Phoebetria fusca*) a spread of 12 days (sample of 40 pairs; Jouventin and Weimer-skirch 1984). In contrast, the Waved Albatross (*Phoebastria irrorata*) spreads laying over 77 days (sample of 495 pairs; Harris 1973) and the Madeiran Storm-petrel (*Oceanodroma castro*) over 115 days (sample of 110 pairs; Allan 1962). Warham (1990) quoted many examples of petrels which lie between these extremes, with 16 out of 40 species exceeding 28 days. Synchrony of breeding well within these ranges occurs in many noncolonial species, e.g., several tits, Pied Flycatcher (*Ficedula hypoleuca*), Meadow Pipit (*Anthus pratensis*) in the Arctic, European Nightjar (*Caprimulgus europaeus*), and several birds of prey studied at restricted sites; also, many examples are given in Cramp and Simmons (1977).

Despite having shown that there is considerable variation in breeding synchrony within the petrels (which also occurs in other seabird taxa), there is, of course, no reason why some colonial seabirds should not use social interactions to produce or enhance the degree of breeding synchrony in situations where this is advantageous. Social interaction could be used in addition to, or in combination with, the proximate methods used in noncolonial species to determine laying date, including photoperiod, temperature-sums, and internal "clocks" (Lehrman 1965, Gwinner 1989, 1996).

4.8.2 AGE EFFECTS AND SYNCHRONY

It is now well established that in many birds, colonial and noncolonial alike, young individuals tend to nest later than older birds. As a result, the degree of synchrony is decreased since all breeding groups typically contain old and young birds. Gochfeld (1980) showed that the laying period in almost all colonial birds is not symmetrical about a mean date, but is skewed toward late laying. Presumably these late layers are mainly young birds (but in some cases they also involve some older birds which have relayed). If synchrony was of major importance, it might be expected that late laying by young birds would be reduced or eliminated by selection. In the kittiwake, first-breeding birds lay an average of 10 days later than older birds (Coulson and White 1958), and while older birds in a group may be highly synchronized, the presence of young birds dramatically reduces the measured degree of synchrony in a colony or subcolony. If synchrony were important, selection would have operated to reduce or prevent the later laying by young birds. In most species this has not occurred.

4.9   IS COLONIAL BREEDING ALWAYS AN ADVANTAGE?

Advantages from colonial breeding may occur intermittently in time, and only be evident when considered over several years. Again, colonial breeding may have been advantageous in the past, but because environmental situations have changed, it may no longer be so in some species (Burger and Gochfeld 1994). This is a possibility which is often overlooked. Some would argue that in this situation, natural selection would have eliminated colonial breeding in these species, but that is not necessarily so. When what had been an adaptive advantage becomes a disadvantage, there are three possibilities: (1) the species changes its behavior and manner of breeding, (2) the species does not change but persists, despite the disadvantage, and (3) the disadvantage becomes so great that the species becomes extinct. Once colonial breeding has evolved in a species, it may well be that the species cannot revert to solitary breeding; the species is trapped because the mechanism needed for solitary breeding cannot be recovered or redeveloped. The situation could be further complicated by the development of secondary advantages developed from colonial breeding, and which now have become more important than the originally selected factors.

4.10 FUTURE RESEARCH

There is still no evidence that colonial breeding in seabirds has arisen for different reasons than in land birds or even in several mammal taxa. Comparative studies of colonial breeding across taxonomic lines are needed. It may be that there is not a common theme to colonial breeding in vertebrates, but it is likely that some clusters of similarity will occur between species which are not taxonomically close.

The points considered in the above sections return several times to unique or characteristic features of seabirds: having vast colonies, most species being colonial, monogamy, and long feeding trips during the breeding season. It seems likely that, ultimately, the functions of colonial breeding in seabirds will be found to lie within the interactions of such characteristics, but the precise mechanism has probably yet to be recognized. Wittenberger and Hunt (1985) concluded that there was not a single cause of colonial breeding in seabirds (let alone birds in general). A wealth of new general hypotheses indicate that this conclusion has not yet been universally accepted.

In recent years, more new theories about the function of colonial breeding have been produced. Some of these ideas (hypotheses) will be rejected and some (as in the past) will persist as unproven because they are virtually impossible to test. Bear in mind the difficulty and length of time it took to critically examine Darling's (1938), Ward and Zahavi's (1973), and Wynne-Edward's (1962) ideas. Current thought has turned toward breeding strategies and individual quality in coloniality, and here is envisaged difficulty in devising critical and satisfactory tests needed to validate the ideas. Recent ideas involve complex concepts, and while these may be stimulating to argue and discuss in armchairs, science will not progress far without well-thought-out experiments to validate them.

We still know little about the social organization of birds in colonies and how they communicate with each other. We need to know more of the effects of age, the extent of philopatry, the factors which cause birds to change colonies, and how birds communicate within colonies. It is not difficult to produce a hypothesis or a model. It is much more difficult to present it in a form which can be adequately tested. Accordingly, a plea is hereby made for more detailed biological studies on colonial animals where simple questions are asked and satisfactorily answered.

LITERATURE CITED

ALLAN, R. G. 1962. The Madeiran Storm Petrel *Oceanodroma castro*. Ibis 103B: 274–295.

AINLEY, D. G., R. E. LeRESCHE, AND W. J. L. SLADEN. 1983. Breeding Biology of the Adelie Penguin. University of California Press, Berkeley.

ANDERSON, D. A. 1993. Masked Booby (*Sula dactylatra*). No. 73 *in* The Birds of North America (A. Poole and F. Gill, Eds.). Academy of Natural Sciences, Washington, D.C.; American Ornithologists' Union, Philadelphia.

ANDREW, D. J., AND K. R. DAY. 1999. Reproductive success in the Great Cormorant *Phalacrocorax carbo carbo* in relation to colony nest position and timing of nesting. Atlantic Seabirds 1: 107–120.

ASHMOLE, N. P. 1963. The biology of the Wideawake or Sooty Tern (*Sterna furcata*) on Ascension Island. Ibis 103B: 297–364.

ASHMOLE, N. P., AND M. J. ASHMOLE. 1967. Comparative feeding ecology of sea birds of a tropical oceanic island. Peabody Museum of Natural History, Yale University Bulletin 24, New Haven, CT.

AUSTIN, O. L. 1929. Contributions to the knowledge of the Cape Cod Sterninae. NE Bird Banding Association 5: 123–140.

AUSTIN, O. L. 1948. Predation by the common rat *Rattus norvegicus* in the Cape Cod colonies of nesting terns. Bird Banding 19:60–65.

BOULINIER, T., E. DANCHIN, J.-T. MONNAT, C. DOUTRELANT, AND B. CADIOU. 1996. Timing of prospecting and the value of information in a colonial breeding bird. Journal of Avian Biology 27: 252–256.

BROWN, C. R., AND M. B. B. BROWN. 1996. Coloniality in the Cliff Swallow. The effects of group size on social behavior. University of Chicago Press, Chicago.

BUCKLEY, F. G., AND P. A. BUCKLEY. 1972. The breeding ecology of Royal Terns *Sterna (Thalasseus) maxima maxima*. Ibis 114: 344–359.

BUCKLEY, P. A., AND F. G. BUCKLEY. 1979. What constitutes a waterbird colony? Reflections from northeastern U.S. Proceedings of the Colonial Waterbird Group 3: 1–15.

BURGER, J. 1984. Colony stability in Least terns. Condor 86: 61–67.

BURGER, J., AND M. GOCHFELD. 1991. The Common Tern: Its Breeding Biology and Social Behavor. Columbia University Press, New York.

BURGER, J., AND M. GOCHFELD. 1994. Seabirds on Islands (D. N. Nettleship, J. Burger, and M. Gochfeld, Eds.). BirdLife International, Cambridge.

BURGER, J., AND J. SHISLER. 1980. The process of colony formation among Herring Gulls (*Larus argentatus*) nesting in New Jersey. Ibis 122: 15–26.

CADIOU, B. 1994. Un evenement rarissime: L'elevage de deux nichees avec succes par un couple de cormorans huppes *Phalalcrocorax aristotelis*. Alauda 62: 134–135.

CHABRZYK, G., AND J. C. COULSON. 1976. Survival and recruitment in the herring gull *Larus argentatus*. Journal of Animal Ecology 45: 187–203.

COULSON, J. C. 1963. The status of the Kittiwake in the British Isles. Bird Study 10: 147–179.

COULSON, J. C. 1968. Differences in the quality of birds nesting in the centre and on the edges of a colony. Nature (London) 217: 478–479.

COULSON, J. C. 1971. Competition for breeding sites causing segregation and reduced young production in colonial animals. Proceedings of the Advanced Study Institute Dynamics of Populations 1970: 257–266 (Pudoc, Wageningen).

COULSON, J. C. 1983. The changing status of the Kittiwake *Rissa tridactyla* in the British Isles, 1969–1979. Bird Study 30: 9–16.

COULSON, J. C. 1985a. Density regulation in colonial seabird colonies. Proceedings of the XVIII International Ornithological Congress, Moscow 783–791.

COULSON, J. C. 1985b. A new hypothesis for the adaptive significance of colonial breeding in the Kittiwake *Rissa tridactyla* and other seabirds. Proceedings of the XVIII International Ornothological Congress, Moscow 892–899.

COULSON, J. C., AND F. DIXON. 1979. Colonial breeding sea birds. Pp. 445–458 *in* Biology and Systematics of Colonial Organisms (G. Larwood and B. R. Rosen, Eds.). Academic Press, London.

COULSON, J. C., N. DUNCAN, AND C. S. THOMAS. 1982. Changes in the breeding biology of the Herring Gull *(Larus argentatus)* induced by reduction in the size and density of the colony. Journal of Animal Ecology 51: 739–756.

COULSON, J. C., AND G. NEVE DE MEVERGNIES. 1992. Where do young Kittiwakes breed, philopatry or dispersal? Ardea 80: 187–197.

COULSON, J. C., AND S. J. RAVEN. 1997. The distribution and abundance of *Larus* gulls nesting on buildings in Britain and Ireland. Bird Study 44: 13–34.

COULSON, J. C., AND C. S. THOMAS. 1983. Mate choice in the Kittiwake Gull *in* Mate Choice (P. Bateson, Ed.). Cambridge University Press, Cambridge.

COULSON, J. C., AND E. WHITE. 1956. A study of colonies of the Kittiwake *Rissa tridactyla* (L.). Ibis 98: 63–79.

COULSON, J. C., AND E. WHITE. 1958. The effect of age on the breeding biology of the Kittiwake *Rissa tridactyla*. Ibis 100: 40–51.

COULSON, J. C., AND E. WHITE. 1960. The effect of age and density of breeding birds on the time of breeding of the kittiwake *Rissa tridactyla*. Ibis 102: 71–86.

COULSON, J. C., AND R. D. WOOLLER. 1976. Differential survival rates among breeding kittiwake gulls *Rissa tridactyla* L. Journal of Animal Ecology 45: 205–213.

CRAIK, J. C. A. 1995. Effects of North American Mink on the breeding success of terns and smaller gulls in west Scotland. Seabirds 17: 3–11.

CRAIK, J. C. A. 1997. Aspects of the biology of the Common Gull *Larus canus* from remains left by predators. Ringing and Migration 18: 84–90.

CRAMP, S., AND K. E. L. SIMMONS. 1977. Birds of the Western Palearctic. Oxford University Press, Oxford.

CRAMP, S. 1985. Birds of the Western Palearctic, Vol. IV. Oxford University Press, Oxford.

CULLEN, E. 1957. Adaptations in the kittiwake to cliff nesting. Ibis 99: 275–302.

DANCHIN, E., AND J.-Y. MONNAT. 1992. Population dynamics modelling of two neighbouring kittiwake *Rissa tridactyla* colonies. Ardea 80: 171–180.

DANCHIN, E., AND R. H. WAGNER. 1997. The evolution of coloniality: the emergence of new perspectives. Trends in Ecology and Evolution 12: 342–347.

DARLING, F. F. 1938. Bird Flocks and the Breeding Cycle. Cambridge University Press, Cambridge.

DIAMOND, A. W. 1972. Sexual dimorphism in breeding cycles and unequal sex ratio in Magnificent Frigatebirds. Ibis 114: 395–398.

DEXHEIMER, M., AND W. E. SOUTHERN. 1974. Breeding success relative to nest location and density in Ring-billed Gull colonies. Wilson Bulletin 85: 288–290.

DUNCAN, N. 1978. The effects of culling Herring Gulls (*Larus argentatus*) on recruitment and population dynamics. Journal of Applied Ecology 15: 697–713.

DUNCAN, N., AND P. MONAGHAN. 1977. Infidelity to the natal colony by breeding Herring Gulls. Ringing and Migration 1: 166–172.

DUNNET, G. M., AND J. C. OLLASON. 1978. The estimation of survival rate in the fulmar, *Fulmarus glacialis*. Journal of Animal Ecology 47: 507–520.

DUNNET, G. M., J. C. OLLASON, AND A. ANDERSON. 1979. A 28 year study of breeding Fulmars *Fulmarus glacialis* in Orkney. Ibis 121: 293–300.

ELTON, C. S. 1963. Self-regulation of animal populations. Nature 97: 634.

FAIRWEATHER, J. A., AND J. C. COULSON. 1995. Mate retention in the kittiwake *Rissa tridactyla* and the significance of nest site tenacity. Animal Behaviour 50: 455–464.

FEARE, C. J. 1976. The breeding of the Sooty Tern *Sterna fuscata* in the Seychelles and the effect of experimentally removal of its eggs. Journal of Zoology, London 179: 317–360.

FISHER, J. 1952. The Fulmar. Collins, London.

FREDERIKSEN, M., AND A. PETERSEN. 2000. The importance of natal dispersion in a colonial seabird, the Black Guillemot *Cepphus grylle*. Ibis 142: 48–57.

FURNESS, R. W., AND T. R. BIRKHEAD. 1984. Seabird colony distributions suggest competition for food supplies during the breeding season. Nature 311: 655–656.

GOCHFELD, M. 1979. Breeding synchrony in Black Skimmers: colony vs. subcolonies. Proceedings Second Conference Colonial Waterbird Group 2: 171–177.

GOCHFELD, M. 1980. Mechanisms and adaptive value of reproductive synchrony in colonial sea-birds. Pp. 207–270 *in* Behavior of Marine Animals. Vol. 4. Marine Birds (J. Burger, B. L. Ollo, and H. E. Winn, Eds.). Plenum Press, New York.

GREENWOOD, P. J. 1980. Mating systems, philopatry and dispersal in birds and mammals. Animal Behaviour 28: 1140–1162.

GWINNER, E. 1989. Photoperiod as a modifying and limiting factor in the expression of avian circannual rhythms. Journal of Biological Rhythms 4: 237–250.

GWINNER, E. 1996. Circannual clocks in avian reproduction and migration. Ibis 138: 47–63.

HAILMAN, J. P. 1964. Breeding synchrony in the equatorial Swallow-tailed gull. American Naturalist 98: 79–83.

HAMILTON, W. D. 1971. Geometry for the selfish herd. Journal of Theoretical Biology 31: 295–311.

HARRIS, M. P. 1973. The biology of the Waved Albatross *Diomedea irrorata* of Hood Island, Galapagos. Ibis 115: 483–510.

HAYS, H. 1984. Common terns raise young from successive broods. Auk 101: 274–280.

HOWELL, T. R., B. ARAYA, AND T. R. MILLIE. 1974. Breeding biology of the Gray Gull, *Larus modestus*. University of California Publications in Zoology 104: 1–57.

JOUVENTIN, P., AND H. WEIMERSKIRCH. 1984. L'Albatross Fuligineux a Dos Sombre *Phoebetria fusca*, example de strategie d'adaptation extreme a la vie pelagique. Review of Ecologie (Terre et Vie) 39: 401–429.

JOUVENTIN, P., J.-L. MOUGIN, J.-C. STAHL, AND H. WEIMERSKIRCH. 1985. Comparative biology of the burrowing petrels of the Crozet Islands. Notornis 32: 157–220.

KHARITONOV, S. P., AND D. SIEGEL-CAUSEY. 1988. Colony formations in seabirds. Current Ornithology 5: 223–269.

KREBS, J. R. 1974. Colonial nesting and social feeding as stategies for exploiting food resources in the Great Blue Heron *Ardea herodias*. Behaviour 51: 99–134.

KUSHLAN, J. A. 1986. Colonies, sites and surveys: the terminology of waterbird studies. Colonial Waterbirds 9: 119–120.

LACK, D. 1954. The Natural Regulation of Animal Numbers. Clarendon Press, Oxford.

LACK, D. 1966. Population Studies on Birds. Clarendon Press, Oxford.

LANGHAM, N. E. P. 1974. Comparative breeding biology of the Sandwich Tern. Auk 91: 255–277.

LEHRMAN, D. S. 1965. Interaction between internal and external environments in the regulation of the reproductive cycle of the Ring Dove, in Sex and Behavior (F. A. Beach, Ed.). Wiley, New York.

MACROBERTS, M. H. AND R. R. MACROBERTS. 1972. Social stimulation of reproduction in Herring and Lesser Black-backed Gulls. Ibis 114: 495–506.

MOCK, D. W., T. C. LAMEY, AND D. B. A. THOMPSON. 1988. Falsifiability and the information center hypothesis. Ornis Scandinavica 19: 231–248.

MØLLER, A. P., AND T. R. BIRKHEAD. 1993. Cuckoldry and sociality: a comparative study of birds. American Naturalist 142: 118–140.

NELSON, J. B. 1978. The Gannet. Poyser, Berkhamsted.

NORTON, R. L., AND E. A. SCHREIBER. In preparation. Brown Booby (Sula leucogaster), in The Birds of North America (A. Poole and F. Gill, Eds.). The Birds of North America, Inc., Philadelphia.

ORIANS, G. H. 1961. Social stimulation within blackbird colonies. Condor 63: 330–337.

PARSONS, J. 1975. Seasonal variation in the breeding success of the Herring Gull: an experimental approach. Journal of Animal Ecology 44: 553–573.

PARSONS, J. 1976. Nesting density and breeding success in the Herring Gull Larus argentatus. Ibis 118: 537–546.

PEARSON, T. H. 1968. The feeding biology of sea-bird species breeding on the Farne Islands, Northumberland. Journal of Animal Ecology 37: 521–552.

PERRINS, C. M. 1970. The timing of birds' breeding seasons. Ibis 112: 242–255.

PORTER, J. M. 1990. Patterns of recruitment to the breeding group in the kittiwake Rissa tridactyla. Animal Behaviour 40: 350–360.

PORTER, J. M., AND J. C. COULSON. 1987. Long-term changes in recruitment to the breeding group, and quality of recruits at a kittiwake Rissa tridactyla colony. Journal of Animal Ecology 56: 675–689.

POTTS, G. R., J. C. COULSON, AND I. R. DEANS. 1980. Population dynamics and breeding success of the Shag (Phalacrocorax aristotelis (L.)), on the Farne Islands, Northumberland. Journal of Animal Ecology 49: 465–484.

RICHNER, H., AND P. HEEB. 1995. Is the information center hypothesis a flop? Advanced Studies in Behaviour 24: 1–45.

RICHNER, H., AND P. HEEB. 1996. Communal life: honest signaling and the recruitment centre hypothesis. Behavioral Ecology 7: 115–118.

ROBERTS, B. 1940. The life-cycle of Wilson's Petrel Oceanites oceanicus (Kuhl). British Graham Land Expedition Scientific Report 1: 141–194.

ROWAN, M. K. 1952. The Greater Shearwater Puffinus gravis at its breeding grounds. Ibis 94: 97–121.

SCHREIBER, E. A., AND R. W. SCHREIBER. 1993. Red-tailed Tropicbird, in The Birds of North America 43 (A. Poole and F. Gill, Eds.). The Birds of North America, Inc., Philadelphia.

SCHREIBER, E. A., R. W. SCHREIBER, AND G. A. SCHENK. 1996. Red-footed Booby, in The Birds of North America 241 (A. Poole and F. Gill, Eds.). The Birds of North America, Inc., Philadelphia.

SIEGEL-CAUSEY, D., AND S. P. KHARITINOV. 1990. The evolution of coloniality. Current Ornithology 7: 285–330.

STENHOUSE, I. J., AND W. A. MONTEVECCHI. 1999. Increasing and expanding populations of breeding Northern Fulmars in Atlantic Canada. Waterbirds 22: 382–391.

SWINGLAND, I. R., AND P. J. GREENWOOD. 1984. The Ecology of Animal Movement, Caradon Press, Oxford.

TAPPER, S. 1992. Game Heritage: An Ecological Review from Shooting and Gamekeeping Records. Game Conservancy, Fordingbridge.

TELLA, J. L., F. HIRALDO, AND J. A. DONAZAR. 1998. The evolution of coloniality: does commodity selection explain all? Trends in Ecology and Evolution 13: 75–76.

TICKELL, L. 2000. Albatrosses. Pica Press, Mountfield, East Sussex.

VERCRUIJSSE, H. J. P. 1999. Zilvermeeuwen uit de duinen van Schouwen; Verspreiding sterfte en broedbiolgoie. IBN-DLO. Published Doctoral thesis. Tilburg, Netherlands.

WANLESS, S., AND M. P. HARRIS. 1997. Successful double-brooding in European Shags. Colonial Waterbirds 20: 291–294.

WARD, P., AND A. ZAHAVI. 1973. The importance of certain assemblages of birds as "information centres" for food finding. Ibis 115: 517–534.

WARHAM, J. 1990. The Petrels; their ecology and breeding systems. Academic Press, London.

WIGGINS, D. A., R. D. MORRIS, I. C. T. NISBET, AND T. W. CUSTER. 1984. Occurrence and timing of second clutches in common terns. Auk 101: 281–287.

WITTENBERGER, J. F., AND G. L. HUNT. 1985. The adaptive significance of coloniality in birds, *in* Avian Biology 8 (D. S. Farner, J. R. King, and K. C. Parkes, Eds.). Academic Press, New York.

WYNNE-EDWARDS, V. C. 1962. Animal Dispersion in Relation to Social Behaviour. Oliver & Boyd, Edinburgh.

WYNNE-EDWARDS, V. C. 1986. Evolution through Group Selection. Blackwell, Oxford.

A Family of European Shags

5 Seabird Demography and Its Relationship with the Marine Environment

Henri Weimerskirch

CONTENTS

5.1 DEMOGRAPHY AND LIFE HISTORY STRATEGIES

Demography is the study of the size and structure of populations and of the process of replacing individuals constituting the population. The study of demography was developed to forecast population growth. The rate at which a population increases or decreases depends basically on the fecundity (number of eggs laid) and survivorship of the individuals that belong to the population (Figure 5.1, bottom), but also to a lesser extent (especially for seabirds) on migration. Because many organisms, and especially seabirds, breed several times in their lives, a population consists of cohorts of individuals of different ages, born in different years. Moreover, mortality and fecundity rates are generally age-specific; life tables represent these birth and death probabilities. The relationship between the rate of increase or decrease and demographic parameters can be translated into more or less complex equations. The basic equation is the Euler–Lotka equation (Euler 1760, Lotka 1907) that specifies the relationships of age at maturity, age at last reproduction, probability of survival to age classes, and number of offspring produced for each age class, to the rate of growth of the population (r).

The demography of organisms is a key to the evolution of life histories because it allows us to examine the strength of selection on life history traits. Although they can achieve similar population growth rates, i.e., being stable, increasing, or declining, each population living in a particular habitat has specific dynamics, with specific age-related survivals and fecundities. The particular values of the demographic traits depend upon the adaptation of individuals and the attributes of the environment in which they live. Therefore, comparing demographic traits of populations allows us to elucidate the ecological and evolutionary responses of populations to their

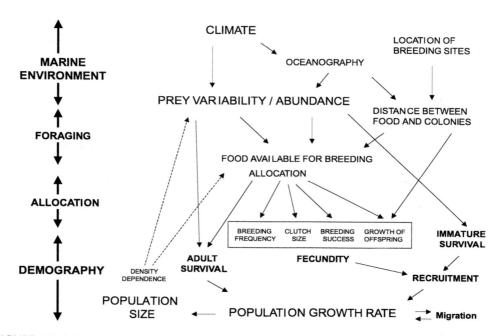

FIGURE 5.1 Schematic representation of the relationships between demographic traits and the marine environment.

environments. The comparison of demographic traits among taxa shows that demographic "tactics" exist; the concept of demographic tactic describes a complex co-adaptation of demographic parameters (Stearns 1976). Basically these co-adaptations result in the existence of a gradient from taxa with high fecundity and a low survival, to species with a high survival and a low fecundity. This fast–slow gradient (fast meaning fast turnover, and slow, slow turnover) or r/K gradient (Pianka 1970) provides a convenient (although not perfect) summary of the patterns linking life histories and habitats.

However, caution must be taken when life histories are compared. First it is possible to compare taxa from an ecological point of view as long as the allometric relationship linking them at a higher taxonomic level is known (Clutton-Brock and Harvey 1979). For example, within a taxa (a genus, for example), individuals of a particular species may live longer or produce fewer offspring than another species, not because they rely on a different habitat, but only because they are larger. Because they are larger they have a lower metabolism and therefore could live longer; they may produce fewer offspring because their offspring are larger and therefore require more energy (Calder 1984). The second constraint is phylogenetic (Harvey and Pagel 1991). Species are prisoners of their evolutionary past and can evolve to only a limited number of options. The single egg clutch of all Procellariiformes, and many other seabird species, has often been taken as an example for this (Stearns 1992). The life histories and habitats of two albatrosses can reasonably be compared, but care has to be taken when an albatross is compared to a species belonging to a different order. Phylogeny sets limits on an organism's life history and habitat but the ecological task of relating life histories to habitats is a fundamental challenge in ecology (Begon et al. 1996). Comparing demographic tactics within taxonomic levels that are closely related (ideally within the same species, see Lack 1947) to habitats or ecology remains a powerful tool to understand the influence of the environment on the evolution of life histories (Figure 5.1).

The aim of this chapter is first to describe the demographic traits of seabirds and compare these traits between taxa to examine whether demographic tactics can be found between and within the four orders of seabirds. Second, the variation in demographic traits will be examined to see whether it can be related to differences in the marine environment or the way seabirds exploit it, when comparing species within the same order, but also by comparing populations within the same species.

5.2 SEABIRDS AND OTHER BIRDS

In this study, a seabird is considered the species breeding along the seashore and relying on marine resources during the breeding season. Therefore several species of Pelecanidae, Laridae, Sternidae, and Phlacrocoracidae breeding inland or relying on freshwater resources are excluded, although they often winter in marine habitats. The data set used here includes 177 species of seabirds, with information on fecundity for 103 species, on age at first breeding for 111 species, and on survival/life expectancy for 76 species. All three parameters were simultaneously available for 62 species, and fecundity and age at first breeding for 84 species. Data were taken from Cramp (1978), Jouventin and Mougin (1981), Cramp and Simmons (1983), Marchant and Higgins (1990), Del Hoyo et al. (1992, 1996), Gaston and Jones (1998), unpublished data from a long-term data base for southern seabirds (CEBC-IFRTP), and unpublished data provided by E. A. Schreiber for tropical Pelecaniformes.

When compared with other birds, seabirds have lower fecundity; they breed at an older age and have higher adult survival. Since age at first breeding, survival, and to a lesser extent clutch size, are explained in part by mass (relationship between log body mass and log of demographic parameters: clutch $y = -0.081x - 1.33$, $r^2 = 0.028$, $p < 0.01$, $n = 362$ species of seabirds and other birds; age at first breeding $y = 0.215x - 0.545$, $r^2 = 0.313$, $p < 0.001$, $n = 261$, survival $y = 0.249x + 0.4859$, $r^2 = 0.394$, $n = 127$, $p < 0.001$), it is important to remove the effect of size. Indeed it could be argued that on average, seabirds are larger than land birds. To remove the variation of demographic traits related to body mass, they were transformed as log (parameter) $- 0.25$ log (mass) (Stearns 1983, Gaillard et al. 1989). Once the effect of body mass has been removed for clutch size and age at first breeding, seabirds still appear to stand at the extreme slow end of the fast–slow gradient that exists for bird species (Figure 5.2), underlying the low reproductive rate of seabirds.

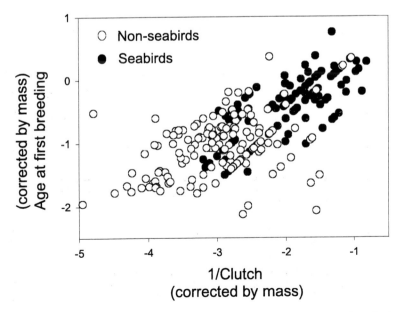

FIGURE 5.2 Relationship between the clutch size and minimum age at first breeding (both corrected for body mass) in 175 species of seabirds (black dots, $y = 0.571x + 0.72$, $r^2 = 0.541$, $p < 0.001$) and 187 species of other birds (white dots, $y = 0.306x - 0.118$, $r^2 = 0.216$, $p < 0.001$) belonging to all the existing orders of birds for which data are available.

5.3 DEMOGRAPHIC PARAMETERS OF SEABIRDS

When examining the demographic parameters of seabirds, extensive differences exist between and within orders, families, and species (Table 5.1). Fecundity is the product of clutch size, breeding frequency, and breeding success. Fecundity of seabirds is generally low, with all Procellariiformes, Phaethontidae, Fregatidae, and several species of Sulidae, Alcidae, Sternidae, Spheniscidae, and even some Laridae having a clutch of one (Table 5.1; see also Appendix 2). Several species of Diomedeidae, one of Procellaridae (the White-headed petrel *Pterodroma lessonii*; Figure 5.3), and probably most Fregatidae (at least females) breed only every second year when successful. On the other hand, some species of Phalacrocoracidae can have clutch sizes reaching five to seven eggs and many species of Laridae have clutch sizes of three and are able to lay a replacement clutch when failing early in the season (Figure 5.4).

The reasons for the low fecundity of seabirds have been much debated, and David Lack used seabirds, especially pelagic seabirds with a very low fecundity, to illustrate his general theory on clutch size (Lack 1948, 1968). Basically, Lack suggested that altricial birds should lay the clutch that fledges the most offspring. The ability to provide enough food to offspring would therefore be the main reason for the low reproductive rate of some seabirds. The development of life history theory and especially the concept of cost of reproduction and residual reproductive value (Williams 1966) later sophisticated this view. The basic idea is that, because resources are assumed to be limited, reproduction can have a negative influence on the probability of survival to the next reproduction, and therefore individuals should balance present and future reproduction (allocation; see Figure 5.1). For a long-lived species, the risk taken, especially during the first years of life, should be limited in order to enhance future reproductive success. Long-lived animals would therefore behave as "prudent parents," trying to limit risks of increased mortality when reproducing.

Therefore the single clutch of albatrosses and many other seabirds may have evolved as the result of the low provisioning rate of chick due to distant foraging zones (Lack 1968), but also of the "prudent" behavior of the parents that would limit energetic investment because of their high reproductive value. However, whether a clutch of one is the best option for other seabirds with a different ecological specialization is not clear (Ricklefs 1990). Indeed the low fecundity of seabirds is generally attributed to the marine environment on which they rely, an environment that is assumed to be poor, patchy, and unpredictable (Ashmole 1971). However, obviously the marine environment is very diverse and heterogeneous, with localized rich feeding areas or areas of low productivity. Therefore we might expect differences in demographic tactics within taxa according to the environment exploited, or to the foraging technique used, or diet. Conversely, convergence might be expected between taxa exploiting the same resources or environment, and divergence within taxa when environments exploited are different.

The minimum age at first breeding ranges from 2 to 4–5 years in most species of seabirds, except for Diomedeidae and Fregatidae and some species of Procellaridae that start breeding later (Table 5.1). Late age at first breeding is generally assumed to be necessary for long-lived species to attain similar foraging skills to those of adults, either because skills are complex to attain (e.g., Orians 1969, Burger 1987) and/or because of the high reproductive value of young birds.

Like age at first breeding, but even more importantly, survival is a parameter that is difficult to estimate accurately because it requires the marking of birds and their recapture over several years. Estimates of adult survival are available for a limited number of species (Table 5.1) and have to be treated with caution. Indeed, the statistical methods to estimate survival are in constant refinement, resulting in an overall increase of the estimates of survival rates within a species as techniques improve (Clobert and Lebreton 1991). Therefore, comparisons of survival are often difficult to perform unless the same method has been used. Average longevity is generally used to illustrate survival but cannot be compared to longevity records that only give maximum age based on isolated recaptures. Most Procellariidae and Diomedeidae have high survival and life expectancy, but also several species within the other orders, for example, several species of Alcidae and one

TABLE 5.1
Range of Demographic Parameters Observed in the Families of Seabirds

Order	Family	Symbol Used in Figures	Number of Species with at Least One Parameter	Average Clutch	Frequency of Breeding	Age at First Breeding	Adult Life Expectancy (number of species with an estimate of survival)	Relationship between Age at First Breeding − 1/Fecundity (both corrected for body mass)	Relationship between 1/Fecundity − Life Expectancy (both corrected for body mass)
Sphenisciformes	Sphenisciidae	Cross	15	1–2	0.7–1	2–5	6.4–20.5 (10)	$y = 0.219x - 0.648$, $r^2 = 0.085$, $p > 0.1$	$y = 0.251x - 1.11$, $r^2 = 0.0262$, $p > 0.1$
Procellariiformes	Diomedeidae	Circle	14	1	0.5–1	5–9	11.6–33.8 (12)		
	Procellariidae	Square	22	1	0.5–1	2–8	6.9–25.5 (20)	$y = 0.165x + 0.087$, $r^2 = 0.112$, $p < 0.05$	$y = 0.97x - 1.8$, $r^2 = 0.359$, $p < 0.01$
	Hydrobatidae	Diamond	5	1	1	2–3	7.6–17.2 (4)		
	Pelecanoididae	Triangle	2	1	1	2	5.7 (1)		
Pelecaniformes	Phaethontidae	Circle	3	1	1	2–5	25.5 (1)		
	Pelecanidae		1	4	1	2	?		
	Sulidae	Diamond	9	1–3	1	2–5	17.2–20.5 (4)	$y = 0.322x - 0.299$, $r^2 = 0.360$, $p < 0.05$	$y = 1.032x - 2.16$, $r^2 = 0.495$, $p < 0.01$
	Phalacrocoracidae	Triangle	8	2–4	1	2–4	6.7–10.4 (3)		
	Fregatidae	Square	4	1	0.5	5–8	?		
Charadriiformes	Stercorariidae	Diamond	4	2	1	3–5	6.7–11.6 (4)		
	Laridae	Square	12	1–3	1	2–4	8.8–19 (4)	$y = 0.07x - 0.268$, $r^2 = 0.0226$, $p > 0.1$	$y = 0.543x - 1.702$, $r^2 = 0.1236$, $p > 0.1$
	Sternidae	Triangle	14	1–2.1	1	2–4	5.7–9.6 (3)		
	Alcidae	Circle	19	1–2	1	2–5	4.7–20.5 (11)		

FIGURE 5.3 A White-headed Petrel. They breed only every other year, incubating their egg for 60 days and spending 112 days raising their single chick. (Photo by H. Weimerskirch.)

Laridae. Unfortunately, no estimate is available for Fregatidae, nor for most tropical Procellariidae, Laridae, and Sternidae, limiting the scope of a general comparison. The low fecundity and late age at first breeding of Fregatidae suggest high survival rate (maximum age recorded 34 years [E.A. Schreiber personal communication]), probably similar to Diomedeidae. One reason for the high survival of seabirds, especially those breeding on oceanic islands, is the absence of terrestrial predators; this is probably true for most large species, but not for the smaller species that can suffer heavy mortality from avian predators. Estimates of survival between fledging and recruitment into the breeding population are more difficult to obtain logistically because of the delayed age at first breeding, and are rare in the literature, limiting the scope for meaningful comparisons between groups.

5.4 COMPARING THE DEMOGRAPHY OF THE FOUR ORDERS OF SEABIRDS

Within seabirds, minimum age at first breeding and life expectancy (log transformed) are somewhat related to the log of mass (y = 0.092x + 0.666, r^2 = 0.0788, $p < 0.01$ and y = 0.1148x + 1.675, r^2 = 0.1532, $p < 0.001$). These relationships express the allometric component of demographic pattern and indicate that body mass is a significant, but not fundamental, determinant of the variation in demographic traits in seabirds. They represent a first-order tactic which expresses the biomechanical constraints of body mass (Western 1979, Gaillard et al. 1989). When parameters are corrected for the effect of body mass, the relationships between demographic traits are still very significant (Figure 5.5), representing a second-order tactic (Western 1979). It indicates that demographic parameters of seabirds covary after correction for the effect of body mass, which suggests the existence of demographic tactics among seabirds. The relationship between fecundity and life expectancy is very significant (Figure 5.5) and highlights the classical balance between clutch size and survival rates. The relationship between fecundity and age at first breeding, and that between age at first breeding and life expectancy, are also highly significant (Figure 5.5). The regression lines for the three relationships each describe a similar gradient within seabirds going from species with a fast turnover (high fecundity, early age at first breeding, and short life expectancy) to species with a slow turnover.

When examining the species within each order, they appear not to be distributed evenly along this fast–slow gradient. Spheniciformes appear to be distributed at the left-hand size of the gradient

(a)

(b)

FIGURE 5.4 Seabird species exhibit a range of fecundities. (a) Some gulls, such as this Herring Gull, may raise three chicks in a year, spending 45 to 50 days feeding them before they fledge. (b) Giant Petrels raise one chick a year and spend 100 to 120 days feeding it before it fledges. (Photos by J. Burger.)

or fast turnover end of the gradient: penguins breed relatively early, have a short life expectancy, and a high fecundity relative to their size. Conversely, many Procellariiformes species are found at the slow turnover extreme (Figure 5.5). Since the relationship considers all seabirds, i.e., four different orders, it is important to examine whether the relationships are a result of taxonomic differences in demography. Controlling for phylogeny (Harvey and Pagel 1991) was not possible because of the lack of a complete phylogeny covering all species of seabirds, and was out of the scope of this study. When investigating the existence of a gradient within orders, it appears that significant relationships persist within Procellariiformes and Pelecaniformes, whereas there is a tendency, yet nonsignificant for Charadriiformes, and no relationship for Spheniciformes (Table 5.1). This suggests the existence of different demographic tactics within Procellariiformes and Pelecaniformes, and perhaps Charadriiformes. We will now examine whether these tactics among taxa tending to show a fast or a slow turnover can be related to different environmental conditions or foraging strategies.

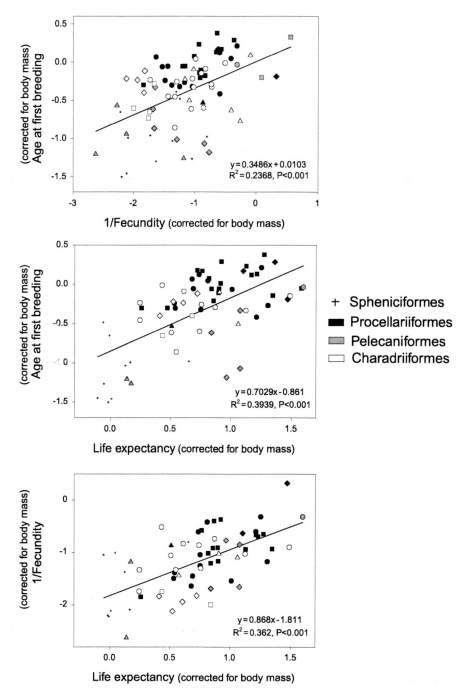

FIGURE 5.5 Relationships between 1/fecundity, age at first breeding, and life expectancy (corrected by body mass) in the four orders of seabirds (Sphenisciformes, crosses; Procellariiformes, symbols filled in black; Pelecaniformes in gray; and Charadriiformes in white). The inverse of fecundity is used for clarity, so that the three variables are positively linked. Correspondences of symbols for families are given in Table 5.1. Fecundity is estimated as the number of young produced per female per year. It is the product of the average clutch size per year by the overall breeding success. Because data on the average age at first breeding are scarce, minimum age at first breeding is used. Adult life expectancy is directly derived from adult survival and is measured as $(0.5 + 1/(1 - s))$ (Seber 1973). When parameters are available for several populations, average values are used.

To allow an easier representation of the ranking of species along this gradient, the species have been plotted along the first component of a principal component analysis (PCA) performed on the demographic parameters. When the three parameters are used, the first principal component explains 71.1% of the total variance (Figure 5.6a). One extreme, the left-hand side, is characterized by a high fecundity, short life expectancy, and early age at first breeding, while the other extreme presents the opposite characteristics. Because of the low number of species for which life expectancy is known, with an absence of data for some families like Fregatidae (see Table 5.1), a PCA was also performed on the fecundity and age at first breeding only, to be able to plot a larger number of species. The first principle component then explains 74.3% of the total variance (Figure 5.6b). Because the two analyses provide very similar ranking (compare Figure 5.6a and 5.6b, Factor 1 (2 parameters) = $0.924 \times$ Factor 1 (3 parameters) + 0.043, r = 0.956, $p < 0.001$). We use the ranking obtained from the PCA performed on fecundity and age at first breeding only, with the larger number of species (Figure 5.6b).

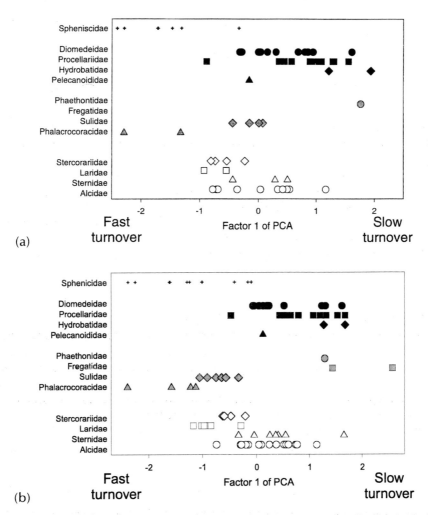

FIGURE 5.6 Ranking of the four orders of seabirds along a slow–fast gradient described by the first principal component of the PCA analyses (see symbols for families in Table 5.1): (a) PCA performed on 1/fecundity, life expectancy, and age at first reproduction, all corrected for body weight (eigenvalues 2.133, 0.546, and 0.321); and (b) PCA performed on 1/fecundity and age at first reproduction, both corrected for body weight (eigenvalues 1.487 and 0.513).

Spheniciformes and Procellariiformes almost do not overlap on the gradient, whereas Pelecaniformes extend throughout the gradient, and Charadriiformes are intermediate (Figure 5.6b). Whereas the species within the four families of Procellariiformes are scattered throughout the gradient, in Pelecaniformes the four families appear to be clearly separated from one another: Phalacrocoracidae, Sulidae, Phaethontidae, and Fregatidae ranking separately on the fast–slow gradient. This ranking probably reflects a strong phylogenetic effect on demographic tactics within this order, with each family having a distinct morphology and feeding specialization. Conversely, within Procellariiformes, Diomedeidae and Procellaridae are very similar in terms of morphology and feeding technique and are ranked similarly. Similarities in demographic traits between some families belonging to different orders suggest convergence. Phalacrocoracidae appear to have equivalent demographic tactics to those of Spheniciformes, having a fast turnover. Diving petrels, Pelecanoididae, also appear have faster turnover than most other petrels. This tendency to be at the fast extreme of the gradient in these three families could be associated with the constraints of diving that make birds poor fliers and therefore reduce foraging range. Convergence in demographic tactics may also be found between Fregatidae and the longest lived albatrosses and petrels. These birds have in common a pelagic life but especially economic flight. In Charadriiformes, a ranking of demographic tactics by families is also apparent, although less clear-cut than in Pelecaniformes, with Laridae (with the exception of one species, the Swallow-tailed Gull *Creagrus furcatus*) and Stercorariidae toward the fast extreme. Conversely, Sternidae and Alcidae are distributed over a wider range, rather at the slow end, suggesting convergence in demographic tactics with species that are well known to be long-lived like Procellariiformes.

5.5 FACTORS RESPONSIBLE FOR DIFFERENCES IN DEMOGRAPHIC TACTICS

Some demographic traits are phylogenetically conservative and fixed at high taxonomic levels. For example, all Procellariiformes have a clutch size of one. Others, like minimum age at first breeding and maximum life expectancy, probably do not vary within populations of a species because they are likely not to be adapted to local environmental conditions. Maximum life expectancy is probably mainly related to allometric pressures or phylogeny. Small birds have a higher energy expenditure and therefore shorter life span than larger birds (Lindstedt and Calder 1976; see Chapter 11). There are negative correlations between survival and vigorous, energy-expensive activity such as flight (Bryant 1999); consequently, birds with a low-energy flight such as albatrosses may live longer compared to birds with a highly expensive flight such as shags. On the other hand, breeding success, breeding frequency, average age at first breeding, and adult and juvenile survival express the interactions between phenotype and environment and are influenced by the environment (Figure 5.1). These demographic traits are likely to be different between closely related species exploiting different marine environments, or even within the same species exploiting different environments. Therefore, families covering a wide range over the fast–slow gradient suggest a broad range of demographic tactics due, for example, to a group of species exploiting a diversity of habitats. Conversely, families with a restricted range along the fast–slow gradient suggest that all species belonging to this group probably face similar environmental conditions. For example, Sulidae rank over a relatively restricted range, but they breed from tropical to sub-Arctic waters.

Seabirds have been classically separated into inshore, offshore, and oceanic or pelagic (Ashmole 1971), and it is generally assumed that pelagic species are the most long-lived, whereas inshore species are shorter lived (Lack 1968). Therefore, we might expect that pelagic species should be found at the slow turnover extreme of the fast–slow gradient. When considering the four orders simultaneously, there is indeed a tendency for oceanic families to stand at the slow end of the gradient (e.g., most Procellaridae, Hydrobatidae, Diomedeidae, or Fregatidae), whereas more inshore families are found at the other extreme. However, this is mainly due to the fact that many

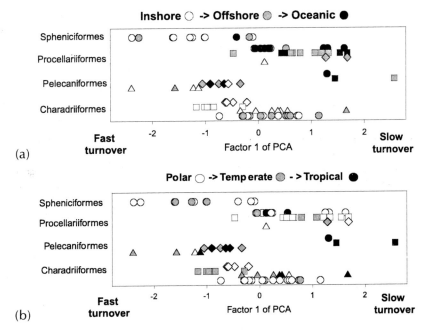

FIGURE 5.7 Ranking of the four orders of seabirds along a slow–fast gradient described by the first principal component of the PCA on 1/fecundity and age at first breeding: (a) Symbols are inshore (white), offshore (gray), and oceanic (black) species. (b) Symbols are polar (white), temperate (gray), and tropical (black) species.

Procellariiformes are pelagic and stand at the slow extreme of the gradient. Examining the distribution of inshore, offshore, or oceanic species within families does not lead to the clustering of inshore or oceanic species at one or the other extreme of the gradient (Figure 5.7a). This suggests that the assumption that pelagic species are more long-lived than inshore species only exists when groups are compared at high taxonomic levels (for example, when comparing Procellariiformes and Charadriiformes). But when the effect of size is controlled and each order examined separately, the data available today do not allow us to conclude that pelagic species have a slower turnover than inshore species.

Polar waters are generally more productive than tropical waters, which may have influenced the evolution of demographic traits. Therefore, we might expect that tropical species might be located at the slow end of the gradient compared to polar species. This tendency is not apparent with the data available (Figure 5.7b). For most families, either some breed only over a narrow range of climates (Alcidae or Fregatidae, for example) or data are not available (tropical Procellaridae, for example), limiting the possibility of making generalizations.

These first examinations indicate that the role of the marine environment in shaping demographic tactics is difficult to determine, and that the conjunction of several factors has probably been involved in shaping the demographic traits of marine birds. Because data are lacking for many groups, comparisons at lower taxonomic levels are impossible at this time.

5.6 INTRASPECIFIC VARIATIONS IN DEMOGRAPHIC TRAITS

Some species that are separated geographically show very homogenous demographic traits between populations. For example, Wandering Albatrosses (*Diomedia exulans*) breeding in the Atlantic, Indian, or Pacific Ocean have very similar demographic traits (Weimerskirch and Jouventin 1987, Croxall et al. 1990, Weimerskirch et al. 1997, DeLamare and Kerry 1992; Figure 5.8), suggesting that each population is relying on similar resources in the three regions. Indeed both in the Atlantic

FIGURE 5.8 A pair of Wandering Albatrosses at their nest. They are one of the most long-lived seabirds and have one of the longest breeding periods, incubating for 75 to 83 days and taking about 280 days to raise their chick. (Photo by H. Weimerskirch.)

and the Indian Ocean, Wandering Albatrosses are pelagic feeders that rely on distant food resources, have similar diets, growth of the chick, and foraging strategies (Weimerskirch et al. 1993, Prince et al. 1998). However, homogeneous demographic traits between populations of the same species are probably exceptions rather than the general rule. This will be highlighted by two examples.

Hatch et al. (1993) and Golet et al. (1998) noticed that Black-legged Kittiwakes (*Rissa tridactyla*) from the Pacific appear to be much longer-lived than those breeding in the north Atlantic. Golet et al. (1998) suggested that differences might be the result of higher winter mortality in Atlantic populations because of lower food availability due to different oceanographic conditions. A close examination of the data and the inclusion of other data indicate that the situation is not so clear-cut, although all Pacific colonies have, on average, lower fecundity and higher survival than Atlantic populations (Table 5.2). In the Atlantic, colonies in northern Norway, and to a lesser extent in Scotland, are longer-lived than colonies in more southern waters (Table 5.2). The common characteristic of the most long-lived populations is that fecundity is not only low on average (fecundity can be fairly high some years in the Pacific), but it is very variable in all localities, with complete breeding failures occurring frequently. Demographic tactics appear to vary extensively between oceans, but also between the different sites in the Atlantic Ocean. The striking feature of the data available for kittiwakes is that the range of average demographic traits (fecundity and adult survival) for a single species covers, in fact, almost that of the whole Charadriiformes order (see Figure 5.6).

Extensive differences exist between three populations of Black-browed Albatross (*Thalassarche melanophris*) for which fecundity, adult survival, and other life history characteristics are known (Table 5.3). The Kerguelen population is characterized by high breeding success that does not vary from year to year, a relative low minimum age at first breeding, and a relatively low adult survival (see references in Table 5.3). On the other end, the South Georgia population has a low and very variable breeding success, years with complete breeding failures, a later minimum age at first breeding, and a fairly high survival. The Campbell population (the smallest birds) is intermediate between the two others, similar in fecundity to Kerguelen and in survival to South Georgia. Birds at three sites rely on similar diets with the same squid species, typical of the Polar frontal zone. At Kerguelen birds forage close to colonies, at an average distance of 250 km, over offshore waters (on the slope of the shelf) where the Polar front passes. At South Georgia, birds forage over the shelf, or slope area, and feed on krill to a large extent during some years, but have to forage farther

TABLE 5.2
Fecundity and Survival of Black-Legged Kittiwake (*Rissa tridactyla*) in Different Sites of the North Atlantic and Pacific Oceans

	Fecundity Chick/Year (Range)	Adult Survival	References
Atlantic			
North Shield — England	1.25 (1–1.4)	0.801	Aebischer and Coulson 1990, Coulson and Thomas 1985
Brittany — France	0.78 (0.23–1.48)	0.808	Danchin and Monnat 1992
Isle of May — Scotland		0.890	Harris and Calladine 1993
Shetlands — Scotland	1.1 (0–1.8)		Monaghan 1996
Hornøya — Norway	0.55	0.922	Erikstad et al. 1995
Eastern Canada	0.74		
Pacific			
Middleton Is. — Alaska	0.31 (0–1.2)	0.926	Hatch et al. 1993
Tatan Is. — Sea of Okhotsk		0.920	Hatch in Golet et al. 1998
St. George — Bering Sea		0.930	Dragoo and Dragoo 1996
Bluff — Alaska	0.55 (0–1.16)		Murphy et al. 1991
Shoup Bay — Alaska	0.35	0.925	Golet et al. 1998

TABLE 5.3
Comparative Life History Traits of Three Populations of Black-Browed Albatrosses

	South Georgia[a]	Kerguelen[b]	Campbell[c]
Ocean	Atlantic	Indian	Pacific
Latitude	55°S	50°S	52°S
Size of population	65000	3300	10–15000
Mass of adult (g)	3560 ± 396	3655 ± 353	2750 ± 161
Culmen length (mm)	119.0 ± 2.4	118.4 ± 3.9	112.5 ± 2.9
Breeding success (%)	34.2 ± 24.0	63.0 ± 10	66.3 ± 12.9
Fecundity	0.27	0.580	0.543
Minimum age at first breeding	8	6	6
Average age at first breeding	10	9.7	10
Adult survival	0.95 ± 0.006	0.906 ± 0.005	0.945 ± 0.007
Juvenile survival	0.240	0.21	0.186
Trend of population (%/year)	0	0	+1.1
Foraging zone	Shelf, Polar Front	Shelf, Polar Front	Shelf, Polar Front
Distance to main feeding zone	100–600 km	250 km	2200 km
Food	Fish, Krill	Fish	Fish
Foraging trips incubation	12 days	4 days	11 days
Foraging trips chick rearing	2.1 (1–12)	2.1 (1–7)	2.0 (1–12)
Fledging period (days)	116	120	130

[a] Before 1990, Prince et al. 1994, Croxall et al. 1998, Prince et al. 1998, Tickell and Pinder 1974.
[b] References: Weimerskirch et al. 1997, Weimerskirch 1998.
[c] References: Waugh et al. 1999a, 1999b, 1999c.

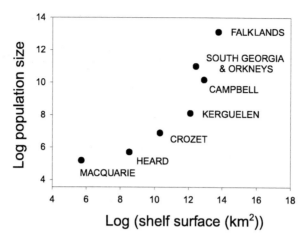

FIGURE 5.9 Relationship between population size and the surface of shelf as an index of food resource in different populations of Black-browed Albatrosses.

away when krill is rare (Veit and Prince 1997). They also exploit the Polar Frontal zone that is on average 500 km north of the breeding site. At Campbell birds forage within a year alternately in the Polar Frontal zone (2000 km south of the island), and on the shelf close to the island. On the three sites, birds are using the same habitats, the Polar Frontal zone and shelf slopes, but the geographic location and trophic situations of these favored habitats are different, leading to varying demographic tactics. At Campbell the extensive distance to the Polar Front area makes provisioning more difficult, with longer fledging periods and small fledglings, possibly the reason for the smaller size of the adult birds. Yet fecundity, as well as survival, is high. The sizes of the populations are also different between sites. Probably because the size of the shelf is related to the amount of resources available to the population, the size of the Black-browed Albatross population for each breeding locality varies directly with the size of the surrounding shelf (Figure 5.9).

Similar divergent evolution in demographic tactics probably exists within many other taxa of seabirds where populations rely on different marine environments, or when the food resource is more or less distant from the breeding grounds. The two examples presented here highlight the importance of the marine environment in shaping demography of seabirds (Figure 5.1). Fecundity is dependent on the amount of resources available in the environment, i.e., on oceanographic processes, and their variability is influenced by climatic variability (see Chapter 6). Seabirds rely on marine resources but have to breed on land. Colonies are often located in proximity to productive ocean zones, but the distances between the colony and the resources put constraints on the amount of energy seabirds are able to invest in reproduction. Seabirds must therefore allocate resources (Figure 5.1). It is not surprising that the diversity in morphology that allows various families of seabirds to efficiently exploit more or less distant resources has resulted in different demographic strategies. For example, within Pelecaniformes, shags are poor fliers but excellent divers that forage close to colonies, whereas frigatebirds are magnificent fliers and probably forage at great distances from colonies. Similarities in demographic tactics are found in diverse families that exploit resources in the same way: penguins and cormorants have restricted foraging range and rely on diving to feed and have similar demographic tactics. Pelagic albatrosses and frigate birds share many life history traits and have a similar demography. Alcids, often compared to penguins in terms of morphology or foraging habits, are in fact closer to Procellariiformes in demographic terms (Figure 5.6, Croxall and Gaston 1988), possibly because several species of alcids forage at long distances from colonies, but also because other factors probably have played a role in the evolution of low fecundity. However, within families, when clutch size is fixed by phylogenetic history, fecundity is greatly influenced by breeding success or frequency (highly dependent on the availability of

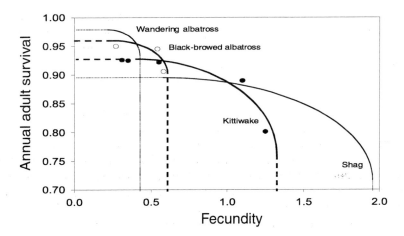

FIGURE 5.10 Relationship between fecundity and survival in kittiwakes (black dots) and Black-browed Albatross (white dots) populations, suggesting a convex curve, representing the optimization between survival and fecundity, specific to each species. Hypothetical relationships for a long-lived species (Wandering Albatross) and a short-lived species (shag) are represented with the dotted lines.

resources). Variability in resources, quantity of resources, and distance to the resource all influence both parameters (Figure 5.1).

The data available on fecundity and adult survival in several populations of kittiwakes (Table 5.2) suggest a negative relationship between the two that has a convex shape (Figure 5.10). The shape of the relationship is similar to the classical figure representing the optimization of the trade-off between survival and fecundity (Cody 1966). The trade-off underlines the idea that when extrinsic factors such as resource availability or predation cause little mortality, evolution could reduce parental investment and therefore fecundity. The curve for kittiwakes could represent the range of possible demographic tactics, with the curve becoming asymptotic at the two extremes (Figure 5.7). The maximum asymptotic values for survival and fecundity could be determined genetically (maximum clutch size or maximum life span), and the intermediate values represent the result of optimization of the trade-off, i.e., the possible phenotypes realized according to the environmental conditions. The curve would be different for each species, as suggested by that for Black-browed Albatrosses, and represents the possible demographic scenarios for a stable population. Points with lower values (inside the convex curve) represent declining populations; those outside, increasing populations. Hypothetical scenarios are shown as extremes for a very long-lived species (Wandering Albatrosses with long life span and low fecundity, maximum one egg every second year) and a shorter-lived species (shag with multiegg clutches, Figure 5.7). These different scenarios are based on the assumption of density-dependent feedback. Although hypothetical, they probably represent more accurately the possible variability in demographic traits found within a species, and contrast, of course, with the single figure or average figure that is often proposed in comparative studies. This representation does not integrate the survival of chicks from fledging to recruitment, which plays a significant role in the population dynamics of long-lived species, but is generally not considered, in part, because of the paucity of empirical data.

5.7 POPULATION REGULATION AND ENVIRONMENTAL VARIABILITY

It has been suggested that populations of seabirds are mainly regulated by food availability, in a density-dependent way (Ashmole 1963, Birkhead and Furness 1985). The sizes of populations in relation to potential food availability around breeding grounds (Figure 5.6), or in relation to the location of other colonies of conspecifics (Furness and Birkhead 1984), are good examples. Also, the degree of density dependence is likely to be different, especially during the breeding season, between species relying on resources close or distant from breeding grounds (Birkhead and Furness 1985). Other factors such as predation or breeding site availability are likely to be important only in isolated species, with the exception of predation by introduced species, such as cats or rats, on islands that were free of terrestrial predators (Burger and Gochfeld 1994). The question remains open: How does density dependence affect populations? For example, density-dependent feedback may affect fecundity during the breeding season, or survival of adults or immatures during or outside the breeding season (Figure 5.1).

Many seabird species live in a variable environment and it is not surprising that when examining long-term trends in demographic traits these parameters can vary extensively over time. The most extreme cases are, for example, populations of seabirds affected by El Niño–Southern Oscillation (ENSO) events (e.g., Schreiber and Schreiber 1984; see Chapter 7). Variability in demographic parameters mainly reflects the variability in food availability, which drops dramatically during en ENSO event. One classical example is three species of "guano" (deposits used as fertilizer) seabirds relying on anchovies off Peru. Huge populations of Guanay Cormorant (*Leucocarbo bougainvilli*), Peruvian Booby (*Sula variegata*), and Brown Pelican (*Pelecanus occidentalis*) collapse regularly (20 times in the last 100 years) with the occurrence of severe ENSO affecting the Humboldt current (see, e.g., Duffy 1990; and Chapter 7).

Fecundity is the demographic trait that is mainly affected by variations in environmental conditions off Peru (when the upwelling disappears), with lowered breeding success and fewer birds breeding. This has also been documented in other areas such as the North Atlantic (for kittiwakes, Aebischer et al. 1990) or in the Southern Ocean (for Blue Petrels, *Halobaena caerulea*, Guinet et al. 1998) through correlations between breeding performances and oceanographic-climatic parameters. There is less evidence that survival, especially adult survival, may be dependent on environmental conditions. Theoretically, because they are long-lived, adult seabirds should be less likely to suffer high mortality when environmental conditions are poor, than shorter lived species. For example in the three species of seabirds from Peru affected by ENSO, it has been suggested that adults die in large numbers from starvation (Duffy 1990). Inshore or sedentary species, such as "guano" seabirds, could be more susceptible to die when resources are scarce, but this has not been proved by either large numbers of birds found dead or by estimates of survival rates.

The lower susceptibility of long-lived species to increased mortality from environmental conditions affecting the food supply might be partly due to the fact that seabirds, especially pelagic species, can move to more favorable waters or migrate when conditions become unfavorable around breeding grounds. But in general, the extent to which adult survival is affected by environmental variability is poorly known due to few field studies of marked birds. Only recently developed techniques in modeling of survival should allow us in the future to relate environmental variability and adult survival in seabirds. Using such techniques, it has been possible to relate the survival of adult Emperor Penguins to oceanographic anomalies related to the Antarctic Circumpolar Wave. During the warm events that occur every 4 to 5 years, adult survival drops to low values, some years to 0.75, whereas in other years survival is 0.92–0.97 (Barbraud and Weimerskirch 2001).

Another aspect of the demography of long-lived seabirds that is still poorly known is the extent to which nonbreeding by adult mature birds affects the dynamics of populations. In addition to the absence of reproduction in populations strongly affected by ENSO (Schreiber and Schreiber 1984, Duffy 1990), it appears that nonbreeding could be a general feature in other populations in response to ENSO. In some petrels in the Southern Ocean, up to 70% of the population refrains from breeding in some years (Chastel et al. 1995), but adult survival is not affected by these poor years when breeding success is low, or when few birds are able to breed (unpublished data). Thus it is important to be able to distinguish between absence due to nonbreeding and absence due to mortality.

Environmental variability has probably had a major influence on the evolution of life history traits of seabirds. It is generally assumed that birds which live in a highly variable environment have increased reproductive rate and therefore reduced survival (Schaffer 1974). However, in our examples with kittiwakes, as well as in Black-browed Albatrosses, populations with highly variable fecundity, i.e., probably living in the most variable environment, are those with the highest survival and the lowest fecundity. This paradox shows that the degree to which environmental variability influences the evolution of life history strategies is not clearly understood (Cooch and Ricklefs 1994).

The possible tendency for some seabirds to be longer-lived when living in a variable environment may be explained by taking into account several important factors specific to seabirds, and especially the ability or inability of species to disperse when conditions are unfavorable. Kittiwakes and Black-browed Albatrosses disperse from the vicinity of breeding grounds, and are thus able to escape from poor environmental conditions, especially outside the breeding season. Low average fecundity due to high variability in breeding success and especially to the occurrence of complete breeding failure has to be balanced by high survival. Conversely, species such as cormorants, boobies, or pelicans in Peru and Ecuador are relying all year round on the same system, the Humboldt current. Therefore collapse in food availability results in breeding failure and possibly in adult mortality. Selection has probably, in this case, resulted in a high potential fecundity to balance the possible regular crashes in adult population. Average adult survival is probably low, but also probably highly variable, possibly being high during favorable years. In addition to the ability or inability of adults to escape from poor trophic conditions, the frequency of extreme events and the extent of variability in resource availability have probably played an important role in the selection of demographic tactics.

It could be suggested that two kinds of demographic tactics might be selected. Highly variable environments can select for boom/bust unstable populations with high fecundities if adults are unable to escape adverse conditions ("guano" seabirds). Population sizes fluctuate extensively in this case. Alternatively, if adults can move to different zones when feeding conditions decline around the breeding area (Black-browed Albatrosses, kittiwakes, frigatebirds), lower fecundities and high survival are selected for in variable environments. In less variable environments, populations do not vary extensively in size from one year to the next, and may be in some cases close to saturation, with high density-dependent feedbacks: higher fecundity is balanced by lower survival, or alternatively lower fecundity by high survival.

5.8 PERSPECTIVES

Understanding the demography of seabirds requires studying them over long periods having populations of marked birds. Short-term studies of seabirds are inadequate to characterize the demographic pattern of seabird populations because seabirds are long-lived and live in variable environments. It is therefore necessary for studies to encompass the exceptional events that punctuate the life of seabirds in order to measure demographic parameters. This is a difficult task, because long-term studies have often been carried out by individual researchers (e.g., Fulmars, G. Dunnet; Kittiwake, J. Coulson; Short-tailed Shearwaters, D. Serventy, R.D. Wooller, and J.S. Bradley) rather than by institutions or governmental agencies, but also because humans have a life span that is not much longer than that of some seabirds. Furthermore much modern research is generally based on short-term projects.

One difficulty in studying seabird populations is that an important part of the population is not accessible to study. Young birds, after they have fledged, remain at sea until they first breed, and it is impossible to obtain information on the factors that affect their survival and maturation. Yet immature birds represent a significant portion of the population, up to 40% in some species. Again the only way to obtain information is to carry out long-term population studies and band large numbers of fledglings. Doing this is vital to understanding the demography of a seabird population. Another aspect that is also poorly known is the dispersal rates of a seabird population. Seabirds, and especially Procellariiformes, are generally assumed to be highly philopatric but there is some evidence that it is not always the rule. For example, the expansion of fulmars (*Fulmarus*) in the Atlantic can only be explained by high emigration rates from large colonies. Most snow petrel (*Pagodroma nivea*) chicks do not return to their birthplace (Chastel et al. 1993). The role of dispersal in the dynamics of seabird populations is technically difficult to study, but it is likely that, in some species at least, it plays a significant role.

There is a paucity in studies on tropical species for most families compared to the large number of long-term studies on the demography of temperate or polar species. There is definitely a need for studies on tropical species such as most Pelecaniformes (especially tropicbirds and frigatebirds), tropical terns, or Procellariiformes. This would allow fruitful comparison and allow tests of hypothesis such as, for example, that related to the lower productivity of tropical waters.

To conclude, it appears that much is still to be learned on the demography of seabirds, and many exciting questions remain unanswered. Problems will only be solved by the development of new ideas and modeling, but there is a striking need for empirical data. Comparative studies between high taxonomic levels are probably not optimal to understand the role of the marine environment in shaping demographic studies. Comparing populations of the same species, living in contrasted environments is probably more promising.

ACKNOWLEDGMENTS

I would like to thank E. A. Schreiber and J. Burger for inviting me to write this chapter and for extensive help in the preparation of the manuscript

LITERATURE CITED

AEBISCHER, N. J., AND J. C. COULSON. 1990. Survival of the kittiwake in relation to sex, year, breeding experience and position in the colony. Journal of Animal Ecology 59: 1063–1071.

AEBISCHER, N. J., J. C. COULSON, AND J. M. COLEBROOK. 1990. Parallel long term trends across four marine trophic levels and weather. Nature 347: 753–755.

ASHMOLE, N. P. 1963. The regulation of numbers of tropical oceanic seabirds. Ibis 103: 458–473.

ASHMOLE, N. P. 1971. Seabird ecology and the marine environment. Pp. 223–286 in Avian Biology, Vol. 1 (D. S. Farner and J. R. King, Eds.). Academic Press, New York.

BARBRAUD, C., AND H. WEIMERSKIRCH. 2001. Emperor penguins and climate change. Nature 411: 160–163.

BEGON, M., J. L. HARPER, AND C. R. TOWNSEND. 1996. Ecology: Individuals, Populations and Communities. Blackwell Scientific, Oxford.

BIRKHEAD, T. R., AND R. W. FURNESS. 1985. Regulation of seabird populations. Pp. 145–167 in Behavioural Ecology: Ecological Consequences of Adaptive Behaviour (Sibly and Smith, Eds.). Blackwell Scientific, Oxford.

BRYANT, D. M. 1999. Energetics and lifespan in birds. Pp. 412–421 in Proceedings of the 22nd International Ornithological Congress, Durban (N. J. Adams and R. H. Slotow, Eds.). BirdLife, Johannesburg, South Africa.

BURGER, J. 1987. Foraging efficiency in gulls: a congeneric comparison of age differences in efficiency and age of maturity. Studies in Avian Biology 10: 83–90.

BURGER, J., AND M. GOCHFELD. 1994. Predation and effect of humans on island-nesting seabirds. Pp. 39–67 in Seabirds on Islands: Threats, Case Studies and Action Plans (D. N. Nettleship, J. Burger, and M. Gochfeld, Eds.). BirdLife International, Cambridge.

CALDER, W. A. 1984. Size, Function and Life-History. Harvard University Press, Cambridge, MA.

CHASTEL, O., H. WEIMERSKIRCH, AND P. JOUVENTIN. 1995. Body condition and seabird reproductive performance: a study of three petrel species. Ecology 76: 2240–2246.

CLOBERT, J., AND J. D. LEBRETON. 1991. Estimation of demographic parameters in bird populations. Pp. 75–104 in Bird Population Studies: Relevance to Conservation and Management (C. M. Perrins, J. D. Lebreton, and G. J. M. Hirons, Eds.). Oxford University Press, Oxford.

CODY, M. L. 1966. A general theory of clutch size. Evolution 20: 174–184.

COOCH, E. G., AND R. E. RICKLEFS. 1994. Do variable environments significantly influence optimal reproductive effort in birds? Oikos 69: 447–459.

COULSON, J. C., AND C. S. THOMAS. 1985. Changes in the biology of the kittiwake (*Rissa tridactyla*): a 31-year study of a breeding colony. Journal of Animal Ecology 54: 9–26.

CRAMP, S. (Ed.). 1977. The birds of the Western Palearctic. Vol. 1. Oxford University Press, Oxford.

CRAMP, S., AND K. E. L. SIMMONS (Eds.). 1983. The Birds of the Western Palearctic. Vol. 3. Oxford University Press, Oxford.

CROXALL, J. P., AND D. R. BRIGGS. 1991. Foraging economics and performance of polar and subpolar Atlantic seabirds. Polar Research 10: 561–578.

CROXALL, J. P., P. A. PRINCE, P. ROTHERY, AND A. G. WOOD. 1998. Population changes in albatrosses at South Georgia. Pp. 69–83 *in* Albatross Biology and Conservation (G. Robertson and R. Gales, Eds.). Surrey Beatty & Sons, Sydney, Australia.

CROXALL, J. P., AND A. J. GASTON. 1988. Patterns of reproduction in high-latitude northern and southern seabirds. Pp. 1176–1194 *in* Proceedings of the XIX International Ornithological Congress (H. Ouellet, Ed.). University of Ottawa, Ottawa.

CROXALL, J. P., P. ROTHERY, et al. 1990. Reproductive performance, recruitment and survival of wandering albatrosses *Diomedea exulans* at Bird Island, South Georgia. Journal of Animal Ecology 59: 775–796.

DANCHIN, E., AND J. Y. MONNAT. 1992. Population dynamics modeling of two neighbouring kittiwake *Rissa tridactyla* colonies. Ardea 80: 171–180.

DEL HOYO, J., A. ELLIOTT, AND J. SARGATAL (Eds.). 1992. Handbook of the Birds of the World. Vol. 1. Lynx Edicion, Barcelona.

DEL HOYO, J., A. ELLIOTT, AND J. SARGATAL (Eds.). 1992. Handbook of the Birds of the World. Vol. 3. Lynx Edicion, Barcelona.

De LaMARE, W. K., AND K. R. KERRY. 1994. Population dynamics of the wandering albatross (*Diomedea exulans*) on Macquarie Island and the effects of mortality from longline fishing. Polar Biology 14: 231–241.

DRAGOO, D. E., AND DRAGOO, B. K. 1996. Results of productivity monitoring of kittwakes and murres at St. George Island, Alaska, *in* 1995. U.S. Fish and Wildlife Service Report, AMNWR 96/01. Homer, Alaska.

DUFFY, D. C. 1990. Seabirds and the 1982–1984 El Nino–Southern Oscillation. Pp. 395–415 *in* Global Ecological Consequences of the 1982–83 El Nino–Southern Oscillation (P. W. Glynn, Ed.). Elsevier Oceanography Series 52. Elsevier, Amsterdam.

ERIKSTAD, K. E., T. TVERAA, AND R. BARRETT. 1995. Adult survival and chick production in long-lived seabirds: a 5-year study of the kittiwake *Rissa tridactyla*. Pp. 471–477 *in* Ecology of Fjords and Coastal Waters (H. R. Skjodal, C. Hopkins, K. E. Erikstad, and H. P. Leinaas, Eds.). Elsevier, Amsterdam.

EULER, L. 1760. Recherches générales sur la mortalité: la multiplicité du genre humain. Mémoires de l'Académie des Sciences, Berlin 16: 144–164.

FURNESS, R. W., AND T. R. BIRKHEAD. 1984. Seabird colony distributions suggest competition for food supplies during the breeding season. Nature 311: 655–656.

GAILLARD, J. M., D. PONTIER, D. ALLAINE, J. D. LEBRETON, J. TROUVILLIEZ, AND J. CLOBERT. 1989. An analysis of demographic tactics in birds and mammals. Oikos 56: 56–76.

GASTON, A. J., AND I. L. JONES. 1998. The Auks. Alcidae. Oxford University Press, Oxford.

GOLET, G. H., D. B. IRONS, AND J. A. ESTES. 1998. Survival costs of chick rearing in black-legged kittiwakes. Journal of Animal Ecology 67: 827–841.

GUINET, C., O. CHASTEL, M. KOUDIL, J. P. DURBEC, AND P. JOUVENTIN. 1998. Effects of warm sea surface temperature anomalies related to El Nino and the Antarctic Circumpolar Wave on the blue petrel at Kerguelen Island. Proceedings of the Royal Society London B 265: 1001–1006.

HARRIS, M. P., AND J. CALLADINE. 1993. A check on the efficiency of finding colour-ringed kittiwakes *Rissa tridactyla*. Ringing and Migration 14: 113–116.

HARVEY, P. H., AND M. PAGEL. 1991. The Comparative Method in Evolutionary Biology. Oxford University Press, Oxford.

HATCH, S. A., B. D. ROBERTS, AND B. S. FADELY. 1993. Adult survival of black-legged kittiwakes *Rissa tridactyla* in a Pacific colony. Ibis 135: 247–254.

JOUVENTIN, P., AND J. L. MOUGIN. 1981. Les stratégies adaptatives des oiseaux de mer. Terre et Vie 35: 217–272.

KARR, J. R., J. N. NICHOLS, M. K. KLIMKIEWICZ, AND J. D. BRAWN. 1990. Survival rates of birds of tropical and temperate forests: will the dogma survive? American Naturalist 136: 277–291.

LACK, D. 1947. The significance of clutch size. Ibis 89: 302–352.

LACK, D. 1968. Ecological Adaptations for Breeding in Birds. Methuen, London.

LOTKA, A. J. 1907. Studies on the mode of growth of material aggregates. American Journal of Sciences 24: 199–216; 375–376.

LINDSTEDT, S. L., AND W. A. CALDER. 1976. Body size, physiological time and longevity of homeothermic animals. Quarterly Review of Biology 56: 1–161.

MARCHANT, S., AND P. J. HIGGINS. 1990. Handbook of Australian, New Zealand and Antarctic Birds. Vol. 1. Oxford University Press, Melbourne.

MONAGHAN, P. 1996. Relevance of the behavior of seabirds to the conservation of marine environments. Oikos 77: 227–237.

MURPHY, E. C., A. M. SPRINGER, AND D. G. ROSENEAU. 1991. High annual variability in reproductive success of kittiwakes (*Rissa tridactyla* L.) at a colony in western Alaska. Journal of Animal Ecology 60: 515–534.

ORIANS, G. H. 1969. Age and hunting success in the brown Pelican (*Pelecanus occidentalis*). Animal Behaviour 17: 316–319.

PIANKA, E. R. 1970. On r- and k-selection. American Naturalist 104: 592–597.

PRINCE, P. A., J. P. CROXALL, P. ROTHERY, AND A. G. WOOD. 1994. Population dynamics of black-browed and grey-headed albatrosses *Diomedea melanophris* and *D. chrysostoma* at Bird Island, South Georgia. Ibis 136: 50–71.

PRINCE, P. A., J. P. CROXALL, P. N. TRATHAN, AND A. G. WOOD. 1997. The pelagic distribution of South Georgia albatrosses and their relationships with fisheries. Pp. 137–167 *in* Albatross Biology and Conservation (G. Robertson and R. Gales, Eds.). Surrey Beatty & Sons, Sydney, Australia.

RICKLEFS, R. E. 1990. Seabirds life histories and the marine environment: some speculations. Colonial Waterbirds 13: 1–6.

SCHAFFER, W. M. 1974. Optimal reproductive effort in fluctuating environments. American Naturalist 108: 783–790.

SCHREIBER, R. W., AND E. A. SCHREIBER. 1984. Central Pacific seabirds and the El Nino Southern Oscillation: 1982 to 1983 perspectives. Science 225: 713–716.

SEBER, G. A. F. 1973. The estimation of animal abundance and related parameters. Griffin, London.

STEARNS, S. 1976. Life history tactics: a review of the ideas. Quarterly Review of Biology 51: 3–47.

STEARNS, S. 1992. The Evolution of Life Histories. Oxford University Press, Oxford.

TICKELL, W. L. N., AND R. PINDER. 1975. Breeding biology of the black-browed albatross, *Diomedea melanophris*, and grey-headed albatross, *D. chrysostoma*, at Bird Island, South Georgia. Ibis 117: 433–452.

VEIT, R. D. R., AND P. A. PRINCE. 1997. Individual and population level dispersal of black-browed albatrosses *Diomedea melanophris* and grey-headed albatrosses *D. chrysostoma* in response to Antarctic krill. Aredea 85: 129–134.

WAUGH, S. M., P. A. PRINCE, AND H. WEIMERSKIRCH. 1999. Geographical variation in morphometry of black-browed and grey-headed albatrosses *Diomedea melanophrys* and *D. chrysostoma*. Polar Biology 22: 189–194.

WAUGH, S., H. WEIMERSKIRCH, Y. CHEREL, U. SHANKAR, P. A. PRINCE, AND P. SAGAR. 1999. The exploitation of the marine environment by two sympatric albatrosses in the Pacific Southern Ocean. Marine Ecology Progress Series 177: 243–254.

WAUGH, S., H. WEIMERSKIRCH, P. MOORE, AND P. SAGAR 1999. Population dynamics of New Zealand Black-browed and Grey-headed Albatross *Diomedea melanophrys impavida* and *D. chrysostoma* at Campbell island, New Zealand, 1942–1996. Ibis 141: 216–225.

WEIMERSKIRCH, H. 1998. Foraging strategies of southern albatrosses and their relationship with fisheries. Pp. 168–179 *in* Albatross Biology and Conservation (G. Robertson and R. Gales, Eds.). Surrey, Beatty & Sons, Sydney, Australia.

WEIMERSKIRCH, H., AND P. JOUVENTIN. 1998. Changes in population size and demographic parameters of 6 albatross species breeding in the French sub-antarctic slands. Pp. 84–91 *in* Albatross Biology and Conservation (G. Robertson and R. Gales, Eds.). Surrey Beatty & Sons, Sydney, Australia.

WEIMERSKIRCH, H., T. MOUGEY, AND X. HINDERMEYER. 1997. Foraging and provisioning strategies of black-browed albatrosses in relation to the requirements of the chick: natural variation and experimental study. Behavioral Ecology 8: 635–643.

WESTERN, D. 1979. Size, life history and ecology in mammals. East Africa Wildlife Journal 17: 185–204.

WILLIAMS, G. C. 1966. Natural selection, the cost of reproduction and a refinement of Lack's principle. American Naturalist 100: 687–690.

Wilson's Storm-petrel Feeding on the Wing

6 Foraging Behavior and Food of Seabirds

David A. Shealer

CONTENTS

6.1 INTRODUCTION

Evolution has produced a vast array of morphological and physiological adaptations in seabirds that enable them to exploit the ocean's food resources in myriad ways. As consumers, seabirds are found at most trophic levels of the marine food web (Figure 6.1), ranging from the storm-petrels that feed on surface zooplankton, to gannets and some penguins that take large pelagic fish and squid, to gulls and albatrosses that scavenge the remains of dead animals.

In terms of its exploitability, the ocean represents a challenge for air-breathing seabirds. The problem of finding enough food to survive and reproduce in this realm has generally been solved in three ways. Coastal or inshore species, such as most gulls and terns, gather to feed in areas where prey are abundant or are forced to the surface. Diving seabirds, including penguins and alcids, can exploit a greater range of depths to obtain food. Pelagic species, such as albatrosses,

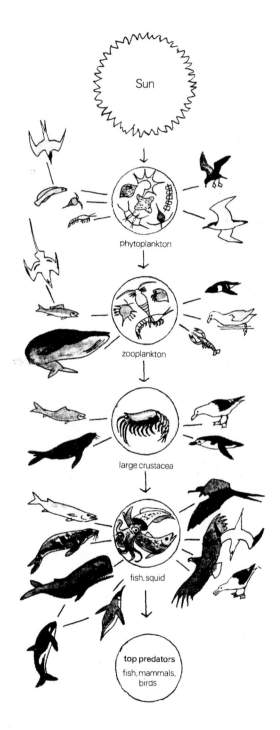

FIGURE 6.1 A simplified marine food web illustrating how seabirds feed at different trophic levels. (From Nelson 1979.)

FIGURE 6.2 Brown Pelicans diving for fish in Florida. They are the only pelican species to plunge dive. (Photo by R. W. Schreiber.)

shearwaters, and some petrels, soar over vast expanses of the ocean surface at a relatively low energy cost in search of widely dispersed surface prey.

Some authors categorize seabirds as inshore, offshore, and pelagic feeders. These categories of human construct are somewhat artificial, however, because most seabirds are opportunistic feeders and will cross such boundaries if necessary to find adequate prey. Solitary foraging in seabirds is rare. Most seabirds feed on localized concentrations of prey in single- or mixed-species aggregations, suggesting that they may rely on social facilitation either to locate or catch prey. Seabirds may feed either during the daylight hours (diurnal) or at night (nocturnal), and presumably have evolved these diel habits in response to the behavior of their primary prey. Some species, such as Red-footed Boobies (*Sula sula*) and Red-tailed Tropicbirds (*Phaethon rubricauda*), may forage both during the day and at night (Schreiber and Schreiber 1993). Seabird diets consist mainly of three types of prey: small pelagic fishes, crustaceans, and mollusks (Montevecchi 1993). Modes of prey acquisition include picking from or diving to the surface, plunging below the surface (Figure 6.2), and diving from the surface to depths of several meters (Table 6.1; see also Nelson 1979 for detailed discussion of feeding methods).

TABLE 6.1
Most Common Foraging Techniques Used by 14 Families of Seabirds

Family	Distribution at Sea	Pursuit Diving	Surface Seizing	Dipping	Pattering	Plunge Diving	Kleptoparasitism	Scavenging
Spheniscidae	Coastal-inshore	X						
Diomedeidae	Pelagic		X				X	?
Procellariidae	Inshore-pelagic	X	X	X	X		X	X
Oceanitidae	Inshore-pelagic		X	X	X	?		
Pelecanoididae	Inshore	X						
Phaethontidae	Pelagic					X		
Pelecanidae	Coastal-inshore		X			X	X	
Sulidae	Inshore-pelagic	X				X		
Phalacrocoracidae	Coastal-inshore	X						
Fregatidae	Inshore-pelagic			X			X	
Chionididae	Coastal						X	X
Stercorariidae	Coastal-inshore			X			X	X
Laridae (gulls)	Coastal-inshore		X	X			X	X
Laridae (terns)	Inshore-pelagic			X		X	X	
Alcidae	Inshore	X						

Note: References: Harris 1977, Croxall and Prince 1980, Ainley and Boekelheide 1983, Duffy 1983, 1989.

Our knowledge of seabird foraging behavior is in its infancy. Past studies of seabird diets, usually based at breeding colonies, and the at-sea distribution of seabirds, determined by ship surveys, have provided a fundamental basis for our understanding of the basic foraging ecology of many species. Recent advances in technology include satellite imagery to correlate habitats of marine birds with ship-based censusing (Haney 1989a); satellite telemetry to track individual birds during foraging trips (Jouventin and Weimerskirch 1990); electronic activity recorders to determine time budgets of birds (Cairns et al. 1987); archival temperature, depth, and compass loggers to determine diving depths and foraging patterns (Burger and Wilson 1988, Garthe et al. 2000); and stable isotope analysis to determine trophic utilization (Hobson et al. 1994). These technological advances are furthering our knowledge and raising new and exciting questions about the way seabirds obtain food.

This chapter reviews the current state of knowledge about seabird foraging behavior and diet, examines various hypotheses about the at-sea distribution and food habits of seabirds, and points out gaps in our understanding of how and why seabirds feed where and when they do. We begin with an overview of the morphology and design of seabirds that enable them to be successful marine predators. This section is followed by a discussion of global circulation patterns of the ocean and some of its physical properties that create feeding opportunities for seabirds. Foraging behavior is addressed after this and covers such topics as daily foraging patterns, foraging range, social interactions, and learning. The chapter concludes with a brief review of diets of seabirds and the various methods that have been used to quantify them.

6.2 FORM AND FUNCTION

The principle of allocation states that a given amount of energy obtained from food can be used in one of three ways: for basic metabolism, growth, or reproduction. The fact that all three of these metrics are generally much lower in seabirds than they are in most landbirds led Ashmole (1963) and Lack (1967, 1968) to suggest that energy is one of the most critical limiting factors in the life of a seabird. Recent work, however, has begun to chip away at this long-held belief, and we are beginning to understand how seabirds can thrive in the ocean.

Evolution has sculpted seabirds into formidable marine predators. The various morphological adaptations seen today reveal the winners of this evolutionary competition and reflect both convergence toward a basic body form and divergence at a finer scale that is related to how individual species obtain food. For example, wing morphology among various tropical seabird species correlates well with their primary foraging mode (Hertel and Ballance 1999), with more-pelagic species having a lower wing loading. Among diving species, such as penguins and alcids, maximum dive depths and dive durations generally scale positively with body size (Watanuki and Burger 1999): larger species dive to greater maximum depths and submerge for longer periods than smaller species do, although there are some exceptions (Figure 6.3a and b). Conversely, among the tube-noses that dive below the surface, it is the smaller species that are able to dive more deeply (Figure 6.3c), presumably because larger wing spans hinder underwater mobility. Relative to body mass, the Common Diving-Petrel (*Pelecanoides urinatrix*) can dive deeper than penguins and possibly alcids (Bocher et al. 2000).

Petrels (Procellariidae) inhabiting tropical and southern polar oceans are all pelagic with similar breeding habits and prey requirements. However, polar species have significantly higher subcutaneous and mesenteric fat levels than tropical petrels (Figure 6.4a), and tropical species have greater wing spans and wing areas than polar species (Figure 6.4b). Spear and Ainley (1998) argued that these differences are independent of phylogenetic relationships and reveal distinct adaptations of these groups to their respective environments.

Some species require wind for efficient long-distance flight. Wandering Albatrosses (*Diomedea exulans*) rely almost entirely on the wind for gliding at high speeds in search of food, because their high-aspect-ratio wings are not suited for flapping flight. If they are caught in a high-pressure system, they rest on the sea surface until the winds return (Jouventin and Weimerskirch 1990).

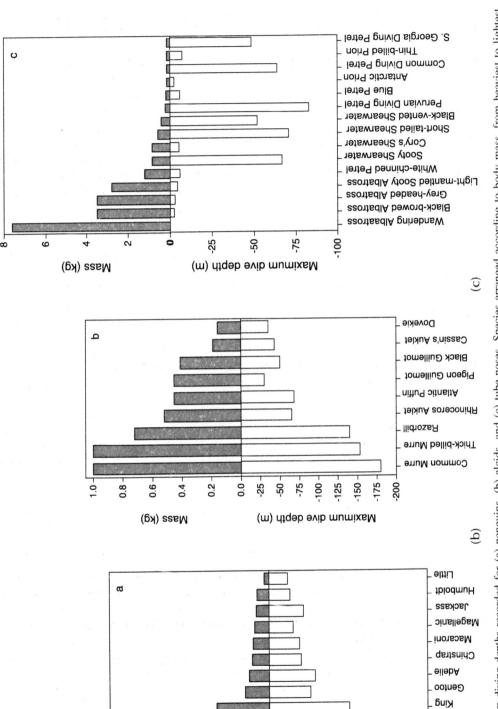

FIGURE 6.3 Maximum diving depths recorded for (a) penguins, (b) alcids, and (c) tube-noses. Species arranged according to body mass, from heaviest to lightest. (Data from a variety of sources, primarily Watanuki and Burger 1999 and Bocher et al. 2000.)

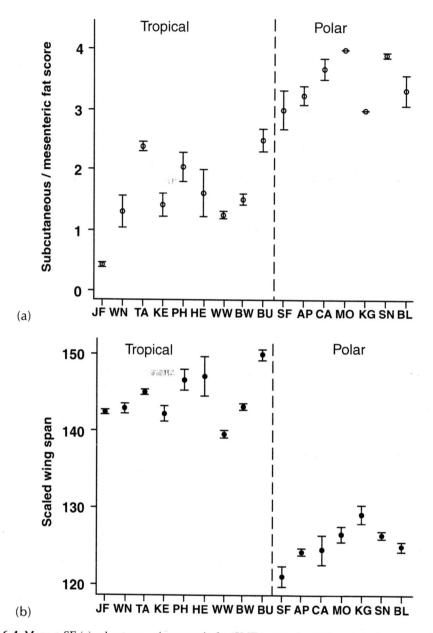

FIGURE 6.4 Mean ± SE (a) subcutaneous/mesenteric fat (SMF) scores, and (b) scaled wing spans of tropical and polar petrels. Species are JF, Juan Fernandez Petrel; WN, White-necked Petrel; PH, Phoenix Petrel; HE, Herald Petrel; WW, White-winged Petrel; BW, Black-winged Petrel; BU, Bulwer's Petrel; SF, Southern Fulmar; AP, Antarctic Petrel; CA, Cape Petrel; MO, Mottled Petrel; KG, Kerguelen Petrel; SN, Snow Petrel; and BL, Blue Petrel. (From Spear and Ainley 1998, used with permission.)

Similarly, Northern Fulmars (*Fulmarus glacialis*) require wind for efficient soaring flight. In calm conditions they are forced to use flapping flight, which is energetically expensive and results in little forward progress (Furness et al. 1993). In the central Pacific, frigatebirds and boobies were more abundant at a roost site on calm than on windy nights and remained at roost longer during calm periods (Schreiber and Chovan 1986). Thus, the high-aspect-ratio wings of many seabirds most likely evolved to take advantage of sustained oceanic winds, which keep them aloft.

Not all seabirds are adapted for foraging in windy conditions. Small species, such as terns and storm-petrels, probably experience considerable difficulty maintaining stability in high winds. Winds also ripple the water surface and obscure visibility for those species that hunt by plunge diving. Although foraging efficiency among terns may improve from calm to moderate wind conditions (Dunn 1973, Reed and Ha 1983), high winds depress foraging success (Sagar and Sagar 1989). Moderate winds may be beneficial for terns either because they do not need to flap so vigorously while hovering or because the rippled surface of the water may reduce the ability of prey to detect a tern overhead (Dunn 1973).

Water clarity also may exert selective pressure on seabird foraging modes and may constrain the distribution of species that require clear water. Turbidity reduces visibility and may impair the foraging proficiency of aerial hunters, such as terns and gannets. Ainley (1977), working in the eastern tropical Pacific Ocean, found that plunge divers were more common in clear than in turbid waters, whereas pursuit divers were significantly correlated with increasingly turbid waters. He proposed that plunge divers require relatively clear water to keep a visual fix on their target prey as they dive and that pursuit divers rely more on surprise to capture prey, a strategy that may be enhanced in turbid water. Subsequent cruises through a more extensive region of the southern Pacific Ocean supported Ainley's hypothesis: moving from polar to tropical regions, pursuit-diving species were replaced by plunge divers (Ainley and Boekelheide 1983). In particular, mainly deep-plunging species, such as most sulids, are found only in clear, warm water. The exception among deep-plunging species appears to be gannets, which occur mostly at high latitudes and also engage in pursuit diving (Garthe et al. 2000).

Across a turbidity gradient off the southeastern United States, both numbers and proportions of plunge-diving species and individuals decreased significantly with increasing water clarity (Haney and Stone 1988), arguing against Ainley's hypothesis. However, Haney and Stone redefined the foraging behavior of terns (*contra* Ainley), and the range of water clarity was much reduced compared to Ainley's study.

Clearly there are differing views on the influence of water clarity on the evolution of seabird foraging methods. Further tests of Ainley's hypothesis need to incorporate or control for the influence of water temperature, which varies with water clarity. For example, pursuit divers are common at high latitudes characterized by cold water, whereas surface feeders dominate in the tropics. This difference in foraging method may relate to the slower burst speeds of fishes in cold water, which would make them easier to catch by homeothermic seabirds (Montevecchi 1993).

6.3 OCEANOGRAPHIC CONSIDERATIONS

Global oceanic circulation patterns are important determinants of where seabirds feed (Hunt 1991a). In the absence of any circulation, many of the nutrients in the ocean would be unavailable to surface-feeding organisms, primarily because of the pycnocline, a vertical density gradient that forms at some depth below the surface and inhibits vertical mixing. The pycnocline forms primarily from the differential warming of the oceanic surface water by the sun, relative to the water at depth. A strong temperature gradient, or thermocline, is thus formed. Warmer surface water is less dense than cooler water at depth and "floats" at the surface in a wind-mixed layer known as the epipelagic zone. All organic material that dies above the pycnocline eventually sinks toward the bottom, which creates a nutrient-rich subsurface layer below.

The depth and temporal stability of the pycnocline varies seasonally and geographically. In tropical and subtropical oceans, the pycnocline is relatively permanent and shallow, and its depth may range from 25 m in eastern tropical oceans to 250 m in the center of subtropical gyres (Longhurst 1999). At higher latitudes the pycnocline is more dynamic, due to stronger seasonality of wind velocity and solar heating. During the winter, surface cooling erodes the pycnocline and allows for vertical mixing, but in the spring a new seasonal mixed layer develops at the surface and deepens progressively throughout the summer. Because tropical pycnoclines are more sharply defined, and because precipitation usually exceeds evaporation in the tropical oceans, these waters have a greater resistance to mixing than waters at higher latitudes. This reduction in mixing results in a much lower resource base for tropical marine biota (Raymont 1980).

Fortunately, there are many areas of the ocean that mix regularly and force this nutrient-rich water toward the surface. These nutrients are used by photosynthetic plankton, which form the basis for almost every marine food web. To understand how and why these regions form and how they are maintained, it is necessary to understand the nature of global circulation patterns of the ocean.

The circulation of ocean currents is either wind-driven or thermohaline. Wind-driven currents are created by moving air masses of the atmosphere and effect horizontal movement of the upper layer of the water. Wind is generated by the differential heating of the earth's atmosphere, which enables a constant transfer of energy from warm equatorial air to colder air at higher latitudes (see Chapter 7, Figures 7.10 to 7.12). Vertical mixing is accomplished primarily by thermohaline circulation, whereby temperature and salinity conditions produce a dense water mass that sinks and spreads below the surface waters.

Gyres are large, closed circular regions of oceanic current that occur in the northern and southern hemispheres of the Atlantic and Pacific Oceans, and in the tropical Indian Ocean (Figure 6.5). Gyres are set in motion by the earth's rotation and the prevailing winds. Along the equator the direction of surface flow is from east to west in each region. Due to the difference in the velocity of an object at the equator compared to higher latitudes, there is a tendency for water to be deflected at an angle to the main flow — toward the right in the northern hemisphere and toward the left in the southern hemisphere. This phenomenon is known as the Coriolis Effect.

The magnitude of the Coriolis Effect depends on the angular velocity of the earth, the velocity of an object, and the latitude at which it is moving, and can be described by the equation: $CD = V \sin\phi$. No Coriolis Effect is apparent at the equator (sine $0° = 0$), and the maximum deflection occurs at the poles (sine $90° = 1$). Most air masses travel in a north-south direction in the Northern Hemisphere. An air mass traveling from 30°N to the equator will also travel several hundred kilometers westward before reaching the equator due to the higher velocity of the earth at the equator. Since the wind is responsible for horizontal surface currents, these too will be deflected to the right. The same reasoning applies to wind and currents in the Southern Hemisphere, except that the deflection is to the left, resulting in a counterclockwise pattern of circulation (see Figure 7.11 in Chapter 7).

Wind-driven surface currents create friction drag on the layers of water below. The effects of the wind can be transmitted to depths of 100 m. Due to the Coriolis Effect, each underlying layer of water also is deflected to the right of the layer directly above. The result is termed an Ekman spiral, the progressive deflection of surface current with increasing depth. The theoretical net deflection of all surface layers is 90°, with the deepest layers sometimes traveling in the opposite direction from the uppermost layer.

Ekman circulation causes the less-dense surface waters to be deflected toward the center of large oceanic gyres. The deflection creates a higher sea surface in the center of the gyres relative to the edges, creating gravitational pull. As water flows down the slope of the elevation gradient, it is deflected to the right until it is balanced between gravity and the Coriolis Effect.

In the Atlantic and Pacific Oceans, major gyres are centered at the subtropical latitudes of 30°N and 30°S in each hemisphere. The rotation of the earth displaces gyres toward the western side, causing western boundary currents (e.g., the Gulf Stream) to be stronger and narrower than eastern boundary currents. Wind transports surface water away from tropical western coasts of continents (due to Ekman transport), allowing the upwelling of cold, nutrient-rich water from the bottom to replace it. Regions of upwelling (Figure 6.6) are important areas of productivity in tropical marine ecosystems because nutrients are brought up into the photic zone. Major upwellings occur along the western coasts of South America (Humboldt Current) and Africa (Benguela Current), where

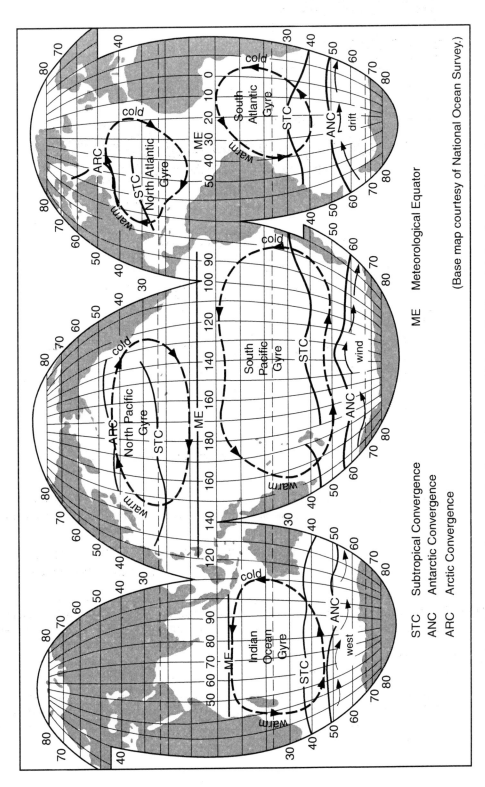

FIGURE 6.5 Global oceanic circulation patterns and currents, showing major gyres of Atlantic, Pacific, and Indian Oceans. Poleward-moving currents are warm, and equatorward-moving currents are cold. (From Thurman 1983, used with permission.)

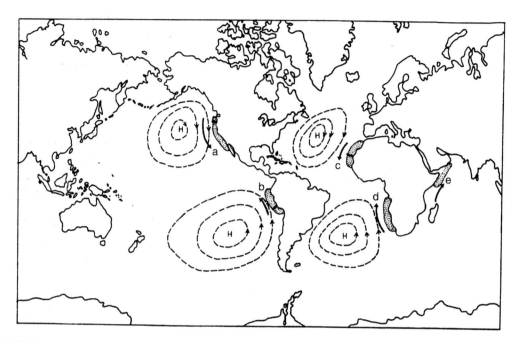

FIGURE 6.6 Major coastal upwelling regions (stippled areas) of the world and the sea-level atmospheric pressure systems (anticyclones) that influence them. The dashed circles represent mean idealized positions of isobars during the season of maximum upwelling in a given region. Arrows indicate the location of the (a) California Current, (b) Peru Current, (c) Canary Current, (d) Benguela Current, and (e) Somali Current. (From Longhurst 1981, used with permission.)

large numbers of a diverse array of seabird species congregate to forage (Duffy 1983, 19   ). However, the unstable nature of these upwelling systems leads to periodic crashes in the marine resource base (see Chapter 7).

Convergence fronts (Figure 6.7a) also are important areas of productivity, particularly in the open ocean where little vertical mixing occurs otherwise. Atmospheric (Hadley) cells in tropical latitudes are controlled by high to low pressure flow and are set in horizontal motion by the trade winds. This interaction produces an extensive oceanic convergence zone called the Intertropical Convergence Zone (ITCZ). The ITCZ is the principal convergence zone in the earth's atmospheric and oceanic circulation, and its location changes seasonally (following the apparent movement of the sun) from about 2°S to about 9°N in the Atlantic and Pacific Oceans (Longhurst 1999). Differences in temperature or salinity between two converging water masses cause sinking, turbulence, and vertical mixing.

Langmuir circulation (Figure 6.7b) is a local phenomenon caused by the displacement of water by the prevailing wind. The surface layer is displaced at an angle to either side of the wind, causing turbulence and potential upwelling of water at depth to replace the water that has been displaced. Eddies and meanderings (Figure 6.8) from strong currents also create localized areas of high productivity. Gulf Stream eddies cause vertical mixing and carry large quantities of nutrients toward the surface. Farther north, these eddies carry warm-water masses to otherwise cold-water regions such as the Gulf of Maine or Grand Banks. All of the above oceanographic phenomena create feeding opportunities for seabirds.

In summary, ocean regions are characterized by a great diversity of temperatures, salinities, currents, and areas of productivity sufficient to support millions of marine animals. As we will see in the next section, the distribution of the world's seabirds is not random, but often intimately associated with these inconsistencies in the ocean environment.

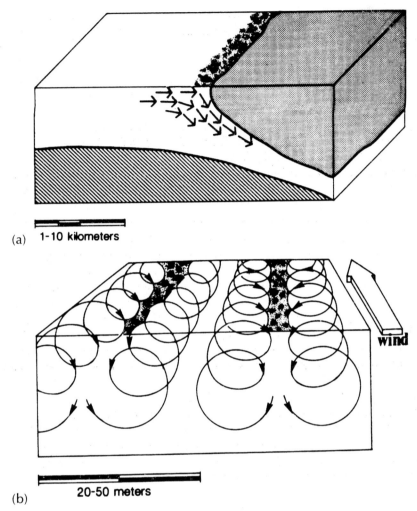

(a)   **1-10 kilometers**

(b)   **20-50 meters**

FIGURE 6.7 Convergence front (a) and Langmuir circulation (b) influencing the spatial distribution of plankton in the ocean. (From Haney 1986, used with permission.)

6.4 RELATIONSHIPS BETWEEN SEABIRDS AND THE PHYSICAL OCEAN ENVIRONMENT

Seabirds should congregate to feed in areas of high prey availability. Studies have correlated seabird abundance with prey abundance at local (Safina and Burger 1985, 1989), regional (Hunt et al. 1990), and multiple (Russell et al. 1992) scales. Recent research has examined factors responsible for concentrating aggregations of prey to determine whether these aggregations are random or predictable. In specific situations, both physical oceanographic phenomena and biotic activity have been shown to concentrate prey for seabirds over a variety of spatial and temporal scales.

Physical processes that cause predictable aggregations of prey include boundary fronts where water masses converge or diverge (Haney 1986), regions of interaction between water currents and bathymetry (underwater topography, like sea mounts) that promote physical forcing (deeper, nutrient-rich water is forced toward the surface; Hunt 1991a), and at ice edges (Ainley et al. 1993). In the vertical dimension, a strong or persistent pycnocline may concentrate prey at a particular depth. These processes are not static, and environmental conditions may induce changes in the physical structure of the ocean to which seabirds must adapt, migrate, or suffer the consequences

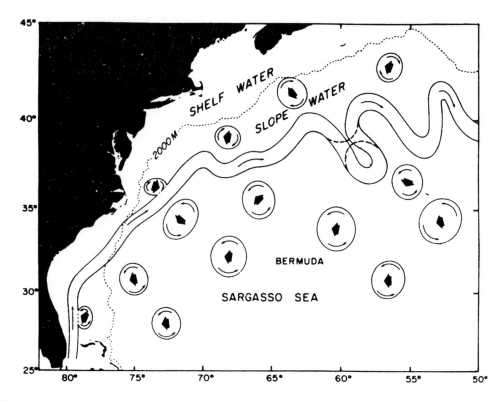

FIGURE 6.8 Schematic representation of the path of the Gulf Stream and the distribution and movement of ring eddies. (From Longhurst 1981, used with permission.)

FIGURE 6.9 Black-browed Albatross feed on krill, fish, and squid found within an average of 250 km from the breeding colony. (Photo by J. Burger.)

of higher mortality or large-scale breeding failure. The most notorious of these environmental changes is the El Niño–Southern Oscillation (ENSO) which impacts seabirds most strongly in the tropical Pacific Ocean, but its effects can be manifest worldwide (see Chapter 7; also Schreiber and Schreiber 1989).

Satellite telemetry studies have provided insights into the foraging patterns of the most pelagic species, which were previously unknown. Studies of three species of albatrosses in the Southern Ocean during the breeding season have revealed distinctly different foraging behaviors (Prince et al. 1999). The Black-browed Albatross (*Thalassarche melanophris*) feeds on krill, fish, and squid found at short distances (average 250 km) from the breeding colony, primarily over shelves and shelf-slope areas, ignoring oceanic waters in-between the shelf habitats (Figure 6.9). The Grey-headed Albatross (*T. chrysostoma*) also forages relatively close to the breeding colony (200 to 300 km), but this species associates largely with the Antarctic Polar Frontal Zone (APFZ) where it feeds on lanternfish and ommastrephid squid, possibly confined to warm core rings. The Wandering Albatross, which feeds on fish and cephalopods, shows a much greater diversity and complexity of foraging patterns. It may move randomly on looping courses (up to 15,000 km), stop for prolonged bouts of foraging in relatively restricted areas, or it may undertake more directed movements of both long and short distance (Prince et al. 1999).

Using combined information from satellite imagery and ship-based censuses, Haney (1989a) found that Black-capped Petrels (*Pterodroma hasitata*) were more common than expected in the cold-core eddies in the Gulf Stream (upwelled water) and less common than expected in other habitats. Iterative techniques using time-series satellite images provide information on how oceanic habitats change over time, allow identification of recurring consistently used habitats by seabirds, and may reveal time-dependent use by seabirds associated with seasonal or annual variation in habitat quality (Figure 6.10; Haney 1989b).

The center of the North Atlantic gyre is characterized by nutrient-poor waters. However, regions of *Sargassum* algae coalesce to provide localized areas of high biomass and concentrated prey for seabirds (Morris and Mogelberg 1973, Haney 1986). *Sargassum* mats are formed by Gulf Stream eddies and warm-core rings, convergence fronts, and Langmuir circulation. As many as 23 species of seabirds use these *Sargassum* "reefs" as foraging or resting sites. Species that occur most commonly in association with *Sargassum* are those that use aerial-dipping and plunge-diving to capture prey, such as boobies, tropicbirds, and Bridled Terns (*Sterna anaethetus*).

The Humboldt and Benguela upwelled ecosystems off Peru and southern Africa, respectively, are important feeding areas for a variety of seabird species that are attracted to the abundance of surface prey. The high overlap in species composition of foraging flocks and diversity of foraging techniques used by seabirds in these regions indicate independent attraction to rich food sources (Duffy 1989).

Strong tidal currents passing over a submerged reef in the Aleutian Islands cause upwelling on the upstream side of the reef and zones of surface convergence on the downstream side. Hunt et al. (1998) found that three species of auklets partitioned foraging habitat along this gradient. Crested Auklets (*Aethia cristatella*) fed on euphausiids carried to the surface by upwelling on the upstream side, Least Auklets (*A. pusilla*) fed on copepods concentrated in the downstream surface convergence zones, and Parakeet Auklets (*Cyclorrhyncus psittacula*) foraged over the reef. The strength of the tidal current was positively related to the number of birds present, and when the tide reversed, so did the sides of the reef occupied by Least and Crested Auklets.

In the eastern tropical Pacific, Ribic and Ainley (1997) found that Black-winged Petrels (*Pterodroma nigripennis*) and Gould's Petrels (*P. leucoptera*) associated primarily with cool waters of the South Equatorial Current. Juan Fernandez Petrels (*P. externa*), Sooty Terns, and Wedge-tailed Shearwaters (*Puffinus pacificus*) were associated with deep thermoclines and low salinities. Storm-petrels (*Oceanodroma leucorhoa* and *O. tethys*) were not linked to any physical characteristic and were found over a wide variety of habitats. Qualitatively similar results were obtained in the same area by Ballance et al. (1997).

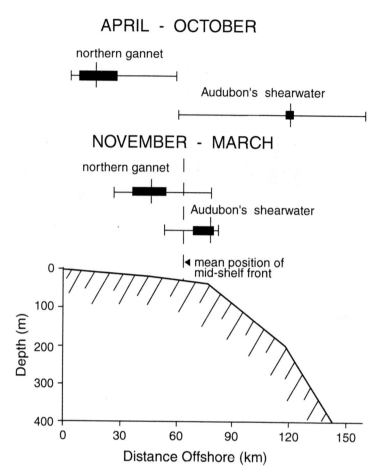

FIGURE 6.10 Changes in cross-shelf distributions of two seabird species compared to seasonal availability of the midshelf front. Distributions are illustrated with means, standard deviations, and ranges during two seasons. (From Haney 1989b, used with permission.)

Around Antarctica, the ocean is relatively uniform over a broad longitudinal range. Nevertheless, certain physical features of the ocean may consistently attract foraging seabirds. Two such features are the seaward edge of the pack ice, which concentrates plankton, and a hydrographic front over the Antarctic continental slope where krill aggregates. Veit and Hunt (1991) found a uniform species composition of seabird assemblages in the Antarctic Ocean, and little cohesion among the distributions of individual species. Although seabird assemblages were associated with the ice edge, abundance there was no higher than other in habitats surveyed, and seabird biomass was only slightly elevated over the continental slope. More than 45% of the total number of birds seen during the survey were confined to three extremely large aggregations (one of which contained >1 million Antarctic Petrels, *Thalassoica antarctica*). Veit and Hunt (1991) suggested that localized patches of prey were more important to Antarctic seabirds than any physical feature per se.

By contrast, Ainley et al. (1993) used three rapid advances and retreats of pack ice within a 200-km-wide band at the Scotia-Weddell Confluence in the Southern Ocean to examine seabird distribution and foraging patterns (Figure 6.11). They found that, despite rapid movement of the ice edge, pack-ice and open-water species remained densest in their preferred foraging habitat and maintained an identical diet regardless of where their preferred habitat was located. Ainley et al. (1993) concluded that physical variables are more important than biotic factors in maintaining species composition among Antarctic seabird assemblages.

FIGURE 6.11 King Penguins foraging off Macquerie Island. (Photo by J. Burger.)

The relationship between seabird predators and their marine prey appears to be scale-dependent (Schneider and Duffy 1985, Mehlum et al. 1999, Hunt et al. 1999). Studies of planktivorous seabirds at fine scales (<5 km) have typically found weak associations between predator and prey (e.g., Obst 1985, Hunt et al. 1990), suggesting that seabirds may sample prey randomly at small spatial scales (Hunt et al. 1991). Stronger correlations have been found at larger scales (Ryan and Cooper 1989) and among piscivorous seabirds (Schneider and Piatt 1986). Because of the difficulty in correlating seabird abundance with prey abundance at a particular scale, Russell et al. (1992) explored fractal geometry as a means to overcome scale-dependent attributes of seabird foraging patterns. Fractal geometry is a relatively new analytical technique that can resolve patterns when variability is autocorrelated across a continuum of spatial scales (Hunt et al. 1999). The spatial patterns of Least Auklets and their planktonic prey were reasonably well described by fractal dimensions at scales of up to 10 km (Russell et al. 1992). Although the study of spatial pattern does not directly reveal the processes underlying the pattern, it can enable the formulation of process-oriented hypotheses (Russell et al. 1992).

Physical processes are important for seabirds to locate prey resources in the ocean and are probably more regular and predictable than biological phenomena over a long time span. Hunt (1991b) argued that, at the smallest scales, physical processes result in greater predictability and availability of prey for seabirds than do behavioral or biotic interactions. Now that we have begun to uncover the "whats" and the "wheres" of the at-sea distribution of seabirds, we need to shift our attention to trying to understand the "whys."

6.5 FORAGING BEHAVIOR

6.5.1 DAILY PATTERN OF FORAGING

It would be naive to assume that seabird foraging behavior is random since prey are not randomly distributed. Just as evolution has sculpted the form and function of seabirds, it has also been an important agent in shaping knowledge-based foraging behavior.

The diurnal pattern of foraging exhibited by most seabirds is probably influenced by extrinsic factors that are related to prey availability. For example, Black-legged Kittiwakes (*Rissa tridactyla*) in Alaska do not forage at the same time every day, but shift their foraging schedule to coincide with daily tidal cycles (Figure 6.12, Irons 1998). Some individuals consistently foraged on a flood tide whereas others tracked the diurnal ebb tide. Most individuals exhibited preferences for

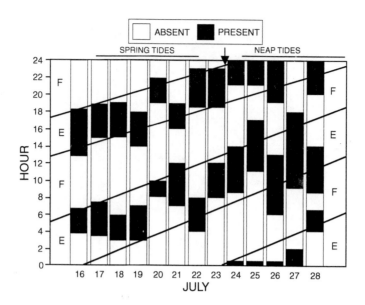

FIGURE 6.12 Foraging trips of a radio-tagged Black-legged Kittiwake during a spring and neap tide series. Black areas correspond to the time bird was on a foraging trip. Diagonal lines approximate the periods of flood (F) and ebb (E) tides. (From Irons 1998, used with permission.)

particular sites and returned to them frequently, suggesting that kittiwakes rely more on "local knowledge" or memory than information transfer to locate prey patches (Irons 1998). A similar pattern was found among individual Common Terns (*Sterna hirundo*) foraging in the Wadden Sea, Germany (Becker et al. 1993).

Breeding Gentoo Penguins (*Pygoscelis papua*) exhibit clear diel patterns in their deep (>30 m) dives, averaging 40 m at dawn and dusk and 80 to 90 m at midday (Williams et al. 1992a). A similar pattern is evident among nonbreeding Gentoo Penguins in the winter (Williams et al. 1992b). This pattern likely reflects the vertical migration of krill at various times of day, but could also relate to the difference in attenuation of light with water depth, which is greatest at midday (Wanless et al. 1999, Garthe et al. 2000).

Most seabird species are visual predators and forage most actively during daylight hours. However, several species may forage regularly at night. In the Southern Ocean, 13 of 20 species from three different orders (Procellariiformes, Pelecaniformes, Charadriiformes) were directly observed feeding at night, and five species were exclusive nocturnal feeders (Harper 1987). Birds that are present at the breeding colony by day, leave at night, and return at dawn are presumed to be feeding mostly at night. Diet studies indicating a prey base that is more likely to be available at or near the surface at night (such as bioluminescent myctophid fish or vertically migrating euphausiids) can also be used to infer nocturnal foraging behavior. Brooke and Prince (1991) listed three general trends related to nocturnal foraging:

1. Nocturnal feeders tend to feed offshore rather than inshore.
2. There is better evidence for nocturnal feeding by tropical and Southern Hemisphere species than for cool-water northern species.
3. Nocturnal foraging is often associated with a diet of squid caught in the open ocean, because most squid remains at depth during the day and surfaces at night.

Squid specialists, such as the albatrosses (*Diomedea*) and gadfly petrels (*Pterodroma*), are unable to dive deeply, which indicates that they would be restricted primarily to feeding at night. However, there remains considerable uncertainty about the means by which these birds catch squid.

Some squid species float when they die and others may be regurgitated by whales. Regurgitations of fresh squid remains by Wandering Albatrosses at the Crozet Islands revealed numerous cephalopod spermatophores (Cherel and Weimerskirch 1999), suggesting that this area is an important mating/spawning region for squids. Since Wandering Albatrosses concentrate their foraging activity on the southeast edge of the Crozet Shelf (Cherel and Weimerskirch 1999), these birds may feed primarily on squid that rise moribund to the surface following spawning.

In addition to albatrosses and gadfly petrels, other procellariids suspected of nocturnal feeding include Northern Fulmar, Bulwer's Petrel (*Bulweria bulwerii*), and Wedge-rumped Storm-Petrel (*Oceanodroma tethys*). Among Pelecaniformes, the Red-footed and Brown (*S. leucogaster*) Boobies consume large quantities of squid, as do Red-tailed Tropicbirds. Among Charadriiformes, the Dovekie (*Alle alle*) feeds on vertically migrating copepods, the Red-legged Kittiwake (*Rissa brevirostris*) on myctophid fish, the Black-legged Kittiwake on vertically migrating euphausiids, and the Swallow-tailed Gull (*Creagrus furcatus*) and White Tern (*Gygis alba*) on squid. Only one sphenisciform, the Macaroni Penguin (*Eudyptes chrysolophus*), has been identified as a nocturnal forager, based on dive recorder data indicating higher dive rates at night than during the day (see Brooke and Prince 1991, and references therein).

Diel patterns of foraging by seabirds may relate more to the behavior and availability of preferred prey species than inherent activity patterns of the predator. For example, Common Murres (*Uria aalge*) breeding in Newfoundland forage primarily during daylight hours on diurnally active capelin (Burger and Piatt 1990). However, in the Northwest Territories, Canada, the closely related Thick-billed Murre (*U. lomvia*) feeds mostly at night on krill that typically undergo a vertical migration toward the surface in the evening (Croll et al. 1992).

6.5.2 Feeding Methods

Seabirds use a variety of feeding methods to obtain prey (Table 6.1, Figure 6.13), and although most species have a primary mode of foraging, they may use one to several alternative methods opportunistically. The most common general foraging method is diving from the surface and pursuing prey while swimming underwater, using either the wings or the feet (Nelson 1979). Pursuit diving is used extensively by penguins, alcids, cormorants, and diving-petrels. Picking prey from the surface, or "dipping," is the second most common foraging method, used by storm-petrels, skuas, gulls, terns, and large petrels. Some species, such as albatrosses and pelicans, may use a combination of dipping while floating on the surface and occasional pursuit diving. The most specialized mode of foraging is plunge diving, which is used by sulids (gannets and boobies), tropicbirds, many terns, and pelicans. Other less common foraging behaviors include aerial pursuit of prey, kleptoparasitism, and scavenging (Figure 6.14; Nelson 1979).

6.5.3 Foraging Range

Studies of the foraging patterns of seabirds have been conducted primarily during the breeding season, when seabirds must return periodically to the nest site to assume incubation duties or to feed nestlings. Because of these obligations, it is generally assumed that foraging ranges of seabirds are constrained during the breeding season. Nevertheless, some species continue to range widely in search of prey. Most notable are several species of Procellariiformes that may travel over 15,000 km of ocean during a single foraging trip (Figure 6.15, Jouventin and Weimerskirch 1990).

In general, foraging ranges of seabirds can be classified as coastal, inshore, offshore, and pelagic (Table 6.1). Coastal-feeding species, such as gulls and cormorants, usually forage within sight of land and sometimes may venture into estuarine habitats. Inshore species, such as penguins, terns, and alcids, may forage out of sight of land but typically remain within the range of the continental shelf. Offshore and pelagic species, such as albatrosses, petrels, shearwaters, and tropicbirds, often travel great distances from land to forage at frontal zones along the shelf-slope break and beyond.

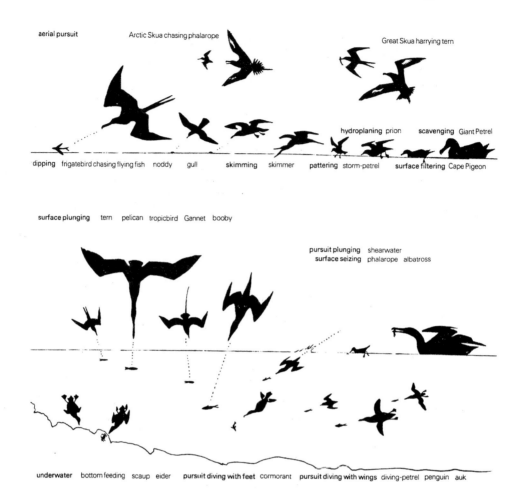

FIGURE 6.13 Various feeding methods used by seabirds to obtain prey. (From Nelson 1979.)

FIGURE 6.14 Western Gulls feeding on a pup elephant seal's placenta (San Nicholas Island, California). They are opportunistic feeders and eat a wide variety of food. (Photo by E. A. and R. W. Schreiber.)

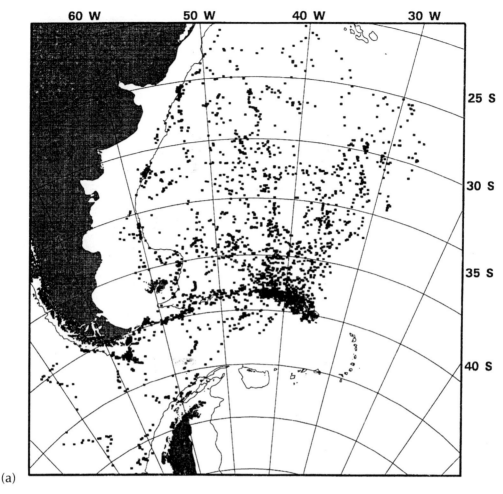

(a)

FIGURE 6.15 Distribution of uplinks from satellite-tracked (a) Wandering Albatrosses and (b) Black-browed Albatrosses breeding at Bird Island, South Georgia. (From Prince et al. 1997, used with permission.)

Foraging ranges can be highly variable within a family. For example, among the terns there are species that are strictly coastal (e.g., Common Tern) and others that are among the most pelagic of any family (e.g., Sooty Tern).

Closely related species that breed sympatrically may differ in foraging ranges, either horizontally or vertically. Penguins breeding in Antarctica partition foraging habitat, a strategy that may enable them to coexist and feed on similar prey (Trivelpiece et al. 1987). Gentoo Penguins are deep-diving inshore feeders, whereas Adélie (*Pygoscelis adeliae*) and Chinstrap (*P. antarctica*) penguins are shallow-diving, offshore foragers; all three species specialize on krill during the chick-rearing period. The breeding chronology of the two offshore penguins is offset, which may reduce competition for food during the time of peak demand.

One of the most exciting recent discoveries, revealed by satellite telemetry studies, is that several species of procellariids alternate or mix long (pelagic) and short (inshore) foraging trips during the chick-rearing period (Weimerskirch et al. 1994). The duration of foraging trips in these species is clearly bimodal (Figure 6.16), with short trips generally lasting from 1 to 5 days and long trips from 7 to a maximum of 20 days in the Wandering Albatross. Body masses of adults returning to the colony suggest that parents use short trips almost exclusively to gather food for their chicks, whereas long trips are a strategy to replenish energy lost during chick-provisioning trips (Table 6.2).

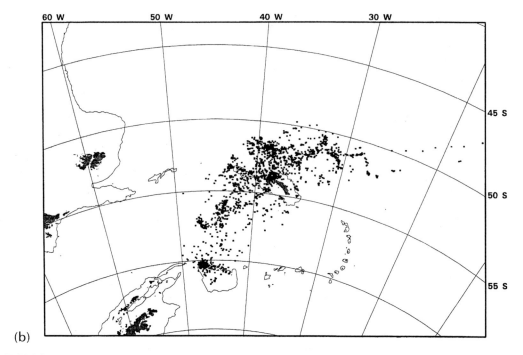

(b)

FIGURE 6.15 *Continued.*

Individual variation in foraging ranges also were reported for alcids breeding on the Isle of May, Scotland (Wanless et al. 1990), and for boobies breeding in the Galapagos (Anderson and Ricklefs 1987). Individuals foraged in different places on consecutive days and sometimes on consecutive trips during the same day. Least Auklets breeding on three islands in the Bering Sea differed with respect to distance flown during foraging trips, ranging from a foraging radius of 5 km (St. Matthew Island) to 56 km (St. Lawrence Island; Obst et al. 1995).

6.5.4 OLFACTION

Although birds are not known for their acute sense of smell, at least 22 species of Procellariiformes are suspected to use olfaction to detect food on the ocean surface (Lequette et al. 1989). By creating cod-liver oil slicks, Verheyden and Jouventin (1994) identified specific search behaviors in birds using olfactory cues. These species approached the slick from downwind, flying close to the surface in a zig-zag pattern that became progressively narrower as the birds approached the oil slick. Species indifferent to the oil slick approached from all directions and at greater heights above the surface. Of the species examined, all members of the Oceanitidae (storm-petrels) showed positive reactions to odors, but results with species from other families were less conclusive. Verheyden and Jouventin (1994) suggested that other species might rely as much or more on visual cues to locate a food source. That smaller species (e.g., storm-petrels) show the most obvious olfactory capacities and tend to arrive first at an oil slick suggest that a compensatory mechanism is operative. Smaller species can be displaced by later-arriving larger species, and greater olfactory abilities may enable them to arrive first at a food source.

6.5.5 SOCIAL ATTRACTION

More than any other group of birds, seabirds are primarily flock foragers (Duffy 1983), and feeding flocks may sometimes exceed 1 million birds (Veit and Hunt 1991). However, seabirds typically do not *search* for food in groups (Hoffman et al. 1981), suggesting that flock formation is a

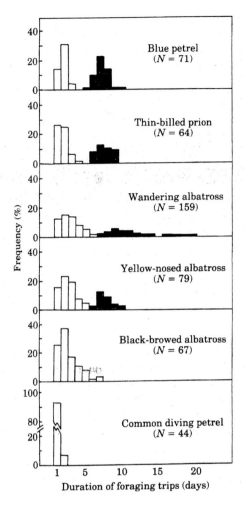

FIGURE 6.16 Frequency distribution of the duration of foraging trips of petrels and albatrosses (sample size indicates number of foraging trips). Open bars, short trips; black bars, long trips. (From Weimerskirch et al. 1994, used with permission.)

TABLE 6.2
Duration of Foraging Trips, Change in Adult Body Mass and Feed Mass after Long and Short Foraging Trips in Four Species of Procellariiformes

	Trip Duration (days)		Percent Body Mass Change after Trip		Feed Mass (% adult body mass)	
	Short	**Long**	**Short**	**Long**	**Short**	**Long**
Blue Petrel	1.8 ± 0.6	7.2 ± 1.0	−7.2 ± 2.3	+5.5 ± 3.4	31.5 ± 6.1	38.3 ± 4.8
Thin-billed Prion	1.7 ± 0.8	7.5 ± 1.1	−6.4 ± 5.1	+3.5 ± 5.6	25.3 ± 8.4	29.4 ± 8.1
Wandering Albatross	2.5 ± 1.3	10.5 ± 2.8	−2.3 ± 8.3	+2.6 ± 7.6	9.2 ± 5.4	11.1 ± 6.0
Yellow-nosed Albatross	2.5 ± 1.2	7.7 ± 1.1	−0.2 ± 4.3	+1.6 ± 7.1	16.7 ± 4.8	19.3 ± 5.8

From Weimerskirch et al. 1994.

consequence of recruitment to prey patches following initial discovery by individual foragers (e.g., Gochfeld and Burger 1982). The white or partially white plumage of many seabirds, combined with typical foraging activity, such as diving (Ostrand 1999), may alert potential recruits to the presence of a distant prey patch (Simmons 1972). In these situations, the food-finder would benefit from other recruits, as long as individual foraging success improves with increasing group size (Götmark et al. 1986). Seabird prey usually consists of schooling fish or invertebrates, and multiple attacks from the surface or below may disrupt the cohesiveness of predator avoidance tactics by the prey and facilitate their capture.

Haney et al. (1992) used chumming experiments to examine recruitment rates and potential attraction distances for mixed-species seabird groups in the western Atlantic Ocean. Based on arrival times and potential flight speeds of recruits, estimated recruitment distances (~4.5 km) were close to lower theoretical limits, derived from information on the earth's curvature, heights of forager, and potential recruits. Although the theoretical maximum recruitment distance exceeded 20 km, most prey patches are ephemeral. Haney et al. (1992) argued that even if potential recruits are able to detect a flock at long distances, the travel time required to arrive at a prey patch is likely to exceed the life of the patch.

Large mixed-species feeding flocks also create the potential for intense foraging competition among the participants. Although consumptive competition has not been demonstrated in such situations, there is anecdotal evidence that interference competition may occur, particularly among species that use different foraging techniques. Sooty Shearwaters (*Puffinus griseus*) displaced Antarctic Terns (*Sterna vittata*) from crustacean swarms in New Zealand, first by flying and diving through the tern flocks and then by covering the foraging surface in dense rafts (Sagar and Sagar 1989). In the tropics, noddies (*Anous* spp.) feed by dipping close to the water surface. Such an arrangement blocks access to prey schools for other birds that plunge dive into the water from greater heights (Hulsman 1989, Shealer and Burger 1993). Dense packing by foraging Common Terns over prey fish may force Roseate Terns (*Sterna dougallii*) to peripheral areas of the flock where prey are more dispersed (Duffy 1986). Thus, both active and passive interference mechanisms may depress the foraging success of the inferior competitor (Safina 1990, Shealer and Burger 1993).

Diamond (1978) argued that avoidance of foraging competition may explain why Sooty Terns fly past inshore feeding flocks of Brown Noddies (*Anous stolidus*) en route to their pelagic feeding areas. Sooty Terns and noddies had very similar diets in Hawaii (Brown 1975), although diet overlap between the two species at Christmas Island was only 64% (Ashmole and Ashmole 1967). Thus, long-distance foraging flights of Sooty Terns may or may not be made to obtain different prey. Because of a low energetic cost of flight, Sooty Terns may be more able than noddies to fly offshore to better feeding areas. Long-distance commuting may not be energetically more expensive than feeding near the breeding colony if prey profitability compensates for the increased flight duration (cf. Obst et al. 1995).

6.5.6 COMMENSAL FORAGING ASSOCIATIONS

The most widely reported commensal foraging associations among seabirds involve interactions with various species of cetaceans (Evans 1982, Pierotti 1988). In New Zealand, White-fronted Terns (*Sterna striata*) and Hutton's Shearwaters (*Puffinus huttoni*) were attracted to the surface feeding of Hector's dolphins (*Cephalorhynchus hectori*) in neritic waters (Brager 1998). The feeding activity of the dolphins caused fish to surface where terns and shearwaters could take them. This association was observed only during a 3-month period coinciding with the terns' breeding season. Presumably the prey fish were present in the area only during this time because dolphins used other feeding behaviors during the remainder of the year.

Along the Lebanese coast in winter, mixed flocks of hundreds of Common Black-headed Gulls (*Larus ridibundus*) and Slender-billed Gulls (*L. genei*) feed in association with bottlenose dolphins (*Tursiops truncatus*; Evans 1987). Groups of three to six dolphins circle a school of fish and drive

FIGURE 6.17 Flock involving Antarctic fur seals, Black-browed Albatrosses, giant-petrels, Cape Petrels, and a Gray-headed Albatross. The spatial arrangement shown here is typical, with the fur seals at the front, followed by the Black-browed Albatrosses, and with the giant-petrels at the rear. (From Harrison et al. 1991, used with permission.)

it to the surface in a concentrated area where the fish are made available to diving gulls. A similar situation exists between predatory fishes and Laughing Gulls (*Larus atricilla*) at St. John, U.S. Virgin Islands (Coblentz 1985). Here predators force schools of dwarf herring (*Jenkinsia lamprotaenia*) to form a roiling "bait ball" at the surface where the herring are easy prey for seabirds.

In the Bering Sea, several species of seabirds feed at mud plumes created by foraging gray whales (*Eschrichtius robustus*; Obst and Hunt 1990). Gray whales are benthic feeders that slurp deep furrows in the sea floor and strain prey through their baleen. Much of the disturbed sediment is carried to the surface as the whales surface to breathe. Large muddy plumes are formed in their wakes and these plumes attract a variety of seabirds that exploit prey in different ways. Black-legged Kittiwakes and Red Phalaropes (*Phalaropus fulicaria*) were among the first to arrive at a surfacing whale. Thick-billed Murres were the only pursuit diving species at the plumes and occurred irregularly, presumably to feed on prey in the settling sediment. Stomach analyses of these participant species indicated that prey were several species of benthic amphipods, which are not normally present in surface waters and not usually accessible to the birds. Obst and Hunt (1990) argued that whale plumes represent an important resource to kittiwakes and murres, whose breeding season coincides with the nearby foraging of gray whale herds. In Argentina, Kelp Gulls (*Larus dominicanus*) fed on small marine invertebrates stirred up by mating and calving right whales (*Eubalaena glacialis*; Verheyden 1993).

Seabirds also are attracted to the feeding activities of seals and sea lions. Multispecies feeding flocks near Bird Island, South Georgia often were initiated by Black-browed Albatrosses that responded to the surfacing of Antarctic fur seals (*Arctocephalus gazella*) and penguins during a feeding bout (Harrison et al. 1991). Other species of petrels and albatrosses apparently followed the lead of Black-browed Albatrosses and converged at a particular location to feed on live zooplankton or to scavenge the remains (Figure 6.17).

Although it is generally accepted that subsurface predators facilitate prey capture for seabirds by increasing prey availability, these associations appear to be ephemeral and facultative (but see Pitman and Ballance 1992). Few studies have determined the benefits to seabirds engaged in these associations, relative to foraging on their own. In Puerto Rico, Roseate Terns forage in a variety of habitats and rely on both physical (reef margins) and biotic (predatory fish) factors to make prey available to them. Shealer (1996) determined that foraging success (captures per minute) was four times higher for individuals feeding over predatory fish schools than in their absence. Since 84% of all flocks were located over predatory fish, and flock size was significantly greater in the presence than in the absence of predatory fish, Roseate Terns foraged primarily in habitats that result in the highest rate of prey capture.

6.5.7 KLEPTOPARASITISM

Facultative kleptoparasitism (food theft) appears to be common among seabirds, particularly among those that regularly participate in mixed-species foraging aggregations. Frigatebirds (Figure 6.18), sheathbills, skuas, jaegers, gulls, and terns are the most notorious pirates of the sea, but many other species engage in this behavior opportunistically. Specialist kleptoparasites exhibit a host of adaptations that are apparently lacking in opportunist species. These include the ability to detect and attack hosts carrying food concealed in the proventriculus, to discriminate among the most suitable hosts, to sustain prolonged aerial chases, to respond to changes in host availability and prey selection, and to adapt their breeding cycle to match that of their host (see review in Furness 1987).

For opportunists, kleptoparasitic behavior appears to be context-dependent and is probably related more to variation in food availability than to size differences among the participants (Furness 1987). Kleptoparasitism becomes increasingly more frequent during the chick-rearing period, when parents are carrying food back to the colony. Juveniles of many species, which are usually less proficient foragers than adults, are more likely to attempt piracy. Duffy (1980) examined piracy among a guild of Peruvian seabirds and found that deeper-diving species, which presumably can exploit more of the resource base, are more likely to be targeted by shallower-diving species. For example, shallow-diving Band-tailed Gulls (*Larus belcheri*) frequently attempted piracy on Brown Pelicans (*Pelecanus occidentalis*), and pelicans in turn commonly attacked deeper-diving Peruvian

FIGURE 6.18 Some frigatebirds wait near a booby colony for birds returning to feed their chick. They then chase the booby (here a Red-footed), often grabbing a feather, until the booby regurgitates, whereupon the frigatebird swoops down and grabs the fish. (Drawing by J. Busby.)

Boobies (*Sula variegata*). A follow-up study (Duffy 1982) supported this hypothesis for seabird foraging assemblages of the Galapagos Islands and southern Africa and suggests that differential access to prey may promote kleptoparasitic behavior in a variety of species.

If kleptoparasitism is more profitable than direct foraging, natural selection should favor this behavior. The aerial agility of frigatebirds (*Fregata* spp.) has been interpreted as an adaptation for a kleptoparasitic habit. However, several studies, including a study of Magnificent Frigatebird (*F. magnificens*) piracy on boobies (*Sula* spp.) in Mexico (Osorno et al. 1992), revealed a low percentage (6%) of successful attacks. In this study, most attempts at piracy were initiated by females and juveniles, but Gochfeld and Burger (1981) observed mostly adult males and juveniles pirating Laughing Gulls at another site in Mexico. Thus, either sex seems capable of successful kleptoparasitism.

Some evidence suggests that even opportunistic kleptoparasites are able to discriminate among the most profitable hosts. Laughing Gulls stealing fish from Brown Pelicans preferentially targeted adult pelicans in both years of a 2-year study (1991–1992), but were less prone to attack adults in 1992, the year that pelican foraging success did not differ between adults and juveniles (Shealer et al. 1997).

6.5.8 Learning

The skills necessary to gain proficiency in foraging at sea presumably are acquired gradually by most seabird species. Many seabirds have a protracted period of parental care following fledging (Burger 1980) and defer the first breeding attempt for 2 to 9 years thereafter (Appendix 2). Seabirds as independent foragers must learn to recognize physical or biotic features of the ocean that are predictable indicators of food, and the times that these features can be exploited. For example, tidal forcing may concentrate prey in a narrow island passage, but particular tidal stages do not occur at the same time each day. Migrating cetaceans may herd and drive prey fish to the surface, but these predators are mobile and occur in a particular area only at certain times of the year. Seabirds may learn the migration patterns of marine mammals and possibly some species may even time their breeding season to coincide with local aggregations of cetaceans or pinnipeds, but this topic has not been investigated in detail.

Flock feeding may facilitate the process of learning in juvenile or naïve foragers. Juvenile gulls of several species did not initiate feeding flocks in British Columbia, but readily joined existing flocks (primarily initiated by adult gulls) and also were attracted to floating gull models (Porter and Sealy 1982). Newly fledged terns travel to feeding flocks with their parents where they may learn both the location of prey and the techniques necessary to capture fish.

Although some degree of information transfer may occur among colonially breeding seabirds, individuals breeding at the same colony often use different foraging areas (e.g., Becker et al. 1993, Irons 1998), suggesting that learning where and when to forage may be an individual process (Hunt et al. 1999). In tropicbirds and some alcids, chicks near fledging must go to sea alone. Although fledgling alcids may join up with large rafts of conspecifics at sea, Red-tailed Tropicbirds are believed to be primarily solitary feeders (Schreiber and Schreiber 1993). Thus the question of how young tropicbirds learn to forage remains a mystery.

6.5.9 Age-Dependent Foraging Proficiency

A wealth of literature supports the generalization that adult birds of a particular species are more proficient foragers than immatures (see review in Wunderle 1991). Seabirds typically require an extended maturation period of two to several years before the first breeding attempt, suggesting that the period of delayed maturation may reflect the length of time young birds require to obtain the foraging skills necessary to sustain them as well as to rear offspring. Studies of gulls that have age-specific plumage patterns have shown that foraging proficiency increases gradually with age until adult status is achieved (MacLean 1986, Burger 1987). Gulls are generalists, and juvenile gulls seem to try a variety of foraging techniques during the maturation period, including attempts at stealing food from other birds (Burger 1988, Hackl and Burger 1988) and frequenting garbage dumps (Burger and Gochfeld 1983).

More specialist foragers, such as terns, require a period of parental care that may extend into the first winter (Ashmole and Tovar 1968). Even after making the transition to independence, juvenile terns are less adept than adults at plunge-diving for prey (Dunn 1972, Kallander 1991) and this deficit may last for a year or more (Shealer and Burger 1995).

6.6 DIET

Diet studies of seabirds abound, and we probably know more about what seabirds eat than any other aspect of their foraging ecology. Diet studies, coupled with information on the marine resource base, can reveal other aspects of seabird foraging behavior, such as where birds go to feed, foraging site fidelity, trip duration, and energetic considerations. The type of prey seabirds select can have important fitness consequences, particularly during the chick-rearing period (Pierotti and Annett 1990). Moreover, diet studies of seabirds can be used as indicators of the health of the marine ecosystem, because prey harvests often are correlated with prey abundance at a variety of scales (Montevecchi and Myers 1995), and changes in diet often reflect changes in the prey base (Montevecchi et al. 1988, Montevecchi 1993; see also Chapter 16). Some evidence suggests that surface feeders are more vulnerable than subsurface feeders to environmental changes (Baird 1990). This idea is supported by large-scale breeding failures among Arctic Terns (*Sterna paradisaea*) following periods of low food availability (Monaghan et al. 1989, 1992), but further study is needed.

Diet studies of entire seabird communities are few, but have provided insight into the manner by which seabirds partition resources at a variety of spatial and temporal scales. Some of the most comprehensive diet studies of seabird communities include 18 species breeding in the northwest Hawaiian Islands (Harrison et al. 1983), 14 species in the Seychelles (Diamond 1983), several species of terns at Christmas Island, Pacific Ocean (Ashmole and Ashmole 1967), 10 species breeding on the Farallon Islands off of California (Ainley and Boekelheide 1990), 8 species breeding off the northeast coast of England at the Farne Islands (Pearson 1968), and 6 species breeding at Bird Island, South Georgia (Croxall et al. 1997). The potential for interspecific competition for food is high among these seabird communities, but competition has never been demonstrated convincingly in any study. Seabirds at Christmas Island exhibited high overlap in diet (Ashmole and Ashmole 1967), suggesting the possibility of intense competition for food. However, Farallon Islands seabirds bred most successfully during years when their feeding niches overlapped the most (Ainley and Boekelheide 1990), presumably because of a superabundant food supply. Ecological segregation among seabirds appears to be maintained either by differences in foraging range and feeding methods, or by differences in the size and type of prey taken (Croxall et al. 1997).

Seabird diets usually consist of a very limited range of taxa: primarily pelagic fishes, squid, and crustaceans, with krill and euphausiids being the most common arthropods (Table 6.3). However, some species, such as skuas and gulls, also prey heavily upon other seabirds, primarily during the breeding season (e.g., Ryan and Moloney 1991). Giant petrels (*Macronectes* spp.) are predators of penguins and burrowing petrels, but also may scavenge carcasses of seabirds and seals (Hunter and Brooke 1992).

The composition of seabird diets usually is not static among years, seasons, or even weeks. Changes in diet probably more reflect changes in the marine prey base (Montevecchi and Myers 1995, 1996) or shifts in foraging habitat (Ainley et al. 1996) than they reflect inherent changes in preference. However, certain species may exhibit flexibility in their diet at certain times of the year. Great Frigatebirds (*Fregata minor*) breeding at Christmas Island (central Pacific) begin eating Sooty Tern chicks after they hatch (Schreiber and Hensley 1976). Some Western Gulls (*Larus occidentalis*)

TABLE 6.3
Major Components of the Diet of Seabirds by Numerical Rank of Importance (% by mass, volume, or number)

Family	Fish	Cephalopods	Crustaceans	Polychaetes	Other	Location
Spheniscidae						
Gentoo Penguin (a)	1	3	2			Antarctica[1]
Gentoo Penguin (c)	2		1			South Georgia[2]
Chinstrap Penguin (c)	2		1			S. Shetland Is.[3]
Macaroni Penguin (c)	2		1			S. Shetland Is.
Royal Penguin (c)	2		1			Macquarie I.[4]
King Penguin (c)	1	2				Crozet Is.[5]
Adelie Penguin (c)			1			S. Shetland Is.[6]
Diomedeidae						
Wandering Albatross (a)	2	1	3			South Georgia[7]
Grey-headed Albatross (a)	2	1	3			South Georgia[7]
Black-browed Albatross (a)	1	2	3			South Georgia[7]
Black-footed Albatross (a, c)	1	2	3		4	Hawaii[8]
Laysan Albatross (c)	2	1	2		4	Hawaii[8]
Procellariidae						
White-chinned Petrel (a)	2		1			South Georgia[9]
Peruvian Diving-petrel	2		1			Peru[10]
S. Georgia Diving-petrel (c)			1		2	South Georgia[11]
Common Diving-petrel (c)			1		2	South Georgia[11]
Antarctic Prion	2	3	1			South Georgia[12]
Fairy Prion	2	3	1			South Georgia[13]
Short-tailed Shearwater (c)			1			Alaska[14]
Wedge-tailed Shearwater (a, c)	1	2	3			Hawaii[8]
Black-vented Shearwater (c)	1	2				Mexico[15]
Oceanitidae						
Leach's Storm-petrel (c)	2		1			British Columbia[16]
Sooty Storm-petrel (a)	2	1	4		3	Hawaii[8]
Phaethontidae						
Red-tailed Tropicbird (a)	1	2				Hawaii[8]

TABLE 6.3 *Continued*

Major Components of the Diet of Seabirds by Numerical Rank of Importance (% by mass, volume, or number)

Family	Fish	Cephalopods	Crustaceans	Polychaetes	Other	Location
Pelecanidae						
Brown Pelican (c)	1					California, U.S.[17]
Sulidae						
Cape Gannet	1					W. South Africa[18]
Northern Gannet (c)	1					Shetland, U.K.[19]
Peruvian Booby (a, c)	1	2				Peru[20]
Blue-footed Booby (a, c)	1					Peru[20]
Masked Booby (a,c)	1	2				Christmas I., Pacific[21]
Red-footed Booby (a,c)	1	2				Christmas I., Pacific[21]
Phalacrocoracidae						
Antarctic Shag (a)	1	2	3			Nelson I., Antarctica[22]
European Shag (a, c)	1		3	4		Scotland[23]
Double-crested Cormorant (c)1	1		2	4	3	Maine, U.S.[24]
Fregatidae						
Great Frigatebird (a,c)	1	2			3	Christmas I., Pacific[21]
Chionididae						
Lesser Sheathbill (c)					1 (birds)	Antarctica[25]
Pale-faced Sheathbill (a)					1 (seals)	Antarctica[26]
Stercorariidae						
Subantarctic Skua	2				1 (birds)	S. Atlantic Ocean[27]
South Polar Skua (a, c)	1		3		2 (birds)	Antarctica[28]
Brown Skua (a, c)	3		2		1 (birds)	Antarctica[28]
Laridae (terns)						
South American Tern (a)	1					Argentina[29]
Arctic Tern (c)	1					Shetland, U.K.[30]
Common Tern (c)	1					Many sites

					Location
Roseate Tern (a)	1				Puerto Rico[31]
Roseate Tern (c)	1				Puerto Rico[31], Azores[32]
Sandwich Tern (a)	1				Puerto Rico[31]
Sooty Tern (c)	2	1			Hawaii[33]
Gray-backed Tern (a, c)	1	2	3		Hawaii[8]
White Tern (a, c)	1	2	3		Hawaii[8]
Brown Noddy (a)	1	2	3		Hawaii[8]
Alcidae					
Rhinoceros Auklet (c)	1				Vancouver I., Canada[34]
Atlantic Puffin (breeding-c)	1				Shetland, U.K.[19]
Atlantic Puffin (winter-a)	1	3	2	5	Faroe Is., Denmark[35]
Common Murre (breeding-a)	1	3	2	4	California, U.S.[36]
Common Murre (winter-a)	1	2	3		California, U.S.[36]
Common Murre (c)	1	2			California, U.S.[36]
Thick-billed Murre (breeding-a)	2	1		3	Canadian Arctic[37]
Thick-billed Murre (winter-a)	1			2	Japan[38]
Least Auklet (c)	1			2	Alaska, U.S.[11]

Note: Adult (a) and chick (c) diets indicated if reported.

References: [1]Adams and Klages 1989, [2]Croxall et al. 1997, [3]Croxall and Furse 1980, [4]Hindell 1988, [5]Cherel and Ridoux 1992, [6]Trivelpiece et al. 1987, [7]Prince et al. 1999, [8]Harrison et al. 1983, [9]Croxall et al. 1995, [10]Jahncke et al. 1999, [11]Roby 1991, [12]Reid et al. 1997, [13]Prince and Copestake 1990, [14]Hunt et al. 1996, [15]Keitt et al. 2000, [16]Vermeer and Devito 1988, [17]Anderson et al. 1982, [18]Berruti et al. 1993, [19]Martin 1989, [20]Jahncke and Goya 2000, [21]Schreiber and Hensley 1976, [22]Favero et al. 1998, [23]Harris and Wanless 1991, [24]Blackwell et al. 1995, [25]Burger 1981, [26]Favero 1996, [27]Ryan and Moloney 1991, [28]Pietz 1987, [29]Favero et al. 2000, [30]Monaghan et al. 1992, [31]Shealer 1998b, [32]Ramos et al. 1998, [33]Brown 1975, [34]Davoren and Burger 1999, [35]Falk et al. 1992, [36]Ainley et al. 1996, [37]Gaston and Bradstreet 1993, [38]Ogi and Shiomi 1991.

switch their diet from garbage to fish once chicks hatch (Annett and Pierotti 1989). Fish have higher nutritional content and result in higher fledging success than garbage (Pierotti and Annett 1990). Roseate Tern adults mostly take larval schooling fish (<30 mm in length) when self-feeding, but deliver increasingly larger fish to their chicks throughout the rearing period (Shealer 1998a). These studies and others demonstrate some plasticity in diet among seabirds.

6.6.1 METHODS FOR STUDYING SEABIRD DIETS

Duffy and Jackson (1986) reviewed the various methods available for determining seabird diets. At that time, the primary means of assessing diet composition were (1) examining stomach contents; (2) catching birds and collecting food samples resulting from either spontaneous or forced regurgitation; (3) collecting prey samples dropped near nest sites; and (4) visually identifying prey carried in the bill, usually during the delivery of prey by adults to chicks. A recent method involving stable isotope analysis (see below) has been used effectively to make inferences regarding trophic positions of seabirds in marine food webs. Each method has its own associated limitations and biases, and methods chosen must depend on the goals of the study.

Results of diet studies are reported in numerous ways, such as frequency of occurrence, mass and volume, or numerical abundance. Frequency of occurrence may be the most appropriate method when a few prey items are taken of approximately the same size, or when samples are well digested. Mass and volume measures are preferred when prey differ in size, to reduce the overall contribution of numerous but tiny prey items. Numerical abundance data are appropriate when prey items are similar in size, but will overestimate the importance of small prey when prey are of different sizes. See Duffy and Jackson (1986) for more detailed discussion of diet methodology.

6.6.1.1 Regurgitations

Regurgitated food samples collected by stomach lavage or other techniques can provide useful information about the diets of many seabirds. However, analysis of regurgitations and stomach samples may be biased because of the differential digestibility of certain prey types (Jackson and Ryan 1986). For example, squid beaks and fish otoliths are not readily digested because they are composed primarily of calcium carbonate. However, soft-bodied prey, such as cnidarians or poly-chaetes, may go undetected in a diet sample if enough time has elapsed between ingestion and the collection of the sample. The diet of Ring-billed Gulls (*Larus delawarensis*) during the nestling period was compared simultaneously using regurgitated pellets cast by adults and direct observations of chick provisioning (Brown and Ewins 1996). Pellets contained largely insects, vegetation, and garbage, but direct observation revealed that mostly earthworms or fish were fed to chicks. Given this result, Brown and Ewins (1996) argued for more systematic, controlled studies to calibrate sampling techniques.

Most diet studies have been conducted at breeding colonies, where adults return to land to feed dependent offspring. Sometimes the assumption is made that nestling diets and adult diets are similar, but some studies have shown that adults feed on prey of different sizes or species than they deliver to offspring (Shealer 1998b), or that their foraging behavior differs between self-feeding and chick provisioning (Davoren and Burger 1999). The Emperor Penguin (*Aptenodytes forsteri*) spends an average of 15 days at sea on a single foraging trip during the chick-rearing period (Kooyman and Kooyman 1995). Time-depth recorders indicate that from 1 to 3 days before a penguin returns to the colony, it alters its diving behavior from prolonged deep dives (>200 m) to a series of dives at midwater depths. If the shift in diving pattern is indicative of a change in prey, regurgitated samples collected at the breeding colony may not be representative of what adult penguins are eating during most of their time at sea (Kooyman and Kooyman 1995).

6.6.1.2 Prey Dropped near Nest Sites

For species in which adults transfer whole or partially digested prey items to chicks during provisioning, analysis of prey dropped near nest sites can provide a qualitative assessment of chick diet (Atwood and Kelly 1984, Ramos et al. 1998). If quantitative determination is required, this method is likely to be biased toward larger, more spiny, or deeper-bodied prey due to difficulties in handling or swallowing prey of this type by chicks. Prey scraps found at nests may also include high proportions of rejecta.

6.6.1.3 Prey Carried in the Bill

A skilled observer may be able to identify the species of prey carried in the bill (by species such as terns, skimmers, and some alcids) in certain areas where diet diversity is limited. Estimates of the size or mass of the prey items can be made by comparing them to the length of the bill and using known mass-length relationships for individual prey species. Whereas this method may be useful for determining prey that adults feed to chicks, it does not necessarily reveal what adults themselves eat. Moreover, this method is prone to have the largest margin of error of any since it relies primarily on remote observation. Bias in prey identification may also be an issue, because an observer will generally have more difficulty identifying a small prey item compared to a larger one. Few calibration studies have been done to assess the validity of this method. The ones that have (e.g., Bayer 1985, Goss-Custard et al. 1987, Cezilly and Wallace 1988, Rodway and Montevecchi 1996) usually show that the method is unreliable.

6.6.1.4 Stable Isotope Analysis

Stable isotope analysis (SIA) is a relatively recent technique used to determine seabird diets and trophic relationships (Hobson et al. 1994). This technique is based on the fact that ratios of stable isotopes of nitrogen ($^{15}N/^{14}N$) and carbon ($^{13}C/^{12}C$) in tissues of the consumer reflect those in their prey in a predictable manner. The nitrogen signature is enriched (by three parts per thousand) at each successive step up the trophic scale in a food web, due to the preferential loss of the lighter ^{14}N isotope during protein synthesis in the consumer. The carbon signature also may be enriched, but to a lesser degree than nitrogen. The ^{13}C value, however, can provide information about the source of carbon, such as discriminating between inshore and pelagic feeding in seabirds (Sydeman et al. 1997). Depending on the question of interest, samples for SIA analysis can be obtained from a variety of sources, including body tissues, feathers, and egg albumin. However, stable isotopic studies using different tissues may not be directly comparable to one another because of differences in turnover rates and, hence, potentially different inferences about feeding ecology.

In the Gulf of the Farallones, California, SIA was used to infer relative trophic position of several seabird species during the early breeding period in conjunction with more conventional dietary studies to infer chick diets later in the season (Sydeman et al. 1997). Analyses of egg albumin separated Brandt's and Pelagic Cormorant (*Compsohalieus penicillatus* and *Stictocarbo pelagicus*), Pigeon Guillemot (*Cepphus columba*), and Rhinoceros Auklet (*Cerorhinca monocerata*) from Cassin's Auklet (*Ptychoramphus aleuticus*), Common Murre, and Western Gull. The former group had higher ^{15}N signatures, indicating a piscivorous feeding habit, whereas the latter group had lower ^{15}N signatures, indicating a more planktivorous diet. Brandt's Cormorant had the highest ^{13}C values, confirming a neritic and benthic feeding habit, whereas Pelagic Cormorant and Cassin's Auklet had relatively low values, consistent with their more pelagic foraging range (Figure 6.19). Studies of chick diets later in the season largely supported the results of the SIA on the eggs, except for Common Murres, which apparently shifted from a planktivorous diet early in the breeding season to a piscivorous one later in the summer when they were provisioning chicks.

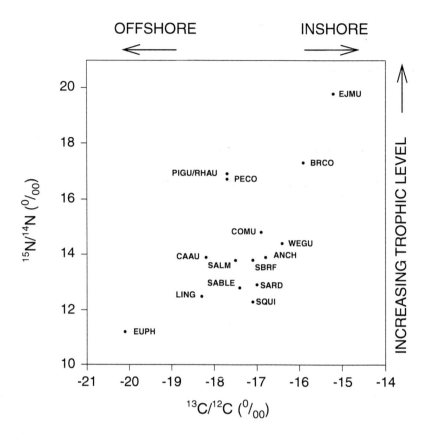

FIGURE 6.19 Trophic structure in the Gulf of the Farallones based on seabird egg albumin, sea lion and fish muscle tissue, and whole bodies of krill. Abbreviations: EUPH (krill = euphausiids), LING (lingcod), SABLE (sablefish), SQUI (squid), SARD (sardine), SALM (salmon), SBRF (short-bellied rockfish), ANCH (anchovy), CAAU (Cassin's Auklet), WEGU (Western Gull), COMU (Common Murre), PIGU (Pigeon Guillemot), RHAU (Rhinoceros Auklet), PECO (Pelagic Cormorant), BRCO (Brandt's Cormorant), EJMU (northern sea lion). (From Sydeman et al. 1997, used with permission.)

SIA was used to infer trophic relationships among six species of seabirds in Iceland, where stomach content analysis had revealed that capelin (*Mallotus villosus*) dominated the diets of birds in north Iceland and sandlance (*Ammodytes* sp.) was more prevalent in south Iceland (Thompson et al. 1999). Overall ^{15}N values tended to be enriched in seabirds from the north and ^{13}C values tended to be enriched in seabirds from the south, suggesting that northern birds fed at a higher trophic level and farther from shore than southern birds. However, there were no consistent patterns in isotope signatures in seabird species between the two locations. The pelagic foraging habit of Thick-billed Murres, indicated by the relatively depleted ^{13}C values, was later confirmed by a transmitter study that tracked birds up to 168 km from the breeding colony (Benvenuti et al. 1998).

Seasonal dietary shifts in Northern Fulmars were detected by applying SIA to feather samples (Thompson and Furness 1995). Comparison of the difference in ^{13}C and ^{15}N signatures between primary 2 and primary 10, which in fulmars are molted at different times (primary molt proceeds outward beginning toward the end of the chick-rearing period), indicated that during the winter fulmars feed at a lower trophic position than they do during the summer. The advantage of using feathers in SIA is that sampling is nondestructive, and analysis of selected feathers permits detection of seasonal variations in diet.

6.6.2 NONBREEDING AND WINTER DIET

Seabirds are most accessible for study when they come ashore to breed. Thus, most information on diet is limited to the breeding season. Our knowledge of what seabirds eat during the nonbreeding season is scant and requires further study.

Stomach samples from Atlantic Puffins (*Fratercula arctica*) wintering in the Northeast Atlantic near the Faroe Islands contained squid and a variety of fish and crustaceans (Falk et al. 1992). Puffins wintering in coastal areas fed primarily on euphausiids and fish, whereas birds collected in a more pelagic realm of the Norwegian Sea ate almost exclusively squid and one species of lanternfish (*Benthosema glaciale*). The pelagic diet suggested a more nocturnal foraging habit for puffins in winter, because lanternfish undergo a vertical migration to the surface at night.

During summer, King Penguins (*Aptenodytes patagonicus*) at South Georgia and elsewhere feed primarily on mesopelagic fish (Olsson and North 1997). However, during winter, cephalopods increase in importance and may comprise up to 64% of the diet (Adams and Klages 1987, Cherel et al. 1996). Gentoo Penguins at Marion Island ate mostly crustaceans from March to June 1984, then switched to a diet primarily consisting of fish from July to March 1985 (Adams and Klages 1989). The switch coincided with peak egg laying, suggesting that changes in diet may reflect changes in seasonal nutrient requirements. By contrast, Gentoo Penguins at South Georgia ate mostly krill during the winter months of 1987–1988 (Williams 1991). During the nonbreeding season, seabirds are not constrained to return frequently to the colony, and therefore diets may reflect a more opportunistic nature of prey acquisition.

6.7 ENERGETIC CONSEQUENCES OF FORAGING AT SEA

Pelagic seabirds share certain life history traits, such as single-chick broods, slow growth rates, and delayed maturation, which have been suggested to indicate that the ocean is an energetically expensive environment relative to terrestrial ecosystems (Ashmole 1963, Lack 1967, 1968). However, new paradigms of seabird growth and energetics are emerging that do not support the energy limitation hypothesis (e.g., Prince and Ricketts 1981, Schreiber 1994). The energetics of reproduction and chick rearing, and of free-ranging seabirds, are discussed in Chapters 11 and 13. See also Croxall and Briggs (1991), who reviewed the pattern of acquisition and use of energy by several pelagic seabird species in polar and subpolar regions of the Atlantic Ocean.

6.8 SUMMARY AND CONCLUSIONS

Seabirds have evolved many morphological, physiological, sensory, and behavioral adaptations that permit them to find adequate food to maintain themselves and rear offspring. The ocean is characterized by physical features, such as upwellings, convergence zones, eddies, tidal forcing, and ice edges, that make prey available to foraging seabirds. Moreover, some species have learned to exploit subsurface predators that drive prey toward the surface. High-latitude species appear to rely more on physical forcing, whereas seabirds in the tropics are more likely to associate with subsurface predators.

The nonrandom distribution of seabirds over the oceans is largely a result of localized concentrations of prey. Earlier characterization of seabirds as coastal, inshore, or pelagic may be valid during the breeding season but not at other times of the year when seabirds are free to roam more widely in search of food. Both diet studies and telemetry support the opportunistic and variable nature of seabird foraging patterns at daily, seasonal, and annual scales.

Our understanding of seabird foraging behavior has increased considerably over the past 20 years, due in part to advances in technology, but we still have much to learn. To date, most studies of seabird foraging behavior have used the inductive approach, whereby inferences about a phenomenon are drawn from a set of observations. Few studies have used the hypothetico-deductive approach to address questions about foraging behavior. This lack of rigorous hypothesis testing may be partly due to the logistic constraints of studying seabirds in the marine environment and the inability to control or constrain such a vast array of parameters to examine a single effect. However, the deductive approach is a more powerful tool in ascribing causation to a particular phenomenon, whereas correlative results are about the best we can muster using the inductive method.

New paradigms are emerging that challenge the dogmatic notion that seabirds operate under severe energetic constraints and that the ocean is a food-limited environment. While building on the pioneering work of Lack and Ashmole, our future work must not be constrained by past theoretical frameworks. This chapter has attempted to summarize our current state of knowledge about seabird foraging behavior and diet, and hopefully has provided some insights to guide future research. The ocean is an alluring siren.

LITERATURE CITED

ADAMS, N. J., AND N. T. KLAGES. 1987. Seasonal variation in the diet of the King Penguin (*Aptenodytes patagonicus*) at sub-Antarctic Marion Island. Journal of Zoology, London 212: 303–324.

ADAMS, N. J., AND N. T. KLAGES. 1989. Temporal variation in the diet of the Gentoo Penguin *Pygoscelis papua* at sub-Antarctic Marion Island. Colonial Waterbirds 12: 30–36.

AINLEY, D. G. 1977. Feeding methods in seabirds: a comparison of polar and tropical nesting communities in the eastern Pacific Ocean. Pp. 669–685 *in* Adaptations within Antarctic Ecosystems (G. A. Llano, Ed.). Smithsonian Institution, Washington, D.C.

AINLEY, D. G., AND R. J. BOEKELHEIDE. 1983. An ecological comparison of oceanic seabird communities of the South Pacific Ocean. Studies in Avian Biology 8: 2–23.

AINLEY, D. G., AND R. J. BOEKELHEIDE. 1990. Seabirds of the Farallon Islands. Ecology, Dynamics, and Structure of an Upwelling-System Community. Stanford University Press, Stanford, CA.

AINLEY, D. G., C. A. RIBIC, AND L. B. SPEAR. 1993. Species-habitat relationships among Antarctic seabirds: a function of physical or biological factors? Condor 95: 806–816.

AINLEY, D. G., L. B. SPEAR, S. G. ALLEN, AND C. A. RIBIC. 1996. Temporal and spatial patterns in the diet of the Common Murre in California waters. Condor 98: 691–705.

ANDERSON, D. J., AND R. E. RICKLEFS. 1987. Radio-tracking Masked and Blue-footed Boobies (*Sula* sp.) in the Galapagos Islands. National Geographic Research 3: 152–163.

ANDERSON, D. W., F. GRESS, AND K. F. MAIS. 1982. Brown Pelicans: influence of food supply on reproduction. Oikos 39: 23–31.

ANNETT, C., AND R. PIEROTTI. 1989. Chick hatching as a trigger for dietary switching in the Western Gull. Colonial Waterbirds 12: 4–11.

ASHMOLE, N. P. 1963. The regulation of numbers of tropical oceanic birds. Ibis 103b: 458–473.

ASHMOLE, N. P., AND M. J. ASHMOLE. 1967. Comparative feeding ecology of sea birds of a tropical oceanic island. Peabody Museum of Natural History, Yale University Bulletin 24.

ASHMOLE, N. P., AND H. TOVAR. 1968. Prolonged parental care in Royal Terns and other birds. Auk 85: 90–100.

ATWOOD, J. L., AND P. R. KELLY. 1984. Fish dropped on breeding colonies as indicators of Least Tern food habits. Wilson Bulletin 96: 34–47.

BAIRD, P. H. 1990. Influence of abiotic factors and prey distribution on diet and reproductive success of three seabird species in Alaska. Ornis Scandinavica 21: 224–235.

BALLANCE, L. T., R. L. PITMAN, AND S. B. REILLY. 1997. Seabird community structure along a productivity gradient: importance of competition and energetic constraint. Ecology 78: 1502–1518.

BAYER, R. D. 1985. Bill length of herons and egrets as an estimator of prey size. Colonial Waterbirds 8: 104–109.

BECKER, P. H., D. FRANK, AND S. R. SUDMANN. 1993. Temporal and spatial pattern of Common Tern (*Sterna hirundo*) foraging in the Wadden Sea. Oecologia 93: 389–393.

BENVENUTI, S., F. BONADONNA, L. DALL'ANTONIA, AND G. A. GUDMUNDSSON. 1998. Foraging flights of breeding Thick-billed Murres (*Uria lomvia*) as revealed by bird-borne direction recorders. Auk 115: 57–66.

BERRUTI, A., L. G. UNDERHILL, P. A. SHELTON, C. MOLONEY, AND R. J. M. CRAWFORD. 1993. Seasonal and interannual variation in the diet of two colonies of the Cape Gannet (*Morus capensis*) between 1977–78 and 1989. Colonial Waterbirds 16: 158–175.

BLACKWELL, B. F., W. B. KROHN, AND R. B. ALLEN. 1995. Foods of nestling Double-crested Cormorants in Penobscot Bay, Maine, USA: temporal and spatial comparisons. Colonial Waterbirds 18: 199–208.

BOCHER, P., B. LABIDOIRE, AND Y. CHEREL. 2000. Maximum dive depths of Common Diving Petrels (*Pelecanoides urinatrix*) during the annual cycle at Mayes Island, Kerguelen. Journal of Zoology, London 251: 517–524.

BRAGER, S. 1998. Feeding associations between White-fronted Terns and Hector's dolphins in New Zealand. Condor 100: 560–562.

BROOKE, M. DE L., AND P. A. PRINCE. 1991. Nocturnality in seabirds. Proceedings of the International Ornithological Congress 20: 1113–1121.

BROWN, K. M., AND P. J. EWINS. 1996. Technique-dependent biases in determination of diet composition: an example with Ring-billed Gulls. Condor 98: 34–41.

BROWN, W. Y. 1975. Parental feeding of young Sooty Terns (*Sterna fuscata* L.) and Brown Noddies (*Anous stolidus* L.) in Hawaii. Journal of Animal Ecology 44: 731–742.

BURGER, A. E. 1981. Food and foraging behavior of Lesser Sheathbills at Marion Island. Ardea 69: 167–180.

BURGER, A. E., AND J. F. PIATT. 1990. Flexible time budgets in breeding Common Murres: buffers against variable prey abundance. Studies in Avian Biology 14: 71–83.

BURGER, A. E., AND R. P. WILSON. 1988. Capillary-tube depth gauges for diving animals: an assessment of their accuracy and applicability. Journal of Field Ornithology 59: 345–354.

BURGER, J. 1980. The transition to independence and postfledging parental care in seabirds. Pp. 367–447 *in* Behavior of Marine Animals, Vol. 4: Marine Birds (J. Burger, B. Olla, and H. E. Winn, Eds.). Plenum Press, New York.

BURGER, J. 1987. Foraging efficiency in gulls: a congeneric comparison of age differences in efficiency and age of maturity. Studies in Avian Biology 10: 83–90.

BURGER, J. 1988. Foraging behavior in gulls: differences in method, prey, and habitat. Colonial Waterbirds 11: 9–23.

BURGER, J., AND M. GOCHFELD. 1983. Behavior of nine avian species at a Florida garbage dump. Colonial Waterbirds 6: 54–63.

CAIRNS, D. K., K. A. BREDIN, AND W. A. MONTEVECCHI. 1987. Activity budgets and foraging ranges of breeding Common Murres. Auk 104: 218–224.

CEZILLY, F., AND J. WALLACE. 1988. The determination of prey captured by birds through direct field observations: a test of the method. Colonial Waterbirds 11: 110–112.

CHEREL, Y., AND V. RIDOUX. 1992. Prey species and nutritive value of food fed during summer to King Penguin *Aptenodytes patagonica* chicks at Possession Island, Crozet Archipelago. Ibis 134: 118–127.

CHEREL, Y., AND H. WEIMERSKIRCH. 1999. Spawning cycle of onychoteuthid squids in the southern Indian Ocean: new information from seabird predators. Marine Ecology Progress Series 188: 93–104.

CHEREL, Y., V. RIDOUX, AND P. G. RODHOUSE. 1996. Fish and squid in the diet of King Penguin chicks, *Aptenodytes patagonicus*, during winter at sub-Antarctic Crozet Islands. Marine Biology 126: 559–570.

COBLENTZ, B. E. 1985. Mutualism between Laughing Gulls *Larus atricilla* and epipelagic fishes. Cormorant 13: 61–63.

CROLL, D. A., A. J. GASTON, A. E. BURGER, AND D. KONNOFF. 1992. Foraging behavior and physiological adaptation for diving in Thick-billed Murres. Ecology 73: 344–356.

CROXALL, J. P., AND J. R. FURSE. 1980. Food of Chinstrap Penguins *Pygoscelis antarctica* and Macaroni Penguins *Eudyptes chrysolophus* at Elephant Island Group, South Shetland Islands. Ibis 122: 237–245.

CROXALL, J. P., AND P. A. PRINCE. 1980. Food, feeding ecology and ecological segregation of seabirds at South Georgia. Biological Journal of the Linnean Society 14: 103–131.

CROXALL, J. P., AND D. R. BRIGGS. 1991. Foraging economics and performance of polar and subpolar Atlantic seabirds. Polar Research 10: 561–578.

CROXALL, J. P., A. J. HALL, H. J. HILL, A. W. NORTH, AND P. G. RODHOUSE. 1995. The food and feeding ecology of the White-chinned Petrel *Procellaria aequinoctialis* at South Georgia. Journal of Zoology, London 237: 133–150.

CROXALL, J. P., P. A. PRINCE, AND K. REID. 1997. Dietary segregation of krill-eating South Georgia seabirds. Journal of Zoology, London 242: 531–556.

DAVOREN, G. K., AND A. E. BURGER. 1999. Differences in prey selection and behaviour during self-feeding and chick provisioning in Rhinoceros Auklets. Animal Behaviour 58: 853–863.

DIAMOND, A. W. 1978. Feeding strategies and population size in tropical seabirds. American Naturalist 112: 215–223.

DIAMOND, A. W. 1983. Feeding overlap in some tropical and temperate seabird communities. Studies in Avian Biology 8: 24–46.

DUFFY, D. C. 1980. Patterns of piracy by Peruvian seabirds: a depth hypothesis. Ibis 122: 521–525.

DUFFY, D. C. 1982. Patterns of piracy in the seabird communities of the Galapagos Islands and southern Africa. Cormorant 10: 71–80.

DUFFY, D. C. 1983. The foraging ecology of Peruvian seabirds. Auk 100: 800–810.

DUFFY, D. C. 1986. Foraging at patches: interactions between Common and Roseate terns. Ornis Scandinavica 17: 47–52.

DUFFY, D. C. 1989. Seabird foraging aggregations: a comparison of two southern upwellings. Colonial Waterbirds 12: 164–175.

DUFFY, D. C., AND S. JACKSON. 1986. Diet studies of seabirds: a review of methods. Colonial Waterbirds 9: 1–17.

DUNN, E. K. 1972. Effect of age on the fishing ability of Sandwich Terns (*Sterna sandvicensis*). Ibis 114: 360–366.

DUNN, E. K. 1973. Changes in fishing ability of terns associated with windspeed and sea surface conditions. Nature 244: 520–521.

EVANS, P. G. H. 1982. Associations between seabirds and cetaceans: a review. Mammal Review 12: 187–206.

EVANS, D. L. 1987. Dolphins as beaters for Gulls? Bird Behaviour 7: 47–48.

FALK, K., J. JENSEN, AND K. KAMPP. 1992. Winter diet of Atlantic Puffins (*Fratercula arctica*) in the northeast Atlantic. Colonial Waterbirds 15: 230–235.

FAVERO, M. 1996. Foraging ecology of Pale-faced Sheathbills in colonies of Southern Elephant Seals at King George Island, Antarctica. Journal of Field Ornithology 67: 292–299.

FAVERO, M., R. CASAUX, P. SILVA, E. BARRERA-ORO, AND N. CORIA. 1998. The diet of the Antarctic Shag during summer at Nelson Island, Antarctica. Condor 100: 112–118.

FAVERO, M., M. S. BO, M. P. SILVA R., AND C. GARCIA-MATA. 2000. Food and feeding biology of the South American Tern during the nonbreeding season. Waterbirds 23: 125–129.

FURNESS, R. W. 1987. Kleptoparasitism in seabirds. Pp. 77–100 *in* Seabirds: Feeding Ecology and Role in Marine Ecosystems (J. P. Croxall, Ed.). Cambridge University Press, Cambridge.

FURNESS, R. W., J. J. D. GREENWOOD, AND P. J. JARVIS. 1993. Can birds be used to monitor the environment? Pp. 1–41 *in* Birds as Monitors of Environmental Change (R. W. Furness and J. J. D. Greenwood, Eds.). Chapman & Hall, London.

GARTHE, S., S. BENVENUTI, AND W. A. MONTEVECCHI. 2000. Pursuit plunging by Northern Gannets (*Sula bassana*) feeding on capelin (*Mallotus villosus*). Proceedings of the Royal Society of London B 267: 1717–1722.

GASTON, A. J., AND M. S. W. BRADSTREET. 1993. Intercolony differences in the summer diet of Thick-billed Murres in the eastern Canadian Arctic. Canadian Journal of Zoology 71: 1831–1840.

GOCHFELD, M., AND J. BURGER. 1981. Age-related differences in piracy of frigatebirds from Laughing Gulls. Condor 83: 79–82.

GOCHFELD, M., AND J. BURGER. 1982. Feeding enhancement by social attraction in the Sandwich Tern. Behavioral Ecology and Sociobiology 10: 15–17.

GOSS-CUSTARD, J. D., J. T. CAYFORD, J. S. BOATES, AND S. E. A. DURELL. 1987. Field tests of the accuracy of estimating prey size from bill length in oystercatchers, *Haematopus ostralegus*, eating mussels, *Mytilus edulis*. Animal Behaviour 35: 1078–1083.

GÖTMARK, F., D. W. WINKLER, AND M. ANDERSSON. 1986. Flock-feeding on fish schools increases individual success in gulls. Nature 319: 589–591.

HACKL, E., AND J. BURGER. 1988. Factors affecting piracy in Herring Gulls at a New Jersey landfill. Wilson Bulletin 100: 424–430.

HANEY, J. C. 1986. Seabird patchiness in tropical oceanic waters: the influence of *Sargassum* "reefs." Auk 103: 141–151.

HANEY, J. C. 1989a. Remote characterization of marine bird habitats with satellite imagery. Colonial Waterbirds 12: 67–77.

HANEY, J. C. 1989b. Iterative techniques for characterizing marine bird habitats with time-series of satellite images. Colonial Waterbirds 12: 78–89.

HANEY, J. C., AND A. E. STONE. 1988. Seabird foraging tactics and water clarity: are plunge divers really in the clear? Marine Ecology Progress Series 49: 1–9.

HANEY, J. C., K. M. FRISTRUP, AND D. S. LEE. 1992. Geometry of visual recruitment by seabirds to ephemeral foraging flocks. Ornis Scandinavica 23: 49–62.

HARPER, P. C. 1987. Feeding behaviour and other notes on 20 species of Procellariiformes at sea. Notornis 34: 169–192.

HARRIS, M. P. 1977. Comparative ecology of seabirds in the Galapagos Archipelago. Pp. 65–76 *in* Evolutionary Ecology (B. Stonehouse and C. Perrins, Eds.). University Park Press, Baltimore.

HARRIS, M. P., AND S. WANLESS. 1991. The importance of the lesser sandeel *Ammodytes marinus* in the diet of the Shag *Phalacrocorax aristotelis*. Ornis Scandinavica 22: 375–382.

HARRISON, C. S., T. S. HIDA, AND M. P. SEKI. 1983. Hawaiian seabird feeding ecology. Wildlife Monographs 85: 1–71.

HARRISON, N. M., M. J. WHITEHOUSE, D. HEINEMANN, P. A. PRINCE, G. L. HUNT, JR., AND R. R. VEIT. 1991. Observations of multispecies seabird flocks around South Georgia. Auk 108: 801–810.

HERTEL, F., AND L. T. BALLANCE. 1999. Wing ecomorphology of seabirds from Johnston Atoll. Condor 101: 549–556.

HINDELL, M. A. 1988. The diet of the Royal Penguin *Eudyptes schlegeli* at Macquarie Island. Emu 88: 219–226.

HOBSON, K. A., J. F. PIATT, AND J. PITOCHELLI. 1994. Using stable isotopes to determine seabird trophic relationships. Journal of Animal Ecology 63: 786–798.

HOFFMAN, W., D. HEINEMANN, AND J. A. WIENS. 1981. The ecology of seabird feeding flocks in Alaska. Auk 98: 437–456.

HULSMAN, K. 1989. The structure of seabird communities: an example from Australian waters. Pp. 59–91 *in* Seabirds and Other Marine Vertebrates: Competition, Predation, and Other Interactions (J. Burger, Ed.). Columbia University Press, New York.

HUNT, G. L., JR. 1991a. Marine ecology of seabirds in polar oceans. American Zoologist 31: 131–142.

HUNT, G. L., JR. 1991b. Occurrence of polar seabirds in relation to prey concentrations and oceanographic factors. Polar Research 10: 553–559.

HUNT, G. L., JR., N. M. HARRISON, AND R. T. COONEY. 1990. The influence of hydrographic structure and prey abundance on foraging of Least Auklets. Studies in Avian Biology 14: 7–22.

HUNT, G. L., JR., J. F. PIATT, AND K. E. ERIKSTAD. 1991. How do foraging seabirds sample their environment? Proceedings of the International Ornithological Congress 20: 2272–2279.

HUNT, G. L., JR., K. O. COYLE, S. HOFFMAN, M. B. DECKER, AND E. N. FLINT. 1996. Foraging ecology of Short-tailed Shearwaters near the Pribilof Islands, Bering Sea. Marine Ecology Progress Series 141: 1–11.

HUNT, G. L., JR., F. MEHLUM, R. W. RUSSELL, D. IRONS, M. B. DECKER, AND P. H. BECKER. 1999. Physical processes, prey abundance, and the foraging ecology of seabirds. Proceedings of the International Ornithological Congress 22: 2040–2056.

HUNT, G. L., JR., R. W. RUSSELL, K. O. COYLE, AND T. WEINGARTNER. 1998. Comparative foraging ecology of planktivorous auklets in relation to ocean physics and prey availability. Marine Ecology Progress Series 167: 241–259.

HUNTER, S., AND M. DE L. BROOKE. 1992. The diet of giant petrels *Macronectes* spp. At Marion Island, Southern Indian Ocean. Colonial Waterbirds 15: 56–65.

IRONS, D. B. 1998. Foraging area fidelity of individual seabirds in relation to tidal cycles and flock feeding. Ecology 79: 647–655.

JACKSON, S., AND P. G. RYAN. 1986. Differential digestion rates of prey by White-chinned Petrels (*Procellaria aequinoctialis*). Auk 103: 617–619.

JAHNCKE, J., AND E. GOYA. 2000. Responses of three booby species to El Niño 1997–1998. Waterbirds 23: 102–108.

JAHNCKE, J., A. GARCIA-GODOS, AND E. GOYA. 1999. The diet of the Peruvian Diving-Petrel at La Vieja and San Gallan, Peru. Journal of Field Ornithology 70: 71–79.

JOUVENTIN, P., AND H. WEIMERSKIRCH. 1990. Satellite tracking of Wandering Albatrosses. Nature 343: 746–748.

KALLANDER, H. 1991. Differences in prey capture efficiency of adult and juvenile Common *Sterna hirundo* and Arctic *S. paradisaea* terns. Ornis Svecica 1: 121–122.

KEITT, B. S., D. A. CROLL, AND B. R. TERSHY. 2000. Dive depth and diet of the Black-vented Shearwater (*Puffinus opisthomelas*). Auk 117: 507–510.

KOOYMAN, G. L., AND T. G. KOOYMAN. 1995. Diving behavior of Emperor Penguins nurturing chicks at Coulman Island, Antarctica. Condor 97: 536–549.

LACK, D. 1967. Interrelationships in breeding adaptations as shown by marine birds. Acta 14th Congressus Internationalis Ornithologici: 3–42. Montreal, Quebec, Canada.

LACK, D. 1968. Ecological Adaptations for Breeding in Birds. Methuen, London.

LEQUETTE, B., C. VERHEYDEN, AND P. JOUVENTIN. 1989. Olfaction in subantarctic seabirds: its phylogenetic and ecological significance. Condor 91: 732–735.

LONGHURST, A. 1981. Analysis of Marine Ecosystems. Academic Press, New York.

LONGHURST, A. 1999. Ecological Geography of the Sea. Academic Press, New York.

MACLEAN, A. A. E. 1986. Age-specific foraging ability and the evolution of deferred breeding in three species of gulls. Wilson Bulletin 98: 267–279.

MARTIN, A. R. 1989. The diet of Atlantic Puffin *Fratercula arctica* and Northern Gannet *Sula bassana* chicks at a Shetland colony during a period of changing prey availability. Bird Study 36: 170–180.

MEHLUM, F., G. L. HUNT, JR., Z. KLUSEK, AND M. B. DECKER. 1999. Scale-dependent correlations between the abundance of Brünnich's Guillemots and their prey. Journal of Animal Ecology 68: 60–72.

MONAGHAN, P., J. D. UTTLEY, M. D. BURNS, C. THAINE, AND J. BLACKWOOD. 1989. The relationship between food supply, reproductive effort and breeding success in Arctic Terns *Sterna paradisaea*. Journal of Animal Ecology 58: 261–274.

MONAGHAN, P., J. D. UTTLEY, AND M. D. BURNS. 1992. Effect of changes in food availability on reproductive effort in Arctic Terns *Sterna paradisaea*. Ardea 80: 71–81.

MONTEVECCHI, W. A. 1993. Birds as indicators of change in marine prey stocks. Pp. 217–266 *in* Birds as Monitors of Environmental Change (R. W. Furness and J. J. D. Greenwood, Eds.). Chapman & Hall, London.

MONTEVECCHI, W. A., AND R. A. MYERS. 1995. Prey harvests of seabirds reflect pelagic fish and squid abundance on multiple spatial and temporal scales. Marine Ecology Progress Series 117: 1–9.

MONTEVECCHI, W. A., AND R. A. MYERS. 1996. Dietary changes of seabirds indicate shifts in pelagic food webs. Sarsia 80: 313–322.

MONTEVECCHI, W. A., V. L. BIRT, AND D. K. CAIRNS. 1988. Dietary changes of seabirds associated with local fisheries failures. Biological Oceanography 5: 153–161.

MORRIS, B. F., AND D. D. MOGELBERG. 1973. Identification manual of the pelagic *Sargassum* fauna. Bermuda Biological Station Special Publication 11: 1–63.

NELSON, B. 1979. Seabirds: Their Biology and Ecology. A and W Publishers, New York.

OBST, B. S. 1985. Densities of Antarctic seabirds at sea and the presence of the krill *Euphausia superba*. Auk 102: 540–549.

OBST, B. S., AND G. L. HUNT, JR. 1990. Marine birds feed at Gray Whale mud plumes in the Bering Sea. Auk 107: 678–688.

OBST, B. S., R. W. RUSSELL, G. L. HUNT, JR., Z. A. EPPLEY, AND N. M. HARRISON. 1995. Foraging radii and energetics of Least Auklets (*Aethia pusilla*) breeding on three Bering Sea islands. Physiological Zoology 68: 647–672.

OGI, H., AND K. SHIOMI. 1991. Diet of murres caught incidentally during winter in northern Japan. Auk 108: 184–185.

OLSSON, O., AND A. W. NORTH. 1997. Diet of the King Penguin *Aptenodytes patagonicus* during three summers at South Georgia. Ibis 139: 504–513.

OSORNO, J. L., R. TORRES, AND C. M. GARCIA. 1992. Kleptoparasitic behavior of the Magnificent Frigatebird: sex bias and success. Condor 94: 692–698.

OSTRAND, W. D. 1999. Marbled Murrelets as initiators of feeding flocks in Prince William Sound, Alaska. Waterbirds 22: 314–318.

PEARSON, T. H. 1968. The feeding biology of sea-bird species breeding on the Farne Islands, Northumberland. Journal of Animal Ecology 37: 521–548.

PIEROTTI, R. 1988. Associations between marine birds and mammals in the northwest Atlantic Ocean. Pp. 31–58 in Seabirds and Other Marine Vertebrates (J. Burger, Ed.). Columbia Univeristy Press, New York.

PIEROTTI, R., AND C. A. ANNETT. 1990. Diet and reproductive output in seabirds. BioScience 40: 568–574.

PIETZ, P. J. 1987. Feeding and nesting ecology of sympatric South Polar and Brown Skuas. Auk 104: 617–627.

PITMAN, R. L., AND L. T. BALLANCE. 1992. Parkinson's Petrel distribution and foraging ecology in the eastern Pacific: aspects of an exclusive feeding relationship with dolphins. Condor 94: 825–835.

PORTER, J. M., AND S. G. SEALY. 1982. Dynamics of seabird multispecies feeding flocks: age-related feeding behaviour. Behaviour 81: 91–109.

PRINCE, P. A., AND C. RICKETTS. 1981. Relationships between food supply and growth in albatrosses: an interspecies chick fostering experiment. Ornis Scandinavica 12: 207–210.

PRINCE, P. A., AND P. G. COPESTAKE. 1990. Diet and aspects of Fairy Prions breeding at South Georgia. Notornis 37: 59–69.

PRINCE, P. A., J. P. CROXALL, P. N. TRATHAN, AND A. G. WOOD. 1997. The pelagic distribution of South Georgia albatrosses and their relationships with fisheries. Pp. 137–167 in Albatross Biology and Conservation (G. Robertson and R. Gales, Eds.). Surrey Beatty and Sons, Chipping Norton, Australia.

PRINCE, P. A., H. WEIMERSKIRCH, A. G. WOOD, AND J. P. CROXALL. 1999. Areas and scales of interactions between albatrosses and the marine environment: species, populations and sexes. Proceedings of the International Ornithological Congress 22: 2001–2020.

RAMOS, J. A., E. SOLA, L. R. MONTEIRO, AND N. RATCLIFFE. 1998. Prey delivered to Roseate Tern chicks in the Azores. Journal of Field Ornithology 69: 419–429.

RAYMONT, J. E. G. 1980. Plankton and Productivity in the Oceans, 2nd ed. Vol. 1: Phytoplankton. Pergamon Press Ltd., Oxford.

REED, J. M., AND S. J. HA. 1983. Enhanced foraging efficiency in Forster's Terns. Wilson Bulletin 95: 479–481.

REID, K., J. P. CROXALL, AND T. M. EDWARDS. 1997. Interannual variation in the diet of the Antarctic Prion Pachyptila desolata at South Georgia. Emu 97: 126–132.

RIBIC, C. A., AND D. G. AINLEY. 1997. The relationships of seabird assemblages to physical habitat features in Pacific equatorial waters during spring 1984–1991. ICES Journal of Marine Science 54: 593–599.

ROBY, D. D. 1991. Diet and postnatal energetics in convergent taxa of plankton-feeding seabirds. Auk 108: 131–146.

RODWAY, M. S., AND W. A. MONTEVECCHI. 1996. Sampling methods for assessing the diets of Atlantic Puffin chicks. Marine Ecology Progress Series 144: 41–55.

RUSSELL, R. W., G. L. HUNT, JR., K. O. COYLE, AND R. T. COONEY. 1992. Foraging in a fractal environment: spatial patterns in a marine predator-prey system. Landscape Ecology 7: 195–209.

RYAN, P. G., AND J. COOPER. 1989. The distribution and abundance of aerial seabirds in relation to Antarctic krill in the Prydz Bay region, Antarctica, during late summer. Polar Biology 10: 199–209.

RYAN, P. G., AND C. L. MOLONEY. 1991. Prey selection and temporal variation in the diet of Subantarctic Skuas at Inaccessible Island, Tristan da Cunha. Ostrich 62: 52–58.

SAFINA, C. 1990. Bluefish mediation of foraging competition between Roseate and Common terns. Ecology 71: 1804–1809.

SAFINA, C., AND J. BURGER. 1985. Common Tern foraging: seasonal trends in prey fish densities, and competition with bluefish. Ecology 66: 1457–1463.

SAFINA, C., AND J. BURGER. 1989. Ecological dynamics among prey fish, bluefish, and foraging Common Terns in an Atlantic coastal system. Pp. 95–173 in Seabirds and Other Marine Vertebrates (J. Burger, Ed.). Columbia University Press, New York.

SAGAR, P. M., AND J. L. SAGAR. 1989. The effects of wind and sea on the feeding of Antarctic Terns at the Snares Islands, New Zealand. Nortornis 36: 171–182.

SCHNEIDER, D. C., AND D. C. DUFFY. 1985. Scale-dependent variability in seabird abundance. Marine Ecology Progress Series 25: 211–218.

SCHNEIDER, D. C., AND J. F. PIATT. 1986. Scale-dependent correlation of seabirds with schooling fish in a coastal ecosystem. Marine Ecology Progress Series 32: 237–246.

SCHREIBER, E. A. 1994. El Niño–Southern Oscillation effects on provisioning and growth in Red-tailed Tropicbirds. Colonial Waterbirds 17: 105–119.

SCHREIBER, E. A., AND R. W. SCHREIBER. 1989. Insights into seabird ecology from a global "natural experiment." National Geographic Research, Winter 1989: 64–81.

SCHREIBER, E. A., AND R. W. SCHREIBER. 1993. Red-tailed Tropicbird (*Phaethon rubricauda*). *In* The Birds of North America, No. 43 (A. Poole and F. Gill, Eds.). Academy of Natural Sciences, Philadelphia; American Ornithologists' Union, Washington, D.C.

SCHREIBER, R. W., AND D. A. HENSLEY. 1976. The diets of *Sula dactylatra*, *Sula sula*, and *Fregata minor* on Christmas Island, Pacific Ocean. Pacific Science 30: 241–248.

SCHREIBER, R. W., AND J. L. CHOVAN. 1986. Roosting by pelagic seabirds: energetic, populational, and social considerations. Condor 88: 487–492.

SHEALER, D. A. 1996. Foraging habitat use and profitability in tropical Roseate Terns and Sandwich Terns. Auk 113: 209–217.

SHEALER, D. A. 1998a. Size-selective predation by a specialist forager, the Roseate Tern. Auk 115: 519–525.

SHEALER, D. A. 1998b. Differences in diet and chick provisioning between adult Roseate and Sandwich terns in Puerto Rico. Condor 100: 131–140.

SHEALER, D. A., AND J. BURGER. 1993. Effects of interference competition on the foraging activity of tropical Roseate Terns. Condor 95: 322–329.

SHEALER, D. A., AND J. BURGER. 1995. Comparative foraging success between adult and one-year-old Roseate and Sandwich terns. Colonial Waterbirds 18: 93–99.

SHEALER, D. A., T. FLOYD, AND J. BURGER. 1997. Host choice and success of gulls and terns kleptoparasitizing Brown Pelicans. Animal Behaviour 53: 655–665.

SIMMONS, K. E. L. 1972. Some adaptive features of seabird plumage types. British Birds 65: 465–479.

SPEAR, L. B., AND D. G. AINLEY. 1998. Morphological differences relative to ecological segregation in petrels (Family: Procellariidae) of the Southern Ocean and tropical Pacific. Auk 115: 1017–1033.

SYDEMAN, W. J., K. A. HOBSON, P. PYLE, AND E. B. MCLAREN. 1997. Trophic relationships among seabirds in central California: combined stable isotope and conventional dietary approach. Condor 99: 327–336.

THOMPSON, D. R., AND R. W. FURNESS. 1995. Stable-isotope ratios of carbon and nitrogen in feathers indicate seasonal dietary shifts in Northern Fulmars. Auk 112: 493–498.

THOMPSON, D. R., K. LILLIENDAHL, J. SOLMUNDSSON, R. W. FURNESS, S. WALDRON, AND R. A. PHILLIPS. 1999. Trophic relationships among six species of Icelandic seabirds as determined through stable isotope analysis. Condor 101: 898–903.

THURMAN, H. V. 1983. Essentials of Oceanography. C. E. Merrill, Columbus, OH.

TRIVELPIECE, W. Z., S. G. TRIVELPIECE, AND N. J. VOLKMAN. 1987. Ecological segregation of Adélie, Gentoo, and Chinstrap penguins at King George Island, Antarctica. Ecology 68: 351–361.

VEIT, R. R., AND G. L. HUNT, JR. 1991. Broadscale density and aggregation of pelagic birds from a circumnavigational survey of the Antarctic Ocean. Auk 108: 790–800.

VERHEYDEN, C. 1993. Kelp Gulls exploit food provided by active right whales. Colonial Waterbirds 16: 88–91.

VERHEYDEN, C., AND P. JOUVENTIN. 1994. Olfactory behavior of foraging Procellariiformes. Auk 111: 285–291.

VERMEER, K., AND K. DEVITO. 1988. The importance of *Paracallisoma coecus* and myctophid fishes to nesting Fork-tailed and Leach's storm-petrels in the Queen Charlotte Islands, British Columbia. Journal of Plankton Research 10: 63–75.

WANLESS, S., M. P. HARRIS, AND J. A. MORRIS. 1990. A comparison of feeding areas used by individual Common Murres (*Uria aalge*), Razorbills (*Alca torda*) and an Atlantic Puffin (*Fratercula arctica*) during the breeding season. Colonial Waterbirds 13: 16–24.

WANLESS, S., S. K. FINNEY, M. P. HARRIS, AND D. J. MCCAFFERTY. 1999. Effect of the diel light cycle on the diving behaviour of two bottom feeding marine birds: the Blue-eyed Shag *Phalacrocorax atriceps* and the European Shag *P. aristotelis*. Marine Ecology Progress Series 188: 219–224.

WATANUKI, Y., AND A. E. BURGER. 1999. Body mass and dive duration in alcids and penguins. Canadian Journal of Zoology 77: 1838–1842.

WEIMERSKIRCH, H., O. CHASTEL, L. ACKERMAN, T. CHAURAND, F. CUENOT-CHAILLET, X. HINDERMEYER, AND J. JUDAS. 1994. Alternate long and short foraging trips in pelagic seabird parents. Animal Behaviour 47: 472–476.

WILLIAMS, T. D. 1991. Foraging ecology and diet of Gentoo Penguins *Pygoscelis papua* at South Georgia during winter and an assessment of their winter prey consumption. Ibis 133: 3–13.

WILLIAMS, T. D., D. R. BRIGGS, J. P. CROXALL, Y. NAITO, AND A. KATO. 1992a. Diving pattern and performance in relation to foraging ecology in the Gentoo Penguin, *Pygoscelis papua*. Journal of Zoology, London 227: 211–230.

WILLIAMS, T. D., A. KATO, J. P. CROXALL, Y. NAITO, D. R. BRIGGS, S. RODWELL, AND T. R. BARTON. 1992b. Diving pattern and performance in nonbreeding Gentoo Penguins (*Pygoscelis papua*) during winter. Auk 109: 223–234.

WUNDERLE, J. M., JR. 1991. Age-specific foraging proficiency in birds. Pp. 273–324 *in* Current Ornithology, Vol. 8 (D. M. Power, Ed.). Plenum Press, New York.

Magnificent Frigatebirds Fly Out before a Storm

7 Climate and Weather Effects on Seabirds

E. A. Schreiber

CONTENTS

7.1 INTRODUCTION

For a group of birds with similar life-history characteristics (deferred onset of breeding, long life, small clutch size, slow growth), seabirds live in a highly diverse variety of environments in their worldwide distribution. They experience the full gamut of weather patterns, whether daily, seasonally, annually, or on greater scales, and these patterns affect their survival, their habitat, their food supply, their ability to feed, and, thus, the continuing evolution of their species.

Effects of weather on birds can be long term, occurring over hundreds of years, or as short as a passing rain storm. The long-term effects of weather on birds undoubtedly have help shaped their particular demography and other life-history characteristics. In the short term, weather effects on seabirds can be seen in more proximate factors: the decision to nest that year or not, where to nest, annual nest success, growth rates of chicks, and survival of adults. Weather can cause the extirpation of a species from an area or only the loss of a few eggs to chilling. It can affect birds directly through increased wind levels or rain causing difficulty in flying, through flooding of nests, and through thermal stress. Effects can also be indirect: weather parameters can alter or destroy nesting habitat, change fish or krill distribution, or cause decreased visibility of prey.

Weather effects on seabirds are easy to observe on land but determining what is occurring while the birds are at sea has proven to be a difficult challenge, partly because our picture of how seabirds sample the ocean is incomplete. Seabirds primarily eat fish, squid, krill, and plankton that they must find and catch in the ocean, a medium that can exhibit dramatic variability or cycles from daily to seasonally to annually (see Chapter 6). Not only must seabirds feed in all these conditions, but during the nesting season, food must be transported back to the colony to feed young. Weather can affect:

1. Cost of catching food (Konyukhov 1997, Finney et al. 1999).
2. Transportation cost of food (Pennycuick 1982, Jouventin and Weimerskirch 1990, Furness and Bryant 1996, Weimerskirch et al. 1997).
3. Ability of birds to find food (Dobinson and Richards 1964, Dunn 1973, Taylor 1983, Schreiber and Schreiber 1989, Cruz and Cruz 1990, Finney et al. 1999).
4. Timing of the breeding season (Nelson 1978, Anderson 1989, Schreiber and Schreiber 1989, Konyukhov 1997, Anderson et al. unpublished).
5. Number of birds that attempt to nest in a given season (Schreiber and Schreiber 1989, Duffy 1990, Warham 1990).
6. Clutch size (Springer et al. 1986, Coulson and Porter 1985).
7. Reproductive success and chick growth (Prince and Ricketts 1981, Gaston et al. 1983, Coulson and Porter 1985, Cruz and Cruz 1985, Schreiber and Schreiber 1993, Schreiber 1996, Arnould et al. 1996, Finney et al. 1999).
8. Thermoregulation (Howell and Bartholomew 1961, 1962, Bartholomew and Dawson 1979, Kildaw 1999).
9. Adult survival (Schreiber and Schreiber 1989, Duffy 1990, Chastel et al. 1993, Montevecchi and Myers 1997).
10. Availability of food (Murphy 1936, Nelson 1978, Gaston et al. 1983, Arntz and Tarazona 1990, Montevecchi and Myers 1996, 1997, Finney et al. 1999).

Unusual or severe weather can impose a burden on seabird populations. In the most severe cases, such as occur during El Niño–Southern Oscillation events (ENSO; Schreiber and Schreiber 1984, 1989, Ainley et al. 1988), many adults may die. ENSO events, a worldwide rebalancing of the heat load of the earth that causes extreme weather patterns, are discussed in detail below. Seabirds respond to anything that affects their food source and often can serve as a good indicator of fish availability (Bailey et al. 1991, Hunt et al. 1991, Montevecchi and Myers 1992, Montevecchi 1993). However, a confounding factor in using birds as indicators of fish availability is that some species of seabirds alter their feeding effort to adjust to changes in fish stocks in order to supply a more constant amount of food to chicks (Drent and Daan 1980, Finney et al. 1999). Other seabird breeding parameters may be useful as an indicator of fish stocks, such as length of feeding trips, growth rates of young, changes in the mass of adults, and reproductive energetics (Cairns et al. 1987, Ainley and Boekelheide 1990, Montevecchi and Myers 1992, Montevecchi 1993, Schreiber 1994, 1996).

FIGURE 7.1 R. W. Schreiber bands a young Masked Booby on Christmas Island (Pacific Ocean). Long-term studies of individually marked birds are necessary to understanding seabird demography. (Photo by E. A. Schreiber.)

Determining the effects of weather on seabirds and understanding the evolutionary implications of the documented short-term changes require long-term monitoring of banded individuals (Figure 7.1). Few studies have accomplished this. The Farallon Islands, one such example, are one of the best-studied seabird communities in the world and provide abundant data on the myriad effects of weather on seabirds. From 1971 through today biologists from the Pt. Reyes Bird Observatory monitored the approximately 300,000 individuals of 11 species nesting there (Ainley and Lewis 1974, Ainley and Boekelheide 1990). This long-term perspective has taught us much about seabird biology, the effects of the environment, and the vital importance of long-term studies. Notably, the population was not found to be stable and in equilibrium, contrary to what we have frequently been taught in the way most populations exist, at least until the 1982–1983 ENSO event (Schreiber and Schreiber 1989). ENSO events were one reason for the instability in population numbers found on the Farallons, but seasonal variation also was not constant from year to year, causing further variation in the population. Overlaid on this variability, photoperiod also had an effect on initiation of laying. Interestingly, when the originally planned 5- to 6-year study began, researchers thought they would find a normal pattern in most years. Thirteen years later, through severe ENSO, mild ENSO, severe droughts, and record rains, no two years were found to be the same (Ainley 1990). In the author's own work in the Pacific, years used to be referred to as ENSO years and normal years. But, as the Farallon researchers found, no one prevailing pattern emerged in normal years! Long ago I began calling the years ENSO years and non-ENSO years, after coming to realize that the environment of seabirds was constant only in its inconsistency.

General effects of weather on fitness in seabirds, weather effects at the nest, and effects on the feeding ability of birds at sea are discussed in this chapter. Finally, weather effects are addressed in terms of the length of the event:

1. Long-term effects (50 years or more) — ice ages, warming trends.
2. Annual and several years — individual ENSO events, La Niña.
3. Seasonal — air temperature, water temperature, wind levels, ocean current strength.
4. Short-term (weekly, daily) — hurricanes, rains, cloud cover, fronts.

While weather events occur over varying lengths of time, the above divisions are not necessarily relevant to the effects experienced by birds in a particular region of the world. For example, ENSO events last from 1 to 2 or more years, but may exhibit only seasonal effects on birds in a particular area (Schreiber and Schreiber 1989). Because ENSO events begin in the tropical Pacific and have some of the most dramatic effects on birds there, the discussions below place particular emphasis on this area.

Seabirds, having evolved over millions of years, have had their life histories shaped by the effects of weather, as well as other pressures of survival, all of which can alter their lifetime reproductive success. However, in many cases today, it is difficult or impossible for us to piece together cause-and-effect relationships in the evolution of seabird life histories in relation to weather or any other selective parameter.

7.2 WEATHER EFFECTS ON FITNESS AND BEHAVIOR

7.2.1 FITNESS

Seabirds have evolved adaptations that enable them to live in the oceanic environment quite successfully: they survive and reproduce in all areas of the world. During periods of "normal" weather patterns, seabirds appear essentially unaffected by weather, or at least are well adapted to tolerate local weather conditions. Yet, every aspect by which we measure fitness in a bird can be affected by severe or unusual weather: clutch size, adult mass, hatching success, chick growth, fledging success, presence of disease, population size, and survival.

Not all seabirds in an area exhibit similar responses to any one weather event, possibly because they use different feeding areas, different feeding methodologies, a different food source, or have different flight capabilities. For instance, during strong ENSO events, most Masked Boobies (*Sula dactylatra*) on Christmas Island (central Pacific) desert the nesting colony at the very beginning of the event, while Red-tailed Tropicbirds (*Phaethon rubricauda*) and Great Frigatebirds (*Fregata minor*) do nest but lose many young to starvation as the event progresses and they cannot find enough food (Figure 7.2; Schreiber and Schreiber 1989). Within a seabird species, sexual differences may have evolved as adaptations to weather conditions and feeding in different areas. Wandering Albatrosses (*Diomedea exulans*) exhibit sexual segregation of foraging zones which can be explained by differences in wing-loading: males, with 12% greater wing-loading, feed in areas of highest wind levels (Shaffer 2000).

Most seabirds are able to avoid some weather effects by flying to a different area, and this, in fact, is how many survive a storm. Albatrosses rely on, and apparently soar effortlessly in, heavy winds, and may even select nesting areas that have high wind levels to conserve energy (which can then be spent raising their young). On Midway Island (northern Pacific) Laysan (*Phoebastria immutabilis*) and Black-footed Albatrosses (*P. nigripes*) nesting inland in calm areas often walk to the edge of the island, where updrafts help them get airborne (Whittow 1993a, b). Field metabolic rates of Northern Fulmars (*Fulmarus glacialis*) were inversely related to wind level, perhaps accounting for birds spending more time at the nest during calm periods (Furness and Bryant 1996). Wandering Albatrosses often sit on the sea and wait for higher wind levels to avoid the cost of flapping flight during calm periods (Jouventin and Weimerskirch 1990). Fulmars apparently could not afford the time to take the "sit and wait" approach and often suffered the increased costs of flight in low winds. Furness and Bryant (1996) suggested this as a factor that limits their breeding range.

Assessing both direct (death) and indirect effects (such as decreased food supply or lost habitat) of dramatic weather patterns can be difficult. If banded populations exist, adult mortality may be

FIGURE 7.2 A young Great Frigatebird on Christmas Island (Pacific Ocean) calls to its arriving parent overhead. Last year's dead chick lies beneath the nest. It was deserted by its parents in 1983, during one of the worst ENSO events on record. (Photo by R. W. and E. A. Schreiber.)

assessed directly, but few studies of seabirds have this luxury. A 27-year study of Snow Petrels (*Pagodroma nivea*) documented reduced adult survival in connection with ENSO events (Chastel et al. 1993). Population level (one of the most frequently measured parameters) may not always be a good direct indicator of the fitness of a local population since it can be affected by many parameters. For instance, changes in vegetation may make a colony site unsuitable for nesting, causing a decline in the number of birds in an area (Schreiber and Schreiber 1989). Dispersal, migration, and changes in food availability can also affect population size (Klomp and Furness 1992). Few studies have tracked a multispecies seabird community over time and attempted to explain why breeding numbers of individual species do not fluctuate in synchrony, and whether or not this reflects a degree of adaptedness. Differing responses of species to stochastic events may explain the proximate effects seen (Ainley and Boekelheide 1990, Schreiber and Schreiber 1989) but does not answer the question of a species fitness.

Chick growth rate and fledging mass are sometimes directly correlated with survival to breeding age (Croxall et al. 1988, Chastel et al. 1993), perhaps the most important measure of a species fitness as a whole. Unusual variation in chick mass is generally a reflection of changes in the adults' ability to feed brought on by weather (changes in oceanographic parameters) or changes in the distribution of the food resource, again, often itself brought on by weather (Konarzewski et al. 1990, Schreiber and Schreiber 1993). Weather is used here to mean either a short-term perturbation such as a storm, or a larger-scale event such as an ENSO, and to include oceanographic as well as atmospheric events.

7.2.2 BEHAVIOR

Seabird adults and chicks use various behavioral methods to thermoregulate when overheated or chilled (Bartholomew et al. 1968, Bartholomew and Dawson 1979, Welty and Baptista 1988). To

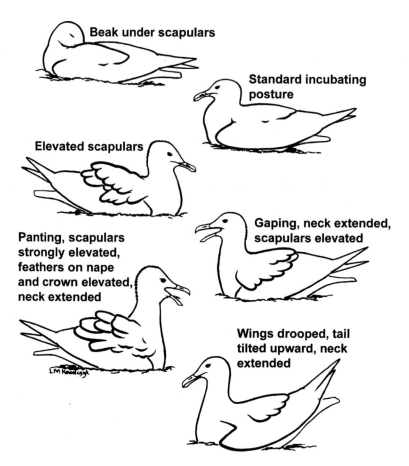

Beak under scapulars

Standard incubating posture

Elevated scapulars

Gaping, neck extended, scapulars elevated

Panting, scapulars strongly elevated, feathers on nape and crown elevated, neck extended

Wings drooped, tail tilted upward, neck extended

FIGURE 7.3 Gulls and other seabirds use various behavioral methods to adjust to temperature. By adjusting its posture and the erection of its feathers, this gull goes from keeping warm in cooler temperatures to "cooling" postures in the heat of the day. (From Bartholomew and Dawson 1979, University of Chicago. Used with permission.)

dispel heat when hot, they can fluff feathers or droop wings to increase circulation around them, pant (gular flutter) to provide evaporative cooling, seek shade, expose or shade certain body parts, or sit with their backs to the sun and hang their heads in their own shade or stand on a rock. Figure 7.3 illustrates the components of behavioral thermoregulation in Hermann's Gulls (*Larus heermanni*), shown in the order in which they appear as heat load increases (Bartholomew and Dawson 1979). Under the severest heat load, all behaviors are used together.

Young of many species seek shade during the heat of the day if it is available (even that offered by their parents' body; Howell et al. 1974, Konyukhov 1997, Schreiber et al. in preparation). Seeking shade may provide a less energetically expensive method of cooling rather than using physiological methods, an energy-conservation technique that may be important in times when food is limited. Wind level and panting rate are inversely correlated in Herring Gulls (*Larus argentatus*), implying an effective cooling function of wind (Baerends and Drent 1970). Frigatebird and booby chicks often raise their rear end toward the sun and tilt their upper body and head down low in their own shade during the heat of the day (Schreiber et al. 1996, Norton and Schreiber in press).

Storks are known to defecate on their legs to provide evaporative cooling (Hancock et al. 1992), a behavior not common in seabirds although it is documented in Cape Gannets (*Morus capensis*; Cooper and Siegfried 1976). Desert birds, such as the Gray Gull (*Larus modestus*), are far more active at night, conserving their energy during the day (Howell et al. 1974). Rather than transferring

FIGURE 7.4 Black Skimmers and other seabirds nesting in hot climates may rise up and stand over their eggs (or young chicks) during the heat of the day, shading them but not applying additional heat by incubating (or brooding) them. (Photo by E. A. Schreiber.)

heat to eggs or small chicks by direct contact, adults may stand over them in the heat of the day (Figure 7.4; Nelson 1965, Bartholomew 1966, Huggins 1941, Howell and Bartholomew 1962), though Baerends and Drent (1970) consider this an adult comfort movement that provides cooling. Frigatebirds, herons, storks, and some other species will stand, facing the sun, wings drooped and twisted so that the ventral surface faces the sun. This is most likely a thermoregulatory behavior although its function is not really understood. Some species drink water when hot (Schreiber et al. in preparation).

During cold weather birds may hide their bills in feathers, shiver, or huddle in groups, and attentiveness by adults to eggs and chicks is increased (Bartholomew and Dawson 1979, Baerends and Drent 1970). Adults may sit tighter on the nest during cold spells or rain to not expose their eggs (Baerends and Drent 1970).

7.3 WEATHER EFFECTS AT THE NEST

7.3.1 EFFECTS ON TIMING OF BREEDING

The ultimate reason for breeding at a particular time may be tied to food availability, but our ability to understand this connection is hampered by our inability to document the actual availability of food to seabirds. Little is known about changes in the abundance of fish populations through the year. In polar areas the breeding period of seabirds approaches the limit of available time to fit in a breeding cycle, and probably commences as soon as conditions allow. Commencement of nesting in polar through temperate areas varies closely with the arrival of spring weather in many species (Sladen 1958, Warham 1963, Gaston and Nettleship 1981, Williams 1995, Konyukhov 1997, Hatch and Nettleship 1998). Late winters cause delayed nesting due to snow and ice on the breeding grounds and to fish availability. In temperate habitats, while climate may allow a longer breeding season than in polar areas, adequate food may only be available during summer months. At lower latitudes, air temperature variations are small and may have less significance for the onset of breeding, but food is frequently still seasonally available which has important effects on timing of laying. Brown Pelicans (*Pelecanus occidentalis*) in Florida lay earlier in warm winters and cease laying when a late cold spell occurs, possibly because of the effects of cold weather on the fish (Schreiber 1980). On tropical Johnston Atoll (central Pacific Ocean; 16°N, 169°W), some species exhibit strict seasonality of nesting (Christmas Shearwaters, *Puffinus nativitatus;* Wedge-tailed Shearwaters, *P. pacificus;* Brown Boobies, *Sula leucogaster;* Brown Noddies, *Anous stolidus;* and

Gray-backed Terns, *Sterna lunata*), while others do not (Red-footed Boobies, *Sula sula*; Sooty Terns; and White Terns, *Gygis alba*; Schreiber 1999). Since we know little about the at-sea feeding behavior of these species, we do not understand why they differ in their flexibility of laying.

Oceanographic factors (most likely because of their affect on the food source) can alter the timing of the nesting season. At both high and low latitudes, unpredictable changes in food availability induced by environmental events cause changes in the onset of breeding and increased mortality of chicks and adults (Schreiber 1980, Duffy et al. 1984, Schreiber and Schreiber 1984, Croxall et al. 1988, Hatch and Hatch 1990, Chastel et al. 1993, Regehr and Montevecchi 1997, Finney et al. 1999). Chick mortality appears to be generally higher than adult mortality but we have few data from marked populations. When changes in local environmental conditions cause a decrease in the available food supply, adult seabirds often desert nests, going elsewhere to find food. Adults deserting young and allowing them to die of starvation during a hurricane or ENSO event permits these long-lived birds to reproduce in another year, and in years after that.

There are a few reports in the literature of subannual breeding by seabirds, laying every 8 to 10 months (Ashmole 1963, Stonehouse 1960, Snow 1965a, Harris 1969, 1970, Diamond 1976, Nelson 1977, 1978). Some of the studies which documented these short breeding cycles were conducted during ENSO events, when we now know that the breeding season can be altered by changes in food availability. While, in areas of normally rich food supply such as the Humboldt Current, subannual breeding may occur (Swallow-tailed Gull, *Creagrus furcatus*; Harris 1970), careful examination of nesting cycles over a period of years is needed to determine periodicity of breeding. The normal cycle for most seabirds probably is annual, with leeway for adjustment according to the food supply.

7.3.2 EFFECTS ON NEST SITES

Many species are probably selecting breeding sites based on weather–climate parameters such as degree of shade (Red-tailed Tropicbird, Clark et al. 1983), wind level (Adelie Penguin, *Pygoscelis adeliae*, Volkman and Trivelpiece 1981; Brown Booby, Schreiber 1999), density of grass (Sooty Tern, Schreiber et al. in preparation), or distance to open water during chick rearing owing to extensive ice pack (Emperor Penguin, *Aptenodytes forsteri*, Williams 1995). The selection of breeding sites that allows birds to save energy may become significant in times of low food availability when adults increase feeding effort in order to successfully raise young. Beyond this, effects of weather on nest sites vary, causing nests to be covered by blowing sand, flooded by tides or rain (Figure 7.5), or being destroyed by unstable substrate (Burger and Gochfeld 1990, Warham 1990, Schreiber 1999). Black Skimmers (*Rhynchops niger*) actively keep their eggs above drifting sand in windstorms (Burger and Gochfeld 1990). Ground nests in some areas get flooded in high spring tides or storms. Laughing Gulls (*Larus atricilla*) nesting in salt marshes build substantial nests, continue to add nest material throughout the breeding season, and repair damaged nests (Burger 1979). Repair and maintenance are often not enough, however, and nests may be lost to floods and storms. Burrow-nesting species (such as petrels and shearwaters) commonly suffer nest loss to rains flooding the nest and to subsequent erosion or collapse (Warham 1990). The largest Adélie Penguin colonies occur in areas where dispersal of fast sea-ice occurs early in the breeding season, allowing birds easier access to feeding areas (Stonehouse 1963, 1975).

Hurricanes can destroy habitat, making it unusable for many years, and may thus cause decreased nesting numbers in philopatric species (birds that return to the same nesting area each year). If nest loss occurs early enough in the season, many species will relay (Dorward 1963, Amerson and Shelton 1975, Gaston and Nettleship 1981, Coulson and Porter 1985, Warham 1990, Schreiber and Schreiber 1993, Casey 1994, Schreiber et al. 1996), although this is apparently least likely in the Procellariiformes. When nest or chick loss occurs late in the season, few to no birds relay, possibly due to insufficient time to complete the cycle or because of energetic constraints, or both.

FIGURE 7.5 Burrowing nesting birds like this Magellanic Penguin chick may have their burrow flooded during severe storms. (Photo by P. D. Boersma.)

7.3.3 EFFECTS ON CARE AND DEVELOPMENT OF EGGS AND YOUNG

Weather effects on breeding seabirds are confounded by factors such as adult age, adult experience, flexible time budgets of adults, and nest location (Drent and Daan 1980, Montevecchi and Porter 1980, Ainley et al. 1983, Cairns et al. 1987, Burger and Piatt 1990, Hamer et al. 1991, 1993, Croxall et al. 1992, Schreiber 1994, Falk and Moller 1997). Some data indicate that less experienced breeders are more affected by weather parameters which cause food shortages (Ainley et al. 1983), but due to a lack of known-age populations of seabirds, this has received little study. Females that lay a multiegg clutch sometimes lay fewer eggs during seasons with unusual weather patterns (Boekelheide et al. 1990). In years with unusual ice conditions in the high Arctic, Northern Fulmars and Black-legged Kittiwakes (*Rissa tridactyla*) may not lay at all (Nettleship 1987, Baird 1994). Individual egg size is thought to be genetically constrained and unable to change significantly in response to climate variability, as has been found in some studies (Monaghan et al. 1989, Schreiber and Schreiber 1993, Schreiber 1999). But egg size does change with adult age, at least in some species, possibly obscuring any potential weather or food availability effects (Coulson and White 1958).

In general, adults must protect eggs and small young (unable to thermoregulate yet) from both hot and cold temperatures (Sladen 1958, Howell et al. 1974, Ainley et al. 1983, Burger and Gochfeld 1990, Warham 1990, Schreiber and Schreiber 1993, Williams 1995, Schreiber et al. 1996), but there are few data on fatal temperatures for eggs or chicks. During very hot weather, adults may stand over eggs, shading them rather than transferring heat to them (Howell 1979). They may soak their belly with water and then sit on the egg (Howell 1979), although this might be done to cool the adult, rather than the egg (Baerends and Drent 1970), or both. In Antarctica, where it reaches −45°C and lower during the Emperor Penguin breeding season, the ability to keep eggs warm is vital to hatching success (Kooyman et al. 1971). Exceptionally high winds in Antarctica can actually blow eggs away during nest relief of penguin pairs, or blow the adults themselves away from their nest (Ainley et al. 1983).

A drop in chick mass and increased mortality in small chicks are often related to precipitation and accompanying chilling (Nye 1964, Dunn 1975, Konarzewski and Taylor 1989). Small Red-tailed Tropicbird chicks on Johnston Atoll (central Pacific Ocean) experience higher mortality during rainy days (Schreiber 1999), chilling of the chick being the probable cause. A severe rainstorm in Newfoundland caused the death of 90% of the Herring Gull chicks present (Threlfall et al. 1974). Small chicks are also susceptible to short-term weather perturbations that cause difficulties for adults catching food since they cannot survive long without food. Year-to-year changes in reproductive success and chick growth rates have been related to changes in sea-surface temperatures worldwide (particularly during ENSO events: Boersma 1978, Springer et al. 1984, Murphy et al. 1986, Ainley et al. 1988, Croxall et al. 1988, Anderson 1989, Schreiber and Schreiber 1989, Duffy 1990, Warham 1990), probably due to changes in the distribution of fish. The effects of longer-term weather patterns on seabirds, such as brought on by ENSO events, are discussed below.

The effects of various wind levels on adults' ability to feed can be determined from measuring chick growth rates, but growth rate is also affected by the chicks' energetic expenditure. Kidlaw (1999) found high wind levels resulted in reduced chick growth rates in Kittiwakes in extremely windy sites, while chicks in less windy sites grew normally. In a case such as this it is difficult to determine whether adults had difficulty feeding chicks in the windier sites or the windier sites increased chicks' energetic expenditure. Persistent pack ice cover in Antarctica (probably due to low winds, such as occurred in 1968–1969) causes desertion of nests by Adelie Penguins (Ainley et al. 1983). In normal years the sea ice disappears at about the time chicks hatch, allowing shorter foraging trips by adults. Here again, however, experience of adults plays a role that overlies the role of the environment: more experienced adults delivered more food to chicks during difficult feeding times (Ainley and Schlatter 1972).

The harsh environment of polar regions often causes the loss of eggs and chicks when nests get snowed in, ice freezes eggs to ledges, or the pack ice does not melt in time (Ainley et al. 1983, Nettleship et al. 1984, Hatch and Nettleship 1998, Warham 1990). Chicks of many species breeding in cold climates undergo long periods of fasting and still survive (Tickell 1968, Wasilewski 1986, Warham 1990, Hatch and Nettleship 1998). Red-tailed Tropicbird chicks on Christmas Island (central Pacific) grow more slowly and do not reach an asymptotic mass during ENSO events, but still fledge successfully at the normal fledging mass, taking longer to do so (Schreiber 1996). Audubon's Shearwater (*Puffinus lherminieri*) chicks in the Galapagos required from 62 days (non-ENSO years) to 100 days (ENSO years) to fledge (Harris 1969). This flexibility in chick growth rates appears to be a common adaptation in seabird chicks of all orders to survive variable weather patterns (Harris 1969, Mougin 1975, Ainley et al. 1983, Warham 1990).

Severe weather or dead calm during the period of fledging can be perilous for young birds as they first learn to fly. Fledgling albatrosses in the north Pacific often get stranded on the beach during days of low wind levels and appear to be dependent on high winds to take their first flights (Fisher and Fisher 1969). Many reports of wrecks of beached birds (mass mortality) are disproportionately young birds (Harris and Wanless 1984, Piatt and van Pelt 1997, Work and Rameyer 1999), possibly reflecting the difficulty young birds have learning to fly and feed themselves.

7.4 WEATHER EFFECTS ON FEEDING

The broad variation in seabird flight style and abilities leads to a wide variety of feeding methods, and while there are data on the effects of weather on birds in their colonies, the data for how it affects birds at sea are scant. Most often, our knowledge is derived from what we can measure on land. Weather affects the ability of seabirds to find food due to: (1) wind speed and direction and precipitation affecting flight; (2) cloud cover, precipitation, clarity of water, and turbidity of water affecting their ability to see and capture their prey; and (3) its effects on prey behavior and distribution (Dunn 1973, Taylor 1983, Sagar and Sagar 1989) .

The energetic cost of flight has not been studied in many seabirds (see Chapter 11) and studies can be difficult to carry out. When a bird is at sea, out of sight of land, it is not easy to determine how much time is spent sitting on the water vs. actively diving or swimming after food vs. flying. Various devices, such as activity recorders, have been developed to help determine amount of time spent on the water and number of dives made during a trip to sea (Cairns et al. 1987, Schreiber 1996, Arnould et al. 1996). While these devices can assist in determining the energetic cost of various activities, there is also a cost to the bird carrying the device which must be considered.

Some of the effects of weather on feeding ability can be deduced from observation and measurement at colonies and roosts (Harrington et al. 1972, Wanless and Harris 1989). More chicks may be fed during the night around the full moon, implying that a species can feed at night given sufficient illumination. More Red-footed Boobies and Great Frigatebirds roost during days of low winds (Schreiber and Chovan 1986), which suggests: (1) it is energetically more costly to fly in low winds, or (2) food is less available on low wind days (upwelling could be reduced). This illustrates the difficulty in determining the ultimate reason for some bird behaviors.

7.4.1 EFFECTS ON FINDING AND TRANSPORTING FOOD

Some meteorological and oceanographic factors have widespread, consistent properties that seabirds have undoubtedly learned to use both in getting from the nesting colony or roost to feeding grounds and in finding food. These features can have a degree of consistency that allows birds to reliably use them in most years and most likely were a selective force on the evolution of seabird distribution and lifestyles (Schreiber and Schreiber 1989, Jouventin and Weimerskirch 1990, Haney and Lee 1994). Seabirds often feed at specific oceanographic features (like oceanic fronts and eddies) because these features concentrate food near the surface (Ashmole 1971, Haney 1986, 1989, Hull et al. 1997; see Chapter 6). Weather disturbances that disrupt these oceanographic features affect seabirds' ability to find food. Thermals also play a role in efficient transportation for some seabirds, and riding up in thermals may be a way for birds to spot feeding flocks or individuals to join (Welty and Baptista 1988). Birds such as pelicans (*Pelecanus* spp.) and frigatebirds (*Fregata* spp.) take off and immediately head for thermals to assist their flight.

Obviously anything that interferes with the birds' ability to see food beneath the water will interfere with finding food, such as rough weather stirring up the surface. Heavy plankton blooms or polluted water may do the same thing. The cost of transporting food may be more dependent on wind direction and speed than any other factor, but few studies have addressed this. Several studies document increased difficulty in feeding during storms (Dobinson and Richards 1964, Ainley and LeResche 1973, Birkhead 1976, Sagar and Sagar 1989, Konyukhov 1997, Finney et al. 1999).

7.4.2 EFFECTS ON CAPTURING FOOD

Seabirds' ability to capture food can be affected by anything that interferes with their flight, their ability to dive, or their ability to sit on the surface and pick up food (Figure 7.6). Terns may have a higher catch rate per dive on days with some wind. When the wind gets too high, however, it becomes more difficult to dive accurately and fish swim deeper, making them less visible and less accessible (Elkins 1995). The ability to determine catch rates in nearshore feeding birds facilitates our understanding of seabird energetics, but in many species, we have not been able to do this since they feed well out of sight of land over the open ocean.

Auks (Alcidae) pursue prey underwater, using their wings to propel them. The cold water of high latitudes has been suggested to slow the swimming speed of fish, allowing their capture by underwater pursuit divers, and accounts for the lack of this method of feeding in the tropics. Cloud cover and wind speed are important variables in determining feeding success for seabirds. Higher wind levels are hypothesized to assist in feeding by making it harder for fish to see the birds above them (Elkins 1995). However, greater turbidity of water (accompanying higher winds) can make it more difficult for birds to see the fish, too (Gaston and Nettleship 1981). Steady winds behind cold fronts give ideal feeding conditions for Manx Shearwaters (*Puffinus puffinus*), Fulmars (*Fulmarus* spp.), and Gannets (*Morus* spp.) to change feeding grounds. Black Terns hawk for insects over land on mild, still days, but in cooler, windier weather they stay over the sea (Elkins 1995).

FIGURE 7.6 A feeding flock of Brown Pelicans and Laughing gulls in Florida. Unusual weather parameters can affect a seabird's ability to find food and to catch it. (Photo by R. W. Schreiber.)

Whether this change in feeding areas is related to the direct effect of wind levels on bird flight or its effect on fish availability is unknown.

Indicators of weather effects on birds' feeding ability can include changes in the amount of food delivered to chicks, the amount of time adults spend with chicks vs. at-sea feeding, and growth rates of chicks. These factors can be difficult to measure, and once measured may require an understanding of the behavior of the birds to interpret. Common Murres (*Uria aalge*) and Red-tailed Tropicbirds have flexible time-budgets that allow adults to spend more time feeding in years when food is less available (Burger and Piatt 1990, Monaghan et al. 1994, Schreiber and Schreiber 1993, Schreiber 1996), but the studies conducted to establish this conclusion are very different. Common Murres breed in colonies on cliff sides which are generally inaccessible. Measurement of growth rates of chicks is essentially impossible, but through observation the length of adult attendance at the nest and the number of chick feedings can be determined. Also, adults carry food to chicks in their bills allowing meal size (and often species of prey) to be estimated. In rough weather chicks receive smaller meals and adults spend less time with them, implying difficulty in catching food (Finney et al. 1999). Researchers also documented that adults make more dives per feeding bout during poor food years (Burger and Piatt 1990, Monaghan et al. 1994). For Common Murres the amount of time spent at the breeding site was a good indicator of foraging effort (Finney et al. 1999, Uttley et al. 1994).

Adult Red-tailed Tropicbirds spend only 30 s to 10 min at the nest when coming in to feed chicks, and this time is unrelated to the amount of food delivered or to feeding frequency (Schreiber 1994, 1996). Additionally, meals fed to chicks are carried in the gullet of adults and regurgitated directly into the chicks' mouths so that food delivered is not visible to a researcher. In order to measure food delivery to chicks, they must be weighed before and after meals (Figure 7.7; Schreiber 1994).

7.4.3 EFFECTS ON PREY DISTRIBUTION

Prey distribution is significantly influenced by atmospheric and oceanographic parameters (Birkhead and Nettleship 1987, Arntz and Tarazona 1990; see Chapter 6), which must be taken into account in any study of seabird ecology and breeding biology. For instance, in a season-long study

FIGURE 7.7 Weighing chicks daily, or even several times a day, is one method ornithologists use to study growth rates and energetics in seabirds. This banded Laughing Gull chick is weighed in a Pringles® can that restrains its movement while being weighed. (Photo by R. W. Schreiber.)

of Common Murres, adults fed chicks food with a lower energy value during stormy weather (although frequency of feeds did not change) indicating a change in fish behavior and availability (Finney et al. 1999). Unfortunately, it is difficult to determine prey availability or changes in it.

7.5 TYPES OF WEATHER EVENTS AND THEIR SPECIFIC EFFECTS

Climate-weather events occur on varying scales from thousands of years to a few hours and have shaped seabird life histories (see Chapter 8). While this discussion is divided into the length of time an event lasts (1 — hundreds of years; 2 — 1 to 2 years; 3 — a season; 4 — a few hours), this may not reflect the length of time an event affects a particular species or area of the world. For instance, a hurricane may pass through an area in a matter of hours, but it can permanently destroy a nesting colony causing an effect on seabirds in an area for many years to come. Mass mortality and failed breeding seasons for seabirds have been reported in the literature many times over the years (Murphy 1936, Ashmole 1963, Dorward 1963, Scott et al. 1975, Nettleship et al. 1984, Schreiber and Schreiber 1984, Lyster 1986a,b, Ainley and Boekelheide 1990). The causes are most likely related to changes in food availability brought on by changes in oceanographic–atmospheric parameters.

7.5.1 LONG-TERM EVENTS

Determining the effects of large-scale weather patterns (lasting hundreds to thousands of years) on seabirds can be difficult for several reasons. There are few good data on bird population sizes prior to about 100 years ago, and even fewer data on long-term trends, making population changes almost impossible to find and track. There are also few good historic data on sea or air temperatures. Additionally, it can be difficult to tease apart the effects of natural environmental factors from those of human-induced factors on birds. The effects of the ice ages and fluctuations in sea level on seabirds can generally only be hypothesized, except for some fossil data. Undoubtedly, changes in fish distribution occurred. The Great Bahamas Bank off southeastern United States and the Puerto Rican Bank in the Caribbean were each a continuous island until the end of the Pleistocene (about 7000 years ago), which brought higher sea level and split them into many islands. In both areas many potential seabird nesting sites were flooded and studies of fossils have determined that extinctions of avifauna occurred (Pregill and Olson 1981, Olson 1981; see Chapter 2). Gray Gulls nesting in the Atacama Desert in Chile may have begun using this extremely harsh nesting area when it had a different climate (Howell et al. 1974). Because their adaptations to nesting in the desert heat are behavioral rather than physiological, and from geologic evidence, Howell et al. (1974) suggested that their nesting sites had once been on the shores of a lake that were much more hospitable.

The potential effects of the predicted global warming can be speculated upon, and smaller warming events give us a clue to the kinds of things that can happen. In the northwest Atlantic a small warming trend in sea surface temperatures from the 1930s to 1950s brought the return of mackerel (*Scomber scombrus*), commonly eaten by Northern Gannets (*Morus bassanus*), to the area, and an increase in the number of nesting Gannets (Montevecchi and Myers 1997). Current warming trends may increase sea levels worldwide, causing the loss of many seabird nesting sites, while creating other potential nesting sites. Since warmer or colder water would affect the distribution of birds' food sources, the distribution of individual species would probably change as they follow their food. In sub-Antarctic waters, scientists are not sure whether a change in the amount of ice means more or less prey for seabirds (Croxall 1992). Changes in sea level will certainly alter habitat availability to and use by seabirds, but there has been no broad-scale analysis of potential alternate habitat (Burger 1990) and thus we cannot predict the loss of species or even changes in population size. The occurrence of major natural weather events, such as ENSO, will further hamper our ability to understand the potential effects of global warming on seabirds.

7.5.2 ONE- TO THREE-YEAR EVENTS

Weather events that last for a season or a year or two (such as ENSO events or droughts) are less likely to cause the extinction of a species than are longer-term events (Vermeij 1990). Seabirds live in an environment that exhibits tremendous variability and may cause the loss of several breeding seasons during their long lifetime, such as occurs during ENSO events (Schreiber and Schreiber 1989, Duffy 1990). They are suggested to have evolved some specific adaptations to stochastic events which help ensure the survival of the species (Schreiber and Schreiber 1989; see discussion below). Also, many seabird species are widely distributed and weather effects may not extend throughout a species' range. Even during ENSO events (see below), which have worldwide ramifications, local effects vary, causing differential mortality rates. For detailed discussions of ENSO effects on plants and animals worldwide see Glynn (1990) and references therein.

One- to three-year weather events have a variety of effects on seabird populations and differing effects on different species: (1) reduced numbers of breeding birds; (2) delayed breeding; (3) increased egg and chick mortality; (4) reduced juvenile survival; (5) increased adult mortality; or (6) changes in vegetation, owing to changes in precipitation level, that physically affect nesting (Ainley et al. 1981, Schreiber and Schreiber 1989, Duffy 1990).

Other weather effects on seabirds can be more difficult to discern. Some researchers found adult seabirds to have flexible time budgets, allowing them to adjust effort to changes in prey availability and supply chicks with a constant amount of food (Cairns et al. 1987, Burger and Piatt 1990). Determining that birds have "spare time" during normal years or that they are, in fact, working harder in a particular year is not always easy and requires multiyear studies. Chick growth rates and fledging success may also not directly reflect food supply if adults are adjusting their effort. The presence of "extra time" in normal or high food years may be what allows some birds to increase effort in poor food years and still raise young (Drent and Daan 1980, Schreiber 1994). Chicks also have flexible growth rates which allow them to survive and fledge in spite of periods of food shortage (Cruz and Cruz 1990, Schreiber 1994).

7.5.2.1 El Niño–Southern Oscillation (ENSO) Events

Biologists have long known that El Niño events affect marine birds along the coast of Ecuador and Peru, but it was not until the 1982–1983 event that they realized these ENSOs affect weather patterns worldwide and affect both marine and land birds (Duffy et al. 1984, Ainley et al. 1988, Hall et al. 1988, Monaghan et al. 1989, Schreiber and Schreiber 1989, Barrett and Rikardsen 1992, Cairns 1992, Saether 2000). Generally, effects of ENSO events on seabirds are seen first in the central Pacific where they develop and are the most severe (Schreiber and Schreiber 1984, 1989), but parallel oceanographic and atmospheric changes occur in the Atlantic and Indian Oceans (Longhurst and Pauly 1987). There are many instances of major seabird wrecks reported in the literature with the cause attributed to storms or starvation, and autopsies revealing underweight birds. Most often these occurrences can be linked to ENSO events (Bailey and Kaiser 1972, Harris and Wanless 1996, Piatt and van Pelt 1997, Work and Rameyer 1999). So, although seabirds may have the ability to fly away from storms, this apparently does not always occur, particularly if the affected area is far reaching as with ENSO events. Interestingly, as Ian Newton (1998) pointed out, for unknown reasons most wrecks of seabirds involved one species of bird. Birds deserting nests or found dying of starvation are often found to carry high parasite loads (Norman et al. 1992, Shealer 1999), which this author believes is something that occurs after birds are weakened by lack of food.

To understand the development and propagation of ENSO events, it is first necessary to understand global atmospheric circulation patterns. Remember: wind is described as where it comes from and currents as where they flow toward. Solar energy drives a wind system created by unequal heating of the earth, a pattern first described by George Hadley in 1735 (Lutgens and Tarbuck 1995). He realized that unequal heating of the earth's surface causes air movement to balance the heat load: hot, equatorial air rises (causing a low pressure system) and moves poleward where it eventually cools and sinks, flowing outward, to the south or north as it hits the ocean surface (high pressure system; Figure 7.8). It wasn't until the 1920s that we came to understand the complexity of the global circulation pattern and the fact that there are actually three circulation cells of air (the ones nearest the equator being named for Hadley). Also, the spinning of the earth deflects wind so that flow is more east–west than north–south. For instance, the trade winds across the central Pacific are from the northeast and southeast (Figure 7.9). Warm air still moves poleward, redistributing the equatorial heat, but not as directly as Hadley proposed. Fronts are created where moving air masses of differing temperatures and directions come together, as occurs at about 60°N in the low pressure system of the Polar Front Zone (Figure 7.9).

Where atmospheric wind systems come into contact with the ocean they drive surface currents through friction (Figure 7.9). The normal trade winds in the tropical Pacific (and in the Atlantic) do two things: (1) they keep a body of warm surface water pushed against the western Pacific where sea level is higher than in the eastern Pacific, and (2) they strip away the warm surface water along the coast in the eastern Pacific allowing the upwelling of the cool, nutrient-rich waters which lie beneath (Figure 7.10). These nutrient-rich upwellings allow increased phytoplankton production which feeds the beginning of a food chain that supports huge populations of fish and seabirds. The area of upwelling along the west coast of South America is called the Humboldt Current and it normally supports millions of marine birds (see Chapter 6).

El Niño events were originally described from the east coasts of Peru and Ecuador. The name, meaning "the child" in Spanish, refers to the normally mild seasonal warming of the ocean along the coast (suppression of the thermocline and rich-upwelled cold waters as described above) that

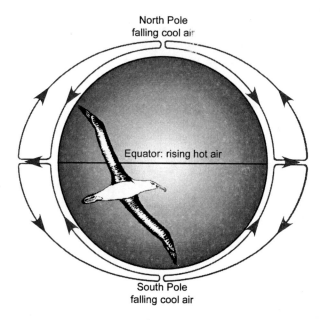

FIGURE 7.8 The global circulation pattern as described by George Hadley in 1735. (Adapted from *Ocean Circulation*, Open University.)

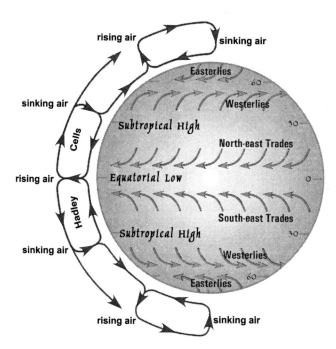

FIGURE 7.9 The actual three-cell global circulation pattern, showing the main wind and pressure systems of the globe. Areas of rising air are low pressure and areas of falling air are high pressure. (Adapted from *Ocean Circulation*, Open University.)

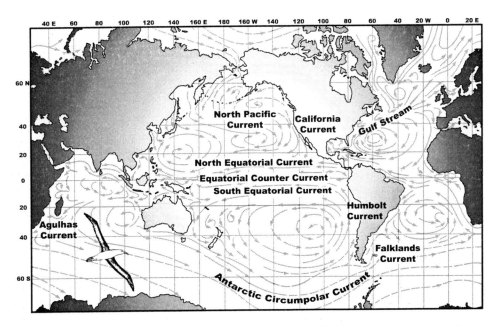

FIGURE 7.10 The main wind-driven current patterns of the globe. Seabirds do not feed randomly over the ocean but feed at oceanographic features that bring increased nutrients to an area and thus have high food availability. (Adapted from *Ocean Circulation*, Open University.)

occurs, to some degree, each winter as winds decrease. The name stemmed from the fact that the events usually begin around Christmas time. The warm water causes the disappearance of the abundant fish and the collapse of the food chain to varying degrees. Every few years the winter warming is more severe, with drastic repercussions in the food chain, which collapses. These events are what is called an El Niño–Southern Oscillation event, or ENSO event.

As an ENSO event develops, the normal South Pacific high pressure system decreases in strength and the Indonesian low pressure system increases (Figure 7.11a; Cane and Zebiak 1985). Changes in these two pressure systems are referred to as the Southern Oscillation and the difference between them (high pressure value minus low) is the Southern Oscillation Index (SOI). A negative SOI is one of the signs of an ENSO event. As the SOI becomes negative, the trade winds relax and actually reverse, allowing the warm western Pacific pool of nutrient-depleted water to flow toward the eastern Pacific, depressing the thermocline as it progresses and eventually halting the eastern ocean upwelling (Figure 7.11b). Sea-surface temperature and sea level rise as the water moves east (Wyrtki 1975, Cane 1983, Rasmusson and Wallace 1983), causing extensive fish mortality and changes in fish distributions. As the warm water hits the coastline of the Americas, it spreads north and south, and the once-rich feeding grounds for seabirds in the Humboldt and California Currents disappear as they are overlaid with warm, nutrient-poor water. The connection between El Niño events and the Southern Oscillation was first recognized by Jacob Bjerknes (1966) during the severe 1957–1958 event. ENSO has a similar development and effect in the Atlantic Ocean where massive mortality is experienced by seabirds that use the Benguela current area as a feeding ground (Crawford and Shelton 1978).

Eventually worldwide wind and pressure regimes (Figure 7.9) are upset, altering weather patterns on a global scale. For instance, warming of the North Pacific sea surface during ENSO causes alterations in weather patterns over northern Canada and the United States, most frequently bringing drought to the prairies (Bonsal et al. 1993). Ethiopia and northern India also experience droughts (Bletrando and Camberlin 1993). In southern India, winter monsoon rains are extreme,

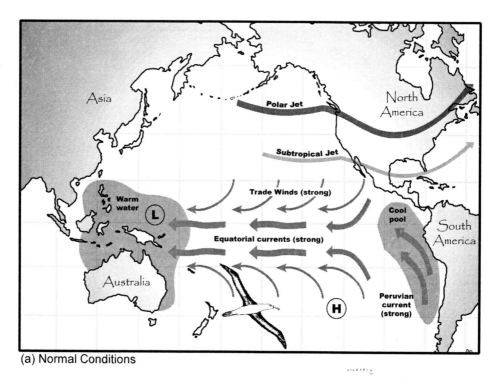

(a) Normal Conditions

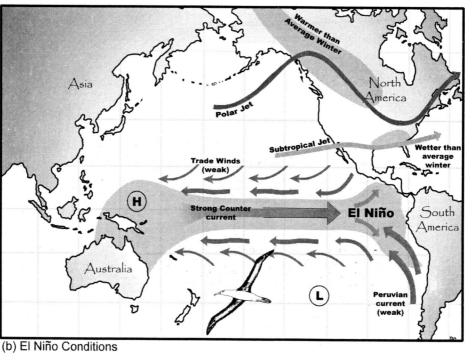

(b) El Niño Conditions

FIGURE 7.11 (a) Normal Pacific Ocean pressure and wind patterns. (b) El Niño (or ENSO) conditions in the Pacific Ocean: as the trade winds die down, the body of warm water they kept pushed toward the western Pacific moves back across the ocean, suppressing the thermocline and nutrient-rich upwellings of the Humboldt and California Current. (From F.K. Lutgens and E.J. Tarbuck 1995, The Atmosphere, Prentice-Hall, Englewood Cliffs, NJ. Used with permission.)

TABLE 7.1
Years of El Niño–Southern Oscillation Events (ENSO) and Strength of Event

Year	Strength	Year	Strength	Year	Strength
1726	Moderate	1857	Weak	1923	Weak
1728	Strong	1862	Weak	1925–1926	Strong
1763	Strong	1864	Weak	1929–1930	Moderate
1770	Strong	1866	Weak	1932	Weak
1791	Strong	1871	Strong	1939–1941	Strong
1803–1804	Strong	1873	Weak	1943–1944	Weak
1814	Strong	1877–1878	Strong	1950–1951	Weak
1817	Moderate	1880	Weak	1953	Moderate
1819	Moderate	1884–1885	Strong	1957–1958	Strong
1821	Moderate	1887–1888	Moderate	1963	Weak
1824	Moderate	1891	Strong	1965	Moderate
1828–1829	Strong	1896	Moderate	1969	Weak
1832	Moderate	1899–1900	Strong	1972–1973	Strong
1837	Moderate	1902	Moderate	1976–1977	Moderate
1844–1846	Strong	1905	Moderate	1982–1983	Strong
1850	Weak	1911–1913	Strong	1986–1987	Moderate
1852	Weak	1914	Moderate	1990–1994	Moderate
1854–1855	Weak	1917–1919	Strong	1997–1998	Strong

Note: Climate Analysis Center 1982–2000; Quinn et al. 1978, Schweigger 1961, Wyrtki et al. 1976.

flooding many regions (Ropelewski and Halpert 1987). Northwestern Europe experiences colder than normal winters during ENSO (Fraedrich and Muller 1992).

These events occur periodically every 2 to 7 years (Table 7.1). Each event varies in strength and timing of onset, and thus the degree to which it affects birds in the Pacific and areas outside the Pacific basin. The teleconnections between wind and pressure regimes, and between the atmosphere and ocean around the world are affected by many factors which interact to make each event somewhat different (Cane and Zebiak 1985, Rasmusson 1985, Hamilton 1988). As wind and pressure systems pass over land masses and mountain ranges, weather patterns are altered; thus while we know that ENSOs affect weather patterns around the world, each event differs as it propagates (Wyrtki 1975, Thompson 1981, Glynn 1990), and effects on sea and land birds differ. ENSO effects on birds often diminish in severity with distance from the tropical Pacific (Prince and Ricketts 1981, Ainley et al. 1988, Schreiber and Schreiber 1989, Duffy 1990, Barrett and Rikardsen 1992). In the most severe cases, adults die when their food source disappears and they cannot find food elsewhere in time (Schreiber and Schreiber 1984, Duffy 1990). Lesser effects include (1) desertion of nests by adults, (2) death of young from starvation when adults cannot find enough food, (3) young fledge underweight and suffer increased mortality during the period when they are learning to feed themselves, (4) delayed breeding season, (5) reduced numbers of adults attempting to nest, and (6) changes in vegetation caused by rain or lack of rain that make the habitat unsuitable for nesting (Ainley et al. 1988, Schreiber and Schreiber 1989, Duffy 1990).

Our lack of knowledge about the behavior of marine birds at sea and about the energetic cost of feeding has hampered our ability to fully understand what is causing the effects we see on land. Seabirds generally feed at oceanographic features where cool water upwellings concentrate nutrients, enriching a local ocean area and providing good feeding grounds (see Chapter 6; Nettleship 1996,

Hull et al. 1997). The changes in wind patterns that occur during ENSO cause the disappearance of these features, and birds, perhaps, must randomly search the ocean for new feeding areas.

In the following sections, the effects of ENSO on seabirds around the world are discussed. This is not a comprehensive summary but is intended to give an overview of the global extent of ENSO effects and the types of responses exhibited by seabirds.

7.5.2.1.1 ENSOs on Christmas Island, Pacific Ocean

One of the first indications that an ENSO event is beginning is seen in the response of seabirds breeding on Christmas Island (2°N, 157°W), Pacific Ocean. From 1979 through 1991 ornithologists studied the breeding biology and ecology of the 18 breeding species of seabirds on Christmas Island: 5 petrels and shearwaters, 1 tropicbird, 3 boobies, 2 frigatebirds, and 7 terns. Two to three months prior to each of the three ENSOs that occurred during that time (1982–1983, 1986–1987, and 1990–1994; Table 7.1), they recorded reduced numbers of breeding birds, increased chick mortality, and lowered growth rates of chicks indicating a shortage of food. The types of food brought to chicks generally change during ENSO, with a greater variety of species appearing in meals. No changes in egg sizes have been recorded during ENSO events or in the years after an event (E. A. Schreiber, unpublished).

Responses to a coming ENSO differ among species for reasons that are not completely understood, but they may be related to their method of catching food or differences in their energy budgets. Immediately prior to each of these three ENSO events, fewer than normal numbers of Masked and Red-footed Boobies attempted to breed and many that did later deserted nests during incubation or the early chick stage. Great Frigatebirds, Red-tailed Tropicbirds, and Sooty Terns courted and laid eggs in normal or slightly decreased numbers, but breeding failure was much higher (Schreiber and Schreiber 1989). Responses were most dramatic with the highest mortality of adults and young during the 1982–1983 ENSO (Schreiber and Schreiber 1989). No young fledged during 1982 as they were left to starve to death in their nests by deserting adults (Figure 7.12).

While some young fledge during most ENSO events, catching food may be more difficult and survival of these birds may be lower. Years of recaptures of known-age cohorts of young, through ENSO and non-ENSO years, are needed to examine differences in postfledging survival. These data are not available for Christmas Island where local human populations consume the birds in

FIGURE 7.12 During severe ENSO events on Christmas Island (Pacific Ocean), adults leave young to starve to death while they go to sea in search of food, most surviving to breed in another year. This Black Noddy chick died during the 1982–1983 ENSO. (Photo by R. W. and E. A. Schreiber.)

unknown numbers (Schreiber and Schreiber 1993, Schreiber et al. 1996). On Johnston Atoll (16°N, 169°W), where ENSO effects are less dramatic, extensive band recoveries of Red-tailed Tropicbirds do not indicate lower survival of young fledged in ENSO years, but adult survival is decreased ($X^2 = 9.90$, df = 1, $p < 0.001$; P. Doherty and E. A. Schreiber, unpublished).

7.5.2.1.2 ENSO in Peru and Ecuador

The effects of ENSO events are perhaps best documented in this region due to historic records from over 200 years of seabird guano harvesting for fertilizer and commercial fishing, both of which are affected by the events (Murphy 1925, 1936, Duffy et al. 1988). Robert Cushman Murphy's visits to the area in the early 1900s (Murphy 1925, 1936) have provided an early record of the effect of ENSO on seabirds as rainfall increases 5 to 10 times normal and sea surface temperatures rise 5 to 10°C above normal. Thousands of seabirds desert their nests, many dying, and fish die or go elsewhere, to cooler waters (Murphy 1925, Harris 1973, Robinson and del Pino 1985, Ainley et al. 1988, Duffy et al. 1988).

As the strength of events varies, so do their effects on birds and fish. Data collected during the 1982–1983 event indicate that the population of 9 million Peruvian Boobies (*Sula variegata*), Brown Pelicans, and Guanay Cormorants (*Leucocarbo bougainvillii*) dropped to one million (most thought to have died; Duffy et al. 1988). An estimated 65% of the Humboldt Penguin (*Spheniscus humboldti*) population died (Hayes 1986). Waved Albatrosses (*Phoebastria irrorata*), Blue-footed Boobies (*Sula nebouxii*), and Swallow-tailed Gulls left the area by the thousands and did not begin reproducing successfully again until 1984 (Rechten 1985). Growth of Dark-rumped Petrels was significantly reduced and fledging took a longer period of time (Cruz and Cruz 1990).

7.5.2.1.3 ENSO in Other Areas of the Pacific

Hawaii: ENSO effects on seabirds are not as severe in Hawaii as they are in the central Pacific, possibly because water temperatures and thus fish distributions around Hawaii change less during ENSO than they do in the central Pacific. Most seabird species experience reduced breeding success in the more southern islands (Schreiber and Schreiber 1989). There are no good data for the northern islands to determine what the effects are, but they appear to be less severe (E. A. Schreiber unpublished).

Gulf of California: Breeding and wintering seabirds in the Gulf of California are strongly affected by ENSO events. During the 1997–1998 event fewer than 5% of normal numbers of Brown Pelicans, Brandt's Cormorants (*Compsohalieus penicillatus*), Double-crested Cormorants (*Hypoleucos auritus*), Yellow-footed Gulls (*Larus livens*), and Elegant Terns (*Sterna elegans*) attempted to breed (Anderson et al. unpublished). Breeding effort was reduced in most other seabird species, also. Nonbreeding species, such as Eared Grebes (*Podiceps nigricollis*), could not find food and experienced high mortality.

Bering Sea and Alaska: In the summer of 1997, as the 1997–1998 ENSO was in its early stages, warm, nutrient-depleted water invaded the area extending to a depth of 60 m. Thousands of Short-tailed Shearwaters (*Puffinus tenuirostris*) died, presumably of starvation (Baduini et al. 1999). Half the breeding Ancient Murrelets (*Synthliboramphus antiquum*) deserted their eggs and breeding success of those that bred fell by 43% (Smith et al. 1999).

Central California: The Farallon Islands are one of the best-studied seabird communities in the world. From 1971 through 1983, biologists from the Point Reyes Bird Observatory monitored the approximately 300,000 individuals of 11 species nesting there (Ainley and Lewis 1974, Ainley and Boekelheide 1990). Nearshore feeders, such as Pigeon Guillemots (*Cepphus columba*) and Pelagic Cormorants (*Strictocarbo pelagicus*), were severely affected by ENSO events, abandoning nesting attempts. Other species that feed farther offshore were less affected. After the 1982–1983 event, the lack of returning Cassin's Auklets (*Ptychoramphus aleuticus*) and Western Gulls (*Larus occidentalis*; Figure 7.13) that had previously bred on the Farallons led biologists to believe that many older breeders were killed by the event.

FIGURE 7.13 Western Gulls experience high nest failures during ENSO events on the Farallon Islands, California, and severe events also cause adult mortality. (Photo by R. W. and E. A. Schreiber.)

Western Pacific: We do not have data for what occurs to seabirds in the western Pacific during ENSOs. Water temperatures become colder than normal during ENSOs and productivity may even increase (NOAA 1984–2000).

7.5.2.1.4   ENSO in Other Areas of the World

Antarctica: Snow Petrels show high annual variability in reproductive success and numbers of breeding birds in relation to some ENSO events (Chastel et al. 1993). Antarctic Petrels (*Thalassoica antarctica*) experience similar population fluctuations to Snow Petrels (Jouventin and Weimerskirch 1991), while Adélie and Chinstrap (*Pygoscelis antarctica*) Penguins show less variability (Trivelpiece et al. 1990, Jouventin and Weimerskirch 1991). Newton (1998) and Chastel et al. (1993) suggested that the differences relate to demographic strategies, with the longer-lived petrels more likely to refrain from breeding in poor food years. The 1976–1977 ENSO caused reduced breeding numbers of Emperor Penguins, Adélie Penguins, and Antarctic Petrels (Jouventin and Weimerskirch 1991). Not all events appear to affect the birds, and during those of 1965, 1968, 1972, and 1987, little effect was detected (Chastel et al. 1993).

Atlantic Ocean: The position of the Gulf Stream shifts northward during ENSO, although the timing of the event lags that in the Pacific. This alters zooplankton abundance and changes the location of good feeding areas for seabirds. Seabird die-offs along the coast of South Africa have been shown to coincide with ENSO events (La Cock 1986, Duffy et al. 1984). Cape Gannets appear to be one of the least-affected species, perhaps because they feed offshore and have a large foraging range (La Cock 1986). Cape Cormorants (*Leucocarbo capensis*), a near-shore feeder, are very severely affected with widespread mortality of eggs and young (Duffy et al. 1984), but the hypotheses proposed for the different responses need investigation. Mortality is not as severe as that occurring in the central Pacific and birds show effects such as shifts in diet and a slight increase in mortality of fledglings (Duffy et al. 1984).

On South Georgia Island, off southeastern South America, most seabird species experience reduced reproductive performance related to ENSO, although the timing of the effect is delayed over that seen in the tropical Pacific (Croxall et al. 1988). Breeding success of the Black-browed Albatross (*Thalassarche melanophris*) is greatly reduced (Prince 1985). Lyster (1986a, b) reported extensive mortality of adult Rockhopper Penguins (*Eudyptes chrysocome*, over 3000 adults) and Gentoo Penguins (*Pygoscelis papua*, over 300 dead adults) around the Falkland Islands during an

FIGURE 7.14 Sooty Terns, a pan-tropical breeding seabird, suffer from the effects of ENSO events throughout their range, from experiencing massive nesting failures on islands in the Pacific to reduced breeding success in the Caribbean. (Photo by R. W. and E. A. Schreiber.)

ENSO. Chick mortality during severe weather conditions is not unusual, but adult mortality is. In eastern Argentina, Imperial Shags (*Notocarbo atriceps*) ceased breeding in 1982–1983 (ENSO 1982–1983), and Rock Cormorants (*Stricticarbo magellanicus*) laid few eggs. Both experienced poor breeding seasons in 1984–1985 and 1985–1986 (ENSO 1982–1983), illustrating the delayed development of ENSO in the Atlantic (Duffy et al. 1988).

During an ENSO, the northwestern Atlantic experiences colder than normal sea surface temperatures that influence the timing, movements, and availability of pelagic fish and squid to seabirds. Herring, associated with cooler waters, become an important prey item for Northern Gannets as the availability of mackerel decreases during some ENSOs (Montevecchi and Myers 1996). Reproductive success of Black-legged Kittiwakes decreases, probably due to changes in the food supply (Casey 1994, Regeher 1995).

Caribbean Sea: ENSO events change rainfall patterns in the Caribbean and during the 1982–1983 event the basin experienced severe drought conditions and enhanced hurricane activity (Gray 1984). Effects on seabirds are somewhat milder than those seen elsewhere, with decreased nesting populations (Sooty Terns; Figure 7.14) and slightly reduced nesting success (Roseate Terns, *Sterna dougallii*, and Sooty Terns; R. Norton unpublished).

Indian Ocean: We know of no data from the Indian Ocean on what happens to seabird populations during ENSO.

7.5.2.1.5 Mass Mortalities and Vagrancy

Many unusual mortality events occur during ENSOs and reports of vagrant sightings of seabirds outside their normal range are common. Massive seabird mortality in the Benguela and Humboldt Current areas during ENSO is well documented (Duffy 1990, Duffy 1988 et al., Schreiber and Schreiber 1989). The deaths of thousands of starving Common Murres in the Gulf of Alaska in 1993 was associated with an ENSO (Piatt and van Pelt 1997). In 1983, 40,000 auks washed ashore in the North Sea during the 1982–1983 ENSO (Harris and Wanless 1984). Researchers working with seabirds need to be aware of the occurrence of ENSO events (Table 7.1) when examining causal relationships in their data.

7.5.2.1.6 Land Birds

Passerine birds are affected by ENSO events both during the breeding season and during winter migrations (Hall et al. 1988, Miskelly 1990, Lindsey et al. 1997, Jasick and Lazo 1999, Saether 2000, Sillett et al. 2000). Changes in rainfall, either droughts or excessive rain, alter the amount of food available (seeds, insects) for birds. Increases in food brought on by wetter than normal conditions allow high reproductive success, and an increase in bird species diversity and density (Gibbs and Grant 1987, Jasick and Lazo 1999). Effects are also seen in timing of the onset of breeding, timing of migration, and survivorship of young and adults during the nonbreeding season. On Christmas Island (Pacific Ocean), the one land bird, the Christmas Island Warbler (*Acrocephalus aequinoctialis*), has higher than normal breeding success during ENSO due to increased production of insects, and their sources of food during heavy rains (R. W. and E. A. Schreiber unpublished). ENSO events reduce productivity in Black-throated Blue Warblers (*Dendroica caerulescens*) at their breeding grounds in New Hampshire and decrease survival at their wintering grounds in Jamaica (Saether 2000).

7.5.2.2 La Niña Events

La Niña, also referred to as a cold-water event, is a more recently described phenomenon which involves colder than normal water across the central and eastern Pacific and Atlantic (Center for Ocean Atmosphere Prediction Studies: www.coaps.fsu.edu, Philander 1990). Their effects are often the opposite of those that occur during an El Niño. La Niña generally follows closely behind an ENSO and can have broad-reaching ramifications in all parts of the world. Winters in the south-eastern United States are warmer, whereas from the Great Lakes to the Pacific northwest they are colder. There are not many data published on the effects of these events on seabirds, but any major change in water temperature can be expected to affect fish distributions and thus seabird breeding parameters. Seabird chicks of several species breeding on Johnston Atoll (16°N, 169°W) experience reduced growth rates during La Niñas, which indicates that food may not be as available (Schreiber 1999). Both Arctic (*Sterna paradisaea*) and Common Terns failed to fledge any young on Mousa Island off Scotland during the 1988 La Niña (Uttley et al. 1989). Adult Cory's Shearwaters (*Calonectris diomedea*) experience reduced survival during La Niñas while they are wintering off the coast of South Africa (Brichetti et al. 2000).

7.5.2.3 ENSO Have Shaped Our Thinking on Seabird Demography

Prior to the 1982–1983 ENSO event, the only seabirds thought to be affected by ENSO events were those in the Galapagos Islands and along the coast of Peru and Ecuador. Ornithologists now recognize that ENSO events affect birds around the world. Schreiber and Schreiber (1989) noted that, strangely enough, most early seabird studies (1900–1965) were conducted during ENSO events (Table 7.1). Each of Robert Cushman Murphy's trips to the equatorial Pacific was made during an ENSO when he witnessed massive starvation of seabird young and adults (Murphy 1936). It must be noted that Murphy was fully aware of the occurrence of El Niño and the fact that a large number of nest failures occurred during the events. The British Centenary Expedition to Ascension Island (1957–1958; ornithologists: N. P. Ashmole, J. M. Cullen, D. F. Dorward, and B. Stonehouse) occurred during a strong ENSO event and researchers documented extensive breeding failures in the seabirds (see Ibis 103b, 1962). K. E. L. Simmons also visited Ascension Island during an ENSO soon thereafter (Simmons 1967, 1968), and D. W. Snow and J. B. Nelson both conducted research on the Galapagos Islands during the 1963–1965 ENSO (Snow 1965a,b, Nelson 1968, 1969).

The fact that these and other ornithologists witnessed extensive bird mortality and nesting failures undoubtedly helped shape the current literature on seabirds and the original hypotheses that seabirds are strictly energy limited (Ashmole 1963, Lack 1968). Recent studies of energetics, food delivery to chicks, and general breeding biology are leading us to believe that (1) seabirds are more adaptable to climate conditions than we had thought, and (2) there may be a combination of factors which account for their life-history characteristics (Shea and Ricklefs 1985, Taylor and Konarzewski 1989, Konarzewski et al. 1993, Schreiber 1994, 1996, Hamer and Hill 1993, 1997, Houston et al. 1996, Hamer et al. 2000; see Chapter 1).

7.5.2.4 ENSO and Weather Websites

The following Websites provide information on weather and the occurrence of ENSO or other unusual weather events. When unusual breeding conditions are noted in any field study, weather factors should be investigated and detailed information is normally available from the following Websites:

National Oceanic and Atmospheric Administration: www.elnino.noaa.gov
Jet Propulsion Laboratory: topex-www.jpl.nasa.gov
International Research Institute for Climate Prediction: www.iri.ldeo.columbia.edu
NOAA Office of Global Programs: www.opg.noaa.gov
Scripps Institute of Oceanography: meteora.ucsd.edu
Center for Ocean-Atmospheric Prediction Studies: www.coaps.fsu.edu
National Center for Environmental Prediction: www.ncep.noaa.gov
Earth Space Research Group: www.creso.ucsb.edu
World Climate Research Program, Climate Variability and Predictability: www.clivar.org

7.5.3 SEASONAL WEATHER PATTERNS

All marine birds are affected by seasonal changes in weather and ocean parameters because these changes physically affect their feeding ability and affect food availability. For instance, seabirds may have timed their nesting season to avoid the hurricane season in Florida and the Caribbean (Schreiber 1980, Lee 1996). Seasonal weather patterns (such as late cold fronts) can affect the timing of laying (Schreiber 1980), though the reasons for this are unclear and may be related to food availability. Since we know little about daily fluctuations in food availability, it is often necessary to measure other parameters as indicators of food availability. Many researchers have used timing of commencement of nesting, mean mass of adults, and growth rates of chicks as indicators of food availability (Birkhead 1976, Nettleship 1977, Dunnet et al. 1979, Schreiber 1980, 1996, Gaston and Nettleship 1981, Hamer et al. 1991, Barrett and Rikardson 1992, Anderson et al. unpublished).

7.5.3.1 Seasonal Oceanographic Changes

While seasonal changes are well known and significant at high latitudes, they also occur in the tropical oceans. Equatorial circulation in tropical oceans exhibits dramatic seasonal changes and these undoubtedly influence nesting seasons of birds in their proximity (see Chapter 6). Many species nesting in the tropics exhibit definite annual nesting cycles, while others appear less constrained and may lay eggs over many months of the year. The patterns are not necessarily consistent throughout the tropics, so that while Brown Boobies on Johnston Atoll (Pacific Ocean) are strictly seasonal nesters (Norton and Schreiber in preparation), in the Caribbean they nest in all months of the year (British Virgin Islands: E. A. Schreiber unpublished). The ultimate reason for the timing of breeding is probably related to food availability.

The wind systems of the Indian Ocean change seasonally as a result of differential heating of the Asian land mass and the ocean. As a result of the change in winds, the Somali Current flows southwest during the northeast monsoon, and becomes a major western boundary current during the southwest monsoon bringing intense upwelling along the Somali coast (Bearman 1989). The periodicity of the California Current upwelling undoubtedly helps set the timing of the summer breeding season of marine birds in the area because it affects fish availability (Ainley and Boekelheide 1990). In Alaska, upwelling off the continental shelf near Kodiak Island fluctuates (Ingraham et al. 1976), causing between-season variability in timing of laying for Fork-tailed Storm Petrels (*Oceanodroma furcata*; Boersma et al. 1980). See Chapter 6 for a full discussion of seasonal changes in oceanography and food availability.

7.5.3.2 Winter

At high latitudes, winters bring on severe weather conditions as well as low temperatures, and both affect seabirds. To some extent mid- and high-latitude marine birds are adapted to the commonly occurring high winds, low temperatures, wind chill, and precipitation. For instance, penguins have a thick coat of dense feathers and fat deposits under the skin, but even these birds can suffer increased mortality in severe winters. Late snow-melt (which may not be the ultimate cause) delays nesting in many high-latitude seabirds (Konyukhov 1997).

Late break-up of the ice pack in Antarctica causes increased egg and chick mortality in Adélie Penguins (Ainley and LeResche 1973). Colder than normal temperatures can affect weight and fat reserves in birds and, historically, several colder-than-normal winters have caused increased mortality in birds (Armitage 1964, Ash 1964, Beer and Boyd 1964, Dobinson and Richards 1964). The offshore dispersal of gulls during the winter may be caused by seasonal changes in ocean wind fields and temperature inversions that facilitate efficient flight, and not by changes in food distribution or availability (Haney and Lee 1994). While the lower latitudes are generally not considered to have winter conditions, cold weather is known to inhibit breeding (Schreiber 1980).

7.5.3.3 Migration

After the breeding season is over, many seabird species disperse from their colonies. Migration, or dispersal, may range from long-distance travelers that circumnavigate the globe, to never leaving the breeding colony vicinity. In the higher latitudes it may be necessary to go to lower latitudes or to the opposite hemisphere to find food as does the Arctic Tern (Burger and Gochfeld 1996). Some species remain in the colder latitudes throughout the year and winter at sea, taking advantage of oceanographic or physical features which make food locally abundant (Nettleship 1996). For tropical and subtropical nesting seabirds, dispersal is not driven by temperature but perhaps by greater availability of resources elsewhere after months of thousands of birds feeding on the resources around a colony. Some coastal breeding seabirds in tropical and temperate regions disperse in multiple directions (up and down the coast) after breeding (Nelson 1978, Schreiber and Mock 1988). Island breeding species may stay at sea near rich feeding areas.

Deferred maturity (not breeding until two to several years of age) allows juvenile seabirds to exploit the widely scattered resources of the open ocean since they may not return to the breeding colony for several years. While little is known about where many seabirds go during the nonbreeding season, there are a few studies showing that juvenile birds go to different areas of the ocean than adults. Band recovery data from the thousands of birds banded by the Pacific Ocean Biological Survey Project of the Smithsonian Institution during the 1960s showed that juveniles were recovered thousands of kilometers from their natal colonies, while adults tended to be recovered closer to home. Greater migration distances of juvenile seabirds may allow them to exploit oceanic areas where food might be more available than it is nearer the breeding colony (Coulson 1966). It may also be the means by which species survive during ENSO or other catastrophic events. During stochastic events, when limited food causes high seabird mortality in an area, birds in other areas may survive. If adults and juveniles are in different oceanic areas, one group may survive to breed in the future (Schreiber and Schreiber 1989).

7.5.4 Short-Term Weather Effects

Passing storms can cause flooding and loss of nests, and the chilling and death of eggs and small chicks (Figure 7.15). Coastal breeding gulls and terns are particularly susceptible to loss of nests or young to flooding during storms (Burger 1979). Newly hatched chicks, unable to thermoregulate, are more susceptible to mortality during storms than are older chicks (Schreiber and Schreiber 1993, Burger and Gochfeld 1990). Storms can also inhibit adult feeding, thus causing mortality in

FIGURE 7.15 Short-term weather events, such as a rainstorm, can kill young unbrooded chicks which do not yet have the ability to thermoregulate well. On Midway Island, some small Laysan Albatross chicks often die in a spring storm. (Photo by R. W. and E. A. Schreiber.)

small chicks that cannot survive for a couple days without food. The behavior of prey during storms may alter their availability to birds. Common Murres breeding on the Isle of May, off Scotland, caught smaller sandeels (*Ammodytes marinus*) during stormy days, possibly because smaller, younger individuals exhibit less vertical migration than do older, larger ones (Finney et al. 1999).

In the Caribbean, terns generally have completed breeding by the hurricane season, thus avoiding nest losses to storms. Pelecaniform species, on the other hand, begin breeding during the hurricane season and lose a season to these storms when one strikes their colony. Their extended nesting season cannot be completed during the period between hurricane seasons. But these long-lived adults normally move out ahead of a hurricane, abandoning their nests and surviving to breed again next year. Around their breeding or wintering areas, birds know where to find food, but when forced to leave by unusual weather patterns or storms, they may have to search to find food. If they don't find it within a few days, they die of starvation. Documenting the extent of storm-caused mortality away from a nesting colony can be difficult. A few studies in colonies of marked birds documented lower recapture rates of birds in the year after a storm (Schreiber and Schreiber 1984).

While the immediate effects of a storm on birds may pass within hours or a few days, changes caused to the habitat by the storm may last for years. If the storm is severe enough to destroy nesting habitat, birds may not be able to use the area again, at least until the vegetation recovers. Hurricanes in the Caribbean often destroy mangrove trees used by nesting Magnificent Frigatebirds (*Fregata magnificens*) and other tree-nesting species (Schreiber 1997, Diamond and Schreiber in press). Since they are philopatric, fewer birds may be able to nest in subsequent years with less habitat available. Storms destroy not only nesting habitat for seabirds, but also can destroy grass beds and reefs which are, directly or indirectly, an important food source for seabirds.

7.6 CONCLUSIONS

Through long-term studies, ornithologists have come to appreciate environmental variability as a fact of life for seabirds and that, perhaps, no years are "normal" (Schreiber and Schreiber 1989, Chastel et al. 1993). Seabirds appear to have evolved to exist in a stochastic system where weather patterns can be severe and unpredictable. Additionally, population levels may be in a constant state of flux brought on by stochastic weather patterns, as suggested by Schreiber and Schreiber (1989)

FIGURE 7.16 Long-term studies of individually marked birds are vital to further our understanding of the ecology and demographics of seabirds. This Masked Booby on Johnston Atoll (Pacific) is getting a worn band replaced so that ornithologists can continue to follow its yearly activities. (Photo by E. A. Schreiber.)

and as found by long-term studies of some seabird colonies (Ainley and Boekelheide 1990, Chastel et al. 1993, Schreiber 1999).

Seabirds have reserves in their time–energy budgets that allow them to increase effort at times when foraging or flying may be more difficult. Studies are needed to relate specific weather parameters to seabird behavior, breeding biology, feeding ecology, demography, or even evolution. With the sophisticated weather data available today, this can be a fruitful line of investigation. For instance, gull distribution at sea was found to be strongly correlated with the occurrence of temperature inversions in the western Atlantic (Haney and Lee 1994). Undoubtedly, an animal dependent on flight to survive and the ocean to feed has evolved to use the meteorological regimes that govern its flight and oceanographic features that govern the distribution of its food source. The evolution of life-history traits in seabirds, as well as their geographic distributions, must have been strongly influenced by weather patterns on all scales.

The variability seen among species in their responses to weather bears further investigation, particularly as these responses affect the evolution of species. More long-term studies of individually banded birds are badly needed (Figure 7.16). And studies of seabird energetics across phylogenetic and geographic lines are needed in order to fully understand the occurrence of periodic breeding failures and the selective pressures they exert on populations. There are still many unanswered questions regarding seabird breeding biology and ecology for the interested student to pursue.

ACKNOWLEDGMENTS

I am deeply indebted to my husbands, Ralph W. Schreiber and Gary A. Schenk, for their support, for endless discussions on seabird biology, for untold hours of labor in the field, and for many other reasons. I thank Klaus Wyrtki for his help and insight into the world of oceanography. J. Burger, D. N. Nettleship, and G. A. Schenk (who missed several weekends of shooting clay pigeons to work on this book) provided insightful reviews of drafts of this manuscript.

LITERATURE CITED

AINLEY, D. G. 1990. Farallon seabirds, patterns at a community level. Pp. 349–380 *in* Seabirds of the Farallon Islands (D. C. Ainley and R. J. Boekelheide, Eds.). Stanford University Press, Stanford, CA.

AINLEY, D. G., AND R. J. BOEKELHEIDE, Eds. 1990. Seabirds of the Farallon Islands. Stanford University Press, Stanford, CA.

AINLEY, D. G., AND R. E. LeRESCHE. 1973. The effects of weather and ice conditions on breeding in Adelie Penguins. Condor 75: 235–239.

AINLEY, D. G., AND T. J. LEWIS. 1974. The history of the Farallon bird populations 1854–1872. Condor 76: 432–446.

AINLEY, D. G., AND R. P. SCHLATTER. 1972. Chick raising ability in Adelie Penguins. Auk 89: 559–566.

AINLEY, D. G., R. E. LeRESCHE, AND W. J. L. SLADEN. 1983. Breeding Biology of the Adelie Penguin. University of California Press, Berkeley.

AINLEY, D. G., H. R. CARTER, D. W. ANDERSON, K. T. BRIGGS, M. C. COULTER, F. CRUZ, J. B. VALLEE, C. A. VALLEE, S. I. HATCH, E. A. SCHREIBER, R. W. SCHREIBER, AND N. G. SMITH. 1988. ENSO effects on Pacific Ocean marine bird populations. Pp. 1747–1758 *in* Acta 19th Congressus Internationalis Ornithologici (H. Ouelette, Ed.). University of Ottawa Press, Ottawa, Canada.

AMERSON, A. B., JR., AND P. C. SHELTON. 1975. The natural history of Johnston Atoll, central Pacific Ocean. Atoll Research Bulletin 192: 1–479.

ANDERSON, D. J. 1989. Differential responses for boobies and seabirds in the Galapagos to the 1986–87 El Niño–Southern Oscillation event. Marine Ecology Progress 52: 209–216.

ANDERSON, D., J. KEITH, E. PALACIOS, E. VELARDE, F. GRESS, AND K. A. KING. 1999. El Niño 1997–98: Seabird Responses from the Southern California Current and Gulf of California. Unpublished.

ARMITAGE, T. S. 1964. The effects on birds of the winter, 1962–63. Naturalist 889: 49–52.

ARNOULD, J. P. Y., D. R. BRIGGS, J. P. CROXALL, P. A. PRINCE, AND A. G. WOOD. 1996. The foraging behaviour and energetics of wandering albatrosses brooding chicks. Antarctic Science 8: 229–236.

ARNTZ, W. E., AND J. TARAZONA. 1990. Effects of El Niño 1982–83 on benthose, fish and fisheries off the South American Pacific coast. Pp. 323–360 *in* Global Ecological Consequences of the 1982–83 El Niño–Southern Oscillation (P. W. Glynn, Ed.). Elsevier Oceanography Series, Amsterdam.

ASH, J. S. 1964. Observations in Hampshire and Dorset during the 1963 cold spell. British Birds 57: 221–239.

ASHMOLE, N. P. 1963. The biology of the wideawake or sooty tern *Sterna fuscata* on Ascension Island. Ibis 103b: 297–364.

ASHMOLE, N. P. 1971. Sea bird ecology and the marine environment. Pp. 224–286 *in* Avian Biology, Vol. 1 (D. S. Farner and J. R. King, Eds.). Academic Press, New York.

BADUINI, C. L., K. D. HYRENBACH, AND G. L. HUNT, JR. 1999. Anomalous weather events in the southeastern Bering Sea: the condition of Short-tailed Shearwaters in 1997 and 1998. Pacific Seabirds 26: 24.

BAERENDS, G. P., AND R. H. DRENT. 1970. The Herring Gull and its egg. Behaviour Supplement 17: 1–312.

BAILEY, E. P., AND G. W. KAISER. 1972. Die-off of Common Murres on the Alaska peninsula and Unimak Island. Condor 74: 215–218.

BAILEY, R. S., R. W. FURNESS, J. A. GAULD, AND P. A. KUNZLIK. 1991. Recent changes in the population of sand eel (*Ammodytes, marinus*) at Shetland in relation to estimates of seabird predation. ICES Marine Science Symposia 193: 209–216.

BAIRD, P. H. 1994. Black-legged Kittiwake *Rissa tridactyla*. No. 92 *in* The Birds of North America (A. Poole and F. Gill, Eds.). Academy of Natural Sciences, Philadelphia; American Ornithologists' Union, Washington, D.C.

BARRETT, R. T., AND F. RIKARDSEN. 1992. Chick growth, fledging periods and adult mass loss of Atlantic Puffins, *Fratercula arctica*, during years of prolonged food stress. Colonial Waterbirds 15: 24–32.

BARTHOLOMEW, G. A. 1966. The role of behavior in temperature regulation in the Masked Booby. Condor 68: 523–535.

BARTHOLOMEW, G. A., AND W. R. DAWSON. 1979. Thermoregulatory behavior during incubation in Heermann's Gulls. Physiological Zoology 52: 422–437.

BARTHOLOMEW, G. A., R. C. LASIEWSKI, AND E. C. CRAWFORD, JR. 1968. Patterns of panting and gular flutter in cormorants, pelicans, owls and doves. Condor 70: 31–34.

BEARMAN, G. 1989. Ocean Circulation. Pergamon Press, Oxford.

BEER, J. V., AND H. BOYD. 1964. Deaths of wild white-fronted geese at Slimbridge in January 1963. Wildfowl Trust Annual Report 15: 40–44.

BIRKHEAD, T. R. 1976. Effects of sea conditions on rates at which Guillemots feed chicks. British Birds 69: 490–492.

BIRKHEAD, T. R., AND D. N. NETTLESHIP 1987. Ecological relationships between Common Murres, *Uria aalge*, and Thick-billed Murres, *Uria lomvia*, at the Gannet Islands, Labrador III feeding ecology of the young. Canadian Journal of Zoology 65: 1638–1649.

BJERKNES, J. 1966. The possible response of the atmospheric Hadley circulation to equatorial anomalies of ocean temperature. Tellus 4: 820–829.

BLETRANDO, G., AND P. CAMBERLIN. 1993. Interannual variability of rainfall in the eastern horn of Africa and indicators of atmospheric circulation. International Journal of Climatology 13: 533–546.

BOEKELHEIDE, R. J., D. G. AINLEY, H. R. HUBER, AND T. J. LEWIS. 1990. Pelagic Cormorant and Double-crested Cormorant. Pp. 195–217 *in* Seabirds of the Farallon Islands (D. G. Ainley and R. J. Boekelheide, Eds.). Stanford University Press, Stanford, CA.

BOERSMA, P. D. 1978. Breeding patterns of Galapagos Penguins as an indicator of oceanographic conditions. Science 200: 1481–1483.

BOERSMA, P. D., N. T. WHEELWRIGHT, M. K. NERINI, AND E. S. WHEELWRIGHT. 1980. The breeding biology of the Fork-tailed Storm-petrel (*Oceanodroma furcata*). Auk 97: 268–282.

BONSAL, B. R., A. K. CHAKRAVARTI, AND R. G. LAWFORD. 1993. Teleconnections between north Pacific SST anomalies and growing season extended dry spells on the Canadian prairies. International Journal of Climatology 13: 865–878.

BRICHETTI, P., U. F. FOSCHI, AND G. BOANO. 2000. Does El Niño affect survival rate of Mediterranean populations of Cory's shearwaters? Waterbirds 23: 147–336.

BURGER, A. E., AND J. F. PIATT. 1990. Flexible time budgets in breeding Common Murres: buffers against variable prey abundance. Studies in Avian Biology 14: 71–83.

BURGER, J. 1979. Nest repair behavior in birds nesting in salt marshes. Journal of Comparative Physiology and Psychology 11: 189–199.

BURGER, J. 1990. Seabirds, tropical biology and global warming: are we missing the ark. Colonial Waterbirds 13: 81–84.

BURGER, J., AND M. GOCHFELD. 1990. The Black Skimmer, Social Dynamics of a Colonial Species. Columbia University Press, New York.

BURGER, J., AND M. GOCHFELD. 1996. Family Laridae (gulls). Pp. 572–622 *in* Handbook of Birds of the World (J. del Hoyo, A. Elliott, and J. Sargatal, Eds.). Lynx Editions, Barcelona.

CAIRNS, D. K. 1992. Population regulation of seabird colonies. Current Ornithology 9: 37–61.

CAIRNS, D. K., K. A. BREDIN, AND W. A. MONTEVECCHI. 1987. Activity budgets and foraging ranges of breeding common murres. Auk 104: 261–271.

CANE, M. A. 1983. Oceanographic events during El Niño. Science 222: 1189–1195.

CANE, M. A., AND S. E. ZEBIAK. 1985. A theory for the El Niño and Southern Oscillation. Science 228: 1085–1086.

CASEY, J. 1994. Reproductive Success of Black-Legged Kittiwakes in the Northwest Atlantic. 1991–1993. B. Sc. Honours thesis (biopsychology), Memorial University Newfoundland, St. John's. 39 pp.

CHASTEL, O., H. WEIMERSKIRCH, AND P. JOUVENTIN. 1993. High annual variability in reproductive success and survival of an Antarctic seabird, the Snow Petrel *Pagodroma nivea*. Oecologia 94: 278–285.

CLARK, L., R. E. RICKLEFS, AND R. W. SCHREIBER. 1983. Nest-site selection by the Red-tailed Tropicbird. Auk 100: 953–959.

CLIMATE ANALYSIS CENTER. 1982–2000. Climate Diagnostics Bulletin NOAA, Washington, D.C. Publ. monthly.

COOPER, J., AND W. R. SIEGFRIED. 1976. Behavioural responses of young Cape Gannets Sula capensis to high ambient temperatures. Marine Behaviour and Physiology 3: 211–220.

COULSON, J.C., AND E. WHITE. 1958. The effect of age on the breeding biology of the Kittiwake *Rissa tridactyla*. Ibis 100: 40–51.

COULSON, J. C. 1966. The movements of the Kittiwake. Bird Study 13: 107–115.

COULSON, J. C., AND J. M. PORTER. 1985. Reproductive success of the Kittiwake *Rissa tridactyla*: the role of clutch size, chick growth rates and parental quality. Ibis 127: 450–466.

CRAWFORD, R. J. M., AND P. A. SHELTON. 1978 Pelagic fish and seabird inter-relationships off the coasts of South West and South Africa. Biological Conservation 14: 85–109.

CROXALL, J. P. 1992. Southern Ocean environmental changes: effects on seabird, seal and whale populations. Philosophical Transactions of the Royal Society of London 338: 319–328.

CROXALL, J. P., T. S. McCANN, P. A. PRINCE, AND P. ROTHERY. 1988. Reproductive performance of seabirds and seals on south Georgia and Signey Island, South Orkney Islands, 1976–1897: implications for Southern Ocean monitoring studies. Pp. 261–285 in Antarctic Ocean and Resources Variability (D. Sahrahage, Ed.). Springer-Verlag, Berlin.

CROXALL, J. P., P. ROTHERY, AND A. CRISP. 1992. The effect of maternal age and experience on egg-size and hatching success in Wandering Albatross Diomedea exulans. Ibis 134: 219–228.

CRUZ, F., AND J. B. CRUZ. 1985. The effect of El Niño on the breeding of the Dark-rumped Petrel on Cerro Pajas, Floreana. Pp. 259–272 in El Niño en Las Islas Galapagos: el Evento de 1982–83 (G. Robinson and E. M. del Pino, Eds.). Fundacion Darwin, Quito.

CRUZ, J. B., AND F. CRUZ. 1990. Effect of El Niño–Southern Oscillation conditions on nestling growth rate in the Dark-rumped Petrel. Condor 92: 160–165.

DIAMOND, A. W. 1976. Subannual breeding and moult cycles in the Bridled Tern Sterna anathetus in the Seychelles. Ibis 118: 414–419.

DIAMOND, A. W., AND E. A. SCHREIBER. In press. Magnificent Frigatebird (Fregata magnificens). In The Birds of North America (A. Poole and F. Gill, Eds.). The Birds of North America, Inc., Philadelphia.

DOBINSON, H. M., AND J. A. RICHARDS. 1964. The effects of the severe winter of 1962/63 on birds in Britain. British Birds 57: 373–434.

DORWARD, D. F. 1963. Comparative biology of white boobies and brown boobies Sula spp. at Ascension Island. Ibis 103b: 174–220.

DRENT, R. H., AND S. DAAN. 1980. The prudent parent: energetic adjustments in avian breeding. Ardea 68: 225–252.

DUFFY, D. C. 1990. Seabirds and the 1982–84 El Niño Southern Oscillation. Pp. 395–415 in Global Ecological Consequences of the 1982–83 El Niño–Southern Oscillation (P. W. Glynn, Ed.). Elsevier, Amsterdam.

DUFFY, D. C., A. BERRUTI, R. M. RANDALL, AND J. COOPER. 1984. Effects of the 1982–83 warm water event on the breeding of South African seabirds. South African Journal of Science 80: 65–69.

DUFFY, D. C., W. E. ARNTZ, H. T. SERPA, P. D. BOERSMA, AND R. L. NORTON. 1988. A comparison of the effects of El Niño and the Southern Oscillation in Peru and the Atlantic Ocean. Pp. 1740–1745 in Acta 19th Congressus Internl. Ornithologici (H. Ouelette, Ed.). University of Ottawa Press, Ottawa, Canada.

DUNN, E. K. 1973. Changes in the fishing ability of terns associated with wind speed and sea surface conditions. Nature 244: 520–521.

DUNN, E. K. 1975. The role of environmental factors in the growth of tern chicks. Journal of Animal Ecology 44: 743–754.

DUNNET, G. M., C. OLLASON, AND A. ANDERSON. 1979. A 28 year study of breeding fulmars Fulmarus glacialis in Orkney. Ibis 121: 293–300.

ELKINS, N. 1995. Weather and bird behaviour. T. & A. D. Poyser, London.

FALK, K., AND S. MOLLER. 1997. Breeding ecology of the Fulmar Fulmarus glacialis and the Kittiwake Rissa tridactyla in high-arctic northeastern Greenland, 1993. Ibis 139: 270–281.

FINNEY, S. K., S. WANLESS, AND M. P. HARRIS. 1999. The effect of weather conditions on the feeding behaviour of a diving bird, the Common Guillemot Uria aalge. Journal of Avian Biology 30: 23–30.

FISHER, H. I., AND M. L. FISHER. 1969. The visits of Laysan Albatrosses to the breeding colony. Micronesia 5: 173–201.

FRAEDRICH, K., AND K. MULLER. 1992. Climate anomalies in Europe associated with ENSO extremes. International Journal of Climatology 12: 25–31.

FURNESS, R. W., AND D. M. BRYANT. 1996. Effect of wind on field metabolic rates of breeding Northern Fulmars. Ecology 77: 1181–1188.

GASTON, A. J., AND D. N. NETTLESHIP. 1982. The Thick-billed Murres of Prince Leopold Island — a study of the breeding biology of a colonial high arctic seabird. Canadian Wildlife Service Monograph Series 6, 350 pp.

GASTON, A. J., G. CHAPDELAINE, AND D. G. NOBLE. 1983. The growth of Thick-billed Murre chicks at colonies in Hudson Strait: inter- and intra-colony variation. Canadian Journal of Zoology 61: 2465–2475.

GIBBS, H. L., AND P. R. GRANT. 1987. Ecological consequences of an exceptionally strong El Niño event on Darwin's finches. Ecology 68: 1735–1746.

GOCHFELD, M., AND J. BURGER. Family Sternidae (terns). Pp. 624–666 in Handbook of Birds of the World (J. del Hoyo, A. Elliott, and J. Sargatal, Eds.). Lynx Editions, Barcelona.

GLYNN, P. W. 1990. Global ecological consequences of the 1982–83 El Niño-Southern Oscillation. Elsevier Oceanography Series No. 52. New York.

GRAY, W. M. 1984. Atlantic seasonal hurricane frequency. Part I: El Niño and 30 mb quasi-biennial oscillation influences. Monthly Weather Review 112: 1649–1668.

HALL, G. A., H. L. GIBBS, P. R. GRANT, L. W. BOTSFORD, AND G. S. BUTCHER. 1988. Effects of El Niño–Southern Oscillation (ENSO) on terrestrial birds. Pp. 1759–1769 in Acta 19th Congressus Internl. Ornithologici (H. Ouelette, Ed.). University of Ottawa Press, Ottawa, Canada.

HAMER, K. C., AND J. K. HILL. 1993. Variation and regulation of meal size and feeding frequency in Cory's Shearwaters Calonectris diomedea. Journal of Animal Ecology 62: 441–450.

HAMER, K. C., AND J. K. HILL. 1997. Nestling obesity and variability of food delivery in Manx Shearwaters, Puffinus puffinus. Functional Ecology 11: 489–497.

HAMER, K. C., R. W. FURNESS, AND R. W. G. CALDOW. 1991. The effects of changes in food availability on the breeding ecology of great skuas Catharacta skua in Shetland. Journal of Zoology, London 223: 75–188.

HAMER, J. A., P. MONAGHAN, J. D. UTTLEY, P. WALTON, AND M. D. BURNS. 1993. The influence of food supply on the breeding success of Kittiwakes Rissa tridactyla in Shetland. Ibis 135: 255–263.

HAMER, K. C., R. A. PHILLIPS, S. WANLESS, M. P. HARRIS, AND A. G. WOOD. 2000. Foraging ranges, diets and feeding locations of gannets Morus bassanus in the North Sea: evidence from satellite telemetry. Marine Ecology Progress Series 200: 257–264.

HAMILTON, K. 1988. A detailed examination of the extratropical response to tropical El Niño/Southern Oscillation events. Journal of Climatology 8: 67–86.

HANCOCK, J. A., J. A. KUSHLAN, AND M. P. KAHL. 1992. Storks, Ibises and Spoonbills of the World. Academic Press, New York.

HANEY, J. C. 1986. Seabird patchiness in tropical oceanic waters: the influence of Sargassum "reefs." Auk 103: 141–151.

HANEY, J. C. 1989. Remote characterization of marine bird habitats with satellite imagery. Colonial Waterbirds 12: 67–77.

HANEY, J. C., AND D. S. LEE. 1994. Air-sea heat flux, ocean wind fields, and offshore dispersal of gulls. Auk 111: 427–440.

HARRINGTON, B. A., R. W. SCHREIBER, AND G. E. WOOLFENDEN. 1972. The distribution of male and female Magnificent Frigatebirds, Fregata magnificens, along the Gulf coast of Florida. American Birds 26: 927–931.

HARRIS, M. P. 1969. Food as a factor controlling the breeding of Puffinus lherminieri. Ibis 111: 139–156.

HARRIS, M. P. 1970. Breeding ecology of the Swallow-tailed Gull (Creagrus furcatus). Auk 87: 215–243.

HARRIS, M. P. 1973. The biology of the waved albatross Diomedea irrorata of Hood Island, Galapagos. Ibis 115: 483–510.

HARRIS, M. P., AND S. WANLESS. 1984. Breeding success of British kittiwakes Rissa tridactyla in 1986–88: evidence for changing conditions in the northern Sea. Journal of Applied Ecology 27: 172–187.

HARRIS, M. P., AND S. WANLESS. 1996. Differential responses of Guillemot (Uria aalge) and Shag (Phalacrocorax aristotelis) to a late winter wreck. Bird Study 43: 220–230.

HATCH, S. A. 1989. Diurnal and seasonal patterns of colony attendance in the Northern Fulmar, Fulmarus glacialis, in Alaska. Canadian Field Naturalist 103: 248–260.

HATCH, S. A., AND M. A. HATCH. 1990. Breeding season of oceanic birds in a sub-arctic colony. Canadian Journal of Zoology 68: 1664–1679.

HATCH, S. A., AND D. N. NETTLESHIP. 1998. Nothern Fulmar, Fulmarus glacialis. No. 361 in The Birds of North America (A. Poole and F. Gill, Eds.). Academy of Natural Sciences, Philadelphia; American Ornithologists' Union, Washington, D.C.

HAYES, C. 1986. Effects of the 1982–83 El Nino on Humboldt Penguin colonies in Peru. Biological Conservation 36: 169–180.

HOUSTON, A. I., W. A. THOMPSON, AND A. J. GASTON. 1996. The use of a time and energy budget model of a parent bird to investigate limits to fledging mass in the thick-billed murre. Functional Ecology 10: 432–439.

HOWELL, T. R. 1979. Breeding biology of the Egyptian Plover. University of California Publications in Zoology 113: 1–76.

HOWELL, T. R., AND G. A. BARTHOLOMEW. 1961. Temperature regulation in Laysan and Black-footed albatrosses. Condor 63: 185–197.

HOWELL, T. R., AND G. A. BARTHOLOMEW. 1962. Temperature regulation in the Sooty Tern (*Sterna fuscata*). Ibis 104: 98–105.

HOWELL, T. R., B. ARAYA, AND W. R. MILLIE. 1974. Breeding biology of the Gray gull, *Larus modestus*. University of California, Publication in Zoology 104: 1–57.

HUGGINS, R. A. 1941. Egg temperatures of wild birds under natural conditions. Ecology 22: 148–157.

HULL, C. L., M. A. HINDELL, AND K. MICHAEL. 1997. Foraging zones of royal penguins during the breeding season, and their association with oceanographic features. Marine Ecology Progress Series 153: 217–228.

HUNT, G. L., JR., J. F. PIATT, AND K. E. ERIKSTAD. 1991. How do foraging seabirds sample their environment? Pp. 2272–2279 *in* Acta 20th Congressus Internalis Ornithologici. New Zealand International Congress Trust Board, Wellington.

INGRAHAM, W. J., A. BAKUN, AND F. FAVORITE. 1976. Physical oceanography of the Gulf of Alaska. Proc. Report, Northwest Fisheries Center, National Marine Fisheries, National Oceanic and Atmospheric Administration, Seattle, WA.

JASICK, F. M., AND I. LAZO. 1999. Response of a bird assemblage in semiarid Chile to the 1997–98 El Niño. Wilson Bulletin 111: 527–535.

JOUVENTIN, P., AND H. WEIMERSKIRCH. 1990. Satellite tracking of Wandering Albatrosses. Nature 343: 746–748.

JOUVENTIN, P., AND H. WEIMERSKIRCH. 1991. Changes in the population size and demography of southern seabirds: management implications. Pp. 297–314 *in* Bird Population Studies, Relevance to Conservation and Management (C. M. Perrins, J. D. Lebreton, and G. J. M. Hirons, Eds.). Oxford University Press, Oxford.

KILDAW, S. D. 1999. Effect of wind on the growth rate of kittiwake chicks. Pacific Seabirds 26: 38, abstract.

KLOMP, N. I., AND R. W. FURNESS. 1992. Non-breeders as a buffer against environmental stress: declines in numbers of great skuas on Foula, Shetland, and prediction of future recruitment. Journal of Applied Ecology 29: 341–348.

KONAWZEWSKI, M., AND J. R. E. TAYLOR. 1989. The influence of weather conditions on growth of the little auk (*Alle alle*) chicks. Ornis Scandinavica 20: 112–116.

KONARZEWSKI, M., C. LILJA, J. KOZLOWSKI, AND B. LEWONCZUK. 1990. On the optimal growth of the alimentary tract in avian postembryonic development. Journal of Zoology, London 222: 89–101.

KONARZEWSKI, M., J. R. E. TAYLOR, AND G. W. GABRIELSEN. 1993. Chick energy requirements and adult energy expenditures of dovekies (*Alle alle*). Auk 110: 343–353.

KONYUKHOV, N. B. 1997. Weather factors of breeding in alcids. Biology Bulletin 24: 177–182.

KOOYMAN, G. L., C. M. DRABEK, R. ELSNER, AND W. B. CAMPBELL. 1971. Diving behavior of the Emperor Penguin, *Aptenodytes forsteri*. Auk 88: 775–795.

LACK, D. 1968. Ecological Adaptations for Breeding in Birds. Methuen, London.

LA COCK, G. D. 1986. The Southern Oscillation, environmental anomalies, and mortality of two southern African seabirds. Climate Change 8: 173–184.

LEE, D. S. 1996. Sex, seabirds, and cyclones: the benefits of planned parenthood. Paper given at the 1996 meeting, Society Caribbean Ornithology.

LINDSEY, G. D., T. K. PRATT, M. H. REYNOLDS, AND J. D. JACOBI. 1997. Response of six species of Hawaiian forest birds to a 1991–1992 El Niño drought. Wilson Bulletin 109: 339–343.

LONGHURST, A. R., AND D. PAULY. 1987. Ecology of Tropical Oceans. Academic Press, New York.

LUTGENS, F. K., AND E. J. TARBUCK. 1995. The Atmosphere: An Introduction to Meteorology. Prentice-Hall, Englewood Cliffs, NJ.

LYSTER, S. 1986a. Penguin deaths worry. Falklands Island Foundation Newsletter 5.

LYSTER, S. 1986b. Penguin mortality in the Falklands. Oryx 20: 206.

MISKELLY, C. M. 1990. Effects of the 1982–83 El Niño on two endemic land birds on the Snares Islands, New Zealand. Emu 90: 24–27.

MONAGHAN, P., J. D. UTTLEY, M. D. BURNS, C. THAINE, AND J. BLACKWOOD. 1989. The relationship between food supply, reproductive effort and breeding success in Arctic Terns, *Sterna paradisaea*. Journal of Animal Ecology 58: 261–274.

MONAGHAN, P., P. WALTON, S. WANLESS, J. D. UTTLEY, AND M. D. BURNS. 1994. Effects of prey abundance on the foraging behaviour, diving efficiency and time allocation of breeding Guillemots Uria aalge. Ibis 136: 214–222.

MONTEVECCHI, W. A. 1993. Birds as indicators of change in marine prey stocks. Pp. 217–266 in Birds as Monitors of Environmental Change (R. W. Furness and J. J. D. Greenwood, Eds.). Chapman & Hall, London.

MONTEVECCHI, W. A., AND R. A. MYERS. 1992. Monitoring population fluctuations in pelagic fish availability with seabirds. Canadian Atlantic Fisheries Advisory Commission Research Document 92/94, 1–22.

MONTEVECCHI, W. A., AND R. A. MYERS. 1996. Dietary changes of seabirds indicate shifts in pelagic food webs. Sarsia 80: 313–322.

MONTEVECCHI, W. A., AND R. A. MYERS. 1997. Centurial and decadal oceanographic influences on changes in northern gannet populations and diets in the north-west Pacific. Journal of Marine Science 54: 608–614.

MONTEVECCHI, W. A., AND J. M. PORTER. 1980. Parental investments by seabirds at the breeding area with emphasis on northern gannets. Pp. 323–365 in Behavior of Marine Animals. Vol. 4 (J. Burger, B. L. Olla, and H. E. Winn, Eds.). Plenum Press, New York.

MOUGIN, J.-L. 1975. Ecologie comparee des Procellariidae Antarctiques et Subantarctiques. Com. Natn. Fr. Rech. Antarctique 36: 1–195.

MURPHY, E. C., A. M. SPRINGER, AND D. G. ROSENAU. 1986. Population status of Uria aalge at colony in western Alaska: results and simulations. Ibis 128: 348–363.

MURPHY, R. C. 1925. Bird Islands of Peru. G. P. Putnam's Sons, New York.

MURPHY, R. C. 1936. Oceanic Birds of South America. Vol. 2. American Museum of Natural History, New York.

NATIONAL OCEANOGRAPHIC AND ATMOSPHERIC ADMINISTRATION. 1982–2000. Climate Diagnostics Bulletin, published monthly, Boulder, CO.

NELSON, J. B. 1965. The behaviour of the Gannet. British Birds 58: 233–288.

NELSON, J. B. 1968. Galapagos, Island of Birds. Longman, Green, London.

NELSON, J. B. 1969. The breeding ecology of the red-footed booby in the Galapagos. Journal of Animal Ecology 38: 181–198.

NELSON, J. B. 1977. Some relationships between food supply and breeding in the marine Pelecaniformes. Pp. 76–87 in Evolutionary Ecology (B. Stonehouse and C. Perrins, Eds.). University Park Press, London.

NELSON, J. B. 1978. The Sulidae: Gannets and Boobies. Oxford University Press, Cambridge.

NETTLESHIP, D. N. 1977. Studies of seabirds at Prince Leopold Island and vicinity, Northwest Territories. Canadian Wildlife Service Progress Notes 73: 1–11.

NETTLESHIP, D. N. 1987. Arctic seabirds: differential responses in breeding to unusual environmental conditions. Pp. 4–7 in Proceedings of "Polar Zeevogels," Arctisch Centrum and Nederlandse Ornithologische Unie, Univeristy of Groningen, Netherlands.

NETTLESHIP, D. N. 1996. Family Alcidae (Auks). Pp. 678–722 in Handbook of Birds of the World (J. del Hoyo, A. Elliott, and J. Sargatal, Eds.). Lynx Editions, Barcelona.

NETTLESHIP, D. N., T. R. BIRKHEAD, AND A. J. GASTON. 1984. Breeding of Arctic seabirds in unusual ice years: the Thick-billed Murre, Uria lomvia in 1978. Pp. 35–38 in BIO Review 1984 (M. P. Latremouille, Ed.). Bedford Institute of Oceanography, Fisheries and Oceans Canada, Dartmouth, Nova Scotia.

NEWTON, I. 1998. Population Limitation in Birds. Academic Press, London.

NORMAN, F. I., P. B. Du GUESCLIN, AND P. DANN. 1992. The 1986 'wreck' of Little Penguins Eudyptula minor in Western Victoria. Emu 91: 369–376.

NORTON, R., AND E. A. SCHREIBER. In press. Brown Booby (Sula leucogaster). In The Birds of North America (A. Poole and F. Gill, Eds.). Academy of Natural Sciences, Philadelphia; American Ornithologists' Union, Washington, D.C.

NYE, P. 1964. Heat loss in wet ducklings and chicks. Ibis 106: 189–197.

OLSON, S. L. 1981. Natural history of vertebrates on the Brazilian Islands of the mid south Atlantic. National Geographic Research Reports 13: 481–492.

PENNYCUICK, C. J. 1982. The flight of petrels and albatrosses (Procellariiformes), observed in South Georgia and its vicinity. Philosophical Transactions of the Royal Society of London Biological Sciences 300: 75–106.

PHILANDER, S. G. H. 1990. El Niño, La Niña and the Southern Oscillation. Academic Press, San Diego, CA. 289 pp.

PIATT, J. F., AND T. I. VAN PELT. 1997. Mass mortality of guillemots (*Uria aalge*) in the Gulf of Alaska in 1963. Marine Pollution Bulletin 34: 656–662.

PREGILL, G. K., AND S. L. OLSON. 1981. Zoogeography of West Indian vertebrates in relation to Pleistocene climatic cycles. Annual Review of Ecological Systematics 12: 75–98.

PRINCE, P. A. 1985. Population and energetic aspects of the relationships between Black-browed and Grey-headed Albatross and the southern ocean marine environment. Pp. 473–477 *in* Antarctic Nutrient Cycles and Food Webs (W. R. Siegfried, P. R. Condy, and R. M. Laws, Eds.). Springer-Verlag, Berlin.

PRINCE, P. A., AND C. RICKETTS. 1981. Relationship between food supply and growth in albatrosses: an interspecies chick fostering experiment. Ornis Scandinavica 12: 207–210.

QUINN, W. H., D. O. ZOPF, AND K. S. SHORT. 1978. Historical trends and statistics of the Southern Oscillation, El Nino and Indonesian droughts. Fisheries Bulletin 76: 663–678.

RASMUSSON, E. M. 1985. El Niño: the ocean/atmosphere connection. Oceanus 27: 4–12.

RASMUSSEN, E. M., AND J. M. WALLACE. 1983. Meteorological aspects of the El Niño/Southern Oscillation. Science 222: 1195–1202.

RECHTEN, C. 1985. The waved Albatross in 1983. El Niño leads to complete breeding failure. Pp. 227–238 *in* El Niño in the Galapagos Islands: The 1982–83 Event (G. Robinson and E. M. del Pino, Eds.). Charles Darwin Foundation for the Galapagos Islands, Quito, Ecuador.

REGEHR, H. M. 1995. Breeding Performance of Black-Legged Kittiwakes on Great Island, Newfoundland, during Periods of Reduced Food Availability. M. Sc. thesis (biology) Memorial University of Newfoundland, St. John's. 199 pp.

REGEHR, H. M., AND W. A. MONTEVECCHI. 1997. Interactive effects of food shortage and predation on breeding failure of Black-legged Kittiwakes: implications for indicator species, seabird interactions and indirect effects of fisheries activities. Marine Ecology Progress Series, 155: 249–260.

ROBINSON, G., AND E. M. DEL PINO. 1985. El Niño in the Galapagos Islands: the 1982–83 event. Charles Darwin Foundation, Galapagos.

ROPELEWSKI, C. F., AND M. S. HALPERT. 1987. Global and regional scale precipitation patterns associated with the El Niño/Southern Oscillation. Monthly Weather Review 115: 1606–1626.

SAETHER, B. 2000. Weather ruins winter vacations. Science 288: 1975–1976.

SAGAR, P. M., AND J. L. SAGAR. 1989. The effects of wind and sea on the feeding of Antarctic Terns at the Snares Islands, New Zealand. Notornis 36: 171–182.

SCHREIBER, E. A. 1989. Unusual seabird breeding parameters on Christmas Island. Climate Diagnostics Bulletin, September 1989. NOAA, Washington, D.C.

SCHREIBER, E. A. 1994. El Niño–Southern Oscillation effects on chick provisioning and growth in red-tailed tropicbirds. Colonial Waterbirds 17: 105–119.

SCHREIBER, E. A. 1996. Experimental manipulation of feeding in Red-tailed Tropicbird chicks. Colonial Waterbirds 19: 45–55.

SCHREIBER, E. A. 1997. The Barbuda Magnificent Frigatebird colony: status report and management recommendations. Report to ENCORE and World Wildlife Fund, Washington, D.C. 20 pp.

SCHREIBER, E. A., AND R. W. SCHREIBER. 1989. Insights into seabird ecology from a global natural experiment. National Geographic Research 5: 64–81.

SCHREIBER, E. A., AND R. W. SCHREIBER. 1993. Red-tailed Tropicbird (*Phaethon rubricauda*). No. 43 *in* The Birds of North America (A. Poole and F. Gill, Eds.). Academy of Natural Sciences, Philadelphia; American Ornithologists' Union, Washington, D.C.

SCHREIBER, E. A., R. W. SCHREIBER, AND G. A. SCHENK. 1996. The Red-footed Booby (*Sula sula*). No. 241. *in* The Birds of North America (A. Poole and F. Gill, Eds.). Academy of Natural Sciences, Philadelphia; American Ornithologists' Union, Washington, D.C.

SCHREIBER, E. A., C. FEARE, B. A. HARRINGTON, B. MURRAY, W. B. ROBERTSON, M. J. ROBERTSON, AND G. E. WOOLFENDEN. In preparation. Sooty Tern (*Sterna fuscata*). *In* The Birds of North America (A. Poole and F. Gill, Eds.). The Birds of North America, Inc., Philadelphia.

SCHREIBER, R. W. 1980. Nesting chronology of the eastern Brown Pelican. Auk 97: 491–508.

SCHREIBER, R. W., AND E. A. SCHREIBER. 1984. Central Pacific seabirds and the El Niño–Southern Oscillation: 1982–1983 perspectives. Science 225: 713–716.

SCHREIBER, R. W., AND J. L. CHOVAN. 1986. Roosting by pelagic seabirds: energetic, populational, and social considerations. Condor 88: 487–492.

SCHREIBER, R. W., AND E. A. SCHREIBER. 1986. Unusual seabird breeding parameters on Christmas Island. Climate Diagnostics Bulletin, March 1986. NOAA, Washington, D.C.

SCHREIBER, R. W., AND P. J. MOCK. 1988. Eastern Brown Pelicans: what does 60 years of banding tell us. Journal of Field Ornithology 59: 171–182.

SCHWEIGGER, E. H. 1961. Temperatures anomalies in the eastern Pacific and their forcasting. Soc. Geogr. Lima, Bol. 78: 3–50.

SCOTT, J. M., J. A. WEINS, AND R. R. CLAEYS. 1975. Organochlorine levels associated with a murre die-off in Oregon. Journal of Wildlife Management 39: 310–320.

SHAFFER, S. A. 2000. Foraging Ecology of the Wandering Albatross (Diomedea exulans): Impacts on Reproduction and Life History. Ph.D. thesis, University of California, Santa Cruz. 112 pp.

SHEA, R. E., AND R. E. RICKLEFS. 1985. An experimental test of the idea that food supply limits growth rate in a tropical pelagic seabird. American Naturalist 125: 116–122.

SHEALER, D. 1999. Sandwich Tern (Sterna sandvicensis) No. 405 in The Birds of North America (A. Poole and F. Gill, Eds.). The Birds of North America, Inc., Philadelphia.

SILLETT, T. A., R. T. HOLMES, AND T. W. SHERRY. 2000. Impacts of a global cycle on population dynamics of a migratory songbird. Science 288: 2040–2042.

SIMMONS, K. E. L. 1967. Ecological adaptations in the life history of the Brown Booby at Ascension Island. The Living Bird 6: 187–212.

SIMMONS, K. E. L. 1968. Occurrence and behaviour of the red-footed booby at Ascension Island. Bulletin of the British Ornithologists' Club 88: 15–20.

SLADEN, W. J. L. 1958. The Pygoscelid penguins. Falklands Islands Depend. Survey Scientific Report 17: 1–97.

SMITH, J. L., C. D. FRENCH, AND A. J. GASTON. 1999. Effects of the 1997–98 El Niño on Ancient Murrelets in Haida Gwaii (Queen Charlotte Islands). Talk given at 26th Annual Meeting of the Pacific Seabird Group.

SNOW, D. W. 1965a. The breeding of the red-tailed tropicbird in the Galapagos. Condor 67: 210–214.

SNOW, D. W. 1965b. The breeding of Audubon's Shearwater in the Galapagos. Auk 82: 591–597.

SPRINGER, A. M., D. G. ROSENEAU, E. C. MURPHY, AND M. I. SPRINGER. 1984. Environmental controls of marine food webs: food habits of seabirds in the eastern Chukchi Sea. Canadian Journal of Fisheries and Aquatic Sciences 41: 1202–1215.

SPRINGER, A. M., D. G. ROSENEAU, D. S. LLOYD, C. P. McROY, AND E. C. MURPHY. 1986. Seabird responses to fluctuating prey availability in the eastern Bering Sea. Marine Ecology Progress Series 32: 1–12.

STONEHOUSE, B. 1960. Wideawake Island, the Story of the B.O.U. Centenary Expedition to Ascension. Hutchinson and Co., Ltd., London.

STONEHOUSE, B. 1963. Observations on Adelie Penguins (Pygocelis adeliae) at Cape Royds, Antarctica. Pp. 766–779 in Proceeding of the 13th International Ornithological Congress, Ithaca, NY.

STONEHOUSE, B. 1975. The Biology of Penguins. Macmillan, London.

TAYLOR, I. R. 1983. Effect of wind on the foraging behavior of Common and Sandwich Terns. Ornis Scandinavica 14: 90–96.

TAYLOR, J. R. E., AND M. KONARZEWSKI. 1989. On the importance of fat reserves for the Little Auk (Alle alle) chicks. Oecologia 81: 551–558.

TICKELL, W. L. N. 1968. The biology of the great albatrosses, Diomedea exulans and Diomedea epomophora. Antarctic Research Series 12: 1–55.

THOMPSON, J. D. 1981. Climate, upwelling, and biological productivity: some preliminary relationships. Pp. 13–34 in Resource Management and Environmental Uncertainty (M. H. Glantz and J. D. Thompson, Eds.). Wiley, New York.

THRELFALL, W., E. EVELEIGH, AND J. E. MAUNDER. 1974. Seabird mortality in a storm. Auk 91: 846–849.

TRIVELPIECE, W. Z., S. G. TRIVELPIECE, G. R. GEUPEL, J. KJELMYR, AND N. J. VOLKMAN. 1990. Adélie and Chinstrap Penguins: their potential as monitors of the southern ocean marine ecosystem. Pp. 191–202 in Antarctic Ecosystem, Ecological Change and Conservation (K. R. Kerry and G. Hempel, Eds.). Springer, Berlin.

UTTLEY, J., P. MONAGHAN, AND S. WHITE. 1989. Differential effects of reduced sandeel availability on two sympatrically breeding species of tern. Ornis Scandinavica 20: 273–277.

UTTLEY, J. P., P. WALTON, P. MONAGHAN, AND G. AUSTIN. 1994. The effects of food abundance on breeding performance and adult time budgets of Guillemots (*Uria aalge*). Ibis 136: 204–213.

VERMEIJ, J. 1990. An ecological crisis in an evolutionary context: El Niño in the eastern Pacific. Pp. 505–517 *in* Global Ecological Consequences of the 1982–83 El Niño–Southern Oscillation (P. W. Glynn, Ed.). Elsevier, Amsterdam.

VOLKMAN, N. J., AND W. Z. TRIVELPIECE. 1981. Nest site selection among Adelie, Chinstrap and Gentoo Penguins in mixed species rookeries. Wilson Bulletin 93: 243–248.

WANLESS, S., AND M. P. HARRIS. 1989. Kittiwake attendance patterns during chick rearing on the Isle of May. Scottish Birds 15: 156–161.

WARHAM, J. 1963. The Rockhopper Penguin, *Eudyptes chrysocome*, at Macquarie Island. Auk 80: 229–256.

WARHAM, J. 1990. The Petrels: Their Ecology and Breeding Systems. Academic Press, London.

WASILEWSKI, A. 1986. Ecological aspects of the breeding cycle in the Wilson's Storm Petrel, *Oceanites oceanicus* (Kuhl), at King George Island (South Shetland Islands, Antarctica). Polish Polar Research 7: 173–216.

WEIMERSKIRCH, H., T. MOUGEY, AND X. HINDERMEYER. 1997. Foraging and provisioning strategies of black-browed albatrosses in relation to the requirements of the chick: natural variation and experimental study. Behavioral Ecology 8: 635–643.

WELTY, J. C., AND L. BAPTISTA. 1988. The Life of Birds. Saunders College Publishing, New York.

WHITTOW, G. C. 1993a. Black-footed Albatross (*Diomedea nigripes*). No. 65 *in* The Birds of North America (A. Poole and F. Gill, Eds.). Academy of Natural Sciences, Philadelphia; American Ornithologists' Union, Washington, D.C.

WHITTOW, G. C. 1993b. Laysan Albatross (*Diomedea immutabilis*). No. 66 *in* The Birds of North America (A. Poole and F. Gill, Eds.). Academy of Natural Sciences, Philadelphia; American Ornithologists' Union, Washington, D.C.

WILLIAMS, T. D. 1995. The Penguins. Oxford University Press, Oxford.

WORK, T. M., AND R. A. RAMEYER. 1999. Mass stranding of wedge-tailed shearwater chicks in Hawaii. Journal of Wildlife Disease 35: 487–495.

WYRTKI, K. 1975. El Niño — the dynamic response of the equatorial Pacific Ocean to atmospheric forcing. Journal of Physical Oceanography 5: 572–584.

WYRTKI, K., E. STROUP, W. PATZERT, R. WILLIAMS, AND W. QUINN. 1976. Predicting and observing El Niño, Science 191: 343–346.

A Brown Pelican Feeds Its Three Hungry Chicks

8 Breeding Biology, Life Histories, and Life History–Environment Interactions in Seabirds

Keith C. Hamer, E. A. Schreiber, and Joanna Burger

CONTENTS

0-8493-9882-7/02/$0.00+$1.50
© 2002 by CRC Press LLC

8.1   INTRODUCTION

Seabirds comprise about 328 species in four orders, the Spenisciformes (penguins; 17 species in one family), Procellariiformes (albatrosses, shearwaters, petrels, diving petrels, storm-petrels; 125 species in four families, here termed petrels), Pelecaniformes (pelicans, tropicbirds, frigatebirds, gannets, and cormorants; 61 species in five families), and Charadriiformes (gulls, terns, skuas, skimmers, and auks; 128 species in four families: see Appendix 1 for a complete list of species). Seabirds range in size from the Least Storm-petrel (*Halocyptena microsoma*; body mass = 20 g) to the Emperor Penguin (*Aptenodytes forsteri*; body mass = 30 kg). They exploit a broad spectrum of marine habitats, from littoral to pelagic and from tropical to polar, breeding at higher latitudes and in colder environments than any other vertebrate on earth. The general characteristics of the different families of seabirds are summarized in Table 8.1 (Family Sternidae is included in the Family Laridae following Croxall et al. 1984, Nelson 1979, Croxall 1991). Seabirds can all be broadly categorized as long-lived species with delayed sexual maturation and breeding and low annual reproductive rates. Many species have a lifespan well in excess of 30 years with fewer than 10% of adults dying each year, and most do not commence breeding until age 3 years or older (over 10 years in some albatrosses; see Appendix 2). Most species lay only one to three eggs per clutch and in some cases rearing offspring takes so long (e.g., 380 days in Wandering Albatrosses, *Diomedea exulans*) that successful parents breed only every second year.

These life history traits are adaptive evolutionary responses to conditions of living in the marine and maritime environment, both at sea and on land. They have been generally assumed to reflect the patchy and unpredictable distribution of marine food resources, although there are additional explanations that have not received sufficient recognition (see Chapter 1 and conclusions below). Some confusion has arisen in the literature because of a failure to distinguish between life histories (comprising sets of evolved traits) and life-table variables such as age-specific fecundity and mortality (that indicate an individual's performance and are the consequence of how life history traits interact with the environment; Charnov 1993, Ricklefs 2000). For instance, all petrels are constrained by their life history evolution to lay a single-egg clutch, no matter how favorable the environment (Figure 8.1). Some other species have the potential to lay larger clutches, with the number of eggs laid varying from one individual to another, and between years within individuals. Life history adaptations determine the potential limits to this variation within each population, whereas variation within individuals is better expressed in life-table variables.

This chapter explores the variation in breeding biology and nesting ecology among seabirds. It examines breeding phenology and habitat in different species and environments, breeding systems and social organization, life history traits (including analysis of comparative data for different species from Appendix 2), the relationships between different life history traits, breeding performance and life history–environment interactions, and postbreeding biology, focusing in particular on postbreeding migration and dispersal.

8.2   BREEDING PHENOLOGY

About 98% of seabirds are colonial and have synchronously timed breeding cycles within colonies. The benefits and costs of breeding synchronously are discussed in Chapter 4. At the beginning of

TABLE 8.1
Range of Demographic Parameters Observed in the Families of Seabirds

Order	Family	No. of Species	Avg. Clutch Size	Breeding Cycle	Age 1st Breed (yr)	Incubation Period	Chick Period (d)	Post-Fledging Care (d)	Nest Location	Hatch Type	Breeding Region	Forage Distance	Annual Survival (%)
Sphenisciformes	Spheniscidae, penguins	17	1–2	A–B	2–5	33–63	54–170	0–50	O-Bu	SA	STr-P	NS,OS	62–95
Procellariiformes	Diomedeidae, albatross	21	1	A–B	5–9	62–79	115–280	0–44	O	SP	Tr-P	OS,NS+	91–96
	Procellariidae, shearwaters	79	1	A–B	2–8	43–62	45–130	0–?	Bu(O)	SP	STr-Tm	OS	72–96
	Pelecanoididae, diving petrels	4	1	A	2	42–58	42–75	0–?	Bu	SP	STr-Tm	OS	75–87
	Hydrobatidae, storm-petrels	21	1	A	2–3	38–55	55–75	0–?	Cr,Bu	SP	STr-P	OS	79–93
Pelecaniformes	Phaethontidae, tropicbirds	3	1	A	2–5	39–51	72–90	0	Un,Cr	SP	STr-Tr	OS	90
	Pelecanidae, pelicans	7	2–3	A	2–3	28–32	71–88	7–20	O,Tr	A	STr-Tr	NS,NS+	?
	Fregatidae, frigatebirds	5	1	A–B	5–8	52–60	150–170	30–200	Tr	A	Tr	OS	?
	Sulidae, boobies	10	1–3	A	2–5	41–58	78–139	0–200	O,Tr	A	STr-Tr	NS,OS	83–96
	Phalacrocoracidae, cormorants	36	2–4	A	2–4	27–35	38–80	20–65	O,Tr	A	Tm-STr	NS	80–91
Charadriiformes	Stercorariidae, skuas	7	2	A	3–7	24–32	24–50	14–24	O	SP	P-Tm	NS,NS+	90–98
	Laridae, gulls	50	1–3	A	2–4	24–30	32–60	7–45	O,Tr	SP	Tm-Tr	NS,OS	74–97
	Laridae, terns	45	1–2	A	2–4	22–37	20–67	5–30	O(Tr)	SP	Tm-Tr	NS,OS	75–93
	Rhynchopidae, skimmers	3	1–5	A	3	21–24	28–30	14–20	O	SP	Tm-STr	NS	?
	Alcidae, auks, murres (total)	23	1–2	A	2–5	28–46	26–50	0–?	Cr,Bu,O	SP	P-Tm	NS,OS	75–95
	Synthliboramphus sp., Endomychura sp.	4	2	A	2–4	31–36	2–4	long	Bu,Cr	P	Tm-STr	OS,NS	77
	Alca torda, Uria sp.	3	1	A	4–5	33–35	20	long	Bu,Cr	SP	P-Tm	NS,OS	75

Note: Breeding cycle: A = annual breeder, B = biennial breeder. Hatchling type: SA = semialtricial, SP = semiprecocial, A = altricial, P = precocial. Breeding region: P = polar, SP = subpolar, Tm = temperate, STr = subtropical, Tr = tropical. Foraging distance: OS = feeds offshore, NS = feeds nearshore, + indicates feeding at a slightly greater distance than nearshore. Phalacrocoracidae includes only subfamily Phalacrocoracinae (cormorants). The four genera of Alcids that have chicks that fledge (leave the nest) before they can fly are listed separately. (See further explanation of codes in Appendix 2.)

FIGURE 8.1 Grey-backed Storm-petrels, like all members of the Order Procellariiformes, lay one egg. (Photo by H. Weimerskirch.)

the breeding season, birds generally arrive back at the colony site over a short period of time, moving into the colony and establishing nesting territories. The majority breed on an annual cycle, although there may be small fluctuations in the commencement of nesting that are related to weather variations and/or food availability. Some albatrosses and petrels, and probably at least female frigatebirds, breed biennially (every other year) due to the length of time it takes chicks to become independent (Appendix 2). Several factors play a role in setting the timing of the breeding cycle: temperature, food availability, age, experience, and length of daylight, and probably others. Temperature is very important in polar, subpolar, and temperate breeding seabirds, while it is probably unimportant in subtropical and tropical areas. Seabird food (fish, squid, krill, etc.) is not uniformly available in space or time in the oceans and fluctuations in it undoubtedly play a significant role in setting the timing of breeding in all areas of the world (see Chapters 1 and 6).

8.2.1 Effects of Age on Breeding Phenology

Older breeders are commonly the first ones to return to the breeding colony at the beginning of the season and have the highest nesting success, suggesting that experience may have an important influence on timing of breeding (Adelie Penguins, *Pygoscelis adeliae*, Ainley et al. 1983; Wandering Albatrosses, Pickering 1989; Northern Fulmars, *Fulmarus glacialis*, Weimerskirch 1990; Manx Shearwaters, *Puffinus puffinus*, Brooke 1990; Northern Gannets, *Morus bassanus*, Nelson 1964; Black-legged Kittiwakes, *Rissa tridactyla*, Coulson and Porter 1985). Young birds may spend one to several seasons around the colony learning how to court and claim a territory before they begin breeding (Fisher and Fisher 1969, Harrington 1974, Nelson 1978, Hudson 1985, Schreiber and Schreiber 1993; Chapter 10). There are some data to indicate that there is an optimum age for first breeding and that birds beginning earlier may have a shorter life span (Ollason and Dunnet 1978, Croxall 1981). This implies a cost to the bird of breeding so that beginning at a younger age does not necessarily mean the pair will raise more offspring in their lifetime.

8.2.2 Effects of Weather

Seabirds are well adapted to their surrounding climate. They have a good insulation of feathers, are endothermic, and have a suite of behaviors that allow further adjustment to local weather patterns. However, any extremes of climate or unusual climatic events can affect the nesting cycle of seabirds and their breeding success. These effects may be due directly to the weather itself, or indirectly to changes in food availability. Direct effects of weather on nesting are discussed in detail in Chapter 7 and only a brief overview is presented here.

FIGURE 8.2 Adelie Penguin chicks in Antarctica wait for their parents to return from sea and feed them. Polar nesting species must have enough thermal insulation to survive cold temperatures. (Photo by P. D. Boersma.)

Polar and subpolar seabirds may have the greatest energetic constraints imposed on them by climate. They have a short time available for breeding, they must cope with low air temperatures (Figure 8.2), prey are available only during a restricted season, and the length of the nesting period approaches the limit of available time. Since they need to begin their breeding season as soon as possible each year, they may arrive on the colony to find snow and ice inhibiting access to burrows or nesting areas, thus late season storms can delay nesting (Procellariiformes, Warham 1990; Adelie Penguins, Ainley and Le Resche 1973; Gentoo Penguins, *P. papua,* and Chinstrap Penguins, *P. Antarctica,* in Antarctica, Williams 1995).

The effects of different weather variables in temperate breeding species are less clear. Wind speed is inversely correlated with site attendance in the early stages of the breeding season in Thick-billed Murres (*Uria lomvia*; Gaston and Nettleship 1981). Several species delay nesting during cold weather, including Brown Pelicans (*Pelecanus occidentalis*; Schreiber 1976), Black Skimmers (*Rynchops niger*), Common Terns (*Sterna hirundo*; Burger and Gochfeld 1990, 1991), and many other species.

Subtropical and tropical species are less confined to a season by weather patterns, but food availability is still generally seasonal (see Chapter 6) and most species nest seasonally, although the season is less constricted than in most higher latitude species. For instance, on Johnston Atoll (central Pacific Ocean) Wedge-tailed Shearwaters (*Puffinus pacificus*), Christmas Shearwaters (*Puffinus nativitatus*), Brown Boobies (*Sula leucogaster*), Brown Noddies (*Anous stolidus*), and Grey-backed Terns (*Sterna lunata*) lay in a strictly confined season over 1 to 2 months. Masked Boobies (*S. dactylatra*), Red-footed Boobies (*S. sula*), Red-tailed Tropicbirds (*Phaethon rubricauda*), and White Terns (*Gygis alba*) lay in most months of the year, although a definite laying peak occurs in the spring (Schreiber 1999). The reasons for these differences among species have not been determined. It could be that social facilitation is more important in some species, resulting in a short laying period. Seasonal changes in food availability may also affect the energetic expenditures of some species more than others. El Niño–Southern Oscillation (ENSO) events have dramatic effects on breeding cycles for species in the tropical Pacific (Schreiber and Schreiber 1984, Duffy 1990; Chapter 7). The ultimate reason for their effect on breeding cycles is probably related to food availability.

8.2.3 EFFECTS OF FOOD AVAILABILITY

Seabirds tolerate almost any degree of cold and heat but are highly sensitive to changes in food availability as documented by their responses to ENSO events (Schreiber and Schreiber 1989, Duffy 1990; Chapter 7). Birds may not attempt to nest at all during such events, or initiation of nesting may be delayed until food supplies increase (Ainley and Boekelheide 1990, Schreiber 1999). Food availability fluctuates seasonally on a global scale (Chapter 6) and therefore it is not equally available throughout the prolonged reproductive periods of seabirds. Even in the tropics, which we associate with a uniform climate, there are seasonal changes that affect the abundance and distribution of food, and these play a regulatory role in seabird nesting cycles (Chapter 6).

The highest seasonality of food availability occurs in polar areas, where some species commence nesting before the great flushes of summer oceanic productivity. Given that adults are more adept foragers than immature birds (Chapter 6), young birds should fledge during the period of highest food availability to help ensure their survival while they learn to feed themselves. Emperor Penguins lay during the Antarctic winter and their chicks fledge 7 to 8 months later during the summer, when food availability is highest (Williams 1995). Wandering Albatrosses also time their nesting season so that chicks fledge when food is most available (Salamolard and Weimerskirch 1993).

We know little about food availability to seabirds, making it difficult to determine why nesting cycles are timed the way they are, or why cycles are altered in some years. In some cases, ornithologists roughly determine changes in food availability by weighing adults, measuring growth rates in chicks, monitoring provisioning rates of chicks, or measuring nest success (Jarvis 1974, Gaston 1985, Chastel et al. 1993, Schreiber 1994, 1996, Phillips and Hamer 2000a).

8.2.4 BIENNIAL BREEDING

Some species with extended nesting seasons are able to breed only every second year (e.g., King Penguins, *Aptenodytes patagonicus;* several of the albatrosses; White-headed Petrels, *Pterodroma lessoni,* of which 13% are annual breeders; Carboneras 1992, Chastel et al. 1995, Williams 1995). Some frigatebirds (*Fregata* sp.) may also breed biennially, particularly females, which continue to feed fledglings for 30 to about 180 days after they fledge, by which time they are 8 to 12 months old or more (Figure 8.3; Diamond 1975, Diamond and Schreiber in press; Appendix 2). The complete nesting cycle in King Penguins takes about 400 days, the longest of all seabirds.

FIGURE 8.3 A female Lesser Frigatebird broods her single small chick on Christmas Island (central Pacific). Chicks hatch naked and take 5 to 6 months to fledge, after which they return to the nest for another 2 to 5 months to be fed. (Photo by R. W. and E. A. Schreiber.)

Biennial breeding in species with a nesting cycle lasting less than a complete year has been attributed to birds being unable to breed and molt at the same time, owing to the energy requirements of each. White-headed Petrels, for instance, breed biennially even though they need only 160 to 180 days for a breeding cycle (seemingly allowing enough time to molt and breed annually, and similar to the breeding period of Great-winged Petrels, *Pterodroma macroptera,* that breed annually). Chastel (1995) suggests they breed biennially because they fledge their young at the end of summer and must molt during the winter when food availability is low, which slows the molt process.

8.2.5 ASEASONAL BREEDING

There are some reports of subannual breeding by seabirds (a cycle of fewer than 12 months; Ashmole 1962, Dorward 1963, Harris 1970, Nelson 1977, 1978, King et al. 1992). Some of these studies were conducted during ENSO events that we now know cause changes in the timing of the nesting season due to changes in food availability (Schreiber 1999; Chapter 7). For some purported aseasonal breeders, breeding is probably annual with some adjustment according to food supply. Among the best-known reports of subannual breeding are those from the British Ornithologists' Union Centenary Expedition to Ascension Island from October 1957 through May 1959 (see Ibis 103b, 1962, 1963). During this period, one of the most pronounced ENSO ever recorded was underway (Glynn 1990). Reports of subannual breeding in several species may have represented delayed breeding in one year because of unusual changes in food availability, and this needs further investigation. Sooty Terns (*Sterna fuscata*) may apparently lay every 10 months (Ashmole 1963), but there are not good data that it is the same birds breeding each time.

Snow and Snow (1967) and Harris (1970) reported a 9- to 10-month cycle in Swallow-tailed Gulls (*Creagrus furcatus*) in the Galapagos, but both studies were during ENSO events. Earlier, Murphy (1936) had found them to breed in all months of the year, although this also was during an ENSO event in 1925. This species may actually have a true subannual cycle. Perhaps because these birds nest in an area of abundant food associated with the Humboldt Current, they are not constrained to an annual cycle by seasonal food availability. They may also have the ability to alter their diet during the year to adjust to seasonality of food resources.

King et al. (1992) documented both annual and subannual breeding in seven seabird species in a 6-year study on Michaelmas Cay (16°S, 145°E), during which two ENSO events occurred (1986–1987, 1990–1994). Interestingly, the two pelagic feeders, Sooty Terns and Brown Noddies, remained at the island year round and experienced the greatest nesting failures and desertions. Diet was not studied in this population, but food was most likely the factor controlling timing of nesting and presence on the island. On Johnston Atoll (central Pacific), Sooty Terns breed in all months of the year during ENSO events, when they have repeated failures and relayings (Schreiber 1999), leading one to wonder if this was the reason they were breeding in all months on Michaelmas Cay. On Christmas Island (central Pacific Ocean) some White Terns are reported to breed on a subannual cycle (Ashmole 1968), although this has not been studied over a multiyear period.

Some seabirds are actually double-brooded and able to raise two broods in a year: Brown Noddies, Black Noddies (*Anous minutus*), White Terns, Cassin's Auklets (*Ptchoramphus aleuticus*; Manuwal and Thoresen 1993, E. A. Schreiber unpublished; Appendix 2). It is interesting that three of these species nest in the supposed "depauperate" tropical waters.

8.3 BREEDING HABITAT

8.3.1 NESTING AND FORAGING HABITATS

Habitat use in seabirds can be divided into nesting habitat and foraging habitat. While many land birds, such as passerines, often use the same habitat for both of these functions, seabirds do not. Instead seabirds nest on land and forage in estuarine or oceanic waters, often far from their nest sites. Further, since many seabirds have delayed breeding, they may spend years at sea, coming to land only occasionally until they begin breeding

Species in the four orders of marine birds fall into three main habitat categories as a broad generalization: (1) species that feed pelagically and nest mainly on oceanic islands, such as albatrosses, petrels, frigatebirds, tropicbirds, boobies, and some terns; (2) species that nest along the coasts and feed in nearshore environments, such as some pelicans, cormorants, gulls, some terns, and alcids; (3) those few species that nest and forage in inland habitats, and come to the coasts during the nonbreeding season (such as some skuas and jaegers, Franklin's Gull (*Larus pipixcan*), Bonaparte's Gull (*L. philadelphia*), Ring-billed Gull (*L. delawarensis*), and Black Tern (*Chlidonias niger*). Grey Gull (*L. modestus*) is unusual in that it breeds in the interior deserts of Chile, but feeds coastally even during the breeding season (Howell et al. 1974).

There are several important issues with habitat use in marine birds: (1) colony and nesting habitats used; (2) habitat selection in high-latitude and low-latitude species; (3) habitat use vs. habitat selection; (4) competition for habitat use and the role of competitive exclusion; and (5) the roles of predation, weather, and other factors in habitat selection.

8.3.2 COLONY AND NESTING HABITATS USED

Seabirds nest in a great variety of habitats from steep cliffs to flat ground, laying their eggs in trees or bushes, in burrows, in crevices, or in the open (Appendix 2; Figures 8.4 and 8.5). They nest on the mainland, in marshes, or on coastal or oceanic islands. Some even nest on roofs (Vermeer et al. 1988). A typical cliff habitat in eastern Canada illustrates habitat use by some species of breeding seabirds (Figure 8.4), from the large surface-nesting Northern Gannets at the top of the cliff to the smaller Black Guillemots (*Cephus grylle*) in crevices in the middle areas of a cliff. The habitat is partitioned to some extent by the size of the birds, with larger birds nesting in the open and toward the top, and smaller birds on ledges and in crevices lower down. In a crowded area, the species on the cliff face tend to be in small subcolony units of their own species, and during courtship there is much competition both between and within species for nesting sites.

A typical tropical coral atoll may have 14 to 18 nesting species of seabirds (Figure 8.5). Some of the largest species nest in the open on the ground, such as Masked and Brown Boobies, although some nest in bushes and trees (Red-footed Boobies and Great Frigatebirds, *Fregata minor*). Terns may nest in bushes or trees (White Terns, Brown and Black Noddies), or on the ground (Sooty and Grey-backed Terns, Brown Noddies). Burrow-nesting birds may include Wedge-tailed Shearwaters and Audubon's Shearwaters (*Puffinus lherminieri*), while crevice-nesting species include White-throated Storm-petrels (*Nesofregatta fuliginosa*). Christmas Shearwaters nest under bunches of grass or other vegetation. There are species-specific preferences for the various available breeding areas, which in some cases overlap and there is competition for nest sites. This occurs more on smaller atolls with less habitat available.

Within each order there is wide diversity of habitat use, and this variability may extend to within some species as well. For instance, Red-footed Boobies nest in trees or on the ground (Schreiber et al. 1996); Sooty Terns nest in the open at some colonies, while in other places they nest under bushes (Schreiber et al. in preparation); and Herring Gulls (*Larus argentatus*) nest in nearly all habitats from flat ground to cliffs and trees (Pierotti and Good 1994). Some species, however, nest in only one habitat; most albatrosses, skuas, and most gulls nest only on the ground in the open. Franklin's Gulls build floating nests in marshes and nest in no other habitat (Burger and Gochfeld 1994a). Many seabird species can be adaptable in the habitat they use, and given varying conditions, may change habitats.

The type of available habitat influences competition for nest sites, both within and between species. The greater the diversity in spatial heterogeneity, the greater niche diversification is possible. Even on an apparently uniform sandy atoll in the tropics, there can be great diversification of nesting sites and birds can make choices about which areas to use. On Johnston Atoll, Christmas

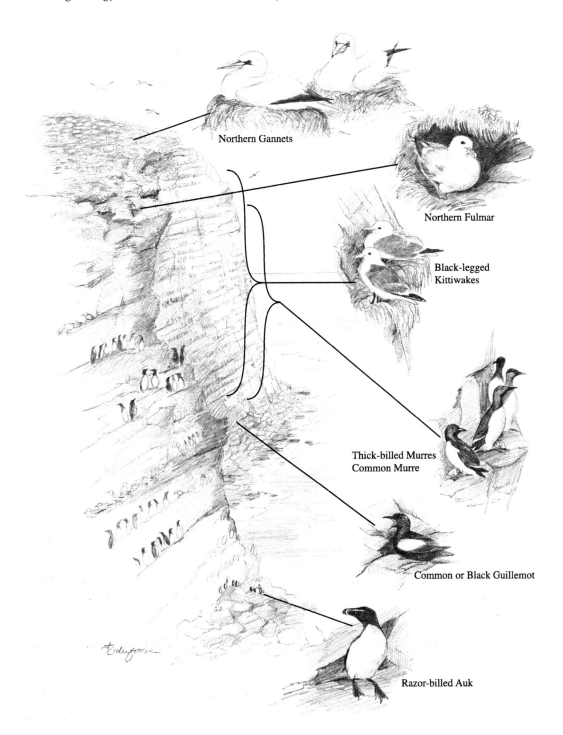

FIGURE 8.4 Typical nesting habitat and location of each species for a colony of seabirds along the coast of eastern Canada. Northern Gannets nest mostly on the flatter areas toward the top of the cliff, building mounded nests of rock or turf. Northern Fulmars nest toward the top of the cliff under dense grasses or vegetation, in crevices or shallow burrows. Black-legged Kittiwakes nest over a broad range of heights along the cliff face on narrow ledges, mounding nest material to hold their eggs. Thick-billed and Common Murres also nest over a broad range of heights on the cliff face. They build no nest laying their single egg on a narrow ledge. Razor-billed Auks nest toward the bottom of the cliff in the rock rubble. (Drawn by J. Zickefoose.)

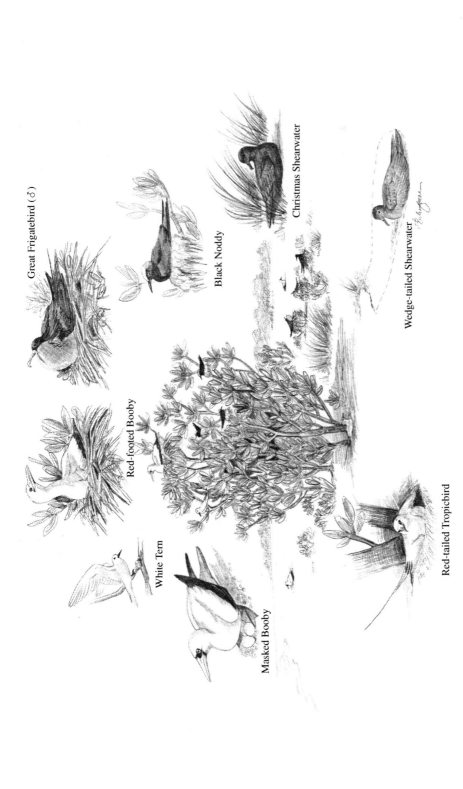

FIGURE 8.5 Typical nesting habitat for tropical Pacific Ocean seabirds. While all species are colonial, nesting densities vary by species and with habitat. Masked Boobies nest on open ground, building no nest, though they may collect a few small pebbles (nests tend to be widely dispersed and rarely occur singly). Red-footed Boobies nest near the tops of trees or bushes (nests from about 0.5 to 10 m apart) building a nest of twigs lined with some vegetation. Great Frigatebirds nest in the tops of bushes, or in or near the tops of trees (nests tend to be quite close together). They build a similar but less substantial and smaller nest than Red-footed Boobies. Black Noddies nest in trees, generally under leafy cover when possible (nests about 0.5 to 1.5 m apart depending on tree structure). Christmas Shearwaters nest under grasses or other vegetation, in crevices or short burrows (about 1 to 8 m apart). Wedge-tailed Shearwaters nest in burrows (0.5 to 3.0 m long and about 1 to 10 m apart, partly depending on substrate). Red-tailed Tropicbirds nest under bushes or other vegetation providing shade and some movement space (individuals are grouped by availability of vegetation, desired space between nests, and desired isolation from neighbor; from 0.5 to 10 m apart).

TABLE 8.2
Number of Species in Each Family of
Seabird Breeding at Different Latitudes

	Tropical	Temperate	Polar
Sphenisciformes			
Penguins	1	11	5
Procellariiformes			
Albatrosses	1	20	0
Shearwaters	21	46	7
Diving-petrels	0	4	0
Storm-petrels	8	11	1
Pelecaniformes			
Tropicbirds	3	0	0
Pelicans	3	4	0
Frigatebirds	5	0	0
Boobies	7	3	0
Cormorants	5	28	1
Charadriiformes			
Skuas	0	2	5
Gulls and terns	28	50	7
Skimmers	3	0	0
Auks	0	17	3

Note: Tropical (between the Tropic of Cancer and Tropic of Capricorn); Temperate (Tropic of Cancer to Arctic Circle [66.3°N] and Tropic of Capricorn to 60°S). Where species breed in more than one zone, that given is where the majority of individuals breed.

Shearwaters and Brown Boobies choose areas of highest wind levels, while Great Frigatebirds tend to nest in areas of lowest wind levels in the lee of the island (Schreiber 1999). The frigatebird choice of nest sites may relate to wing loading and body mass, since a black bird might not be expected to nest in a windless area if avoiding heat stress was its main consideration. High winds may make it difficult for frigatebirds to remain on their nests, given their light weight and extremely long wings (they have the lowest wing loading of any flying bird; Diamond and Schreiber in press).

In all habitats there is the opportunity for both species preferences and competition to influence nest-site selection (see below). In some cases, larger species may obtain breeding sites by virtue of their size and ability to defend these sites. However, where there is great heterogeneity, there can be separation by habitat type. As early as the 1950s, Fisher and Lockley (1954) noted differences in habitat use by a range of species nesting on rocky cliff faces.

8.3.3 HIGH-LATITUDE VS. LOW-LATITUDE SPECIES

All four orders of seabirds breed over a wide range of latitudes from tropical to polar environments, although all four do not have species equally distributed across all latitudes (Table 8.2). Species particularly associated with high latitudes and cold water include penguins, most albatrosses, diving petrels, skuas, some gulls and terns, and auks. Those nesting in warm water areas include one penguin, one albatross, some petrels and shearwaters, frigatebirds, boobies, tropicbirds, and some gulls and terns. These environments affect the choice of both nesting and foraging habitat. Some habitat choices faced by seabirds are a function of the habitats available within these regions. For example, in high-latitude environments, there are no trees. Temperate region birds may have marshes, trees, rocky or grassy slopes, and rocky islands, giving way at lower latitudes to salt marshes, sandy beaches, coral rubble, and bushes. At high latitudes, cold temperatures influence nest-site locations, while in tropical regions, heat stress may play a more critical role.

8.3.4 Habitat Use vs. Habitat Selection

The fact that seabirds forage in marine habitats, often pelagically, defines their habitat use to some extent. The factors that affect foraging habitat selection are complex, and include fish and invertebrate distribution, weather patterns and ocean currents, and upwellings. These factors are discussed in Chapter 6. Choice of breeding habitat for seabirds is somewhat constrained by their foraging behavior (type of foraging habitat, daily flight distances, and other energetic considerations). However, within these constraints, the question of habitat use vs. habitat selection is critical. Do seabirds use the nearest available habitat to breed or are they selecting specific habitats from the range available? Additionally, are their breeding habitat preferences influencing their choice of foraging habitats?

Habitat selection can be defined as the choice of a place to live (Partridge 1978), and it implies selection from a range of available characteristics (Burger 1985). Selection of breeding habitat, however, includes at least two distinct choices: colony site location and nest site location. In the few solitary-nesting seabird species, the two are functionally the same, but in most seabirds, the two types of selection are distinct. Colony-site selection generally occurs when the colony is first established, and subsequent birds simply nest within the colony. It is difficult to examine colony-site selection in species that use the same nesting place for many years, although frequently, scientists can measure microclimate of nest sites to determine what birds are selecting, such as sun exposure, wind level, or temperature. A new colony may form from a roost (birds loafing in areas other than their nesting colony; Brown Pelicans, Schreiber and Schreiber 1982; Red-footed Boobies, Schreiber 1999). There are not good data on why this happens or what causes these birds not to return to their natal colony.

For some species, environmental conditions vary from year to year altering the habitat of the colony and making colony-site selection more frequent. Heavy rains during ENSO events cause dense vegetation growth in Sooty Tern colony sites on Christmas Island, making them unsuitable for nesting (Schreiber and Schreiber 1989). During the following breeding cycle, birds select an area of less-dense vegetation in which to lay. Franklin's Gulls and Forster's Terns (*Sterna forsterii*) that nest in marshes frequently change sites from year to year (McNicholl 1975, Burger 1974; Figure 8.6). Black-billed Gulls (*Larus bulleri*) that nest in braided rivers shift colony sites annually (Beer 1966), and skimmers and terns nesting on sand bars in tributaries of the Amazon shift sites as new islands are created following floods (Krannitz 1989).

Seabirds select specific colony sites based on a range of abiotic and biotic characteristics (Buckley and Buckley 1980). While many studies describe the nesting habitat of seabirds, implying that the birds have selected these sites, in order to demonstrate habitat selection it is necessary to compare the habitat characteristics used by the species with the characteristics that are available, such as substrate, wind levels, and vegetation density. Burger and Lesser (1978) compared the characteristics of 34 Common Tern colonies in Barnegat Bay, New Jersey, with those of 225 other salt marsh islands. They found that the nesting islands differed significantly from the 225 other islands in size, distance to nearest other island and to shore, exposure to open water, and vegetation characteristics. The characteristics that terns selected were islands that were sufficiently high to avoid tidal flooding during the nesting season, but sufficiently low to lack mammalian predators (such as rats and foxes; Burger and Lesser 1978).

Similar selection of colony sites occurs in places with numerous potential nesting islands, such as in the Caribbean (Schreiber and Lee 2000) and in the Galapagos Islands (Cepeda and Cruz 1984). However, for some species that nest on isolated oceanic islands, there may be no other available island nearby and these birds tend to exhibit high philopatry (Great Frigatebirds, Schreiber

FIGURE 8.6 Franklin's Gulls often build floating nests in marshes as a way to avoid predation, but this unstable habitat changes from year to year causing frequent changes in colony site. (Photo by J. Burger.)

and Schreiber 1988; Red-tailed Tropicbirds, Schreiber and Schreiber 1993; Red-footed Boobies, Schreiber et al. 1996). To examine nest-site selection, researchers compare the specific characteristics at nest sites with those available within the colony. For examples of the methodology, see Squibb and Hunt (1983), Duffy (1984), Clark et al. (1983), Burger and Gochfeld (1990, 1991), Fasola and Canova (1992), and Hagelin and Miller (1997).

8.3.5 COMPETITION FOR HABITAT: IS THERE COMPETITIVE EXCLUSION?

Many seabird species are commonly found nesting together, whether on oceanic islands in the tropics (Figure 8.5) or on cliffs at higher latitudes (Figure 8.4). The occurrence of many species nesting together suggests that competition for specific sites may occur, and certainly in any seabird colony there is a plethora of aggressive encounters occurring among neighbors. Competition for nesting sites is often difficult to prove, however, even when there is apparent separation between two species nesting near each other (see Duffy 1984). Neighboring birds may be fighting over a nesting site, or simply defending a chosen nesting territory. Moreover, in many instances the availability of breeding sites appears sufficient, and birds seem not to be limited by appropriate sites. Olsthoorn and Nelson (1990) found that there were many unused, but apparently adequate, breeding sites for European Shags (*Strictocarbo aristotelis*), Black-legged Kittiwakes, Common Murres (*Uria aalge*), Razor-billed Auks (*Alca torda*), and Northern Fulmars on sea cliffs in Britain; there was even some exchange among species using specific sites in different years.

There are examples where there is a shortage of nesting spaces and the potential for intense competition. For example, Duffy (1983) reported that when Peruvian managers greatly increased the nesting space for three surface-nesting species (Guanay Cormorant, *Phalacrocorax bougain-villa;* Peruvian Booby, *Sula variegata;* Peruvian Brown Pelican), their numbers increased from 8 million to 20 million birds. These three species have overlapping nesting preferences, with the pelican being dominant in aggressive interactions and nest usurpations (Duffy 1983). In Alaska, there is habitat separation on nesting cliffs partly as a function of ledge size and bird size (Squibb and Hunt 1983). Nonetheless, discriminant analysis revealed greater overlap than expected between some species pairs, and nest usurpations occurred most frequently between the pairs that overlapped the most (Squibb and Hunt 1983).

In the Mediterranean, eight species of gulls and terns coexist and use different nesting habitats, based on vegetation cover and height (Fasola and Canova 1992). However, within mixed-species colonies, there is often some aggression, with the subordinate species losing territorial clashes, suggesting competition for space. For instance, Black Skimmers displace Common Terns on salt marsh islands (Burger and Gochfeld 1990). Trivelpiece and Volkman (1979) found that Chinstrap

Penguins displaced Adelie Penguins on King George Island, Antarctica, thereby lowering reproductive success. On the Farallon Islands, seabirds are strongly influenced by space limitations and interspecific competition for nest sites (Ainley and Boekelheide 1990). The two largest species (Pigeon Guillemot, *Cepphus columba,* and Rhinoceros Auklet, *Cerorhinca monocerata*) regularly usurp nests of the smaller Cassin's Auklet, often destroying eggs or killing chicks (Wallace et al. 1992). In general, the larger species usually wins among seabirds (Fasola and Canova 1992, Wallace et al. 1992).

8.3.6 Role of Predation, Weather, and Other Factors

Predation, weather, and other factors have shaped both colony-site and nest-site selection in seabirds.

8.3.6.1 Predators

Lack (1968) proposed that seabirds nested on inaccessible oceanic islands to avoid mammalian predators and to be far enough from the mainland to avoid many avian predators (see Figure 8.5). Since then, the role of nesting on cliffs and in trees was similarly ascribed an anti-predator function (Cullen 1957; Figure 8.4). We do not necessarily know that either of these breeding situations arose because of predators, and this is an interesting topic in need of further investigation. On flat ground, nest location can influence predation rates (Emms and Verbeek 1989). And, while nesting on remote islands may have evolved as an anti-predator mechanism, the lack of anti-predator behavior places seabirds at particular risk when predators (such as rats or cats) are accidentally introduced to these islands (see Moors and Atkinson 1984, Burger and Gochfeld 1994b, Schreiber 2000, Thompson and Hamer 2000).

8.3.6.2 Weather

Weather can influence nest-site choice and this is discussed primarily in Chapter 7, Climate and Weather Effects on Seabirds. In colder nesting areas, choosing nesting sites with some protection from the weather can save energy, a factor that may make a great difference in years of poor food availability. In the hot tropics, some seabirds select nest sites that are under vegetation (Red-tailed Tropicbirds, Schreiber and Schreiber 1993) or are in trees where shade is provided (Black Noddies, Buttemer and Astheimer 1990), or they may breed in burrows (Wedge-tailed Shearwaters, Whittow 1997). For instance, adult body temperatures can average 6°C higher on exposed compared to protected sites, creating a thermal stress for Black Noddies (Buttemer and Astheimer 1990). However, while thermal stress can be a problem, other factors may also influence nesting location. Jehl and Mahoney (1987) demonstrated that although thermal stress can kill embryos and small chicks of California Gulls (*Larus californicus*), their choice of nest sites near the shores of islands relates to the early detection of predators, providing hiding places for small chicks, and offering escape routes for larger young and parents (Jehl and Mahoney 1987).

8.3.6.3 Other Factors Affecting Nest-Site Selection

Other factors can affect nest-site choice. The birds themselves can render habitat less suitable with time. Species such as Great Cormorants (*Phalacrocorax carbo*; Grieco 1999), Great Frigatebirds (E. A. Schreiber unpublished), and Red-footed Boobies (Schreiber 1999) that nest in trees kill the tree with guano deposition after a few years, forcing the birds to seek other sites in subsequent years. Guano deposition or greatly increased rains from ENSO events can cause greatly increased vegetation growth in Sooty Tern colonies preventing the terns from being able to get to the ground to nest in the next year (Schreiber 1999, Schreiber and Schreiber 1989). Over many years of use, a dense colony of burrow-nesting seabirds can undermine a colony, leading to burrow collapse and mortality of adults and young (Stokes and Boersma 1991).

8.3.7 Nest-Site Selection and Reproductive Success

Presumably colony- and nest-site selection have been strongly shaped by differences in reproductive success in different habitats as has been shown in Magellanic Penguins (*Spheniscus magellanicus*, Stokes and Boersma 1991, Frere et al. 1992), Northern Gannets (Montevecchi and Wells 1984), Common Murres (Harris et al. 1997), Pigeon Guillemots (Emms and Verbeek 1989), Razor-billed Auks (Rowe and Jones 2000), Black Skimmers (Burger and Gochfeld 1990), and Common Terns (Burger and Gochfeld 1991). Where breeding success is influenced by characteristics of the breeding site, competition may occur, and higher-quality birds should obtain higher-quality sites (Burger and Gochfeld 1991, Rowe and Jones 2000). In summary, researchers have shown that (1) nesting habitat is limited for some species; (2) there is interspecific habitat partitioning in some species, although in other colonies there is a high degree of overlap; and (3) reproductive success varies in some colonies as a function of habitat choice.

8.4 BREEDING SYSTEMS AND SOCIAL ORGANIZATION

8.4.1 Coloniality and Dispersal

This topic is discussed in detail by Coulson (Chapter 6) and is only briefly mentioned here. The high frequency of coloniality among seabirds is perhaps surprising given the potential fitness costs of breeding at high densities, which include increased competition for food, nest sites, and mates, increased transmission of parasites and diseases, cuckoldry, cannibalism, and infanticide (Brown et al. 1990, Burger and Gochfeld 1990, 1991, Møller and Birkhead 1993, Danchin and Wagner 1997). Coloniality could result simply from there being a limited number of suitable breeding sites relative to the large foraging areas of seabirds (Forbes et al. 2000). However, this cannot explain why, in many species, nests are clumped together while apparently suitable neighboring areas remain empty (Siegel-Causey and Kharitonov 1990, Danchin and Wagner 1997). One possible advantage of coloniality is the avoidance of predators, through dilution effects or social mobbing (Kruuk 1964, Birkhead 1977, Burger and Gochfeld 1991, Anderson and Hodum 1993), although many colonies have no natural predators. Research on colony- and nest-site selection in seabirds is needed.

Coloniality may result from conspecific-based habitat selection, where individuals use the reproductive success of conspecifics to assess and select nesting sites. Or it may have evolved in the context of sexual selection and competition for breeding partners, by increasing the opportunities for individuals to assess the secondary sexual characteristics of potential mates (Danchin and Wagner 1997). This seems unlikely in seabirds given their high degree of social and genetic monogamy (see below), although it could facilitate the identification and selection of potential alternative mates for subsequent breeding seasons (Dubois et al. 1998). See Chapter 6 for details.

While there are fairly good data on migration in some species of seabirds, we know much less about dispersal (birds fledged from one colony and breeding in another, or birds breeding in one colony and then moving to another to breed). Learning about dispersal patterns requires banded populations of birds, and then recoveries of those birds in other colonies — often a difficult and expensive task. With the advent of molecular techniques that allow us to determine individual and populational relationships, we may soon understand better the extent of dispersal in seabird colonies. It is easy to assume for species such as seabirds that dispersal is common, since seabirds do travel great distances easily, there are vast oceans in which they can live and feed, and many species are widely distributed. But, in fact, many populations are highly philopatric with little dispersal. For example, recent evidence suggests that two currently, sympatrically nesting populations of Masked Boobies are separate species (Friesen et al. submitted). However, genetic studies, which indicate gene flow, do not tell us the current story of dispersal in a colony. For that, recaptures of marked individuals are needed.

Dispersal is difficult to study, especially long-distance dispersal, since few people can visit many potential colony sites a great distance from the one in which they work to search for banded birds. One method by which ornithologists can assume dispersal is to examine survival in a local colony. Frederiksen (1998) found survival to age 2 years was 0.116 in a population of Black Guillemots, unusually low for a species with 87% annual adult survival. From this he assumed a high degree of emigration from the colony. Degree of dispersal is related to distance to the nearest other potential nesting area, colony size, and potential competition for nesting space. Intercolony dispersal is an important aspect of population dynamics for species with high dispersal rates (Fulmars 90%, Dunnet and Ollason 1978a; Herring Gulls 70%, Coulson 1991; Common *Murres* 25 to 33%, Harris et al. 1996; Black Guillemots 50%, Frederiksen and Petersen 2000; Atlantic Puffins, *Fratercula arctica*, 46%, Gaston and Jones 1998).

8.4.2 MATING SYSTEMS

Seabirds are predominantly socially monogamous, with little evidence of polygyny, polyandry, communal, or cooperative nesting. This monogamy may be related to the need for biparental care, although each can occur without the other (Mock and Fujioka 1990). Exceptions to social monogamy occur in local populations of several species of gulls and terns that have uneven sex ratios with more females than males of breeding age (see Chapters 4 and 5). In some cases this has resulted in a small proportion of females (usually <10%) pairing with other females, laying eggs in a single nest and either sharing a male mate or obtaining extra-pair copulations to fertilize their eggs (Conover 1984, Nisbet and Hatch 1999, Bried et al. 1999). A much rarer situation is for more than one male to pair with a single female. Such polyandrous trios have been recorded in several populations of Brown Skuas (*Catharacta antarctica;* Young 1978).

No seabirds are strictly cooperative breeders, where young from a previous generation assist in raising their siblings, but the adoption of nonfilial young by foster parents does occur in a variety of gulls and terns (Saino et al. 1994, Oro and Genovart 1999, Bukacinski et al. 2000), in Black Skimmers (Quinn et al. 1994), Thick-billed Murres (Gaston et al. 1995), and Emperor Penguins (Jouventin et al. 1995). The frequency of foster parenting varies between species and years, usually ranging from 5 to 35% (Brown 1998), with up to 50% of broods affected in poor food years (Oro and Genovart 1999). It is commonest in ground-nesting species but has also been recorded in cliff-nesting Black-legged Kittiwakes, where 8% of chicks departed their nests before fledging and were adopted by foster parents (Roberts and Hatch 1994). Adoption is usually permanent until fledging, but in Emperor Penguins, most last fewer than 10 days (Jouventin et al. 1995). A poorly fed chick departing its own brood, thus replacing both its parents, usually initiates the adoption process. Foster parents often raise fewer of their own chicks to fledging than pairs that do not adopt (Salino et al. 1994, Brown 1998). This suggests that adoption may be the outcome of an evolutionary arms raise between poorly fed chicks that benefit through foster care, and foster parents that can avoid providing foster care through infanticide, but only at the risk of mistakenly killing their own offspring on some occasions (Brown 1998). Fostering may also increase inclusive fitness if foster parents and foster chicks are closely related, as appears to be the case in Thick-billed Murres (Friesen et al. 1996) and Common Gulls (*Larus canus;* Bukacinski et al. 2000).

8.4.3 OBTAINING A MATE

There are several ways seabirds obtain a mate, the most common being that a male claims a territory within the colony and displays or courts females (most penguins, all Pelecaniformes except tropicbirds, most Charadriiformes). Prospecting females come to the territories and interact with the male (see Chapter 10). Frigatebird males will select another site if they are not successful in obtaining a mate at one (Diamond and Schreiber in press). Pairs may form away from the colony site and move to it together (some albatrosses, some terns). Males may select burrows and defend

FIGURE 8.7 A pair of Royal Albatrosses at their nest on Campbell Island, New Zealand. They mate for life and show high nest philopatry. (Photo by J. Burger.)

them, then court nearby on the surface (petrels and shearwaters). It is not understood exactly how site selection and obtaining a mate occurs in tropicbirds, which court in the air (Schreiber and Schreiber 1993). Males may already have chosen a site, to which they lead a female to inspect. Previously mated birds frequently return to the nest site and wait for their mate to return (Figure 8.7).

8.4.4 DURATION OF PAIR BONDS

Most seabirds have long-term partnerships that endure from one breeding season to the next, and there are not any apparent trends by order of seabird. In Short-tailed Shearwaters (*Puffinus tenuirostris*; Wooller and Bradley 1996) and Red-billed Gulls (*Larus novaehollandiae*; Mills et al. 1996), 50% of adults had one partner during their lifetime and the mean number of partners was less than two, with 30 to 40% of mate changes being due to the death of one partner. Some seabirds do not retain mates between years, but these are either species where the sexes differ in the time required between breeding attempts (e.g., frigatebirds; Orsono 1999, Diamond and Schreiber in press) or nomadic species that lack a mechanism for reestablishing contact with a previous partner (e.g., Pomarine Jaegers, *Catharacta pomarinus*; Furness 1987). Less-extreme cases of low-rate mate retention occur in species that tend to change nesting localities from one year to the next (e.g., Caspian Terns, *Sterna caspia*, where 75% of individuals change mates between years; Cuthbert 1985) and in species that have low within-pair synchrony of arrival at the breeding colony at the start of the season (e.g., Emperor Penguins, where 85% of birds change mates between years; Williams 1996).

The long duration of pair bonds in seabirds probably reflects the benefits of more efficient and better coordinated breeding in an established pair. For instance, delayed egg laying and lower clutch sizes in gulls that changed partners between years were probably due to their need to spend more time establishing a pair bond, reducing the time available to forage (Mills 1973, Chardine 1987). Established pairs may also have more regular and better coordinated patterns of activity during incubation and chick rearing (Schreiber and Schreiber 1993, Coulson and Wooller 1984, Mills et al. 1996, Wooller and Bradley 1996). In Herring Gulls, pairs that had equal investment in incubation, chick feeding, and chick defense raised more young than did pairs with unequal investment (Burger 1984).

Changes in pair bond duration can easily be confounded by changes in adult age and experience, but in Cassin's Auklets (Sydeman et al. 1996) and Short-tailed Shearwaters (Wooller and Bradley 1996), reproductive success increased with the duration of the pair bond between partners independently of age and experience. A further problem of interpretation arises because more successful pairs tend to stay together longer (Mills et al. 1996). Thus an apparent increase in breeding success with increasing pair bond duration could be due simply to the different duration of successful and unsuccessful partnerships. To overcome this problem would require longitudinal study of serial changes in the reproductive success of individual pairs, controlled for age and experience.

A further advantage of long-term partnerships is that they reduce the potential costs of mate sampling, which include injury or predation, delays in finding a mate, or missing a breeding opportunity altogether (see Chapter 9). In many species, a high proportion of individuals that change mates fail to breed the next season (e.g., 26% in Great Skuas, *Catharacta skua*, Catry et al. 1997; 30 to 60% in Red-billed Gulls, Mills et al. 1996), and some never breed again. However, ending a partnership may nonetheless be advantageous in some circumstances, especially for young birds following one or more seasons with poor reproductive success (Schreiber and Schreiber 1983). The causes and consequences of divorce in seabirds are discussed in more detail by Bried and Jouventin (Chapter 9).

8.5 BREEDING BIOLOGY AND LIFE HISTORIES

8.5.1 EGGS

The sizes of eggs laid by seabirds are discussed in Chapter 12 by Whittow. The majority of seabirds lay clutches of one to two eggs, and 54% of species lay single-egg clutches. Some species of cormorant lay up to six and skimmers up to seven eggs per clutch (four to five in most). The small clutch sizes of seabirds are hypothesized to reflect the relative scarcity of food resources in marine ecosystems compared to terrestrial ecosystems (Ashmole 1963, Lack 1968; see Chapter 1). Modal clutch sizes might be expected to be lower in pelagic species, that travel farther from the nest to obtain food than nearshore foragers, and lower in tropical species reflecting the presumed relatively low productivity of tropical waters. Across all species (in Appendix 2), modal clutch size is higher in nearshore feeders (mean = 2.1, n = 113, SD = 0.9) than in pelagic feeders (mean = 1.1, n = 80, SD = 0.4; Mann-Whitney $Z = 7.83$, n = 174, $p < 0.001$). This difference partly reflects phylogeny, since petrels, frigatebirds, and tropicbirds all lay one egg and predominantly feed in pelagic waters, whereas cormorants and Laridae have larger clutches and predominantly feed inshore. The difference appears to hold within some individual families: in the Sulidae, pelagic foragers have smaller clutches than inshore foragers (mean = 1.17 and 1.88, respectively; $Z = 2.2$, n = 10, $p = 0.025$). However, there are few data on actual distance boobies feed from colonies, and broad generalizations were used in this analysis. In terns, those that nest along coasts frequently lay three eggs, while more pelagic species lay fewer (Gochfeld and Burger 1996). However, in other families, clutch size is not related to foraging distance (e.g., in the Alcidae, $Z = 0.3$, n = 16, $p = 0.8$). Understanding differences in clutch size as they relate to foraging distances of seabirds is somewhat confounded by the fact that we do not know where or how far many seabird species travel to feed, nor the energetic costs of foraging.

Across all species, modal clutch size is not related to latitude (Kruskal-Wallis $\chi^2_2 = 1.3$, n = 244, $p = 0.5$). However, if Procellariiformes, which invariably lay a single egg, are removed from the analysis, then clutch size is lower in tropical species (1.8, n = 51, SD = 1.0) than in temperate species (2.3, n = 100, SD = 0.6) or polar species (2.0, n = 22, SD = 0.6; K-W $\chi^2_2 = 12.3$, $p = 0.002$). This difference could reflect phylogeny, since the different orders are not distributed evenly across latitudes (Table 8.2; $\chi^2_6 = 14.3$, n = 175, $p < 0.01$). However, the difference holds within the

Laridae (tropical mean = 1.6, n = 25, SD = 0.7; temperate mean = 2.5, n = 45, SD = 0.6; polar mean = 2.2, n = 7, SD = 0.8; K-W χ^2_2 = 21.0, p <0.001). The reasons for this are not known at this time but could be related to food availability, foraging techniques, or differences in thermal environment and its effect on energetics. See Section 8.8 below for further discussion of this point.

In multiparous species (those that lay more than one egg per clutch), eggs are normally laid at intervals of 1 to 2 days. The interval may be up to 4 days in penguins, and longer in some coastal nesting species such as Franklin's Gulls and Skimmers (Burger 1974, Burger and Gochfeld 1990). In Crested Penguins (*Eudyptes* spp.), the first-laid egg is much smaller than the second (e.g., 60% as large in Macaroni Penguins, *E. chrysolophus*), but the second-laid egg hatches first, because it is more often incubated in the posterior part of the brood pouch where it is maintained at a higher and less variable temperature (Burger and Williams 1979). In Brown Boobies, either of the two first laid eggs may be larger, but a third egg (rarely laid) is always smaller (Schreiber 1999). In gulls and terns with three-egg clutches, the last-laid egg is typically about 10% smaller than the first two and produces a smaller chick that has a lower probability of survival to fledging (Schreiber et al. 1979, Sydeman and Emslie 1992, Royle and Hamer 1998). This size hierarchy could simply reflect a progressive decline in the female's nutritional reserves, or it might have adaptive value in allowing parents to more efficiently tailor their brood size to prevailing environmental conditions (Lack 1947, Temme and Charnov 1987, see Incubation below).

Some seabirds are determinate layers that do not replace lost eggs, even if this means the termination of the current breeding attempt. Other species (indeterminate layers) often replace eggs that are lost soon after laying (Haywood 1993). The current record is held by a Lesser Black-backed Gull (*Larus fuscus*) that responded to repeated removal of her clutch by laying a total of 16 eggs (Nager et al. 2000). Replacement eggs had a similar mass to those of last-laid eggs in normal clutches, but contained relatively less lipid and more water (Nager et al. 2000). Male offspring had particularly high mortality and as the body condition of female parents declined, the sex ratio of their eggs was progressively skewed toward females (Nager et al. 1999).

8.5.2 INCUBATION

Once laid, eggs are incubated more or less continuously. This maintains an adequate temperature for embryonic development and protects the eggs from thermal stress and potential predators. Incubation generally starts as soon as the first egg is laid, with the result that in multiparous species, the eggs hatch asynchronously. This, coupled with the fact that the last-laid egg in a clutch is often smaller, results in hatching asynchrony and the early establishment of a hierarchy of sizes and survival among chicks in the brood (Figure 8.8). The first chick to hatch, being larger when its siblings hatch, is able to monopolize the food and has a higher chance of survival. In Crested Penguins, however, there is a gradual change in the behavior of the adult over the 4-day laying period from partial protection to complete incubation of the egg (Williams 1995). In conjunction with the higher and less variable incubation temperature of the last-laid egg, this may ensure that the smaller (in this case first-laid) egg hatches last, again promoting the early establishment of a brood size hierarchy.

Larger, older chicks are able to compete more effectively for food provided by the parents, with the result that the youngest chicks are adequately fed only when food supply is plentiful, and so in many cases they die within a few days of hatching (Schreiber 1976, Norton and Schreiber in press). This process is termed brood reduction and may be adaptive in allowing parents to lay an optimistic clutch size, which can then be quickly and efficiently reduced if necessary to fit the prevailing conditions (Lack 1947, Mock and Forbes 1994). Alternatively, parents may create extra offspring as backup for members of the core brood that chance to die early (Forbes and Mock 2000). Brood reduction was challenged by Mock and Schwagmeyer (1990), who argued that in some species, hatching asynchrony could serve to ensure that chicks in a brood do not all reach their maximum food requirements at the same time, so reducing the maximum daily food demand

FIGURE 8.8 Brown Pelican eggs (generally three are laid) hatch asynchronously. The first chick to hatch is larger than its siblings when they hatch and generally monopolizes the food. Only in very good food years do Brown Pelicans raise all three young. (Photo by R. W. Schreiber.)

at the nest during chick-rearing. However, in most multiparous species, hatching asynchrony is only 2 to 5 days, which does not separate peak food requirement significantly among chicks.

Both sexes share in incubation except in Emperor Penguins, where the male is alone for the entire 62- to 64-day incubation period. In most species (especially petrels), the first long incubation shift is undertaken by the male, allowing the female to go to sea to replenish body reserves used in egg formation. Parents then alternate between incubating and foraging, at average intervals of about 5 h to 20 days (generally longer in larger species). Some penguins stay together at the nest and share incubation duties for the first half of the incubation period. The two sexes usually spend about the same amount of time incubating, although in some species the male may spend longer (e.g., Hatch 1990). In both cases, incubation shifts by both sexes may shorten toward the end of incubation, increasing the probability that the incubating bird has food for the chick when it hatches or that its partner returns relatively soon after (Warham 1990).

If for whatever reason the foraging bird does not return sufficiently quickly, the incubating bird may head to sea before its partner returns, leaving the egg unattended. Such eggs are usually vulnerable to thermal stress and predators, although in burrow-nesting petrels, they can remain viable for many days without incubation (e.g., up to 23 days in Madeiran Storm-petrel, *Oceanodroma castro*, Galapagos Islands; Harris 1969). Cooling of the embryo during periods it is not incubated does, however, retard growth, resulting in a delay in hatching. Across species, incubation period increases significantly and allometrically with body mass (Figure 8.9; linear regression of log-transformed data: $F_{1,176} = 11.6$, $p < 0.001$), according to the equation:

$$\text{incubation period (days)} = 26.67 \ (\text{SE} \pm 1.12) \ \text{body mass (g)}^{0.0584 \ (\text{SE} \pm 1.05)} \qquad (8.1)$$

However, the relationship is weak: log body mass explains only 6% of the variation in log incubation period. Deviations from this relationship could be explained by differences among species in vulnerability to predators, which might be expected to produce selection for shorter incubation periods. To test this hypothesis, we divided species into those with nests most vulnerable to predators (surface nesting on relatively level ground), those with nests vulnerable only to aerial predators (nesting in trees, on cliffs, or with floating nests), and those with nests relatively well protected against predators (in burrows or concealed crevices). We partly controlled for nonindependence of

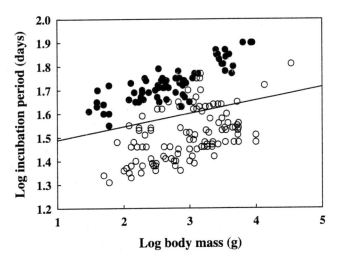

FIGURE 8.9 The relationship between log body mass and log incubation period in seabirds. Solid circles are procellariiformes; open circles are other species.

data due to common phylogeny using a two-way analysis of variance of residuals from the above relationship (Equation 8.1), with nest type and order (penguins, petrels, pelecaniformes, or charadriiformes) as factors. Mass-adjusted incubation period was much longer in petrels than in all other seabirds (Figure 8.9; $F_{3,167} = 13.8$, $p = 0.01$) and there was a significant interaction between order and nest location ($F_{4,167} = 7.5$, $p < 0.001$). Mass-adjusted incubation period increased with presumed vulnerability of nest sites to predation among the charadriiformes ($F_{2,48} = 24.3$, $p < 0.001$). However, this relationship was due almost entirely to longer incubation periods among the Alcidae (which almost all nest in burrows or other protected locations) than in other species (most of which have surface nests on relatively level ground). There was no relationship with nest site independently of phylogeny (Figure 8.10).

Vulnerability to predation may nonetheless have affected incubation periods in some species through the influence of body size. Among the Procellariiformes, the largest species are all surface

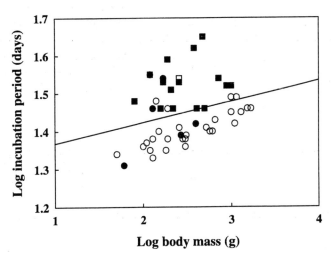

FIGURE 8.10 The relationship between log body mass and log incubation period in Charadriiformes. Squares = alcids, circles = other species; open symbols = surface nesters, closed symbols = species with protected nest sites (in burrows, crevices or trees, on cliffs, or floating).

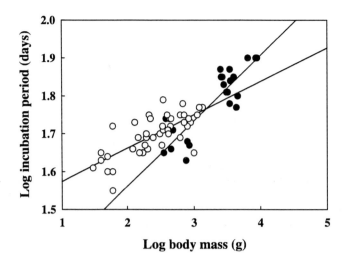

FIGURE 8.11 The relationship between log body mass and log incubation period in procellariiform seabirds. Solid circles = surface-nesters, open circles = burrow-nesters.

nesters. This is probably due to physical constraints on burrowing by large individuals, but it could also reflect the relative invulnerability of large species to predators. Incubation period decreased more rapidly with decreasing body mass in surface-nesting petrels than in burrow nesters (Figure 8.11 analysis of covariance; for effect of body mass $F_{1,64} = 128.8$, $p < 0.001$; for difference in slopes, $F_{1,64} = 13.0$, $p = 0.001$) with the result that smaller surface-nesting petrels had shorter incubation periods than burrow nesters of the same size. This may mean there was selection for shorter incubation periods among species most vulnerable to predation (small surface nesters). However, as discussed in Section 8.3.6 above, many seabirds nest in locations relatively free from predators. It may be that since eggs of burrow nesters are somewhat protected, adults are able to leave them unattended, thus lengthening the apparent incubation period. Also, natural selection may have operated on chick growth rate (not incubation period), which embryo growth merely reflects.

8.5.3 CHICKS AND CHICK-REARING

Growth rates of chicks are closely related to adult body mass (Figure 8.12; $F_{1,68} = 547.4$, $p < 0.001$) by the following allometric equation that explains 88% of the variance in growth rate:

$$\text{growth rate (g/day)} = 0.19 \ (\text{SE} \pm 1.09) \ \text{body mass (g)}^{0.690 \ (\text{SE} \pm 0.029)} \tag{8.2}$$

There is no difference among the four orders in residuals from this relationship ($F_{3,59} = 1.7$, $p = 0.3$) but pelagic foragers have significantly lower growth rates for their size than nearshore foragers (Figure 8.12; $F_{3,59} = 26.4$, $p = 0.006$). There is no interaction between order and mass-adjusted growth rate ($F_{3,59} = 0.4$, $p = 0.8$) and no relationship with latitude ($F_{2,60} = 1.6$, $p = 0.3$).

The length of the chick-rearing period is determined in part by chick growth rate but also by the pattern of development and the stage of development at which chicks leave the nest. Among birds there is a broad range, from altricial species (e.g., passerines and parrots) that hatch in an almost embryo-like state and are fed by their parents at the nest, to precocial species that hatch at a more advanced stage of development and can quickly move around and feed themselves (Starck and Ricklefs 1998). Most seabirds are intermediate and are variously termed semiprecocial or semialtricial. Pelecaniformes (except perhaps the Phaethontidae) are altricial and murrelets of the genera *Endomychura* and *Synthliboramphus* are precocial (young are not fed at the nest site, they leave when only a few days old and are fed entirely at sea; Houston et al. 1996; Appendix 2). Chick development patterns are discussed in detail by Visser (Chapter 13).

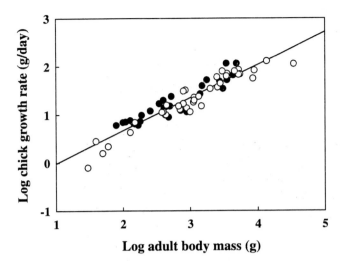

FIGURE 8.12 The relationship between log adult body mass and log chick growth rate in seabirds. Solid circles = nearshore foragers and open circles = pelagic foragers.

Regardless of their position on the altricial-precocial spectrum, in most seabirds body mass increases almost monotonically during growth before reaching an asymptotic mass similar to that of adults at or around fledging. However, there are a number of exceptions to this pattern. In some penguins, asymptotic masses of chicks are only 70 to 90% of adult mass (50% in Emperor Penguins), which means that chicks fledge earlier, allowing adults more time to complete their postnuptial molt while food availability is still high (Williams 1995). In other species of seabirds, chicks accumulate large quantities of nonstructural body fat during their development (e.g., up to 30% of body mass in Northern Fulmars, Phillips and Hamer 1999). This results in chicks attaining peak body masses far in excess of adult body mass (up to 170% of adult mass in Yellow-nosed Albatross *Thalassarche chlororhynchos*; Weimerskirch et al. 1986) before losing mass prior to fledging. This pattern is most pronounced in the procellariiformes but also occurs among tropicbirds, gannets, and boobies and some auks, while King Penguin chicks also accumulate large quantities of body fat during the first 3 to 4 months of posthatching development (Cherel et al. 1994).

The traditional explanation for nestling obesity was that chicks needed large fat reserves to sustain them over long intervals between feeds (Lack 1968, Ashmole 1971). This is certainly true of King Penguins (see below), but for other species this view is not supported. Recent evidence indicates that chicks do not encounter intervals between feeds of sufficient duration to account for such large fat stores (Hamer and Hill 1993, Schreiber and Schreiber 1993, Lorentsen 1996, Hamer and Hill 1997).

Five alternative hypotheses have been proposed to explain the evolution of nestling obesity.

1. Lipid reserves may provide a buffer against the cumulative effects of stochastic variation in food availability and/or foraging success by individual parents, rather than against the acute effects of long intervals between feeds. This hypothesis is supported by a simulation model of provisioning and growth in Leach's Storm-petrel (*Oceanodroma leucorhoa*; Ricklefs and Schew 1994) and by empirical evidence for other species (Hamer et al. 2000).
2. Chicks may accumulate lipid when they are young and their energy requirements are relatively low, and use these stores to subsidize greater metabolic energy requirements later in development.
3. Fat reserves may act as an energy sink if food provided by parents is energy rich but nutrient poor. For instance, Dovekies (*Alle alle*) feed their chicks on lipid-rich

zooplankton, and to obtain sufficient nutrients to sustain normal growth, chicks need to consume an excess of lipids (Taylor and Konarzewski 1989).

4. Fat stores may provide fuel for chicks at the end of their development and so allow parents to commence postbreeding migration sooner than they could otherwise (Brooke 1990).

5. Fat accumulated during the nestling period may be crucial for survival after fledging, while chicks develop foraging skills (Perrins et al. 1973, Chapter 6).

The decline in body mass of chicks at the end of their development has generally been taken as evidence that chicks use up their fat stores prior to fledging. However, in Northern Fulmars, prefledging mass recession was due entirely to loss of water from tissues as they attained functional maturity, and chicks continued to accumulate fat until fledging (Phillips and Hamer 1999). These data are not compatible with hypotheses 2 or 4 above, and strongly support the idea that fat deposits in Fulmars serve to fuel chicks during the initial critical period after fledging while they learn to forage for themselves. Fat reserves could also ensure against energy demands and/or nutritional stresses affecting the quality of flight feathers, which continue to grow up to or beyond the end of the nestling period (Reid et al. 2000).

Apart from the precocial murrelets, most seabirds do not fledge until they have more or less completed their development and are capable of flight. However, three species of semiprecocial auk (Common Murre, Thick-billed Murre, and Razor-billed Auk) follow an intermediate strategy where the young leave the nest site accompanied by one parent, usually the male, at less than one third of adult mass and unable to fly (Sealy 1973). This strategy may result from poor food provisioning ability of parents, because the relative load-carrying capacity of large auks is lower than that of smaller species, and the ratio of egg size to adult size is lower in larger species (Birkhead and Harris 1985). Hence, chicks of large auks are too small at hatching to be precocial. However, Tufted Puffins (*Fratercula cirrhata*) are larger than Razor-billed Auks (Appendix 2), yet have a semiprecocial mode of development. Moreover, mathematical modeling of parent time and energy budgets suggests that the departure of chicks before reaching typical semiprecocial size was not a prerequisite for successful chick rearing (Houston et al. 1996). In contrast to Tufted Puffins, none of the three species with intermediate strategies are burrow nesters, and the age at which chicks fledge may represent an optimum balance between the relative risks of chick starvation and predation at the nest and at sea (Ydenberg 1989, Byrd et al. 1991).

From data in Appendix 2, excluding the precocial murrelets, the length of the chick-rearing period in seabirds was significantly and independently related to both body mass and chick growth rate (stepwise multiple regression; $F_{2,67} = 26.7$, $p < 0.001$) according to the following allometric equation:

$$\text{chick-rearing period (days)} = 6.34 \ (\text{SE} \pm 1.43) \ \{\text{body mass (g)}^{0.537(\text{SE} \pm 0.151)}$$
$$\times \ 1/\text{growth rate (g/day)}^{0.419(\text{SE} \pm 0.151)}\} \tag{8.3}$$

Body mass and growth rate accounted for 38 and 6%, respectively, of the variance in chick-rearing period. Because growth rate differed between pelagic and nearshore feeders, the residuals from Equation 8.3 effectively estimated differences in chick-rearing period adjusted for body mass and foraging range.

Adjusted chick-rearing period was significantly longer in Procellariiformes and Pelecaniformes than in Sphenisciformes and Charadriiformes (Table 8.3; two-way ANOVA followed by post-hoc range tests; $F_{3,60} = 14.4$, $p < 0.001$). It was also significantly longer in tropical species (mean = 0.13, n = 9, SD = 0.09) than in polar species (mean = –0.08, n = 17, SD = 0.20) with temperate species intermediate (mean = 0.004, n = 14, SD = 0.20; $F_{2,60} = 6.3$, $p = 0.02$; Table 8.3). There was no interaction between order and latitude ($F_{4,60} = 0.4$, $p = 0.8$). The relationship between chick-rearing period and latitude probably reflects an advantage to polar species in leaving the nest before conditions deteriorate at the end of the breeding season (Williams 1995). Latitudinal differences

TABLE 8.3
Adjusted Chick-Rearing Periods of Seabirds by Order and Latitude

	Latitude								
	Tropical			Temperate			Polar		
Order	mean	SD	(n)	mean	SD	(n)	mean	SD	(n)
Sphenisciformes	—	—	—	0.02	0.21	(6)	−0.17	0.07	(5)
Procellariiformes	0.24	0.01	(2)	0.13	0.10	(23)	0.09	0.14	(7)
Pelecaniformes	0.13	0.05	(6)	−0.01	0.13	(2)	—	—	—
Charadriiformes	−0.02	—	(1)	−0.02	0.17	(13)	−0.23	0.21	(5)
Total	**0.13**	**0.09**	**(9)**	**0.01**	**0.20**	**(44)**	**−0.08**	**0.20**	**(17)**

Note: Data are residuals from a multiple linear regression of log chick-rearing period on log body mass and log chick growth rate (see text for further details).

in food supply appear to impact on clutch size rather than chick growth rate (see above), and so the shorter chick-rearing periods of polar species compared to tropical species are probably not related to food availability.

With the exception of precocial species, newly hatched chicks are unable to regulate their own body temperatures and so they are brooded more or less continually until they become thermally independent. This takes a few days in most species but up to 3 weeks or longer in the altricial Pelecaniformes (Montevecchi and Vaughan 1989). Newly hatched chicks are able to draw upon energy reserves retained from the egg and can survive 1 to 3 days without being fed by the brooding adult. Emperor Penguin males feed the chick on a protein-rich esophageal secretion for the first few days after hatching, after which they are relieved by the females that have been at sea feeding.

Small chicks have limited gut capacity, and so during the brooding period they are fed small meals several times a day by the attending parent. Once the adult's presence is not required for brooding, adults show a variety of responses. In petrels, penguins, burrow-nesting auks, and tropicbirds, chicks are left unattended for most of the time. In many cases both parents are foraging at sea. In Red-tailed Tropicbirds, much of the time away from the nest is spent sitting at sea, not foraging, and this may simply be the preferred roosting location (Schreiber and Schreiber 1993). This change in parental attendance is accompanied by an increase in meal size and a decrease in feeding frequency. For most of the nestling period parents deliver meals weighing 5 to 35% of adult body mass (proportionately smaller in heavier species; Birkhead and Harris 1985, Schreiber 1994, Phillips and Hamer 2000a) to the chick. Overall feeding frequency, from provisioning by both parents, is generally between one meal every 2 to 3 days and two to three meals per day (higher in auks, fulmars, and diving petrels than in other species; higher in penguins that forage relatively close to the colony than in those that forage further afield; Figure 8.13). King Penguin chicks are fed on average only every 39 days overwinter and may not be fed for up to 5 months during this period (Williams 1995). In most species, feeding frequency then declines at the end of the nestling period. In most cases, it is not known if feeding frequency is directly related to the adults' ability to find food. Further studies of seabird foraging behavior at sea are needed.

With the exception of diving petrels, most petrels alter the chemical composition of captured prey during transport to the nest by differential retention of the aqueous and lipid fractions of the digesta within the proventriculus. Following liquefaction of the food, the denser aqueous fraction passes into the duodenum first, leaving a lipid-rich liquid termed stomach oil in the proventriculus. Stomach oil has a caloric density up to 30 times greater than the prey and may be essential in some species to allow adults to carry enough energy back to the chick (Roby et al. 1997). The extent to which adults form stomach oil increases with foraging range and trip duration.

FIGURE 8.13 These Lesser Frigatebird chicks on Christmas Island (central Pacific) are normally fed about 0.7 to 1.0 times per day. (Photo by R. W. and E. A. Schreiber.)

In surface-nesting species, unattended chicks are vulnerable to predation in some cases. Some petrels such as fulmars can deter potential predators by spitting stomach oil at them. In most penguin species, chicks form into creches, which may comprise loose aggregations of a few individuals or, in King and Emperor Penguins, a single large very dense aggregation of all chicks in the colony (Williams 1995). Creches provide protection against aerial predators (particularly skuas) and may also protect against adverse weather conditions.

In Pelecaniformes, chicks are normally attended by at least one parent for the first quarter to third of the nestling period, probably reflecting their altricial mode of development, which results in a relatively long period before chicks become endothermic. Alternatively, this pattern could simply reflect no need for both adults to be foraging at once (Figure 8.14). In species such as gannets, that form dense colonies, unattended chicks are frequently attacked by neighboring adults and nonbreeders, and so they are seldom left alone at the nest (Nelson 1978). In other species, chicks are increasingly left unattended as they grow. Some data indicate that adults are not foraging the whole time they are away from the nest. Brown and Red-footed Booby adults are often roosting around the colony during the day (Schreiber et al. 1996, Norton and Schreiber in press). Red-tailed Tropicbirds spend the majority of their time away from the nest sitting on the water, not foraging (Schreiber and Schreiber 1993).

In the Charadriiformes (with the exception of burrow-nesting auks), chicks are usually attended at all times by at least one adult (Burger 1984), although species such as Kittiwakes with relatively well-protected nests may sometimes leave older chicks unattended, especially in conditions of poor food supply (Hamer et al. 1993). In some species, the two parents alternate between foraging at sea and guarding the chick; in others, males and females have specialized roles during chick rearing. For instance, in Great Skuas, most food is normally provided by males while the females defend the brood. In conditions of poor food availability, both parents forage simultaneously, but this leaves unattended chicks vulnerable to predation by neighboring conspecifics (Hamer et al. 1991).

During chick-rearing, all seabirds need to judge how much energy to invest in their chicks without impairing their own ability to breed again. In general the rate of food provisioning is regulated by interactions between the chicks' nutritional requirements, their begging behavior, and

FIGURE 8.14 Red-tailed Tropicbird adults brood a chick past the time when the chick can regulate its own body temperature, possibly because the adults do not need to spend more time at sea feeding. (Photo by E. A. Schreiber.)

the response of adults (Kilner and Johnstone 1997, Hamer et al. 1999, Weimerskirch et al. 2000). An exception to this pattern occurs in some petrels that make long foraging trips and feed their chick comparatively infrequently, where adults deliver food according to an intrinsic rhythm that is insensitive to changes in chick requirements (Ricklefs 1992, Hamer and Hill 1993, Saether et al. 1993). In either case, adults may limit the potential costs of food provisioning by monitoring their own body condition, and foraging for themselves or for their offspring accordingly (Chaurand and Weimerskirch 1994, Tveera et al. 1998). In some species of petrel, this results in adults adopting a dual foraging strategy, in which comparatively short trips undertaken to obtain food for the chick are interspersed with less frequent long trips during which adults obtain food for themselves (Weimerskirch et al. 1997, Granadeiro et al. 1998, Weimerskirch et al. 2000). The switch between short and long trips seems to be determined by a threshold body condition, below which adults forage only for themselves. In some cases the areas of ocean visited on short and long trips are far apart. For instance, Short-tailed Shearwaters breeding in southeast Australia forage comparatively close to the colony to obtain food for their chick but obtain food for themselves in the highly productive waters of the Polar Frontal Zone at least 1000 km away (Weimerskirch and Cherel 1998).

8.5.4 POSTFLEDGING CARE

Parental care in seabirds is a continuum from the intense care most recently hatched chicks receive, which may include brooding, feeding, and protection from predators, to a gradual tapering off that may occur when the chicks leave the nest, or fledge (are able to fly), and then become fully independent (Burger 1980). Normally, postfledging care refers to the feeding of chicks after they are able to fly (or have left the nest site in the case of penguins).

Major issues in the transition from dependence to independence include protection by parents from predators or conspecifics, protection from inclement weather or thermal stress, provisioning by parents, and learning or observing foraging techniques or areas by young birds (Burger 1980; see Chapters 6, 7, 11, 13). Regardless of the ambiguity of the term "fledging," all seabirds

FIGURE 8.15 Fledgling Red-footed Boobies spend their days exploring their environment, learning how to use their bills well, and perfecting their flying skills. They return to the nest at night to be fed by their parents but soon will be on their own. (Photo by R. W. and E. A. Schreiber.)

eventually are completely independent of their parents. The period of postfledging care is suggested to allow young birds time to learn to fly efficiently and feed themselves while still returning to the nest to get fed by their parents. For instance, Red-footed Booby fledglings associate in groups and explore their surroundings during the day. They may pick up sticks and play with them, land in trees, and poke at things to investigate them (Schreiber et al. 1996; Figure 8.15). This innate curiosity may be a behavior that enables finding food at sea, where any disturbance in the surface could be worth investigating.

While it is relatively easy to determine the duration of prefledging care (see Appendix 2; Burger 1980), it is often difficult to determine how long young are dependent after fledging because parents and young leave the colony site and disperse far along coasts or out to sea. Further, young may be fed only once a day, or much less often, making it difficult to observe these interactions, even for species that remain along coasts (Burger 1981). For species that travel far from their breeding colonies or wander at sea (Nelson 1975), following them to ascertain postfledging care is nearly impossible. Another difficulty with understanding postfledging care is that many authors often note that parental care occurs for many days or weeks following fledging, but do not give specific time periods. Short-term postfledging care is possible to observe (Phillips et al. 1998), often because some seabirds remain near the natal colony. With more detailed study of the period following fledging in a range of seabirds, it will be possible to correlate other aspects of seabird breeding biology with the degree and intensity of postfledging care.

It is worthwhile considering how the presence or absence of postfledging parental care, in most cases limited to feeding, varies in seabirds (Table 8.4). In this table we show only those species with care, not the duration of that care, largely because of the difficulties discussed above. None-theless, data on duration of postfledging care, given in Appendix 2, indicate great variability among closely related species. Most penguins have no postfledging care, yet Gentoo Penguins have extended care of 50 days. While most procellariiformes have no postfledging care, Black-footed Albatross (*Phoebastria nigripes*) have about 40 days.

Postfledging parental care is commonest and most extended in species in the Pelecaniformes and Charadriiformes (Table 8.4), with Great Frigatebirds reported to have a period of over 400 days in some cases, and Black-naped Terns (*Sterna sumatrana*) having a period of 180 days. In total, 98% of the gulls, terns, and skimmers have postfledging care, while only 23% of the alcids are reported to have such care. While many gulls have diverse feeding methods (Chapter 6), terns specialize in diving which may be a difficult skill to perfect and is perhaps the reason for their longer period of postfledging care.

A detailed analysis of postfledging parental care can be found in Burger (1980). Remarkably, relatively little attention has been devoted to this topic since that time, largely because of the difficulties of obtaining this information. Perhaps with satellite and radio-telemetry, more data will

TABLE 8.4
Postfledging Care in the Four Main Seabird Orders

Order	With Information (No.)	With Care (No.)	With Care (%)
Sphenisciformes	11	2	18
Procellariiformes	36	3	8
Pelecaniformes	30	24	80
Charadriiformes	69	57	83

Note: Shown are (1) the number of species with information, (2) the number of those that have postfledging care, and (3) the percent of species within each order with postfledging care. Data from Appendix 2.

be gathered on this topic, such as determining whether parents and young remain together after leaving the colony, as well as whether members of a pair remain together during migration and over-wintering. Detailed behavioral observations of interactions between parents and young in the weeks and months following fledging are necessary to understand the nature and importance of this care to different aspects of foraging and, ultimately, survival.

8.5.5 SURVIVAL

There are two types of postfledging survival that ornithologists monitor: (1) survival of young birds to breeding age, and (2) annual survival of adults. There are few good data on either type of survival from populations of marked birds. Mortality is highest in young birds, possibly because many do not learn how to feed themselves (Orians 1969, Dunn 1972). For instance, survival to breeding age in Red-tailed Tropicbirds ranges from 40 to 45% (17-year study; Schreiber 1999) and in Roseate Terns it is estimated at 20% (12-year study; Nisbet and Spendelow 1999). In Appendix 2 there are estimates for annual survival of adults in 64 species ranging from 61.7% in the Jackass Penguin (*Spheniscus demersus*) to 97.3% in the Light-mantled Sooty *Albatross (Phoebetria palpebrata)*. The majority of the estimates in Appendix 2 are based on 2- to 4-year studies and more long-term data are needed. Yet, the available data consistently show that seabirds have high annual survival: 90% of the species for which we have data have annual survival of 80% or higher.

Annual survival can be affected by factors such as age, number of years in which a bird has bred, climate conditions, and predation rates. Younger breeders were found to have lower survival than more experienced breeders in some species (Black-legged Kittiwakes, Coulson and Wooller 1976; Northern Fulmars, Dunnet and Ollason 1978b; Adelie Penguins, Ainley and DeMaster 1980) and survival can decline with age (Dunnet and Ollason 1978b, Wooller et al. 1989).

Survivorship is highest in procellariiformes and lowest in penguins (Table 8.5; factorial ANOVA of arcsine transformed data; $F_{1,71} = 23.4$, $p = 0.01$), and in all four orders, survivorship is higher in pelagic foragers than inshore foragers (Table 8.5; $F_{1,71} = 16.8$, $p = 0.03$), with no interaction between order and foraging distance ($F_{3,71} = 0.3$, $p = 0.9$). Possible reasons for these differences are discussed in Section 8.8 below, but commitments to long-term studies in seabirds are badly needed if we are to come to understand the relationships between demographic parameters and seabird life histories.

8.5.6 AGE AT FIRST BREEDING

Seabirds do not attain sexual maturity until they are several years old (Table 8.1; see Appendix 2 for data on individual species and references). Age at first breeding is related to annual adult survival (Figure 8.16; Pearson correlation of arcsine and log-transformed data; $r = 0.57$, $n = 68$, $p < 0.01$)

TABLE 8.5
Annual Adult Survivorship of Seabirds by Order and Foraging Range

Order	Pelagic Foragers			Nearshore Foragers		
	Mean	SD	n	Mean	SD	n
Sphenisciformes	0.81	0.11	5	0.84	0.06	6
Procellariiformes	0.91	0.02	3	0.92	0.05	25
Pelecaniformes	0.85	0.02	8	0.91	0.02	4
Charadriiformes	0.85	0.06	25	0.89	0.07	3

Note: SD = standard deviation and n = number of species for which there are data. Data from Appendix 2.

and varies from 2 to 3 years (tropicbirds, Schreiber and Schreiber 1993; gulls and some terns, Burger and Gochfeld 1996) to 8 to 11 years (albatross, some petrels, and possibly frigatebirds; Marchant and Higgins 1990). Only 3% of the variation among species in age at first breeding is explained by body mass, although the regression is significant ($F_{1,114} = 4.5$, $p = 0.037$). Age at first breeding is significantly higher in procellariiformes and lower in pelecaniformes than in the remaining two orders (Table 8.6; multifactorial ANOVA with log-transformed data; $F_{3,103} = 30.6$, $p < 0.001$). There is no effect of latitude, foraging distance, or nest type and there are no interactions among factors ($p > 0.1$ in all cases).

The reasons for the long period of immaturity in seabirds are not known but may be related to: (1) time required to perfect feeding methods and become efficient at locating prey (Orians 1969, Irons 1998); (2) time required to attain physiological maturity (Ainley 1978); or (3) time required to learn the behaviors related to mating and claiming a territory (Sooty Terns, Harrington 1974; boobies, Nelson 1978; alcids, Hudson 1985). There is some evidence to suggest that frigatebirds have unusually high ages at first breeding among the Pelecaniformes, which Nelson (1975, 1977) suggests may have resulted from their difficult feeding technique (snatching fish from the air as they escape from predatory fish beneath the surface). However, we have few data on actual age at first breeding in frigatebirds (which may be as young as 3 years in some cases; Diamond 1975), nor on the degree of difficulty in learning to catch prey by this technique.

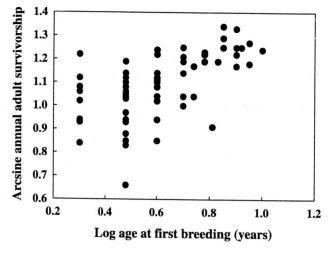

FIGURE 8.16 The relationship between log age at sexual maturity and arcsine annual adult survivorship in seabirds.

TABLE 8.6
Mean Age at First Breeding in Different
Orders of Seabirds

Order	Mean	SD	n
Sphenisciformes	3.49[a]	1.59	12
Procellariiformes	5.30[b]	1.52	40
Pelecaniformes	2.39[c]	1.19	23
Charadriiformes	3.27[a]	1.30	41

Note: Mean age is followed by standard deviation and number of species for which there are data. Means followed by a different letter are significantly different from each other. Data from Appendix 2.

8.5.7 RELATIONSHIPS AMONG LIFE HISTORY TRAITS

To examine the relationships among life history traits, we used a principal components analysis (Picozzi et al. 1992) for 48 species (8 penguins, 19 petrels, 7 Pelecaniformes, 14 Charadriiformes) for which we had values for all of the above seven life history variables (Table 8.7), with annual adult survival arcsine transformed and all other variables log transformed. The analysis extracted two components (Principal Component 1, PC 1; Principal Component 2, PC 2), which accounted for 48 and 24% of the variance in the data, respectively (Figure 8.17). The PC 1 score increased (in order from greatest to least importance) with increasing body mass, chick growth rate, and chick-rearing period. The PC 2 score increased with decreasing clutch size and with increasing incubation period, survivorship, and age at sexual maturity (Table 8.7). Thus PC 1 primarily measured body size, whereas PC 2 primarily measured survivorship and maximum reproductive output from each breeding attempt.

PC 1 was higher in penguins and pelecaniformes than in the other two orders (Figure 8.17: $F_{3,44} = 15.2$, $p < 0.001$), reflecting their large body masses (Table 8.8), but did not vary with latitude,

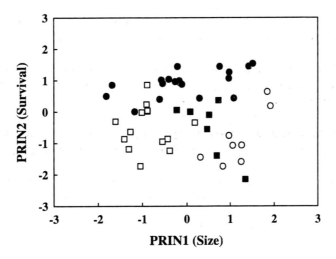

FIGURE 8.17 Principal component scores for SIZE and SURVIVAL components of life history variation in 48 species of seabirds, by order. Sphenisciformes, open circles; procellariiformes, solid circles; pelecaniformes, solid squares; charadriiformes, open squares. Size described 48% of the variability and survival described 24%. See text for discussion.

TABLE 8.7
A Principal Components Analysis of the Relationship among
Seven Life History Traits

	Principal Component 1	Principal Component 2
Body mass	**0.964**	0.067
Clutch size	0.150	**–0.858**
Incubation period	0.503	**0.764**
Chick-rearing period	**0.750**	0.438
Chick growth rate	**0.952**	–0.059
Age at sexual maturity	0.164	**0.741**
Annual adult survivorship	0.095	**0.755**

Note: Traits for which we had values for 48 species (8 penguins, 19 petrels, 7 Pelecaniformes, 14 Charadriiformes; annual adult survival arcsine transformed and all other variables log transformed). Principal Component 1 and 2 accounted for 48 and 24% of the variance in the data, respectively. Variables contributing most to each component are in bold type. Data from Appendix 2.

foraging range, or nest type, with no interactions among factors ($p > 0.3$ in all cases). PC 2 was consistently higher in pelagic foragers (mean = 0.49, n = 30, SD = 0.83) than nearshore foragers (mean = –0.81, n = 18, SD = 0.70; two-way ANOVA by order and foraging range; $F_{1,40} = 22.6$, $p = 0.01$), reflecting smaller clutches and higher adult survivorship in the former (see above). PC 2 also differed among orders (Figure 8.17; $F_{3,40} = 11.3$, $p = 0.038$) being highest in petrels and lowest in penguins (Table 8.8), with no interaction between factors ($F_{3,40} = 0.8$, $p = 0.5$) and no influence of latitude or nest type. The implications of these data are discussed in Section 8.8 below.

8.6 BREEDING PERFORMANCE AND LIFE HISTORY–ENVIRONMENT INTERACTIONS

8.6.1 AGE-SPECIFIC SURVIVAL AND FECUNDITY

Many aspects of reproduction vary as a function of adult age, improving steadily from first breeding up to the onset of senescence and declining thereafter. For instance, the proportion of two-egg clutches in Great Skuas increased from 0.73 in 5-year-olds to a peak of 0.93 in 18-year-olds before declining to 0.82 in 29-year-olds (Ratcliffe et al. 1998). The volume of two-egg clutches had a similar relationship with age, reaching a peak at 14 years and declining beyond 15 years (Hamer and Furness 1991). Overall breeding success has also been shown to increase with age in many species (see Wooller et al. 1992, Forslund and Pärt 1995 for reviews), although this depends to some extent on environmental conditions. For instance, in Great Skuas, hatching and fledging success are uniformly low when food supply is very poor, increasing with age when food supply is intermediate, and uniformly high when food is abundant (Furness 1984, Hamer and Furness 1991, Ratcliffe et al. 1998).

Age-specific increases in reproductive output have been attributed to a lack of breeding and/or foraging experience in young breeders and to an increased reproductive effort in old birds to offset declining reproductive opportunities. There also may be differential survival of individuals, resulting in a higher proportion of more successful breeders in older cohorts. There is evidence in support of all of these processes and at present it appears that they may all operate in different species, without a single, identifiable common pattern (Wooller et al. 1992).

Seabirds have often been cited as examples of long-lived animals with constant adult survival rates throughout life. However, with increasingly long-term data sets, decreases in annual survival

FIGURE 8.18 A White Tern adult carries fish in its bill to feed its chick. Some pairs are able to raise two broods (one chick each) per year. (Photo by E. A. and R. W. Schreiber.)

after many years of breeding have now been recorded in a number of species including Short-tailed Shearwaters (Bradley et al. 1989), Brandt's Cormorants (Nur and Sydeman 1999), and Great Skuas (Ratcliffe et al. in press). Both reproductive success and adult survival may be heavily dependent on environmental conditions, particularly during catastrophic events associated with climatic and oceanographic anomalies (Schreiber and Schreiber 1984, 1989, Duffy 1990). These changes may have important repercussions for the regulation of seabird population sizes and are discussed in more detail by Weimerskirch in Chapter 5.

8.6.2 Breeding Frequency

Most species breed annually but some take occasional years off from breeding and a few species raise two broods per year. Both sabbatical years and raising two broods in a year may be more common than we currently realize but few marked populations of seabirds have been studied. One study of Short-tailed Shearwaters found that 28% of individuals known to be alive and to have bred before were either not present at the colony or present but not breeding in any one year (Bradley et al. 2000). The birds most likely to defer breeding are either young and inexperienced or old and senescent or lower quality individuals (Calladine and Harris 1997, Mougin et al. 1997, Catry et al. 1998). The frequency of such sabbatical years possibly increases as a function of adult survivorship (Chastel et al. 1995, Catry et al. 1998). Nonbreeding is associated with poor body condition at the start of the breeding season (Chastel et al. 1995, Osorno 1999) and may represent a life history decision to avoid the costs of reproduction at a stage when these costs might be too high (but see Bradley et al. 2000). Hence the incidence of nonbreeding at a colony is correlated with environmental conditions and is higher in years of poor breeding success (Schreiber and Schreiber 1989, Crawford and Dyer 1995, Catry et al. 1998, Nur and Sydeman 1999). Double-brooding occurs most often in seabirds with shorter breeding cycles (see discussion above; Figure 8.18).

8.6.3 Adult Quality

In seabirds and other species, there are often significant differences among individuals in their lifetime reproductive success, with the result that a small proportion of individuals is responsible for most of the next generation (see Partridge 1989 for a review; Wendeln and Becker 1999). For seabirds, differences in quality among individuals appear to be linked to factors that influence the probability of having a long breeding life span, and in Red-billed Gulls some of this variation is linked to body size (Mills 1989). However, in Short-tailed Shearwaters, no associations were found with size or other morphological characteristics (Wooller et al. 1989). Moreover, in Great Skuas, a number of indices of adult quality were highly repeatable in consecutive years but declined rapidly over longer periods, suggesting that apparent individual quality may be only a transient attribute (Catry et al. 1999). We do not yet know whether factors that influence lifetime reproductive success are heritable or whether contingency is largely responsible, and critical evaluation will require better data on the characteristics of high-quality individuals and the importance of cohort effects (Wooller et al. 1992).

8.7 POSTBREEDING BIOLOGY

After the breeding season, the migration patterns of seabirds range from the long-distance travels of Arctic Terns (*Sterna paradisaea*) and South Polar Skuas (*Catharacta maccormicki*) that traverse northern and southern hemisphere polar regions (Salomonsen 1967, Furness 1987), to not leaving the colony site at all (e.g., Brown Boobies on Ascension, Dorward 1962). Coastal nesting species of gulls and terns may not migrate, but just disperse from the colony area. Greater migration distances occur in polar and subpolar nesting species than in species that can continue to find food near the colony throughout the year. The main reason for leaving the colony area is probably related to food availability in many cases, although some cases are less clear.

There are some data to indicate that migration routes are dictated by prevailing winds, with birds choosing the routes of least energy expenditure by taking advantage of global wind patterns. However, we actually know little about where many species of seabirds go once they leave the breeding colony. Most seabirds undergo their major molt cycle during the nonbreeding season, probably as a way to avoid overlapping the energetic cost of breeding and that of molting. The timing of molt varies among orders and species, as does migration distance and timing, and each can affect the other (Marshall 1956, Mean and Watmough 1976). Penguins do not undergo their major molt and migrate at the same time, possibly to avoid being in the water during molt. After breeding they go to sea, build up their body reserves (10 to 70 days), then return to land again to molt (Croxall 1982). Energy expenditure during molt is about twice that during incubation, and Emperor Penguins lose up to 45% of their initial mass during this period (Williams 1995). Most penguins leave the breeding colony and go to sea during the nonbreeding season and adults may not travel as far as do juveniles (Williams 1995).

Waved Albatrosses (*Phoebastria irrorata*) in the Galapagos Islands probably have the shortest migration, heading to the rich feeding waters of the neighboring Humboldt Current. Most of the southern ocean albatrosses disperse over the southern hemisphere ocean and some circumnavigate the globe (Carboneras 1992). When not breeding, some of the more oceanic species range over the open ocean, not needing to go to land, as they can roost on the water at night or stay in the air.

Long transequatorial migrations occur in several petrel species: Manx Shearwaters (Thompson 1965), Sooty Shearwaters (*P. griseus*, Carboneras 1992), Short-tailed Shearwaters (Serventy 1967). Great Shearwaters (*Puffinus gravis*) migrate in a figure eight, from the southern ocean to the north Atlantic and back again. Several species of storm-petrel migrate long distances from high latitudes in the northern and southern hemispheres to high latitudes in the opposite hemisphere (Wilson's Storm-petrels, *Oceanites oceanicus*; Leach's Storm-petrels, White-faced Storm-petrels, *Pelagodroma marina;* and Black-bellied Storm-petrels, *Fregetta tropica*; Carboneras 1992). Others are basically sedentary (Ashy Storm-petrel, *Oceanodroma homochroa*; Tristram's Storm-petrel, *O. tristrami;* Grey-backed Storm-petrel, *Garrodia nereis*; Carboneras 1992). Little is known about migration in the diving petrels but it is thought that they do not go far from the nesting colonies.

Migration in the Pelecaniformes ranges from circumnavigating the north Pacific (Lesser Frigatebirds, *Fregata ariel*; Orta 1992) to remaining on the breeding territories (Brown Boobies on

Ascension, Dorward 1962). Tropicbirds go to sea and are not seen on land until the next nesting season (Schreiber and Schreiber 1993). The American White Pelican (*Pelecanus erythrorhynchos*) migrates from northern and central Canada and the U.S. to the southern states, Baja, and the Gulf of Mexico (Anderson 1984). Brown Pelicans in Florida move both north and south from the breeding sites, while those nesting farther north move to the south (Schreiber and Mock 1988). Among boobies, Northern Gannets all migrate south to warmer climes (Carboneras 1992). There are few data on boobies, which tend to spread out widely over tropical waters, not necessarily needing to roost on land at night. Hundreds of immature Red-footed Boobies from French Frigate Shoals roost on Johnston Atoll, 850 km to the south (returning home to nest; Schreiber et al. 1996), but we do not know where the adults go when not breeding.

Cormorants are mostly sedentary, dispersing small distances from colony sites. Those nesting in harsher climes move to ice-free areas to find food. Double-crested Cormorants (*Hypoleucos auritus*) may migrate the farthest, going from Alaska to Baja, or from Maine to the Caribbean (Harrison 1983, Orta 1992). It is thought that adult frigatebirds are fairly sedentary, remaining near nesting areas, while immatures go vast distances during the 3 to 7 years it takes them to mature.

Gulls and terns also present considerable variety in their nonbreeding season movements. Laughing Gulls (*Larus atricilla*) and Franklin's Gulls may travel thousands of kilometers to the south of their breeding grounds (Southern 1980). Franklin's Gull, the gull that migrates the farthest, even has a complete molt twice a year (Burger and Gochfeld 1994a). Many gull species simply disperse from the breeding area, not migrating any great distance, and most stay in coastal waters. Even within a species, such as Herring Gulls and Laughing Gulls, some individuals nesting in lower latitudes fail to migrate at all, while those nesting in higher latitudes may migrate hundreds of miles. Other species that nest entirely in tropical regions, such as Lava Gull (*Larus fuliginosus*) and Swallow-tailed Gull, remain relatively close to their nesting islands.

There are data for a few species to suggest that juvenile birds migrate to different areas than adults, although many of them eventually return to their natal colony to breed. Coulson (1966) found that young Black-legged Kittiwakes winter farther south than do adults. Lesser Frigatebirds from the central Pacific follow the trades winds west and then north, with young birds going farther than adults (Sibley and Clapp 1967). Juvenile Brown Boobies, but not adults, from Johnston Atoll are recovered in and around French Caledonia (E. A. Schreiber unpublished). Band recoveries of juvenile Sooty Terns from Florida showed that they go to the coast of west Africa and offshore (Robertson 1969). Juveniles of several gull species range much farther during their maturing years than do adults in the postbreeding season (Southern 1980). We do not know the reasons for juvenile birds moving to different areas from adults, but it was suggested by Schreiber and Schreiber (1989) that this behavior may have evolved as a result of living in environments with stochastic weather patterns which can cause extensive mortality on unpredictable, varying scales. A population of juveniles away from the event may survive to repopulate a devastated colony.

8.8 THE EVOLUTION OF SEABIRD LIFE HISTORIES

Discussion of seabird life histories has focused on their low annual reproductive rates, which have been taken as suggesting severe limits to the rates at which adults can provide food for their young (see Ydenberg and Bertram 1989, Ricklefs 1990 for reviews). However, low annual reproductive output should not be viewed in isolation, as indicated in Section 8.5.7 above, where in a principal-components analysis, one quarter of the variation in seabird life history traits was explained by a single component combining high annual adult survival, delayed sexual maturation, and low annual clutch size (Table 8.8).

Optimum annual reproductive output is that which maximizes the difference between breeding offspring raised in the current attempt and future offspring lost as a result of parental mortality (Charnov and Krebs 1974, Partridge 1989, Charnov 1993). Thus whenever extrinsic mortality (independent of reproduction) is low, natural selection will favor a reduction in parental investment

TABLE 8.8
Mean Values of Seven Life History Variables

	Sphenisciformes		Procellariiformes		Pelecaniformes		Charadriiformes	
	mean	SD	mean	SD	mean	SD	mean	SD
Body mass (kg)	6.2	7.6	1.3	1.9	2.3	2.1	0.5	0.4
Clutch size	1.9	0.3	1.0	0.0	2.5	1.2	2.0	0.8
Incubation period	39.7	8.3	54.1	10.2	36.8	9.4	27.2	4.9
Chick-rearing period	95.1	69.1	98.7	47.4	83.5	37.4	35.7	12.5
Chick growth rate	75.0	34.6	25.7	25.4	44.9	35.9	18.1	20.2
Age at sexual maturity	3.8	1.6	5.0	2.3	2.6	1.0	3.4	1.1
Annual adult survivorship	0.82	0.08	0.91	0.06	0.86	0.06	0.88	0.07
Size	1.18	0.52	−0.04	0.97	0.51	0.49	−0.88	0.47
Survival	−0.85	0.85	0.93	0.42	−0.54	0.91	−0.50	0.09

Note: From the principal components analysis presented in Table 8.7, and mean scores for the SIZE and SURVIVAL components of variation (PC 1 and PC 2, respectively) in the four orders of seabirds.

to reduce mortality related to reproduction and preserve the potentially long life span of individuals (Saether et al. 1996, Ricklefs 2000). Seabirds have high annual survivorship and so selection would be expected to lead to small clutch sizes independently of any constraints on the maximum rate of food delivery to the nest. Protracted breeding seasons could also result not from energy limitation but from physiological constraints on the rate of maturation of embryos and chicks (Konarzewski et al. 1990, Ricklefs et al. 1998, Phillips and Hamer 2000b), which may be particularly important for offspring with a long potential life span (Ricklefs 1990).

Most studies of optimum annual reproductive output have assumed that the environment stays constant, so that the costs and benefits of a given level of reproductive investment do not vary between breeding attempts. For seabirds this has long been viewed as a reasonable assumption. However, we now know that the marine environment is highly variable on all scales such that individuals are likely to encounter several years of very low breeding success during their lifetimes (Schreiber and Schreiber 1989, Hamer et al. 1993, Schreiber 1994, Nur and Sydeman 1999). Under these circumstances, individuals that lay large clutches may do well in good years, but they will do particularly badly in poor years and may additionally suffer higher mortality as a result of their greater reproductive effort. Long-term fitness is then increased by always laying the optimum clutch size for bad years, especially because each offspring produced in a bad year forms a higher proportion of its cohort than does an offspring from a good year (Partridge 1989). If bad years are more frequent in tropical environments, for instance due to El Niño events, then it could be this rather than lower average food availability that has led to lower clutches in tropical species (see Section 8.5.1 above).

In conclusion, seabird life histories are undoubtedly related to the conditions of the marine environment, but the traditional explanation that they are entirely due to the patchy and unpredictable nature of marine food resources needs to be revised in the light of available evidence. In some species, adults do appear to have little capacity to increase their food provisioning rates, but this limitation is due primarily to adults meeting their own nutritional requirements before those of their offspring (Weimerskirch et al. 2000). Moreover, other species do not appear to be limited in this way (e.g., Schreiber 1996). Thus greater attention should be given to a range of explanations for the entire suite of attributes that characterize seabird life histories.

LITERATURE CITED

AINLEY, D. G., AND R. J. BOEKELHEIDE (Eds.). 1990. Seabirds of the Farallon Islands. Ecology, Dynamics and Structure of an Upwelling-System Community. Stanford University Press, Stanford, CA.

AINLEY, D. G., AND D. P. DEMASTER. 1980. Survival and mortality in a population of Adelie Penguins. Ecology 61: 522–530.

AINLEY, D. G., AND R. E. LERESCHE. 1973. The effects of weather and ice conditions on breeding in Adelie Penguins. Condor 75: 235–239.

AINLEY, D. G., R. E. LERESCHE, AND W. J. L. SLADEN. 1983. Breeding biology of the Adelie Penguin. University of California Press, Berkeley.

ANDERSON, D. J., AND P. J. HODUM. 1993. Predator behavior favors clumped nesting in an oceanic seabird. Ecology 74: 2462–2464.

ANDERSON, D. W. 1984. Pelicans. Family Pelecanidae. Pp. 84–91 in Seabirds of North Pacific and Arctic Waters (D. Haley, Ed.). Pacific Search Press, Washington.

ASHMOLE, N. P. 1962. The Black Noddy Anous tenuirostris on Ascension Island. I. General biology. Ibis 103b: 235–273.

ASHMOLE, N. P. 1963. The biology of the Wide-awake or Sooty Tern Sterna fuscata on Ascension Island. Ibis 103b: 297–364.

ASHMOLE, N. P. 1968. Breeding and molt in the White Tern (Gygis alba) on Christmas Island, Pacific Ocean. Condor 70: 35–55.

ASHMOLE, N. P. 1971. Seabird ecology and the marine environment. Pp. 223–286 in Avian Biology No. 1 (D. S. Farner and J. R. King, Eds.). Academic Press, New York.

BEER, C. G. 1966. Adaptations to nesting habitat in the reproductive behaviour of the Black-billed Gull, Larus bulleri. Ibis 108: 394–410.

BIRKHEAD, T. R. 1977. The effect of habitat and density on breeding success in the Common Guillemot (Uria aalge). Journal of Animal Ecology 46: 751–764.

BIRKHEAD, T. R., AND M. P. HARRIS. 1985. Ecological adaptations for breeding in the Atlantic Alcidae. Pp. 205–231 in The Atlantic Alcidae (D. N. Nettleship and T. R. Birkhead, Eds.). Academic Press, London.

BRADLEY, J. S., R. D. WOLLER, AND I. J. SKIRA. 2000. Intermittent breeding in the Short-tailed Shearwater Puffinus tenuirostris. Journal of Animal Ecology 69: 639–650.

BRADLEY, J. S., R. D. WOOLLER, I. J. SKIRA, AND D. L. SERVENTY. 1989. Age-dependent survival of Short-tailed Shearwaters Puffinus tenuirostris. Journal of Animal Ecology 58: 175–188.

BRIED, J., O. DURIEZ, AND G. JUIN. 1999. A first case of female-female pairing in the Black-faced Sheathbill Chionis minor. Emu 99: 292–295.

BROOKE, M. DE L. 1990. The Manx Shearwater. T. and A. D. Poyser, London.

BROWN, C. R., B. J. STUTCHBURY, AND P. D. WALSH. 1990. Choice of colony size in birds. Trends in Ecology and Evolution 5: 398–403.

BROWN, K. 1998. Proximate and ultimate causes of adoption in Ring-billed Gulls. Animal Behavior 56: 1529–1543.

BUCKLEY, F. C., AND P. A. BUCKLEY. 1980. Habitat selection and marine birds. Pp. 69–112 in Behavior of Marine Animals: Current Perspectives in Research, Vol. 4: Marine Birds. Plenum Press, New York.

BUKACINSKI, D., M. BUKACINSKA, AND T. LUBJUHN. 2000. Adoption of chicks and the level of relatedness in Common Gull, Larus canus, colonies: DNA fingerprinting analyses. Animal Behavior 59: 289–299.

BURGER, A. E., AND A. J. WILLIAMS. 1979. Egg temperatures of the Rockhopper Penguin and some other penguins. Auk 96: 100–105.

BURGER, J. 1974. Breeding adaptations of the Franklin's Gull (Larus pipixcan) to a marsh habitat. Animal Behavior 22: 521–567.

BURGER, J. 1980. The transition to independence and post fledging parental care in seabirds. Pp. 367–447 in Behavior of Marine Organisms: Perspectives in Research, Vol. 4: Marine Birds (J. Burger, B. Olla, and H. Winn, Eds.). Plenum Press, New York.

BURGER, J. 1981. On becoming independent in Herring Gulls: parent-young conflict. American Naturalist 117: 444–456.

BURGER, J. 1984. Pattern, mechanism, and adaptive significance of territoriality in Herring Gulls (Larus argentatus). American Ornithologists' Union Monograph No. 34.

BURGER, J. 1985. Habitat selection in temperate marshes. Pp. 253–281 *in* Nest Site Selection in Birds (M. Cody, Ed.). Academic Press, New York.

BURGER, J., AND M. GOCHFELD. 1990. The Black Skimmer: Social Dynamics of a Colonial Species. Columbia University Press, New York.

BURGER, J., AND M. GOCHFELD. 1991. The Common Tern: Its Breeding Biology and Social Behavior. Columbia University Press, New York.

BURGER, J., AND M. GOCHFELD. 1994a. Franklin Gull. Pp. 1–28 *in* Birds of North America. No. 116 (A. Poole and F. Gill, Eds.). Academy of Natural Sciences, Philadelphia.

BURGER, J., AND M. GOCHFELD. 1994b. Predation and effects of humans on island-nesting seabirds. Pp. 39–67 in Seabirds on Islands: Threats, Case Studies and Action Plans (D. N. Nettleship, J. Burger, and M. Gochfeld, Eds.). BirdLife International, Cambridge, U.K.

BURGER, J., AND M. GOCHFELD. 1996. Seasonal changes in use of space by nesting Black-billed Gulls (*Larus bulleri*): behavioral constraints of unpredictable environments. Emu 96: 73–80.

BURGER, J., AND F. LESSER. 1978. Colony and nest site selection in Common Terns. Ibis 120: 443–449.

BUTTEMER, W., AND L. B. ASTHEIMER. 1990. Thermal and behavioural correlates of nest site location in Black Noddies. Emu 90: 114–118.

BYRD, J., A. I. HOUSTON, AND P. SOZOU. 1991. An examination of optimal fledging times for the Common Murre. Ecology 72: 1893–1896.

CALLADINE, J., AND M. P. HARRIS. 1997. Intermittent breeding in the herring gull *Larus argentatus* and the Lesser Black-backed Gull *Larus fuscus*. Ibis 139: 259–263.

CARBONERAS, C. 1992. Order Procellariiformes, Family Diomedeidae. Pp. 198–214 *in* Handbook of Birds of the World, Vol. 1 (J. del Hoyo, A. Elliott, and J. Sargatoa, Eds.). Lynx Edicions, Barcelona.

CATRY, P., R. A. PHILLIPS, K. C. HAMER, N. RATCLIFFE, AND R. W. FURNESS. 1998. The incidence of nonbreeding by adult Great Skuas and Parasitic Jaegers from Foula, Shetland. Condor 100: 448–455.

CATRY, P., N. RATCLIFFE, AND R. W. FURNESS. 1997. Partnerships and mechanisms of divorce in the Great Skua, *Catharacta skua*. Animal Behavior 54: 1475–1482.

CATRY, P., G. D. RUXTON, N. RATCLIFFE, K. C. HAMER, AND R. W. FURNESS. 1999. Short-lived repeatabilities in long-lived Great Skuas: implications for the study of individual quality. Oikos 84: 473–479.

CEPEDA, F., AND J. B. CRUZ. 1994. Status and management of seabirds on the Galapagos Islands, Ecuador. Pp 268–278 *in* Seabirds on Islands: Threats, Case Studies and Action Plans (D. N. Nettleship, J. Burger, and M. Gochfeld, Eds.). BirdLife International, Cambridge, U.K.

CHARDINE, J. W. 1987. The influence of pair status on the breeding behaviour of the Kittiwake *Rissa tridactyla* before laying. Ibis 129: 515–526.

CHARNOV, E. L. 1993. Life History Invariants. Some Explorations of Symmetry in Evolutionary Ecology. Oxford University Press, Oxford.

CHARNOV, E. L., AND J. R. KREBS.1974. On clutch size and fitness. Ibis 116: 217–219.

CHASTEL, O. 1995. Influence of reproductive success on breeding frequency in four southern petrels. Ibis 137: 360–363.

CHASTEL, O., H. WEIMERSKIRCH, AND P. JOUVENTIN. 1993. High annual variability in reproductive success and survival of the Antarctic seabird, the Snow Petrel *Pagodroma nivea*. Oecologia 94: 278–285.

CHASTEL, O., H. WEIMERSKIRCH, AND P. JOUVENTIN. 1995. Body condition and seabird reproductive performance — a study of three petrel species. Ecology 76: 2240–2246.

CHAURAND, T., AND H. WEIMERSKIRCH. 1994. The regular alternation of short and long trips in the Blue Petrel *Halobaena caerulea*: a previously undescribed strategy of food provisioning in a pelagic seabird. Journal of Animal Ecology 63: 275–282.

CHEREL, Y., J. GILLES, Y. HANDRICH, AND Y. Le MAHO. 1994. Nutrient reserve dynamics and energetics during long-term fasting in the King Penguin (*Aptenodytes patagonicus*). Journal of Zoology London 234: 1–12.

CLARK, L., R. E. RICKLEFS, AND R. W. SCHREIBER. 1983. Nest-site selection by the red-tailed tropicbird. Auk 100: 953–959.

CONOVER, M. R. 1984. Occurrence of supernormal clutches among the Laridae. Wilson Bulletin 96: 249–267.

COULSON, J. C. 1966. The movements of the Kittiwake. Bird Study 13: 107–115.

COULSON, J. C. 1991. The population dynamics of culling Herring and Lesser Black-backed Gulls. Pp. 479–497 *in* Bird Population Studies: Relevance to Conservation and Management (C. M. Perrins, J.-D. Lebreton, and G. J. M. Hirons, Eds.). Oxford University Press, Oxford.

COULSON, J. C., AND J. M. PORTER. 1985. Reproductive success of the Kittiwake *Rissa tridactyla*: the role of clutch size, chick growth rates and parental quality. Ibis 127: 450–466.

COULSON, J. C., AND R. D. WOOLLER. 1976. Differential survival rates among breeding kittiwake gulls. Journal of Animal Ecology 45: 205–213.

COULSON, J. C., AND R. D. WOOLLER. 1984. Incubation under natural conditions in the Kittiwake Gull, *Rissa tridactyla*. Animal Behavior 32: 1204–1215.

CRAWFORD, R. J. M., AND B. M. DYER. 1995. Responses by four seabird species to a fluctuating availability of Cape Anchovy *Engraulis capensis* off South Africa. Ibis 137: 329–339.

CROXALL, J. P. 1981. Aspects of the population demography of Antarctic and Subantarctic seabirds. Colloque sur les ecosystèmes subantarctiques, Paimpont, Committee Nat. French Res. Antarctica 51: 479–488.

CROXALL, J. P. 1982. Energy costs of incubation and moult in petrels and penguins. Journal of Animal Ecology 51: 177–194.

CROXALL, J. P. 1991. Seabird Status and Conservation: A Supplement. International Council for Bird Preservation, Technical Publication No. 11.

CROXALL, J. P., P. G. H. EVANS, AND R. W. SCHREIBER (Eds.). 1984. Status and Conservation of the World's Seabirds. International Council for Bird Preservation, Technical Publication No. 2.

CULLEN, E. 1957. Adaptation in the kittiwake to cliff-nesting. Ibis 99: 275–303.

CUTHBERT, F. J. 1985. Mate retention in Caspian Terns. Condor 87: 74–78.

DANCHIN, E., AND R. H. WAGNER. 1997. The evolution of coloniality: the emergence of new perspectives. Trends in Ecology and Evolution 12: 342–347.

DIAMOND, A. W. 1975. Biology and behaviour of frigatebirds *Fregata* spp. on Aldabra Atoll. Ibis 117: 302–323.

DIAMOND, A. W., AND E. A. SCHREIBER. In press. The Magnificent Frigatebird Fregata magnificens. *In* The Birds of North America (A. Poole and F. Gill, Eds.). Academy of Natural Sciences, Philadelphia; American Ornithologists' Union, Washington, D.C.

DORWARD, D. F. 1962. Comparative biology of the White Booby and the Brown Booby *Sula* sp. at Ascension. Ibis 103b: 174–220.

DORWARD, D. F. 1963. The fairy tern *Gygis alba* on Ascension Island. Ibis 103b: 365–389.

DUBOIS, F., F. CEZILLY, AND M. PAGEL. 1998. Mate fidelity and coloniality in waterbirds: a comparative analysis. Oecologia 116: 433–440.

DUFFY, D. C. 1983. Competition for nesting space among Peruvian guano birds. Auk 100: 680–688.

DUFFY, D. C. 1984. Nest site selection by Masked and Blue-footed Boobies on Isla Espanola, Galapagos. Condor 86: 301–304.

DUFFY, D. C. 1990. Seabirds and the 1982–84 El Niño–Southern Oscillation. Pp. 395–415 *in* Global Ecological Consequences of the 1982–83 El Niño–Southern Oscillation (P. W. Glynn, Ed.). Elsevier, Amsterdam.

DUNN, E. K. 1972. Effect of age on the fishing ability of Sandwich Terns *Sterna sandvicensis*. Ibis 114: 360–366.

DUNNET, G. M., AND J. C. OLLASON. 1978a. Survival and longevity in the Fulmar. Ibis 120: 124–125.

DUNNET, G. M., AND J. C. OLLASON. 1978b. The estimation of survival rate in the fulmar, *Fulmarus glacialis*. Journal of Animal Ecology 47: 507–520.

EMMS, S. K., AND N. M. VERBEEK. 1989. Significance of the pattern of nest distribution in the Pigeon Guillemot (*Cepphus columba*). Auk 106: 193–202.

FASOLA, M., AND L. CANOVA. 1992. Nest habitat selection by eight synoptic species of Mediterranean gulls and terns. Colonial Waterbirds 15: 169–191.

FISHER, H. I., AND M. L. FISHER. 1969. The visits of Laysan Albatross to the breeding colony. Micronesia 5: 173–221.

FISHER, J., AND R. M. LOCKLEY. 1954. Seabirds. Collins, London.

FORBES, L. S., AND D. W. MOCK. 2000. A tale of two strategies: life-history aspects of family strife. Condor 102: 23–34.

FORBES, L. S., M. JAJAM, AND G. W. KAISER. 2000. Habitat constraints and spatial bias in seabird colony distributions. Ecography 23: 575–578.

FORSLUND, P., AND T. PÄRT. 1995. Age and reproduction in birds — hypotheses and tests. Trends in Ecology and Evolution 10: 374–378.

FREDERIKSEN, M. 1998. Population Dynamics of a Colonial Seabird. Analysis of a Long-Term Study of Survival, Recruitment and Dispersal in a Black Guillemot *Cepphus grylle* Population. Ph.D. thesis, University of Copenhagen.

FREDERICKSEN, M., AND A. PETERSEN. 2000. The importance of natal dispersal in a colonial seabird, the Black Guillemot *Cepphus grylle*. Ibis 142: 48–57.

FRERE, E., P. GANDINI, AND P. D. BOERSMA. 1992. Effects of nest type and location on reproductive success of the Magellanic Penguin *Spheniscus magellanicus*. Marine Ornithology 20: 1–6.

FRIESEN, V. L., D. J. ANDERSON, T. E. STEEVES, H. JONES, AND E. A. SCHREIBER. Submitted. The Nazca Booby: a challenge to the classical model of speciation.

FRIESEN, V. L., W. A. MONTEVECCHI, R. T. BARRETT, AND W. S. DAVIDSON. 1996. Molecular evidence for kin groups in the absence of large-scale genetic differentiation in a migratory seabird. Evolution 50: 924–930.

FURNESS, R. W. 1984. Influences of adult age and experience, nest location, clutch size and laying sequence on the breeding success of the Great Skua *Catharacta skua*. Journal of Zoology London 202: 565–576.

FURNESS, R. W. 1987. The Skuas. T. and A. D. Poyser, London.

GASTON, A. J. 1985. Development of young in the Atlantic Alcidae. Pp. 319–354 *in* The Atlantic Alcidae (T. R. Birkhead and D. N. Nettleship, Eds.). Academic Press, London.

GASTON, A. J., AND D. N. NETTLESHIP. 1981. The Thick-billed Murres of Prince Leopold Island, Canadian Wildlife Service Monograph Series, No. 6. Canadian Wildlife Service, Canada.

GASTON, A. J., AND I. JONES 1998. The Auks. Birds of the World. Oxford University Press, Oxford.

GASTON, A. J., C. EBERL, M. HIPFNER, AND K. LEFEVRE. 1995. Adoption of chicks among Thick-billed Murres. Auk 112: 508–510.

GLYNN, P. W. 1990. Global ecological consequences of the 1982–83 El Niño–Southern Oscillation. Elsevier Oceanography Series No. 52. New York.

GOCHFELD, M., AND J. BURGER. 1996. Sternidae (terns). Pp. 572–623 *in* Handbook of Birds of the World (J. del Hoya, A. Elliott, and J. Sargatal, Eds.). Lynx Edicions, Barcelona.

GRANADEIRO, J. P., M. UNES, M. C. SILVA, AND R. W. FURNESS. 1998. Flexible foraging strategy of Cory's Shearwater, *Calonectris diomedea*, during the chick-rearing period. Animal Behaviour 56: 1169–1176.

GRIECO, F. 1999. Nest-site limitations and colony development in tree-nesting Great Cormorants. Waterbirds 22: 417–423.

HAGELIN, J. C., AND G. D. MILLER. 1997. Nest-site selection in South Polar Skuas: balancing nest safety and access to resources. Auk 114: 638–645.

HAMER, K. C., AND R. W. FURNESS. 1991. Age-specific breeding performance and reproductive effort in Great Skuas *Catharacta skua*. Journal of Animal Ecology 60: 693–704.

HAMER, K. C., R. W. FURNESS, AND R. W. G. CALDOW. 1991. The effects of changes in food supply on the breeding ecology of Great Skuas *Catharacta skua* in Shetland. Journal of Zoology London 223: 175–188.

HAMER, K. C., AND J. K. HILL. 1993. Variation and regulation of meal size and feeding frequency in Cory's Shearwater *Calonectris diomedea*. Journal of Animal Ecology 62: 441–450.

HAMER, K. C., AND J. K. HILL. 1997. Nestling obesity and variability of food delivery in Manx Shearwaters *Puffinus puffinus*. Functional Ecology 11: 489–497.

HAMER, K. C., J. K. HILL, J. S. BRADLEY, AND R. D. WOOLLER. 2000. Contrasting patterns of nestling obesity in three species of *Puffinus* shearwaters: the role of predictability. Ibis 142: 139–158.

HAMER, K. C., A. S. LYNNES, AND J. K. HILL. 1999. Parent-offspring interactions in food provisioning of Manx Shearwaters: implications for nestling obesity. Animal Behaviour 57: 627–631.

HAMER, K. C., P. MONAGAN, J. D. UTTLEY, P. WALTON, AND M. D. BURNS. 1993. The influence of food supply on the breeding ecology of kittiwakes *Rissa tridactyla* in Shetland. Ibis 135: 255–263.

HARRINGTON, B. A. 1974. Colony visitation behavior and breeding ages of Sooty Terns (Sterna fuscata). Bird-Banding 45: 115–144.

HARRIS, M. P. 1969. The biology of storm petrels in the Galapagos Islands. Proceedings of the California Academy of Science (4th Series) 37: 95–166.

HARRIS, M. P. 1970. Breeding ecology of the swallow-tailed gull, *Creagrus furcatus*. Auk 87: 215–243.

HARRIS, M. P., D. J. HALLEY, AND S. WANLESS. 1996. Philopatry in the Common Guillemot *Uria aalge*. Bird Study 43: 134–137.

HARRIS, M. P., S. WANLESS, T. R. BARTON, AND D. A. ELSTON. 1997. Nest site characteristics, duration of use and breeding success in the Guillemot *Uria aalge*. Ibis 139: 468–476.

HARRISON, P. 1983. Seabirds: an identification guide. Houghton Mifflin Co., Boston.

HATCH, S. A. 1990. Time allocation by Northern Fulmars *Fulmarus glacialis* during the breeding season. Ornis Scandinavica 21: 89–98.

HAYWOOD, S. 1993. Sensory and hormonal control of clutch size in birds. Quarterly Review of Biology 68: 33–60.

HOUSTON, A. I., W. A. THOMPSON, AND A. J. GASTON. 1996. The use of a time and energy budget model of a parent bird to investigate limits to fledging mass in the Thick-billed Murre. Functional Ecology 10: 432–439.

HOWELL, T. R., B. ARAYA, AND W. R. MILLIE. 1974. Breeding biology of the Grey Gull, *Larus modestus*. University of California Publications in Zoology 104: 1–57.

HUDSON, P. J. 1985. Population parameters for the Atlantic Alcidae. Pp. 233–261 *in* The Atlantic Alcidae (D. N. Nettleship and T. R. Birkhead, Eds.). Academic Press, London.

IRONS, D. B. 1998. Foraging area fidelity of individual seabirds in relation to tidal cycles and flock feeding. Ecology 79: 647–655.

JARVIS, M. J. F. 1974. The ecological significance of clutch size in the South African Gannet (*Sula capensis*) Journal of Animal Ecology 43: 1–17.

JEHL, J. R., JR., AND S. A. MAHONEY. 1987. The roles of thermal environment and predation in habitat choice in the California Gull. Condor 89: 850–862.

JOUVENTIN, P., C. BARBRAUD, AND M. RUBIN. 1995. Adoption in the Emperor Penguin, *Aptenodytes forsteri*. Animal Behavior 50: 1023–1029.

KILNER, R., AND R. A. JOHNSTONE. 1997. Begging the question: are offspring solicitation behaviours signals of need? Trends in Ecology and Evolution 12: 11–15.

KING, B. R., J. T. HICKS, AND J. CORNELIUS. 1992. Population changes, breeding cycles and breeding success over six years in a seabird colony at Michaelmas Cay, Queensland. Emu 92: 1–10.

KONARZEWSKI, M. C., C. LILJA, J. KOZLOWSKI, AND B. LEWONCZUK. 1990. On the optimal growth of the alimentary tract in avian postembryonic development. Journal of Zoology, London 222: 89–101.

KRANNITZ, P. G. 1989. Nesting biology of Black Skimmers, Large-billed Terns, and Yellow–billed Terns in Amazonian Brazil. Journal of Field Ornithology 60: 216–223.

KRUUK, H. 1964. Predators and anti-predator behaviour of the Black-headed Gull, *Larus ridibundus*. Behaviour Supplement 11: 1–129.

LACK, D. 1947. The significance of clutch size. Ibis 89: 302–352.

LACK, D. 1968. Ecological adaptations for breeding in birds. Methuen, London.

LORENTSEN, S.-H. 1996. Regulation of food provisioning in the Antarctic Petrel *Thalassoica antarctica*. Journal of Animal Ecology 65: 381–388.

MANUWAL, D. A., AND A. C. THORESEN. 1993. Cassin's Auklet: *Ptychoramphus aleuticus*. No. 50 *in* The Birds of North America (A. Poole and F. Gill, Eds.). Academy of Natural Sciences, Philadelphia; American Ornithologists' Union, Washington, D.C.

MARCHANT, S., AND P. J. HIGGINS. 1990. Handbook of Australian, New Zealand and Antarctic Birds, Vol. 1. Oxford University Press, Australia.

MARSHALL, A. J. 1956. Moult adaptation in relation to long distance migration in Petrels. Nature 177: 943.

MCNICHOLL, M. K. 1975. Larid site-tenacity and group adherence in relation to habitat. Auk 92: 98–104.

MILLS, J. A. 1973. The influence of age and pair bond on the breeding biology of the Red-billed Gull *Larus novaehollandiae scopulinus*. Journal of Animal Ecology 42: 147–162.

MILLS, J. A. 1989. Red-billed Gull. Pp. 387–404 *in* Lifetime Reproduction in Birds (I. Newton, Ed.). Academic Press, London.

MILLS, J. A., J. W. YARRALL, AND D. A. MILLS. 1996. Causes and consequences of mate fidelity in Red-billed Gulls. Pp. 286–304 *in* Partnerships in Birds. The Study of Monogamy (J. M. Black, Ed.). Oxford University Press, Oxford.

MOCK, D. W., AND L. S. FORBES. 1994. Life-history consequences of brood reduction. Auk 111: 115–123.

MOCK, D., AND M. FUJIOKA. 1990. Monogamy and long-term pair-bonding in vertebrates. Trends in Ecology and Evolution 5: 39–43.

MOCK, D. W., AND P. L. SCHWAGMEYER. 1990. The peak load reduction hypothesis for avian hatching asynchrony. Evolutionary Ecology 4: 249–260.

MØLLER, A. P., AND T. R. BIRKHEAD. 1993. Cuckoldry and sociality: a comparative study of birds. American Naturalist 142: 118–140.

MONTEVECCHI, W. A. AND R. B. VAUGHAN. 1989. The ontogeny of thermal independence in nestling Gannets. Ornis Scandinavica 20: 161–168.

MONTEVECCHI, W. A., AND J. WELLS. 1984. Fledging success of Northern Gannets from different nest-sites. Bird Behaviour 5: 90–95.

MOORS, P. J., AND I. A. E. ATKINSON. 1984. Predation on seabirds by introduced animals, and factors affecting its severity. Pp 667–690 *in* Status and Conservation of the World's Seabirds (J. P. Croxall, P. G. H. Evans, and R. W. Schreiber, Eds.). International Council for Bird Preservation (Technical Publication 2), Cambridge, U.K.

MOUGIN, J.-L., C. H. R. JOUANIN, AND F. ROUX. 1997. Intermittent breeding in Cory's Shearwater *Calonectris diomedea* of Selvagem grande, North Atlantic. Ibis 139: 40–44.

MURPHY, R. C. 1936. Oceanic Birds of South America. American Museum of Natural History, New York.

NAGER, R., P. MONAGHAN, R. GRIFFITHS, D. C. HOUSTON, AND R. DAWSON. 1999. Experimental demonstration that offspring sex ratio varies with maternal condition. Proceedings of the National Academy of Sciences of the United States of America 96: 570–573.

NAGER, R., P. MONAGHAN, AND D. C. HOUSTON. 2000. Within-clutch trade-offs between the number and quality of eggs: experimental manipulations in gulls. Ecology 81: 1339–1350.

NELSON, J. B. 1964. Factors influencing clutch-size and chick growth in the Northern Gannet *Sula bassana*. Ibis 106: 63–77.

NELSON, J. B. 1975. The breeding biology of frigatebirds: a comparative review. The Living Bird 14: 113–156.

NELSON, J. B. 1977. Some relationships between food and breeding in the marine Pelecaniformes. Pp. 77–87 *in* Evolutionary Ecology (B. Stonehouse and C. Perrins, Eds.). University Park Press, London.

NELSON, J. B. 1978. The Sulidae: Gannets and Boobies. Oxford University Press, Cambridge.

NELSON, J. B. 1979. Seabirds, Their Biology and Ecology. A & W Publishers, Inc., New York.

NISBET, I. C. T., AND J. J. HATCH. 1999. Consequences of a female-biased sex-ration in a socially monogamous bird: female-female pairs in the Roseate Tern *Sterna dougallii*. Ibis 141: 307–320.

NISBET, I. C. T., AND J. A. SPENDELOW. 1999. Contribution of research to management and recovery of the Roseate Tern: review of a twelve year project. Waterbirds 22: 239–252.

NORTON, R., AND E. A. SCHREIBER. In press. The Brown Booby (*Sula leucogaster*). *In* The Birds of North America (A. Poole and F. Gill, Eds.). The Birds of North American, Inc., Philadelphia.

NUR, N., AND W. J. SYDEMAN. 1999. Survival, breeding probability and reproductive success in relation to population dynamics of Brandt's Cormorants *Phalacrocorax penicillatus*. Bird Study 46 (suppl.): S92–103.

OLLASON, J. C., AND G. M. DUNNET. 1978. Age, experience and other factors affecting the breeding success of the Fulmar *Fulmaris glacialis* in Orkney. Journal of Animal Ecology 47: 961–976.

ORO, D., AND M. GENOVART. 1999. Testing the intergenerational conflict hypothesis: factors affecting adoptions in Audouin's Gulls, *Larus audouinii*. Canadian Journal of Zoology 77: 433–439.

OLSTHOORN, J. C. M., AND J. B. NELSON. 1990. The availability of breeding sites for some British Birds. Bird Study 37: 145–164.

ORIANS, G. H. 1969. Age and hunting success in the Brown Pelican (*Pelecanus occidentalis*). Animal Behaviour 17: 316–319.

ORTA, J. 1992. Family Phalacrocoracidae. Pp. 326–353 *in* Handbook of Birds of the World Vol. 1 (J. del Hoyo, A. Elliott, and J. Sargatoa, Eds.). Lynx Edicions, Barcelona.

OSORNO, J. L. 1999. Offspring desertion in the Magnificent Frigatebird: are males facing a trade-off between current and future reproduction? Journal of Avian Biology 30: 335–341.

PARTRIDGE, L. 1978. Habitat selection. Pp 351–376 *in* Behavioural Ecology: An Evolutionary Approach (J. R. Krebs and N. B. Davis, Eds.). Sinauer Associates, Inc., Sunderland, MA.

PARTRIDGE, L. 1989. Lifetime reproductive success and life-history evolution. Pp. 421–440 *in* Lifetime Reproduction in Birds (I. Newton, Ed.). Academic Press, London.

PERRINS, C. M., M. P. HARRIS, AND C. K. BRITTON. 1973. Survival of Manx Shearwaters *Puffinus puffinus*. Ibis 115: 535–548.

PHILLIPS, R. A., R. W. FURNESS, AND F. M. STEWART. 1998. The influence of territory density on the vulnerability of Arctic Skua *Stercorarius parasiticus* to predation. Biological Conservation 86: 21–31.

PHILLIPS, R. A., AND K. C. HAMER. 1999. Lipid reserves, fasting capability and the evolution of nestling obesity in procellariiform seabirds. Proceedings of the Royal Society of London B 266: 1329–1334.

PHILLIPS, R. A., AND K. C. HAMER. 2000a. Growth and provisioning strategies of Northern Fulmars *Fulmarus glacialis*. Ibis 142: 435–445.

PHILLIPS, R. A., AND K. C. HAMER. 2000b. Postnatal development of northern fulmar chicks, *Fulmarus glacialis*. Physiological and Biochemical Zoology 73: 597–604.

PICKERING, S. P. C. 1989. Attendance patterns and behaviour in relation to experience and pair-bond formation in the Wandering Albatross *Diomedea exulans* at South Georgia. Ibis 131: 183–195.

PICOZZI, N., D. C. CATT, AND R. MOSS. 1992. Evaluation of Capercaillie habitat. Journal of Applied Ecology 29: 751–762.

PIEROTTI, R. J., AND T. P. GOOD. 1994. Herring Gull, *Larus argentatus*. Birds of North America 124: 1–28.

QUINN, J. S., L. A. WHITTINGHAM, AND R. D. MORRIS. 1994. Infanticide in skimmer and terns: side effects of territorial attacks or inter-generational conflict? Animal Behavior 47: 363–367.

RATCLIFFE, N., P. CATRY, K. C. HAMER, N. I. KLOMP, AND R. W. FURNESS. In press. The effect of age and year on the survival of breeding adult Great Skuas *Catharacta skua* in Shetland. Ibis.

RATCLIFFE, N., R. W. FURNESS, AND K. C. HAMER. 1998. The interactive effects of age and food supply on the breeding ecology of Great Skuas. Journal of Animal Ecology 67: 853–862.

REID, K., P. A. PRINCE, AND J. P. CROXALL. 2000. Fly or die: the role of fat stores in the growth and development of Grey-headed Albatross *Diomedea chrysostoma* chicks. Ibis 142: 188–198.

RICKLEFS, R. E. 1990. Seabird life histories and the marine environment: some speculations. Colonial Waterbirds 13: 1–6.

RICKLEFS, R. E. 1992. The role of parent and chick in determining feeding rates in Leach's Storm-petrel. Animal Behaviour 43: 895–906.

RICKLEFS, R. E. 2000. Density dependence, evolutionary optimization, and the diversification of avian life histories. Condor 102: 9–22.

RICKLEFS, R. E., AND W. A. SCHEW. 1994. Foraging stochasticity and lipid accumulation by nestling petrels. Functional Ecology 8: 159–170.

RICKLEFS, R. E., J. M. STARCK, AND M. KONARZEWSKI. 1998. Internal constraints on growth in birds. Pp. 266–287 *in* Avian Growth and Development (J. M. Stark and R. E. Ricklefs, Eds.). Oxford University Press, Oxford.

ROBERTS, B. D., AND S. A. HATCH. 1994. Chick movements and adoption in a colony of Black-legged Kittiwakes. Wilson Bulletin 106: 289–298.

ROBERTSON, W. B., JR. 1969. Transatlantic migration of juvenile Sooty Terns. Nature 223: 632–634.

ROBY, D. D., J. R. E. TAYLOR, AND A. R. PLACE. 1997. Significance of stomach oil for reproduction in seabirds: an interspecies cross-fostering experiment. Auk 114: 725–736.

ROWE, S., AND I. L. JONES. 2000. The enigma of Razorbill *Alca torda* breeding site selection: adaptation to a variable environment. Ibis 142: 324–327.

ROYLE, N. J., AND K. C. HAMER. 1998. Hatching asynchrony and sibling size hierarchies in gulls: effects on parental investment decisions, brood reduction and reproductive success. Journal of Avian Biology 29: 266–272.

SAETHER, B.-E., R. ANDERSEN, AND C. PEDERSEN. 1993. Regulation of parental effort in a long-lived seabird: an experimental manipulation of the cost of reproduction in the Antarctic Petrel, *Thalassoica antarctica*. Behavioural Ecology and Sociobiology 33: 147–150.

SAETHER, B.-E., T. H. RINGSBY, AND E. RØSKAFT. 1996. Life history variation, population processes and priorities in species conservation: towards a reunion of research paradigms. Oikos 77: 217–226.

SALAMOLARD, M., AND H. WEIMERSKIRCH. 1993. Relationship between foraging effort and energy requirement throughout the breeding season in the Wandering Albatross. Functional Ecology 7: 643–652.

SALINO, N., M. FASOLA, AND E. CROCICCHIA. 1994. Adoption behavior in Little and Common Terns (Aves, Sternidae) — chick benefits and parents' fitness costs. Ethology 97: 294–309.

SALOMONSEN, F. 1967. Migratory movements of the Arctic Tern (*Sterna paradisaea*) in the Southern Ocean. Biol. Medd., Kobenhavn 24: 1–42.

SCHREIBER, E. A. 1994. El Niño–Southern Oscillation effects on provisioning and growth in Red-tailed Tropicbirds. Colonial Waterbirds 17: 105–119.

SCHREIBER, E. A. 1996. Experimental manipulation of feeding in Red-tailed Tropicbird chicks. Colonial Waterbirds 19: 45–55.

SCHREIBER, E. A. 1999. Breeding Biology and Ecology of the Seabirds of Johnston Atoll, Central Pacific Ocean. Report to the Department of Defense.

SCHREIBER, E. A. 2000. Action plan for conservation of West Indian Seabirds. Pp. 182–191 *in* Status and Conservation of West Indian Seabirds (E. A. Schreiber and D. S. Lee, Eds.). Society of Caribbean Ornithology, Special Publication No. 1, Ruston, LA.

SCHREIBER, E. A., AND D. S. LEE (Eds.). 2000. Status and Conservation of West Indian Seabirds. Society of Caribbean Ornithology, Special Publication No. 1, Ruston, LA.

SCHREIBER, E. A., AND R. W. SCHREIBER. 1989. Insights into seabird ecology from a global natural experiment. National Geographic Research 5: 64–81.

SCHREIBER, E. A., AND R. W. SCHREIBER. 1993. The Red-tailed Tropicbird (*Phaethon rubricauda*). No. 43 *in* The Birds of North America (A. Poole and F. Gill, Eds.). Academy of Natural Sciences, Philadelphia; American Ornithologists' Union, Washington, D.C.

SCHREIBER, E. A., R. W. SCHREIBER, AND J. J. DINSMORE. 1979. Breeding biology of Laughing Gulls in Florida. Part I: nesting, egg, and incubation parameters. Bird-Banding 50: 304–321.

SCHREIBER, E. A., R. W. SCHREIBER, AND G. A. SCHENK. 1996. The Red-footed Booby *Sula sula*. No. 241 *in* The Birds of North America (A. Poole and F. Gill, Eds.). The Birds of North America, Inc., Philadelphia.

SCHREIBER, E. A., C. FEARE, B. A. HARRINGTON, B. MURRAY, W. B. ROBERTSON, M. J. ROBERT-SON, AND G. E. WOOLFENDEN. In preparation. The Sooty Tern (*Sterna fuscata*). *In* The Birds of North America (A. Poole and F. Gill, Eds.). The Birds of North America, Inc., Philadelphia.

SCHREIBER, R. W. 1976. Growth and development of nestling Brown Pelicans. Bird-Banding 47: 19–39.

SCHREIBER, R. W., AND P. J. MOCK. 1988. Eastern Brown Pelicans, what does 60 years of banding tell us? Journal of Field Ornithology 59: 171–182.

SCHREIBER, R. W., AND E. A. SCHREIBER. 1982. Essential habitat of the Brown Pelican in Florida. Florida Field Naturalist 10: 9–17.

SCHREIBER, R. W., AND E. A. SCHREIBER. 1984. Central Pacific seabirds and the El Niño–Southern Oscillation: 1982–83 perspectives. Science 225: 713–716.

SEALY, S. G. 1973. Adaptive significance of posthatching development patterns and growth rates in the Alcidae. Ornis Scandinavica 4: 113–121.

SERVENTY, D. L. 1967. Aspects of the population ecology of the Short-tailed Shearwater *Puffinus tenuirostris*. Proceedings 14th International Ornithology Congressus, 1966, pp. 165–190.

SIBLEY, F. C., AND R. B. CLAPP. 1967. Distribution and dispersal of central Pacific Lesser Frigatebirds *Fregata ariel*. Ibis 109: 328–337.

SIEGEL-CAUSEY, D., AND S. P. KHARITONOV. 1990. The evolution of coloniality. Pp. 285–330 *in* Current Ornithology Vol. 7 (D. M. Power, Ed.). Plenum Press, New York.

SNOW, D. W., AND B. K. SNOW. 1967. The breeding cycle of the swallow-tailed gull *Creagrus furcatus*. Ibis 109: 14–24.

SOUTHERN, W. E. 1980. Comparative distribution and orientation of North American gulls. Pp. 449–498 *in* Behavior of Marine Animals (J. Burger, B. L. Olla, and H. E. Winn, Eds.). Plenum Press, New York.

SQUIBB, R. C., AND G. L. HUNT, JR. 1983. A comparison of nesting-ledges used by seabirds on St. George Island. Ecology 64: 727–734.

STARCK, J. M., AND R. E. RICKLEFS. 1998. Patterns of development: the altricial-precocial spectrum. Pp. 3–30 *in* Avian Growth and Development (J. M. Stark and R. E. Ricklefs, Eds.). Oxford University Press, Oxford.

STOKES, D. L., AND P. D. BOERSMA. 1991. Effects of substrate on the distribution of Magellanic Penguin (*Spheniscus magellanicus*) burrows. Auk 108: 923–933.

SYDEMAN, W. J., AND S. D. EMSLIE. 1992. Effects of parental age on hatching asynchrony, egg-size and third-chick disadvantage in Western Gulls. Auk 109: 242–248.

SYDEMAN, W. J., P. PYLE, S. D. EMSLIE, AND E. B. MCLAREN. 1996. Causes and consequences of long-term partnerships in Cassin's Auklets. Pp. 211–222 *in* Partnerships in Birds. The Study of Monogamy (J. M. Black, Ed.). Oxford University Press, Oxford.

TAYLOR, J. R. E., AND M. KONARZEWSKI. 1989. On the importance of fat reserves for the Little Auk (*Alle alle*) chicks. Oecologia 81: 551–558.

TEMME, D. H., AND E. L. CHARNOV. 1987. Brood size adjustments in birds: economical tracking in a temporally varying environment. Journal of Theoretical Biology 126: 137–147.

THOMPSON, D. R., AND K. C. HAMER. 2000. Stress in seabirds: causes, consequences and diagnostic value. Journal of Aquatic Ecosystems Stress and Recovery 7: 91–110.

TVEERA, T., B. E. SAETHER, R. AANES, AND K. E. ERIKSTAD. 1998. Regulation of food provisioning in the Antarctic Petrel: the importance of parental body condition and chick body mass. Journal of Animal Ecology 67: 699–704.

TRIVELPIECE, W., AND N. J. VOLKMAN. 1979. Nest-site competition between Adelie and Chinstrap Penguins: an ecological interpretation. Auk 96: 675–681.

VERMEER, K., D. POWER, AND G. E. JOHN SMITH. 1988. Habitat selection and nesting biology of roof-nesting Glaucous–winged Gulls. Colonial Waterbirds 11: 189–201.

WALLACE, G. E., B. COLLIER, AND W. J. SYDEMAN. 1992. Interspecific nest-site competition among cavity-nesting alcids on southeast Farallon Island, California. Colonial Waterbirds 15: 241–244.

WARHAM, J. 1990. The petrels. Their ecology and breeding systems. Academic Press, London.

WEIMERSKIRCH, H. 1990. The influence of age and experience on breeding performance of the Antarctic fulmar, *Fulmarus glacialoides*. Journal of Animal Ecology 59: 867–875.

WEIMERSKIRCH, H., AND Y. CHEREL. 1998. Feeding ecology of short-tailed shearwaters: breeding in Tasmania and foraging in the Antarctic? Marine Ecology Progress Series 167: 261–274.

WEIMERSKIRCH, H., Y. CHEREL, F. CUENOT-CHAILLET, AND V. RIDOUX. 1997. Alternative foraging strategies and resource allocation by male and female Wandering Albatrosses. Ecology 78: 2051–2063.

WEIMERSKIRCH, H., P. JOUVENTIN, AND J. C. STAHL. 1986. Comparative ecology of six albatross species breeding on the Crozet Islands. Ibis 128: 195–213.

WEIMERSKIRCH, H., P. A. PRINCE, AND L. ZIMMERMANN. 2000. Chick provisioning in the Yellow-nosed Albatross: response of foraging effort to experimentally increased costs and demands. Ibis 142: 103–110.

WENDELN, H., AND P. BECKER. 1999. Effects of parental quality and effort on the reproduction of Common Terns. Journal of Animal Ecology 68: 205–214.

WHITTOW, G. C. 1997. The Wedge-tailed Shearwater, *Puffinus pacificus*. No. 305 *in* The Birds of North America (A. Poole and F. Gill, Eds.). The Birds of North America, Inc., Philadelphia.

WILLIAMS, T. D. 1995. The Penguins: Spheniscidae. Oxford University Press, Oxford.

WILLIAMS, T. D. 1996. Mate fidelity in penguins. Pp. 268–285 *in* Partnerships in Birds. The Study of Monogamy (J. M. Black, Ed.). Oxford University Press, Oxford.

WOOLLER, R., AND S. BRADLEY. 1996. Monogamy in a long-lived seabird: the Short-tailed Shearwater. Pp. 223–234 *in* Partnerships in Birds. The Study of Monogamy (J. M. Black, Ed.). Oxford University Press, Oxford.

WOOLLER, R. D., J. S. BRADLEY, AND J. P. CROXALL. 1992. Long-term population studies of seabirds. Trends in Ecology and Evolution 7: 111–114.

WOOLLER, R. D., J. S. BRADLEY, I. J. SKIRA, AND D. L. SERVENTY. 1989. Short-tailed Shearwater. Pp. 405–417 *in* Lifetime Reproduction in Birds (I. Newton, Ed.). Academic Press, London.

YDENBERG, R. C. 1989. Growth-mortality trade-offs and the evolution of juvenile life histories in the Alcidae. Ecology 70: 1494–1506.

YDENBERG, R. C., AND D. F. BERTRAM. 1989. Lack's clutch size hypothesis and brood enlargement studies on colonial seabirds. Colonial Waterbirds 12: 134–137.

YOUNG, E. C. 1978. Behavioural ecology of *lönnbergi* skuas in relation to environment on the Chatham Islands, New Zealand. New Zealand Journal of Zoology 5: 401–416.

A Black-Legged Kittiwake Pair with Chicks

9 Site and Mate Choice in Seabirds: An Evolutionary Approach

Joël Bried and Pierre Jouventin

CONTENTS

9.1 INTRODUCTION

In more than 90% of avian species, monogamy is the mating system (Lack 1968) but it still remains "the neglected mating system" (Mock 1985), because many studies on mating systems focus on the evolution and the maintenance of alternative mating systems (i.e., polygamy and promiscuity) rather than on the reasons why monogamy evolved. The concept of monogamy has been debated by (among others) Wickler and Seibt (1983) and Gowaty (1996). Distinguished from genetic monogamy, social monogamy can be defined as the association of one male and one female usually with some level of biparental care. In birds, this partnership, exclusive for incubation and chick-rearing, can be maintained during an entire lifetime.

FIGURE 9.1 A Short-tailed Albatross incubating its egg on Torishima Island, Japan. The incubation period is about 60 days and it takes the pair about 180 days to raise their single chick. (Photo by E. A. and R. W. Schreiber.)

The choice of a breeding place (Cody 1985, Ens et al. 1995) and a sexual partner (Orians 1969, Hunt 1980, Ligon 1999) has important consequences for reproduction. In birds, males classically compete over sites and/or females, whereas females perform mate choice (Darwin 1871, Orians 1969, Trivers 1972). Whether individuals should retain their site and/or mate from year to year, or change, is ultimately determined by breeding success, considering both previous and expected future reproductive performances (Greenwood and Harvey 1982, Switzer 1993, McNamara and Forslund 1996). According to Hinde (1956) and Rowley (1983), individuals of long-lived species should be able to retain both site and mate from year to year, because of their high adult survival rates. Moreover, life history theory predicts that high longevity should be associated with reduced fecundity or low reproductive effort (Stearns 1992). Therefore, individuals of long-lived species should optimize their reproductive outputs, while minimizing the costs of breeding not to jeopardize their future survival and residual reproductive value (Drent and Daan 1980, Partridge 1989, Ricklefs 1990, Stearns 1992; but see also Erikstad et al. 1998). Maximizing their chances to replace themselves (by producing at least one chick that will recruit into the breeding population) can be achieved through a high number of breeding attempts (iteroparity), and hence a long reproductive life span. Because site and mate fidelity are known to enhance reproductive performances in many avian species (Domjan 1992, Ens et al. 1996), long-lived species classically are expected to show high site and mate fidelity, with fidelity rates and life expectancy being positively correlated (Rowley 1983; Figure 9.1). However, very few studies have so far examined the relationships between fidelity and survival (e.g., Ens et al. 1996, Bried, Pontier, and Jouventin in preparation), considering fidelity rates as demographic parameters.

Seabirds appear as a choice model for these studies, being particularly long-lived, laying small clutches, and having a deferred sexual maturity (Jouventin and Mougin 1981; see also Chapter 5 by H. Weimerskirch; Table 9.1). Furthermore, the probability for seabird young to recruit into the breeding population is low (review in Nelson 1980; see also Ollason and Dunnet 1988, Wooller et al. 1989, Weimerskirch et al. 1992, Prince et al. 1994, Weimerskirch and Jouventin 1997), because a high proportion of seabird fledglings die from starvation during their first year of independence, presumably lacking sufficient foraging skills (Nelson 1980, Nur 1984). Due to biparental care, all seabird species are socially monogamous (Lack 1968). However, genetic monogamy may not always occur, promiscuous matings and polygyny having been observed in

TABLE 9.1
Nest and Mate Fidelity in Seabirds

Taxon[a]	Locality	Nest Fidelity[b] (%)	Mate Fidelity[b] (%)	S (ALE)	Body Mass (g)	References[c]
Sphenisciformes						
Aptenodytes patagonicus	Iles Crozet	39.4 (site fidelity)	22.4	0.952 (21.33)	13,400	1, 2, 3, 4
A. patagonicus	South Georgia	—	18.8	—	13,800	5
A. forsteri	Terre Adélie	—	14.5	0.91 (11.61)	30,000	2, 6, 4
Pygoscelis adeliae	Cape Crozier	59.4	18–50	0.696 (3.79)	3,900	7, 7, 8, 4
P.adeliae	Cape Bird	98.2	56.5	0.736 (4.29)	4,200	9, 9, 9, 4
P. adeliae	Wilkes Land	76.8	84.0	0.77 (4.85)	4,490	10, 10, 10, 4
P. adeliae	Iles Crozet	—	76.0	0.865 (7.91)	6,740	11, 8, 4
P. papua papua	South Georgia	93.0	90.2	ca. 0.8 (ca. 5.5)	6,800	12, 12, 13, 4
P. p. ellsworthi	King George Is.	61.4	90.0		5,300	14, 14, 14, 15
P. antarctica	King George Is.	87.9	82.0		4,000	14, 14, 14, 4
Eudyptes (chrysolophus) chrysolophus	South Georgia	83	90.8		4,120	12, 12, 16
E. (c.) schlegeli	Macquarie Is.	—	80.0	0.86 (7.64)	5,000	16, 17, 18
E. chrysocome filholi	Iles Kerguelen	53.0	78.6	—	2,500	19, 19, 19
E. c. moseleyi	Amsterdam Is.	34.9	46.3	0.84 (6.75)	2,400	19, 19, 20, 21
E. robustus	The Snares	—	no case reported	—	3,100	4, 4
Megadyptes antipodes	New Zealand	30.0	82.0	0.87/0.86 (ca. 8.5)	5,300	22, 23, 24, 4
Spheniscus mendiculus	Galápagos Is.	—	>89	0.844 (6.91)	2,030	25, 16, 16
S. demersus	South Africa	59.8	86.2	0.617 (3.11)	3,100	26, 26, 26, 16
S. magellanicus	Punta Tombo	80/70	90.4	0.85 (7.17)	4,440	27, 16, 16, 16
Eudyptula minor	Philip Is.	43.9	82.0	0.858 (7.54)	1,110	28, 28, 28, 4
Procellariiformes						
Diomedea exulans	South Georgia	20.0	no case reported	0.94 (17.17)	8,700	29, 30, 4
D. exulans	Iles Crozet	28.9	95.1	0.931 (14.99)	9,600	31, 19, 32, 33
D. amsterdamensis	Amsterdam Is.	—	97.9	0.966 (29.91)	6,270	19, 19, 34
D. epomophora epomophora	Campbell Is.	—	no case reported	—	9,280	35, 4
D. e. sanfordi	Taiaroa Is.	—	1 case reported	0.946 (19.02)	6,500	36, 36, 4
Diomedea (Phoebastria) irrorata	Galápagos Is.	—	no case reported	0.95 (20.50)	3,500	37, 37, 37
D. immutabilis	Midway Atoll	—	97.9	0.947/0.946 (ca. 19.19)	2,900	38, 39, 40

TABLE 9.1 (Continued)
Nest and Mate Fidelity in Seabirds

Taxon[a]	Locality	Nest Fidelity[b] (%)	Mate Fidelity[b] (%)	S (ALE)	Body Mass (g)	References[c]
Phoebetria fusca	Iles Crozet	41.1	94.8	0.95 (20.50)	2,600	19, 19, 41, 42
Diomedea (Thalassarche) chlororhynchos	Amsterdam Is.	92.6	90.6	0.912 (11.86)	2,100	43, 19, 41, 42
D. bulleri	The Snares	67.0	96.2	0.913 (11.99)	2,700	44, 44, 45, 4
D. chrysostoma	Campbell Is.	—	96.3	0.953 (21.78)	3,180	46, 47, 4
D. melanophris melanophris	Iles Kerguelen	74.1	92.3	0.914 (12.13)	3,740	19, 19, 19, 4
D. m. melanophris	South Georgia	93.5	—	0.934 (15.65)	3,600	48, 48, 4
D. m. impavida	Campbell Is.	—	95.5	0.945 (18.68)	2,900	46, 47, 4
Pagodroma nivea	Terre Adélie	89.8	88.3	0.934 (15.65)	380	49, 49, 50, 51
Daption capense capense	Terre Adélie	88.0	85.0	—	472	52, 52, 53
D. c. capense	South Orkney Is.	84.0	73.0	0.942 (17.74)	425	54, 54, 54, 55
D. c. australe	The Snares	97.5	97.3	0.892 (9.76)	435	56, 56, 56, 57
Fulmarus glacialoides	Terre Adélie	82.5	77.1	0.916 (12.40)	800	19, 19, 19, 4
F. glacialis	Orkney Is.	93.4	96.9	0.968 (31.75)	813	58, 58, 59, 60
Macronectes giganteus	Terre Adélie	59.0	80.8	0.902 (10.70)	4,500	19, 19, 51, 51
M. giganteus	South Orkney Is.	92.9	no case reported	—	4,360	61, 61, 61
Pelecanoides urinator	Iles Kerguelen	81.6	92.8	0.807 (5.68)	140	19, 19, 19, 19
Pterodroma lessonii	Iles Kerguelen	96.6	91.2	0.921 (13.16)	708	19, 19, 51, 62
P. macroptera	Iles Kerguelen	80.2	87.5	—	581	19, 19, 63
P. inexpectata	The Snares	96.9	>83	—	320	64, 64, 64
P. phaeopygia	Galápagos Is.	96.7	—	—	410	65, 65
Calonectris diomedea borealis	Salvages Is.	91.4	94.0	0.956 (23.23)	890	66, 66, 67, 4
C. d. diomedea	Crete	95.9	96.4	0.89 (9.59)	552	68, 68, 68, 69
Puffinus puffinus	Skokholm Is.	93.3	90.3	0.905 (11.03)	450	70, 70, 70, 71
P. tenuirostris	Bass Strait	—	82.2	0.897/0.899 (ca. 10.30)	590	72, 73, 4
Procellaria aequinoctialis	Iles Crozet	80.5	93.7	—	1,300	74, 74, 75
P. parkinsoni	New Zealand	—	88.0	0.94 (17.17)	700	76, 76, 71
P. cinerea	Iles Kerguelen	90.2	95.9	0.924 (13.66)	1,131	19, 19, 51, 77
Bulweria bulwerii	Salvages Is.	63.0	78.5	0.947 (19.37)	95	78, 78, 79, 80
Halobaena caerulea	Iles Kerguelen	88.3	80.0	0.88 (8.83)	190	19, 19, 51, 81
Pachyptila belcheri	Iles Kerguelen	87.5	79.2	0.852 (7.26)	145	32, 32, 51, 63

Species	Location					References
P. desolata	Iles Kerguelen	86.5	88.0	—	150	19, 19, 19
P. turtur	Whero Is.	87.0	—	0.844 6.91	130	80, 82, 4
Oceanites oceanicus	South Orkney Is.	—	80.0	0.908 (11.37)	40	83, 83, 83, 83
Hydrobates pelagicus	Skokholm Is.	—	77.3	0.88 (8.83)	28	84, 71, 71
Oceanodroma leucorhoa	Maine, U.S.A.	95.0	—	0.86 (7.64)	45	85, 86, 71
Pelecaniformes						
Phaethon rubricauda	Kure Atoll	25.0	—	—	612	87, 88
P. lepturus	Seychelles Is.	—	97.0	—	341	89, 89
Morus bassanus	Bass Rock	89.8	83.5	0.89 (9.59)	3,000	90, 90
S. dactylatra personata	Kure Atoll	10.0	54.8	0.895 (10.02)	2,030	90, 90, 90, 90
Sula (d.) granti	Galápagos Is.	87.1	—	0.832 (6.45)	1,750	90, 87
S. leucogaster	Kure Atoll	—	≥97.7	0.92—0.955 (ca. 16.50)	1,110	90, 90, 90, 90
S. abbotti	Christmas Is.	—	≥90	—	1,480	90, 4
Phalacrocorax aristotelis	Isle of May	49.2	69.0	0.84 (6.75)	2,000	91, 91, 92, 71
P. penicillatus	Farallon Is.	62.3	—	0.80 (5.50)	2,450	93, 93, 71
P. atriceps	South Orkney Is.	—	40.3	—	2,880	94, 71
Nannopterum harrisi	Galápagos Is.	35.9	11.9	0.876 (8.56)	3,200	95, 95, 95, 71
Charadriiformes						
Catharacta skua skua	Foula Is.	—	93.6	0.93 (14.78)	1,418	96, 97, 97
C. s. lönnbergi	Iles Kerguelen	98.3	96.5	0.925 (13.83)	1,835	19, 19, 19, 19
C. s. lönnbergi	Anvers Is.	—	>89	0.95 (20.50)	1,700	98, 98, 99
C. maccormicki	Terre Adélie	89.0	90.9	0.912 (11.86)	1,405	19, 19, 19, 100
C. maccormicki	Anvers Is.	—	>85	0.95 (20.50)	1,200	98, 98, 99
C. maccormicki	Cape Crozier	87.3	98.5	0.938 (16.63)	1,300	101, 102, 99
L. (novaehollandiae) scopulinus	New Zealand	—	89.5	0.856 (7.44)	280	103, 103, 99
Rissa tridactyla tridactyla	Great Britain	—	71.9	0.81/0.86 (ca. 6.56)	408	104, 105, 106
R. t. pollicaris	Alaska	—	80.7	0.93 (14.78)	408	107, 107, 106
Sterna anaethetus	Western Australia	82.3	—	0.78 (5.04)	127	108, 108, 99
Sterna hirundo	Germany	—	81.1	0.89 (9.59)	134	109, 110, 109
Anous minutus	Ascension Is.	81.8	—	—	118	111, 111
Uria lomvia	Prince Leopold Is.	73.0	88.3	0.91 (11.61)	945	112, 113, 113
U. aalge	Isle of May	85.7	94.3	0.949 (20.11)	862	114, 115, 116, 117
Alca torda	Isle of May	93.0	92.7	0.888 (9.43)	710	118, 118, 118, 113
Ptychoramphus aleuticus	Farallon Is.	—	—	0.75 (4.50)	170	119, 119, 113
Cepphus grylle	Iceland	90.0	95.5	0.87 (8.19)	500	120, 120, 120, 113
Aethia cristatella	Buldir Is.	75/62	64.5	0.89 (9.59)	260	121, 121, 121, 113

TABLE 9.1
Nest and Mate Fidelity in Seabirds

Taxon[a]	Locality	Nest Fidelity[b] (%)	Mate Fidelity[b] (%)	S (ALE)	Body Mass (g)	References[c]
A. pusilla	Pribilof Is.	—	63.6	0.808 (5.71)	85	122, 122, 113
Fratercula arctica	Skomer Is.	92.2	92.2	0.942 (17.74)	460	123, 123 123, 113
F. arctica	Unknown locality	—	84.0	0.87 (8.19)	460	124, 124, 113

[a] Only adult individuals (i.e., known to have bred in the past) were considered. Data from populations known to live in unstable environments, or not to be in equilibrium, were excluded.

[b] Studies involving less than 25 individual-years or 25 pair-years for site fidelity and mate fidelity, respectively, were excluded. Only adult individuals were considered. Site fidelity rates were calculated as 1 minus (number of site changes/number of adult-years). Mate fidelity was calculated as 1 minus the probability of divorce when both previous partners survive, following Black (1996). When two values separated by a slash (/) are given for the same parameter (e.g., 80/70), the former is for males, the latter for females.

[c] Numbers refer to the source of nest fidelity, mate fidelity, adult survival rate, respectively, and when data were available. Although some of these sources did not express fidelity rates in the same manner as ours, they provided the data that enabled us to calculate them as described above.

1, Barrat (1976); 2, Bried et al. (1999); 3, Weimerskirch et al. (1992); 4, Marchant and Higgins (1990); 5, Olsson (1998); 6, Jouventin and Weimerskirch (1991); 7, Ainley et al. (1983); 8, Ainley and DeMaster (1980); 9, Davis (1988); 10, Penney (1968); 11. Bost and Jouventin (1991); 12, Williams and Rodwell (1992); 13, Croxall and Rothery (1994); 14, Trivelpiece and Trivelpiece (1990); 15, Volkman et al. (1980); 16, Williams (1995); 17, Carrick (1972); 18, Carrick and Ingham (1970); 19, this study; 20, Guinard et al. (1998); 21, E. Guinard (unpublished data); 22, Richdale (1949); 23, Richdale (1947); 24, Jouventin and Mougin (1981); 25, Boersma (1976); 26, LaCock et al. (1987); 27, Scolaro (1990); 28, Reilly and Cullen (1981); 29, Tickell (1968); 30, Croxall et al. (1990); 31, Fressanges du Bost and Ségonzac (1976); 32, Weimerskirch and Jouventin (1997); 33, Rice and Kenyon (1992); 34, Jouventin et al. (1989); 35, Waugh et al. (1997); 36, Robertson (1993); 37, Harris (1973); 38, Rice and Kenyon (1962); 39, Fisher (1975); 40, Frings and Frings (1961); 41, Weimerskirch et al. (1987); 42, Weimerskirch and Jouventin (1897); 43, Jouventin et al. (1983); 44, Sagar and Warham (1997); 45, P. M. Sagar, J. Molloy, H. Weiberskirch, and J. Warham (unpublished data); 46, S. M. Waugh and J. Bried (unpublished data); 47, Waugh et al. (1999); 48, Prince et al. (1994); 49, Jouventin and Bried (in press); 50, Chastel et al. (1993); 51, Chastel (1995); 52, Mougin (1975); 53, Isenmann (1970); 54, Hudson (1966); 55, Pinder (1966); 56, Sagar et al. (1996); 57, Sagar (1986); 58, Ollason and Dunnet (1978); 59, Dunnet and Ollason (1978); 60, Ollason and Dunnet (1988); 61, Conroy (1972); 62, Zotier (1990b); 63, Weimerskirch et al. (1989); 64, Warham et al. (1977); 65, Cruz and Cruz (1990); 66, Mougin et al. (1987a); 67, Mougin et al. (1987b); 68, Swatschek et al. (1994); 69, Ristow and Wink (1980); 70, Brooke (1990); 71, del Hoyo et al. (1992); 72, Bradley et al. (1990); 73, Wooller and Bradley (1996); 74, Bried and Jouventin (1999); 75, A. Catard (unpublished data); 76, Imber (1987); 77, Zotier (1990a); 78, Mougin (1989); 79, Mougin (1990); 80, Warham (1990); 81, Chastel et al. (1995a); 82, Richdale (1963); 83, Beck and Brown (1972); 84, Scott (1970); 85, Morse and Buchheister (1979); 86, Warham (1996); 87, Harris (1979b); 88, Fleet (1974); 89, Phillips (1987); 90, Nelson (1978); 91, Aebischer et al. (1995); 92, Potts (1969); 93, Boekelheide and Ainley (1989); 94, Shaw (1986); 95, Harris (1979a); 96, Catry et al. (1997); 97, Furness (1987); 98, Pietz and Parmelee (1994); 99, Higgins and Davis (1996); 100, Jouventin and Guillotin (1979); 101, Ainley et al. (1990); 102, Wood (1971); 103, Mills et al. (1996); 104, Coulson (1966); 105, Coulson and Wooller (1976); 106, Burger and Gochfeld (1996); 107, Hatch et al. (1993); 108, Dunlop and Jenkins (1992); 109, González-Solís et al. (1999); 110, Wendeln and Becker (1998); 111, Ashmole (1962); 112, Gaston and Nettleship (1981); 113, Nettleship (1996); 114, Harris et al. (1996); 115, Ens et al. (1996); 116, Harris and Wanless (1995); 117, Cramp (1985); 118, Harris and Wanless (1989); 119, Sydeman et al. (1996); 120, Petersen (1981); 121, Gaston and Jones (1998); 122, Jones and Montgomerie (1991); 123, Ashcroft (1979); 124, Davidson (unpublished data, in Ens et al. (1996).

gulls (Burger and Gochfeld 1996). Because living organisms tend to optimize their own fitness, but also that of their offspring (Maynard-Smith 1978), do the long-lived seabirds choose their breeding places and their partners carefully?

In this chapter, we provide a new insight into the relationships between site fidelity, mate fidelity, and longevity, by using an evolutionary approach with seabirds as a model. In order to achieve this purpose, we will (1) identify the constraints on reproduction faced by seabirds, and (2) test the classical predictions concerning site, mate choice, and fidelity (see above) by assessing the effects of these selective pressures on the reproductive strategy of seabirds.

9.2 THE MAJOR EVOLUTIONARY CONSTRAINT IN SEABIRDS: BREEDING ON LAND AND FEEDING AT SEA

9.2.1 THE KEY FACTOR

Seabirds face an important constraint during reproduction, which appears as the key factor in the evolution of their life history traits: they exclusively rely on marine resources for feeding and yet they need to come ashore for breeding (Jouventin and Mougin 1981). For "inshore" and "offshore" feeders, nesting and feeding areas are not only distinct, but there is a continuous gradient from the more coastal seabirds to the most pelagic ones: foraging trips can range from a few hundred meters from the nest in terns to several thousands kilometers in albatrosses and petrels (Jouventin and Weimerskirch 1991, Weimerskirch 1997, Weimerskirch et al. 1999). Consequently, trips can last up to several days and sometimes several weeks, and their duration has affected the evolution of seabird life histories (see Figures 9.2 and 9.3).

These foraging trips represent a constraint in seabirds, for demography but also for morphology (wing shape) and metabolism, because seabirds must fly long distances and fast when on land (Warham 1975, 1990, Schreiber and Schreiber 1993, Chaurand and Weimerskirch 1994a). A breeding adult that undertakes a long foraging trip while its partner incubates or broods will have to return to its nest before its mate has exhausted its body reserves and abandoned the nest, implying a good synchronization between mates (e.g., Jouventin et al. 1983 for albatrosses, Davis 1988 for Adélie Penguins [*Pygoscelis adeliae*], Schreiber and Schreiber 1993 for Red-tailed Tropicbirds [*Phaethon rubricauda*]). Nevertheless, successful breeding in seabirds does not only depend on food and synchronization between parents; two other conditions must be met on land. The first one is the ownership of a breeding territory to have a place to incubate the clutch (Newton 1992). The second condition is obtaining the sexual partner.

In many seabird species, males return ashore earlier than females at the onset of breeding and settle on their nests before attracting a mate (Hunt 1980; for examples of taxonomic groups, see e.g., Warham 1990 for albatrosses and petrels, and Nelson 1983 for sulids and some cormorants). However, in Emperor Penguins (*Aptenodytes forsteri*), females generally return earlier than males (Bried et al. 1999), and both sexes can return simultaneously in frigatebirds (Nelson 1983), but also in some terns and alcids (Hunt 1980). However, site quality and mate quality vary in birds, including seabirds (e.g., Adélie Penguins, Carrick and Ingham 1967; Caspian Terns [*Hydroprogne caspia*], Cuthbert 1985; sulids, Nelson 1988; Snow Petrels [*Pagodroma nivea*], Chastel et al. 1993). Therefore, individuals should settle on the most suitable sites available (this implies proximity to foraging area, concealment from weather and potential predators, and easy access and departure for breeders) and should obtain as high quality mates as possible ("ideal" choice, Fretwell and Lucas 1970). Because of the constraints of oviparity and the duration of their foraging trips, seabirds have evolved obligate biparental care during both incubation and chick rearing, which has led to social monogamy (Lack 1968, Jouventin and Cornet 1980, Wittenberger and Tilson 1980, Ligon 1999). An optimal mate choice should enable each mate to assume its parental duties successfully during incubation and chick rearing and to optimize its reproductive output. However, other

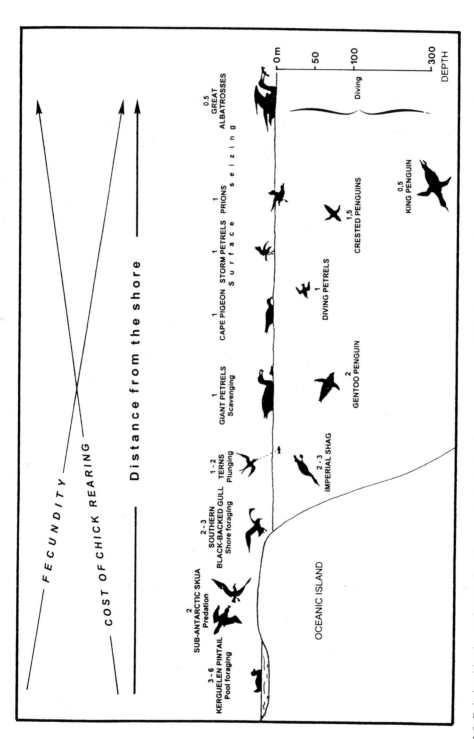

FIGURE 9.2 Relationships in a subantarctic avian community between foraging range and life history traits such as clutch size (above species name). The key factor in seabirds is the distance between feeding and breeding grounds. Foraging trips, both flying and diving, represent an energetic cost that prevents the most pelagic seabirds from rearing more than one chick per year (or every other year). Are site and mate fidelity a consequence of this low fecundity and high longevity? (Modified from Jouventin and Mougin 1981.)

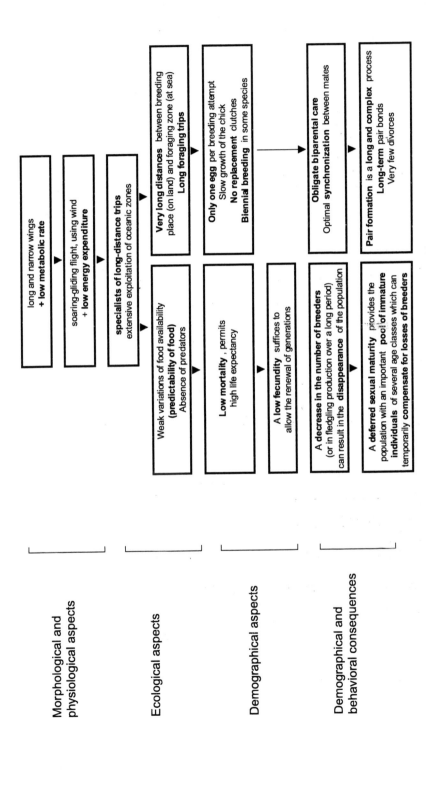

FIGURE 9.3 Demographic characteristics and reproductive strategy of albatrosses. Delayed maturity results in the presence of high numbers of immature individuals belonging to several age classes. If survival of breeders decreases, immatures recruit into the breeding population at a younger age than if the population were at equilibrium, acting as a buffer (at least temporarily) against a decline in the breeding population. (Modified from Jouventin and Weimerskirch 1984b.)

constraints than the distance between the breeding grounds and the feeding area may influence the breeding distribution on land and the breeding schedule of seabirds.

9.2.2 PHYLOGENETIC CONSTRAINTS

Phylogeny is likely to play a major role in breeding distribution and reproduction in seabirds; however, there are considerable variations of the breeding range within the same group (Figure 9.4; see Chapter 3). Phylogeny also has a strong influence on fecundity. Seabirds lay small clutches (most no more than three eggs; see Appendix 2) and/or have a low fecundity. The 2 *Aptenodytes* penguins, all procellariiforms, frigatebirds, and tropicbirds, 5 sulids out of 10, 1 gull out of 51, 12 terns out of 44, and 11 alcids out of 22 lay one egg only (see Table 9.2 for within-group variations of clutch size). In addition, many species do not lay replacement clutches (Nelson 1978, Jouventin and Mougin 1981, Warham 1990, Gaston and Jones 1998). Moreover, some breed only every other year (Nelson 1978, Warham 1990, Zotier 1990b). However, some species can raise successfully two broods in succession during the same year (del Hoyo et al. 1992, Williams 1995).

Seabirds have extended periods of parental care. Incubation varies between *circa* 20 days in the smallest terns to 79 days in the largest albatrosses (see Appendix 2). The nestling period (i.e., from hatching until departure from the colony) ranges from 2 days in *Synthliboramphus* alcids (Gaston and Jones 1998) to almost 1 year in King Penguins (*Aptenodytes patagonicus;* Marchant and Higgins 1990). Moreover, conditions at sea can affect chick growth, which takes longer in years of poor food availability, due to, e.g., El Niño events (see Schreiber and Schreiber 1993, Red-tailed Tropicbirds). In tropical sulids, frigatebirds, skuas, many gulls, terns, and alcids, parents provide postfledging care (from 1 month in some alcids, Gaston and Jones 1998; to several months in frigatebirds and Abbott's Booby [*Sula abbotti*], Nelson 1972, 1976). Conversely, chicks of penguins (del Hoyo et al. 1992), petrels, and albatrosses (Warham 1990), tropicbirds (del Hoyo et al. 1992), gannets (*Morus* sp., Nelson 1978), and puffins (*Fratercula* sp., Gaston and Jones 1998) must fend for themselves after leaving their natal colony. High adult life expectancy, however, may enable seabirds to compensate for the long duration of their breeding cycles, low fecundity, and mortality of juveniles at sea (see Introduction). Although adult life expectancy is between 12 and 15 years in many seabird species (Table 9.1), some individuals attain very old ages: Northern Royal Albatross (*Diomedea epomophora*) 61 years (Robertson 1993, see Appendix 2). Some Emperor Penguins and Snow Petrels that the authors banded as breeders in the mid-1960s at the French station of Dumont d'Urville, Terre Adélie (Antarctica), are still alive at over 35 years of age.

9.2.3 PHENOLOGICAL CONSTRAINTS

Food is classically considered the ultimate factor that determines the breeding period in most avian species (Lack 1968, Daan et al. 1988). Because energetic demands of birds are highest during breeding, birds generally breed during periods of highest food availability (Perrins 1970, Martin 1987, Harrison 1990). Marine productivity increases with latitude, but undergoes marked seasonal changes (Nelson 1970, Jouventin and Mougin 1981, Harrison 1990) and breeding synchrony is higher in temperate and polar areas (although some exceptions may occur, Croxall 1984) and chick growth becomes faster as latitude increases (Ashmole 1971, Nelson 1983). Accordingly, we checked for a negative correlation between latitude and the duration of chick growth (i.e., until chicks fledge). For each species, latitude was determined by calculating the average value (accuracy: 1°) between the northernmost and the southernmost locality in its breeding area. We excluded the Emperor Penguin from our analyses because chicks of this species depart to sea at only half of adult body mass (Isenmann 1971). Life history theory predicts that large-sized organisms should have a slower growth than small ones (Stearns 1992); therefore, we divided the chick growth period by adult body mass (after assuming that hatchling body mass was negligible compared to adult body mass). We obtained the amount of time necessary to produce a unit of mass (TUM), which

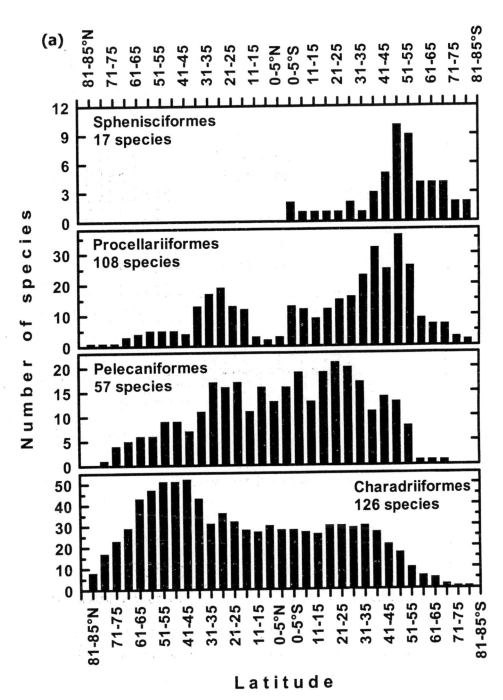

FIGURE 9.4 Breeding distribution of seabirds, with respect to latitude. (a) Considering the four orders of seabirds; however, two orders are very heterogeneous, and we considered the different families for each of them; (b) within the order Pelecaniformes; (c) within the order Charadriiformes. Histograms were drawn using data in Nelson (1978), del Hoyo et al. (1992), Lequette et al. (1995), Burger and Gochfeld (1996), Gochfeld and Burger (1996), Furness (1996), Nettleship (1996), and Zusi (1996).

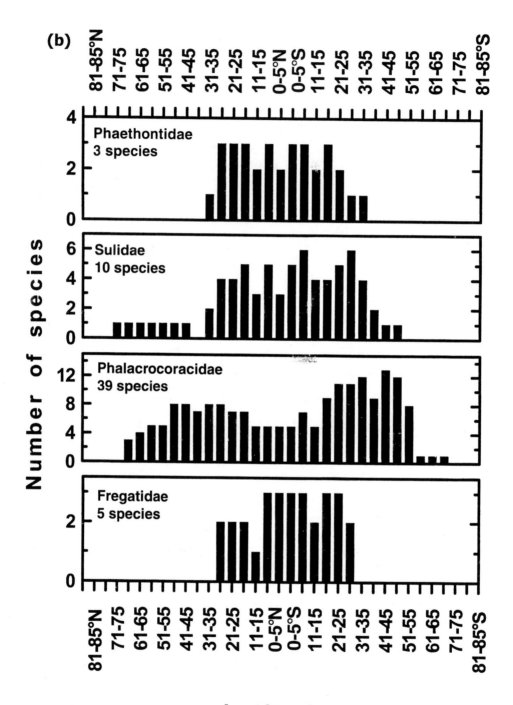

FIGURE 9.4 *Continued.*

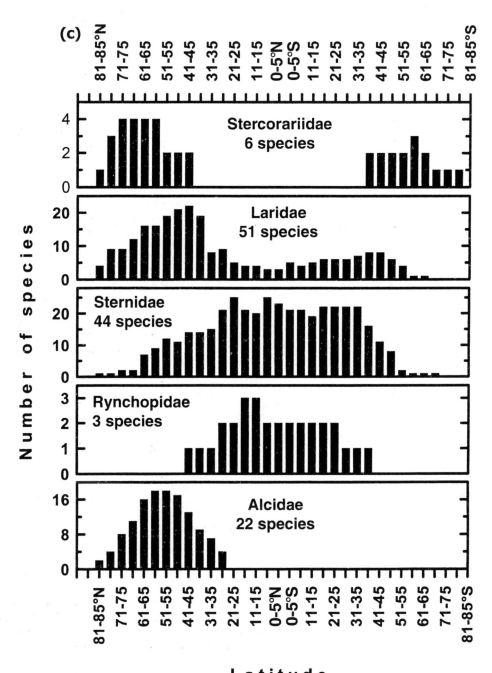

FIGURE 9.4 *Continued.*

TABLE 9.2

Duration of Parental Investment (from Laying until Chicks Become Independent) in Seabirds

Family	Clutch Size Range	Incubation Duration	Chick-Rearing Period	Postfledging Care
Sphenisciformes (one family only)				
Spheniscidae (penguins)	1–2 (3)	33–64 days	52 days to 13 months	No
Procellariiformes				
Diomedeidae (albatrosses)	1	60–79 days	120–278 days	No
Pelecanoididae (diving petrels)	1	48–55 days	ca. 52 days	No
Procellariidae (gadfly petrels, fulmars, prions, shearwaters)	1	45–61 days	47–130 days	No
Hydrobatidae (storm petrels)	1	40–53 days	58–84 days	No
Pelecaniformes				
Phaethontidae (tropicbirds)	1	41–43 days	75–85 days	No
Phalacrocoracidae (cormorants)	(1) 3–4 (7)	24–35 days	35–70 days	10–120 days
Sulidae (gannets, boobies)	1–3 (4)	42–57 days	90–160 days	0–280 days
Fregatidae (frigatebirds)	1	45–55 days	20–29 weeks	5–18 months
Charadriiformes				
Stercorariidae (skuas, jaegers)	(1) 2 (3)	24–30 days	30–50 days	Exists
Laridae (gulls)	1–3 (4)	20–30 days	3–7 weeks	30–70 days
Sternidae (terns)	1–3	19–35 days	18–60 days	Up to 5 months
Rynchopidae (skimmers)	2–4	21–26 days	28–30 days	ca. 2 weeks
Alcidae (auks)	1–2	28–45 days	27–52 days	0 to 12 weeks

Data in del Hoyo et al. (1992) for Sphenisciformes, Procellariiformes, and Pelecaniformes; Furness (1996) for skuas and jaegers; Burger and Gochfeld (1996) for gulls; Gochfeld and Burger (1996) for terns; Zusi (1996) for skimmers; Gaston and Jones (1998) for auks. Values in brackets represent extreme (albeit normal) clutch sizes.

seemed us to be a more reliable parameter. We found a significant negative relationship between TUM and latitude (Spearman's rank correlation: $r_s = -0.27$, $p = 0.0002$; TUM was not normally distributed) when considering 187 seabird species for which data were available (i.e., 14 penguins, 13 albatrosses, 48 petrels, all sulids, tropicbirds, and frigatebirds, 14 cormorants, 6 skuas, 29 gulls, 28 terns, 1 skimmer, and 16 alcids; data in Nelson 1978, del Hoyo et al. 1992, Lequette et al. 1995, Burger and Gochfeld 1996, Furness 1996, Gochfeld and Burger 1996, Nettleship 1996, Zusi 1996). This relationship remained significantly negative if we considered each taxonomic order separately, except for Procellariiformes (Figure 9.5). However, unpredictable short-term decreases in food availability, due to local oceanographic changes (such as El Niño events), can occur at any latitude and affect chick growth and/or breeding success (Ashmole and Ashmole 1967, Boersma 1978, Schreiber and Schreiber 1989, 1993, Guinet et al. 1998).

The duration of the breeding cycle varies according to species (due partly to body size, Stearns 1992). Thus, the largest species (*Aptenodytes* penguins, albatrosses, Figure 9.6) face the strongest temporal constraints at high latitudes (Croxall 1984, Croxall and Gaston 1988). In polar areas, the breeding cycle must be completed during summer (e.g., Nelson 1980, Harrison 1990). An exception exists, however: the largest penguin, the 30-kg Emperor Penguin, breeds during the antarctic winter. Breeding colonies are established on the sea ice. Chicks achieve sufficient growth and complete their molt before ice break-up during the austral summer, so that they can depart successfully to sea, even though their body mass is half of that of adults (Isenmann 1971).

In all latitudinal areas, both sexes participate in incubation and chick rearing. The duration of parental investment over the entire chick-rearing period also may be sex dependent. Chicks are fed

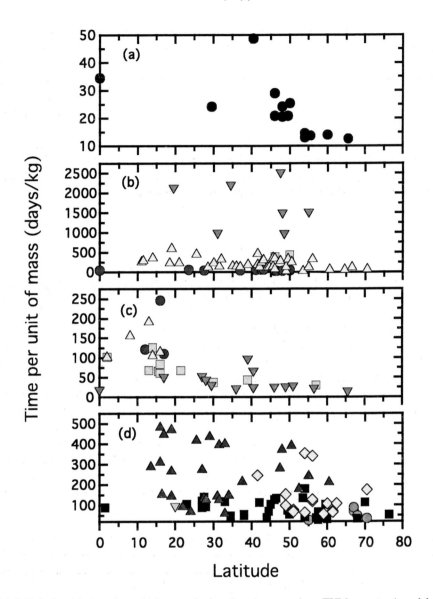

FIGURE 9.5 Relationship between chick growth duration (expressed as TUM, see text) and latitude in seabirds. TUM not normally distributed in either taxonomic order. (a) Sphenisciformes: $r_s = -0.84$, $p = 0.0002$, $n = 14$. (b) Procellariiformes: $r_s = -0.20$, $p = 0.11$, $n = 61$; filled circles, *Diomedeidae*; squares, *Pelecanoididae*; light triangles, *Procellariidae*; black triangles, *Hydrobatidae*. (c) Pelecaniformes: $r_s = -0.66$, $p = 0.0001$, $n = 32$; filled circles, *Phaethontidae*; squares, *Sulidae*; light triangles, *Fregatidae*; black triangles, *Phalacrocoracidae*. (d) Charadriiformes: $r_s = -0.45$, $p = 0.0001$, $n = 80$; filled circles, *Stercorariidae*; squares, *Laridae*; triangles, *Sternidae*; inverted triangle, *Rynchops niger*; rhombuses, *Alcidae*.

longer by male King Penguins at Iles Crozet (F. Jiguet and P. Jouventin unpublished data), as in Flightless Cormorants (*Nannopterum harrisi*, Harris 1979a) and some alcids (Razorbill [*Alca torda*] and murres [*Uria* sp.], Gaston and Jones 1998). Conversely, females provide parental care longer than males (sometimes up to 14 months after fledging) in Greater Frigatebirds (*Fregata minor*) and Magnificent Frigatebirds (*F. magnificens;* Schreiber and Ashmole 1970, Nelson 1976, Trivelpiece and Ferraris 1987).

It appears important for seabirds to minimize the effects of energetic and phenologic constraints by choosing carefully their breeding places and their partners. In seabirds, breeding failures are

FIGURE 9.6 Wandering Albatross courting. Adults provision chicks for 9 months. (Photo by P. Jouventin.)

due not only to low food availability some years, but also to late breeding or poor synchronization between mates, and to predation (by native or introduced predators) if the eggs (or the small young) are left unattended, as occurs in petrels and albatrosses (Warham 1990), tropicbirds (Schreiber and Schreiber 1993), frigatebirds, and some boobies (Nelson 1980).

9.3 HABITAT SELECTION

9.3.1 CHOICE OF THE BREEDING PLACE

The ultimate factors that determine choice of the breeding place are food and shelter from predators (Lack 1968, Nelson 1980, Warham 1996). Because seabirds leave their young (and sometimes their eggs) unattended during long periods (see above), these two factors play a major role in their breeding strategies. As expected, seabirds generally settle in breeding localities that are as close to their feeding areas as possible, so that the benefits of feeding are not outweighed by the costs of flying, and that have (if possible) no predators (Lack 1968, Buckley and Buckley 1980, Warham 1996). A scarcity of suitable breeding localities can result in aggregations of large numbers of individuals, sometimes in colonies harboring several species (e.g., Bayer 1982). Consequently, coloniality is widespread among seabirds, at least 93% of seabird species being colonial (Lack 1968).

Colony size and density vary greatly between species, and even for a given species (Nelson 1980, Marchant and Higgins 1990, Higgins and Davis 1996). Some species can either nest solitarily or colonially (e.g., Caspian Tern, Gochfeld and Burger 1996; Black Guillemot [*Cepphus grylle*], Gaston and Jones 1998). Availability of or competition for both nesting sites and food might influence colony size (Ashmole and Ashmole 1967, Furness and Birkhead 1984, Harrison 1990; and see Chapter 4). The importance of each of these factors still needs to be assessed more accurately.

9.3.2 NEST-SITE SELECTION

Depending on species, seabirds nest on the surface, dig burrows, or use crevices or tree holes. *Aptenodytes*, *Pygoscelis*, and most *Eudyptes* penguins, albatrosses, and large petrels, most pelecaniforms, skuas, gulls, terns, and three alcids (the two murres and the Dovekie [*Alle alle*]) nest on the surface. Some of these species build no nest, but incubate their egg on the ground (e.g., Masked Booby [*Sula dactylatra*], murres, and skimmers) or on their feet (King and Emperor Penguins; individuals of the latter species can also walk in the colony while incubating their eggs, whereas displacements of incubating King Penguins do not exceed a few meters from the laying site). Most smaller petrels, the Little Penguin, the four *Spheniscus* penguins, tropicbirds (sometimes), and 18 alcids nest in burrows or cavities (see Chapter 8). Among surface nesters, some species build their nests on the ground (penguins, gannets, many gulls, and terns), whereas others like frigatebirds, the Red-footed Booby (*Sula sula*), Abbott's Booby, Bonaparte's Gull (*Larus philadelphia*), and the Black Noddy usually build them in trees (see Chapter 8). Some species, like murres and kittiwakes (*Rissa* sp.), nest on cliffs. The White Tern (*Gygis alba*) builds no nest, laying its single egg at the fork of a tree (Neithammer and Patrick 1998). Table 9.3 shows the different types of nesting sites utilized by each order of seabirds.

However, nesting site quality can vary. Nests situated at the periphery of colonies generally are less productive than those situated in the central part of the colony (review in Rowley 1983), suffering highest predation rates and being the most exposed to the consequences from agonistic interactions (Carrick and Ingham 1967, Tenaza 1971, Nelson 1988). Furthermore, coloniality can create competition for nesting sites (reviews in Forbes and Kaiser 1994, Rolland et al. 1998), so that some individuals may nest in suboptimal areas (e.g., Rowan 1965, Aebischer et al. 1995). Moreover, the occurrence of several species breeding in the colony at the same time may have led to a partitioning in the selection of nesting sites (Buckley and Buckley 1980, Nelson 1980; see also Figure 9.7).

9.4 MATE CHOICE

In order to optimize reproduction, birds must choose a mate that will enable them to produce as many high-quality offspring as possible, i.e., a mate whose genotype will enable the offspring to inherit the best combination of genes possible ("good genes hypothesis," see Andersson 1994). All other things being equal, females should seek a male that provides good parental care (Trivers 1972, Halliday 1983, Qvarnström and Forsgren 1998). Foraging skills and resource provisioning (both qualitative and quantitative) are essential to breeding success in seabirds (Hunt 1980, Nur 1984; see also Section 9.1 of this chapter), as in other long-lived species with biparental care like raptors (Simmons 1988, Bildstein 1992). Consequently, foraging abilities should be an essential proximate criterion of mate choice in seabirds. This hypothesis is supported, for example, by the existence of courtship-feeding, which may help females to evaluate the foraging abilities of their

TABLE 9.3
Nesting Sites Used by Seabirds

	Burrows	Crevices	Boulders or Rock Cavities	Tree Holes	Cliffs, Ledges	Trees	Flat Ground or Smooth Slopes	Steep Slopes	No Nesting Site
Spheniscifomes	+		+				+	+	+
Procellariiformes	+	+	+		+		+	+	
Pelecaniformes									
Tropicbirds		+	+	+	+		+		
Sulids					+	+	+		
Frigatebirds						+	+		
Cormorants					+	+	+	+	
Charadriiformes									
Skuas and jaegers			+				+		
Gulls and terns					+	+	+		
Alcids	+	+	+	+	+				

For references, see Nelson 1978, 1980, Warham 1990, del Hoyo et al. 1992, Furness 1996, Burger and Gochfeld 1996, Gochfeld and Burger 1996, Nettleship 1996, and Gaston and Jones 1998.

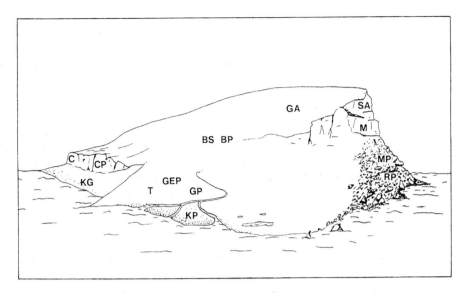

FIGURE 9.7 Partitioning of breeding habitats in a seabird community on a sub-Antarctic island: Kelp Gulls (KG) nest on the shore; cormorants (C) and Cape Pigeons (CP) on coastal cliffs; King Penguins (KP) on sandy beaches and estuaries; terns (T) on gravelled coastal plateaus; Gentoo Penguins (GEP) and giant petrels (GP) on grassy slopes; great albatrosses (GA) on grassy plateaus; Brown Skuas (BS) on grassy slopes and plateaus; small petrels (BP) dig burrows; mollymawks (M) nest on steep slopes; sooty albatrosses (SA) on ledges in cliffs; Macaroni Penguins (MP) in boulders on the surface; and Rockhopper Penguins (RP) under boulders.

prospective males. Amongst marine birds *sensu lato*, courtship-feeding of the female by the male occurs in skuas, gulls, terns (Figure 9.8), and skimmers (Hunt 1980, Zusi 1996). Courtship-feeding yields nutrients and energy to females (Hunt 1980, Halliday 1983), enabling them to lay larger clutches (Andersson 1994).

Body condition or body mass upon pair formation also may reflect foraging abilities (Chastel et al. 1995b) and may be used as quality indices by some species (Moorhen [*Gallinula chloropus*], Petrie 1983; Black-tailed Godwit [*Limosa limosa*], Hegyi and Sasvari 1998; Cooper's Hawk [*Accipiter cooperi*], Rosenfield and Bielefeldt 1999), including seabirds. Thus, body mass appears to be

FIGURE 9.8 A Common Tern brings food to its mate at the nest site as part of their courtship ritual. He may continue bringing food to her (pair-bond maintenance behavior) even after the chick hatches. (Photo by J. Burger.)

an important criterion of mate choice in Brown Noddies ([*Anous stolidus*], Chardine and Morris 1989). Male Blue Petrels (*Halobaena caerulea*) may give some information on their body condition when calling from their burrows (Genevois and Bretagnolle 1994); yet, it remains unknown whether petrel females do take this information into account (Bretagnolle 1996). Indeed, body condition may be a predictor of the quality of parental care during incubation and chick-rearing (Mock and Fujikoa 1990, Hegyi and Sasvari 1998, Buchanan et al. 1999). In seabirds, body condition may indicate how long body reserves enable each parent to fast when incubating (Davis 1988, Chaurand and Weimerskirch 1994b). Accordingly, female Adélie Penguins seem to choose large males as mates, presumably because fat storage capacities increase with body size (Davis and Speirs 1990).

Quality also may be assessed from the date of return to the breeding grounds, high quality individuals returning earlier than poor quality ones (Rowley 1983, Ens et al. 1996, Bried et al. 1999), particularly those in good body condition (Blue Petrel [*Halobaena caerulea*], Chastel et al. 1995b) and/or being the most experienced (Royal Penguin [*Eudyptes (chrysolophus) schlegeli*] and Adélie Penguin, Carrick and Ingham 1970). However, individual quality may vary in the course of life (e.g., Ens et al. 1996, Catry et al. 1999), especially in long-lived species such as seabirds, in which breeding success and/or adult survival rates may be affected by senescence (see, e.g., Coulson and Horobin 1976 for Arctic Terns [*Sterna paradisea*], Bradley et al. 1989 for Short-tailed Shearwaters [*Puffinus tenuirostris*], Weimerskirch 1992 for Wandering Albatrosses [*Diomedea exulans*]).

In species with biparental care like seabirds, both sexes are expected to perform active mate choice, with the sex that invests the most in reproduction being the choosiest (Trivers 1972, Hunt 1980, Parker 1983, Jones and Hunter 1993, Johnstone et al. 1996). Simultaneously, males and females should also be equally likely to initiate divorce to improve their reproductive performances (Birkhead and Møller 1996). The existence of mutual mate choice implies that sexual selection (i.e., the selection of characters giving greater chances to achieve successful matings to their owners than to other conspecifics of the same sex, Darwin 1871, Partridge and Halliday 1984) exists for both sexes (Jones and Hunter 1993, Andersson 1994). The existence of biparental care also implies that each individual strives to obtain a partner that, ultimately, maximizes the efficiency of the pair as a unit, during incubation and chick-rearing, but possibly also when performing territorial defense (see Ens and Haverkort in Ens 1992; for seabirds, see Isenmann 1970, Warham 1990). Consequently, an "ideal" pair should be formed by two compatible mates (Coulson 1972) and the "ideal" mate should have genes (or qualities and abilities) which can complement those of the other mate (Halliday 1983, Black et al. 1996, Qvarnström and Forsgren 1998).

Mate choice appears to be very important in seabirds and takes several years, particularly in biennial albatrosses, which seem to be extremely choosy (Jouventin and Weimerskirch 1984a, Jouventin et al. 1999a). Conversely and surprisingly, some studies failed to find reliable criteria of mate choice in petrels (Mougin et al. 1988a, Jouventin and Bried in press; but see Brooke 1978, Bradley et al. 1995). This result may be explained by the fact that in many cases individuals fail to obtain the best partner available (Mougin et al. 1988a, Olsson 1998, Bried et al. 1999), probably because of time and energy constraints during the courtship period, partial information on mate quality, and/or intrasexual competition for mates (Johnston and Ryder 1987, Real 1990, Sullivan 1994). Despite the possible occurrence of suboptimal matings, seabird pairs involving experienced mates often have higher breeding success than pairs in which both mates are inexperienced (Ainley et al. 1983, Weimerskirch 1990, Schreiber and Schreiber 1993, Burger and Gochfeld 1996, Wooller and Bradley 1996). Therefore, individuals should mate with experienced partners (Forslund and Larsson 1991, Jouventin et al. 1999a), partly because foraging skills may increase with age and/or experience in seabirds (Lack 1968, Nur 1984, Weimerskirch 1992). However, the optimal age of a partner depends on the reproductive potential (in terms of residual reproductive value) of the individual that exerts mate choice (Hunt 1980). This may explain why individuals, and especially seabirds, tend to mate with partners of similar age or experience (Coulson 1966, Hunt 1980, Mougin et al. 1988a, Reid 1988, Bradley et al. 1995), so that the duration of the pair bond can be maximized.

In species with biparental care, assortative mating is expected to occur if the male has a high investment and also may result from the combined effects of mate choice and competition between males (Parker 1983). However, Reid (1988) suggested that it could be attributed to the contemporary recruitment of birds from the same age cohort into the breeding population. He also argued that in some cases, assortative mating by age in seabirds (e.g., Adélie Penguin) might be no more than a by-product of an age-dependent date of arrival at the colony, suggesting passive mate choice within an age cohort (see also Ens et al. 1996). Yet, at least two studies on seabirds (Shaw 1985, Jouventin et al. 1999a) do not support Reid's (1988) explanations, but are consistent with Hunt's (1980) "optimal age of the mate" hypothesis, which states that the optimal age of the partner depends on the reproductive potential (in terms of residual reproductive value) of the individual that exerts mate choice. Shaw (1985) provided evidence that assortative mating by age actually was an active process in Blue-eyed Shags (*Phalacrocorax atriceps* spp.). Jouventin et al. (1999a) also showed that Wandering Albatrosses selected their partners on the basis of age at Iles Crozet, young females rejecting old widowed males which were much more numerous at this locality. However, mate choice in Wandering Albatrosses still appeared more complex than in Blue-eyed Shags: not only the ages of mated pairs were correlated, but birds selected experienced individuals as partners within a given age cohort. If females chose the oldest (i.e., most experienced) males, they would mate with those individuals that are the most likely to die during subsequent years. Because (1) the time spent by widowed individuals before breeding with a new mate represents on average 15% of reproductive life span in the Wandering Albatross (Jouventin et al. 1999a) and (2) this species breeds biennially, Wandering Albatrosses cannot afford missing too many breeding seasons. Choosing experienced males of similar ages represents the best solution for females. Thus, the Wandering Albatross appears as one of the choosiest species that have been so far studied. Other types of assortative mating also may occur in seabirds. For example, a tendency to assortative mating by size has been observed in the Snow Petrel (Barbraud and Jouventin 1998) and in the Razorbill (Wagner 1999), whereas Brown Noddies mate assortatively by body mass, irrespective of size (Chardine and Morris 1989). Assortative mating by color morph might occur in some populations of the Parasitic Jaeger (*Stercorarius parasiticus*, Furness 1987).

Yet individual characteristics may not suffice to explain mate choice in birds. The females of some territorial species assess and choose males not on the basis of quality, but on the quality of their territory (e.g., Pied Flycatcher [*Ficedula hypoleuca*], Alatalo et al. 1986), or on both male and territory quality (see, for example, Davies 1978 for Dunnocks [*Prunella modularis*], or Searcy 1979 for Red-winged Blackbirds [*Agelaius phoeniceus*]). In these studies, territory quality was assessed in terms of nest and food availability. Male body condition also may be related to territory quality (e.g., Willow Warbler [*Phylloscopus trochilus*], Nyström 1997), suggesting that the quality of a site may reflect that of its owner (Hunt 1980, Ligon 1999).

Mate choice based (at least partially) on territory quality might occur in seabird species where males return earlier than females and establish the nest (Hunt 1980). Thus, a long-term study (Jouventin and Bried in press) showed that nest-site quality appeared to be prevailing in the Snow Petrels from Terre Adélie. Average breeding success varied greatly amongst nests at this locality. Some nests experienced very low occupancy rates and breeding success due to the presence of snow or ice that obturated them in certain years, making occupation or successful breeding impossible. Therefore, individuals refrained from breeding during these years, showing on average low breeding frequency (Chastel et al. 1993). Some evidence was found that the Snow Petrel colonies from Terre Adélie had already attained their carrying capacity and that the ownership of a nest site had a greater influence than mate choice on breeding success in this species.

Individuals prospecting for mates should minimize the costs for mate assessment and acquisition. The former comprise a greater risk of predation for smallest species (Andersson 1994, McNamara and Forslund 1996). Thus in seabirds, prospecting Thin-billed Prions (*Pachyptila belcheri*) and Blue Petrels seem to suffer heavier predation by Brown Skuas (*Catharacta skua lönnbergi*) than breeding conspecifics (Mougeot et al. 1998). Costs of mate assessment also involve the use of time and body reserves that would otherwise have been available for reproduction *sensu stricto* when prospecting for mates (McNamara and Forslund 1996; for seabirds, see Hunt 1980 and Olsson 1998). Moreover, the number of prospective mates that can be sampled, and hence the amount of information available, decreases as time elapses, creating additional costs in case of late mate sampling (Sullivan 1994). Similarly, individuals (males in a majority of species) that seek to attract prospectors may signal their quality through ornaments (e.g., crests, tails, plumage), the production of which requires great amounts of energy, and/or through energy-consuming acoustic and/or visual displays (Andersson 1994; for seabirds, see Genevois and Bretagnolle 1994 for the Blue Petrel; Harrison 1990 for frigatebirds). Most of the costs of obtaining a mate are due to intrasexual competition, which may limit opportunities for mate sampling (Andersson 1994, Johnstone 1995, Reynolds 1996). These latter costs also may lead to reduced fecundity (Andersson 1994). In seabirds, for example, intrasexual competition may increase if the sex ratio is biased and time constraints for breeding are strong, sometimes resulting in missed breeding years for the individuals of the most represented sex. Thus, female Emperor Penguins monopolize each male that returns ashore at the onset of the breeding season, due to the female-biased sex ratio in this species, combined with the urge to breed early because of incubation and chick-rearing constraints (see Isenmann 1971, Jouventin 1971, and Bried et al. 1999 for more details). As a consequence, the latest-arriving females have greater difficulties to obtain a mate, and some of them may find themselves unpaired (see also Isenmann and Jouventin 1970).

Other costs of intrasexual competition even involve reduced survival, because individuals may be exposed to diseases and parasites when fighting for mates and/or for territories, or may be severely injured or even killed by conspecifics (Andersson 1994, Gustafsson et al. 1994). Such fights to death can occur in some seabird groups. For example, petrels (e.g., White-chinned Petrels [*Procellaria aequinoctialis*], Mougin 1970), tropicbirds (del Hoyo et al. 1992), gannets (Nelson 1978), and skuas [*Catharacta* spp.], Furness 1987) are known to defend fiercely their burrows or their nests, and takeover of territory and males by female skuas sometimes causes the death of the ancient resident female (Furness 1987). Conversely, there seems to be no evidence for sexual competition in frigatebirds (Nelson 1976, Harrison 1990). Moreover, when research costs become too important, the threshold value for mate acceptability then should decrease (Real 1990), leading to possible mismatch between the two partners (Johnstone et al. 1996). Such imperfect matings probably occur in *Aptenodytes* penguins, in which search costs are *a priori* low but can increase rapidly because of strong time constraints for breeding and expensive fat storing (Olsson 1998, Bried et al. 1999). This hypothesis of a lowered acceptability threshold strongly suggests the existence of a trade-off during mate selection. Consequently, it may be better to mate with a partner whose quality (or complementarity) is greater than a threshold value ("threshold-based" choice) than with the highest-quality partner available ("best of *n*" hypothesis) when search costs exist (Janetos 1980, Real 1990, Valone et al. 1996).

Remating with familiar individuals (for example, neighbors, whose quality has been assessed readily during the previous partnership) after the loss of the previous partner may represent a way of achieving this trade-off between choosing the best mate available and limiting the costs of mate selection. Thus, risks and time-wasting (McNamara and Forslund 1996) are minimized, and individuals can benefit from familiarity with the new mate and the new site (Black and Owen 1995). In seabirds, remating amongst neighbors has been observed, for instance, in Adélie Penguins (Davis and Speirs 1990), Cory's Shearwaters (*Calonectris diomedea*) (Mougin et al. 1988b), and some larids (Johnston and Ryder 1987). Such rematings also may be favored because synchrony is generally higher amongst neighbors than amongst more distant individuals in the colony (Smith 1975, Mougin et al. 1988a, b).

FIGURE 9.9 Sooty Tern adults call from the air as they fly in to feed their chick. The chick recognizes its parents' voices and calls back. They continue calling to each other while the chick walks out from the shade or from a creche to meet its parent and get fed. (Photo by E. A. Schreiber.)

9.5 SITE AND/OR MATE TENACITY, OR SWITCHING?

A prerequisite to nest/mate fidelity is nest/mate recognition. Numerous birds (e.g., Stoddard 1996), including seabirds, can recognize vocally an individual (neighbor, partner, parent, or chick) among many conspecifics. Pioneering studies showed that individual recognition between mates, or between parents and young, existed in larids (Beer 1969, Evans 1970; Figure 9.9). Individual acoustic recognition has ever since been proved in penguins (Jouventin 1982, Davis and Speirs 1990, Jouventin et al. 1999b, Lengagne et al. 1999), procellariiforms (Bretagnolle 1996, Jouventin et al. 1999a), sulids (Nelson 1978), and terns (McNicholl 1975, Møller 1982). According to Bretagnolle (1996), individual recognition would occur more frequently in colonial or highly mate-faithful species. Yet, the nesting habits of seabirds should induce different constraints on the localization and the recognition of the previous nest and mate. Indeed, surface-nesting seabirds are diurnal on land, using visual cues to locate their previous nest/mate and achieving specific and sexual recognition through visual and acoustic displays. Amongst burrow-nesting and cavity-nesting species, the Little Penguin, most small petrels, and eight auks are predominantly or strictly nocturnal when ashore. These species probably cannot utilize vision to locate their nesting sites except during moonlit nights, and whether they utilize olfaction remains to be proved (Grubb 1974, Brooke 1979). Thus, they use vocalizations essentially to communicate when on land (Nelson 1980, Jouventin and Mougin 1981, Davis and Speirs 1990, del Hoyo et al. 1992, Warham 1996, Gaston and Jones 1998). When they return ashore at the onset of a new breeding season, seabirds have the possibility either to retain their previous site and mates, or to switch. Switches are expected to occur only if their benefits outweigh their costs (Choudhury 1995). Therefore, we will examine the benefits and the costs of fidelity, divorce, and site changes before determining the factors that elicit changes of site and/or mate in seabirds.

9.5.1 BENEFITS OF SITE AND MATE FIDELITY

The most obvious and direct benefit of site and mate fidelity is an increase in breeding success (Greenwood and Harvey 1982, Domjan 1992; Table 9.4). This improvement of reproductive performances may be achieved through better coordination between mates of incubation shifts or chick feeding schedules (Choudhury 1995) and/or greater familiarity with the territory. Individuals exploit

TABLE 9.4

Costs and Benefits of Site Fidelity, Mate Fidelity, Site Change, and Divorce in Seabirds

Benefits	Costs
Site Fidelity	
Increased breeding success (1)	Low breeding success when remaining on a poor quality site (18, 10)
Better knowledge of neighbors and potential mates (2, 3, 4)	
Dominance in territorial contests (5, 6)	Failure to breed (number of years spent without breeding) when remaining on the territory after loss of mate (15)
Utilization of the site as a meeting point for pairs to reunite, hence all the benefits from mate retention (7, 8, 9)	Costs of territorial defense (20, 21, 6)
Mate Fidelity	
Increased breeding success (10, 5, 1, 11, 12)	Low breeding success when remaining with a poor quality mate (19)
Limitation of the costs of divorce (13, 14, 15)	
Reduced costs of reproduction, enhanced future survival, and fecundity (16)	Costs of late breeding, or not breeding at all, when waiting for previous mate (22, 23)
Site change	
Acquisition of a higher-quality site (17)	Costs of prospecting (increased mortality, number of years missed: 17, 21, 10, 24)
Increased breeding success upon the first breeding attempt on the new site or in the long term (1, 18)	Costs of settlement on the new site (2, 17, 21, 10)
	Lower breeding success (17, but see 25)
	Risk of finding oneself without a territory (26, 27)
Divorce	
Acquisition of a higher-quality site and/or mate (19)	Costs of prospecting (see above; see also 28, 29)
Increased breeding success upon the first breeding attempt with the new mate or in the long term (7)	Costs of intrasexual competition for access to the new mate (30)
	Time spent displaying with the new mate (31)
	Remating with a poor quality partner (19, 32)
	Lower breeding success (25)

Note: Numbers in brackets refer to the following references: 1, Ollason and Dunnet (1988); 2, McNicholl (1975); 3, Møller (1982); 4, Mougin et al. (1988b); 5, Greenwood and Harvey (1982); 6, Warham (1990); 7, Jouventin (1982); 8, Morse and Kress (1984); 9, Bried, Pontier, and Jouventin (in preparation); 10, Nelson (1978); 11, Bradley et al. (1990); 12, Scott (1988); 13, Ens et al. (1996); 14, Jouventin et al. (1999a); 15, Jouventin and Bried (in press); 16, Pianka and Parker (1975); 17, Switzer (1993); 18, Carrick and Ingham (1967); 19, Choudhury (1995); 20, del Hoyo et al. (1992); 21, Mougin (1970); 22, Bried et al. (1999); 23, Davis (1988); 24, Mougeot et al. (1998); 25, Ollason and Dunnet (1978); 26, Ens et al. (1995); 27, Bried and Jouventin (1998); 28, Gochfeld (1980); 29, Jouventin and Weimerskirch (1984a); 30, Trivers (1972); 31, Van Ryzin and Fisher (1976); 32, Ainley et al. (1983); and 33, Jouventin et al. (1999a).

a familiar territory more efficiently because they have a better knowledge of food resources, potential mates, neighbors, and refuges against predators (Hinde 1956, Greenwood and Harvey 1982). Moreover, the ownership of a breeding site confers dominance in aggressive encounters in many species (Hinde 1956; for seabirds, see, e.g., Nelson 1978 for sulids, Burger and Gochfeld 1996 for gulls). The advantages of dominance also may be particularly important in species that compete intensively for sites like burrowing procellariiforms (Warham 1990). In species where competition for sites is severe, fidelity to the territory may help reduce competition and the risks of usurpation. For example, White-headed Petrels (*Pterodroma lessonii*) show very high burrow fidelity (Bried, Pontier, and Jouventin in preparation) and individuals visit their burrows every year (Warham 1967), although this species breeds biennially (Zotier 1990b). In many species, males return to the breeding grounds earlier than females (Greenwood 1980, 1983, Ens et al. 1996; for seabirds, see references quoted in Section 9.2). Greater familiarity with the site may give males an advantage in male–male competition and territorial contests. Also, aggressive interactions with familiar neighbors are generally less numerous and less severe than those with total strangers (Falls and Mc Nichol 1979, Stamps 1987).

Finally, nests or territories may serve as meeting points for pairs to reunite in species that have part-time pair bonds (i.e., species where pairs do not spend the nonbreeding period together), so that mate retention is facilitated, being classically considered as a by-product of site fidelity (Hinde 1956). Most seabirds have part-time pair bonds, and the role of nesting sites as meeting points has already been documented, for example, in petrels (Morse and Kress 1984), skuas (Pietz and Parmelee 1994), and gulls (see Burger 1974). Yet, it deserves further examination (Jouventin 1982, Bried, Pontier, and Jouventin in preparation). Although they do not reunite at the nest site, pairs of many gull species reunite in "clubs" at proximity from breeding colonies (Burger and Gochfeld 1996). Partners that reunite do not waste time prospecting for new mates, and they spend less time displaying. Consequently, they often lay earlier and have higher fledging success than newly formed pairs (reviews in Rowley 1983, Domjan 1992). Early breeding has been shown to be advantageous for reproduction in several species, including seabirds (e.g., Isenmann 1971, Jouventin and Lagarde 1995 for *Aptenodytes* penguins; Ollason and Dunnet 1978 for the Atlantic Fulmar [*Fulmarus glacialis*]; Spear and Nur 1994 for the Western Gull [*Larus occidentalis*]). Experienced seabird pairs not only do better than newly formed pairs (e.g., Nelson 1978, Ollason and Dunnet 1988, Scott 1988, Burger and Gochfeld 1996), but breeding success may be positively correlated with pair experience in some species, at least during the first few years (e.g., Bradley et al. 1990, Ens et al. 1996).

Therefore, the reproductive advantages of mate fidelity are very important in seabirds (Lack 1968). Moreover, early breeding may be advantageous not only for breeding success, but also for offspring fitness, early-hatched chicks having higher postfledging survival and sometimes being more successful upon their first breeding attempt than later-hatched young (Perrins 1970, Nelson 1980, Visser and Verboven 1999). In some seabirds, however, breeding success does not significantly increase with pair experience (Shaw 1986, Williams and Rodwell 1992, Bried and Jouventin 1999), but the costs of divorce (in terms of years spent without breeding, see below) are high. In these species, however, high mate fidelity rates might enable individuals to limit the costs of divorce (Ens et al. 1996, Dubois et al. 1998). Monogamy and mate fidelity also may increase the residual reproductive value (and therefore survival) of the parents (Pianka and Parker 1975) in another way, because (1) in most species, a female mated with a monogamous male shares parental duties (see Jouventin and Cornet 1980 or Ligon 1999), and (2) in birds, a monogamous male may be less involved in male–male competition than if he had to inseminate other females (Trivers 1972).

9.5.2 COSTS OF SITE AND MATE FIDELITY

Site fidelity involves the costs of maintaining a territory, for which competition may be severe or which is of poor quality. In the latter case, individuals do the best of a bad situation. These costs may range from egg or chick losses resulting from disturbance by intruders, to death if fights occur (e.g., Mougin 1970, Nelson 1980). Another cost of site fidelity is represented by the years spent without breeding by widowers and divorcees which retain their breeding site after the loss of their previous mate, but fail to attract a new partner (e.g., Bried and Jouventin 1998). Similarly, it may be costly for individuals to remain with a low quality mate or with a mate whose abilities do not complement their own qualities (Choudhury 1995). Independent of quality, the facility with which pairs can reunite may depend on whether the species to which they belong are migratory or sedentary, and also on whether mates remain together during the nonbreeding season. Because they have part-time pair bonds, seabirds may incur the costs of waiting for their previous mate at the onset of the next breeding cycle. These costs are likely to become higher (i.e., late breeding or not breeding at all) as asynchrony of return between the previous mates increases, especially if time constraints for breeding are strong. Accordingly, the probability of divorce increases as asynchrony of arrival of previous partners increases in Adélie Penguins (Davis 1988) and *Aptenodytes* penguins (Bried et al. 1999), which face predictable and marked seasonal environmental changes.

9.5.3 Benefits and Costs of Divorce and Site Changes

Benefits of site changes and divorce involve the acquisition of a better territory and/or a better mate, and higher reproductive performances. The increase in fitness is likely to be dependent on the amount of variation in territory quality or mate quality, but also on the quality of the individual that initiates divorce or site change (Choudhury 1995, McNamara and Forslund 1996).

Recent reviews or studies of the costs of site changes and divorce have been published by Switzer (1993) for the former, Choudhury (1995), Ens et al. (1996), and McNamara and Forslund (1996) for the latter. To summarize, individuals that change site or divorce must devote some time to prospection for a new site and/or mate. So, they use reserves that would otherwise be available for reproduction, at the risk of breeding late or missing one or several breeding seasons. Moreover, vulnerability to predators increases for some species when prospecting. Costs of fighting due to intraspecific competition also may occur when obtaining a new breeding site or choosing a new mate. New owners of a territory often bear some costs of familiarization with their territory (lower feeding efficiency, predation when exploring the new territory, temporary insufficient knowledge of shelters against predators, potential nesting sites, and local food resources). Suboptimal matings also may occur if divorcees remate with a poor quality partner or one with whom timing does not work — as for incubation shifts. Yet, other important costs may exist. Thus, individuals that move toward a new breeding site may find themselves in an environment to which they are not genetically adapted (Oring 1988); however, this hypothesis seems to be doubtful for seabirds, which show extremely strong fidelity to their breeding locality. Other risks when switching territory are obtaining a poorer quality territory, or finding oneself without a territory (Ens et al. 1995, Bried and Jouventin 1998). Moreover, courtship-displaying with the new partner, even if the latter is of high quality, is time- and energy-consuming and often implies later breeding (or lower breeding success) than in experienced pairs (but see Dann and Cullen 1990 for the Little Penguin). Therefore, divorcing or changing site may affect reproductive output upon the first breeding attempt with the new mate or on the new site. In the long term, changes may affect residual reproductive value or survival (Switzer 1993, Choudhury 1995, McNamara and Forslund 1996).

However, the costs of divorce also may depend on the quality of divorcees and that of their new mates. Low breeding success in newly formed pairs is most likely to occur in species where reproductive performance increases with pair experience (Choudhury 1995, Ens et al. 1996). It may also occur if divorcees re-mate with an inexperienced or young partner (Black et al. 1996). Young individuals often have poor breeding success (Curio 1983). In some cases, however, reproductive performance may be higher than during the year preceding divorce (Rowley 1983), especially if no year has been missed between divorce and re-mating (see Ollason and Dunnet 1978).

In seabirds, re-pairing often is prolonged and can take several years (Gochfeld 1980, Jouventin and Weimerskirch 1984a, Bried and Jouventin 1999 and unpublished data), and pair formation may take longer if new mates are experienced, as occurs with experienced female Laysan Albatrosses (*Diomedea [Phoebastria] immutabilis*). These females appear to have difficulties getting synchronized with their new males in the early stages of courtship-displaying and may take longer to mate than younger ones (Van Ryzin and Fisher 1976).

Finally, costs should be less important in pairs divorcing at the end of the breeding season than in pairs splitting at the beginning, because the former may save time by prospecting for new mates before the postbreeding dispersal of adults at sea occurs (Johnston and Ryder 1987).

9.5.4 Changing Site and/or Mate

Birds may choose their breeding territories on criteria (food availability, shelters from bad weather and predators, breeding success of conspecifics established in the same area) that should enable them to maximize their reproductive outputs (Hildén 1965, Cody 1985, Boulinier and Danchin 1997). Therefore, breeding dispersal (i.e., the movement of individuals between two successive breeding sites, Greenwood 1980) is usually interpreted as a means for birds to improve their reproductive performances through establishment on a more suitable territory and/or with a higher quality mate (Rowley 1983, Switzer 1993).

Accordingly, site changes are most likely to occur after a breeding failure (review and theoretical developments in Greenwood 1980 and Switzer 1993; see also an experimental study by Haas 1998), although the role of breeding failures may be confounded by that of conditions on land, for example, if predators regularly visit the colony. Thus, the presence of predators (which have an impact on eggs and young) at a seabird colony has been shown to elicit colony site desertion (Austin 1940, Burger 1982) or burrow switching (see Bried and Jouventin 1999 for an experimental study on the White-chinned Petrel). Territory or nest switches also can occur even when individuals produce young. Such moves generally result in an improvement of breeding performances (Greenwood and Harvey 1982, Switzer 1993), or are caused by deteriorating conditions on the previous site. For instance, it is well known that Greater Frigatebirds, Lesser Frigatebirds (*Fregata ariel*), and Red-footed Boobies change sites when their nesting bushes or trees get killed from too much guano and plucking (Schreiber et al. 1996, M. Le Corre unpublished data from Europa Island). However, individuals may not take into account their own reproductive performance before remaining faithful or changing, but the breeding success of the colony or that of the patch to which they belong if the environment is patchy (see Boulinier and Danchin 1997 for a theoretical study on terrestrial and marine birds). High breeding success of the whole population at a given site enhances fidelity or makes the colony or the subcolony more attractive to prospectors (e.g., Thibault 1994, 1995 for Cory's Shearwater; Burger 1982 for the Black Skimmer; Sydeman et al. 1996 for Cassin's Auklet [*Ptychoramphus aleuticus*]).

Some characteristics of the nesting habitat also may affect site fidelity. Thus, site fidelity is lower in unstable, unpredictable, or heterogeneous habitat (Switzer 1993). Most seabirds exhibit quasiabsolute fidelity to their breeding area (del Hoyo et al. 1992, Burger and Gochfeld 1996, Furness 1996, Gochfeld and Burger 1996, Nettleship 1996). However, the Flightless Cormorant regularly changes breeding locality (Harris 1979a). Some seabird species also can shift breeding location in case of environmental instability caused by floodings (e.g., Laughing Gull [*Larus atricilla*], Burger and Shisler 1980; Franklin's Gull, Burger 1974; Caspian Tern, Cuthbert 1985). The extreme situation is that of the Pomarine Jaeger (*Stercorarius pomarinus*), which shows almost no fidelity to its previous territory in reponse to interannual variations of prey (lemmings) numbers (Furness 1987). In unpredictable habitats, however, individuals should retain their nests if average site quality does not vary between the different parts of the habitat; moreover, individuals should not take their previous breeding performance into account when settling at the onset of the new breeding cycle (Switzer 1993). In predictable habitats, individuals should switch nests after a breeding failure, especially if they have an opportunity to settle on a site where chicks fledged the previous year; individuals that move after a successful breeding attempt are expected to enhance their reproductive output on their new territory (Switzer 1993). In some colonial terns, an interesting pattern is observed: for example, in the Common Tern (*Sterna hirundo*) and the Gull-billed Tern (*Gelochelidon nilotica*), nest fidelity is high in stable habitats. However, the members of the colony can shift together to a more suitable area if environmental disasters such as floodings occur at the usual locality, and individuals can retain their mates and settle with the same neighbors at the new breeding locality. This phenomenon is called "group adherence" (McNicholl 1975, Møller 1982).

Similarly, breeding success appears to be the ultimate factor of divorce in seabirds, which is generally promoted by breeding failures (e.g., Richdale 1957, Nelson 1978, Ollason and Dunnet 1978, Guillotin and Jouventin 1980, Harrison 1990, Thibault 1994; but see Emslie et al. 1992). Some pairs may divorce, not because they have failed to fledge offspring, but because (1) one of the partners has an opportunity to improve its reproductive performance with a new partner and/or on a new territory (see below, the "better option" hypothesis in Oystercatchers [*Haematopus ostralegus*], Ens et al. 1993), or (2) the cost of mate retention is late breeding or not breeding at all, outweighing all potential benefits (see Davis 1988 for Adélie Penguins; Olsson 1998 and Bried et al. 1999 for *Aptenodytes* penguins). A confounding effect of reproductive performance and individual quality on divorce remains possible, i.e., poor-quality individuals not only have low breeding success, but they also fail to retain their mates (Perrins and McCleery 1985, Ens et al. 1996; but see Lindén 1991).

Other factors, albeit of lesser importance, may promote divorce in seabirds. For example, intense competition for mates at the onset of the new breeding season also can increase divorce rates. Thus, the female-biased sex-ratio in the Emperor Penguin creates a competition between females for males, which is part of the combination of factors responsible for high divorce rates in this species (Bried et al. 1999). An increase in adult mortality also can lead to a higher incidence of divorce, because mortality creates a surplus of widowers seeking new mates (Johnston and Ryder 1987, Black et al. 1996). Pair experience also may be a factor of divorce in seabirds, the incidence of divorces being highest amongst inexperienced pairs (i.e., that bred together for the first time), especially if they have experienced a breeding failure (Richdale 1957, Ollason and Dunnet 1978, Schreiber and Schreiber 1993). Inexperienced pairs often have lower reproductive performances than more experienced pairs; but care must be taken when studying the effects of reproduction and experience on divorce and site changes, because inexperienced birds are generally young ones, so that a confounding effect of age and poor breeding success (Curio 1983) on divorce is possible.

Proximate factors of divorce may be the active decision to break the pair bond or the failure to meet one's previous mate. Active decisions to break the pair bond may be explained by the "incompatibility" hypothesis (Coulson 1966, 1972), which predicts that divorce should occur after a breeding failure or low breeding success in pairs with no or little previous common experience and that each divorcee should improve its breeding success with its new partner. The rather similar "error of mate choice" hypothesis (Johnston and Ryder 1987) and the "better option" hypothesis (Ens et al. 1993) also may explain active pair bond disruptions, the latter hypothesis stating that one member of the pair initiates divorce if it has the opportunity to improve breeding success on another territory with a higher-quality partner (by filling a vacancy, for instance). Choudhury (1995) hypothetized that the incidence of divorce in a population should be determined (at least partly) by the degree of assortative mating by quality.

Failures to find the previous mate can be explained by the "accidental loss" hypothesis (which explains that some divorces occur because the previous partner fails to return or returns too late), or the "musical chairs" hypothesis, which states that divorce occurs because the latest returning previous partners find their territory occupied (Dhondt and Adriaensen 1994). A situation similar to the "accidental loss" is represented by the "sabbatical" years in procellariiforms. The "sabbatical" partner does not return ashore at the onset of the breeding season, remaining at sea. Sabbatical years may represent an important cause of divorce in Cory's Shearwater (Mougin et al. 1987a, b). Some of these divorces, however, are temporary: when the partner that took a year off returns to the colony at the onset of the next breeding season, the previous pair bond can be re-formed. Such temporary divorces have been observed in the Northern Rockhopper Penguin (*Eudyptes chrysocome moseleyi*) from Amsterdam Island and in 15 out of the 18 procellariiform species that we have long-term monitored annually in the French Austral and Antarctic Territories (up to 35 consecutive years for some species). The absence of cases for the three remaining species must be due to the small number of years of study rather than to behavioral particularities. Another extreme situation is that of frigatebirds and Flightless Cormorants. In Greater and Magnificent Frigatebirds, the male stops feeding the chick after *circa* 3 months, molts, and can start a new breeding attempt with another female the next year, leaving his previous female still taking care of the young for several months (Nelson 1976, Trivelpiece and Ferraris 1987, Osorno 1999). Conversely, male Flightless Cormorants keep on feeding the young several weeks after fledging while their previous females undertake a new breeding cycle with another mate (Harris 1979a).

9.6 CONCLUSIONS AND PERSPECTIVES

9.6.1 WHICH STRATEGY SEEMS THE MOST ADAPTIVE FOR SEABIRDS?

Because of the importance of the benefits of fidelity (in terms of breeding success) and the existence of potentially high costs of divorce and site changes in many long-lived species (reduced reproductive life span; see references quoted above and Table 9.4), one should expect seabirds to be highly site and mate tenacious. Most of the data presented in Table 9.1 support this assumption. Moreover, the existence of long-term pair bonds may be favored by assortative mating by age, which limits the costs of widowhood; the latter are sometimes higher than those of divorce (Johnston and Ryder 1987). Under these conditions, divorce should occur only if the reproductive success with the old mate the next year is less than the average residual reproductive value with the new mate (McNamara and Forslund 1996). This prediction also seems to be relevant for site changes. Switzer (1993) predicted that site fidelity should be positively correlated to the costs of territory switching. Therefore, retaining one's previous nest upon a divorce also may enable individuals to avoid potentially important costs of territory switching (Tenaza 1971, Jouventin and Bried in press).

Similarly, small costs may suffice to inhibit divorce if the scope for improvement is small (Choudhury 1995). However, seabirds may minimize the costs of changing by utilizing familiarity as a factor of site/mate choice. For example, divorced Cory's Shearwaters and female Adélie Penguins that have lost their partner often remate with neighbors at the beginning of the next breeding season (Mougin et al. 1988b, Davis and Speirs 1990), benefitting from prior knowledge of their new mate, the greater familiarity of the latter with the environment and its higher level of synchrony. Likewise, the absence of nests and territorial behavior in King and Emperor Penguins would allow these species to assess potential mates rapidly and with low costs, by walking across the colony (Olsson 1998).

9.6.2 IS FIDELITY POSITIVELY RELATED TO LONGEVITY?

Although they are long-lived, seabirds can switch nesting site and divorce (Table 9.1), even though the frequency of occurrence of changes is low. Changes enable seabirds to improve their breeding success or just to breed in the most extreme cases (*Aptenodytes* penguins).

More surprisingly, some very long-lived and uniparous species (*Aptenodytes* penguins, Bried et al. 1999; frigatebirds, Nelson 1976) have very high divorce rates despite the advantages of fidelity and high costs of divorce in seabirds. This phenomenon suggests that mate fidelity may not be related to longevity only in seabirds. To test this hypothesis, we checked for a correlation between site fidelity, mate fidelity, and adult life expectancy (ALE), the latter parameter being defined by the formula (Seber 1973):

$$ALE = 0.5 + 1/(1 - S)$$

S being the adult survival rate.

In a first step, we performed a regression of each parameter on body mass, using Log-transformed variables (Calder 1984). If the correlation was significant (i.e., $p < 0.05$), we used the residuals from the regression for subsequent analyses; otherwise we used raw variables. If data were available for several populations from the same taxon, we performed analyses on several data sets, choosing one population at random. Results are presented in Table 9.5 and for one data set only, analyses conducted on the other data sets yielding similar results. We failed to find significant correlations between fidelity rates and ALE if we lumped all seabird species of our data sets together, or if we considered penguins only or charadriiforms only. Amongst pelecaniforms, we also failed to find significant relationships between nest fidelity and adult life expectancy, and between mate fidelity and adult life expectancy (both $p > 0.3$). However, our results on pelecaniforms may not be

TABLE 9.5
Relationships between Mate Fidelity and Adult Life Expectancy (ALE) in Seabirds

All Seabirds	Sphenisciformes Aptenodytes included	Charadriiformes
Relative mate fidelity vs. ALE $r^2 = 0.02$, $F_{1, 55} = 1.26$, $p = 0.27$	Relative mate fidelity vs. ALE $r^2 = 0.26$, $F_{1, 9} = 3.09$, $p = 0.11$	Relative mate fidelity vs. relative ALE $r^2 = 0.01$, $F_{1, 12} = 0.16$, $p = 0.70$
Relative nest fidelity vs. ALE $r^2 = 0.009$, $F_{1, 40} = 0.37$, $p = 0.55$	Nest fidelity vs. ALE $r^2 = 0.34$, $F_{1, 4} = 2.05$, $p = 0.22$	Nest fidelity vs. relative ALE $r^2 = 0.003$, $F_{1, 6} = 0.017$, $p = 0.9$
Relative mate fidelity vs. relative nest fidelity $r^2 = 0.18$, $F_{1, 34} = 7.58$, $p = 0.009$ (Regression slope = 0.35)	Relative mate fidelity vs. nest fidelity $r^2 = 0.39$, $F_{1,4} = 2.61$, $p = 0.18$	Relative mate fidelity vs. nest fidelity $r^2 = 0.02$, $F_{1,4} = 0.10$, $p = 0.77$
	Aptenodytes excluded	
Mate fidelity vs. ALE $r^2 = 0.12$, $F_{1, 53} = 7.18$, $p = 0.01$ (Regression slope = 0.85)	Mate fidelity vs. ALE $r^2 = 0.03$, $F_{1, 7} = 0.23$, $p = 0.64$	
	Nest fidelity vs. ALE $r^2 = 0.34$, $F_{1, 4} = 2.05$, $p = 0.22$	
Relative nest fidelity vs. ALE $r^2 = 0.009$, $F_{1, 40} = 0.37$, $p = 0.55$	Mate fidelity vs. nest fidelity $r^2 = 0.32$, $F_{1, 4} = 1.86$, $p = 0.24$	
Mate fidelity corrected for ALE vs. relative nest fidelity $r^2 = 0.31$, $F_{1, 34} = 15.49$, $p < 0.001$ (Regression slope = 20.65)		

Data were taken from Table 9.1. Species in which parameters were given for each sex and not for the entire study population were not considered. The term "relative" was employed whenever we used the residuals from the regression of one parameter on body mass (see text for methodology).

reliable due to the paucity of data concerning this order (see Table 9.1). Therefore, we have not presented them in Table 9.5.

However, King (Figure 9.10) and Emperor Penguins show low mate fidelity despite very high longevity (Bried et al. 1999). When these two nonnest building species were removed from our data sets, we found a positive significant correlation between mate fidelity and adult life expectancy when considering all seabirds together (Table 9.5, Figure 9.11), but no correlation between these two parameters in penguins.

Performing the same analyses using independent contrasts would enable us to control for phylogeny (Harvey and Purvis 1991, Garland et al. 1992). However, this method may not provide reliable results if sample size is too small (Björklund 1997), as is the case for some groups (e.g., cormorants, Figure 9.12). Within the order Procellariiformes, Bried, Pontier, and Jouventin (in preparation) found a significant positive correlation between mate fidelity and ALE after controlling for phylogeny. Conversely, Ens et al. (1996) found negative correlations between the same parameters in both penguins and procellariiforms (but significant in procellariiforms only); yet, they used much smaller data sets than ours for their analyses. They controlled for phylogeny, but it remains unknown whether they did so for body mass. Thus, our results seem to indicate that mate fidelity and adult life expectancy are positively correlated in seabirds, consistent with Rowley's (1983) prediction, but only if we exclude King and Emperor Penguins which are "aberrant" species, in that (1) they have no nesting site, incubating their egg on their feet, and (2) they face almost no costs of divorce. Similarly, some monogamous and very long-lived colonial waterbirds like the Greater Flamingo (*Phoenicopterus ruber*) show almost no mate fidelity (Cézilly and Johnson 1995),

FIGURE 9.10 King Penguins exhibit low mate fidelity, possibly because they have no specific nest site. (Photo by P. Jouventin.)

but this species breeds opportunistically in unstable habitats and may have evolved without mate fidelity as did large penguins. The presence of such aberrant species, amongst seabirds and non-marine waterbirds, therefore suggests that high longevity alone seems unlikely to have promoted mate fidelity. We suggest that these species show low mate fidelity because (amongst other factors) they have no meeting point for previous partners to reunite (Bried et al. 1999), having no nest site (King and Emperor Penguins) or being highly nomadic in their breeding because of instability of their breeding habitat (Greater Flamingo).

9.6.3 Is Mate Fidelity Just a By-Product of Nest Fidelity?

Most seabirds have part-time pair bonds (partners do not remain together during the nonbreeding period). Therefore, nesting sites serve as meeting points for the previous partners at the onset of the next breeding cycle. However, some seabird pairs can both switch nest or burrow and remain together. Such moves are known in terns (see "group adherence" phenomenon above), but also, for instance, in penguins, petrels, and skuas (Ollason and Dunnet 1978, Ainley et al. 1990, Bried and Jouventin 1999 and unpublished data). More surprisingly, Dark-mantled Sooty Albatrosses (*Phoebetria fusca*) and Royal Albatrosses show high mate fidelity (Jouventin and Weimerskirch 1984a, Robertson 1993) despite low nest fidelity (Sorensen 1950, Jouventin et al. 1983). Biennial albatrosses are the exception that confirms the rule, being unable to keep every year the same nest, yet showing a very high fidelity to the subcolony (almost 100%, J. Bried and P. Jouventin unpublished data). The hypothesis that mate fidelity merely results only from site fidelity has recently been questioned in a study on procellariiforms, although the existence of a meeting point seems to us to be of major importance for pairs to reunite from one breeding cycle to the next (Bried, Pontier, and Jouventin in preparation).

As a general rule, when considering all seabirds for which data are available, mate fidelity seemed to be correlated with site fidelity, even if we kept the two nonnesting *Aptenodyte*s in our

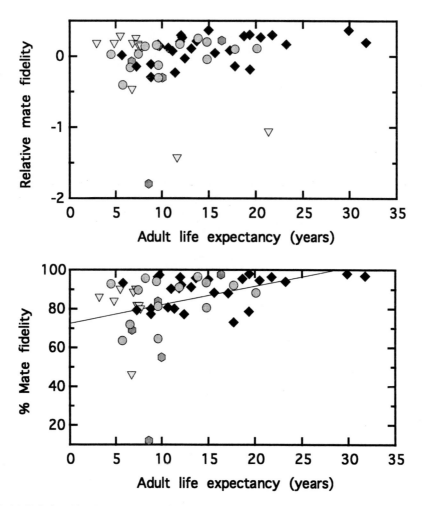

FIGURE 9.11 Relationships between mate fidelity and adult life expectancy in seabirds. Top: including *Aptenodytes* penguins. Bottom: *Aptenodytes* penguins excluded. The term "relative" was employed whenever we used the residuals from the regression of one parameter on body mass (see text for methodology). Triangles, Sphenisciformes; rhombuses, Procellariiformes; hexagons, Pelecaniformes; and filled circles, Charadriiformes.

data sets (Table 9.5). As a consequence, it would be interesting to examine more carefully the determinants of nest site selection and fidelity, and those of mate choice and mate fidelity (ultimate and proximate factors) in seabirds. Such studies are all the more desirable in petrels, since no reliable criterion of mate choice (age, experience) has been found for the individuals that retain their burrow after the loss of their previous partner (e.g., Cory's Shearwater, Mougin et al. 1988b; Snow Petrel, Jouventin and Bried in press). The ownership of a nest might be more important than mate choice in this group where the availability of burrows often is the limiting factor.

9.6.4 INFLUENCE OF BREEDING SUCCESS ON FIDELITY: RELEVANCE TO CONSERVATION

Finally, special attention should be paid when studying the relationships between reproductive performance and fidelity in seabirds. Such studies may have implications, not only for life history studies, but also for conservation biology (see Thibault 1994, 1995 on Cory's Shearwater; Bried and Jouventin 1999 on the White-chinned Petrel). For example, if introduced predators are present, not only breeding success but also site fidelity may be affected (see above). Therefore, alien

FIGURE 9.12 A pair of Kerguelen Cormorants on their nest. They lay 2 to 4 eggs. (Photo by J. Bried.)

predators might be indirectly responsible for a decrease of adult survival rate in the long term, by increasing the level of predation on the prospecting adults of the smallest species by native predators (Mougeot et al. 1998). Population trends of long-lived species depend mostly on the variations of adult survival rate (e.g., Lebreton and Isenmann 1976, Jouventin and Weimerskirch 1988). Because petrels are amongst the most long-lived birds (Warham 1996), a slight increase of adult mortality would result in a steeper decrease of population size than in shorter-lived species (see Weimerskirch et al. 1987, Croxall and Rothery 1991). Introduced mammals might therefore have more dramatic effects on small species than previously expected. Yet, controlling the density of introduced predators in a locality where eradication is impossible has proved to be an efficient means of increasing both breeding success and fidelity rates (Bried and Jouventin 1999).

ACKNOWLEDGMENTS

Support for this work was provided by the Institut Français pour la Recherche et la Technologie Polaires. We also are grateful to J.-L. Mougin, J. Burger, E. A. Schreiber, and S. Benhamou for comments and help with the literature, and to L. Jouventin for drawing figures.

LITERATURE CITED

AEBISCHER, N. J., G. R. POTTS, AND J. C. COULSON. 1995. Site and mate fidelity of Shags *Phalacrocorax aristotelis* at two British colonies. Ibis 137: 19–28.

AINLEY, D. G., AND D. P. DEMASTER. 1980. Survival and mortality in a population of Adélie Penguins. Ecology 61: 522–530.

AINLEY, D. G., R. E. LERESCHE, AND W. J. L. SLADEN. 1983. Breeding Biology of the Adélie Penguin. University of California Press, Berkeley.

AINLEY, D. G., C. A. RIBIC, AND R. C. WOOD. 1990. A demographic study of the South Polar Skua *Catharacta maccormicki* at Cape Crozier. Journal of Animal Ecology 59: 1–20.

ALATALO, R. V., A. LUNDBERG, AND C. GLYNN. 1986. Female Pied Flycatchers choose territory quality and not male characteristics. Nature 323: 152–153.

ANDERSSON, M. 1994. Sexual Selection. Princeton University Press, Princeton, NJ.

ASHCROFT, R. E. 1979. Survival rates and breeding biology of Puffins on Skomer Island, Wales. Ornis Scandinavica 10: 100–110.

ASHMOLE, N. P. 1962. The Black Noddy *Anous tenuirostris* on Ascension Island. I. General biology. Ibis 103b: 235–273.

ASHMOLE, N. P. 1971. Seabird ecology and the marine environment. Pp. 223–286 *in* Avian Biology, Vol. 1 (D. S. Farner, J. R. King, and K. C. Parkes, Eds.). Academic Press, New York.

ASHMOLE, N. P., AND M. J. ASHMOLE. 1967. Comparative feeding ecology of seabirds of a tropical oceanic island. Peabody Museum of Natural History, Yale University Bulletin 24: 1–131.

AUSTIN, O. L. 1940. Some aspects of individual distribution in the Cape Cod Tern colonies. Bird-Banding 11: 155–159.

BARBRAUD, C., AND P. JOUVENTIN. 1998. What causes body size variation in the Snow Petrel *Pagodroma nivea*? Journal of Avian Biology 29: 161–171.

BARRAT, A. 1976. Quelques aspects de la biologie et de l'écologie du Manchot royal (*Aptenodytes patagonicus*) des îles Crozet. Comité National Français de la Recherche en Antarctique 40: 9–51.

BAYER, R. D. 1982. How important are bird colonies as information centers? Auk 99: 31–40.

BECK, J. R., AND D. W. BROWN. 1972. The biology of Wilson's Storm Petrel, *Oceanites oceanicus* (Kuhl), at Signy Island, South Orkney Islands. British Antarctic Survey Science Reports 69: 1–54.

BEER, C. G. 1969. Laughing Gull chicks: recognition of their parents' voices. Science 166: 1030–1032.

BILDSTEIN, K. L. 1992. Causes and consequences of reversed sexual dimorphism in raptors: the head start hypothesis. Journal of Raptor Research 26: 115–123.

BIRKHEAD, T. R., AND A. P. MØLLER. 1996. Monogamy and sperm competition in birds. Pp. 323–343 *in* Partnerships in Birds — The Study of Monogamy (J. M. Black, Ed.). Oxford University Press, Oxford.

BJÖRKLUND, M. 1997. Are "comparative methods" always necessary? Oikos 80: 607–612.

BLACK, J. M. 1996. Introduction: pair bonds and partnerships. Pp. 3–20 *in* Partnerships in Birds — The Study of Monogamy (J. M. Black, Ed.). Oxford University Press, Oxford.

BLACK, J. M., S. CHOUDHURY, AND M. OWEN. 1996. Do Barnacle Geese benefit from lifelong monogamy? Pp. 91–117 *in* Partnerships in Birds — The Study of Monogamy (J. M. Black, Ed.). Oxford University Press, Oxford.

BLACK, J. M., AND M. OWEN. 1995. Reproductive performance and assortative pairing in relation to age in Barnacle Geese. Journal of Animal Ecology 64: 234–244.

BOERSMA, P. D. 1976. An ecological and behavioral study of the Galapagos Penguin. The Living Bird 15: 43–93.

BOERSMA, P. D. 1978. Breeding patterns of Galápagos Penguins as an indicator of oceanographic conditions. Science 200: 1481–1483.

BOEKELHEIDE, R. J., AND D. G. AINLEY. 1989. Age, resource availability, and breeding effort in Brandt's Cormorant. Auk 106: 389–401.

BOST, C.-A., AND P. JOUVENTIN. 1991. The breeding biology of the Gentoo Penguin *Pygoscelis papua* in the northern edge of its range. Ibis 133: 14–27.

BOULINIER, T., AND E. DANCHIN. 1997. The use of conspecific reproductive success for breeding patch selection in terrestrial migratory species. Evolutionary Ecology 11: 505–517.

BRADLEY, J. S., R. D. WOOLLER, AND I. J. SKIRA. 1995. The relationship of pair-bond formation and duration to reproductive success in Short-tailed Shearwaters *Puffinus tenuirostris*. Journal of Animal Ecology 64: 31–38.

BRADLEY, J. S., R. D. WOOLLER, I. J. SKIRA, AND D. L. SERVENTY. 1989. Age-dependent survival of breeding Short-tailed Shearwaters *Puffinus tenuirostris*. Journal of Animal Ecology 58: 175–188.

BRADLEY, J. S., R. D. WOOLLER, I. J. SKIRA, AND D. L. SERVENTY. 1990. The influence of mate retention and divorce upon reproductive success in Short-tailed Shearwaters *Puffinus tenuirostris*. Journal of Animal Ecology 59: 487–496.

BRETAGNOLLE, V. 1996. Acoustic communication in a group of nonpasserine birds, the petrels. Pp. 160–177 *in* Ecology and Evolution of Acoustic Communication in Birds (D. E. Kroodsma and E. H. Miller, Eds.). Comstock, Cornell University Press, Ithaca, NY.

BRIED, J., F. JIGUET, AND P. JOUVENTIN. 1999. Why do *Aptenodytes* penguins have high divorce rates? Auk 116: 504–512.

BRIED, J., AND P. JOUVENTIN. 1998. Why do Lesser Sheathbills *Chionis minor* switch territory? Journal of Avian Biology 29: 257–265.

BRIED, J., AND P. JOUVENTIN. 1999. Influence of breeding success on fidelity in long-lived birds: an experimental study. Journal of Avian Biology 30: 392–398.

BROOKE, M. DE L. 1978. Some factors affecting the laying date, incubation and breeding success in the Manx Shearwater, *Puffinus puffinus*. Journal of Animal Ecology 47: 477–495.

BROOKE, M. DE L. 1979. Determination of the absolute visual threshold of a nocturnal seabird, the Common Diving Petrel *Pelecanoides urinatrix*. Ibis 29: 2.

BROOKE, M. DE L. 1990. The Manx Shearwater. T. & A. D. Poyser, London.

BUCHANAN, K. L., C. K. CATCHPOLE, J. W. LEWIS, AND A. LODGE. 1999. Song as an indicator of parasitism in the Sedge Warbler. Animal Behaviour 57: 307–314.

BUCKLEY, F. G., AND P. A. BUCKLEY. 1980. Habitat selection and marine birds. Pp. 69–112 *in* Behavior of Marine Animals, Vol. 4: Marine Birds (J. Burger, B. L. Olla, and H. E. Winn, Eds.). Plenum, New York.

BURGER, J. 1974. Breeding adaptations of Franklin's Gull (*Larus pipixcan*) to a marsh habitat. Animal Behaviour 22: 521–567.

BURGER, J. 1981. Sexual differences in parental activities of breeding Black Skimmers. American Naturalist 117: 975–984.

BURGER, J. 1982. The role of reproductive success in colony-site selection and abandonment in Black Skimmers (*Rynchops niger*). Auk 99: 109–115.

BURGER, J., AND M. GOCHFELD. 1988. Habitat selection in Mew Gulls: small colonies and site plasticity. Wilson Bulletin 100: 395–410.

BURGER, J., AND M. GOCHFELD. 1996. Family *Laridae* (gulls). Pp. 572–623 *in* Handbook of the Birds of the World, Vol. 3 (J. del Hoyo, A. Elliott, and J. Sargatal, Eds.). Lynx Edicions, Barcelona, Spain.

BURGER, J., AND J. SHISLER. 1980. Colony and nest site selection in Laughing Gulls in response to tidal flooding. Condor 82: 251–258.

CALDER, W. A., III. 1984. Size, Function and Life History. Harvard University Press, Cambridge, MA.

CARRICK, R. 1972. Population ecology of the Australian Black-backed Magpie, Royal Penguin and Silver Gull. Pp. 41–99 *in* Population ecology of migratory birds: a symposium. U.S. Department of the Interior Wildlife Research Report, No 2.

CARRICK, R. E., AND S. E. INGHAM. 1967. Antarctic sea-birds as subjects for ecological research. Proceedings of the symposium on Pacific-Antarctic sciences. JARE Scientific Reports, Special Issue 1: 151–184.

CARRICK, R. E., AND S. E. INGHAM. 1970. Ecology and population dynamics of Antarctic sea birds. Pp. 505–525 *in* Antarctic Ecology (M. Holdgate, Ed.). Academic Press, London.

CATRY, P., N. RATCLIFFE, AND R. W. FURNESS. 1997. Partnerships and mechanisms of divorce in the Great Skua. Animal Behaviour 54: 1475–1482.

CATRY, P., G. D. RUXTON, N. RATCLIFFE, K. C. HAMER, AND R. W. FURNESS. 1999. Short-lived repeatabilities in long-lived Great Skuas: implications for the study of individual quality. Oikos 84: 473–479.

CÉZILLY, F., AND A. R. JOHNSON. 1995. Re-mating between and within breeding seasons in the Greater Flamingo *Phoenicopterus ruber.* Ibis 137: 543–546.

CHARDINE, J. W., AND R. D. MORRIS. 1989. Sexual size dimorphism and assortative mating in the Brown Noddy. Condor 91: 868–874.

CHASTEL, O. 1995. Effort de reproduction chez les oiseaux longévifs: fréquence de reproduction et condition physique chez les pétrels. Unpublished Ph.D thesis. Université François-Rabelais, Tours.

CHASTEL, O., H. WEIMERSKIRCH, AND P. JOUVENTIN. 1993. High annual variability in reproductive success and survival of an Antarctic seabird, the Snow Petrel *Pagodroma nivea*. Oecologia 94: 278–285.

CHASTEL, O., H. WEIMERSKIRCH, AND P. JOUVENTIN. 1995a. Body condition and seabird reproductive performance: a study of three petrel species. Ecology 76: 2240–2246.

CHASTEL, O., H. WEIMERSKIRCH, AND P. JOUVENTIN. 1995b. Influence of body condition on reproductive decision and reproductive success in the Blue Petrel. Auk: 112: 964–972.

CHAURAND, T., AND H. WEIMERSKIRCH. 1994a. The regular alternation of short and long foraging trips in the Blue Petrel *Halobaena caerulea*: a previously undescribed strategy of food provisioning in a pelagic seabird. Journal of Animal Ecology 63: 275–282.

CHAURAND, T., AND H. WEIMERSKIRCH. 1994b. Incubation routine, body mass regulation and egg neglect in the Blue Petrel *Halobaena caerulea*. Ibis 136: 285–290.

CHOUDHURY, S. 1995. Divorce in birds: a review of the hypotheses. Animal Bahaviour 50: 413–429.

CODY, M. L. 1985. An introduction to habitat selection. Pp. 3–56 *in* Habitat Selection in Birds (M. L. Cody, Ed.). Academic Press, San Diego, CA.

CONROY, J. W. H. 1972. Ecological aspects of the biology of the Giant Petrel *Macronectes giganteus* (Gmelin), in the maritime Antarctic. British Antarctic Survey Science Reports 75: 1–74.

COULSON, J. C. 1966. The influence of the pair-bond and age on the breeding biology of the Kittiwake Gull *Rissa tridactyla*. Journal of Animal Ecology 35: 269–279.

COULSON, J. C. 1972. The significance of the pair-bond in the Kittiwake. Proceedings of the 15th International Congress of Ornithology: 5: 424–433.

COULSON, J. C., AND J. HOROBIN. 1976. The influence of age on the breeding biology and survival of the Arctic Tern *Sterna paradisea*. Journal of Zoology of London 178: 247–260.

COULSON, J. C., AND R. D. WOOLLER. 1976. Differential survival rates among breeding Kittiwake Gulls *Rissa tridactyla*. Journal of Animal Ecology 45: 205–215.

CRAMP, S. 1985. Handbook of the Birds of Europe, the Middle East and North Africa. The Birds of the Western Palearctic. Vol. 4: Terns to Woodpeckers. Oxford University Press, Oxford.

CROXALL, J. P. 1984. Seabirds. Pp. 533–619 *in* Antarctic Ecology, Vol. 2 (R. M. Laws, Ed.). Academic Press, London.

CROXALL, J. P., AND A. J. GASTON. 1988. Patterns of reproduction in high-latitude northern and southern hemisphere seabirds. Proceedings of the 19th International Congress of Ornithology 1: 1176–1194.

CROXALL, J. P., AND P. ROTHERY. 1991. Population regulation of seabirds: implications of their demography for conservation. Pp. 272–296 *in* Bird Population Studies: Relevance to Conservation and Management (C. M. Perrins, J.-D. Lebreton, and G. M. Hirons, Eds.). Oxford University Press, Oxford.

CROXALL, J. P., AND P. ROTHERY. 1994. Population change in Gentoo Penguins *Pygoscelis papua* at Bird Island, South Georgia: potential roles of adult survival, recruitment and deferred breeding. Pp. 26–38 *in* The Penguins (P. Dann, I. Norman, and P. Reilly, Eds.). Surrey Beatty and Sons, Chipping Norton.

CROXALL, J. P., P. ROTHERY, S. P. C. PICKERING, AND P. A. PRINCE. 1990. Reproductive performance, recruitment and survival of Wandering Albatrosses *Diomedea exulans* at Bird Island, South Georgia. Journal of Animal Ecology 59: 775–796.

CRUZ, F., AND J. B. CRUZ. 1990. Breeding, morphology, and growth of the endangered Dark-rumped Petrel. Auk 107: 317–326.

CURIO, E. 1983. Why do young birds reproduce less well? Ibis 125: 400–404.

CUTHBERT, F. J. 1985. Mate retention in Caspian Terns. Condor 87: 74–78.

DAAN, S., C. DIJKSTRA, R. H. DRENT, AND T. MEIJER. 1988. Food supply, and the annual timing of avian reproduction. Proceedings of the 19th International Congress of Ornithology (Ottawa) 1: 392–407.

DANN, P. J., AND J. M. CULLEN. 1990. Survival, patterns of reproduction and lifetime reproductive output in Little Penguins (*Eudyptula minor*) on Philips Island, Victoria, Australia. Pp. 63–84 *in* Penguin Biology (L. S. Davis, and J. T. Darby, Eds.). Academic Press, San Diego, CA.

DARWIN, C. 1871. The descent of man, and selection in relation to sex. John Murray, London.

DAVIES, N. B. 1978. Ecological questions about territorial behavior. Pp. 317–350 *in* Behavioral Ecology: An Evolutionary Approach (J. R. Krebs and N. B. Davies, Eds.). Blackwell, Oxford.

DAVIS, L. S. 1988. Coordination of incubation routines and mate choice in Adélie Penguins (*Pygoscelis adeliae*). Auk 105: 428–432.

DAVIS, L. S., AND E. A. H. SPEIRS. 1990. Mate choice in penguins. Pp. 377–397 *in* Penguin Biology (L. S. Davis and J. T. Darby, Eds.). Academic Press, San Diego, CA.

DEL HOYO, J., A. ELLIOTT, AND J. SARGATAL. 1992. Handbook of the Birds of the World. Vol 1: Ostrich to Ducks. Lynx Edicions, Barcelona, Spain.

DHONDT, A. A., AND F. ADRIAENSEN. 1994. Causes and effects of divorce in the Blue Tit *Parus caeruleus*. Journal of Animal Ecology 63: 979–987.

DOMJAN, M. 1992. Adult learning and mate choice: possibilities and experimental evidence. American Zoologist 32: 48–61.

DRENT, R. H., AND S. DAAN. 1980. The prudent parent: energetic adjustments in avian breeding. Ardea 68: 225–252.

DUBOIS, F., F. CÉZILLY, AND M. PAGEL. 1998. Mate fidelity and coloniality in waterbirds: a comparative analysis. Oecologia 116: 433–440.

DUNLOP, J. N., AND J. JENKINS. 1992. Known-age birds at a subtropical breeding colony of the Bridled Tern (*Sterna anaethetus*): a comparison with the Sooty Tern. Colonial Waterbirds 15: 75–82.

DUNNET, G. M., AND J. C. OLLASON. 1978. Survival and longevity in the Fulmar. Ibis 120: 124–125.

EMSLIE, S. D., W. J. SYDEMAN, AND P. PYLE. 1992. The importance of mate retention and experience on breeding success in Cassin's Auklet (*Ptychoramphus aleuticus*). Behavioral Ecology 3: 189–195.

ENS, B. J. 1992. The social prisoner. Causes of natural variation in reproductive success of the Oystercatcher. Ph.D thesis, University of Groningen, The Netherlands.

ENS, B. J., S. CHOUDHURY, AND J. M. BLACK. 1996. Mate fidelity and divorce in monogamous birds. Pp. 344–401 *in* Partnerships in Birds — The Study of Monogamy (J. M. Black, Ed.). Oxford University Press, Oxford.

ENS, B. J., U. N. SAFRIEL, AND M. P. HARRIS. 1993. Divorce in the long-lived and monogamous Oystercatcher *Haematopus ostralegus*: incompatibility or choosing the better option? Animal Behaviour 45: 1193–1217.

ENS, B. J., F. J. WEISSING, AND R. H. DRENT. 1995. The despotic distribution and deferred maturity: two sides of the same coin. American Naturalist 146: 625–650.

ERIKSTAD, K. E., P. FAUCHALD, T. TVERAA, AND H. STEEN. 1998. On the cost of reproduction in long-lived birds: the influence of environmental variability. Ecology 79: 1781–1788.

EVANS, R. M. 1970. Parental recognition and the "Mew Call" in Black-billed Gulls (*Larus bulleri*). Auk 87: 503–513.

FALLS, J. B., AND M. K. MCNICHOL. 1979. Neighbour-stranger discrimination in male Blue Grouse. Canadian Journal of Zoology 57: 457–462.

FISHER, H. I. 1975. Mortality and survival in the Laysan Albatross *Diomedea immutabilis*. Pacific Science 29: 279–300.

FLEET, R. R. 1974. The Red-tailed Tropicbird on Kure Atoll. Ornithological Monographs 16: 1–64.

FORBES, L. S., AND G. W. KAISER. 1994. Habitat choice in breeding seabirds: when to cross the information barrier. Oikos 70: 377–384.

FORSLUND, P., AND K. LARSSON. 1991. The effect of mate change and new partner's age on reproductive success in the Barnacle Goose, *Branta leucopsis*. Behavioral Ecology 2: 116–122.

FRESSANGES DU BOST, D., AND M. SÉGONZAC. 1976. Note complémentaire sur le cycle reproducteur du Grand Albatros (*Diomedea exulans*) de l'île de la Possession, archipel Crozet. Comité National Français de la Recherche en Antarctique 40: 53–60.

FRETWELL, S. D., AND H. L. LUCAS. 1970. On territorial behavior and other factors affecting habitat distribution in birds. I. Theoretical development. Acta Biotheoretica 19: 16–36.

FRINGS, H., AND M. FRINGS. 1961. Some biometric studies on the albatrosses of Midway Atoll. Condor 63: 304–312.

FURNESS, R. W. 1987. The Skuas. T. & A. D. Poyser, London.

FURNESS, R. W. 1996. Family *Stercorariidae* (skuas). Pp. 556–571 in Handbook of the Birds of the World, Vol. 3 (J. del Hoyo, A. Elliott, and J. Sargatal, Eds.). Lynx Edicions, Barcelona, Spain.

FURNESS, R. W., AND T. R. BIRKHEAD. 1984. Seabird colony distributions suggest competition for food supplies during the breeding season. Nature 311: 655–656.

GARLAND, T., JR., P. H. HARVEY, AND A. R. IVES. 1992. Procedures for the analysis of comparative data using phylogenetically independent contrasts. Systematic Biology 41: 18–32.

GASTON, A. J., AND I. L. JONES. 1998. The Auks. Oxford University Press, New York.

GASTON, A. J., AND I. L. NETTLESHIP. 1981. The Thick-billed Murres of Prince Leopold Island — a study of the breeding ecology of a colonial High Arctic seabird. Canadian Wildlife Service Monographs, Series 6. Canadian Wildlife Service, Ottawa.

GENEVOIS, F., AND V. BRETAGNOLLE. 1994. Male Blue Petrels reveal their body mass when calling. Ethology, Ecology and Evolution 6: 377–383.

GOCHFELD, M. 1980. Mechanisms and adaptive value of reproductive synchrony in colonial seabirds. Pp. 207–270 in Behavior of Marine Animals, Vol. 4, Marine Birds (J. Burger, B. L. Olla, and H. E. Winn, Eds.). Plenum, New York.

GOCHFELD, M., AND J. BURGER. 1996. Family *Sternidae* (terns). Pp. 624–667 in Handbook of the Birds of the World, Vol. 3 (J. del Hoyo, A. Elliott, and J. Sargatal, Eds.). Lynx Edicions, Barcelona, Spain.

GONZÁLEZ-SOLÍS, J., P. H. BECKER, AND H. WENDELN. 1999. Divorce and asynchronous arrival in Common Terns, *Sterna hirundo*. Animal Behaviour 58: 1123–1129.

GOWATY, P. A. 1996. Battles of the sexes and origins of monogamy. Pp. 21–52 in Partnerships in Birds — The Study of Monogamy (J. M. Black, Ed.). Oxford University Press, Oxford.

GREENWOOD, P. J. 1980. Mating systems, philopatry and dispersal in birds and mammals. Animal Behaviour 28: 1140–1162.

GREENWOOD, P. J., AND P. H. HARVEY. 1982. The natal and breeding dispersal of birds. Annual Review of Ecology and Systematics 13: 1–21.

GRUBB, T. C., JR. 1974. Olfactory navigation to the nesting burrow in Leach's Petrel (*Oceanodroma leucorhoa*). Animal Behaviour 22: 192–202.

GUILLOTIN, M., AND P. JOUVENTIN. 1980. Le Pétrel des neiges à Pointe Géologie. Gerfaut 70: 51–72.

GUINARD, E., H. WEIMERSKIRCH, AND P. JOUVENTIN. 1998. Population changes and demography of the Northern Rockhopper Penguin on Amsterdam and Saint Paul Islands. Colonial Waterbirds 21: 222–228.

GUINET, C., O. CHASTEL, M. KOUDIL, J.-P. DURBEC, AND P. JOUVENTIN. 1998. Effects of warm sea-surface temperature anomalies on the Blue Petrel at the Kerguelen islands. Proceedings of the Royal Society of London (B) 265: 1001–1006.

GUSTAFSSON, L., D. NORDLING, M. S. ANDERSSON, B. C. SHELDON, AND A. QVARNSTRÖM. 1994. Infectious diseases, reproductive effort, and the cost of reproduction in birds. Philosophical Transactions of the Royal Society of London (B) 346: 323–331.

HAAS, C. A. 1998. Effects of prior nesting success on site fidelity and breeding dispersal: an experimental approach. Auk 115: 929–936.

HALLIDAY, T. R. 1983. The study of mate choice. Pp. 3–32 in Mate choice (P. Bateson, Ed.). Cambridge University Press, Cambridge, MA.

HARRIS, M. P. 1973. The biology of the Waved Albatross *Diomedea irrorata* of Hood Island, Galapagos. Ibis 115: 483–510.

HARRIS, M. P. 1979a. Population dynamics of the Flightless Cormorant *Nannopterum harrisi*. Ibis 121: 135–146.

HARRIS, M. P. 1979b. Survival and ages of first breeding of Galapagos seabirds. Bird-Banding 50: 56–61.

HARRIS, M. P., AND S. WANLESS. 1989. The breeding biology of Razorbills *Alca torda* on the Isle of May. Bird Study 36: 105–114.

HARRIS, M. P., AND S. WANLESS. 1995. Survival and non-breeding of adult Common Guillemots *Uria aalge*. Ibis 137: 192–197.

HARRIS, M. P., S. WANLESS, AND T. R. BARTON. 1996. Site use and fidelity in the Common Guillemot *Uria aalge*. Ibis 138: 399–404.

HARRISON, C. S. 1990. Seabirds of Hawaii. Natural history and conservation. Cornell University Press, Ithaca, NY.

HARVEY, P. H., AND A. PURVIS. 1991. Comparative methods for explaining adaptations. Nature 351: 619–624.

HATCH, S. A., B. D. ROBERTS, AND B. S. FADELY. 1993. Adult survival of Black-legged Kittiwakes *Rissa tridactyla* in a Pacific colony. Ibis 135: 247–254.

HEGYI, Z., AND L. SASVARI. 1998. Parental condition and breeding effort in waders. Journal of Animal Ecology 67: 41–53.

HIGGINS, P. J., AND S. J. J. F. DAVIS. 1996. Handbook of Australian, New Zealand and Antarctic Birds, Vol. 3. Oxford University Press, Melbourne.

HILDÉN, O. 1965. Habitat selection in birds: a review. Annales Zoologici Fennici 2: 53–75.

HINDE, R. A. 1956. The biological significance of the territories in birds. Ibis 98: 340–369.

HUDSON, R. 1966. Adult survival estimates for two antarctic petrels. British Antarctic Survey Bulletin 8: 63–73.

HUNT, G. L. 1980. Mate selection and mating systems in seabirds. Pp. 113–151 *in* Behavior of Marine Animals, Vol. 4, Marine Birds (J. Burger, B. L. Olla, and H. E. Winn, Eds.). Plenum, New York.

IMBER, M. J. 1987. Breeding ecology and conservation of the Black Petrel *Procellaria parkinsoni*. Notornis 34: 19–39.

ISENMANN, P. 1970. Note sur la biologie de reproduction comparée des Damiers du Cap *Daption capensis* aux Orcades du Sud et en Terre Adélie. L'Oiseau et la R. F. O. 40 (special issue): 135–141.

ISENMANN, P. 1971. Contribution à l'éthologie et à l'écologie du Manchot empereur (*Aptenodytes forsteri* Gray) à la colonie de Pointe Géologie (Terre Adélie). L'Oiseau et la R. F. O. 41: 9–64.

ISENMANN, P., AND P. JOUVENTIN. 1970. Eco-éthologie du Manchot empereur (*Aptenodytes forsteri*) en comparaison avec le Manchot Adélie (*Pygoscelis adeliae*) et le Manchot royal (*Aptenodytes patagonica*). L'Oiseau et la R. F. O. 41: 9–64.

JANETOS, A. C. 1980. Strategies of female mate choice: a theoretical analysis. Behavioral Ecology and Sociobiology 7: 107–112.

JOHNSTON, V. H., AND J. P. RYDER. 1987. Divorce in larids: a review. Colonial Waterbirds 10: 16–26.

JOHNSTONE, R. A. 1995. Sexual selection, honest advertisement and the handicap principle: reviewing the evidence. Biological Review 70: 1–65.

JOHNSTONE, R. A., REYNOLDS, J. D., AND J. D. DEUTSCH. 1996. Mutual mate choice and sex differences in choosiness. Evolution 50: 1382–1391.

JONES, I. L., AND F. L. HUNTER. 1993. Mutual sexual selection in a monogamous seabird. Nature 362: 238–239.

JONES, I. L., AND R. MONTGOMERIE. 1991. Mating and remating of Least Auklets (*Aethia pusilla*) relative to ornamental traits. Behavioral Ecology 2: 249–257.

JOUVENTIN, P. 1971. Comportement et structure sociale chez le Manchot empereur. La Terre et la Vie 25: 510–586.

JOUVENTIN, P. 1982. Visual and vocal signals in penguins, their evolution and adaptive characters. Advances in Ethology 24: 1–149.

JOUVENTIN, P., T. AUBIN, AND T. LENGAGNE. 1999b. Finding a parent in a King Penguin colony: the acoustic system of individual recognition. Animal Behaviour 57: 1175–1183.

JOUVENTIN, P., AND J. BRIED. In press. The effect of mate choice on speciation in Snow Petrels. Animal Behaviour 61.

JOUVENTIN, P., AND A. CORNET. 1980. The sociobiology of Pinnipeds. Advances in the Study of Behaviour 11: 121–141.

JOUVENTIN, P., AND M. GUILLOTIN. 1979. Socio-écologie du Skua antarctique à Pointe Géologie. Revue d'Ecologie (Terre & Vie) 33: 109–127.

JOUVENTIN, P., AND F. LAGARDE. 1995. Evolutionary ecology of the King Penguin *Aptenodytes patagonicus*: the self-regulation of the breeding cycle. Pp. 80–95 *in* The Penguins (P. Dann, I. Norman, and P. Reilly, Eds.). Surrey Beatty & Sons, Chipping Norton.

JOUVENTIN, P., B. LEQUETTE, AND F. S. DOBSON. 1999a. Age-related mate choice in the Wandering Albatross. Animal Behaviour 57: 1099–1106.

JOUVENTIN, P., J. MARTINEZ, AND J.-P. ROUX. 1989. Breeding biology and current status of the Amsterdam Island Albatross *Diomedea amsterdamensis*. Ibis 131: 171–182.

JOUVENTIN, P., AND J.-L. MOUGIN. 1981. Les stratégies adaptatives des oiseaux de mer. Revue d'Ecologie (Terre & Vie) 35: 217–272.

JOUVENTIN, P., J.-L. MOUGIN, J.-C. STAHL, AND H. WEIMERSKIRCH. 1985. Comparative biology of the burrowing petrels of the Crozet Islands. Notornis 32: 157–220.

JOUVENTIN, P., J.-P. ROUX, J.-C. STAHL, AND H. WEIMERSKIRCH. 1983. Biologie et fréquence de reproduction chez l'Albatros à bec jaune (*Diomedea chlororhynchos*). Gerfaut 73: 161–171.

JOUVENTIN, P., AND H. WEIMERSKIRCH. 1984a. L'Albatros fuligineux à dos sombre *Phoebetria fusca*, exemple de stratégie d'adaptation extrême à la vie pélagique. Revue d'Ecologie (Terre & Vie) 39: 401–429.

JOUVENTIN, P., AND H. WEIMERSKIRCH. 1984b. Les albatros. La Recherche 159: 1228–1240.

JOUVENTIN, P., AND H. WEIMERSKIRCH. 1988. Demographic strategies of southern albatrosses. Proceedings of the 19th International Congress of Ornithology (Ottawa 1986): 857–865.

JOUVENTIN, P., AND H. WEIMERSKIRCH. 1991. Changes in the population size and demography of southern seabirds: management implications. Pp. 297–314 *in* Bird Populations Studies, Relevance to Conservation and Management (C. M. Perrins, J. D. Lebreton, and G. J. M. Hirons, Eds.). Oxford University Press, Oxford.

LACK, D. 1968. Ecological Adaptations for Breeding in Birds. Methuen, London.

LACOCK, G. D., D. C. DUFFY, AND J. COOPER. 1987. Population dynamics of the African Penguin *Spheniscus demersus* at Marcus Island in the Benguela upwelling Ecosystem: 1979–1985. Biological Conservation 40: 117–126.

LEBRETON, J.-D., AND P. ISENMANN. 1976. Dynamique de la population camarguaise de Mouettes rieuses *Larus ridibundus* L.: un modèle mathématique. Revue d'Ecologie (Terre & Vie) 30: 529–549.

LENGAGNE, T., P. JOUVENTIN, AND T. AUBIN. 1999. Finding one's mate in a King Penguin colony: efficiency of acoustic communication. Behaviour 136: 833–846.

LEQUETTE, B., D. BERTEAUX, AND J. JUDAS. 1995. Presence and first breeding attempts of southern gannets *Morus capensis* and *M. serrator* at Saint Paul Island, southern Indian Ocean. Emu 95: 134–137.

LIGON, J. D. 1999. The Evolution of Avian Breeding Systems. Oxford University Press, Oxford.

LINDÉN, M. 1991. Divorce in Great Tits: chance or choice? An experimental approach. American Naturalist 138: 1039–1048.

MARCHANT, S., AND P. J. HIGGINS. 1990. Handbook of Australian, New Zealand and Antarctic Birds, Vol. 1. Oxford University Press, Melbourne.

MARTIN, T. E. 1987. Food as a limit on breeding birds: a life-history perspective. Annual Review of Ecology and Systematics 18: 453–467.

MAYNARD-SMITH, J. 1978. Optimization theory in evolution. Annual Review of Ecology and Systematics 9: 31–56.

MCNAMARA, J. M., AND P. FORSLUND. 1996. Divorce rates in birds: predictions from an optimization model. American Naturalist 14: 609–640.

MCNICHOLL, M. K. 1975. Larid site tenacity and group adherence in relation to habitat. Auk 92: 98–104.

MILLS, J. A., J. W. YARRALL, AND D. A. MILLS. 1996. Causes and consequences of mate fidelity in Red-billed Gulls. Pp. 286–304 *in* Partnerships in Birds — The Study of Monogamy (J. M. Black, Ed.). Oxford University Press, Oxford.

MOCK, D. W. 1985. An introduction to the neglected mating system. Pp. 1–10 *in* Avian Monogamy (P. A. Gowaty and D. Mock, Eds.). Ornithological Monographs 37. American Ornithologists' Union, Washington, D.C.

MOCK, D. W., AND M. FUJIOKA. 1990. Monogamy and long-term pair bonding in vertebrates. Trends in Ecology and Evolution 5: 39–43.

MØLLER, A. P. 1982. Coloniality and colony structure in Gull-billed Terns *Gelochelidon nilotica*. Journal of Ornithology 12: 41–53.

MORSE, D. H., AND C. W. BUCHHEISTER. 1979. Nesting patterns of Leach's Storm-petrels on Matinicus Rock, Maine. Bird-Banding 50: 145–158.

MORSE, D. H., AND S. W. KRESS. 1984. The effect of burrow losses on mate choice in Leach's Storm-Petrel. Auk 101: 158–160.

MOUGEOT, F., F. GENEVOIS, AND V. BRETAGNOLLE. 1998. Predation on burrowing petrels by the Brown Skua at Mayes Island, Kerguelen. Journal of Zoology of London 244: 429–438.

MOUGIN, J.-L. 1970. Le Pétrel à menton blanc *Procellaria aequinoctialis* de l'île de la Possession (archipel Crozet). L'Oiseau et la R. F. O. 40 (special volume): 62–96.

MOUGIN, J.-L. 1975. Ecologie comparée des *Procellariidae* antarctiques et subantarctiques. Comité National Français de la Recherche en Antarctique 36: 1–195.

MOUGIN, J.-L. 1989. Données préliminaires sur la structure et la dynamique de la population de Pétrels de Bulwer *Bulweria bulwerii* de l'île Selvagem Grande (30°09'N, 15°52'W). Compte-rendus de l'Académie des Sciences de Paris 308: 103–106.

MOUGIN, J.-L. 1990. La fidélité au partenaire et au nid chez le Pétrel de Bulwer *Bulweria bulwerii* de l'île Selvagem Grande (30°09'N, 15°52'W). L'Oiseau et la R. F. O. 60: 224–232.

MOUGIN, J.-L., B. DESPIN, C. JOUANIN, AND F. ROUX. 1987a. La fidélité au partenaire et au nid chez le Puffin cendré, *Calonectris diomedea borealis*, de l'île Selvagem Grande. Gerfaut 77: 353–369.

MOUGIN, J.-L., C. JOUANIN, AND F. ROUX. 1987b. Structure et dynamique de la population de Puffins cendrés *Calonectris diomedea borealis* de l'île Selvagem Grande (30°09'N, 15°52'W). L'Oiseau et la R. F. O. 57: 201–225.

MOUGIN, J.-L., C. JOUANIN, AND F. ROUX. 1988a. Les différences d'âge et d'expérience entre partenaires chez le Puffin cendré *Calonectris diomedea borealis* de l'île Selvagem Grande (30°09'N, 15°52'W). L'Oiseau et la R. F. O. 58: 113–119.

MOUGIN, J.-L., C. JOUANIN, AND F. ROUX. 1988b. L'influence des voisins dans la nidification du Puffin cendré *Calonectris diomedea*. Comptes-rendus de l'Académie des Sciences de Paris (III) 307: 195–198.

NEITHAMMER, K. R., AND L. B. PATRICK. 1998. The White Tern (*Gygis alba*) No. 371. *In* The Birds of North America (A. Poole and F. Gill, Eds.). Academy of Natural Sciences, Philadelphia; American Ornithologists' Union, Washington, D.C.

NELSON, J. B. 1972. The biology of the seabirds of the Indian Ocean Christmas Island. Journal of the Marine Biological Association of India 14: 643–662.

NELSON, J. B. 1976. The breeding biology of frigatebirds — a comparative review. Living Bird 14: 113–155.

NELSON, J. B. 1978. The Sulidae — Gannets and Boobies. University of Aberdeen. Oxford University Press, Oxford.

NELSON, J. B. 1980. Seabirds. Their Biology and Ecology. Hamlyn, London.

NELSON, J. B. 1983. Contrasts in breeding strategies between some tropical and temperate marine pelecaniformes. Studies in Avian Biology 8: 95–114.

NELSON, J. B. 1988. Age and breeding in seabirds. Proceedings of the 19th International Congress of Ornithology 1: 1081–1097.

NETTLESHIP, D. N. 1996. Family *Alcidae* (auks). Pp. 678–722 *in* Handbook of the Birds of the World, Vol. 3 (J. del Hoyo, A. Elliott, and J. Sargatal, Eds.). Lynx Edicions, Barcelona, Spain.

NEWTON, I. 1989. Lifetime Reproduction in Birds. Academic Press, London.

NEWTON, I. 1992. Experiments on the limitation of bird numbers by territorial behaviour. Biological Review 67: 129–173.

NUR, N. 1984. Increased reproductive success with age in the California Gull: due to increased effort or improvement of skill? Oikos 43: 407–408.

NYSTRÖM, K. G. K. 1997. Food density, song rate, and body condition in territory establishing Willow Warblers (*Phylloscopus tochilus*). Canadian Journal of Zoology 75: 47–58.

OLLASON, J. C., AND G. M. DUNNET. 1978. Age, experience and other factors affecting the breeding success of the Fulmar, *Fulmarus glacialis*, in Orkney. Journal of Animal Ecology 47: 961–976.

OLLASON, J. C., AND G. M. DUNNET. 1988. Variation in breeding success in fulmars. Pp. 268–278 *in* Reproductive Success (T. H. Clutton-Brock, Ed.). University of Chicago Press, Chicago.

OLSSON, O. 1998. Divorce in King Penguins: asynchrony, expensive fat storing and ideal free mate choice. Oikos 83: 574–581.

ORIANS, G. H. 1969. On the evolution of mating systems in birds and mammals. American Naturalist 103: 589–602.

ORING, L. W. 1988. Philopatry and breeding site fidelity: an introduction. Proceedings of the 19th International Congress of Ornithology (Ottawa): 561–562.

OSORNO, J.-L. 1999. Offspring desertion in the Magnificent Frigatebird: are males facing a trade-off between current and future reproduction? Journal of Avian Biology 30: 335–341.

PARKER, G. A. 1983. Mate quality and mating decisions. Pp. 141–166 *in* Mate Choice (P. Bateson, Ed.). Cambridge University Press, Cambridge.

PARTRIDGE, L. 1989. Lifetime reproductive success and life-history evolution. Pp. 421–440 *in* Lifetime Reproduction in Birds (I. Newton, Ed.). Academic Press, London.

PARTRIDGE, L., AND T. R. HALLIDAY. 1984. Mating patterns and mate choice. Pp. 222–250 *in* Behavioural Ecology. An Evolutionary Approach (J. R. Krebs and N. B. Davies, Eds.). Blackwell Scientific, Oxford.

PENNEY, R. L. 1968. Territorial and social behaviour in the Adélie Penguin. Antarctic Research Series 12: 83–131.

PERRINS, C. M. 1970. The timing of birds' breeding seasons. Ibis 112: 242–255.

PERRINS, C. M., AND R. H. MCCLEERY. 1985. The effect of age and pair bond on the breeding success of Great Tits *Parus major.* Ibis 127: 306–315.

PETERSEN, Æ. 1981. Breeding Biology and Feeding Ecology of Black Guillemots. Unpublished Ph.D thesis. University of Oxford, Oxford.

PETRIE, M. 1983. Female Moorhens compete for small fat males. Science 220: 413–415.

PHILLIPS, N. P. 1987. The breeding biology of White-tailed Tropicbirds *Phaethon lepturus* at Cousin Island, Seychelles. Ibis 129: 10–24.

PIANKA, E. R., AND W. S. PARKER. 1975. Age-specific reproductive tactics. American Naturalist 109: 453–464.

PIETZ, P. J., AND D. F. PARMELEE. 1994. Survival, site and mate fidelity in South Polar Skuas *Catharacta maccormicki* at Anvers Island, Antarctica. Ibis 136: 12–18.

PINDER, R. 1966. The Cape Pigeon, *Daption capensis* Linnaeus, at Signy Island, South Orkney Islands. British Antarctic Survey Bulletin 8: 19–47.

POTTS, G. R. 1969. The influence of eruptive movements, age, population size and other factors on the survival of the Shag *Phalacrocorax aristotelis* (L.). Journal of Animal Ecology 38: 53–102.

PRINCE, P. A., P. ROTHERY, J. P. CROXALL, AND A. G. WOOD. 1994. Population dynamics of Black-browed and Grey-headed Albatrosses *Diomedea melanophris* and *D. chrysostoma* at Bird Island, South Georgia. Ibis 136: 50–71.

QVARNSTRÖM, A., AND E. FORSGREN. 1998. Should females prefer dominant males? Trends in Ecology and Evolution 13: 498–501.

REAL, L. 1990. Search theory and mate choice. I. Models of single-sex discrimination. American Naturalist 136: 376–405.

REID, W. V. 1988. Age correlations between pairs of breeding birds. Auk 105: 278–285.

REILLY, P. N., AND J. M. CULLEN. 1981. The Little Penguin *Eudyptula minor* in Victoria. II. Breeding. Emu 81: 1–19.

REYNOLDS, J. D. 1996. Animal breeding systems. Trends in Ecology and Evolution 11: 68–72.

RICE, D. W., AND K. W. KENYON. 1962. Breeding cycles and behavior of Laysan and Black-footed Albatrosses. Auk 79: 517–567

RICHDALE, L. E. 1947. The pair bond in penguins and petrels: a banding study. Bird-Banding 18: 107–117.

RICHDALE, L. E. 1949. A study of a group of penguins of known age. Otago Daily Times and Witness Newspapers Co., Ltd., Dunedin.

RICHDALE, L. E. 1957. A Population Study of Penguins. Clarendon Press, Oxford.

RICHDALE, L. E. 1963. Breeding behaviour of the Narrow-billed and the Broad-billed Prion on Whero Island. Transmissions from the Zoologist Society of London 31: 87–155.

RICKLEFS, R. E. 1990. Seabird life histories and the marine environment: some speculations. Colonial Waterbirds 13: 1–6.

RISTOW, D., AND M. WINK. 1980. Sexual dimorphism in Cory's Shearwater. Il-Merill 21: 9–12.

ROBERTSON, C. J. R. 1993. Survival and longevity of the Northern Royal Albatross *Diomedea epomophora sanfordi* at Taiaroa Head 1937–93. Emu 93: 269–276.

ROLLAND, C., E. DANCHIN, AND M. DE FRAIPONT. 1998. The evolution of coloniality in birds in relation to food, habitat, predation, and life-history traits: a comparative analysis. American Naturalist 151: 514–529.

ROSENFIELD, R. N., AND J. BIELEFELDT. 1999. Mass, reproductive biology, and nonrandom pairing in Cooper's Hawks. Auk 116: 830–835.

ROWAN, M. K. 1965. Regulation of sea-bird numbers. Ibis 107: 54–59.

ROWLEY, I. 1983. Re-mating in birds. Pp. 331–360 *in* Mate choice (P. Bateson, Ed.). Cambridge University Press, London.

SAGAR, P. M. 1986. The sexual dimorphism of Snares Cape Pigeons. Notornis 33: 259–263.

SAGAR, P. M., A. J. D. TENNYSON, AND C. M. MISKELLY. 1996. Breeding and survival of Snares Cape Pigeons *Daption capense australe* at the Snares, New Zealand. Notornis 43: 197–207.

SAGAR, P. M., AND J. WARHAM. 1997. Breeding biology of Southern Buller's Albatrosses at The Snares, New Zealand. Pp. 92–98 *in* The Albatross Biology and Conservation (G. Robertson and R. Gales, Eds.). Surrey Beatty and Sons, Chipping Norton.

SCHREIBER, R. W., AND N. P. ASHMOLE. 1970. Seabird breeding seasons on Christmas Island, Pacific Ocean. Ibis 112: 363–394.

SCHREIBER, E. A., AND R. W. SCHREIBER. 1993. Red-tailed Tropicbird. *In* The Birds of North America, Vol. 43 (A. Poole, and F. Gill, Eds.). Academy of Natural Sciences, Philadelphia; American Ornitholgists' Union, Washington, D.C.

SCHREIBER, E. A., R. W. SCHREIBER, AND G. A. SCHENK. 1996. Red-footed Booby. *In* The Birds of North America, Vol. 241 (A. Poole and F. Gill, Eds.). Academy of Natural Sciences, Philadelphia; American Ornitholgists' Union, Washington, D.C.

SCOLARO, J. A. 1990. Effects of nest density on breeding success in a colony of Magellanic Penguins (Spheniscus magellanicus). Colonial Waterbirds 13: 41–49.

SCOTT, D. A. 1970. Breeding biology of the Storm Petrel. Ph.D thesis, Edward Grey Institute of Field Ornithology, Botanic Garden, Oxford.

SCOTT, D. K. 1988. Breeding success in Bewick's Swans. Pp. 220–236 in Reproductive Success (T. H. Clutton-Brock, Ed.). University of Chicago Press, Chicago.

SEARCY, W. A. 1979. Male characteristics and pairing success in Red-winged Blackbirds. Auk 96: 353–363.

SEBER, G. A. F. 1973. The Estimation of Animal Abundance and Related Parameters. Griffin, London.

SHAW, P. 1985. Age-differences within breeding pairs of Blue-eyed Shags Phalacrocorax atriceps. Ibis 127: 537–543.

SHAW, P. 1986. Factors affecting the breeding performance of Antarctic Blue-eyed Shags Phalacrocorax atriceps. Ornis Scandinavica 17: 141–150.

SIMMONS, R. 1988. Honest advertising, sexual selection, courtship displays, and body condition of polygynous male harriers. Auk 105: 303–307.

SMITH, A. J. M. 1975. Studies of breeding Sandwich Terns. British Birds 68: 142–156.

SORENSEN, J. H. 1950. The Royal Albatross. Cape Expeditions Series Bulletin 2: 5–37.

SPEAR, L., AND N. NUR. 1994. Brood size, hatching order and hatching date: effects on four life-history stages from hatching to recruitment in Western Gulls. Journal of Animal Ecology 63: 283–298.

STAMPS, J. A. 1987. The effect of familiarity with a neighborhood on territory acquisition. Behavioral Ecology and Sociobiology 21: 273–277.

STEARNS, S. C. 1992. The Evolution of Life Histories. Oxford University Press, Oxford.

STODDARD, P. K. 1996. Vocal recognition of neighbors by territorial passerines. Pp. 365–374 in Ecology and Evolution of Acoustic Communication in Birds (D. E. Kroodsma and H. E. Miller, Eds.). Comstock, Cornell University Press, Ithaca, NY.

SULLIVAN, M. S. 1994. Mate choice as an information gathering process under time constraint: implications for behaviour and signal design. Animal Behaviour 47: 141–151.

SWATSCHEK, I., D. RISTOW, AND M. WINK. 1994. Mate fidelity and parentage in Cory's Shearwater Calonectris diomedea-field studies and DNA fingerprinting. Molecular Ecology 3: 259–262.

SWITZER, P. V. 1993. Site fidelity in predictable and unpredictable habitats. Evolutionary Ecology 7: 533–535.

SYDEMAN, W. J., P. PYLE, S. D. EMSLIE, AND E. B. MCLAREN. 1996. Causes and consequences of long-term partnerships in Cassin's Auklets. Pp. 211–222 in Partnerships in Birds — The Study of Monogamy (J. M. Black, Ed.). Oxford University Press, Oxford.

TENAZA, R. 1971. Behavior and nesting success relative to nest location in Adélie Penguins Pygoscelis adeliae. Condor 73: 81–92.

THIBAULT, J.-C. 1994. Nest-site tenacity and mate fidelity in relation to breeding success in Cory's Shearwater Calonectris diomedea. Bird Study 41: 25–28.

THIBAULT, J.-C. 1995. Effect of predation by the Black Rat Rattus rattus on the breeding success of Cory's Shearwater Calonectris diomedea in Corsica. Marine Ornithology 23: 1–10.

TICKELL, W. L. N. 1968. The biology of the great albatrosses, Diomedea exulans and Diomedea epomophora. Antarctic Birds Studies 12: 191–212.

TRIVELPIECE, W. Z., AND J. D. FERRARIS. 1987. Note on the behavioural ecology of the Magnificent Frigatebird Fregata magnificens. Ibis 129: 168–174.

TRIVELPIECE, W. Z., AND S. G. TRIVELPIECE. 1990. Courtship period of Adélie, Gentoo, and Chinstrap Penguins. Pp. 113–127 in Penguin Biology (L. S. Davis and J. T. Darby, Eds.). Academic Press, San Diego.

TRIVERS, R. L. 1972. Parental investment and sexual selection. Pp. 136–179 in Sexual Selection and the Descent of Man 1871–1971 (B. Campbell, Ed.). Aldine, Chicago.

VALONE, T. J., S. E. NORDELL, L. A. GIRALDEAU, AND J. J. TEMPLETON. 1996. The empirical question of thresholds and mechanisms of mate choice. Evolutionary Ecology 10: 447–455.

VAN RYZIN, M., AND H. I. FISHER. 1976. The age of Laysan Albatrosses, Diomedea immutabilis, at first breeding. Condor 78: 1–9.

VISSER, M. E., AND N. VERBOVEN. 1999. Long-term fitness effects of fledging date in Great Tits. Oikos 85: 445–450.

VOLKMAN, N. J., P. PRESLER, AND W. TRIVELPIECE. 1980. Diets of Pygoscelid penguins at King George Island, Antarctica. Condor 82: 373–378.

WAGNER, R. H. 1999. Sexual size dimorphism and assortative mating in Razorbills (Alca torda). Auk 116: 542–544.

WARHAM, J. 1967. The White-headed Petrel Pterodroma lessonii at Macquarie Island. Emu 67: 1–22.

WARHAM, J. 1975. The crested penguins. Pp. 189–269 in The Biology of Penguins (B. Stonehouse, Ed.). Macmillan, London.

WARHAM, J. 1990. The Petrels. Their Ecology and Breeding Systems. Academic Press, London.

WARHAM, J. 1996. The Behaviour, Population Biology and Physiology of the Petrels. Academic Press, London.

WARHAM, J., B. R. KEELEY, AND G. J. WILSON. 1977. Breeding of the Mottled Petrel. Auk 94: 1–17.

WAUGH, S. M., P. M. SAGAR, AND D. PAULL. 1997. Laying dates, breeding success and annual breeding of Southern Royal Albatrosses *Diomedea epomophora epomophora* at Campbell Island during 1964–69. Emu 97: 194–199.

WAUGH, S. M., H. WEIMERSKIRCH, P. J. MOORE, AND P. M. SAGAR. 1999. Population dynamics of Black-browed and Grey-headed Albatrosses *Diomedea melanophrys* and *D. chrysostoma* at Campbell Island, New Zealand, 1942–96. Ibis 141: 216–225.

WEIMERSKIRCH, H. 1990. The influence of age and experience on breeding performances of the Antarctic Fulmar, *Fulmarus glacialoides*. Journal of Animal Ecology 59: 867–875.

WEIMERSKIRCH, H. 1992. Reproductive effort in long-lived birds: age-specific patterns of condition, reproduction and survival in the Wandering Albatross. Oikos 64: 464–473.

WEIMERSKIRCH, H. 1997. Foraging strategies of Indian Ocean albatrosses and their relationships with fisheries. Pp. 168–179 *in* Albatross Biology and Conservation (G. Robertson and R. Gales, Eds.). Surren Beatty & Sons, Chipping Norton.

WEIMERSKIRCH, H., A. CATARD, P. A. PRINCE, Y. CHEREL, AND J. P. CROXALL. 1999. Foraging White-chinned Petrels *Procellaria aequinoctialis* at risk: from the tropics to Antarctica. Biological Conservation 87: 273–275.

WEIMERSKIRCH, H., J. CLOBERT, AND P. JOUVENTIN. 1987. Survival in five southern albatrosses and its relationship with their life history. Journal of Animal Ecology 56: 1043–1055.

WEIMERSKIRCH, H., AND P. JOUVENTIN. 1987. Population dynamics of the Wandering Albatross, *Diomedea exulans*, of the Crozet Islands: causes and consequences of the population decline. Oikos 49: 315–322.

WEIMERSKIRCH, H., AND P. JOUVENTIN. 1997. Changes in population sizes and demographic parameters of six albatross species breeding on the French sub-Antarctic islands. Pp. 84–91 *in* The Albatross Biology and Conservation (G. Robertson and R. Gales, Eds.). Surrey Beatty & Sons, Chipping Norton.

WEIMERSKIRCH, H., J.-C. STAHL, AND P. JOUVENTIN. 1992. The breeding biology and population dynamics of King Penguins *Aptenodytes patagonica* on the Crozet Islands. Ibis 134: 107–117.

WEIMERSKIRCH, H., R. ZOTIER, AND P. JOUVENTIN. 1989. The avifauna of the Kerguelen Islands. Emu 89: 15–29.

WENDELN, H., AND P. H. BECKER. 1998. Populationbiologische Untersuchungen an einer Kolonie der Flußseeschwalbe, *Sterna hirundo*. Vogelwelt 119: 209–213.

WICKLER, W., AND U. SEIBT. 1983. Monogamy: an ambiguous concept. Pp. 33–50 *in* Mate Choice (P. Bateson, Ed.). Cambridge University Press, London.

WILLIAMS, T. D. 1995. The Penguins. Oxford University Press, Oxford.

WILLIAMS, T. D., AND S. RODWELL. 1992. Annual variation in return rate, mate and nest-site fidelity in breeding Gentoo and Macaroni Penguins. Condor 94: 636–645.

WITTENBERGER, J. F., AND R. L. TILSON. 1980. The evolution of monogamy: hypotheses and evidences. Annual Review of Ecology and Systematics 11: 197-232.

WOOD, R. C. 1971. Population dynamics of breeding South Polar Skuas of unknown age. Auk 88: 805–814.

WOOLLER, R. D., AND J. S. BRADLEY. 1996. Monogamy in a long-lived seabird: the Short-tailed Shearwater. Pp. 223–234 *in* Partnerships in Birds — The Study of Monogamy (J. M. Black, Ed.). Oxford University Press, Oxford.

WOOLLER, R. D., J. S. BRADLEY, I. J. SKIRA, AND D. L. SERVENTY. 1989. Short-tailed Shearwater. Pp. 405–417 *in* Lifetime Reproduction in Birds (I. Newton, Ed.). Academic Press, London.

ZOTIER, R. 1990a. Breeding ecology of a subantarctic winter breeder: the Grey Petrel *Procellaria cinerea* on Kerguelen Islands. Emu 90: 180–184.

ZOTIER, R. 1990b. Breeding ecology of the White-headed Petrel *Pterodroma lessonii* on the Kerguelen Islands. Ibis 132: 525–534.

ZUSI, R. L. 1996. Family *Rynchopidae* (skimmers). Pp. 668–677 *in* Handbook of the Birds of the World, Vol. 3 (J. del Hoyo, A. Elliott, and J. Sargatal, Eds.). Lynx Edicions, Barcelona, Spain.

Northern Gannets Courting

10 Seabird Communication and Displays

J. Bryan Nelson and Patricia Herron Baird

CONTENTS

0-8493-9882-7/02/$0.00+$1.50
© 2002 by CRC Press LLC

10.1 INTRODUCTION

Owing to their colonial habit, communication in seabirds is an important part of their existence and most seabird colonies are noisy and active. They use a combination of calls and postures to find a mate, defend a territory, and communicate their intentions to neighbors. Communication means expressing a motivation (such as aggression) via a posture, movement, or vocalization, or a combination of these, in response to a stimulus from another individual, usually a conspecific. Often it is ritualized (sensu Tinbergen 1952), defined as stereotyped, exaggerated, and repetitive, although information can be conveyed via nonritualized, involuntary behavior such as intention movements (see Beer 1980 for comments on communication in gulls). The behavior expressed depends on context. For instance, male sexual advertising occurs only in relation to a territory or nest site. Communication behaviors are species-specific and are far from equally developed in the various seabird families.

Seabird displays, or signals, are largely visual (often with allied sound) and ground based. They convey information, adaptive both for the sender and receiver, via a visual feature (color, adornment), sound, or movement (rarely a scent), or a combination of these. Signals and vocalizations vary in complexity from simple threat to highly abstract displays. Often, ritualized communication shows off specific external features to which conspecifics respond instinctively. These features include vividly colored facial and orbital skin, crests, colorful and/or ornamented bills and feet, and specialized feather tracts. The ritualized wing-raising of the Great Skua (*Catharacta skua*) displays a bold white wing; the long call of the Lava Gull (*Larus fuliginosus*) shows the flame-colored gape; the flank-touching of Abbott's Booby (*Papasula abbotti*) is precisely directed to the patch of black feathers. These features figure prominently in display and are lost or diminished soon after egg-laying, when the associated behavior also declines. For example, the Great Frigate-bird's (*Fregata minor*) spectacular crimson sac contracts, the American White Pelican's (*Pelecanus erythrorhynchos*) bill horn is shed (Figure 10.1), and the Great Cormorant's (*Phalacrocorax carbo*) white thigh patch disappears.

The contexts for the occurrence of ritualized behavior in seabirds include: acquiring and maintaining a nest site; attracting and bonding with a partner (Figure 10.2); after landing or leaving the nest site; partners meeting at the nest site; before and after copulation; when interacting with neighbors and with offspring. Communication operates within relationships by indicating status and motivation during interactions between individuals, including parent/offspring. We can recognize that communication has occurred only by observing its effect on the behavior of the recipient.

FIGURE 10.1 An American White Pelican with nuptial adornment. The "horn" is grown prior to courtship and is shed after incubation begins. (Photo by R. W. and E. A. Schreiber.)

FIGURE 10.2 Head-turning behavior in the Brown Pelican. As part of finding a mate and courtship, a male and female display ritually by turning their head and bill (the weapon) away from each other. (Photo by R. W. Schreiber.)

Overt behavior such as fighting, fleeing, or feeding differs from ritualized display insofar as the latter's motivation and function are more obscure and usually can be revealed only by analysis of context. Ritualized behavior derives from displacement activities and from simple, everyday acts such as walking, preening, preparing for flight, or picking up nest material. By elaboration and becoming stereotyped, these acts have acquired a communicative function (Tinbergen 1959a). The stereotype confers predictability, which is essential for reliable communication. The elaboration may be simple or complex. Merely by spreading and raising its webs into the air, the Blue-footed Booby (*Sula nebouxii*) transforms ordinary walking into ritualized parading; the male Waved Albatross (*Phoebastria irrorata*) uses exaggerated head movements (once merely a part of balancing as it walked) and a bizarre gait to lead the female to a nest site. By contrast, the complex and stereotyped site-ownership display of the North Atlantic (or just "Atlantic") Gannet (*Morus bassanus*) has combined exaggerated versions of redirected aggression (ground biting) with the resultant bill-cleaning head shake.

Appeasement postures (hunched or lowered body orientation, bill-hiding, turning bill away) contrast strongly with aggressive behavior such as threat (bill pointed at opponent, bill gaping, body forward). A gannet defending its site thrusts its gaping bill directly toward an opponent, with a ritualized twist-cum-withdrawal component. A defensive individual adopts a withdrawn posture, does not initiate threat, is hesitant and often silent, and may turn its bill away. Comparably, the aggressive low forward posture of the Common Black-headed Gull (*Larus ridibundus*, Tinbergen 1959a) involves bill pointing at the opponent and contrasts with the defensive upright, head drawn back. Aggression and fear, so evident in interactions between rivals, is present also in male–female interactions but modified by sexual motivation. Conflict between these primary motivations may lead to contextually irrelevant acts (displacement activities, Tinbergen 1952). These may themselves become ritualized, as in flank-touching (ritualized preening, derived from a normal preening movement) in displaying Waved Albatrosses. In a point of balance between attack and fleeing, Herring Gulls (*Larus argentatus*) may redirect their aggression and attack a substitute object as in grass pulling (Tinbergen 1959a).

Aerial display is common among seabirds. It occurs in all pelecaniform families, but is especially prominent in tropicbirds where it is both communal and pair-based. It is prominent among petrels, it is both communal and nocturnal, but it is extremely difficult to analyze. Aerial display is particularly well developed in terns and occurs in many alcids. Aquatic display is common in some, but is rare in pelecaniforms and procellariiforms, although some pelicans may take to the water during their largely land-based communal display.

Group display occurs in some pelicans, penguins, gulls, auks, and tropicbirds (Figure 10.3). Group display in frigatebirds is distinctive in that as several males display in close proximity, each directs its display to an overflying female without reacting to other males. Among procellariiforms, some albatrosses display in a group, but typically one male displays to one female.

Seabird colonies are usually noisy. Visual signals can be greatly enhanced by sound and most displays employ it. Nevertheless, seabird vocabularies are comparatively limited. In pelicans the syrinx is apparently unmodified and there are no flexible tympanic membranes to facilitate sound production. Yet Greenewalt (1968) concluded that in birds the same physical mechanisms could produce an extremely wide range of sounds, including melodious song and harsh, gutteral noises. Almost all seabird vocalizations are staccato, growling, barking, gargling, hissing, croaking, wailing, yodeling, and other simple sounds. Mechanical sound, such as mandible clattering, occurs in pelicans, frigatebirds, and albatrosses. Parameters that are individually recognized include sound frequencies, amplitude differences, and the spacing of component sounds (White 1971, White and White 1970, Warham 1990, 1996). Parents widely recognize their offsprings' voices, an ability that enables parents to feed only their own young. Indeed, chicks call from the egg even before hatching (Evans 1988, Burger and Gochfeld 1993) and may develop complex vocalizations (Warham 1990). Where young form creches (some pelicans, cormorants, albatrosses, gulls, and terns), adults clearly respond selectively to their offsprings' voices. Others feed strange young if these are "on-site"

FIGURE 10.3 Great White Pelicans participate in group parading as part of the courtship process. (Photo by R. W. and E. A. Schreiber.)

(Red-tailed Tropicbirds, *Phaethon rubricauda*; Northern Gannets, Great Frigatebirds, cormorant, and many others). In such species, where chicks normally do not wander from the nest site, there is little likelihood that they could be usurpers.

The type of ritualized behavior which a species evolves depends in part on its nesting environment. A good example is Cullen's (1957) demonstration of the link between the Black-legged Kittiwake's (*Rissa tridactyla*) behavior (which differs from but is still homologous with that of ground nesting gulls) and its cliff-nesting habitat. Kittiwakes show specialized fighting technique (twisting an opponent off the ledge), lack upright threat posture, hide their beaks in an appeasement posture, and use a special nest-building technique to cement their nest to the ledge. Narrow ledges, trees, and other specialized nesting habitats impose obvious limits on displays such as "parading." Excluding burrow nesters, there seems to be no clear correlation between the type of environment and the complexity of communication behavior. Despite habitat constraints, cliff and tree nesters seem no more or less likely than ground or slope nesters to have evolved complex rituals.

Behavior has evolved its many forms by a process comparable to the adaptive radiation that has given rise to a range of morphologies, yet within every seabird family there are convincing examples of behavioral homology. Table 10.1 demonstrates how a particular display, which may be complex and have various differing parts to it that make it distinct in each species, still retains a basic similarity. In most Pelecaniforms the male sexual advertising display includes a form of raising the head and neck or throwing it back over the back, except in the tropicbirds, where display is arial. Enhancements to this advertising display (see Table 10.1) in the various families are fascinating, particularly when considering how they might have evolved. The Tropicbirds are the most dissimilar, and essentially have no sexual displays homologous with those of other families. A comparison of tropicbird behavior with all other pelecaniforms exemplifies how behavioral characteristics can be used as a taxonomic character: the relationship of the tropicbirds to other pelecaniform groups is also uncertain genetically and morphologically (Nelson in press, Sibley and Ahlquist 1990).

Sometimes one can suggest how the prototype display, from which extant versions have been derived, arose from simpler "building blocks" such as displacement activities. For example, displacement preening probably gave rise to ritualized flank-touching in the Waved Albatross and Abbott's Booby. The precise area touched may have evolved a special feather structure (Waved Albatross) or a plumage mark (Abbott's Booby). Displays are sometimes more stable in evolution than are other inherited traits and can be useful in comparing relationships (Moynihan 1959).

TABLE 10.1
Behavioral Homology within the Pelecaniformes as Illustrated by the Male Sexual Advertising Display

Family, Species, and Description of Display	Illustration	Comment
Sulidae: Blue footed Booby. "Skypointing" bill pointed vertically upward, wings spread and swiveled, dorsal surface forward; bird vocalizes. Unilateral and mutual.		Clear homologues in all boobies except Abbott's. Gannet homologue emancipated and serves different (nonsexual) function. Probably homologous with sexual advertising display of cormorants and pelicans. Funtionally equivalent but probably not homologous with sexual advertising of frigatebirds. Tropicbirds aberrant.
Phalacrocoracidae: Great Cormorant. Neck retracted, head couched backward, bill pointed upward. Loosened wings flicked rhythmically so that tips elevated. Head periodically thrown back to base of tail and rotated. Vocalizes. Unilateral (male to female).		Clear homologues in many other cormorants/shags. Strong similarities to contextual and functional equivalents in boobies, and homology very probable. Variants of this display in this family conform to principles widely applicable to homologous behavior in general (e.g., there are differences in amplitude of components, speed of movement, etc.).
Pelecanidae: Pink backed Pelican. Head thrown back, bill open, mandibles clapped. Wings raised and "thrashed." Vocalizes. Unilateral (male to female). Usually referred to by motivation or function, which is unsatisfactory; should be descriptive.		Sexual advertising in pelicans is less homogeneous than in boobies and cormorants. The Pink back's display is clearly similar in form to its contextual equivalent in cormorants but some pelican spp. are less similar to cormorants. Brown Pelicans (both sexes) "bow" and jerk wings rhythmically. Conclusion about homology remains provisional and detailed comparative study required.
Fregatidae: Great Frigatebird. Display communal. Male presents gular sac to over-flying females, head back, bill pointed upward, wings spread and rotated with underside facing forward to upward. Vocalizes. Unilateral (male to female).		Entirely convincing homology throughout family (5 species). Clearly functionally equivalent to sexual advertising in other Pelecaniform families but no evident homology with them. Upward pointing bill may seem equivalent to booby skypointing, but is merely an orientation movement toward over-flying females. Origin differs in the two cases (boobies and frigatebirds).

TABLE 10.1 *(Continued)*
Behavioral Homology within the Pelecaniformes as Illustrated by the Male Sexual Advertising Display

Family, Species, and Description of Display	Illustration	Comment
Phaethontidae: Red-tailed Tropicbird. Sexual display aerial and communal, using ritualized flight. No obvious one-to-one male-to-female display flight, but sexes impossible to differentiate. Birds vocalize.		Homologous throughout the family (3 species). No evident relationship to pair-formation procedures of other pelecaniforms. Extreme aerialness of feeding/foraging behavior may have coevolved with, or predisposed to, aerial display. Fundamental differences in pair display between tropicbirds and other Pelecaniforms may reflect phylogenetic disparity, i.e., tropicbirds may not belong within the Pelecaniformes.

Note: Other functionally equivalent sets of displays, such as site ownership and greeting, could be treated similarly. Homology in behavior, as in morphology, implies derivation from a common source. Functional equivalence means that two or more behaviors or structures perform equivalent tasks but may be differently derived (bird and butterfly wings subserve flight but are not homologous).

There are other examples of homology in displays among relatives. The "stare-around" of the Eudyptid penguins (a mutual recognition, threat, or appeasement while walking through the colony) is similar to the "salute" of Yellow-eyed Penguins (*Megadyptes antipodes*), their closest relative (Richdale 1951, O'Hara 1989). Vocalizations can also indicate phylogeny, for instance, the Rockhopper Penguin (*Eudyptes chrysocome*) is very different morphologically from the rest of the Eudyptid penguins and, in turn, its vocalizations are the most different from the others (e.g., its "vertical trumpeting" is of shorter duration and higher pitched; Warham 1975).

Comparison of displays between, rather than within, families is much more speculative, while comparing orders becomes little more than guesswork. Therefore, the main comparative emphasis of this chapter is on the nature of communication in breeding behavior within families. The Sphenisciformes and Charadriiformes sections were written by P. H. Baird, and the above section and the Procellariiformes and Pelecaniformes by J. B. Nelson.

10.2 COMMUNICATION AND COLONIES

The orderliness of seabird colonies is mediated by communication between individuals and not by any other means. There are no general conventions, and there are no social hierarchies. To take a simple example, skypointing, the display which in the Atlantic Gannet precedes departure from the nest site, does not ensure for it an unmolested passage through territorial neighbors (which would be a "convention"; Nelson 1978a). On the contrary, it elicits aggression. Its function is to ensure that the partner accepts the posturing bird's imminent departure and remains on the site, thus obviating the danger to egg or chick of an unguarded nest.

While colonial breeding confers significant advantages (Kharitonov and Siegel-Causey 1988, Nelson 1970, 1983; see Chapter 4), it incurs costs, such as intraspecific interference (frigatebirds, Nelson 1976) and competition for resources such as sites, mates, and food (see Chapter 8). Also, it may attract predators and require the use of large amounts of energy in nest-site attendance, defense, and display. In addition to physical features, a colony possesses a "behavior profile" (Nelson 1989), which communicates information about the frequency and intensity of social interactions (comings and goings, fights, and displays) and the state of breeding in the colony.

In this chapter, we concentrate on what behavior does. For example, territorial behavior, which delimits a breeding space, typically differs in form and intensity according to whether sites are limited. Gannets and Peruvian Boobies (*Sula variegata*) show intense threat and overt fighting in their huge, dense colonies, whereas Abbott's Booby rarely shows overt aggression in its sparsely occupied arboreal habitat (Nelson 1971). Pair formation-and-maintenance behavior tends to be most highly developed in long-lived seabirds with near-permanent bonds, although there are significant exceptions. Fulmarine petrels, for example, have enduring pair bonds but an extremely simple display, which appears to serve both territorial defense and pair bonding.

10.3 SPHENISCIFORMES

Ritualized and nonritualized behavior of penguins is similar among species whether they inhabit the parched deserts of Patagonia or the ice-bound coasts of Antarctica. All penguins except the Yellow-eyed and the Fjordland Penguin (*Eudyptes pachyrhynchus*) are colonial. Characteristically, most penguin colonies are large. Colonies of Gentoo Penguins (*Pygoscelis papua*) can number 100 or fewer, but others like Macaroni (*Eudyptes chrysolophus*), King (*Aptenodytes patagonicus*), and Chinstrap (*Pygoscelis antarctica*) can have hundreds of thousands of pairs. This promotes a high degree of social interaction and has led to the evolution of a varied and complex set of communal behaviors (Williams 1995).

Functional categories of displays in penguins are similar to those of other seabirds. Agonistic (aggressive) displays are associated with territorial and pairing behavior. There are advertising displays to attract a mate, various pair-maintenance displays, appeasement displays, and displays between parents and offspring. Some are ritualized, some are not.

10.3.1 TERRITORIAL BEHAVIOR

10.3.1.1 Site Prospecting by Males and Site Defense

Little is known about the cues used in the initial stages of prospecting and how an actual site is selected. Colonies require accessible sites with ice-free substrates (except those of the Emperor Penguin, *Aptenodytes forsteri,* which breeds on stable fast-ice), and suitable habitat for nests (burrows or open ground; Williams 1995). Many penguins show marked natal and breeding philopatry, as in other seabird groups (LeResche and Sladen 1970, Reilly and Balmford 1975, Williams 1995) and some colonies have been used for several hundred years (Williams 1995). Others move their colony site each year, especially those which nest in tussock grass (*Poa* spp.), e.g., Gentoo and Snares (*Eudyptes robustus*) Penguins because the grass gets trampled and dies (Fineran 1964, Boost and Jouventin 1990). Early layers are sometimes displaced by more aggressive or larger species after the arrival of the latter (Warham 1975, Williams 1995). Failed breeders or pre-breeders may sometimes be displaced by other species (e.g., Short-tailed Shearwaters, *Puffinus tenuirostris*; Blue Penguins, *Eudyptula minor*), although penguins are usually dominant (Reilly and Balmford 1975).

10.3.1.2 Defense of Territory

Defense of territories occurs almost constantly in a penguin colony because of the natural crowding and the constant movement on foot through territories. Not surprisingly, there are more interactions

FIGURE 10.4 Nonritualized threat is universal in penguins and involves simple pointing or gape menacing, as in these Adelie Penguin. (From Stonehouse 1975.)

and a greater expenditure of energy in denser colonies (Oelke 1975), as well as a greater diversity of displays. The type of nest site (open or burrow) may dictate how frequently agonistic behaviors are expressed because of the greater encounter rate among open nesting individuals (Waas 1990b). Aggression seems to be less in Emperor Penguins who group very closely together during incubation as an adaptation to cold (Williams 1995).

Most territorial defense involves threat, both ritualized and nonritualized, and appeasement and displacement movements in an effort to maintain personal distance (Williams 1995). Nonritualized threat is universal in penguins and involves simple pointing or gape-menacing (Figure 10.4), often accompanied by vocalizations (Spurr 1975). Each call component has its own meaning and some species have "punctuation marks" between vocalizations, indicating when one communication is finished. Ritualized threat ranges from loud calling, stand-offs, and aggression redirected to inanimate objects, to elaborate displays with stereotyped components such as circling, bowing, head-up or -down positions, and other movements (Ainley 1975a, Oelke 1975, Spurr 1975, Warham 1975, Williams 1995).

Some aggressive displays are similar across species within the order. For example, displacement preening, alarm "wing-flapping," and low-intensity aggressive posturing like "head-circling" (e.g., King Penguins) or "bill-to-axilla" movements (e.g., Adelie Penguins, *Pygoscelis adeliae*) relay an agonistic message (Figure 10.5; Spurr, 1975, Stonehouse 1975, Williams 1995). In contrast, courtship messages are more species-specific. Intermediate aggression which falls short of physical contact is indicated by a "sideways stare" or "twisting postures." Many postures are composed of various levels of aggression. For example, in the Fjordland Penguin, one type of hostile display

FIGURE 10.5 The "bill-to-axilla" is a low-level aggressive posture in the Adelie Penguin. (From Stonehouse 1975.)

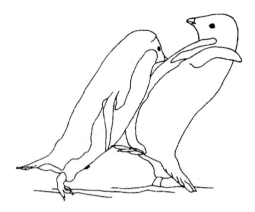

FIGURE 10.6 At the height of aggression, a bird may be physically attacked. An Adelie Penguin shoves an opponent. (From Stonehouse 1975.)

begins with a low-level hiss and ends with a high-intensity growl, often accompanied by a lunge (Warham 1975). Ritualized threat displays and posturing can easily lead to action. The closer is the intruder, the more aggressive is the display in form, rate, and/or duration (Stonehouse 1975).

Demarcation and defense of territory involves damaging fighting in some species (e.g., Jackass Penguins, *Spheniscus demersus;* Blue Penguins), although all species peck and bite one another (Waas 1990a, Williams 1995). Overt aggression involves all possible weapons: bills, feet, and flippers (Figure 10.6; Spurr 1975). During fights there are often growls and interlocked bills and many penguins become bloodied from bites. Erect feathers, rolled eyes, and the body in a forward position are the most aggressive postures used toward neighbors, and the head forward, body upright, flippers out, and bill open were the most aggressive postures toward strangers (Spurr 1975). If aggression erupts at one nest, it will often travel to other nearby pairs. During bouts of territorial defense, a pair may attack each other or even themselves, and then neighboring pairs may start attacking them (Warham 1971).

Many stereotyped movements are linked to special markers such as crests, vividly colored bills, feet, eyes, and facial skin or special tracts of plumage (Figure 10.7). These can increase the effectiveness of the display, e.g., position of eyes or piloerection, in contrast to the gross motor movements which can vary tremendously, depending on morphology. Changes in eye and head positioning are common (Ainley 1975b, Warham 1975, Williams 1995). Penguins in threat postures

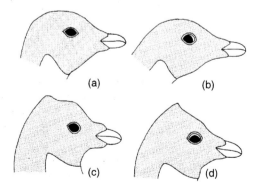

FIGURE 10.7 Many stereotyped movements use special markers such as crests or special tracts of plumage. Penguins in threat postures slick their face feathers around their eyes and eyelids which also has the effect of increasing the white area. Exposed large eyes tend to be more threatening. (a) normal, (b) sleeked feathers, (c) ruffled feathers, (d) crest erected, eyes directed down showing whites. (From Stonehouse 1975.)

FIGURE 10.8 The sideways stare is an aggressive posture given by penguins on their territories. (From Stonehouse 1975.)

slick their face feathers around their eyes and eyelids which also has the effect of increasing the white area, and exposed large eyes tend to be more threatening (Figure 10.7). Vocalizations during aggressive displays may help to mitigate aggression. During silent, agonistic displays, conflicts often escalate. Wass (1991a, b) believes that vocalization may give information about the vocalizing penguin and thus information about potential success in a conflict.

Appeasement communication is a clear signal that the initiator of the signal means no harm. These postures are the antithesis of aggressive postures and are similar across taxonomic lines, with the weapons concealed or placed in a nonthreatening position. The displays are often oriented away from the aggressor, e.g., the "face-away" (Figure 10.8) or the "slender walk" in many other species (Warham 1963). Penguins who live in crowded colonies often crouch or lie flat and withdraw their beaks, not moving off the nest in response to aggressive displays, covering and protecting eggs and chicks (Spurr 1975, Warham 1975). Appeasement behavior may maintain the distance between two individuals or may be distance-increasing, and is usually silent (Williams 1995). In sexual behavior, weapons are concealed or averted, e.g., in the "oblique stare" (Roberts 1940, Penney 1968).

10.3.2 PAIR RELATIONSHIPS

10.3.2.1 Pair Formation: Female Prospecting, Male Advertising

Penguins are serially monogamous (Williams 1995) and reinforcement of the pair bond through ritualized communication may determine the year-to-year maintenance of the pair bond. An average of about 60 to 90% of the pairs keep the same mate over a period of years, although this can be as low as 15% due to death or divorce or separation (Williams 1995). Ainley et al. (1983) believe that mate retention in Adelie Penguins is related to the amount of elaborate pair-formation behavior. Additionally, pair fidelity may be a result of birds returning to the same nest site, which both sexes tend to do, although this relationship between pair and site fidelity remains unclear.

Little is known about female choice or prospecting, and what factors in a male's appearance or behavior cause a female to choose him. Penguins often occupy sites long before the breeding season, sometimes spending the day at the colony but the night at sea. There are two types of pair displays similar in all species of penguin: "bowing" or "quivering," and "ecstatic" or "mutual" displays (Ainley 1975a, Spurr 1975, Warham 1975, Williams 1995). "Mutual displays" help maintain the pair bond (Williams 1995) and can be vocal or silent. Both of these displays are discussed further in the "Pair Maintenance" section below.

Allopreening occurs only in the smaller species and may be both sexual behavior and appeasement behavior (Williams 1995). In addition, King and Emperor Penguins also have the "face-to-face" or "high-pointing" display (a mate-mimicking behavior, Figure 10.9) and the "attraction walk" or "waddling gait" where the pair walks through the colony together. These displays often have components of appeasement, aggression, and attraction (Williams 1995).

When a stranger approaches a territory holder after an "ecstatic display," the territorial penguin may respond with a "sideways stare" and bill at a pre-attack position (Spurr 1975). Males are deterred by this, but females may advance. Subsequent responses, aggressive or sexual, depend on

FIGURE 10.9 A King Penguin pair performs a "high-pointing" display (a mate-mimicking behavior). This display has components of appeasement and aggression. (From Williams 1995.)

FIGURE 10.10 "Vertical trumpeting" by a pair or Erect-crested Penguins (a courtship ritual) helps them to learn each other's voices and reinforces the pair bond. (From Warham 1975.)

the response of the approaching penguin (Spurr 1975). Close proximity of pair members may elicit both escape and aggressive behaviors, thus the incorporation of elements of aggression and appeasement into sexual displays. Much of pair-formation behavior is incorporated into pair-maintenance behavior. One of the main functions of courtship rituals may be for mates to learn each other's voices and ways of displaying (i.e., to reinforce each bird's memory of the other; Warham 1975). "Vertical trumpeting" (Figure 10.10) and "forward trumpeting" as well as "bowing" (Figure 10.11) and "mutual displays" may serve this purpose in penguins (Warham 1975).

10.3.2.2 Pair Maintenance

Season-long, ritualized greeting ceremonies, both auditory and visual, occur in penguins. Greeting ceremonies, generally called "mutual displays," are most commonly seen during nest-relief (Figure

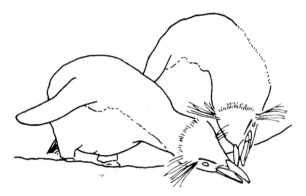

FIGURE 10.11 "Bowing" is another courtship and pair-bonding ritual, performed here by a pair of Erect-crested Penguins. (From Warham 1975.)

10.12; Williams 1995). Bowing helps mitigate aggression because the bill is tucked (Spurr 1975). Females may sleek feathers on the head to show nonaggression. Soft vocalizations or "bill-quivering" often accompany "bowing" as part of the nest-relief ritual, which can include arranging nest material. During mate greetings at the nest, neighbors often mimic the "forward trumpet" recognition display (Warham 1971).

"Ecstatic displays," more often performed by males during pair formation, may help a returning female to recognize her mate at a distance. In some penguins (e.g., Jackass Penguin) females also perform this display, lending support to the "mutual recognition" theory (Williams 1995). Emperor Penguins, which have no fixed nest sites, entirely rely on individual voice recognition to find both chicks and mates (Jouventin 1982). Nest scraping by Adelie Penguins, now a ritualized courtship display, derived from scraping out a hollow in the ground for the egg. Mutual recognition displays

FIGURE 10.12 The ecstatic mutual display is typical of crested penguins (here, Marconi Penguins); it is a season-long, ritualized greeting ceremony, and is most commonly seen during nest relief. (From Warham 1975.)

frequently involve the raising and lowering of the bill, often in response to a potentially agonistic "sideways stare" which is given when a mate may be about to move (Spurr 1975).

10.3.2.3 Copulation

As in other groups, copulation serves the dual purpose of insemination and maintenance of the pair bond. Since both adults are needed to raise a chick, repeated copulations before the egg is laid may help to cement the pair bond. Extra-pair copulations have been observed in Adelie Penguins (Davis and Speirs 1990). "Mutual displays," "bowing," nest-circling, or nest building may precede copulation (Spurr 1975, Warham 1975, Williams 1995). Sexual displays are often most intense as the female approaches, and they show a conflict of motivational states. "Shoulder-hunching" around the nest with a partner (common in *Eudyptes* penguins) is incorporated in the dance and is a territorial and appeasement posture (Warham 1975). The male may then send an appeasement signal to the female just before copulation commences. In some species there are postcopulation displays.

10.3.2.4 Nest Building

Nest building in the surface-nesting penguins is often symbolic, although usually pebble nests or simple scrapes are made. Some species burrow or hollow out nests under bushes and others nest in caves or occasionally under forest canopies or in clearings. Nest building by both sexes helps to strengthen the pair bond. Touching, biting, and handling of nest material are probably displacement activities for aggressive behavior (see above). Emperor and King Penguins do not build nests, nor do Emperor Penguins have a fixed nest site, since they carry the egg and the chick on their feet (Figure 10.13).

10.3.3 INCUBATION AND CARE OF YOUNG

Incubation lasts 30 to about 60 days in penguins; longer in the larger species (Appendix 2). Successful breeding requires efficient communication between offspring and parents, as much as it does between parents. Parent recognition of their offsprings' voices is widespread with obvious benefit. There are calls which beckon the chick out of the nest or away from the colony; calls when parents bring back food, food-begging calls, and alarm calls (Warham 1975). There are general

FIGURE 10.13 The head-down mutual display in Emperor Penguins. Notice the small chick being carried on the feet, as is the egg during incubation. (From Williams 1995.)

calls to which individuals unrelated to the caller respond. For example, in response to aggressive calling by adults, chicks, related or not, often freeze, crouch, or assume the classic horizontal appeasement posture with tucked beaks. Chicks are semi-altricial and poikilothermic at hatching and are completely dependent on their parents for 2 to 6 weeks for brooding, warmth, and guarding. After this, chicks of some species, especially in large colonies, form crèches while both parents forage. Creching may help in thermoregulation and in protection from predators (Davis 1982).

10.3.4 EVOLUTION AND COMMUNICATION: ONTOGENY OF BEHAVIOR

In some species vocalizations are fairly completely matured and highly individualistic at hatching (e.g., King and Emperor), while in others (e.g., Adelie) they are simple. There may be elaborate recognition postures between parents and chicks, always accompanied by calling. For example, chicks of Erect-crested Penguins (*Eudyptes sclateri*) not only call but also flick their flippers to get the attention of their parents (Warham 1975). Many penguin adults call their chicks with some of the same displays that they use in pair-bonding, e.g., bowing and then displaying in a "forward trumpet." Jouventin (1982) altered the appearance of some chicks and adults with paint and these birds were still accepted as relatives, emphasizing the importance of voice in penguins. Most adults call loudly as they approach the nest or crèche with food. Sometimes unrelated chicks will surround an adult returning with food, pecking the adult and trying to steal food from the adults' offspring, but the adult will provision only its own chick (Warham 1975, Jouventin 1982, Williams 1995).

Food begging by chicks is varied in form, but usually is somewhat ritualized and includes vocalization. Chicks peck the adult's bill, breast, and head as they stand (Figure 10.14). They may also beat the adult with their flippers. Violent begging produces avoidance or head-turning away by adults. Chicks take food directly from the bill or throat of the adult. In poor food years especially,

FIGURE 10.14 A King Penguin feeding its chick. The chick pecks at the adult's bill, breast, or head while begging to stimulate feeding. (From Williams 1995.)

FIGURE 10.15 An adult penguin (here, Gentoo Penguin) coming in to feed its chick may lead the chick away from the créche in a feeding chase before feeding, partly as a way to ensure that it feeds only its own chick. (From Williams 1995.)

great numbers of chicks surround and peck a returning adult who may or may not have brought back food. Parent and chicks will sometimes run off together from the melee (Williams 1995). Feeding chases (Figure 10.15) help to eliminate would-be usurpers. In poor food years, adults with two chicks often feed only the stronger one.

Agonistic displays, similar to those of adults, are commonly used by chicks against predators, interlopers, and siblings. In Jackass Penguins, which hatch chicks asynchronously, the larger is usually more aggressive and there are many fights over food. Yellow-eyed Penguins hatch both chicks synchronously, thus they are equally matched, and there is no fighting over food (Seddon and van Heezik 1991). Development of displays involves learning from adults, as has been shown in interspecific egg-swapping experiments (Waas 1990b).

10.3.5 Nonbreeding Behavior

Little is known about penguin communication off the breeding colonies but they are social at sea (Broni 1985). Depending on the species, penguins may swim in small or large groups (Broni 1985, Williams 1995). Remaining in groups may "swamp" predators, decreasing an individual's risk of depredation. When at sea, mates do not necessarily associate with each other, with the result that nonmigratory penguins often have a higher mate retention than do migratory ones (Williams 1995). Courtship displays, nest-building, and sometimes copulation behavior can occur during the nonbreeding season (e.g., Chinstrap and Gentoo Penguins, Warham 1962, Conroy 1972, Conroy et al. 1975).

10.4 PROCELLARIIFORMES

This large order comprises albatrosses (Diomedeidae), fulmars, prions, shearwaters and gadfly petrels (genera *Lugensa* and *Pterodroma*), the storm petrels (Hydrobatidae), and the diving petrels (Pelecanoididae). The Procellariiformes is a much more coherent order than the Pelecaniformes. Nevertheless, during its long evolutionary history (some 50 million years), it has undergone great adaptive radiation in size, feeding behavior, nesting-habitat, and breeding phenology, which has necessarily led to diverse communication behaviors. This chapter has benefitted incalculably from Warham's (1990, 1996) extraordinarily comprehensive accounts, which will not be cited repetitively.

10.4.1 Territorial Behavior

All albatrosses are diurnal and have over-dispersed spacing, facilitating their complex and often perambulatory displays. Among the petrels and shearwaters (family *Procellariidae*), comparatively open habitat is chosen by the diurnal fulmarine petrels, none of which burrow. The bulk of this family, however, visits colonies at night and uses crevices, gaps among boulders, and burrows, thus rendering display based on vision and movement impracticable. Sounds, touch, and possibly smell are important.

10.4.1.1 Site Prospecting by Males

In general, pre-breeders are markedly drawn to their birth colony (Warham 1990, 1996). Most procellariiform colonies are noisy and/or visually conspicuous. Response to playback has shown that male Manx Shearwaters (*Puffinus puffinus*) in burrows increased the frequency of calling in response to vocalizations of both males and females, but females increased their calling only to male calls (Storey 1984). He concluded that male calls provoke competing calls from other males, who were attracting females to their burrows, and from prospecting females who may land and begin "duetting." Similarly, male Blue Petrels (*Halobaena caerulea*) singing near their burrows attracted overflying females who called before landing (Bretagnolle 1990).

Aerial prospecting for a site is conspicuous in fulmars, other diurnal petrels, and ledge-nesting albatrosses. Dispersed, ground-nesting albatrosses, such as Waved and Wandering (*Diomedea exulans)*, move around on foot and visit a breeding colony for 2 or 3 years before the male settles on a definitive site. By their third season ashore, male Wandering Albatrosses aged 8 to 10 were attending a nest site (Tickell 1968), although there is no information about how they selected it, nor whether they had to compete for it. Nor do they perform a specific site-ownership display on it.

Nocturnal burrow- and crevice-nesting shearwaters wander in and out of burrows early in the nesting cycle. Some burrows without breeding owners held up to ten different Sooty Shearwaters (*Puffinus griseus*) during a single season (Richdale 1963). During the breeding season, nonbreeding storm-petrels range widely (hundreds of kilometers) over short time periods; all birds captured away from their birth colony were pre-breeders or non-breeders (Fowler et al. 1982). Diurnal petrels such as the Wedge-rumped Storm-petrel (*Oceanodroma tethys*) inspect breeding areas visually, sometimes in dense crowds. Where it is necessary to locate suitable crevices, they may avoid occupied ones if the owner vocalizes. Or they may use smell. Shearwaters may even visit uncolonized areas such as headlands, where they vocalize, but do not land.

10.4.1.2 Defense of Territory

High breeding density occurs mainly among crevice and burrow nesters, and in these species, site competition can be injurious and occasionally lethal (Nelson 1970, Olsthoorn and Nelson 1990). However, Snow Petrels (*Pagodroma nivea*), which nest in crevices but not densely, nevertheless defend territories aggressively; fighting is common, with oil spitting, interlocked bills, head twisting, and screeching (Brown 1966). They also pursue each other in the air. The Wedge-rumped Storm-petrel has been roughly estimated at 15 pairs per square meter (Harris 1969) and at the other extreme nearly 50% of *Phoebetria* albatrosses nest solitarily (Weimerskirch et al. 1986). Petrels often nest in mixed colonies, but species usually segregate.

Among most albatrosses there is little competition for nesting space, and none have evolved a ritualized display directed solely toward preemptive declaration of ownership, as occurs in many sulids and gulls. Male–male aggression during courtship is common, but is usually in defense of a movable area around a female and not in defense of a site to which the female has been attracted (as occurs in many pelecaniforms). Nevertheless, in many species, a precise site is eventually defended, as in the Black-footed (*Phoebastria nigripes*) and Laysan (*P. immutabilis*) Albatrosses

(Rice and Kenyon 1962). In the colony, aggression in albatrosses takes the form of overt attack biting, threat with wide gape and vocalization, and rapid bill-clapping (in some albatrosses this is used in feeding competition at sea). In many albatrosses, the so-called "gawky-look," a peculiar facial expression caused by flattening head feathers and accentuating eyebrow ridges, is used in an aggressive territorial-cum-sexual context.

Threat vocalizations, or mechanical sounds, are universal in procellariiforms. Gadfly petrels hiss with violent head movements, jab, and grapple. In the *Puffinus* shearwaters, established birds vocalize to repel territorial intruders, but also fight using bill and claws; wounds or even death may ensue. In crevice- and burrow-nesting procellariiforms, vocalizations are used in both territorial and sexual contexts. There are both gender and individual differences in petrel vocalizations, and voice plays a part in isolating sympatrically nesting species. However, the function of much of the aerial calling in petrels is largely unexplained (Warham 1990). Fulmars use oil-spitting, cackling, and head waving in site defense. Spitting is widespread in tuberoses, except diving petrels, but in most species it is used mainly by chicks and, except for fulmars, is badly aimed. Apparently aggressive aerial behavior "upright-threat" in mid-air by Southern Giant Petrels (*Macronectes giganteus*) may be directed against other flying birds or at birds on the ground (Warham 1996).

10.4.2 PAIR RELATIONSHIPS

10.4.2.1 Pair Formation: Female Prospecting, Male Advertising

Before pair formation there is a period of reconnaissance during which females may visit different parts of a colony and even different colonies, as in Northern Fulmars (*Fulmarus glacialis*) and storm-petrels. Among albatrosses, the period during which a pair bond is initially formed is the only time when their complicated courtship repertoire is performed extensively. Thereafter, even when partners reunite for a new breeding season, display is relatively perfunctory. Courtship display is highly ritualized and the various postures are accompanied by specific vocalizations. Most display is ground based, although some aerial display occurs in the Southern Royal Albatross (*Diomedea epomophora*), some mollymawks (five medium-sized dark-backed albatross species), and the giant petrels. In fulmars, pair interactions involve little more than comparatively undifferentiated head waving plus vocalization, during which bills may make contact. This raises issues concerning the adaptive significance of complex vs. simple display. Even the giant petrel's display, though more complex than the fulmar's, is considerably simpler than that of any albatross. There is no obvious relationship between display complexity and the duration of the pair bond.

Understandably, pair formation in burrow- and crevice-nesting petrels and shearwaters is largely based on vocalizations and tactile stimuli, such as billing and preening (Figure 10.16). In some petrels it is preceded by mass aerial maneuvers. Fundamentals of albatross display include (a) its complexity, (b) the pattern in which postures and movements are strung together (these patterns seem less evident in some), and (c) the evident homology of many of the discrete postures-cum-movements, despite modifications in different species. Most behavior patterns may be performed

FIGURE 10.16 A pair of shearwaters billing in a burrow as part of courtship. (From Nelson 1979.)

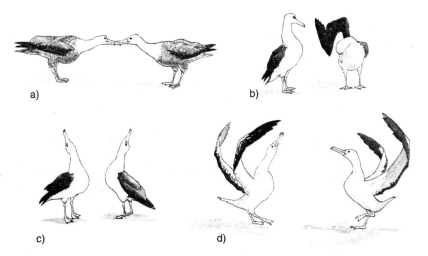

FIGURE 10.17 Courtship in albatrosses: (a) billing, (b) scapular action (touching the side of the breast with the wing raised), (c) skypointing, and (d) a full-intensity dance with wings spread (generally used only during early pair formation). (From Nelson 1979.)

by either sex, but not equally, nor at the same typical intensity. Before a definitive pair bond is formed, variable sequences of display may be performed with different partners, which may lead to aggressive interactions between potential partners. Currently there is no generally agreed terminology for the many discrete postures/movements in albatrosses, and no critical evaluation of homologies. Thus far the interpretations which have been offered for the motivation and function of the various displays of the albatrosses are largely unsubstantiated. However, Warham (1996) performed a seminal service in amalgamating and ordering the existing information, and Jouventin et al. (1981) have analyzed the relationships of the display elements in the Sooty and Light-mantled Sooty Albatrosses (*Phoebetria fusca* and *P. palpebrata*).

Billing ("bill clashing"). Species vary in the amount of bill contact, from mere apposition of part of the bill to vigorous half-circling contact, producing loud rattling (Figure 10.17). Billing is interspersed with other courtship activities, and in most albatrosses is the commonest interaction and is used in greeting. Billing is the main courtship element in fulmar display.

Scapular action. Depending on species, one or both wings are partly lifted and the distal feathers fanned from the wrist. The Black-footed Albatross fans both wings simultaneously, while at the same time touching its side with its bill. The Laysan fans only one wing and touches the bill to that same side (Figure 10.17b). The Waved Albatross does not fan either wing, but still touches its side. In the great albatrosses the bill is hidden in the scapulars, but the wing(s) are not fanned and the action is preceded by closing the widely parted bill with a snap. Scapular action occurs in many species as one of the courtship acts while the partner simultaneously performs another component such as "bill-clappering" or "skypointing."

Sky pointing (probably the same as "sky-call"). This is another prominent courtship component in which the bird stretches its neck and points its bill vertically upward or even backward (Figure 10.17c). It is accompanied by vocalization described as "whining" or "moo-ing." Sky-pointing is widespread in albatrosses, but absent in mollymawks.

Bill gaping. Wide bill gaping with neck stretched is common both in courtship and threat; the latter is accompanied by loud vocalization. Bill gaping in courtship terminates with a resonant "clunk" as the bill closes. Since it is part of the display, this sound has become amplified well beyond that normally produced by forcibly closing the bill.

Sway walking. The Waved Albatross (both sexes) uses a ritualized "sway walk," swinging the head and neck far over to the side opposite to that on which the foot is raised, to move to a potential nest site after a bout of display. Usually the male leads.

FIGURE 10.18 A mutual display in fulmars. They head sway and cackle, bills opposed, and often nostrils seep oil. (From Nelson 1979.)

Bill clappering. This is an extremely rapid clapping together, almost a blur-like vibration, of the mandibles with neck stretched horizontally. It is a common component of Waved and Laysan Albatross display.

Dance or ecstatic ritual. In this display, the great albatrosses arc their wings widely in a curve with the undersurfaces facing forward, meanwhile swinging the head vertically (sky pointing) and vocalizing ("whining"; Figure 10.17d). They intersperse downward sweeps of the head and bursts of bill clappering before swinging it back to vertical. Meanwhile the male steps ostentatiously around the female.

Forward bowing, bobbing, and tail tilt. In "forward bowing" or "pointing" the neck is lengthened and the bill is held more or less horizontal or inclined downward. "Forward bobbing" occurs in a crouched position with the head retracted and bill pointing down or backward between the legs and is sometimes called "yapping" or "nest-pointing" (other features may be added such as the "gawky look"). Whether it is the "croaking and nodding" of mollymawks seems unclear. It appears to be linked to the nest site, perhaps initially indicating its position. In general, aerial display plays a relatively small part in albatross courtship. Dual flights feature prominently in the *Phoebetria* albatrosses, the following bird matching every twist and wing-beat of the other (Warham 1996). They are cliff-ledge nesters with little ground space on which to display.

It is remarkable that compared with albatrosses, Giant Petrels have such a simple repertoire, consisting of little more than bill touching or bill clashing, and mutual and reciprocal allo-preening. They concur with other procellariids. However, they do visit the colony in winter and display together, which may help to maintain the pair bond. Fulmars merely cackle and head wave, with bill contact, during nuptial display (Figure 10.18). This, with mutual and reciprocal allo-preening, serves also for pair maintenance. Antarctic Petrels (*Thalassoica antarctica*), Cape Pigeons (*Daption capense*), and Snow Petrels have not evolved elaborate courtship. Courting pairs perform versions of head waving with bill contact, exposure of mouth cavity, and a range of vocalizations as duets (cackling, chattering, staccato "caws" or "clucks," trills, or chirrups). Warham (1996) points out that the calls of the smaller petrels are all built from very similar pulsed notes with few harmonics.

The large group of gadfly petrels are burrow nesters, have few visual displays, and are largely silent although some have a complex repertoire of calls (Warham 1996). Because of the difficulties in observing identifiable individuals, the initial stages of pair formation have not been described. Once non-breeders (of whatever category) visiting birds leave the colony, partway through the breeding season, there is little activity among breeders except for care of young. Breeders come and go silently, except for a brief call when entering their nesting tunnels. Among tropical gadfly petrels, an aerial chorus begins in late afternoon, peaking just after dark. The high speed, daytime, "dual flighting" of gadfly petrels in which two or more birds pursue each other (at times high in the air) seems largely unexplained in terms of function, though presumably it is most likely sexual in nature.

Of the 19 species of *Puffinus* shearwaters, some may assemble on bare, unburrowed ground to "sing" and court, which Warham (1996) compares to albatross gatherings. Dueting and billing may be intense. Mutual and reciprocal allo-preening may be prolonged. In dueting in a colony, the participants may vocalize while pointing their bills down a burrow. Some storm-petrels (*Hydrobatidae*) are notable for their nocturnal flights over the nesting area. However, Wedge-rumped Storm-petrels gather by the thousands during daylight to fly silently around the colony, at intervals descending to patter over the ground or alight. It is not clear whether they are prospecting for sites, mates, or both. Pair formation in the diving-petrels, though little studied, seems typically procellariiform, with much vocalization, including dueting and pair billing.

10.4.2.2 Pair Formation

It is unusual for display between well-established partners to reach the intensity or duration of that shown in pair formation. At the beginning of a new breeding season, established albatross pairs spend comparatively little time together before the female departs to sea to feed in preparation for egg production. Paired Laysan Albatrosses may mate within an hour of reuniting on the site (Rice and Kenyon 1962) after wintering at sea, presumably separately. At change-over there may be little ritualized interaction, the incomer merely taking over. However, nest relief in some albatrosses may be accompanied by brief display reminiscent of that used when the nest site is first selected ("forward bobbing" in Waved Albatrosses). Vocalizations are usually involved, sometimes persistently, as the relieving bird approaches (albatrosses, fulmars, Snow Petrels, shearwaters), and in some shearwaters, dueting forms part of the change-over ceremony. Mutual allo-preening is widespread in procellariiforms in the prelaying period, during nest relief and as pair-bond maintenance (Warham 1996).

10.4.2.3 Copulation

Procellariiforms do not precede copulation with elaborate display. Among surface nesters (albatrosses and fulmars), most observed displays did not lead to coition (Warham 1996, p.196), although this act is often prefaced by the male preening his mate. The female albatross's precopulation posture, squatting and spreading or drooping scapulars and primaries, is a necessary signal before the male will mount (Fisher 1971, Laysan Albatross). In some albatrosses, shearwaters, and fulmars, the male, while on his mate's back, strops his bill across the top of hers and vocalizes. In mollymawks, fulmars, and shearwaters, copulation is prolonged and involves several cloacal applications. In this latter feature it resembles some Laridae, and differs from pelecaniforms.

Extra-pair copulations are commonplace in many procellariiforms. Tickell (1968) records that fewer than half the copulations of Wandering Albatrosses were by mated or prospective pairs and 2.4% of 1823 copulations of Northern Fulmars were extra pair (Hunter et al. 1992), although they suggest that none of these fertilized an egg.

10.4.2.4 Nest Building

Most procellariiforms do not build a substantial nest, although some albatrosses build a pedestal which is functional, may endure for years, and is repaired annually by the female. Most procellariiforms do not collect and bring in nest material, with the attendant ritualization so marked in pelecaniforms. However, even in species that build a merely symbolic nest, adults show strong fidelity to the site.

10.4.3 INCUBATION AND CARE OF YOUNG

After the first two shifts, which can be rather variable, petrels generally share incubation about equally. The female usually departs 1 to 2 days after laying. Mean shift length varies with species.

Even in the same species and locality it also varies with year, presumably reflecting variations in food. The longest mean shifts are in albatrosses (17.5 days, Laysan Albatross, Rice and Kenyon 1962), medium length shifts in the giant petrels (5 to 6 days, Conroy 1972), Northern Fulmars (3 to 5 days, Mougin 1967), and short ones in storm-petrels (2.6 days, Davis 1957) and diving-petrels (1 day, Richdale 1965). Long shifts allow far-foraging: a breeding Black-footed Albatross has been recovered 3700 km away from its colony (Rice and Kenyon 1962). Typically, breeding petrels are either on nest duty or away at sea. Nest relief occurs quickly and with little ceremony. If a partner is overdue the incubating bird may leave the egg, but chilled petrel embryos can survive many days. Intermittent incubation is common and is not restricted to burrow nesters (Warham 1990). In the absence of avian scavengers, unattended eggs of surface-nesting petrels may remain viable for many days, although the incubation period is lengthened.

Continuing the trend set in late incubation, attendance shifts on small chicks are short, ensuring frequent feeds at a stage when the chicks cannot fast for long. Small petrels such as the European Storm-petrel (*Hydrobates pelagicus*) may change over every night. Chicks vocalize for some time before hatching, thereby cueing the adults, which then shortens their incubation stints; and adults having fed more recently, are more able to feed the hatchling. In species which undertake very long foraging trips, Weimerskirch et al. (1995) suggest that adults adjust their foraging distance to their body condition, thus avoiding debilitation during breeding.

Parents share brooding, guarding, and feeding the chick about equally, and relieve each other with little or no ceremony apart from, maybe, some allo-preening. Petrels leave their young unattended at an early age, thus freeing both adults for foraging. The semiprecocial, mobile young of open-nesting species can seek shade if necessary. Albatrosses and the giant petrels may brood or guard their chicks for 2 to 5 weeks continuously, and then intermittently. Nevertheless, many surface-nesting tropical petrels, such as the Christmas Shearwater (*Puffinus nativitatus*) and *Pterodroma* spp., may leave very young chicks unguarded. All procellariiforms lay a single egg and sibling rivalry is therefore not an issue.

Petrel chicks beg by pecking at the base of the adult's bill from the side, and they feed by placing their partly open bills across the back of the adult's gape and gulping down the regurgitate which spurts from the parent's throat. In the albatrosses a semicontinuous stream of oil and fragments, forcefully ejected, hoses into the chick's lower mandible (Figure 10.19). In some species, large chicks take food from the parent's bill, from the front. Before feeding, the adult may preen its chick which, when well-grown, reciprocates. Once past the guard stage there is seldom much delay between the adult returning, feeding the chick, and departing. Thus mates do not often meet at the nest.

In some albatross species, adults do not discriminate against nonoffspring when the chick is younger than about 10 days, but do so later. However, Black-browed (*Thalassarche melanophris*)

FIGURE 10.19 A Laysan Albatross feeds its chick by squirting oil into the chick's open bill. (From Nelson 1979.)

and Grey-headed (*T. chrysostoma*) Albatrosses will accept strange chicks even at 80+ days when they are almost fully grown. Shearwaters will accept strange chicks even if this greatly extends the care period. This accords with their failure to respond preferentially to the calls of their own chicks (Warham 1996). Generally, however, seabirds refuse or repulse strange young, particularly once the chicks are a few days of age (pelicans, frigatebirds, boobies, cormorants, gulls, terns, auks, and penguins). Intruding adults sometimes attack chicks, and many young seabirds exhibit appeasement behavior such as bill-hiding or lying prone, in some cases enhanced by plumage. Albatross chicks react by tucking the bill beneath their body. However, the young of some procellariiforms spit oil at intruders rather than appease them. Creching does not occur to any great extent in procellariiforms, although some albatross chicks (Waved Albatross) may wander from the nest to seek shade, and in so doing congregate loosely.

There are few details about the developmental sequence, frequency, and precise form of chick behavior. Preening, head scratching (under wing), and yawning are among the earliest behaviors to appear. Later, many shearwater chicks gather nest material; postguard albatrosses may even build nests near the parental one. Little is recorded about the ontogeny of ritualized behavior, but bill clapping and snapping in albatross chicks and oil spitting in fulmars occur early. The chicks of tropical albatrosses reduce heat stress by shivering and using their webs as heat radiators. Petrel chicks spend about 90% of their time sleeping or resting quietly, during which metabolic rates are 20 to 30% below those of the alert state (Warham 1990, 1996; see Chapter 13).

Once they have flown from the nest, fledglings are unaccompanied and do not return to the area, whereas most pelecaniforms, gulls, and terns do. Petrel chicks typically refuse food for some time before fledging. Some Galapagos Petrels (*Pterodroma phaeopygia*) starved for 3 weeks before fledging, while others were fed right up to fledging (Simons 1985). Seemingly late in the growth period, petrel chicks regulate their own intake (Ricklefs et al. 1987, Hamer and Thompson 1997). The precise relationship of parental food delivery rates to chick begging behavior and the regulatory mechanism needs further study.

Before fledging, chicks exercise their wings. In forest-nesting petrels, the adults climb trees from which to launch themselves, leaving a detectable scent which fledglings may follow. In many petrels fledging is hazardous. The young birds may clamber over obstacles and push through vegetation on their way to the sea. Night-fledging petrels are fatally attracted to lights; offshore congregation has been noted in Northern Fulmars, though its function is not known.

10.4.4 Behavior Outside Breeding Season

Seventy percent of petrels are southern hemisphere birds (Warham 1990) and after the breeding season they disperse or migrate. Very few are truly sedentary. Since some procellariiforms, especially albatrosses, forage thousands of kilometers from the colony when breeding, it is not surprising that their dispersals are among the most extreme of any seabird. Transequatorial migrations, especially south to north, form the largest category (e.g., Manx Shearwater, north to south; Short-tailed Shearwaters and Mottled Petrels, *Pterodroma inexpectata,* south to north). The Shy Albatross (*Thalassarche cauta*) migrates east to west; it breeds near southeast Australia and southern New Zealand, wintering near South Africa and western South America (Lofgren 1984). The Wandering Albatross also moves east to west, from the Indian and South Atlantic oceans to coastal New South Wales. One equipped with a satellite transmitter traveled about 20,000 km in 103 days (Nicholls et al. 1995).

10.5 PELECANIFORMES

The pelecaniforms are more diverse, phylogenetically and ecologically, than any other seabird order, and this means that their communication behavior is extremely diverse. In this account, we refer to gannets and boobies together as "sulids." The "core" pelecaniforms (those families widely considered to be closely related) comprise the pelicans, sulids, and cormorants. The other two families (frigatebirds and tropicbirds) differ in several important physical, ecological, and behavioral respects from the core three, and some authors favor removing them from the pelecaniforms. Where generalizations are made without reference to a particular family, they refer to the whole order.

10.5.1 TERRITORIAL BEHAVIOR

10.5.1.1 Site Prospecting by Males

Little is known about the cues used in the initial stages of site prospecting, usually carried out by males. Some pelecaniforms show marked natal and breeding philopatry; many new breeders return to the colony in which they were born, and established breeders remain in the colony in which they first bred (for sulids see Nelson 1978b, Schreiber et al. 1996; pelicans, Evans and Knopf 1993; cormorants/shags, Potts 1967, Aebischer 1995; tropicbirds, Stonehouse 1963, Snow 1965, Schreiber and Schreiber 1993). But even the Northern Gannet, intensely faithful to its site, once established, is quite likely to visit and settle far from the natal colony (Nelson 1978a). The proximate causes of such a move where space is still available in the natal colony are not known. Clearly a high proportion of subadults in a colony increases the chances of a young recruit finding a partner since there is a large pool of unmated individuals, and probably a degree of age-related mating.

Site prospecting involves aerial survey and/or inspection from a point near to or within the group. Aerial prospecting occurs in sulids, in some cormorants, probably in pelicans and frigatebirds, and certainly in some tropicbirds. Stationary prospecting from a perch, usually after aerial search, is found in all sulids, cormorants, and probably in pelicans. Frigatebirds are a special case in that males do not prospect for nest sites as such (Nelson 1976, Reville 1980), but join a display group in which they may or may not succeed in attracting a partner and then breeding. Unsuccessful suitors move to another group. Tropicbirds examine potential nest sites as they fly by, periodically landing to examine them more closely (Schreiber and Schreiber 1993).

FIGURE 10.20 Aggressive territorial displays in the boobies that are used to repel intruders. Head waving, head jerking, the yes/no head shake, and bowing all indicate aggression. (From Nelson 1978.)

FIGURE 10.21 Gannets are the only sulid to have frequent, physically damaging fights. (From Nelson 1979.)

10.5.1.2 Defense of Territory

Territorial displays are designed to repel intruders with a warning posture or call (Figure 10.20), avoiding physical conflict. Most territorial defense involves threat, both ritualized and nonritualized. The displays in sulids, pelicans, and cormorants may be derived from redirected aggression, such as nest biting, and may be quite elaborate. Defense of territory involves physically damaging fighting only in gannet and tropicbirds (Figure 10.21). All overt aggression in pelecaniforms involves bill-grappling or stabbing rather than wing-buffeting or feet-grappling. Simple "threat" in birds involves exhibiting the ability to inflict damage, usually by biting, and an accompanying vocalization may indicate the intensity of the threat. In some species overt behaviors (such as moving toward an intruder) grade into fully ritualized threat. In the Atlantic Gannet unritualized threat is erratic (vocal, bill wide open) and likely to be followed by attack. Ritualized threat is predictable and stereotyped, employing a preset twist-cum-withdrawal motion and highly unlikely to be followed by attack. Ritualized threat correlates with dense nesting in which neighbors are within reach. It facilitates the maintenance of distance without the damage and waste of energy involved in repeated fracas. Nonritualized threat is universal in pelecaniforms and ritualized threat is less common.

Notably, only the three gannets have evolved highly ritualized site ownership display, "bowing," which does not depend for its elicitation on the threat of intrusion, but occurs endogenously* and at high frequency throughout the breeding season (Nelson 1978b; Figure 10.20). Other sulids, some pelicans, and cormorants perform aggressively motivated displays after landing at the site or territory boundaries, manifesting aggression by touching or biting

---

* Here meaning apparently independent of a specific external stimulus.

the substrate. The movements involved in this displacement activity have become stereotyped and the resulting site-ownership display may show little formal resemblance to aggression. Some sulids (e.g., Blue-footed Booby) and cormorants (e.g., Red-legged *Stictocarbo gaimardi*) with conspicuous webs flaunt them in a ritualized prelanding gesture as a signal of ownership. The flaunting of webs during courtship is a different action altogether.

10.5.2 Pair Relationships

In most pelecaniforms the fundamental pair relationship is serial monogamy, although this has been little studied in this order. The male Magnificent Frigatebird (*Fregata magnificens*) is an exception, abandoning one breeding cycle partway through (Diamond 1975). It is theoretically free to start a new one with another female the next year while the previous female continues to raise its chick. The male Flightless Cormorant (*Compsohalieus harrisi*) similarly allows the female to complete rearing of the chicks while he pairs with another female (Harris 1979). Gannets, Abbott's Boobies, and some tropicbirds do exhibit long-term fidelity (Nelson 1978b, Schreiber and Schreiber 1993). Several sulids keep the same mate for two or more successive breeding cycles (Nelson 1978b, Schreiber et al. 1996, Schreiber 1999). No frigatebird has been shown to do so, but it may occasionally happen, although the unusual breeding cycle makes this unlikely (Nelson 1976).

10.5.2.1 Pair Formation

Males generally own a site and potential female breeders look around or "prospect." Little is known about this potentially important behavior and whether females choose a quality site (physical and social) or a quality male. Initially, prospecting may be aerial. Prospecting sulids fly over and may hover at particular spots; tropicbirds repeatedly fly low past potential locations; female Lesser Frigatebirds (*Fregata ariel*) adopt a specific investigative posture (goose neck, Diamond 1975) when inspecting displaying males. On the ground, prospecting female sulids and cormorants are recognizable by their alert, forward-leaning posture, and they elicit the "advertising" display from receptive males.

Pelicans are unique within the order, in that their pre-pairing procedures are communal, which renders the process of pair formation difficult to follow. Moreover, unlike other families, the site on which ground-nesting pelicans subsequently breed is not the spot on which pair formation occurs, but is selected afterward (Schaller 1964, Brown and Urban 1969, Vestjens 1977). In tree-nesting pelicans (Brown; Pink-backed, *P. rufescens;* Spot-billed, *P. philippensis*), the pair forms at the place in which the male advertises (Din and Eltringham 1974, Schreiber 1977, Nagulu 1984; Figure 10.22). The complex communal interactions, chases, and processions on land, water, or in the air are homologous in the Great White, Dalmatian (*Pelecanus crispus*), American White, and Australian Pelicans. There is much male–male interaction (pointing, gaping, thrusting, pouch rippling, pouch swinging, bill clapping), but its function is in need of study, particularly with individually marked birds. Eventually, for example in the Australian (Vestjens 1977) and the Great White Pelicans (Brown and Urban 1969), a particular female may attract a number of males who follow her on foot, swimming, or flying. Even in the arboreal Pink-backed Pelican, parties may fly from tree to tree (Din and Eltringham 1974). In Pink-backed and Brown Pelicans, individual males advertise to prospecting females from a perch near a potential nest site (Din and Eltringham 1974, Schreiber 1977). Thus these males have selected the site, whereas in ground nesters it is the female that selects the site. Ground-nesting pelicans are the only pelecaniforms in which the female selects the site.

In sulids, females prospect for a partner. In the dense-nesting gannets, territories are small and colonies crowded. Males often respond aggressively to a female, even though at the same time they are sexually motivated. Such females show marked appeasement behavior by turning the bill away and presenting the nape. Boobies also show appeasement behavior toward males. Where territory size allows, ground-nesting sulids engage in elaborate parading rituals in which some of the postures strongly resemble those found in ground-nesting pelicans (Figure 10.23). Both aerial and ground

FIGURE 10.22 A pair of Pink-backed Pelicans court, giving the head-up display. (Drawn by J. Busby.)

prospecting occurs in cormorants (e.g., Great Cormorant, Kortlandt 1940; European Shag, *Sticto-carbo aristotelis*, Snow 1963).

Overflying females of all five frigatebird species inspect clusters of frenziedly displaying males (Figure 10.24). Here, too, females probably choose male-plus-site, since successful suitors then

FIGURE 10.23 Blue-footed Boobies parade with exaggerated lifting of the bright blue feet during courtship, displaying a morphological marker. (From Nelson 1979.)

FIGURE 10.24 Male frigates display their enlarged, red gular sac and silver underwings to females flying overhead as part of their courtship display. (Drawn by J. Busby.)

build on their display site; unsuccessful ones move elsewhere. Like terns, with which they show considerable convergent evolution, tropicbirds perform complicated aerial courtship, involving 2 to 20 birds in ritualized maneuvers which differ in detail in the three species (Fleet 1974, Schreiber and Schreiber 1993). This courtship appears to involve both paired and unpaired individuals and may be a prelude to re-occupying a previously held site as well as to prospecting for a new one. Aerial display seems to be primarily sexual in motivation; it attracts peers to the group and synchronizes subgroup breeding.

10.5.2.2 Pair Maintenance

Within the order there is great variety in the degree and intensity of the interactions between partners. Undoubtedly the most complex, highly ritualized, high-intensity, and seasonally prolonged displays (greeting ceremonies) occur in some sulids, particularly the gannets and Abbott's Booby, when they reunite at the nest. In the Atlantic Gannet the greeting ceremony is a modified form of the site-ownership display, which itself derives from redirected aggression (Nelson 1965, 1978a). The greeting ceremony persists at high intensity throughout incubation, chick rearing, and post-chick attendance. Indeed, after the chick has fledged, site ownership and greeting ceremonies are as intense as in the pre-laying period.

Nest material may also be brought after the egg(s) is laid as pair maintenance behavior. In all except tropicbirds, even in species such as the Masked Booby (*Sula dactylatra*) where the nest is merely symbolic, it may involve scores of collecting trips, almost wholly by the male, and the ensuing delivery, acceptance, and building of a nest (often a mutual activity; Figure 10.25). Early in incubation, pelicans display conspicuously at change-over (head up, bill tucking, handling nest material), but later there is little ceremony. Cormorant pairs exhibit a wide range of head movements (snaking, pointing, stretching) and conspicuous alterations in head shape due to depression of the hyoid bone (kink-throating) and lateral expansion of the buccal cavity (discoidal head; Kortlandt 1940, Van Tets 1965). These behaviors appear to be behavioral modifiers; they occur as adjuncts to other displays depending on the situation, perhaps reflecting subtle tensions and the balance between different motivations (sexual, aggressive, fearful) rather than imparting a specific message.

White

Red-foot

FIGURE 10.25 Nest building in the Masked Booby is mostly symbolic as the eggs are laid on bare ground. Red-footed Boobies build a nest of twigs; generally the male brings nest material to the female who places it in the nest. (From Nelson 1978.)

In some cormorants, mutual display has become rigidly stereotyped; partners stand parallel and synchronize their forward stretching (Figure 10.26). These behaviors soon wane after the clutch is complete (Marchant and Higgins 1990).

At the other end of the scale, frigatebirds change-over on egg or chick with perfunctory interaction or none. Tropicbirds interact minimally. Unlike all other pelecaniforms, tropicbirds lack face-to-face display, ground-based courtship, and, at the nest site, any ritualized behavior except mutual preening. Despite this, pair bonds persist for many years (Schreiber and Schreiber 1993). Mutual preening is an important consolidator of the pair bond in pelecaniforms, but it varies in form and extent (Nelson 1978b; Figure 10.27). More than other pelecaniforms, cormorants and some sulids have evolved highly ritualized versions of pre-take-off and post-landing behavior (Van Tets 1965). Some sulids, the three gannets and Abbott's Booby, employ ritualized pre-flight postures which appear to coordinate the pair's activity, thereby preventing simultaneous absence from the nest with its attendant danger of intraspecific interference or predation (details in Nelson 1978b).

10.5.2.3 Copulation

Copulation in pelecaniforms may be regarded as both inseminating and pair bonding; it occurs over a period longer than necessary for mere fertilization. Brown Pelicans may mate within hours of coming together, although these may be mates from a previous season (Schreiber 1977). The first egg may be laid between 3 and 10 days after mating begins. Sulids may regularly mate from 2 to 12 weeks before the egg is laid; the longest period occurs in the Atlantic Gannet. In most cormorants it is 3 to 4 weeks. In Great Frigatebirds it may begin within a few hours of the female settling alongside a displaying male (Reville 1980), and continues for 1 to 2 weeks before egg laying. Tropicbirds may attend the nest site only sporadically before laying, and the length of the period over which copulation occurs, or where it occurs, is not recorded (Schreiber and Schreiber 1993).

All pelecaniform females show distinctive, albeit often inconspicuous, precopulatory behavior. They may crouch, gape, touch, or handle nest material, erect head feathers, shake the head, raise the tail, and dilate the cloaca. Male precopulatory behavior may be inconspicuous: a mere elongation and a particular orientation of the body, a fixed gaze, or an exaggerated locomotion. Some male pelicans

FIGURE 10.26 A pair of Pied-Cormorants in the head-up display. This mutual display has become rigidly stereotyped; partners stand parallel and synchronize their forward stretching. (Drawn by J. Busby.)

FIGURE 10.27 A pair of gannets performs pair-bonding mutual preening; a ritualized form of aggressive behavior. (From Nelson 1978.)

and cormorants perform the sexual "advertising" display as a precursor to copulation. It is somewhat similar in form to that performed by all boobies except Abbott's and is presumably homologous in these three core families, thus providing further evidence of their relationship (Van Tets 1965; see Table 10.1). Boobies, although employing this display in the context of sexual advertising, do not use it precopulatory. Among sulids, only male gannets bite the female's nape during copulation; the others merely lay their bills alongside (though they may nibble), thus discounting the idea that biting is only for maintaining balance. Cormorants vary with regard to nape biting. Some, for instance, Brandt's Cormorant (*Compsohalieus penicillatus*), Cape Cormorant (*Leucocarbo capensis*), Pied Cormorant (*Hypoleucos varius*), and European Shag, nape-bite during mating, while others lay their bills alongside and may nibble (Williams 1942, Snow 1963, Johnsgard 1993).

Postcopulation behavior includes various ritualized elements. Male frigatebirds do not precede or follow copulation with a special display, though they do extend their wings and may rub the bill alongside the female's head and nibble. The female hunches horizontally. Some body maintenance behavior such as gaping, stretching, and preening universally follows mating.

In pelicans, extra-pair copulations are apparently commonplace (Pink-backed, Burke and Brown 1970; Great White, Brown and Urban 1969; Spot-billed, Nagulu 1984). They are reportedly rare in American White and not studied in the Australian. In sulids, extra-pair copulations have been little studied, but are certainly rare in the Atlantic Gannet and never noted in the Blue-footed Booby (Nelson unpublished). They were not noted in Flightless Cormorants (Snow 1966), but are common in the European Shag. In this species up to 18% of chicks resulted from extra-pair copulations (Graves et al. 1992, 1993, Graves and Ortega-Ruano 1994). Extra-pair copulations occur in the Great Cormorant and have been recorded in the Great Frigatebird (Reville 1980). Reverse copulations are apparently extremely rare in pelecaniforms.

10.5.2.4 Nest Building

Nest building in pelecaniforms may produce a functional structure or be merely symbolic; in both cases it involves intimate interactions between mates that probably serve to strengthen the pair bond. The following are some brief generalizations:

1. The touching, biting or handling of material during or after copulation, probably as displacement activities (not functional nest-building), is universal in pelecaniforms.
2. Except tropicbirds, males (rarely females) of all families bring nest material, either in-between closely successive matings, or soon after. This may cost significant energy, or the male may merely take a few steps from the nest site, as in Masked Boobies that return with fragments for their symbolic nest.
3. All families contain some species that build symbolic nests that serve no architectural function (the only frigatebirds that do so are the ground-nesting Ascension Frigatebird, *Fregata aquila*, Stonehouse and Stonehouse 1963; and the Great frigatebird where there are no bushes, Schreiber personal communiction).
4. The core families contain species in which males "show" or "present" material in a ritualized manner.
5. In all species that build structural nests, the female does most (usually all) of the building.
6. All pelecaniforms that build architecturally functional structures and nest densely, pilfer from near neighbors with consequent need to guard nests.

10.5.3 INCUBATION AND CARE OF YOUNG

During incubation, coordination between the sexes at change-over is important in ensuring that disparate motivational states do not lead to eggs being unattended and thus vulnerable to predation, weather, or interference by conspecifics. Commonly in the early stages of incubation in the core pelecaniforms, change-over is preceded and accompanied by ritual. In this respect, as in many others, frigates and tropicbirds differ from other pelecaniforms in that they do not show ritualized greeting behavior during change-over.

Among pelecaniforms, five sulids, all frigatebirds, and all tropicbirds lay clutches of one egg (see Appendix 2). The other four sulids lay two or three, depending on species. Pelicans lay one to three, mostly two or three; cormorants lay two to five, usually three or four (see Appendix 2). Incubation begins with the laying of the first egg, which is then first to hatch. This chick then has a marked size advantage when its sibling hatches. In some species the larger chick actively pushes its sibling out of the nest or kills it (siblicide). In other species, the parents are not able to bring enough food to feed all the chicks and the smaller generally is excluded from food by older, larger sibling(s). The value of the second egg is probably as an insurance policy in case the first chick dies soon after hatching. Thus mean brood size in multiparous pelecaniforms, once the chicks are well grown, is generally less than the mean clutch size.

The reduction in brood size may be achieved by siblicide: obligative (in Masked and Brown Boobies [*Sula leucogaster*]) or facultative (Blue-footed Boobies, Nelson 1978b, Drummond 1987, Drummond and Chavelas 1989, Anderson 1989). Some populations of Great White Pelicans practice obligative reduction (Cooper 1980) but in most pelicans and in cormorants reduction is by unequal treatment of chicks (see also Mock 1984, Shaw 1985, and Drummond et al. 1991). There is dispute about the ultimate causal factors involved in siblicide (Anderson 1989, 1990a, b, Anderson and Ricklefs 1992) and further study is warranted.

Successful breeding requires efficient communication between offspring and parents, although the mechanisms remain poorly understood. As hatching approaches, incubation shifts shorten (Nelson 1966, Evans 1988), partly in response to auditory stimuli from the chick. In pelecaniforms, such as frigatebirds and tropicbirds, with long foraging trips, this change increases the probability that the attendant parent will be able to feed the chick. Pelecaniform hatchlings are weak and uncoordinated and most chick mortality is at this stage. Adult pelicans, sulids, and cormorants stimulate their new chicks to feed by grasping their heads, a trait that has developed to such an extent that in pelicans with large young, it resembles violent attack (Brown and Urban 1969, White 1971). In Red-tailed Tropicbirds (*Phaethon rubricauda*) the chick's gaping response is triggered by the adult, who touches its bill (Fleet 1974).

Chick begging is a normal prerequisite for feeding. Large pelican chicks violently peck the adult's bill and breast and flail their wings; some may attack other young, or bite their own wings and legs. Several studies of pelicans have noted chicks falling into convulsions when begging (Brown and Urban 1969, Burke and Brown 1970, Vestjens 1977, Nagalu 1984, Cash and Evans 1987). Ritualized begging using stereotyped movements is found in all sulids (Nelson 1978b) and can involve bobbing and feinting, repetitive vocalization maintained for long periods without any contact, rapid nodding or head-bobbing movements, touching or aggressively pestering the adult's bill, wing-flailing, or loud calling (Figure 10.28). Violent or consistent begging produces evasive action from adults, and in some species, such as the European Shag, ritualized appeasement. In many pelecaniforms the adult moves out of reach immediately after feeding its chick to escape unwelcome solicitation. In all species except tropicbirds chicks insert the bill into the throat of the adult. In tropicbirds the adult puts its bill into the mouth of the chick to transfer food.

In those pelecaniforms whose chicks form créches (ground-nesting pelicans and some cormorants), adults recognize and feed their own young. Ground-nesting sulids repel strange young as do frigatebirds (Nelson 1968, 1976), and Abbott's Boobies feed only their own juveniles even if others beg (Nelson 1971, 1978b). Gannet chicks recognize and respond to the calls of the incoming parent before it lands (Nelson 1965). The adaptive value of recognition lies in channeling potentially scarce resources to one's own offspring. Discrimination seems to develop gradually, and in sulids, substituted young are widely accepted before they are capable of wandering. Tropicbird adults do not recognize their own young, but feed whatever chick is in the nest at any age (Schreiber 1994, 1996).

FIGURE 10.28 A Red-tailed Tropicbird adult feeds its downy chick. Chicks are fed a diet of fish and squid. (Drawn by J. Busby.)

Ritualized appeasement behavior (in response to attack) is not universally present in young pelecaniforms; young cormorants lack it, but young pelicans bill-tuck (Van Tets 1965). Chicks of the Atlantic Gannet and Abbott's Booby hide their bills for long periods when attacked, but Red-footed Boobies do not, and may be killed by an intruding Masked Booby (Nelson 1968). Why such a basic communication pattern evolved so unequally is unclear. Appeasement in the female and young gannet may relate to the unusually low threshold for adult male aggression, evidenced by vigorous ritualized attack on his mate each time they reunite on the nest.

When the mobile young of ground nesting pelecaniforms cluster together in "pods," or crèches, they conserve energy and possibly deter some predators. While pods are said to help young to conserve heat at night, Brown and Urban (1969) noted that pods of the Great White Pelican increased in size in response to midday heat. Adult and young pelicans recognize each other's voice and chicks rush out of a crèche to be fed when their parents call (Schaller 1964, Hatzilacou 1992). In the Brown Pelican, which nests both in trees and on the ground, pods form only on the ground. Many young cormorants form groups, but these are much less organized and looser than in pelicans. Crèches are not found in other pelecaniforms.

In pelicans, boobies, and cormorants, play, prey-handling movements, handling nest material, nest building, body care, and fairly complete ritualized territorial and sexual display develop in a predictable sequence while the young are still partly downy. Interactions between young birds presumably help to perfect motor responses and communication behavior, thereby facilitating subsequent adult communication. Prey-handling is practiced, using inanimate objects such as twigs. The well-grown young of ground-nesting pelicans, boobies, and cormorants practice flying before achieving full flight, but tree and sheer cliff nesters fly adequately the first time, as do frigatebirds and tropicbirds (Stonehouse 1963, Schreiber and Schreiber 1993). Both of the latter groups are extremely awkward on the ground.

No pelecaniform starves its young prior to fledging, as has been stated in some older literature. In many (Atlantic Gannet, some pelicans, tropicbirds), the prefledgling stops begging, and may go without food for 2 days or more (Nelson 1978b, Schaller 1964, Brown and Urban 1969, Vestjens 1977, Stonehouse 1963, Schreiber and Schreiber 1993). Although the prefledgling exhibits clear overt signals of its shifting motivation toward fledging, this behavior elicits no parental reaction; none is necessary since no adult pelecaniform accompanies its free-flying juvenile on its dispersal/migration.

Postfledging parental investment varies from none (gannets, tropicbirds, some pelicans) to more than 6 months (some boobies, all frigatebirds, some cormorants). This wide disparity must be interpreted as part of each species' overall energy strategy in relation to the increased probability of offspring survival. Pelecaniforms have evolved widely disparate strategies in this regard, whereas the procellariiforms, without exception, have failed to evolve postfledging parental care at all. This reflects the heterogeneity within the pelecaniforms. It may also illustrate the evolutionary paradigm that a species' breeding strategy evolves in the species' typical habitat and circumstances, and that subsequent radiation into a different habitat may fail to change the species' main strategy.

10.5.4 BEHAVIOR OUTSIDE BREEDING

Mortality rates of immature seabirds, particularly in their first year, are typically on the order of 50 to 80+%, compared with an adult mortality which can be as low as 3% (see Appendix 2). For example, estimates of prebreeding mortality rates are Atlantic Gannet 60 to 70% (Nelson 1978a); Masked Booby 67% in the first 2 years (Woodward 1972); Great Cormorant around 40% (Koshelev et al. 1997); Brown Pelican 70% in the first 15 months (Schreiber and Mock 1988); American White Pelican 41% in the first year (Evans and Knopf 1993); Red-tailed Tropicbird 53% prebreeding (Schreiber and Schreiber 1993). Since nomadism offers time and energy savings (no commuting) and safety from land-based dangers (predation, accident), it is not surprising that outside the breeding season many pelecaniforms are nomadic/oceanic (gannets, several boobies, tropicbirds), nomadic/coastal (some boobies, some cormorants), or nomadic/island hopping (some boobies, some frigatebirds). All endemic cormorants and the Ascension Frigatebird, together with some populations of more widespread cormorants, are largely or entirely sedentary. True migration is found in some sulids (especially the three gannets, perhaps in Abbott's Booby) and in some pelicans (some populations of Great White and American White).

Except for tropicbirds, Abbott's Booby, and rare endemics, pelecaniforms gather in large roosting/loafing aggregations. This may be due to the combination of innate gregariousness, communal feeding, and a relative scarcity of adequate roosting sites (whether islands, headlands, cliffs, trees) in relation to feeding area. Apart from defending space, the individuals in such groups show little if any ritualized communication, devoting their time to preening, sleeping, and resting.

10.6 CHARADRIIFORMES

10.6.1 TERRITORIAL BEHAVIOR

10.6.1.1 Site Prospecting

Prior to breeding, individuals of all seabird species may visit potential breeding areas repeatedly to identify suitable sites, a behavior know as prospecting. A site may be visited for 1 to 2 years before breeding occurs. Physical features, absence of predators, proximity to food, as well as social relationships and presence of other species determine where a bird may nest, and Charadriiformes are no exception (see Buckley and Buckley 1980 for a review). The important physical features in nest-site selection include: visibility, protective vegetation or rocks, height above water, likelihood of flooding, distance to foraging area, and exposure to prevailing winds (e.g., Buckley and Buckley 1972a, b, Burger and Lesser 1978, Baird and Moe 1978, Baird 1994a, Gaston and Jones 1998). Besides physical or social preferences, habitat may be chosen for predator avoidance, with birds competing more for crowded sites in the center of colonies where mobbing and predator swamping may afford them higher reproductive success as well as longevity (Coulson 1968, Burger 1979, Pierrotti 1987).

Both intra- and interspecific competition for nest sites occurs, whether direct (involving agonistic behavior) or indirect (by earlier arrival; Hay 1974, Ashcroft 1976, Baird and Gould 1983, Gaston and Jones 1998). During early prospecting, males and females may arrive at different times, and many species spend time on the colony by day but roost off-site on land or at sea at night, weeks before nest-building begins. Birds whose mate does not return may not be able to hold a preferred location in the colony as they court a new mate (e.g., Coulson 1968) and may be displaced during this time.

Alcids may spend 2 to 3 days at the colony and then 2 to 3 days at sea during this prospecting phase and even into the nesting phase (Gaston and Jones 1998). Razor-billed Auk (*Alca torda*) arrive near their colonies 1 to 6 months prior to egg laying, and spend time there displaying and defending potential nest sites by day, returning to the sea at night (Nettleship and Birkhead 1985). In most species, attendance at the colony commences 2 to 4 weeks before first laying. Alcids often raft and display off the colony before entering it (Gaston and Jones 1998). Unlike other alcids and all larids, Rhinoceros (*Cerorhinca monocerata*) and Whiskered Auklets (*Aethia pygmaea*) prospect at night and thus not much is known about their behavior (Gaston and Jones 1998). Rhinoceros Auklets are completely nocturnal on colonies where Bald Eagles (*Haliaetus leucocepahlus*) are present, and diurnal where they are absent or where the auklets nest in caves (Gaston and Jones 1998). Kittlitz's and Marbled Murrelets (*Brachyramphus brevirostris* and *B. marmoratus*) are crepuscular. These nocturnal species display and call to each other during prospecting, unlike some alcids (puffins, murres) that have minimal nocturnal vocal displays.

The majority of seabirds return to their natal colony (philopatry), but some move to different colonies (e.g., Ashcroft 1976, Birkhead 1976, Gaston and Jones 1998). Larids which nest in unstable habitats are the most likely to change colony sites (e.g., many crested terns such as Caspian [*Sterna caspia*], Royal [*S. maxima*], Crested [*S. bergii*], Sandwich [*S. sandvicensis*]; Least Terns [*S. antillarum*]; and Black-billed Gulls [*Larus bulleri*]; Austin 1949, McNicholl 1975, Southern 1977, Baird 1991). In many species, immature birds may attend the colony after the start of incubation, although young birds generally do not remain long on the colony. Nonbreeding Least Auklets (*Aethia pusilla*) may remain in a colony throughout the entire breeding period and court other birds, sometimes forming pair bonds for the next breeding season (Jones 1989, Jones and Montgomerie 1991, 1992, Gaston and Jones 1998). Immature Atlantic Puffins (*Fratercula arctica*) may just inspect burrows and first-year birds usually only raft offshore (Gaston and Jones 1998). Many species of alcids and larids show marked natal and breeding philopatry.

10.6.1.2 Defense of Territory

Territorial aggression is used by males to claim a site and males and females to defend it, which increases reproductive success. Territorial defense usually is composed of two parts: (1) agonistic (aggressive, used to obtain and retain a territory) and (2) appeasement displays (used generally with a mate or potential mate). Aggressive displays have become ritualized in some cases, possibly to avoid physically damaging the birds involved (Veen 1977). Appeasement displays tend to mitigate aggression of the aggressive bird. There are more interactions and thus more stereotyped displays and aggression in denser colonies, or more in gregarious species such as in murres, auklets, some gulls, and terns (Burger 1984, Baird 1985, Pierotti 1987, Gaston and Jones 1998).

The amount of aggression may vary between sexes and among habitat types because preferred habitats are entered more often by prospecting birds, resulting in increased encounter rates (Baird 1977, Burger 1977, 1984, Pierotti 1987). In order to safeguard preferred sites, some species may return to a colony during the winter to attend nest sites. The types of agonistic displays used by seabirds at colonies can be dictated by the physical features of the colony itself, (e.g., its openness, amount of vegetation, steepness). For example, in the Black-legged Kittiwake (nests on shallow ledges with little room), there are more appeasement postures than in other seabirds, and they are simpler than in other gulls (e.g.,"beak hiding," "facing away"; Cullen 1957, Beer 1980). The same holds true for murres, which nest in crowded colonies and whose appeasement displays take up little room. "Nape-presenting" (a turning away and beak-hiding display) is an appeasement posture typical of birds with restricted nesting sites (Cullen 1957, Nelson 1978b).

Levels of aggression change throughout the breeding period in larids, with few species having higher levels of preincubation (e.g., Arctic Tern, *Sterna paradisaea*; Laughing Gull, *Larus atricilla*;

FIGURE 10.29 A Pomarine Jaeger pair defends their nesting territory against a bird that has approached too close, using postures (wings up) and calling (left hand bird is giving the long call). (From Nelson 1979.)

Black-headed Gull; Lesser Black-backed Gull, *L. fuscus*; Kelp Gull, *L. dominicanus*; Fordham 1964, Lemmetyinen 1971, MacRoberts and MacRoberts 1972, Hutson 1977, Burger 1984), or at hatching (Arctic Tern, Lemmetyinen 1971; Black Skimmer, *Rhynchops niger*, Burger and Gochfeld 1990; Common Tern, *Sterna hirundo*, Burger and Gochfeld 1991). Ritualized hostile behavior is assumed to have evolved by modification of an originally non-ritualized action. For example, the common hostile "upright" posture of gulls and skuas is but one component of the initial attack movement in a fight ("an intention movement"): the head is lifted to peck downward and the wings are out with the carpals lifted to be ready to beat the wings (Figure 10.29; Tinbergen 1952). Intention movements are preparatory movements for a certain behavior and reveal the underlying drive and predict what kind of action can be expected next (Tinbergen 1960).

Hostile displays of gulls, dark noddies, and large terns are very similar. The alcid "forward threat" (a higher intensity threat with gaping bill and lowered head) is similar to the gull "long call." Typical agonistic displays in gulls, skuas, and noddies involve holding the body upright, or forward, sometimes "choking" and "gaping," as well as redirected hostile displays such as "grass-pulling" (Tinbergen 1959a, b). The upright threat is an intention movement of fighting (Tinbergen 1960) and the grass-pulling display of gulls, skuas, or noddies can erupt into an overt fight with pulling of wings, tail, leg. Posturing and vocalization often occur simultaneously.

Species which differ morphologically from the large white-headed gulls (where these postures were first described) have correspondingly different behavior (e.g., skimmers; fork-tailed gulls; Sabine's Gull, *Xema sabini*; Swallow-tailed Gull, *Creagrus furcatus*; Silver Gulls, *Larus novae-hollandiae*; Grey-headed Gulls, *L. cirrocephalus*; White Terns, *Gygis alba*; skuas). Black-legged Kittiwakes have lost the "aggressive upright" probably as a result of their nesting habitat (small ledges). The hostile (and sexual) displays of skuas are simple and are more similar to those of the primitive hooded gulls than to the typical displays of the large white-headed gulls. Some noddies and skuas have more overt aggression than ritualized displays. Skimmers, morphologically quite different from terns and gulls, have very simple hostile displays.

Tern hostile displays are homologous to those of gulls. Overall, noddies have more gull-like displays than do other terns. Aggressive communication among alcids is less similar than that among larids, skuas, and noddies. Typical alcid hostile postures include "forward threats," "lunge pecks," "hunched with bill gaping," "skypointing," and "strutting." The "pelican walk" is unique to Atlantic Puffins: head tucked, body erect and stiff, the movements of the feet exaggerated. It is used in a variety of contexts, e.g., defense of burrow, after it lands, or as a precursor to sexual "billing" (see below). As in other seabirds, the intent and interpretation of the message depend upon the context in which it is sent.

Postlanding wing-up postures are very common and ritualized in many alcids (Danchin 1983). The closer a landing bird is to a conspecific, the longer it leaves its wings up (Danchin 1983). Species which do not have these displays nest only in loose colonies where ownership and advertising are not as important. In larids, vocalizations also differ the most among those that are the least phylogenetically related, especially those least related to the large white-headed gulls (Cullen 1957, Moynihan 1959). Uprights of primitive hooded gulls are usually silent, and in Fork-tailed Gulls even the alarm call is absent. Within the Family Alcidae, the auk–murre–guillemot–murrelet group has complex calls while the puffin-auklet group has simple calls (Tinbergen 1953, Jones et al. 1987, 1989, Gaston and Jones 1998). The larger and colonial-nesting species generally have louder vocalizations than do smaller species and those that nest more dispersed (Marbled Murrelets are an exception). Additionally, nocturnally active birds tend to have more complex calls, although there are exceptions to this, also (Gaston and Jones 1998).

Repeated neighbor-to-neighbor interactions eventually result in the waning of aggression (Shugart and Fitch 1987, Pierrotti 1987). In gulls and alcids, opponents may be dragged, and chases can occur over long distances (Tinbergen 1960, Gaston and Jones 1998). Jones (1993b) noted that in Least Auklets, winners of fights have lighter plumage, and in Crested Auklets (*Aethia cristatella*), dominance is correlated with crest size (Jones and Hunter 1999). Exaggerated movements and use of piloerectors which raise crests, whiskers, or eyebrows, or puffing up or sleeking feathers are common in charadriiforms and also in penguins, but these small motor movements are used more often in pairing than in aggression (see below). In Atlantic Puffins, the "gape" display, performed by both sexes in territory defense, where the neck feathers are erect and the bill open, reveals the yellow gape, usually not seen.

As in other seabirds, charadriiforms' appeasement postures are the opposite of aggressive ones, and their role is in conflict avoidance. Weapons, like bills, feet, and wings, are hidden or placed in a non-threatening way (e.g., beak held up or turned away, aggressive markings like masks turned away, wings close to the body). Appeasement walks are diametrically opposite to the hostile walking postures (Asbirk 1979, Evans 1981). In alcids the "low profile walk" is similar in hunched appearance to the horizontal posture or "chick walk" of gulls (Figure 10.30), and these are comparable to the "slender walk" in penguins. These postures mitigate aggression from any species, and since many colonies are mixed-species, they lessen overall aggression.

The "erect" or "sky-pointing" posture of Common and Arctic Terns is used as an appeasement posture at or near the nest and in pairing (Cullen 1956, Burger and Gochfeld 1991). This behavior is similar to the upright appeasement posture, with bill raised, of some alcids (e.g., Common Murres, *Uria Aalge*) used by prospectors in a colony. Some postures like "choking" in gulls and "sky pointing" in alcids and black-capped terns can occur during pairing, threat, and mitigation of aggression (Cullen 1956, Danchin 1983, Baird 1994b). Social facilitation (behavior which elicits

FIGURE 10.30 An Atlantic Puffin doing an appeasement walk as means of deflecting aggression. (From Harris 1984.)

similar behavior in neighbors) often follows aggressive–appeasement postures, especially in dense colonies, with neighbors entering into appeasement postures, also. Incubating Common Murres often turn away and flatten themselves against a cliff to avoid a fight (Gaston and Jones 1998). Even in charadriiforms, which are not highly colonial, this kind of appeasement may occur.

10.6.2 PAIR RELATIONSHIPS

Communication between mates and a well-developed pair bond are needed at the nest site to coordinate site defense, nest building, incubation, foraging for food, and for protection of young. Pairing behavior includes elements of aggression, and appeasement, as such, is more complex than typical territorial interactions. In larids, skuas, noddies, and alcids, the fundamental pair relationship is usually serial monogamy. Polygamy and extra-pair copulations occur at varying rates, depending on the species (rare in larids).

Season-long greeting ceremonies are common in alcids, skuas, and larids, some of which also show high mate fidelity (65 to 95%) over their lifetime (Coulson 1966, Gaston and Jones 1998). A longer-lasting pair bond improves breeding success in the black-legged kittiwake (Coulson 1966a, Coulson and Thomas 1980). Cooperative breeding is unknown in most species, although in California Least Terns, Roseate Terns (*Sterna dougallii*), skuas, and Black-legged Kittiwakes, three-bird nests are known and are sometimes more numerous in years of low food (Baird 1990, J. Chardine personal communication). Laridae have very similar sexual displays, in contrast to their hostile displays which tend to be different among species (Moynihan 1959).

10.6.2.1 Pair Formation

In a few species of alcids (e.g., Least Auklet, Crested Auklet, possibly others), mate choice may be based in part on facial ornaments (Jones and Montgomerie 1992, Jones and Hunter 1993). Ritualized group display, including male–male, occurs in some alcids (e.g., clubs of murres). Terns and gulls may have social gatherings in the precourtship phase but these are not group displays. For example, female Laughing Gulls may food-beg among a group of males, or in the case of terns, males offer fish to birds of either sex who act like females and beg (Noble and Wurm 1943, I. Nisbet personal communication). Courtship of most alcids can occur on land or in the water (1 to 3 km from or adjacent to the colony where large flocks gather; Williams 1982, Jones 1993a, b, Gaston and Jones 1998). Aquatic display is common in most of the alcids (Figure 10.31), which gather at sea to perform elaborate displays (Harris 1986).

Greeting ceremonies are performed when a pair meets after being separated or when they return to the nesting site. These postures, as in other courtship displays, have elements of aggression and appeasement. Greeting ceremonies can range from simple to highly ritualized (Baird 1994b). During greetings, as in pair formation, the female (larids, noddies, skuas) may walk around the male and may also display the appeasement "look away" (Noble and Wurm 1943, Tinbergen 1959a, b, Moynihan 1962). The male may react aggressively at first (Tinbergen 1960). Circular parading in terns is not a greeting but rather a behavior resulting from a conflict between the desires to approach a possible mate and avoidance of conflict (Cullen 1956).

All black-capped terns use begging calls (Moynihan 1959). The Common and Arctic Tern, and presumably other black-capped terns, have a "black-cap-hiding display" (the same as appeasement postures in hostile encounters; Cullen 1956). In contrast, the crested terns, morphologically very different from other black-capped terns, have more gull-like behaviors. White Terns, which build no nest and lay their single egg on a limb, have few display postures. Noddies have sexual postures and vocalizations which are very gull-like (Moynihan 1962). Chilean Skuas (*Catharacta chilensis*) have some of the most hostile intrapair behavior of any charadriid (Moynihan 1962), frequently interrupting a greeting or food-begging display or attacking a mate before or after copulation. While

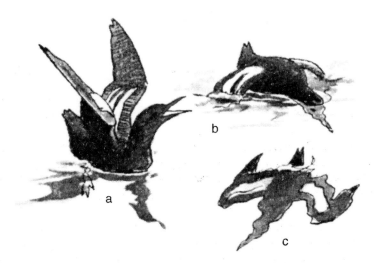

FIGURE 10.31 Pigeon Guillemots performing courtship displays at sea: (a) calling and wing-raising, (b) head dipping below the surface, and (c) flipping below the surface (skittering). (From Nelson 1979.)

most gulls redirect intermate hostility to a neighbor, Chilean Skuas simply attack their mate (Moynihan 1962).

Aerial courtship display and pair flights occur in many terns, noddies, skuas, and some alcids. Group flight displays occur in Tufted (*Fratercula cirrhata*) and Atlantic Puffins and in the sometimes chaotic yet synchronized social flight of the Crested Auklet (Taylor 1984, Jones 1993a). "Wheeling" (flying over the colony) may help puffins locate their burrows, or it may help synchronize colony occupation. It is common in other alcids and larids as a predator-swamping technique at a colony (Skokova 1962). Many larids posture as they fly, sometimes hunching and inclining the beak downward, and at other times, elongating the neck and pointing it skyward. Courting pairs of terns and dark noddies often fly very high (Common Tern, 165 to 300 m; Sandwich Terns, 300 to 1000 m; Hay 1974). Courtship flights of terns may include fish carrying by the male which then segues into courtship feeding once the pair has landed (see section on copulation below). Low flights, with many birds involved, most likely function in sex identification, indicating who is unmated (Cullen 1960).

10.6.2.2 Pair Maintenance

Vocalizations and ritualized displays including threat slowly decrease in frequency at the colony as the breeding season progresses, although nonbreeders may continue to display (Hudson 1979). Most alcids are vocally noisy with their mates. Duetting is common and can last up to 30 min in a bout (e.g., Ancient Murrelet, *Synthliboramphus antiquum*; Gaston and Jones 1998). Many species vocalize during nest relief. The nest-relief greeting helps maintain the pair bond.

The social signals of alcids are less ritualized and more diverse than are those of gulls. Only a few displays are common to all species, unlike gulls and penguins which have similar displays within their respective groups. One of their most intense pairing displays is "mutual fencing," used as a greeting, elicited by an aggressive encounter, or when a nearby bird returns to its site with fish. Displays may or not include vocalizations and have different degrees of bill contact (Evans 1981, Nettleship and Birkhead 1985). During the "billing" display of pair maintenance (Figure 10.32), both partners may walk around on land or, when at sea, swim in circles. Neighbors often approach the displaying pair, even attempting to join in. Mutual bowing or some kind of "head-down" behavior is also common.

FIGURE 10.32 An Atlantic Puffin pair courtship billing. (From Harris 1984.)

10.6.2.3 Copulation

Courtship displays and courtship feeding in gulls and terns are frequently followed by copulation. The male begins with a precopulatory display and the female performs nonhostile, appeasement behavior to the male while they are in close proximity, frequently assuming a chick-like horizontal posture, often food begging when she is receptive. The male, to show nonaggression, often will also become horizontal. Copulation is usually preceded by head bobbing (except in most terns) which is derived from food begging (Tinbergen 1959a.) Female food-begging behavior may serve to decrease aggression or fear in the male, or signal subordination. The male regurgitates or presents fish to the female (Figure 10.33). Courtship feeding is widespread in skuas (Stonehouse 1956, Burton 1968, Andersson 1971), gulls (Lack 1940, Tinbergen 1953, Brown 1967), and terns (Cullen and Ashmole 1963, Nisbet 1973), and probably serves to help the females determine mate quality (Nisbet 1972, 1973). The Chilean Skua, consistent with its repertoire of simplified or perhaps primitive behaviors, does not courtship feed.

In alcids, copulation usually takes place at sea although some species also copulate on land (e.g., Least Auklet, Dovekie [*Alle alle*], murres, guillemot; Gaston and Jones 1998). Copulation in

FIGURE 10.33 A female tern begs from a male in a submissive, chick-like posture during courtship. The male offers a fish to the female. (From Nelson 1979.)

larids generally takes place at the colony. Social facilitation in copulation is common in large groups of birds at the colony.

10.6.2.4 Nest Building

Alcids usually nest in crevices (e.g., Pigeon Guillemot, *Cepphus columba*; auklets), trees (e.g., Marbled Murrelet), burrows (e.g., Tufted Puffins), or on rock ledges with little or no nest material (e.g., murres). Larids nest on bare substrate (e.g., many gulls and terns), on grassy areas or slopes (e.g., Glaucous-winged Gulls, *Larus glaucescens*), in trees (e.g., Bonaparte's Gull, *L. philadelphia*; White Tern), on cliffs (e.g., Black-legged Kittiwake), or on artificial nesting areas such as rooftops (Least Tern, White Tern, Black-legged Kittiwakes). Gathering nest material and building and defending the nest all help to strengthen the pair bond (Tinbergen 1959a, b) and protect the nest against the elements (Burger 1978, Burger and Gochfeld 1991).

Nests in larids range from simple scrapes on the sand or rocks (many terns and gulls) and simply constructed nests (e.g., California, *Larus californicus*; Ring-billed Gulls, *L. delawarensis*), to elaborate mud-vegetation nests on cliffs (kittiwakes). Some burrowing alcids build simple nests or line their burrows with feathers. Some species decorate their scrapes with shells or other items found at the colony. Nest building or passing of, arranging, or presenting nest material or feathers is incorporated in courtship rituals of many alcids and larids. Manipulation of nest material, dropped food, or objects around the nest, such as pebbles and feathers, is common.

10.6.3 INCUBATION AND CARE OF YOUNG

10.6.3.1 Nonfeeding Displays at the Nest

As in other seabirds, successful breeding by larids and alcids requires efficient communication between offspring and parents. When potential danger threatens, alarm calls are given by adults to protect chicks. Chicks respond, depending on the species, by crouching in place (particularly ledge-nesting or tree-nesting species: e.g., Black-legged Kittiwakes; some Swallow-tailed Gulls; Black Noddies, *Anous minutus*), moving to a nearby hiding place (ground-nesting species: e.g., Tufted Puffins, Glaucous-winged Gulls), or running (e.g., Elegant and Least Terns; Cullen and Ashmole 1963, Hailman 1965, Baird and Gould 1983, Baird et al. 1998). Small larid chicks which stray from their nest sites are often killed by other birds. If chicks who violate territorial boundaries crouch instead of exhibiting escape behavior, they may have a greater chance of reducing attack by the adult. Familiarity among neighbors helps ameliorate aggression and, therefore, threats to chicks (Shugart and Fitch 1987).

Parent and chick recognition of each other's voice is widespread in Larids with obvious benefit (Tschanz 1968, Evans 1970a, b, Stevenson et al. 1970, Buckley and Buckley 1972b, Dinsmore 1972, Ingold 1973, Hahn 1977, Shugart 1977, Howell 1978, Burger 1980). However, in the Black-legged Kittiwake (cliff nester) adults do not recognize chicks fewer than 4 weeks old (Cullen 1957), and perhaps that is why foster chicks are accepted (Pierotti and Murphy 1987). Wandering chicks may sometimes be adopted (Holley 1981, 1984).

The threat of predation and decreased foraging competition are hypothesized to be the reasons for the intermediate-age nest departure of semi-precocial Common Murre, Thick-billed Murre (*Uria lomvia*), and Razor-billed Auk chicks (15 to 30 days of age) and the early nest departure of precocial chicks (*Synthliboramphus* and *Endomychura* murrelets, 1 to 3 days; Gaston and Jones 1998). In both of these alcid groups, voice recognition is essential as chicks follow the male to sea (Tschanz 1968, Jones et al. 1987). Puffins nest in burrows (therefore with low predation risk) and do not recognize their chicks. They feed any chick that is placed in their burrow (Harris 1983) and do not accompany it to sea.

FIGURE 10.34 An adult tern presents a small fish to its chick. (From Nelson 1979.)

10.6.3.2 Communication during Chick Provisioning

The type of nest site as well as the method by which parents bring food back to their young (internal or external) may shape the type of communication between adults and chicks at food delivery. Many open nesting species call to the chick as they fly in with food and are answered by the chick. Burrow nesters tend to be quiet, except at the burrow terminus. For example, Atlantic Puffins make soft clicking calls once in the burrow, which alert the chick to presence of a meal.

Chicks display an appeasement posture while food begging, similar to precopulatory food-begging posture used by females. Food, regurgitated or carried in the bill, is taken from the adult's bill, but may be dropped in front of the chick (Figure 10.34; Corkhill 1973, Baird 1986). Gulls, some terns, skuas, jaegers, auklets, and Dovekies bring back food internally. Other alcids and many terns are external transporters, holding food in the bill. Gulls carry food back to chicks in the proventriculus and chicks peck at the bill to elicit regurgitation. The red, orange, striped, or spotted bills of gulls and terns serve as releasers for chick pecking (Figure 10.35; Tinbergen 1960).

10.6.3.3 Predators

Communication is important to avoid predators. There are different gull calls for aerial vs. ground predators, and these calls usually elicit the mobbing response. Gulls alternate between the "charge" call and the "alarm" call (Tinbergen 1953). The alarm call expresses the conflicting desires of flight

FIGURE 10.35 Herring Gulls, showing plumage at three ages: small downy chick, fledgling (chick at the time it first begins flying), and adult plumage. The red tip on the bill of the adult is a releaser for chick pecking to elicit adult feeding. (From Nelson 1979.)

and nest defense. Ground predators usually elicit much mobbing; aerial raptor predators (e.g., eagles, kestrels, falcons) can elicit temporary abandonment of the colony (e.g., Black-legged Kittiwakes, Least Terns; Baird et al. 1998). Mobbing, when it occurs, may be accompanied by alarm calls or may be silent. Most alcids and larids are diurnal. However, members of each are active at night, and nocturnal entrance onto colonies is one means of predator avoidance (see discussion earlier). Nocturnality in some species of alcids may have produced a greater emphasis on vocalizations than on postures (Jones et al. 1989).

10.6.4 BEHAVIOR OUTSIDE BREEDING

As in some other species, some charadriiforms remain together, either in a loose flock or in pairs or family groups, during the nonbreeding season. Since nomadism offers time and energy savings, and safety from land-based dangers, many larids and alcids are nomadic/oceanic, or nomadic/coastal, becoming more widely distributed in winter. The more oceanic larids, e.g., kittiwakes, Arctic, and Aleutian Terns (*Sterna aleutica*), tend to spread out pelagically although larger gulls often group together for foraging and roosting on beaches. In contrast to pelecaniforms, these groups are not large.

10.7 CONCLUSIONS

Seabirds are notable in having evolved complex ritualized communication in the service of establishing a site, pair formation, and pair bonding. This communication employs elaborate interactions; aerial, ground, and aquatic display, striking visual adornments, and mechanical and vocal sounds. Moreover, in many seabirds, elaborate ceremonies attend change-over on egg and chick. These ceremonies communicate the "intentions" and "moods" of the partners, thereby facilitating cooperation.

Seabirds are remarkable in their extreme development of pair-bonding behavior. Other avian groups have evolved complex, colorful, and innovative behavior to serve pair formation, but seabirds are exceptional in their face-to-face billing, complex displays, including ritualized head postures, modified locomotion, and prolonged mutual preening. They have a strong tendency to form a near-permanent nest site and pair bond throughout their long life and to nest in colonies. The latter may lead to strong competition for a physically or socially adequate site, which may favor the permanent nest site and pair bond, which in turn are served by elaborate behavior.

While it is axiomatic that eggs need fertilizing and therefore that sexual contact is necessary, and that all seabirds require space in which to deposit eggs and incubate them, it is by no means obvious why some species have evolved complex communication behavior while others have not. This kind of variability lies at the heart of adaptedness. If one assumes that evolution is nondirected, such variability is inevitable and is not susceptible to post-hoc rationalization. Currently "whole bird" etho-ecological studies tend to be neglected in favor of laboratory-based molecular genetics and physiological work. Ideally these latter aspects of seabird biological study would proceed alongside holistic field studies to provide a fuller picture of the reasons for the evolution of the particular life-history characteristics of seabirds. The study of evolution, adaptation, and the relationship of species with their environment leads not only to an understanding of the species, but provides the information needed for wise conservation management. Long-term studies of these long-lived species, particularly comparative studies among related groups, are still badly needed to provide a more complete understanding of their biology.

LITERATURE CITED

AEBISCHER, N. J. 1995. Philopatry and colony fidelity of shags *Phalacrocorax aristotelis* on the east coast of Britain. Ibis 137: 11–18.

AINLEY, D. 1975a. Development and reproductive maturity in Adélie Penguins *in* The Biology of Penguins (B. Stonehouse, Ed.). Macmillan, London.

AINLEY, D. 1975b. Flocking in Adélie penguins. Ibis 114: 388–390.

AINLEY, D., R. LE RESCHE, AND W. SLADEN. 1983. Breeding Biology of the Adélie Penguin. University of California Press, Berkeley.

ANDERSON, D. J. 1989. The role of hatching asynchrony in siblicidal brood reduction of two booby species. Behavioral Ecology and Sociobiology 25: 363–368.

ANDERSON, D. J. 1990a. Evolution of obligate siblicide in boobies. 1. A test of the insurance-egg hypothesis. American Naturalist 135: 334–350.

ANDERSON, D. J. 1990b. Evolution of obligate siblicide in boobies. 2. Food limitation and parent-offspring conflict. Evolution 44: 2069–2082.

ANDERSON, D. J., AND R. E. RICKLEFS. 1992. Brood size and food provisioning in masked and Blue-footed Boobies (*Sula* spp.) Ecology 73: 1363–1374.

ANDERSSON, M. 1971. Breeding behavior of the Long-tailed Skua *Stercorarius longicaudus* (Viellot). Ornis Scandinavica 2: 35.

ASBIRK, S. 1979. Behavior of the Black Guillemot. Dansk Ornithologisk Forenings Tidsskrift 73: 287–296.

ASHCROFT, R. E. 1976. Breeding Biology and Survival of Puffins. Ph.D. thesis, Oxford University, England.

AUSTIN, O. 1949. Site Tenacity and Behavior Traits of the Common Tern (*Sterna hirundo* L.) Bird Banding 20: 1–39.

BAIRD, P. 1977. Comparative Ecology of California and Ring-billed Gulls, *Larus californicus* and *L. delawarensis*. Ph.D. dissertation, University of Montana, Missoula.

BAIRD, P. 1986. A new method for collecting prey delivered to Tufted Puffin chicks. Wilson Bulletin 98: 169–170.

BAIRD, P. 1991. Seabirds as indicators of the oceanic environment. Pp. 91–104 *in* Perspectives on the Marine Environment (P. Grifman and S. Yoder, Eds.). Proceedings from a Symposium of the 100th Anniversary of the Southern California Academy of Sciences. Sea Grant Program. Publication USCG TR-01-92, University of Southern California, Los Angeles.

BAIRD, P. 1994a. Habitat parameters and nesting success of Black-legged Kittiwakes, *Rissa tridactyla*. Proceedings of the Xth International Congress of Ornithology, Vienna, Austria.

BAIRD, P. 1994b. Black-legged Kittiwake (*Rissa tridactyla*). No. 92 *in* The Birds of North America (A. Poole and F. Gill, Eds.). Academy of Natural Sciences, Philadelphia; American Ornithologists' Union, Washington, D.C.

BAIRD, P., AND A. MOE, 1978. The breeding biology and feeding ecology of marine birds in the Sitkalidak Strait area, Kodiak Island, 1977. Pp. 313–524 *in* Environmental Assessment of the Alaska Continental Shelf. Annual Report of Principal Investigators, Vol. 3, NOAA Environmental Research Laboratory, Boulder, CO.

BAIRD, P., AND P. GOULD (Eds.). 1986. The breeding biology and feeding ecology of marine birds in the Gulf of Alaska. USDC NOAA OCSEAP Final Report 45 (1986): 121–505.

BAIRD, P., S. HINK, AND D. ROBINETTE, 1998. Monitoring of the Least Tern Colony during the X Games in Mission Bay Park, 1998. Final Report submitted to ESPN, San Francisco, CA.

BEER, C. 1980. The communication behaviour of gulls and other seabirds. Pp. 169–205 *in* Behaviour of Marine Animals, Vol. 4 (J. Burger, B. L. Olla, and H. E. Winn, Eds.). Plenum Press, New York.

BIRKHEAD, T. 1976. Breeding biology and survival of Guillemots *Uria aalge*. Ph.D. thesis, University of Oxford, p. 204.

BOOST, C., AND P. JOUVENTIN. 1990. The breeding performance of the Gentoo Penguin, *Pygoscelis papua*, at the northern edge of its range. Ibis 133: 14–25.

BRETAGNOLLE, V. 1990. Behavioural affinities of the Blue Petrel *Halobaena caerulea*. Ibis 132: 102–105.

BRONI, S. 1985. Social and spatial foraging patterns of the Jackass Penguin, *Spheniscus demersus*. South African Journal of Zoology 20: 241–245.

BROWN, D. A. 1966. Breeding biology of the Snow Petrel *Pagodroma nivea* (Forster). Australian National Antarctic Research Expedition Science Report. B(1) Zoology 89: 1–63.

BROWN, L. H., AND E. X. URBAN. 1969. The breeding biology of the Great White Pelican *Pelecanus onocrotalus* roseus at Lake Shala, Ethiopia. Ibis 111: 199–237.

BROWN, L. H., E. X. URBAN, AND K. NEWMAN. 1982. The Birds of Africa, Vol. l. Academic Press, London.

BROWN, R. G. B. 1967. Courtship behavior in Lesser Black-backed Gulls. Behavior 29: 122–153.

BUCKLEY, F., AND P. BUCKLEY. 1972a. The breeding ecology of Royal Terns (*Thalasseus maxima maxima*). Ibis 114: 344–359.

BUCKLEY, F., AND P. BUCKLEY. 1972b. Individual egg and chick recognition by adult Royal Terns (*Sterna maxima maxima*). Animal Behaviour 20: 457–462.

BUCKLEY, F., AND P. BUCKLEY. 1980. Habitat selection in marine birds. Pp. 69–113 *in* Behavior of Marine Animals, Vol. 4 (J. Burger, B. Olla, and H. Winn, Eds.). Plenum Press, New York.

BURGER, J. 1977. The role of visibility in the nesting behavior of five species of *Larus* gulls. Journal of Comparative Physiology and Psychology 9: 1347–1351.

BURGER, J. 1978. Determinants of nest repair in Laughing Gulls. Animal Behaviour 26: 856–861.

BURGER, J. 1979. Competition and predation: Herring Gulls versus Laughing Gulls. Condor 81: 269–277.

BURGER, J. 1980. The transition to independence and post-fledging parental care in seabirds. Pp. 367–448 *in* Behavior of Marine Animals, Vol. 4 (J. Burger, B. L. Olla, and H. E. Winn, Eds.). Plenum Press, New York.

BURGER, J. 1984. Pattern, mechanism and adaptive significance of territoriality in Herring Gulls. Ornithological Monographs No. 34. American Ornithologists' Union, Washington, D.C.

BURGER, J., AND M. GOCHFELD. 1990. The Black Skimmer: Social Dynamics of a Colonial Species. Columbia University Press, New York.

BURGER, J., AND M. GOCHFELD. 1991. The Common Tern. Columbia University Press, New York.

BURGER J., AND M. GOCHFELD. 1993. Lead and behavioral development in young Herring Gulls: effects of timing individual. Fundamental and Applied Toxicology 21:187–195.

BURGER, J., AND F. LESSER. 1978. Selection of colony sites and nest sites by Common Terns Sterna hirundo in Ocean County, New Jersey. Ibis 120: 443–449.

BURKE, V. E. M., AND L. J. BROWN. 1970. Observations on the breeding of the Pink-backed Pelican *Pelecanus rufescens*. Ibis 112: 499–512.

BURTON, R. 1968. Breeding biology of the Brown Skua *Catharacta skua lonnbergi* (Mathews) at Signy Island, South Orkney Islands. British Antarctic Survey Bulletin 15: 9.

CASH, K. J., AND R. M. EVANS. 1987. The occurrence, context and functional significance of aggressive begging behaviours in young American White Pelicans. Behaviour 102: 119–128.

CONROY, J. W. H. 1972. Ecological aspects of the histology of the Giant Petrel, *Macronectes giganteus* (Guelin). *In* the Maritime Antarctic. British Antarctic Survey Science Report No. 75, 74 pp.

CONROY, J., O. DARLING, AND H. SMITH. 1975. The annual cycle of the Chinstrap Penguin, *Pygoscelis antarctica*, on Signey Islands, South Orkney Islands. Pp. 353–362 *in* The Biology of Penguins (B. Stonehouse, Ed.). University Park Press, London.

COOPER, J. 1980. Fatal sibling aggression in pelicans — a review. Ostrich 51: 183– 86.

CORKHILL, P. 1973. Food and feeding ecology of puffins. Bird Study 20: 207–220.

COULSON, J. 1966. The influence of the pair-bond and age on the breeding biology of the Kittiwake Gull. Journal of Animal Ecology 35: 269–279.

COULSON, J. 1968. Differences in the quality of birds nesting in the centre and on the edges of a colony. Nature 217: 478–479.

COULSON, J., AND C. THOMAS. 1980. A study of the factors influencing the pair-bond in the kittiwake gull. Proceedings of International Ornithological Congress 17: 823–233.

CRIVELLI, A. J. 1987. The ecology and behaviour of the Dalmatian pelican, *Pelecanus crispus* Bruch, a world-endangered species. Internal Report, Station Biologique de la Tour du Valat, France.

CULLEN, J. 1956. A Study of the Behavior of the Arctic Tern (*Sterna macrura*). Thesis, Oxford University, England.

CULLEN, E. 1957. Adaptation of kittiwakes to cliff nesting. Ibis 99: 275–302.

CULLEN, J. 1960. Some adaptations in the nesting behavior of terns. Pp. 153–157 *in* Proceedings of the 12th International Congress of Ornithology, Helsinki.

CULLEN, J. M., AND N. P. ASHMOLE. 1963. The Black Noddy, *Anous tenurostris,* on Ascension Island. Part 2. Behaviour. Ibis 103: 423–446.

DANCHIN, E. 1983. La posture de post-attervissage chez le macareux marine, *Fratercula arctica* [The post-bowing posture in the Atlantic Puffin]. Biology of Behavior 8: 3–10.

DAVIS, L. 1982. Creching behavior of Adélie Penguin chicks (*Pygoscelis adeliae*). New Zealand Journal of Zoology 9: 279–286.

DAVIS, L., AND E. SPEIRS. 1990. Mate choice in penguins. Pp. 377–397 *in* Penguin Biology (L. Davis and J. Darby, Eds.). Academic Press, San Diego.

DAVIS, P. 1957. The breeding of the Storm Petrel. British Birds 50: 85–101; 371–384.

DIAMOND, A. W. 1975. The biology and behaviour of frigatebirds *Fregata* spp. on Aldabra Atoll. Ibis 117: 302–323.

DIN, N.A. AND K. ELTRINGHAM. 1974. Breeding of the pink-backed pelican *Pelecanus rufescens* in Rwenzori National Park, Uganda. Ibis 116: 477–493.

DINSMORE, J. 1972. Sooty tern behavior. Bulletin of Florida State Museum of Biological Science 16: 129–179.

DRUMMOND, H. 1987. A review of parent-offspring conflict and brood reduction in the Pelecaniformes. Colonial Waterbirds 10: 1–15.

DRUMMOND, H. 1989. Parent-offspring conflict and siblicidal brood reduction in boobies. Pp. 1244–1253 *in* Proceedings of the XIX International Ornithological Congress.

DRUMMOND, H., AND C. G. CHAVELAS. 1989. Food shortage influences sibling aggression in the Blue-footed Booby. Animal Behaviour 37: 806–819.

DRUMMOND, H., E. GONZALEZ, AND J. L. OSORNO. 1986. Parent-offspring cooperation in the blue-footed booby (*Sula nebouxii*): social roles in infanticidal brood reduction. Behavioral Ecology and Sociobiology 19: 365–372.

DRUMMOND, H., J. L. OSORNO, R. TORRES, C. G. CHAVELAS, AND H. M. LARIOS. 1991. Sexual size dimorphism and sibling competition: implications for avian sex ratios. American Naturalist 138: 623–641.

EVANS, R. M. 1970a. Imprinting and mobility in young Ring-billed Gulls, *Larus delawarensis*. Animal Behaviour Monographs 3: 193–248.

EVANS, R. M. 1970b. Parental recognition and mew calling in Black-billed Gulls. Auk 87: 503–513.

EVANS, P. G. 1981. Ecology and behavior of Little Auk. Ibis 123: 1–18.

EVANS, R. M., AND F. L. KNOPF. 1993. American White Pelican (*Pelecanus erythrorhynchos*). No. 57 *in* The Birds of North America (A. Poole and F. Gill, Eds.). Academy of Natural Sciences, Philadelphia; American Ornithologists' Union, Washington, D.C.

FINERAN, B. 1964. An outline of the vegetation of the Snares Islands. Transactions of the Royal Society of New Zealand (Botany) 17: 229–235.

FISHER, H. I., AND M. L. FISHER. 1969. The visits of Laysan Albatrosses to the breeding colony. Micronesica 5: 173–201.

FLEET, R. R. 1974. The Red-tailed Tropicbird on Kure Atoll. American Ornithologists' Union Monographs 16: 1–64

FORDHAM, R. 1964. Breeding biology of the Southern Black-backed Gull. I. Pre-egg and egg stage. Notornis 11: 3–34.

FOWLER, J. A., J. D. OKILL, AND B. MARSHALL. 1982. A retrap analysis of Storm Petrels tape-lured in Shetland. Ringing Migration 4: 1–7.

GASTON, A., AND I. JONES. 1998. The Auks. Birds of the World. Oxford University Press, Oxford.

GRAVES, J. A., AND J. ORTEGA-RUANO. 1994. Patterns of interaction in the courtship behaviour of shags (*Phalacrocorax aristotelis*). Ecologia 4: 1–9.

GRAVES, J., R. T. HAY, M. SCALLAN, AND S. ROWE. 1992. Extra-pair paternity in the shag *Phalacrocorax aristotelis* as determined by DNA fingerprinting. Journal of the Zoological Society of London 226: 399–408.

GRAVES, J., J. ORTEGA-RUANO, AND R. J. B. SLATER. 1993. Extra-pair copulations and paternity in shags: do females choose better males? Proceedings of the Royal Society of London 253: 3–7.

GREEN, K. 1997. Biology of the Heard Island Shag *Phalacrocorax nivalis*. 1. Breeding behaviour. Emu 97: 60–66.

HAHN, D. 1977. Parent-offspring relations in the Laughing Gull (*Larus atricilla*). American Ornithologists' Union Abstracts No. 115.

HAILMAN, J. P. 1965. Cliff nesting adaptation in the Galapagos Swallow-tailed Gull. Wilson Bulletin 77: 346–362.

HAMER, K. C., AND D. R. THOMPSON 1997. Provisioning and growth rates of nestling fulmars *Fulmarus glacialis*: stochastic variation or regulation? Ibis 139: 31–39.

HARRIS, M. P. 1969. The biology of Storm Petrels in the Galapagos Islands. Proceedings of the California Academy of Science (4th Series) 37: 95–166.

HARRIS, M. P. 1979. Population dynamics of the Flightless Cormorant (*Nannopterum harrisi*). Ibis 121: 135–146.

HARRIS, M. P. 1983. Parent-young communication in the puffin, *Fratercula arctica*. Ibis 125: 109–114.

HATZILACOU, D. 1992. The breeding biology and the feeding ecology of the great white pelican *Pelecanus onocrotalus* L. 1758, at Lake Mikri Prespa (North Western Greece) Ph.D. dissertation. University of Athens, 183 pp. (in Greek).

HAY, J. 1974. Spirit of Survival. E. P. Dutton, New York.

HOLLEY, A. 1981. Naturally missing adoption in the Herring Gull. Animal Behaviour 29: 302–303.

HOLLEY, A. 1984. Adoption, parent-chick recognition and maladaption in the Herring Gull (*Larus argentatus*). Zeitschrift für Tierpsychologie 64: 9–14.

HOWELL, T. 1978. Ecology and reproductive behavior of the Grey Gull of Chile and of the Red-tailed tropicbird and White Tern of Midway Island. National Geography Society Research Report 1969: 251–284.

HUDSON, P. 1979. Survival Rates and Behavior of British Auks. Ph.D. thesis, Oxford University.

HUNTER, F. M., T. BURKE, AND S. E. WATTS. 1992. Frequent copulation as a method of paternity assurance in the Northern Fulmar. Animal Behaviour 44: 149–156.

HUTSON, G. 1977. Agonistic displays and spacing in the Black-headed Gull, *Larus ridibundus*. Animal Behaviour 25: 763–773.

INGOLD, P. 1973. Zur lautlichen Bezietung des Eleters zu seinem Kueken bei Tordalken (*Alca torda*). Behaviour 45: 154–190.

JOHNSGARD, P. A. 1993. Cormorants, Darters and Pelicans of the World. Smithsonian Institution Press, Washington, D.C.

JONES, I. L. 1989. Status Signaling, Sexual Selection and the Social Signals of Least Auklet (*Aethia pusilla*). Ph.D. thesis, Queen's University, Kingston, Ontario, Canada.

JONES, I. L. 1993a. Crested Auklet (*Aethia cristatella*). No. 70 *in* The Birds of North America (A. Poole and F. Gill, Eds.). Academy of Natural Sciences, Philadelphia; American Onithologists' Union, Washington, D.C.

JONES, I. L. 1993b. Least auklet (*Athia pusilla*). No. 69 *in* The Birds of North America (A. Poole and F. Gill, Eds.). Academy of Natural Sciences, Philadelphia; American Ornithologists' Union, Washington, D.C.

JONES, I. L., AND F. M. HUNTER. 1993. Experimental evidence for mutual sexual selection in a monogamous seabird. Nature 362: 238–239.

JONES, I. L., AND F. M. HUNTER. 1999. Experimental evidence for mutual inter- and intrasexual selection favouring a crested auklet ornament. Animal Behaviour 57: 521–528.

JONES, I. L., AND R. D. MONTGOMERIE. 1991. Mating and re-mating patterns of least auklets (*Aethia pusilla*) in relation to plumage and ornamentation. Behavioural Ecology 2: 249–257.

JONES, I. L., AND R. D. MONTGOMERIE. 1992. Least Auklet ornaments: do they function as quality indicators? Behavioral Ecology and Sociobiology 30: 43–52.

JONES, I. L., J. B. FALLS, AND A. J. GASTON. 1987. Vocal recognition between parents and young of Ancient Murrelets (*Synthliboramphus antiquus*). Animal Behaviour 35: 1405–1415.

JONES, I. L., J. B. FALLS, AND A. GASTON. 1989. The vocal repertoire of the Ancient Murrelet. Condor 91: 699–710.

JOUVENTIN, P. 1982. Visual and Vocal Signals in Penguins: Their Evolution and Adaptive Characters. Parey, Berlin.

JOUVENTIN, P., G. D. E. MONICAULT, AND J. M. BLOSSEVILLE. 1981. La danse de l'Albatross *Phoebetria fusca*. Behaviour 78: 43–80.

KHARITONOV, S. R., AND D. SIEGEL-CAUSEY. 1988. Colony formation in seabirds. Current Ornithology 5: 223–272.

KNOPF, F. L. 1979. Spatial and temporal aspects of colonial nesting of white pelicans. Condor 81: 353–363.

KORTLANDT, A. 1940. Eine Übersicht der angeborenen Verhaltungsweisen des Mittel-Europaischen Kormorans (*Phalacrocorax carbo sinensis* (Shaw and Nodd 1)), Funktion, Ontogenetische Entwicklung and Phylogenatische Herkunft. Extrait des Archives Neerlandaises de Zoologie 4: 401–442.

KOSHELEV, A., B. CHABAN, AND R. POKUSA. 1997. Seasonal distribution and mortality of cormorants from the northern Azov Sea. Suppl. Ricerche di Biologia della Selvaggina 26: 153–157.

LACK, D. 1940. Courtship Feeding in Birds. Auk 57: 169–178.

LEMMETYIEN, R. 1971. Nest defense behavior of Common and Arctic Terns and its effects in the success achieved by predators. Ornis Fennica 48: 13–24.

LERESCHE, R., AND W. SLADEN. 1970. Establishment of pair and breeding site bonds by young known-age Adélie Penguins (*Pygoscelis adeliae*). Animal Behavior 18: 517–526.

LOFGREN, L. 1984. Ocean Birds: Their Breeding Biology and Behavior. Groom Helm, London.

MACROBERTS, M., AND B. MacROBERTS. 1972. Social stimulation of reproduction in Herring and Lesser Black-backed Gulls. Ibis 114: 495–506.

MARCHANT, S., AND P. J. HIGGINS. 1990. Handbook of Australian, New Zealand and Antarctic Birds. Oxford University Press, Melbourne.

MCNICHOLL, M. 1975. Larid site tenacity and group adherence in relation to habitat. Auk 92: 98–104.

MOCK, D. W. 1984. Infanticide, siblicide and avian nestling mortality. Pp. 3–30 *in* Infanticide: Comparative and Evolutionary Perspectives (G. O. Hausfater and S. B. Hardy, Ed.). Aldine Publishing Company, New York.

MOUGIN, J. L. 1967. Etude ecologique des deux especes de fulmars le Fulmar Atlantique (*F. glacialis*) et le Fulmar Antarctique (*F. glacialoides*). Oiseau Rapport de Francais Ornithologiste 37: 57–103.

MOYNIHAN, M. 1959. A revision of the Family Laridae (Aves). American Museum Novitates 1928: 1–42.

MOYNIHAN, M. 1962. Hostile and sexual behavior patterns of South American and Pacific Laridae. Behaviour Supplement No. 8. E. J. Brill, Leiden.

NAGULU, V. 1984. Biology of Spot-Billed Pelican, *Pelecanus philippensis*, at Nelapattu, India. Ph.D. thesis.

NELSON, J. B. 1965. The behavior of the gannet. British Birds 58: 233–288; 313–336.

NELSON, J. B. 1966. The breeding biology of the gannet *Sula bassana* on the Bass Rock, Scotland. Ibis 108: 584–626.

NELSON, J. B. 1967. Breeding behavior of the white booby *Sula dactylatra*. Ibis 109: 194–231.

NELSON, J. B. 1968. Galapagos: Islands of Birds. Longman, Green, London.

NELSON, J. B. 1970. The relationship between behavior and ecology in the Sulidae with reference to other seabirds. Oceanographic and Marine Biology Annual Review 8: 501–574.

NELSON, J. B. 1971. The biology of Abbott's Booby *Sula abbotti*. Ibis 113: 429–67.

NELSON, J. B. 1976. The breeding biology of frigatebirds — a comparative review. Living Bird 14: 113–155.

NELSON, J. B. 1978a. The Gannet. T. and A. D. Poyser, Berkhamstead.

NELSON, J. B. 1978b. The Sulidae: Gannets and Boobies. Oxford University Press, London, 1012 pp.

NELSON, J. B. 1979. Seabirds: Their Biology and Ecology. Hamlyn's, London.

NELSON, J. B. 1983. Contrasts in breeding strategies between some tropical and temperate marine pelecaniformes. Studies in Avian Biology 8: 95–114.

NELSON, J. B. 1989. Problems in seabird social behavior. Colonial Waterbirds 12: 13.

NISBET, I. 1972. Courtship feeding and clutch size in Common Terns, *Sterna hirundo*. Pp. 101–109 *in* Evolutionary Ecology (B. Stonehouse and C. Perrins, Eds.). University Park Press, Baltimore.

NISBET, I. 1973. Courtship feeding, egg-size and breeding success in Common Terns, *Sterna hirundo*. Nature 241: 141.

NETTLESHIP, D. N., AND T. BIRKHEAD. 1985. The Atlantic Alcidae. Academic Press, London.

NOBLE, G., AND M. WURM. 1943. The social behavior of the Laughing Gull. Annals of the New York Academy of Science 45: 179–220.

OELKE, H. 1975. Breeding behavior and success in a colony of Adélie Penguins *Pygoscelis adeliae* at Cape Crozier, Antarctica. Pp. 363–396 *in* The Biology of Penguins (B. Stonehouse, Ed.). Macmillan, London.

O'HARA, R. 1989. An estimate of the phylogeny of the living penguins. American Zoologist 29, 11A.

PENNEY, R. L. 1968. Territorial and social behavior in the Adélie Penguin. Antarctic Research Series 12: 83–131 (O. L. Austin Jr., Ed.) American Geophysical Union Publication No. 1686.

PIEROTTI, R. 1987. Behavioral consequences of habitat selection in the Herring Gull. Pp. 119–129 in Ecology and Behavior of Gulls (J. Hand, W. Southern, and K. Vermeer, Eds.). Studies in Avian Biology No. 10. Cooper Ornithological Society.

PIEROTTI, R., AND E. MURPHY. 1987. International conflicts in gull. Animal Behaviour 35: 435–444.

POTTS, G. R. 1967. The influence of eruptive movements, age population size and other factors on the survival of the shag (Phalacrocorax aristotelis L.). Journal Animal Ecology 38: 53–102.

REILLY, P., AND P. BALMFORD. 1975. A breeding study of the Little Penguin, Eudyptula minor. Pp. 161–187 in The Biology of Penguins (B. Stonehouse, Ed.). University Park Press, Baltimore.

REVILLE, B. J. 1980. Spatial and Temporal Aspects of Breeding in the Frigatebirds Fregata minor and F. aerial. Ph.D. thesis, University of Aberdeen, Scotland.

RICE, D. W., AND K. W. KENYON. 1962. Breeding cycles and behavior of Laysan and Black-footed Albatrosses. Auk 79: 517–567.

RICHDALE, L. E. 1951. Sexual behavior of penguins. University of Kansas Press, Lawrence.

RICHDALE, L. E. 1963. Biology of the sooty shearwater Puffinus griseus. Proceedings of the Zoology Society of London 141: 1–117.

RICHDALE, L. E. 1965. Biology of the birds of Whero Island, New Zealand, with special reference to the Diving Petrel and the White-faced Storm Petrel. Transactions of the Zoological Society of London 31: 1–86.

RICKLEFS, R. E., A. R. PLACE, AND D. J. ANDERSON. 1987. An experimental investigation of the influence of diet quality on growth rate in Leach's Storm-petrel. American Naturalist 130: 300–305.

ROBERTS, B. 1940. The breeding behavior of penguins. In British Graham Land Expedition 1934–1937, Science Report 1: 195–254.

SCHALLER, G. B. 1964. Breeding behavior of the white pelican at Yellowstone Lake, Wyoming. Condor 63: 3–23.

SCHREIBER, E. A. 1994. El Niño–Southern Oscillation effects on provisioning and growth in Red-tailed Tropicbirds. Colonial Waterbirds 17: 105–119.

SCHREIBER, E. A. 1996. Experimental manipulation of feeding in Red-tailed Tropicbird chicks. Colonial Waterbirds 19: 45–55.

SCHREIBER, E. A., AND R. W. SCHREIBER. 1993. Red-tailed tropicbird (Phaethon rubricauda). No. 43 in The Birds of North America (A. Poole and F. Gill, Eds.). Academy of Natural Sciences, Philadelphia; American Ornithologists' Union, Washington, D.C.

SCHREIBER, R. W. 1976. Growth and development of nestling Brown Pelicans. Bird-Banding 47: 19–39.

SCHREIBER, R. W. 1977. Maintenance behavior and communication in the Brown Pelican. Ornithological Monographs No. 22. American Ornithologists' Union, Washington, D.C.

SCHREIBER, R. W. 1979. Reproductive performance of the eastern brown pelican, Pelecanus occidentalis. Contributions in Science of the Natural History Museum of Los Angeles County 317: 1–43.

SCHREIBER, R. W., AND P. J. MOCK. 1988. Eastern Brown Pelicans: what does 60 years of banding tell us? Journal of Field Ornithology 59: 171–182.

SEDDON, P., AND Y. VAN HEEZIK. 1991. Hatching asynchrony and brood reduction in the Jackass Penguin: an experimental study. Animal Behaviour 42: 347–356.

SHAW, P. 1985. Brood reduction in the Blue-eyed Shag Phalacrocorax atriceps. Ibis 127: 476–494.

SHUGART, G. 1977. The development of chick recognition by adult Caspian Terns. Proceedings of the Conference of the Colonial Waterbird Group 1977: 110–117.

SHUGART, G., AND M. FITCH. 1987. Neighbor interactions and cooperation among breeding Herring Gulls: an alternative interpretation of gull territoriality. Studies in Avian Biology No. 10. Cooper Ornithological Society.

SIMONS, T. R. 1985. Biology and behavior of the endangered Hawaiian Dark-rumped Petrel. Condor 87: 229–245.

SKOKOVA, N. 1962. Tupik na Ainovych ostrovach (The puffin on the Ainov Islands). Ornitologiya 5: 7–12

SNOW, B. K. 1963. The behavior of the shag. British Birds 56: 77–186.

SNOW, B. K. 1966. Observations on the behavior and ecology of the flightless cormorant *Nannopterum harrisi*. Ibis 108: 265–280.

SNOW, D. 1965. The breeding of the Red-billed Tropicbird in the Galapagos Islands. Condor 67: 210–214.

SOUTHERN, W. 1977. Colony selection and site tenacity in Ring-billed Gulls at a stable colony. Auk 94: 469–478.

SPURR, E. 1975. Communication in the Adélie Penguin. Pp. 449–501 *in* The Biology of Penguins (B. Stonehouse, Ed.). University Park Press, London.

STEVENSON, J., R. HUTCHINGSON, J. HUTCHINGSON, B. BERTRAM, AND W. THORPE. 1970. Individual recognition by auditory cues in the Common Tern (*Sterna hirundo*). Nature (London) 226: 562–563.

STONEHOUSE, B. 1956. The Brown Skua: *Catharacta skua evinbergi* (Matthews) of South Georgia. Falkland Islands Department of Survey Science Report 14: 1–25.

STONEHOUSE, B. 1963. The tropic birds genus (*Phaethon*) of Ascension Island. Ibis 103: 124–161.

STONEHOUSE, B. (Ed.). 1975. The Biology of Penguins. University Park Press, London, 555 pages.

STONEHOUSE, B., AND S. STONEHOUSE. 1963. The frigatebird *Fregata aquila* of Ascension Island. Ibis 103: 409–422.

STOREY, A. E. 1984. Function of Manx Shearwater calls in mate attraction. Behaviour 89: 73–89.

TAYLOR, K. 1984. Puffin Behavior. Pp. 96–105 *in* The Puffin (M. P. Harris, Ed.). T. and A. D. Poyser Ltd., Staffordshire, England.

TICKELL, W. L. N. 1968. The biology of the great albatrosses, *Diomedea exulans* and *Diomedea epomophora*. Antarctic Research Series 12: 1–55.

TINBERGEN, N. 1952. "Derived" activities; their causation, biological significance, origin and emancipation during evolution. Quarterly Review Biology 27: 1–32.

TINBERGEN, N. 1953. The Herring Gull's World. Collins, London.

TINBERGEN, N. 1959a. Comparative study of the behavior of gulls. Behavior 15: 1–70.

TINBERGEN, N. 1959b. "Derived" activities: their causation, biological significance, origin and emancipation during evolution. Quarterly Review of Biology 27: 1–32.

TINBERGEN, N. 1960. The Herring Gull's World. Collins. London.

TSCHANZ, B. 1968. Trotte Ilummen die Entstehung der personlichen Beziehungen zwischen Jungvogelund Eltern (The development of the relationship between young and parents in the Common Murre). Zeitschrift für Tierpsychologie Beiheft 4: 1–100.

VAN TETS, G. F. 1965. A comparative study of some social communication patterns in the Pelecaniformes. Ornithological Monographs 2: 1–88.

VESTJENS, W. J. M. 1977. Breeding behavior and ecology of the Australian Pelican, *Pelecanus conspicillatus*, in New South Wales. Australian Wildlife Research 4: 37–58.

WAAS, J. 1990a. An analysis of communication during the aggressive actions of Little Blue Penguins. Pp. 345–376 *in* Penguin Biology (L. Dain and J. Dark, Eds.). Academic Press, San Diego.

WAAS, J. 1990b. Intraspecific variation in social repertoires: evidence from cave and burrow dwelling Little Blue Penguins. Behaviour 115: 63–99.

WAAS, J. 1991a. Do Little Penguins signal their intentions during aggressive interactions with strangers? Animal Behaviour 41: 375–382.

WAAS, J. 1991b. The risks and benefits of signaling aggressive motivation: a study of cave-dwelling Little Blue Penguins. Behavioral Ecology and Sociobiology 29: 39–46.

WARHAM, J. 1962. The biology of the Giant Petrel, *Macronectes giganteus*. Auk 79: 139–160.

WARHAM, J. 1963. The Rock-hopper Penguin, *Eudyptes chrysocome* at Macquarie Island. Auk 80: 229–256.

WARHAM, J. 1971. Aspects of breeding behavior in the Royal Penguin *Eudyptes chrysolophus schlegeli*. Notornis 18: 91–115.

WARHAM, J. 1975. The Crested Penguins. Pp. 189–269 *in* The Biology of Penguins (B. Stonehouse, Ed.). University Park Press, London.

WARHAM, J. 1990. The Petrels. Their Ecology and Breeding Systems. Academic Press, London.

WARHAM, J. 1996. The Behavior, Population Biology and Physiology of the Petrels. Academic Press. London.

WEIMERSKIRCH, H., P. JOUVENTIN, AND J. C. STAHL. 1986. Comparative ecology of six albatross species breeding on the Crozet Islands. Ibis 128: 195–213.

WEIMERSKIRGH, H., O. CHASTEL, AND L. ACKERMAN. 1995. Adjustment of parental effort to manipulated foraging ability in a pelagic seabird, the Thin-billed Prion *Pachyptila belcheri*. Behavior Ecology and Sociobiology 36: 11–16.

WHITE, S. J. 1971. Selective responsiveness by the gannet (*Sula bassana*) to played-back calls. Animal Behaviour 19: 126–131.

WHITE, S. J., AND R. E. WHITE. 1970. Individual voice production in gannets. Behaviour 37: 40–64.

WILLIAMS, A. 1982. Chick-feeding rates of Macaroni and Rock-hopper Penguins at Marion Island. Ostrich 53: 129–134.

WILLIAMS, L. 1942. Display and sexual behavior of the Brandt Cormorant. Condor 44: 85–104.

WILLIAMS, T. 1995. The Penguins. Birds of the World. Oxford University Press, London.

WOODWARD, P. W. 1972. The natural history of Kure Atoll, earth-western Hawaiian Islands. Atoll Research Bulletin No. 164.

Shy Albatross and White-chinned Petrel Squabbling over Food

11 Energetics of Free-Ranging Seabirds

Hugh I. Ellis and Geir W. Gabrielsen

CONTENTS

11.1 INTRODUCTION

Nearly 30 years ago, Calder and King (1974), noting that metabolic rates on 38 species of passerine and 34 species of nonpasserine birds had been measured since 1950 and recognizing the predictive power of allometric equations, asked whether it was better to add more birds to the list or to ask new questions. Of course, both happened. In fact, adding more species to the list in part led to new questions. Among these developments has been the ability to look at groups of birds in terms of both their phylogeny and their ecology. One such approach has been to single out seabirds as an ecological group (Ellis 1984, Nagy 1987). In the more than 15 years since a comprehensive review of seabird energetics has appeared (Ellis 1984), the information on basal metabolic rates (BMR) in this group has doubled and the reports on field metabolic rates (FMR, using doubly labeled water) have more than tripled. New analyses using both of these measurements have appeared during that time. It is the goal of this chapter to summarize our current knowledge of seabird energetics, provide a comprehensive review of BMR and FMR measurements, and examine many correlates of both. The relationships of BMR with color and activity pattern (Ellis 1984) need no further development. However, unlike the earlier review, we treat thermoregulation and provide information on thermal conductance and lower critical limits of thermoneutrality. For a comprehensive treatment of avian thermoregulation, refer to Dawson and Whittow (2000). Lustick (1984) remains the best source on seabird thermoregulation generally. Ellis (1984) demonstrated a latitudinal gradient for BMR in Charadriiformes. We reevaluate that gradient and consider whether such an analysis can be extended outside that order. We examine a variety of metabolic costs, including locomotion, and survey information on community energetics, critiquing old models and suggesting new ones.

In this chapter, we limit ourselves mainly to adults in the four orders of seabirds: Sphenisciformes, Procellariiformes, Pelecaniformes, and Charadriiformes. Where feasible, we also include available information on sea ducks (Anseriformes). References to shorebirds or other birds are made only when necessary. But because the energetics of seabird migration is so poorly known, we direct the reader to those publications, relevant for shorebirds, which may provide useful insights (e.g., Alerstam and Hedenström 1998).

11.2 BASAL METABOLIC RATE IN SEABIRDS

Basal metabolic rate is a unique parameter (McNab 1997). It represents a limit, the minimal rate of energy expenditure in an endotherm under prescribed conditions (see below) and otherwise subject only to variations in time of day or season. Because it is replicable under those conditions, comparisons across a variety of species are possible. McNab (1997) cites seven conditions for BMR, some of which we view as too restrictive. We believe that BMR should be defined as the rate found in a thermoregulating, postabsorptive, adult animal at rest in its thermoneutral zone. This is fairly close to the definition given by Bligh and Johnson (1973), except that it does not demand measurement in the dark (although in actual practice it is typically measured in the dark or in dim light), and, like McNab (1997), requires the measurement be of adults to remove potential costs of growth. However, we believe that BMR is a statistic, not a constant because of circadian and seasonal effects. For example, Aschoff and Pohl (1970) demonstrated that for many birds that period of activity affects BMR; namely, BMR may be lower in the inactive (ρ) period and higher in the active (α) period. BMR may also change with season as found for a gull (Davydov 1972), sea duck (Jenssen et al. 1989, Gabrielsen et al. 1991a), and shorebird (Piersma et al. 1995); this is also known in terrestrial birds (Gavrilov 1985) and mammals (Fuglei and Ørietsland 1999). Fyhn et al. (2001) have even shown in Black-legged Kittiwakes (*Rissa tridactyla*) that BMR may change from one stage of the breeding season to another (although different individuals were used in the two periods chosen). Consequently, it is essential to note the circumstances under which BMR was measured (i.e., time of day, season) in addition to the complete experimental protocols urged by McNab (1997). The repeatability of BMR measurements within individuals, sometimes assumed by researchers, has now been demonstrated in Black-legged Kittiwakes over relatively long periods of time (1 year; Bech et al. 1999).

There are areas where there is contention over whether measured metabolic rates can be considered basal. McNab (1997) warns against the measurement of endotherms in a reproductive condition; he includes incubating birds. Indeed, King (1973) and Walsberg and King (1978) report incubation metabolic rates (IMR) above BMR, although there may be no appreciable differences between IMR and BMR in other species (cf. Williams 1996). Values for IMR in seabirds are reported in this volume by Whittow (see Chapter 12), who discusses this problem. Whereas the effect of incubation on metabolism is varied, changes in body composition (e.g., liver mass) during chick-rearing can affect metabolic rate (Langseth et al. 2000). In fact, changes in body composition in a variety of contexts, such as migration (Weber and Piersma 1996), can affect metabolic rate. We are undecided on whether these metabolic rates should be considered BMR. Although body composition may change during long-term fasting, metabolic rate may drop in Phase I of the fast before those changes become apparent; Cherel et al. (1988) consider this to be a change in BMR. Long-term fasting is further discussed in Section 11.2.5 below. Is metabolism during sleep BMR? Most metabolic experiments are done in the dark or in dim light, but the bird is thought to be awake. That often is not verifiable. However, Stahel et al. (1984) argue that for Blue Penguins (*Eudyptula minor*) the reduction in BMR (≤8%) due to sleep is minor.

The literature has many measurements reported as SMR (standard metabolic rate) or RMR (resting metabolic rate). Generally, SMR in endotherms can be considered equivalent to BMR. That is not necessarily the case with RMR. Resting rates may not be measured in the zone of thermoneutrality nor on birds that are postabsorptive. The RMR reported for Common (*Uria aalge*) and Thick-billed Murres (*U. lomvia*) were measured under the conditions specified for BMR (Croll and McLaren 1993). On the other hand, insufficient information exists to draw that conclusion in the case of Tufted Ducks (*Aythya fuligula*; Woakes and Butler 1983) used in comparisons with seabirds in Section 11.4.3.1 below. In fact, the ducks' RMRs were measured in water; in most cases RMR of a floating bird is higher than BMR (Prange and Schmidt-Nielsen 1970, Hui 1988a, Luna-Jorquera and Culik 2000, H. Ellis unpublished, in Eared Grebes, *Podiceps nigricollis*). Similar problems are reported in penguins by Culik and Wilson (1991a).

The use of BMR and other physiological parameters has recently come under scrutiny by those who argue that phylogenetic relationships must be considered in all such comparisons, especially across broad taxonomic groups (Garland and Carter 1994, Reynolds and Lee 1996). However, this presumes knowledge of phylogenetic relationships that may be unknown or disputed, and it is not without its detractors (Mangum and Hochachka 1998). In this paper, we have chosen to provide metabolic data in a straightforward manner. However, there are differences among the orders; for example, sphenisciform birds have generally a lower BMR (see Section 11.2.2).

Our allometric equations below are given both for seabirds as a group and for each of the four orders of seabirds. It is our intention to provide as much information as possible, but we recommend that workers interested in making seabird comparisons use the all-seabird equation unless they have specific reasons for doing otherwise. Other, more serious problems affect the validity of the data themselves. These occur during both the measurement of metabolism and the conversion of units in metabolic studies and are discussed below.

11.2.1 METHODS AND ERRORS IN METABOLIC MEASUREMENTS

Direct and indirect calorimetry are the two main methods used to determine BMR in birds. The origins of both go back to Lavoisier; they are compared in Brody (1945). The indirect method has been used in most metabolic studies, including all those cited in this chapter. It is based on determinations of the quantities of oxygen consumed or carbon dioxide produced or food assimilated. In fact, for reasons discussed in most introductory physiology texts, oxygen consumption is the primary means by which such information is obtained.

Two methods have been used to measure oxygen consumption in animals: closed- and open-circuit respirometry. In open-circuit respirometry, a constant flow of air goes to an animal and then to some analytical device. In closed-circuit respirometry, gas pressure is measured as it decreases due to the consumption of oxygen; carbon dioxide production does not compensate for such reductions because it is absorbed by some chemical (NaOH, Ascarite®, soda lime, etc.). Although not essential, closed-circuit respirometry often reduces metabolic chamber size to increase the pressure change signal. These experiments typically have shorter equilibration times and are of shorter duration than open-circuit experiments. All of these introduce sources of error likely to raise metabolic rate. We think that is likely to be the case for the study by Ricklefs and White (1981) on Sooty Terns (*Sterna fuscata*). This study is cited in Table 11.1, which compares data collected in open circuitry with those collected in closed circuitry for the same species but in different studies.

An opposite problem that may occur in closed-circuit respirometry is an apparently reduced metabolic rate due to a buildup of carbon dioxide. This would occur if the CO_2 absorbent failed, was depleted, or was ineffective (this last may occur because, unlike open systems where the absorbent is in columns through which the air passes, in closed systems it is often on the bottom of the chamber providing limited surface area). This may have occurred in the studies by Cairns et al. (1990) on the Common Murre and Birt-Friesen et al. (1989) on the Northern Gannet (*Morus bassana*), as shown in Table 11.1. Not only may the buildup of CO_2 reduce apparent metabolic rate by giving false readings of pressure changes in a closed system, but it may, in extreme cases, actually reduce the metabolic rate of a bird directly. The situation is complicated in the Northern Gannets because while the closed system of Birt-Friesen et al. (1989) may have allowed a buildup of CO_2, the experiment by Bryant and Furness (1995) actually did result in CO_2 levels as high as 2.8%.

Although we tend to trust open-circuit respirometry over closed-circuit respirometry when the results are as different as they often are in Table 11.1, we recognize that other errors may make the results of open systems suspect. The study by Kooyman et al. (1976) on Adélie Penguins (*Pygoscelis adeliae*) probably gives an inflated value for BMR because the birds were restrained. This practice, almost entirely abandoned today, may be necessary in unusual cases; but its consequences are likely to compromise results.

Another problem that can create problems for open- as well as closed-circuit respirometry involves the respiratory quotient. Respiratory quotient (RQ) is the ratio of the volume of CO_2 produced to the volume of O_2 consumed. It varies with the food substrate being metabolized by the subject. A carbohydrate diet yields an RQ of 1.0; a diet based on lipids yields an RQ of 0.71; protein substrates (Elliott and Davison 1975) and mixed substrates are intermediate (Schmidt-Nielsen 1990). An animal that is postabsorptive, a condition of BMR, would typically be sustaining itself on stored fat. Consequently, RQs measured during studies of BMR should be around 0.71. In fact, reported RQs measured in fasting birds, usually during metabolic experiments, show values at or close to 0.71 (King 1957, Drent and Stonehouse 1971). This is equally true for seabirds (Pettit et al. 1985, Gabrielsen et al. 1988, Chappell et al. 1989). Higher values suggest that birds were not postabsorptive or that CO_2 built up during the experiment. This may be illustrated by Iversen and Krog (1972) whose open-circuit BMR for Leach's Storm-petrels (*Oceanodroma leucorhoa*) is about 30% higher than was found in two closed-circuit studies (Table 11.1). Iversen and Krog did not remove CO_2 before measuring oxygen and reported RQ = 0.83. The buildup of CO_2 explains the high RQ, although not the high BMR. That high value may be a function of the very small (0.5 L) chamber used. Small chambers, often used in closed systems (see above) may cause inflated levels of oxygen consumption (H. Ellis unpublished). Here, we prefer the comparable closed-circuit experiments which used much larger chambers. A high RQ may also reflect a nonpostabsorptive condition.

Open and closed systems, when used with care, can give similar results. The nearly identical results coming from the independent studies on Southern Giant Fulmars (*Macronectes giganteus*) by Ricklefs and Matthew (1983) using a closed system and Morgan et al. (1992) using an open one underscore that (see Table 11.1). Overall, while we recognize that a closed system is sometimes

TABLE 11.1
Open- vs. Closed-Circuit Respirometry in Independent Studies

Species	N[a]	Mass[b]	BMR: Open[c]	BMR: Closed[c]	% Open	Reference
Sooty Tern (*Sterna fuscata*)	4	150.4 ± 13.0	0.97 ± 0.14	—	—	MacMillen et al. 1977
	5	156.6 ± 8.4	0.93 ± 0.14	—	—	Ellis, Pettit, and Whittow unpublished in 1982
	4	170.4	—	1.75	80.4	Ricklefs and White 1981
Common Murre (*Uria aalge*)	11	913 ± 53	1.20 ± 0.03	—	—	Gabrielsen 1996
	3	972 ± 24	—	0.77 ± 0.15	-35.8	Cairns et al. 1990
Northern Gannet (*Morus bassana*)	4	2574 ± 289	0.89 ± 0.16	—	—	Bryant and Furness 1995
	4	3030 ± 140	—	0.48 ± 0.10	-46.1	Birt-Friesen et al. 1989
Southern Giant Fulmar (*Macronectes giganteus*)	6	3929	0.92	—	—	Morgan et al. 1992
	8	3460	—	0.89	-3.3	Ricklefs and Matthew 1983
Leach's Storm-petrel (*Oceanodroma leucorhoa*)	2	42	2.77[d]	—	—	Iversen and Krog 1972
	4	47	—	1.92 ± 0.37	-30.7	Ricklefs et al. 1986
	7	46.6	—	2.02 ± 1.01	-27.1	Montevecchi et al. 1992
Adélie Penguin (*Pygoscelis adeliae*)	13	3970	1.20[e]	—	—	Kooyman et al. 1976
	8	3500 ± 60	—	0.92 ± 0.06	-23.3	Ricklefs and Matthew 1983

[a] Number of experimental birds.
[b] Mass in g.
[c] mL O_2 g^{-1} h^{-1}.
[d] RQ = 0.83.
[e] Restrained animals.

FIGURE 11.1 Conducting physiological studies under field conditions is often difficult: catching and confining the animal, working without electricity, dealing with weather conditions. All of these can add error to measurements. (Photo by R. W. and E. A. Schreiber.)

the only practical method under often difficult field conditions, and that it can give reliable results, we think caution should be exercised in choosing it when both options are available (Figure 11.1).

The conversion of metabolic data from units actually measured (typically oxygen consumption) to derivative units of energy (kJ, W, or previously kcal), invariably used in allometric studies (Lasiewski and Dawson 1967, Aschoff and Pohl 1970, Ellis 1984), may also be a source of error. The conversion of oxygen consumption to energy is a function of RQ, for which caloric equivalents of oxygen are provided by Bartholomew (1982). Scattered throughout the metabolic literature is the equivalency of 20.8 kJ/L O_2. This is based on an RQ of 0.79. The more reasonable RQ of 0.71 for a postabsorptive bird gives an equivalency of 19.8 kJ/L O_2. So a common misunderstanding of RQ introduces a 5% overestimate in many metabolic papers. We suggest that authors provide measured data (e.g., mL O_2 h^{-1}) or conversion factors used.

Other problems may affect the data base for seabirds. For instance, it is possible that some values presented in this chapter do not represent true values of BMR because they were not measured within the thermoneutral zone (TNZ, that range of environmental temperatures across which resting metabolic rates are lowest and independent of temperature). McNab (1997) provides examples of this. We have found far fewer data in the seabird literature on thermal conductance and lower limits of thermoneutrality than BMR. This suggests that full metabolic profiles may not always have been done and that the actual TNZ may not always have been known (e.g., Roby and Ricklefs 1986, Bryant and Furness 1995).

Not all differences in BMR can be attributed to obvious sources of error, however. The BMR of Blue Penguins (*Eudyptula minor*) reported by Stahel and Nicol (1982) is 69% higher than the value reported by Baudinette et al. (1986). We cannot explain this difference but it can have implications beyond the BMR value itself, as noted in Section 11.4.2 below. Table 11.2 includes all the measurements of BMR we found in the literature.

11.2.2 ALLOMETRY OF BMR

King and Farner (1961) reviewed previous allometric analyses and provided the best equation then possible. But they noted an incongruity between small birds and those exceeding 125 g. In 1967, Lasiewski and Dawson argued that passerines and nonpasserines required separate allometric analyses. Their nonpasserine equation is given below:

$$BMR = 327.8 \ m^{0.723}$$

(11.1)

TABLE 11.2
Body Mass, Basal Metabolic Rates (BMR; in kJ d⁻¹ and kJ g⁻¹ h⁻¹), and Breeding Region in Seabirds

Order/Species	Body Mass (g)	n	BMR (kJ d⁻¹)	BMR (kJ g⁻¹ h⁻¹)	Latitude/ Region (degree)	Source
Sphenisciformes						
Adelie Penguin	3970	14	1060	0.0111	64 S	Kooyman et al. 1976
Pygoscelis adeliae						
Adelie Penguin	3500	8	1552	0.0185	64 S	Ricklefs and Matthew 1983
P. adeliae						
Emperor Penguin	23370	5	3704	0.0066	78 S	Pinshow et al. 1976
Aptenodytes forsteri						
Emperor Penguin	24800	11	4239	0.0071	46 S	Le Maho et al. 1976
A. forsteri						
Fjordland Penguin	2600	4	599	0.0096	40 S	In Drent and Stonehouse 1971
Eudyptes pachyrhynchus						B. Stonehouse unpublished
Yellow-eyed Penguin	4800	1	996	0.0086	40 S	In Drent and Stonehouse 1971
Megadyptes antipodes						B. Stonehouse unpublished
Humboldt Penguin	3870	3	821	0.0088	49 N	Drent and Stonehouse 1971
Spheniscus humboldti						
Blue Penguin	900	6	384	0.0178	42 S	Stahel and Nicol 1982
Eudyptula minor						
Blue Penguin	1106	8	298	0.0112	36 S	Baudinette et al. 1986
E. minor						
Blue Penguin	1082	14	308	0.0119	42 S	Stahel and Nicol 1988
E. minor						
Procellariiformes						
Wandering Albatross	8130	4	1755	0.0090	47 S	Adams and Brown 1984
Diomedea exulans						
Laysan Albatross	3103	5	637	0.0086	24 N	Grant and Whittow 1983
Phoebastria immutabilis						
Grey-headed Albatross	3753	3	735	0.0082	47 S	Adams and Brown 1984
Thalassarche chrysostoma						
Sooty Albatross	2875	4	715	0.0104	47 S	Adams and Brown 1984
Phoebetria fusca						
Southern Giant Petrel	3460	8	1466	0.0177	64 S	Ricklefs and Matthew 1983
M. giganteus						
Southern Giant Petrel	4780	6	1154	0.0101	47 S	Adams and Brown 1984
M. giganteus						
Southern Giant Petrel	3929	6	1735	0.0184	64 S	Morgan et al. 1992
Macronectes giganteus						
Southern Fulmar	780	5	437	0.0233	69 S	Weathers et al. 2000
Fulmarus glacialoides						
Northern Fulmar	651	16	314	0.0201	79 N	Gabrielsen et al. 1988
F. glacialis						
Northern Fulmar	728	4	330	0.0189	56 N	Bryant and Furness 1995
F. glacialis						
Antarctic Petrel	718	6	408	0.0237	69 S	Weathers et al. 2000
Thalassoica antarctica						
Cape Pigeon	420	7	317	0.0314	69 S	Weathers et al. 2000
Daption capense						
Snow Petrel	292	6	199	0.0284	69 S	Weathers et al. 2000
Pagodroma nivea						

TABLE 11.2 *(Continued)*

Body Mass, Basal Metabolic Rates (BMR; in kJ d⁻¹ and kJ g⁻¹ h⁻¹), and Breeding Region in Seabirds

Order/Species	Body Mass (g)	n	BMR (kJ d⁻¹)	BMR (kJ g⁻¹ h⁻¹)	Latitude/ Region (degree)	Source
Kerguelen Petrel *Leugensa brevirostris*	315	2	153	0.0202	47 S	Adams and Brown 1984
Soft-plumaged Petrel *Pterodroma mollis*	274	2	151	0.0230	47 S	Adams and Brown 1984
Bonin Petrel *Pterodroma hypoleuca*	180	2	89	0.0206	24 N	Grant and Whittow 1983
Bonin Petrel *P. hypoleuca*	167	7	72	0.0181	24 N	Pettit et al. 1985
Salvin's Prion *Pachyptila salvini*	165	3	134	0.0338	47 S	Adams and Brown 1984
Bulwer's Petrel *Bulweria bulwerii*	87	6	44	0.0211	24 N	Pettit et al. 1985
White-chinned Petrel *Procellaria aequinoctialis*	1287	3	545	0.0176	47 S	Adams and Brown 1984
Grey Petrel *P. cinerea*	1014	2	433	0.0178	47 S	Adams and Brown 1984
Wedge-tailed Shearwater *Puffinus pacificus*	332	18	121	0.0152	24 N	Pettit et al. 1985
Sooty Shearwater *P. griseus*	740	3	249	0.0140	37 N	Krasnow 1979
Christmas Shearwater *P. nativitatis*	308	6	127	0.0172	24 N	Pettit et al. 1985
Manx Shearwater *P. puffinus*	413	10	195	0.0197	62 N	Bech et al. 1982
Manx Shearwater *P. puffinus*	367	4	201	0.0228	57 N	Bryant and Furness 1995
Georgian Diving-petrel *Pelecanoides georgicus*	127	2	85	0.0279	47 S	Adams and Brown 1984
Georgian Diving-petrel *P. georgicus*	119	5	122	0.0427	54 S	Roby and Ricklefs 1986
Common Diving-petrel *P. urinatrix*	132	4	126	0.0398	54 S	Roby and Ricklefs 1986
Wilson's Storm-petrel *Oceanites oceanicus*	42	9	37	0.0367	64 S	Obst et al. 1987
Wilson's Storm-petrel *O. oceanicus*	34	6	35	0.0429	64 S	Morgan et al. 1992
Leach's Storm-petrel *Oceanodroma leucorhoa*	47	7	45	0.0399	47 N	Montevecchi et al. 1991
Leach's Storm-petrel *O. leucorhoa*	45	4	43	0.0402	45 N	Ricklefs et al. 1986
Leach's Storm-petrel *O. leucorhoa*	44	6	59	0.0565	48 N	Ricklefs et al. 1980
Leach's Storm-petrel *O. leucorhoa*	42	2	55	0.0548	54 N	Iversen and Krog 1972
Fork-tailed Storm-petrel *O. furcata*	49	16	56	0.0476	54 N	Iversen and Krog 1972
Fork-tailed Storm-petrel *O. furcata*	45	1	39	0.0361	59 N	Vleck and Kenagy 1980

TABLE 11.2 *(Continued)*
Body Mass, Basal Metabolic Rates (BMR; in kJ d^{-1} and kJ g^{-1} h^{-1}), and Breeding Region in Seabirds

Order/Species	Body Mass (g)	n	BMR (kJ d^{-1})	BMR (kJ g^{-1} h^{-1})	Latitude/ Region (degree)	Source
Pelecaniformes						
Red-tailed Tropicbird	593	5	288	0.0202	24 N	Pettit et al. 1985
Phaethon rubricauda						
Australian Pelican	5090	1	1566	0.0128	41 N	Benedict and Fox 1927
Pelecanus conspicillatus						
Brown Pelican	3510	1	1105	0.0131	41 N	Benedict and Fox 1927
P. occidentalis						
Brown Pelican	3038	3	896	0.0123	29 N	H. Ellis and W. Hennemann unpublished data
P. occidentalis						
Magnificent Frigatebird	1078	4	240	0.0093	9 N	Enger 1957
Fregata magnifiscens						
Cape Gannet	2660	5	856	0.0134	32 S	Adams et al. 1991
Morus capensis						
Northern Gannet	3030	4	701	0.0096	47 N	Birt-Friesen et al. 1989
M. bassanus						
Northern Gannet	2574	4	1079	0.0175	55 N	Bryant and Furness 1995
M. bassanus						
Masked Booby	1289	1	476	0.0154	28 N	H. Ellis unpublished data
Sula dactylatra						
Red-footed Booby	1017	8	376	0.0154	21 N	Ellis et al. 1982a
S. sula						
Double-crested Cormorant	1330	5	537	0.0168	28 N	Hennemann 1983a
Hypoleucos auritus						
Great Cormorant	1950	3	721	0.0154	35 N	Sato et al. 1988
Phalacrocorax carbo						
Imperial Shag	2660	6	1317	0.0206	64 S	Ricklefs and Matthew 1983
Notocarbo atriceps						
European Shag	1619	4	739	0.0190	56 N	Bryant and Furness 1995
Stictocarbo arstotelis						
Charadriiformes						
Parasitic Jaeger	351	4	199	0.0236	60 N	Bryant and Furness 1995
Stercorarius parasiticus						
Great Skua	970	1	410	0.0176	41 N	Benedict and Fox 1927
S. skua						
Great Skua	1159	4	538	0.0193	60 N	Bryant and Furness 1995
S. skua						
South Polar Skua	1130	9	705	0.0260	64 S	Ricklefs and Matthew 1983
Catharcta maccormicki						
South Polar Skua	1250	6	708	0.0236	64 S	Morgan et al. 1992
C. maccormicki						
Pacific Gull	1210	1	532	0.0183	41 N	Benedict and Fox 1927
Larus pacificus						
Common Gull	428	1	201	0.0196	55 N	Gavrilov 1985
L. canus						
Ring-billed Gull	439	3	250	0.0237	29 N	Ellis 1980a
L. delawarensis						
Kelp Gull	980	4	610	0.0259	64 S	Morgan et al. 1992
L. dominicanus						

TABLE 11.2 *(Continued)*
Body Mass, Basal Metabolic Rates (BMR; in kJ d⁻¹ and kJ g⁻¹ h⁻¹), and Breeding Region in Seabirds

Order/Species	Body Mass (g)	n	BMR (kJ d⁻¹)	BMR (kJ g⁻¹ h⁻¹)	Latitude/ Region (degree)	Source
Western Gull *L. occidentalis*	761	7	294	0.0161	34 N	Obst unpublished data
Glaucous Gull *L. hyperboreus*	1210	2	754	0.0260	71 N	Scholander et al. 1950b
Glaucous Gull *L. hyperboreus*	1326	9	562	0.0177	79 N	Gabrielsen and Mehlum 1989
Herring Gull *L. argentatus*	1000	6	415	0.0173	45 N	Lustick et al. 1978
Herring Gull *L. argentatus*	924	6	428	0.0193	56 N	Bryant and Furness 1995
Common Black-headed Gull *L. ridibundus*	285	1	173	0.0253	55 N	Gavrilov 1985
Common Black-headed Gull *L. ridibundus*	252	10	188	0.0311	60 N	Davydov 1972
Laughing Gull *L. atricilla*	276	4	162	0.0250	29 N	Ellis 1980a
Black-legged Kittiwake *Rissa tridactyla*	407	11	242	0.0248	57 N	Gabrielsen et al. submitted
Black-legged Kittiwake *R. tridactyla*	420	17	304	0.0302	70 N	G. Gabrielsen unpublished
Black-legged Kittiwake *R. tridactyla*	365	16	289	0.0330	79 N	Gabrielsen et al. 1988
Black-legged Kittiwake *R. tridactyla*	305	4	237	0.0324	56 N	Bryant and Furness 1995
Red-legged Kittiwake *R. brevirostris*	333	7	230	0.0288	57 N	Gabrielsen et al. submitted
Ivory Gull *Pagophila eburnea*	508	2	443	0.0363	79 N	Gabrielsen and Mehlum 1989
Royal Tern *Sterna maxima*	373	3	217	0.0242	29 N	Ellis 1980a
Arctic Tern *S. paradisaea*	85	3	79	0.0386	79 N	Klaassen et al. 1989
Grey-backed Tern *S. lunata*	131	2	61	0.0192	24 N	Pettit et al. 1985
Sooty Tern *S. fuscata*	148	6	69	0.0194	21 N	MacMillen et al. 1977
Brown Noddy *Anous stolidus*	139	16	67	0.0201	21 N	Ellis et al. 1995
Black Noddy *A. tenuirostris*	90	4	55	0.0260	24 N	Pettit et al. 1985
White Tern *Gygis alba*	98	6	70	0.0299	24 N	Pettit et al. 1985
Dovekie *Alle alle*	153	23	178	0.0490	79 N	Gabrielsen et al. 1991b
Razor-billed Auk *Alca torda*	589	2	311	0.0220	56 N	Bryant and Furness 1995
Common Murre *Uria aalge*	836	8	517	0.0258	57 N	Croll and McLaren 1993
Common Murre *U. aalge*	803	10	461	0.0239	57 N	Gabrielsen et al. submitted

TABLE 11.2 *(Continued)*
Body Mass, Basal Metabolic Rates (BMR; in kJ d⁻¹ and kJ g⁻¹ h⁻¹), and Breeding Region in Seabirds

Order/Species	Body Mass (g)	n	BMR (kJ d⁻¹)	BMR (kJ g⁻¹ h⁻¹)	Latitude/ Region (degree)	Source
Common Murre	956	4	588	0.0256	65 N	Johnson and West 1975
U. aalge						
Common Murre	913	11	580	0.0270	70 N	Gabrielsen 1996
U. aalge						
Common Murre	771	4	390	0.0211	56 N	Bryant and Furness 1995
U. aalge						
Thick-billed Murre	803	6	595	0.0309	57 N	Croll and McLaren 1993
U. lomvia						
Thick-billed Murre	1094	11	619	0.0236	57 N	Gabrielsen et al. submitted
U. lomvia						
Thick-billed Murre	989	5	588	0.0248	65 N	Johnson and West 1975
U. lomvia						
Thick-billed Murre	819	11	438	0.0223	79 N	Gabrielsen et al. 1988
U. lomvia						
Black Guillemot	342	13	262	0.0319	79 N	Gabrielsen et al. 1988
Cepphus grylle						
Parakeet Auklet	243	3	172	0.0300	57 N	Gabrielsen et al. submitted
Cyclorrhynchus psittacula						
Least Auklet	83	5	116	0.0582	56 N	Roby and Ricklefs 1986
Aethia pusilla						
Atlantic Puffin	329	4	313	0.0396	56 N	Bryant and Furness 1995
Fratercula arctica						
Atlantic Puffin	470	22	335	0.0300	70 N	Barrett et al. 1995
F. arctica						
Horned Puffin	452	5	296	0.0273	57 N	Gabrielsen et al. submitted
F. corniculata						
Anseriformes						
Common Eider	1600	12	649	0.0169	79 N	Gabrielsen et al. 1991a
Somateria mollissima						
Oldsquaw	490	5	237	0.0202	63 N	Jenssen and Ekker 1989
Clangula hyemalis						

where BMR is in kJ d⁻¹ and m is mass in kg. Unfortunately, Lasiewski and Dawson (1967) assumed a caloric equivalency of 4.8 kcal/L O₂, which represents an RQ of about 0.79, for all data given in original gaseous units. Aschoff and Pohl (1970) proposed separate allometric relationships for passerines and nonpasserines based on activity pattern (anticipated earlier by King and Farner 1961). Their equations were used for most studies that thereafter noted the time that experiments were done, and most experiments were conducted at night from that time on. Their equations for nonpasserines are

$$BMR_\alpha = 381.0 \ m^{0.729} \tag{11.2}$$

$$BMR_\rho = 307.7 \ m^{0.734} \tag{11.3}$$

where α refers to the active phase and ρ the resting phase; the units are as in Equation 11.1. None of these studies included many seabirds. Ellis (1984) provided a comparison of seabird BMR with Aschoff and Pohl (1970) predictions where possible, but relied on the Lasiewski and Dawson (1967) model, which used data collected both in the day and at night, for several reasons: (1) some of the older literature did not give the time of the experiment; (2) it was unclear at very high latitudes, where summers lacked nights and winters days, that the α/ρ differences of Aschoff and Pohl (1970) would hold; and (3) it seemed that not all seabirds followed those activity differences. Ellis (1984) then constructed an allometric relationship exclusively for seabirds:

$$BMR = 381.8 \ m^{0.721} \tag{11.4}$$

where the units are the same as in Equations 11.1 to 11.3. Ellis' equation is very close to the α Equation 11.2 of Aschoff and Pohl (1970), but because he did not distinguish between active and resting phases, it is probably not directly comparable. Ellis meant for the equation to be descriptive only, but in fact it has been used in a predictive manner as well.

While we acknowledged above that BMR may vary with activity phase (Aschoff and Pohl 1970), we suspect that activity phase may not be as important as is often considered. Differences due to activity phase were not found in several high-latitude seabirds (Gabrielsen et al. 1988, Bryant and Furness 1995) or in three tropical or temperate seabirds (H. Ellis unpublished). Brown (1984) found no activity phase difference in either Macaroni Penguins (*Eudyptes chrysolophus*) or Rock-hopper Penguins (*E. chrysocome*), and although Baudinette et al. (1986) did find one in Blue (= Little) Penguins, it was not significant. Because of the difficulty in ascertaining a metabolic difference between activity phases in some seabirds and because not all studies report the time at which measurements were made, our allometric equation for BMR in seabirds includes all measurements without respect to phase. For ease of comparison, our equation, like Equations 11.1 to 11.4 above, employs units of kJ d^{-1}. However, if there are circadian differences, those units are inappropriate; so Table 11.2 also provides units of kJ g^{-1} h^{-1}. But in many instances these mass-specific units are inferred from an average body mass and an average BMR. Readers should consult original papers where possible. Finally, several species in Table 11.2 are represented by multiple studies. We averaged multiple studies, weighting them with the number of individuals (n) used in each.

Our overall equation for BMR in all seabirds of the four main orders, based on 110 studies on 77 species (Table 11.2) and irrespective of any possible circadian influence, is

$$BMR = 3.201 \ m^{0.719} \tag{11.5}$$

with mass in g (intercept s.e. = 1.143; exponent s.e. = 0.021; R^2 = 0.919). The exponent is close to that of Ellis (1984; Equation 11.4 above).

Table 11.3 provides the BMR equations for each order. Based on our analysis, Sphenisciformes and all but the largest Pelecaniformes have the lowest BMRs. The lower body temperatures, longer incubation times, and longer times to raise chicks in procellariiform birds generally are not reflected

TABLE 11.3
Comparison of Allometric Equations for BMR in All Seabirds, including Two Sea Ducks, and by Order

Taxon	Total	N	R^2	s.e. intercept	s.e. exponent
All Seabirds	BMR = 3.201 m$^{0.719}$	77	0.919	1.143	0.021
Charadriiformes	BMR = 2.149 m$^{0.804}$	31	0.842	1.374	0.052
Pelecaniformes	BMR = 1.392 m$^{0.823}$	12	0.756	2.729	0.135
Procellariiformes	BMR = 2.763 m$^{0.726}$	26	0.954	1.176	0.027
Sphenisciformes	BMR = 1.775 m$^{0.768}$	6	0.944	1.721	0.066

Note: BMR is in units of kJ d^{-1} and mass (m) is in g. N refers to number of species; for the number of studies, see Table 11.2. N for all the seabird equations includes two sea ducks, which explains the apparent discrepancy between the values in the table.

in a lower BMR except when compared to charadriiform species. However, at larger body sizes (>1 kg), pelecaniform BMR exceeds that of the procellariiforms. The number of pelecaniform species in our analysis is relatively small (12) and there is a greater variance in both the intercept and the exponent of that equation (reflected also in the low R^2 value). More data on a variety of pelecaniform birds would be useful.

Finally, we would like to address the predictive value of allometric equations. We feel that enough birds fall away from allometric predictions that allometric equations must be used with care. Using an equation to predict BMR and then treating it as fact remains risky, a point also noted by Bryant and Furness (1995). In spite of our hesitancy about using allometric equations for prediction, we know they will inevitably be used that way (e.g., Ellis 1984). If that be the case, we urge readers to pay close attention to the standard errors and R^2 values we provide; only Equation 11.5 and the procellariiform equation (Table 11.3) should even be considered for such use. Given that caveat, we present in Table 11.2 every value for BMR that we know.

11.2.3 ANTICIPATED CORRELATES OF BMR

We tested BMR as a function of: (1) taxonomic order, (2) latitude/region, (3) ocean regime, (4) season, (5) activity mode, and (6) body mass. Of these parameters, only order and latitude increase the ability of body mass to predict BMR. Of those two, latitude was the more important. Using N = 107 studies on 76 species, we found

$$BMR = 1.865 \ (mass^{0.712})[exp_{10} \ (latitude)]^{0.0047} \tag{11.6}$$

where BMR remains in kJ d^{-1}, mass in g, and latitude in degrees (intercept s.e. = 1.120; body mass s.e. = 0.015; and latitude exponent s.e. = 0.001; R^2 = 0.958). The inclusion of order does not increase the predictive value much (R^2 = 0.966). This confirms the importance of latitude in seabird BMR first noted by Ellis (1984) for charadriiforms and extended to other seabird taxa by Bryant and Furness (1995).

A correlate of BMR found in birds (McNab 1988) and mammals (McNab 1986a, b) is food habits. We failed to find such a relationship among seabirds, probably owing to the lack of variety in diet among these carnivores. Whether some relationship may eventually be found that allows, for example, filter-feeders (of plankton) to be separated from feeders of whole fish or squid by BMR awaits a more comprehensive data set.

Ellis (1984) suggested a correlation between activity mode, in terms of flight or feeding, and BMR. That was not verified statistically in this study, when looking at all seabirds as a group. Whether it exists within specific taxa is currently unknown and may also, for some taxa, require a larger data set.

11.2.4 UNUSUAL CORRELATES OF BMR

Basal metabolic rate can be invoked as a correlate of several characters in the life histories and demographics of birds. One of these is life span, since life span in birds scales positively with body size (Lindstedt and Calder 1976), which is the major predictor of BMR as noted above (Figure 11.2; see Chapters 5 and 8). Similarly, mass-specific BMR can be inferred to vary inversely with life span. For example, long-lived Laysan Albatrosses (*Phoebastria immutabilis*) have a low BMR (Grant and Whittow 1983) based on the predictions of Equation 11.5 or even the procellariiform equation (Table 11.3). However, there has not yet been a systematic study of the relationship of BMR and life span in seabirds or any other birds in spite of Calder's (1985) hypothesis. A particularly interesting correlate of BMR is the intrinsic rate of reproduction (r). McNab (1980a, 1987) and Hennemann (1983b) suggested a positive correlation between BMR and r, both factors under the control of natural selection. Though Hennemann's formulation has been challenged

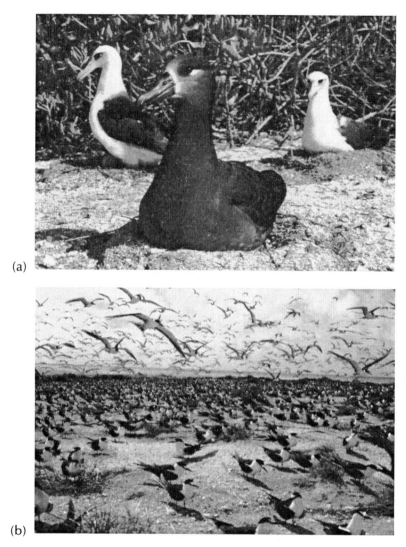

FIGURE 11.2 Body size scales directly with BMR: (a) BMR in albatrosses ranges from 637 to 1755 kJ d^{-1}, here Laysan and Black-footed Albatrosses weighing 3000 g; (b) BMR of Sooty Terns is 69 kJ d^{-1}, body mass 150 g. (Photos by R. W. and E. A. Schreiber.)

(Hayssen 1984), testing this imputed association may be of great value to seabird biologists looking for relationships between reproductive effort and energy costs.

Another interesting correlate of BMR is the cost of feather production. Lindström et al. (1993) demonstrated that the cost of feather production (C$_f$ in kJ g^{-1} of dry feathers) is a function of mass-specific BMR. They found

$$C_f = 270 \text{ BMR m}^{-1} \tag{11.7}$$

where BMR is in units of kJ g^{-1} d^{-1}. They further inferred an inverse relationship between body mass and molt efficiency. Recent work on penguin molting (Cherel et al. 1994) seems to confirm this relationship and therefore suggests confirmation of Equation 11.7 for seabirds as well (see Section 11.4.2).

Once it was recognized that different taxa have different evolutionary molecular clocks (see Nunn and Stanley 1998), efforts were made to determine the factor or factors that set that rate.

Martin and Palumbi (1993) suggested that metabolic rate was the key determinant because it was related to higher mutation rates. Nunn and Stanley (1998), recognizing the close correspondence of FMR and especially BMR with body mass, used body mass as a surrogate in their analysis of 85 species of procellariiform seabirds. They concluded that in these seabirds, metabolic rate was the most likely factor setting the rate of change in the mitochondrial gene for cytochrome b. Stanley and Harrison (1999) subsequently explained why molecular clocks in birds were slower than those of mammals, despite higher metabolic rates in birds, by reconciling the avian constraint hypothesis, which argues that increased functional constraint in birds limits substitutions of mutations, with the metabolic rate hypothesis. This work is likely to stimulate new areas of research for birds generally and may lead to the justification of many more BMR measurements. One question that might be addressed is how very different metabolic rates in closely related birds (e.g., *Egretta*; see Ellis 1980b) may affect this analysis.

11.2.5 LONG-TERM FASTING METABOLISM

While the measurement of BMR is dependent upon the animal being postabsorptive, this involves a fast of only 8 to 14 h. However, several seabirds are deprived of food for longer periods during incubation. The best known of these are the penguins, albatrosses, and eiders which can go from several days to weeks without food (e.g., Croxall 1982, Gabrielsen et al. 1991a). During these long-term fasts, the metabolic substrates can change from a largely lipid form to include more protein (Groscolas 1990), which may be reflected in an increase in the RQ of the animal. A description of the physiology and biochemistry of this kind of fast may be found in Le Maho (1993) and Cherel et al. (1988) who describe the three phases of fasting. Briefly, Phase I is a period of adaptation and lipid mobilization; body mass decreases with BMR decreasing even faster. Phase II is a period of reduced activity and slow loss of body mass; mass-specific BMR reaches an equilibrium, and 90% or more of the metabolic substrate is lipids. It is in Phase III that proteins may be mobilized; daily body mass loss increases rapidly, and various behaviors, including locomotor activity, return, perhaps as a hormonal "refeeding signal" to improve the bird's chances of survival (Robin et al. 1998). These changes in metabolic activity should be noted, because many studies on the costs of molt (Section 11.4.2) and incubation (see Chapter 12 and Section 11.5.1.1 below) have been done on birds during long-term fasting.

11.3 SEABIRD THERMOREGULATION

When physiological studies of thermoregulation were still relatively new, Scholander et al. (1950a, b, c) argued that birds and mammals in cold climates could evolve higher metabolic rates (BMR) or lower thermal conductance (that is, better insulation). They demonstrated the latter, but not the former. However, Weathers (1979) and Hails (1983) showed some effect of climate on BMR in birds. Ellis (1984), using latitude as a general proxy for climate, also demonstrated a BMR correlation for charadriiform seabirds. Reducing thermal conductance would reduce the lower critical limit of an endotherm's thermoneutral zone (TNZ), thus effectively extending downward the range of temperatures at which its metabolism could remain basal. In this section, we address both thermal conductance and the lower critical temperature.

Seabirds have metabolic rates that are somewhat higher than would be expected from an analysis of all nonpasserine birds. Climate might be one reason for this. Due to sea-surface temperatures (SST), tropical seabirds often have cooler environments than their terrestrial counterparts. Polar seabirds may actually benefit in winter from the moderating temperatures of the sea when compared to their terrestrial counterparts. Unlike the majority of polar land birds, many seabird species do not migrate to warmer climates during winter. Whether higher metabolic rates are accompanied by increases in insulation or reductions in the lower critical limit of the thermoneutral zone has not been analyzed in a comprehensive way for seabirds. We present a preliminary analysis here but studies of the thermal biology of seabirds at different latitudes and under different conditions are needed. Aside from a study on the influence of wind speed on thermal conductance in Adélie Penguins and Imperial Shags (*Notocarbo atriceps*) by Chappell et al. (1989), these are not yet available.

11.3.1 THERMAL CONDUCTANCE

Thermal conductance (C) is a coefficient of heat transfer (Calder and King 1974) and is inversely related to insulation. It is the sum of many processes, including radiation, conduction, and convection. Whether it should also include the evaporative process is the subject of some debate. McNab (1980b) distinguishes between "wet" conductance, which includes the evaporative factor, and "dry" conductance, which does not. Drent and Stonehouse (1971) compared the wet and dry thermal conductances of many species, and the difference decreased with increasing size. Of the 16 species in their study exceeding 100 g, wet conductance averaged 15.5% higher than dry. In the only two seabirds in that analysis, the Common (Mew or Short-billed) Gull (*Larus canus*) and Humboldt Penguin (*Spheniscus humboldti*) both showed a difference of 11%. The difference between wet and dry thermal conductance in Double-crested Cormorants (*Hypoleucos auritus*) was also small (13%, which was not significant), though in the same study (Mahoney 1979) a large and significant difference of 31.5% was found in Anhingas (*Anhinga anhinga*).

We have found 37 values for C in seabirds (see Table 11.4), a mix of wet and dry values. Because the differences are likely to be small (≤15%), we do not distinguish between them in our analysis. Most are "wet." It should be noted, however, that these differences often become exacerbated when the correction of Dawson and Whittow (1994) is applied to one set of the data. Using the same data set, Ellis et al. (1982b) referred to a wet thermal conductance 25% higher than the dry, "corrected" values later reported for Brown Noddies (*Anous stolidus*) by Ellis et al. (1995) and cited in Table 11.4.

A more fundamental difference involves the nature of the measurement. Originally, thermal conductance was measured as a function of body surface area. This made sense, since heat exchange is across the surface; it also conforms to the definition provided by Bligh and Johnson (1973). But beginning with Morrison and Ryser (1951), McNab and Morrison (1963), and Lasiewski et al. (1967), conductance was reported as a function of body mass. In our review, we favor the use of body mass since surface area is not measurable, varies with posture, erection of feathers, etc., and is approximated by (Meeh's) equation. Prosser (1973) viewed this approximation as a source of error. McNab (1980b) also noted that having surface area in the units for thermal conductance makes them inconsistent with the units typically reported for metabolism. Luna-Jorquera et al. (1997), analyzing the use of Meeh's equation in penguins, argued that surface area is too problematic a measure and urged the use of body mass in the reporting of thermal conductance. Consequently, we use a modified Meeh's equation to back calculate all values of thermal conductance in surface area units to body mass units (kJ g^{-1} h^{-1} °C^{-1} rather than kJ cm^{-2} h^{-1} °C^{-1}). As with BMR, these are derived units, so wherever possible we began with the original units for oxygen consumption, and converted assuming RQ = 0.71 and a conversion of 19.8 kJ/L O$_2$. Where the original data were already in heat or caloric equivalents, there exists the possibility of a 5% overestimate, as noted above. Finally, because avian conductance often drops with decreasing ambient temperatures (Drent and Stonehouse 1971), wherever possible we follow the convention of McNab (1980b) in using the lowest values of C at which the birds are still thermoregulating. This is the minimal thermal conductance.

Allometric relationships for thermal conductance in birds have been reported by Herreid and Kessel (1967) using cooling curves, Lasiewski et al. (1967) using metabolic data, Calder and King (1974) combining both 1967 data sets, and Aschoff (1981) who distinguished between active and resting phases. Seabirds barely contributed to any of those curves. Weathers et al. (2000) presented thermal conductances for 17 species of seabirds, but all were from high latitudes. The data set

TABLE 11.4
Body Mass, Thermal Conductance (C), and Lower Critical Temperatures (LCT) in Seabirds, by Breeding Region

Order/Species	Body Mass (g)	n	C (mL O_2 g^{-1} h^{-1} $^{\circ}C^{-1}$)	LCT ($^{\circ}C$)	Latitude/ Region (degree)	Source
Sphenisciformes						
Gentoo Penguin	5850		0.0222			Scholander et al. 1940
Pygoscelis papua						
Adelie Penguin	3980	5	0.0132	10	65 S	Chappell et al. 1989
P. adeliae						
Emperor Penguin	23370	5	0.007	–7	78 S	Pinshow et al. 1977
Aptenodytes forsteri						
Blue Penguin	900	6	0.0346	10	41 S	Stahel and Nicol 1982
Eudyptula minor						
Procellariiformes						
Southern Fulmar	780	5	0.036	5.6	69 S	Weathers et al. 2000
Fulmarus glacialoides						
Northern Fulmar	651	16	0.0336	9	79 N	Gabrielsen et al. 1988
F. glacialis						
Antarctic Petrel	718	6	0.037	6.4	69 S	Weathers et al. 2000
Thalassoica antarctica						
Cape Pigeon	420	7	0.058	10.8	69 S	Weathers et al. 2000
Daption capense						
Snow Petrel	292	6	0.058	13.6	69 S	Weathers et al. 2000
Pagodroma nivea						
Wedge-tailed Shearwater	321		0.0625	22.5	20 N	Whittow et al. 1987
Puffinus pacificus						
Manx Shearwater	413	8	0.0513		62 N	Bech et al. 1982
P. puffinus						
Georgian Diving-petrel	119	5	0.070	20	54 S	Roby and Ricklefs 1986
Pelecanoides georgicus						
Common Diving-petrel	132	4	0.070	20	54 S	Roby and Ricklefs 1986
P. urinatrix						
Wilson's Storm-petrel	36		0.117	16	64 S	Obst 1986
Oceanites oceanicus						
Leach's Storm-petrel	45	4	0.0318	14	45 N	Ricklefs et al. 1986
Oceanodroma leucorhoa						
Leach's Storm-petrel	47	7	0.0222		47 N	Montevecchi et al. 1991
O. leucorhoa						
Pelecaniformes						
Magnificent Frigatebird	1100	4	0.023	20	9 N	Enger 1957
Fregata magnificens						
Red-footed Booby	994	4	0.0394	19	21 N	H. Ellis unpublished
Sula sula						
Double-crested Cormorant	1500	12	0.0492		26 N	Mahoney 1979
Hypoleucos aristotolis						
Imperial Shag	2630	6	0.0278	0	65 S	Chappell et al. 1989
Notocarbo atriceps						
Charadriiformes						
Heerman's Gull	383	5	0.0506	23	32 N	H. Ellis unpublished
Larus heermanni						

TABLE 11.4 *(Continued)*
Body Mass, Thermal Conductance (C), and Lower Critical Temperatures (LCT) in Seabirds, by Breeding Region

Order/Species	Body Mass (g)	n	C (mL O_2 g^{-1} h^{-1} $°C^{-1}$)	LCT (°C)	Latitude/ Region (degree)	Source
Ring-billed Gull	470	2	0.0443	16	29 N	Ellis 1980a
L. delawarensis						
California Gull	565	5	0.0412	20	38 N	H. Ellis unpublished
L. californicus						
Glaucous Gull	1326	9	0.0248	2	79 N	Gabrielsen and Mehlum 1989
L. hyperboreus						
Herring Gull	1000	6	0.0385	10	45 N	Lustick et al. 1978
L. argentatus						
Laughing Gull	278	4	0.0559	22	29 N	Ellis 1980a
L. atricilla						
Black-legged Kittiwake	365	16	0.0466	4.5	79 N	Gabrielsen et al. 1988
Rissa tridactyla						
Ivory Gull	508	2	0.0488	0.5	79 N	Gabrielsen and Mehlum 1989
Pagophila eburnea						
Royal Tern	386	3	0.0612	23	29 N	Ellis 1980a
S. maxima						
Sooty Tern	150	4	0.084	30	21 N	MacMillen et al. 1977
Sterna fuscata						
Brown Noddy	140	15	0.0513	20	21 N	Ellis et al. 1995
Anous stolidus						
Dovekie	153	23	0.063	4.5	79 N	Gabrielsen et al. 1991b
Alle alle						
Common Murre	956	4	0.0492	5	65 N	Johnson and West 1975
U. aalge						
Thick-billed Murre	819	11	0.0282	2	79 N	Gabrielsen et al. 1988
Uria lomvia						
Black Guillemot	342	13	0.0475	7	79 N	Gabrielsen et al. 1988
Cepphus grylle						
Least Auklet	83	5	0.084	15	56 N	Roby and Ricklefs 1986
Aethia pusilla						
Anseriformes						
Common Eider	1661	12	0.024	7	79 N	Gabrielsen et al. 1991a
Somateria mollissima						

provided in Table 11.4 is the first comprehensive compilation of thermal conductances for seabirds from a variety of latitudes. It includes 37 measurements on 35 species. Unlike Aschoff (1981) or the restricted set of thermal conductances presented by Weathers et al. (2000), it does not separate these values into active and passive activity categories. This is because that information was not always available in the studies we cited and because of the absence of a clear activity dichotomy in the BMR data of many birds (see Section 11.2 above). Two of these measurements, both for Leach's Storm-petrel, represent significant outliers. Without them, we found the following relationship for all seabirds:

$$C = 0.435 \ m^{-0.374}$$

(11.8)

where m is mass in g and C in mL O_2 g^{-1} h^{-1} $°C^{-1}$ (N = 35; intercept s.e. = 1.225; exponent s.e. = 0.032; R^2 = 0.806). If the outliers were included, R^2 would drop dramatically to 0.511 and the equation would become 0.231 $m^{-0.281}$ (N = 37 measurements on 36 species; intercept s.e. = 1.337; exponent s.e. = 0.046). Equation 11.8 differs considerably from earlier equations. Compared to the equation of Lasiewski et al. (1967), which like ours also avoids circadian phase, our equation predicts higher values of thermal conductance at all body masses above 150 g.

Thermal conductance varies among seabirds. In accordance with the analysis of Scholander et al. (1950a, c), low thermal conductance (i.e., good insulation) is one adaptation which might be expected in cold climates. On the other hand, high values of C (i.e., poor insulation) would promote convective heat loss and might be expected in warm climates (Yarborough 1971). In a hot climate, forced convection (wind) might be advantageous to a bird, but in a cold climate it represents a real threat, lowering effective operative temperatures (T_e). This must be the case for seabirds nesting in polar areas where a combination of wind and cold temperatures leads to substantial increases in metabolic rates, especially in adults (Chappell et al. 1989).

Avian insulation can derive from either the tissues or the feathers. Drent and Stonehouse (1971) reported that about 20% of the total insulation of the Humboldt Penguin comes from body tissues, including subcutaneous fat, the remainder being from the feathers. That being the case, it is likely that molt should be important in certain seasonal adjustments. The winter acclimatized Common Eider (*Somateria mollissima*) has a C which is 25% lower than the summer acclimatized eider (Jenssen et al. 1989, Gabrielsen et al. 1991a). This is also seen in land birds in the Arctic and sub-Arctic (West 1972, Bech 1980, Rintamäki et al. 1983, Barre 1984, Mortensen and Blix 1986). Mortensen and Blix found that ptarmigans (*Lagopus* spp.) reduced C in the winter by 8 to 32% by increasing subcutaneous fat and plumage thickness. Common Eiders probably reduce insulation in the summer by molting their down (which is then used as nest material) and producing naked brood patches (Gabrielsen et al. 1991a). Females also reduce insulation by losing fat during incubation (Korschgen 1977, Parker and Holm 1990, Gabrielsen et al. 1991a).

Thermal conductance does not seem to vary in a predictable way with latitude (Gabrielsen et al. 1988, 1991a, b). This may be because evolution may modify metabolic rate as well as thermal conductance in cold climates (Scholander et al. 1950a, c). But comparing seabirds as a group with land birds does indicate some connection between thermal conductance and climate. As was noted above, polar seabirds may actually be at a thermal advantage compared to polar land birds because of the high heat capacity of water and its moderating effect on climate. In fact, many land birds have a better insulation. Both arctic-breeding ravens (Schwann and Williams 1978) and ptarmigan (West 1972, Mortensen and Blix 1986) have lower values of C than do seabirds, indicating that these permanent residents may be better cold adapted than seabirds.

11.3.2 Lower Limit of Thermoneutrality

The lower critical temperature (LCT) or lower limit of thermoneutrality is an indicator of thermoregulatory ability since below that level metabolism must increase. Scholander et al. (1950b) demonstrated the value of a reduced LCT in the metabolic economy of endotherms. Table 11.4 shows that, as expected, seabirds show an inverse relationship between size and LCT. We also find that there is an influence between LCT and latitude, with Arctic and Antarctic birds having a lower LCT than birds of similar mass from warmer climates. These relationships can be expressed by the equation

$$LCT = 43.15 - 6.58 \log mass - 0.26 \ latitude \qquad (11.9)$$

where LCT is in degrees Celsius; mass is in g; latitude in degrees (N = 33; intercept s.e. = 3.94; log mass coefficient s.e. = 1.43; latitude coefficient s.e. = 0.03; R^2 = 0.779).

11.3.3 BODY TEMPERATURE

Deep body temperature (T_b) is dependent on metabolic rate and insulation (Irving 1972). There is no evidence that body temperature varies with climate or latitude across a range from the Arctic through temperate and tropical to Antarctic regimes (Scholander et al. 1950c, Irving and Krog 1954, Drent 1965, Irving 1972, Barrett 1978, Prinzinger et al. 1991, Morgan et al. 1992). Body temperatures in seabirds are typical of birds generally, though Prinzinger et al. (1991) found T_b to be lower in Procellariiformes and Sphenisciformes than the average for all birds. The earliest measurements were by Eydoux and Souleyet (1838; cited in Warham 1996) on procellariiforms and Martins (1845) who measured T_b at 40.6°C in ten species of "webfooted" birds during summer expeditions to Svalbard in 1838 and 1840. We do not know the species in the Martins study, but they probably included Common Eider, Glaucous Gull (*Larus hyperboreus*), kittiwakes, and alcids. His value is very close to those presented in later studies of Arctic and sub-Arctic seabirds (Irving 1972). In the Antarctic, body temperature remains at expected avian levels (Chappell et al. 1989, Weathers et al. 2000). On the other hand, some tropical species allow T_b to show some lability under different conditions and even fall somewhat (Red-footed Boobies, *Sula sula* [Shallenberger et al. 1974]; Great Frigatebirds, *Fregata minor* [Whittow et al. 1978]).

While T_b is resistant to climate, it is linked tightly to metabolic rate. If metabolism drops for any reason, T_b may drop as well. This is the case with the Atlantic Puffin (*Fratercula arctica*) which can lower its RMR while incubating to conserve its energy reserves. Consequently, T_b drops and incubation times are lengthened (Barrett et al. 1995). There seems to be some linkage to BMR as well: procellariiform birds as a group have somewhat lower BMR than other seabirds (see Section 11.2.2) and their body temperatures are also lower (Warham 1971, 1996).

Body temperature may vary as a function of activity phase. Typically, birds that show a reduction in metabolic rate during the ρ-phase also show a depression in T_b (cf. Warham 1996). Great Frigatebirds drop T_b by 3 to 4°C during the night (Whittow et al. 1978). The linkage between T_b and metabolism is not dependent only on activity phase. Regel and Pütz (1997) found that Emperor Penguins (*Aptenodytes forsteri*) showed increases in body temperature as a function of human disturbance as mediated by metabolic rate.

Body temperature may also be affected by the water which, because of its high heat capacity, can represent an enormous heat sink when cold. Dumonteil et al. (1994) found T_b to remain very constant in water, although it was slightly (0.3°C) depressed below measurements in air. Bank Cormorants (*Compsohalieus neglectus*) show a more pronounced T_b depression in the water, either because of poor insulation or insufficient heat production from swimming activity. These birds may allow T_b to drop as much as 5°C while diving to save energy (Wilson and Grémillet 1996), regaining it quickly through sunning behavior out of the water (Grémillet 1995). On the other hand, Great Cormorants (*Phalacrocorax carbo*), which do not experience as much solar radiation as Bank Cormorants, show smaller depressions of T_b and have better insulation (Grémillet et al. 1998). Imperial Shags (Bevan et al. 1995a, Grémillet et al. 1998) and South Georgia Shags (*Notocarbo georgianus*, Bevan et al. 1997) in Antarctic seas face such cold waters and dive so deeply they cannot prevent T_b from dropping. The T_b of South Georgia Shags may drop by 5°C or more during diving. Abdominal temperature in King Penguins (*Aptenodytes patagonicus*) may fall to as low as 11°C, 10 to 20° below the normal stomach temperature. A slowing of metabolism in certain anatomical areas when diving may help explain why penguins can dive for such long durations (Handrich et al. 1997). Similar studies on diving birds in warm water do not exist.

Deep core temperatures monitored by implants in or near the stomach are likely to be distorted by feeding in free-ranging birds. The ingestion of food in petrels (Obst et al. 1987), boobies (Shallenberger et al. 1974), and cormorants (Ancel et al. 1997) is known to drop stomach temperature by 5°C or more. While there are obvious advantages to knowing when a diving bird ingests prey, the effect that event has on T_b needs to be understood better. Handrich et al. (1997) reported that low abdominal temperatures may preserve food until the bird reaches its chicks in the colony.

11.4 OTHER COSTS

BMR is defined for very specific sets of conditions, as noted above. If any of the restrictions are violated, metabolism is not basal. However, the metabolic rates then measured may convey additional information. Metabolism in nonpostabsorptive birds, for example, may provide information on the costs of digestion. Similarly, the costs of molt and locomotion have been quantified. Croll and McLaren (1993) provided one such measure which is otherwise rare in the seabird literature. They found the cost of preening in murres to be 2.5 to 3 × RMR which was the most expensive activity these birds engaged in. Earlier Butler and Woakes (1984) had reported a preening cost in Humboldt Penguins of just over twice resting rates. Croll and McLaren (1993) suggested that the high increase in metabolic rate in preening murres might be linked to producing more heat for thermoregulation in cold water.

11.4.1 DIGESTION

The cost of digestion is often referred to as specific dynamic action (SDA) in the older literature, and today is more often referred to as the heat increment of feeding (HIF). The heat produced by digestion is transient, but it may aid thermoregulation (Hawkins et al. 1997), though Dawson and O'Connor (1996) did not find such a connection for most birds in their review. Baudinette et al. (1986) found metabolic rate in Blue Penguins increased by 87% as a result of feeding. The increment is smaller, though still appreciable (36 to 49%) in Common and Thick-billed Murres according to two studies (Croll and McLaren 1993, Hawkins et al. 1997). Hawkins et al. suggested that this increment could be responsible for nearly 6% of the daily energy expenditure of either murre species. However, caution is urged because Wilson and Culik (1991) found the increase in metabolic rate associated with feeding in Adélie Penguins to result from heating cold food to body temperature rather than actual SDA. Weathers et al. (2000) discussed the effect of HIF on nestling metabolic rates in four Antarctic fulmarine petrels. They do not attribute a thermoregulatory role to HIF in these birds.

11.4.2 MOLT

The metabolic cost of molt in birds was not known in any detail until late in the 20th century (King 1974, 1981). Murphy (1996) provides an excellent summary of the energetics of molt, but provides no information about seabirds. Among seabirds, molt has been best studied in penguins and was reviewed by Adams and Brown (1990). This section supplements that work with some more recent information and some slightly different perspectives. Readers concerned with the mechanisms of molt in penguins are referred to Groscolas (1990).

Adams and Brown (1990) evaluate the use of mass loss in estimating the energetic cost of molt in penguins. Based on mass loss, Williams et al. (1977) estimated the cost of molt to be 1.6 and 2.1 × BMR for Macaroni Penguins and Rockhopper Penguins, respectively. However, these multiples were based on predictions from the Lasiewski-Dawson (1967) allometric equation, and the mass losses assumed a large component of fat during molt. Relying primarily on studies using mass loss, Croxall (1982) estimated the cost of molt at twice BMR and established that only about half the material lost was fat, which had clear energy implications. Brown (1985) underscored this by comparing the cost of molt in Macaroni and Rockhopper Penguins using both mass loss and oxygen consumption. Using mass loss, he estimated the cost to be 1.96 and 1.79 × IMR (incubation metabolic rate, a value Brown felt was close to BMR; see Whittow on IMR, Chapter 12), respectively; but with oxygen consumption the multiples were 1.81 and 1.50. These two sets of figures could be partially reconciled by reducing the proportion of fat in the mass loss below the level assumed by Williams et al. (1977). Groscolas and Cherel (1992) reported the daily rate of mass specific weight loss to double in King Penguins and increase fivefold in Emperor Penguins during molt compared to breeding, suggesting a high associated cost of molt. Cherel et al. (1994) used

FIGURE 11.3 In King Penguins (Crozet Island), adults during the breeding season (here incubating eggs on their feet) have a significantly lower metabolic rate of fasting than when fasting during molt, implying a high cost of molt. (Photo by H. Weimerskirch.)

mass loss to estimate the cost of molt in King Penguins; it agreed with a value determined by indirect calorimetry. They found the metabolic rate of fasting King Penguins in molt to be 21% higher than in birds that were fasting during the breeding season (Figure 11.3). Their value for cost of molt as a multiple of BMR depends on the value for BMR used. It is 1.30 × BMR as determined by Le Maho and Despin (1976) but 1.67 × BMR (Adams and Brown 1990). These values bracket the 50% increase in Blue Penguins (Baudinette et al. 1986, Gales et al. 1988). Both Baudinette et al. (1986), using oxygen consumption in confined birds, and Gales et al. (1988), using doubly labeled water in free-ranging penguins, found the cost of molt to be 1.5 × BMR. However, they used different values for BMR (see Section 11.2.1). If Gales et al. had used the average value reported by Baudinette et al. (1986), or Stahel and Nicol (1988) instead of Stahel and Nicol (1982), their multiple would have been 2.6 × BMR.

Murphy (1996) reported that the energy content of feathers and other associated keratinous structures is 22 kJ g⁻¹ of dry mass and argued that the cost of depositing these structures should be minimal, perhaps <6% of BMR. However, the actual energy costs of molt are higher because of associated costs including the processing and utilizing of nutrients for feather growth, specific nutritional costs associated with molt, etc. (King 1981, Lindström et al. 1993, Murphy 1996). These associated costs may not include additional thermogenesis, which Murphy (1996) discounted as a problem in most birds (but see Groscolas and Cherel 1992 for a different view regarding penguins). She cites a total cost of molt between 109 and 211% of nonmolt (BMR?) levels. Values for penguins, which have a more intense molt than most other birds, tend toward the upper end of that range. Lindström et al. (1993) looked at energetic efficiencies (energy deposited as feathers and associated structures divided by the feather mass specific cost of molt) of several avian species (none seabirds). They found efficiencies to increase with increasing body mass because the cost of feather production was inversely related to mass. This is validated by Cherel et al. (1994) who found the lowest cost of feather production (85 kJ g⁻¹) and one of the highest efficiencies (25%) in King Penguins, which began their molting fasts at 18 kg and ended them at a still quite large 10 kg.

11.4.3 Locomotion

Seabirds move by flight, swimming, and walking, though several species are incapable of at least one such form (e.g., some of the better diving birds such as tropicbirds, loons, and grebes have legs so far back that they cannot walk; penguins cannot fly; frigatebirds and skimmers do not swim).

The energetics of flight in birds generally was reviewed recently (Norberg 1996, Butler and Bishop 2000). Two papers (Pennycuick 1987a, b) missed in those reviews add to our understanding of flight in seabirds. Pennycuick (1987b) noted that in spite of the great variety of feeding methods and provisioning frequencies found in seabirds, the only factor that has had a "drastic" effect on flight adaptations is the use of wings under water. That is obvious in penguins and will be noted below for alcids. Those interested in the full range of physiological trade-offs between flight and diving should consult Lavvorn and Jones (1994).

The costs of flight in particular species of seabirds was noted in Ellis (1984). Wind seems to be a major environmental factor. Sooty Terns have a low cost of flight due to their partial reliance on soaring (Flint and Nagy 1984). Red-footed Boobies also take advantage of the wind during flight and show considerably lower costs than would otherwise be expected (Ballance 1995). This was also inferred for Gray-headed Albatrosses (*Thalassarche chrysostoma*); the indirect measure of their flight costs was compared also to those of other seabirds known at that time (Costa and Prince 1987). The geographic distribution of the Wandering Albatross (*Diomedea exulans;* Jouventin and Weimerskirch 1990) and Northern Fulmar (*Fulmarus glacialis*; Furness and Bryant 1996) may be limited by the absence of wind. Boobies and frigatebirds roost in greater numbers during low or no-wind days implying a greater cost of flight on those days (Schreiber and Chovan 1986, Schreiber 1999). On the other hand, wind has been reported to increase the cost of flight (Black-legged Kittiwakes and Dovekies, *Alle alle*; Gabrielsen et al. 1987, 1991b).

11.4.3.1 Swimming

Large numbers of species of seabirds swim on the surface of the water; fewer swim under the surface. Of those that do, penguins, alcids (auks and their relatives), sulids (gannets and boobies), and some shearwaters propel themselves under water with their wings, whereas tropicbirds, diving petrels, and cormorants use their feet, as do the seasonally marine grebes and loons. Some of the larger procellariiforms (albatrosses and shearwaters) use both modes. The fact that many albatrosses dive at all was not well known until recently (Prince et al. 1994). In this section, the terms *diving* and *subsurface* or *underwater swimming* are used synonymously.

The earliest examination of the energetics of surface swimming was on ducks (Prange and Schmidt-Nielsen 1970). Most of the information developed recently on the energetics of diving has been for the wing-propelled groups. Baudinette and Gill (1985) compared surface and underwater swimming in Blue Penguins and found a 40% reduction in the cost of a penguin swimming below the surface compared to one swimming at the surface. Several studies have shown that as speed increases, birds that have a choice switch from surface to underwater swimming which can be accomplished more cheaply at higher speeds (Baudinette and Gill 1985, Hui 1988a). The greater efficiency of penguins may be gauged in a comparison of the metabolic costs of wing-propelled Humboldt Penguins at 1.26 × RMR (Butler and Woakes 1984) with wing-propelled Common Murres at 1.8 × RMR and Thick-billed Murres at 2.4 × RMR (Croll and McLaren 1993) and foot-propelled divers (Tufted Ducks at 3.5 × RMR; Woakes and Butler 1983). Schmid et al. (1995) reported a multiple nearly 12 × BMR (daytime) and 2.6 × RMR (in water) in the Great Cormorant (foot-propelled). Given the paucity of data in foot-propelled divers, this very high value cannot be easily evaluated.

Cormorant feathers are more wettable than other diving birds, so buoyancy is a relatively small problem for them (Schmid et al. 1995, Grémillet et al. 1998). That suggests that one reason given for the poorer performance of ducks and alcids (greater costs of overcoming buoyancy; Woakes and Butler 1983, Croll and McLaren 1993) may not be as important as previously thought (but see Ancel et al. 2000). However, thermoregulatory costs may add to the high expense of diving in cormorants (Schmid et al. 1995, Grémillet and Wilson 1999, Ancel et al. 2000; but see also Section 11.3.3 above). Potential thermoregulatory costs may be countered by more fat insulation, but that may confer additional costs for flight (Butler 2000). A more fundamental difference may be that wing-propelled diving is cheaper than foot-propelled diving, and that wings uncompromised by the demands of flight confer an additional advantage.

Total efficiency of swimming is the ratio of power input (the product of drag and speed) to metabolic power output. In surface swimming, the efficiencies of Mallards (*Anas platyrhynchos*; Prange and Schmidt-Nielsen 1970), Black Ducks (*A. superciliosa*; Baudinette and Gill 1985), Blue Penguins (Baudinette and Gill 1985), and Humboldt Penguins (Hui 1988a) are remarkably similar: 4 to 5%. However, maximal efficiency for Humboldt Penguins is achieved when swimming under water; it is 19.2% (Hui 1988a). Hui attributes the increased efficiency to the greater proportion of wing muscles to body mass in penguins compared to the proportion of leg muscles in ducks. Efficiencies can often be reflected in the cost of transport (COT), which is the metabolic expenditure needed to move a unit of mass a unit distance (usually oxygen consumption or SI units of energy times kg^{-1} m^{-1}). Typically, it is the minimal COT which is reported. Blue Penguins swimming underwater have lower costs of transport than surface-swimming birds (Baudinette and Gill 1985); their costs are comparable to those found for Humboldt Penguins (Hui 1988a) and Jackass Penguins (*Spheniscus demersus*; Nagy et al. 1984), 13.5 to 15.5 J kg^{-1} m^{-1}. More recent studies that use birds that dive voluntarily and do not carry external devices indicate that COT values may be much lower in diving penguins. Culik et al. (1994) report values of 7.1, 6.3, and 8.9 J kg^{-1} m^{-1} for Adélie, Chinstrap (*Pygoscelis antarctica*), and Gentoo (*P. papua*) Penguins, respectively. Using a similar analysis, Luna-Jorquera and Culik (2000) found a comparably low cost of transport, 6.8 J kg^{-1} m^{-1} in Humboldt Penguins. A still lower value of 4.7 J kg^{-1} m^{-1} has been reported for King Penguins (Culik et al. 1996). This lower COT increases still further the difference between surface and underwater swimming. By contrast, minimal COT = 19 J kg^{-1} m^{-1} in foot-propelled Great Cormorants (Schmid et al. 1995) and Brandt's Cormorants (*Compsohalieus penicillatus*; Ancel et al. 2000).

The effect of using external devices on birds for which either swimming metabolism or dive performance is measured has been questioned. In a swim channel, Adélie Penguins (Culik and Wilson 1991b) and Great Cormorants (Schmid et al. 1995) carrying external packs had higher costs of transport largely due to increases in drag; the penguins even had higher RMR values than controls. Culik and Wilson.(1991b) predicted that penguins and alcids so instrumented would show reduced speeds, smaller foraging ranges, and lower food acquisition. Ropert-Coudert et al. (2000), using free-ranging animals, confirmed this with King Penguins carrying external packs. Their proportion of consecutive deep dives was reduced compared to birds with internal instrumentation. Ropert-Coudert et al. join Culik and Wilson (1991b) in recommending internal instrumentation in studies of free-living diving birds. However, the implanting of such devices requires a level of surgical skill not necessary with external devices.

The multiples of BMR or RMR noted above are all low, with the possible exception of the Great Cormorant, compared to the maximum multiples we see in birds for aerial or cursorial locomotion. It is reasonable to infer that maximal metabolic rates were never achieved in these studies. In the case of the surface swimmers, the reason was first proposed by Prange and Schmidt-Nielsen (1970), later confirmed by Baudinette and Gill (1985): surface-swimming birds cannot exceed a particular "hull speed" dictated by forces of drag even if they have more metabolic capacity available. In the case of diving birds, it is likely that maximal speeds and thus power output were not achieved under experimental conditions. However, Kooyman and Ponganis (1994) attempted to achieve such a power output by attaching loads to swimming Emperor Penguins. Although they did not find a maximum metabolic rate, they felt that the 7.8 × RMR was close to it. Because they were hesitant to accept RMR as true BMR (for reasons noted also above; Kooyman personal communication), they also provide a multiple of 9.1 × the value predicted by Aschoff and Pohl (1970) for a 20.8-kg bird. Either multiple is smaller than found in running or flying birds, which Kooyman and Ponganis (1994) attribute to a higher anaerobic capacity of (Emperor) penguin muscles and the ability to conserve oxygen for longer periods while diving (see also Kooyman et al. 1992a). It is widely thought that diving birds, especially penguins, will attempt to remain within their aerobic dive limit (ADL), which is the dive duration that produces no increased lactate levels after a dive. Since ADL is rarely measured, a calculated version (cADL) is often used. Analyzing these data for three penguin species, Butler (2000) concluded that the cost of normal dives may be very close to RMR values in the water. This surely is not true for cormorants (Ancel et al. 2000) and warrants additional testing.

The energetics of swimming in penguins is treated in several reviews (Oehme and Bannasch 1989, Croxall and Davis 1990, Kooyman and Ponganis 1990). Croxall and Davis (1990) also presented a valuable analysis and critique of methods used. One concern raised by Butler and Woakes (1984) was that attempts to quantify swimming costs using isotopes (doubly labeled or tritiated water; Kooyman et al. 1982) might confound the costs associated with locomotion and those reflecting thermoregulation. This is only a problem where water temperatures are considerably below the TNZ. An attempt to model the metabolic costs of (underwater) swimming in marine homeotherms, based on pinnipeds, but purportedly applicable to birds as well, is presented by Hind and Gurney (1997). Although it is ancillary to a discussion on metabolic costs, the mechanics of swimming in penguins (Hui 1988b, Oehme and Bannasch 1989) and in foot-propelled swimmers (Lavvorn 1991, Lavvorn et al. 1991) is available. A general review of the hydrodynamics and power requirements of all divers is provided by Kooyman (1989), and Butler and Jones (1997) reviews of the physiology of diving.

11.4.3.2 Walking

LeMaho and Dewasmes (1984) reviewed walking in penguins. In fact, all the work on seabird walking continues exclusively in this group. Although the cost of transport for walking has long been known to be higher than for other modes of locomotion (Baudinette and Gill 1985), the multiple of active metabolic rate to BMR in an extremely cursorial species (Rheas, *Rhea americana;* 35 × BMR) may be the highest locomotion multiple reported in vertebrates (Bundle et al. 1998). To the extent that walking represents a major part of a species' time-activity budget, its energetics is of some importance. The Emperor Penguin has been documented to walk as far as 300 km to get to foraging areas (Ancel et al. 1992).

Pinshow et al. (1977) compared the metabolic rates and costs of transport of Emperor, Adélie, and White-flippered Penguins (*Eudyptula minor albosignata*) with those of other walking birds. They found penguin COT values to be quite high. But Wilson et al. (1999), observing that Magellanic Penguins (*Spheniscus magellanicus*) walked up the slope of a shore from the water's edge at a 39° angle, instead of the shorter 90° angle, concluded that COT in walking penguins may have been overestimated by as much as two times and that waddling walk might not be so expensive as suggested by Pinshow et al. (1977). Griffen and Kram (2000) concluded that the high cost of walking in Emperor Penguins is not due to waddling, which they found actually to conserve energy, but to their short legs which require them to generate muscular force more rapidly. Wilson et al. (1991) showed that tobogganing in Adélie Penguins was less expensive than walking under most conditions, but the savings were countered by feather wear, consequential reduced diving performance, and the added costs of feather maintenance.

11.5 DAILY ENERGY EXPENDITURE AND FIELD METABOLIC RATE IN SEABIRDS

Daily energy expenditure (DEE) is the energetic cost for an animal to live throughout a day during its normal routine. DEE may vary somewhat from day to day and more across seasons. It includes all those general maintenance functions necessary to stay alive and included in measurements of BMR; also included are the cost of thermoregulation and all other activities from feeding to locomotion to reproduction appropriate to the particular part of the annual cycle being studied.

11.5.1 TYPES OF DEE MEASUREMENTS

The development of a daily energy budget was long a goal of those working in the field of energetics. King (1974) explained several ways to estimate energy budgets: extrapolating laboratory measurements of metabolism with or without time-activity budgets, often with activities reported as multiples of BMR; estimating energy consumption by changes in body mass or composition or by feeding activity; comparing activity to heart rate in telemetered birds; and use of doubly labeled water. He considered but rejected use of existence metabolism (see Section 11.5.1.3). Most of these methods have been used with seabirds since that time. Nagy (1989) evaluated some of those methods in a general way. Below, we briefly critique most of the methods commonly found in the seabird literature and then discuss in more detail one of the most direct measures, the field metabolic rate (FMR).

11.5.1.1 BMR Multiples and Mass Loss

Ellis (1984) analyzed the use of time-activity budgets with multiples of BMR in Lesser Sheathbills (*Chionis minor*; Burger 1981). In this study both BMR and individual activities were based on unmeasured estimates — a dangerous decision. Even when BMR and the cost of individual activities are based on actual measurements, there is enormous reliance on the accuracy of the time-activity budgets. For species that stay within sight and whose activities can thus be determined, this may be acceptable. This is rarely the case. Bernstein and Maxson (1985) estimated a DEE for Imperial Shags based on the metabolic measurements of Ricklefs and Matthew (1983), multiples of BMR found in the literature for numerous activities, and time-activity budgets. They reported a DEE/SMR ratio of less than two for both sexes in all phases of the breeding cycle; this is among the lowest ratios in the literature (see Section 11.5.2.3). Variations of this method have been proposed by Blake (1985) whose model included flight parameters for Black Skimmers (*Rynchops niger*) and by Carter and Morrison (1997) whose model for sandpipers was based in part on weather data.

The use of mass loss to estimate DEE, or some component of DEE, was pioneered in the Antarctic on fasting, usually incubating seabirds. It has been used in procellariiform birds (Prince et al. 1981, Croxall and Ricketts 1983, Grant and Whittow 1983, Mougin 1989) and Common Eiders (Gabrielsen et al. 1991a) to measure the costs of incubation; in penguins (Williams et al. 1977, Brown 1985, Adams and Brown 1990, Chercl et al. 1994) and petrels (Croxall 1982) to study the costs of molt. Several studies have noted that substrate use during fasting can vary across species and time (Croxall 1982, Gales et al. 1988, Groscolas 1990); the implication is that assumptions about substrate use in fasting birds can greatly affect the outcome of mass loss studies. This may be illustrated by studies on incubation costs in Wandering Albatrosses (Brown and Adams 1984) and Common Eiders (Gabrielsen et al. 1991a), as well as molt costs in Macaroni and Rockhopper Penguins (Brown 1985) where estimates from mass loss were lower than estimates from oxygen consumption in the first case and higher in the second. These studies cited possible errors in the proportion of fat oxidized during fasting as one possible reason for the discrepancy. The use of mass loss studies in long-term fasting birds may yield different results during different phases of the fast (see Section 11.2.5 above). The use of the correct energy equivalents for fat and proteins is critical when making calculations of the energy cost for incubation and molt in seabirds.

11.5.1.2 Heart Rate

Heart rate (f_H) is known to increase with exercise in birds (cf. Bevan et al. 1994). So if heart rate can somehow be calibrated to metabolic rate, implantable data loggers (Woakes et al. 1995) could provide illuminating data about the cost of particular activities simply by monitoring heart rate. That is exactly the contention of recent studies on Black-browed Albatrosses (*Thalassarche melanophris*; Bevan et al. 1994) and Gentoo Penguins (Bevan et al. 1995b). In these studies, f_H was calibrated to oxygen consumption in treadmill and, in the latter case, swimming channel experiments. In both instances, heart rate was found to be an excellent predictor of metabolism, both in resting and active birds. The advantage of this method is that the cost of individual activities can be monitored in free-ranging birds and that DEE could be partitioned by activity. In addition, the studies can be of longer duration than those using other methods, such as isotopic water. However, this method requires very careful calibration for each species, which may limit its usefulness. In addition, if heart rate is affected by the classic diving bradycardia, as Bevan et al. (1997) found in South Georgia Shags, it may not reflect actual oxygen consumption under certain circumstances. This needs to be tested, especially given the assertion of Kanwisher et al. (1981) that bradycardia is not found in nonstressed diving Double-crested Cormorants. In any event, while f_H allows estimation of metabolic rate during dive bouts (periods under water and at the surface), it cannot measure the cost of diving alone (Butler 2000). At the moment, this method is not yet widely accepted, but its potential, especially for annual energy budgets and when combined with other measures of DEE, is enormous.

11.5.1.3 Existence Metabolism and Metabolizable Energy

Existence metabolism (EM) is the metabolic rate of birds confined to small cages. It includes those costs that go into BMR, as well as the costs of thermoregulation, specific dynamic action, and a small amount of caged activity (Kendeigh 1970). It is typically estimated by measuring the amount of food ingested, the changes in body mass, and the eliminated products of digestion and metabolism. King (1974) discussed some of the problems of estimating the caloric equivalent of weight change; they are similar to the discussion in Sections 11.2.5 and 11.5.1.1 above. Estimations of EM often involve use of metabolizable energy (ME) coefficients that relate ingested food to energy budgets. An equation for the calculation of ME coefficients is provided by Davis et al. (1989). Problems with making certain assumptions about ME are addressed by Miller and Reinecke (1984). Related issues concerning methods using feces production or ecological assimilation are assessed by Nagy (1989).

Allometric predictions of EM exist for passerines and nonpasserines in summer (long photoperiod) and winter (short photoperiod) and at various temperatures (Kendeigh et al. 1977). In spite of the fact that the nonpasserine equations were not based on seabirds, they were often used to predict DEE in seabirds or to model population or community energetics in seabirds; Ellis (1984) reviewed some of the EM equations most often found in the seabird literature. One of the problems with EM, however, is that unlike BMR, it does not represent a limit, so is not easily replicable. Even if temperature were always held constant to control for thermoregulatory variation, the limited locomotion allowed is very difficult to regularize. Especially in seabirds, where swimming or flight may be the normal mode of locomotion, caged activity is often meaningless. We believe King (1974) was correct to avoid the use of EM in estimating DEE. It is an indirect estimate that presumes that all seabirds will follow a model based on very few ecologically equivalent or phylogenetically related species. In spite of its occasional appearance still in the literature (e.g., Gavrilov 1999), and given the direct measurements now available (doubly labeled water, oxygen consumption, perhaps heart rate), its use should be abandoned.

11.5.1.4 FMR and DEE

Field metabolic rate (FMR) has become the expression signifying DEE measurements based on doubly labeled water. First demonstrated by LeFebvre (1964) for pigeons, it has become the most common method for measuring DEE. The method utilizes the turnover of isotopes of hydrogen (either ^2H or more often ^3H) and oxygen (^{18}O) to determine CO_2 production. Its theory was described by Lifson and McClintock (1966) and has been assessed by Nagy (1980, 1989) and Nagy and Costa (1980).

11.5.2 Field Metabolic Rate

Without question, the method most commonly used today in the acquisition of DEE is the field metabolic rate. Nagy and his collaborators have used FMR to partition a variety of components of the DEE (e.g., swimming, brooding, foraging, Nagy et al. 1984; and flight, Flint and Nagy 1984) and even to extrapolate to food requirements (e.g., Nagy et al. 1984; see Section 11.5.2.5 below). Its wide use, however, demands that its liabilities as well as its potential be understood.

11.5.2.1 Conditions and Errors in FMR Studies

Nagy (1980, 1989) addressed problems with errors in isotope concentration in the calculation of FMR, and Nagy and Costa (1980) first considered problems of water turnover. In seabirds, a high water content in food is likely to cause a high water turnover rate in the birds. Both of these require validation studies which are difficult to do in the field. However, there have been comparisons of different methods which, until proper validation studies are done, can be useful. For example, Bevan et al. (1994), in their study of Black-browed Albatrosses, used doubly labeled water, oxygen consumption, and heart rate, finding a strong correspondence in all three.

A very different kind of concern comes from looking at the conditions attendant upon the measurement of FMR. The use of doubly labeled water requires sufficient time for differences in the oxygen and hydrogen isotopes to develop but is limited by the inevitable dilution of those isotopes to immeasurable levels over time. In addition, since animal activity is often tied to a circadian rhythm, the exact length of time between samples needs to be a day or some multiple of days. In most seabird studies, the maximum multiple is likely to be two (this multiple can be increased by switching to a more expensive form of analysis which can detect very low levels of isotope; e.g., Pettit et al. 1988). Measurements that miss the 24- or 48-h interval by more than an hour require a back calculation, introducing a new level of uncertainty (K. Nagy, personal communication). In studies of free-ranging animals, these limitations have led investigators to try to maximize their chance of recovering an injected animal by using animals tied to nest sites. This has created an important bias in the DEE literature for seabirds: almost all the studies have been conducted during the breeding season (a notable exception is Gales and Green 1990). Measurements of FMR tell us much about the DEE during reproduction, but little about most other parts of the annual cycle.

Despite the large number of FMR studies on seabirds during the breeding season, not all of them are comparable. Studies done on incubating birds with long periods of unrelieved incubation (e.g., some penguins and procellariiforms, some tropicbirds) are not equivalent to those done on species that exchange incubation duties with their mates every few hours (e.g., Common Terns, *Sterna hirundo*; Ricklefs et al. 1986; Figure 11.4). Similarly, care must be taken in those species where the period spent at the nest changes during the course of incubation or brooding, or changes between incubation and brooding. Differences within a species may occur due to sex, especially during or just previous to egg-laying when the female costs are almost always elevated (Cary 1996; see Chapter 12).

Another criticism of the use of doubly labeled water, especially by the proponents of the use of heart rate, is that it often integrates many activities. However, as long as an animal's activities can be monitored, FMR can be partitioned (e.g., Nagy et al. 1984). If the actual manipulation of the animal, for example, by the injection of labeled water, affects its subsequent behavior, the value of the study is compromised. This is the contention of Wilson and Culik (1995) who reported that Gentoo Penguins given 5 mL injections in their pectoralis muscles changed their behavior considerably for the next two days, reducing their activity, especially at sea. However, their injection volume is larger than the 1 mL that would be expected with highly enriched water. In fact, this paper may be interpreted as a caution against using less-enriched isotopic water (which requires larger quantities). Alternatively, special methods might be employed such as using water isotonic with the tissue fluids or putting large amounts of water in noncritical muscles (Wilson and Culik 1995, K. Nagy, personal communication). Birt-Friesen et al. (1989) reported that injected Northern Gannets behaved differently than control birds after a 1-mL injection; but they argued that FMR was unaffected. Similarly, injected Arctic Terns (*Sterna paradisaea*; Uttley et al. 1994) reduced feeding of chicks and Black-legged Kittiwakes (Fyhn et al. 2001) showed a reduction in nest

(a)

(b)

FIGURE 11.4 Differences in incubation shift length pose problems in comparing FMR studies among species: (a) A Common Tern pair exchanges incubation duties several times a day, while (b) Red-tailed Tropicbirds incubate for 4 to 8 days in a row. (Photos by J. Burger, R. W. and E. A. Schreiber.)

attentiveness compared to controls, though in the latter case, differences disappeared after one day. Comparing the behavior of injected birds to control animals seems a wise precaution.

11.5.2.2 Allometry of FMR

The first allometric treatment of FMR in seabirds was provided by Nagy (1987). Looking at 15 species, he found

$$FMR = 8.02 \ m^{0.704} \tag{11.10}$$

where FMR is in kJ d^{-1} and m is mass in g. Birt-Friesen et al. (1989) expanded this analysis by looking at 23 species of seabirds. They found

$$FMR = 12.02 \ m^{0.667} \tag{11.11}$$

with units converted to those in Equation 11.10. They further analyzed these birds by water temperature and activity (see Section 11.5.2.4 below). Nagy et al. (1999) increased the sample size to 36 species of "marine" birds (including four species of shorebirds) and showed a similar relationship:

$$FMR = 14.25 \ m^{0.659} \tag{11.12}$$

with units as in Equation 11.10. They also presented FMR equations for four seabird orders separately. We compare their equations to ours in Table 11.5. Nagy et al. (1999) report no significant differences based on order, which is in agreement with the findings of Birt-Friesen et al. (1989) and our analysis below. Nagy et al. (1999) also found no scaling effect separating marine and nonmarine birds, but they did find a significant difference in the intercept: marine birds' FMR averaged 60% higher than that of nonmarine birds.

Our analysis, based on 45 studies on 37 species of seabirds (no shorebirds included), provides the following relationship:

$$FMR = 16.69 \ m^{0.651} \tag{11.13}$$

with FMR in kJ d^{-1} and mass (m) in g (intercept s.e. = 1.2719; exponent s.e. = 0.0360; R^2 = 0.910). The exponent is nearly identical to that of Nagy et al. (1999), but the coefficient is 17% higher. It appears from the last four equations that with the progressive inclusion of more data, the higher the coefficient and therefore the higher the prediction for a seabird of a particular size. This may be because of an increasing proportion of species from very high latitudes. We address this in Section 11.5.2.4 below.

11.5.2.3 FMR/BMR Ratios

For reasons discussed above, most of the FMR data collected in birds has been during the breeding season, which is a time of high energy demand both for parents and offspring. Lack (1954) and Drent and Daan (1980) viewed reproduction as so energy demanding that adult birds had to work at near maximum capacity to produce young successfully; this is no different for seabirds (Ricklefs 1983). However, other authors have found the reproductive effort to be less demanding (e.g., Masman et al. 1989, Weathers and Sullivan 1989). This effort may be represented best as a multiple of BMR, that is, as the FMR/BMR ratio (cf. Drent and Daan 1980, Nagy 1987). Table 11.5 provides FMR/BMR ratios for all seabirds for which both measurements were available. Although the ratio for some species is not particularly high (≤3.0), several reach the predicted maximum of 4.0 (Drent and Daan 1980), and others exceed it.

Since Nagy's (1987) analysis of FMR in a variety of taxa in which comparisons were made with BMR, there has been some tendency to look for parallels in the scaling of FMR and BMR. Koteja (1991) incorrectly attributed that tendency to some authors (see p. 58) but correctly criticized that imputed relationship and, in a reanalysis of BMR and FMR for several groups, concluded that there was no general case for FMR scaling with BMR, although specific cases varied. When we compare our scaling equations for BMR (Table 11.3) and FMR (Table 11.6), it is clear that some of the equations for seabirds, including the overall equation, scale differently. However, that is mainly because of the disparity in the allometric exponents for penguins and especially petrels, because there is some obvious correspondence among charadriiforms and especially pelecaniforms. FMR/BMR ratios shed no light on this question. Our calculations show great variation in all orders. Since both BMR and FMR equations are influenced by latitude, biases in data sets by latitude may affect the results of a comparison. Even a single species can show different FMR/BMR ratios as a function of latitude, as Castro et al. (1992) demonstrated for a shorebird, the Sanderling (*Calidris alba*). Whether the bias in FMR due to reproductive season also affects comparisons cannot yet be assessed.

TABLE 11.5
Body Mass and Field (FMR) and Basal Metabolic Rates (BMR) in Seabirds by Breeding Region

Order/Species	Body Mass (g)	N	FMR (kJ d⁻¹)	BMR (kJ d⁻¹)	FMR/BMR	Latitude/ Region (degree)	Source
Sphenisciformes							
Gentoo Penguin	6100	5	3925	1406	2.8	54 S	Davis et al. 1989
Pygoscelis papua							BMR from Brown 1984
Gentoo Penguin	6170	17	5263	1418	3.7	53 S	Gales et al. 1993
P. papua							BMR from Ellis 1984
Chintrap Penguin	3806	22	4720	1045	4.5	63 S	Moreno and Sanz 1996
P. antarctica							BMR from Brown 1984
Adélie Penguin	3868	18	4002	1039	3.9	64 S	Nagy and Obst 1992
P. adeliae							BMR from Kooyman et al. 1976
Adélie Penguin	3940	24	3787	1233	3.1	64 S	Chappell et al. 1993
P. adeliae							BMR from Chappell and Souza 1988
King Penguin	12900	14	7518	2427	3.1	54 S	Kooyman et al. 1992a
Aptenodytes patagonicus							BMR from Ellis 1984
Macaroni Penguin	3870	6	4380	1188	3.7	54 S	Davis et al. 1989
Eudyptes chrysolophus							BMR from Brown 1984
Jackass Penguin	3170	10	1945	877	2.2	33 S	Nagy et al. 1984
Spheniscus demersus							BMR from Ellis 1984
Blue Penguin	1076	4	986	465	2.1	38 S	Costa et al. 1986
Eudyptula minor							BMR from Stahl and Nicol 1982
Blue Penguin	1050	4	2662	577	4.6	40 S	Gales and Green 1990
E. minor							BMR from B. Green unpublished
Procellariiformes							
Wandering Albatross	10465	17	4485	2260	2.0	46 S	Shaffer 2000
Diomedea exulans							BMR from Brown and Adams 1984
Wandering Albatross	8305	11	3288	1794	1.8	46 S	Adams et al. 1986
D. exulans							BMR from Brown and Adams 1984
Laysan Albatross	3066	8	1803	689	2.6	24 N	Pettit et al. 1988
Phoebastria immutabilis							BMR from Grant and Whittow 1983
Gray-headed Albatross	3890	6	2393	718	3.3	54 S	Costa and Prince 1987
Thalassarche chrysostoma							BMR from Adams and Brown 1984
Southern Giant Petrel	3885	8	4330	1110	3.9	64 S	Obst and Nagy 1992
Macronectes giganteus							BMR from Morgan et al. 1992
Northern Fulmar	728	14	1444	312	4.6	60 N	Furness and Bryant 1996
Fulmarus glacialis							BMR from Bryant and Furness 1995
Antarctic Petrel	618	2	1302	368	3.5	69 S	P. J. Hodum and W. W. Weathers
Thalassoica antarctica							unpublished
Cape Pigeon	440	26	1196	338	3.5	69 S	P. J. Hodum and W. W. Weathers
Daption capense							unpublished
Snow Petrel	245	11	793	174	4.6	69 S	P. J. Hodum and W. W. Weathers
Pagodroma nivea							unpublished

TABLE 11.5
Body Mass and Field (FMR) and Basal Metabolic Rates (BMR) in Seabirds by Breeding Region

Order/Species	Body Mass (g)	N	FMR (kJ d⁻¹)	BMR (kJ d⁻¹)	FMR/BMR	Latitude/ Region (degree)	Source
Antarctic Prion	149	8	391	97	4.0	54 S	Taylor et al. 1997
Pachyptila desolata							BMR from Ellis 1984
Wedge-tailed Shearwater	384	10	614	191	3.2	24 N	Ellis et al. 1983, Ellis 1984
Puffinus pacificus							BMR from Ellis 1984
Georgian Diving Petrel	109	10	464	112	4.1	54 S	Roby and Ricklefs 1986
Pelecanoides georgicus							
Common Diving Petrel	137	13	557	130	4.3	54 S	Roby and Ricklefs 1987
P. urinatrix							
Wilson's Storm Petrel	42	15	157	37	4.2	64 S	Obst et al. 1987
Oceanites oceanicus							
Leach's Storm Petrel	45	8	123	43	2.9	45 N	Ricklefs et al. 1986
Oceanodroma leucorhoa							
Leach's Storm Petrel	47	12	142	45	3.2	47 N	Montevecchi et al. 1992
O. leucorhoa							
Pelecaniformes							
White-tailed Tropicbird	370	10	777	186	4.2	18 N	Pennycuick et al. 1990
Phaethon lepturus							BMR from Ellis 1984
Cape Gannet	2580	10	3380	756	4.5	32 S	Adams et al. 1991
Morus capensis							BMR from Ellis 1989
Northern Gannet	3210	20	4865	1377	3.5	49 N	Birt-Friesen et al. 1989
M. bassanus							BMR from Bryant and Furness 1995
Red-footed Booby	1070	9	1246	401	3.1	16 N	Ballance 1995
Sula sula							BMR from Ellis 1984
Charadriiformes							
Black-legged Kittiwake	392	17	795	310	2.6	76 N	Gabrielsen et al. 1987
Rissa tridactyla							BMR from Gabrielsen et al. 1988
Black-legged Kittiwake	386	15	786	305	2.6	61 N	Golet et al. 2000
R. tridactyla							BMR from Gabrielsen et al. 1988
Common Tern	127	7	343	86	4.0	53 N	Klaassen et al. 1992
Sterna hirundo							BMR from Ellis 1984
Arctic Tern	101	8	335	70	4.8	55 N	Uttley et al. 1994
S. paradisaea							
Sooty Tern	186	14	241	87	2.8	24 N	Flint and Nagy 1984
S. fuscata							BMR from MacMillen et al. 1977
Brown Noddy	195	9	352	95	3.7	24 N	Ellis et al. 1983, Ellis 1984
Anous stolidus							BMR from Ellis et al. 1995
Dovekie	164	13	696	191	3.6	79 N	Gabrielsen et al. 1991b
Alle alle							
Common Murre	1025	11	2198	593	3.7	70 N	Gabrielsen 1996
Uria aalge							
Common Murre	940	4	1790	544	3.3	47 N	Cairns et al. 1990
U. aalge							BMR from Gabrielsen 1996

TABLE 11.5
Body Mass and Field (FMR) and Basal Metabolic Rates (BMR) in Seabirds by Breeding Region

Order/Species	Body Mass (g)	N	FMR (kJ d⁻¹)	BMR (kJ d⁻¹)	FMR/BMR	Latitude/ Region (degree)	Source
Thick-billed Murre *U. lomvia*	1078	12	1783	577	3.1	57 N	E. N. Flint and G. Hunt unpublished BMR from Gabrielsen et al. 1988
Thick-billed Murre *U. lomvia*	980	5	1860	552	3.4	67 N	Croll 1990 BMR from G. Gabrielsen unpublished
Black Guillemot *Cepphus grylle*	380	10	860	291	3.0	79 N	Mehlum et al. 1993 BMR from Gabrielsen et al. 1988
Cassin's Auklet *Ptychoramphus aleuticus*	174	9	413	108	3.8	37 N	Hodum et al. 1998 BMR from Ellis 1984
Least Auklet *Aethia pusilla*	83	24	358	116	3.1	56 N	Roby and Ricklefs 1986
Atlantic Puffin *Fratercula arctica*	460	9	848	309	2.7	70 N	G. Gabrielsen unpublished BMR from Barrett et al. 1995

TABLE 11.6
Comparison of Allometric Equations for FMR by Order

Taxon	This Study	R^2	s.e. Intercept	s.e. Exponent	Nagy et al. 1999
All Seabirds	$16.69\ m^{0.651}$ (N = 37)	0.910	1.210	0.028	$14.25\ mass^{0.659}$ (N = 36)[a]
Charadriiformes	$11.49\ m^{0.718}$ (N = 12)	0.814	1.716	0.095	$8.13\ mass^{0.77}$ (N =13)[a,b]
Pelecaniformes	$3.90\ m^{0.871}$ (N = 4)	0.953	4.209	0.196	$4.54\ mass^{0.844}$ (N = 4)
Procellariiformes	$22.06\ m^{0.594}$ (N = 14)	0.921	1.350	0.047	$18.4\ mass^{0.599}$ (N = 11)[c]
Sphenisciformes	$21.33\ m^{0.626}$ (N=7)	0.681	3.908	0.162	$4.53\ mass^{0.795}$ (N = 7)

Note: The general form of the equation is FMR = a mass^b, with FMR in units of kJ d⁻¹ and mass (m) in g. N refers to number of species; in this study, the number of sources is typically larger and may be found in Table 11.5.

[a] This equation includes four species of shorebirds; shorebirds are not included in our equations.
[b] This equation has been corrected to: $8.49\ mass^{0.77}$ for N = 15 (K. Nagy, pers. comm.).
[c] This equation has been corrected to: $17.9\ mass^{0.600}$ for N = 10 (K. Nagy, pers. comm.).

11.5.2.4 Correlates and Influences on FMR

Ellis (1984) showed a relationship between BMR and various activities. In general, he found that more active life styles were associated with higher values of BMR. Birt-Friesen et al. (1989) did a similar analysis for FMR values in seabirds. They found FMR to be higher in birds living in colder waters and having flapping flight. We also tested FMR as a function of latitude or region, ocean regime, season, activity mode, as well as body mass. Of all those parameters, only mode and latitude increased the ability of body mass to predict FMR. Of those two, latitude was most important, so the relationships can be expressed by the equation

$$FMR = 9.014 \ m^{0.655} \cdot [\exp_{10} (\text{latitude})]^{0.0048} \qquad (11.14)$$

with FMR in kJ d^{-1} and mass (m) in g (intercept s.e. = 1.233; body mass exponent s.e. = 0.023; latitude exponent s.e. = 0.001; R^2 = 0.951). This seems to validate the possibility that the equation describing FMR as a simple allometric equation (11.13) is affected by a sample with a geographic bias.

Another correlate of FMR is activity mode (e.g., flapping vs. gliding flight; plunge diving; etc.). When that parameter is included in our analysis, however, we failed to find a statistically significant relationship. Thus, we are unable to confirm statistically the assertion of Birt-Friesen et al. (1989) that mode of activity is related to FMR. Additional data may change that in the future and taxon-specific data might also show relationships not found in our analysis.

Other factors may play a role in the value of FMR. Growth of chick(s) and number of chicks have a great influence on FMR in Black-legged Kittiwakes: adults raising larger chicks and several chicks have a higher FMR than those raising smaller chicks and broods of one (Gabrielsen et al. 1987, Gabrielsen 1996, Fyhn et al. 2001). In most seabird studies, time away from the colony (and distance) is the most important factor associated with higher FMR (Gabrielsen et al. 1987, 1991b, Birt-Friesen et al. 1989, Gabrielsen 1996, Shaffer 2000). Wind also has an influence on FMR in seabirds. Gabrielsen et al. (1987, 1991b) reported that FMR increased with wind speed in Black-legged Kittiwakes and Dovekies, respectively. Furness and Bryant (1996) found an inverse relationship for Northern Fulmars. The difference is due to differences in the modes of flight: Kittiwakes and auks are flap flyers, while most petrels use substantial gliding. Wind is beneficial to the flight of the latter, but not necessarily the former. This influence on flight costs (also discussed in Section 11.4.3) affects FMR.

11.5.2.5 Partitioning FMR

FMR, in conjunction with time-activity budgets, leads to a partitioning of DEE into specific activities. Birt-Friesen et al. (1989) reported the cost of flying and pursuit diving in six seabirds to average 5.25 × BMR, but that included their own multiple for Northern Gannets (11.3 × BMR). If the BMR value from Bryant and Furness (1995) is used instead, their multiple becomes 6.4 × BMR. Flight costs estimated this way in two other seabirds (where flight has a wind-assisted component) are Red-footed Boobies (4.5 × BMR; Ballance 1995) and Northern Fulmars (4.5 × BMR; Furness and Bryant 1996). Arnould et al. (1996) reported that flight was not a limiting activity for Wandering Albatrosses during foraging trips. Schaffer (2000) confirmed that, finding a very low FMR in Wandering Albatrosses due to a low cost of flight. The only factor which had an influence on FMR was number of landings. For most albatrosses the highest cost is probably associated with getting airborne, particularly in the absence of wind. Shaffer (2000) compared FMR in several albatross species. All of them had a cost of flight between 2 and 3 × BMR. Costs of swimming and diving, occasionally calculated as FMR partitions, are discussed in Section 11.4.3.1.

If greater activity can be correlated with higher values of DEE, can sleep be an energy-reducing mechanism? Stahel et al. (1984) addressed that question for Blue Penguins; they found it to represent a trivial savings (2.4% of DEE) in that species. A logical partition for an FMR study might seem to be the cost of molt. However, using FMR to estimate the cost of molt may be problematic, since molting birds sometimes compensate by reducing activity (Groscolas 1990, Murphy 1996).

One of the more common partitions presented for free-ranging seabirds has been the cost of being at sea. In some sense, this is the real cost of foraging: it includes costs associated with flying, swimming and diving, and thermoregulation; but it may also include the costs of social interactions and other undescribed activities. Because it is an integrated value, it is likely to be less than locomotion, but more than nest attendance. Birt-Friesen et al. (1989) found the cost of being at sea in 11 species of seabirds to be 3.78 × BMR. Furness and Bryant (1996) found that FMR at sea varied greatly among individuals, ranging from 1.40 to 7.85 × BMR; this probably reflected the effects of variable winds, as discussed above. Montevecchi et al. (1992) found at-sea FMR to increase with time spent at sea. At sea FMRs, expressed in ratio with BMR, are often high compared to the overall FMR/BMR ratios reported in Table 11.5, especially among penguins (Nagy et al. 1984, Davis et al. 1989, Kooyman et al. 1992b, Nagy and Obst 1992, Moreno and Sanz 1996), but also among gannets (Adams et al. 1991), ranging from about 4.5 to over 8.0 × BMR. The data set is biased toward high-latitude species, however, and comparable work on low-latitude seabirds would be useful in developing generalizations.

Studies of locomotion, especially diving, have often been done on seabirds where extrapolations to foraging costs and even total food requirements were easily made (e.g., Culik and Wilson 1991a, Grémillet and Wilson 1999, Luna-Jorquera 2000). These extrapolations lend themselves easily to considerations of population, and ultimately community, energetics.

11.6 COMMUNITY ENERGETICS

The ability to estimate the DEE of seabirds has always held out the possibility of converting energy costs to food requirements. It has also promised the ability to extrapolate from individuals to populations and ultimately to communities. The literature on seabird population and community energetics is too broad to treat comprehensively here. Instead, using examples, we will indicate areas where there have been problems and areas which have proven especially fruitful. It should be noted that many DEE studies have extrapolated their energy costs to the numbers of fish, squid, krill, etc. required by an individual seabird or a population (e.g., Nagy et al. 1984, Fitzpatrick et al. 1988, Cairns et al. 1990, Adams et al. 1991, Gabrielsen et al. 1991b, Montevecchi et al. 1992, Nagy and Obst 1992, Mehlum et al. 1993, Moreno and Sanz 1996). Far fewer studies of seabird community energetics exist.

Extrapolations from individuals to populations and communities has always required very specific information: (1) the food eaten, (2) caloric value of the food, (3) an estimate of the efficiency of assimilation, (4) the size of the population(s), (5) the period of time of an event (e.g., the breeding season), and (6) the number of chicks, if any. But pivotal to such an extrapolation is an estimate of the energy demands of individuals. Many of the early studies (e.g., Wiens and Scott 1975, Furness 1978, Croxall and Prince 1982, Furness and Cooper 1982, Pettit et al. 1984) used allometric predictions of existence metabolism to estimate energy requirements at the individual level. The problems inherent in such an approach were discussed in Section 11.5.1.3 and would be extended to those populations and communities.

Although we are particularly critical of the use of EM in energetics studies, the use of any allometric model carries with it some uncertainty. One equation so used in the past was for BMR (discussed above in Section 11.5.1.1), although Schneider et al. (1986) used alternative equations to model community energetics involving seabirds of the Bering Sea. If the energy costs of different activities such as flight, gliding, swimming, and walking are known absolutely or as multiples of BMR, and the time spent in each activity is known, it is possible to calculate energy budgets for free-living seabirds (Croxall and Briggs 1991). However, the use of this method to determine the food harvest of a seabird community requires not only accurate measurements of the cost of different activities, but detailed time budgets for different periods in many seabird species. Good data on costs are especially lacking, except for a few species and groups. The same can be said about time budgets. While the activities of birds in the colony are easy to document, good studies of time-activity budgets of seabirds once they leave the colony (during or outside the breeding period) are few.

The advent and accessibility of doubly labeled water studies, providing FMR values for the birds actually studied, have ended much of the uncertainty associated with individual energy budgets in the construction of larger-scale studies. Furthermore, if diet, chemical composition, and assimilation efficiency (Brekke and Gabrielsen 1994) are known, it is also possible to calculate the food consumption using this method. This has become the basis of more recent studies on population and community energetics. However, other problems remain.

Most data on seabird FMR have been collected during the breeding season, with the majority of bioenergetic models being based on data obtained during incubation and chick rearing. With the exception of the study by Gales and Green (1990), virtually no FMR data have been collected outside the breeding season. That is, bioenergetics models which cover the whole year were based on extrapolations of breeding season data. Gales and Green (1990) found that although chick-rearing in Blue Penguins takes up only 16% of their annual cycle, it accounts for 31% of the annual energy budget. Only recently have models integrated data from breeding colonies with extensive data from birds at sea (e.g., Diamond et al. 1993, from the Canadian Arctic and a report, Anonymous 1994, from the North Sea). Although we have good energetics data from adult breeding birds, we have very little information concerning the energetics of chicks, juveniles, and nonbreeding birds. Most seabirds do not start to breed before they are 4 to 8 years old. In some extreme cases, as in many Procellariiformes and Pelecaniformes, breeding may not begin before 8 to 10 years. Surely, the activity budgets of breeding and nonbreeding birds must differ. Finally, most studies are from high latitudes, especially in the northern hemisphere (Wiens 1984, Furness and Barrett 1985, Croxall 1987, Duffy et al. 1987, Barrett et al. 1990, Furness 1990, Bailey et al. 1991, Diamond et al. 1993, Mehlum and Gabrielsen 1995), and may introduce geographic bias.

Based on the published literature from different parts of the world, the potential impact by seabird communities on fish stocks has been estimated to vary between 5 and 30% of the local annual fish production (Wiens and Scott 1975, Furness and Cooper 1982). Two Norwegian studies suggest that, under very special circumstances, seabird predation has the potential to negatively influence recruitment into some commercial fisheries (Barrett et al. 1990, Anker-Nilsen 1992). However, these estimates were crude and need to be validated. To quantify the impact of seabird populations or communities on fish stocks, it is necessary to estimate the proportion of fish consumed by a seabird population or community in a defined area. That brings us back to the relationship between energetics and demographic variables (e.g., population size and age structure). However, some of these variables will change daily and seasonally, so that an annual figure can be computed only by using values throughout the year. Once again, extrapolations from known periods to unknown periods are likely to add new sources of error of unknown magnitude.

The overexploitation of fisheries by humans may leave seabirds without adequate resources. That being the case, it is important to move from models to actual measurements in order to minimize extrapolations.

11.7 SPECULATIONS AND FUTURE RESEARCH DIRECTIONS

We return to Calder and King's (1974) challenge to their readers: is it better to add more species to such a mass of BMR data now or to ask new questions? Having spent many years in search of more data, we see now the possibility of asking intriguing and new questions. In particular, in what ways is BMR a product of natural selection? Is it a function of ecological considerations such as reproductive output, as McNab (1980a) and Hennemann (1983b) proposed, but perhaps in a more complex way? Is it a response to climate as originally suggested by Scholander et al. (1950b)? Is it phylogenetic baggage set in each order by the exigencies of an ancient and more selective regime and modified now only in minor ways? Are BMR and FMR linked as originally suggested by Nagy (1987) or is that truly a spurious relationship (Koteja 1991)? Is BMR linked to summit metabolism (a concept not really addressed in this review) or to any kind of upper end metabolic measurement?

Seabirds often live in extreme environments, where air and sea can put conflicting demands on their thermoregulatory abilities. We have indicated areas where some very interesting work has begun; future work should amplify this. We suggest that such work needs to be done in warm waters as well as cold and should look at temperature regulation during a variety of activities such as diving and flight as well as land-based activities.

The challenge of Calder and King (1974) is very much with us as we assess our knowledge of energetics based on FMR values. We have more data than we did 20 years ago on FMR in seabirds, but those data now appear to be both narrow and diverse. Our FMR data are still almost entirely based on breeding birds; we lack a real database for the evaluation of DEE in seabirds measured outside the breeding period. At the same time, in the FMR data we do have, studies cover varying aspects of birds' lives (incubation, brooding, life at sea, life on land) and are not all comparable. Future studies should add to the database so that analyses of each of these facets of seabird life can be reviewed and compared. Studies which investigate the factors that affect FMR (growth, age, distance to foraging areas, cost of flight/diving) will be important to address in future research on energetics in seabirds. Additionally, future energetic studies should involve whole-year energy expenditure using on-board data storage of heart rate, body temperature, diving depths, and time in water whenever possible. This will surely require addressing the use and efficacy of external and internal devices. One of the challenges of the future will be to further partition DEE into particular activities. This work has already begun and is summarized above, but it is unclear if that will be the province of further FMR work or the domain of new methods such as heart rate.

The dilemma of Calder and King is not particularly Gordonian. It will be resolved by both data collection and the formulation of new questions, just as it has been for the last quarter of a century. We expect that the latter will lead the former, however, and that should provide some fascinating work for all of us.

ACKNOWLEDGMENTS

We are appreciative of the opportunity given to us by the editors to produce this review, an undertaking far more daunting than we originally anticipated. They have been particularly support-ive of the inclusion of our views, especially given the great breadth of this subject. We are grateful for the invaluable assistance of Espen O. Henriksen and Per Fauchald in developing the statistical analyses. We would like to thank Claus Bech and Gerald Kooyman for their insightful comments on the manuscript. Any remaining errors of commission or omission are our responsibility alone. We are particularly indebted to certain individuals whose professional help opened the paths to our seabird work. For HIE, Brian McNab provided the initial mentoring in avian energetics; Causey Whittow and Robert Loftin provided the original opportunities to work with seabirds; Harvey Fisher donated a part of his seabird library. For GWG, Johan B. Steen imparted the introduction to avian physiology and Fridtjof Mehlum to Arctic seabirds. To all these people we are grateful and indebted.

LITERATURE CITED

ADAMS, N. J., AND C. R. BROWN. 1984. Metabolic rates of sub-Antarctic Procellariiformes: a comparative study. Comparative Biochemistry and Physiology 77A: 169–173.

ADAMS, N. J., AND C. R. BROWN. 1990. Energetics of molt in penguins. Pp. 297–315 in Penguin Biology (L. S. Davis and J. T. Darby, Eds.). Academic Press, San Diego.

ADAMS, N. J., C. R. BROWN, AND K. A. NAGY. 1986. Energy expenditure of free-ranging Wandering Albatrosses (Diomedea exulans). Physiological Zoology 59: 583–591.

ADAMS, N. J., R. W. ABRAMS, W. R. SIEGFRIED, K. A. NAGY, AND I. R. KAPLAN. 1991. Energy expenditure and food consumption by breeding Cape Gannets Morus capensis. Marine Ecology Progress Series 70: 1–9.

ALERSTAM, T., AND A. HEDENSTRÖM (Eds.). 1998. Optimal Migration. Journal of Avian Biology 29: 337–636.

ANCEL, A., G. L. KOOYMAN, P. J. PONGANIS, J.-P. GENDNER, J. LIGNON, X. MESTRE, N. HUIN, P. H. THORSON, P. ROBBISON, AND Y. LE MAHO. 1992. Foraging behaviour of emperor penguins as a resource detector in winter and summer. Nature 360: 336–338.

ANCEL, A., M. HORNING, AND G. L. KOOYMAN. 1997. Prey ingestion revealed by oesophagus and stomach temperature recordings in cormorants. Journal of Experimental Biology 200: 149–154.

ANCEL, A., L. N. STARKE, P. J. PONGANIS, R. VAN DAM, AND G. L. KOOYMAN. 2000. Energetics of surface swimming in Brandt's cormorants (Phalacrocorax penicillatus Brandt). Journal of Experimental Biology 203: 3727–3731.

ANKER-NILSEN, T. 1992. Food Supply as a Determinant of Reproduction and Population Development in Norwegian Puffins *Fratercula arctica*. D.Sc. thesis, University of Trondheim, Norway.

ANONYMOUS. 1994. Report of the Working Group on Seabird Ecology. ICES CM 1994/L: 3.

ARNOULD, J. P. Y., D. R. BRIGGS, J. P. CROXALL, P. A. PRINCE, AND A. G. WOOD. 1996. The foraging behaviour and energetics of wandering albatrosses brooding chicks. Antarctic Science 8: 229–236.

ASCHOFF, J. 1981. Thermal conductance in mammals and birds: its dependence on body size and circadian phase. Comparative Biochemistry and Physiology 69A: 611–619.

ASCHOFF, J., AND H. POHL. 1970. Rhythmic variations in energy metabolism. Federation Proceedings 29: 1541–1552.

BAILEY, R. S., R. W. FURNESS, J. A. GAULD, AND P. A. KUNZLIK. 1991. Recent changes in the population of the sandeel (*Ammodytes marinus* Raitt) at Shetland in relation to estimates of seabird predation. ICES Marine Science Symposium 193: 209–216.

BALLANCE, L. T. 1995. Flight energetics of free-ranging red-footed boobies (*Sula sula*). Physiological Zoology 68: 887–914.

BARRE, H. 1984. Metabolic and insulative changes in winter- and summer-acclimatized king penguin chicks. Journal of Comparative Physiology B 154: 317–324.

BARRETT, R. T. 1978. Adult body temperatures and the development of endothermy in kittiwake (*Rissa tridactyla*). Astarte 11: 113–116.

BARRETT, R. T., N. RØV, J. LOEN, AND W. A. MONTEVECCHI. 1990. Diets of shags *Phalacrocorax aristotelis* and cormorants *P. carbo* in Norway and possible implications for gadoid stock recruitment. Marine Ecology Progress Series 66: 205–218.

BARRETT, R. T., G. W. GABRIELSEN, AND P. FAUCHALD. 1995. Prolonged incubation in the Atlantic puffin (*Fratercula arctica*) and evidence of mild hypothermia as an energy-saving mechanism. Pp. 479–488 *in* Ecology of Fjords and Coastal Waters (H. R. Skjoldal, C. Hopkins, K. E. Erikstad, and H. P. Leinaas, Eds.). Elsevier Science, Amsterdam.

BARTHOLOMEW, G. A. 1982. Energy metabolism. Chapter 3 *in* Animal Physiology, 4th ed. (M. Gordon, ed.). Macmillan, New York.

BAUDINETTE, R. V., AND P. GILL. 1985. The energetics of 'flying' and 'paddling' in water: locomotion in penguins and ducks. Journal of Comparative Physiology B 155: 373–380.

BAUDINETTE, R. V., P. GILL, AND M. O'DRISCOLL. 1986. Energetics of the Little Penguin, *Eudyptula minor*: temperature regulation, the calorigenic effect of food, and moulting. Australian Journal of Zoology 34: 35–45.

BECH, C. 1980. Body temperature, metabolic rate and insulation in winter and summer acclimatized Mute swans (*Cygnus olor*). Journal of Comparative Physiology B 136: 61–66.

BECH, C., R. BRENT, P. F. PEDERSEN, J. G. RASMUSSEN, AND K. JOHANSEN. 1982. Temperature regulation in chicks of the Manx Shearwater *Puffinus puffinus*. Ornis Scandinavica 13: 206–210.

BECH, C., I. LANGSETH, AND G. W. GABRIELSEN. 1999. Repeatability of basal metabolism in breeding female kittiwakes *Rissa tridactyla*. Proceedings of the Royal Society of London B 266: 2161–2167.

BENEDICT, F. G., AND E. L. FOX. 1927. The gaseous metabolism of large wild birds under aviary conditions. Proceedings of the American Philosophical Society 66: 511–534.

BERNSTEIN, N. P., AND S. J. MAXSON. 1985. Reproductive energetics in Blue-eyed Shags in Antarctica. Wilson Bulletin 97: 450–462.

BEVAN, R. M., A. J. WOAKES, P. J. BUTLER, AND I. L. BOYD. 1994. The use of heart rate to estimate oxygen consumption of free-ranging black-browed albatrosses *Diomedea melanophrys*. Journal of Experimental Biology 193: 119–137.

BEVAN, R. M., I. L. BOYD, P. J. BUTLER, K. R. REID, AND A. J. WOAKES. 1995a. Cardiovascular and thermoregulatory adjustments associated with flying and diving in the free-ranging blue-eyed shag, *Phalacrocorax atriceps*. Journal of Physiology 483: 193–194.

BEVAN, R. M., A. J. WOAKES, P. J. BUTLER, AND J. P. CROXALL. 1995b. Heart rate and oxygen consumption of exercising gentoo penguins. Physiological Zoology 68: 855–877.

BEVAN, R. M., I. L. BOYD, P. J. BUTLER, K. REID, A. J. WOAKES, AND J. P. CROXALL. 1997. Heart rates and abdominal temperatures of free-ranging South Georgia Shags. Journal of Experimental Biology 200: 661–675.

BIRT-FRIESEN, V. L., W. A. MONTEVECCHI, D. K. CAIRNS, AND S. A. MACKO. 1989. Activity-specific metabolic rates of free-living northern gannets and other seabirds. Ecology 70: 357–367.

BLAKE, R. W. 1985. A model of foraging efficiency and daily energy budget in the Black Skimmer (*Rynchops niger*). Canadian Journal of Zoology 63: 42–48.

BLIGH, J., AND K. G. JOHNSON. 1973. Glossary of terms for thermal physiology. Journal of Applied Physiology 35: 941–961.

BREKKE, B., AND G. W. GABRIELSEN. 1994. Assimilation efficiency of adult kittiwakes and Brunnich's guillemots fed capelin and arctic cod. Polar Biology 14: 279–284.

BRODY, S. 1945. Bioenergetics and Growth. Reinhold Publishing Corporation, New York. 1023 pp.

BROWN, C. R. 1984. Resting metabolic rate and energetic cost of incubation in macaroni penguins (*Eudyptes chrysolophus*) and rockhopper penguins (*E. chrysocome*). Comparative Biochemistry and Physiology 77A: 345–350.

BROWN, C. R. 1985. Energetic cost of moult in macaroni penguins (*Eudyptes chrysolophyus*) and rockhopper penguins (*E. chrysocome*). Journal of Comparative Physiology B 155: 515–520.

BROWN, C. R., AND N. J. ADAMS. 1984. Basal metabolic rate and energy expenditure during incubation in the Wandering Albatross (*Diomedea exulans*). Condor 86: 182–186.

BRYANT, D. M., AND R. W. FURNESS. 1995. Basal metabolic rates of North Atlantic seabirds. Ibis 137: 219–226.

BUNDLE, M. W., J. H. HOPPELER, R. VOCK, J. M. TESTER, AND P. G. WEYLAND. 1998. High metabolic rates in running birds. Nature 397: 31–32.

BURGER, A. E. 1981. Time budgets, energy needs and kleptoparasitism in breeding Lesser Sheathbills (*Chionis minor*). Condor 83: 106–112.

BUTLER, P. J. 2000. Energetic costs of surface swimming and diving of birds. Physiological and Biochemical Zoology 73: 699–705.

BUTLER, P. J., AND A. J. WOAKES. 1984. Heart rate and aerobic metabolism in Humboldt penguins, *Speniscus humboldti*, during voluntary dives. Journal of Experimental Biology 108: 419–428.

BUTLER, P. J., AND D. R. JONES. 1997. Physiology of diving of birds and mammals. Physiological Review 77: 837–899.

BUTLER, P. J., AND C. M. BISHOP. 2000. Flight. Pp. 391–435 *in* Sturkie's Avian Physiology, 5th ed. (G. C. Whittow, Ed.). Academic Press, San Diego.

CAIRNS, D. K., W. A. MONTEVECCHI, V. L. BIRT-FRIESEN, AND S. A. MACKO. 1990. Energy expenditures, activity budgets, and prey harvest of breeding Common Murres. Studies in Avian Biology 14: 84–92.

CALDER, W. A. 1985. The comparative biology of longevity and lifetime energetics. Experimental Gerontology 20: 161–170.

CALDER, W. A., AND J. R. KING. 1974. Thermal and caloric relations of birds. Pp. 259–413 *in* Avian Biology, Vol. 4 (D. S. Farner and J. R. King, Eds.). Academic Press, New York.

CARTER, R. V., AND R. I. G. MORRISON. 1997. Estimating metabolic costs for homeotherms from weather data and morphology: an example using calidridine sandpipers. Canadian Journal of Zoology 75: 94–101.

CARY, C. 1996. Female reproductive energetics. Pp. 324–374 *in* Avian Energetics and Nutritional Ecology (C. Carey, Ed.). Chapman & Hall, New York.

CASTRO, G., J. P. MYERS, AND R. E. RICKLEFS. 1992. Ecology and energetics of Sanderlings migrating to four latitudes. Ecology 73: 833–844.

CHAPPELL, M. A., AND S. L. SOUZA. 1988. Thermoregulation, gas exchange, and ventilation in Adelie penguins (*Pygoscelis adeliae*). Journal of Comparative Physiology B 157: 783–790.

CHAPPELL, M. A., K. R. MORGAN, S. L. SOUZA, AND T. L. BUCHER. 1989. Convection and thermoregulation in two Antarctic seabirds. Journal of Comparative Physiology B 159: 313–322.

CHAPPELL, M. A., V. H. SHOEMAKER, D. N. JANES, S. K. MALONEY, AND T. L. BUCHER. 1993. Energetics of foraging in breeding Adelie penguins. Ecology 74: 2450–2461.

CHEREL, Y., J.-P. ROBIN, AND Y. LE MAHO. 1988. Physiology and biochemistry of long-term fasting in birds. Canadian Journal of Zoology 66: 159–166.

CHEREL, Y., J.-B. CHARRASSIN, AND E. CHALLET. 1994. Energy and protein requirements for molt in the king penguin *Aptenodytes patagonicus*. American Journal of Physiology 266: 1182–1188.

COSTA, D. P., P. DANN, AND W. DISHER. 1986. Energy requirements of free ranging little penguin, *Eudyptula minor*. Comparative Biochemistry and Physiology 85A: 135–138.

COSTA, D. P., AND P. A. PRINCE. 1987. Foraging energetics of Grey-headed Albatrosses *Diomedea chrysostoma* at Bird Island, South Georgia. Ibis 129: 149–158.

CROLL, D. A. 1990. Diving and Energetics of the Thick-billed Murre, *Uria lomvia*. Ph.D. dissertation, University of California, San Diego.

CROLL, D. A., AND E. MCLAREN. 1993. Diving metabolism and thermoregulation in common and thick-billed murres. Journal of Comparative Physiology B 163: 160–166.

CROXALL, J. P. 1982. Energy costs of incubation and moult in petrels and penguins. Journal of Animal Ecology 51: 177–194.

CROXALL, J. P. (Ed.). 1987. Seabirds: Feeding Ecology and Role in Marine Ecosystems. Cambridge University Press, Cambridge, U.K.

CROXALL, J. P., AND P. A. PRINCE. 1982. A preliminary assessment of the impact of seabirds on marine resources at South Georgia. Le Comité National Français des Recherches Antarctiques 51: 501–509.

CROXALL, J. P., AND C. RICKETTS. 1983. Energy costs of incubation in the Wandering Albatross *Diomedea exulans*. Ibis 125: 33–39.

CROXALL, J. P., AND R. W. DAVIS. 1990. Metabolic rate and foraging behavior of *Pygoscelis* and *Eudyptes* penguins at sea. Pp. 207–228 *in* Penguin Biology (L. S. Davis and J. T. Darby, Eds.). Academic Press, San Diego.

CROXALL, J. P., AND D. R. BRIGGS. 1991. Foraging economics and performance of polar and subpolar Atlantic seabirds. Polar Research 10: 561–578.

CULIK, B., AND R. P. WILSON. 1991a. Energetics of under-water swimming in Adélie penguins (*Pygoscelis adeliae*). Journal of Comparative Physiology B 161: 285–291.

CULIK, B., AND R. P. WILSON. 1991b. Swimming energetics and performance of instrumented Adélie penguins (*Pygoscelis adeliae*). Journal of Experimental Biology 158: 355–368.

CULIK, B. M., R. P. WILSON, AND R. BANNASCH. 1994. Underwater swimming at low energetic cost by pygoscelid penguins. Journal of Experimental Biology 197: 65–78.

CULIK, B. M., K. PÜTZ, R. P. WILSON, D. ALLERS, J. LAGE, C. A. BOST, AND Y. LE MAHO. 1996. Diving energetics in king penguins (*Aptenodytes patagonicus*). Journal of Experimental Biology 199: 973–983.

DAVIS, R. W., J. P. CROXALL, AND M. J. O'CONNELL. 1989. The reproductive energetics of gentoo *Pygoscelis papua* and macaroni *Eudyptes chrysolophus* penguins at South Georgia. Journal of Animal Ecology 58: 59–74.

DAVYDOV, A. F. 1972. Seasonal variations in the energy metabolism and thermoregulation at rest in the black-headed gull. Soviet Journal of Ecology 2: 436–439.

DAWSON, W. R., AND G. C. WHITTOW. 1994. The emergence of endothermy in the Laysan and black-footed albatross. Journal of Comparative Physiology B 164: 292–298.

DAWSON, W. R., AND T. P. O'CONNOR. 1996. Energetic features of avian thermoregulatory responses. Pp. 85–124 *in* Avian Energetics and Nutritional Ecology (C. Carey, Ed.). Chapman & Hall, New York.

DAWSON, W. R., AND G. C. WHITTOW. 2000. Regulation of body temperature. Pp. 344–390 *in* Sturkie's Avian Physiology, 5th ed. (G. C. Whittow, Ed.). Academic Press, San Diego.

DIAMOND, A. W., A. J. GASTON, AND R. G. B. BROWN. 1993. Studies of high-latitude seabirds. 3. A model of the energy demands of the seabirds of eastern and Arctic Canada. Canadian Wildlife Service Occasional Paper 77: 9–39.

DRENT, R. H. 1965. Breeding biology of the Pigeon Guillemot, *Cepphus columba*. Ardea 53: 99–160.

DRENT, R. H., AND B. STONEHOUSE. 1971. Thermoregulatory responses of the Peruvian penguin (*Spheniscus humbolti*). Comparative Biochemistry and Physiology 40A: 689–710.

DRENT, R. H., AND S. DAAN. 1980. The prudent parent: energetic adjustments in avian breeding. Ardea 68: 225–252.

DUFFY, D. C, W. R. SIEGFRIED, AND S. JACKSON. 1987. Seabirds as consumers in the southern Benguela region. South Africa Journal of Marine Science 5: 771–790.

DUMONTEIL, E., H. BARRÉ, J.-L. ROUANET, M. DIARRA, AND J. BOUVIER. 1994. Dual core and shell temperature regulation during sea acclimatization in Gentoo penguins (*Pygoscelis papua*). American Journal of Physiology 266: R1319–R1326.

ELLIOTT, J. M., AND W. DAVISON. 1975. Energy equivalents of oxygen consumption in animal energetics. Oecologia 19: 195–201.

ELLIS, H. I. 1980a. Metabolism and evaporative water loss in three seabirds (Laridae). Federation Proceedings 39: 1165.

ELLIS, H. I. 1980b. Metabolism and solar radiation in dark and white herons in hot climates. Physiological Zoology 53: 358–372.

ELLIS, H. I. 1984. Energetics of free-ranging seabirds. Pp. 203–234 *in* Seabird Energetics (G. C. Whittow and H. Rahn, Eds.). Plenum Press, New York.

ELLIS, H. I., M. MASKREY, T. N. PETTIT, AND G. C. WHITTOW. 1982a. Temperature regulation in Hawaiian Red-footed Boobies. American Zoologist 22: 916.

ELLIS, H. I., M. MASKREY, T. N. PETTIT, AND G. C. WHITTOW. 1982b. Temperature regulation in Hawaiian Brown Noddies (*Anous stolidus*). Physiologist 25: 279.

ELLIS, H. I., T. N. PETTIT, AND G. C. WHITTOW. 1983. Field metabolic rates and water turnover in two Hawaiian seabirds. American Zoologist 23: 980.

ELLIS, H. I., M. MASKREY, T. N. PETTIT, AND G. C. WHITTOW. 1995. Thermoregulation in the Brown Noddy (*Anous stolidus*). Journal of Thermal Biology 20: 307–313.

ENGER, P. S. 1957. Heat regulation and metabolism in some tropical mammals and birds. Acta Physiologica Scandinavica 40: 161–166.

FITZPATRICK, L. C., C. G. GUERRA, AND R. E. AGUILAR. 1988. Energetics of reproduction in the desert nesting seagull *Larus modestus*. Estudios Oceanolaogicos 7: 33–39.

FLINT, E. N., AND K. A. NAGY. 1984. Flight energetics of free-living sooty terns. Auk 101: 288–294.

FUGLEI, E., AND N. A. ØRITSLAND. 1999. Seasonal trends in body mass, food intake and resting metabolic rate, and induction of metabolic depression in arctic foxes (*Alopex lagopus*) at Svalbard. Journal of Comparative Physiology B 169: 361–369.

FURNESS, R. W. 1978. Energy requirements of seabird communities: a bioenergetics model. Journal of Animal Ecology 47: 39–53.

FURNESS, R. W. 1990. A preliminary assessment of the quantities of Shetland sandeels taken by seabirds, seals, predatory fish and the industrial fishery in 1981–1983. Ibis 132: 205–217.

FURNESS, R. W., AND J. COOPER. 1982. Interactions between breeding seabird and pelagic fish populations in the southern Benguela region. Marine Ecology Progress Series 8: 243–250.

FURNESS, R. W., AND R. T. BARRETT. 1985. The food requirements and ecological relationships of a seabird community in north Norway. Ornis Scandinavica 16: 305–313.

FURNESS, R. W., AND D. M. BRYANT. 1996. Effect of wind on field metabolic rates of breeding northern fulmars. Ecology 77: 1181–1188.

FYHN, M., G. W. GABRIELSEN, E. S. NORDØY, B. MOE, I. LANGSETH, AND C. BECH. 2001. Individual variation in field metabolic rate of kittiwake (*Rissa tridactyla*) during the chick rearing period. Physiological and Biochemical Zoology 74: 343–355.

GABRIELSEN, G. W. 1996. Energy expenditure of breeding Common Murres. Canadian Wildlife Service Occasional Paper 91: 49–58.

GABRIELSEN, G. W., AND F. MEHLUM. 1989. Thermoregulation and energetics of arctic seabirds. Pp. 137–146 *in* Physiology of Cold Adaptations in Birds (C. Bech and R.E. Reintersen, Eds.). Pergamon Press, New York.

GABRIELSEN, G. W., F. MEHLUM, AND K. A. NAGY. 1987. Daily energy expenditure and energy utilization of free-ranging Black-legged Kittiwakes. Condor 89: 126–132.

GABRIELSEN, G. W., F. MEHLUM, AND H. E. KARLSEN. 1988. Thermoregulation in four species of arctic seabirds. Journal of Comparative Physiology B 157: 703–708.

GABRIELSEN, G. W., F. MEHLUM, H. E. KARLSEN, Ø. ANDRESEN, AND H. PARKER. 1991a. Energy cost during incubation and thermoregulation in the female Common Eider (*Somateria mollissima*). Norsk Polarinstitut Skrifter 195: 51–62.

GABRIELSEN, G. W., J. R. E. TAYLOR, M. KONARZEWSKI, AND F. MEHLUM. 1991b. Field and laboratory metabolism and thermoregulation in Dovekies (*Alle alle*). Auk 108: 71–78.

GALES, R., AND B. GREEN. 1990. The annual energetics cycle of little penguins *Eudyptula minor*. Ecology 71: 2297–2312.

GALES, R., B. GREEN, AND C. STAHEL. 1988. The energetics of free-living little penguins *Eudyptula minor* (Spheniscidae) during moult. Australian Journal of Zoology 36: 159–167.

GALES, R., B. GREEN, J. LIBKE, K. NEWGRAIN, AND D. PEMBERTON. 1993. Breeding energetics and food requirements of gentoo penguins (*Pygoscelis papua*) at Heard and Macquarie Islands. Journal of Zoology, London 231: 125–139.

GARLAND, T., JR., AND P. A. CARTER. 1994. Evolutionary physiology. Annual Review of Physiology 56: 579–621.

GAVRILOV, V. M. 1985. Seasonal and circadian changes of thermoregulation in passerine and non-passerine birds; which is more important? Proceedings of the International Ornithological Congress 18: 1254–1263.

GAVRILOV, V. M. 1999. Comparative energetics of passerine and non-passerine birds: differences in maximal, potential productive and normal levels of existence metabolism and their ecological implication. Proceedings of the International Ornithological Congress 22: 338–369.

GOLET, G. H., D. B. IRONS, AND D. P. COSTA. 2000. Energy costs of chick rearing in black-legged kittiwakes (*Rissa tridactyla*). Canadian Journal of Zoology 78: 982–991.

GRANT, G. S., AND G. C. WHITTOW. 1983. Metabolic cost of incubation in the Laysan albatross and the Bonin petrel. Comparative Biochemistry and Physiology 74A: 77–82.

GRÉMILLET, D. 1995. "Wing-drying" in cormorants. Journal of Avian Biology 26: 2.

GRÉMILLET, D., AND R. P. WILSON. 1999. A life in the fast lane: energetics and foraging strategies of the great cormorant. Behavioral Ecology 10: 516–524.

GRÉMILLET, D., I. TUSCHY, AND M. KIERSPEL. 1998. Body temperature and insulation in diving Great Cormorants and European Shags. Functional Ecology 12: 386–394.

GRIFFIN, T. M., AND R. KRAM. 2000. Penguin waddling is not wasteful. Nature 408: 929.

GROSCOLAS, R. 1990. Metabolic adaptations to fasting in emperor and king Penguins. Pp. 269–296 *in* Penguin Biology (L. S. Davis and J. T. Darby, Eds.). Academic Press, San Diego.

GROSCOLAS, R., AND Y. CHEREL. 1992. How to molt while fasting in the cold: the metabolic and hormonal adaptations of Emperor and King Penguins. Ornis Scandinavica 23: 328–334.

HAILS, C. J. 1983. The metabolic rate of tropical birds. Condor 85: 61–65.

HANDRICH, Y., R. M. BEVAN, J. B. CHARRASSIN, P. J. BUTLER, K. PUTZ, A. J. WOAKES, J. LAGE, AND Y. LE MAHO. 1997. Hypothermia in foraging king penguins. Nature 388: 64–67.

HAWKINS, P. A. J., P. J. BUTLER, A. J. WOAKES, AND G. W. GABRIELSEN. 1997. Heat increment of feeding in Brünnich's guillemot *Uria lomvia*. Journal of Experimental Biology 200: 1757–1763.

HAYSSEN, V. 1984. Basal metabolic rate and the intrinsic rate of increase: an empirical and theoretical reexamination. Oecologia 64: 419–424.

HENNEMANN, W. W., III. 1983a. Environmental influences on the energetics and behaviour of anhingas and double-crested cormorants. Physiological Zoology 56: 201–216.

HENNEMANN, W. W., III. 1983b. Relationship between body mass, metabolic rate and the intrinsic rate of natural increase in mammals. Oecologia 56: 104–108.

HERREID, C. F., II, AND B. KESSEL. 1967. Thermal conductance in birds and mammals. Comparative Biochemistry and Physiology 21:405–414.

HIND, A. T., AND W. S. C. GURNEY. 1997. The metabolic cost of swimming in marine homeotherms. Journal of Experimental Biology 200: 531–542.

HODUM, P. J., W. J. SYDEMAN, H. VISSER, AND W. W. WEATHERS. 1998. Energy expenditure and food requirements of Cassin's auklets provisioning nestlings. Condor 100: 546–550.

HUI, C.A. 1988a. Penguin swimming. II. Energetics and behavior. Physiological Zoology 61: 344–350.

HUI, C.A. 1988b. Penguin swimming. I. Hydrodynamics. Physiological Zoology 61: 333–343.

IRVING, L. 1972. Zoophysiology and Ecology. Vol. 2 of Arctic Life of Birds and Mammals, including Man (D. S. Farner, Ed.). Springer-Verlag, Berlin.

IRVING, L., AND J. KROG. 1954. Body temperature of arctic and subarctic birds and mammals. Journal of Applied Physiology 6: 667–680.

IVERSEN, J. A., AND J. KROG. 1972. Body temperatures and resting metabolic rates in small petrels. Norwegian Journal of Zoology 20: 141–144.

JENSSEN, B. M., AND M. EKKER. 1989. Thermoregulatory adaptations to cold in winter-acclimatized Long-tailed Ducks (*Clangula hyemalis*). Pp. 147–152 in Physiology of Cold Adaptation in Birds (C. Bech and R. E. Reinertsen, Eds.). Plenum Press, New York.

JENSSEN, B. M., M. EKKER, AND C. BECH. 1989. Thermoregulation in winter-acclimatized common eiders (*Somateria mollissima*) in air and water. Canadian Journal of Zoology 67: 669–673.

JOHNSON, S. R., AND G. C. WEST. 1975. Growth and development of heat regulation in nestling and metabolism in adult Common Murre and Thick-billed Murre. Ornis Scandinavica 6: 109–115.

JOUVENTIN, P., AND H. WEIMERSKIRCH. 1990. Satellite tracking of wandering albatrosses. Nature 343: 746–748.

KANWISHER, J., G. W. GABRIELSEN, AND N. KANWISHER. 1981. Free and forced diving in birds. Science 211: 717–719.

KENDEIGH, S. C. 1970. Energy requirements for existence in relation to size of bird. Condor 72: 60–65.

KENDEIGH, S. C., V. R. DOL'NIK, AND V. M. GAVRILOV. 1977. Pp. 127–205 and 363–378 *in* Granivorous Birds in Ecosystems (J. Pinowski and S. C. Kendeigh, Eds.). Cambridge University Press, London.

KING, J. R. 1957. Comments on the theory of indirect calorimetry as applied to birds. Northwest Science 31: 155–170.

KING, J. R. 1973. Energetics of reproduction in birds. Pp. 78–120 *in* Breeding Biology of Birds (D. S. Farner, Ed.). National Academy of Sciences, Washington, D.C.

KING, J. R. 1974. Seasonal allocation of time and energy resources in birds. Pp. 4–85 *in* Avian Energetics (R. A. Paynter, Jr., Ed.). Publications of the Nuttall Ornithological Club, No. 15.

KING, J. R. 1981. Energetics of avian moult. Proceedings of the International Ornithological Congress 17: 312–317.

KING, J. R., AND D. S. FARNER. 1961. Energy metabolism, thermoregulation, and body temperature. Pp. 215–305 *in* Biology and Comparative Physiology of Birds, Vol. II (A. J. Marshall, Ed.). Academic Press, New York.

KLAASSEN, M., C. BECH, D. MASMAN, AND G. SLAGSVOLD. 1989. Growth and energetics of Arctic Tern chicks (*Sterna paradisaea*). Auk 106: 240–248.

KLAASSEN, M., P. H. BECKER, AND M. WAGENER. 1992. Transmitter loads do not affect the daily energy expenditure of nestling Common Terns. Journal of Field Ornithology 63: 181–185.

KOOYMAN, G. L. 1989. Diverse Divers. Springer-Verlag, Berlin.

KOOYMAN, G. L., AND P. J. PONGANIS. 1990. Behavior and physiology of diving in emperor and king penguins. Pp. 229–242 *in* Penguin Biology (L. S. Davis and J. T. Darby, Eds.). Academic Press, San Diego.

KOOYMAN, G. L., AND P. J. PONGANIS. 1994. Emperor penguin oxygen consumption, heart rate and plasma lactate levels during graded swimming exercise. Journal of Experimental Biology 195: 199–209.

KOOYMAN, G. L., R. L. GENTRY, W. P. BERGMANN, AND H. T. HAMMEL. 1976. Heat loss in penguins during immersion and compression. Comparative Biochemistry and Physiology 54A: 75–80.

KOOYMAN, G. L., R. W. DAVIS, J. P. CROXALL, AND D. P. COSTA. 1982. Diving depths and energy requirements of king penguins. Science 217: 726–727.

KOOYMAN, G. L., P. J. PONGANIS, M. A. CASTELLINI, E. P. PONGANIS, K. V. PONGANIS, P. H. THORSON, S. A. ECKERT, AND Y. LE MAHO. 1992a. Heart rates and swim speeds of emperor penguins diving under sea ice. Journal of Experimental Biology 165: 161–180.

KOOYMAN, G. L., Y. CHEREL, Y. LE MAHO, J. P. CROXALL, AND P. H. THORSON. 1992b. Diving behavior and energetics during foraging cycles in king penguins. Ecological Monographs 62: 143–163.

KORSCHGEN, C. E. 1977. Breeding stress of female eiders in Maine. Journal of Wildlife Management 41: 360–373.

KOTEJA, P. 1991. On the relation between basal and field metabolic rates in birds and mammals. Functional Ecology 5: 56–64.

KRASNOW, L. 1979. Feeding energetics of the Sooty Shearwater *Puffinus griseus* in Monterey Bay. Unpublished M.S. thesis, California State University, Sacramento.

LACK, D. 1954. Natural regulation of animal numbers. Clarendon, Oxford.

LANGSETH, I., B. MOE, M. FYHN, G. W. GABRIELSEN, AND C. BECH. 2000. Flexibility of BMR in an arctic breeding seabird. Pp. 471–477 *in* Life in the Cold (G. Heldmaier, S. Klaus, and M. Klinenspor, Eds.). Springer-Verlag, Heidelberg.

LASIEWSKI, R. C., AND W. R. DAWSON. 1967. A re-examination of the relation between standard metabolic rate and body weight in birds. Condor 69: 13–23.

LASIEWSKI, R. C., W. W. WEATHERS, AND M. H. BERNSTEIN. 1967. Physiological responses of the giant hummingbird, *Patagona gigas*. Comparative Biochemistry and Physiology 23: 797–813.

LAVVORN, J. R. 1991. Mechanics of underwater swimming in foot-propelled diving birds. Proceedings of the International Ornithological Congress 20: 1868–1874.

LAVVORN, J. R., AND D. R. JONES. 1994. Biomechanical conflicts between adaptations for diving and aerial flight in estuarine birds. Estuaries 17: 62–75.

LAVVORN, J. R., D. R. JONES, AND R. W. BLAKE. 1991. Mechanics of underwater locomotion in diving ducks: drag, buoyancy and acceleration in a size gradient of species. Journal of Experimental Biology 159: 89–108.

LeFEBVRE, E. A. 1964. The use of D_2O^{18} for measuring energy metabolism in *Columba livia* at rest and in flight. Auk 81: 403–416.

LE MAHO, Y. 1993. Metabolic adaptations to long-term fasting in Antarctic penguins and domestic geese. Journal of Thermal Biology 8: 91–96.

LE MAHO, Y., AND B. DESPIN. 1976. Réduction de la dépense énergétique au cours du jeûne chez le Manchot royal. Comptes Rendus Académie des Sciences Paris D283: 979–982.

LE MAHO, Y., AND G. DEWASMES. 1984. Energetics of walking in penguins. Pp. 235–243 in Seabird Energetics (G. C. Whittow and H. Rahn, Eds.). Plenum Press, New York.

LE MAHO, Y., P. DELCLITTE, AND J. CHATONNET. 1976. Thermoregulation in fasting emperor penguins under natural conditions. American Journal of Physiology 231: 913–922.

LIFSON, N., AND R. MCCLINTOCK. 1966. The theory of use of the turnover rates of body water for measuring energy and material balance. Journal of Theoretical Biology 12: 46–74.

LINDSTEDT, S. L., AND W. A. CALDER. 1976. Body size and longevity in birds. Condor 78: 91–94.

LINDSTRÖM, A., G. H. VISSER, AND S. DAAN. 1993. The energetic cost of feather synthesis is proportional to basal metabolic rate. Physiological Zoology 66: 490–510.

LUNA-JORQUERA, G., R. P. WILSON, B. M. CULIK, R. AGUILAR, AND C. GUERRA. 1997. Observations on the thermal conductance of Adélie (*Pygoscelis adeliae*) and Humboldt (*Spheniscus humboldti*) penguins. Polar Biology 17: 69–73.

LUNA-JORQUERA, G., AND B. M. CULIK. 2000. Metabolic rates of swimming Humboldt penguins. Marine Ecology Progress Series 203: 301–309.

LUSTICK, S. 1984. Thermoregulation in adult seabirds. Pp. 183–201 in Seabird Energetics (G. C. Whittow and H. Rahn, Eds.). Plenum Press, New York.

LUSTICK, S., B. BATTERSBY, AND M. KELTY. 1978. Behavioral thermoregulation: orientation toward the sun in herring gulls. Science 200: 881–883.

MACMILLEN, R. E., G. C. WHITTOW, E. A. CHRISTOPHER, AND R. J. EBISU. 1977. Oxygen consumption, evaporative water loss, and body temperature in the Sooty Tern. Auk 94: 72–79.

MAHONEY, S. A. 1979. Some aspects of the thermal physiology of Anhingas Anhinga anhinga and Double-crested Cormorants Phalacrocorax auritus. Pp. 461–470 in Proceedings of the Symposium of Birds of the Sea and Shore (J. Cooper, Ed.). African Seabird Group, Cape Town, South Africa.

MANGUM, C. P., AND P. W. HOCHACHKA. 1998. New directions in comparative physiology and biochemistry: mechanisms, adaptations, and evolution. Physiological Zoology 71: 471–484.

MARTIN, A. P., AND S. R. PALUMBI. 1993. Body size, metabolic rate, generation time, and the molecular clock. Proceedings of the National Academy of Sciences, U.S.A. 90: 4087–4091.

MARTINS, C. H. 1845. Memoire sur la temperature des oiseaux palmipedes du nord de l'Europe. Pp. 10–44 in Memoirs Originaux. [Publishing house unknown], Paris.

MASMAN, D., C. DIJKSTRA, S. DAAN, AND A. BULT. 1989. Energetic limitation of avian parental effort: field experiments in the kestrel (*Falco tinnunculus*). Journal of Evolutionary Biology 2: 435–455.

MCNAB, B. K. 1980a. Food habits, energetics, and the population biology of mammals. American Naturalist 116: 106–124.

MCNAB, B. K. 1980b. On estimating thermal conductance in endotherms. Physiological Zoology 53: 145–156.

MCNAB, B. K. 1986a. The influence of food habits on the energetics of eutherian mammals. Ecological Monographs 56: 1–19.

MCNAB, B. K. 1986b. Food habits, energetics, and the reproduction of marsupials. Journal of Zoology 208: 595–614.

MCNAB, B. K. 1987. The reproduction of marsupial and eutherian mammals in relation to energy expenditure. Symposium of the Zoological Society of London, No. 57: 29–39.

MCNAB, B. K. 1988. Food habits and the basal rate of metabolism in birds. Oecologia 77: 343–349.

MCNAB, B. K. 1997. On the utility of uniformity in the definition of basal rate of metabolism. Physiological Zoology 70: 718–720.

MCNAB, B. K., AND P. MORRISON. 1963. Body temperature and metabolism in subspecies of *Peromyscus* from arid and mesic environments. Ecological Monographs 33: 63–82.

MEHLUM, F., AND G. W. GABRIELSEN. 1995. Energy expenditure and food consumption by seabird population in the Barents Sea region. Pp. 457–470 in Ecology of Fjords and Coastal Waters (H. R. Skjoldal, C. Hopkins, K. E. Erikstad, and H. P. Leinaas, Eds.). Elsevier Science, Amsterdam.

MEHLUM, F., G. W. GABRIELSEN, AND K. A. NAGY. 1993. Energy expenditure by Black Guillemots (*Cepphus grylle*) during chick-rearing. Colonial Waterbirds 16: 45–52.

MILLER, M. R., AND K. J. REINECKE. 1984. Proper expression of metabolizable energy in avian energetics. Condor 86: 396–400.

MONTEVECCHI, W. A., V. L. BIRT-FRIESEN, AND D. K. CAIRNS. 1992. Reproductive energetics and prey-harvest in Leach's storm-petrels in the Northwest Atlantic. Ecology 73: 823–832.

MORENO, J., AND J. J. SANZ. 1996. Field metabolic rates of breeding chinstrap penguins (*Pygoscelis antarctica*) in the south Shetlands. Physiological Zoology 69: 586–598.

MORGAN, K. R., M. A. CHAPPELL, AND T. L. BUCHER. 1992. Ventilatory oxygen extraction in relation to ambient temperature in four antarctic seabirds. Physiological Zoology 65: 1092–1113.

MORRISON, P. R., AND F. A. RYSER. 1951. Temperature and metabolism in some Wisconsin mammals. Federation Proceedings 10: 93–94.

MORTENSEN, A., AND A. S. BLIX. 1986. Seasonal changes in resting metabolic rate and mass-specific conductance in Svalbard Ptarmigan, Norwegian Rock Ptarmigan and Norwegian Willow Ptarmigan. Ornis Scandinavica 17: 8–13.

MOUGIN, J.-L. 1989. Evaluation de la dépense énergétique et de la consommation alimentaire du Pétrel de Bulwer *Bulweria bulwerii* d'après l'étude de la décroissance pondérale au cours du jeûne. Boletim do Museu Municipal do Funchal 41: 25–39.

MURPHY, M. E. 1996. Energetics and nutrition of molt. Pp. 158–198 *in* Avian Energetics and Nutritional Ecology (C. Carey, Ed.). Chapman & Hall, New York.

NAGY, K. A. 1980. CO_2 production in animals: analysis of potential errors in the doubly labeled water method. American Journal of Physiology 238: R466–R473.

NAGY, K. A. 1987. Field metabolic rate and food requirement scaling in mammals and birds. Ecological Monographs 57: 111–128.

NAGY, K. A. 1989. Field bioenergetics: accuracy of models and methods. Physiological Zoology 62: 237–252.

NAGY, K. A., AND D. P. COSTA. 1980. Water influx in animals: analysis of the potential errors in the tritiated water method. American Journal of Physiology 238: R454–R465.

NAGY, K. A., AND B. S. OBST. 1992. Food and energy requirements of Adelie penguins *Pygoscelis adeliae* on the Antarctic peninsula. Physiological Zoology 65: 1271–1284.

NAGY, K. A., W. R. SIEGFRIED, AND R. P. WILSON. 1984. Energy utilization by free-ranging jackass penguins, *Spheniscus demersus*. Ecology 65: 1648–1655.

NAGY, K. A., I. A. GIRARD, AND T. K. BROWN. 1999. Energetics of free-ranging mammals, birds, and reptiles. Annual Review of Nutrition 19: 247–277.

NORBERG, U. M. 1996. Energetics of flight. Pp. 199–249 *in* Avian Energetics and Nutritional Ecology (C. Carey, Ed.). Chapman & Hall, New York.

NUNN, G. B., AND S. E. STANLEY. 1998. Body size effects and rates of cytochrome *b* evolution in tube-nosed seabirds. Molecular Biology and Evolution 15: 1360–1371.

OBST, B. S. 1986. The Energetics of Wilson's Storm-petrel *Oceanites oceanicus* nesting at Palmer Station, Antarctica. Ph.D. thesis, University of California, Los Angeles.

OBST, B. S., AND K. A. NAGY. 1992. Field energy expenditures of the Southern Giant-petrel. Condor 94: 801–810.

OBST, B. S., K. A. NAGY, AND R. E. RICKLEFS. 1987. Energy utilization by Wilson's storm-petrel (*Oceanites oceanicus*). Physiological Zoology 60: 200–210.

OEHME, H., AND R. BANNASCH. 1989. Energetics of locomotion in penguins. Pp. 230–240 *in* Energy Transformations in Cells and Organisms (W. Wieser and E. Gnaiger, Eds.). Proceedings of the 10th Conference of the European Society for Comparative Physiology and Biochemistry. Georg Theim Verlag, New York.

PARKER, H., AND H. HOLM. 1990. Patterns of nutrient and energy expenditure in female Common Eiders nesting in Svalbard. Auk 107: 660–668.

PENNYCUICK, C. J. 1987a. Flight of auks (Alcidae) and other northern seabirds compared with southern Procellariiformes: ornithodolite observations. Journal of Experimental Biology 128: 335–347.

PENNYCUICK, C. J. 1987b. Flight of seabirds. Pp. 43–62 *in* Seabirds: Feeding Ecology and Role in Marine Ecosystems (J. P. Croxall, Ed.). Cambridge University Press, Cambridge, U.K.

PENNYCUICK, C. J., F. C. SCHAFFNER, M. R. FULLER, H. H. OBRECHT, III, AND L. STERNBERG. 1990. Foraging flights of the White-tailed Tropicbird (*Phaethon lepturus*): Radiotracking and doubly-labelled water. Colonial Waterbirds 13: 96–102.

PETTIT, T. N., G. C. WHITTOW, AND H. I. ELLIS. 1984. Food and energetic requirements of seabirds at French Frigate Shoals, Hawaii. Pp. 265–282 *in* Proceedings of the Second Symposium on Resource Investigations in the Northwestern Hawaiian Islands (R. W. Grigg and K. W. Tanoue, Eds.). Sea Grant Miscellaneous Report, University of Hawaii.

PETTIT, T. N., H. I. ELLIS, AND G. C. WHITTOW. 1985. Basal metabolic rate in tropical seabirds. Auk 102: 172–174.

PETTIT, T. N., K. A. NAGY, H. I. ELLIS, AND G. C. WHITTOW. 1988. Incubation energetics of the Laysan Albatross. Oecologia 74: 546–550.

PINSHOW, B., M. A. FEDAK, D. R. BATTLES, AND K. SCHMIDT-NIELSEN. 1976. Energy expenditure for thermoregulation and locomotion in emperor penguins. American Journal of Physiology 231: 902–912.

PINSHOW, B., M. A. FEDAK, AND K. SCHMIDT-NIELSEN. 1977. Terrestrial locomotion in penguins: it costs more to waddle. Science 195: 592–594.

PIERSMA, T., N. CADÉE, AND S. DAAN. 1995. Seasonality in basal metabolic rate and thermal conductance in a long-distance migrant shorebird, the knot (*Calidris canutus*). Journal of Comparative Physiology B 165: 37–45.

PRANGE, H. D., AND K. SCHMIDT-NIELSEN. 1970. The metabolic cost of swimming in ducks. Journal of Experimental Biology 53: 763–777.

PRINCE, P. A., C. RICKETTS, AND G. THOMAS. 1981. Weight loss in incubating albatrosses and its implications for their energy and food requirements. Condor 83: 238–242.

PRINCE, P. A., N. HUIN, AND H. WEIMERSKIRCH. 1994. Diving depths of albatrosses. Antarctic Science 6: 353–354.

PRINZINGER, R., A. PREßMAR, AND E. SCHLEUCHER. 1991. Body temperature in birds. Comparative Biochemistry and Physiology 99A: 499–506.

PROSSER, C. L. 1973. Comparative Animal Physiology, 3rd ed. W.B. Saunders Company, Philadelphia.

REGEL, J., AND K. PÜTZ. 1997. Effect of human disturbance on body temperature and energy expenditure in penguins. Polar Biology 18: 246–253.

REYNOLDS, P. S., AND R. M. LEE III. 1996. Phylogenetic analysis of avian energetics: passerines and nonpasserines do not differ. American Naturalist 147: 735–759.

RICKLEFS, R. E. 1983. Some considerations on the reproductive energetics of pelagic seabirds. Studies in Avian Biology 8: 84–94.

RICKLEFS, R. E., AND S. C. WHITE. 1981. Growth and energetics of chicks of the Sooty Tern (*Sterna fuscata*) and Common Tern (*S. hirundo*). Auk 98: 361–378.

RICKLEFS, R. E., AND K. K. MATTHEW. 1983. Rates of oxygen consumption in four species of seabird at Palmer Station, Antarctic Peninsula. Comparative Biochemistry and Physiology 74A: 885–888.

RICKLEFS, R. E., S. C. WHITE, AND J. CULLEN. 1980. Energetics of post-natal growth in Leach's Storm-petrel. Auk 97: 566–575.

RICKLEFS, R. E., D. D. ROBY, AND J. B. WILLIAMS. 1986. Daily energy expenditure by adult Leach's storm-petrels during the nestling cycle. Physiological Zoology 59: 649–660.

RINTAMÄKI, H., S. SAARELA, A. MARJAKANGAS, AND R. HISSA. 1983. Summer and winter temperature regulation in the black grouse *Lyrurus tetrix*. Physiological Zoology 56: 152–159.

ROBIN, J.-P., L. BOUCONTET, P. CHILLET, AND R. GROSCOLAS. 1998. Behavioral changes in fasting emperor penguins: evidence for a "refeeding signal" linked to a metabolic shift. American Journal of Physiology 274: R746–R753.

ROBY, D. D., AND R. E. RICKLEFS. 1986. Energy expenditure in adult least auklets and diving petrels during the chick–rearing period. Physiological Zoology 59: 661–678.

ROPERT-COUDERT, Y., C.-A. BOST, Y. HANDRICH, R. M. BEVAN, P. J. BUTLER, A. J. WOAKES, AND Y. LE MAHO. 2000. Impact of externally attached loggers on the diving behaviour of the king penguin. Physiological and Biochemical Zoology 73: 438–445.

SATO, K., J. HWANG-BO, AND J. OKUMURA. 1988. Food consumption and basal metabolic rate in Common Cormorants (*Phalacrocorax carbo*). Ouyou Tyougaku Shuho 8: 56–62 (in Japanese).

SCHMID, D., D. J. H. GRÉMILLET, AND B. M. CULIK. 1995. Energetics of underwater swimming in the great cormorant (*Phalacrocorax carbo sinensis*). Marine Biology 123: 875–881.

SCHMIDT-NIELSEN, K. 1990. Animal Physiology, 4th ed. Cambridge University Press, Cambridge, U.K.

SCHNEIDER, D. C., G. L. HUNT, JR., AND N. M. HARRISON. 1986. Mass and energy transfer to seabirds in the southeastern Bering Sea. Continental Shelf Research 5: 241–257.

SCHOLANDER, P. F. 1940. Experimental investigations on the respiratory function in diving mammals and birds. Hvalrådets Skrifter 22: 1–131.

SCHOLANDER, P. F., V. WALTERS, R. HOCK, AND L. IRVING. 1950a. Body insulation of some arctic and tropical mammals and birds. Biological Bulletin 99: 225–236.

SCHOLANDER, P. F., R. HOCK, V. WALTERS, F. JOHNSON, AND L. IRVING. 1950b. Heat regulation in some arctic and tropical mammals and birds. Biological Bulletin 99: 237–258.

SCHOLANDER, P. F., R. HOCK, V. WALTERS, AND L. IRVING. 1950c. Adaptation to cold in arctic and tropical mammals and birds in relation to body temperature, insulation, and basal metabolic rate. Biological Bulletin 99: 259–271.

SCHREIBER, E. A. 1999. Breeding Biology and Ecology of the Seabirds of Johnston Atoll, Central Pacific Ocean. Report to the Department of Defense, Aberdeen Proving Ground, MD.

SCHREIBER, R. W., AND J. L. CHOVAN. 1986. Roosting by pelagic seabirds: energetic, populational, and social considerations. Condor 88: 487–492.

SCHWANN, M. W., AND D. D. WILLIAMS. 1978. Temperature regulation in the common raven in interior Alaska. Comparative Biochemistry and Physiology 60A: 31–36.

SHAFFER, S. A. 2000. Foraging Ecology of Wandering Albatrosses (Diomedea exulans): Impact on Reproduction and Life History. Ph.D. dissertation, University of California, Santa Cruz.

SHALLENBERGER, R. J., G. C. WHITTOW, AND R. M. SMITH. 1974. Body temperature of the nesting Red-footed Booby (Sula sula). Condor 76: 476–478.

STAHEL, C. D., AND S. C. NICOL. 1982. Temperature regulation in the little penguin Eudyptula minor, in air and water. Journal of Comparative Physiology B 148: 93–100.

STAHEL, C. D., AND S. C. NICOL. 1988. Ventilation and oxygen extraction in the little penguin (Eudyptula minor), at different temperatures in air and water. Respiration Physiology 71: 387–398.

STAHEL, C. D., D. MEGIRIAN, AND S. C. NICOL. 1984. Sleep and metabolic rate in the little penguin, Eudyptula minor. Journal of Comparative Physiology B 154: 487–494.

STANLEY, S. E., AND R. G. HARRISON. 1999. Cytochrome b evolution in birds and mammals: an evaluation of the avian constraint hypothesis. Molecular Biology and Evolution 16: 1575–1585.

TAYLOR, J. R. E., A. R. PLACE, AND D. D. ROBY. 1997. Stomach oil and reproductive energetics in Antarctic Prions, Pachyptila desolata. Canadian Journal of Zoology 75: 490–500.

UTTLEY, J., P. TATNER, AND P. MONAGHAN. 1994. Measuring the daily energy expenditure of free-living Arctic Terns (Sterna paradisaea). Auk 111: 453–459.

VLECK, C. M., AND G. J. KENAGY. 1980. Embryonic metabolism of the fork-tailed storm petrel: physiological patterns during prolonged and interrupted incubation. Physiological Zoology 53: 32–42.

WALSBERG, G. E., AND J. R. KING. 1978. The heat budget of incubating mountain White-crowned Sparrows (Zonotrichia leucophrys) in Oregon. Physiological Zoology 51: 92–103.

WARHAM, J. 1971. Body temperatures of petrels. Condor 73: 214–219.

WARHAM, J. 1996. The Behaviour, Population Biology and Physiology of the Petrels. Academic Press, London.

WEATHERS, W. W. 1979. Climate adaptation in avian standard metabolic rate. Oecologia 42: 81–89.

WEATHERS, W. W., AND K. A. SULLIVAN. 1989. Juvenile foraging proficiency, parental effort and avian reproductive success. Ecological Monographs 59: 223–246.

WEATHERS, W. W., K. L. GERHART, AND P. J. HODUM. 2000. Thermoregulation in Antarctic fulmarine petrels. Journal of Comparative Physiology B 170: 561–572.

WEBER, T. P., AND T. PIERSMA. 1996. Basal metabolic rate and the mass of tissues differing in metabolic scope: migration-related covariation between individual Knots Calidris canutus. Journal of Avian Biology 27: 215–224.

WEST, G. C. 1972. Seasonal differences in resting metabolic rate in Alaskan ptarmigan. Comparative Biochemistry and Physiology 42A: 867–876.

WHITTOW, G. C., C. T. ARAKI, AND R. L. PEPPER. 1978. Body temperature of the Great Frigatebird (Fregata minor). Ibis 120: 358–360.

WHITTOW, G. C., T. N. PETTIT, R. A. ACKERMAN, AND C. V. PAGANELLI. 1987. Temperature regulation in a burrow-nesting tropical seabird, the wedge-tailed shearwater (Puffinus pacificus). Journal of Comparative Physiology B 157: 607–614.

WIENS, J. A. 1984. Modelling the energy requirements of seabird populations. Pp. 255–284 *in* Seabird Energetics (G. C. Whittow and H. Rahn, Eds.). Plenum Press, New York.

WIENS, J. A., AND J. M. SCOTT. 1975. Model estimation of energy flow in Oregon coastal seabird populations. Condor 77: 439–452.

WILLIAMS, A. J., W. R. SIEGFRIED, A. E. BURGER, AND A. BERRUTI. 1977. Body composition and energy metabolism of moulting eudyptid penguins. Comparative Biochemistry and Physiology 56A: 27–30.

WILLIAMS, J. B. 1996. Energetics of avian incubation. Pp. 375–415 *in* Avian Energetics and Nutritional Ecology (C. Carey, Ed.). Chapman & Hall, New York.

WILSON, R. P., AND B. M. CULIK. 1991. The cost of a hot meal: facultative specific dynamic action may ensure temperature homeostasis in post-ingestive endotherms. Comparative Biochemistry and Physiology 100A: 151–154.

WILSON, R. P., AND B. M. CULIK. 1995. Energy studies of free-living seabirds: do injections of doubly-labeled water affect Gentoo Penguin behavior? Journal of Field Ornithology 66: 484–491.

WILSON, R. P., AND D. GRÉMILLET. 1996. Body temperatures of free-living African penguins (*Spheniscus demersus*) and bank cormorants (*Phalacrocorax neglectus*). Journal of Experimental Biology 199: 2215–2223.

WILSON, R. P., B. CULIK, D. ADELUNG, N. R. CORIA, AND H. J. SPAIRANI. 1991. To slide or stride: when should Adélie penguins (*Pygoscelis adeliae*) toboggan? Canadian Journal of Zoology 69: 221–225.

WILSON, R. P., M. A. M. KIERSPEL, J. A. SCOLARO, S. LAURENTI, J. UPTON, H. GALLELLI, E. FRERE, AND P. GANDINI. 1999. To think or swim: does it really cost penguins more to waddle? Journal of Avian Biology 30: 221–224.

WOAKES, A. J., AND P. J. BUTLER. 1983. Swimming and diving in tufted ducks, *Aythya fuligula*, with particular reference to heart rate and gas exchange. Journal of Experimental Biology 107: 311–329.

WOAKES, A. J., P. J. BUTLER, AND R. M. BEVAN. 1995. Implantable data logging system for heart rate and body temperature: its application to the estimation of field metabolic rates in Antarctic predators. Medical and Biological Engineering and Computing 33: 145–151.

YARBOROUGH, C. G. 1971. The influence of distribution and ecology on the thermoregulation of small birds. Comparative Biochemistry and Physiology 39A: 235–266.

APPENDIX 11.1 TERMS USED IN THIS CHAPTER

ADL   Aerobic dive limit — the maximal dive duration without measurable postdive lactate levels.

BMR   Basal metabolic rate — the minimal metabolic rate measured under specified conditions, including being at rest, post-absorptive, and in the thermoneutral zone (TNZ).

C   Thermal conductance — the coefficient of heat transfer that describes the increase in metabolism that accompanies decreases in temperature in an endotherm; wherever possible values of C reported here are minimal.

C_f Cost of feather production.

COT   Cost of transport — the metabolic expenditure needed to move a unit of mass a unit distance.

DEE   Daily energy expenditure — the daily energy budget.

EM   Existence metabolism — the energy it takes to stay alive, thermoregulating at a particular temperature, with minimal activity; as developed for birds, it does not include any of the BMR restrictions.

FMR   Field metabolic rate — the DEE typically measured in free-ranging animals using isotopically (usually doubly) labeled water.

f_H Heart rate.

HIF   Heat increment of feeding (see SDA).

LCT   Lower critical temperature — the lower limit of the zone of thermoneutrality (TNZ).

ME   Metabolizable energy — ingested energy which is assimilated and used in the DEE.

T_b Deep body temperature.

RMR Resting metabolic rate — the rate of metabolism of an animal at rest; the postabsorptive condition is not always specified and it may not be in its TNZ or in air, hence it may be the same or different from BMR.

RQ   Respiratory quotient — the ratio of the volume of CO_2 produced to the volume of O_2 consumed in an animal.

SDA Specific dynamic action — the cost of digestion; typically measured as an increase in the heat budget of an animal.

SMR Standard metabolic rate (see BMR).

T_b Body temperature.

T_e Operative temperature — an environmental temperature that integrates air temperature, insolation, wind speed, etc.

TNZ Thermoneutral zone — the environmental temperature range to which metabolic rate is insensitive; a condition of BMR.

Atlantic Puffin Entering Burrow with Sand Eels for Its Chick

12 Seabird Reproductive Physiology and Energetics

G. Causey Whittow

CONTENTS

12.1 INTRODUCTION

Seabirds tend to lay large eggs and, in Procellariiformes and tropical seabirds particularly, a single-egg clutch is the rule. Many seabirds feed far offshore during the breeding season, greatly increasing the time away from the nest and the distance over which food has to be transported back to the nestling (Lack 1968, Nelson 1979, Rahn and Whittow 1984). Some tropical seabirds nest under extremely hot conditions; others (penguins) endure colder conditions at the nest site than those to which any other warm-blooded animal is exposed. Thus, the reproductive period may place great demands on the physiological resources of seabirds and the adaptations of seabirds to these stresses invariably have energetic consequences. Nevertheless, the geographical distribution of seabird breeding sites and seabirds' exploitation of the full altricial–precocial developmental spectrum attest to the success of their breeding strategies.

The aim of this chapter is to summarize what is known about the incubation and hatchling physiology of seabirds without repeating information on other birds. For recent comprehensive reviews of reproduction in birds in general, the reader is referred to Cogburn et al. (2000), Johnson (2000), Kirby and Froman (2000), and Tazawa and Whittow (2000).

12.2 ROLE OF THE PARENT BIRDS

The cyclic reproductive physiology of birds effectively begins with the seasonal increased secretion of the hormone gonadotropic releasing hormone (GnRH), and other releasing hormones, from the hypothalamus. The releasing hormones, in turn, stimulate the secretion of hormones by the pituitary gland, leading to the development of the gonads and the full spectrum of reproductive behavior. In many seabirds, the initial event in this behavioral spectrum is the return of the birds at sea to their breeding colonies. Display (Figure 12.1a), mate selection, construction and defense of the nest, copulation (Figure 12.1b), egg-laying, and incubation are also part of this spectrum, but there is little direct information on the underlying physiology and the energetic cost of these reproductive activities in many seabirds.

There is evidence in female albatross that high levels of the hormone progesterone inhibit egg production when the birds are not breeding (Hector 1988). In the Cape Cormorant (*Leucocarbo capensis*), there is a clearly demarcated summer breeding season accompanied by a sharply defined increase in the mass of the pituitary gland, testes, and ovaries, and in the plasma testosterone levels (Berry et al. 1979).

12.2.1 FERTILIZATION OF THE OVUM

Copulation (Figure 12.1b) takes place ashore in all seabirds, and in Gentoo Penguins (*Pygoscelis papua*), plasma levels of luteinizing hormone (LH), estradiol, and testosterone are all high during the copulation period (Williams 1992). Procellariiformes leave on a "pre-laying exodus" after copulation has occurred (Warham 1990); the pre-laying exodus commonly lasts 2 to 3 weeks, but as long as 80 days in the Great-winged Petrel (*Pterodroma macroptera*). The exodus allows the birds to feed at sea while the egg is being formed and before having to endure the long fasts during incubation of the egg. Fertilization of the ovum occurs at sea; the delayed fertilization is possible because birds have sperm-storage glands, described in a number of seabirds (Horned Puffins, *Fratercula corniculata*; Leach's Storm Petrel, *Oceanodroma leucorhoa*; and Cassin's Auklet, *Ptychoramphus aleuticus*; Grau 1984) and identified in the Northern Fulmar (*Fulmarus glacialis*) as microtubules in the uterovaginal area (Hatch and Nettleship 1998).

12.2.2 EGG FORMATION

Yolk formation — In Procellariiformes, yolk formation occurs at sea while the birds are on their pre-laying exodus (Grau 1984). Astheimer and Grau (1990) present data on the time required for

FIGURE 12.1 (a) Three Laysan Albatross engage in ritual courtship display on Tern Island, French Frigate Shoals, Northwestern Hawaiian Islands. The birds illustrate three different component postures of the dance. (b) Two Laysan Albatross copulating on Trig Island, French Frigate Shoals. (Photos by G. C. Whittow.)

the rapid phase of yolk formation in 40 species of seabirds, which varies from 5 to 8 days in the Common Gull (*Larus canus*) to 30 days in the Southern Royal Albatross (*Diomedea epomophora*). It takes significantly longer to lay down the yolk in Procellariiformes than it does in other seabirds. The highest rates of yolk deposition are in alcids such as the Common Murre (*Uria aalge*) and Tufted Puffin (*Lunda cirrhata*). In Cassin's Auklet, daily yolk deposition increases rapidly to a sharp peak; in Buller's Shearwater (*Puffinus bulleri*), the increase is gradual, over a prolonged period, and the peak rate of deposition broad. Among Procellariiformes, yolk is deposited most rapidly in albatross and least rapidly in the smaller petrels.

Ovulation — Completion of yolk deposition is followed by ovulation (the release of the ovum into the body cavity) and fertilization of the ovum. The fertilized egg moves down the oviduct and in the magnum it is invested in a thick protein gel — the albumen. The albumen proteins have been characterized in two penguins (Adeliae Penguin, *Pygoscelis adeliae*; Blue Penguin, *Eudyptula minor*; Grau 1984). Formation of the shell in the shell gland and finally the laying of the egg complete the process. As many as 10 days (Royal Albatross, *Diomedea* spp.) may elapse between completion of yolk deposition and laying (Grau 1984). Observations on Cassin's Auklet suggest that the delay in laying is largely the result of a delay in ovulation after yolk formation is complete.

Egg laying — The laying period, the period over which eggs are laid by individual members of a particular species, varies considerably; even within a single order (Procellariiformes), the laying period may vary from 9 to 110 days (Warham 1990). Some seabirds — the Herring Gull (*Larus argentatus*) is an example (Parsons 1976) — can compensate for the loss of the first egg by laying an additional egg so that the customary clutch size of three eggs is maintained. The female may lose body mass during the egg-laying period: female Common Terns (*Sterna hirundo*) lose 16.8% of their body mass between laying the first and second eggs of a two-egg clutch (Nisbet 1977). Laying is influenced by availability of food in some species: Pelagic Cormorants (*Strictocarbo pelagicus*) on the Farallon Islands do not lay when food supplies are less plentiful in warm-water years (Hobson 1997).

Investment in the egg — Some idea of the investment that the female bird makes in the egg may be gleaned from the amount of protein, lipid, and energy incorporated in the two eggs of the Fiordland Penguin (*Eudyptes pachyrhynchus*), a species that lays two eggs. The peak in the rate of lipid deposition exceeds 1.5 g/day; it occurs earlier than that in protein synthesis, reflecting the fact that most of the lipid is incorporated in the yolk. Yolk formation precedes albumen deposition, the latter accounting for much of the protein in the egg. The maximal rate of energy deposition is greater than 104.6 kJ/day. Fiordland Penguins do not eat during the formation of the eggs — the lipid, protein, and energy are derived from the parent bird's tissues (Grau 1984). The energy cost of producing the yolk varies from 2 to 50% of the standard metabolic rate of the bird in different seabirds; the energy cost is lowest in Pelecaniformes and highest in the smaller Charadriiformes (Astheimer and Grau 1990) which tend to have large eggs in relation to the size of the adults.

12.2.3 Incubation

Incubation patch — Pelecaniformes do not develop an incubation patch, transmitting heat to their eggs possibly from their feet, but more likely through the abdominal feathers in contact with the egg (Howell and Bartholomew 1962a; Figure 12.2). The temperature of the bare incubation patch of seabirds that do have incubation patches, in comparison with egg temperature, is shown in a small sample of seabirds in Figure 12.3. The mean difference in temperature amounts to 2.75°C in Procellariiformes, and 2.72°C in Charadriiformes.

FIGURE 12.2 Brown Boobies do not have a brood patch and incubate by wrapping their large webbed feet around their eggs. (Photo by R. W. and E. A. Schreiber.)

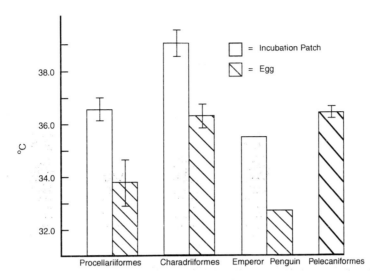

FIGURE 12.3 Mean (± S.E.) incubation-patch and egg temperatures of six species of Procellariiformes, six species of Charadriiformes, and the Emperor Penguin. Also shown is the mean (± S.E.) egg temperature of three species of Pelecaniformes, which either do not have incubation patches, or for which data are not available. References: Barrett 1980, Bucher et al. 1986, Drent 1970, Grant et al. 1982a, b, Howell and Bartholomew 1961, 1962a, b, Pettit et al. 1981, Rahn and Huntington 1988, Spellerberger 1969, Wheelwright and Boersma 1979, Whittow and Grant 1985, Whittow et al. 1982, 1989, G. C. Whittow, G. S. Grant and T. N. Pettit unpublished data.

Hormonal changes — In the Great Black-backed Gull (*Larus marinus*) and Wandering Albatross (*Diomedea exulans*), gonads regress during incubation (Harris 1964, Hector et al. 1986), which is characterized in the Wandering Albatross by high prolactin levels and a rapid decrease in progesterone and estradiol levels. In the Gray-headed (*Thalassarche chrysostoma*) and Black-browed (*T. melanophris*) Albatross, there is also a diminution in LH during incubation (Hector et al. 1986). In Gentoo Penguins, also, plasma levels of LH, estradiol, and testosterone diminish in incubating birds (Williams 1992), while prolactin concentrations increase during incubation (Williams and Sharp 1993). Evidence is accumulating in seabirds that elevated prolactin levels are strongly associated with incubation behavior in both male and female birds (Schradin and Anzenberger 2000).

Incubation energetics — Incubation shifts are generally longest in Procellariiformes, offshore-feeding Pelecaniformes, and in some penguins that nest on the ice far from their feeding grounds at sea. Many seabirds endure long incubation spells during which they neither feed nor drink. The first incubation shift of the male Laysan Albatross (*Phoebastria immutabilis*), for example, lasts 24.1 days (Fisher 1971, Whittow 1993b). Grant and Whittow (1983) calculate that after a 10-day fast, a Laysan Albatross loses 243 g (9.6%) of body mass; 144 g of this loss is due to loss of body fat. The birds do not become dehydrated because water loss from the body is balanced by metabolic water production. The fasting of the male Emperor Penguin (*Aptenodytes forsteri*), 115 days during pairing, incubation, and brooding of the nestling, results in the loss of 40.5% of its body mass. The male incubates the egg for the entire 68.5-day incubation period. Plasma lipids, the main source of energy during fasting, are high during courtship, decline during incubation, and then increase to maximal values at hatching (Groscolas 1982). In Macaroni Penguins (*Eudyptes chrysolophus*), fasting during incubation results in a 30.9 to 34.1% decrease in body mass over a 34- to 36-day period; body lipid reserves are the main source of energy (Williams et al. 1992). In Leach's Storm Petrel, which has one of the highest daily rates of mass loss during incubation, stomach oils provide some energy to the incubating bird (Ricklefs et al. 1986).

The energy cost of incubation has been assessed by comparing the measured energy expenditure of incubating birds with that of resting birds that are not incubating (Grant 1984), and, more recently

TABLE 12.1
Energy Expenditure of Incubating Seabirds (IMR, Joules/hour × gram) Compared with That of Resting Seabirds That Are Not Incubating (RMR, Joules/hour × gram)

Order/Species		IMR (J/h × g)	RMR (J/h × g)	IMR/RMR	Reference
Sphenisciformes					
Rockhopper Penguin	*Eudyptes chrysocome*	10.56[a]	14.35[a]	0.74	Brown 1984
Macaroni Penguin	*E. chrysolophus*	8.88[a]	12.79[a]	0.69	Brown 1984
Procellariiformes					
Wandering Albatross	*Diomedea exulans*	12.71	9.00[c]	1.41	Brown and Adams 1984
Laysan Albatross	*Phoebastria immutabilis*	8.23[d]	10.64[d]	0.77	Grant and Whittow 1983
Kerguelen Petrel	*Lugensa brevirostris*	17.67	—[e]	0.88	Brown 1988
Great-winged Petrel	*Pterodroma macroptera*	16.58	—[e]	0.82	Brown 1988
Bonin Petrel	*P. hypoleuca*	20.88[d]	26.91[d]	0.78	Grant and Whittow et al. 1983
Blue Petrel	*Halobaena caerulea*	31.25	—[e]	0.94	Brown 1988
White-chinned Petrel	*Procellaria aequinoctialis*	14.83	—[e]	0.84	Brown 1988
Wilson's Storm-petrel	*Oceanites oceanicus*	79.80	43.17[b]	1.85	Obst et al. 1987
Leach's Storm-petrel	*Oceanodroma leucorhoa*	72.80	40.56[c]	1.79	Montevecchi et al. 1992
Pelecaniformes					
Cape Gannet	*Morus capensis*	33.75	11.60[c]	2.91	Adams et al. 1991
Northern Gannet	*M. bassanus*	44.81	9.63[c]	4.65	Birt-Friesen et al. 1989
Charadriiformes					
Sooty Tern	*Sterna fuscata*	31.29	31.56[b]	0.99	Flint and Nagy 1984, MacMillen et al. 1977

[a] Measured under similar conditions in the laboratory; not known if the birds were within their thermoneutral zone.
[b] Within their thermoneutral zone.
[c] Not known if the birds were within their thermoneutral zone.
[d] Measured under similar conditions; not known if the birds were within their thermoneutral zone.
[e] Value not given in reference; not known if the birds were within their thermoneutral zone.

(Williams 1996), by comparing calculated and measured incubation metabolic rates with basal metabolic rates (BMR). Table 12.1 presents data for the *measured* metabolic rates of incubating birds. The incubating metabolism varies from 69 to 465% of the metabolic rate of birds that are not incubating. In incubating Wilson's Storm-Petrel (*Oceanites oceanicus*), the smallest seabird nesting in Antarctica, the metabolic rate during incubation is 2.2 × the basal metabolic rate (Obst et al. 1987). However, in 9 of the 14 species included in Table 12.1, the metabolic rate is lower in incubating than in nonincubating birds. Care is needed in the interpretation of the data in Table 12.1; in many cases different methods have been used to measure the metabolic rates of the incubating and nonincubating birds in the same species. The most valid comparisons are between incubating and nonincubating birds tested under similar conditions in the field (e.g., Laysan Albatross) or between incubating birds and the BMR of the birds (e.g., Sooty Tern, *Sterna fuscata*). The measurement of BMR requires that the bird be in its thermoneutral zone, and there is little information available on the thermoneutral zones of seabirds (Whittow et al. 1987).

With few exceptions (one is the Emperor Penguin cited above), male and female seabirds alternate incubation duties, although *Eudyptes* penguins literally share the first shift, i.e., they are at the nest together. The use of the doubly labeled water technique has made it possible to partition energy expenditure of incubating birds between incubation duties and other activities, such as

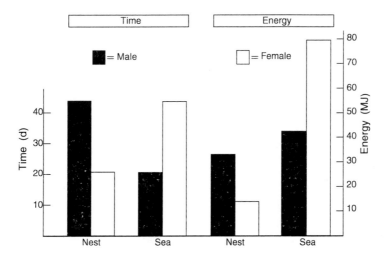

FIGURE 12.4 Partition of the time (days) and energy expenditure (megajoules) of an incubating pair of Laysan Albatross, between the nest and foraging at sea. Based on Pettit et al. 1988.

foraging at sea. The allocation of time and energy to incubation and foraging at sea in an incubating pair of Laysan Albatross is illustrated in Figure 12.4. Females are at sea longer than they are ashore incubating the egg, and their energy expenditure is correspondingly greater. This pattern is not universal. In Wandering Albatross the energy expenditure of males is greater than that of females, reflecting the disparity in body size between the two sexes (male, 9.5 kg; female, 7.4 kg; Adams et al. 1986).

12.2.4 BREEDING AGE AND PERIODICITY

Some seabirds do not breed until they are 11 years old (Royal Albatross, Skutch 1976), and not all seabirds are able to breed every year (see Appendix 2). At Tern Island, French Frigate Shoals, in the Northwestern Hawaiian Islands, but not elsewhere, Black Noddies (*Anous minutus*) regularly lay two clutches each year, approximately 5 months apart (Gauger 1999). Abundance of food may explain this phenomenon; in addition, Black Noddies on Tern Island have low metabolic rates which may permit them to divert more energy to reproductive activities than is possible in birds that have a higher minimum metabolic energy expenditure (Pettit et al. 1985).

12.2.5 HEAT STRESS

Many tropical seabirds endure extremely demanding conditions during incubation. Air and ground temperatures may be very high; there is usually no shade and very often little wind, and solar radiation is intense. The incubating bird relies on physiological mechanisms of thermoregulation and a repertoire of behavioral responses to moderate the elevation of body temperature and maintain egg temperature within the bounds permissible for successful incubation (Figures 12.3 and 12.5). Plumage color may not be an infallible guide to the radiant heat load on incubating seabirds, based on studies of terrestrial species (Whittow 1986). Thus, some dark plumages may re-radiate to the environment heat gained from solar radiation, while some white plumages may transmit the gained heat inward to the bird's skin, thus adding to its heat stress. Information is needed on the radiative properties of seabirds' plumages. An important part of incubation in hot climates is the prevention of overheating of eggs (Bennett and Dawson 1979, Figure 12.5). All seven species of Charadriiformes that nest under desert conditions at the Salton Sea in California keep their eggs cool with water conveyed to the nest in their abdominal plumage ("belly soaking," Grant 1982).

FIGURE 12.5 An incubating Masked Booby rises from its eggs, shading them and not providing further heat to the eggs (Kure Atoll, Northwestern Hawaiian Islands). Note the gular flutter (bill agape), raised scapulars, and drooping wings, measures promoting heat loss from the incubating bird. The eggs and the feet of the parent bird are in the shade of the bird's body. (Photo by G. C. Whittow.)

12.3 NESTS

Seabirds as a group show exceptional versatility in the selection of nest sites, from a scrape in the desert sand devoid of all trace of vegetation (Grey Gull, *Larus modestus*, Howell et al. 1974) to the substantial nest of twigs of Abbott's Booby (*Papasula abbotti*) in dense jungle on Christmas Island, Indian Ocean (Nelson 1971). Some species exhibit plasticity in the selection of a nest site: in the Hawaiian Islands, Black Noddies usually build substantial nests in trees or bushes, but they may also nest on cliffs. The White Tern (*Gygis alba*) makes no pretence at constructing a nest, usually balancing its single egg precariously on the branch of a tree, but it will, in fact, lay its egg anywhere (Figure 12.6). In addition to the obvious purpose of providing protection for the eggs and nestlings, nests can, potentially, alter the rate of transfer of carbon dioxide away from the egg, and diffusion of oxygen (needed for development) from the ambient air into the egg. Depletion of oxygen and accumulation of carbon dioxide in the nest are most likely to occur in the burrow nests favored by puffins, petrels, and shearwaters. The Bonin Petrel (*Pterodroma hypoleuca)*, for instance,

FIGURE 12.6 The White Tern does not build a nest, laying its single egg in a depression on a branch or in a fork between two branches in a bush or tree. (Photo by R. W. and E. A. Schreiber.)

may nest in burrows as long as 3 m; the partial pressures of oxygen and carbon dioxide in occupied burrows average 148 and 4.4 torr, respectively, in comparison with values of 154 and 0.6 torr in empty burrows (Pettit et al. 1982a). These levels of carbon dioxide and oxygen are not likely to be inimical to either the developing embryo in the egg or the adult bird (Pettit and Whittow 1982a, b).

12.4   EGGS

Clutches are smallest in tropical seabirds, offshore feeders, and in Procellariiformes; they are largest — up to six eggs — in some cormorants and terns. Procellariiformes invest a great deal of energy (see below), in the form of yolk (Table 12.2), in a single egg, and incubate it for a long time (Figures 12.7 and 12.8).

12.4.1   MASS AND SIZE

The mass of the freshly laid egg is the only value for which valid allometric relationships may be made because the fresh-egg mass represents the end product of the parent birds' efforts. Once incubation proceeds, the egg loses mass, and materials are consumed by the embryo. Unfortunately, many studies of the breeding biology of seabirds report egg masses taken throughout incubation. These data have little value unless accompanied by accurate measurements of the daily mass loss of the egg and the elapsed number of days of incubation.

In general, seabirds are larger than terrestrial birds and, as larger birds lay larger eggs, it follows that the average mass of seabird eggs is greater than that of land birds. However, in all birds the mass of the freshly laid egg relative to mass of the adult bird is less in large than in small birds (Rahn and Whittow 1984). Seabird eggs range in size from 5.5 g (European Storm-Petrel, *Hydrobates pelagicus*; Rahn et al. 1984) to 638 g (Emperor Penguin, Williams et al. 1982). Procellariiformes and Alcidae have the largest eggs in relation to their body size (mean value = 11.2% of body mass); the relatively large eggs of Procellariiformes are the result of greater contents rather than heavier shells (Rahn et al. 1984, Table 12.2). There are large differences in egg size between families of the order Pelecaniformes (Table 12.2): the smallest eggs relative to the size of the adults belong to the cormorants (3% of body mass), and the largest are those of the uniparous tropicbirds (9.3%). In Brown Boobies (*Sula leucogaster*), cormorants, terns, and gulls, the first-laid egg is usually larger than subsequent ones, but in Eudyptes penguins the second egg is markedly heavier than the first egg, which rarely produces a viable hatchling.

12.4.2   SHELL CHARACTERISTICS

Shells of Sphenisciformes and Pelecaniformes are appreciably heavier and thicker than those of Procellariiformes and Charadriiformes (Rahn et al. 1984; Table 12.2). Penguins also have the densest shells. There may be a relation between the thicker shells of penguins, gannets, and boobies and their mode of incubating the eggs with their feet. Thinning of the shells since the introduction of pesticides has been documented in Brown Pelicans, Brown and Masked (*Sula dactylatra*) Boobies (Morrison 1979), and in Brandt's Cormorant (*Compsohalieus penicillatus*, Wallace and Wallace 1998).

The eggshell provides a rigid container for the contents of the egg which contain all of the materials necessary for the development of the embryo, with one exception, oxygen. Oxygen is essential for the release of energy stored in the yolk and albumen, and for many of the embryo's metabolic reactions. It is derived from surrounding air and it passes into the egg through microscopic pores in the shell. Oxygen transport occurs by simple diffusion through the pores and the number, shape, and dimensions of the pores determine to a large extent the rate at which oxygen diffuses into the egg. The same pores provide a conduit for the removal of carbon dioxide produced by the embryo. Water vapor also passes through the pores from the warm, moist interior of the egg to outside. Thus, the pores are the "lungs" of the egg (Ar and Rahn 1985, Rahn et al. 1987).

TABLE 12.2
Composition of Seabird Eggs

Order/Species		Egg Mass (g)	Shell (% of egg mass)	Yolk (% of egg contents)	Water (% of egg contents)
Sphenisciformes					
Blue Penguin	Eudyptula minor	60.5	10.1	26.0	—
Rockhopper Penguin	Eudyptes chrysocome	86.3	14.8	29.6	77.2
Jackass Penguin	Spheniscus demersus	97.5	14.8	29.0	78.8
Fiordland Penguin	Eudyptes pachyrhynchus	108.6	11.0	27.3	—
Adeliae Penguin	Pygoscelis adeliae	114.3	12.8	25.3	78.3
Gentoo Penguin	P. papua	117.1	13.5	28.5	80.4
Macaroni Penguin	Eudyptes chrysolophus	119.5	14.4	32.2	78.4
King Penguin	Aptenodytes patagonicus	322.4	14.4	27.0	77.8
Emperor Penguin	A. forsteri	469.4	15.7	30.4	—
Mean		**166.2**	**13.5**	**28.4**	**78.5**
Procellariiformes					
Leach's Storm-petrel	Oceanodroma leucorhoa	9.3	6.8	39.7	74.4
White-faced Storm-petrel	Pelagodroma marina	13.4	5.7	38.6	—
Fairy Prion	Pachyptila turtur	25.2	6.7	40.0	—
Bulwer's Petrel	Bulweria bulwerii	27.1	5.9	38.0	72.2
Broad-billed Prion	Pachyptila vittata	32.3	9.9	42.6	76.5
Bonin Petrel	Pterodroma hypoleuca	39.2	5.0	40.0	70.6
Soft-plumaged Petrel	P. mollis	49.3	11.4	39.0	78.5
Snow Petrel	Pagodroma nivea	56.8	9.7	37.8	—
Wedge-tailed Shearwater	Puffinus pacificus	60.0	6.2	40.0	73.2
Buller's Shearwater	P. bulleri	66.0	6.7	38.1	—
Cape Pigeon	Daption capense	67.3	11.3	35.8	—
Great-winged Petrel	Pterodroma macroptera	78.9	9.3	42.0	77.6
Southern Fulmar	Fulmarus glacialoides	103.4	10.9	33.1	—
Greater Shearwater	Puffinus gravis	107.2	7.3	38.3	—
White-chinned Petrel	Procellaria aequinoctialis	125.2	9.9	44.0	78.7
Sooty Albatross	Phoebetria fusca	241.9	9.1	31.0	75.5
Southern Giant Petrel	Macronectes giganteus	248.4	11.4	35.0	78.3
Northern Giant Petrel	M. halli	253.3	12.2	39.0	78.2
Light-mantled Albatross	Phoebetria palpebrata	254.3	9.9	34.0	78.3
Laysan Albatross	Phoebastria immutabilis	284.8	7.2	34.8	74.2
Black-footed Albatross	P. nigripes	304.9	7.6	33.0	74.9
Southern Royal Albatross	Diomedea epomophora	440.5	7.2	26.2	—
Wandering Albatross	D. exulans	504.2	9.2	30.0	76.5
Mean		**147.5**	**8.5**	**37.0**	**75.9**
Pelecaniformes					
Crowned Cormorant	Microcarbo coronatus	22.6	14.2	28.4	83.5
Little Pied Cormorant	M. melanoleucos	26.0	8.1	17.2	—
Cape Cormorant	Leucocarbo capensis	38.0	14.8	20.0	82.8
Spotted Shag	Strictocarbo punctatus	40.4	9.2	17.4	—
Bank Cormorant	Compsohalieus neglectus	46.7	13.2	20.2	83.3
Great Cormorant	Phalacrocorax carbo	47.2	16.4	19.4	83.1
Imperial Shag	Notocarbo atriceps	50.5	13.2	23.9	87.9
Red-billed Tropicbird	Phaethon aethereus	56.0	9.8	35.6	—
Brown Booby	Sula leucogaster	57.8	12.5	18.3	—
Red-footed Booby	S. sula	58.3	8.1	15.8	83.6
Red-tailed Tropicbird	Phaeton rubricauda	71.6	7.9	24.4	74.8
Great Frigatebird	Fregata minor	89.1	7.4	25.7	77.5

TABLE 12.2 *(Continued)*
Composition of Seabird Eggs

Order/Species		Egg Mass (g)	Shell (% of egg mass)	Yolk (% of egg contents)	Water (% of egg contents)
Brown Pelican	*Pelecanus occidentalis*	92.1	12.2	28.1	76.7
Cape Gannet	*Morus capensis*	97.8	15.5	21.2	82.8
Northern Gannet	*M. bassanus*	117.7	12.1	17.6	83.4
Eurasian White Pelican	*Pelecanus onocrotalus*	203.8	14.6	14.2	83.2
Mean		**69.7**	**11.8**	**21.7**	**81.9**
Charadriiformes					
Black Tern	*Chlidoniasis niger*	11.4	7.0	29.9	—
White-winged Black Tern	*C. leucopterus*	14.0	7.2	30.0	78.1
Whiskered Tern	*C. hybridus*	15.0	5.3	31.7	—
Arctic Tern	*Sterna paradisaea*	18.4	6.5	30.8	—
Roseate Tern	*S. dougallii*	20.6	10.7	33.2	—
Common Tern	*S. hirundo*	21.5	7.5	28.6	74.8
White Tern	*Gygis alba*	23.3	5.2	38.0	71.1
Black Noddy	*Anous minutus*	24.8	5.5	34.9	74.9
White-fronted Tern	*Sterna striata*	26.4	6.6	30.0	—
Gray-backed Tern	*S. lunata*	28.7	5.8	36.5	74.6
Cassin's Auklet	*Ptychoramphus aleuticus*	29.1	6.6	—	74.7
Sandwich Tern	*Sterna sandvicensis*	34.6	—	33.0	76.4
Sooty Tern	*S. fuscata*	35.7	5.8	37.7	71.7
Common Black-headed Gull	*Larus ridibundus*	36.9	6.6	29.1	—
Silver Gull	*L. novaehollandiae*	39.2	7.0	31.1	—
Brown Noddy	*Anous stolidus*	40.1	6.6	35.3	74.7
Laughing Gull	*Larus atricilla*	40.6	9.1	36.5	76.2
Ancient Murrelet	*Synthliboramphus antiquum*	44.9	7.7	46.6	—
Hartlaub's Gull	*Larus hartlaubii*	49.8	11.6	39.1	80.0
Common Gull	*L. canus*	52.3	7.8	28.4	—
Crested Tern	*Sterna bergii*	56.9	10.0	33.4	75.6
Caspian Tern	*S. caspia*	65.0	6.9	27.0	—
Atlantic Puffin	*Fratercula arctica*	65.0	—	37.2	70.5
Black-tailed Gull	*Larus crassirostris*	66.0	8.4	29.2	76.9
Royal Tern	*Sterna maxima*	70.2	—	30.0	77.6
Lesser Black-backed Gull	*Larus fuscus*	78.1	6.3	26.9	—
Rhinocerus Auklet	*Cerorhinca monocerata*	82.0	7.7	34.7	74.5
Kelp Gull	*Larus dominicanus*	88.1	9.0	30.7	75.7
Razor-billed Auk	*Alca torda*	90.0	9.3	39.8	71.9
Herring Gull	*Larus argentatus*	93.0	9.3	26.5	78.1
Western Gull	*L. occidentalis*	97.7	6.7	—	77.8
South Polar Skua	*Catharacta maccormicki*	99.3	8.6	28.7	76.4
Brown Skua	*C. antarctica*	102.1	11.1	33.2	80.7
Great Black-backed Gull	*Larus marinus*	116.1	7.4	27.9	—
Common Murre	*Uria aalge*	120.0	14.4	37.0	73.1
Mean		**54.2**	**7.9**	**32.8**	**75.5**

Note: References: Ar and Rahn 1985, Carey et al. 1980, Gaston 1994, Grant et al. 1982a, b, Grau 1982, Mehlum et al. 1986, Montevecchi et al. 1983, Morgan et al 1978, Pettit and Whittow 1983, 1985, Pettit et al. 1981, Rahn and Dawson, 1979, Rahn and Hammel 1982, Rahn and Huntington 1988, Rahn et al. 1976, 1982, 1984, Ricklefs 1977, Ricklefs and Montevecchi 1979, Roudybush et al. 1980, Sotherland and Rahn 1987, Wallace and Wallace 1998, Whittow and Grant 1985, Whittow and Pettit 2000, Whittow et al. 1985, Williams et al. 1982.

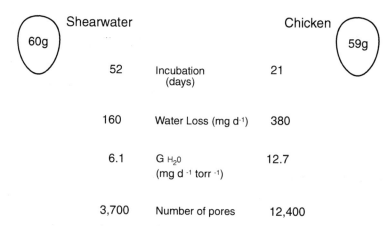

FIGURE 12.7 Prolonged incubation in a tropical, procellariiform seabird (Wedge-tailed Shearwater). Comparison of the egg mass, incubation period, the daily water loss from the egg, the water-vapor conductance (G_{H_2O}), and the number of pores in the eggshell, in the chicken's egg and in that of the Wedge-tailed Shearwater. Note that the initial mass of the two eggs is almost identical. Drawing courtesy of Charles V. Paganelli.

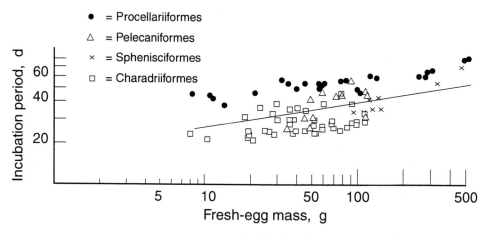

FIGURE 12.8 Relationship between the incubation period in days (d) and the mass of the freshly laid egg, in grams (g) in 99 species of seabirds. The regression line for the relationship ($r = 0.458$, $p < 0.001$): incubation period (d) = $16.448 \times$ fresh-egg mass (g)$^{0.186}$. References: Adams 1990, Ar and Rahn 1978, Ainley 1995, Ar et al. 1974, Brown and Adams 1988, Bucher 1986, 1987, Carey et al. 1980, Drost and Lewis 1995, Gaston 1994, Gauger 1999, Gochfeld and Burger 1994, Grant et al. 1982a, 1984, Grau 1982, Hatch and Nettleship, 1998, Hobson 1997, Howell et al. 1974, Kepler 1969, Lishman 1985, Morgan et al. 1978, Nelson 1971, 1976, Paganelli et al. 1974, Parnell et al. 1995, Pettit and Whittow 1985, Pettit et al. 1981, Rahn and Dawson 1979, Rahn and Hammel 1982, Rahn et al. 1976, 1982, 1984, Ricklefs and Montevecchi 1979, Roudybush et al. 1980, Vleck and Kenagy 1980, Vleck et al. 1983, Wallace and Wallace 1998, Warham 1974, Whittow 1993a, b, 1994, 1997, Whittow and Grant 1985, Whittow and Naughton 1999, Whittow and Pettit 2000, Whittow et al. 1979, 1984, 1985, 1989, Wiley and Lee 1998; also, Birds of North American Species accounts (Birds of North America, Inc., Philadelphia, PA) for individual species.

Pores of Royal Tern (*Sterna maxima*) eggs are funnel-shaped, a common configuration in birds. In both the Royal Tern and the Royal Penguin (*Eudyptes schlegeli*), the pores are rather more numerous at the blunt pole of the egg than at the pointed end. The relationship between the total number of pores in a seabird egg (N) and the mass of the freshly laid egg (M) is linear when plotted on logarithmic coordinates and the relationship may be represented by the equation (Whittow 1984):

$$N = 278.6 \times M^{0.704}$$

12.4.3 Egg Contents

Freshly laid eggs contain everything (except oxygen) that the embryo needs to grow and develop into a hatchling (Table 12.2).

Yolk and energy — The most conspicuous components of eggs are the yolk and albumen. Seabird eggs vary in the proportions of yolk and albumen in their eggs (Table 12.2) and these differences are reflected in variations in the energy content because the yolk is rich in lipids which, in turn, are rich in energy. The yolk may make up as much as 46.6% or as little as 14.2% (by mass) of the egg contents (Table 12.2). The energy content of the Northern Gannet's (*Morus bassanus*) egg is smaller, in relation to egg mass, than that of any other avian species (Ricklefs and Montevecchi 1979). Many of the differences in egg yolk content between eggs of different species are related to variations in hatchling maturity (see below). There is more yolk in the eggs of semiprecocial puffins and terns and petrels than in those of altricial boobies and gannets (Table 12.2, Sotherland and Rahn 1987); the production of a relatively mature hatchling requires more energy than does that of a helpless, immature hatchling (Vleck and Vleck 1987). However, other factors have determined the yolk content of seabird eggs, including the proximity of feeding grounds and the length of the incubation period (discussed below). Offshore feeders visit their young less frequently, necessitating a larger hatchling yolk reserve and, in turn, requiring a greater amount of yolk in the egg. Offshore feeders also have long incubation periods (see below) which are also associated with a higher egg yolk content (Sotherland and Rahn 1987).

Water and albumen — While the yolk provides lipids, the albumen in the egg is an important source of protein for the developing embryo. The albumen also has a high water content (Sotherland and Rahn 1987). Water and albumen content in a freshly laid egg are inversely proportional to the yolk content. Consequently, in the semiprecocial eggs of Procellariiformes and Charadriiformes, the water and albumen content are relatively less than in the eggs of altricial Pelecaniformes. For example, water makes up 87.9 and 70.5% of the contents of the altricial Imperial Shag's (*Notocarbo atriceps*) and precocial Atlantic Puffin's eggs (*Fratercula arctica*), respectively (Table 12.2). When the water present in the egg is removed by drying the contents of the egg, solids remain. The solids are largely made up of organic compounds — fats, proteins, and carbohydrates, but also minerals.

The water regulation of the egg is complex: water is *lost* from the egg through microscopic pores in the shell throughout incubation (see next section). However, water is also *produced* as a result of embryonic metabolism. The *net* effect is that the percentage water content of the hatchling is the same as that of the freshly laid egg (Rahn 1984).

12.4.4 Water Loss from the Egg

The decrease in mass of eggs during incubation is entirely attributable to the loss of water (Rahn and Ar 1974). Water is replaced by air, giving rise to the aircell. The daily mass loss ($\dot{M}_{H_2O}$) of unpipped eggs is defined by two factors (Rahn and Ar 1974): (1) the difference in water-vapor pressure ($\Delta P_{H_2O} = P_{H_2O,egg} - P_{H_2O,nest}$) between the contents of the egg ($P_{H_2O,egg}$) and the microclimate of the incubated egg ($P_{H_2O,nest}$), and (2) the water–vapor conductance of the shell and shell membranes (G_{H_2O}):

$$\dot{M}_{H_2O} = \Delta P_{H_2O} \cdot G_{H_2O}$$

The water loss from the eggs of Procellariiformes and tropical Charadriiformes and Pelecaniformes tends to be low in relation to the size of their eggs (Whittow 1984) as a result of low water-vapor conductance. With few exceptions, the low G_{H_2O} is due to fewer pores in the shell (Ar and Rahn 1985; Figure 12.5). A reduced rate of decrease in the mass of the egg is part of the suite of characteristics of birds that feed far offshore, including prolonged incubation and slow embryonic growth (Lack 1968, Rahn et al. 1974, 1984, Whittow 1980, 1984, Furness and Monaghan 1987, Sotherland and Rahn 1987).

Water loss may also be influenced by factors that affect ΔP_{H_2O}. One of these, the water-vapor pressure of egg contents ($P_{H_2O,egg}$), is largely a function of egg temperature. The mean egg temperature of 36 species of seabirds is 35.4°C ± 1.6; mean egg temperature (and, therefore, $P_{H_2O,egg}$) is lowest in Procellariiformes (Figure 12.3). Egg temperature is not constant throughout incubation and may vary diurnally as well as through the course of incubation (Whittow et al. 1982, Grant et al. 1982b).

The other factor that influences the ΔP_{H2O}, water-vapor pressure in the microclimate of the egg ($P_{H_2O,nest}$), can be affected by ambient air in the nesting colony. For example, the water-vapor pressure in the egg's microclimate of the Adelie Penguin is quite low (10.2 torr), reflecting the very cold, dry Antarctic air (Rahn and Hammel 1982). The low egg microclimate water-vapor pressure might have led to excessive water loss from the egg; however, the water-vapor conductance of the shell (G_{H_2O} — see above equation) is relatively low (13.1 mg/d.torr), thus curtailing water loss and counteracting the effects of the dry environment. A similar problem is faced by the Hawaiian Dark-rumped Petrel (*Pterodroma sandwichensis*) which nests in the very dry air at 3000 m on the rim of Haleakala Crater, Hawaiian Islands (Whittow et al. 1984). When an incubating bird rises from the egg (Figure 12.5), turns the egg, or applies its incubation patch less closely to the egg, it "ventilates" the nest, permitting the replacement of moist air in the nest microclimate with drier ambient air (Rahn and Dawson 1979).

Water loss from the egg increases when the egg is pipped. The pipping process is prolonged in many seabirds (see below) and accounts for a significant fraction of the total water loss from the egg. In 12 species of tropical seabirds, water loss during pipping amounts to an average of 28.3% of the total water loss during incubation, although the duration of pipping is only 10.5% of the incubation period (Table 12.3). Total water loss from the egg during incubation averages 17.1% of the mass of the freshly laid egg. Similar information is needed for seabirds breeding in higher latitudes.

12.4.5 INCUBATION PERIOD

The relationship between the incubation period and egg mass in 99 species of seabirds is shown in Figure 12.8. The incubation period is longer in a large than in a small egg, as might be expected. Incubation periods of Procellariiformes tend to be longer than those of other seabirds of comparable egg size, over the entire range of egg mass (Rahn et al. 1984). Embryonic growth is slow in Procellariiformes but the hatchlings are relatively mature, factors conducive to prolongation of incubation.

Egg neglect occurs in a few seabird species. In Fork-tailed Storm Petrels (*Oceanodroma furcata*) the mean cumulative duration of egg neglect is 11 days and, as a result of the variable degree of neglect, the incubation period ranges from 37 to 68 days. The neglected eggs cool to approximately 10°C, a temperature at which embryonic development is arrested. Hence, incubation is prolonged (Vleck and Kenagy 1980). Egg neglect is a feature of some offshore feeders, e.g., Xantu's Murrelets (*Endomychura hypoleuca*), which produce, on the average, two eggs representing a large percentage of their body mass (Drost and Lewis 1995). Egg neglect allows birds more time to forage (Murray et al. 1980). Unattended Heermann's Gulls' eggs at night may chill to temperatures at which the heart beat ceases, but the exposure is not lethal if the eggs are rewarmed (Bennett and Dawson 1979).

TABLE 12.3

Decrease in the Mass of the Egg during Incubation and Partition of the Decrease between the Pre-pipping and Pip-hatch Periods, in 12 Species of Tropical Seabirds

Order/Species		Total Decrease in Egg Mass during Incubation as % of Initial Egg Mass	Duration of Pip-hatch Interval as % of Incubation Period	Decrease in Egg Mass in Pipped Eggs as % of Total Decrease in Egg Mass
Procellariiformes				
Black-footed Albatross	*Phoebastria nigripes*	16.1	5.6	11.1
Laysan Albatross	*P. immutabilis*	16.0	4.9	8.8
Bonin Petrel	*Pterodroma hypoleuca*	18.6	12.1	34.9[a]
Bulwer's Petrel	*Bulweria bulwerii*	20.6	10.0	36.3
Wedge-tailed Shearwater	*Puffinus pacificus*	17.2	9.2	29.0
Pelecaniformes				
Red-tailed Tropicbird	*Phaethon rubricauda*	12.9	8.0	17.1
Red-footed Booby	*Sula sula*	14.5	3.1	9.6
Charadriiformes				
Gray-backed Tern	*Sterna lunata*	14.7	13.5	30.8
Sooty Tern	*S. fuscata*	18.5	15.9	41.0
Brown Noddy	*Anous stolidus*	16.9	12.9	36.5
Black Noddy	*A. minutus*	21.7	15.9	42.7
White Tern	*Gygis alba*	17.6	14.6	42.2
		17.1	10.5	28.3

[a] Star-fracture of the shell.

Note: References: Grant et al. 1982a, b, Pettit and Whittow 1983, 1985, Pettit et al. 1981, Whittow 1994, Whittow and Grant 1985, Whittow and Whittow 1988, Whittow et al. 1982, 1985.

FIGURE 12.9 Pipped Brown Pelican eggs: rate of water loss during the pipped stage of incubation is higher than in unpipped eggs. See text. (Photo by R. W. Schreiber.)

12.4.6 Pipping

Pipping is an important part of the incubation period in many seabirds because it is long (up to 6 days in the Bonin Petrel), and substantial embryonic growth and physiological development may occur in pipped eggs. Unfortunately, the pipping process has been inadequately described in most seabirds. In many species, the initial event in the pipping process is fracture of the shell ("star fracture"), permitting an increased rate of diffusion of water vapor out of the egg, and the opportunity for an augmented diffusion of oxygen into the egg. This is followed by penetration of the aircell of the egg by the embryo's beak, allowing the embryo to begin to use its lungs to rebreathe aircell air. This "internal pipping" has little effect on the water loss from the egg but, by using up some of the oxygen in the aircell, results in a reduction in the oxygen pressure in the aircell gas. The reduction in aircell oxygen pressure increases the oxygen pressure difference between ambient air outside the egg and aircell air, resulting in an increased rate of oxygen diffusion into the egg. Subsequently, a pip hole is formed in the shell, allowing the embryo to breathe fresh air (Figure 12.9). Pip-hole formation is followed by hatching. This sequence of events during pipping is seen in petrels, shearwaters, tropicbirds, and most, if not all, Charadriiformes. It is associated with an acceleration of the heart beat (Tazawa et al. 1991, Tazawa and Whittow 1994). In albatross and boobies, and probably in gannets also, the initial event during pipping is penetration of the aircell before the shell is cracked.

12.4.7 Hatching Success

Severe predation by natural or introduced predators, and natural calamities such as hurricanes, flooding, and ENSO events (Schreiber and Schreiber 1983, 1984) or man-made catastrophes (e.g., wars), can reduce hatching success (eggs hatched/eggs laid × 100) in all seabirds. It is important, therefore, to specify the year that studies of reproductive success are carried out, and to record any unusual associated events. Failure to do this may explain variation in hatching success from 4.8 to 75.0% in an earlier summary of 14 species of seabirds (Skutch 1976). More recent reports (e.g., Gaston 1994) place hatching success as high as 96% in some species.

12.5 EMBRYONIC FUNCTION

12.5.1 Embryonic Growth

An indication of the mean embryonic growth rate of a species may be obtained by dividing hatchling mass by the incubation period. Procellariiformes have the lowest mean embryonic growth rates, in keeping with their long incubation periods, and possibly also with their relatively low egg temperature (Figure 12.3). Embryonic growth (g/day) in the Laridae (Charadriiformes) is twice as rapid as in procellariiformes, but, even among the Laridae, the rate of growth in offshore feeders is lower than that of inshore feeding gulls and terns (Rahn et al. 1984). Penguin growth rates are intermediate between those of Laridae and Procellariiformes. The low growth rates of the altricial Sulidae partly reflect the immaturity of the hatchling.

The mean embryonic growth rate gives no indication of the pattern of growth, i.e., the shape of the growth curve during incubation. In the Wedge-tailed Shearwater, peak growth rate occurs 60% of the way through incubation (Ackerman et al. 1980). Similarly shaped logistic curves describe the embryonic growth of many seabirds for which information is available. A comparison of growth rate constants derived from such curves reveals that, in four tropical Procellariiformes, the smaller species have lower growth rates and the maximal growth rate occurs earlier in incubation than in larger species (Pettit et al. 1982a).

The increase in embryonic mass during incubation is the result of the growth of individual tissues and organs. There is little information available on the differential growth rates of embryonic organs and body parts in seabirds. Zhang and Whittow (1993a) conclude that embryonic organ growth rates in the Sooty Tern and Wedge-tailed Shearwater — two species with substantially different incubation periods (29 and 52 days, respectively) and degrees of maturity at hatching (albeit both nominally semiprecocial) are very similar. The most rapidly growing organs are the liver and stomach; the slowest growth is seen in the heart and lungs. A similar pattern of organ growth is evident in the semiprecocial Laysan Albatross (Zhang and Whittow 1992, 1993b).

Little information is available on defective embryonic growth in seabirds; Pettit and Whittow (1981) reported an embryonic double monster in a naturally incubated Wedge-tailed Shearwater's egg but the etiology of the "Siamese Twin" is not known.

12.5.2 Embryonic Oxygen Consumption

Oxygen consumption of the embryo increases as the embryo grows. The character of the increase differs in altricial and semiprecocial seabirds (Tazawa and Whittow 2000). In the altricial Brown Pelican (*Pelecanus occidentalis*) oxygen consumption increases continuously (Bartholomew and Goldstein 1984), but in the semiprecocial Laysan Albatross and White-chinned Petrel (*Procellaria aequinoctialis*), a plateau in oxygen consumption is evident just prior to pipping, with an accelerated increase after pipping of the egg (Pettit et al. 1982b, Brown and Adams 1988). Prior to pipping, oxygen demands of the embryo may exceed the rate at which oxygen can diffuse into the egg through the unpipped shell (Whittow and Tazawa 1991). A substantial percentage of the total oxygen consumed by the embryo occurs after pipping. Thus, in nine procellariiform seabirds, 41.4 to 50.0% of the total volume of oxygen consumed by the embryo is accounted for by pipped eggs (Pettit et al. 1982a, Brown and Adams 1988). Similarly, in the Black and Brown (*Anous stolidus*) Noddies and the White Tern, 42.0 to 51.8% of the total oxygen consumed occurs in pipped eggs (Pettit and Whittow 1983).

Some investigators have used the time immediately prior to internal pipping as a reference point for measurements of embryonic oxygen consumption. Unfortunately, in practice, the beginning of internal pipping can be detected with certainty only by opening the egg to verify that the embryonic beak is in the aircell, which precludes the possibility of making further measurements on that egg. Other investigators have used the beginning of internal pipping as a reference point in species in which the initial event in the pipping process is external pipping. In the latter species, external pipping (the initial pipping event) is the appropriate demarcation between the unpipped and pipped states.

Energy released from organic substrates in the egg (mostly yolk lipids) is utilized in the synthesis of new tissue, and also to support "maintenance" activities of the embryo, such as the pumping of blood by the heart. The longer the incubation period, and the longer an energy-consuming embryo is present in the egg, the greater the amount of oxygen consumed for maintenance activities over and above that needed for growth (Pettit et al. 1982a, Pettit and Whittow 1983).

12.5.3 EMBRYONIC ENERGETICS

To provide for the greater energy requirement contingent upon prolonged incubation, the freshly laid eggs of birds with long incubation periods have a greater energy content, and, as most of an egg's energy is contained in the yolk, this implies a greater yolk content, also (Carey et al. 1980, Rahn et al. 1984, Sotherland and Rahn 1987). In Procellariiformes, the relative yolk content increases as mass of the egg diminishes. This phenomenon parallels the increase in incubation time, relatively speaking, in the smaller eggs of the Procellariiformes. Consequently, in the small eggs of petrels, the energy and yolk content, as well as total energy cost of incubation, are relatively greater, and incubation time relatively longer, than in the large eggs of albatrosses (Pettit et al. 1982b, Rahn et al. 1984). The efficiency ("efficiency of biosynthesis") of conversion of the energy contained in a freshly laid egg to hatchling tissue energy is correspondingly less for eggs with a long incubation period, because a greater proportion of energy is allocated to maintenance of the embryo rather than to growth. In seven tropical species from two orders, the efficiency of biosynthesis varies from 39.3 to 47.6% (Pettit et al. 1984).

As indicated in a previous section, a greater adult investment of energy in the egg is necessary if the end product of incubation is a relatively mature, independent, precocial, rather than a helpless, naked, altricial hatchling. A precocial hatchling requires a more energy-dense egg with a higher yolk fraction than does an altricial hatchling (Vleck and Vleck 1987). This is partly due to the greater maintenance energy requirements over the longer incubation period necessary to produce a relatively mature hatchling. It is also attributable to the greater amount of energy incorporated into tissues of the hatchling (Pettit et al. 1984). Not all energy contained in the newly laid egg is allocated to hatchling tissue or energy costs of tissue synthesis and maintenance. A small amount is incorporated in residual egg tissues (extra-embryonic membranes), and a larger fraction in the yolk reserve of the hatchling (see below).

Figure 12.10 shows the relationship between total energy expenditure of the embryo during incubation ("energy cost") and the mass of the freshly laid egg in 21 species of seabirds. The relationship is linear, on logarithmic coordinates. The regression line in Figure 12.10 predicts that a seabird hatchling from an egg with a fresh-egg mass of 50 g would cost 127 kJ. When birds as a whole class are considered, altricial hatchlings require 52 kJ, and precocial hatchlings need 93 kJ (Vleck et al. 1980). Thus, the energy cost of embryonic development in seabirds is considerably higher than that of other birds. It is likely that prolonged incubation in procellariiform and tropical seabirds contributes to the high embryonic energy costs, a conclusion that conforms with that of Vleck et al. (1980). The high energy cost of producing a hatchling is another feature of offshore feeders (see above). Although much smaller, in all birds, than the energy required by both a nestling and an adult, embryonic energy requirements may represent a relatively greater percentage of the total lifetime energy expenditure in seabirds than in other birds because of the prevalence of prolonged incubation in seabirds.

12.5.4 EMBRYONIC TEMPERATURE REGULATION

A study of five semiprecocial seabirds (Black-footed, *Phoebastria nigripes*, and Laysan Albatross; Wedge-tailed Shearwater, Sooty Tern, and the Brown Noddy) reveals that not even late-incubation embryos in eggs with pip-holes are able to respond to cooling of the egg with an increased production of metabolic heat (Matsunaga et al. 1989, Mathiu et al. 1991, 1992, 1994, Dawson and Whittow 1994, Whittow 1993a, b, 1997). However, some seabird embryos (American White Pelican, *Pelecanus erythrorhynchos*; Herring Gull; and the Ring-billed Gull, *Larus delawarensis*) are known to be able to elicit parental incubation behavior by vocalizing. The vocalizations are induced by

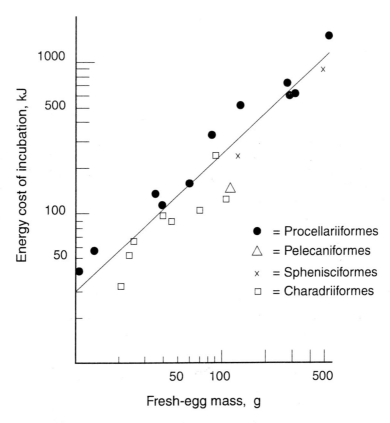

FIGURE 12.10 Relationship between the total energy expenditure (cost) during incubation, in kilojoules (kJ), and the mass of the freshly laid egg, in grams (g), in 21 species of seabirds. The regression ($r = 0.956$; $p < 0.001$) line represents the relationship: energy cost of incubation (kJ) = 3.452 × fresh-egg mass (g)$^{0.922}$. References: Bartholomew and Goldstein 1984, Brown and Adams 1988, Bucher and Bartholomew 1984, Bucher et al. 1986, Drent 1970, Pettit and Whittow 1983, Pettit et al. 1981, 1982a, b, Rahn and Ar 1974, Rahn and Huntington 1988, Vleck et al. 1980.

cooling of the egg and incubation behavior is effective in raising egg temperature to typical incubation levels (Evans 1990, Evans et al. 1994).

12.6 HATCHLING

12.6.1 HATCHLING MATURITY

Hatchling size varies from 6.3 g (Leach's Petrel) to 326 g in the Wandering Albatross, and hatchling maturity from the helpless altricial booby to the precocial alcid (Figure 12.11). The maturity of avian hatchlings has been traditionally classified according to obvious morphological and behavioral characteristics such as mobility and the presence of down (Nice 1962). More recently, Starck and Ricklefs (1998) advocated the use of the dry-matter content of hatchling tissues as an index of hatchling functional maturity, based on the diminution in the water content of tissues as they age and mature. Their index suggests that there is a continuous spectrum of hatchling maturity rather than the discrete categories implicit in Nice's (1962) classification. Other indices use physiological criteria for assessment of hatchling maturity (discussed below). Indices based on the dry mass of the hatchling or physiological criteria assign a quantitative value to the maturity of each species. Hatchling maturity has also been correlated with properties of the egg (e.g., yolk content; see above) rather than the hatchling.

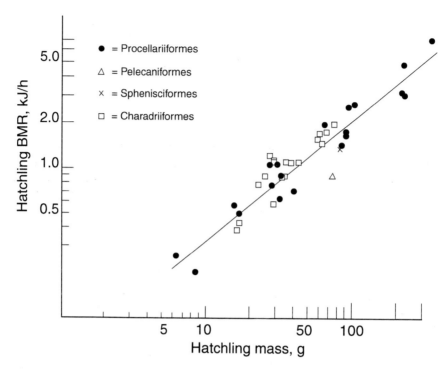

FIGURE 12.11 Relationship between hatchling basal metabolic rate (BMR), in kilojoules (kJ)/hour, and body mass, in grams (g), in 38 species of seabirds. The equation for the regression line is BMR (kJ) = 0.057 × body mass (g)$^{0.795}$ ($r = 0.938$, $p < 0.001$). References: Ackerman et al. 1980, Bartholomew and Goldstein 1984, Bech et al 1984, 1987, 1991, Brown 1988, Brown and Prys-Jones 1988, Bucher et al. 1986, Dawson and Bennett 1980, 1981, Dawson et al. 1972, 1976, Eppley 1984, Paganelli and Rahn 1982, Pettit and Whittow 1983, Pettit et al. 1981, 1982a, b, Vleck and Kenagy 1980.

12.6.2 BASAL METABOLIC RATE

In Black-legged Kittiwakes (*Rissa tridactyla*) there is evidence that hatchling metabolic rate is elevated by nutrients from the yolk sac (Klassen et al. 1987). Strictly speaking, therefore, basal metabolic rate of fasting hatchlings is not comparable to that of adults, which do not have a yolk sac. With this proviso, the basal metabolic rate (BMR) of a hatchling is linearly related to its mass, in a logarithmic manner (Figure 12.11). The relationship may be used to predict BMR of hatchlings weighing between 6.3 and 326 g. However, the species represented in Figure 12.11 are almost entirely semiprecocial Procellariiformes and semiprecocial/precocial Charadriiformes. There are few data on penguin and altricial pelecaniform hatchlings. The BMR of the altricial Brown Pelican hatchling is 48% lower than the value predicted by the regression line for a hatchling weighing 72.3 g. BMR of the hatchling may be compared with that of the adult, as a physiological measure of the maturity of the hatchling, provided that differences in the body mass of the hatchling and adult are taken into account (Vleck and Bucher 1998).

12.6.3 BODY TEMPERATURE

The mean hatchling body temperature for 11 species of seabirds is 38.2 ± 1.1°C (all semiprecocial except the precocial Xantu's Murrelet). In 8 of the 10 species for which adult body temperatures are available, hatchling body temperature is lower; in the two species of albatross, however, hatchling body temperatures are slightly higher than those of incubating adults.

TABLE 12.4
Ranking of Hatchling Maturity of 28 Seabirds according to Physiological Criteria

Order/Species		Ratio PMR/RMR	Hatchling Maturity	Reference
Gentoo Penguin	*Pygoscelis papua*	1.05	SA	Visser 1998
Adeliae Penguin	*P. adeliae*	1.12	SA	Visser 1998
Chinstrap Penguin	*P. antarctica*	1.13	SA	Visser 1998
Common Diving Petrel	*Pelecanoides urinatrix*	1.21	SP	Visser 1998
Brown Noddy	*Anous stolidus*	1.28	SP	Mathiu et al. 1991
Black-footed Albatross	*Phoebastria nigripes*	1.40	SP	Dawson and Whittow 1994
Laysan Albatross	*P. immutabilis*	1.41	SP	Dawson and Whittow 1994
Southern Giant Petrel	*Macronectes giganteus*	1.45	SP	Visser 1998
Kelp Gull	*Larus dominicanus*	1.45	SP	Visser 1998
Pigeon Guillemot	*Cepphus columba*	1.47	SP	Visser 1998
Laughing Gull	*Larus atricilla*	1.5	SP	Visser 1998
Antarctic Petrel	*Thalassoica antarctica*	1.60	SP	Visser 1998
Thick-billed Murre	*Uria lomvia*	1.60	SP-P	Visser 1998
Common Black-headed Gull	*Larus ridibundus*	1.62	SP	Visser 1998
Sooty Tern	*Sterna fuscata*	1.64	SP	Mathiu et al. 1991
Common Tern	*S. hirundo*	1.69	SP	Visser 1998
South Polar Skua	*Catharacta maccormicki*	1.71	SP	Visser 1998
Black-legged Kittiwake	*Rissa tridactyla*	1.72	SP	Visser 1998
Wedge-tailed Shearwater	*Puffinus pacificus*	1.74	SP	Mathiu et al. 1991
Western Gull	*Larus occidentalis*	1.85	SP	Visser 1998
Great-winged Petrel	*Pterodroma macroptera*	1.96	SP	Brown and Prys-Jones 1988
Arctic Tern	*Sterna paradisaea*	1.98	SP	Visser 1998
Grey Petrel	*Procellaria cinerea*	1.99	SP	Brown and Prys-Jones 1998
Ring-billed Gull	*Larus delawarensis*	2.04	SP	Visser 1998
Atlantic Puffin	*Fratercula arctica*	2.10	SP	Visser 1998
Georgian Diving Petrel	*Pelecanoides georgicus*	2.19	SP	Visser 1998
Antarctic Prion	*Pachyptila desolata*	2.61	SP	Visser 1998
Xantu's Murrelet	*Endomychura hypoleuca*	3.51	SP	Visser 1998

Note: Ratio: peak metabolic rate during exposure to cold ([PMR]/metabolic rate within the thermoneutral zone [RMR]). The species are ranked in order of increasing PMR/RMR. Also shown is the hatchlings' maturity according to traditional criteria: P = precocial, SP = semiprecocial, SA = semialtricial, A = altricial.

12.6.4 THERMOREGULATION

Semiprecocial hatchlings of the Laysan and Black-footed Albatross, Wedge-tailed Shearwater, Sooty Tern, and Brown Noddy have a demonstrable ability to regulate their body temperature in the face of exposure to heat or cold, in contrast to their failure to do so a matter of hours earlier while they are still in the egg. In these species, therefore, thermoregulatory ability develops abruptly after the hatchling has emerged from the egg (Mathiu et al. 1991, 1992, 1994, Dawson and Whittow 1994, 2000). The magnitude of the increase in metabolic rate during exposure to cold has been used as a criterion of hatchling physiological maturity in birds (Visser 1998). It is the most physiologically valid of all indices of hatchling maturity because it measures a physiological *response*. Table 12.4 presents the ratios of peak metabolic rate during exposure to cold/metabolic rate in a thermoneutral environment, in 28 species of seabirds; the metabolic rate of hatchlings during exposure to cold increased 1.05- to 3.51-fold, the greatest increase occurring in the precocial Xantu's Murrelet (Eppley 1984) and the smallest increase in the semialtricial Gentoo Penguin. It is also apparent in Table 12.4 that, among birds that are all traditionally semiprecocial, (1) there are considerable differences in physiological maturity and (2) there is little separation of the species along taxonomic lines. Data for a small sample of Charadriiformes reveal that the thermal insulation of the plumage and tissues of six out of the seven species is less than that of adults, while in the seventh species, they are comparable.

12.6.5 YOLK RESERVES

Not all of the yolk is consumed by the embryo; the remaining yolk is withdrawn into the hatchling's abdomen. Hatchling yolk reserves of seven species of semiprecocial tropical seabirds varies from 6.5 to 17.4% of the mass of the hatchlings, and amounts to 15.8 to 37.8% of the hatchling caloric content. It may be calculated that the yolk reserves can sustain the hatchlings for 1.5 to 6.2 days. The yolk reserves of the four procellariiform species is double that of the three Charadriiformes (Pettit et al. 1984).

12.7 SUMMARY AND CONCLUSIONS

Seabirds usually nest in large colonies and are relatively old when they start to breed, but seabirds are long lived. The incubation biology of seabirds is related to their foraging behavior on the one hand, and to the hatchling maturity on the other. The array of incubation characteristics associated with exploitation of distant, offshore feeding grounds has much in common with that correlated with the production of a precocial or semiprecocial hatchling. Nevertheless, there are altricial seabirds that feed far offshore, suggesting that there is considerable plasticity in the ways in which incubation can proceed in seabirds.

DEDICATION

This chapter is dedicated to the memory of my colleague, Hermann Rahn, who showed what a physiologist can contribute to ornithology.

LITERATURE CITED

ACKERMAN, R. A., G. C. WHITTOW, C. V. PAGANELLI, AND T. N. PETTIT. 1980. Oxygen consumption, gas exchange, and growth of embryonic Wedge-tailed Shearwaters (*Puffinus pacificus chlororhynchus*). Physiological Zoology 53: 210–221.

ADAMS, N. J. 1990. Feeding Biology and Energetics of King *Aptenodytes patagonicus* and Gentoo *Pygoscelis papua* Penguins at Subantarctic Marion Island. Ph.D. thesis, University of Cape Town, 246 pp.

ADAMS, N. J., R. W. ABRAMS, W. R. SIEGFRIED, K. A. NAGY, AND I. R. KAPLAN. 1991. Energy expenditure and food consumption by breeding Cape Gannets *Morus capensis*. Marine Ecology Progress Series 70: 1–9.

ADAMS, N. J., C. R. BROWN, AND K. A. NAGY. 1986. Energy expenditure of free-ranging Wandering Albatrosses *Diomedea exulans*. Physiological Zoology 59: 583–591.

AINLEY, D. 1995. Ashy Storm-Petrel *Oceanodroma homochroa*. No. 185 *in* The Birds of North America (A. Poole and F. Gill, Eds.). Academy of Natural Sciences, Philadelphia; American Ornithologists' Union, Washington, D.C.

AR, A., AND H. RAHN. 1978. Interdependence of gas conductance, incubation length, and weight of the avian egg. Pp. 227–236 *in* Respiratory Function in Birds: Adult and Embryonic (J. Piiper, Ed.). Springer-Verlag, New York.

AR, A., AND H. RAHN. 1985. Pores in avian eggshells: gas conductance, gas exchange and embryonic growth rate. Respiration Physiology 61: 1–20.

AR, A., C. V. PAGANELLI, R. B. REEVES, D. G. GREENE, AND H. RAHN. 1974. The avian egg: water vapor conductance, shell thickness, and functional pore area. Condor 76: 153–158.

ASTHEIMER, L. B., AND C. R. GRAU. 1990. A comparison of yolk growth rates in seabird eggs. Ibis 132: 380–394.

BARRETT, R. T. 1980. Temperature of Kittiwake *Rissa tridactyla* eggs and nests during incubation. Ornis Scandinavica 11: 50–59.

BARTHOLOMWE, G. A., AND D. L. GOLDSTEIN. 1984. The energetics of development in a very large altricial bird, the Brown Pelican. Pp. 347–357 *in* Respiration and Metabolism of Embryonic Vertebrates (R. S. Seymour, Ed.). Dr. W. Junk Publishers, Dordrecht.

BECH, C., F. J. AARVIK, AND D. VONGRAVEN. 1987. Temperature regulation in hatchling Puffins (*Fratercula arctica*). Journal für Ornithologie 128: 163–170.

BECH, C., S. MARTINI, R. BRENT, AND J. RASMUSSEN. 1984. Thermoregulation in newly hatched Black-legged Kittiwakes. Condor 86: 339–341.

BECH, C., F. MEHLUM, AND S. HAFTORN. 1991. Thermoregulatory abilities in chicks of the Antarctic Petrel (*Thalassoica antarctica*). Polar Biology 11: 233–238.

BENNETT, A. F., AND W. R. DAWSON. 1979. Physiological responses of embryonic Heermann's Gulls to temperature. Physiological Zoology 52: 413–421.

BERRY, H. H., R. P. MILLAR, AND G. N. LOUW. 1979. Environmental clues influencing the breeding biology and circulating levels of various hormones and triglycerides in the Cape Cormorant. Comparative Biochemistry and Physiology 62A: 879–884.

BIRT-FRIESEN, V. L., W. A. MONTEVECCHI, D. K. CAIRNS, AND S. A. MACKO. 1989. Activity-specific metabolic rates of free-living Northern Gannets and other seabirds. Ecology 70: 357–367.

BROWN, C. R. 1984. Resting metabolic rate and energetic cost of incubation in Macaroni Penguins (*Eudyptes chrysolophus*) and Rockhopper Penguins (*E. Chrysocome*). Comparative Biochemistry and Physiology 77A: 345–350.

BROWN, C. R. 1988. Energy expenditure during incubation in four species of sub-antarctic burrowing petrels. Ostrich 59: 67–70.

BROWN, C. R., AND N. J. ADAMS. 1984. Basal metabolic rate and energy expenditure during incubation in the Wandering Albatross (*Diomedea exulan*s). Condor 86: 182–186.

BROWN, C. R., AND N. J. ADAMS. 1988. Egg temperature, embryonic metabolism, and water loss from the eggs of subantarctic Procellariiformes. Physiological Zoology 61: 126–136.

BROWN, C. R., AND R. P. PRYS-JONES. 1988. Development of homeothermy in chicks of sub-Antarctic burrowing petrels. Suid-Afrikaanse Tydskrif vir Dierkunde 23: 288–294.

BUCHER, T. L. 1986. Ratios of hatchling and adult mass-independent metabolism: a physiological index to the altricial-precocial continuum. Respiration Physiology 65: 69–83.

BUCHER, T. L. 1987. Patterns in the mass-independent energetics of avian development. Journal of Experimental Biology, Suppl. 1: 139–150.

BUCHER, T. L., AND G. A. Bartholomew. 1984. Analysis of variation in gas exchange, growth patterns, and energy utilization in a parrot and other avian embryos. Pp. 359–372 *in* Respiration and Metabolism of Embryonic Vertebrates (R. S. Seymour, Ed.). Dr. W. Junk Publishers, Dordrecht.

BUCHER, T. L., G. A. BARTHOLOMEW, W. Z. TRIVELPIECE, AND N. J. VOLKMAN. 1986. Metabolism, growth, and activity in Adelie and Emperor Penguin embryos. Auk 103: 485–493.

CAREY, C., H. RAHN, AND P. PARISI. 1980. Calories, water, lipid and yolk in avian eggs. Condor 82: 335–343.

COGBURN, L. A., J. BURNSIDE, AND C. G. SCANES. 2000. Physiology of growth and development. Pp. 635–656 *in* Sturkie's Avian Physiology (G. C. Whittow, Ed.), 5th ed. Academic Press, San Diego, CA.

DAWSON, W. R., AND A. F. BENNETT. 1980. Metabolism and thermoregulation in hatchling Western Gulls. Condor 82: 103–105.

DAWSON, W. R., AND A. F. BENNETT. 1981. Field and laboratory studies of the thermal relations of hatchling Western Gulls. Physiological Zoology 54: 155–164.

DAWSON, W. R., AND G. C. WHITTOW. 1994. The emergence of endothermy in the Black-footed and Laysan Albatrosses. Journal of Comparative Physiology B. 164: 292–298.

DAWSON, W. R., AND G. C. WHITTOW. 2000. Regulation of body temperature. Pp. 344–390 *in* Sturkie's Avian Physiology, 5th ed. (G. C. Whittow, Ed.). Academic Press, San Diego, CA.

DAWSON, W. R., A. F. BENNETT, AND J. W. HUDSON. 1976. Metabolism and thermoregulation in hatchling Ring-billed Gulls. Condor 78: 49–60.

DAWSON, W. R., J. W. HUDSON, AND R. W. HILL. 1972. Temperature regulation in newly hatched Laughing Gulls (*Larus atricilla*). Condor 74: 177–184.

DRENT, R. H. 1970. Functional aspects of incubation in the Herring Gull. Pp. 1–132 *in* The Herring Gull and Its Egg (G. P. Baerends and R. H. Drent, Eds.). E. J. Brill, Leiden.

DROST, C. A., AND D. B. LEWIS. 1995. Xantus' Murrelet (*Synthliramphus hypoleucus*). No. 164 *in* The Birds of North America (A. Poole, and F. Gill, Eds.). Academy of Natural Sciences, Philadelphia; American Ornithologists' Union, Washington, D.C.

EPPLEY, Z. A. 1984. Development of thermoregulatory abilities in Xantus' Murrelet chicks *Synthliboramphus hypoleucus*. Physiological Zoology 57: 307–317.

EVANS, R. M. 1990. Vocal regulation of temperature by avian embryos: a laboratory study with pipped eggs of the American White Pelican. Animal Behavior 40: 969–979.

EVANS, R. M., A. WHITAKER, AND M. O. WIEBE. 1994. Development of vocal regulation of temperature by embryos in pipped eggs of Ring-billed Gulls. Auk 11: 596–604.

FISHER, H. I. 1971. The Laysan Albatross: its incubation, hatching, and associated behaviors. Living Bird, 10: 19–78.

FLINT, E. N., AND K. A. NAGY. (1984). Flight energetics of free-living Sooty Terns. Auk 101: 288–294.

FURNESS, R. W., AND P. MONAGHAN. 1987. Seabird ecology. Chapman & Hall, New York.

GASTON, A. J. 1994. Ancient Murrelet (*Synthliramphus antiquus*). No. 132 *in* The Birds of North America (A. Poole and F. Gill, Eds.). Academy of Natural Sciences, Philadelphia; American Ornithologists' Union, Washington, D.C.

GAUGER, V. H. 1999. Black Noddy *Anous minutus*. No. 412 *in* The Birds of North America (A. Poole and F. Gill, Eds.). Academy of Natural Sciences, Philadelphia; American Ornithologists' Union, Washington, D.C.

GRANT, G. S. 1982. Avian incubation: egg temperature, nest humidity, and behavioral thermoregulation in a hot environment. Ornithological Monographs No. 30, American Ornithologists' Union, Washington, D.C., 75 pp.

GRANT, G. S. 1984. Energy cost of incubation to the parent seabird. Pp. 59–71 *in* Seabird Energetics (G. C. Whittow and H. Rahn, Eds.). Plenum Press, New York.

GRANT, G. S., AND G. C. WHITTOW. 1983. Metabolic cost of incubation in the Laysan Albatross and Bonin Petrel. Comparative Biochemistry and Physiology 74A: 77–82.

GRANT, G. S., C. V. PAGANELLI, AND H. RAHN. 1984. Microclimate of Gull-billed Tern and Black Skimmer nests. Condor 86: 337–338.

GRANT, G. S., T. N. PETTIT, H. RAHN, G. C. WHITTOW, AND C. V. PAGANELLI. 1982a. Regulation of water loss from Bonin Petrel (*Pterodroma hypoleuca*) eggs. Auk 99: 236–242.

GRANT, G. S., T. N. PETTIT, H. RAHN, G. C. WHITTOW, AND C. V. PAGANELLI. 1982b. Water loss from Laysan and Black-footed Albatross eggs. Physiological Zoology 55: 405–414.

GRAU, G. S. 1982. Egg formation in Fiordland Crested Penguins (*Eudyptes pachyrhynchus*). Condor 84: 172–177.

GRAU, C. R. 1984. Egg formation. Pp. 35–57 *in* Seabird Energetics (G. C. Whittow and H. Rahn, Eds.). Plenum Press, New York.

GROSCOLAS, R. 1982. Changes in plasma lipids during breeding. molting, and starvation in male and female Emperor Penguins (*Aptenodytes forsteri*). Physiological Zoology 55: 45–55.

HARRIS, M. P. 1964. Aspects of the breeding biology of the gulls *Larus argentatus, L. fuscus* and *L. marinus*. Ibis 106: 432– 56.

HATCH, S. A., AND D. N. NETTLESHIP. 1998. Northern Fulmar *Fulmarus glacialis*. No. 361 *in* The Birds of North America (A. Poole and F. Gill, Eds.). Academy of Natural Sciences, Philadelphia; American Ornithologists' Union, Washington, D.C.

HECTOR, J. A. L. 1988. Reproductive endocrinology of albatrosses. Proceedings of the XIX International Ornithological Congress, pp. 1702–1709.

HECTOR, J. A. L., B. K. FOLLETT, AND P. A. PRINCE. 1986. Reproductive endocrinology of the Black-browed Albatross *Diomedea melanophris* and the Grey-headed Albatross *D. chrysostoma*. Journal of Zoology (London) A 208: 237–253.

HOBSON, K. A. 1997. Pelagic Cormorant *Phalacrocorax pelagicus*. No. 282 *in* The Birds of North America (A. Poole, and F. Gill, Eds.). Academy of Natural Sciences, Philadelphia; American Ornithologists' Union, Washington, D.C.

HOWELL, T. R., AND G. A BARTHOLOMEW. 1961. Temperature regulation in Laysan and Black-footed Albatrosses. Condor 63: 185–197.

HOWELL, T. R., AND G. A. BARTHOLOMEW. 1962a. Temperature regulation in the Red-tailed Tropicbird and the Red-footed Booby. Condor 64: 6–18.

HOWELL, T. R., AND G. A. BARTHOLOMEW. 1962b. Temperature regulation in the Sooty Tern *Sterna fuscata*. Ibis 104: 98–105.

HOWELL, T. R., B. ARAYA, AND W. R. MILLIE. 1974. Breeding biology of the Gray Gull, *Larus modestus*. University of California Publications in Zoology 104: 1–57.

JOHNSON, A. L. 2000. Reproduction in the female. Pp. 569–596 *in* Sturkie's avian physiology, 5th ed. (G. C. Whittow, Ed.). Academic Press, San Diego, CA.

KEPLER, C. B. 1969. Breeding biology of the Blue-faced Booby *Sula dactylatra personata* on Green Island, Kure Atoll. Publ. Nutall Ornithological Club No. 8.

KIRBY, J. D., AND D. P. FROMAN. 2000. Reproduction in male birds. Pp. 597–616 *in* Sturkie's Avian Physiology, 5th ed. (G. C. Whittow, Ed.). Academic Press, San Diego, CA.

KLASSEN, M., G. SLAGSVOLD, AND C. BECH. 1987. Metabolic rate and thermostability in relation to availability of yolk in hatchlings of Black-legged Kittiwake and domestic chicken. Auk 104: 787–789.

LACK, D. 1968. Ecological Adaptations for Breeding in Birds. Methuen, London.

LISHMAN, G. S. 1985. The comparative breeding biology of Adelie and Chinstrap Penguins *Pygoscelis adeliae* and *P. antarctica* at Signy Island, South Orkney Islands. Ibis 127: 84–89.

MACMILLEN, R. E., G. C. WHITTOW, E. A. CHRISTOPHER, AND R. J. EBISU. 1977. Oxygen consumption, evaporative water loss, and body temperature in the Sooty Tern. Auk 94: 72–79.

MATHIU, P. M., W. R. DAWSON, AND G. C. WHITTOW. 1991. Development of thermoregulation in Hawaiian Brown Noddies (*Anous stolidus pileatus*). Journal of Thermal Biology 16: 317–325.

MATHIU, P. M., W. R. DAWSON, AND G. C. WHITTOW. 1994. Thermal responses of late embryos and hatchlings of the Sooty Tern. Condor 96: 280–294.

MATHIU, P. M., G. C. WHITTOW, AND W. R. DAWSON. 1992. Hatching and the establishment of thermoregulation in the Wedge-tailed Shearwater (*Puffinus pacificus*). Physiological Zoology 65: 583–603.

MATSUNAGA, C., P. M. MATHIU, G. C. WHITTOW, AND H. TAZAWA. 1989. Oxygen consumption of Brown Noddy (*Anous stolidus*) embryos in a quasiequilibrium state at lowered ambient temperatures. Comparative Biochemistry and Physiology 93A: 707–710.

MEHLUM, F., H. RAHN, AND S. HAFTORN. 1986. Interrelationships between egg dimensions, pore numbers, incubation time, and adult body mass in Procellariiformes with special reference to the Antarctic Petrel *Thalassoica antarctica*. Polar Research 5: 53–58.

MONTEVECCHI, W. A., V. L. BIRT-FRIESEN, AND D. K. CAIRNS. 1992. Reproductive energetics and prey harvest of Leach's Storm-Petrels in the Northwest Atlantic. Ecology 73: 823–832.

MONTEVECCHI, W. A., I. R. KIRKHAM, D. D. ROBY, AND K. L. BRINK. 1983. Size, organic composition, and energy content of Leach's Storm-Petrel (*Oceanodroma leucorhoa*) eggs with reference to position in the precocial-altricial spectrum and breeding ecology. Canadian Journal of Zoology 61: 1456–1463.

MORGAN, K. R., C. V. PAGANELLI, AND H. RAHN. 1978. Egg weight loss and nest humidity during incubation in two Alaskan gulls. Condor 80: 272–275.

MORRISON, M. L. 1979. Eggshell thickness changes in Pacific Ocean Phaethontidae and Sulidae. Condor 81: 209.

MURRAY, K. G., K. WINNETT-MURRAY, AND G. L. HUNT. 1980. Egg neglect in Xantus' Murrelet. Proceedings. Colonial Waterbirds 3: 186–195.

NELSON, J. B. 1971. The biology of Abbott's Booby *Sula abbotti*. Ibis 113: 429–467.

NELSON, J. B. 1976. The breeding biology of frigatebirds — a comparative review. Living Bird 14: 113–156.

NELSON, J. B. 1979. Seabirds: Their Biology and Ecology. A & W. Publishers, Inc., New York.

NICE, M. M. 1962. Development of behavior in precocial birds. Transactions of the Linnean Society N.Y. 8: 1–211.

NISBET, I. C. T. 1977. Courtship-feeding and clutch size in Common Terns *Sterna hirundo*. Pp. 101–109 *in* Evolutionary Ecology (B. Stonehouse and C. M. Perrins, Eds.). University Park Press, Baltimore.

OBST, B. S., K. A. NAGY, AND R. E. RICKLEFS. 1987. Energy utilization by Wilson's Storm Petrel (*Oceanites oceanicus*). Physiological Zoology 60: 200–210.

PAGANELLI, C. V., AND H. RAHN. 1982. Adult and embryonic metabolism in birds and the role of shell conductance. Pp. 193–204 *in* Respiration and Metabolism of Embryonic Vertebrates (R. S. Seymour, Ed.). Martinus Nijhoff/Dr. W. Junk, The Hague.

PAGANELLI, C. V., A. OLSZOWKA, AND A. AR. 1974. The avian egg: surface area, volume and density. Condor 76: 319–325.

PARNELL, J. F., R. M. ERWIN, AND K. C. MOLINA. 1995. Gull-billed Tern *Sterna nilotica*. No. 140 *in* The Birds of North America (A. Poole and F. Gill, Eds.). Academy of Natural Sciences, Philadelphia; American Ornithologists' Union, Washington, D.C.

PARSONS, J. 1976. Factors determining the number and size of eggs laid by the Herring Gull. Condor 78: 481–492.

PETTIT, T. N., AND G. C. WHITTOW. 1981. Embryonic double monster in the Wedge-tailed Shearwater. Condor 83: 91.

PETTIT, T. N., AND G. C. WHITTOW. 1982a. The initiation of pulmonary respiration in a bird embryo: blood and air cell gas tensions. Respiration Physiology 48: 199–208.

PETTIT, T. N., AND G. C. WHITTOW. 1982b. The initiation of pulmonary respiration in a bird embryo: tidal volume and frequency. Respiration Physiology 48: 209–218.

PETTIT, T. N., AND G. C. WHITTOW. 1983. Embryonic respiration and growth in two species of noddy terns. Physiological Zoology 56: 455–464.

PETTIT, T. N., AND G. C. WHITTOW. 1985. Water loss from pipped eggs of two species of noddies. Journal of Field Ornithology 56: 277–280.

PETTIT, T. N., H. I. ELLIS, AND G. C. WHITTOW. 1985. Basal metabolic rate in tropical seabirds. Auk 102: 172–174.

PETTIT, T. N., G. S. GRANT, G. C. WHITTOW, H. RAHN, AND C. V. PAGANELLI. 1981. Respiratory gas exchange and growth of White Tern embryos. Condor 83: 455–464.

PETTIT, T. N., G. S. GRANT, G. C. WHITTOW, H. RAHN, AND C. V. PAGANELLI. 1982a. Respiratory gas exchange and growth of Bonin Petrel embryos. Physiological Zoology 55: 162–170.

PETTIT, T. N., G. S. GRANT, G. C. WHITTOW, H. RAHN, AND C. V. PAGANELLI. 1982b. Embryonic oxygen consumption and growth of Laysan and black-footed albatross. American Journal of Physiology 242: R121–R128.

PETTIT, T. N., K. A. NAGY, H. I. ELLIS, AND G. C. WHITTOW. 1988. Incubation energetics of the Laysan Albatross. Oecologia 74: 546–550.

PETTIT, T. N., G. C. WHITTOW, AND G. S. GRANT. 1984. Caloric content and energetic budget of tropical seabird eggs. Pp. 113–137 *in* Seabird Energetics (G. C. Whittow and H. Rahn, Eds.). Plenum Press, New York.

RAHN, H. 1984. Factors controlling the rate of incubation water loss in bird eggs. Pp. 271–288 *in* Respiration and Metabolism of Embryonic Vertebrates (R. S. Seymour, Ed.). Martinus Nijhoff/Dr. W. Junk, The Hague.

RAHN, H., AND A. AR. 1974. The avian egg: incubation time and water loss. Condor 76: 147–152.

RAHN, H., AND W. R. DAWSON. 1979. Incubation water loss in eggs of Heermann's and Western Gulls. Physiological Zoology 52: 451–460.

RAHN, H., AND H. T. HAMMEL. 1982. Incubation water loss, shell conductance, and pore dimensions in Adelie Penguin eggs. Polar Biology 1: 91–97.

RAHN, H., AND C. E. HUNTINGTON. 1988. Eggs of Leach's Storm Petrel: O_2 uptake, water loss, and microclimate of the nest. Comparative Biochemistry and Physiology 91A: 519–521.

RAHN, H., AND G. C. WHITTOW. 1984. Introduction. Pp. 1–32 *in* Seabird energetics (G. C. Whittow and H. Rahn, Eds.). Plenum Press, New York.

RAHN, H., R. A. ACKERMAN, AND C. V. PAGANELLI. 1984. Eggs, yolk and embryonic growth rate. Pp. 89–112 *in* Seabird Energetics (G. C. Whittow and H. Rahn, Eds.). Plenum Press, New York.

RAHN, H., C. V. PAGANELLI, AND A. AR. 1974. The avian egg: air-cell gas tension, metabolism and incubation time. Respiration Physiology 22: 297–309.

RAHN, H., C. V. PAGANELLI, AND A. AR. 1987. Pores and gas exchange of avian eggs: a review. Journal of Experimental Zoology (Suppl.) 1: 165–172.

RAHN, H., C. V. PAGANELLI, I. C. T. NISBET, AND G. C. WHITTOW. 1976. Regulation of incubation water loss in eggs of seven species of terns. Physiological Zoology 49: 245–259.

RAHN, H., P. PARISI, AND C. V. PAGANELLI. 1982. Estimating the initial density of birds' eggs. Condor 84: 339–341.

RICKLEFS, R. E. 1977. Composition of eggs of several bird species. Auk 94: 350–356.

RICKLEFS, R. E., AND W. A. MONTEVECCHI. 1979. Size, organic composition and energy content of North Atlantic Gannet *Morus bassanus* eggs. Comparative Biochemistry and Physiology 64A: 161–165.

RICKLEFS, R. E., D. D. ROBY, AND J. B. WILLIAMS. 1986. Daily energy expenditure by adult Leach's Storm Petrels during the nesting cycle. Physiological Zoology 59: 649–660.

ROUDYBUSH, T., L. HOFFMAN, AND H. RAHN. 1980. Conductance, pore geometry, and water loss of eggs of Cassin's Auklet. Condor 82: 105–106.

SCHRADIN, C., AND G. ANZENBERGER. 2000. Prolactin, the hormone of paternity. News in Physiological Sciences 14: 223–231.

SCHREIBER, R. W., AND E. A. SCHREIBER. 1983. Reproductive failure of marine birds on Christmas Island, Fall 1982. Tropical Ocean Atmosphere Newsletter, February 10.

SCHREIBER, R. W., AND E. A. SCHREIBER. 1984. Central Pacific seabirds and the El Niño–Southern Oscillation: 1982–1983 perspectives. Science 225: 713–716.

SKUTCH, A. F. 1976. Parent Birds and Their Young. University of Texas Press, Austin.

SOTHERLAND, P. R., AND H. RAHN. 1987. On the composition of bird eggs. Condor 89: 48–65.

SPELLERBERGER, I. F. 1969. Incubation temperatures and thermoregulation in the McCormick Skua. Condor 71: 59–67.

STARCK, J. M., AND R. E. RICKLEFS. 1998. Patterns of development: the altricial-precocial spectrum. Pp. 3–30 *in* Avian Growth and Development (J. M. Starck and R. E. Ricklefs, Eds.). Oxford University Press, New York.

TAZAWA, H., AND G. C. WHITTOW. 1994. Embryonic heart rate and oxygen pulse in two procellariiform seabirds, *Diomedea immutabilis* and *Puffinus pacificus*. J. Comp. Physiol. B 163: 642–648.

TAZAWA, H., AND G. C. WHITTOW. 2000. Incubation physiology. Pp. 617–634 *in* Sturkie's Avian Physiology, 5th ed. (G. C. Whittow, Ed.). Academic Press, San Diego, CA.

TAZAWA, H., O. KURODA, AND G. C. WHITTOW. 1991. Noninvasive determination of embryonic heart rate during hatching in the Brown Noddy (*Anous stolidus*). Auk 108: 594–601.

VISSER, G. H. 1998. Development of temperature regulation. Pp. 117–156 *in* Avian Growth and Development (J. M. Starck and R. E. Ricklefs, Eds.). Oxford University Press, New York.

VLECK, C. M., AND T. L. BUCHER. 1998. Energy metabolism, gas exchange and ventilation. Pp. 89–116 *in* Avian Growth and Development (J. M. Starck and R. E. Ricklefs, Eds.). Oxford University Press, New York.

VLECK, C. M., AND G. J. KENAGY. 1980. Embryonic metabolism of the Fork-tailed Storm Petrel: physiological patterns during prolonged and interrupted incubation. Physiological Zoology 53: 32–42.

VLECK, C. M., AND D. VLECK. 1987. Metabolism and energetics of avian embryos. Journal of Experimental Zoology Supplement 1: 111–125.

VLECK, C. M., D. VLECK, AND D. F. HOYT. 1980. Patterns of metabolism and growth in avian embryos. American Zoologist 20: 405–416.

VLECK, C. M., D. VLECK, H. RAHN, AND C. V. PAGANELLI. 1993. Nest microclimate, water-vapor conductance, and water loss in heron and tern eggs. Auk 100: 76–83.

WALLACE, E. A. H., AND G. E. WALLACE. 1998. Brandt's Cormorant (*Phalacrocorax penicillatus*). No. 362 *in* The Birds of North America (A. Poole and F. Gill, Eds.). Academy of Natural Sciences, Philadelphia; American Ornithologists' Union, Washington, D.C.

WARHAM, J. 1974. The Fiordland Crested Penguin *Eudyptes pachyrhynchus*. Ibis 116: 1–27.

WARHAM, J. 1990. The Petrels: Their Ecology and Breeding Systems. Academic Press, London.

WHEELWRIGHT, N. T., AND P. D. BOERSMA. 1979. Egg chilling and the thermal environment of the Fork-tailed Storm Petrel (*Oceanodroma furcata*) nest. Physiological Zoology 52: 231–239.

WHITTOW, G. C. 1980. Physiological and ecological correlates of prolonged incubation in seabirds. American Zoologist 20: 427–436.

WHITTOW, G. C. 1984. Physiological ecology of incubation in tropical seabirds. Studies in Avian Biol. No. 8: 47–72.

WHITTOW, G. C. 1985.Partition of water loss from the eggs of the Sooty Tern between the pre-pipping and pipped periods. Wilson Bull. 97: 240–241.

WHITTOW, G. C. 1986. Regulation of body temperature. Pp. 221–252 *in* Avian Physiology, 4th ed. (P. D. Sturkie, Ed.). Springer-Verlag, New York.

WHITTOW, G. C. 1993a. Black-footed Albatross *Diomedea nigripes*. No. 65 *in* The Birds of North America (A. Poole and F. Gill, Eds.). Academy of Natural Sciences, Philadelphia; American Ornithologists' Union, Washington, D.C.

WHITTOW, G. C. 1993b. Laysan Albatross *Diomedea immutabilis*. No. 66 *in* The Birds of North America (A. Poole and F. Gill, Eds.). Academy of Natural Sciences, Philadelphia; American Ornithologists' Union, Washington, D.C.

WHITTOW, G. C. 1994. Incubation biology and nestling growth of Bulwer's Petrels on Manana Island, Oahu, Hawaii. Pacific Science 48: 136–144.

WHITTOW, G. C. 1997. Wedge-tailed Shearwater *Puffinus pacificus*. No. 305 *in* The Birds of North America (A. Poole and F. Gill, Eds.). Academy of Natural Sciences, Philadelphia; American Ornithologists' Union, Washington, D.C.

WHITTOW, G. C., AND G. S. GRANT. 1985. Water loss and pipping sequence in the eggs of the Red-tailed Tropicbird (*Phaethon rubricauda*). Auk 102: 749–753.

WHITTOW, G. C., AND M. B. NAUGHTON. 1999. Christmas Shearwater egg dimensions and shell characteristics on Laysan Island, Northwestern Hawaiian Islands. Wilson Bulletin 111: 421–422.

WHITTOW, G. C., AND T. N. PETTIT. 2000. Egg dimensions and shell characteristics of Bulwer's Petrels, *Bulweria bulwerii*, on Laysan Island, Northwestern Hawaiian Islands. Pacific Science 54: 183–188.

WHITTOW, G. C., AND H. TAZAWA. 1991. The early development of temperature regulation in birds. Physiological Zoology 64: 1371–1390.

WHITTOW, G. C., AND C. S. K. WHITTOW. 1988. Pipping in the eggs of the Red-footed Booby (*Sula sula*) Elepaio 48: 45–46.

WHITTOW, G. C., R. A. ACKERMAN, C. V. PAGANELLI, AND T. N. PETTIT. 1982. Pre-pipping water loss from the eggs of the Wedge-tailed Shearwater. Comparative Biochemistry and Physiology 72A: 29–34.

WHITTOW, G. C., G. S. GRANT, AND E. N. FLINT. 1985. Egg water loss, shell water-vapor conductance, and the incubation period of the Gray-backed Tern (*Sterna lunata*). Condor 87: 269–272.

WHITTOW, G. C., T. N. PETTIT, R. A. ACKERMAN, AND C. V. PAGANELLI. 1987. Temperature regulation in a burrow-nesting tropical seabird, the Wedge-tailed Shearwater (*Puffinus pacificus*). Journal of Comparative Physiology B 157: 607–614.

WHITTOW, G. C., T. N. PETTIT, R. A. ACKERMAN, AND C. V. PAGANELLI. 1989. The regulation of water loss from the eggs of the Red-footed Booby (*Sula sula*). Comparative Biochemistry and Physiology 93A: 807–810.

WHITTOW, G. C., T. R. SIMONS, AND T. N. PETTIT. 1984. Water loss from the eggs of a tropical seabird (*Pterodroma phaeopygia*) at high altitude. Comparative Biochemistry and Physiology 78A: 537–540.

WILEY, R. H., AND D. S. LEE. 1998. Long-tailed Jaeger *Stercorarius longicaudus*. No. 365 *in* The Birds of North America (A. Poole and F. Gill, Eds.). Academy of Natural Sciences, Philadelphia; American Ornithologists' Union, Washington, D.C.

WILLIAMS, J. B. 1996. Energetics of avian incubation. Pp. 375–415 *in* Avian Energetics and Nutritional Ecology (C. Carey, Ed.). Chapman & Hall, New York.

WILLIAMS, T. D. 1992. Reproductive endocrinology of Macaroni (*Eudyptes chrysolophus*) and Gentoo (*Pygoscelis papua*) Penguins. II. Plasma levels of gonadal steroids and LH in immature birds in relation to deferred sexual maturity. General and Comparative Endocrinology 85: 241–247.

WILLIAMS, T. D., AND P. J. SHARP. 1993. Plasma prolactin during the breeding season in adult and immature Macaroni (*Eudyptes chrysolophus*) and Gentoo (*Pygoscelis papua*) Penguins. General and Comparative Endocrinology 92: 339–346.

WILLIAMS, A. J., W. R. SIEGFRIED, AND J. COOPER. 1982. Egg composition and hatchling precocity in seabirds. Ibis 124: 456–470.

WILLIAMS, T. D., K. GHEBREMESKEL, G. WILLIAMS, AND M. A. CRAWFORD. 1992. Breeding and moulting fasts in Macaroni Penguins: do birds exhaust their fat reserves? Comparative Biochemistry and Physiology 103A: 783–785.

ZHANG, Q., AND G. C. WHITTOW. 1992. Embryonic oxygen consumption and organ growth in the Wedge-tailed Shearwater. Growth, Development and Aging 56: 205–214.

ZHANG, Q., AND G. C. WHITTOW. 1993a. Oxygen consumption and organ growth in Sooty Tern (*Sterna fuscata*) embryos. Auk 110: 939–943.

ZHANG, Q., AND G. C. WHITTOW. 1993b. Organ growth and oxygen consumption in Laysan Albatross embryos. Wilson Bulletin 105: 657–665.

Black-browed Albatross Chicks on Their Nest, Waiting for Parents to Return with Food

13 Chick Growth and Development in Seabirds

G. Henk Visser

CONTENTS

13.1 INTRODUCTION

Chicks of most seabird species grow up on land situated in close proximity to the sea. It is presumed that the nature of their food supply has not allowed the evolution of the self-feeding precocial mode of development in seabirds (Lack 1968). Although there are marked interspecific differences with respect to developmental mode, in the majority of seabird species, chicks stay in or close to their nest until fledging, being parentally fed and brooded. For example, chicks of pelicans, frigatebirds, gannets, and boobies are born naked with their eyes closed, being totally dependent on parental food and warmth (Figure 13.1). Chicks that hatch in this developmental state have been classified

FIGURE 13.1 A newly hatched altricial Lesser Frigatebird chick (*Fregata ariel*). (Photo by R. W. and E. A. Schreiber.)

as being altricial by Nice (1962; see Table 13.1). Chicks of tropicbirds (Figure 13.2) hatch with their eyes closed, but are covered in down (being classified as being semialtricial-2; Nice 1962), whereas tern, auk, murre, and jaeger chicks hatch with a downy plumage with their eyes open, and are able to walk (semiprecocial: Figure 13.3). In contrast, chicks of some murrelet species (*Synthliboramphus* spp. and *Brachyrhamphus* spp.) leave the nest shortly after hatching, being fed at sea by their parent (precocial-4; Nice 1962, Eppley 1984, Gaston 1992, Starck and Ricklefs 1998a). Chicks of Common Murre (*Uria aalge*), Thick-billed Murre (*Uria lomvia*), and Razor-billed Auk (*Alca torda*) do so after having attained about 25% of adult body mass (Daan and Tinbergen 1979, Gaston 1985, Starck and Ricklefs 1998a). Obviously, early nest desertion by the chick potentially reduces parental traveling time and enables exploitation of remote feeding areas (Ydenberg 1989). However, this strategy can only be achieved with the co-evolution of some specific physiological adaptations of the chick to minimize and compensate for its heat loss (e.g., Eppley 1984).

TABLE 13.1
Criteria for Classification of Neonates

Type of Neonate	Plumage	Eyes	Nest Attendance	Parental Brooding	Parental Feeding
Precocial-1	Contour feathers	Open	Leave nest	None	None
Precocial-2	Down	Open	Leave nest	Yes	None
Precocial-3	Down	Open	Leave nest	Yes	Showing food
Precocial-4	Down	Open	Leave nest	Yes	Yes
Semiprecocial	Down	Open	Around nest	Yes	Yes
Semialtricial-1	Down	Open	In nest	Yes	Yes
Semialtricial-2	Down	Closed	In nest	Yes	Yes
Altricial	None	Closed	In nest	Yes	Yes

From Nice 1962.

FIGURE 13.2 A 1-day-old Red-tailed Tropicbird chick. They are the only Pelecaniform chicks to hatch with a full coat of down. (Photo by E. A. Schreiber.)

FIGURE 13.3 A Sooty Tern chick hatches with its eyes open and able to walk. The down on this just-hatched chick has not dried out yet. (Photo by E. A. Schreiber.)

Seabirds have developed different feeding strategies, ranging from in-shore feeding to off-shore feeding (see Chapter 6). In species that feed in-shore (e.g., some pelicans, cormorants, gulls, and some terns), the chick can be fed several times a day and one parent can remain at the nest to brood it. This foraging mode may not necessitate chicks developing homeothermy at an early age (Klaassen 1994). Small seabird chicks frequently receive food as whole particles, such as fish (e.g., in most tern species), or as a predigested food mash (e.g., in young chicks of the Black-legged Kittiwake [*Rissa tridactyla*], cormorants, boobies). In pelagic seabirds (e.g., albatrosses, petrels, fulmars, boobies, and some terns), however, due to the long travel distances to their food source, parents are often gone for one or more days on a foraging trip. Therefore, until the achievement of homeothermy by the chick, feeding rates of the chick may be somewhat reduced over those of near-shore feeding birds. After the chick(s) achievement of homeothermy, both parents can leave the nest, potentially resulting in a doubling of the amount of food brought to the brood (Ricklefs and Roby 1983). While foraging, most pelagic seabirds store the food in their stomach to carry it back to the colony, although a few species carry fish in the bill (e.g., White Terns, *Gygis alba*; puffins). Procellariiform birds are unique in the sense that parents partly concentrate the food caught into stomach oil. This substance mainly consists of wax esters with a very high energy density, and stomach oil formation has been interpreted as a strategy to increase the amount of energy per feeding trip (Ricklefs et al. 1985, Roby 1991, Roby et al. 1997). This physiological development has enabled procellariiform birds to exploit more remote feeding areas.

Recent reviews in the literature include growth patterns of birds in general (Starck and Ricklefs 1998b), developmental plasticity (Schew and Ricklefs 1998), energy budgets during growth (Weathers 1992, 1996), and development of temperature regulation (Visser 1998). It is the aim of this chapter to partly update the information presented in these reviews with special emphasis on seabird chicks. In addition, current knowledge of several aspects of postnatal development on seabird chicks is integrated into the text to provide insight into patterns of evolutionary and geographical diversification. Some recent methodological developments are evaluated in an attempt to provide guidelines for the standardization of future work on the construction of energy budgets in seabird chicks and adults.

13.2 GROWTH PATTERNS OF SEABIRD CHICKS IN RELATION TO TAXON AND PARENTAL FEEDING STRATEGY

13.2.1 INTERSPECIFIC VARIATION IN GROWTH RATES

Starck and Ricklefs (1998b) present growth parameter estimates based on logistic growth curves for a mixed data set of altricial and precocial land and seabird species (n = 557 species). The logistic growth curve, which has a sigmoid shape, enables the description of the development of body masses (M, g) of chicks as a function of age (t, d):

$$M = A/(1 + \exp(-k_l(t - t_i))) \tag{13.1}$$

where A represents the asymptotic body mass of chicks (g), t_i the time of inflection point of the curve (d), and k_l the logistic growth rate constant (d^{-1}). For the seabird subset (Figure 13.4),

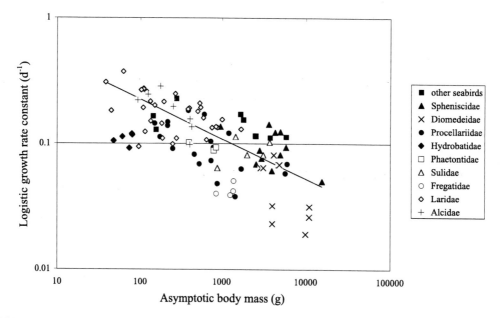

FIGURE 13.4 Relationship between the logistic growth constant (k_l, d^{-1}) and asymptotic body mass (A, g) in seabird chicks. Drawn diagonal line represents the general relationship between A and k_l in birds (Equation 13.2 of this chapter, Starck and Ricklefs 1998b).

asymptotic body masses range between 38.5 g for the Least Tern (*Sterna antillarum*, Schew 1990) and 15,500 g for the Emperor Penguin (*Aptenodytes forsteri*, Stonehouse 1953), and the logistic growth constants range between 0.019 d^{-1} for the Amsterdam Albatross (*Diomedea amsterdamensis*, Jouventin et al. 1989) and 0.38 d^{-1} for the fastest growing species, the Black Tern (*Chlidonias niger*, Schew 1990). Assuming independence of the data points for all 557 species, the relationship between both parameters was described by:

$$k_1 = 0.962A^{-0.31} \tag{13.2}$$

(Figure 13.4; Starck and Ricklefs 1998b). Equation 13.2 was used to predict k_1 for each seabird species listed by Starck and Ricklefs (1998b). Next, its residual value was calculated with the general equation:

$$\text{Residual value} = 100 \cdot (\text{observed value} - \text{predicted value})/\text{predicted value} \tag{13.3}$$

The analysis revealed that, after correction for differences in asymptotic body masses, most seabird families exhibit relatively low growth rate constants, which is particularly the case for the Fregatidae (average level relative to prediction: –58.0%, which differs significantly from zero, after comparison of the standard error of the residuals, $t_3 = -14.4$; $p < 0.001$), Hydrobatidae (–56.9%; $t_4 = -19.0$; $p < 0.0001$), Diomedeidae (–28.3%; $t_9 = -2.81$; $p < 0.02$), Phaethontidae (–4.2%; $t_2 = -6.1$; $p < 0.03$), and Procellariidae (–20.9%; $t_{17} = -3.0$; $p < 0.01$). In the Spheniscidae, average relative growth rates were significantly above prediction (+32%; $t_{10} = 2.5$; $p < 0.03$); and in the Sulidae (+2.9%; $t_4 = 0.18$; $p = 0.86$), Alcidae (+6.9%; $t_{11} = 1.0$; $p = 0.33$), and Laridae (+8.2%; $t_{31} = 1.4$; $p = 0.16$), residual values were higher than prediction, but these differences were not significant. These values indicate that growth rates are particularly low for many pelagic seabird species, and tend to be higher in species that feed in-shore such as the Laridae. Highest relative growth rates are observed in the Spheniscidae. This has been interpreted to be an adaptation to the short Antarctic summer enabling the chicks to leave the colony before the onset of the winter (Volkman and Trivelpiece 1980). In Section 13.3.4 we explore the energetic consequences of variations in growth rate for free-living chicks and their parents.

13.2.2 INTRASPECIFIC VARIATION IN GROWTH RATES

Chicks of most seabird species are generally fed a high-quality diet rich in protein and energy, with the possible exception of chicks of some petrel and albatross species that are mainly fed squid with a relatively low energy content (Prince and Ricketts 1981, see also Section 13.3.2). However, the quantity of food delivered by parents can be less predictable (see Chapters 6 and 7). In some species there is tremendous variation in postnatal growth rates owing to changes in the available food supply. For example, residual values for the logistic growth constants of Wedge-tailed Shearwaters (*Puffinus pacificus*) range between –72 and –9% (n = 13 studies), Black-legged Kittiwakes between –6 and +52%, Common Terns (*Sterna hirundo*) between –27 and +74% (n = 21 studies), and Atlantic Puffins (*Fratercula arctica*) between –45 and +22% (n = 27 studies). In addition, these four species exhibit marked intraspecific differences in calculated asymptotic body mass values which vary between 424 and 750 g, 335 and 421 g, 100 and 133 g, and 265 and 400 g, respectively. These large differences may reflect differences in body condition of the fledglings at this stage.

There is considerable evidence that growth retardation results from reduced food availability. For example, in a year with poor food availability, parental foraging trips of the Magellanic Penguin (*Spheniscus magellanicus*) lasted 20% longer than normal, chick feeding rates were reduced, and average body mass in 5-day-old chicks was 30% lower than in years with normal food availability (Boersma et al. 1990). Consequently, there were large differences between years with respect to the number of chicks fledged per nest, and in a 5-year study values ranged from 0.02 chicks per nest to 0.60 (Boersma et al. 1990). In some cases, variation in food availability is related to El Niño–Southern Oscillation (ENSO) events (e.g., Schreiber and Schreiber 1993, Schreiber 1994). In other cases, intraspecific variations in chick growth and nesting success have been attributed to individual differences in timing of egg laying within a particular year (Brooke 1986, Catry et al. 1998), age and breeding experience of the parents (Brooke 1986, Coulson and Porter 1985), individual quality, genetically determined (Brooke 1986), weather conditions (especially wind speed; Konarzewski and Taylor 1989), differences in food availability between colonies within a season (Frank 1992, Frank and Becker 1992), and between years (Boersma et al. 1990, Danchin 1992, Crawford and Dyer 1995). One of the key parameters for the interpretation of inter- and intraspecific variations in growth rates in seabird chicks seems to be the amount of energy collected by parents during chick rearing per unit of energy spent (see Section 13.6).

On the longer term, food restrictions and the subsequent growth retardation can potentially result in reduced survival (e.g., in the Common Murre, Harris et al. 1992), tolerance to starvation (e.g., in the Lesser Black-backed Gull [*Larus fuscus*], Griffiths 1992), and reduced recruitment rate (e.g., in the Black-legged Kittiwake, Coulson and Porter 1985).

13.3 ENERGETICS OF GROWTH

13.3.1 INTRODUCTION

One of the key factors needed for interpreting seabird life histories is the construction of energy budgets of free-living chicks and their parents (Drent et al. 1992). It is assumed that during the evolution of avian life histories, chicks have developed an array of adaptive responses, for instance:

1. In single-chick broods, the total amount of food required until independence (TME, kJ). However, in multi-chick broods, sibling competition may select for rapid growth and active food solicitation, potentially resulting in an increase of the TME.
2. Peak level of daily metabolized energy (peak-DME, kJ d^{-1}).
3. Duration of the growth period (t_{fl}, d) in order to reduce the risk of predation (Lack 1968), and (in polar environments) to complete the reproductive cycle before the onset of winter (Obst and Nagy 1993).

It has to be noted that minimizing the duration of the development period may require increasing growth rate and therefore daily energy requirement (Weathers 1992, and Section 13.3.4). These adaptations may enable parents to maximize their lifetime reproductive output.

13.3.2 COMPONENTS OF THE CHICKS' ENERGY BUDGET

Of all energy ingested by a chick (gross energy intake: GEI, kJ d^{-1}) only a part can be metabolized (metabolizable energy intake: MEI, kJ d^{-1}); the remainder is excreted in the form of feces and urine (FU, kJ d^{-1}). The assimilation coefficient (Q, dimensionless) is defined as:

$$Q = (GEI - FU)/GEI \qquad (13.4)$$

Once gross energy intake and assimilation coefficient are known, metabolizable energy intake can be calculated by

$$MEI = Q \cdot GEI \qquad (13.5)$$

The gross energy content of chick food varies with the type of diet, and is reported to be about 2.9 to 4.5 kJ g^{-1} fresh mass for krill (depending on its reproductive status; Clarke and Prince 1980,

TABLE 13.2
Assimilation Coefficients for Seabird Chicks in Relation to Food Type

Family/Species	Food Type	Coefficient	Source
Spheniscidae			
Jackass Penguin (*Spheniscus demersus*)	Fish	0.76	Cooper 1978
Jackass Penguin (*S. demersus*)	Fish	0.83	Heath and Randall 1985
Jackass Penguin (*S. demersus*)	Zooplankton	0.71	Heath and Randall 1985
Gentoo Penguin (*Pygoscelis papua*)	Krill	0.74	Davis et al. 1989
Procellariidae			
White-chinned Petrel (*Procellaria aequinoctialis*)	Fish	0.78	Jackson 1986
White-chinned Petrel (*P. aequinoctialis*)	Zooplankton	0.74	Jackson 1986
White-chinned Petrel (*P. aequinoctialis*)	Krill	0.76	Jackson 1986
Sulidae			
Cape Gannet (*Morus capensis*)	Fish	0.74	Batchelor 1982
Cape Gannet (*M. capensis*)	Fish	0.76	Cooper 1978
Phalacrocoracidae			
Double-crested Cormorant (*Hypoleucos auritus*)	Fish	0.85	Dunn 1975
Laridae			
Common Tern (*Sterna hirundo*)	Fish	0.81	Klaassen et al. 1992
Arctic Tern (*S. paradisaea*)	Fish	0.80	Drent et al. 1992
Sandwich Tern (*S. sandvicensis*)	Fish	0.82	Klaassen et al. 1992
Black-legged Kittiwake (*Rissa tridactyla*)	Fish	0.80	Gabrielsen et al. 1992
Alcidae			
Dovekie (*Alle alle*)	Zooplankton	0.79	Taylor and Konarzewski 1992

Note: In chicks of some diving petrels, prions, and storm petrels, the digestion efficiencies of the dietary wax component was near 0.99 (Roby et al. 1986).

Davis et al. 1989), 2.9 to 4.9 kJ g^{-1} for zooplankton (Clarke and Prince 1980, Montevecchi et al. 1984, Simons and Whittow 1984, Clarke et al. 1985), 4.2 to 10.3 kJ g^{-1} for fish (depending on its fat content; Montevecchi et al. 1984), and 39 to 41.7 kJ g^{-1} for the oil component of procellariiform diets (Warham et al. 1976, Simons and Whittow 1984, Obst and Nagy 1993). Some seabird species are known for having highly specialized diets (e.g., feeding exclusively on fish [terns] or krill [some penguins]), whereas other species (like most Procellariiformes) exhibit a nonspecialized aquatic diet (squid and other zooplankton, krill, fish, and trawler offal).

Assimilation coefficients have been determined in seabird chicks of several species and for different diets (Table 13.2). Average values for fish, krill, and zooplankton diets are 0.80 (SD = 0.035, n = 10), 0.75 (SD = 0.014, n = 2), and 0.75 (SD = 0.040, n = 3), respectively. It is interesting to note that the measured values in chicks are in close agreement with those reported for adult birds fed fish or invertebrates (0.77 and 0.74, respectively; Castro et al. 1989), which suggests that digestion efficiency in seabird chicks is high. Little information is available on the development of the assimilation efficiency as a function of the chicks' ages. An increase in assimilation coefficient from 0.8 at 11 to 12 days of age to a value of 0.88 at 20 to 21 days of age was reported in Double-crested Cormorant (*Hypoleucos auritus*) chicks (Dunn 1975). A similar trend with age was found in Jackass Penguin (*Spheniscus demersus*) chicks (Heath and Randall 1985), but not in chicks of the Cape Gannet (*Morus capensis*, Cooper 1978) and Common and Sandwich Tern (*Sterna sandvicensis*, Klaassen et al. 1992).

It appears as if there are marked differences between species with respect to assimilation efficiency as a function of diet type. For example, assimilation coefficients in White-chinned Petrel (*Procellaria aequinoctialis*) chicks were relatively insensitive to diet type, and values ranged from 0.74 for a squid diet to 0.78 for a fish diet (Jackson 1986). In contrast, assimilation coefficients of Jackass Penguin chicks varied from 0.68 and 0.87 for these two diets (Heath and Randall 1985). The high flexibility of the digestive system of White-chinned Petrel chicks is interpreted to be an adaptation to their nonspecialized diets (squid, krill, fish, and trawler offal; Jackson 1986; see also Brown 1988).

Metabolized energy can be allocated to the following components of the chicks' energy budget: (1) resting metabolism at thermoneutrality (i.e., the energy required for maintaining some basal physiological functions within the chicks' body; RMR, units kJ d^{-1}), (2) heat increment of feeding (i.e., the energy required to warm and digest the food; HIF [also referred to as specific dynamic action of food; SDA], units kJ d^{-1}), (3) temperature regulation (to compensate for heat losses from the chick to its environment: TR, units kJ d^{-1}), (4) activity (e.g., walking, preening, calling, and begging; A, units kJ d^{-1}), (5) biosynthesis-related heat production (the energy required for synthesizing new tissue such as fat and protein; S, units kJ d^{-1}), and (6) tissue energy (energy deposited as protein and fat; TE, units kJ d^{-1}):

$$MEI = RMR + HIF + TR + A + S + TE \qquad (13.6)$$

At a given level of metabolizable energy intake of a chick (e.g., the maximum level that can be provided by the parents), growth is highest at low levels of energy expenditure. Growth is zero if MEI equals energy expenditure, and growth is negative if MEI is lower than the level of energy expenditure. Under the latter conditions, body tissue (e.g., fat or protein) is used to produce energy for supporting other physiological functions.

13.3.3 METHODS TO DETERMINE ENERGY BUDGETS IN FREE-LIVING CHICKS

Four different methods have been used to determine the chick's level of MEI under free-living conditions:

1. Determination of gross energy intake based on periodic weighing of the chick in the field (e.g., Prince and Walton 1984, Ricklefs et al. 1985, Obst and Nagy 1993)
2. Determination of energy expenditure of the chick based on the extrapolation of laboratory measurements to field conditions, with an added component for energy deposited in tissues (see Ricklefs and White 1981)
3. Measurement of water influx rates and subsequent conversion to gross energy intake (Gabrielsen et al. 1992, Konarzewski et al. 1993)
4. Measurement of the level of energy expenditure directly in the field with doubly labeled water method, with an added component for growth energy (see Klaassen et al. 1989, Klaassen 1994, Visser and Schekkerman 1999)

13.3.3.1 Periodic Chick Weighing

A method used frequently to assess levels of food intake in seabird chicks is based on periodic chick weighing (expressed in grams per unit of time; e.g., Ricklefs 1984, Ricklefs et al. 1985, Schreiber 1994, 1996, Philips and Hamer 2000). Chicks are weighed regularly (e.g., at 2- to 12-h intervals) to monitor changes in their body mass. It is assumed that body mass decreases with time, a process that can be approached mathematically (e.g., by taking initial body mass, age, body size index, and time into account; Philips and Hamer 2000). If the chick exhibits a positive change in body mass, it is assumed that it was fed exactly between two weighings. The food intake level at the assumed feeding time is calculated as the difference between backward and forward extrapolation of the mass loss curves of a recently fed and fasting chick, respectively. The value obtained represents the amount of food eaten by the chick (in grams per unit of time). Next, to convert this value to gross energy intake (GEI), an assumption must be made with respect to the mass-specific energy content of the food (see Section 13.3.2). Finally, MEI can be calculated on the basis of Equation 13.5, after assuming a specific value for the assimilation coefficient of the diet (see Section 13.3.2 and Table 13.2).

Although this method is very easy to apply, it can only be used in chicks that are fed meals that are heavy relative to their body mass. In addition, apart from weighing and extrapolation errors, the calculated MEI level is subject to several other accumulating methodological errors. The first potential error is caused by the uncertainty with respect to the exact feeding time. This error is larger if weighings are done with a lower frequency. However, a high weighing frequency may, in some cases, interfere with chick begging, or parental feeding behavior, although some species are not bothered by it. In some sedentary seabird chicks, this problem can be circumvented by continuous weighing on an electronic balance, as employed in albatross chicks (Prince and Walton 1984, Huin et al. 2000).

The second potential error relates to the conversion of mass change to GEI. This error is probably smallest in species with a specialized diet, facilitating accurate estimation of the mass-specific energy content of the diet. The error is probably largest in procellariiform chicks because of the large difference in energy density of separate components of their diet, which ranges from about 4 kJ g^{-1} for predigested food to about 40 kJ g^{-1} for the stomach-oil component (see Section 13.3.2). Another complication is the large variation in the relative quantity of the oil component between meals within a species (e.g., values determined for different birds from the same colony during one observation day ranged between 20 and 83% in Wilson's Storm-Petrel [*Oceanites oceanicus*]; Obst and Nagy 1993), between species (e.g., see Roby 1991), and the relative difficulty to estimate the fraction of the oil component in a chick's diet (Roby et al. 1997).

The third potential error of this method is the conversion of GEI to MEI (Equation 13.5), after assuming a specific value for the assimilation coefficient. As discussed in Section 13.3.2, these average values range from 0.77 for krill and zooplankton to about 0.99 for stomach oil (Roby et al. 1986). Because of its high mass-specific gross energy content and its high assimilation coefficient, stomach oil is the most important component in energy budgets of procellariiform chicks, and it may contribute up to about 80% of their energy budgets (Roby 1991, Obst and Nagy 1993). The overall error of using the "chick weighing" method to estimate MEI can be as high as about 25%, depending on the number of assumptions made (Weathers 1992, 1996).

13.3.3.2 The Time-Energy Budget

The "time energy budget" method differs fundamentally from the "chick weighing" method in the sense that in the former method, metabolizable energy intake is estimated on the basis of measurements on energy expenditure (the components RMR, HIF, TR, A, and S; Equation 13.6) with an added component of the tissue energy (TE; Equation 13.6). As a first step, levels of oxygen consumption (and carbon dioxide production) are determined in resting chicks while housed in a small respiration chamber (indirect calorimetry; Weathers 1996). Next, metabolic rate (MR) can be calculated after assuming a specific energy equivalent per unit oxygen consumed or carbon dioxide produced. Typically, levels of energy expenditure are determined at different ambient temperatures to reveal the lowest level of energy expenditure at thermoneutrality (RMR), and the thermoregulatory costs (TR) at temperatures below the lower critical temperature (LCT, units °C). At each temperature, thermal conductance can be calculated being the metabolic level per degree temperature difference between the chick's body and its environment (see Visser 1998). The thermal conductance is assumed to be minimal at ambient temperatures below LCT.

To facilitate extrapolation of laboratory measurements to field conditions, the thermal environment of a chick must be characterized in its habitat (Bakken 1976, Klaassen 1994). This is most easily accomplished in chicks that live in deep burrows (e.g., procellariiform chicks, by measuring burrow-air temperatures), and it is most difficult in mobile chicks that live in sparsely vegetated colonies. When fully exposed, a chick experiences cooling effects of wind, compensated for by the chick elevating its metabolism. These effects can be strongly diminished by the chick positioning itself in vegetation (i.e., an energy-saving mechanism). In contrast, when fully exposed, a chick may experience the heating effects of solar radiation (enabling the chick to reduce thermoregulatory costs). Both effects can be integrated when employing heated taxidermic mounts, or (partly) with temperature measurements using black spheres (Gabrielsen et al. 1992, Klaassen 1994).

To estimate the tissue energy component of the energy budget in relation to the chicks' age, it is necessary to determine their growth curve in the field (see Section 13.2.1), as well as their mass-specific energy content of the body. The latter component is often estimated with the general equation:

$$E = a + b \cdot (M/A) \tag{13.7}$$

where E represents the mass-specific energy density of the whole body (kJ g^{-1}); a the intercept value at zero body mass (g); b the slope of the relationship; M the body mass (g) of the chick at a particular stage; and A its asymptotic body mass (g; Ricklefs 1974, Weathers 1996). The mass-specific energy density increases from about 3 to 4 kJ g^{-1} in young Double-crested Cormorant chicks (Ricklefs 1974) to about 22 kJ g^{-1} in heavy Wilson's Storm-Petrel chicks 1 week prior to fledging (Obst and Nagy 1993). For some species-specific estimates, see Table 13.3 (*Note:* a steeper slope indicates that as birds get heavier, they have a higher energy density per gram of body mass).

As can be seen, seabirds exhibit large differences in developmental patterns, and energy densities are particularly high in some pelagic seabird species (Ricklefs et al. 1980, Obst and Nagy 1993, Ricklefs and Schew 1994). Therefore, the use of group-specific estimates of the regressions to estimate energy accumulation is suggested, instead of the use of Weathers' (1996) Equation 13.10 for birds in general. To estimate the biosynthesis-related heat production in chicks, a synthesis efficiency value of 0.75 has traditionally been assumed (Ricklefs 1974), which has been used for the construction of energy budgets of most species listed in Table 13.3. More recently, Weathers (1996) advocated the separation of tissue growth into a fat component (with high synthesis efficiency) and a protein component (with low synthesis efficiency). Although this approach is more correct, it appears as if little error is made when employing a value of 0.75 (Konarzewski 1994, Ricklefs et al. 1998).

There are a number of potential methodological errors involved in this time energy budget method that merit attention. First, it is virtually impossible to account for the costs associated with locomotion or activity of the chick (e.g., see Dunn 1980). Probably these costs are lowest in individual procellariiform chicks that spend "about 90% of their time resting and sleeping" (Simons and Whittow 1984, Brown 1988), but the costs can be much higher for chicks growing up in dense colonies where frequent social interactions occur (Figure 13.5). Second, the extrapolation of laboratory measurements to field conditions for estimating the costs for temperature regulation is relatively difficult, especially because of the difficulty in accounting quantitatively for energy-saving mechanisms such as huddling, sheltering, or exposure to solar radiation. Third, poikilothermic chicks are frequently brooded by a parent, which considerably reduces its energy expenditure level (Klaassen 1994). Fourth, the extrapolation procedure is very sensitive to the assumed body temperature of the chick under field conditions. Although chicks of most species tend to keep up a body temperature of about 40°C, chicks of some species enter into torpor periodically between feeding bouts (Pettit et al. 1982, Boersma 1986). Occurrence of torpor has been interpreted to be an energy-saving mechanism to minimize costs of temperature regulation.

TABLE 13.3

Development of the Mass-Specific Energy Content of Chicks during Growth of Some Seabird Species

Family/Species	Intercept (a)	Slope (b)	Source
Spheniscidae			
Gentoo Penguin (*Pygoscelis papua*)	3.26	7.57	Myrcha and Kaminski 1982
Chinstrap Penguin (*P. antarctica*)	3.18	7.08	Myrcha and Kaminski 1982
Pelecanoididae			
Georgian Diving-Petrel (*Pelecanoides georgicus*)	4.28	5.11	Roby 1991
Hydrobatidae			
Wilson's Storm-Petrel (*Oceanites oceanicus*)	7.55	12.26	Obst and Nagy 1993
Phalacrocoracidae			
Double-crested Cormorant (*Hypoleucos auritus*)	2.97	6.40	Ricklefs 1974
Laridae			
Common Tern (*Sterna hirundo*)	3.86	4.94	Drent et al. 1992
Sandwich Tern (*S. sandvicensis*)	3.97	4.68	Drent et al. 1992
Arctic Tern (*S. paradisaea*)	4.13	3.73	Drent et al. 1992
Black-legged Kittiwake (*Rissa tridactyla*)	4.24	4.51	Gabrielsen et al. 1992
Alcidae			
Least Auklet (*Aethia pusilla*)	3.69	5.04	Roby 1991

Note: Model used $E = a + b \cdot (M/A)$, where E represents the mass-specific energy density of the body (kJ g^{-1}), M the actual body mass of the chick (g), and A the asymptotic or fledging body mass (g).

Some of the aforementioned difficulties with the extrapolation of laboratory-based measurements to field conditions also apply to adult birds. However, the energetic importance of energy-saving mechanisms in chicks is magnified because (1) chicks exhibit a larger surface area per unit body mass than adults, and (2) per unit of body surface, minimal thermal conductances are higher in chicks than in adults (Visser 1998). The average error of the time-energy budget method has been estimated to be on the order of 25% (Weathers 1992).

FIGURE 13.5 In a colony of Sooty Terns, nests are very close together. Here adult and large chicks are constantly interacting, possibly raising the energy budget of birds in these colonies. (Photo by R. W. Schreiber.)

13.3.3.3 The Measurement of Water Influx Rates and Subsequent Conversion to Energy Intake

Apart from metabolic water formation, food is the only other water source for most seabird chicks (Gabrielsen et al. 1992, Konarzewski et al. 1993). Water fluxes can be measured in free-living chicks following labeling with a heavy hydrogen isotope (i.e., water enriched with respect to ^{2}H or ^{3}H). Its concentration decreases in relation to the water-influx level (Nagy and Costa 1980) and can be measured on the basis of small blood samples (see Section 13.3.3.4). Food-water intake can be calculated after correction for metabolic water formation (Gabrielsen et al. 1992, Konarzewski et al. 1993). Finally, the amount of food eaten (in g) can be calculated after assuming a specific water percentage in the chicks' diet. This method has been validated in chicks of the Black-legged Kittiwake (Gabrielsen et al. 1992) by comparing the quantity of gross energy provided to chicks with concomitant determination of water influx with stable isotopes. On average, the validation revealed an excellent agreement of both methods (the isotope method tended to underestimate the true GEI level by only 2%). However, determinations on individual birds could differ by 34% at maximum, which suggests that a considerable sample size is required for the construction of energy budgets. The most critical steps of this method seem to be the estimation of the amount of metabolic-water formation, and the conversion of water fluxes to levels of food intake, the latter conversion being very sensitive to the hydration level of the food (plus attached water). The method cannot be applied in chicks that have access to other sources of water.

13.3.3.3.1 The Doubly Labeled Water Method: Some General Principles

The doubly labeled water (DLW) method has frequently been used to measure the rate of energy expenditure of adult and young seabirds under free-living conditions. Its principle is based on labeling the bird's body water pool with the heavy isotopes ^{2}H and ^{18}O, and the subsequent determination of their fractional turnover rates (k_d and k_o, respectively, units d^{-1}; Lifson and McClintock 1966, Nagy 1980, Speakman 1997, Visser et al. 2000a). In the past, the radioactive isotope ^{3}H has also been used instead of the stable ^{2}H isotope, but due to permit restrictions, nowadays most DLW experiments on free-living seabirds are performed with ^{2}H. It is assumed that, following labeling, ^{2}H leaves from the body water pool as water only, and ^{18}O both as water and as carbon dioxide. Thus, the difference between ^{18}O and ^{2}H can be converted to a rate of carbon dioxide production (rCO$_2$, moles d^{-1}):

$$rCO_2 = N/2 \cdot (k_o - k_d) \qquad (13.8)$$

where N represents the size of the body-water pool (moles), which can be determined on the basis of isotope dilution (Speakman 1997). The rate of carbon dioxide production can subsequently be converted to a level of energy expenditure, after assuming a diet-specific energy equivalent of carbon dioxide. For example, for seabird species with diets rich in proteins, it can be assumed that the production of 1 l of carbon-dioxide per hour is equivalent to a level of energy expenditure of 27.3 kJ h^{-1} (see Gessaman and Nagy 1988).

Concentrations of the heavy isotopes ^{2}H and ^{18}O in the body-water pool of animals are often determined on the basis of three blood samples, stored in flame-sealed capillaries or vacutainers. The first sample is taken prior to administration of the dose, to determine the natural abundance of both heavy isotopes in the bird's body-water pool (typically 0.015 and 0.20 atom percent for ^{2}H and ^{18}O, respectively). The second sample is taken after equilibration of the pulse dose (often referred to as "initial blood sample"), and the animal can be released. It needs to be recaptured after 24 to 72 h to take the third blood sample (often referred to as "final blood sample") to determine the isotope concentrations at the end of the measurement period. Thus, the calculated rate of CO$_2$ production is related to the level of energy expenditure by the bird between the taking of the "initial" and "final" blood sample.

Another important assumption of the DLW method for its application in seabirds is that the heavy ^2H and ^{18}O isotopes in the water molecule exhibit the same physical kinetics as the normal ^1H and ^{16}O isotopes (Lifson and McClintock 1966, Speakman 1997). This is true with respect to *fecal and urine water loss* (Lifson and McClintock 1966, Visser et al. 2000b), but not with respect to *evaporative water loss*: water molecules with the heavy ^2H or ^{18}O are less likely to evaporate than those with the lighter isotopes ^1H or ^{16}O. This process is called fractionation. As a result, the estimated fractional turnover rates of both isotopes (based on samples of the body-water pool) are too low. Theoretically, this fractionation effect also affects the calculation of water fluxes based on determinations of ^2H turnover (see Section 13.3.3.3), but an error analysis has revealed that a maximum error of about 1% is made if a fractional evaporative water loss value of 0.25 is used (Visser et al. 2000b). In contrast, the DLW method to estimate the rate of CO_2 production is much more sensitive to the effects of fractionation. Because this fractionation effect is larger for the ^2H isotope than for the ^{18}O isotope, this process also affects the difference between the ^{18}O and ^2H turnover rate (i.e., the calculated level of CO_2 production). Therefore, to solve this fractionation issue mathematically, the water efflux is considered to consist of one route subject to fractionation (i.e., evaporative water loss), and another route not subject to fractionation (i.e., fecal and urine water loss). After taking this effect into account, in combination with fractionation of the ^{18}O isotope in the CO_2 molecule (Speakman 1997), Equation 13.8 can be rewritten as:

$$rCO_2 = N/2.078 \cdot (k_o - k_d) - (r_G \cdot 0.0249 \cdot N \cdot k_d) \qquad (13.9)$$

where r_G (dimensionless) represents the fraction of the water efflux lost through evaporative pathways.

13.3.3.3.2   Applications of the DLW Method in Adult Seabirds: The Need for Standardization

After having collected and analyzed blood samples to determine N, k_o, and k_d, we only need to estimate r_G in order to calculate the rate of carbon-dioxide production (Equation 13.9). After studying small mammals in the laboratory, Lifson and McClintock (1966) assumed that an r_G value of 0.5 was appropriate (i.e., 50% of the total water efflux is lost through evaporative pathways). This value has been used for many terrestrial and aquatic species (see Speakman 1997). However, for other seabird species, rates of carbon dioxide production were calculated with Equation 13.8, thereby assuming that fractionation did not play a role (e.g., Nagy 1980). Speakman (1997) recognized the fact that in many free-living species, the assumption of a fractional evaporative water loss value of 0.5 was too high, and he proposed to use a value of 0.25 to be applied in Equation 13.9.

To illustrate the importance of assumptions concerning fractional evaporative water loss in adult seabirds, the levels of energy expenditure in adult Black-legged Kittiwakes were calculated at two different types of behavior. In both cases fractional ^2H and ^{18}O turnover rates were measured based on small blood samples. If the bird was foraging to feed its young, the rate of water efflux was high (about 450 g d^{-1}), and the calculated level of energy expenditure using Equation 13.8 was 1654 kJ d^{-1}. However, if Equation 13.9 were applied with the same ^2H and ^{18}O turnover rates, but after assuming that (1) fractional evaporative water loss was zero ($r_G = 0$), (2) fractional evaporative water loss was 0.25, and (3) fractional evaporative water loss was 0.5, calculated levels of energy expenditure were 1584 kJ d^{-1}, 1487 kJ d^{-1}, and 1389 kJ d^{-1}, respectively. This indicates that at high water fluxes, the DLW method is very sensitive to the assumptions made, and calculated rates of energy expenditure could differ by about 19%. In contrast, during incubation, water efflux was much lower (about 55 g d^{-1}), and the maximum difference in calculated levels of energy expenditure was only 9%. This clearly indicates that the DLW method needs to be validated in adult seabirds under different feeding conditions. Pending these results, the author proposes the use of Equation 13.9 in adult seabirds, with a fractional evaporative water loss value of 0.25.

13.3.3.3.3 Applications of the DLW Method in Seabird Chicks

Application of the DLW method in growing chicks has been hampered by uncertainties with respect to the routes of ^{2}H (or ^{3}H) and ^{18}O loss from the chick's body-water pool. Both isotopes may not leave from the body-water pool exclusively as water (^{2}H) or as water and carbon-dioxide gas (^{18}O), but they may also be incorporated in growing tissues (Williams and Nagy 1985, Weathers and Sullivan 1991). Differential rates of isotope incorporation could potentially result in an underestimation of the true rate of CO_2 production in the order of 10 to 25%.

In only two seabird species, the Arctic Tern (*Sterna paradisaea*; Klaassen et al. 1989) and Kittiwake (Gabrielsen et al. 1992), has the DLW method been validated to estimate the rate of CO_2 production in growing chicks. After assuming a fractional evaporative water loss value of 0.5, it was found that the DLW method underestimated the "true" rate of CO_2 production by about 10% on average (Klaassen et al. 1989), which suggests that differential rates of isotope incorporation did play a role. However, at a later stage it was demonstrated that the error of the DLW method could be significantly reduced after assuming a fractional evaporative water loss value of 0.25 (Visser and Schekkerman 1999). This value was also found to yield the best results in growing shorebird chicks (i.e., Northern Lapwing [*Vanellus vanellus*] and Black-tailed Godwit [*Limosa limosa*], both belonging to the Charadriiformes; Visser and Schekkerman 1999) and Japanese Quail (*Coturnix c. japonica*; Visser et al. 2000a). Therefore, to calculate energy expenditure in growing seabird chicks, the author recommends the use of Equation 13.9, with a fractional evaporative water loss value of 0.25. Because meals are often rich in fat and proteins, an energy equivalent of 27.33 kJ should be used per liter CO_2 produced (Gessaman and Nagy 1988). Unfortunately, in the past, little effort has been made to standardize this conversion, and in many studies a lower value has been employed.

As a last step in the construction of an energy budget, the level of metabolizable energy intake of a chick is calculated from its level of energy expenditure (as measured with DLW) plus an added component of tissue growth (see above). Of all methods described here, the DLW method is assumed to be the most accurate (Weathers 1992, 1996).

13.3.4 ENERGY BUDGETS OF GROWING SEABIRD CHICKS: THE IMPORTANCE OF ASYMPTOTIC BODY MASS, DURATION OF THE NESTLING PERIOD, AND LATITUDE

Weathers (1992, 1996) identified two important components of energy budgets of chicks: (1) the total amount of energy metabolized until fledging (TME, kJ) and (2) the peak level of daily metabolized energy during the growth period (peak-DME, kJ d^{-1}). His analysis for birds in general revealed significant effects of body mass at fledging (M, g) and the duration of the fledging period (t_{fl}, d) on the TME level

$$\text{TME} = 6.65 \ M^{0.852} \cdot t_{fl}^{0.710} \qquad (13.10)$$

and also on the peak-DME level

$$\text{peak-DME} = 11.69 \ M^{0.9082} \cdot t_{fl}^{-0.428} \qquad (13.11)$$

In other words, TME is highest in chicks with a high asymptotic body mass and a long nestling period, and peak-DME is highest in chicks with a high asymptotic body mass, but with a short nestling period. A compilation was made of published energy budgets of seabird chicks (Table 13.4). The analysis concentrated on the following questions: (1) Do energy budgets for seabird chicks differ from patterns observed for birds in general? and (2) Are there other factors than asymptotic body mass and duration of the fledging period that can explain variation in TME and peak-DME?

TABLE 13.4
Energy Budgets of Seabird Chicks

Species	A (g)	T_{fl} (d)	TME (kJ)	Peak-DME (kJ d⁻¹)	Latitude (deg)	Source
Northern Gannet (*Morus bassanus*)	3700	91	144990	2760	48	Montevecchi et al. 1984
Cape Gannet (*M. capensis*)	3240	97	200346	2758	34	Cooper 1978
Double-crested Cormorant (*Hypoleucos auritus*)	1900	46	59517	1987	43	Dunn 1975
Herring Gull (*Larus argentatus*)	1016	42	43839	1620	54	Drent et al. 1992
Glaucus-winged Gull (*L. glaucescens*)	940	39	31390		49	Drent et al. 1992
Herring Gull (*L. argentatus*)	850	45	29240	920	43	Dunn 1980
Southern Fulmar (*Fulmaris glacialoides*)	808	52	50692	1480	69	Hodum and Weathers, manuscript
Antarctic Petrel (*Thassaloica antarctica*)	590	48	31657	854	69	Hodum and Weathers, manuscript
Great-winged Petrel (*Pterodroma macroptera*)	537	108	34498		47	Brown 1988
Cape Pigeon (*Daption capense*)	441	47	31728	942	69	Hodum and Weathers, manuscript
Pigeon Guillemot (*Cepphus columba*)	430	33	11390	904	49	Dunn 1980
Hawaiian Dark-rumped Petrel (*Pterodroma sandwichensis*)	426	111	54232	608	25	Simons and Whittow 1984, Brown 1988
Black-legged Kittiwake (*Rissa tridactyla*)	410	35	18400	852	79	Gabrielsen et al. 1992
Snow Petrel (*Pagodroma nivea*)	246	47	17637	455	69	Hodum and Weathers, manuscript
Sooty Tern (*Sterna fuscata*)	198	60	6882	135	25	Ricklefs and White 1981
Blue Petrel (*Halobaena caerulea*)	177	53	13273		47	Brown 1988
Salvin's Prion (*Pachyptila salvini*)	154	52	9267		47	Brown 1988
Georgian Diving-Petrel (*Pelecanoides georgicus*)	148	43	8680	204	53	Roby 1991
Common Diving-Petrel (*P. urinatrix*)	139	53.5	7400	178	53	Roby 1991
Antarctic Tern (*S. vittata*)	133	27	7150	398	62	Klaassen 1994
Arctic tern (*S. paradisaea*)	115	22	4628	277	79	Klaassen 1994
Dovekie (*Alle alle*)	115	27	5293	284	77	Konarzweski et al. 1993
Common Tern (*S. hirundo*)	114	25	4852	239	53	Klaassen 1994
Common Tern (*S. hirundo*)	110	30	4412	199	25	Ricklefs and White 1981
Arctic tern (*S. paradisaea*)	107	22	3996	233	53	Klaassen 1994
Least Auklet (*Aethia pusilla*)	96	26	3334	176	57	Roby 1991
Leach's Storm-Petrel (*Oceanodroma leucorhoa*)	67	60	4798	97	45	Ricklefs et al. 1980
Wilson's Storm-Petrel (*Oceanites oceanicus*)	54	60	8820	217	65	Obst and Nagy 1993

Note: A: Asymptotic body mass (g); T_{fl}: duration of the fledging period (d); TME: total amount of energy metabolized until fledging (kJ); Peak-DME: peak level of daily metabolized energy during the growth period (kJ d⁻¹).

Using Equation 13.10 for the different seabird species, calculated residual TME values range from -37.5% in the Sooty Tern (*Sterna fuscata*, Ricklefs and White 1981) to 142.2% in the Wilson's Storm-Petrel (Obst and Nagy 1993), with an average value of 25.1% (SD = 36.01, n = 28 cases) which differs significantly from zero ($t_{27} = 3.62$, $p < 0.001$). In addition, values for peak-DME range from -45.3% in the Sooty Tern (Ricklefs and White 1981) to 186.0% in the Wilson's Storm-Petrel (Obst and Nagy 1993), with an average value of 24.7% (SD = 44.45%, n = 24 cases) which also differed significantly from zero ($t_{23} = 2.67$, $p < 0.014$). In conclusion, after correction for body mass and duration of the fledging period, in seabird chicks both TME and peak-DME appear to be significantly higher than values reported for birds in general. Thus, compared to most land birds, seabird chicks exhibit both high TME and peak-DME levels, especially at high latitudes. Since these chicks are parentally fed, adult seabirds need to collect more food than most land birds to satisfy the energy requirements of each chick.

Apparently, seabird chicks need more energy to achieve the fledging stage than do landbirds. Therefore, the data set presented in Table 13.4 was used by this author to derive predictive equations for TME and peak-DME in seabirds:

$$TME = 11.09 \, A^{0.771} \cdot t_{fl}^{0.747} \tag{13.12}$$

($F_{2, 25} = 211$, $r^2 = 0.939$, $p < 0.001$, standard error of the exponent for A: 0.053, standard error of the exponent for t_{fl}: 0.134), and

$$peak\text{-}DME = 14.06 \, A^{0.848} \cdot t_{fl}^{-0.341} \tag{13.13}$$

($F_{2, 21} = 211$, $r^2 = 0.894$, $p < 0.001$, standard error of the exponent for A: 0.066, standard error of the exponent for t_{fl}: 0.177).

After having established these relationships for seabirds, for each case listed in Table 13.4, the residual value was calculated using the measured value, and predicted values from Equations 13.12 and 13.13. Further analysis revealed a significant relationship between residual TME and latitude (degrees):

$$Res\text{-}TME = -39.28 + 0.796 \cdot latitude \tag{13.14}$$

($r^2 = 0.21$, $p < 0.015$, n = 28 cases, standard error of the slope $[s_b] = 0.411$; Figure 13.6), and also between residual peak-DME and latitude:

$$Res\text{-}peak\text{-}DME = -45.30 + 0.927 \cdot latitude \tag{13.15}$$

($r^2 = 0.208$, $p < 0.025$, n = 24 cases, $s_b = 0.286$, Figure 13.7). Thus, after correcting for differences in asymptotic body masses and duration of the nestling period, TME and peak-DME are positively related with latitude, and increase by about 1% per degree. Thus, rearing chicks at higher latitudes is more costly than at lower latitudes. Interestingly, these trends seem to be also applicable at the intraspecific level. For example, in Common and Arctic Tern chicks relative levels of TME and peak-DME were highest at high latitudes (see Table 13.2; Ricklefs and White 1981, Klaassen 1994).

As can be seen from Figures 13.6 and 13.7, the highest residual value is observed in the energy budget of the Wilson's Storm-Petrel determined at Palmer station, Antarctica, with the "chick weighing" method and "time-energy-budget" method (Obst and Nagy 1993). High TME and peak-DME levels at high latitudes can be explained by the low ambient temperatures that prevail even during the breeding season (e.g., see Obst and Nagy 1993), resulting in high thermoregulatory costs for the chicks. This effect may be magnified in species that rear only one chick, resulting in a permanent full exposure of the chick to low ambient temperatures.

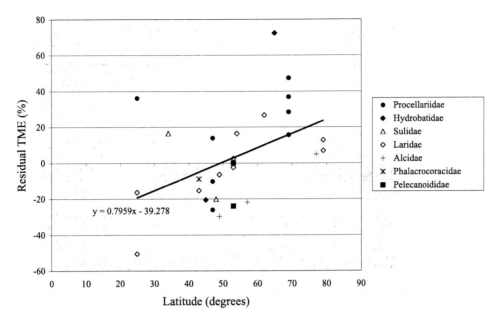

FIGURE 13.6 Relationship between residual total metabolizable energy intake during growth (residual-TME, %) and degrees latitude in seabird chicks. Drawn diagonal line represents the relationship between residual-TME (%) and degrees-latitude in seabird chicks (Equation 13.14).

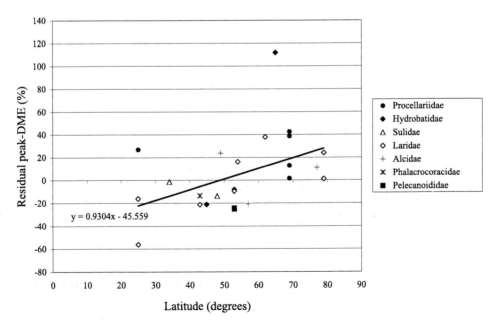

FIGURE 13.7 Relationship between residual peak-daily metabolizable energy intake during growth (residual-peak-DME, %) and degrees latitude in seabird chicks. Drawn diagonal line represents the relationship between residual-peak-DME (%) and degrees-latitude in seabird chicks (Equation 13.15).

FIGURE 13.8 Three Brown Pelican siblings huddle in their nest, possibly reducing thermoregulatory costs over that of single chicks in a nest. (Photo by R. W. Schreiber.)

In species that rear more chicks in a single nest, chick may reduce thermoregulatory costs due to huddling (Visser 1998, Figure 13.8). In young chicks, thermoregulatory costs are relatively low because of parental brooding, but older, unbrooded chicks are permanently exposed to lower ambient temperatures (Klaassen 1994). Due to these high thermoregulatory costs the overall growth efficiency is low in chicks that grow at high latitudes. For example, in chicks of the Wilson's Storm-Petrel, only about 12% of the TME is converted to tissue, which is much lower than the average value of 23.1% (SD = 4.45) reported for nine other seabird species (Drent et al. 1992).

13.4 DEVELOPMENT OF TEMPERATURE REGULATION

There is an intimate relationship between the chicks' development of homeothermy and parental time budgets. Until achievement of homeothermy of the chick(s) one parent needs to be at the nest site to provide warmth (Coulson and Johnson 1993, Visser 1998). The time that adults spend brooding considerably reduces their foraging time, which possibly affects the upper limit of the amount of food that can be brought to the nest. After achievement of homeothermy of the chick(s), both parents can leave resulting in an increase in the amount of food collected (Ricklefs and Roby 1983). Therefore, it seems advantageous for the chick to develop homeothermy at an early age. However, in some tropical seabird species, such as the Red-tailed Tropicbird and Red-footed Booby, adults brood the chick for a long time beyond its achievement of homeothermy (Figure 13.9; Schreiber and Schreiber 1993, Schreiber et al. 1996), which suggests that at these latitudes parental time budgets are not as tight as in birds breeding in polar regions. It has to be noted that the achievement of homeothermy of the chick and being left unbrooded incur energetic costs for temperature regulation, which affects the chicks' growth efficiency (e.g., Gabrielsen et al. 1992, Klaassen 1994; see below). Furthermore, it has been hypothesized that the development of tissues with high levels of functional maturity occurs at the expense of a high possible growth rate (i.e., the growth rate–maturity trade-off, Ricklefs 1979). In addition, costs of chicks being unattended can include an elevated predation risk from natural or human-introduced predators in some colonies. Given these costs and benefits of early or late acquisition of homeothermy, it is not surprising that in seabirds a range of developmental strategies exists with respect to the timing of the chicks' homeothermy.

The body temperature that can be maintained by a seabird chick (T_b, °C) above ambient temperature (T_a, °C) can be considered to be a function of its maximum level of thermogenic heat production (PMR, W) and its minimum thermal conductance (K, mW °C^{-1}):

FIGURE 13.9 Red-tailed Tropicbird chicks are sometimes brooded well past attaining homeothermy, as is this 3 1/2-week-old chick. Brooding still may offer some energetic savings to the chick. (Photo by E. A. Schreiber.)

$$(T_b - T_a) = PMR/K \qquad (13.16)$$

Thus, at a given ambient temperature, a large temperature gradient can be maintained by a chick if it exhibits a high peak metabolic rate level and/or a low minimal thermal conductance. Visser (1998) reviewed the literature with respect to RMR, PMR, and K in neonates of bird species. For birds, in general, allometric relationships were derived to predict the values for these parameters on the basis of neonatal body mass (M_n, g):

$$RMR = 0.0112 \cdot M_n^{0.861} \qquad (13.17)$$

$$PMR = 0.0139 \cdot M_n^{1.010} \qquad (13.18)$$

$$K = 6.57 \cdot M_n^{0.501} \qquad (13.19)$$

Thus, because PMR allometrically scales with the body mass to the power 1.01, and minimal thermal conductance with the power 0.501, the ratio of PMR/K allometrically scales with the power 0.509. Consequently, larger neonates exhibit a more favorable ratio of heat production to heat loss, resulting in a larger temperature difference that can be maintained by the chick. Therefore, the production of relatively large eggs as observed in seabirds (see Chapter 12) gives their chicks considerable thermal advantage.

To facilitate comparison between seabirds and other bird species, for each seabird species listed in Visser (1998), residual values were calculated for RMR, PMR, and K, using the predictive Equations 13.17, 13.18, and 13.19, respectively, and Equation 13.3.

In neonates of the Laridae, after correction for body mass as outlined above, average residual values for RMR, PMR, and K were +18.0% (SD = 23.7, n = 19 species, being significantly higher than zero, $p < 0.004$), +4.1% (SD = 19.1, n = 12 species, not being significantly different from zero, $p < 0.49$), and +4.9% (SD = 20.7, n = 11 species, not being significantly different from zero, $p < 0.50$).

In neonates of the Alcidae, average residual values for RMR, PMR, and K were +26.3% (SD = 29.5, n = six species), 27.9% (SD = 79.2, n = four species), and +23.9 (SD = 48.9, n = five species), none of these differences being significantly different from zero. To some extent, the large standard error for PMR residuals is caused by the value of the Xantus' Murrelet (*Endomychura hypoleuca*, Eppley 1984) which exhibits very high residual PMR value (145% above prediction), resulting in a relatively high degree of homeothermy (Epply 1984). This physiological adaptation enables these chicks to leave the nest at a very early age to join their parents at sea (Eppley 1984).

No PMR values are available for neonates of cormorants, pelicans, or boobies. However, because of their altricial mode of development, it can be assumed that newly hatched chicks of these taxonomic groups lack the ability of thermogenic heat production and require parental brooding. Cormorant and pelican chicks appear to achieve homeothermic status when weighing approximately 30 and 18% of their asymptotic body mass, respectively (Visser 1998). However, chicks of these species are being parentally brooded beyond this stage (Schreiber 1976, Hatch and Weseloh 1999).

In neonates of the Spheniscidae, average residual values for RMR and PMR were +0.8% (SD = 21.5, n = four species) and –47.2% (SD = 6.9, n = three species), only the latter being significantly different from zero (t_2 = 9.6, p <0.01). Apparently, penguin neonates exhibit a low level of thermogenic heat production and require parental brooding. Penguin chicks attain the homeothermic state after having achieved about 7% of the asymptotic body mass, but they are being frequently brooded at higher body masses (Visser 1998).

In neonates of the Procellariidae, average residual values for RMR, PMR, and K were +10.9% (SD = 26.0, n = 23 species), +15.1% (SD = 63.0, n = 14 species), and –20.2% (SD = 27.4, n = 12 species), none of these differences being significantly different from zero. Antarctic Prion (*Pachyptila desolata*) neonates exhibit a high level of thermogenic heat production (PMR = 2.6 × RMR) compared to chicks of Common Diving-petrels (*Pelecanoides urinatrix*: PMR = 1.2 × RMR). In addition, chicks of Antarctic Prions achieve the homeothermic state at an earlier age than do those of the Common Diving-petrel; as a result Prion chicks are left unattended at an earlier age (Ricklefs and Roby 1983). As chicks grow, PMR levels increase rapidly due to the accumulation of more muscle tissue, as well as the increase of metabolic heat production per unit of muscle tissue (Ricklefs 1979). In addition, whole-body minimal thermal conductance values increase at a lower rate than the increase of the surface area of the chicks. This is to some extent due to the development of vasomotor control in the legs, resulting in a reduction of heat loss (e.g., Eppley 1984). In general, chicks of larger species achieve homeothermy at an earlier age, permitting both parents to leave the nest unattended. Homeothermy is achieved at a later stage in chicks of pelicans (at achievement of approximately 18% of their asymptotic body mass), cormorants (at 30% of A), gannets (at 16% of A), and in penguins (at about 7% of A; see Visser 1998).

As mentioned before, thermoregulatory costs of chicks in the field can be offset by parental brooding, through which a chick can save energy for growth. To estimate the costs for temperature regulation in the field, uncorrected for parental brooding, heated taxidermic mounts were positioned in Black-legged Kittiwake (Gabrielsen et al. 1992) and tern colonies at different latitudes (Klaassen 1994). In addition, at these study sites the DLW method was also applied to construct energy budgets in growing chicks under free-living conditions. The comparison revealed energy savings from parental brooding to be on the order of 80% in the Kittiwake at Spitsbergen (latitude 79°N), 67 and 38% for the Arctic Tern in The Netherlands (53°N) and Spitsbergen, respectively, 38% for the Common Tern in The Netherlands, and 46 to 81% for the Antarctic tern on King George Island (63°S). Clearly, these energy-saving mechanisms will also hold for chicks of other species that are brooded, particularly at higher latitudes. A research topic that needs further investigation is the contribution of parental brooding in low-latitude colonies to protect chicks from overexposure to sunny conditions. Some species have circumvented the problem of overheating by digging burrows or nesting in the shade.

13.5 PHYSIOLOGICAL EFFECTS OF FOOD RESTRICTION

As shown above, if less food is provided by parents, the growth rate is lower than normal, which can be reflected by achievement of a lower value for the logistic growth constant and/or the achievement of a lower asymptotic body mass. In the preceding section it was demonstrated that achievement of homeothermy is to a large extent the product of an increase of muscle mass as well as an increase in muscle-mass-specific thermogenic heat production (Ricklefs 1979). Little is known about differences in carcass composition between well-fed and underfed chicks. However, Klaassen and Bech (1992) found strong reductions in levels of thermogenic heat production in Arctic Tern chicks weighing 25% or more below normal. Apparently, at these reduced growth levels, development of normal physiological function is impaired. As a consequence of stunted thermoregulatory development, chicks may require more parental brooding, reducing potential parental foraging time and, consequently, the potential amount of food delivered to the nest (Schew and Ricklefs 1998). In the longer term, food restriction and the subsequent growth retardation can potentially result in reduced survival (e.g., in the Common Murre, Harris et al. 1992), reduced tolerance to starvation (e.g., in the Lesser Black-backed Gull, Griffiths 1992), and reduced recruitment rate (e.g., in the Black-legged Kittiwake, Coulson and Porter 1985).

13.6 TOWARD THE CONSTRUCTION OF ENERGY BUDGETS OF ENTIRE FAMILY UNITS DURING THE PEAK DEMAND OF THE BROOD

In Section 13.3.4, it was shown that, after correction for asymptotic body mass and the duration of the nestling period, peak-DME levels of growing seabird chicks exceed average values for land birds by about 25%. As a consequence, adult seabirds need to collect more food per chick than land birds, which potentially has a major impact on the planning of the optimal brood size in seabirds. In addition, after correcting for asymptotic body mass and the duration of the nestling period, peak-DME appeared to increase by about 0.9% per degree latitude (Equation 13.15). It was postulated that during peak demand of the brood, adults work at a maximum physiological capacity of about four times their basal metabolic rate (Drent and Daan 1980). At this specific "work load," parents are supposed to collect sufficient food to cover their own expenses plus that of their brood.

Tieleman and Williams (2000) report values for parental daily energy expenditure during chick-rearing (DEE_{par}, kJ d^{-1} individual^{-1}) for a range of bird species, including ten seabird and eight landbird species for which energy budgets are available for growing chicks (Table 13.5). Using their predictive equation for the level of parental energy expenditure for all birds (their Table 13.2 and Equation 13.7, prediction based on body mass), and Equation 13.3 of this chapter, this author calculated residual levels of parental energy expenditure for each species (Table 13.5). As can be seen, seabird parents exhibit levels of daily energy expenditure which are on average 54.4% higher than predicted. In addition, for the landbird subset, the average residual value was only 3.7%. Based on this data set it can be concluded that after correction for body mass, seabird parents exhibit much higher levels of energy expenditure than landbirds.

To break down the energy budgets of entire family units during the peak demand of the brood, the following components can be identified: (1) the level of parental metabolizable energy intake, which must be equal to the level of parental energy expenditure if parental body mass remains constant (DEE_{par}, kJ d^{-1} individual^{-1}), and (2) the peak level of metabolizable energy intake of the entire brood (peak-DME_{br}, kJ d^{-1} brood^{-1}). Thus at this particular stage of maximum energy demand of the brood, the overall foraging efficiency of the parents (FE_{par}, units: kJ d^{-1} of food collected per kJ d^{-1} of energy spent, i.e., dimensionless) for species with biparental care (assuming equal DEE_{par} for both parents) can be calculated as:

$$FE_{par} = (2 \cdot DEE_{par} + \text{peak-}DME_{br}) / (2 \cdot DEE_{par}) \tag{13.20}$$

Thus, if $FE_{par} = 1$ (for example, during periods with high wind [short-term effect] or during El Niño–Southern Oscillation events [longer-term effect], making it difficult for the parents to collect

TABLE 13.5
Comparison of Energy Budgets of Family Units between Seabirds and Landbirds

Species	A (g)	Peak-DME (kJ d⁻¹ per chick)	Parental Mass (g)	Parental DEE (kJ d⁻¹)	Res-parental DEE (%)	Average No. of Chicks	Peak Demand Brood (kJ/d)	Parental Foraging Efficiency	Source
Seabirds									
Northern Gannet (*Morus bassanus*)	3700	2760	3120	4865	76.8	1	2760	1.28	Montevecchi et al. 1984, Birt-Friesen et al. 1989
Cape Gannet (*M. capensis*)	3240	2758	2620	4670	91.9	1	2758	1.30	Cooper 1978, Adams et al. 1991
Black-legged Kittiwake (*Rissa tridactyla*)	410	852	391.6	995	55.6	1	852	1.43	Gabrielsen et al. 1987, Gabrielsen et al. 1991
Sooty Tern (*Sterna fuscata*)	198	135	184	340.4	-9.5	1	135	1.20	Ricklefs and White 1981, Flint and Nagy 1984
Georgian Diving-Petrel (*Pelecanoides georgicus*)	148	204	119.2	463.5	67.2	1	204	1.22	Roby and Ricklefs 1986, Roby 1991
Common Diving-Petrel (*P. urinatrix*)	139	178	132.3	556.6	86.6	1	178	1.16	Roby and Ricklefs 1986, Roby 1991
Dovekie (*Alle alle*)	115	284	163.7	696.1	100.9	1	284	1.20	Gabrielsen et al. 1991, Konarzewski et al. 1993
Least Auklet (*Aethia pusilla*)	96	176	83.5	357.9	65.8	1	176	1.25	Roby and Ricklefs 1986, Roby 1991
Leach's Storm-Petrel (*Oceanodroma leucorhoa*)	67	97	46.8	131.8	-8.3	1	97	1.37	Ricklefs et al. 1980, Ricklefs et al. 1986
Wilson's Storm-Petrel (*Oceanites oceanicus*)	54	217	42.2	157	17.5	1	217	1.69	Obst et al. 1987, Obst and Nagy 1993
Average					54.5			1.31	
Landbirds									
Eurasian Kestrel (*Falco tinnunculus*)	238	351	215.5	328	-22.0	5.1	1790.1	3.73	Masman et al. 1988, Visser et al. unpublished
Common Starling (*Sturnus vulgaris*)	65	135	75.5	299.5	48.9	5	675	2.13	Westerterp 1973, Ricklefs and Williams 1984
Blue-throated Bee-eater (*Merops viridis*)	33	57.5	33.8	77.4	-32.3	3.9	224.25	2.45	Bryant and Hails 1983, Bryant et al. 1984
Western Bluebird (*Sialia mexicana*)	27.5	66.1	27.5	95	-3.9	5	330.5	2.74	Mock et al. 1991, Mock 1991
Yellow-eyed Junco (*Junco phaeonotus*)	18.5	54.1	19.6	71.7	-8.0	4	216.4	2.51	Weathers and Sullivan 1991, Weathers and Sullivan 1993
House Martin (*Delichon urbica*)	17.4	45.3	18.1	87.2	18.4	4	181.2	2.04	Bryant and Gardiner 1979, Westerterp and Bryant 1984
Savannah Sparrow (*Passerculus sandwichensis*)	15.7	46.6	18.3	77.8	4.8	3.2	149.12	1.96	Williams and Prints 1986, Williams 1987
Pacific Swallow (*Hirundo tahitica*)	14.6	31.7	14.1	76.6	23.9	3	95.1	1.62	Bryant and Hails 1983, Bryant et al. 1984
Average					3.7			2.40	

Note: A: Asymptotic body mass (g); Peak-DME: peak level of daily metabolized energy during the growth period (kJ d⁻¹ per chick); Parental DEE: parental level of daily energy expenditure (kJ d⁻¹); Res-parental DEE: residual value of parental daily energy expenditure (%).

food), no energy is available for chicks and, consequently, they lose mass. Another short-term option under these conditions is that parents bring some food to chicks at the expense of their own consumption, resulting in parental body mass loss. If the FE_{par} value exceeds 1, some food is left to cover the demand of the brood, and the higher this value the more food is available for the brood.

For ten species of seabirds, DEE_{par}, peak-DME_{br}, and FE_{par} values were compiled from literature data (Table 13.5). As can be seen, values for parental foraging efficiency range from a low 1.16 in the Common Diving-petrel, to a high 1.69 in the Wilson's Storm-petrel, with an average value of 1.31 for all seabird species. This average value for seabirds is particularly low compared to the average value of 2.40 found for eight species of land birds (Table 13.5). Thus, in seabirds, per unit of parental energy spent during foraging, the amount of energy obtained is only 55% of that for land birds. This low foraging efficiency observed in seabirds may result from (1) relatively long parental flying distances to forage during the breeding season, and (2) from low food densities in the sea (Ricklefs et al. 1986, Roby and Ricklefs 1986). Possibly as a result of the low foraging efficiencies experienced by adult seabirds, in many species sufficient energy is collected to cover the energetic needs of only one chick. In other inshore feeding species, foraging efficiencies may be higher permitting parents to rear more than one chick (such as the Black-legged Kittiwake; Gabrielsen et al. 1992; and gull, tern, cormorant, and pelican species). Clearly, more research is needed to quantify the effects of flying distance, food availability, and weather on parental foraging efficiency and, thus, on the overall growth rate of the chicks.

ACKNOWLEDGMENTS

David Goldstein, Erpur Hansen, and Bob Ricklefs gave many valuable comments on earlier drafts of this manuscript. Peter Hodum and Wes Weathers generously made available data on energy budgets of some Antarctic petrels.

LITERATURE CITED

ADAMS, N. J., R. W. ABRAMS, W. R. SIEGFRIED, K. A. NAGY, AND I. R. KAPLAN. 1991. Energy expenditure and food consumption by breeding Cape Gannets (*Morus capensis*). Marine Ecology Progress Series 70: 1–9.

BAKKEN, G. S. 1976. A heat transfer analysis of animals: unifying concepts and the application of metabolism chamber data to field ecology. Journal of Theoretical Biology 60: 337–384.

BATCHELOR, A. L. 1982. The Diet of the Cape Gannet (*Sula capensis*) Breeding on Bird Island, Algoa Bay. M.Sc. thesis, University of Port Elisabeth, Port Elisabeth, South Africa.

BIRT-FRIESEN, V. L., W. A. MONTEVECCHI, D. K. CAIRNS, AND S. A. MACKO. 1989. Activity-specific metabolic rates of free-living Northern Gannets and other seabirds. Ecology 70: 357–367.

BOERSMA, P. D. 1986. Body temperature, torpor, and growth of chicks of Fork-tailed Storm-Petrels (*Oceanodroma furcata*). Physiological Zoology 59: 10–19.

BOERSMA, P. D., D. L. STOKES, AND P. M. YORIO. 1990. Reproductive variability and historical change of Magellanic Penguins (*Spheniscus magellanicus*) at Punta Tombo, Argentina. Pp. 15–43 in Penguin Biology (L. S. Davis, and J. T. Darvy, Eds.). Academic Press, New York.

BROOKE, M. DE L. 1986. Manx Shearwater chicks: seasonal, parental, and genetic influences on the chick's age and weight at fledging. Condor 88: 324–327.

BROWN, C. R. 1988. Energy requirements for growth of Salvin's *Prions Pachyptila* vittata salvini, Blue Petrels *Halobaena caerulea* and Great-winged Petrels *Pterodroma macroptera*. Ibis 130: 527–534.

BRYANT, D. M., AND A. GARDINER. 1979. Energetics of growth in House Martins (*Delichon urbica*). Journal of Zoology, London 189: 275–304.

BRYANT, D. M., AND C. J. HAILS. 1983. Energetics and growth patterns of three tropical bird species. Auk 100: 425–439.

BRYANT, D. M., C. J. HAILS, AND P. TATNER. 1984. Reproductive energetics of two tropical bird species. Auk 101: 25–37.

CASTRO, G., N. STOYAN, AND J. P. MYERS. 1989. Assimilation efficiency in birds: a function of taxon or food type? Comparative Biochemistry and Physiology 92A: 271–278.

CATRY, P., N. RATCLIFFE, AND R. W. FURNESS. 1998. The influence of hatching date on different life-history stages of Great Skuas *Catharacta skua*. Journal of Avian Biology 29: 299–304.

CLARKE, A., AND P. A. PRINCE. 1980. Chemical composition and calorific value of food fed to Mollymauk chicks *Diomeda melanophris* and *D. chrysostoma* at Bird Island, South Georgia. Ibis 122: 488–494.

CLARKE, A., M. R. CLARKE, J. L. HOLMES, AND T. D. WATERS. 1985. Calorific values and elemental analysis of eleven species of oceanic squids (Mollusca: Cephalopoda). Journal of Marine Biology Assessment U.K. 65: 983–986.

COOPER, J. 1978. Energetic requirements for growth of the Jackass Penguin. Zoologica Africana 12: 201–213.

COULSON, J. C., AND J. M. PORTER. 1985. Reproductive success of the Kittiwake *Rissa tridactyla*: The role of clutch size, chick growth rates, and parental quality. Ibis 127: 450–466.

COULSON, J. C., AND M. P. JOHNSON. 1993. The attendance and absence of adult Kittiwakes *Rissa tridactyla* from the nest site during the chicks stage. Ibis 135: 372–378.

CRAWFORD, R. J. M., AND B. M. DYER. 1995. Responses by four seabird species to a fluctuating availability of Cape Anchovy *Engraulis capensis* off South Africa. Ibis 137: 329–339.

DAAN, S., AND J. M. TINBERGEN. 1979. Young Guillemots (*Uria lomvia*) leaving their arctic breeding cliffs: a daily rhythm in numbers and risk. Ardea 67: 96–100.

DANCHIN, E. 1992. Food shortage as a factor in the 1988 Kittiwake *Rissa tridactyla* breeding failure in Shetland. Ardea 80: 93–98.

DAVIS, R. W., J. P. CROXALL, AND M. J. O'CONNELL. 1989. The reproductive energetics of Gentoo (*Pygoscelis papua*) and Macaroni (*Eudyptes chrysolophus*) Penguins at South Georgia. Journal of Animal Ecology 58: 59–74.

DRENT, R., AND S. DAAN. 1980. The prudent parent: energetic adjustments in avian breeding. Ardea 68: 225–252.

DRENT, R. H., M. KLAASSEN, AND B. ZWAAN. 1992. Predictive growth budgets in terns and gulls. Ardea 80: 5–17.

DUNN, E. H. 1975. Caloric intake of nestling Double-crested Cormorants. Auk 92: 553–565.

DUNN, E. H. 1980. On the variability in energy allocation of nestling birds. Auk 97: 19–27.

EPPLEY, Z. A. 1984. Development of thermoregulatory abilities in Xantus's Murrelet chicks (*Synthliboramphus hypoleucus*). Physiological Zoology 57: 307–317.

FLINT, E. N., AND K. A. NAGY. 1984. Flight energetics of free-living Sooty Terns. Auk 101: 288–294.

FRANK, D. The influence of feeding conditions on food provisioning of chicks in Common Terns *Sterna hirundo* nesting in the German Wadden Sea. Ardea 80: 45–56.

FRANK, D., AND P. H. BECKER. 1992. Body mass and nest reliefs in Common Terns *Sterna hirundo* exposed to different feeding conditions. Ardea 80: 57–70.

GABRIELSEN, G. W., F. MEHLUM, AND K. A. NAGY. 1987. Daily energy expenditure and energy utilization of free-ranging Black-legged Kittiwakes. Condor 89: 126–132.

GABRIELSEN, G. W., J. R. E. TAYLOR, M. KONARZEWSKI, AND F. MEHLUM. 1991. Field and laboratory metabolism and thermoregulation in Dovekies (*Alle alle*). Auk 108: 71–78.

GABRIELSEN, G. W., M. KLAASSEN, AND F. MEHLUM. 1992. Energetics of Black-legged Kittiwake (*Rissa tridactyla*) chicks. Ardea 80: 29–40.

GASTON, A. J. 1985. Development of the young in the Atlantic Alcidae. Pp. 319–354 *in* The Atlantic Alcidae (D. N. Nettleship and T. R. Birkhead, Eds.). Academic Press, London.

GASTON, A. J. 1992. The Ancient Murrelet. A Natural History in the Queen Charlotte Islands. Poyser, London.

GESSAMAN, J. A., AND K. A. NAGY. 1988. Energy metabolism: errors in gas-exchange conversion factors. Physiological Zoology 61: 507–513.

GRIFFITHS, R. 1992. Sex-biased mortality in the Lesser Backed Gull *Larus fuscus* during the nestling stage. Ibis 134: 237–244.

HARRIS, M. P., D. J. HALLEY, AND S. WANLESS. 1992. The post-fledging survival of young Guillemots *Uria aalge* in relation to hatching date and growth. Ibis 134: 335–339.

HATCH, J. H., AND D. V. WESELOH. 1999. The Double-crested Cormorant (*Phalacrocorax auritus*). No. 441 *in* The Birds of North America (A. Poole and F. Gill, Eds.). The Birds of North America, Inc., Philadelphia.

HEATH, R. G. M., AND R. M. RANDALL. 1985. Growth of Jackass Penguin chicks (*Spheniscus demersus*) hand reared on different diets. Journal of Zoology, London 205: 91–105.

HODUM, P. J., AND W. W. WEATHERS. Submitted. Ecological energetics and prey consumption of an Antarctic seabird community. Submitted to Journal of Animal Ecology.

HUIN, N., P. A. PRINCE, AND D. R. BRIGGS. 2000. Chick provisioning rates and growth in Black-browed Albatross *Diomedea melanophris* and Grey-headed Albatross *D. chrysostoma* at Bird Island, South Georgia. Ibis 142: 550–565.

JACKSON, S. 1986. Assimilation efficiencies of White-chinned Petrels (*Procellaria aequinoctialis*) fed different prey. Comparative Biochemistry and Physiology 85A: 301–303.

JOUVENTIN, P., J. MARTINEZ, AND J. P. ROUX. 1989. Breeding biology and current status of the Amsterdam Island Albatross, *Diomedea amsterdamensis*. Ibis 131: 171–182.

KLAASSEN, M. 1994. Growth and energetics of tern chicks from temperate and polar environments. Auk 111: 525–544.

KLAASSEN, M., AND C. BECH. 1992. Resting and peak metabolic rates of arctic Tern nestlings and their relations to growth rate. Physiological Zoology 65: 803–814.

KLAASSEN, M., C. BECH, D. MASMAN, AND G. SLAGSVOLD. 1989. Growth and energetics of Arctic Tern chicks (*Sterna paradisaea*). Auk 106: 240–248.

KLAASSEN, M., B. ZWAAN, P. HESLENFELD, P. LUCAS, AND B. LUIJCKX. 1992. Growth rate associated changes in the energy requirements of tern chicks. Ardea 80: 19–28.

KONARZEWSKI, M. 1994. Allocation of energy to growth and respiration in avian postembryonic development. Ecology 76: 8–19.

KONARZEWSKI, M., AND J. R. E. TAYLOR. 1989. The influence of weather conditions on growth of Little Auk (*Alle alle*) chicks. Ornis Scandinavica 20: 112–115.

KONARZEWSKI, M., J. R. E. TAYLOR, AND G. W. GABRIELSEN. 1993. Chick energy requirements of Dovekies (*Alle alle*). Auk 110: 343–353.

LACK, D. 1968. Ecological adaptations for breeding in birds. Methuen, London.

LIFSON, N. A., AND R. M. McCLINTOCK 1966. Theory of use of the turnover rates of body water for measuring water and energy balance. Journal of Theoretical Biology 12: 46–74.

MASMAN, D., S. DAAN, AND H. J. A. BELDHUIS. 1988. Ecological energetics of the kestrel: daily energy expenditure throughout the year based on time-energy budget, food intake and doubly labeled water methods. Ardea 76: 64–81.

MOCK, P. J. 1991. Daily allocation of time and energy of Western Bluebirds feeding nestlings. Condor 93: 598–611.

MOCK, P. J., M. KHUBESRIAN, AND D. M. LARCHEVEQUE. 1991. Energetics of growth and maturation in sympatric passerines that fledge at different ages. Auk 108: 34–41.

MONTEVECCHI, W. A., R. E. RICKLEFS, I. R. KIRKHAM, AND D. GABALDON. 1984. Growth energetics of nestling Northern Gannets (*Sula bassanus*). Auk 101: 334–341.

MYRCHA, A., AND P. KAMINSKI. 1982. Changes in body calorific values during nestling development of penguins of the genus Pygoscelis. Polish Polar Research 3: 81–88.

NAGY, K. A. 1980. CO_2 production in animals: an analysis of potential errors of the doubly labeled water technique. American Journal of Physiology 238: R466–R473.

NAGY, K. A., AND D. P. COSTA. 1980. Water flux in animals: analyses of potential errors in the tritiated water method. American Journal of Physiology 238: R454–R465.

NICE, M. M. 1962. Development of behavior in precocial birds. Transactions of the Linnean Society, New York 8: 1–211.

OBST, B. S., K. A. NAGY, AND R. E. RICKLEFS. 1987. Energy utilization by Wilson's Storm-Petrel (*Oceanites oceanicus*). Physiological Zoology 60: 200–210.

OBST, B. S., AND K. A. NAGY. 1993. Stomach oil and the energy budget of Wilson's Storm-Petrel nestlings. Condor 95: 792–805.

PETTIT, T. N., G. S. GRANT, AND G. C. WHITTOW. 1982. Body temperatures and growth of Bonin Petrel chicks. Wilson Bulletin 94: 358–361.

PHILLIPS, R. A., AND K. C. HAMER. 2000. Periodic weighing and the assessment of meal mass and feeding frequency in seabirds. Journal of Avian Biology 31: 75–80.

PRINCE, P. A., AND C. RICKETTS. 1981 Relationships between food supply and growth in albatrosses: an interspecies chick fostering experiment. Ornis Scandinavica 12: 207–210.

PRINCE, P. A., AND D. W. H. WALTON. 1984. Automated measurement of meal sizes and feeding frequency in albatrosses. Journal of Applied Ecology 21: 789–794.

RICKLEFS, R. E. 1974. Energetics of reproduction in birds. Pp. 152–292 *in* Avian Energetics (J. R. Paynter, Jr., Ed.). Publications of the Nuttall Ornithology Club, No. 15, Cambridge, MA.

RICKLEFS, R. E. 1979. Adaptation, constraint, and compromise in avian postnatal development. Biological Reviews 54: 269–290.

RICKLEFS, R. E. 1984. Meal sizes and feeding rates of Christmas Shearwaters and Phoenix Petrels on Christmas Island, Central Pacific Ocean. Ornis Scandinavica 15: 16–22.

RICKLEFS, R. E., AND S. C. WHITE. 1981. Growth and energetics of chicks of the Sooty Tern (*Sterna fuscata*) and Common Tern (*Sterna hirundo*). Auk 98: 361–378.

RICKLEFS, R. E., S. WHITE, AND J. CULLEN. 1980. Energetics of postnatal growth in Leach's Storm-Petrel. Auk 97: 566–575.

RICKLEFS, R. E., AND D. D. ROBY. 1983. Development of homeothermy in the diving petrels *Pelecanoides urinatrix exsul* and *P. georgicus*, and the Antarctic Prion *Pachyptila desolata*. Comparative Biochemistry and Physiology 75A: 307–311.

RICKLEFS, R. E., AND J. B. WILLIAMS. 1984. Daily energy expenditure and water-turnover rate of adult European Starlings (*Sturnus vulgaris*) during the nestling cycle. Auk 101: 707–716.

RICKLEFS, R. E., C. H. DAY, C. E. HUNTINGTON, AND J. B. WILLIAMS. 1985. Variability in feeding rate and meal size of Leach's Storm-Petrel at Kent Island, New Brunswick. Journal of Animal Ecology 54: 883–898.

RICKLEFS, R. E., D. D. ROBY, AND J. B. WILLIAMS. 1986. Daily energy expenditure of adult Leach's Storm-Petrels during the nesting cycle. Physiological Zoology 59: 649–660.

RICKLEFS, R. E., AND W. A. SCHEW. 1994. Foraging stochasticity and lipid accumulation by nestling petrels. Functional Ecology 8: 159–170.

RICKLEFS, R. E., J. M. STARCK, AND M. KONARZEWSKI. 1998. Internal constraints on growth in birds. Pp. 266–287 *in* Avian Growth and Development. Evolution within the Altricial-Precocial Spectrum (J. M. Starck and R. E. Ricklefs, Eds.). Oxford Ornithology Series. Oxford University Press, Oxford.

ROBY, D. D. 1991. Diet and postnatal energetics in convergent taxa of plankton-feeding seabirds. Auk 108: 131–146.

ROBY, D. D., AND R. E. RICKLEFS. 1986. Energy expenditure in adult Least Auklets and diving petrels during the chick-rearing period. Physiological Zoology 59: 661–678.

ROBY, D. D., A. R. PLACE, AND R. E. RICKLEFS. 1986. Assimilation and deposition of wax esters in planktivorous seabirds. Journal of Experimental Zoology 238: 29–41.

ROBY, D. D., J. R. E. TAYLOR, AND A. R. PLACE. 1997. Significance of stomach oil for reproduction in seabirds: an interspecies cross-fostering experiment. Auk 114: 725–736.

SCHEW, W. A. 1990. Ecological and Environmental Determinants of Growth Patterns in Terns. Ph.D. thesis, University of California, Long Beach.

SCHEW, W. A., AND R. E. RICKLEFS. 1998. Developmental plasticity. Pp. 288–304 *in* Avian Growth and Development. Evolution within the Altricial-Precocial Spectrum (J. M. Starck and R. E. Ricklefs, Eds.). Oxford Ornithology Series. Oxford University Press, Oxford.

SCHREIBER, E. A. 1994. El Niño–Southern Oscillation effects on chick provisioning and growth in Red-tailed Tropicbirds. Colonial Waterbirds 17: 105–119.

SCHREIBER, E. A. 1996. Experimental manipulation of feeding in Red-tailed Tropicbird chicks. Colonial Waterbirds 19: 45–55.

SCHREIBER, E. A., AND R. W. SCHREIBER. 1993. The Red-tailed Tropicbird (*Phaeton rubricauda*). No. 43 *in* The Birds of North America (A. Poole and F. Gill, Eds.). Academy of Natural Sciences, Philadelphia; American Ornithologists' Union, Washington, D.C.

SCHREIBER, E. A., R. W. SCHREIBER, AND G. A. SCHENK. 1996. The Red-footed Booby (*Sula sula*). No. 241 *in* The Birds of North America (A. Poole and F. Gill, Eds.). Academy of Natural Sciences, Philadelphia; American Ornithologists' Union, Washington, D.C.

SCHREIBER, R. W. 1976. Growth and development of nestling Brown Pelicans. Bird-Banding 47: 19–39.

SIMONS, T. R., AND G. C. WHITTOW. 1984. Energetics of breeding Dark-rumped Petrels. Pp. 159–181 *in* Seabird Energetics (G. C. Whittow and H. Rahn, Eds.). Plenum Press, New York.

SPEAKMAN, J. R. 1997. Doubly Labeled Water. Theory and Practice. Chapman & Hall, London.

STARCK, J. M., AND R. E. RICKLEFS. 1998a. Patterns of development: the altricial-precocial spectrum. Pp. 3–30 *in* Avian Growth and Development. Evolution within the Altricial-Precocial Spectrum (J. M. Starck and R. E. Ricklefs, Eds.). Oxford Ornithology Series. Oxford University Press, Oxford.

STARCK, J. M., AND R. E. RICKLEFS. 1998b. Data set of avian growth parameters. Pp. 381–423 *in* Avian Growth and Development. Evolution within the Altricial-Precocial Spectrum (J. M. Starck and R. E. Ricklefs, Eds.). Oxford Ornithology Series. Oxford University Press, Oxford.

STONEHOUSE, B. 1953. The Emperor Penguin, *Aptenodytes forsteri*. I. Breeding biology and development. Falkland Islands Dependencies Survey Scientific Report 6: 1–33.

TAYLOR, J. R. E., AND M. KONARZEWSKI. 1992. Budget of elements in Little Auk (*Alle alle*) chicks. Functional Ecology 6: 137–144.

TIELEMAN, B. I., AND J. B. WILLIAMS. 2000. The adjustment of avian metabolic rates and water fluxes to desert environments. Physiological and Biochemical Zoology 73: 461–479.

VISSER, G. H. 1998. Development of temperature regulation. Pp. 117–156 *in* Avian Growth and Development. Evolution within the Altricial-Precocial Spectrum (J. M. Starck and R. E. Ricklefs, Eds.). Oxford Ornithology Series. Oxford University Press, Oxford.

VISSER, G. H., AND H. SCHEKKERMAN. 1999. Validation of the doubly labeled water method in growing precocial birds: the importance of assumptions concerning evaporative water loss. Physiological and Biochemical Zoology 72: 740–749.

VISSER, G. H., P. E. BOON, AND H. A. J. MEIJER. 2000a. Validation of the doubly labeled water method in Japanese Quail (*Coturnix c. japonica*): is there an effect of growth rate? Journal of Comparative Physiology B 170: 365–372.

VISSER, G. H., A. DEKINGA, B. ACHTERKAMP, AND T. PIERSMA. 2000b. Ingested water equilibrates isotopically with the body water pool of a shorebird with unrivaled water fluxes. American Journal of Physiology 209: R1795–R1804.

VOLKMAN, N. J., AND W. TRIVELPIECE. 1980. Growth in pygoscelid penguin chicks. Journal of Zoology, London 191: 521–530.

WARHAM, J., J. WATTS, AND R. J. DAINTY. 1976. The composition, energy content and function of the stomach oils of petrels (Order Procellariiformes). Journal of Experimental Marine Biology and Ecology 23: 1–13.

WEATHERS, W. W. 1992. Scaling nestling energy requirements. Ibis 134: 142–153.

WEATHERS, W. W. 1996. Energetics of postnatal growth. Pp. 461–496 *in* Avian Energetics and Nutritional Ecology (C. Carey, Ed.). Chapman and Hall, New York.

WEATHERS, W. W., AND K. A. SULLIVAN. 1991. Growth and energetics of nestling Yellow-eyed Juncos. Condor 93: 138–146.

WEATHERS, W. W., AND K. A. SULLIVAN. 1993. Seasonal patterns of time and energy allocation by birds. Physiological Zoology 66: 511–536.

WESTERTERP, K. R. 1973. The energy budget of the nestling Starling (*Sturnus vulgaris*): a field study. Ardea 61: 137–158.

WESTERTERP, K. R., AND D. M. BRYANT. 1984. Energetics of free existence in swallows and martins (Hirundinidae) during breeding: a comparative study using doubly labeled water. Oecologia 62: 376–381.

WILLIAMS, J. B. 1987. Field metabolism and food consumption of the Savannah Sparrow during the breeding season. Auk 104: 277–289.

WILLIAMS, J. B., AND K. A. NAGY. 1985. Water flux and energetics of nestling Savannah Sparrows in the field. Physiological Zoology 58: 515–525.

WILLIAMS, J. B., AND A. PRINTS. 1986. Energetics of growth in nestling Savannah Sparrows: a comparison of doubly labeled water and laboratory estimates. Condor 88: 74–83.

YDENBERG, R. C. 1989. Growth-mortality trade-offs and the evolution of juvenile life histories in the Alcidae. Ecology 70: 1494–1506.

Sooty Terns Flying out from Colony to Drink Seawater

14 Water and Salt Balance in Seabirds

David L. Goldstein

CONTENTS

14.1 INTRODUCTION

The regulation of a normal balance of body water and salts (osmoregulation, or osmotic homeostasis, Table 14.1) is one of the fundamental physiological demands on vertebrates. The volume of body fluids contributes to the maintenance of blood pressure and flow, and the concentrations of salts contribute to the proper ionic environment for biomolecules like enzymes, as well as to the regulation of cellular volume and electrical balance. The mechanisms by which marine birds achieved this balance remained unresolved for many years, based on inadequate knowledge concerning routes of input (e.g., do marine birds drink?) and output (e.g., what are the excretory organs and their capacities?). Much of this historical background is summarized in Schmidt-Nielsen's seminal paper describing for the first time the function of the supra-orbital salt-secreting glands (Schmidt-Nielsen et al. 1958).

Today, we have a much more complete understanding about the roles played by the various organs of osmoregulation, including many details of physiological control mechanisms. We also

TABLE 14.1
Typical Values of Some Regulated Osmoregulatory Variables in Adult Marine Birds

Variable	Value	Reference
Total body water	60–65% of body mass[*]	Mahoney and Jehl 1984
Exchangeable sodium pool	40–50 mmol/kg body mass	Green and Gales 1990
Hematocrit	45–50%	Work 1996
Plasma osmolality	315 mmol/kg	Gray and Erasmus 1988
Plasma Na[+]	150 mmol/l	Gray and Erasmus 1988
Plasma uric acid	5–10 mg/dl	Work 1996

[*] See text for exceptions.

have quantitative data on many aspects of osmoregulation in the field, including rates of water turnover, estimates of drinking rate, and urine composition for selected species. In the present overview I will summarize, for each of the regulatory pathways, what we know of the system in birds generally and how marine birds might compare with this general model. I will also summarize what is known of function and integration of the osmoregulatory system in marine birds in the wild.

14.2   INPUTS

Water can become available to marine birds via one of three routes (Figure 14.1). First, birds gain pre-formed water in their food. Second, birds may drink, either intentionally or incidentally, along with feeding. Last, water is produced as an end product of oxidative metabolism. For the body, it does not matter which of these routes is the source of its water. Ions enter a marine bird either through food or drinking water.

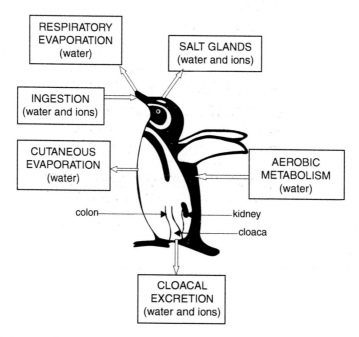

FIGURE 14.1 Pathways of water and ion influx and efflux in marine birds.

14.2.1 Food

Marine birds generally consume fresh foods, either vertebrate (fish) or invertebrate (Figure 14.2). In either case, the water content of the food is relatively high, 70 to 80% (Table 14.2; Green and Gales 1990). The mix of these two types of food items (fish and invertebrates) may vary within a species or population over time (e.g., Gales and Green 1990) or between populations (Cooper et al. 1990). The choice of food has implications for the relative intakes of water and sodium. In general, fish are osmoregulators; body fluid osmotic and ionic concentrations are similar to those of terrestrial animals, with osmolarity about 300 to 400 mmol/l and extracellular Na^+ about 150 mmol/l. In contrast, marine invertebrates are typically osmoconformers, with body fluid concentrations similar to seawater, and two to three times those of fish (Withers 1992). Because of the regulation of specific ions and the varying contributions of body components (e.g., skeleton, intracellular and extracellular fluids) to fresh mass, the sodium content of marine invertebrates may not differ markedly from that of fish when expressed per unit fresh prey mass. Nevertheless, the intake of sodium in the food relative to energy intake is probably two to three times higher for birds consuming invertebrates (on the order of 25 mmol Na^+ per MJ) than for those consuming fish (about 10 to 15 mmol Na^+ per MJ; Table 14.2; Green and Gales 1990). Feeding on more salty foods may be reflected in adjustments of physiology such as enlargement of the salt glands (Ensor and Phillips 1972).

The food fed to chicks may differ significantly in water and salt content from that consumed by adults. Even in birds that regurgitate relatively whole food to their chicks, as opposed to stomach oils, the composition of regurgitated foods may be altered while in the adult gut, so that chicks may receive foods with reduced concentrations of Na^+ (Janes 1997).

14.2.2 Drink

Because of the high water content of their foods, it is unlikely that marine birds require drinking water for survival. Indeed, penguins fed fish in the laboratory can maintain body mass for at least a week, and presumably longer, without access to drinking water (Adams 1984, Gales 1989). In the field, too, a variety of birds that encounter saline waters may have mechanisms to avoid ingestion of that water. In Wilson's Phalaropes (*Phalaropus tricolor*) and American Avocets (*Recurvirostra americana*), for example, the morphology of the tongue and palate may provide a mechanism for filtering water out of the mouth before ingestion of prey (Mahoney and Jehl 1985a). In Adelie Penguins (*Pygoscelis adeliae*), a comparison of the composition of krill fed to chicks vs. fresh samples of the same food items suggests that adults might be able to remove superficial adherent water from prey items prior to swallowing (Green and Gales 1990). Similarly, California Gulls (*Larus californicus*) feeding at Mono Lake had stomach contents with osmolalities of about 635 mosmol/kg, similar to their brine shrimp prey and just one quarter as concentrated as the lake water, suggesting minimal intake of the hypersaline waters (Mahoney and Jehl 1985b).

Nevertheless, marine birds do ingest seawater. In 1958, Schmidt-Nielsen postulated that "It would be extremely difficult to prove whether oceanic birds regularly use sea water as a normal water supply" (p. 101). More recently, techniques have become available which do allow estimation of seawater drinking rates by free-living marine birds. Green and colleagues (Green and Brothers 1989, Gales and Green 1990) have accomplished this by simultaneously measuring the fluxes of water and of Na^+, based on the turnover of injected isotopic compounds (^{3}H-labeled water and ^{22}Na). By an iterative procedure that accounts for intakes of sodium and water from food as well as drink, it is possible to calculate the volume of seawater ingested. Seawater contributed approximately 8.5% of total water influx in Common Diving Petrels (*Pelecanoides urinatrix*), 17% in Fairy Prions (*Pachyptila turtur*), and between 6 and 30% (though usually less than 10%) in Blue Penguins (*Eudyptula minor*). Seawater ingestion also contributed 20 to 35% of total sodium intake.

(a)

(b)

FIGURE 14.2 (a) Brown Pelican chicks eat fish which they take from the parents gullet as the adult regurgitates. (b) This Laysan Albatross chick is being fed a rich oil which the adult squirts directly into its throat. Chicks also receive pieces of fish and squid. (c) Common Tern chicks are fed whole fish which their parents offer to them in the bill. This adult is waiting for its chick to come out of hiding and feed. (Photos a and b by E. A. Schreiber; photo c by J. Burger.)

Birds in saline environments may make use of fresh water if it is available. California Gulls at Mono Lake regularly visited freshwater sources along the lake shore, though they may not have needed to do so to maintain osmotic balance (Mahoney and Jehl 1985b). Even when inhabiting sites without access to standing fresh water, Laysan Albatrosses (*Phoebastria immutabilis*) acquire

(c)

FIGURE 14.2 *Continued.*

TABLE 14.2
Composition of Some Foods Consumed by Marine Birds

Food	Water (ml/kg)	Sodium (mmol/kg)	Energy (MJ/kg)	mmol Na/MJ
Fish	697	110	8.6	12.8
Squid	788	122	4.6	26.5
Krill	791	124	4.3	28.7

Note: Values are means taken from data compiled by Green and Gales (1990).

fresh water by ingesting rainwater as it falls or runs off of their beaks (Rice and Kenyon 1962, Pettit et al. 1988).

14.2.3 METABOLISM

The amount of water produced as an end product of metabolism varies with the nutrient composition of the food being oxidized, being highest for fat (about 1 ml/g) and lower (about 0.5 ml/g) for protein or carbohydrate (Withers 1992). For seabirds eating fish or invertebrates, metabolic water may provide 10 to 15% of the total water available from ingested food (Gales 1989).

14.3 OUTPUTS

14.3.1 EVAPORATION

Birds lose evaporative water by two pathways, "insensible" evaporation through the skin (birds lack sweat glands and so have no cutaneous route of ion loss) and water vapor contained in exhaled air (Figure 14.1). The general principles involved in these avenues of water loss are well understood.

For cutaneous evaporation, the driving force for water vapor flux is a gradient of vapor density, and such a gradient will essentially always exist. Ambient vapor densities can approach those of body fluids only if the ambient air is warm and fully humidified. In addition to the vapor density gradient, the actual rate of evaporative water loss depends on the surface area exposed and on the combined resistance to vapor diffusion of the surface tissues, feathers, and air-boundary layer (see Withers 1992).

Respiratory evaporative water loss occurs because air is warmed and humidified as it is inspired. Exhaled air is usually saturated with water vapor (Withers et al. 1981), and so even after it passes over the cooled nasal surfaces during expiration the result is a net loss of water.

FIGURE 14.3 A Masked Booby excretes prior to flying from nest. (Photo by B. A. Schreiber.)

Although we understand these principles, there is much about evaporative water loss that remains unquantified. We know little about the extent to which rates of either cutaneous or respiratory evaporation are regulated in birds. We also have little idea about the way in which water loss is partitioned between excretion and evaporation in the field (Giladi and Pinshow 1999), or how evaporative water loss is partitioned between cutaneous and respiratory components. It is highly likely that such partitioning varies under changing circumstances. Certainly birds that are experiencing thermal stress, as may occur for marine birds during incubation or while nestlings, can enhance evaporation by panting (rapid, shallow breathing) or gular flutter (fluttering of the throat region so as to create air flow over the moist, vascular tissues) (e.g., Lasiewski and Snyder 1969, Bartholomew and Dawson 1979). However, the pattern of evaporative water loss in relation to ambient temperature does vary among species (Morgan et al. 1992). A bird that has demands for water conservation (an albatross flying over long distances, or a penguin on an extended incubation shift, for example) may well have strategies to minimize evaporative water loss. The mechanisms by which these responses may occur — whether by changing patterns of respiration or changes in skin permeability to water vapor — and the extent to which evaporation can be reduced remain largely unstudied in birds generally, and entirely so in marine birds.

14.3.2 EXCRETION

Birds possess several excretory organs. In all birds, a pair of kidneys produces urine that serves to remove wastes, including excess water and ions, from the plasma. The urine drains from the kidneys into the cloaca, the common opening of the digestive and reproductive systems (Figure 14.3). From there it may move by reverse peristalsis into the colon, which has the potential to substantially modify the urine prior to excretion. An additional set of excretory organs, the paired salt glands, are present in a subset of birds, including all marine families. These glands are primarily responsible for excreting NaCl in a concentrated solution, and so with relatively little water loss. All three of these organs are regulated by a combination of neural and humoral factors that respond to changes in osmoregulatory balance.

14.3.2.1 Kidneys

The basic functional unit and principles of operation of the avian kidney are similar to those of other vertebrates. That is, birds have nephrons consisting of a filtering apparatus (the glomerular capillaries) and a tubule that can modify the filtered fluid by a combination of reclaiming urinary constituents back to the extracellular fluid (reabsorption) and adding materials to the tubular urine (secretion). Birds generally have a large number of nephrons with small glomeruli. However, within a single kidney the morphology of these nephrons is quite heterogeneous; some resemble mammalian nephrons in that they possess loops of Henle, whereas others lack these loops and more closely resemble reptilian nephrons. The proportions of looped and unlooped nephrons may vary among species (Goldstein and Braun 1989). The loops of Henle confer upon birds the capacity, as in mammals, to produce a urine more concentrated than the plasma.

Overall, the morphology of kidneys from marine species resembles that of other birds. The kidney has three divisions, though, as in other groups of birds, the distinctness of the middle division varies among species, even within orders (e.g., some procellariiforms have distinct middle divisions, whereas others do not; Johnson 1968). Marine birds and others with salt glands generally have larger kidneys (>1% of body mass) than do birds without salt glands (<1% of body mass; Hughes 1970).

Within the kidney, the avian renal medulla, containing the loops of Henle along with the collecting ducts, is divided into subunits known as medullary cones. The number of medullary cones relative to kidney size is not obviously unusual in marine birds, but the proportion of kidney mass accounted for by medullary tissue and the lengths of the medullary cones both appear high in the marine species (Goldstein and Braun 1989, Goldstein 1993).

In principle, one might expect that a higher proportion of medullary tissue and longer medullary cones would translate into an enhanced urine concentrating ability. However, it has not been demonstrated that this is the case. It does appear that marine birds often produce urine significantly hyperosmotic to the plasma (Table 14.3). For example, hydrated Jackass Penguins (*Spheniscus demersus*) infused in the laboratory with physiological saline produced urine with a concentration of 650 mmol/kg (Oelofsen 1973), and Adelie Penguin chicks and adults in the field had urine osmolalities averaging about 800 mmol/kg (Janes 1997). Likewise, chicks of Leach's Storm-petrel (*Oceanodroma leucorhoa*) induced into antidiuresis in the laboratory by infusion of hypertonic saline had urine concentrations of 840 mmol/kg. In the field these chicks, which are fed sporadically in their burrows and so are likely to be conserving water, had urine osmolality of 735 mmol/kg (Goldstein 1993). Although all of these values are two to three times the plasma osmolality, it is not clear that the urine concentrating ability per se — i.e., the maximum urine osmolality — is higher among marine birds than is expected based on the relation between maximal urine concentration and body mass in birds generally (Goldstein and Braun 1989).

One of the mechanisms by which birds can effect a reduced urine flow and increased concentration is to modulate the rate of initial urine filtration. Birds, including marine birds (Oelofsen 1973), have a renal portal system that can deliver blood flow to the renal cortex without its first passing through the glomeruli. Consequently, birds can reduce the rate of glomerular filtration (GFR) while maintaining perfusion of the renal tubules and continued delivery of urate for secretion. The reduced GFR in turn reduces the flow of urine. Marine birds, with varying sensitivity, respond to water deprivation or saline acclimation with increases in circulating levels of the antidiuretic hormone, arginine vasotocin (Figure 14.4, Gray and Erasmus 1988, 1989). Saline acclimation itself does not necessarily induce any changes in GFR (Hughes 1980, 1995), which can induce diminished GFR. However, chicks of Leach's Storm-petrels can reduce the GFR by 85% when infused with hypertonic saline (compared with the condition when infused with hypotonic saline). In combination with an enhanced tubular reabsorption of water, they can reduce urine flow by more than 98% (Goldstein 1993). GFR and urine flow are variable in terrestrial birds under natural conditions, too, being reduced during the overnight inactive period (Goldstein and Rothschild 1993) and during flight (Giladi et al. 1997). It seems likely that such strategies would be used by marine birds during times of water scarcity (incubation, flight, intermeal intervals of nestlings), but studies are lacking.

Other than the osmolality, we have little information on urine composition in wild marine birds (Table 14.3). Only a small proportion of urinary osmolality was accounted for by Na^+ and K^+ in Leach's Storm-petrels and Adelie Penguins in the field (Goldstein 1993, Janes 1997).

TABLE 14.3
Urine Composition of Marine Birds in the Field

Species	Body Mass (g)	Age	Osmolality (mmol/kg)	Na+ (mmol/l)	K+ (mmol/l)	Cl- (mmol/l)	NH4+ (mmol/l)	Reference
Oceanodroma leucorhoa	54	Chick	735	72	19			Goldstein 1993
Larus californicus	~650	Adult	492	86	7			Mahoney and Jehl 1985
L. californicus	?	Chick	475	89	27			Mahoney and Jehl 1985
Pygoscelis adeliae	~4000	Adult	777	81	30	235	270	Janes 1997
P. adeliae	?	Chick	805	20	98	68	237	Janes 1997

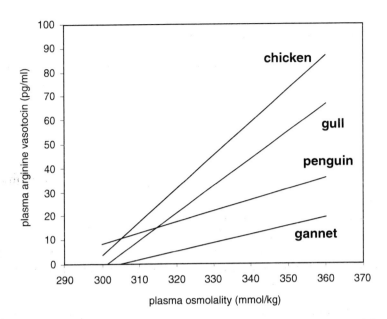

FIGURE 14.4 Relation between circulating concentration of the antidiuretic hormone arginine vasotocin and plasma osmolality in three marine birds (the Kelp Gull, *Larus dominicanus;* Cape Gannet, *Morus capensis;* and Jackass Penguin, *Spheniscus demersus*) and the domestic chicken *Gallus gallus*. Note that neither the circulating concentrations of hormone nor the sensitivity of hormone release to changes in plasma osmolality are the same among species, or among marine species. Data taken from Gray and Erasmus (1988) and Stallone and Braun (1986).

In addition to ion excretion, another important function of the avian kidney is elimination of nitrogenous waste. In marine birds as in others (Janes 1997), nitrogen is excreted primarily as uric acid and its salts (urates). Marine birds have diets with high protein content, and so must presumably excrete large loads of urate. In avian urine, urates form a colloidal mass that protects against uric acid crystallization within the renal tubules. This colloid incorporates substantial amounts of protein (Janes and Braun 1997), and Oelofsen (1973) noted copious amounts of mucin in the urine of Jackass Penguins. Moreover, terrestrial birds placed on high protein diets typically have high rates of urine flow (McNabb et al. 1972, Ward et al. 1975). Renal (and postrenal; see below) handling of nitrogen, and the ability to limit the losses of urinary protein and water, might be especially critical in special circumstances like the molting fast of penguins (Cherel and Freby 1994). The implications of the high-protein diet for osmoregulation in marine birds have not been studied.

14.3.2.2 Intestines

The avian lower intestine (colon) receives inputs from both the small intestine and from the kidneys, which empty via the ureters into the coprodeum of the cloaca. As such, the lower intestine serves as an integrator of excretory loss of water and ions. Marine birds have just vestigial digestive ceca (McLelland 1989) that are probably unimportant in osmoregulation. This integration has been studied most intensively in the chicken, and has been reviewed several times (e.g., Skadhauge 1981, Goldstein 1989). Schmidt-Nielsen et al. (1963) postulated that in birds with salt glands (see below) the lower intestine should maintain a high capacity for salt reabsorption even during times of high salt intake, and thereby cycle the ions to the salt glands for excretion. A few studies of intestinal osmoregulation have been made in gulls and ducks, both of which have functional salt glands and so may illustrate patterns that pertain in other marine species. In gulls, the lower intestine does have a relatively high area-specific capacity for salt absorption, even in birds drinking saline (Goldstein et al. 1986). However, this organ has a small surface area, and it is likely that the small

FIGURE 14.5 Salt gland secretions dripping from the beak of a Northern Fulmar. The glands, whose position is shown in outline on the head of the bird, are largest in birds like fulmars that feed at sea. They secrete a saline solution up to twice the concentration of seawater, thereby eliminating NaCl that is absorbed from the diet. Ducts from the glands empty into the anterior nasal cavity, from where the secretion flows out through the nares and is shaken off or drips from the beak. (Drawings by J. Zickefoose.)

intestine (Roberts and Hughes 1984) and kidneys (Goldstein et al. 1986) absorb most of the salt, which is excreted by the salt glands without the ions having reached the lower intestinal epithelium. Domestic ducks have a larger lower intestine, but its capacity for salt uptake is reduced in birds with high salt intake (Skadhauge et al. 1984). The only data available for true marine species appear to be the observation that in Adelie Penguin chicks, the ion concentrations in cloacal fluid were not different from those in ureteral urine (Janes 1997), again suggesting perhaps a minor role for postrenal modification of the urine.

14.3.2.3 Salt Glands

Supra-orbital salt-secreting glands (Figure 14.5) are present in approximately ten of the avian orders. Biologists knew for many years of the presence of these glands, that they are largest in marine species, and that marine birds secrete a fluid from the orbital or nasal region (see historical summary in Peaker and Linzell 1976). However, not until the elucidation of the function of these glands in the late 1950s (Schmidt-Nielsen et al. 1958) was it evident that they helped solve the problem of how marine birds cope with their high intake of salt (Krogh 1939). Since that discovery, the anatomy and physiology of the salt glands has been studied in great detail (see Peaker and Linzell 1976, Gerstberger and Gray 1993 for thorough reviews). The most common model for these studies has been the domestic Mallard Duck, *Anas platyrhynchos*. However, there may be variation among species in the details of salt gland function, and little of the recent work on salt glands has been accomplished using wild species that are habitually exposed to seawater.

Avian salt glands secrete solutions that are nearly pure NaCl, with small amounts of other ions (e.g., K^+, HCO_3^-; see Schmidt-Nielsen 1960). The stimulation for salt-gland secretion involves some combination of rises in extracellular fluid volume and osmolality (see Hughes 1989); salt-gland secretion may (Stewart 1972) or may not (e.g., Schmidt-Nielsen et al. 1958) occur in response to water deprivation leading to a rise in osmolality with a contraction of volume. The concentration and flow rate of secretions from the salt glands vary between species and increase as the glands hypertrophy in response to exposure to elevated salt intake. Similar gland hypertrophy may occur even in the absence of an increasing salt ingestion as part of an annual cycle of physiological adjustment (Burger and Gochfeld 1984). The most concentrated salt gland secretions (about 1000 mM Na^+, twice the concentration of seawater) have been measured in procellariiforms (Schmidt-Nielsen 1960). Following a salt load, the salt glands may excrete the great majority of the NaCl (e.g., Bøkenes and Mercer 1995), while the kidneys are responsible for the elimination of other components of the load (other ions, water).

14.4 INTEGRATION

14.4.1 WATER

14.4.1.1 Total Body Water

The object of osmoregulation is to maintain a body-water volume and composition within some normal range. The homeostatic body-water volume is that required to maintain normal intracellular and extracellular volumes. Total body water is measured either from the dilution of an injected volume of labeled water or from desiccating a carcass to constant mass. For the purposes of comparison, these volumes are commonly expressed in terms of the fraction of body mass consisting of water (percent body water). This percent varies not only as a function of the amount of water in the body, but also as a function of the amount of nonwater components. In particular, adipose tissue, bone, and feathers are tissues with low water content; a high proportion of any of these tissues will result in low values of percent body water. Thus, for example, even if the absolute total body water does not change, the percent body water may vary seasonally as body fat is gained and used (Williams et al. 1977, Gales et al. 1988).

For terrestrial birds, total body water (TBW) averages approximately 60% of body mass (Skadhauge 1981). Mahoney and Jehl (1984) compiled data for 22 aquatic and marine bird species (mostly charadriiforms) and found that for these species, too, TBW was typically near 60% (range 57 to 70%). Penguins (summarized by Green and Gales 1990) and some procellariiforms (the Northern Fulmar, *Fulmarus glacialis;* Mahoney and Jehl 1984) and the Gray-headed Albatross (*Thalassarche chrysostoma*; Costa and Prince 1987) also have TBW about 60% of body mass. In contrast, TBW in a variety of other procellariiform birds, ranging in size from Wilson's Storm-petrel (*Oceanites oceanicus*) to Wandering Albatross (*Diomedea exulans*), is substantially lower, 44 to 52% (Adams et al. 1986, Pettit et al. 1988, Arnauld et al. 1996). The explanations offered for these low TBW values include a high proportion of body mass in feathers, a high fat content, and perhaps older age (birds tend to become less "watery" with aging, and albatrosses are particularly long lived). It has not been determined which of these explanations is actually correct.

14.4.1.2 Water Turnover Rate

The total turnover rate (influx and efflux) of water through a bird can be evaluated by injecting isotopic water (water enriched in either deuterium [^2H] or tritium [^3H]) and then following the decline in isotope specific activity as the heavy water is lost to the environment and is replaced by "normal" water. The water turnover rate (flux) integrates all of the avenues of water entry and loss. The water flux is not itself a regulated variable, but integrates the function of the other physiological systems and provides insight into the overall balance of these systems.

Water flux of free-living birds increases as a function of body mass (Nagy and Peterson 1988, Williams 1993). The slope (exponent) of this relation in seabirds (0.65; Figure 14.6) is similar to that for birds generally (0.698; Williams 1993). However, it is difficult at this point to say how the magnitude of the water fluxes (the y-intercept in Figure 14.6) compares in seabirds with other birds because there are too few data for the proper comparison group (nonpasserine, nondesert terrestrial species between 100 and 10,000 g body mass). It is interesting to note that all of the charadriiform species for which there are data have water fluxes slightly on the high side (Figure 14.6); these data include species that inhabit a combination of marine and freshwater habitats. Seabirds that

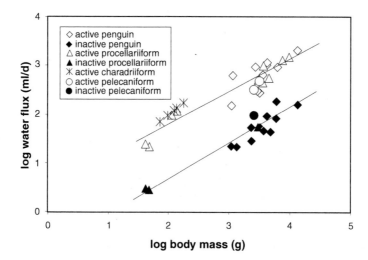

FIGURE 14.6 Water efflux (ml/d) in a variety of active and inactive (brooding or incubating) marine birds. Equations for regression lines are: log Y = 0.51 + 0.65 log X, N = 29, r^2 = 0.88 (active); and log Y = –0.77 + 0.72 log X, N = 16, r^2 = 0.91 (inactive). The charadriiform data include two species (American Avocet, Ruddy Turnstone) that are not truly marine and nestlings of another species (Glaucous-winged Gull). Sources for data are Adams et al. (1986), Adams et al. (1991), Birt-Friesen et al. (1989), Costa and Prince (1987), Costa et al. (1986), Davis et al. (1989), Gabrielsen et al. (1987, 1991), Gales et al. (1988), Green and Gales (1990), Green et al. (1988), Hotker et al. (1996), Hughes (1984), Kooyman et al. (1982), Nagy et al. (1984), Obst et al. (1987), Pettit et al. (1988), Piersma and Morrison (1994), Ricklefs et al. (1986), and Roby and Ricklefs (1986).

are relatively inactive while confined to land (during incubation or brooding) have water fluxes less than one tenth those of active animals (Figure 14.6).

14.4.2 SODIUM

14.4.2.1 Exchangeable Sodium Pool Size

As with the regulation of total body water, the total pool of exchangeable sodium ions in an animal is maintained at a homeostatic level that provides a proper concentration of sodium in the intracellular and extracellular fluids. The exchangeable-sodium pool size can be measured from the dilution of an injected amount of labeled sodium (^{22}Na).

In terrestrial birds, the exchangeable Na pool size appears to be generally 40 to 50 mmol/kg body mass (Goldstein and Bradshaw 1998). In marine species, the values are similar (Roberts and Hughes 1984, Green and Brothers 1989, Green and Gales 1990). The Na pool may be somewhat higher in penguin chicks than in adults (summarized in Green and Gales 1990), probably associated with the higher body water fraction in younger animals. No data are available for the procellariiform species cited above as having low total-body waters; given that these species maintain concentrations of Na$^+$ in their body fluids that are similar to those of other birds, they presumably should have low Na$^+$ pool sizes as well.

14.4.2.2 Sodium Flux

Sodium enters a bird by two routes, food and drink. The sodium influx is thus determined by the intake rate and sodium content of food and drink. In general, food intake rates will be greater for larger species, with their larger energy demands, but this is not necessarily the case for drinking. As noted above, the sodium content of marine invertebrate food is likely to be higher than that of vertebrate prey items.

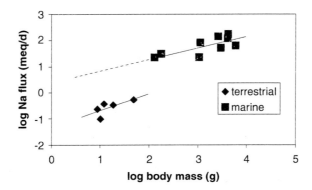

FIGURE 14.7 Sodium influx (meq/d) as a function of body mass in a variety of terrestrial and marine birds. The data include Na taken in through both food and water for marine birds. Solid lines are least-squares linear regressions: $\log Y = -1.36 + 0.66 \log X$, $N = 5$, $r^2 = .51$ ($p = 0.18$ for this small set of data points) for terrestrial species, and $\log Y = 0.49 + 0.41 \log X$, $N = 9$, $r^2 = 0.54$ ($p = 0.02$) for marine species. The dashed line extends the marine regression to the range of body masses for which data are available for nonmarine species. Sources of data are Goldstein and Bradshaw (1998), Ambrose and Bradshaw (1988), Green and Brothers (1989), Green and Gales (1990), and Green et al. (1988).

The available data on sodium fluxes in the field are limited, at least in part because of constraints on using ^{22}Na in field situations. Nevertheless, some patterns do emerge from the available data (Goldstein and Bradshaw 1998; Figure 14.7). First, Na flux increases with body mass. Second, marine species appear to have higher Na fluxes than terrestrial species, though the data come from species with nonoverlapping body masses. Last, when Na fluxes are compared with water fluxes (with correction for drinking water), consumers of marine invertebrates appear to have Na influxes about twice those of piscivores (Goldstein and Bradshaw 1998).

14.4.3 VARIATIONS WITHIN AND BETWEEN SPECIES

The above discussion has focused principally on general patterns in osmoregulation. It is worth making explicit note, too, of variations within these themes. First, it is clear that "marine birds" do not form a homogeneous group of species with respect to osmoregulatory biology. Certainly some of the challenges facing these species are quite widespread (high levels of salt intake or extended periods without fresh water, for example), and some of the response systems (like the presence of salt-secreting glands) are ubiquitous. Nevertheless, quite basic aspects of morphology and physiology — what is the stimulus for initiating salt gland secretion, for example, or how does the lower intestine respond to salt intake, or what is the sensitivity for release of osmoregulatory hormones — may differ among species. The domestic duck is commonly used as a laboratory model for studying the physiology of "birds with salt glands." Many physiological systems remain poorly studied in most groups of marine birds, and these systems might be poorly modeled by the Mallard Duck.

Osmoregulatory challenges and responses may also vary substantially through time for marine birds, as seasons and ontogeny progress. For adults, alternating periods of inactivity (incubation, brooding) and activity, and changing activity levels associated with caring for growing young, induce marked variation in throughputs of food and water (Figure 14.1; see also summary in Green and Gales 1990, and Adams et al. 1991). For example, incubating Red-tailed Tropicbirds (*Phaethon rubricauda*) lose 30 to 40 g body mass per day and incur rises in plasma osmolarity and hematocrit as high as 16 to 25% (S. Mahoney and E. A. Schreiber unpublished). It is possible that the most severe challenges to osmoregulation may come from dehydration of nest-bound individuals, rather than from dietary salt intake.

Chicks and nestlings of marine birds have been studied much less than adults. Nestlings are likely to have fewer behavioral options than adults, requiring them to respond to osmoregulatory challenge with a physiological response. Yet, the sizes and capacities of the organs of osmoregu-lation, such as the salt glands, kidneys, and adrenals, may change with age, such that the ability to excrete a salt load or to tolerate water deprivation changes, too (e.g., Douglas 1968, Hughes 1984). A limited ability to excrete salt loads may render nestlings particularly sensitive to the salt contents of their diets (Dosch 1997), and for some species this may force parents to forage far afield to procure low-salt foods (Johnston and Bildstein 1990).

14.4.4 APPLICATIONS

Water and sodium fluxes integrate many of the homeostatic demands on marine birds in their natural habitats. Food consumption, temperature regulation, activity level, and microhabitat selection all contribute to determining the magnitude of osmoregulatory fluxes. Conversely, quantifying water and sodium balance can provide insight into the ecology and health of the animal.

As noted above, the combination of water and sodium fluxes has provided a means of assessing rates of seawater consumption by foraging marine birds. If the diet items contain different amounts of sodium, as is likely for marine vertebrate and invertebrate prey items, then the combination of sodium and water fluxes may also provide an estimate of the relative contributions of the two prey types (Goldstein and Bradshaw 1998).

In addition to indicating aspects of normal physiological ecology, osmoregulatory balance can also reflect the health of animals in potentially pathological situations. Marine birds that come in contact with spilled oil may acquire the oils on their surface or by ingestion (Boersma 1986), which could influence salt and water balance (though, depending on the exposure, some organs might continue to function normally; see McEwan 1978). Externally acquired oil has the potential to disrupt the oil layers on the feathers and the skin, thereby altering the potential for wetting or evaporation (Hughes et al. 1990). Ingested oils can alter the morphology and/or physiology of the intestinal mucosa in such a way as to modify the capacities for salt absorption (Crocker et al. 1974, 1975). In either case, normal osmotic regulation may be disturbed. Other contaminants, such as heavy metals, may also be excreted by and accumulated in the kidneys and salt glands (Burger et al. 2000), though studies are lacking as to the effect of these metals on the function of the excretory organs.

Marine birds face many situations that seem extreme: prolonged fasts, extended nestling duration with intermittent feeding, long-distance migrations, high salt intake. In the face of these demands, the balance of water and ions may be challenged or compromised, and net losses of body fluids may constrain behavior and physiology. Marine birds have a diversity of osmoregulatory strategies available under these circumstances. Future studies will help define their implementation under varying circumstances, the limits to their physiological capacities, and the extent to which osmoregulatory physiology informs bird behavioral and ecological decisions.

ACKNOWLEDGMENTS

Thanks to those who have made my work with seabirds possible, especially to Maryanne Hughes and Allen Place. Some of my work was supported by the National Science Foundation (IBN-9982985).

LITERATURE CITED

ADAMS, N. J. 1984. Utilization efficiency of a squid diet by adult king penguins (*Aptenodytes patagonicus*). Auk. 101: 884–886.

ADAMS, N. J., C. R. BROWN, AND K. A. NAGY. 1986. Energy expenditure of free-ranging wandering albatrosses *Diomedea exulans*. Physiological Zoology 59: 583–591.

ADAMS, N. J., R.W. ABRAMS, W. R. SIEGFRIED, K. A. NAGY, AND I. R. KAPLAN. 1991. Energy expenditure and food consumption by breeding Cape gannets *Morus capensis*. Marine Ecology Progress Series 70: 1–9.

AMBROSE, S. J., AND S. D. BRADSHAW. 1988. The water and electrolyte metabolism of free-ranging and captive white-browed scrubwrens, *Sericornis frontalis* (Acanthizidae), from arid, semi-arid and mesic environments. Australian Journal of Zoology 36: 26–51.

ARNOULD, J. P. Y., D. R. BRIGGS, J. P. CROXALL, P. A. PRINCE, AND A. G. WOOD. 1996. The foraging behaviour and energetics of wandering albatrosses brooding chicks. Antarctic Science 8: 229–236.

BARTHOLOMEW, G. A., AND W. R. DAWSON. 1979. Thermoregulatory behavior during incubation in Heerman's gulls. Physiological Zoology 52: 422–437.

BIRT-FRIESEN, V. L., W. A. MONTEVECCHI, D. K. CAIRNS, AND S. A. MACKO. 1989. Activity-specific metabolic rates of free-living northern gannets and other seabirds. Ecology. 70: 357–367.

BrKENES, L., AND J. B. MERCER. 1995. Salt gland function in the common eider duck (*Somateria mollissima*). Journal of Comparative Physiology B 165: 255–267.

BOERSMA, P. D. (1986). Ingestion of petroleum by seabirds can serve as a monitor of water quality. Science 231: 373–376.

BURGER, J., AND M. GOCHFELD. 1984. Seasonal variation in size and function of the nasal salt gland of the Franklin's gull (*Larus pipixcan*). Comparative Biochemistry and Physiology 77A: 103–110.

BURGER, J., C. D. TRIVEDI, AND M. GOCHFELD. 2000. Metals in herring and great black-backed gulls from the New York Bight: the role of salt gland in excretion. Environmental Monitoring and Assessment 64: 569–581.

CHEREL, Y., AND F. FREBY. 1994. Daily body-mass loss and nitrogen excretion during molting fast of Macaroni penguins. Auk 111: 492–495.

COOPER, J., C. R. BROWN, AND R. P. GALES. 1990. Diets and dietary segregation of crested penguins (*Eudyptes*). Pp. 131– 147 *in* Penguin Biology (L. S. Davis and J. T. Darby, Eds.). Academic Press, San Diego.

COSTA, D. P., AND P. A. PRINCE. 1986. Foraging energetics of grey-headed Albatrosses *Diomedea chrysostoma* at Bird Island, South Georgia. Ibis 129: 149–158.

COSTA, D. P., P. DANN, AND W. DISHER. 1986. Energy requirements of free-ranging little penguin, *Eudyptula minor*. Comparative Biochemistry and Physiology 85A: 135–138.

CROCKER, A. D., J. CRONSHAW, AND W. N. HOLMES. 1974. The effect of crude oil on intestinal absorption in ducklings (*Anas platyrhynchos*). Environmental Pollution 7: 165–177.

CROCKER, A. D., J. CRONSHAW, AND W. N. HOLMES. 1975. The effect of several crude oils and some petroleum distillation fractions on intestinal absorption in ducklings (*Anas platyrhynchos*). Environmental Physiology and Biochemistry 5: 92–106.

DAVIS, R. W., J. P. CROXALL, AND M. J. O'CONNELL. 1989. The reproductive energetics of gentoo *Pygoscelis papua* and macaroni *Eudyptes chrysolophus* penguins at South Georgia. Journal of Animal Ecology 58: 59–74.

DOSCH, J. J. 1997. Salt tolerance of nestling laughing gulls: an experimental field investigation. Colonial Waterbirds 20: 449–457.

DOUGLAS, D. S. 1968. Salt and water metabolism of Adelie penguin. Antarctic Research Series 12: 167–190.

ENSOR, D. M., AND J. G. PHILLIPS. 1972. The effect of age and environment on extrarenal salt excretion in juvenile gulls (*Larus argentatus* and *L. fuscus*). Journal of Zoology 168: 119–126.

GABRIELSEN, G. W., F. MEHLUM, AND K. A. NAGY. 1987. Daily energy expenditure and energy utilization of free-ranging black-legged kittiwakes. Condor 89: 126–132.

GABRIELSEN, G. W., M. KONARZEWSKI, F. MEHLOUM, AND J. R. E. TAYLOR. 1991. Field and laboratory metabolism and thermoregulation in dovekies (*Alle alle*). Auk 108: 71–78.

GALES, R. P. 1989. Validation of the use of tritiated water, doubly labeled water and ^{22}Na for estimating food, energy and water intake in little penguins *Eudyptula minor*. Physiological Zoology 62: 147–169.

GALES, R., AND B. GREEN. 1990. The annual energetics cycle of little penguins (*Eudyptula minor*). Ecology 71: 2297–2312.

GALES, R., GREEN, B., AND C. STAHEL. 1988. The energetics of free-living little penguins *Eudyptula minor* (Spheniscidae) during moult. Australian Journal of Zoology 36: 159–167.

GERSTBERGER, R., AND D. A. GRAY. 1993. Fine structure, innervation, and functional control of avian salt glands. International Review of Cytology 144: 129–215.

GILADI, I., AND B. PINSHOW. 1999. Evaporative and excretory water loss during free flight in pigeons. Journal of Comparative Physiology B 169: 311–318.

GILADI, I., D. L. GOLDSTEIN, B. PINSHOW, AND R. GERSTBERGER. 1997. Renal function and plasma levels of arginine vasotocin during free flight in pigeons. Journal of Experimental Biology 200: 3203–3211.

GOLDSTEIN, D. L. 1989. Transport of water and electrolytes by the lower intestine and its contribution to avian osmoregulation. Pp. 271–294 *in* Progress in Avian Osmoregulation (M. R. Hughes and A. Chadwick, Eds.). Leeds Philosophical and Literary Society, Leeds, U.K.

GOLDSTEIN, D. L. 1993. Renal response to saline infusion in chicks of Leach's storm petrel (*Oceanodroma leucorhoa*). Journal of Comparative Physiology B 163: 167–173.

GOLDSTEIN, D. L., AND S. D. BRADSHAW. 1998. Regulation of water and sodium balance in the field by Australian honeyeaters (Aves: Meliphagidae). Physiological Zoology. 71: 214–225.

GOLDSTEIN, D. L., AND E. J. BRAUN. 1989. Structure and concentrating ability in the avian kidney. American Journal of Physiology 256: R501–R509.

GOLDSTEIN, D. L., M. R. HUGHES, AND E. J. BRAUN. 1986. Role of the lower intestine in the adaptation of gulls (*Larus glaucescens)* to sea water. Journal of Experimental Biology 123: 345–357.

GOLDSTEIN, D. L., AND E. L. ROTHSCHILD. 1993. Daily rhythms in rates of glomerular filtration and cloacal excretion in captive and wild song sparrows (*Melospiza melodia*). Physiological Zoology 66: 708–719.

GRAY, D. A., AND T. ERASMUS. 1988. Plasma arginine vasotocin and angiotensin II in the water deprived kelp gull (*Larus dominicanus*), Cape gannet (*Sula capensis*) and jackass penguin (*Spheniscus demersus*). Comparative Biochemistry and Physiology 91A: 727–732.

GRAY, D. A., AND T. ERASMUS. 1989. Plasma arginine vasotocin, angiotensin II, and salt gland function in freshwater- and seawater-adapted kelp gulls (*Larus dominicanus*). Journal of Experimental Zoology 249: 138–143.

GREEN, B., AND N. BROTHERS. 1989. Water and sodium turnover and estimated food consumption rates in free-living fairy prions (*Pachyptila turtur*) and common diving petrels (*Pelecanoides urinatrix*). Physiological Zoology 62: 702–705.

GREEN, B., N. BROTHERS, AND R. P. GALES. 1988. Water, sodium, and energy turnover in free-living little penguins, *Eudyptula minor.* Australian Journal of Zoology 36: 429–440.

GREEN, B., AND R. P. GALES. 1990. Water, sodium, and energy turnover in free-living penguins. Pp. 245–268 *in* Penguin Biology (L. S. Davis and J. T. Darby, Eds.). Academic Press, San Diego, CA.

HÖTKER, H., G. KÖLSCH, AND G. H. VISSER. 1996. Der Energieumsatz brütender Säbelschnäbler *Recurvisrostra avosetta*. Journal für Ornithologie 137: 203–212.

HUGHES, M. R. 1970. Relative kidney size in nonpasserine birds with functional salt glands. Condor 72: 164–168.

HUGHES, M. R. 1980. Glomerular filtration rate in saline acclimated ducks, gulls, and geese. Comparative Biochemistry and Physiology 65: 211–213.

HUGHES, M. R. 1984. Osmoregulation in nestling glaucous-winged gulls. Condor 86: 390–395.

HUGHES, M. R. 1989. Stimulus for avian salt gland secretion. Pp. 143–161 *in* Progress in Avian Osmoregulation (M. R. Hughes and A. Chadwick, Eds.). Leeds Philosophical and Literary Society, Leeds, U.K.

HUGHES, M. R., C. KASSERRA, AND B. R. THOMAS. 1990. Effect of externally applied bunker fuel on body mass and temperature, plasma concentration, and water flux of glaucous–winged gulls, *Larus glaucescens*. Canadian Journal of Zoology 68:716–721.

JANES, D. N. 1997. Osmoregulation by Adelie penguin chicks on the Antarctic peninsula. Auk 114: 485–495.

JANES, D. N., AND E. J. BRAUN. 1997. Urinary protein excretion in red jungle fowl (*Gallus gallus*). Comparative Biochemistry and Physiology 118A: 1273–1275.

JOHNSON, O. W. 1968. Some morphological features of avian kidneys. Auk 85: 216–228.

JOHNSTON, J. W., AND K. L. BILDSTEIN. 1990. Dietary salt as a physiological constraint in white ibis breeding in an estuary. Physiological Zoology 63: 190–207.

KOOYMAN G. L., R. W. DAVIS, J. P. CROXALL, AND D. P. COSTA. 1982. Diving depths and energy requirements of king penguins. Science 217: 726–728.

KROGH, A. 1939. Osmotic regulation in aquatic animals. Cambridge University Press, London.

LASIEWSKI, R. C., AND G. K. SNYDER. 1969. Responses to high temperature in nestling double-crested and pelagic cormorants. Auk 86: 529–540.

MAHONEY, S. A., AND JEHL, J. R. 1984. Body water content in marine birds. Condor 86: 208–209.

MAHONEY, S. A., AND JEHL, J. R. 1985a. Adaptations of migratory shorebirds to highly saline and alkaline lakes: Wilson's phalarope and American avocet. Condor 87: 520–527.

MAHONEY, S. A., AND JEHL, J. R. 1985b. Physiological ecology and salt loading of California gulls at an alkaline, hypersaline lake. Physiological Zoology 58: 553–563.

McEWAN, E. H. 1978. The effect of crude oils on salt gland sodium secretion of orally imposed salt loads in glaucous-winged gulls, *Larus glaucescens*. Canadian Journal of Zoology 56: 1212–1213.

McLELLAND, J. 1989. Anatomy of the avian cecum. Journal of Experimental Zoology S3: 2–9.

McNABB, F. M. A., R. A. McNABB, AND J. M. WARD. 1972. The effects of dietary protein content on water requirements and ammonia excretion in pigeons *Columba livia*. Comparative Biochemistry and Physiology 43A: 181–185.

MORGAN, K. R., M. A. CHAPPELL, AND T. L. BUCHER. 1992. Ventilatory oxygen extraction in relation to ambient temperature in four Antarctic seabirds. Physiological Zoology 65: 1092–1113.

NAGY, K. A., W. R. SIEGFRIED, AND R. P. WILSON. 1984. Energy utilization by free-ranging jackass penguins *Spheniscus demersus*. Ecology 65: 1648–1655.

NAGY, K. A., AND C. C. PETERSON. 1988. Scaling of Water Flux Rate in Animals. University of California Publications in Zoology 120: 1–172.

OBST, B. S., K. A. NAGY, AND R. E. RICKLEFS. 1987. Energy utilization by Wilson's storm-petrel (*Oceanites oceanicus*). Physiological Zoology 60: 200–210.

OELOFSEN, B. W. 1973. Renal function in the penguin (*Spheniscus demersus*) with special reference to the role of the renal portal system and renal portal valves. Zoologica Africana 8: 41–62.

PEAKER, M., AND J. L. LINZELL. 1976. Salt Glands in Birds and Reptiles. Cambridge University Press, Cambridge, U.K.

PETTIT, T. N., K. A. NAGY, H. I. ELLIS, AND G. C. WHITTOW. 1988. Incubation energetics of the Laysan albatross. Oecologia 74: 546–550.

PIERSMA, T., AND R. I. G. MORRISON. 1994. Energy expenditure and water turnover of incubating ruddy turnstones: high costs under high arctic climatic conditions. Auk 111: 366–376.

RICE, D. W., AND K. W. KENYON. 1962. Breeding cycles and behavior of Laysan and black-footed albatrosses. Auk 79: 517–567.

RICKLEFS, R. E., D. D. ROBY, AND J. B. WILLIAMS. 1986. Daily energy expenditure of adult Leach's storm-petrel during the nesting cycle. Physiological Zoology 59: 649–660.

ROBERTS, J. R., AND M. R. HUGHES. 1984. Exchangeable sodium pool size and sodium turnover in freshwater- and saltwater-acclimated ducks and gulls. Canadian Journal of Zoology. 62: 2142–2145.

ROBY, D. D., AND R. E. RICKLEFS. 1986. Energy expenditure in adult least auklets and diving petrels during the chick-rearing period. Physiological Zoology 59: 661–678.

SCHMIDT-NIELSEN, K. 1960. The salt-secreting gland of marine birds. Circulation 21: 955–967.

SCHMIDT-NIELSEN, K., C. B. JÖRGENSEN, AND H. OSAKI. 1958. Extrarenal salt excretion in birds. American Journal of Physiology 193: 101–107.

SCHMIDT-NIELSEN, K., A. BORUT, P. LEE, AND E. CRAWFORD, JR. 1963. Nasal salt excretion and the possible function of the cloaca in water conservation. Science 142: 1300–1301.

SKADHAUGE, E. 1981. Osmoregulation in Birds. Springer-Verlag, New York.

SKADHAUGE, E., B. G. MUNCK, AND G. E. RICE. 1984. Regulation of NaCl and water absorption in duck intestine. Pp. 131–142 *in* Lecture Notes in Coastal and Estuarine Studies, Vol. 9, Osmoregulation in Estuarine and Marine Animals (A. Pequeux, R. Gilles, and L. Bolis, Eds.). Springer-Verlag, New York.

STALLONE, J. N., AND E. J. BRAUN. 1986. Regulation of plasma arginine vasotocin in conscious water-deprived domestic fowl. American Journal of Physiology 250: R658–R664.

STEWART, D. J. 1972. Secretion by salt gland during water deprivation the duck. American Journal of Physiology 223: 384–386.

WARD, J. M., JR., McNABB, R. A., AND McNABB, F. M. A. 1975. The effects of changes in dietary protein and water availability on urinary nitrogen compounds in the rooster, *Gallus domesticus* — I. Urine flow and the excretion of uric acid and ammonia. Comparative Biochemistry and Physiology 51A: 165–169.

WILLIAMS, A. J., W. R. SIEGFRIED, A. E. BURGER, AND A. BERRUTI. 1977. Body composition and energy metabolism of moulting eudyptid penguins. Comparative Biochemistry and Physiology 56A: 27–30.

WILLIAMS, J. B., W. R. SIEGFRIED, S. J. MILTON, N. J. ADAMS, W. R. J. DEAN, M. A. DUPLESSIS, AND S. JACKSON. 1993. Field metabolism, water requirements, and foraging behavior of wild ostriches in the Namib. Ecology 74: 390–404.

WITHERS, P. C. 1992. Comparative Animal Physiology. Saunders, New York.

WITHERS, P. C., W. R. SIEGFRIED, AND G. N. LOUW. 1981. Desert ostrich exhales unsaturated air. South Africa Journal of Science 77: 569–570.

WORK, T. M. 1996. Weights, hematology, and serum chemistry of seven species of free-ranging tropical pelagic seabirds. Journal of Wildlife Diseases 32: 643–657.

Brown Pelican with Thin-shelled Eggs

15 Effects of Chemicals and Pollution on Seabirds

Joanna Burger and Michael Gochfeld

CONTENTS

15.1 INTRODUCTION

In a world where the use of chemicals is increasing daily, in industry, on farms, and in homes, levels of many chemicals are elevated in marine and coastal environments. There remain many threats from local point-source polluters such as industries, water treatment plants, and sewage outfalls, as well as from nonpoint sources (pollution arising from many locations). Moreover, the threat from long-range atmospheric transport and deposition is increasing as many chemicals from power plants and industries are transported to all regions, including the Arctic and the Antarctic (Houghton et al. 1992). Aquatic and marine environments are particularly at risk because of the rapid movement of contaminants in water, compared to movement in terrestrial environments.

Marine birds are exposed to a wide range of chemicals and other forms of pollution because they spend most of their time in aquatic environments where they are exposed by external contact, by inhalation, and particularly by ingestion of food and water (Figure 15.1). The major groups of pollutants of concern are chlorinated hydrocarbons, metals, petroleum products, plastic particles, and artefacts. Recently attention has focused on a much wider range of industrial and agricultural compounds which may be bioactive, including those that interact with the endocrine system.

The potential impact of a pollutant occurs both at the individual and the population levels. Whether a pollutant causes an effect depends on intrinsic toxicity and exposure. For exposure to occur, there must be contact to a substance that is readily bioavailable, which must gain access from the external environment to target organ systems, which usually requires absorption into the blood stream. The amount absorbed and the intrinsic toxicity of the substance determine the toxic impact on target organs, and this is in turn modified by the susceptibility of individuals to toxic effects. We distinguish susceptibility (an intrinsic property of the receptor organism based on genetics, nutritional status, and state of health) from vulnerability (whether it is likely to be exposed to a significant dose based on location, ecology, and behavior). However, these terms are often used interchangeably. Since different families of seabirds, and different species within these families, have different life cycles, behavior, ecologies, and habitat uses, their vulnerability varies. Further, as with other animals, susceptibility varies with age, reproductive stage, and gender.

In this chapter, we review why seabirds are particularly vulnerable, examine why some families are more vulnerable than others, describe the methods of assessing potential effects of pollution,

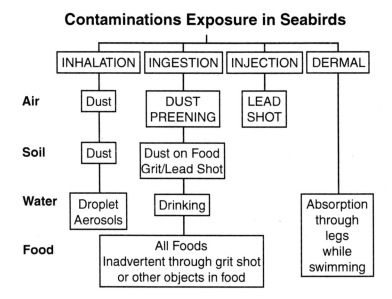

FIGURE 15.1 Pathways of exposure for seabirds in air, soil, water, and food.

describe the types of pollutants with their major effects, discuss exposure and uptake, and examine cases where pollutants affect individual reproductive success, survival, and population levels. Finally, we discuss future research needs and data gaps. Major advances in understanding the concentrations, distribution, and effects of pollutants have occurred since the mid-1970s (Burton and Statham 1990, Beyer et al. 1996). There is a rich literature on pollutants in birds which can be roughly assigned to four major categories: laboratory studies, residue measurements in sick or dead birds, surveys of contaminants in a species, and finally, recently emerging studies in a risk assessment framework.

15.1.1 EXPOSURE ASSESSMENT

An important aspect of pollutant effects on seabirds lies in exposure assessment. The pathway from source to environmental fate and transport, food chain bioamplification, contact, intake, bioavailability and absorption, metabolism, transport, and excretion and distribution within the body ultimately determines the dose delivered to a target organ. Since many of the contaminants discussed in this chapter are taken up and stored in tissues, tissue levels can be used as biomarkers of exposure and of possible effects on the seabirds themselves (Peakall 1992, Nisbet 1994). The time frame of exposure is important. Exposures can be acute or chronic. Acute exposure to a contaminant will have a different impact than chronic ingestion of small quantities, even when the same total dose is achieved.

The effects of contaminants may also be acute and short-lived such that once exposure has ended, there is no further risk (e.g., organophosphates). Or the substance may accumulate or produce a cumulative effect so that the impact may not be apparent until long after the exposure has begun, or in some cases, even after it has terminated (e.g., organochlorines, some heavy metals). Once exposure has ended, and there are no effects apparent, the likelihood of subsequent effects begins to decline (see Eaton and Klaasen 1996, Gochfeld 1998).

15.1.2 STATISTICAL POWER

Most studies that consider statistically significant differences in contaminant residues from among localities, species, tissues, age classes, or sexes, rely on the traditional alpha = 0.05 level. There is no *a priori* basis for relying on this particular value. In many cases, studies involving a few individuals lack the statistical power to identify differences that may be real. Conversely, differences that are statistically different may represent sampling artefacts. Both phenomena should be considered in interpreting research or planning new studies.

The National Research Council (NRC 1993) has encouraged reliance on a weight-of-evidence approach, which recognizes that although each study may have some problems, it is prudent to examine the totality of evidence from a meta-analysis approach. For example, if a dozen studies of a substance all show an excess of a particular endpoint, the weight of evidence approach supports a relationship even if none achieved "statistical significance."

15.2 SEABIRDS AS BIOINDICATORS

A few groups of birds, raptors, waterfowl, and seabirds, dominate the contaminant literature. Seabirds offer the advantage of being large, wide ranging, conspicuous, long lived, easily observed, and important to people. They are often at the top of the food chain where they can be exposed to relatively high levels of contaminants in their prey. Since many species of seabirds are philopatric, returning to the same nest site and colony site for years, contaminant loads of individuals can be studied (Burger 1993). Although many seabird populations are already threatened or endangered through habitat loss, exploitation, overfishing, and other anthropogenic impacts (Croxall et al. 1984; Chapters 8, 16, 17), populations of many species are robust, and the collecting of limited individuals does not pose a conservation problem.

While contaminant levels can be examined in seabirds as an indication of potential harm to the seabirds themselves, seabirds have also been used as bioindicators of coastal and marine pollution (Hays and Risebrough 1972, Gochfeld 1980b, Walsh 1990, Peakall 1992, Furness 1993, Furness and Camphuysen 1997). They have been used to assess pollution over local, regional, or wide-scale geographical areas as well to determine whether levels of contaminants have changed over time (Walsh 1990). Feathers in museum collections have been used to examine changes in mercury levels over centuries (Berge et al. 1966, Thompson et al. 1992). Seabirds are bioindicators for local, regional, and global scales, and can integrate over both spatial and temporal scales. Seabirds have proven particularly useful as bioindicators for contamination in the Great Lakes (Fox 1976, Mineau et al. 1984, Weseloh et al. 1995, Pekarik and Weseloh 1998).

Like any bioindicator, there are advantages and disadvantages of using seabirds. Seabirds are excellent bioindicators because they are sensitive to chemical and radiological hazards and are widespread over the world in coastal and marine habitats where pollution is often great and where contaminants are transported rapidly through aquatic systems and within food chains. They "integrate" contamination over time and space (Walsh 1990, Burger 1993). Since seabirds travel over substantial distances to obtain food, they sample prey from different regions, and the resultant levels in their tissues are an indication of contamination over that area. Sampling contaminants in seabirds is often more cost effective than sampling water, sediment, or invertebrates, because those samples represent only the small number of points or locations sampled. To sample a large bay or estuary, many points are required to obtain a picture of pollutant levels, with serial samples needed to capture seasonal fluctuations. However, by sampling only a few seabirds, it is possible to determine whether there is a problem in the bay generally.

The advantage of using seabirds to integrate over space and time, however, is also a disadvantage. If high levels of any contaminant are discovered in seabirds, then it is necessary to understand the life cycle, migration routes, prey base, foraging range, and habitat of the seabird. Knowing contaminant loads in a seabird will not normally identify the exact location of point-source pollution; further sampling of other bioindicators is required. With an understanding of the prey consumed by seabirds, it is possible to determine where they might have foraged, thus identifying potential sites of high contamination. Finally, it is important to understand the migratory behavior of seabirds before interpreting contaminant levels. Sedentary species reflect local levels of pollution, but for migratory species it is essential to know how long the seabirds have been in the local area.

Some of the disadvantages discussed above can be ameliorated by using eggs or young seabirds as bioindicators. Coastal-nesting species of seabirds often arrive at the breeding colony a month or more before laying eggs, and the contaminant loads in eggs largely reflect local exposure. Species that nest on oceanic islands, however, may arrive only a few days before egg laying, and thus levels in their eggs do not reflect local exposure. Young birds that have not yet fledged have obtained all of their food from their parents, who usually obtained it from the local area. Exceptions are albatrosses and some petrels that might have traveled several hundred kilometers to obtain food (Fisher and Fisher 1969, Weimerskirch 1997).

Studies of contaminants in seabirds have examined the internal tissues (liver, brain, kidney, and muscle) of adult and young birds, eggs (both viable and nonviable), and young chicks. Each kind of tissue addresses different questions. Since feathers have been used so often to examine levels of metals, we will describe briefly why they work for metals and not for other substances. Feathers are rich in disulfide bonds that are readily reduced to sulfhydryl groups that bind to metals. As feather protein is laid down, it becomes a chelator that binds and removes metals from the blood supply. Metal levels in a feather reflect circulating blood levels during the 3- to 4-week period when a feather is forming (Bearhop et al. 2000). Thereafter, the blood supply atrophies, leaving the feather as a permanent record of blood levels, for many years or centuries if specimens are in permanent collections.

The blood levels of heavy metals are a result of current exposure and metals mobilized from other internal tissues (Burger 1993). Thus the molt cycle and the location of seabirds during feather formation must be known before feather levels can be interpreted. Using feathers from prefledging seabirds is a useful method of ascertaining local levels since parents obtained the food within foraging distance of the colony.

The utility of feathers hinges on the high affinity of metals for the sulfhydryl group of the structural protein melanin. Organic pollutants do not have this same affinity, and do not concentrate in feathers. An issue with feathers is whether the metals in the feather have been delivered by the blood supply (a reflection of internal exposure) or deposited superficially from atmospherically transported contamination. Vigorous washing will remove loosely adherent contamination but not necessarily metals bound to the protein (regardless of their origin). When individuals in the same population exposed to similar atmospheric deposition show great differences in feather levels of a metal, we infer that the difference is largely due to internal rather than external deposition.

15.3 SEABIRD VULNERABILITY AND SUSCEPTIBILITY

Different seabirds are affected by pollutants in different ways depending upon breeding schedules, foraging methods, geographical ranges, and life history strategies (see Chapter 8). Species, such as seabirds, that are long lived have longer to accumulate toxics than do shorter-lived species. Further, seabirds that lay fewer eggs may well deposit higher levels in their one egg. All seabirds are not equally vulnerable to contaminants even when exposed to the same levels in their food or water because they do not eat the same proportions of any given prey and they have varying abilities to excrete, metabolize, or sequester xenobiotics. Understanding the relative role of each of these differences requires controlled laboratory experiments on toxicodynamics (the movement of chemicals between and among organs and compartments of an animal), as well as extensive field studies. Toxicodynamic studies have been conducted for mercury (Braune 1987, Lewis and Furness 1991), organochlorines (Clark et al. 1987), and plastic particles (Ryan 1988a, b). Burger (1993) provides a table of the ratio of metal levels among tissues for seabirds, which can be used to assess which tissues concentrate which metals. There are other vulnerabilities that include differences in exposure, location on the food chain, age, or gender.

15.3.1 EXPOSURE AND FOOD CHAIN VULNERABILITIES

Since seabirds have a patchy distribution over a wide range of spatial and temporal scales (Schneider et al. 1988), exposures can vary widely. Levels in tissues are a function of uptake and absorption and how and where each pollutant is stored in different tissues. Uptake is a function of exposure and intake rates. For contaminants to be taken in, they have to be bioavailable to the organisms, otherwise they are excreted and are not absorbed into the bloodstream nor distributed to the tissues. If the contaminant is not bioavailable it will not be incorporated into the tissues, and thus high levels in soil or water may not be biologically relevant.

Once contaminants are in aquatic systems, they enter the food chain where some are biomagnified at each transfer from prey to predator (Hahn et al. 1993). At every step, organisms take in more of a substance than they excrete, resulting in a net increase in the concentration of that substance in their tissues during their lifetime. Top-level carnivores and piscivores can have much higher levels of contaminants than organisms that are lower on the food chain (Hunter and Johnson 1982, van Strallen and Ernst 1991). Ideally, food-chain effects should be examined by evaluating the levels of contaminants in known food chains, which might include water, invertebrates, small fish, squid, large fish, and seabirds. Alternatively, food chain effects can be examined by measuring contaminant levels in a range of seabirds that represent different trophic levels.

15.3.2 AGE- AND GENDER-RELATED VULNERABILITIES

Young seabirds usually have lower levels of contaminants than adults. A summary of metal levels in feathers (Burger 1993) showed that adults had significantly higher concentrations than young for mercury (20 of 21 studies), lead (4 of 7), cadmium (3 of 5), manganese (5 of 5), and selenium (3 of 3), with chromium showing less of a difference (only 1 of 4 studies). Since then, age-related differences in some metals were found for other species (Thompson et al. 1993, Gochfeld et al. 1996, 1999, Burger 1996, Stewart et al. 1997, Burger and Gochfeld 1997a, b, Burger and Gochfeld 2000a, b). Differences between adults and young depend on the contaminant and the species being studied. Age-related differences are not consistent for cadmium or manganese, and generally do not occur for chromium (Burger 1993).

Few studies have examined differences in metal levels of internal tissues as a function of age. Furness and Hutton (1980) reported that cadmium levels in liver increased with age in Great Skua (*Catharacta skua*). In Laughing Gulls (*Larus atricilla*) from the New York City area, Gochfeld et al. (1996) reported that selenium and mercury decreased with adult age, and cadmium levels increased with age. In Franklin's Gull (*Larus pipixcan*) from northern Minnesota, chicks generally had lower levels of metals in tissues than adults (Burger and Gochfeld 1999). Young might be exposed to higher levels of certain metals if adults feed different food to their offspring than they eat themselves.

Adult Laysan Albatrosses (*Diomedea immutabilis*) from Midway Atoll in the northern Pacific Ocean had higher levels of cadmium, selenium, and mercury in most tissues (Burger and Gochfeld 2000a). However, chicks had higher concentrations of manganese in liver and arsenic in salt glands, than did adults. Lock et al. (1992) examined cadmium, lead, and mercury in the feathers, liver, kidney, and bone of adults and juveniles of some seabirds, including several albatrosses, in New Zealand. There were significant age differences, with adults having higher levels of cadmium and mercury in the liver than did young birds. For the metals and tissues examined at both New Zealand and Midway, the concentrations of cadmium were similar, but mercury levels were up to three times higher in the New Zealand albatrosses. The New Zealand and Midway data suggest that albatrosses may be less sensitive to mercury than smaller species of birds that show reproductive effects at liver concentrations of 2 ppm in laboratory studies (Scheuhammer 1987). These two studies on albatrosses indicate the value of data on contaminants in the same species from different parts of the world.

Less is understood about gender-related differences in contaminant levels and effects, largely because birds collected outside the breeding season are difficult to sex (gonads have recrudesced); sexually monomorphic species (i.e., most seabirds) are often not possible to sex. Comparing contaminant levels in females and males is very important, however, since females have an additional route of excretion (to the eggs) that males do not have, which constitutes a major reproductive vulnerability. There does not seem to be any clear pattern, at least in metal levels in feathers, although this requires more study with more species (Burger 1993).

15.3.3 FAMILY VULNERABILITIES

Some families of seabirds are more vulnerable to pollution than others because of their foraging method, prey, or nesting habitat. Most gulls, most cormorants, and some terns and alcids are exposed to high levels of pollutants because they nest near shore in close proximity to sources of industrial or agricultural pollution (Mailman 1980, Fowler 1990). Within families, species may differ in their ability to rid themselves of contaminants, as Henriksen et al.(2000) suggested for Glaucous Gulls (*Larus hyperboreus*) compared with Herring Gulls (*Larus argentatus*).

15.3.4 INDIVIDUALS VS. POPULATIONS

The focus of early studies of the effects of pollutants on birds centered on direct mortality (Bellrose 1959), although recent work has demonstrated a wide range of sublethal effects on development, physiology, and behavior of individuals. Sublethal effects of pollutants on seabirds include reproductive deficits (Ashley et al. 1981), teratogenicity and embryotoxicity (Hoffman 1990), eggshell

Links to Determine Effects

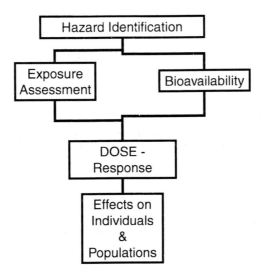

FIGURE 15.2 Methods to establish cause-and-effect relationship of chemicals and adverse outcomes in seabirds. This is an ecological risk-assessment methodology.

thinning (Risebrough 1986), enzyme induction (Fossi et al. 1989, Ronis et al. 1989), effects on endocrine function (Peakall et al. 1973, Peakall 1992), and behavioral abnormalities of adults and young (Burger and Gochfeld 1985, 2000c, Burger 1990). These sublethal effects on overall reproduction, survival, and population dynamics are not well understood, and effects, particularly if localized, do not necessarily lead to population declines.

It is difficult to assess the toxic effects of contaminants on seabird populations because seabirds are long-lived and a population is made up of many overlapping generations. Even the dramatic losses due to a massive oil spill that might eliminate an entire age cohort of young birds may not be obvious if such losses are compensated by improved reproduction and survival of remaining birds, enhanced recruitment from a pool of nonbreeders, or immigration from birds nesting in nearby colonies. Establishing cause-and-effect requires a series of discrete steps in a chain involving both laboratory tests and field observations (Gilbertson 1990, Fox 1991; Figure 15.2). It involves identifying the hazard (types of effects or endpoints), determining exposure and bioavailability of the chemical, estimating dose–response relationship for each endpoint, and examining overall effects on individuals and populations. Establishing these links cannot be done without both laboratory and field experimentation.

In 1991 Fox applied the Bradford Hill postulates (Hill 1965) used by epidemiologists to establish causal relationships for humans to ecotoxicology. These criteria for evaluating the relationship between a contaminant and an observed health effect include the strength and consistency of the association between an outcome and its putative cause, the temporal relationship (exposure must precede effect), the biological plausibility based on knowledge of toxicology and biology, the ability to replicate the relationship, and its predictability (does the endpoint occur in other situations where the exposure occurs).

Toxicologists establish the links between cause and effect, but seldom examine the overall ecological relevance of these effects. Seabird biologists, on the other hand, must examine a wide range of sublethal effects on reproduction and survival of populations (Figure 15.3). This model, developed for lead (Burger 1995), shows how reproduction and survival can be affected by a substance, leading to declines in populations. While it is possible to establish an effect of pollutants on local populations, it is more difficult to demonstrate that these effects have led to regional or

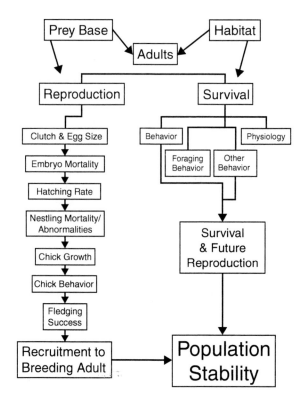

FIGURE 15.3 Model for establishing toxicity for contaminants. Shown are links (arrows) where sublethal and lethal effects can be demonstrated, leading to population declines if the effects are severe enough. (After Burger 1995.)

worldwide declines in a species. We do not, however, believe that it is necessary to prove this last link because seabirds, like other animals, have evolved mechanisms to deal with such perturbations, and unless the level of pollution is similar worldwide, worldwide effects would not be expected.

15.4 CHEMICALS AND THEIR EFFECTS ON SEABIRDS

The major categories of pollutants that we deal with in this chapter are metals and metalloids, organochlorine compounds, polyaromatic hydrocarbons and petroleum products, plastics, and floatables (Table 15.1). We do not deal with substances that are primarily acutely toxic such as the organophosphate pesticides. Space also precludes our dealing with radionuclides, although there is a growing literature on various radioisotopes in seabirds as analytic techniques become available. Seabirds can acquire radionuclides from discharges from fuel reprocessing plants (Woodhead 1986) or from nuclear testing fallout (Noshkin et al. 1994). For a review of the effects of radionuclides on birds see Brisbin (1991).

15.5 METALS

Cadmium, lead, and mercury are the primary metals of concern for oceanic and estuarine environments (Fowler 1990), and thus for seabirds, while selenium is of concern for those seabirds that nest inland (Ohlendorf et al. 1986). Other elements such as arsenic bioaccumulate as organic compounds with apparently relatively low toxicity. Metals are present naturally in the earth's crust and in seawater (Wong et al. 1983), but the contributions from anthropogenic sources are increasing (Schaule and Patterson 1981). For seabirds that breed along coasts, local anthropogenic sources of

TABLE 15.1
Major Chemicals and Pollutants of Concern for Seabirds

Metals and metalloids	Many metals have potent effects on development and the nervous system, including mercury, lead, cadmium, manganese, and selenium.
Organochlorine insecticides	Many of the chlorinated pesticides or their breakdown products are highly persistent in the environment and in the body (DDT).
Polychlorinated di-aromatic compounds (PCB, dioxins)	These are highly persistent chemicals, which vary greatly in their toxicity. Effects on the nervous system of some of these compounds are secondary.
Organophosphates	Organophosphates exert mainly acute nervous-system toxicity by interfering with acetylcholinesterase. There is some evidence of prolonged and even delayed neurotoxicity in survivors; they may break down quickly in the environment and the body.
Petroleum products	These are complex mixtures of aliphatic and organic compounds.
Solvents	Of particular concern are short-chain chlorinated aliphatics such as trichloroethylene, tetrachloroethylene, and formerly carbon tetrachloride. Also of concern are aromatic solvents such as toluene and xylene.
Plastics and floatables	Plastic material and others that float on the ocean surface are of concern.

lead, cadmium, and mercury are a substantial part of their exposure. While other metals, such as chromium (Eisler 1986), are potentially problematic for seabirds, we discuss only cadmium, lead, mercury, and selenium in detail here.

15.5.1 CADMIUM

Cadmium is a nonessential metal that can come from a variety of anthropogenic sources such as smelters and from the manufacture and disposal of commercial products such as batteries, paints, and plastic stabilizers (Burger 1993, Furness 1996). It is a relatively rare element in the environment (Wren et al. 1995), and in most of the earth's crust it is present at levels below 1 ppm (usually less than 0.2 ppm, Farnsworth 1980). Volcanic action is the major natural source of atmospheric cadmium; other natural sources include ocean spray, forest fires, and the releases of particles from terrestrial vegetation (Hutton 1987).

Compared to other organisms, cadmium levels are often relatively high in marine organisms, including seabirds (Bull et al. 1977, Furness 1996). Levels seem to be higher among squid-eating seabirds than among those that eat primarily fish (Muirhead and Furness 1988, Thompson 1990) or crustaceans (Monteiro et al. 1998), and this will probably apply to consumption of other molluscs as well (Furness 1996). Cadmium causes sublethal and behavioral effects at lower concentrations than mercury or lead, and causes kidney toxicity in vertebrates and is an animal carcinogen (Eisler 1985a), although little work has been done on seabirds. Effects also include altered behavior, suppression of egg production, egg-shell thinning, and testicular damage (Furness 1996). Stock et al. (1989) suggested that cadmium is regulated metabolically in adult birds, thus cadmium levels do not increase with age. Eisler (1985a) estimated that a kidney concentration of about 10 ppm (wet weight) was associated with adverse effects, based on laboratory studies (Table 15.2). Unlike mercury and most metals where the feather concentration exceeds the kidney concentration, virtually all studies of cadmium have shown kidney:feather ratios substantially greater than 1. The ratios exceed 100:1 in some species of shearwaters (Osborn et al. 1979), while levels in terns range from 5:1 to 10:1 (Burger 1993). Cadmium levels are usually undetectable or very low in seabird eggs, while relatively high cadmium levels have been reported in kidneys and livers of pelagic species, such as petrels, fulmars, prions, albatrosses, penguins, skuas, and alcids, compared to coastal and inshore species (Nisbet 1994). This suggests a natural, oceanic source of cadmium. Furness (1996) suggested that the threshold level above which adverse effects occur in pelagic

TABLE 15.2
Levels (ppm, dry weight) of Metals Associated with
Adverse and Toxic Effects

	Liver	Kidney	Feathers	Source
Cadmium	>5	10	?[b]	Eisler 1985
Lead	>5[a]	>15	4[a]	Custer and Hohman 1994
				Burger and Gochfeld 2000c
				Ohlendorf 1993
Mercury	>6	>6	5	Ohlendorf 1993
				Eisler 1987
Selenium	9	>10	?	Heinz 1996
				Ohlendorf 1993

[a] For seabirds.
[b] Unknown.

seabirds may be higher than for other birds, and that no adverse cadmium effects have been documented in wild seabirds.

15.5.2 LEAD

Lead average concentration in the Earth's crust is 19 ppm, making it a relatively rare metal (EPA 1980, Pain 1995). Lead also comes from industrial processes, burning of leaded gasoline, stormwater runoff, agricultural practices, eroded lead paint, and to some degree from natural processes such as erosion and volcanism (Eisler 1988, Prater 1995). Lead contamination is ubiquitous; there are no longer natural environmental concentrations because of widespread atmospheric deposition (Pain 1995) and runoff, with contamination of nearshore environments.

Lead affects all body systems; organolead compounds are more toxic than inorganic lead compounds, and young animals are more sensitive than older animals (Eisler 1988). In vertebrates, lead poisoning can be chronic or acute, and there is no "no effect" level since the lowest measurable levels affect some biological systems (Franson 1996), although specific effects on seabirds have been studied in only a few species. Lead levels are considered elevated if liver levels are above about 7 ppm (dry weight, Eisler 1988).

Lead exposure can cause direct mortality, as well as sublethal effects (Eisler 1988). Early studies focused on waterfowl exposed directly by shooting or indirectly from ingesting lead shot as grit or with food items (Bellrose 1959). Symptoms of lead poisoning include drooped wings, loss of appetite, lethargy, weakness, tremors, impaired locomotion, balance and depth perception, and other neurobehavioral effects (Sileo and Fefer 1987, Eisler 1988, Burger and Gochfeld 1994, 1997a).

15.5.2.1 Lead on Midway

In the mid-1980s, lead poisoning due to ingestion of lead paint from buildings was reported for Laysan Albatross chicks from Midway Atoll (Sileo and Fefer 1987, Sileo et al. 1990, Work and Smith 1996). Some chicks that hatched near buildings exhibited symptoms that included drooping wings, weight loss, and death (Figure 15.4). Sileo and Fefer (1987) reported that paint chips with up to 144,000 ppm lead were found in the proventriculus of affected chicks. Acid-fast intranuclear inclusion bodies were present in the kidneys, and degenerative lesions were present in the myelin of some brachial nerves in affected chicks. Further, in 1997, albatross chicks near buildings that exhibited droop-wings (some of which died), had mean lead levels of 4.7 ppm wet weight in the

(a)

(b)

FIGURE 15.4 Laysan Albatross chick on Midway Atoll with droop-wing, indicative of lead toxicity (top), and building with lead paint flaking off (bottom). Chicks are unable to hold their wing against the body, and they fall to the ground. (Photos by J. Burger.)

liver; non-droop-wing albatross chicks away from the buildings averaged 0.7 ppm wet weight in the liver (Burger and Gochfeld 2000b).

15.5.2.2 Lead Effects in Larids in the New York–New Jersey Harbor

One of the difficulties with contaminants work is that almost no studies examine both fate and effects in the same species. For three decades we examined contaminant levels in seabirds nesting in the New York–New Jersey region, and studied effects in the laboratory and the field in larids

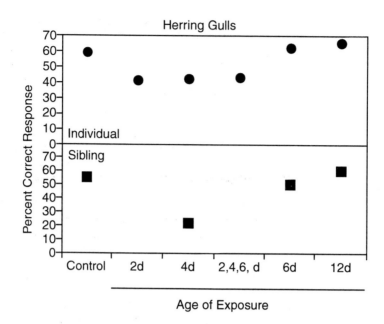

FIGURE 15.5 Effect of age of lead exposure on individual and sibling recognition in Herring Gulls. (After Burger and Gochfeld 1993, 2000c, Burger 1998.)

(Herring Gulls and Common Terns *Sterna hirundo*), subsequently shown to be sensitive to PCB and other chemicals in the Great Lakes (Mineau et al. 1984, Pekarik and Weseloh 1998, Grasman et al. 1998). Our overall protocol was to examine levels of lead in species in the wild and use these levels to determine exposure for laboratory experiments to determine the sublethal effects of lead on neurobehavioral development. We did this by examining the levels of lead in the feathers of Herring Gulls and Common Terns in the wild, and dosing them in the laboratory until their feathers had the same levels as occurred in the field (Burger 1990, 1998, Burger and Gochfeld 1985, 1994, 1995a, 1995b, 1996, 1997a). There is usually a significant correlation between concentrations of lead in feathers and those in internal tissues, including blood, and concentrations of lead in feathers are a good predictor of internal dose (Burger 1993).

This research showed several sublethal neurobehavioral effects from lead levels (although at the high end of exposure; Burger 1990, Burger et al. 1994, Burger and Gochfeld 1997a). Lead affects a wide range of behaviors, including locomotion, balance, begging, feeding, growth, and cognitive abilities, that in turn affect survival in nature.

Effects vary depending upon dose and age of exposure (Burger and Gochfeld 1995a, b). For example, recognition is more severely affected when chicks are exposed at 2 to 4 days than when exposed at 12 days (Figure 15.5), not surprising since individual recognition develops by this age. Delayed recognition can be lethal in nature because once chicks begin to move away from the nests they can be killed by neighbors if they approach a gull other than their parent (Burger 1984). Similar effects occur in the laboratory and the field, although the intensity may vary (Figure 15.6; Burger and Gochfeld 1994). Without continued exposure, there is recovery in some behaviors (Burger and Gochfeld 1995a, b, 1996). The levels of lead that cause lead toxicosis in nonseabird laboratory birds (Mallards) are similar to those that caused lead poisoning in Laysan Albatross chicks on Midway.

Some of the behavioral deficits demonstrated with lead also occurred with chromium and manganese (Figure 15.7; Burger and Gochfeld 1995c). It is important to note that although we know much less about these metals, they have a significant potential for toxicity in seabirds. Similarly, tin, used in organotin compounds, is an important potential toxicant for seabirds.

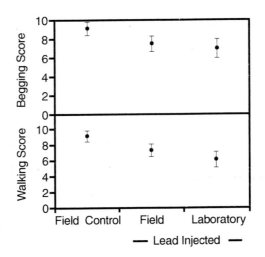

FIGURE 15.6 Comparison of begging scores and walking scores of control, and laboratory and field-exposed Herring Gulls. (After Burger and Gochfeld 2000c.)

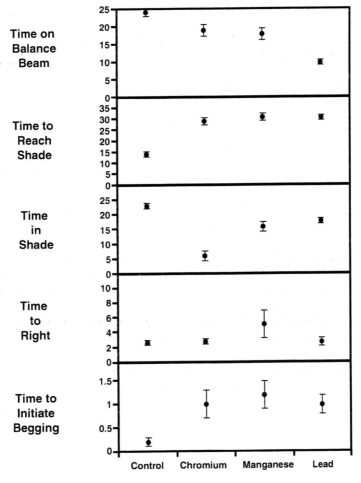

FIGURE 15.7 Comparison of the neurobehavioral deficits caused by chromium, manganese, and lead in Herring Gulls. (After Burger and Gochfeld 1995c.)

15.5.3 MERCURY

Natural sources of mercury include erosion, natural flooding, volcanism, and upwellings (Thompson 1996), but these are dwarfed by anthropogenic sources (WHO 1990, Wren et al. 1995). Local levels of mercury in soil and water are a result of natural levels, local anthropogenic emissions, and global atmospheric transport (Porcella 1994). Global transport of mercury emitted primarily from coal-fired power plants is emerging as a major source of mercury pollution and bioaccumulation. The half time to circumnavigate the earth is about 14 days, with perhaps 10% transfer to the Southern Hemisphere (Porcella 1994).

Mercury is present in elemental, inorganic, and organic forms. Methylmercury is the most toxic form, and most exposure for seabirds is from methylmercury because it is preferentially accumulated in tissues of fish and other prey (Nisbet 1994). Inorganic mercury can be converted into methyl-mercury by some organisms, particularly anaerobic bacteria, and higher organisms can both produce and demethylate methylmercury (Jensen and Jernelov 1969, Ohlendorf et al. 1978). Most studies measure only total mercury. However, in studies that speciate the mercury (analyze methyl and total separately), methylmercury makes up almost 100% of the total mercury in liver, kidney, muscle, and feathers of birds in some studies (Norheim and Froslie 1978, Thompson and Furness 1989a, but see Thompson et al. 1991). Some seabirds seem able to demethylate mercury and store inorganic mercury in the liver, but almost all the mercury in feathers is methylmercury (Thompson and Furness 1989a, 1989b).

The relative percent of methyl to total mercury in tissues may not be similar among seabirds. Thompson and Furness (1989b) reported that the percentage of methylmercury in livers ranged from 2.6% in Wandering Albatrosses (*Diomedia exulans*) to 93% in Little Shearwaters (*Puffinus assimilis*). Also, Furness et al. (1995) noted that Common Tern chicks had nearly 100% methyl-mercury in their livers, perhaps indicating that, if there is a demethylation mechanism, it might not function in young birds. This raises the possibility that the presence of inorganic mercury is simply due to the small accumulation each year, coupled with their inability to eliminate inorganic mercury.

Some seabirds that nest along coasts, such as cormorants and gulls, may not have evolved the demethylating abilities of more pelagic species, and may be more sensitive to mercury intoxication. For example, Double-crested Cormorants (*Phalacrocorax auritus*) from the Everglades have mean levels of mercury of 41 mg/kg (ppm wet weight) in their liver, a level which is associated with mercury poisoning in some species (Sepulveda et al. 1998).

Feathers are the major excretory pathway for mercury (Honda et al. 1985, Braune 1987, Furness et al.1986); from 70 to 93% of the body burden of mercury is in the plumage (Burger 1993), although Kim et al. (1996) reported much lower levels. It is because such a high percentage of the body burden is in feathers, as well as the fact that they can be collected noninvasively, that has led to their use to assess mercury levels in seabirds (Furness et al. 1986, Burger 1993). There is a strong need for effects studies with mercury in seabirds before it is possible to interpret the levels found in nature. In general, pelagic seabirds have higher mercury levels than coastal birds, and those that feed on mesopelagic prey are the highest due to the patterns of methylation of mercury in low-oxygen, deep water.

Mercury has no known metabolic function and causes a wide range of teratogenic and mutagenic effects, as well as causing embryocidal, cytochemical, histopathological, and behavioral effects (Eisler 1987). Unlike other metals, mercury both bioconcentrates and is bioamplified through the food chain. In laboratory experiments, mercury causes a wide range of reproductive effects, includ-ing lowered egg weight and shell-less eggs (Fimreite 1979), embryo malformations (Heinz 1975, 1976), reduced hatchability (Fimreite 1979, Spann et al. 1972, Heinz 1974, 1976, Finley and Stendell 1978), reduced growth (Hoffman and Moore 1979), altered behavior (Heinz 1976), and reduced chick survival (Spann et al. 1972, Finley and Stendell 1978), as well as neural shrinkage, neural lesions, and demyelination (Stendell 1978) and sterility (Solonen and Lodenius 1984). The levels associated with these effects are 5 to 65 ppm (dry weight) in feathers, and 1 to 5 ppm dry

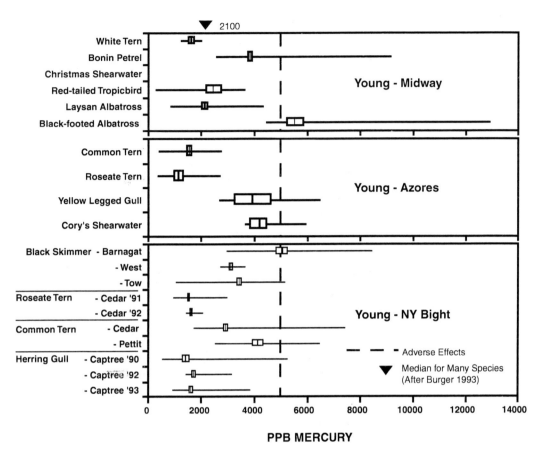

FIGURE 15.8 Levels of mercury in feathers of young seabirds (at fledging) from Midway (north Pacific Ocean, from Burger and Gochfeld 2000d), the Azores (north Atlantic, Monteiro et al. 1995), and along coastal North America (Atlantic, Burger and Gochfeld 1997).

(0.05 to 5.53 ppm wet weight in different species) in eggs (Eisler 1987, Burger and Gochfeld 1997). One difficulty is that toxicity depends upon the form, dose, route of exposure, species, age, gender, and physiological condition (Eisler 1987), as it does with most contaminants. Further, the presence of other metals, such as selenium, can reduce the adverse effects of mercury (Satoh et al. 1985).

Readers are referred to Burger (1993) for a summary of levels in feathers, and to Eisler (1987) for mercury in other tissues. Levels of mercury in the feathers of young are very variable both among species and between locations (Figure 15.8). Levels of mercury from Bonin Petrel (*Pterodroma hypoleuca*) and Black-footed Albatross (*Diomedea nigripes*) were all above the levels known to be associated with adverse effects in nonseabird species (Eisler 1987). Remarkably, mercury levels in Laysan Albatrosses were far lower, despite their similar diet compared with Black-footed Albatrosses (Whittow 1993a, b). There were interspecific differences on the Azores, but mean levels did not exceed the effects level (Monteiro et al. 1998). Terns had some of the lowest levels. The levels of mercury in young seabirds from the east coast of North America (bottom of Figure 15.8) are not as high generally as those from the more pelagic sites.

15.5.6 Selenium

Relatively high concentrations of selenium in the kidneys and liver of dying waterbirds are associated with symptoms such as hepatic lesions, liver changes, and congenital malformations, leading to decreased survival and lowered reproductive success (Ohlendorf et al. 1988, 1990), as well as adult mortality (Ohlendorf et al. 1986, Skorupa and Ohlendorf 1991, King et al. 1978, 1994). Similar reproductive effects were obtained in controlled laboratory conditions (Eisler 1985b, Heinz and Fitzgerald 1993, Heinz 1996). The wide-ranging effects of selenium on reproductive success suggests that there might be subtle behavioral effects from selenium in seabirds.

Selenium has a protective effect on mercury toxicosis (Ganther et al. 1972), but at high levels it can cause behavioral abnormalities, reproductive deficits, and ultimately mortality (Eisler 1985b, Ohlendorf et al. 1986, 1989, 1990, Heinz 1996). Concentrations of 19 to 130 ppm in livers of birds were associated with 40% of the nests having one dead embryo (Ohlendorf et al. 1986, 1989). Using 19 to 130 ppm as the levels associated with adverse effects, and a feather:liver ratio of 1:5 (Burger 1993), indicates that feather levels of 3.8 to 26 ppm would be associated with severe adverse effects (mortality of eggs). However, more recently, Heinz (1996) gives 9 ppm in the liver as the level of concern for embryonic deformities.

High levels of selenium have been reported in eggs and tissues of seabirds in the North Pacific (Stoneburner and Harrison 1981, Honda et al. 1990, Burger and Gochfeld 2000d), in the Antarctic (Norheim 1987), and in Common Murres (*Uria aalge*) from Puget Sound (Ohlendorf 1993), but levels of selenium are often not measured. Levels of selenium in 80% of the eggs of Least Terns (*Sterna antillarum*) from interior regions of North America were above those considered safe (Allen et al. 1998).

Toxicity of selenium in seabirds has not been studied, although it has been suggested that selenium may also be subject to a detoxification mechanism, much like mercury (Hutton 1981, Norheim 1987). The effects of selenium can be ameliorated by arsenic (Hoffman et al. 1992). In Table 15.3 we present a summary of metal levels in feathers in different groups of seabirds (compiled by Burger 1993, and other papers by Burger and Gochfeld, and Furness).

15.6 ORGANOCHLORINE COMPOUNDS

The chlorinated hydrocarbons or organochlorines (OC) groups include many organochlorine insecticides (typified by dichlorodiphenyltrichloroethane or DDT) as well as the 209 isomers of polychlorinated biphenyls (PCB), and the isomers of the polychlorinated dibenzofurans (PCDF) and dioxins (PCDD), of which 2,3,7,8-tetrachloro-dibenzo-*p*-dioxin (TCDD, dioxin) is the best known and most toxic. Many of the effects of these compounds have been reviewed by Gilbertson (1988, 1989).

Although it was the acute lethality of the early chlorinated pesticides that prompted Rachel Carson to publish *Silent Spring* in 1962, it is the cumulative exposure and chronic effects that are the main ecological concern. These chemicals are highly persistent in the environment and in the body, which account for their relatively high levels and sometimes severe adverse effects on seabirds.

By virtue of their lipophilia, these compounds can accumulate at high concentrations in predators at the top of a food chain (Hoffman et al. 1996) and they persist in tissues for months to decades (Peakall 1986). Recent reviews of the levels of these compounds in the tissues of birds can be found in Blus (1996), Peakall (1996), Wiemeyer (1996), Custer et al. (1983), Peakall (1986), and Nisbet (1994). Concentrations are generally low in seabirds from remote oceanic islands and are higher in those that feed near industrialized or agricultural areas in the Northern Hemisphere (Nisbet 1994).

During the past few decades, links have been demonstrated between the accumulation of organochlorine compounds in bird tissues and a variety of effects observed in raptorial and fish-eating birds, including seabirds (Fox 1982, Gilbertson et al. 1991, Giesy et al. 1994a, b, Bosveld and Van den Berg 1994). In the 1960s and 1970s, there were major population declines in some seabirds reported from areas with point-source pollution from manufacturing plants as well as nonpoint-source pollution (Koeman 1972, Blus et al. 1979, Cress et al. 1973, Anderson and Gress 1983, Riseborough 1986), and several species from the Great Lakes (Weseloh et al. 1983, 1995,

TABLE 15.3
Metal Levels by Major Taxa of Marine Birds

	Mercury			Cadmium			Lead		
	Range of Means	Median of Means	No. of Data Sets	Range of Means	Median of Means	No. of Data Sets	Range of Means	Median of Means	No. of Data Sets
Loons and Grebes	9.7–10.9	10.4	3			0	0.5		1
Penguins	0.2–1722.5	2.5	4	0.1–0.4	0.2	3	nd–1.7	0.28	3
Albatrosses	1.6–40[a]	6.7	30	0.05–2.5	0.47	22	0.2–40.2	0.97	22
Shearwaters, Petrels	0.2–30.7	2.6	35	0.07–0.95	0.4	21	0.1–40.8	1.4	19
Storm Petrels	0.2–12.9	6.5	2	1.2	—	1	19.2		1
Gannets, Boobies	2.9–4.5	3.8	6	0.05–0.22	0.13	4	0.82–3.13	1.3	7
Cormorants, Shags	0.4–22	3.2	6	0.2–0.96	0.3	3	1.0–2.2	1.9	3
Frigatebirds and Tropicbirds	1.7–6.4	2.5	3	1.7–3.5	2.4	3	0.63–1.5	0.68	3
Herons and Egrets	0.3–6.1	3	19	0.08–2.0	0.1	4	0.1–9.7	0.59	13
Storks	0.1–2.6	0.8	9	0.03–0.21	0.16	4	0.7–3.6	1.6	4
Gulls	0.2–32	1.7	73	0.08–1.2	0.22	35	0.17–25.8	2.15	40
Terns	0.1–12.9	2.2	42	0.03–1.25	0.13	33	0.1–4.35	1.38	36
Skimmers	0.1–14.4	0.2	4	0.06–0.18	0.13	3	0.8–4.1	1.6	4
Skuas	1.3–8.1	6.8	7			0			
Puffins, Guillemots	1.2–9.2	3.8	9			1			0

[a] Albatross lead levels include individuals with evidence of lead poisoning. Otherwise highest mean is 3.1.

Gilbertson 1989, Gilbertson et al. 1991). Recent studies have demonstrated that some isomers of PCB, PCDF, and PCDD are up to 1000 times more toxic than others (Safe 1990).

In addition to the well-documented effect on egg-shell formation, these compounds have many effects on the nervous system, delayed growth, decreased parental attentiveness, impaired courtship behavior, brought about cessation of nest building and incubation behavior, impaired avoidance behavior and brought about destruction of eggs (Ratcliff 1970, Dahlgren and Linder 1974, Tori and Peterle 1983). Levels of the DDT metabolite DDD in brain exceeding 150 µg/g are associated with lethality (Prouty et al. 1975).

15.6.1 DDT AND EGG-SHELL THINNING

DDT and its breakdown products DDD and DDE have been extensively studied. Field studies have shown that exposure to great concentrations of DDT (212 mg/kg in brain, 838 mg/kg in liver) in wild Bald Eagles (*Haliaeetus leucocephalus)* causes tremors prior to death (Garcelon and Thomas 1997), and these studies were used to determine a threshold of 8 mg/kg, wet weight, DDE in eggs. Abnormal nest defense behavior can result (Fox and Donald 1980).

The classic example of the effect of DDT on seabirds involved egg-shell thinning that occurred in the 1960s with raptors and fish-eating birds. Eggs became so thin shelled that when the birds sat on them to incubate, they broke. Significant egg-shell thinning was shown in Brown Pelicans (*Pelecanus occidentalis*) and White Pelicans (*P. erythrorhynchos*; Blus et al. 1971, Anderson et al. 1975), Northern Gannets (*Sula bassanus*) in Quebec, Double-crested Cormorants in Canada, murres (Gress et al. 1971), petrels (Coulter and Risebrough 1973), and many other seabirds (Nisbet 1994). The correlation between DDE residues and percent thinning varies from species to species, and from study to study, such that the percent of eggshell thinning can indicate elevated DDE levels, but cannot quantitatively predict DDE residues (Blus 1996).

Clear population declines occurred in Brown Pelicans in southern California (Blus 1982) and breeding pelicans disappeared from most of the southeastern United States. Northern Gannets in the Gulf of St. Lawrence (Chapdelaine et al. 1987) also declined, as did cormorants (Weseloh et al. 1995). It was the decline of fish-eating birds that ultimately led to the general ban of DDT for use in the United States. In terns many of the same effects were noted in the 1970s, including thin eggshells (Figure 15.9; Hays and Risebrough 1972, Gochfeld 1971).

The mechanism of eggshell thinning involved disruptions in calcium metabolism, which affected calcium deposition in the eggs and their subsequent thickness (Peakall 1970, 1985, 1986,

FIGURE 15.9 Thin eggshell from a Common Tern nesting on Long Island in the 1970s, illustrating thin eggshells resulting in broken eggs. (Photo by M. Gochfeld.)

Peakall et al. 1973). Among the mechanisms proposed was the efficient induction of liver enzymes by xenobiotics, which in turn increased the breakdown of estrogenic hormones. This represents the first and one of the best-documented examples of endocrine disruption.

Fox (1976) provided another mechanism whereby DDE could affect eggshells to induce embryonic mortality independent of shell thinning. Abnormalities in shell structure and composition were responsible for damage, which resulted in egg disappearance or embryonic death through hypoxia (Fox 1976). Organochlorine-induced estrogenic effects have been suggested based on field observation of increases in the incidence of female–female pairings in gull populations in regions contaminated by DDT (Fry et al. 1987, Fox 1992). Female–female pairs may be due to a shortage of eligible males resulting from a skewed sex ratio (Fry et al. 1987), resulting from increased mortality of males or feminization (Fry and Toone 1981).

15.6.2 OTHER CYCLODIENE PESTICIDES

One group of chlorinated hydrocarbon pesticides are represented by dieldrin, aldrin, endrin, and related compounds (reviewed by Peakall 1996). These compounds are interconverted to some extent and are readily metabolized with 12-ketoendrin being the environmentally important form. They were used extensively as a coating on seeds, and acute avian mortality was widely reported among seed-eating passerines. Metabolic pathways and relative toxicity varies among organisms. A die-off of terns in Holland was putatively attributed to exposure to telodrin. Eggshell thickness and hatchability decreased and chick mortality increased with increasing levels (Peakall 1996).

Other cyclodienes reviewed by Wiemeyer (1996) include chlordane (widely used for termite control and known to be carcinogenic), heptachlor and heptachlor epoxide, methoxychlor, toxaphene, and mirex (used extensively for Fire Ant, *Solenopsis invicta* control). Aside from laboratory studies, relatively little is known of these compounds in birds. Since they are little known, they are often not analyzed, which perpetuates the lack of information.

15.6.3 PCB

Gilbertson (1989) summarized episodes of apparent PCB or PCDD effects on fish-eating birds in the Great Lakes: embryo mortality, subcutaneous and pericardial edema, growth retardation, liver damage, aberrant breeding behavior, and developmental defects in Herring Gulls, Forster's (*Sterna fosterii*) and Common Terns, and Double-crested Cormorants (see Table 15.4). Abnormal porphyrin metabolism, correlated with both TCDD and PCB was reported in the gulls (Fox et al. 1988). Adult mortality was associated with PCB. The death of more than 15,000 Common Murres (= Guillemots) in the Irish Sea in 1969 was associated with a twofold increase in PCB in liver, although the PCB were considered only contributory and not the primary cause (Parslow and Jeffries 1973).

In the Great Lakes, concentrations of certain PCB congeners (co-planar isomers) are associated with both embryo lethality and greater rates of congenital deformities (Giesy et al. 1994a, b; Table 15.4), including chicks born with extra legs, a variety of craniofacial abnormalities such as cross-bill (Hoffman et al. 1987, Figures 15.10 and 15.11). Similar deformities, as well as feather abnormalities, eggshell thinning, cross-bills, extranumerary limbs, microcephalia, anophthalmia and microphthalmia, and cyclopia, were noted among Common Terns on Long Island (Gochfeld 1975, Figures 15.12 and 15.13), possibly due to PCB (Hays and Risebrough, 1972), or to a combination of PCB and mercury (Gochfeld 1971). These abnormalities in terns disappeared by the mid-1970s.

Inadequate parental care was implicated as the cause of poorer hatching success of Herring Gulls and Forster's terns breeding on the Great Lakes (Fox et al. 1978, Kubiak et al. 1989), the mechanism being disruption of adult behavior, embryotoxicity, or a combination of the two. Mora et al. (1993) reported that nest-site tenacity of Caspian Terns (*Hydroprogne caspia*) in the North American Great Lakes was inversely associated with concentrations of PCB in the blood of the

TABLE 15.4
Seabirds in Which Certain Effects of Endocrine-Disrupting Chemicals such as PCB from the Great Lakes Ecosystem Have Been Reported (see also NRC 1999)

Species	Reproductive Effects	Behavioral Deficits	Population Declines	Community Effects	Sources
Double-crested Cormorant	X		X	X	Anderson and Hickey 1972
					Fox et al. 1991a
					Giesy et al. 1994a
					Larson et al. 1996
Herring Gull	X	X	X		Fox et al. 1978, 1991a
					Peakall and Fox 1987
					Grasman et al. 1996
Forster's Gull	X	X	X		Hoffman et al. 1987
					Kubiak et al. 1989
					Allan et al. 1991
					Fox et al. 1991a, b
					Giesy et al. 1994
Caspian Tern	X	X	X		Mora et al. 1993

FIGURE 15.10 Double-crested Cormorant chick with deformed upper mandible. This individual was found in a Lake Huron colony in 1989. (Photo by Birgit Braune, Canadian Wildlife Service.)

FIGURE 15.11 Herring Gull chick with extra legs, hatched in a Lake Huron colony in 1972. (Photo by W. Southern, courtesy of Canadian Wildlife Service.)

FIGURE 15.12 Abnormality of down production in Common Tern chick from Long Island in the 1970s. (Photo by M. Gochfeld.)

FIGURE 15.13 Feather loss in Common Tern chicks from Long Island in the 1970s. Chicks were missing wing and tail feathers. (Photo by M. Gochfeld.)

terns. This may have been due to direct neurobehavioral effects, or to depressed reproduction at the locations where concentrations of PCB were greater. The establishment of a cause–effect relationship based on field studies of pollutant-related behavioral changes is difficult because of other confounding factors such as weather, changes in food supply, and human disturbance.

15.6.4 DIOXINS AND DIELDRIN

Laboratory studies with nonseabirds have shown a variety of effects with dioxin and dieldrin, including lethality, chick edema, decreased growth rates (Hoffman et al. 1996), decreases in locomotory responses, deficits in body motions and balance (Gesell et al. 1979), aggressive behavior (Dahlgren and Linder 1974, Kreitzer and Heinz 1974), and changes in brain neurotransmitters such as serotonin, norepinephrine, and dopamine (Sharma et al. 1976).

Fish-eating birds inhabit areas contaminated with TCDD are chronically exposed during embryonic development via the yolk. TCDD, dieldrin, and some related chemicals have antiestrogenic effects (Janz and Bellward 1996), and *in ovo* exposure to these compounds during the perinatal period may be responsible for certain behavioral characteristics and reproductive dysfunction. Organochlorines may cause behavioral effects through mechanisms such as endocrine disruption (effects on steroid or thyroid hormone metabolism) or by disrupting vitamin A homeostasis. The effects of organochlorines on thyroid hormone have profound effects on neurological function, which are similar to the endocrine disruption mechanism for the observed behavioral changes in birds (Porterfield 1994). However, little experimental evidence is present for birds, and none exists for seabirds (Janz and Bellward 1996).

Finally, the difficulty of assessing the causes of physiological, behavioral, or reproductive failures in the field are made more difficult because wild seabirds are exposed to mixtures. For example, Renzoni et al. (1986) noted that chlorinated hydrocarbon and mercury levels were higher in tissues and eggs from Cory's Shearwaters collected in the Mediterranean Sea than from those collected from the Atlantic. Eggshells were thinner in the shearwaters from the Mediterranean Sea as well, but because of the presence of both contaminants, no single cause could be ascribed, or synergism may occur.

15.6.5 Selected Syndromes

In addition to shell-thinning, chlorinated hydrocarbons and metals have been implicated in several clearly defined syndromes which are mentioned below. However, many of the effects of these chemicals are nonspecific, including failure to thrive, loss of appetite, and listlessness, which are difficult to ascribe to a specific contaminant. Risebrough and Hays (1972) called attention to chick edema disorder that was previously demonstrated in laboratory birds exposed to PCB. Gilbertson (1983) provided evidence that chick edema disease was associated with a TCDD concentration in eggs of about 1 ppb. Many of the congeners of PCB, PCDF, and PCDD are capable of inducing this condition, although their potency varies.

Increased incidence rates of craniofacial defects are associated with pollutants, but cross-bill can occur in uncontaminated populations as well as in laboratory-raised chicks with low contamination levels (Kuiken et al. 1999). From 1988 to 1996 there were 31 Double-crested Cormorant chicks with bill defects at 21% of the Great Lakes colonies surveyed for a prevalence rate of 0.28 per thousand, while no such abnormalities occurred at uncontaminated reference sites (Ryckman et al. 1998). The evidence from both locations supported a causal relationship of PCB (Hays and Risebrough 1972).

A variety of feather abnormalities have been reported in seabirds. A dramatic condition labeled premature feather loss involved the breakage of feather shafts of wing and tail feathers at about 2 weeks of age, resulting in chicks with fully feathered bodies but no flight feathers. Common Tern chicks with feather loss had significantly higher mercury levels than normal chicks (Gochfeld 1980a), but OC were not analyzed. The disease was traced to a rather uniform fragility or fault bar in the feathers, produced by anything that interferes with nutrition of the feather during a critical period. A role for mercury was suggested as well as a synergistic interaction of mercury and PCB (Gochfeld 1975).

15.6.6 Toxic Equivalency Factors

Safe (1990) and others have developed a series of toxic equivalency factors (TEF) for these compounds, based on their potency in inducing ethoxyresorufin deoxylase (EROD) with TCDD having a potency of 1. The relative potencies vary by several orders of magnitude. The advantage of the TEF system is that it is possible to determine a single total dioxin equivalency value for an organism by summing the concentration of each PCB, furan, and dioxin isomer multiplied by its TEF and expressed as toxic equivalents (TEQ) in picograms per gram of sample (pg/g).

This creates at least an estimate of the total toxic burden from these compounds. The weakness of the approach, however, is the uncertainty over whether many of the effects of concern are directly related to the ability to induce EROD. The World Health Organization recently developed TEF for wild birds (Van den Berg et al. 1998). These yield lower TEQ values. Herring Gull chicks in the Great Lakes, showed an inverse relationship between plasma corticosterone and TEQ (Lorenzen et al. 1999). Although the effect of PCB was realized by the late 1960s and of dioxins by the late

1970s, it required improvements in analytic techniques and isomer-specific analysis before meaningful progress could be made in understanding mechanisms. Among PCB it is the co-planar (flat) isomers that have the greatest potency relative to TCDD. In Forster's Terns at Green Bay, over 90% of the TCDD equivalency was contributed by the 2,3,4,3',4' and 3,4,5,3',4' pentachlorobiphenyls (Kubiak et al. 1989). In general, PCB are the greatest contributor to TEQ everywhere.

15.7 PETROLEUM PRODUCTS

Petroleum in the form of crude or processed oil and its polycyclic aromatic hydrocarbon (PAH) constituents are considered together because petroleum products are a major source of PAH in the environment (Albers 1995). This class of products consists of crude oils and a variety of refined products, such as diesel oil, fuel oil, gasoline, kerosene, solvents, jet fuels, and home-heating oils, among others (Burger 1997). PAH are hydrocarbons with two or more fused aromatic 6-carbon rings, and range from naphthalene to coronene. Crude oil contains up to 7% PAH (Albers 1995).

15.7.1 POLYCYCLIC AROMATIC HYDROCARBONS

PAH are ubiquitous and are derived mainly from human activities including transportation accidents, oil seeps, road runoff, incineration, and industrial sources. They are released from the combustion of wood, coal, and oil and are transported in the atmosphere. The concentrations of PAH are highest in sediment, and surface water has higher concentrations than most of the water column (Hellou 1996). PAH are widespread in marine environments and have been implicated as carcinogens in fish (Malins et al. 1988). Total PAH concentrations exceeding 100 ppm are not unusual in fish in the region of oil spills (Hellou 1996). A major difficulty in dealing with PAH is that not all PAH are analyzed. Some are known to be carcinogenic; others are not.

15.7.2 OIL SPILLS AND OILING

The effects of oil on seabirds have been extensively demonstrated. Oil in coastal and marine waters is derived from natural ocean seeps and from anthropogenic activities (Figure 15.14). Natural seeps account for 11% of the oil that flows into the oceans. Since natural seeps emanate from bedrock

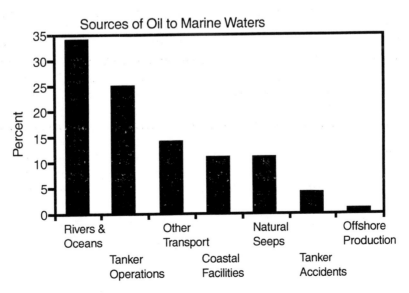

FIGURE 15.14 Sources of oil into marine waters. Note that the highest percentage of oil comes from runoff from rivers into oceans, but tanker operations are second. (After Burger 1997.)

FIGURE 15.15 Laysan Albatross on Midway with a small patch of oil on its breast. Such low-level chronic oil exposure can prove a problem for seabirds that continually bring back oil to their eggs or chicks, and who ingest it chronically when they preen. (Photo by J. Burger.)

at the bottom of the ocean, the oil is dispersed over large areas and does not form massive oil slicks that follow blowouts and tanker wrecks. The pumping of bilge water is a widespread practice that contributes great amounts of oil.

Seabirds are particularly vulnerable to oil spills because many species nest in large colonies of hundreds or thousands near coasts and on small offshore islands (Croxall 1977, Burger 1997). Differences in vulnerability among seabirds reflect differences in breeding schedules, foraging methods, and geographical ranges (Brown 1982, Burger 1997). Species that swim on the surface have the greatest exposure. Major oil spills can cause massive mortality to seabirds. Less severe oil releases can still cause reproductive losses back at their nests from oiling of the eggs. Further, there is more oil spilled in small chronic spills than in the major oil spills (Burger 1997). Thus in evaluating the effect of oil on seabirds it is also critical to understand the effect of chronic, low-level oil exposure (Figure 15.15). In addition to the usual exposures (inhalation, ingestion), oil coats the feathers of seabirds, thereby decreasing their insulation abilities. When they remove oil by preening, they ingest it. Adults can transfer oil from their plumage to eggs during incubation, thereby decreasing their hatching success (Lewis and Malecki 1984). Oil on the plumage of seabirds can be fingerprinted and used to identify the source of the oil spill (Furness and Camphuysen 1997).

Burger and Fry (1993) reviewed a number of beached bird surveys and found that proportion of corpses found that were oiled ranged from about 5% (open shores of Washington, 1981–1984) to 68% in the Netherlands (1969–1990) and 83% in California (1971–1985). Even with information from beached-bird surveys, it is difficult to determine the ratio between oiled birds found along beaches and the number of oiled birds that actually died. Models used in the North Pacific show that ocean currents, wind, seabird distribution, and the persistence of oiled carcasses affect the assessment of mortality; few birds killed by oil spills at sea are likely to come ashore (Burger and Fry 1993).

Estimates of mortality from major oil spills are now routinely made because of the need for damage assessment (Burger 1997). For example, the estimated mortality from the *Apex Houston* accident was 10,577, while the estimated mortality from the *Exxon Valdez* accident in Prince William Sound, Alaska (1989) was between 250,000 to 645,000 (see Burger 1997 for detailed methodology). Nearly 75% of the mortality from the latter spill was Common Murres, 7% were other alcids, and 5% were various seaducks (see Piatt et al. 1990). Nine years after the spill, seabird populations were still depressed for diving species, including cormorants, Pigeon Guillemot (*Cepphus columba*), murres, and ducks; surface-feeding species were not (Irons et al. 2000), nor were Black Oyster-catchers (*Haematopus bachmani*; Murphy and Mabee 2000). They attributed the effects to persistent oil remaining in the environment that reduced forage fish abundance.

Hoffman (1990) reported that external application of oil reduced hatchability in all 34 studies reviewed, and also retarded growth, caused developmental defects, and induced other sublethal effects in chicks. The aromatic fraction of oil was the most toxic. Lewis and Maleck (1984), in an experiment with Great Black-backed Gulls (*Larus marinus*), found that only 10 µl of No.2 fuel oil caused 50% embryonic mortality. Similar effects have been reported for Great Black-backed Gulls (Birkhead et al. 1973), Sandwich Terns (*Thalasseus sandvicensis*; Rittinghaus 1956), and Brown Pelicans (Parnell et al. 1984).

Other sublethal effects on seabirds include lowered breeding rates and breeding success (Wedge-tailed Shearwaters (*Puffinus pacificus*; Fry et al. 1986), reduced weight gain and survival of chicks (Leach's Storm-petrels, *Oceanodroma leucorhoa*; Trivelpiece et al. 1984), reduced breeding rates and reduced clutch size (Cassin's Auklets, *Ptychoramphus aleuticus*; Ainley et al. 1981), and reduced feeding rates (Sanderlings, *Calidris alba;* Burger and Tsipoura 1998). Exposure to oil also causes a number of growth and developmental defects (Hunt 1987), osmoregulation deficits and changes in corticosterone levels (Peakall 1992), induction of mixed function oxidases (Lee et al. 1985), and induction of Heinz-body hemolytic anaemia (Fry and Addiego 1987). Although some of these latter effects have been demonstrated only in the laboratory, they might account for the lowered reproductive success and survival of rehabilitated, oiled birds (Morant et al. 1981).

The best-studied example of the long-term effects of oil is the *Exxon Valdez*. Analyses of marine bird surveys conducted in Prince William Sound in 1972 before this spill, and in 1989, 1990, 1991, and 1993 indicated that several marine birds that eat fish declined, while those that fed on benthic invertebrates did not decline following the spill (Agler et al. 1999). The declines were over 50% for Bonaparte's Gull (*Larus philadelphia*), cormorant (*Phalacrocorax* spp.), Pigeon Guillemot (*Cepphus columba*), murrelets (*Brachyramphus* spp.), Parakeet Auklet (*Cyclorrhynchus psittacula*), Tufted Puffin (*Fratercula cirrhata*), and Horned Puffin (*F. corniculata*; Agler et al. 1999). Agler et al. (1999) point out the difficulties of assigning cause-and-effect since there have also been population declines of some of these species in the Gulf of Alaska, Bering Sea, and along the coast of California, partly attributed to climatic shift and overfishing that have changed forage-fish abundance.

Wiens et al. (1996) suggested that seabird community structure had recovered by 1991, indicating that seabird populations have some resiliency to severe short-term perturbations. Recovery is largely dependent upon having nearby source populations that can supply breeders for populations depleted by oil.

15.8 PLASTICS, FLOATABLES, AND ARTEFACTS

Plastics and floatables have increased in the ocean and along coasts over the last 50 years. The primary objects of concern are industrial pellets (raw material for manufacturing plastics) and manufactured items (fishing nets, kite lines, balloons, cigarette lighters, and other disposable plastic

TABLE 15.5
Plastic in the Gizzards of Selected Seabirds

Species	Body Weight (%)	Relaxed Gizzard Volume (%)	Source
Northern Fulmars	1	59	Furness 1985
Wandering Albatross	2	82	Ryan 1988b
Laysan Albatross	1–8	—	Fry et al. 1987a

items). The sources are from recreational and commercial fishing and transport vessels, and discharges from coastal plants near waterways. Seabirds encounter plastic when they forage, and become entangled or ingest them. Plastics appear to be increasing in the world, although Annex V of the International Convention for the Prevention of Pollution from Ships (1973, modified in 1978) that came into force in 1988, should have resulted in declines worldwide. For more extensive reviews of this topic, see Laist 1987, Pruter 1987, Ryan 1988a, Nisbet 1994.

Ingestion is the major cause of mortality and injury, although most seabirds regurgitate indigestible plastic and other floatables. Procellariformes (other than albatrosses) do not regurgitate and may retain these objects for months or years. In Procellariformes, including albatrosses, most exposure is through the feeding of chicks, who swallow the plastic and then are unable to regurgitate it (Ryan 1988a). Plastics have been found in the guts of a majority of seabirds examined, but in all surveys, Procellariformes have both the highest loads per bird, and the highest incidence (Nisbet 1994, Table 15.5). Furness (1985) found evidence for a negative relationship between plastic loads and weight. Fry et al. (1987) found higher plastic loads in dead than live Laysan Albatrosses from Midway Island, resulting in impacted proventriculi and ulcerations (Figure 15.16). Eventually, if complete obstruction does not occur, the gut may become perforated by sharp plastic fragments. Sievert and Sileo (1993) found that Laysan chicks on Midway Island with large volumes of

FIGURE 15.16 Contents from the stomach of one young Laysan Albatross chick found dead on Midway Island in 1997. Note the squid beaks (black), plastic (including buttons), and a cigarette lighter from Japan. (Photo by J. Burger.)

proventricular plastic had asymptotic fledging weights that were significantly lower than did chicks with smaller amounts. However, plastics had no effect on growth in Black-footed Albatross, and ingested plastic pellets did not reduce meal size and food consumption in White-chinned Petrels (*Procellaria aequinoctalis*; Ryan and Jackson 1987).

Entanglement in discarded fishing nets, kite strings, plastic six-pack rings, lures, packing materials, and other debris is a substantial cause of mortality for seabirds (Gochfeld 1983, Laist 1987). Discarded nets, fishing line, string, and plastic six-pack holders continue to be a problem (Figure 15.17). More information is needed on the relative effect of plastic ingestion on young, juvenile, and adult seabirds before the importance of plastics on survival and population dynamics can be ascertained.

15.9 INVESTIGATING CONTAMINANT EFFECTS

Although it is tempting to attribute deaths, abnormalities, reproductive failure, and population declines to contaminants, documenting such a relationship is difficult, time consuming, and often expensive. Even finding a high level of a contaminant does not assure that it has produced a toxic effect. One of the important aspects of field biomonitoring of pollutant levels in seabirds is to be able to interpret whether the levels found in the field are unusual for a given species or place, whether there is cause for concern, and whether they have caused the observed problems. The toxic levels of mercury in seabirds might be expected to differ from land birds because mercury in the open ocean is mainly natural, and seabirds would have developed mechanism to deal with it. Furness (1996) has offered a similar argument for cadmium in seabirds. There are also problems in understanding the effects of the chlorinated hydrocarbons because of correlations. In epidemiology this problem is called confounding and requires new studies to control for co-linearity among potentially independent variables.

There are two methods of interpreting field levels: (1) comparison of levels with known levels in other similar species, and (2) comparison of field levels known to cause adverse effects in the laboratory, albeit in nonseabirds. In Table 15.3 we give the mean levels of several metals in the feathers of several seabirds, and compare them to those in some other nonseabirds (Burger 1993). Although such a table could be produced for other tissues, there are far fewer data because of the difficulties of collecting seabirds for tissue analysis. Feather collection has thus been a useful tool for both assessing levels in specific seabirds and for biomonitoring (Goede and deBruin 1984, Furness et al. 1986, Burger 1993, Hahn et al. 1993, Gochfeld and Burger 1998). In some cases, however, other tissues must be collected for analyses.

Figure 15.8 can also be used to compare levels to those known to cause adverse effects in laboratory animals (dashed line). This analysis shows that populations of some species have higher mercury and lead levels than those known to cause adverse effects for mercury. It is possible that species with such levels are more tolerant of the metals, but it is also possible that birds at the higher end of the distribution have already died or failed to return to breeding or other sampling areas, so that distributions may under-represent metal levels (compared to studies of moribund and dead birds that over-estimate contaminant levels).

When epidemics of defects, disease, reproductive failure, or population declines are detected, it is reasonable to consider the possible contribution of contaminants. In initial studies, abnormal or dead birds were collected and residues of metals or pesticides were analyzed, usually in only one or two tissues and often in only a few individuals. The concept of controlled studies evolved slowly. With the advent of epidemics of developmental defects in the Great Lakes, coupled with advances in toxicology in general, a more comprehensive protocol developed (Gilbertson 1988), employing the measurement of toxicant residues or their metabolites and biochemical markers.

Today rather elaborate protocols can be conducted, as illustrated by Rattner et al. (2000) who investigated the population decline of herons at Pea Patch Island in Delaware Bay, using a

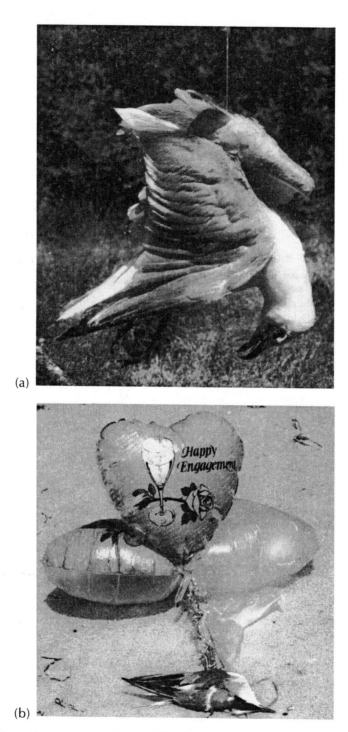

FIGURE 15.17 Young Herring Gull caught in a fishing line at Captree, Long Island (top) and adult Common Tern (bottom) killed by balloons and string from an engagement celebration. (Photos by J. Burger.)

combination of residue analysis (metals, 21 organochlorine pesticides); isomer-specific PCB, PCDF, and PCDD analysis; histopathology; organ weights; and biomarkers of tissue damage, enzyme induction, and oxidative stress, including reduced and oxidized glutathione. Benzyloxyresorufin-O-dealkylase (BROD) and ethoxyresorufin-O-dealkylase (EROD) are enzymes readily induced by organochlorines, and elevated levels reflect this important toxic activity. EROD induction has been used to rank the potency of different compounds (Safe 1990).

15.10   TEMPORAL TRENDS

In the 50 years that contaminants in wildlife has been a serious topic for study, the levels of many contaminants have declined in environmental media and in biota, but the frequency of oil spills and the atmospheric emission of pollutants from power plants can be expected to increase as the world population grows and increases its reliance on fossil fuel. PCB and organochlorine residues declined in some Lake Michigan species from the late 1970s to 1990 (Heinz et al. 1994). From 1970 to 1995, OC levels in cormorants declined in Lake Ontario, Lake Huron, and Lake Superior but not in Lake Erie (Ryckman et al. 1998), and for a suite of 11 organochlorines from 1974 to 1995 in Herring Gull eggs (Pekarik and Weseloh 1998). Between 1983 and 1993, PCB, DDE, and other OC declined in the eggs of six seabird species in north Norway and northwest Russia, while mercury remained unchanged (Barrett et al. 1996). Elliott et al. (1988) showed not only a decline in DDE levels in fresh eggs from 1968 to 1984, but an associated decline in eggshell thinning in Northern Gannets in Eastern Canada.

15.11   FUTURE RESEARCH NEEDS AND CONCLUSIONS

The research needs for chemicals and pollution fall into six main categories: (1) methodologies for more accurate analysis and comparability of measurements, (2) biomarkers of exposure, (3) comparative studies within and among species including sampling strategies, (4) comparative studies within and among trophic levels and other ecological considerations, (5) studies of mixtures, and (6) detailed effects studies in the field and the laboratory to identify thresholds (if any) and dose–response curves.

Clearly a wide range of inorganic and organic chemicals, plastics, and other floatables can result in direct mortality of seabirds, as well as causing sublethal behavioral, reproductive, and physiological effects. It is more difficult to show how this direct mortality and sublethal effects affect population dynamics. Seabirds have evolved mechanisms for coping with a wide range of natural events (such as storm tides, food shortages, changes in habitat) that cause mortality and sublethal effects, and these same mechanisms confer a certain resiliency to pollution events.

However, there are clear examples where chemicals have caused temporary population declines, including several species following the excessive use of DDT (pelicans and cormorants; King et al. 1978, Risebrough 1986), and PCB and other contaminants (gulls and terns in the Great Lakes; Fox et al. 1978, Gilbertson 1988, Fox 1992), as well as oil spills (seabirds in the Gulf of Alaska; Agler et al. 1999, Irons et al. 2000). These declines were largely due to point-source pollution, and not to wide-scale chronic pollution over large geographic areas.

These cases illustrate the potential for wide-spatial-scale effects on seabirds under several conditions: (1) multiple point-source events, (2) overlapping events in the same areas, (3) increased atmospheric transport and subsequent deposition, and (4) lack of human intervention. The last one is particularly critical since if the ban on DDT had not come about, there could have been catastrophic effects on several seabirds species, and the potential existed for extinction, particularly of species that nested coastally (rather than on distant offshore islands). There is no cause for complacency, and every reason to continue with vigorous research programs aimed at understanding the fate and effects of chemicals on seabirds at a local, regional, and global scale.

ACKNOWLEDGMENTS

We acknowledge fruitful discussions with many colleagues over the years, including K. Cooper, B. D. Goldstein, G. Fox, R. Furness, M. Gilbertson, H. Hayes, J. Jehl, Jr., K. King, B. G. Murray, I. Nisbet, C. Powers, C. Safina, J. Saliva, B. A. Schreiber, and J. Spendelow. We thank G. Fox for helpful comments on the manuscript and for providing some photographs from the Great Lakes; T. Benson, C. Dixon, T. Shukla, S. Shukla, and M. McMahon for laboratory assistance; and R. Ramos for graphics. Over the years our research has been funded by NIMH, NIEHS (ESO 5022), EPA, the Consortium for Risk Evaluation with Stakeholder Participation (CRESP) through the Department of Energy (AI # DE-FC01-95EW55084, DE-FG 26-OONT 40938), the New Jersey Department of Environmental Protection (Endangered and Nongame Species Program, Office of Science and Research), Penn Foundation, and the Environmental and Occupational Health Sciences Institute.

LITERATURE CITED

AINLEY, D. G., C. R. GRAU, T. E. ROUDYBUSSH, S. H. MORRELL, AND J. M. UTTS. 1981. Petroleum ingestion reduces reproduction in Cassin's Auklets. Marine Pollution Bulletin 12: 314–317.

ALBERS, P. H. 1995. Petroleum and individual polycyclic aromatic hydrocarbons. Pp. 330–355 in Handbook of Ecotoxicology (D. J. Hoffman, B. A. Rattner, G. A. Burton, Jr., and J. Cairns, Jr., Eds.). Lewis Publishers, Boca Raton, FL.

AGLER, B. A., S. J. KENDALL, D. B. IRONS, AND S. P. KLOSIEWSKI. 1999. Declines in marine bird populations in Prince William Sound, Alaska coincident with a climatic regime shift. Waterbirds 22: 98–103.

ALLAN, R. J., A. J. BALL, V. W. CAIRNS, G. A. FOX, A. P. GILMAN, D. B. PEAKALL, D. A. PIEKARZ, J. C. VANOOSDAM, D. C. VILLENEUVE, AND D. T. WILLIAMS. 1991. Toxic chemicals in the Great Lakes and associated effects. Volume II. Effects. Environment Canada, Department of Fisheries and Oceans, Health and Welfare, Ottawa, Canada.

ALLEN, G. T., S. H. BLACKFORD, AND D. WELSH. 1998. Arsenic, mercury, selenium, and organochlorines and reproduction of interior Least Terns in the northern Great Plains, 1992–1994. Colonial Waterbirds 21: 356–366.

ANDERSON, D. W., AND F. GRESS. 1983. Status of a northern population of California Brown Pelican. Condor 83: 79–88.

ANDERSON, D. W., J. R. JEHL, R. W. RISEBROUGH, L. A. WOODS, L. R. DEWEESE, AND W. G. EDGECOMB. 1975. Brown Pelicans: improved reproduction off the southern California coast. Science 190: 806–808.

ASHLEY, D. G., C. R. GRAU, T. E. ROUDYBUSH, S. H. MORRELL, AND J. M. UTTS. 1981. Petroleum ingestion reduces reproduction in Cassin's Auklets. Marine Pollution Bulletin 12: 314–317.

BARRETT, R. T., J. U. SKAARE, AND G. W. GABRIELSEN. 1996. Recent changes in levels of persistent organochlorines and mercury in eggs of seabirds from the Barents Sea. Environmental Pollution 92: 13–18.

BEARHOP, S., G. D. RUSTON, AND R. W. FURNESS. 2000. Dynamics of mercury in blood and feathers of Great Skuas. Environmental Toxicology and Chemistry 19: 1638–1643.

BELLROSE, F. C. 1959. Lead poisoning as a mortality factor in waterfowl populations. Illinois Natural History Survey Bulletin 27: 235–288.

BERG, W., A. JOHNELS, B. SJOSTRAND, AND T. WESTERMARK. 1966. Mercury content in Swedish birds from the past 100 years. Oikos 17: 71–83.

BEYER, W. N., G. H. HEINZ, AND A. W. REDMON-NORWOOD. 1996. Environmental Contaminants in Wildlife: Interpreting Tissue Concentrations. SETAC Spec. Publ., Lewis Publishers, Boca Raton, FL.

BIRKHEAD, T. R., C. LLOYD, AND P. CORKHILL. 1973. Oiled seabirds successfully cleaning their plumage. British Birds 66: 535–537.

BLUS, L. J. 1982. Further interpretation of the relation of organochlorine residues in Brown Pelican eggs to reproductive success. Environmental Pollution A 28: 15–33.

BLUS, L. J. 1996. DDT, DDD, and DDE in birds. Pp. 49–73 *in* Environmental Contaminants in Wildlife: Interpreting Tissue Concentrations (W. N. Beyer, G. H. Heinz, and A. W. Redmon-Norwood, Eds.). SETAC Spec. Publ., Lewis Publishers, Boca Raton, FL.

BLUS, L. J., E. CROMARTIE, L. McNEASE, AND T. JOANEN. 1979. Brown Pelican: population status, reproductive success, and organochlorine residues in Louisiana, 1971–1976. Bulletin of Environmental Contamination and Toxicology 22: 128–135.

BLUS, L. J., R. G. HEATH, C. D. GISH, A. A. BELISLE, AND R. M. PROUTY. 1971. Eggshell thinning in the Brown Pelican: implications of DDE. BioScience: 1213–1215.

BOSVELD, A. T. C., AND M. VAN DEN BERG. 1994. Effects of polychlorinated biphenyls, dibenzo-*p*-dioxins and dibenzofurans on fish-eating birds. Environmental Reviews 2: 147–166.

BRAUNE, B. W. 1987. Comparison of total mercury levels in relation to diet and molt for nine species of marine birds. Archives of Environmental Contamination and Toxicology 16: 217–224.

BRISBIN, I. L., JR. 1991. Avian radioecology. Pp. 69–140 *in* Current Ornithology, Vol. 8 (D. M. Power, Ed.). Plenum Press, New York.

BROWN, R. C. G. 1982. Birds, oil and the Canadian environment. Pp. 105–112 in Oil and Dispersants in Canadian Seas: Research Appraisals and Recommendations (J. B. Sprague, J. H. Vandermuelen, P. G. Wells, Eds.). Environ. Protect. Service, Environment Canada (Envir. Tech. Rev. Report EPS-3-EC-82-L), Ottawa.

BULL, K. R., R. K. MURTON, D. OSBORN, AND P. WARD. 1977. High levels of cadmium in Atlantic seabirds and sea-skaters. Nature 269: 507–509.

BURGER, A. E., AND D. M. FRY. 1993. Effects of oil pollution on seabirds in the northeast Pacific. Pp. 254–263 *in* The Status, Ecology, and Conservation of Marine Birds of the North Pacific (K. Vermeer, K. T. Briggs, K. H. Morgan, and D. Siegel-Causey, Eds.). Special Publ. Canadian Wildlife Service, Ottawa.

BURGER, J. 1984. Pattern, mechanism, and adaptive significance of territoriality in Herring Gulls (*Larus argentatus*). Ornithological Monograph 34: 1–91.

BURGER, J. 1990. Behavioral effects of early postnatal lead exposure in Herring Gulls (*Larus argentatus*) chicks. Pharmacology, Biochemistry and Behavior 35: 7–13.

BURGER, J. 1993. Metals in avian feathers: bioindicators of environmental pollution. Reviews in Environmental Toxicology 5: 203–311.

BURGER, J. 1995. A risk assessment for lead in birds. Journal of Toxicology and Environmental Health 45: 369–396.

BURGER, J. 1996. Heavy metal and selenium levels in feathers of Franklin's gulls in interior North America. Auk 113: 399–407.

BURGER, J. 1997. Oil spills. Rutgers University Press, New Brunswick, NJ.

BURGER, J. 1998. Effects of lead on sibling recognition in young Herring Gulls. Toxicological Sciences 43: 155–160.

BURGER, J., AND M. GOCHFELD. 1985. Early postnatal lead exposure: behavioral effects in Common Tern chicks (*Sterna hirundo*). Journal of Toxicology and Environmental Health 16: 869–886.

BURGER, J., AND M. GOCHFELD. 1994. Behavioral impairments of lead-injected young Herring Gulls in nature. Fundamentals of Applied Toxicology 23: 553–561.

BURGER, J., AND M. GOCHFELD. 1995a. Behavior effects of lead exposure on different days for gull (*Larus argentatus*) chicks. Pharmacology, Biochemistry and Behavior 50: 97–105.

BURGER, J., AND M. GOCHFELD. 1995b. Effects of varying temporal exposure to lead on behavioral development in Herring Gull (*Larus argentatus*) chicks. Pharmacology, Biochemistry and Behavior 52: 601–608.

BURGER, J., AND M. GOCHFELD. 1995c. Growth and behavioral effects of early postnatal chromium and manganese exposure in Herring Gull (*Larus argentatus*) chicks. Pharmacology, Biochemistry and Behavior 50: 607–612.

BURGER, J., AND M. GOCHFELD. 1996. Lead and behavioral development: parental compensation for behaviorally impaired chicks. Pharmacology, Biochemistry and Behavior 55: 339–349.

BURGER, J., AND M. GOCHFELD. 1997a. Lead and neurobehavioral development in gulls: a model for understanding effects in the laboratory and the field. NeuroToxicology 18: 279–287.

BURGER, J., AND M. GOCHFELD. 1997b. Risk, mercury levels, and birds: relating adverse laboratory effects to field biomonitoring. Environmental Research 75: 160–172.

BURGER, J., AND M. GOCHFELD. 1999. Heavy metals in Franklin's Gull tissues: age and tissue differences. Environmental Toxicology and Chemistry 18: 673–678.

BURGER, J., AND M. GOCHFELD. 2000a. Metals in albatross feathers from Midway Atoll: influence of species, age and nest location. Environmental Research 82: 207–221.

BURGER, J., AND M. GOCHFELD. 2000b. Metals in Laysan Albatrosses from Midway Atoll. Archives of Environmental Contamination and Toxicology 38: 254–259.

BURGER, J., AND M. GOCHFELD. 2000c. Effects of lead on birds (Laridae): a review of laboratory and field studies. Journal of Toxicology and Environmental Health, Part B 3: 59–78.

BURGER, J., AND M. GOCHFELD. 2000d. Metal levels in feathers of 12 species of seabirds from Midway Atoll in the northern Pacific Ocean. The Science of the Total Environment 257: 37–52.

BURGER, J., AND N. TSIPOURA. 1998. Experimental oiling of Sanderlings (*Calidris alba*): behavior and weight changes. Environmental Toxicology and Chemistry 17: 1154–1158.

BURGER, J., T. SHUKLA, T. BENSON, AND M. GOCHFELD. 1994. Lead levels in exposed herring gulls: differences in the field and laboratory. Pp. 115–123 *in* Hazardous Waste: Impacts on Human and Ecological Health (B. L. Johnson, C. Xintaras, and J. S. Andrews, Jr., Eds.). Princeton Scientific Publishing Co., Princeton, NJ.

BURGER, J., T. SHUKLA, C. DIXON, S. SHUKLA, M. McMAHON, R. RAMOS, AND M. GOCHFELD. In press. Metals in feathers of Sooty Tern, White Tern, Grey-backed Tern and Brown Noddy from islands in the North Pacific. Environmental Monitoring and Assessment.

BURTON, J. D., AND P. J. STATHAM. 1990. Trace metals in seawater. Pp. 5–25 *in* Heavy Metals in the Marine Environment (R. W. Furness and P. S. Rainbow, Eds.). CRC Press, Inc., Boca Raton, FL.

CHAPDELAINE, G., P. LAPORTE, AND D. N. NETTLESHIP. 1987. Population, productivity, and DDT contamination trends of Northern Gannets (*Sula bassanus*) at Bonaventure Island, Quebec, 1967–1984. Canadian Journal of Zoology 65: 2922–2926.

CLARK, T. P., R. J. NORSTROM, G. A. FOX, AND H. T. WON. 1987. Dynamics of organochlorine compounds in Herring Gulls (*Larus argentatus*). II. A two-compartment model and data for ten compounds. Environmental Toxicology and Chemistry 6: 547–555.

COULTER, M. C., AND R. W. RISEBROUGH. 1973. Shell-thinning in eggs of the Ashy Petrel (*Oceanodroma homochroa*) from the Farallon Islands. Condor 75: 254–255.

CROXALL, J. P. 1977. The effects of oil on seabirds. Rapports es Proces-verbaux des Reunions. Conseil Intern. pour l'Exploration de la Mer 171: 191–195.

CROXALL, J. P., P. G. H. EVANS, AND R. W. SCHREIBER. 1984. Status and conservation of the World's seabirds. ICBP Technical Publication No. 2. Cambridge, England.

CUSTER, T. W., AND W. L. HOHMAN. 1994. Trace elements in Canvasback (*Aythya valisineria*) wintering in Louisiana, USA, 1987–1988. Environmental Pollution 84: 253–259.

CUSTER, T. W., R. W. ERWIN, AND C. STAFFORD. 1983. Organochlorine residues in Common Terns from nine Atlantic coast colonies, 1980. Colonial Waterbirds 6: 197–204.

DAHLGREN, R. B., AND R. L. LINDER. 1974. Effects of dieldrin in penned pheasants through the third generation. Journal of Wildlife Management 34: 320–330.

DEPARTMENT OF ENERGY (DOE). 1994. Products from the refining of crude oil. U.S. Department of Energy, Washington, D.C.

EATON, D. L., AND C. D. KLAASSEN. 1996. Principles of toxicology. Pp. 13–33 in Casarett and Doull's Toxicology: The Basic Science of Poisons (C. D. Klaassen, Ed.). McGraw-Hill, New York.

EISLER, R. 1985a. Cadmium Hazards to Fish, Wildlife, and Invertebrates: A Synoptic Review. U.S. Fish and Wildlife Service: Biological Report 85 (#1.2).

EISLER, R. 1985b. Selenium Hazards to Fish, Wildlife, and Invertebrates: A Synoptic Review. U.S. Fish and Wildlife Service: Biological Report 85 (#1.5).

EISLER, R. 1986. Chromium Hazards to Fish, Wildlife, and Invertebrates: A Synoptic Review. U.S. Fish and Wildlife Service: Biological Report 85 (#1.6).

EISLER, R. 1987. Mercury Hazards to Fish, Wildlife, and Invertebrates: A Synoptic Review. U.S. Fish and Wildlife Service: Biological Report 85 (#1.1).

EISLER, R. 1988. Lead Hazards to Fish, Wildlife, and Invertebrates: A Synoptic Review. U.S. Fish and Wildlife Service: Biological Report 85 (#1.14).

ELLIOTT, J. E., R. J. NORSTROM, AND J. A. KEITH. 1988. Organochlorines and eggshell thinning in Northern Gannets (*Sula bassanus*) from Eastern Canada, 1968–1984. Environmental Pollution 52: 81–102.

ENVIRONMENTAL PROTECTION AGENCY (EPA). 1980. Ambient water quality criteria for lead. Rep. 440/5-80-057, U.S. EPA, Washington, D.C.

FARNSWORTH, M. 1980. Cadmium chemicals. International Lead Zinc Research Organization, Inc., New York.

FIMREITE, N. 1979. Accumulation and effects of mercury in birds. Pp. 601–627 *in* The Biogeochemistry of Mercury in the Environment (J. O. Nriagu, Ed.). Elsevier Press, New York.

FINLEY, M. T., AND R. C. STENDELL. 1978. Survival and reproductive success of black ducks fed methylmercury. Environmental Pollution 16: 51–64.

FISHER, H. I., AND M. L. FISHER. 1969. The visits of Laysan albatrosses to the breeding colony. Micronesica 5: 173–221.

FOSSI, G., C. LEONZIO, AND S. FOCARDI. 1989. Seasonal variation of mixed-function oxidase activity in a population of Yellow-legged Herring Gulls: relationship to sexual cycle and pollution. Marine Environmental Research 28: 35–37.

FOWLER, S. W. 1990. Critical review of selected heavy metal and chlorinated hydrocarbon concentrations in the marine environment. Marine Environmental Research 29: 1–64.

FOX, G. A. 1976. Eggshell quality: its ecological and physiological significance in a DDE-contaminated Common Tern populations. Wilson Bulletin 88: 459–477.

FOX, G. A. 1991. Practical causal inference for ecoepidemiologists. Journal of Toxicology and Environmental Health 33: 359–373.

FOX, G. A. 1992. Epidemiological and pathobiological evidence of contaminant-induced alterations in sexual development in free-living wildlife. Pp. 147–158 *in* Chemically-Induced Alterations in Sexual and Functional Development: The Wildlife/Human Connection (T. Colborn and C. Clement, Eds.). Princeton Scientific Publishing Co., Princeton, NJ.

FOX, G.A., AND T. DONALD. 1980. Organochlorine pollutants, nest defense behavior and reproductive success in Merlins. Condor 82: 81–84.

FOX, G.A., A. P. GILMAN, D. B. PEAKALL, AND F. W. ANDERKA. 1978. Behavioral abnormalities of nesting Lake Ontario herring gulls. Journal of Wildlife Management 42: 477–483.

FOX, G. A., S. W. KENNEDY, R.J. NORSTROM, AND L. WIGFIELD. 1988. Porphyria in herring gulls: a biochemical response to chemical contamination of Great Lakes food chains. Journal of Environmental Toxicology and Chemistry 7: 831–839

FOX, G. A., D. V. WESELOH, T. J. KUBIAK, AND T. C. ERDMAN. 1991. Reproductive outcomes in colonial fish-eating birds: a biomarker for developmental toxicants in Great Lakes food chains. I. Historical and ecotoxicological perspectives. Journal of Great Lakes Research 17: 153–157.

FOX, G. S., B. COLLINS, E. HAYAKAWA, D.V. WESELOH, J. P. LUDWIG, T. J. KUBIAK, AND T. C. ERDMAN. 1991b. Reproductive outcomes in colonial fish-eating birds: a biomarker for developmental toxicants in Great Lakes food chains. II. Spatial variation in the occurrence and prevalence of bill defects in young double-crested cormorants in the Great Lakes, 1979–1987. Journal of Great Lakes Research 17: 158–167.

FRANSON, J. C. 1996. Interpretation of tissue lead residues in birds other than waterfowl. Pp. 265–280 *in* Environmental Contaminants in Wildlife: Interpreting Tissue Concentrations (W. N. Beyer, G. H. Heinz, and A. W. Redmon-Norwood, Eds.). Lewis Publishers, Boca Raton, FL.

FRY, D. M., AND L. ADDIEGO. 1987. Hemolytic anemia complicates the cleaning of oiled seabirds. Wildlife Journal of Diseases 10: 1–8.

FRY, D. M., AND C. K. TOONE. 1981. DDT-induced feminization of gull embryos. Science 213: 933–934.

FRY, D. M., S. I. FEFER, AND L. SILEO. 1987. Ingestion of plastic debris by Laysan Albatrosses and Wedge-tailed Shearwaters in the Hawaiian Islands. Marine Pollution Bulletin 18: 339–343.

FRY, D. M., C. K. TOONE, S. M. SPEICH, AND R. J. PEARD. 1987. Sex ratio skew and breeding patterns of gulls: demographic and toxicological considerations. Studies in Avian Biology 10: 26–43.

FRY, D. M., J. SWENSON, L. A. ADDIEGO, C. R. GRAU, AND A. KANG. 1986. Reduced reproduction in Wedge-tailed Shearwaters exposed to weathered Santa Barbara crude oil. Archives of Environmental Contamination and Toxicology 15: 453–463.

FURNESS, R. W. 1985. Plastic particle pollution: accumulation by Procellariform seabirds Scottish colonies. Marine Pollution Bulletin 16: 103–106.

FURNESS, R. W. 1993. Birds as monitors of pollutants. Pp. 86–143 *in* Birds as Monitors of Environmental Change (R. W. Furness and J. J. D. Greenwood, Eds.). Chapman & Hall, London.

FURNESS, R. W. 1996. Cadmium in birds. Pp. 389–404 *in* Environmental Contaminants in Wildlife: Interpreting Tissue Concentrations (W. N. Beyer, G. H. Heinz, and A. W. Redmon-Norwood, Eds.). SETAC Spec. Publ., Lewis Publishers, Boca Raton, FL.

FURNESS, R. W., AND K. C. J. CAMPHUYSEN. 1997. Seabirds as monitors of the marine environment. ICES Journal of Marine Science 54: 726–23.

FURNESS, R. W., AND M. HUTTON. 1980. Pollutants and impaired breeding of Great Skua *Catharacta skua* in Britain. Ibis 122: 88–94.

FURNESS, R. W., S. J. MUIRHEAD, AND M. WOODBURN. 1986. Using bird feathers to measure mercury in the environment: relationship between mercury content and moult. Marine Pollution Bulletin 17: 27–37.

FURNESS, R. W., D. R. THOMPSON, AND P. H. BECKER. 1995. Spatial and temporal variation in mercury contamination of seabirds in the North Sea. Helgolander Meeresuntersuchungen 49: 605–615.

GANTHER, H. E., C. GOUDIE, M. L. SUNDE, M. J. KOPECKY, R. WAGNER, O. H. SANG-HWANG AND W. G. HOEKSTRA. 1972. Selenium relation to decreased toxicity of methylmercury added to diets containing Tuna. Science 72: 1122–1124.

GARCELON, D. K., AND N. J. THOMAS. 1997. DDE poisoning in an adult bald eagle. Journal of Wildlife Diseases 33: 299–303.

GESSELL, G. G., R. J. ROBEL, A. D. DAYTON, AND J. FRIEMAN. 1979. Effects of dieldrin on operant behavior of bobwhites. Journal of Environmental Science and Health, B14: 153–170.

GIESY, J. P., J. P. LUDWIG, AND D. E. TILLITT. 1994a. Dioxins, dibenzofurans, PCBs and colonial fish-eating water birds. Pp. 249–307 *in* Dioxins and Health (A. Schecter, Ed.). Plenum Press, New York.

GIESY, J. P., J. P. LUDWIG, AND D. E. TILLITT. 1994b. Deformities in birds of the Great Lakes region: assigning causality. Environmental Science and Technology 28: 128A–135A.

GILBERTSON, M. 1988. Epidemics in birds and mammals caused by chemicals in the Great Lakes. Pp. 133–152 *in* Toxic Contaminants and Ecosystem Health: A Great Lakes Focus (M. S. Evans, Ed.). Wiley, New York.

GILBERTSON, M. 1989. Effects on fish and wildlife populations. Pp. 103–127 *in* Halogenated Biphenyls, Terphenyls, Naphthalense, Dibenzodioxins and Related Products (R. Kimbrough and A. A. Jensen, Eds.), Elsevier, New York.

GILBERTSON, M. (Ed.). 1990. Proceedings of the workshop on cause-effect linkages. International Joint Commission, Windsor, Ontario.

GILBERTSON, M., T. J. KUBIAK, J. P. LUDWIG, AND G. FOX. 1991. Great Lakes embryo mortality, edema and deformity syndrome (GLEMEDS) in colonial fish-eating birds: similarity to chick edema disease. Journal of Toxicology and Environmental Health 33: 455–520.

GOCHFELD, M. 1971. Premature feather-loss — a new disease of Common Terns on Long Island, New York. Kingbird 21: 206–211.

GOCHFELD, M. 1973. Effect of artefact pollution on the viability of seabird colonies on Long Island, New York. Environmental Pollution 4: 1–6.

GOCHFELD, M. 1975. Developmental defects in Common Terns of western Long Island, New York. Auk 92: 58–65.

GOCHFELD, M. 1980a. Tissue distribution of mercury in normal and abnormal young Common Terns. Marine Pollution Bulletin 11: 362–366.

GOCHFELD, M. 1980b. Mercury levels in some seabirds of the Humboldt Current, Peru. Environmental Pollution A22: 197–205.

GOCHFELD, M. 1998. Principles of toxicology. Pp. 415–457 *in* Maxcy-Rosenau-Last: Public Health and Preventive Medicine (R. B. Wallace, Ed.). Appleton & Lange, Stamford, CT.

GOCHFELD, M., and J. BURGER. 1998. Temporal trends in metal levels in eggs of the endangered Roseate Tern (*Sterna dougallii*) in New York. Environmental Research 77: 36–42.

GOCHFELD, M., J. L. BELANT, T. SHUKLA, T. BENSON, and J. BURGER. 1996. Heavy metals in laughing gulls: gender, age and tissue differences. Environmental Toxicology and Chemistry 15: 2275–2283.

GOCHFELD, M., D. J. GOCHFELD, D. MINTON, B. G. MURRAY, JR., P. PYLE, N. SETO, D. SMITH, AND J. BURGER. 1999. Metals in feathers of Bonin Petrel, Christmas Shearwater, Wedge-tailed Shearwater, and Red-tailed Tropicbird in the Hawaiian Islands, Northern Pacific. Environmental Monitoring and Assessment 59: 343–358.

GOEDE, A. A., AND M. deBRUIN. 1984. The use of bird feather parts as a monitor for metal pollution. Environmental Pollution 8: 281–289.

GRASMAN, A., P. F. SCANLON, AND G. A. FOX. 1998. Reproductive and physiological effects of environmental contaminants in fish-eating birds of the Great Lakes: a review of historical trends. Environmental Monitoring and Assessment 52: 117–145.

GRASMAN, A., G. FOX, P. SCANLON, AND J. LUDWIG. 1996. Organochlorine-associated immunosuppression in pre-fledging Caspian Terns and Herring Gulls from the Great Lakes: an ecoepidemiological study. Environmental Health Perspectives 104 (Suppl. 4): 829–842.

GRESS, F., R. W. RISEBROUGH, AND F. C. SIBLEY. 1971. Shell-thinning in eggs of Common Murres (*Uria aalge*) from the Farallon Islands. Condor 73: 368–369.

GRESS, F., R. W. RISEBROUGH, D. W. ANDERSON, L. F. KIFF, AND J. R. JEHL, JR. 1973. Reproductive failures of Double-crested Cormorants in Southern California and Baja California. Wilson Bulletin 85: 197–208.

HAHN, E., K. HAHN, AND M. STOEPPLER. 1993. Bird feathers as bioindicators in areas of the German Environmental Specimen Bank — bioaccumulation of mercury in food chains and exogenous deposition of atmospheric pollution with lead and cadmium. Science of the Total Environment 139: 259–270.

HAYS, H., AND R. W. RISEBROUGH. 1972. Pollutant concentrations in abnormal young terns from Long Island Sound. Auk 89: 19–35.

HEINZ, G. H. 1974. Effects of low dietary levels of methyl mercury on mallard reproduction. Bulletin of Environmental Contamination and Toxicology 11: 386–392.

HEINZ, G. H. 1976. Methylmercury: second generation reproductive and behavioral effects on Mallard ducks. Journal of Wildlife Management 40: 710–715.

HEINZ, G. H. 1996. Selenium in birds. Pp. 447–458 *in* Environmental Contaminants in Wildlife: Interpreting Tissues Concentrations (W. N. Beyer, G. H. Heinz, and A. W. Redmon-Norwood, Eds.). Lewis Publishers, Boca Raton, FL.

HEINZ, G., AND M. A. FITZGERALD. 1993. Reproduction of mallards following overwintering exposure to selenium. Environmental Pollution 81: 117–122.

HEINZ, G. H., D. S. MILLER, B. J. EBERT, AND K. L. STROMBORG. 1994. Declines in organochlorines in eggs of Red-breasted Mergansers from Lake Michigan, 1977–1978 versus 1990. Environmental Monitoring and Assessment 33: 175–182.

HELLOU, J. 1996. Polycyclic aromatic hydrocarbons in marine mammals, finfish, and molluscs. Pp. 229–250 *in* Environmental Contaminants in Wildlife: Interpreting Tissue Concentrations (W. N. Beyer, G. H. Heinz, and A. W. Redmon-Norwood, Eds.). Lewis Publishers, Boca Raton, FL.

HENRIKSEN, E. O., G. W. GABRIELSEN, S. TRUDEAU, J. WOLKERSM, K. SAGERU, AND J. U. SKAARE. 2000. Organochlorines and possible biochemical effects in Glaucous Gulls (*Larus hyperboreus*) from Bjornoya, the Barents Sea. Archives of Environmental Contamination and Toxicology 38: 234–243.

HILL, A. B. 1965. The environment and disease: association or causation? Proceedings of the Royal Society of Medicine 58: 295–300.

HOFFMAN, D. J. 1990. Embryotoxicity and teratogenicity of environmental contaminants to bird eggs. Reviews of Environmental Contamination and Toxicology 115: 40–89.

HOFFMAN, D. J., AND J. M. MOORE. 1979. Teratogenic effects of external egg applications of methylmercury in the mallard, *Anas platyrhynchos*. Teratology 20: 453–462.

HOFFMAN, D. J., B. A. RATTNER, L. SILEO, D. DOCHERTY, AND T. J. KUBIAK. 1987. Embryotoxicity, teratogenicity, and aryl hydrocarbon hydroxylase activity in Forster's Terns on Green Bay, Lake Michigan. Environmental Research 42: 176–184.

HOFFMAN, D. J., C. P. RICE, AND T. J. KUBIAK. 1996. PCBs and dioxins in birds. Pp. 165–208 *in* Environmental Contaminants in Wildlife: Interpreting Tissue Concentrations (W. N. Beyer, G. H. Heinz, and A. W. Redmon-Norwood, Eds.). Lewis Publishers, Boca Raton, FL.

HOFFMAN, D. J., C. J. SANDERSON, L. J. LeCAPTAIN, E. CROMARTIE, AND G. W. PENDLETON. 1992. Interactive effects of arsenate, selenium, and dietary protein on survival, growth, and physiology in Mallard ducklings. Archives of Environmental Contamination and Toxicology 22: 55–62.

HONDA, K., B. Y. MIN, B., Y. MIN, AND R. TATSUKAWA. 1985. Heavy metal distribution in organs and tissues of the Eastern Great White Egret, *Egretta alba modesta*. Bulletin of Environmental Contamination and Toxicology 35: 781–789.

HONDA, K., J. E. MARCOVECCHIO, S. KAN, R. TATSUKAWA, AND H. OGI. 1990. Metal concentrations in pelagic seabirds from the North Pacific Ocean. Archives of Environmental Contamination and Toxicology 19: 704–711.

HOUGHTON, J. T., B. A. CALLANDER, AND S. K. VARNEY. 1992. Climate Change 1992. Cambridge University Press, Cambridge, U.K.

HUNT, G. L. 1987. Offshore oil development and seabirds: the present status of knowledge and long-term research needs. Pp. 539–586 in Long-Term Environmental Effects of Offshore Oil and Gas Development (D. F. Boesch and N. N. Rabalais, Eds.). Elsevier, London.

HUNTER, R. A., AND J. G. JOHNSON. 1982. Food chain relationships of copper and cadmium in contaminated grassland ecosystems. Oikos 38: 108–117.

HUTTON, M. 1981. Accumulation of heavy metals and selenium in three seabird species from the United Kingdom. Environmental Pollution 26: 129–145.

HUTTON, M. 1987. Cadmium. Pp. 35–42 in Lead, Mercury, Cadmium and Arsenic in the Environment (T. C. Hutchinson and K. M. Meema, Eds.). John Wiley & Sons, Chichester, U.K.

IRONS, D.B., S. J. KENDALL, W. P. ERICKSON, L. L. McDONALD, AND B. K. LANCE. 2000. Nine years after the Exxon Valdez oil spill: effects on marine bird populations in Prince William Sound, Alaska. Condor 104: 723–737.

JANZ, D. M., AND G. D. BELLWARD. 1996. In ovo 2,3,7,8-tetrachlorodibenzo-p-dioxin exposure in three avian species. 2. Effects on estrogen receptor and plasma sex steroid hormones during the perinatal period. Toxicology and Applied Pharmacology 139: 292–300.

JENSEN, S., AND A. JERNELOV. 1969. Biological methylation of mercury in aquatic organisms. Nature 223: 753–754.

KIM, E. Y., T. MURAKAMI, K. SAEKI, AND R. TATSUKAWA. 1996. Mercury levels and its chemical form in tissues and organs of seabirds. Archives Environmental Contamination and Toxicology 30: 259–266.

KING, K. A., D. R. BLANKINSHIP, E. PAYNE, A. J. KRYNITSKY, AND G. L. HENSLER. 1985. Brown Pelican populations and pollutants in Texas 1975–1981. Wilson Bulletin 97: 210–214.

KOEMAN, J. H. 1972. Side-effects of persistent pesticides and other chemicals on birds and mammals in the Netherlands. TNO News 27: 527–632.

KREITZER, J. F., AND G. HEINZ. 1974. The effects of sublethal dosages of five pesticides and a polychlorinated biphenyl on the avoidance response of Coturnix quail chicks. Environmental Pollution 6: 21–29.

KUBIAK, T. J., H. J. HARRIS, L. M. SMITH, T. R. SCHWARTZ, D. L. STALLING, J. A. TRICK, L. SILEO, D. E. DOCHERTY, AND T. C. ERDMAN. 1989. Microcontaminants and reproductive impairment of the Forster's tern on Green Bay, Lake Michigan — 1983. Archives of Environmental Contamination and Toxicology 18: 706–727.

KUIKEN, T., G. A. FOX, AND K. L. DANESIK. 1999. Bill malformations in Double-crested Cormorants with low exposure to organochlorines. Environmental Toxicology and Chemistry 18: 2908–2913.

LAIST, D. W. 1987. Overview of the biological effects of lost and discarded plastic debris in the marine environment. Marine Pollution Bulletin 18: 310–326.

LANGE, T. R., H. E. ROYALS, AND L. L. CONNOR. 1994. Mercury accumulation in largemouth bass (Micropterus salmoides) in a Florida lake. Archives of Environmental Contamination and Toxicology 27: 466–471.

LARSON, J. M., W. H. KARASOV, L. SILEO, K. L. STROMBORG, B. A. HANBIDGE, J. P. GIESY, P. D. JONES, D. E. TILLIT, AND D. A. VERBRUGGE. 1996. Reproductive success, developmental anomalies, and environmental contaminants in Double-crested Cormorant (Phalacrocorax auritus). Environmental Toxicology and Chemistry 15: 553–559.

LEE, Y. Z., F. A. LEIGHTON, D. B. PEAKALL, R. NORSTROM, C. J. O'BRIAN, J. F. PAYNE, AND A. D. RAHIMTULA. 1985. The effects of ingestion of Hibernia and Prudhoe Bay crude oils on hepatic and renal mixed function oxidases in nestling Herring Gulls. Environmental Research 36: 248–255.

LEWIS, S. A., AND R. W. FURNESS. 1991. Mercury accumulation and excretion by laboratory reared Black-headed Gull (Larus ridibundus) chicks. Archives of Environmental Contamination and Toxicology 21: 316–320.

LEWIS, S. J., AND R. A. MALECKI. 1984. Effects of egg oiling on larid productivity and population dynamics. Auk 101: 584–592.

LOCK, J. W., D. R. THOMPSON, R. W. FURNESS, AND J. A. BARTLE. 1992. Metal concentrations in seabirds of the New Zealand region. Environmental Pollution 75: 289–300.

LONGCORE, J. R., R. ANDREWS, L. N. LOCKE, G. E. BAGLEY, AND L. T. YOUNG. 1974. Toxicity of Lead and Proposed Substitute Shot to Mallards. U.S. Fish and Wildlife Service Special Science Report No. 182.

LORENZEN, A., T. W. MOON, S. W. KENNEDY, AND G. A. FOX. 1999. Relationships between environmental organochlorine contaminant residues, plasma corticosterone concentrations, and intermediary metabolic enzyme activities in Great Lakes herring gull embryos. Environmental Health Perspectives 107: 179–186.

LOWE, T. P., T. W. MAY, W. G. BRUMBAUGH, AND D. A. KANE. 1985. National contaminant biomonitoring program: concentrations of seven elements in freshwater fish, 1978–1981. Archives of Environmental Contamination and Toxicology 14: 363–388.

MAILMAN, R. B. 1980. Heavy metals. Pp. 34–43 in Introduction to Environmental Toxicology (F. E. Gunthrie and J. J. Perry, Eds.). Elsevier, New York.

MALINS, D. C., B. B. McCAIN, M. S. LANDAHL, M. M. KRAHN, D. W. BROWN, S. L. CHEN AND W. T. ROUBAL. 1988. Neoplastic and other diseases in fish in relation to toxic chemicals: an overview. Aquatic Toxicology 11: 43–67.

MINEAU, P., G. A. FOX, R. J. NORSTROM, D. V. WESELOH, D. J. HALLET, AND J. A. ELLENTON. 1984. Using the Herring Gull to monitor levels and effects of organochlorine contamination in the Canadian Great Lakes. Advances in Environmental Science and Technology 14: 425–452.

MONTEIRO, L. R., AND R. W. FURNESS. 1995. Seabirds as monitors of mercury in the marine environment. Water, Air, Soil Pollution. 80: 831–870.

MONTEIRO, L. B., J. P. GRANADEIRO, AND R. W. FURNESS. 1998. Relationship between mercury levels and diet in Azores seabirds. Marine Ecological Progress Series 166: 259–265.

MORA, M. A., H. J. AUMAN, J. P. LUDWIG, J. P. GIESY, D. A. VERBRUGGE, AND M. E. LUDWIG. 1993. PCB's and chlorinated insecticides in plasma of Caspian Terns: relationships with age, productivity and colony-site tenacity in the Great Lakes. Archives of Environmental Toxicology and Chemistry 24: 320–331.

MORANT, P. D., J. COOPER, AND R. M. RANDALL. 1981. The rehabilitation of oiled Jackass Penguins (Spheniscus demersus), 1970–1980. Pp. 267–301 in Proceedings of the Symposium on Birds of the Sea and Shore (J. Cooper, Ed.). African Seabird Group, Cape Town, South Africa.

MUIRHEAD, S. J., AND R. W. FURNESS. 1988. Heavy metals concentrations in the tissues of seabirds from Gough Island, South Atlantic Ocean. Marine Pollution Bulletin 19: 278–283.

MURPHY, S. M., AND T. J. MABEE. 2000. Status of Black Oystercatchers in Prince William Sound, Alaska, nine years after the Exxon Valdez oil spill. Waterbirds 23: 204–213.

NATIONAL RESEARCH COUNCIL (NRC). 1993. Issues in Risk Assessment. National Academy of Sciences, Washington, D.C.

NATIONAL RESEARCH COUNCIL (NRC). 1999. Hormonally Active Agents in the Environment. National Academy of Sciences, Washington, D.C.

NISBET, I. C. T. 1994. Effects of pollution on marine birds. Pp. 8–25 in Seabirds on Islands: Threats, Case Studies, and Action Plans (D. N. Nettleship, J. Burger, and M. Gochfeld, Eds.). BirdLife International, Cambridge, U.K.

NORHEIM, G. 1987. Levels and interactions of heavy metals in seabirds from Svalbard and the Antarctic. Environmental Pollution 47: 83–94.

NORHEIM, G., AND A. FROSLIE. 1978. The degree of methylation and organ distribution of mercury in some birds of prey in Norway. Acta Pharmacogica Toxicologica 54: 196–204.

NOSHKIN, V. E., V. L. ROBISON, AND K. M. WONG. 1994. Concentration of 210PO and 210PB in the diet at the Marshall Islands. Science of the Total Environment 155: 87–107.

OHLENDORF, H. 1993. Marine birds and trace elements in the temperate North Pacific. In The Status, Ecology, and Conservation of Marine Birds of the North Pacific. Canadian Wildlife Service Special Publication, Ottawa.

OHLENDORF, H., R. L. HOTHEM, C. M. BUNCK, T. W. ALDRICH, AND J. R. MOORE. 1986. Relationship between selenium concentrations and avian reproduction. Transactions of the 51st North American Wildlife Research Conference 51: 330–342.

OHLENDORF, H., R. L. HOTHEM, AND D. WALSH. 1989. Nest success, cause-specific nest failures and hatchability of aquatic birds at selenium contaminated Kesterson Reservoir and a reference site. Condor 91: 787–796.

OHLENDORF, H. M., R. W. RISEBROUGH, AND K. VERMEER. 1978. Exposure of marine birds to environmental pollutants. U. S. Fish and Wildlife Service Research Report, 9: 1–40.

OHLENDORF, H. M., R. L. HOTHEM, C. M. BUNCK, AND K. C. MAROIS. 1990. Bioaccumulation of selenium in birds at Kesterson Reservoir, California. Archives of Environmental Contamination and Toxicology 19: 495–507.

OHLENDORF, H. M., A. W. KILNESS, J. L. SIMMONS, R. D. STROUD, D. J. HOFFMAN, AND J. F. MOORE. 1988. Selenium toxicosis in wild aquatic birds. Journal of Toxicology and Environmental Health 24: 67–92.

OSBORN, D., M. P. HARRIS, AND J. K. NICHOLSON. 1979. Comparative tissue distribution of mercury, cadmium, and zinc in three species of pelagic seabirds. Comparative Biochemistry and Physiology 64C: 61–67.

PAIN, D. J. 1995. Lead in the environment. Pp. 356–391 in Handbook of Ecotoxicology (D. J. Hoffman, B. A. Rattner, G. A. Burton, Jr., and J. Cairns, Jr., Eds.). Lewis Publishers, Boca Raton, FL.

PARNELL, J. F., M. A. SHIELDS, AND D. FRIERSON. 1984. Hatching success of Brown Pelican eggs after contamination with oil. Colonial Waterbirds 7: 22–24.

PARSLOW, J. L. F., AND D. J. JEFFRIES. 1973. Environmental Pollution 5: 87–101.

PEAKALL, D. B. 1970. p,p′-DDT: effect on calcium metabolism and concentration of estradiol in the blood. Science 168: 592–594.

PEAKALL, D. B. 1985. Behavioral responses of birds to pesticides and other contaminants. Research Reviews 96: 45–77.

PEAKALL, D. B. 1986. Accumulation and effects on birds. Pp. 31–47 in PCBs in the Environment, Vol. II (Chap. 3) (J. S. Aaid, Ed.). CRC Press, Boca Raton, FL.

PEAKALL, D. B. 1992. Animal Biomarkers as Pollution Indicators. Chapman & Hall, London, UK.

PEAKALL, D. B. 1996. Dieldrin and other cyclodiene pesticides in wildlife. Pp. 73–98 in Environmental Contaminants in Wildlife: Interpreting Tissue Concentrations (W. N. Beyer, G. H. Heinz, and A. W. Redmon-Norwood, Eds.). SETAC Special Publication Series, Lewis Publishers, Boca Raton, FL.

PEAKALL, D. B., AND G. A. FOX. 1987. Toxicological investigations of pollutant-related effects in Great Lakes gulls. Environmental Health Perspectives 71: 187–193.

PEAKALL, D. B., J. TREMBLAY, W. B. KENTER, AND D. S. MILLER. 1981. Endocrine dysfunction in seabirds caused by ingested oil. Environmental Research 24: 6–14.

PEAKALL, D. B., J. L. LINCER, R. W. RISEBROUGH, J. B. PRITCHARD, AND W. B. KINTER. 1973. DDE-induced eggshell thinning: structural and physiological effects in three species. Comparative General Pharmacology 4: 305–315.

PEKARIK, C., AND D. V. WESELOH. 1998. Organochlorine contaminants in Herring Gull eggs from the Great Lakes 1974–1995: change point regression analysis and short-term regression. Environmental Monitoring and Assessment 53: 77–115.

PIATT, J. F., C. J. LENSINK, W. BUTLER, M. KENDIZIOREK, AND D. NYSEWANDER. 1990. Immediate impacts of the Exxon Valdez oil spill on marine birds. Auk 107: 387–397.

PORCELLA, D. B. (1994). Mercury in the environment: biogeochemistry. Pp. 3–19 in Mercury Pollution: Integration and Synthesis (C. Watras and J. W. Huckabee, Eds.). Lewis Publishers, Boca Raton, FL.

PORTERFIELD, S. P. 1994. Vulnerability of the developing brain to thyroid abnormalities: environmental insults to the thyroid system. Environmental Health Perspectives 102: 125–130.

PRATER, J. C. 1995. Environmental Contaminant Reference Data Book, vol. 1. Van Nostrand Reinhold, New York.

PROUTY, R. M., J. E. PETERSON, L. N. LOCKE, AND B. M. MULHERN. 1975. DDD poisoning in a Loon and the identification of the hydroxylated form of DDD. Bulletin of Environmental Contamination and Toxicology 14: 385–388.

PRUTER, A. T. 1987. Sources, quantities and distribution of persistent plastics in the marine environment. Marine Pollution Bulletin 18: 305–310.

RATCLIFFE, D. A. 1970. Changes attributable to pesticides in egg breakage frequency and egg shell thickness in some British birds. Journal of Applied Ecology 7: 67–115.

RATTNER, B. A., D. J. HOFFMAN, M. J. MELANCON, G. H. OLSEN, S. R. SCHMIDT, AND K. C. PARSONS. 2000. Organochlorine and metal contaminant exposure and effects in hatching Black-crowned Night Herons (Nycticorax nycticorax) in Delaware Bay. Archives of Environmental Contamination and Toxicology 39: 38–45.

RISEBROUGH, R. W. 1986. Pesticides and bird populations. Current Ornithology 3: 397–427.

RITTINGHAUS, H. 1956. On the indirect spread of oil in a seabird sanctuary. Ornithologischen Mitteilungen 8: 43–46.

RONIS, M. J., J. BORLAKOGLU, C. H. WALKER, T. HANSEN, AND J. J. STEGEMANN. 1989. Expression of orthologues in rat P-450AI and IIBI in seabirds from the Irish Sea 1978–88, evidence for environmental induction. Marine Environmental Research 28: 123–130.

RYAN, P. G. 1988a. Intraspecific variation in plastic ingestion by seabirds and the flux of plastic through seabird populations. Condor 90: 446–452.

RYAN, P. G. 1988b. The incidence and characteristics of plastic particles ingested by birds. Marine Pollution Bulletin 19: 125–128.

RYAN, P. G., AND S. JACKSON. 1987. The lifespan of ingested plastic particles in seabirds and their effect on digestive efficiency. Marine Pollution Bulletin 18: 217–219.

RYCKMAN, D. P., D. V. WESELOH, P. HAMR, G. A. FOX, B. COLLINS, P. J. EWINS, AND R. J. NORSTROM. 1998. Spatial and temporal trends in organochlorine contamination and bill deformities in Double-crested Cormorants (Phalacrocorax auritus) from the Canadian Great Lakes. Environmental Monitoring and Assessment 53: 169–195.

SAFE, S. 1990. Polychlorinated biphenyls (PCBs), dibenzo-p-dioxins (PCDDs), dibenzofurans (PCDFs), and related compounds: environmental and mechanistic considerations which support the development of toxic equivalency factors (TEFs). Critical Reviews Toxicology 21: 51–88.

SATOH, H., N. YASUDA, AND S. SHIMAI. 1985. Development of reflexes in neonatal mice prenatally exposed to methylmercury and selenite. Toxicology Letters 25: 199–203.

SCHAULE, B. K., AND G. C. PATTERSON. 1981. Lead concentrations in the north-east Pacific: evidence for global anthropogenic perturbations. Science Letters 54: 97–116.

SCHEUHAMMER, A. M. 1987. The chronic toxicity of aluminum, cadmium, mercury, and lead in birds: a review. Environmental Pollution 46: 265–295.

SCHNEIDER, D. C., D. C. DUFFY, AND G. L. HUNT. 1988. Cross-shelf gradients in the abundance of pelagic birds. International Ornithological Congress 19: 976–981.

SEPULVEDA, M. S., R. H. POPPENGA, J. J. ARRECIS, AND L. B. QUINN. 1998. Concentrations of mercury and selenium in tissues of Double-crested Cormorants (Phalacrocoras auritus) from Southern Florida. Colonial Waterbirds 21: 35–42.

SHARMA, R. P., D. S. WINN, AND J. B. LOW. 1976. Toxic, neurochemical and behavioral effects of dieldrin exposure in mallard ducks. Archives of Environmental Contamination and Toxicology 5: 43–53.

SIEVERT, P. R., AND L. SILEO. 1993. The effects of ingested plastic on growth and survival of albatross chicks. Pp. 212–217 in The Status, Ecology, and Conservation of Marine Birds of the North Pacific (K. Vermeer, K. T. Briggs, K. H. Morgan, and D. Siegel-Causey, Eds.). Special Publication, Canadian Wildlife Service, Ottawa.

SILEO, L., AND S. I. FEFER. 1987. Paint chip poisoning of Laysan Albatross at Midway Atoll. Journal of Wildlife Diseases 23: 432–437.

SILEO, L., P. R. SIEVERT, and M. D. SAMUEL. 1990. Causes of mortality of albatross chicks at Midway Atoll. Journal of Wildlife Diseases 23: 432–437.

SKORUPA, J. P., AND H. M. OHLENDORF. 1991. Contaminants in drainage water and avian risk thresholds. Pp. 345–368 in The Economics and Management of Water and Drainage in Agriculture (A. Dinar and D. Zilberman, Eds.). Kluwer Academic Publishers, Norwell, MA.

SOLONEN, T., AND M. LODENIUS. 1984. Mercury in Finnish sparrow hawks Accipiter nisus. Ornis Fennica 61: 58–63.

SPANN, J. W., R. G. HEATH, J. F. KREITZER, AND L. N. LOCKE. 1972. Ethyl mercury p-toluene sulfonanilide: lethal and reproductive effects on pheasants. Science 175: 128–131.

STEWART, F. M., R. A. PHILLIPS, P. CATRY, AND R. W. FURNESS. 1997. Influence of species, age and diet on mercury concentrations in Shetland seabirds. Marine Ecology Progress Series 151: 237–244.

STOCK, M., R. F. M. HERBER, AND H. M. A. GERON. 1989. Cadmium levels in Oystercatcher Haematopus ostralegus from the German Wadden Sea. Marine Ecology Progress Series 53: 227–234.

STONEBURNER, D. L., AND C. S. HARRISON. 1981. Heavy metal residues in Sooty Tern tissues from the Gulf of Mexico and North Central Pacific Ocean. The Science of the Total Environment 17: 51–58.

THOMPSON, D. R. 1990. Metal levels in marine vertebrates. Pp. 143–182 in Heavy Metals in the Marine Environment (R. W. Furness and P. S. Rainbow, Eds.). CRC Press, Boca Raton, FL.

THOMPSON, D. R. 1996. Mercury in birds and terrestrial mammals. Pp. 341–356 *in* Environmental Contaminants in Wildlife: Interpreting Tissue Concentrations (W. N. Beyer, G. H. Heinz, and A. W. Redmon-Norwood, Eds.). Lewis Publishers, Boca Raton, FL.

THOMPSON, D. R., AND R. W. FURNESS. 1989a. Comparison of the levels of total and organic mercury in seabird feathers. Marine Pollution Bulletin 20: 577–579.

THOMPSON, D. R., AND R. W. FURNESS. 1989b. The chemical form of mercury stores in South Atlantic seabirds. Environmental Pollution 60: 305–317.

THOMPSON, D. R., P. H. BECKER, AND R. W. FURNESS. 1993. Long-term changes in mercury concentrations in herring gulls *Larus argentatus* and common terns *Sterna hirundo* from the German North Sea coast. Journal of Applied Ecology 30: 316–320.

THOMPSON, D. R., R. W. FURNESS, AND P. M. WALSH. 1992. Historical changes in mercury concentrations in the marine ecosystem of the north and north-east Atlantic Ocean as indicated by seabird feathers. Journal of Applied Ecology 29: 79–84.

THOMPSON, D. R., K. C. HAMER, AND R. W. FURNESS. 1991. Mercury accumulation in great skuas *Catharacta skua* of known age and sex and its effects upon breeding and survival. Journal of Applied Ecology 28: 672–684.

TORI, G. M., AND T. J. PETERLE. 1983. Effects of PCBs on mourning dove courtship behavior. Bulletin of Environmental Contamination and Toxicology 30: 44–49.

TRIVELPIECE, W., R. G. BUTLER, D. S. MILLER, AND D. B. PEAKALL. 1984. Reduced survival of chicks of oil-dosed Leach's Storm-petrels. Condor 86: 81–82.

U.S. FISH AND WILDLIFE SERVICE. 1996. Seabirds and Shorebirds of Midway Atoll National Wildlife Refuge. Midway Island: U.S. Fish and Wildlife Service. 12 pgs.

VAN DEN BERG, M. L., L. BIRNBAUM, A. T. C. BOSVELD, B. BRUNSTROM, AND P. COOK. 1998. Toxic equivalency factors (TEFs) for PCBs, PCDDs, PCDFs, for humans and wildlife. Environmental Health Perspectives 106: 775–792.

VAN STRALLEN, N. M., AND E. ERNST. 1991. Metal biomagnification may endanger species in critical pathways. Oikos 62: 255–256.

WALSH, P. M. 1990. The use of seabirds as monitors of heavy metals in the marine environment. Pp. 183–294 *in* Heavy Metals in the Marine Environment (R. W. Furness and P. S. Rainbow, Eds.). CRC Press, Boca Raton, FL.

WEIMERSKIRCH, H. 1997. Foraging strategies of Indian Ocean albatrosses and their relationships with fisheries. Pp. 168–179 *in* Albatross Biology and Conservation (G. Robertson and R. Gales, Eds.) Surren Beatty and Sons, Chipping Norton.

WESELOH, D. V., P. J. EWINS, J. STRUGER, P. MINEAU, C. A. BISHOP, S. POSTUPALSKY, AND J. P. LUDWIG. 1995. Double-crested Cormorants of the Great Lakes: changes in population size, breeding distribution and reproductive output between 1913 and 1991. Colonial Waterbirds 18 (Special Publication 1) 48–59.

WESELOH, D. V., S. M. TEEPLE, AND M. GILBERTSON. 1983. Double-crested Cormorants on the Great Lakes: egg-laying parameters, reproductive failure, and contaminant residues in eggs, Lake Huron 1972–1973. Canadian Journal of Zoology 61: 427–436.

WHITTOW, G. C. 1993A. Laysan Albatross (*Diomedea immutabilis*). Birds of North America 66: 1–20.

WHITTOW, G. C. 1993b. Black-footed Albatross (*Diomedea nigripes*). Birds of North America 66: 1–16.

WIEMEYER, S. N. 1996. Other organochlorine pesticides in birds. Pp. 99–116 *in* Environmental Contaminants in Wildlife: Interpreting Tissue Concentrations (W. N. Beyer, G. H. Heinz, and A. W. Redmon-Norwood, Eds.). SETAC Spec. Publ., Lewis Publishers, Boca Raton, FL.

WIENS, J. A., T. O. CHRIST, R. H. DAY, S. M. MURPHY, AND G. D. HAYWARD. 1996. Effects of the Exxon Valdez oil spill on marine bird communities in Prince William Sound, Alaska. Ecological Applications 6: 828–841.

WONG, C. S., E. BOYLE, K. W. BOYLAND, J. D. BURTON, AND E. D. GOLDBERG (Eds.). 1983. Trace metals in seawater. Plenum Press, New York.

WOODHEAD, D. S. 1986. The radiation exposure of Black-headed Gulls (*Larus ridibundus*) in the Ravenglass Estuary, Cumbria, U. K.: a preliminary assessment. Science of the Total Environment 58: 273–281.

WORLD HEALTH ORGANIZATION (WHO). 1990. Methylmercury. World Health Organization, Geneva, Switzerland.

WORK, T. M., AND M. R. SMITH. 1996. Lead exposure in Laysan albatross adults and chicks in Hawaii: prevalence, risk factors, and biochemical effects. Archives of Environmental Contamination and Toxicology 31: 115–119.

WREN, C. D., S. HARRIS, AND N. HARTTRUP. 1995. Ecotoxicology of mercury and cadmium. Pp. 392–423 *in* Handbook of Ecotoxicology (D. J. Hoffman, B. A. Rattner, G. A. Burton, Jr., and J. Cairns, Jr., Eds.). Lewis Publishers, Boca Raton, FL.

Seabirds Following Fishing Boat

16 Interactions between Fisheries and Seabirds

William A. Montevecchi

CONTENTS

16.1 INTRODUCTION

Since humans first inhabited coastal margins and ventured out to sea, they have exploited marine birds. Seabirds provided sources of food and bait (Collins 1882), fishermen used marine birds for navigational information about the locations of fishing banks and landfalls and followed birds at sea to find schools of fishes (Nelson 1978, Montevecchi and Tuck 1987). Over millennia and perhaps most rapidly in the present century, human populations and their technological capabilities at sea have increased many fold, and so have their demands for marine prey. Human harvests have moved consistently from exploitive to over-exploitive levels with marine birds (e.g., Burger and Gochfeld 1994, Montevecchi and Kirk 1996), mammals (Laws 1985), fishes (Harris 1990), crustaceans (Pauly et al. 1998), cephalopods (Montevecchi 1993b), and shellfish (Dahl 1992). Clearly, these and other human harvests influence seabirds and other marine animals in many ways.

The rapid enhancement of fishing capabilities and overexploitations of fish stocks in the 19th and 20th centuries inevitably led to questions about the influences of marine predators, such as seals (Harwood and Croxall 1988), whales (Harwood 1983), and seabirds (Milton et al. 1995, Cairns 1998) on commercial fishery stocks (Nettleship 1990, Tasker et al. 2000). Large-scale energetics and trophic models of prey consumption by seabirds demonstrated that marine birds consume huge tonnages of prey (Furness 1978, Furness and Cooper 1982, Croxall et al. 1985, Cairns et al. 1990, Montevecchi 2000), mostly small pelagic fishes and crustaceans (Montevecchi 1993a). These levels of consumption are matched or well exceeded by marine mammals (e.g., Furness 1990) and dwarfed by orders of magnitude by the consumption levels of large predatory fishes (Bundy et al. 2000). For example, Table 16.1 shows estimates of prey consumption by large predators in the northwest Atlantic.

Interactions between seabirds and fisheries are dominated by influences of fisheries on birds (Montevecchi 1993a, 1993b; Tasker et al. 2000). These influences may be direct or indirect and either negative or positive (Table 16.2). Direct effects include entrapment in fishing gear, disturbance, and food provisioning with fishery discards and offal. Indirect effects include prey depletion, increases in scavenging and predatory seabirds, decreases in large fish competitors, and increases

---

TABLE 16.1

Consumption Estimates (tons) of Capelin (*Mallotus villosus*), a Small Pelagic Fish, by Large Predators in the Northwest Atlantic

Taxa	Capelin Consumption (tons)	Source
Birds	250,000	Montevecchi 2000
Seals	800,000	Stenson and Lawson 2000
Whales	700,000	Stenson et al. 2000
Cod	1,000,000–3,000,000	Lilly et al. 2000
Human Quota	40,000	Carscadden et al. 2001

in the availability of small fishes. Owing to the small biomass of birds compared to fishes in the world's oceans, influences of seabirds on fisheries tend to be localized, small-scale events, often occurring in artificial situations involving aquaculture (Kirby et al. 1996, Cairns 1998, Tasker et al. 2000) or the stocking of commercial or game fishes (Blackwell et al. 1995, Roby et al. 1999).

The life history attributes of seabirds are such that their populations are relatively robust to interannual variation in breeding success, but highly sensitive to slight changes in adult mortality. Seabirds are long lived, have delayed maturity (often 5 to 10 years) and recruitment to breeding populations, and exhibit low fecundity and high annual adult survival (on the order of 80 to 90% or more; Furness and Monaghan 1987). Hence, poor reproduction must be long term and extensive to decrease populations. When such effects do occur they often lag well behind the environmental factors that caused them. Seabird populations are therefore buffered from environmental perturbations that influence annual production (Montevecchi and Berutti 1991). Yet even slight changes in adult mortality can have profound effects on seabird populations (Furness 2000). Hence, throughout this review an attempt is made to differentiate potential fishery influences on reproduction from those on adult survival.

The present chapter reviews the influences of fisheries on marine birds and also reviews the influences of seabirds on fisheries. Interactions and cumulative effects among fisheries, oceanographic perturbations, pollution, and hunting are also considered. Research and management recommendations to protect seabirds and the large-scale natural ecosystem processes that sustain them are also provided.

TABLE 16.2
Influences of Fisheries on Marine Birds

	Negative	Positive
Direct	Entrapment in fishing gear	Provide food via fisheries discards and offal
	Disturbance	
Indirect	Prey depletion	Remove competitors
	Increase populations of scavengers/predators	Increase abundances of small fishes
	Increase predation by removing artificial food sources of scavengers	

16.2 NEGATIVE EFFECTS OF FISHERIES ON SEABIRDS

16.2.1 DIRECT EFFECTS

16.2.1.1 Entrapment in Fishing Gear

By-catches of seabirds in fishing gear have resulted in negative population effects on birds on a global scale (Tasker et al. 2000).

Nets — Pursuit divers, such as auks and shearwaters, are the seabirds most commonly killed in gill nets in the North Atlantic and North Pacific (Tull et al. 1972, Ainley et al. 1981, King 1984, Ogi 1984, Piatt and Nettleship 1987, Petersen 1994, Artukhin et al. 2000). Loons, cormorants, and gannets are also caught in high numbers with surface-feeding gulls and storm-petrels being caught to a much lesser extent (Piatt and Nettleship 1987). As well as seabirds, seaducks, marine mammals, sharks, and sea turtles also become entrapped in fishing gear (e.g., Harwood 1983).

Before their banning in 1993, high-seas drift nets set for salmon and squid entrapped millions of birds including shallow divers and surface-feeders. More deeply set gill nets catch birds that dive below the foraging range of these species. Among the pursuit divers, birds that densely aggregate (e.g., alcids, shearwaters) are most vulnerable to mortality in nets (e.g., Artukhin et al. 2000), especially nets set near breeding colonies and migratory concentrations (e.g., Piatt and

FIGURE 16.1 Common Murre holding a capelin to be delivered to a chick. (Photo by W. A. Montevecchi.)

Nettleship 1987). Entrapments are often most frequent during periods of fish movements near fishing gear (Christensen and Lear 1977, Piatt and Nettleship 1987), when both birds and targeted fish are pursuing forage fishes.

Common Murre (*Uria aalge*) is the species most widely affected on a global basis by mortality in fishing nets (Melvin et al. 1999). Net mortality has been implicated in population declines of Common Murres in northern Norway (Vader at al. 1990a, Strann et al. 1991) and of Thick-billed Murres (*U. lomvia*) in western Greenland (Evans and Waterston 1976, Evans and Nettleship 1985; Figure 16.1). Net mortality has been implicated in negative population effects on murres in the western Bering Sea and on the Farallon Islands as well as on Red-legged Kittiwakes (*Rissa brevirostris*) on the Commander Islands (Artukhin et al. 2000) and on Sooty Shearwaters (*Puffinus griseus*) and Short-tailed Shearwaters (*P. tenuirostris;* DeGange et al. 1993, Veit et al. 1996). Net mortality has also been associated with population declines of endangered Marbled Murrelet (*Brachyramphus marmoratus;* Carter and Sealey 1984, Grettenberger et al. 2000) and endangered Japanese Murrelet (*Sythliboramphus antiquus*; Piatt and Gould 1994). Northern Gannets (*Morus bassanus*), Atlantic Puffins (*Fratercula arctica*), and nonbreeding Dovekies (*Alle alle*) are also killed in gill nets. Relationships that show that entrapments decrease with increasing distance from colonies (Ainley et al. 1981, Piatt and Nettleship 1987) indicate that no-fishing zones around breeding sites could in some circumstances benefit some seabird populations. Nets set inshore for lumpfish also catch high numbers of marine birds, especially Black Guillemots (*Cepphus grylle*) and Common Eiders (*Somateria mollissima*; Petersen 1998). Some evidence indicates that juvenile and immature murres may be more vulnerable to net mortality than older birds, suggesting that birds may learn to avoid nets (Strann et al. 1991, Brothers 1999), as marine mammals do (Lien et al. 1988).

During the 1990s, the Japanese set about 150,000 km of salmon drift nets (Artukhin et al. 2000) and almost 2,000,000 km of squid drift nets in the North Pacific (DeGange et al. 1993). Concurrent increases in the frequency of free-traveling, unattended nets have increased the mortality of birds and other marine animals, and continue to do so as fixed gear is lost or discarded. Many seabirds, especially gannets and cormorants, scavenge bits of nets, rope, line, etc. from the sea surface for nest material that may in turn entangle adults and chicks at nests (Montevecchi 1991). Alcids, Northern Gannets, and Great Cormorants (*Phalacrocorax carbo*) collected during beach surveys are often entangled in fishing gear (Tasker et al. 2000; see Figure 16.2).

FIGURE 16.2 Drowned murre in a fragment of net washed up on a beach on the south coast of Newfoundland, Canada. (Photo by W. A. Montevecchi.)

The mortality imposed by gill nets is evident when nets are removed. For instance, the closure of the Atlantic salmon fishery (Potter and Crozier 2000) and ground-fisheries in eastern Canada during the 1990s removed gill nets in waters off eastern Newfoundland. Concordantly, numbers of breeding murres, Atlantic Puffins, Razorbills, and gannets appear to be responding positively. Increases in murre populations have also been reported following closure of the drift-net salmon fishery in western Greenland (Piatt and Reddin 1984).

Even though high-seas gill nets have been banned globally, much of the fishing effort that used gill nets subsequently shifted focus to long-lining (see Brothers et al. 1999; see previous section). Thus, while populations of pursuit-diving seabirds benefited from this ban, populations of surface-feeding birds have suffered from the consequences.

Long-lines — Long-line fishing (i.e., setting extensive lines of more than 100 km in length with hundreds of thousands of baited hooks) is an old technique that is used in all of the world's oceans (Bjordal and Løkkeborg 1996). Long-line fishing is generally conservative, in that it catches mainly target species and causes little disturbance to habitat (Løkkeborg 1998). Pelagic long-lining fisheries are directed at tuna, swordfishes, and sharks, primarily in tropical and temperate oceans, and demersal long lining is directed at deep-water fishes like cod, halibut, hakes, toothfish, and snappers in colder waters. Fisheries for large pelagic fishes operate near ocean fronts and continental shelf breaks where marine birds forage (Croxall and Prince 1996, Robertson 1998, Brothers et al. 1999). The major pelagic long-line fisheries for tuna are Japan, Taiwan, and Korea, primarily in the Pacific Ocean (Figure 16.3). Pelagic long-line fisheries for swordfish are smaller and carried out by Spain, the U.S.A., Canada, Portugal, Italy, Greece, and Brazil, mostly in the Atlantic (Figure 16.4).

Seabirds vulnerable to long-line fisheries include those that feed at or near the surface, scavenge, and attempt to steal bait from hooks. These include petrels (e.g., Northern Fulmars, *Fulmarus*

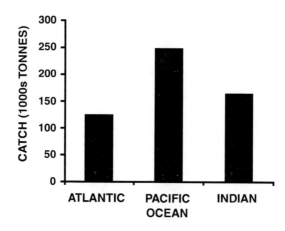

FIGURE 16.3 Long-line catches of tuna. (Based on data in Brothers et al. 1999.)

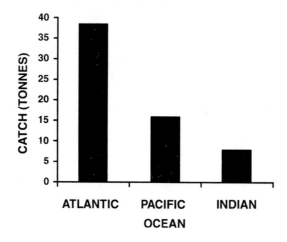

FIGURE 16.4 Long-line catches of swordfish. (Based on data in Brothers et al. 1999.)

glacialis; White-chinned Petrel, *Procellaria acquinoctialis*; Giant Petrels, *Macronectes* spp.), albatrosses (e.g., Gray-headed Albatross, *Diomedea chrysostoma*; Black-browed Albatross, *D. melanophris*; Wandering Albatross, *D. exulans*; Black-footed Albatross, *Phoebastria nigripes*; Laysan
Albatross, *P. immutabilis*; Mollymawk Albatrosses, *Thalassarche* spp.), gulls, and skuas (Cherel et
al. 1995, Croxall and Prince 1996, Brothers et al. 1999, Tasker et al. 2000; Figure 16.5). Major
by-catches of seabirds are documented in the Southern and Pacific Oceans where Brothers (1991)
estimated that approximately 108,000,000 hooks are set by the Japanese tuna fisheries with an
estimated annual mortality of 44,000 albatrosses. Clearly, by-catches of this magnitude hold serious,
nonsustainable consequences for long-lived albatrosses and petrels (Brothers, 1991, Moloney et al.
1993, Robertson and Gales 1998, Brothers et al. 1999). These consequences are intensified by
seabird by-catches that are both adult- and sex-biased (Brothers et al. 1999). Short-tailed Albatrosses
(*Phoebastria albatrus*) and endangered Spectacled Petrels (*Pteraldroma conspicillata*) are also
caught (Table 16.3; Ryan 1998, Brothers et al. 1999).

Many hundreds of thousands and possibly millions of seabirds are killed by long-line fisheries.
Table 16.4 summarizes the available information on fishing effort and avian mortality in the world's
long-line fisheries. Much information still needs to be collected in order to assess fisheries effects
on birds, e.g., Indian Ocean, northwest Atlantic. Additionally, there is little information on unregulated and illegal long-line fisheries that operate in many regions (Brothers et al. 1999). Birds are

FIGURE 16.5 Long-liner setting hooks with Laysan and Black-footed Albatrosses taking the bait. (Drawing by J. Zickefoose.)

TABLE 16.3

Endangered and Critically Endangered Seabird Species That Are Killed by Long-Line Fishing Activities

Species	IUCN Status	Ocean
Tristan Albatross, *Diomedea dabbenena*	Endangered	South Atlantic
Northern Royal Albatross, *D. sanfordi*	Endangered	Southern Ocean
Amsterdam Albatross, *D. amsterdamensis*	Critically Endangered	Southern Ocean
Chatham Albatross, *Thalassarche eremita*	Critically Endangered	Southern Ocean
Spectacled Petrel, *Pterodroma conspicillata*	Endangered	Southern Ocean

Note: International Union for the Conservation of Nature (IUCN) Criteria.

also killed by ingesting hooks in discarded offal and by-catch, as well as by sport fishers (Figure 16.6). Moreover, long-lining crews commonly shoot birds to discourage bait-stealing (Brothers et al. 1999).

Seabird mortality can be reduced by using streamers trailed on lines behind vessels to scare birds, by releasing baited hooks under water line and at night, by increasing their sinking rates, and by avoiding discards (Brothers 1991, Cherel et al. 1995, Løkkeborg 1998, Robertson 1998, Brothers et al. 1999). However, mitigative procedures are not currently widely used (Croxall and Prince 1996) and require the cooperation of fishers for effective implementation (Robertson 1998). Interestingly, Løkkeborg (2000) showed that streamers on lines trailed behind vessels (see Figure 2 in Tasker et al. 2000) significantly reduce bird by-catch and bait-loss *and* increase target fish catch. This effect could motivate fishers to incorporate these techniques.

16.2.1.2 Disturbance

Shellfish aquaculture sites remove potential habitat from use by seaducks, while their dense cultured food sources also attract them. Cormorants, gulls, diving ducks, and other birds are attracted to sites where marine fishes are held in cages or holding pens (Kirby et al. 1996), and to rivers and estuaries where hatchery-reared fishes are released (e.g., Wood 1985, Kalas et al. 1993, Cairns 1998). Many cormorants (*Phalacrocorax* spp.), Shags (*P. aristotelis*), and herons are shot in these situations (e.g., Carss 1994). Birds also are disturbed by fishers and fishing vessels working near concentrations of nonbreeding birds and near colonies where they at times store gear (e.g., lobster pots).

16.2.2 INDIRECT EFFECTS

16.2.2.1 Prey Depletion

Negative effects of fisheries on seabirds are expected when fisheries target the same species and size-classes that birds consume. In contrast, when fisheries take fishes larger than those that birds prey on, the effects of fisheries on seabirds can be positive (see below). Instances of the former are associated with industrial fisheries that exploit abundant, highly aggregative species for fish meal and oil that are used for animal feeds, aquaculture, and other industrial uses (Aikman 1997). Catches by industrial fisheries doubled in the last 30 to 40 years (Aikman 1997), consistent with patterns of overfishing stocks ("raw material") to commercial extinction. Industrial fisheries account for about a third of the world catch of marine fish (Coull 1993, Aikman 1997). As inappropriate as it seems, sandlance catches in the North Sea are essentially unregulated (Aikman 1997).

Another complication of fishery effects on the depletion of seabird prey involves the by-catch of nontarget species. For example, the by-catches of larval and forage fishes in small-mesh shrimp trawls are very substantial (Alverson and Hughes 1996, Aikman 1997), at times exceeding shrimp

TABLE 16.4
Estimated Numbers of Hooks Set, Catch Rates and Estimated Seabird Mortality by Different Long-Line Fisheries in Different Oceanographic Regions (based primarily on Brothers et al. 1999)

Fishery (years)	Oceanographic Region	Estimated No. of Hooks Set ($\times 10^6$)	Catch Rates/1000 Hooks (range [median])	Estimated Mortality	Species killed
Patagonian toothfish; king klip hake, tung, shark (1988–1997)	Southern	125	<0.01–1.90 [0.32] 0.043 (mitigation) 0.02 (night)	32,268–40,200	White-chinned Petrel, Albatrosses (Grey-headed, Black-browed, Wandering, Shy, Antipodean, Chatham, White-capped, Flesh-footed, Buller's, Yellow-nosed, Sooty, Campbell), Petrels (Giant, Grey, Giant-winged, Westland, Black, Pintado), Cape Gannet, Subantarctic Skua, Penguins (Gentoo, Macaroni), Shearwaters (Short-tailed, Wedge-tailed, Flesh-footed, Sooty)
Tuna, swordfish	Indian	~154	?	?	Albatrosses
Cod, ling, haddock, redfish (1980s)	NE Atlantic	~1000	[1.75] 0.04 (scaring)	1,750,000 ?	Fulmar, Gannet
Wolffish, swordfish, tuna (1996)	NE Atlantic		0.49 (underwater sets)		Skua, Gulls (Glaucous, Great Black-backed, Lesser Black-backed, Herring)
Cod, tusk, haddock, halibut, plaice, saithe, hake, tuna, swordfish (1996)	NW Atlantic	200+	?	?	Fulmar, Shearwaters, Gulls
Tuna, swordfish (1987–1994)	Atlantic	?	0.8–15 [7.6]	?	Petrels (White-chinned, Spectacled), Albatrosses (Wandering Black-browed, Yellow-nosed), Shearwaters
Halibut, pollock, cod, sablefish, turbot, rockfish, flounder, tuna, sharks, swordfish (1996)	NE Pacific	510	0.059–0.087 [0.073]	13,042–37,230	Fulmar, Gulls, Shearwaters, Albatrosses (Laysan, Black-footed, Short-tailed)
Pollock, cod, halibut	NW Pacific	270+	?	?	
Tuna, swordfish, sharks, snappers, hake, king klip, skate (1994–1995)	Central Pacific	7+	0.083–0.41 [0.214]	1,898	Albatrosses (Yellow-nosed, Laysan, Chatham, Black-browed), Petrels (White-chinned, Spectacled), Shearwaters (Great, Cory's)

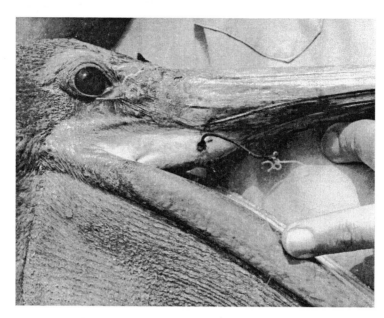

FIGURE 16.6 Brown Pelican hooked by a sports fisher in Florida. (Photo by E. A. Schreiber.)

catches by an order of magnitude or more (e.g., Pender et al. 1992; see Pauly et al. 1998). Surface-feeding seabirds (e.g., gulls, terns) and shallow-diving species (e.g., puffins) are generally considered the most vulnerable to the over-fishing of small pelagic fishes (see Furness and Ainley 1984, Monaghan et al. 1992), though deep divers such as murres can also be negatively affected (e.g., Vader et al. 1990a).

There are many demonstrations of the negative effects of intense and over-exploitive fishing pressures on the reproduction and populations of seabirds (Table 16.5). For example, the breeding

TABLE 16.5
Associations between Intense Fishing Pressures and Breeding Failures or Population Declines of Seabirds

Bird	Fish	Location	Date	Sources
Jackass Penguin, Cape Gannet	Pilchard	Benguela	1956–1980	Burger and Cooper 1984, Crawford et al. 1985
Peruvian Brown Pelican, Guanay Cormorant, Peruvian Booby	Anchoveta	Humboldt Current	1950s–1970s	Duffy 1983
Brown Pelican, Elegant Tern	Anchovy	S. California	1969–1980	Anderson et al. 1982, Anderson and Gress 1984, Schaffner 1986
Shag, Great Skua, Black-legged Kittiwake, Arctic Tern, Common Tern, Common Murre	Sandlance Herring	Shetland North Sea	1986–1990	Furness 1990, Monaghan et al. 1989, 1992, Uttley et al. 1989, Huebeck 1989; Hamer et al. 1991; Bailey et al. 1991, Klomp and Furness 1992, Monaghan 1992
Atlantic Puffin	Herring	Norway North Sea		Anker-Nilssen 1987, 1992, Barrett et al. 1987, Vader et al. 1990b; Anker-Nilssen and Røstad 1993
Atlantic Puffin	Capelin	NW Atlantic	1981	Brown and Nettleship 1984
Common Murre	Capelin	N. Norway	1985–1987	Vader et al. 1990a, b

population of Cape Gannets (*Sula capensis*) in southern Africa decreased by about half from 1956 to 1980 and has been attributed to a fishery-induced collapse of pilchard (*Sardinops ocellata*) and reduced availability of anchovies (*Engraulis capensis*; Crawford et al. 1980, 1983). The over-exploitation and crash of anchoveta fishery and the resultant population declines of Peruvian guano birds have been well documented (Schaefer 1970, Paulik 1981). The fishery-induced collapse of Atlanto-Scandian herring stock along the Norwegian coast resulted in 14% per annum declines in the breeding population of Atlantic Puffins in Lofoten (Anker-Nilssen and Røstad 1993). Breeding populations of Common Murres in the Barents Sea were reduced by 80% in the mid-1980s due to a drastic reduction in capelin stocks induced at least in part by industrial fishery catches (Vader et al. 1990a, b). Changes in the diets of seabirds are also associated with fishery catches. For example, due to economic incentives from Asian markets, there was intense fishing pressure on short-finned squid (*Illex illecebrosus*) during the late 1970s in the northwest Atlantic (Figure 16.7). These squid subsequently disappeared from both seabird diets and commercial catches for at least two decades (Montevecchi and Myers 1995, 1996, unpublished).

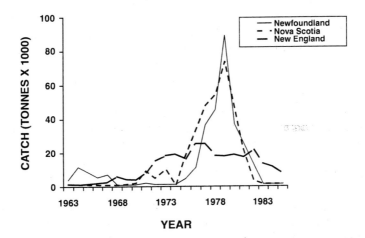

FIGURE 16.7 Catches of short-finned squid from different areas in eastern Canada. (After Montevecchi 1993b.)

Some of the complexities of potential effects of fisheries on seabirds were pointed out by Burger and Cooper (1984). They suggested that purse-seine fisheries targeting pelagic pilchards off southern Africa negatively affected pursuit-diving penguins through prey depletion but also positively affected surface-foraging gannets by providing fishery discards (see also Duffy et al. 1987; see below). Furness (2000) suggested that the huge harvests of sandeel fisheries in the North Sea may not have negatively affected seabird populations because fishery catches and seabird consumption occur in different regions and because the population of major predators of forage fish (mackerel, *Scomber scombrus*) was greatly reduced (Camphuysen and Garthe 2000). Over-fishing large pelagic fishes in tropical oceans can have an opposite effect on marine birds dependent on these fishes to drive small fishes to the surface where birds can access them (Au and Pitman 1988, Safina and Burger 1985).

In considerations of seabird population fluctuations and their associations with fisheries, it is notable that both marine bird and fish populations often change significantly in the absence of fisheries (Cushing 1982, Hatch et al. 1993, Carscadden et al. 2001). Extreme population fluctuations occurred well before fisheries were initiated (e.g., Soutar and Isaacs 1974). Climatic and oceano-graphic changes can induce significant fluctuations in avian populations (e.g., Schreiber and Schreiber 1984, 1989, Montevecchi and Myers 1997; see Chapter 7), fishes (Welch et al. 1998, 2000, Carscadden et al. 2001), and other marine animals. Both physical perturbations and fisheries impacts can at times induce regime shifts, i.e., ecosystem-wide changes in community structures and food webs (Steele 1996, 1998). Because physical oceanographic influences interact with fishing and pollution, it is often very difficult to partition effects attributable to either natural or anthropogenic factors (Duffy and Schneider 1994, Steele 1996, Tasker et al. 2000; cf. Hutchings and Myers 1994).

16.2.2.2 Competition and Predation by Scavenging Seabirds

In many locations, populations of scavenging gulls, kittiwakes, skuas, fulmars, and gannets have become dependent on food sources associated with fishery discards and offal (see next section). Many of these scavenging birds include species that are often highly predatory on other seabirds. Owing in large part to their exploitation of discards, populations of scavenging gulls and skuas have increased sharply during the 20th century (Furness et al. 1992, Howes and Montevecchi 1992) and are having negative effects on other seabirds. For example, gulls displaced tern colonies in the Gulf of St. Lawrence (Howes and Montevecchi 1993), the mid-Atlantic U.S. coast (Burger and Gochfeld 1991), and the Wadden Sea (Becker and Erdelen 1987).

The potentially negative, indirect effects of fishery waste generation can be intensified when disposal is reduced or eliminated (Furness 2000). The use of larger mesh sizes permits the escape of smaller target species, reduces by-catch, and decreases the number of discards available to scavenging birds (Hudson and Furness 1988). The closure of the eastern Canadian ground-fishery from 1992 through 1999 essentially eliminated discarding and offal production in the northwest Atlantic and imposed severe food-stress on populations of Herring Gulls (*Larus argentatus*) and Great Black-backed Gulls (*L. marinus*; Regehr and Montevecchi 1997). Consequently, these gulls greatly increased predation pressure on Leach's Storm-Petrels (*Oceanodroma leuchoroa*), Black-legged Kittiwakes (*Rissa tridactyla*), and Atlantic Puffins (Russell and Montevecchi 1996, Regehr and Montevecchi 1997, Stenhouse and Montevecchi 1999). Similar effects of discard removal were reported for Great Skuas (*Catharacta skua;* Phillips et al. 1999), Yellow-legged Gulls (*Larus cachinnans*), and Audouin's Gulls (*L. audouinii*; Oro and Martinez-Vilalta 1994, Oro et al. 1995).

If the elimination of discards and offal continues in the longer term, then scavenger (e.g., gull) populations are expected to follow them in a density-dependent manner (Figure 16.8). However, as scavenger/predator populations decline to supportable levels, extreme pressure will be exerted on prey species by food-stressed predators.

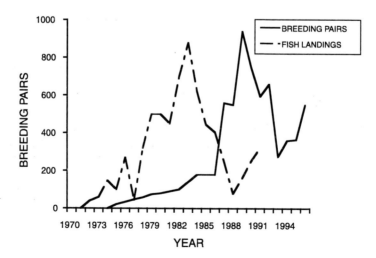

FIGURE 16.8 Populations of Herring and Great Black-backed Gulls in the Gulf of St. Lawrence off western Newfoundland, Canada, and local fishery landings used as indices of fishery discards and offal production. (W. A. Montevecchi, unpublished.)

16.3 POSITIVE EFFECTS OF FISHERIES ON SEABIRDS

16.3.1 DIRECT EFFECTS

16.3.2 Provisioning of Fisheries Discards and Offal

Offshore fishing vessels generate huge tonnages of discarded fish, scraps, and waste from demersal fishes and invertebrates that would otherwise be unavailable to marine birds (Table 16.6), creating a global feeder-at-sea program for avian scavengers. Otter trawlers and stern trawlers attract higher numbers of birds compared to beam trawlers and purse seiners, because they produce the higher levels of discards (Camphuysen et al. 1995).

Many avian species scavenge at trawlers (Hudson and Furness 1988, Furness et al. 1992, Blaber et al. 1995). Fulmars and kittiwakes are the most common and can aggregate so densely at trawlers that their concentrations have been referred to as "blizzards" (R.G.B. Brown pers. comm.). The utilization of fish discards and offal from vessels and from fish plants influences populations of scavenging birds (Oro et al. 1995, Oro 1996, Garthe et al. 1996, Hüppop and Wurm 2000). For example, during the 20th century, populations of fulmars, large *Larus* gulls, kittiwakes, skuas, and gannets have increased greatly and expanded in association with increasing levels of fishery discards throughout the North Atlantic (Fisher 1952, Kadlec and Drury 1968, Drury and Kadlec 1974, Harris 1970, Howes and Montevecchi 1993, Camphuysen et al. 1995, Stenhouse and Montevecchi 2000), the South Atlantic (Burger and Cooper 1984), and elsewhere.

Discards and offal comprise about 30% of the food of seabirds in the North Sea (Tasker and Furness 1996). Up to 70% of the diet of adult Great Skuas and about 30% of the food fed to their chicks in Shetland consists of discards (Furness and Hislop 1981). These percentages increased when the abundance of sandlance declined (Hamer et al. 1991). However, when the proportion of discards increased at the expense of sandlance in the chicks' diet, their growth was reduced (Furness 1987). The tonnages of discards and offal produced (Table 16.6) could potentially support more than 6 million birds in the North Sea (Garthe et al. 1996).

Most seabirds attending fishing vessels are adults, and scavenging levels are highest in winter when there is more competition for scraps. Discards are partitioned by size and shape, and by feeding technique among seabirds (Garthe and Hüppop 1998). In food-stealing interactions at fishing vessels, the largest species (gannets, Great Black-backed Gulls, skuas) fare best (Hudson and Furness 1988, Garthe and Hüppop 1998). Hence, if discards are reduced and competition increases, smaller species are expected to fare worse (Tasker et al. 2000).

Multiple influences of fisheries on seabirds are common. For example, purse-seine fisheries for pelagic fishes (e.g., pilchards) off southern Africa appear to have positively affected surface-foraging

---

TABLE 16.6
Estimated Discards and Offal Produced by Fishing Vessels in the North Sea and Estimated Consumption by Marine Birds

Item Discarded	Tonnage	Energy Density (kJ/g)	Estimated Consumption by Birds	
			%	Tonnes
Offal	70,000	10	95	66,500
Roundfish	273,000	5	80	96,000
Flatfish	307,300	4	20	40,000
Benthic Invertebrate	302,500	2.5	6	10,800
Total	**945,600**		**37**	**213,300**

Based on Camphuysen et al. 1995, Garthe et al. 1996, 1999.

gannets by providing discards while at the same time negatively affecting pursuit-diving penguins through reduction of pelagic fish stocks (Burger and Cooper 1984). For wide-ranging species such as albatrosses, petrels, and fulmars, discarding can change their distributions at sea (Abrams 1983, Ryan and Moloney 1988). The consequences of such changes may be either positive as in the breeding range expansions of fulmars (Fisher 1952, Stenhouse and Montevecchi 1999) or potentially negative as in the case of large numbers of Black-browed and Shy Albatrosses (*Diomedea cauta*) being attracted to trawling sites well outside of their "normal" foraging areas in the southern Benguela region (Abrams 1983). Some birds including penguins, cormorants, and petrels often avoid feeding aggregations at trawlers (Ryan and Moloney 1988).

16.3.2 INDIRECT EFFECTS

16.3.2.1 Removal of Competitors — Multispecies Interactions

Ecosystem interactions are often less straightforward than they appear initially, and concepts of surplus production and predator release are tenuous (May et al. 1979, Lavigne 1996). The over-harvesting of large predators has, however, been associated with increases in the abundance of forage fishes used by seabirds. The depletion of herring (*Clupea harengus*) and mackerel in the North Sea resulted in increases in the abundances of sandlance and sprats (Sherman et al. 1981; see also Springer et al. 1986, Hatch and Sanger 1992). Over-fishing Atlantic Cod (*Gadus morhua*) in the northwest Atlantic (Hutchings and Myers 1994) removed a major predator of the primary prey, capelin (*Mallotus villosus*), of marine birds and mammals (Table 1). Hence, the depletion of cod can be expected to enhance food conditions for birds in the northwest Atlantic during the next decade (see Table 16.1). Factor in the removal of inshore fishing gear during the eastern Canadian ground-fishery closure (1992–1999), and the circumstances for seabirds in the northwest Atlantic appear even better. Perhaps the greatest ecosystem concern is that there is no indication of population increases by capelin (Carscadden and Nakashima 1997), though situations may well be beneficial for birds.

Increases in the population of Chinstrap Penguins (*Pygoscelis antarctica*) in the Southern Ocean since about 1950 were linked to the depletion of baleen whales by commercial whaling and a subsequent increase in krill abundance (Conroy 1975, Coxall et al. 1984). This relationship was questioned, however, and growing penguin populations are attributed to climate warming and less extensive sea ice cover that in turn gave breeding penguins easier access to foraging sites (Fraser et al. 1992).

16.3.2.2 Increase Abundance of Small Fishes

As indicated above, over-fishing large predatory fish has at times resulted in increased abundance of their prey, including small plantivorous fishes and crustaceans (Sherman et al. 1981, Hamre 1988). Increased numbers of juveniles (small fish) of commercially exploited species can be marked among species such as cod and pollock that are cannibalistic. When such events occur they can benefit seabirds and be reflected in increases in their populations.

16.4 NEGATIVE EFFECTS OF SEABIRDS ON COMMERCIAL FISHERIES

16.4.1 DIRECT INFLUENCES

16.4.1.1 Interactions with Aquaculture

Aquaculture is a major global growth industry that accounts for about 15% of current world fisheries production (FAO 1995). Successful aquaculture ventures can lead to complacency about the state of wild fisheries. Predation by cormorants on channel catfish (*Ictalurus punctatus*) farms in the

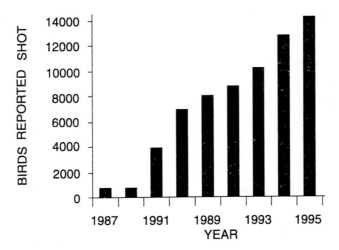

FIGURE 16.9 Number of waterbirds reported killed at fish farms in the southeastern U.S. (From data in Belant et al. 1998.)

southeastern U.S. is a problem (Glahn and Stickney 1995, Glahn and Brugger 1995). However, even though their predation effects may be negligible on the whole (e.g., Kalas et al. 1993, Wooten and Dupree 2000), they can impact individual farming operations. Permits are issued to control avian predators (such as cormorants, shags, anhingas, grebes, seaducks, herons, and kingfishers) feeding at fish pens and shellfish farms in the U.S. (Belant et al. 1998, Kirby et al. 1996). The predation tends to be size-selective, and in situations where avian predators take many fish from holding pens, the vast bulk tend to be small fingerlings (10 to 20 cm length; Glahn and Stickney, 1995). There are some means to mitigate predator effects (see below). About five times the number of waterbirds were killed at fish farms in Arkansas compared to Mississippi, even though there is almost twice as much area taken up by catfish farms in Mississippi (Belant et al. 1998). This is because there are many more baitfish farms in Arkansas (Belant et al. 1998). The permitted killing of waterbirds at aquaculture sites is increasing (Figure 16.9), and with rapidly expanding aquacultural industries, fish farming and bird conservation are clearly proceeding on a head-on collision course. Belant et al. (1998) argue that the levels of bird killing at catfish farms in the southeastern U.S. do not influence local avian populations as indicated by Christmas Bird Counts (CBC). They do not, however, point out that the number of Great Egrets killed by catfish farmers (1000s) is an order of magnitude higher than number of birds sighted on Arkansas CBC (100s). They report that farmers shot only about 60% of the number of waterbirds that they were permitted to kill. This suggests that kill permits may be excessive, though these are also circumstances in which under-reporting would be expected.

16.4.1.2 Bait Stealing

Bait stealing by seabirds can pose problems for long-lining fisheries (Brothers 1991, Løkkeborg 1998). Techniques for scaring birds, for releasing baited hooks at night and underwater, as well as for increasing the sinking rates of baited hooks may benefit the fishery directly and help minimize seabird mortality.

16.4.2 Indirect Effects

16.4.2.1 Prey Depletion

Assessments of seabird interactions with commercial fish stocks come from two general orientations. First, bioenergetic models of prey consumption indicate that birds, like other marine

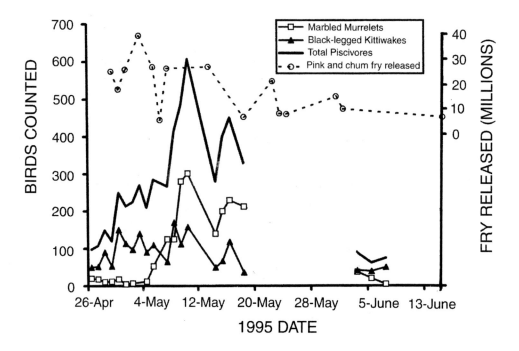

FIGURE 16.10 Aggregative responses of surface-feeders (Black-legged Kittiwakes) and pursuit-divers (Marbled Murrelets *Brachyramphus marmoratus*) and all piscivorous birds to concentrations of hatchery-released pink salmon (*Oncorhynchus gorbuscha*) and chum salmon (*O. keta*) in Prince William Sound, Alaska. (After Scheel and Hough 1997.)

predators, consume substantial tonnages of prey (Table 16.1), at times representing up to 30% of the estimated production in a localized area (see Montevecchi 1993a). The question then becomes, what consequences, if any, do these consumption levels hold for commercial fisheries? Second, estimates of the consumption of fresh-water fishes, especially those that spend significant parts of their life cycle at sea, are used in considerations of avian "impacts" on both commercial and sport fisheries.

A recent bioenergetic modeling exercise led to the suggestion that predation by Northern Gannets could negatively impact the population of Atlantic salmon (*Salmo salar*) in the northwest Atlantic (D. Cairns and W. Montevecchi unpublished). Interestingly, the salmon are not an important prey for gannets, comprising less than 3% of their diet on average (W. Montevecchi and D. Cairns unpublished). Hence, while gannets may negatively influence the population dynamics of salmon, there are no ecologically responsible management options to address this interaction.

Avian predators have been considered to limit salmon production (Elson 1962). Double-crested Cormorants eat salmon and trout, and also damage fish that they do not kill (Kirby et al. 1996, Cairns 1998). Levels of avian consumption of commercially exploited fish are greater in restricted freshwater environments (rivers, streams, lakes) than in the open ocean, though local consumption can be inappropriately generalized to larger scales and populations (Scheel and Hough 1997). Birds often show aggregative responses to hatchery-released fishes (Figure 16.10), but not to natural smolt runs (Wood 1985, Bayer 1986, Mullins et al. 1999). This is possibly related to differences in the behavior of hatchery- and wild-reared smolts. Many seabirds, including cormorants, gulls, murrelets, and terns, prey on concentrations of hatchery-released salmonids. While birds consume many hatchery-reared fishes, they do little to negatively impact populations (Scheel and Hough 1997), except in localized situations (see Cairns 1998). Assessments of predator consumption need to be evaluated in the context of broad-scale multispecies and eco-system models rather than as simple linear predator–prey interactions (e.g., May et al. 1979, Lavigne 1996).

One study has indicated that, on a local scale, seabirds can deplete prey. Double-crested Cormorants (*Phalacrocorax auritus*) deplete benthic fishes near their colonies (Birt et al. 1987). Their prey are mostly noncommercial species, some of which prey on commercial species (e.g., cunners on cod). No other direct evidence of prey depletion by seabirds is available, and due to the small biomass of birds in marine ecosystems, none is expected on spatial scales relevant to commercial fisheries.

16.5 POSITIVE EFFECTS OF SEABIRDS ON COMMERCIAL FISHERIES

16.5.1 DIRECT EFFECTS

16.5.1.1 Birds as Fishing Devices

There are a few instances of fishermen exploiting birds directly to obtain food. For instance, some artisanal fishers have used tethered cormorants to catch fish. The cormorants are fitted with restraints around their necks that do not allow them to swallow the fish they capture and the fishermen retrieve them.

16.5.1.2 Birds as Indicators of Prey Location

Fishermen pursuing pelagic fishes and crustaceans often use sightings of seabirds to help locate schools and concentrations of prey (Montevecchi 1993a). Mobile gear fisheries pursuing pelagic prey are cases in point.

16.5.2 INDIRECT EFFECTS

16.5.2.1 Predation on Predators, Competitors, and Diseased and Parasitized Fish

Interrelationships within marine food webs are complex and often indirect (Lavigne 1996). Marine birds could benefit commercially exploited species by preying on their predators, e.g., eels (see Birt et al. 1987, Cairns 1998) or via the removal of weak, parasitized, or diseased fishes (Feare 1988).

16.5.2.2 Guano and Nutrient Recycling

Seabird guano is collected as agricultural fertilizer in a few locations. Guano that is not collected fertilizes both terrestrial and marine environments (Threlfall 1980). Dense rich growths of seaweed in the vicinities of seabird colonies indicate the enriching effects of seabird excretions on marine plants. These growths also benefit invertebrates and fish that associate with them. The positive effects of this nutrient enrichment tend to be localized in scale (Bédard et al. 1980, Bosman and Hockey 1986).

16.5.2.3 Prey Information

Of the many ways that marine birds might benefit commercial fishers, perhaps the most useful is through systematic indices of biological and ecological information. Knowledge of avian ecology can enhance understanding of fish stock conditions, availabilities, movements, spatial and temporal distributions, natural mortality, and of changing ecosystem and oceanographic conditions more generally (e.g., Cairns 1987, 1992, Montevecchi et al. 1988, Barrett et al. 1990, Barrett 1991, Montevecchi and Berutti 1991, Hatch and Sanger 1992, Montevecchi 1993, Montevecchi and Myers 1995, 1996, 1997, Bunce 2001). Records of guano harvests have been used to indicate fluctuations in fish populations including historic ones before commercial exploitation (e.g., Crawford and Shelton 1978).

16.6 INTERACTIONS OF FISHERIES AND OTHER PERTURBATIONS ON SEABIRDS

Considerations of fisheries mortality have to include other additive and synergistic cumulative effects. These can involve oceanographic influences, pollution, and hunting.

16.6.1 OCEANOGRAPHIC FLUCTUATIONS

Oceanographic events involving cold- and warm-water events can have pervasive effects on fish distributions, recruitment, and population dynamics. Moreover, population resiliency to natural perturbations can be greatly reduced when populations are at low levels and fragmented (Myers et al. 1999, Stephens and Sutherland 1999). Oceanographic influences can cause regime shifts (Steele 1998) or trophic cascades (Pace et al. 1999) that can radically change food web and ecosystems dynamics (e.g., Wooton 1995).

16.6.2 POLLUTION

The many influences of contaminant pollution on marine birds have been well considered (Wiens et al. 1996, Furness and Camphuysen 1997; Chapter 15). Pesticides, herbicides, heavy metals (Furness 1993, Focardi et al. 1996, Jones et al. 1996, Joiris et al. 1997, Van Den Brink et al. 1998, Burger and Gochfeld 2000, Montevecchi 2001), radionuclides (Brisbin 1993), plastics (Montevecchi 1991, Spear et al. 1995, Blight and Burger 1997) and hydrocarbons (Wiese and Ryan 1999) have negative effects on seabirds. Beach-bird surveys indicate that mortality associated with oiling at sea has increased through the 1980s and 1990s in the northwest Atlantic (Wiese and Ryan 1999), but not in the northeast Atlantic (Camphuysen 1998).

16.6.3 HUNTING

The hunting of marine birds is likely decreasing on a global basis, though both aboriginal and traditional hunting is ongoing in many places (e.g., Faroes, Greenland, Newfoundland, New Zealand, etc., Burger and Gochfeld 1994). The influences of hunting on populations can be profound (e.g., Montevecchi and Tuck 1987, Elliot 1991).

16.6.4 CUMULATIVE EFFECTS

To adequately evaluate the range of human effects including fisheries on seabirds, it is essential to consider all additive mortality effects in the context of cumulative effects. For example, the population of African Penguins (*Spheniscus demersus*) decreased sharply in recent decades due in part to prey depletion (e.g., Burger and Cooper 1984). During 2000, their South African breeding area was the site of a large oil spill. While a very successful campaign to save birds was mounted (Underhill 2000), the effects of oil-induced mortality on the population level of African Penguins are yet to be determined.

16.7 MANAGEMENT AND MITIGATION

16.7.1 MISGUIDED MANAGEMENT

16.7.1.1 Culls

Culling strategies are based on assumptions that killing predators will leave "surplus" production from prey populations that would have been consumed by predators and that the "surplus" in turn can be taken by a commercial fishery. In North America and Europe, cormorants are considered a competitor for fish by many sport and commercial fishers. In recent years there have been both authorized culls and illegal mass killings at cormorant colonies. These activities proceed in the absence of evidence that these birds are having an impact on commercial species and that culls can benefit fishers (Tasker et al. 2000).

16.7.1.2 Colony Displacements

Questionable efforts have been made to displace the world's largest colony of Caspian Terns (*Sterna caspia*) on the Columbia River (Henson 1999, Roby et al. 1999, 2000). This drastic action follows from findings that the terns eat substantial numbers of hatchery-raised salmon smolts on the river, though this contention remains to be fully substantiated (Harrison 1999; cf. Scheel and Hough, 1997). Whether or not this consumption of hatchery-reared fishes has any influence on salmon populations also appears to be an unresolved issue.

16.7.2 MITIGATION

Birds are the useful indicators for monitoring the global conditions and health of the marine environment (Furness 1993, Montevecchi 1993a). Besides being the most conspicuous organisms in marine ecosystems, birds are also easily studied. The monitoring of multispecies complexes is maximally efficient by directing attention to top predators whose populations and processes are changing on the slowest time and largest spatial scales in the system (May et al. 1979). Particular species are especially useful for monitoring different system components, including pollution (Furness 1993, Burger 1993, Montevecchi 2001).

Indications that many birds are trapped in fixed fishing gear at dawn and dusk and during inclement weather (Strann et al. 1991, Melvin et al. 1999) suggest that net visibility is an important aspect of avoidance. Manipulations of features of nets that increase visual and auditory detectability to birds have proven useful in facilitating gear avoidance by birds as it has with marine mammals (e.g., Kraus et al. 1997). Visual and acoustic enhancements to salmon gill nets reduced seabird by-catch by 40% or more (Melvin et al. 1999).

Fisheries management options can be used to reduce seabird by-catch in fishing gear. For example, in California where murre populations declined by 50% or more, regulation of fishing depth and large closure areas were used by fisheries managers to reduce seabird by-catch (Salzman 1989, Martin 2000; see also Bryant and Martin 1996). Simple multispecies considerations that link the seasonal or diurnal timing of net fisheries to periods when target species are available, and nontarget, by-catch species are at minimal or low abundance can also be helpful in some circumstances (Melvin et al. 1999). By-catch quotas could be established and regulated (Piatt and Nettleship 1987). Working with fishers to use logbooks to help solve the by-catch problem is another possibility (see Neis et al. 1999).

Scaring birds with streamers and nocturnal and subsurface release of baited hooks should be used to minimize avian mortality from long-line fisheries. Nocturnal line-settings can reduce avian by-catch by 60 to almost 100%, though nocturnal species like White-chinned Petrels are often hooked during night sets (Cherel et al. 1995, Brothers et al. 1999). Nocturnal settings are less effective on moonlit nights, and are often not possible at high latitudes during summer when darkness is brief or nonexistent (Barnes et al. 1997, Brothers et al. 1999). Greater educational efforts and more cooperation and input from the fishing industry are needed (Robertson 1998, Brothers et al. 1999, Neis et al. 1999). An effective approach to solving by-catch problems would be to involve fishers to help create international legislative protections, to monitor their proper execution with fishery observers on vessels, and to soundly prosecute violators. It is extremely difficult to realize adequate conservation legislation in international water (Duffy and Schneider 1994), as evident from fisheries overexploitation and marine oil pollution. The implementation of temporal no-fishing zones (Croxall and Prince 1996) and consumer lobbies could play constructive roles in promoting ecologically conservative fishing practices. These may be essential because the mitigative measures available are not widely used (Croxall and Prince 1996), though fishers may be motivated to use bird-scaring techniques that reduce bait-stealing and increase the harvests of target species (Løkkenborg 2000).

With regard to avian predation at fish farms, scaring birds from farms and roosts has been effective (Mott and Boyd 1995). Moreover, the bulk of the predation that is on fingerling fishes allows farmers opportunities to release fish in larger holding ponds and/or when bird numbers are lower due to migratory movements (Glahn and Stickney 1995, Stickney et al. 1995). "Green" consumer lobbies that pressure for the least invasive predator "control" techniques and for the registering of ecologically conservative fish farming could also prove useful. Nocturnal releases of hatchery-reared fish reduce avian predation (Bayer 1986, Kalas et al. 1993). These fish take many hours to adjust to environmental conditions in the wild, and night releases allow adaptations before daylight when avian predators prey on them.

Well-considered approaches to fishing practices in the face of uncertainty (Ludwig et al. 1993) and the setting of "precautionary quotas" (Aikman 1997) are essential for effective ecosytem-based fisheries management. Minimizing fisheries by-catch is an integral component of such an approach, although it is not always easy to achieve (Alverson and Hughes 1996, Lugten 1997, Melvin et al. 1999).

16.7.2.1 Observer Programs

Independent observer programs on fisheries vessels are ongoing in many jurisdictions. However, vessel coverage and information on seabird (and fish) by-catch are often so limited as to make them ineffectual (Brothers et al. 1999). Formalization of observer programs and systematic data gathering are needed to produce effective fisheries management and conservation (see FAO 1999).

16.7.3 Marine Protected Areas

The extensive and continuous over-exploitation of marine fishes requires that all potential solutions for ecosystem management be considered (Boersma and Parrish 1999). Positive influences of fisheries closures on commercial stocks were clearly indicated during both world wars when fishing was curtailed in large sectors of the North Sea and after which fish populations exhibited substantial increases (Smith 1994). Marine protected areas, or harvest-free zones, offer options for preserving ecosystem processes. The few existing marine protected areas tend to be small and located in tropical regions. There is some evidence that these reserves are helping to create increased biodiversity and biomass that support large, long-lived predators (Williams 1998). To date, marine reserves and their design concepts are most applicable to relatively sedentary, tropical species (e.g., coral reef communities, Rakitin and Kramer 1996, Chapman and Kramer 1999). Dispersal patterns of protected species are the key to determining if reserves might act as sources for surrounding areas (Fogarty 1999, Chapman and Kramer 2000). Experimental tests are needed (Dugan and Davis 1993). Fishing pressure is intense on reserve boundaries, and designs and management programs need to include buffering features (Chapman and Kramer 1999, Fogarty 1999, Day and Roff 2000). Defining and designing marine protected areas on a hierarchical basis of physical features offers methodological promise for high-latitude oceans (Day and Roff 2000). However, there are major challenges in developing marine reserves at higher latitude where both pelagic and demersal fish species move over vast areas. The planning and establishment of marine protected areas require grass-roots community support and input (Lien 1999) and fishers' knowledge in resource and ecosystem management (Neis et al. 1999).

ACKNOWLEDGMENTS

My long-term research program with marine birds has been supported by the Natural Sciences and Engineering Research Council of Canada (NSERC) and at times supplemented by Fisheries and Oceans Canada and the Canadian Parks Service. I thank Betty Anne Schreiber and Joanna Burger for encouragement, manuscript reviews, and inputs; Dave Robichaud for information and references about research on marine protected areas; Cynthia Mercer for help with figure preparations; and Eileen Ryan, Marilyn Hicks, Peggy Ann Parsons, and Sharon Wall for word-processing many less-than-immaculate iterations of the manuscript.

LITERATURE CITED

ABRAMS, R. W. 1983. Pelagic seabirds and trawl-fisheries in the southern Benguela Current region. Marine Ecology Progress Series 11: 151–156.

AIKMAN, P. 1997. Industrial 'Hoover' Fishing: A Policy Vacuum. Greenpeace, Amsterdam.

AINLEY, D. G., A. R. DEGANGE, L. L. JONES, AND R. J. BEACH. 1981. Mortality of seabirds in high seas salmon gill-nets. Fisheries Bulletin 79: 800–806.

ALVERSON, D. L., AND S. E. HUGHES. 1996. Bycatch: from emotion to effective natural resource management. Review of Fishing and Biology of Fisheries 6: 443–462.

ANDERSON, D. W., AND F. GRESS. 1984. Brown Pelicans and the anchovy fishery off southern California. Pp. 128–135 in Marine Birds: Their Feeding Ecology and Commercial Fishery Relationships (D. N. Nettleship, G. A. Sanger, and P. F. Springer, Eds.). Canadian Wildlife Service Special Publication, Ottawa.

ANDERSON, D. W., F. GRESS, AND F. KAIS. 1982. Brown Pelicans: influence of food supply on reproduction. Oikos 39: 23–31.

ANKER-NILSSEN, T. 1987. The breeding performance of puffins Fratercula arctica on Røst, northern Norway in 1979–1985. Fauna Norvegica Series C, Cinclus 10: 21–38.

ANKER-NILSSEN, T. 1992. Food Supply as a Determinant of Reproduction and Population Development in Norwegian Puffins Fratercula arctica. D.Sc. thesis, University of Trondheim, Norway.

ANKER-NILSSEN, T., AND O. W. RØSTAD. 1993. Census and monitoring of puffins Fratercula arctica on Røst, North Norway, 1979–1988. Ornis Scandinavica 24: 1–9.

ARTUKHIN, Y. B., V. N. BURKANOV, AND P. S. VYATKIN. 2000. Incidental mortality of seabirds in the salmon gillnet fishery on the Russian Far East EEZ, 1993–98. (Abstract) Workshop Seabird Bycatch Waters Arctic Countries. Dartmouth, Nova Scotia.

AU, D. W. K., AND R. L. PITMAN. 1988. Seabird relationships with tropical tunas and dolphins. Pp. 174–212 in Seabirds and Other Marine Vertebrates (J. Burger, Ed.). Columbia University, New York.

BAILEY, R. S., R. W. FURNESS, J. A. GAULD, AND P. A. KUNZLIK. 1991. Recent changes in the population of the sandeel (Ammodytes marinus Raitt) at Shetland in relation to estimates of seabird predation. ICES Marine Science Symposium 193: 209–216.

BARNES, K. N., P. G. RYAN, AND C. BOIX-HINZEN. 1997. The impact the hake Merluccius spp. longline fishery off South Africa on procellariiform seabirds. Biological Conservation 82: 227–234.

BARRETT, R. T. 1991. Shags (Phalacrocorax aristotelis) as potential samplers of juvenile saithe (Pollachius virens (L.)) stocks in northern Norway. Sarsia 76: 153–156.

BARRETT, R. T., T. ANKER-NILSSEN, F. RIKARDSEN, K. VALDE, N. RØV, AND W. VADER. 1987. The food, growth and fledging success of Norwegian puffin chicks on *Fratercula arctica* in 1980–1983. Ornis Scandinavica 18: 73–83.

BARRETT, R. T., N. RØV, J. LOEN, AND W. A. MONTEVECCHI. 1990. Diets of Shags *Phalacrocorax aristotelis* and Cormorants *P. carbo* in Norway and implications for gadoid stock recruitment. Marine Ecology Progress Series 66: 205–218.

BAYER, R. D. 1986. Seabirds near an Oregon estuarine salmon hatchery in 1982 and during the 1983 El Niño. Fisheries Bulletin 84: 279–286.

BECKER, P. H., AND M. ERDELEN. 1987. Die Bestandsentwicklung von Brutvogeln der deutschen Nordseeküste 1959–1979. Journal für Ornithologie 128: 1–32.

BÉDARD, J., J. C. THERRIAULT, AND J. BÉRUBÉ. 1980. Assessment of the importance of nutrient recycling by seabirds in the St. Lawrence Estuary. Canadian Journal of Fisheries and Aquatic Sciences 37: 583–588.

BELANT, J. L., L. A. TYSON, AND P. M. MASTRANGELO. 1998. Effects of lethal control at aquaculture facilities on populations of piscivorous birds. Wildlife Society Bulletin 26.

BIRT, V. L., T. P. BIRT, D. GOULET, D. K. CAIRNS, AND W. A. MONTEVECCHI. 1987. Ashmole's halo: evidence for prey depletion by a seabird. Marine Ecology Progress Series 40: 205–208.

BJORDAL, Å., AND S. LØKKEBORG. 1996. Longlining. Fishing News Books, Oxford.

BLABER, S. J., D. A. MILTON, G. C. SMITH, AND M. J. FARMER. 1995. Trawl discards in the diets of tropical seabirds of the Northern Great Barrier Reef, Australia. Marine Ecology Progress Series 127: 1–13.

BLACKWELL, B. F., W. B. KROHN, N. R. DUBE, AND A. J. GODIN. 1995. Spring prey use by Double-crested Cormorants on the Penobscot River, Maine USA. Colonial Waterbirds 20: 77–86.

BLIGHT, L. K., AND A. E. BURGER. 1997. Occurrence of plastic particles in seabirds from the eastern North Pacific. Marine Pollution Bulletin 34: 323–325.

BOERSMA, P. D., AND J. K. PARRISH. 1999. Limiting abuse: marine protected areas, a limited solution. Ecological Economics 31: 287–304.

BOSMAN, A. L., AND P. A. R. HOCKEY. 1986. Seabird guano as a determinant of rocky intertidal community structure. Marine Ecology Progress Series 32: 247–257.

BRISBIN, I. L., JR. 1993. Birds as monitors of radionuclide contamination. Pp. 144–278 in Birds as Monitors of Environmental Change (R. W. Furness and J. J. D. Greenwood, Eds.). Chapman & Hall, London.

BROTHERS, N. 1991. Albatross mortality and the associated bait loss in the Japanese longline fishery in the Southern Ocean. Biological Conservation 55: 255–268.

BROTHERS, N. 1999. Understanding the fisherman's perspective. World Birdwatch 21(1): 28.

BROTHERS, N., J. COOPER, AND S. LOKEBORG. 1999. The incidental catch of seabirds by longline fisheries: worldwide review and technical guidelines for mitigation. Preliminary version. FAO Fisheries Circular 937, 99 pp.

BROWN, R. G. B., AND D. N. NETTLESHIP. 1984. Capelin and seabirds in the Northwest Atlantic. Pp. 184–195 in Marine Birds: Their Feeding Ecology and Commercial Fishery Relationships (D. N. Nettleship, G. A. Sanger, and P. F. Springer, Eds.). Canadian Wildlife Service Special Publication, Ottawa.

BRYANT, S., AND B. MARTIN. 1996. Ancient Rights: The Protected Fishing Area of Petty Harbour — Maddox Cove. Protected Areas Association of Newfoundland and Labrador, St. John's.

BUNCE, A. 2001. Population Dynamics of Australasian Gannets (*Morus Serrator*) Breeding in Port Phillip Bay, Victoria: Competition with Fisheries and the Potential Use of Seabirds in Managing Marine Resources. Ph.D. thesis, University of Melbourne.

BUNDY, A., G. R. LILLY, P. A. SHELTON, AND E. DALLEY. 2000. Cod. Pp. 18–22 in A Mass Balance Model the Newfoundland-Labrador Shelf (A. Bundy, G. R. Lilly, and P. A. Shelton, Eds.). Canadian Technical Report of Fisheries and Aquatic Sciences 2310.

BUNDY, A., G. R. LILLY, AND P. A. SHELTON (Eds.). 2000. A Mass Balance Model of the Newfoundland–Labrador Shelf. Canadian Technical Report of Fisheries and Aquatic Sciences 2310.

BURGER, A. E., AND J. COOPER. 1984. The effects of fisheries on seabirds in South Africa and Namibia. Pp. 150–161 in Marine Birds: Their Feeding Ecology and Commercial Fishery Relationships (D. N. Nettleship, G. A. Sanger, and P. F. Springer, Eds.). Canadian Wildlife Service Special Publication, Ottawa.

BURGER, J. 1993. Metals in avian feathers: bioindicators of environmental pollution. Review of Environmental Toxicology 5: 203–311.

BURGER, J., AND M. GOCHFELD. 1991. The Common Tern. Columbia University Press, New York.

BURGER, J., AND M. GOCHFELD. 1994. Predation and effects of humans on island nesting seabirds. Pp. 39–67 in Seabirds on Islands: Threats, Case Studies and Action Plans (D. N. Nettleship, J. Burger, and M. Gochfeld, Eds.). BirdLife International, Cambridge, U.K.

BURGER, J., AND M. GOCHFELD. 2000. Effects of lead on larids: a review of laboratory and field studies. Toxicology and Environmental Health Part B 3: 59–78.

CAIRNS, D. K. 1987. Seabirds as indicators of marine food supplies. Biological Oceanography 5: 261–271.

CAIRNS, D. K. 1992. Bridging the gap between ornithology and fisheries biology: use of seabird data in stock assessment models. Condor 94: 811–824.

CAIRNS, D. K. 1998. Diets of cormorants, mergansers, and king-fishers in northeastern North America. Canadian Technical Report of Fisheries and Aquatic Sciences No. 2225.

CAIRNS, D. K., G. CHAPDELAINE, AND W. A. MONTEVECCHI. 1990. Prey harvest by seabirds in the Gulf of St. Lawrence. Special Publication of Canadian Journal of Fisheries and Aquatic Sciences 113: 277–291.

CAMPHUYSEN, K. 1998. Beached bird surveys indicate decline in chronic oil pollution in the North Sea. Marine Pollution Bulletin 36: 519–526.

CAMPHUYSEN, K., B. CALVO, J. DURINCK, K. ENSOR, A. FOLKESTAD, R. W. FURNESS, G. GARTHE, G. LEAPER, H. SKOV, M. L. TASKER, AND C. J. N. WINTER. 1995. Consumption of discards by seabirds in the North Sea. Final Rep. EC DG XIV Res. Con. BIOECO/93/10. NIOZ-Report 1995–5. Netherlands Institute for Sea Research, Texel, 202 pp.

CAMPHUYSEN, C. J., AND S. GARTHE. 2000. Seabirds and commercial fisheries: population trends of piscivorous seabirds explained? Pp. 163–184 in The Effects of Fishing on Non-Target Species and Habitats (M. J. Kaiser and S. J. de Groot, Eds.). Blackwell Scientific, Oxford.

CARSS, D. N. 1993. Cormorants Phalacrocorax carbo at cage fish farms in Agyll western Scotland. Seabirds 15: 38–44.

CARSS, D. N. 1994. Killing of piscivorous birds at Scottish finfish farms, 1984–87. Biological Conservation 68: 181–188.

CARSS, D. N., AND M. MARQUISS. 1992. Avian predation at farmed and natural fisheries. Proceedings of the Institute of Fisheries Management Study Course, Aberdeen, 1991.

CARSCADDEN, J. E., AND B. S. NAKASHIMA. 1997. Abundance and changes in distribution, biology, and behavior of capelin in response to cooler water of the 1990s. Pp. 457–486 in Forage Fishes in Marine Ecosystems. Proceedings of the International Symposium on the Role of Forage Fishes in Marine Ecosystems, Alaska Sea Grant College Program Report 97–01.

CARSCADDEN, J. E., K. T. FRANK, AND W. C. LEGGETT. In press. Ecosystem changes and the effects on capelin (Mallotus villosus), a major forage species. Canadian Journal of Fisheries and Aquatic Sciences.

CARTER, H. R., AND S. G. SEALEY. 1984. Marbled Murrelet mortality gill-net fishing in Berkeley Sound, British Columbia. Pp. 212–220 in Marine Birds: Their Feeding Ecology and Commercial Fishery Relationships (D. N. Nettleship, G. A. Sanger, and P. F. Springer, Eds.). Canadian Wildlife Service Special Publication, Ottawa.

CHAPMAN, M. R., AND D. L. KRAMER. 1999. Gradients in coral reef fish density and size across the Barbados Marine Reserve boundary: effects of reserve protection and habitat characteristics. Marine Ecology Progress Series 181: 81–96.

CHAPMAN, M. R., AND D. L. KRAMER. 2000. Movements of fishes within and among fringing coral reefs in Barbados. Environmental Biology of Fishes 57: 11–24.

CHEREL, Y., H. WEIMERSKIRCH, AND G. DUHAMEL. 1995. Interactions between longline vessels and seabirds in Kerguelen waters and a method to reduce mortality. Biological Conservation 75: 63–70.

CHRISTENSEN, O., AND W. H. LEAR. 1977. By-catches in salmon drift nets at west Greenland in 1972. Meddelelser om Grønland 205(5): 1–83.

COLLINS, J. W. 1882. Notes on the habits and methods of capture of various species of sea birds that occur on the fishing banks off the eastern coast of North America, and which are used as bait for catching codfish by New England fishermen. Smithsonian Miscellaneous Collections 46: 311–338.

CONROY, W. H. 1975. Recent increases a penguin populations in the Antarctic and Subantarctic. Pp. 321–336 *in* The Biology of Penguins (B. Stonehouse, Ed.). Macmillan, London.

COULL, J. R. 1993. World fishery resources. Pp. 217–218 *in* Oceans Management and Policy Series (H. D. Smith, Ed.). Routledge, London.

CRAWFORD, R. J. M., AND P. A. SHELTON. 1978. Pelagic fish and seabird interrelationships off the coasts of South West Africa. Biological Conservation 14: 85–109.

CRAWFORD, R. J. M., R. A. CRUICKSHANK, P. A. SHELTON, AND I. KRUGER. 1985. Partitioning of a goby resource among four avian predators and evidence for altered trophic flow in the pelagic community of an intense, perennial upwelling system. South African Journal of Marine Science 3: 215–228.

CRAWFORD, R. J. M., P. A. SHELTON, A. L. BATCHELOR, AND C. F. CLINNING. 1980. Observations on the mortality of juvenile Cape Gannets *Phalacrocorax capensis* during 1975 and 1979. Fisheries Bulletin of South Africa 13: 69–75.

CRAWFORD, R. J. M., P. A. SHELTON, J. COOPER, AND R. K. BROOKE. 1983. Distribution, population size and conservation of the Cape Gannet *Morus capensis*. South African Journal of Marine Science 1: 153–174.

CROXALL, J. P., AND P. A. PRINCE. 1996. Potential interactions between Wandering Albatrosses and longline fisheries Patagonia toothfish at South Georgia. CCAMLR 3: 101–110.

CROXALL, J. P., P. A. PRINCE, AND C. RICKETTS. 1985. Relationships between prey life-cycles and the extent, nature and timing of seal and seabird predation in the Scotia Sea. Pp. 516–534 *in* Antarctic Nutrient Cycles and Food Webs (W. R. Siegfried, P. R. Condy, and R. M. Laws, Eds.). Springer, Berlin.

CROXALL, J. P., P. A. PRINCE, I. HUNTER, S. J. MCINNES, AND P. G. COPESTAKE. 1984. The seabirds of the Antarctic Peninsula, islands of the Scotia Sea and Antarctic Continent between 80°W and 20°W: their status and conservation. Pp. 635–644 *in* Status and Conservation of the World's Seabirds (J. P. Croxall, P. G. H. Evans, and R. W. Schreiber, Eds.). ICBP, Cambridge, U.K.

CUSHING, D. H. 1982. Climate and Fisheries. Academic, New York.

DAHL, K. 1992. Mussel fishery in the Danish Wadden Sea. Schriftenreihe der Schutzgemeinschaft Deutsche Nordseeküste 1: 71.

DAY, J. C., AND J. C. ROFF. 2000. Planning for Representative Marine Protected Areas: A Framework for Canada's Oceans. World Wildlife Fund Report, Toronto.

DEGANGE, A. R., R. H. DAY, J. E. TAKEKAWA, AND V. M. MENDENHALL. 1993. Losses of seabirds in gill nets in the North Pacific. Pp. 204–211 *in* The Status, Ecology, and Conservation of Marine Birds of the North Pacific (K. Vermeer, K. T. Briggs, K. H. Morgan, and D. Diegel-Causey, Eds.). Canadian Wildlife Service Special Publication, Ottawa.

DRURY, W. H., AND J. A. KADLEC. 1974. The current status of the Herring Gull population in the Northeastern United States. Bird-Banding 45: 297–306.

DUFFY, D. C. 1983. Environmental uncertainty and commercial fishing: effects on Peruvian guano birds. Biological Conservation 26: 227–238.

DUFFY, D. C., AND D. C. SCHNEIDER. 1994. Seabird-fishery interactions: a manager's guide. Pp. 26–38 in Seabirds on Islands: Threats, Case Studies and Action Plans (D. N. Nettleship, J. Burger, and M. Gochfeld, Eds.). BirdLife International, Cambridge, U.K.

DUFFY, D. C., R. P. WILSON, R. E. RICKLEFS, S. C. BRONI, AND H. VELDHUS. 1987. Penguins and purse seiners: competition or coexistence? National Geographic Research 3: 480–488.

DUGAN, J. E., AND G.E. DAVIS. 1993. Applications of marine refugia to coastal fisheries management. Canadian Journal of Fisheries and Aquatic Sciences 50: 2029–2042.

ELSON, P. F. 1962. Predator-prey relationships between fish-eating birds and Atlantic salmon. Bulletin Fisheries Research Board of Canada 47: 39–53.

ELLIOT, R. D. 1991. The management of the Newfoundland turr hunt. Pp. 29–35 *in* Studies of High-Latitude Seabirds. 2. Conservation Biology of Thick-billed Murres in the Northwest Atlantic (A. J. Gaston and R. D. Elliott, Eds.). Canadian Wildlife Service Occasional Paper 69.

EVANS, P. G. H., AND D. N. NETTLESHIP. 1985. Conservation of the Atlantic Alcidae. Pp. 428–488 *in* The Atlantic Alcidae (D. N. Nettleship and T. R. Birkhead, Eds.). Academic, Orlando.

EVANS, P. G. H., AND G. WATERSTON. 1976. The decline of the Thick-billed Murre in Greenland. Polar Record 18: 283–286.

FAO (Food and Agricultural Organization of the United Nations). 1995. Review of the State of the World Fishery Resources: Aquaculture. FAO Circular 886 (Rome).

FAO (Food and Agricultural Organization of the United Nations). 1999. International Plan of Action for Reducing Incidental Catch of Birds in Longline Fisheries. International Plan of Action for the Conservation and Management of Sharks. International Plan of Action for the Management of Fishing Capacity. FAO, Rome.

FEARE, C. J. 1988. Cormorants as predators at freshwater fisheries. Institute of Fisheries Management Annual Study Course 18: 18–42.

FISHER, J. 1952. The Fulmar. Collins, London.

FOCARDI, S., C. FOSSI, C. LEONZID, S. CORSOLINI, AND O. PARRA. 1996. Persistent organochlorine residues in fish and water birds from the Biobio River, Chile. Environmental Monitoring and Assessment 43: 73–92.

FOGARTY, M. J. 1999. Essential habitat marine reserves and fishery management. Trends in Ecology and Evolution 14: 133–134.

FRASER, W. R., W. Z. TRIVELPIECE, D. G. AINLEY, AND S. G. TRIVELPIECE. 1992. Increases in Antarctic penguin populations: reduced competition with whales or a loss of sea ice due to environmental warming. Polar Biology 11: 525–531.

FURNESS, R. W. 1978. Energy requirements of seabird communities: a bioenergetics model. Journal of Animal Ecology 47: 39–53.

FURNESS, R. W. 1990. A preliminary assessment of the quantities of Shetland sandeels taken by seabirds, seals, predatory fish and the industrial fishery in 1981–1983. Ibis 132: 205–217.

FURNESS, R. W. 1993. Birds as monitors of pollution. Pp. 84–143 in Birds as Monitors of Environmental Change (R. W. Furness and J. J. D. Greenwood, Eds.). Chapman & Hall, London.

FURNESS, R. W. 2000. Impacts of fisheries on seabird community stability. ICES 88th Statutory Meeting (Brugges) CM 2000/ Q: 03.

FURNESS, R. W., AND D. G. AINLEY. 1984. Threats to seabird populations presented by commercial fisheries. ICBP Technical Report 2: 701–708.

FURNESS, R. W., AND C. J. CAMPHUYSEN. 1997. Seabirds as monitors of the marine environment. ICES Journal of Marine Science 54: 726–737.

FURNESS, R. W., AND J. COOPER. 1982. Interactions between breeding seabird and pelagic fish populations in the southern Benguela region. Marine Ecology Progress Series 8: 243–250.

FURNESS, R. W., AND J. R. G. HISLOP. 1981. Diets and feeding ecology of Great Skuas Catharacta Skua during the breeding season in Shetland. Journal of Zoology, London 195: 1–23.

FURNESS, R. W., AND P. MONAGHAN. 1987. Seabird Ecology. Blackie, London.

FURNESS, R. W., K. ENSOR, AND A. V. HUDSON. 1992. The use of fishery waste by gull populations around the British Isles. Ardea 80: 105–113.

GARTHE, S., AND O. HÜPPOP. 1998. Foraging success, kleptoparasitism and feeding techniques in scavenging seabirds: does crime pay? Helgoländer Meeresuntersuchungen 52: 187–196.

GARTHE, S., C. J. CAMPHUYSEN, AND R. W. FURNESS. 1996. Amounts of discards by commercial fishes and their significance as food for seabirds in the North Sea. Marine Ecology Progress Series 136: 1–11.

GARTHE, S., U. WALTER, M. L. TASKER, P. H. BECKER, G. CHAPDELAINE, AND R. W. FURNESS. 1999. Evaluation of the role of discards in supporting bird population and their effects on the species composition of seabirds in the North Sea. ICES Cooperative Research Report 232: 29–41.

GLAHN, J. F., AND K. E. BRUGGER. 1995. The impact of the Double-crested Cormorants on the Mississippi Delta Catfish Industry: a bioenergetics model. Colonial Waterbirds 18: 165–175.

GLAHN, J. F., AND A. R. STICKNEY, JR. 1995. Wintering Double-crested Cormorants in the Delta Region of Mississippi: population levels and their impacts on the catfish industry. Colonial Waterbirds 18: 137–142.

GRETTENBERGER, J., E. MELVIN, AND J. PARRISH. 2000. Seabird bycatch in salmon gillnet fisheries in Puget Sound, Washington — A case study. Workshop (Abstract) Seabird Bycatch Waters Arctic Countries. Dartmouth, Nova Scotia.

HAMER, K. C., R.W. FURNESS, AND R. W. G. CALDOW. 1991. The effects of changes in food availability on the breeding ecology of Great Skuas *Catharacta skua* in Shetland. Journal of Zoology, London 223: 175–183.

HAMRE, J. 1988. Some Aspects of the Interrelationship between the Herring in the Norwegian Sea and the Stocks of Capelin and Cod in the Barrent Sea. ICES CM 1988/H42.

HARRIS, L. 1990. Independent Review of the State of the Northern Cod Stock. Supply and Services Canada, Ottawa.

HARRIS, M. P. 1970. Rates and causes of increase of some British gull populations. Bird Study 17: 325–335.

HARRISON, C. 1999. Caspian Tern may be removed from Columbia River system. Pacific Seabirds 26: 62.

HARWOOD, J. 1983. Interactions between marine mammals and fisheries. Advances in Applied Biology 8: 189–214.

HARWOOD, J., AND J. P. CROXALL, 1988. The assessment of competition between seals and commercial fisheries in the North Sea and the Antarctic. Marine Mammal Science 4: 13–33.

HATCH, S. A., AND G. A. SANGER. 1992. Puffins as predators on juvenile Pollock and other forage fish in the Gulf of Alaska. Marine Ecology Progress Series 80: 1–14.

HATCH, S. A., G. V. BYRD, D. B. IRONS, AND G. L. HUNT, JR. 1993. Status and ecology of kittiwakes (*Rissa tridactyla* and *R. brevirostris*) in the North Pacific. Pp. 140–153 *in* The Status, Ecology, and Conservation of Marine Birds in the North Pacific (K. Vermeer, K. T. Briggs, K. H. Morgan, and D. Siegel-Causey, Eds.). Canadian Wildlife Service Special Publication, Ottawa.

HENSON, C. 1999. Caspian Tern predation on juvenile salmonids. Pacific Seabirds 26: 62.

HEUBECK, M. 1989. Breeding success of Shetland's seabirds: Arctic Skua, kittiwake, guillemot, Razorbill and puffin. Pp. 11–18 *in* Seabirds and Sandeels (M. Heubeck, Ed.). Shetland Bird Club, Lewick.

HOWES, L. A., AND W. A. MONTEVECCHI. 1993. Population trends of gulls and terns in Gros Morne National Park, Newfoundland. Canadian Journal of Zoology 71: 1516–1520.

HUDSON, A. V., AND R. W. FURNESS. 1988. Utilization of discarded fish by scavenging seabirds behind whitefish trawlers in Shetland. Journal of Zoology, London 215: 151–166.

HUTCHINGS, J. A., AND R. A. MYERS. 1994. What can be learned from the collapse of a renewable resource? Atlantic cod, *Gadus morhua*, of Newfoundland and Labrador. Canadian Journal of Fisheries and Aquatic Sciences 51: 2126–2146.

HÜPPOP, O., AND S. WURM. 2000. Effects of winter fishery activities on resting numbers, food and body condition of large gulls *Larus argentatus* and *L. marinus* in the south-eastern North Sea. Marine Ecology Progress Series 194: 241–247.

JOIRIS, C. R., G. TAPIA, AND L. HOLSBEEK. 1997. Increase of organochlorines and mercury levels in Common Guillemots during winter in the southern North Sea. Marine Pollution Bulletin 34: 1049–1057.

JONES, P. D., D. J. HANNAH, S. J. BACKLAND, P. J. DAY, S. V. LEATHEM, L. J. PORTER, H. J. AUMAN, J. T. SANDERSON, C. SUMMER, J. P. LUDWIG, T. L. COLBORN, AND J. P. GIESY. 1996. Persistent synthetic chlorinated hydrocarbons in albatross tissue samples from Midway Atoll. Environmental Toxicology and Chemistry 15: 1793–1800.

KADLEC, J. A., AND W. H. DRURY. 1968. Structure of the New England Herring Gull population. Ecology 49: 644–676.

KALAS, J. A., T. G. HEGGEBERGET, P. A. BJØRNE, AND O. REITAN. 1993. Feeding behaviour diet of goosanders (*Mergus serrator*) in relation to salmonid seaward migration. Aquatic Living Resources 6: 31–38.

KING, W. B. 1984. Accidental mortality of seabirds in gillnets in the North Pacific. Pp. 709–715 *in* Marine Birds: Their Feeding Ecology and Commercial Fishery Relationships (D. N. Nettleship, G. A. Sanger, and P. F. Springer, Eds.). Canadian Wildlife Service Special Publication, Ottawa.

KIRBY, J. S., J. S. HOLMES, AND R. M. SELLERS. 1996. Cormorants *Phalacrocorax carbo* as fish predators: an appraisal of their conservation and management in Great Britain. Biological Conservation 75: 191–199.

KLOMP, N. I., AND R.W. FURNESS. 1992. Non-breeders as a buffer against environmental stress: declines in numbers of Skuas on Foula, Shetland, and prediction of future recruitment. Journal of Applied Ecology 29: 341–348.

KRAUS, S., A. READ, E. ANDERSON, A. SOLOW, K. BALDWIN, T. SPRADLIN, AND J. WILLIAMSON. 1997. Acoustic alarms reduce porpoise mortality. Nature 338: 525.

LAVIGNE, D. M. 1996. Ecological interactions between marine mammals, commercial fisheries, and their prey: unravelling the tangled web. Pp. 59–71 *in* Studies of High-Latitude Seabirds. 4. Trophic Relationships and Energetics of Endotherms in Cold Ocean Systems (W. A. Montevecchi, Ed.). Canadian Wildlife Service Occasional Paper 91: 59–71.

LAWS, R. M. 1985. The ecology of the Southern Ocean. American Scientist 73: 26–40.

LIEN, J. 1999. When marine conservation efforts sink: what can be learned from the abandoned effort to examine the feasibility of a National Marine Conservation Area on the NE Coast of Newfoundland. Proceedings Canadian Council of Ecological Areas 16th Conference.

LIEN, J., G. B. STENSON, AND I. H. NI. 1988. A review of incidental entrapment of seabirds, seals and whales in inshore fishing gear in Newfoundland and Labrador: a problem for fishermen and fishing gear designers. Proceedings World Symposium on Fishing Gear and Vessel Design (St. John's): 67–71.

LUDWIG, D., R. HILBORNE, AND C. WALTERS. 1993. Uncertainty, resource exploitation, and conservation: lessons from history. Science 260: 17, 36.

LUGTEN, G. L. 1997. The rise and fall of the Patagonian toothfish — food for thought. Environmental Policy and Law 27: 401–407.

LØKKEBORG, S. 1998. Seabird by-catch and bait loss in long-lining using different setting methods. ICES Journal of Marine Science 55: 145–149.

LØKKEBORG, S. 2000. Review and evaluation of three mitigation measures — bird-scaring line, under water setting and line shooter — to reduce seabird bycatch in the Norwegian long-line fishery. ICES 88th Statutory Meeting (Brugges) CM 2000/J: 10.

MARTIN, G. 2000. Commercial gill-net ban tightened in California move to protect wildlife. San Francisco Chronicle: September 23.

MAY, R. M., J. R. BEDDINGTON, C. W. CLARK, S. J. HOLT, AND R. M. LAWS. 1979. Management of multispecies fisheries. Science 205: 267–277.

MELVIN, E. F., J.K. PARRISH, AND L. L. CONQUEST. 1999. Novel tools to reduce seabird bycatch in coastal gillnet fisheries. Conservation Biology 13: 1386–1397.

MILTON, G. R., P. J. AUSTIN-SMITH, AND G. J. FARMER. 1995. Shouting at shags: a case study of cormorant management in Nova Scotia. Colonial Waterbirds 18: 91–98.

MOLONEY, C. L., J. COOPER, P. G. RYAN, AND W. R. SIEGFRIED. 1993. Use of a population model to assess the impact of longline fishing on Wandering Albatross *Diomedea exulans* populations. Biological Conservation 70: 195–203.

MONAGHAN, P. 1992. Seabirds and sandeels: the conflict between exploitation and conservation in the northern North Sea. Biodiversity Conservation 1: 98–111.

MONAGHAN, P., J. D. UTTLEY, AND M. D. BURNS. 1989. The relationship between food supply, reproductive effort and breeding and success in Arctic Terns *Sterna paradisaea.* Journal of Animal Ecology 58: 261–274.

MONAGHAN, P., J. D. UTTLEY, AND M. D. BURNS. 1992. Effect of changes in food availability on reproduction effort in Arctic Terns *Sterna paradisaea.* Ardea 80: 71–81.

MONTEVECCHI, W. A. 1991. Incidence and types of plastic in gannet nests in the northwest Atlantic. Canadian Journal of Zoology 69: 295–297.

MONTEVECCHI, W. A. 1993a. Birds as indicators of change in marine prey stocks. Pp. 217–266 *in* Birds as Monitors of Environmental Change (R. W. Furness and J. J. D. Greenwood, Eds.). Chapman & Hall, London.

MONTEVECCHI, W. A. 1993b. Seabird indication of squid stocks in the Northwest Atlantic. Journal of Cephalopod Biology 2: 57–63.

MONTEVECCHI, W. A. 1996. Introduction. Pp. 7–9 *in* Studies of High-Latitude Seabirds. 4. Trophic Relationships and Energetics of Endotherms in Cold Ocean Systems (W. A. Montevecchi, Ed.). Canadian Wildlife Service Occasional Paper 9l.

MONTEVECCHI, W. A. 2000. Seabirds. Pp. 15–18 *in* A Mass Balance Model of the Newfoundland and Labrador Shelf (A. Bundy, G. R. Lilly, and P. A. Shelton, Eds.). Canadian Technical Report of Fisheries and Aquatic Sciences 2310.

MONTEVECCHI, W. A. 2001. Seabirds as indicators of ocean pollution. *In* Encyclopedia of Ocean Sciences (J. Steele, S. Thorpe, and K. Turekian, Eds.). Academic Press, London.

MONTEVECCHI, W. A., AND A. BERUTTI. 1991. Avian indication of pelagic fishery conditions on the southeast and northwest Atlantic. Acta International Ornithological Congress 20: 2246–2256.

MONTEVECCHI, W. A., AND D. KIRK. 1996. Great Auk (*Penguinus impennis*). *In* Birds of North America (A. Poole and F. Gill, Eds.). No. 260. American Ornithologists' Union and Allen Press, Lawrence, KS.

MONTEVECCHI, W. A., AND R. A. MYERS. 1995. Prey harvests of seabirds reflect pelagic fish and squid abundance on multiple spatial and temporal scales. Marine Ecology Progress Series 117: 1–9.

MONTEVECCHI, W. A., AND R. A. MYERS. 1996. Dietary changes of seabirds reflect with shifts in pelagic food webs. Sarsia 80: 313–322.

MONTEVECCHI, W. A., AND R. A. MYERS. 1997. Centurial and decadal oceanographic influences on changes in the Northern Gannet populations and diets in the north-west Atlantic: implications for climate change. ICES Journal of Marine Science 54: 608–614.

MONTEVECCHI, W. A., AND L. M. TUCK. 1987. Newfoundland Birds: Exploitation, Study, Conservation. Nuttall Ornithological Club, Cambridge, MA.

MONTEVECCHI, W. A., D. K. CAIRNS, AND V. L. BIRT. 1998. Migration of post-smolt Atlantic salmon (*Salmo salar*) off northeastern Newfoundland, as inferred from tag recoveries in a seabird colony. Canadian Journal of Aquatic and Fisheries Science 45: 568–571.

MOTT, D. F., AND F. L. BOYD. 1995. A review of techniques for preventing cormorant depredations at aquaculture facilities in the southeastern United States. Colonial Waterbirds 18: 176–180.

MULLINS, C. C., D. CAINES, AND S. L. LOWE. 1999. Status of Atlantic salmon (*Salmo salar* L.) stocks of three selected rivers in salmon fishing area 14A, 1998. Canadian Stock Assessment Secretariat Research Document 99/101.

MYERS, R. A., N. J. BARROWMAN, J. A. HUTCHINGS, AND A. A. ROSENBERG. 1999. Population dynamics of exploited fish stocks at low population levels. Science 269: 1106–1108.

NEIS, B., L. F. FELT, R. C. HAEDRICH, AND D. C. SCHNEIDER, 1999. An interdisciplinary method for collecting and integrating fishers' ecological knowledge into resource management. Pp. 217–238 *in* Fishing Places, Fishing People (D. Newell and R. Ommer, Eds.). University of Toronto Press, Toronto.

NELSON, J. B. 1978. The Gannet. Buteo, Vermillion, South Dakota.

NETTLESHIP, D. N. 1990. The diet of Atlantic Puffin chicks in Newfoundland before and after the initiation of an international capelin fishery. Acta International Ornithological Congress 20: 2263–2271.

OGI, H. 1984. Seabird mortality incidental to the Japanese gill-net fishery. Pp. 717–721 *in* Status and Conservation of the World's Seabirds (J. Croxall, P. G. H. Evans, and R. W. Schreiber, Eds.). ICBP Technical Publication No. 2.

ORO, D. 1996. Effects of trawler discard availability on egg laying and breeding success in the Lesser Black-backed Gull *Larus Fuscus* in the western Mediterranean. Marine Ecology Progress Series 132: 43–46.

ORO, D., AND A. MARTINEZ-VILALTA. 1994. Factors affecting kleptoparasitism and predation rates upon a colony of Audouin's Gull (*Larus audouinii*) by Yellow-legged Gulls (*Larus achinnans*) in Spain. Colonial Waterbirds 17: 35–41.

ORO, D., M. BOSCH, AND X. RUIZ. 1995. Effects of a trawling moratorium on the breeding success of the Yellow-legged Gull *Larus cachinnans*. Ibis 137: 547–549.

PACE, M. L., J. J. COLE, S. R. CARPENTER, AND J. F. KITCHELL. 1999. Trophic cascades revealed in diverse ecosystems. Trends in Ecology and Evolution 14: 483–488.

PAULIK, G. J. 1981. Anchovies, birds, and fishermen in the Peru Current. Pp. 35–79 *in* Resource Management and Environmental Uncertainty: Lessons from Coastal Upwelling Fisheries (M. H. Glanz and J. D. Thompson, Eds.). Wiley, New York.

PAULY, D., V. CHRISTENSEN, J. DALSGAARD, R. FROESE, AND F. TORRES, JR. 1998. Fishing down marine food webs. Science 279: 860–863.

PENDER, P. J., R. S. WILLING, AND B. CANN. 1992. Prawn fishery by catch: a valuable resource. Australian Fisheries 51: 30–31.

PETERSEN, A. 1994. Potential mortality factors of auks in Iceland. Circumpolar Seabird Bulletin 1: 3–4.

PETERSEN, A. 1998. Incidental take of seabirds in Iceland. Circumpolar Seabird Working Group 1: 23–27.

PHILLIPS, R. A., D. R. THOMPSON, AND K. C. HAMER. 1999. The impact of Great Skua predation on seabird population at St. Kilda: a bioenergetics model. Journal of Animal Ecology 36: 218–232.

PIATT, J. F., AND P. J. GOULD. 1994. Endangered Japanese Murrelets: incidental catch in high seas driftnets and post-breeding dispersal. Auk 111: 953–961.

PIATT, J. F., AND D. N. NETTLESHIP. 1987. Incidental catch of marine birds and mammals in fishing nets off Newfoundland, Canada. Marine Pollution Bulletin 18: 344–349.

PIATT, J. F., AND D. G. REDDIN. 1984. Recent trends in the west Greenland salmon fishery, and implications for Thick-billed Murres. Pp. 208–210 in Marine Birds: Their Feeding Ecology and Commercial Fishery Relationships (D. N. Nettleship, G. A. Sanger, and P. F. Springer, Eds.). Canadian Wildlife Service Special Publication, Ottawa.

PIATT, J. F., D. N. NETTLESHIP, AND W. THRELFALL. 1984. Net mortality of Common Murres and Atlantic Puffins in Newfoundland, 1951–1981. Pp. 196–206 in Marine Birds: Their Feeding Ecology and Commercial Fishery Relationships (D. N. Nettleship, G. A. Sanger, and P. F. Springer, Eds.). Canadian Wildlife Service Special Publication, Ottawa.

POTTER, E. C. E., AND W. W. CROZIER. 2000. A perspective on the marine survival of Atlantic salmon. Pp. 19–36 in The Ocean Life of Atlantic Salmon: Environmental and Biological Factors Influencing Survival (D. Mills, Ed.). Fishing News Books, Oxford.

RAKITIN, A., AND D. L. KRAMER. 1996. Effect of a marine reserve on the distribution of coral reef fishes in Barbados. Marine Ecology Progress Series 131: 97–113.

REGEHR, H. M., AND W. A. MONTEVECCHI. 1997. Interactive effects of food shortage and predation on breeding failure of Black-legged Kittiwakes: indirect effects of fisheries activities and implications for indicator species. Marine Ecology Progress Series 155: 249–260.

ROBERTSON, G. 1998. The culture and practice of longline tuna fishing: implications for seabird by-catch mitigation. Bird Conservation International 8: 211–221.

ROBERTSON, G., AND R. GALES (Eds.). 1998. Albatross Biology and Conservation. Surrey Beatty, Chipping Norton.

ROBY, D. D., K. COLLIS, D. P. CRAIG, S. L. ADAMANY, AND D. E. LYONS. 1999. What to do with the world's largest Caspian Tern colony: When ESA collides with MBTA and NEPA in the Columbia River estuary. Waterbirds Society Meeting (Blaine, Washington).

ROBY, D. D., D. P. CRAIG, D. LYONS, AND K. COLLIS. 2000. Caspian Tern conservation and management in the Pacific Northwest: Who's tern is next. American Ornitholologists' Union Meeting (St. John's, Newfoundland, Canada).

RUSSELL, J. O., AND W. A. MONTEVECCHI. 1996. Predation on adult puffins Fratercula arctica by Great Black-backed Gulls Larus marinus at a Newfoundland colony. Ibis 138: 791–794.

RYAN, P. G. 1998. The taxonomic and conservation status of the Spectacled Petrel Pterodroma conspicillata. Bird Conservation International 8: 223–235.

RYAN, P. G., AND C. L. MOLONEY. 1988. Effect of trawling on bird and seal distributions in the southern Benguela region. Marine Ecology Progress Series 45: 1–11.

SAFINA, C., AND J. BURGER. 1985. Common Terns (Sterna hirundo) foraging: seasonal trends in prey fish densities and competition with bluefish (Pomatomus saltretrix). Ecology 66: 1457–1463.

SALZMAN, J. C. 1989. Scientists as advocates: the Point Reyes Bird Observatory and gill-netting in central California. Biological Conservation 3: 170–180.

SCHAEFER, M. R. 1970. Men, birds and anchovies in the Peru current dynamic interaction. Transactions of the American Fisheries Society 9: 461–467.

SCHAFFNER, F. L. 1986. Trends in Elegant Tern and anchovy populations in California. Condor 88: 347–354.

SCHEEL, D., AND HOUGH, K. R. 1997. Salmon fry predation by seabirds near an Alaskan fishery. Marine Ecology Progress Series 150: 35–48.

SCHREIBER, E. A., AND R. W. SCHREIBER. 1989. Insight into seabird ecology from a global 'natural experiment.' National Geographic Research 5: 64–81.

SCHREIBER, R. W., AND E. A. SCHREIBER. 1984. Central Pacific seabirds and the El Niño Southern Oscillation: 1982 to 1983 retrospectives. Science 225: 713–716.

SHERMAN, K., C. JONES, L. SULLIVAN, W. SMITH, P. BERRIEN, AND L. EJYSMONT. 1981. Congruent shifts in sandeel abundance in western and eastern North Atlantic ecosystems. Nature 291: 486–489.

SMITH, T. D. 1994. Scaling Fisheries: The Science of Measuring the Effects of Fishing, 1855–1955. Cambridge University Press, Cambridge.

SOUTAR, A., AND J. D. ISAACS. 1974. Abundances of pelagic fish during the 19th and 20th centuries as recorded in anaerobic sediment off the Californias. Fisheries Bulletin 72: 257–273.

SPEAR, L. B., D. G. AINLEY, AND C. A. RIBIC. 1995. Incidence of plastic in seabirds from the tropical Pacific, 1984–91: relation with distribution of species, sex, age, season, year and body weight. Marine Environmental Research 40: 123–146.

SPRINGER, A. M., D. G. ROSENEAU, D. S. LLOYD, C. P. MCROY, AND E. C. MURPHY. 1986. Seabird responses to fluctuating prey availability in the eastern Bering Sea. Marine Ecology Progress Series 32: 1–12.

STEELE, J. H. 1996. Regime shifts in fisheries management. Fisheries Research 25: 19–23.

STEELE, J. H. 1998. Regime shifts in marine ecosystems. Ecological Applications 8: S33–S36.

STENHOUSE, I., AND W. A. MONTEVECCHI. 1999. Indirect effects of the availability of forage fish and fisheries discards: gull predation on breeding storm-petrels. Marine Ecology Progress Series 184: 303–307.

STENHOUSE, I., AND W. A. MONTEVECCHI. 2000. Increasing and expanding populations of breeding Northern Fulmars in Atlantic Canada. Colonial Waterbirds 22: 382–391.

STENSON, G., AND J. LAWSON. 2000. Harp seals. Pp. 13–14 in A Mass Balance Model of the Newfoundland-Labrador Shelf (A. Bundy, G. R. Lilly, and P. A. Shelton, Eds.). Canadian Technical Report Fisheries and Aquatic Sciences 2310.

STENSON, G., J. LAWSON, AND A. BUNDY. 2000. Whales. Pp. 11–13 in A Mass Balance Model of the Newfoundland-Labrador Shelf (A. Bundy, G. R. Lilly, and P. A. Shelton, Eds.). Canadian Technical Report Fisheries and Aquatic Sciences 2310.

STEPHENS, P. A., AND W. J. SUTHERLAND. 1999. Consequences of the Allee effect for behaviour, ecology and conservation. Trends in Ecology and Evolution 14: 401–405.

STICKNEY, P. A., JR., J. F. GLAHN, J. O. KING, AND D. T. KING. 1995. Impact of Great Blue Herons depredations on channel catfish farms. Journal of the World Aquacultural Society 26: 194–199.

STRANN, K.-B., W. VADER, AND R. T. BARRETT. 1991. Auk mortality in fishing nets in north Norway. Seabird 13: 22–29.

TASKER, M. L., K. CAMPHUYSEN, J. COOPER, S. GARTHE, M. LEOPOLD, W. A. MONTEVECCHI, AND S. BLABER. 2000. The impacts of fisheries on marine birds. ICES Journal of Marine Science 57: 531–547.

TASKER, M. L., AND R. W. FURNESS. 1996. Estimation of food consumption by seabirds in the North Sea. Pp. 6–42 in Seabird/Fish Interactions with Particular Reference to Seabirds in the North Sea (G. L. Hunt, Jr. and R. W. Furness, Eds.). ICES Cooperative Research Report 216.

THRELFALL, W. 1980. Seabirds. Pp. 467–508 in The Biogeography and Ecology of the Island of Newfoundland (R. South, Ed.). Dr. W. Junk, The Hague.

TULL, C. E., P. GERMAIN, AND A. W. MAY. 1972. Mortality of Thick-billed Murres in the west Greenland salmon fishery. Nature 237: 42–44.

UNDERHILL, L. 2000. A slick operation. World Birdwatch 22(3): 6–7.

UTTLEY, J., P. MONAGHAN, AND W. WHITE. 1989. Differential effects of reduced sandeel availability on two sympatrically breeding species of tern. Ornis Scandinavica 20: 273–277.

VADER, W., R. T. BARRETT, K. E. ERICSTAD, AND K.-B. STRANN. 1990a. Differential responses of Common and Thick-billed Murres to a crash in the capelin stock in the southern Baerents Sea. Studies in Avian Biology 14: 175–180.

VADER, W., T. ANKER-NILSSEN, V. BAKKEN, R. BARRETT, AND K.-B. STRANN. 1990b. Regional and temporal differences in breeding success and population development of fish-eating seabirds in Norway after collapses of herring and capelin stocks. Transactions of the 19th IUGB Congress (Trondheim): 143–150.

VAN DEN BRINK, N. W., J. A. VAN FRANEKER, AND E. M. DE RUITER-DIJKMAN. 1998. Fluctuating concentration of organochlorine pollutants during a breeding season in two Antarctic seabirds: Adelie Penguins and Southern Fulmar. Environmental Toxicology and Chemistry 17: 702–709.

VEIT, R. R., P. PYLE, AND J. A. MCGOWAN. 1996. Ocean warming and long term change in pelagic bird abundance within the California Current system. Marine Ecology Progress Series 139: 11–18.

WELCH, D. W., Y. ISHIDA, AND K. NAGASAWA. 1996. Thermal limits and ocean migrations of sockeye salmon (*Oncorhychus nerka*): long-term consequences of global warming. Canadian Journal of Fisheries and Aquatic Sciences 55: 937–948.

WELCH, D. W., B. R. WARD, B. D. SMITH, AND J. P. EVESON. 2000. Temporal and spatial responses of British Columbia steelhead (*Oncorhychus mykiss*) populations to ocean climate shifts. Fisheries Oceanography 9: 17–32.

WIENS, J. A., T. O. CRIST, R. H. DAY, S. M. MURPHY, AND G. D. HAYWARD. 1996. Effects of the Exxon Valdez oil spill on marine bird communities in Prince William Sound, Alaska. Ecological Applications 6: 828–841.

WIESE, F. K., AND P. C. RYAN. 1999. Trends of chronic oil pollution in southeast Newfoundland assessed through beached-bird surveys. Bird Trends 7: 36–40.

WILLIAMS, N. 1998. Overfishing disrupts entire ecosystems. Science 270: 809.

WOOD, C. C. 1985. Aggregative response of Common Merganser (*Mergus mergansr*): Predicting flock size and abundance on Vancouver Island salmon streams. Canadian Journal of Fisheries and Aquatic Sciences 42: 1259–1271.

WOOTON, D. E., AND H. K. DUPREE. 2000. Food habits of Lesser Scaup occupying baitfish facilities in central Arkansas. Paper presented at Waterbirds Society Meeting, Plymouth, MA.

WOOTON, J. T. 1995. Effects of birds on sea urchins and algae: a lower-intertidal trophic cascade. Ecoscience 2: 321–328.

Least Terns and Semipalmated Plovers Nesting on an Unprotected Beach

17 Seabird Conservation

P. Dee Boersma, J. Alan Clark, and Nigella Hillgarth

CONTENTS

17.1 INTRODUCTION

The growth of human population and human resource consumption are probably the major factors affecting seabirds today. Between one third and one half of the Earth's terrestrial surface has been modified by humans (Vitousek et al. 1997), and about one fourth of bird species have been driven to extinction by humans in the last 2000 years (Steadman 1995). The world's population currently exceeds 6.1 billion people, and more than 1 billion people are added to the planet every 13 years (PRB 2000). The exponentially increasing human population is correlated strongly with species declines (Soulé 1991). Seabirds are no exception, and humans have had significant impacts on many seabird populations.

Many of the same traits that make seabirds well adapted to their environment also make them particularly susceptible to population declines and extinction. They frequently aggregate in colonies to which they return each breeding season, even if the habitat is degraded or destroyed. During the breeding season, seabirds nest in coastal areas or on islands, both habitats that humans have developed extensively, destroying many seabird nesting sites. Two hundred and seventeen taxa (species or races) of birds have become extinct in the last 400 years, and over 200 of these taxa nested on islands (Rodda et al. 1998). Two thirds of all currently threatened birds are threatened on islands (Collar and Andrew 1992). Seabirds face intense threats to their survival both on islands and in coastal areas (see Nettleship et al. 1994 for a more in-depth treatment of seabirds on islands).

Any effort to conserve seabirds needs to start with a determination and understanding of the major threats these species face. The primary reasons species become endangered are habitat loss, over-harvest, invasive species, pollution, and disease (Wilcove et al. 1998). These same factors are primarily responsible for the decline of seabirds. In this chapter, we give a brief overview of some of these major threats to seabirds. These threat categories are somewhat artificial and ignore the potential interactive and synergistic effects of multiple impacts. Nonetheless, understanding threats provides a foundation for informed discussion of seabird conservation. In addition to threats, we also discuss the role of legal systems in seabird protection as well as recent progress in seabird conservation efforts.

17.2 EFFECTS OF HABITAT MODIFICATION ON SEABIRDS

In much of the world, habitat modification is the single most prevalent cause of species becoming endangered (studies summarized in Czech et al. 2000, Meffee and Carroll 1997). Seabird habitat includes three primary components: (1) nesting habitat, (2) foraging habitat during the breeding season, and (3) at-sea habitat during the nonbreeding season (Boersma and Parrish 1998). Foraging and at-sea habitat are most affected by commercial fishing and pollution, and may become highly affected by climate change. Because these impacts are addressed in detail elsewhere in this book, we will not focus on them here.

Although some human activity, such as guano extraction (Duffy 1994a), destroyed seabird nesting habitat as long as 200 years ago (Figure 17.1), wide-scale destruction and modification

FIGURE 17.1 Historically guano was removed from many seabird colonies. Unfortunately, guano (in this case from cormorants) was once removed when birds were breeding, causing colony-wide reproductive failure for that year. Guano mining is now more controlled and usually occurs after the breeding season. Nonetheless, guano mining remains a problem for some seabirds. (Photo by P. D. Boersma.)

FIGURE 17.2 A small sandbar in St. Petersburg, Florida provides the only remaining local habitat where these birds (pelicans, cormorants, gulls, terns, skimmers, and shorebirds) can roost to rest and preen their feathers. (Photo by E. A. Schreiber.)

have taken place only in the last 50 to 70 years. Coastal and island real estate are highly valued and are being increasingly developed for human use, particularly in highly populous regions (Figure 17.2). Approximately 50% of mangrove ecosystems have been modified or destroyed by human activity (WRI 1996). Bryant (1995) calculated that half of the remaining coastal ecosystems in the world are at a moderate to high risk to development-related threats. In Europe, 86% of the remaining undeveloped coastline is at moderate to high risk (Bryant 1995). Other causes of habitat loss and modification include the interrelated threats of logging, farming, and grazing. The current trend of habitat loss for seabirds is a major threat to their survival.

17.3 INTRODUCED SPECIES IN SEABIRD COLONIES

Another of the most prevalent causes of species endangerment is the introduction of nonnative species (Czech et al. 2000). The scientific literature is replete with examples of extinctions, extirpations, and drastic reductions in seabird populations caused by the introduction of nonnative species into seabird nesting habitat (see summaries in Jones and Byrd 1979, Moors and Atkinson 1984, Burger and Gochfeld 1994). Introduced species can be divided into three main categories: (1) escaped pets, such as cats and dogs; (2) accidental introduction, such as mice, rats, and snakes; and (3) intentional releases for food, sport, fur, and greenery, and as biological control agents (Boersma and Parrish 1998). Not all introduced species have had a detrimental impact on seabirds, but several introduced mammalian predators, grazers, and plants have had significant negative impacts.

17.3.1 PREDATORS

Many seabird colonies are naturally free from mammalian predators and as a consequence, seabirds evolved without appropriate behavioral, ecological, and reproductive defenses against them (Loope and Mueller-Dombois 1989). The impact of introduced predators on seabirds is well documented (see compilation in Burger and Gochfeld 1994). Introduced predators with well-documented negative impacts include cats (e.g., Ashmole et al. 1994), dogs (e.g., Everett 1988), rats (e.g., Hobson et al. 1999), mice (e.g., Drost and Lewis 1995), stoats (e.g., Taylor and Tilly 1984), ferrets (e.g., Moors and Atkinson 1984), hedgehogs (e.g., Monteiro et al. 1996), raccoons (e.g., Hartman and Eastman 1999), monkeys (e.g., Gochfeld et al. 1994), and fox (e.g., Bailey 1993). Recognition of the detrimental impacts of introduced predators led to eradication programs in many seabird

FIGURE 17.3 Sheep grazing in a Magellanic Penguin colony reduces habitat quality through trampling, overgrazing, and erosion. (Photo by P. D. Boersma.)

breeding sites and the subsequent recovery of the seabirds (e.g., see discussions in Moors and Atkinson 1984, Drost and Lewis 1995, Taylor et al. 2000). However, eradication programs should be carefully planned and implemented, or they may have inadvertent negative impacts on the species of concern (Howald et al. 1999).

17.3.2 GRAZERS

Unregulated introductions of grazing animals cause habitat destruction through trampling of nests, overgrazing, and erosion resulting from overgrazing (Jones and Byrd 1979, Schreiber and Lee 2000). In addition, some smaller grazers, such as rabbits, may also compete for nest space (Ainley and Lewis 1974). Feral mammals, including grazers, are a widespread problem at many seabird colonies. Introduced grazers significantly alter vegetation structure (Kirk and Racey 1991) and destroy habitat for use by seabirds (Figure 17.3). In the Seychelles, grazing hares appear to prevent the regeneration of *Cauarina equisetifolia*, a tree important for breeding seabirds (Kirk and Racey 1991). The impact of introduced grazers on seabirds, although less obvious than that for introduced predators, has been documented for many species, including rabbits (e.g., Monteiro et al. 1996), hares (e.g., Kirk and Racey 1991), goats (e.g., Keegan et al. 1994), and sheep (e.g., Schwartz 1994). In addition, Jones and Byrd (1979) found impacts on seabirds from introduced cattle, caribou, deer, elk, and musk oxen. Removing introduced grazers from colonies allows seabird populations to regenerate (e.g., McChesney and Tershy 1998).

17.3.3 PLANTS

For some seabirds, specific vegetative communities or individual plant species are important, or even critical, elements of nesting habitat (Feare et al. 1997). Introduced plants often dramatically change terrestrial landscapes and make them unsuitable for use by seabirds. Plants may colonize areas that previously contained few to no plants, interfering with nesting by species that require open ground. Introduced plants may also crowd out native species used by seabirds for nesting. For example, in the Seychelles, the spread of epi bleu (*Stachytarpheta jamaicensis*), an introduced plant, has reduced suitable nesting habitat for Sooty Terns (*Sterna fuscata*, Feare et al. 1997). An introduced cane grass (*Arundo donax*) in the Azores archipelago is blamed for major losses of suitable burrowing ground for the Cory's Shearwater (*Calonectris diomedea;* Hamer, cited in Monteiro et al. 1996).

Humans can spread detrimental species to seabird breeding areas (Van Driesche and Van Driesche 2000). Increased travel to, and settlement in, remote areas is also increasing the threat of unwanted introductions of such species as mice and rats. Furthermore, seabird populations may also be threatened by diseases carried by introduced species (de Lisle et al. 1990). Efforts to prevent the further introduction of nonnative species into seabird colonies and efforts to control and eliminate species from colonies where they have been introduced are valuable conservation measures.

17.4 HUMAN HARVEST OF SEABIRDS

Seabirds have been exploited directly and indirectly throughout human history, and avian extinctions have followed in the wake of human exploration and settlement for at least a millennium (Steadman 1995, 1997). The demise of the Great Auk (*Pinguinis impennis*) is an early example. This flightless seabird flourished in massive numbers in the North Atlantic from the Arctic Circle to Massachusetts. Because they were flightless, Great Auks and their eggs were easy to harvest (Wilcove 1999). Early explorers and sailors began harvesting Great Auks in the late 1400s, and the last two great auks were killed in 1844 (Allen 1876). Although human consumption has driven few other seabirds into extinction in the last 200 years, human exploitation continues to be a significant factor in the decline of many seabird populations (Steadman 1997, Schreiber and Lee 2000). Humans have harvested seabirds for food (commercial, subsistence, and recreational), ornamentation (e.g., feathers), clothing (e.g., gloves), and oil.

17.4.1 HUNTING

At the turn of the century, many seabirds were still heavily harvested by humans, and it is surprising that more species were not lost due to hunting for food, feathers, and oil. During the late 19th and early 20th centuries, ornamental feathers on women's hats were highly fashionable in both Europe and North America, and many birds, including seabirds, were killed to supply the millinery trade. Between 1897 and 1914, over 3.5 million seabirds were killed for their feathers in the central Pacific Ocean alone (Spennemann 1998). Some seabird species are rich in fat deposits, making them valuable for their oil. During the 19th and early 20th centuries, production of penguin oil resulted in the killing of millions of adults. A single company in the Falkland Islands rendered 405,000 birds for their oil in 1867 alone (Sparks and Soper 1987). Although there is a more responsible and sustainable attitude to harvesting seabirds today, there are regions where hunting continues to threaten seabird populations. In Newfoundland, 300,000 to 725,000 murres are shot annually, which may be more than the murre population can sustain (Elliot et al. 1991). In West Greenland, hunters kill 283,000 to 386,000 murres annually (Falk and Durinck 1992), and murre populations may have declined by 80 to 90% (Kampp et al. 1994).

17.4.2 EGGING

The collection of eggs as a food source has probably occurred everywhere humans have come into contact with seabird colonies (Cott 1953, Boersma and Parrish 1998). Eggs were consumed by sailors on long voyages, and local settlers used eggs as an important protein source (Spennemann 1998). In 1897, over 700,000 eggs were taken from penguin colonies along the coast of South Africa, and during a 30-year period, over 13,000,000 eggs were collected from the Cape Islands of South Africa (Frost et al. 1976, Shelton et al. 1984). Egging can have community-level effects. For example, harvest of Jackass Penguin (*Spheniscus demersus*) eggs has been identified as the primary factor that initiated the replacement of this species by the Cape Gannet (*Morus capensis*; Crawford 1987). While egging is no longer common nor commercial for most seabirds (Yorio et

FIGURE 17.4 Well-controlled ecotourism can often coexist with successful nesting by seabirds and is an important part of promoting conservation. Here a tourist (M. Gochfeld) to Antarctica learns about King Penguins (*Aptenodytes patagonicus*). (Photo by J. Burger.)

al. 1999), egging continues to have a significant impact on others (e.g., Canada, Blanchard 1994; the Caribbean, Schreiber and Lee 2000; Greenland, P. D. Boersma unpublished).

17.5 HUMAN INTRUSIONS IN SEABIRD COLONIES

While not as obvious as the impacts of hunting or direct habitat destruction, the negative impact of human visitors to seabird colonies, particularly tourists, is extensively documented (e.g., Manuwal 1978, Anderson and Keith 1980, van Halewyn and Norton 1984, Rodway et al. 1996). These impacts commonly include nest desertion, temporary nest abandonment, increased risk of predation, and, ultimately, reduced breeding success (see Burger and Gochfeld 1983, 1993).

17.5.1 TOURISM

Humans are traveling in ever-increasing numbers to previously remote areas throughout the globe. The World Tourism Organization estimates that nature tourism generates 7% of all international travel expenditure (noted in Lindberg et al. 1997). Nature-based tourism is also one of the fastest growing segments of the tourist industry, growing at an annual rate between 10 and 30% (Lindberg et al. 1997). A positive aspect of this trend is that people who have close encounters with nature are more likely to support conservation measures.

Tourism in the Antarctic increased from under 300 people/year in the 1950s, to over 5,500 people/year in the early 1990s (Enzenbacher 1993, Kenchington 1989). In the Galapagos Islands, tourist numbers grew from negligible levels in 1970 to over 60,000 in 1998 (Damsgard 1999). At Punta Tombo, Argentina, the number of tourists at a penguin colony grew from a few hundred per season (September through April) in the early 1970s to over 55,000 in the late 1990s (Boersma unpublished). The presence of tourists in or near colonies can decrease bird numbers and must be carefully managed. Tourists trampled approximately 28% of all burrows at a penguin colony in the Punihuil Islands, Chile (Simeone and Schlatter 1998). Tourists may also cause adult seabirds to abandon nests, making eggs and chicks susceptible to predation (DesGranges and Reed 1981) or to temperature extremes and other inclement weather (Hunt 1972). However, tourist impacts can be reduced through thoughtful management (Tershey et al. 1997) and tourism is often compatible with seabird colonies when proper management practices are in place (Figure 17.4).

FIGURE 17.5 Research is a vital part of conservation because it provides the basis for sound management. Here R. W. Schreiber replaces a worn band on a Laysan Albatross on Midway Island in order to continue following birds banded 30 years before. (Photo by E. A. Schreiber.)

17.5.2 SCIENTIFIC RESEARCH

Scientific research programs may have short- or long-term impacts on seabird populations if they are not implemented carefully (see summaries in Rodway et al. 1996, Carney and Sydeman 1999, Nisbet 2000, Carney and Sydeman 2000). Seabirds have varying responses to researcher disturbance, and while some are unaffected by it, others are susceptible to being disturbed. Some populations of Atlantic Puffins (*Fratercula arctica*) readily desert their eggs when disturbed by researchers (Rodway et al. 1996), and Adélie Penguin (*Pygoscelis papua*) populations decreased in one colony as a result of disturbance associated with scientific studies (Woehler et al. 1991). Studies on Red-tailed Tropicbirds (*Phaethon rubricauda*) and Brown (*Sula leucogaster*) and Red-footed Boobies (*S. sula*) indicate that these species are not particularly susceptible to human disturbance (Schreiber 1994, 1999). With appropriate precautions, researchers can often reduce their impact and conduct research without decreasing nesting success (see discussions in Burger and Gochfeld 1993, Nisbet 2000).

Appropriate and constructive management decisions cannot be made without quality data on the species involved (Figure 17.5). We are dependent on researchers to provide these data (Schreiber 2000). It is impossible to effectively manage seabird populations without knowing such basic biological information as clutch size, incubation period, fledging period, energetic constraints, and the threats seabirds face during each of their reproductive phases. Furthermore, without understanding the ways in which different threats, such as predators, pollutants, or humans, affect seabirds, it is impossible to design effective management plans.

In most cases, the risks to individual birds from research activity can be drastically reduced by careful research design and implementation. This may involve limiting time in the colony, visiting colonies only during specific periods during the day or reproductive season, or altering investigator behavior. For example, some birds are less disturbed by slow movement through a colony (as opposed to running), by indirect approaches, and by not making eye contact (Burger and Gochfeld 1994). In some cases, noninvasive methods can be used, such as collecting feathers for heavy metal analysis or collecting nonviable eggs, and taking some nest and colony site measurements after the birds are no longer present. In other cases, the conservation problem may be best addressed by collecting individuals to assess physiological parameters, determining contaminant effects, and providing voucher specimens for historical archives (Remsen 1995, Schreiber 2000). Furthermore, many techniques, such as banding and physiological studies, are critical to obtain data on reproductive success, long-term survival, and population dynamics, which in turn aid in managing populations.

17.5.3 Other Disturbances

Impacts to seabirds from disturbance can be subtle. For example, some seabirds are negatively affected by the presence of artificial lights (Reed et al. 1985). Negative impacts on seabirds have also been attributed to noise and disturbance from helicopters (McKnight and Knoder 1979), motorboats (Burger 1998), and personal watercraft (Burger and Leonard 2000). Recreational activities can reduce reproductive success if unmanaged (Burger 1995). However, recreation does not necessarily present a direct conflict with seabird conservation if such activities are carefully managed (Burger et al. 1995, Knight and Gutzwiller 1995).

Human recreational activity at nesting colonies has the potential to cause mortality, reduce reproductive success, or degrade nesting areas. Some seabirds are adversely affected by any visitation or disturbance (Burger and Gochfeld 1993, Yorio and Quintana 1996), but others habituate to the presence of humans (Yorio and Boersma 1992, Burger and Gochfeld 1999). If well managed, human presence appears to have little impact on the reproductive success of many habituated birds (Burger and Gochfeld 1983, Yorio and Boersma 1992).

17.6 OTHER THREATS TO SEABIRDS

17.6.1 Climate Change, Pollution, and Commercial Fishing

Several other major threats face seabird populations. Three of these are so significant that they are treated in separate chapters: climate change (Chapter 7), pollution (Chapter 15), and commercial fishing (Chapter 16). For many seabird populations, these threats, individually or in combination, present the greatest danger to their persistence (Figure 17.6).

17.6.2 Interspecific Competition and Threat Interactions

Interspecific competition is a major factor in the decline of some seabird populations. For example, at some seabird colonies, competition from increasing numbers of gulls is the primary cause of population declines (Anderson and Devlin 1999). But many gull species are increasing because of human activity, such as the presence of garbage dumps (Hunt 1972) and offal from fishing operations.

Often, more than one threat faces a seabird population and these can act synergistically. For example, Williams (1995) concluded that the Yellow-eyed Penguin (*Megadyptes antipodes*) suffered steep population declines because of a combination of human disturbance, habitat modification, introduced predators and grazers, and fisheries impacts. Multiple physiological stresses from several sources can produce severe problems (Livingstone et al. 1992). It is not difficult to imagine a population already affected by pollution, predators, and/or disease might be more heavily impacted by climate change or fishing impacts than would an otherwise healthy population.

FIGURE 17.6 Oil pollution kills a large number of seabirds. Here, an oiled adult Laysan Albatross transfers oil to its chick. (Photo by E. A. Schreiber.)

17.7 LEGAL PROTECTION

Laws protecting seabirds differ dramatically from country to country and from state to state. However, the principles of legal protection are similar in most jurisdictions, and we focus primarily on U.S. law for illustrative purposes. In the U.S., legal protection for seabirds occurs almost entirely at the state, federal, and international levels. No major state or federal law focuses exclusively on seabirds. Rather, legal protection of seabirds is found within laws protecting wildlife generally or in laws regulating fishing. State and federal law is a combination of statutes passed by elected officials, regulations promulgated by implementing agencies, agency policies, and decisions by courts. International law, however, consists primarily of treaties between countries (Figure 17.7).

17.7.1 FEDERAL PROTECTIONS

Federal law provides the primary protective regime for most wildlife, including seabirds (for discussions of U.S. wildlife law, see Bean and Rowland 1997, Musgrave et al. 1998). The first statute passed by Congress to protect wildlife was the Lacey Act of 1900 (current version at 16 U.S.C. §§ 701 and 3371–3378 and 18 U.S.C. § 42), which was passed because of growing national concern over the demise of the Passenger Pigeon (*Ectopistes migratorius*) and the decimation of heron and egret populations for the millinery trade. Although the Lacey Act did not provide explicit protection for seabirds, it provided for federal enforcement of state laws protecting wildlife by prohibiting interstate transfer of wildlife killed in violation of state laws.

   The first major U.S. wildlife statute that included protection of seabirds was the Migratory Bird Act of 1913 (Ch. 145, 37 Stat. 828, 847 [*repealed* 1918]). Passed because of continuing concern over reductions in bird populations, this ground-breaking law prohibited the hunting of migratory

FIGURE 17.7 King Penguins nest by the thousands on Maquerie Island around the massive drums (digestive tanks) used to boil penguins in the late 1800's to make oil. (Photo by J. Burger.)

game birds, insectivorous birds, and other migratory nongame birds (which included seabirds) except in compliance with federal regulations. This Act was quickly challenged as unconstitutional (*U.S. v. Shauer*, 214 F. 154, E.D. Ark. 1914). Recognizing the weakness of its legal arguments, the federal government abandoned its legislative approach. Instead, the federal government entered into an international treaty in 1916 with Great Britain (on behalf of Canada) to provide these same protections (Convention for the Protection of Migratory Birds). The U.S. eventually entered into treaties protecting migratory birds with Mexico (1936), Japan (1972), and the former Soviet Union (1976) as well. The 1916 treaty was implemented by Congress' passage of the Migratory Bird Treaty Act of 1917 (MBTA: current version at 16 U.S.C. §§ 703–711). Ruling on a legal challenge to the constitutionality of the MBTA, the U.S. Supreme Court held that Congress' treaty gave it power to protect migratory birds (*Missouri v. Holland*, 252 U.S. 416, 1920).

Another well-known U.S. statute protecting several seabirds is the Endangered Species Act of 1973 (ESA; 16 U.S.C. §§ 1531–1543). Under the ESA, listed species are protected and conserved, habitat critical to the species is to be preserved, and federal actions are not to jeopardize the species (see Clark 1994 and Bean and Rowland 1997 for discussions of the ESA's basic provisions). The ESA also implements the Convention on International Trade in Species of Wild Fauna and Flora, an international treaty (discussed in more detail below). The list of endangered species protected under the ESA includes several seabirds threatened both nationally and internationally (see Table 17.1). However, the ESA's protections apply only to the boundaries of the U.S., U.S. territorial waters, and persons under U.S. jurisdiction. Despite these limitations, the ESA has provided significant benefits for some seabirds. For example, U.S. courts have affirmed regulations promulgated under the ESA that protect Marbled Murrelet nesting habitat (*Marbled Murrelet v. Babbitt*, 83 F.3d 1060, 9th Cir. 1996).

Indirect protection of seabirds is found in federal regulation of various fishing methods (e.g., gill nets, long lines, driftnets, etc.), such as the Magnuson-Stevens Fishery Conservation and Management Act of 1976 (16 U.S.C. §§ 1801–1882), Fish and Wildlife Conservation Act of 1980 (16 U.S.C. §§ 2901–2911), Driftnet Amendments of 1990 (16 U.S.C. §§ 1826, 1857–1859), and High Seas Driftnet Fisheries Enforcement Act of 1992 (16 U.S.C. §§ 1826a–1826c). Because seabirds are often killed during fishing efforts, regulation of fishing methods provides significant protection to seabirds.

TABLE 17.1
Seabirds Listed under the U.S. Endangered Species Act

Listed Species	Found within U.S.	Found outside U.S.
Galapagos Penguin (*Spheniscus mendiculus*)	No	Yes
Amsterdam Albatross (*Diomedea amsterdamensis*)	No	Yes
Short-tailed Albatross (*Phoebastria albatrus*)	No	Yes
Madeira Petrel (*Pterodroma madeira*)	No	Yes
Cahow (Bermuda Petrel) (*P. cahow*)	No	Yes
Hawaiian Dark-rumped Petrel (*P. phaeopygia sandwichensis*)	Yes	No
Mascarene Black Petrel (*P. aterrima*)	No	Yes
Newell's Townsend's Shearwater (*Puffinus auricularis newelli*)	Yes	No
Brown Pelican (*Pelecanus occidentalis*)	Yes	Yes
Christmas Island Frigatebird (*Fregata andrewsi*)	No	Yes
Audouin's Gull (*Larus audouinii*)	No	Yes
Relict Gull (*L. relictus*)	No	Yes
Roseate Tern (*Sterna dougallii dougallii*)	Yes	Yes
Least Tern (*S. antillarum*)	Yes	Yes
California Least Tern (*S. antillarum browni*)	Yes	No
Marbled Murrelet (*Brachyramphus marmoratus marmoratus*)	Yes	Yes

Note: Species are listed in order listed within ESA regulations.

17.7.2 STATE PROTECTIONS

Section 6 of the ESA requires federal cooperation with states and creates a process through which individual states can enter into management and cooperative agreements for listed species. In addition to having authority to administer aspects of the ESA, many states prepare their own list of threatened species, and legal protections often accompany those lists (see Musgrave and Stein 1993). Some states provide separate and supplemental legal protection of seabirds, though laws differ from state to state (see generally, Musgrave and Stein 1993). For example, Washington State statutes provide that wildlife designated by a state commission as protected cannot be hunted (RCW 77.08.010 (19)). Through state regulations, the commission designated all seabirds as protected wildlife (WAC 232-12-011). Therefore, no seabird can be killed in Washington State. In addition, Washington State law prohibits the taking of protected wildlife (including seabirds). Here, "taking" includes the destruction of eggs or nests as well as killing (RCW 77.15.130). Unlike the ESA, these state protections apply to all seabird species, not just those listed as threatened or endangered. However, most of these state protections overlap with protections contained in the MBTA.

17.7.3 INTERNATIONAL PROTECTIONS

Even for most lawyers, international law is a source of much confusion, and the concept of international law is difficult to grasp (for more in-depth discussions of international environmental and wildlife law, see van Heijnsbergen 1997, UNEP 1997, and Kiss 2000). These conceptual difficulties arise because, unlike federal law, international law has no constitutional foundation (e.g., the U.S. Constitution), elected legislative body, or implementing or enforcing agencies. Nor is there an equivalent of the U.S. court system. International law is more of a shaking of hands, a mere agreement between independent entities that depends almost entirely on goodwill and cooperation for success. An "international law" is usually just an agreement signed between two or more nations, and is usually referred as a "treaty" or "convention."

A useful way to think about international agreements is to consider them contracts, in which two or more entities agree on certain language and indicate their agreement by signature. A further complication is that these "contracts" are generally not effective without implementing legislation by the signatory countries which the governments back home must authorize and endorse. For example, the Convention on the Conservation of Antarctic Marine Living Resources (CCAMLR) was adopted by representatives of several nations at a meeting in the U.S. in 1980. But the treaty did not come into effect in until 1982, after a requisite number of signatory countries adopted implementing legislation. The U.S. did not officially adopt such implementing legislation until 1984 (codified at 16 U.S.C. §§ 2431–2442) at which time CCAMLR applied to the U.S. Because of the complex nature of international treaties and the lack of traditional police and courts, enforcement of international law is often extremely difficult to secure.

Early efforts to provide legal protection to birds focused exclusively on hunting restrictions. Migratory bird treaties between Canada and the U.S., and Mexico and the U.S., however, prohibit not only hunting, but also the capture, possession, sale, or transport of listed birds. And the more recent U.S. migratory bird treaties with Japan and the Soviet Union attempt to protect habitat as well. Other international treaties and conventions providing some measure of protection for seabirds include the Convention on International Trade in Species of Wild Fauna and Flora (CITES) (1973), the Convention on the Conservation of Migratory Species of Wild Animals (1979), CCAMLR (1980), the UN Convention on the Law of the Sea (1982), and the Convention for the Prohibition of Fishing with Long Driftnets in the South Pacific (1989).

CITES prohibits international trade by signatory countries in species listed on one of its three Appendices, except in accordance with CITES' provisions (for discussions of CITES, see Favre 1992 and Hemley 1994). The Appendices provide a hierarchy of threat: Appendix I species are those threatened with extinction and which are or may be affected by trade; Appendix II species include somewhat less threatened species which are or may be affected by trade; and Appendix III species are placed unilaterally by any party to CITES and are species protected in that party's home country but needing international cooperation. CITES was originally enacted over concern about trade in a few charismatic animals such as tigers, elephants, and crocodiles, but has expanded its protections to thousands of plants and animals. Its primary weakness is its reliance on each member state for implementation and enforcement (Hemley 1994). Nevertheless, CITES has retained a high public profile and had a significant impact on worldwide awareness and protection of wildlife, and more than 120 countries are signatories to it. Although CITES is often considered a successful international treaty, the trade it regulates is, unfortunately, only a minor component of the major threats to seabirds, few of which are listed under CITES (see Table 17.2).

The Convention on the Conservation of Antarctic Marine Living Resources (CCAMLR) was enacted because of the fear of negative consequences from unregulated harvest of krill in Antarctica and the loss of Antarctica's rich biological diversity. Unique among international treaties is its

TABLE 17.2
Seabirds Listed under CITES Appendices

Listed Species	Found in U.S.	CITES Appendix
Jackass Penguin (*Spheniscus demersus*)	No	II
Humboldt Penguin (*S. humboldti*)	No	I
Short-tailed Albatross (*Phoebastria albatrus*)	Yes	I
Dalmatian Pelican (*Pelecanus crispus*)	No	I
Christmas Island Frigatebird (*Fregata andrewsi*)	No	I
Abbott's Booby (*Papasula abbotti*)	No	I
Relict Gull (*Larus relictus*)	No	I

ecosystem approach to environmental protection, rather than a focus on individual species or commercial practices. In addition to a permit system to regulate fishing, CCAMLR also contains catch and gear restrictions, area and seasonal closures, and anti-pollution provisions. CCAMLR applies to all living marine resources, including seabirds, and has been used to reduce bycatch of seabirds and modify longline fisheries methods.

The World Trade Organization (WTO) may have an impact on international treaties protecting wildlife, particularly CITES. Determining the exact relationship between the WTO and these international treaties has become highly controversial and remains unclear (Charnovitz 2000, Wofford 2000). One distinct possibility is that the WTO may nullify provisions of these treaties deemed to be in conflict with WTO regulations and decisions on trade. International law remains an enigmatic but essential aspect of legal protection for seabirds.

17.8 PROGRESS IN SEABIRD CONSERVATION

Although isolated concerns were raised about seabird populations earlier this century (see McChesney and Tershy 1998), most seabird conservation efforts began only within the last 20 to 30 years. Because legal mechanisms provide inadequate protection for seabirds and almost no affirmative actions to recover populations, most conservation efforts take place outside the legal arena. Scientists are often at the forefront in crafting, recommending, and implementing conservation efforts. Recommendations for seabird conservation measures have been proposed at the species level (Jackass Penguin, *Spheniscus demersus;* Frost et al. 1976), national level (Yorio et al. 1999, Argentina), regional level (Croxall et al. 1984, multiple regions; Schreiber and Lee 2000, West Indies; Kondratyev et al. 2000, Russian Far East), and global level (Duffy 1994b). More of these concrete, substantive directives are needed, as well as coordinated efforts to implement the recommended conservation measures. In addition, more examples of successful approaches to seabird conservation need to be entered in the literature as a reference tool for further management efforts (e.g., Gochfeld et al. 1994). Below, we briefly discuss selected areas in which significant progress in seabird conservation has been achieved.

17.8.1 POLICY APPROACHES

As we have learned more about threats to seabirds, many individuals, organizations, and national and international bodies have made concerted efforts to protect, preserve, and recover seabird populations. Apart from the very few laws that provide legal protection to listed seabirds (e.g., ESA, CITES), many national and international organizations publish policy or position statements addressing seabird conservation. Some entities publish lists of birds, including seabirds, that are endangered, threatened, or are of special concern. For example, the U.S. Fish and Wildlife Act of 1980 created a list of species of concern that includes over a dozen seabirds (16 U.S.C. §§ 7901–12). In North America, a Colonial Waterbird Conservation Plan and conservation prioritization protocol are currently being developed by scientists, government agencies, and environmental organizations. BirdLife International (2000) recently published a list of endangered and threatened seabirds worldwide (see Appendix 1). The IUCN (International Union for the Conservation of Nature) published a recently updated worldwide list of threatened species (Hilton-Taylor 2000). Although these lists provide no legal protection in and of themselves, they are valuable education and reference tools. But they are insufficient for successful seabird protection and recovery.

To achieve effective seabird conservation, biologists, institutions, organizations, and policy makers must work at unprecedented levels of cooperation across national and international boundaries. Interest groups with divergent and sometimes conflicting priorities must work together educating the public and political decision-makers on seabird problems and solutions. Scientific research is badly needed to identify problems and threats to seabirds, identify conservation and management options, and assess the success of management programs. Education programs are also needed, so that people can learn to care about the natural world around them and understand the importance (economic and cultural) of preserving biodiversity.

17.8.2 RESEARCH

If seabird populations are to remain viable in an increasingly modified world, science has to play a key role. For managers to make quality, workable decisions about seabird conservation, basic research on both breeding biology and ecology is required (Schreiber 2000). We still know very little about many seabird species, and, therefore, we cannot make effective management decisions. Basic research on single problems at a local level is a critical component of any effort to protect, preserve, or recover seabird populations. We have identified four main research objectives for seabird conservation: (1) fill knowledge gaps about species biology, (2) work to understand how marine systems function and change over time, (3) determine how humans are changing systems that seabirds use, and (4) test how modifications in law, policy, human behavior, and use of the environment effect seabird survival and abundance. One of the main gaps in our knowledge, which directly impacts conservation efforts, is our meager understanding of how seabirds use the marine environment. We need to know where they are, when they are there, what resources they use, and what they are doing. Although we know where many key seabird breeding colonies are located, we know little about foraging areas, particularly outside the breeding season (see Chapter 6). More basic biological information is needed on nearly all seabird species to allow development of appropriate seabird conservation measures.

17.8.3 MANAGEMENT AND RESTORATION

Remoteness and isolation are rapidly becoming descriptors of seabird colonies of the past. To reduce people pressure on seabirds, large-scale changes in how the marine environment (including islands, estuaries, coastlines, and sea cliffs) is used are needed. Areas must be protected for seabird use, and human activities in these areas may also need to be proscribed. Seabirds require protection of their breeding colonies and key feeding areas both inside and outside the breeding season. Ocean zoning and marine protected areas offer opportunities to regulate human use (Hyrenbach et al. 2000).

In response to the extirpation and decline of many seabird populations, an increasing number of seabird restoration efforts have been undertaken (e.g., Towns et al. 1997). The populations of three extant species of seabirds numbered fewer than 50 individuals at one time: Amsterdam Albatross (*Diomedea amsterdamensis*), Short-tailed Albatross (*Phoebastria albatrus*), and Cahow Petrel (*Pterdroma cahow*: Wingate 1978, Collar and Andrew 1992). Through the efforts of ornithologists and conservationists, these species are now well on the road to recovery, although continued monitoring is necessary. Other efforts that have had dramatic results include the removal of introduced predators from seabird breeding islands. Removal of foxes, rats, cats, and pigs have had dramatic positive results for many seabirds (e.g., Bailey 1993). In addition to predator removal, some restoration attempts include habitat improvement. Nesting boxes or artificial burrows have been used successfully for several seabird species (Wingate 1978, Podolsky and Kress 1989, Boersma and Silva 2001).

Many restoration efforts have moved beyond predator removal and habitat restoration to efforts to rebuild entire seabird colonies (Anderson and Devlin 1999). Social attraction, for example, has been used successfully for alcids and procellariiformes (Podolsky and Kress 1989, 1991). To attract birds, decoys of breeding birds are placed in an area where the birds bred historically or where habitat seems suitable (Kress 1983, Kotliar and Burger 1984). In addition to decoys, sometimes mirror boxes are used to increase the perceived density of attending birds, and sound systems may be installed to play the breeding calls of the birds, giving the impression of an established breeding colony (Kress 1983).

In 1986, the *Apex Houston* oil spill off California's coast killed approximately 9,900 seabirds (Page et al. 1990). Following the spill, Common Murres (*Uria aalae*) were absent from their nesting sites at Devil's Slide Rock and researchers used social attraction to lure murres back. However, restoration efforts may be considerably more expensive than extirpation prevention, habitat improvement, or predator removal. The cost of a species conservation program increases from 10- to 10,000-fold at each of three levels of intervention from (1) managing a species where it occurs naturally, (2) to those that depend on parks or reserves for survival, (3) to species that are dependent on zoos or botanical gardens for care (Conway 1986, Woodruff 1989). It is far cheaper and more economically and ecologically desirable to conserve species in nature than in captivity and to prevent or reduce the decline of a species or population than to restore a species or population.

17.8.4 EDUCATION

Education and public support are necessary pillars to support seabird conservation while fostering changes in human behavior and use of the environment. Education is inexorably linked to public support for conservation (Pearl 1989: Figure 17.8). Evidence exists that education may be able to reduce certain human behavior, such as the illegal harvest of seabirds and their eggs (Blanchard and Monroe 1990). Educating the public on the impact of killing seabirds in the United States not only led to the founding of national groups such as the National Audubon Society, but also to the enactment of laws protecting seabird breeding areas, eggs, and young. More and more valuable information on seabird conservation is available on the internet (see, e.g., BirdLife International Seabird Conservation Program's Website at: http: //www.uct.ac.za/depts/stats/adu/seabirds).

Successful seabird conservation requires cooperative and interdisciplinary efforts. A good example showing the benefits of linking scientists, environmental advocates, and international decision-makers is apparent in the reductions in seabird mortality resulting from implementing recommendations of CCAMLR (see discussion above). Part of the outcome was the U.S. published regulations for U.S. flagged vessels in subantarctic seas that implemented setting of longlines at night and deployment of streamers to reduce seabird bycatch (61 Fed. Reg. 8483, March 5, 1996). These measures came about because of strong scientific and public support. Changes in how longlines should be set did not end with CCAMLR. Scientists and public interest groups using the

FIGURE 17.8 Through interactions with seabirds, people begin to value them. This, in turn, leads to increased public support of measures to conserve seabird populations. (Photo by R. W. Schreiber.)

internet pushed for global actions. As a result, IUCN Resolutions CGR1.69 and CGR2.PRG049 have been adopted, outlining multiple measures to reduce or eliminate the deaths of seabirds as a result of longline fisheries in the world's oceans.

People are both the ultimate problem and solution for seabirds. Our ability to exploit resources on a global scale, and the emergence of a global economy leave no place on earth untouched. Ultimately, control of our resource consumption and population will be necessary if we hope to protect seabirds and their habitats, along with the rest of the richness and diversity of life on our planet.

LITERATURE CITED

AINLEY, D. G., AND T. J. LEWIS. 1974. The history of Farallon Island marine bird populations 1843–1972. Condor 76: 432–446.

ALLEN, J. J. 1876. The extinction of the Great Auk at the Funk Islands. American Naturalist 10: 48.

ANDERSON, D. W., AND J. O. KEITH. 1980. The human influence on seabird nesting success: conservation implications. Biological Conservation 18: 65–80.

ANDERSON, J. G. T., AND C. M. DEVLIN. 1999. Restoration of a multi-species seabird colony. Biological Conservation 90: 175–181.

ASHMOLE, N. P., ASHMOLE, M. J., AND SIMMONS, K. E. L. 1994. Seabird conservation and feral cats on Ascension Island, South Atlantic. BirdLife Conservation Series 1: 94–121.

BAILEY, E. P. 1993. Introduction of foxes to Alaskan islands: history, effects on avifauna, and eradication. U.S. Department of Interior, FWS/Resource Publication 193: 1–53.

BEAN, M. J., AND M. ROWLAND. 1997. The Evolution of National Wildlife Law, 3d ed. Praeger, Westport, CT.

BIRDLIFE INTERNATIONAL. 2000. Threatened Birds of the World. Lynx Edicions and BirdLife International. Barcelona, Spain and Cambridge, U.K.

BLANCHARD, K. A. 1994. Culture and seabird conservation: the north shore of the Gulf of St Lawrence, Canada. Pp. 294–310 in Seabirds on Islands: Threats, Case Studies and Action Plans (D. N. Nettleship, J. Burger, and M. Gochfeld, Eds.). BirdLife International, Cambridge, U.K.

BLANCHARD, K. A., AND M. C. MONROE. 1990. Effective educational strategies for reversing population declines in seabirds. Pp. 108–118 in Transactions of the Fifty-Fifth North American Wildlife and Natural Resources Conference (R. E. McCabe, Ed.).

BOERSMA, P. D., AND M. C. SILVA. 2001. Fork-tailed Storm-Petrel (*Oceanodrom furcata*) in The Birds of North America No. 569 (A. Poole and F. Gill, Eds.). The Birds of North America, Inc., Philadelphia.

BOERSMA, P. D., AND J. K. PARRISH. 1998. Threats to seabirds: Research, education, and societal approaches to conservation. Pp. 237–259 in Avian Conservation: Research and Management (J. M. Marzluff and R. Sallabanks, Eds.). Island Press, Washington, D.C.

BRYANT, D. 1995. Coastlines at risk: an index of potential development-related threats to coastal ecosystems. World Resources Institute Indicator Brief. World Resources Institute, Washington, D.C.

BURGER, J. 1995. Beach recreation and nesting birds. Pp. 281–295 in Wildlife and Recreationists: Coexistence through Management and Research (R. L. Knight and K. J. Gutzwiller, Eds.). Island Press, Washington, D.C.

BURGER, J. 1998. Effects of motorboats and personal watercraft on flight behavior over a colony of Common Terns. Condor 100: 528–534.

BURGER, J., AND M. GOCHFELD. 1983. Behavioural responses to human intruders of Herring Gulls (*Larus argentatus*) and Great Black-backed Gulls (*L. marinus*) with varying exposure to human disturbance. Behavioural Processes 8: 327–344.

BURGER, J., AND M. GOCHFELD. 1993. Tourism and short-term behavioral responses of nesting Masked, Red-footed, and Blue-footed Boobies in the Galapagos. Environmental Conservation 20: 255–259.

BURGER, J. AND M. GOCHFELD. 1994. Predation and effects of humans on island-nesting seabirds. Pp. 39–67 in Seabirds on Islands: Threats, Case Studies and Action Plans (D. N. Nettleship, J. Burger, and M. Gochfeld, Eds.). BirdLife International, Cambridge, England.

BURGER, J., AND M. GOCHFELD. 1999. Role of human disturbance in response behavior of Laysan Albatrosses (*Diomedea immutabilis*). Bird Behavior 13: 23–30.

BURGER, J., AND J. LEONARD. 2000. Conflict resolution in coastal waters: The case of personal watercraft. Marine Policy 24: 61–67.

BURGER, J., M. GOCHFELD, AND L. J. NILES. 1995. Ecotourism and birds in coastal New Jersey: contrasting responses of birds, tourists, and managers. Environmental Conservation 22: 56–65.

CARNEY, K. M., AND W. J. SYDEMAN. 1999. A review of human disturbance effects on nesting colonial waterbirds. Waterbirds 22: 68–79.

CARNEY, K. M., AND W. J. SYDEMAN. 2000. Response: disturbance, habituation and management of waterbirds. Waterbirds 23: 333–334.

CHARNOVITZ, S. 2000. World trade and the environment: A review of the new WTO report. Georgetown International Environmental Law Review 12: 523–541.

CLARK, J. A. 1994. The Endangered Species Act: its history, provisions, and effectiveness. Pp. 19–43 *in* Endangered Species Recovery: Finding the Lessons, Improving the Process (T. M. Clark, R. P. Reading, and A. L. Clarke, Eds.). Island Press, Washington, D.C.

COLLAR, N. J., AND P. ANDREW (Eds.). 1992. Birds to watch: the ICBP world checklist of threatened species. International Council for Bird Protection, Technical Bulletin No. 8. Smithsonian Institution Press, Washington, D.C.

CONWAY, W. 1986. The practical difficulties and financial implications of endangered species breeding programs. International Zoo Yearbook 24/25: 210–219.

COTT, H. B. 1953. The exploitation of wild birds for their eggs. Ibis 95: 409–449.

CRAWFORD, R. J. M. 1987. Food and population variability in five regions supporting large stocks of anchovy, sardine and horse mackerel. South African Journal of Marine Science 5: 735–757.

CROXALL, J. P., P. G. H. EVANS, AND R. W. SCHREIBER (Eds.). 1984. Status and Conservation of the World's Seabirds. International Council for Bird Preservation (Technical Publication 2). Cambridge, England.

CZECH, B., P. R. KRAUSMAN, AND P. K. DEVERS. 2000. Economic associations among causes of species endangerment in the United States. BioScience 50: 593–601.

DAMSGARD, B. 1999. Galapagos: as a tourist in Darwin's footsteps. Fauna Oslo 52: 136–145.

DE LISLE, G. W., W. L. STANISLAWEK, AND P. J. MOORS. 1990. *Pasteurella multocida* infections in Rockhopper Penguins (*Eudyptes chrysocome*) from Campbell Island, New Zealand. Journal of Wildlife Disease 26: 283–285.

DESGRANGES, J. L., AND A. REED. 1981. Disturbances and control of Double-crested Cormorants in Quebec. Colonial Waterbirds 4: 12–19.

DROST, C. A., AND D. B. LEWIS. 1995. Xantu's Murrelet (*Synthliboramphus hypoleucus*) *in* The Birds of North America, No. 164 (A. Poole and F. Gill, Eds.). Academy of Natural Sciences, Philadelphia; American Ornithologists' Union, Washington, D.C.

DUFFY, D. C. 1994a. The guano islands of Peru: The once and future management of a renewable resource. Pp. 68–76 *in* Seabirds on Islands: Threats, Case Studies and Action Plans (D. N. Nettleship, J. Burger, and M. Gochfeld, Eds.). BirdLife International, Cambridge, England.

DUFFY, D. C. 1994b. Toward a world strategy for seabird sanctuaries. Colonial Waterbirds 17: 200–206.

ELLIOT, R. D., B. T. COLLINS, E. G. HAYAKAWA, AND L. METRAS. 1991. The harvest of murres in Newfoundland from 1977–78 to 1987–88. Pp. 36–44 *in* Studies of High Latitude Seabirds 2. Conservation Biology of Thick-Billed Murres in the Northwest Atlantic (W. A. Montevecchi and W. A. Gaston, Eds.). Canadian Wildlife Service, Ottawa, Ontario.

Enzenbacher, D. J. 1993. Antarctic tourism: 1991/1992 season activity. Polar Record 29: 240–244.

EVERETT, W. T. 1988. Biology of the Black-vented Shearwater. Western Birds 19: 89–104.

FALK, K., AND J. DURINCK. 1992. Thick-billed Murre hunting in West Greenland, 1988–89. Arctic 45: 167–178.

FAVRE, D. S. 1992. International trade in endangered species: a guide to CITES. Martinus Nijhoff, Dordrecht.

FEARE, C. J., E. L. GILL, P. CARTY, H. CARTY, AND V. J. AYRTON. 1997. Habitat use by Seychelles Sooty Terns *Sterna fuscata* and implications for colony management. Biological Conservation 81: 69–76.

FROST, P. G. H., W. R. SIEGFRIED, AND J. COOPER. 1976. The conservation of the Jackass Penguin (*Spheniscus demersus*). Biological Conservation 9: 79–99.

GOCHFELD, M., J. BURGER, A. HAYNES-SUTTON, R. VAN HALEWYN, AND J. SALIVA. 1994. Successful approaches to seabird protection in the West Indies. Pp. 186–209 *in* Seabirds on Islands: Threats, Case Studies and Action Plans (D. N. Nettleship, J. Burger, and M. Gochfeld, Eds.). BirdLife International, Cambridge, England.

HARTMAN, L. M., AND D. S. EASTMAN. 1999. Distribution of introduced raccoons *Procyon lotor* on the Queen Charlotte Islands: implications for burrow-nesting seabirds. Biological Conservation 88: 1–13.

HEMLEY, G. (Ed.). 1994. International Trade: CITES sourcebook. World Wildlife Fund and Island Press, Washington, D.C.

HILTON-TAYLOR, C. (Compiler). 2000. 2000 IUCN red list of threatened species. IUCN, Gland, Switzerland and Cambridge, England.

HOBSON, K. A., M. C. DREVER, AND G. W. KAISER. 1999. Norway rats as predators of burrow-nesting seabirds: insights from stable isotope analyses. Journal of Wildlife Management 63: 14–25.

HOWALD, G. R., P. MINEAU, J. E. ELLIOT, AND K. M. CHENG. 1999. Brodifacoum poisoning of avian scavengers during rat control on a seabird colony. Ecotoxicology 8: 431–447.

HUNT, G. L. 1972. Influence of food distribution and human disturbance on the reproductive success of Herring Gulls. Ecology 53: 1051–1061.

HYRENBACH, K. D., K. A. FORNEY, AND P. K. DAYTON. 2000. Marine protected areas and ocean basin management. Aquatic Conservation: Marine and Freshwater Ecosystems 10: 437–458.

JONES, R. D., AND G. V. BYRD. 1979. Interrelations between seabirds and introduced animals. Pp. 221–216 *in* Conservation of Marine Birds of Northern North America (J. C. Bartonek and D. N. Nettleship, Eds.). Wildlife Research Report 11. U.S. Fish and Wildlife Service, Washington, D.C.

KAMPP, K., D. N. NETTLESHIP, AND P. G. H. EVANS. 1994. Thick-billed murres of Greenland: status and prospects. Pp. 133–154 *in* Seabirds on Islands: Threats, Case Studies and Action Plans (D. N. Nettleship, J. Burger, and M. Gochfeld, Eds.). BirdLife International, Cambridge, England.

KEEGAN, D. R., B. E. COBLENTZ, AND C. S. WINCHELL. 1994. Ecology of feral goats eradicated on San Clemente Island, California. Pp. 321–330 *in* The Fourth California Islands Symposium: Update on the Status of Resources (W. L. Halvorson and G. J. Maender, Eds.). Santa Barbara Museum of Natural History, Santa Barbara, CA.

KENCHINGTON, R. A. 1989. Tourism in the Galapagos Islands: the dilemma of conservation. Environmental Conservation 16: 227–236.

KIRK, D. A., AND P. A. RACEY. 1991. Effects of the introduced black-naped hare *Lepus nigricollis nigricollis* on the vegetation of Cousin Island, Seychelles and possible implications for avifauna. Biological Conservation 61: 171–179.

KISS, A. 2000. International Environmental Law, 2nd ed. Transnational Publishers, Ardsley, NY.

KNIGHT, R. L., AND K. J. GUTZWILLER. (EDS.) 1995. Wildlife and recreationists: coexistence through management and research. Island Press, Washington, D.C.

KONDRATYEV, A. Y., P. S. VYATKIN, AND Y. V. SHIBAEV. 2000. Conservation and protection of seabirds and their habitat. Pp. 117–129 *in* Seabirds of the Russian Far East (A. Y. Kondratyev, N. M. Litvinenko, and G. W. Kaiser, Eds.). Canadian Wildlife Service Special Publication, Ottawa.

KOTLIAR, N. B., AND J. BURGER. 1984. The use of decoys to attract least terns (*Sterna antillarum*) to abandoned colony sites in New Jersey. Colonial Waterbirds 7: 134–138.

KRESS, S. W. 1983. The use of decoys, sound recordings, and gull control for re-establishment of a tern colony in Maine. Colonial Waterbirds 6: 185–196.

LINDBERG, K., B. FURZE, M. STAFF, AND R. BLACK. 1997. Ecotourism in the Asia-Pacific Region: Issues and Outlook. The International Ecotourism Society, Bennington, VT.

LIVINGSTONE, D. R., P. DONKIN, AND C. H. WALKER. 1992. Pollutants in marine ecosystems: an overview. Pp. 235–263 *in* Persistent Pollutants in Marine Ecosystems. Pergamon Press, New York.

LOOPE, L. L., AND D. MUELLER-DOMBOIS. 1989. Characteristics of invaded islands, with special reference to Hawaii. SCOPE 37: 257–280.

MANUWAL, D. A. 1978. Effect of man on marine birds: a review. Pp. 140–160 *in* Wildlife and People. Purdue University Press, West Lafayette, IN.

MCCHESNEY, G. J., AND B. R. TERSHY. 1998. History and status of introduced mammals and impacts to breeding seabirds on the California Channel and Northwestern Baja California Islands. Colonial Waterbirds 21: 335–347.

MCKNIGHT, D. E., AND D. E. KNODER. 1979. Resource development along coasts and on the ocean floor: Potential conflicts with marine bird conservation. Pp. 183–194 *in* Conservation of Marine Birds of Northern North America (J. C. Bartonek and D. N. Nettleship, Eds.). Wildlife Research Report 11. U.S. Fish and Wildlife Service, Washington, D.C.

MEFFE, G. K., AND R. C. CARROLL. 1997. Principles of Conservation Biology, 2nd ed. Sinauer Associates, Sunderland, MA.

MONTEIRO, L. R., J. A. RAMOS, AND R. W. FURNESS. 1996. Past and present status and conservation of the seabirds breeding in the Azores Archipelago. Biological Conservation 78: 319–328.

MOORS, P. J., AND I. A. E. ATKINSON. 1984. Predation on seabirds by introduced animals and factors affecting its severity. Pp. 667–690 *in* Status and Conservation of the World's Seabirds (J. P. Croxall, P. G. H. Evans, and R. W. Schreiber, Eds.). International Council for Bird Preservation (Tech. Publ. 2). Cambridge, England.

MUSGRAVE, P. S., J. A. FLYNN-O'BRIEN, P. A. LAMBERT, A. A. SMITH, AND Y. D. MARINAKIS. 1998. Federal Wildlife Laws Handbook with Related Laws. Government Institutes, Rockville, MD.

MUSGRAVE, P. S., AND M. A. STEIN. 1993. State Wildlife Laws Handbook. Government Institutes, Rockville, MD.

NETTLESHIP, D. N., J. BURGER, AND M. GOCHFELD (Eds.). 1994. Seabirds on Islands: Threats, Case Studies and Action Plans. BirdLife International, Cambridge, England.

NISBET, I. C. T. 2000. Disturbance, habituation, and management of waterbird colonies. Waterbirds 23: 312–332.

PAGE, G. W., H. R. CARTER, AND R. G. FORD. 1990. Numbers of seabirds killed or debilitated in the 1986 *Apex Houston* oil spill in central California. Studies in Avian Biology 14: 164–174.

PEARL, M. C. 1989. The human side of conservation. Pp. 221–231 *in* Conservation for the Twenty-First Century (D. Western and M. Pearl, Eds.). Wildlife Conservation International, New York.

PODOLSKY, R. H., AND S. KRESS. 1989. Factors affecting colony formation in Leach's Storm-Petrel. Auk 106: 332–336.

PODOLSKY, R. H., AND S. KRESS. 1991. Attraction of the endangered dark-rumped petrel to recorded vocalizations in the Galapagos Islands. Condor 94: 448–453.

POPULATION REFERENCE BUREAU (PRB). 2000. 2000 world population data sheet. Population Reference Bureau, Washington, D.C.

REED, J. R., J. L. SINCOCK, AND J. P. HAILMAN. 1985. Light attraction in endangered procellariiform birds: reduction by shielding upward radiation. Auk 102: 377–383.

REMSEN, J. V., JR. 1995. The importance of continued collecting of bird specimens to ornithology and bird conservation. Bird Conservation International 5: 177–212.

RODDA, G. H., T. H. FRITTS, M. J. MCCOID, AND E. W. CAMPBELL, III. 1998. An overview of the biology of the Brown Treesnake, *Boiga irregularis*, a costly introduced pest on Pacific islands *in* Problem Snake Management: Habu and Brown Treesnake Examples (G. H. Rodda, Y. Sawani, D. Chiszar, and H. Tanaka, Eds.). Cornell University Press, Ithaca, NY.

RODWAY, M. S., W. A. MONTEVECCHI, AND J. W. CHARDINE. 1996. Effects of investigator disturbance on breeding success of Atlantic Puffins *Fratercula arctica*. Biological Conservation 76: 311–319.

SCHREIBER, E. A. 1994. El Niño–Southern Oscillation effects on chick provisioning and growth in red-tailed tropicbirds. Colonial Waterbirds 17: 105–119.

SCHREIBER, E. A.1996. Experimental manipulation of feeding in Red-tailed Tropicbird chicks. Colonial Waterbirds 19: 45–55.

SCHREIBER, E. A. 1999. Breeding biology and ecology of the seabirds of Johnston Atoll, Central Pacific Ocean. Report to the Dept. of Defense, Aberdeen, MD.

SCHREIBER, E. A. 2000. The vital role of research and museum collections in the conservation of seabirds. Pp. 126–133 *in* Status and Conservation of West Indian Seabirds (E. A. Schreiber and D. S. Lee, Eds.). Society of Caribbean Ornithology, Special Publication No. 1, Ruston, LA.

SCHREIBER, E. A., AND D. S. LEE. (EDS). 2000. Status and Conservation of West Indian Seabirds. Society of Caribbean Ornithology, Special Publication No. 1, Ruston, LA.

SCHWARTZ, S. J. 1994. Ecological ramifications of historic occupation of San Nicolas Island. Pp. 171–180 *in* The Fourth California Islands Symposium: Update on the Status of Resources (W. L. Halvorson and G. J. Maender, Eds.). Santa Barbara Museum of Natural History, Santa Barbara, CA.

SHELTON, P. A., R. J. M. CRAWFORD, J. COOPER, AND R. K. BROOKE. 1984. Distribution, population size and conservation of the Jackass Penguin (*Spheniscus demersus*). South African Journal of Marine Science 2: 217–257.

SIMEONE, A., AND R. P. SCHLATTER. 1998. Threats to a mixed species colony of *Spheniscus* penguins in Southern Chile. Colonial Waterbirds 21: 418–421.

SOULÉ, M. E. 1991. Conservation: Tactics for a constant crisis. Science 253: 744–750.

SPARKS, J., AND T. SOPER. 1987. Penguins. Facts on File Publications, New York.

SPENNEMANN, D. H. R. 1998. Excessive exploration of central Pacific seabird populations at the turn of the 20th century. Marine Ornithology 26: 49–57.

STEADMAN, D. W. 1995. Prehistoric extinctions of Pacific Island birds: biodiversity meets zooarchaeology. Science 267: 1123–1130.

STEADMAN, D. W. 1997. Human caused extinction of birds. Pp. 139–161 *in* Biodiversity II: Understanding and Protecting Our Biological Resources (M. L. Reaka-Kudla, D. E. Wilson, and E. O. Wilson, Eds.). Joseph Henry Press, Washington, D.C.

TAYLOR, R. H., AND J. A. V. TILLY. 1984. Stoats (*Mustela erminea*) on Adele and Fisherman Island, Abel Tasman National Park, and other offshore islands in New Zealand. New Zealand Journal of Wildlife Management 34: 372–382.

TAYLOR, R. H., G. W. KAISER, AND M. C. DREVER. 2000. Eradication of Norway rats for recovery of seabird habitat on Langara Island, British Columbia. Restoration Ecology 8: 151–160.

TERSHEY, B. R., D. BREESE, AND D. A. CROLL. 1997. Human perturbations and conservation strategies for San Pedro Martir Island, Islas del Golfo de California Reserve, Mexico. Environmental Conservation 24: 261–270.

TOWNS, D. R., D. SIMBERLOFF, AND I. A. E. ATKINSON. 1997. Restoration of New Zealand islands: redressing the effects of introduced species. Pacific Conservation Biology 3: 99–124.

UNITED NATIONS ENVIRONMENT PROGRAMME. 1997. UNEP Environmental Law Training Manual. United Nations Environment Programme, Nairobi.

VAN DRIESCHE, J., AND R. VAN DRIESCHE. 2000. Nature Out of Place: Biological Invasions in the Global Age. Island Press, Washington, D.C.

VAN HALEWYN, R., AND R. L. NORTON. 1984. The status and conservation of seabirds in the Caribbean. ICBGP Technical Publication 2: 169–222.

VAN HEIJNSBERGEN, P. 1997. International Legal Protection of Wild Fauna and Flora. IOS Press, Amsterdam.

VITOUSEK, D. M., H. A. MOONEY, J. LUBCHENCO, AND J. M. MELILLO. 1997. Human domination of Earth's ecosystem. Science 277: 494–499.

WILCOVE, D. S. 1999. The Condor's Shadow: The Loss and Recovery of Wildlife in America. W. H. Freeman and Co., New York.

WILCOVE, D. S., D. ROTHSTEIN, J. DUBOW, A. PHILLIPS, AND E. LOSOS. 1998. Quantifying threats to imperiled species in the United States. BioScience 48: 607–615.

WILLIAMS, T. D. 1995. The Penguins: *Spheniscidae*. Oxford University Press, New York.

WINGATE, D. B. 1978. Excluding competitors from Bermuda Petrel nesting burrows. *In* Endangered Birds: Management Techniques for Preserving Threatened Species (S. A. Temple, Ed.), University of Wisconsin Press, Madison.

WOEHLER, E. J., D. J. SLIP, L. M. ROBERTSON, P. J. FULLAGAR, AND H. R. BURTON. 1991. The distribution, abundance and status of Adélie Penguins *Pygoscelis adeliae* at the Windmill Islands, Wilkes Land, Antarctica. Marine Ornithology. 19: 1–18.

WOFFORD, C. 2000. A greener future at the WTO: the refinement of WTO jurisprudence on environmental exceptions to GATT. Harvard Environmental Law Review 24: 563–591.

WOODRUFF, D. S. 1989. The problems of conserving genes and species. Pp. 76–88 *in* Conservation for the Twenty-First Century (D. Western and M. C. Pearl, Eds.). Oxford University Press, Oxford.

WORLD RESOURCES INSTITUTE (WRI). 1996. World Resources: A Guide to the Global Environment: The Urban Environment 1996–1997. Oxford University Press, Oxford.

YORIO, P., AND P. D. BOERSMA. 1992. The effects of human disturbance on Magellanic Penguin behavior and breeding success. Bird Conservation International 2: 161–173.

YORIO, P., AND F. QUINTANA. 1996. Efectos del disturbio humano sobre una colonia mixta de aves marinas en Patagonia. El Hornero 14: 89–96.

YORIO, P., E. FRERE, P. GANDINI, AND W. CONWAY. 1999. Status and conservation of seabirds breeding in Argentina. Bird Conservation International 9: 299–314.

Sanderlings on a Windy Shore

18 Shorebirds in the Marine Environment

Nils Warnock, Chris Elphick, and Margaret A. Rubega

CONTENTS

18.1 INTRODUCTION

The purpose of this chapter is to review the ecology of shorebirds in the context of their relationship to the marine environment. The shorebirds are a group of families usually placed in the order Charadriiformes along with gulls, skuas, terns, skimmers, and auks (e.g., del Hoyo et al. 1996, American Ornithologists' Union 1998). An alternative view, based on studies of DNA-DNA hybridization, is to treat this entire group as a suborder within the order Ciconiiformes (Sibley and Monroe 1990). The shorebirds are traditionally thought of as a monophyletic group, although this may not be the case and the relationships among the families within the order remain uncertain (American Ornithologists' Union 1998). Depending on taxonomic source, shorebirds are variously divided into about a dozen families. Three of these are of particular significance to this discussion: the Scolopacidae (sandpipers, snipes, and phalaropes) and Charadriidae (plovers) which include 71% of all shorebird species, and the Haematopodidae (oystercatchers), with 11 species found largely in marine environments (Table 18.1). Even though their name conveys an affinity to water, shorebirds are not traditionally considered marine birds (Burger 1984a). However, the taxonomic, ecological, and behavioral characteristics of the two groups indicate much in common, and shorebirds (especially the phalaropes) have many traits that suit them for a life on or near salt water.

Overall, 58% of shorebird species are known to use marine habitats regularly, either during breeding or nonbreeding seasons (Table 18.1). Thirty-nine percent of breeding shorebirds sometimes or always nest along the coast, while 66% of nonbreeding shorebirds use the coast for stopovers or nonbreeding grounds (Burger 1984a). The majority of shorebirds migrate (62%, Table 18.1), and most of these birds cross marine bodies. Two shorebird species, the Red-necked Phalarope (*Phalaropus lobatus*) and the Red Phalarope (*P. fulicaria*), spend up to 75% of their time directly on the open ocean, more than many species traditionally referred to as seabirds.

18.2 GENERAL FEATURES OF SHOREBIRD BIOLOGY

18.2.1 FORAGING

On land and at sea, shorebirds tend to be omnivorous, although invertebrates are their dominant prey. In a survey of shorebird diets in the Western Hemisphere, the most common taxonomic classes of invertebrate prey eaten were the Insecta, Malacostraca, Gastropoda, Polychaeta, and Bivalvia (Skagen and Oman 1996), but other important prey for shorebirds include small amphibians, fishes, seeds, and fruit. The four species of seedsnipes apparently only eat plant matter (Fjeldså 1996), while at the other extreme sheathbills often eat carrion and small penguin chicks (Burger 1996).

Shorebirds typically obtain their prey by locating it visually and plucking it from the water column, ground, or other surfaces, or by probing in mud. There is considerable interspecific and occasional intraspecific variability in bill morphology (Burton 1974, Sutherland et al. 1996, Rubega 1997) that results in a wide variety of feeding habits. Many species have straight bills for making rapid thrusts through soft substrates or firm soils or for picking prey off the water's surface (see also Pelagic Feeding Biology of Phalaropes). The curved bills of species like the Long-billed Curlew (*Numenius americanus*) and the Whimbrel (*N. phaeopus*) are similar in shape to the burrows of invertebrates such as ghost shrimps (*Callianassa californiensis*) and probably facilitate capture of these prey species. Many scolopacids have bills with tactile and chemosensitive receptors at their

TABLE 18.1
Shorebird Families of the World, Whether They Migrate, and Their Use of Marine Habitat

Common Name		Migratory	Marine
Thinocoridae			
Rufous-bellied Seedsnipe	*Attagis gayi*	No	No
White-bellied Seedsnipe	*A. malouinus*	No	No
Grey-breasted Seedsnipe	*Thinocorus orbignyianus*	Unknown	No
Least Seedsnipe	*T. rumicivorus*	Yes	No
Pedionomidae			
Plains-wanderer	*Pedionomus torquatus*	No	No
Scolopacidae			
Eurasian Woodcock	*Scolopax rusticola*	Yes	No
Amami Woodcock	*S. mira*	No	No
Rufous Woodcock	*S. saturata*	No	No
Sulawesi Woodcock	*S. celebensis*	No	No
Moluccan Woodcock	*S. rochussenii*	No	No
American Woodcock	*S. minor*	Yes	No
Solitary Snipe	*Gallinago solitaria*	Yes	Unknown
Latham's Snipe	*G. hardwickii*	Yes	Unknown
Wood Snipe	*G. nemoricola*	Yes	No
Pintail Snipe	*G. stenura*	Yes	Unknown
Swinhoe's Snipe	*G. megala*	Yes	No
Great Snipe	*G. media*	Yes	Unknown
Common Snipe	*G. gallinago*	Yes	Yes
African Snipe	*G. nigripennis*	No	Unknown
Madagascar Snipe	*G. macrodactyla*	No	Unknown
South American Snipe	*G. paraguaiae*	Yes	No
Noble Snipe	*G. nobilis*	No	No
Giant Snipe	*G. undulata*	Unknown	No
Andean Snipe	*G. jamesoni*	No	No
Fuegian Snipe	*G. stricklandii*	Unknown	Unknown
Imperial Snipe	*G. imperialis*	No	No
Jack Snipe	*Lymnocryptes minimus*	Yes	No
Chatham Snipe	*Coenocorypha pusilla*	No	No
Subantarctic Snipe	*C. aucklandica*	No	Unknown
Black-tailed Godwit	*Limosa limosa*	Yes	Yes
Hudsonian Godwit	*L. haemastica*	Yes	Yes
Bar-tailed Godwit	*L. lapponica*	Yes	Yes
Marbled Godwit	*L. fedoa*	Yes	Yes
Little Curlew	*Numenius minutus*	Yes	Yes
Eskimo Curlew	*N. borealis*	Yes	Yes
Whimbrel	*N. phaeopus*	Yes	Yes
Bristle-thighed Curlew	*N. tahitiensis*	Yes	Yes
Slender-billed Curlew	*N. tenuirostris*	Yes	Yes
Eurasian Curlew	*N. arquata*	Yes	Yes
Long-billed Curlew	*N. americanus*	Yes	Yes
Far Eastern Curlew	*N. madagascariensis*	Yes	Yes
Upland Sandpiper	*Bartramia longicauda*	Yes	No
Spotted Redshank	*Tringa erythropus*	Yes	Yes
Common Redshank	*T. totanus*	Yes	Yes
Marsh Sandpiper	*T. stagnatilis*	Yes	Yes
Common Greenshank	*T. nebularia*	Yes	Yes

TABLE 18.1 *(Continued)*
Shorebird Families of the World, Whether They Migrate, and Their Use of Marine Habitat

Common Name		Migratory	Marine
Nordmann's Greenshank	*T. guttifer*	Yes	Yes
Greater Yellowlegs	*T. melanoleuca*	Yes	Yes
Lesser Yellowlegs	*T. flavipes*	Yes	Yes
Solitary Sandpiper	*T. solitaria*	Yes	No
Green Sandpiper	*T. ochropus*	Yes	No
Wood Sandpiper	*T. glareola*	Yes	No
Terek Sandpiper	*Xenus cinereus*	Yes	Yes
Common Sandpiper	*Actitis hypoleucos*	Yes	Yes
Spotted Sandpiper	*A. macularia*	Yes	Yes
Grey-tailed Tattler	*Heteroscelus brevipes*	Yes	Yes
Wandering Tattler	*H. incanus*	Yes	Yes
Willet	*Catoptrophorus semipalmatus*	Yes	Yes
Tuamotu Sandpiper	*Prosobonia cancellata*	No	Yes
Ruddy Turnstone	*Arenaria interpres*	Yes	Yes
Black Turnstone	*A. melanocephala*	Yes	Yes
Short-billed Dowitcher	*Limnodromus griseus*	Yes	Yes
Long-billed Dowitcher	*L. scolopaceus*	Yes	Yes
Asian Dowitcher	*L. semipalmatus*	Yes	Yes
Surfbird	*Aphriza virgata*	Yes	Yes
Great Knot	*Calidris tenuirostris*	Yes	Yes
Red Knot	*C. canutus*	Yes	Yes
Sanderling	*C. alba*	Yes	Yes
Semipalmated Sandpiper	*C. pusilla*	Yes	Yes
Western Sandpiper	*C. mauri*	Yes	Yes
Little Stint	*C. minuta*	Yes	Yes
Red-necked Stint	*C. ruficollis*	Yes	Yes
Temminck's Stint	*C. temminckii*	Yes	Yes
Long-toed Stint	*C. subminuta*	Yes	Yes
Least Sandpiper	*C. minutilla*	Yes	Yes
White-rumped Sandpiper	*C. fuscicollis*	Yes	Yes
Baird's Sandpiper	*C. bairdii*	Yes	Yes
Pectoral Sandpiper	*C. melanotos*	Yes	Yes
Sharp-tailed Sandpiper	*C. acuminata*	Yes	Yes
Purple Sandpiper	*C. maritima*	Yes	Yes
Rock Sandpiper	*C. ptilocnemis*	Yes	Yes
Dunlin	*C. alpina*	Yes	Yes
Curlew Sandpiper	*C. ferruginea*	Yes	Yes
Stilt Sandpiper	*Micropalama himantopus*	Yes	No
Buff-breasted Sandpiper	*Tryngites subruficollis*	Yes	No
Spoon-billed Sandpiper	*Eurynorhynchus pygmeus*	Yes	Yes
Broad-billed Sandpiper	*Limicola falcinellus*	Yes	Yes
Ruff	*Philomachus pugnax*	Yes	Yes
Wilson's Phalarope	*Phalaropus tricolor*	Yes	No
Red-necked Phalarope	*P. lobatus*	Yes	Yes
Red Phalarope	*P. fulicaria*	Yes	Yes
Rostratulidae			
Greater Painted-snipe	*Rostratula benghalensis*	Yes	Unknown
South American Painted-snipe	*Nycticryphes semicollaris*	Unknown	No
Jacanidae			
African Jacana	*Actophilornis africanus*	No	No

TABLE 18.1 *(Continued)*
Shorebird Families of the World, Whether They Migrate, and Their Use of Marine Habitat

Common Name		Migratory	Marine
Madagascar Jacana	*A. albinucha*	No	No
Lesser Jacana	*Microparra capensis*	No	No
Comb-crested Jacana	*Irediparra gallinacea*	Unknown	No
Pheasant-tailed Jacana	*Hydrophasianus chirurgus*	Yes	No
Bronze-winged Jacana	*Metopidius indicus*	No	No
Northern Jacana	*Jacana spinosa*	No	No
Wattled Jacana	*J. jacana*	No	No
Chionidae			
Pale-faced Sheathbill	*Chionis alba*	Yes	Yes
Black-faced Sheathbill	*C. minor*	No	Yes
Burhinidae			
Stone Curlew	*Burhinus oedicnemus*	Yes	No
Senegal Thick-knee	*B. senegalensis*	No	No
Water Dikkop	*B. vermiculatus*	No	Yes
Spotted Dikkop	*B. capensis*	No	No
Double-striped Thick-knee	*B. bistriatus*	No	No
Peruvian Thick-knee	*B. superciliaris*	No	No
Bush Thick-knee	*B. grallarius*	No	Yes
Great Thick-knee	*Esacus recurvirostris*	No	Yes
Beach Thick-knee	*E. giganteus*	No	Yes
Haematopodidae			
Eurasian Oystercatcher	*Haematopus ostralegus*	Yes	Yes
Canarian Black Oystercatcher	*H. meadewaldoi*	Unknown	Yes
African Black Oystercatcher	*H. moquini*	No	Yes
American Black Oystercatcher	*H. bachmani*	Yes	Yes
American Oystercatcher	*H. palliatus*	Yes	Yes
Australian Pied Oystercatcher	*H. longirostris*	No	Yes
Variable Oystercatcher	*H. unicolor*	No	Yes
Sooty Oystercatcher	*H. fuliginosus*	No	Yes
Blackish Oystercatcher	*H. ater*	No	Yes
Magellanic Oystercatcher	*H. leucopodus*	Unknown	Yes
Chatham Oystercatcher	*H. chathamensis*	No	Yes
Ibidorhynchidae			
Ibisbill	*Ibidorhyncha struthersii*	No	No
Recurvirostridae			
Black-winged Stilt	*Himantopus himantopus*	Yes	Yes
Black Stilt	*H. novaezelandiae*	No	Unknown
Banded Stilt	*Cladorhynchus leucocephalus*	Yes	Yes
Pied Avocet	*Recurvirostra avosetta*	Yes	Yes
American Avocet	*R. americana*	Yes	Yes
Red-necked Avocet	*R. novaehollandiae*	No	Yes
Andean Avocet	*R. andina*	Unknown	Yes
Charadriidae			
Eurasian Golden Plover	*Pluvialis apricaria*	Yes	Yes
Pacific Golden Plover	*P. fulva*	Yes	Yes
American Golden Plover	*P. dominica*	Yes	Yes
Grey (Black-bellied) Plover	*P. squatarola*	Yes	Yes

TABLE 18.1 *(Continued)*
Shorebird Families of the World, Whether They Migrate, and Their Use of Marine Habitat

Common Name		Migratory	Marine
Red-breasted Plover	*Charadrius obscurus*	Yes	Yes
Common Ringed Plover	*C. hiaticula*	Yes	Yes
Semipalmated Plover	*C. semipalmatus*	Yes	Yes
Long-billed Plover	*C. placidus*	Yes	Yes
Little Ringed Plover	*C. dubius*	Yes	Yes
Wilson's Plover	*C. wilsonia*	Yes	Yes
Killdeer	*C. vociferus*	Yes	Yes
Black-banded Plover	*C. thoracicus*	No	Yes
St Helena Plover	*C. sanctaehelenae*	No	No
Kittlitz's Plover	*C. pecuarius*	Unknown	Yes
Three-banded Plover	*C. tricollaris*	Unknown	Yes
Forbes's Plover	*C. forbesi*	Yes	No
Piping Plover	*C. melodus*	Yes	Yes
Chestnut-banded Plover	*C. pallidus*	Yes	Yes
Kentish (Snowy) Plover	*C. alexandrinus*	Yes	Yes
White-fronted Plover	*C. marginatus*	Yes	Yes
Red-capped Plover	*C. ruficapillus*	Unknown	Yes
Malaysian Plover	*C. peronii*	No	Yes
Javan Plover	*C. javanicus*	No	Unknown
Collared Plover	*C. collaris*	No	Yes
Double-banded Plover	*C. bicinctus*	Yes	Yes
Puna Plover	*C. alticola*	Unknown	Yes
Two-banded Plover	*C. falklandicus*	Yes	Yes
Lesser Sandplover	*C. mongolus*	Yes	Yes
Greater Sandplover	*C. leschenaultii*	Yes	Yes
Caspian Plover	*C. asiaticus*	Yes	No
Oriental Plover	*C. veredus*	Yes	Yes
Eurasian Dotterel	*C. morinellus*	Yes	No
Mountain Plover	*C. montanus*	Yes	No
Rufous-chested Plover	*C. modestus*	Yes	Yes
Hooded Plover	*C. rubricollis*	No	Yes
Shore Plover	*C. novaeseelandiae*	No	Yes
Red-kneed Dotterel	*Erythrogonys cinctus*	Unknown	No
Tawny-throated Dotterel	*Oreopholus ruficollis*	Yes	Unknown
Wrybill	*Anarhynchus frontalis*	Yes	Yes
Diademed Plover	*Phegornis mitchellii*	Unknown	No
Inland Dotterel	*Peltohyas australis*	Unknown	Unknown
Black-fronted Dotterel	*Elseyornis melanops*	Unknown	No
Magellanic Plover	*Pluvianellus socialis*	Yes	Yes
Northern Lapwing	*Vanellus vanellus*	Yes	Yes
Long-toed Lapwing	*V. crassirostris*	No	No
Yellow-wattled Lapwing	*V. malarbaricus*	No	No
Javanese Wattled Lapwing	*V. macropterus*	Unknown	No
Banded Lapwing	*V. tricolor*	Unknown	No
Masked Lapwing	*V. miles*	Yes	Yes
Blacksmith Lapwing	*V. armatus*	No	Yes
Spur-winged Lapwing	*V. spinosus*	Yes	Yes
River Lapwing	*V. duvaucelii*	No	No
Black-headed Lapwing	*V. tectus*	No	No
Spot-breasted Lapwing	*V. melanocephalus*	No	No

TABLE 18.1 *(Continued)*
Shorebird Families of the World, Whether They Migrate, and Their Use of Marine Habitat

Common Name		Migratory	Marine
Grey-headed Lapwing	*V. cinereus*	Yes	No
Red-wattled Lapwing	*V. indicus*	Yes	No
White-headed Lapwing	*V. albiceps*	Yes	Unknown
African Wattled Lapwing	*V. senegallus*	Unknown	No
Lesser Black-winged Lapwing	*V. lugubris*	Yes	No
Greater Black-winged Lapwing	*V. melanopterus*	Yes	Yes
Crowned Lapwing	*V. coronatus*	No	No
Brown-chested Lapwing	*V. superciliosus*	Yes	No
Sociable Lapwing	*V. gregarius*	Yes	Unknown
White-tailed Lapwing	*V. leucurus*	Yes	Unknown
Pied Lapwing	*V. cayanus*	No	Yes
Southern Lapwing	*V. chilensis*	Unknown	Unknown
Andean Lapwing	*V. resplendens*	No	No
Dromadidae			
Crab Plover	*Dromas ardeola*	Yes	Yes
Glareolidae			
Egyptian Plover	*Pluvianus aegyptius*	Yes	Yes
Double-banded Courser	*Smutsornis africanus*	No	No
Bronze-winged Courser	*Rhinoptilus chalcopterus*	Yes	No
Three-banded Courser	*R. cinctus*	No	No
Jerdon's Courser	*R. bitorquatus*	No	No
Cream-colored Courser	*Cursorius cursor*	Yes	No
Burchell's Courser	*C. rufus*	No	No
Temminck's Courser	*C. temminckii*	Yes	No
Indian Courser	*C. coromandelicus*	No	No
Collared Pratincole	*Glareola pratincola*	Yes	Yes
Oriental Pratincole	*G. maldivarum*	Yes	Yes
Black-winged Pratincole	*G. nordmanni*	Yes	Unknown
Madagascar Pratincole	*G. ocularis*	Yes	Yes
Rock Pratincole	*G. nuchalis*	Unknown	Yes
Grey Pratincole	*G. cinerea*	No	Yes
Small Pratincole	*G. lactea*	Yes	Yes
Australian Pratincole	*Stiltia isabella*	Yes	Yes

Note: Under Migrate, Yes = species known to migrate regularly (note that if part of a species migrates and part does not, the species would be listed under Yes), No = species known not to migrate, and Unknown = species for which it is unclear if the species is migratory. Under Marine, Yes = species known to use marine habitat regularly (marine habitat defined as beginning with coastal estuaries, mudflats, and other types of marine shoreline extending into the pelagic zone), No = species not known to use marine habitat regularly, and Unknown = species for which it is unclear if the species regularly uses marine habitat.

Information summarized from del Hoyo et al. (1996) including accounts by Baker-Gabb (1996), Burger (1996), Fjeldså (1996), Hockey (1996), Hume (1996), Jenni (1996), Kirwan (1996), Knystautas (1996), Maclean (1996), Pierce (1996), Rands (1996), van Gils and Wiersma (1996), and Wiersma (1996).

tips — thus prey can be found by touch and smell (von Bolze 1968, Heezik et al. 1983), and even pressure gradients (Piersma et al. 1998). Species with these adaptations tend to feed both during the day and at night (Warnock and Gill 1996, van Gils and Piersma 1999).

18.2.2 SOCIALITY

Generally, shorebirds are gregarious when not breeding and territorial during the breeding season. However, gregarious nonbreeding shorebirds will often vigorously defend small feeding territories, abandoning them to rejoin flocks when tides cover feeding areas or predators appear (Myers et al. 1979). Some species are colonial breeders, especially members of the Recurvirostridae (avocets and stilts), Dromadidae (crab-plovers), Chionidae (sheathbills), and Glareolidae (coursers and pratincoles) families (Burger 1996, Maclean 1996, Pierce 1996, Rands 1996). In an extreme example, perhaps the entire population of Banded Stilts (*Cladorhynchus leucocephalus*) in eastern Australia, up to 100,000 birds, will attempt to breed at one inland lake (Alcorn and Alcorn 2000). This event happens when the normally dry interior alkali lakes of the region receive rain, creating breeding habitat and stimulating blooms of invertebrates.

18.2.3 BREEDING SYSTEMS

Breeding systems vary considerably among shorebird species. Most shorebirds are monogamous, with individuals forming a pair bond with just one individual each breeding season and both parents caring for the young (Emlen and Oring 1977, Oring and Lank 1984). In many monogamous species, pair bonds are strong and often persist from year to year (e.g., Eurasian Oystercatchers *Haematopus ostralegus*, Ens et al. 1996; Semipalmated Sandpipers *Calidris pusilla*, Sandercock 1997).

Polygyny, in which some males mate with more than one female within a single breeding season, is found in at least 25 species, most of which are sandpipers, snipes, and woodcocks (Emlen and Oring 1977, Oring and Lank 1984). In a few of these species, birds gather at leks, where males display to females from small, vigorously defended territories (Hoglünd and Alatalo 1995). Copulation does not involve pair-bonding and males play no role in parental care. Of shorebirds with lek behavior, the best known are the Buff-breasted Sandpiper (*Tryngites subruficollis*, Lanctot and Laredo 1994), the Great Snipe (*Gallinago media*, Hoglünd and Alatalo 1995), and the Ruff (*Philomachus pugnax*, Van Rhijn 1991), all northern latitude breeders.

Polyandry, in which a female mates with multiple males (Emlen and Oring 1977), occurs in the phalaropes, all jacanas except for the Lesser Jacana (*Microparra capensis*), and in some plovers, painted-snipes, and sandpipers (Baker-Gabb 1996, Jenni 1996, van Gils and Wiersma 1996, Wiersma 1996). The males of these species generally incubate the eggs and raise the young. A few species practice rapid multiple clutch polygamy in which males and females have access to multiple mates within a season, and each may simultaneously incubate separate clutches. Perhaps the best-documented case of this occurs with the Temminck's Stint (*Calidris temminckii*), where both males and females exhibit multiclutch behavior with multiple mates within the breeding season (Hildén 1975, Breiehagen 1989). Often the breeding system varies among individuals within a species. For example, many individuals of nominally "polygamous" species may mate monogamously, and polygamy may occur in some species that are generally monogamous. Males and females of the monogamous Eurasian Oystercatcher often engage in extra-pair copulations, although DNA fingerprinting has shown that few chicks (1 of 65 chicks) are not actually fathered by the dominant male partner (Heg et al. 1993, in Ens et al. 1996).

18.2.4 NESTS, EGGS, AND YOUNG

The typical shorebird nest is a bowl-like scrape in the ground (often near water) that is lined with pebbles, shells, grasses, or leaves. A few species, such as the Solitary Sandpiper (*Tringa solitaria*) in North America and the Wood Sandpiper (*T. glareola*) in Eurasia, are tree-nesters. These birds

FIGURE 18.1 An adult American Oystercatcher pries a limpet off a rock while its chick waits to be fed. Note, the oystercatcher has broken a piece off the limpet shell to insert the blade-like, laterally flattened bill tip. (Drawing by J. Zickefoose.)

do not build nests, but instead use abandoned nests of passerines. Some species of plovers that breed in hot environments such as Africa's White-fronted Plover (*Charadrius marginatus*) and Australia's Inland Dotterel (*Peltohyas australis*) cover their nests with sand, probably to regulate temperatures and hide them from predators (Wiersma 1996). Sheathbills also often lay nests in caves, crevices, and petrel burrows to avoid having their nests depredated by skuas or trampled by penguins (Burger 1996).

Shorebirds lay one to four pyriform eggs (most lay four-egg clutches) that are extremely well camouflaged in shades of off-white, buff, and olive, marked with black or brown splotches (Harrison 1978, Cramp and Simmons 1983). Incubation periods of nesting shorebirds last 15 to 40 days depending on the species and location. Hatching usually takes 12 to 48 h from hole-pipping to actual hatch, with all chicks leaving their eggs within 24 h of each other. Most shorebird chicks are precocial (sheathbills hatch asynchronously and are semiprecocial; Burger 1996) and covered with down at hatching. Generally, they leave the nest within a day or two (sometimes hours) of hatching to forage with at least one parent (often the male), and are brooded by the parents for at least the first few days after hatching. Most shorebird chicks feed on their own after hatching, but there are exceptions. Oystercatcher and sheathbill chicks rely on their parents for food until they fledge (Burger 1996, Hockey 1996, Safriel et al. 1996; Figure 18.1). The Magellanic Plover (*Pluvianellus socialis*) is unique among shorebirds in that parents apparently regurgitate food from their crop to chicks until after the chicks fledge (Jehl 1975). The Burhinidae (thick-knees, Hume 1996), Glareolidae (coursers and pratincoles, MacLean 1996), Dromadidae (crab plovers, Rands 1996), and some species — snipes and woodcocks — of Scolopacidae (sandpipers, snipes, and phalaropes, Piersma 1996b) provide food for their young for about the first week of life.

As with incubation, the timing of fledging varies among shorebird species. Smaller sandpipers and plovers fledge at 14 to 26 days; larger sandpipers and plovers fledge at 28 to 45 days, while some jacanas, oystercatchers, thick-knees, and stilts may take 50 days or more (del Hoyo et al. 1996). Many species breed in their first spring (at approximately 1 year of age); some (especially larger species) do not mature sexually until 2 to 5 years of age. In species where nesting habitat is limited, individuals may have to wait several years before acquiring high-quality breeding sites. In some Eurasian Oystercatchers, it may take up to 10 years before a bird is able to successfully get a mate and a breeding territory (Ens et al. 1996).

18.2.5 SURVIVAL AND LONGEVITY

Annual adult survival rates of shorebirds typically range from 60 to 70% in small species and 85 to 95% in larger species; survivorship of shorebirds in their first year is often less than 50% (Evans and Pienkowski 1984, Evans 1991, Jackson 1994, Sandercock and Gratto-Trevor 1997, Warnock et al. 1997, Reed et al. 1998). Shorebirds are relatively long lived at 4 to 10 years, with some individuals surviving for 20 years or more. In one amazing example, a Eurasian Oystercatcher that was banded as a nestling in 1949 was killed by a Eurasian Sparrowhawk (*Accipiter nisus*) in 1992 at the age of 43 years and 6 months (Exo 1993). Even small sandpipers can be long lived as evidenced by a female Least Sandpiper (*Calidris minutilla*, the world's smallest shorebird with a mass of 19 to 25 g) that was observed breeding at a minimum age of 16 years (Miller and McNeil 1988, Cooper 1994).

18.3 SHOREBIRDS AT THE OCEAN–CONTINENT INTERFACE

The lives of many shorebird species are intimately connected to the ocean, especially at the boundary between land and sea. Shorebirds use a wide variety of coastal habitats, ranging from rocky surf-battered shorelines, to mangrove swamps and sheltered coastal bays. Some species breed in coastal areas, but the majority use these habitats primarily during the nonbreeding season. Because of the relatively low freezing point of salt water, littoral environments often provide accessible food despite cold weather. Also, ambient temperatures in coastal areas frequently are warmer than sites farther inland. Consequently, coastal wetlands and the ocean shore continue to be desirable wintering grounds long after interior wetlands at similar latitudes have become unsuitable for foraging shorebirds.

18.3.1 COASTAL HABITATS

18.3.1.1 Coastal Wetlands

Coastal wetlands include some of the most productive habitats in the world, and shorebirds are found in virtually every kind of coastal wetland. The nontidal portions of saltmarshes and coastal lagoons provide breeding and foraging habitat for many species. Tidal mudflats are home to foraging flocks, which can number into the hundreds of thousands of birds. Tidal marsh breeders include a few species of oystercatchers and some tringine sandpipers, such as Common Redshank (*Tringa totanus*) and Willet (*Catoptrophorus semipalmatus*). Coastal lagoons support a wider variety of breeding species, including avocets, stilts, phalaropes, and plovers.

Shorebirds commonly interact with the marine environment on estuarine mudflats, where vast numbers gather during migration and winter in some parts of the world. For example, 3 to 4 million Western Sandpipers (*Calidris mauri*) may stop at the Copper River Delta in Alaska during a 4-week period in early spring (Bishop et al. 2000). These birds are en route to their tundra nesting grounds and join millions of other shorebirds also stopping at the delta (Isleib 1979). During the winter, over 2 million shorebirds use the vast tidal flats of the Banc d'Arguin in Mauritania, western Africa (Wolff and Smit 1990). In Asia, West Africa, Central and South America, and other tropical areas, mangroves and their associated mudflats are important habitats for shorebirds (Hepburn 1987, Parish et al. 1987, Morrison et al. 1998). Estuaries are highly productive and shorebirds take advantage of the abundance of soft sediment invertebrate prey that they can find by probing in the mud.

Different shorebird species can be found in subtly different parts of marshes. Larger species, such as curlews and godwits, with their long legs and bills, are capable of feeding in deeper water than small sandpipers and plovers. Some species tend to feed in small flocks at the edge of a marsh where they can pick at the base of small clumps of vegetation, whereas others are found in larger flocks on exposed mudflats. Despite these differences, there is much overlap in habitat use and many birds form mixed species foraging flocks. Birds of different species also come together at high-tide roosts. The latter are often located within the marsh or in nearby adjacent agricultural fields or wetland lagoons. Large foraging and roosting flocks are vulnerable to avian predators, such as falcons, accipiters, and owls (Page and Whiteacre 1975, Cresswell 1996). The amazing sight of a tightly bunched, wheeling flock of *Calidris* sandpipers attempting to evade a hunting Peregrine Falcon (*Falco peregrinus*) can been seen on estuaries throughout the world.

18.3.1.2 Beaches

Sandy beaches provide important breeding habitats for certain *Charadrius* plovers in many areas of the world. The most ubiquitous species, the Snowy Plover (*Charadrius alexandrinus*, known as Kentish Plover outside the Americas), will lay its eggs in little more than a shallow depression in the sand, perhaps lined with a few pebbles, bits of shells, or pieces of vegetation (Page et al. 1995). These nests are exceptionally vulnerable (see Conservation of Marine Shorebirds, below). A number of closely related species, including Wilson's Plover (*C. wilsonia*) in the Americas, White-fronted Plover in Africa, and Malaysian Plover (*C. peronii*) in Southeast Asia, also nest on sandy beaches (Wiersma 1996).

Pebble beaches are used as nesting habitat by Common Ringed Plovers (*C. hiaticula*) in Eurasia and Semipalmated Plovers (*C. semipalmatus*) in North America (Cramp and Simmons 1983, Nol and Blanken 1999). Several oystercatcher species, such as Blackish Oystercatchers (*Haematopus ater*) of South America, will also nest in this habitat (Hockey 1996).

Many beach-nesting species regularly place nests near seaweed, driftwood, or other beach debris (e.g., Piping Plover *Charadrius melodus*, Haig 1990). In this respect, they are similar to some tundra- and taiga-nesting shorebirds, which often build nests near a small tree or shrub (e.g., Greater Yellowlegs *Tringa melanoleuca*, Elphick and Tibbitts 1998). Such locations may provide partial protection from predators by placing the nests in an area with some visual variation, or they may simply serve to help the parents find their nest in a relatively featureless landscape.

Various shorebird species will nest close to colonies of beach-nesting seabirds (usually small terns), sometimes even placing nests amidst the colony (Burger 1987, Burger and Gochfeld 1990, Alleng and Whyte-Alleng 1993). This behavior may offer protection in that terns mob potential predators (Burger 1987). Individuals may gain indirect benefits from nesting in a colony: if their nest is surrounded by other nests there is a higher probability that a predator will encounter a different bird's nest first and become satiated than if their nest were isolated (also see J. C. Coulson, Chapter 4, this book). Colonial nesting, however, is not without disadvantages. Groups typically are more conspicuous than singletons, making them vulnerable to predators. Raising the stakes further, gulls, which often nest near terns, will prey upon shorebird eggs. Some shorebirds turn the tables and prey on the eggs of colonial larids (Crossin and Huber 1970, Burger and Gochfeld 1990, Alberico et al. 1991); Ruddy Turnstones (*Arenaria interpres*) can destroy entire colonies (Loftin and Sutton 1979) and will even steal fish brought in to young Arctic Terns (*Sterna paradisaea*) by their parents (Brearey and Hildén 1985).

Other beach-nesting shorebirds include the Beach Thick-knee (*Esacus magnirostris*) of Australia and Southeast Asia, which feeds by stalking crabs like a heron (Hume 1996), and the unusual Crab Plover (*Dromas ardeola*) of the Middle East. Crab Plovers are the only shorebird to dig nest burrows, which they build in coastal sand dunes (Rands 1996). Presumably, burrowing offers protection from predators and the sun. This species, which is unrelated to other plovers and placed in a monotypic family (Dromadidae), breeds colonially and lays white eggs just like many other hole-nesting birds (Rands 1996).

During the nonbreeding season, various inland-breeding shorebirds occur in beach habitats. These are often tundra-nesting birds, and most do not use beaches in large numbers. Sanderlings (*Calidris alba*) are one exception and are usually found in small groups at the water's edge where they move back and forth with the wave front as they feed on small invertebrates (Cramp and Simmons 1983).

18.3.1.3 Rocky Shores and Coral Reefs

Rocky-shore specialists are particularly concentrated along the highly productive Pacific coast of northwestern North America. This group consists of the American Black Oystercatcher (*Haematopus bachmani*), Rock Sandpiper (*Calidris ptilocnemis*), Wandering Tattler (*Heteroscelus incanus*), Black Turnstone (*Arenaria melanocephala*), and Surfbird (*Aphriza virgata*). Tattlers also are found on tropical islands throughout the Pacific, where they frequent coral reefs and the shores of volcanic islands, and Surfbirds range to southern Chile (van Gils and Wiersma 1996). These species rarely wander away from rocky coastal substrate. Purple Sandpipers (*Calidris maritima*) are a North Atlantic counterpart to Rock Sandpipers, which they resemble closely (Paulson 1993), and rocky shorelines are an important habitat for several oystercatcher species in the Southern Hemisphere (Hockey 1996).

Other species that regularly use rocky shores during the nonbreeding season, but which are by no means restricted to them, include Whimbrel, Ruddy Turnstone, and Grey-tailed Tattler (*Heteroscelus brevipes*, van Gils and Wiersma 1996). In many areas, rocky promontories and islands are used as roosting areas by shorebirds that feed on estuarine mudflats that become inundated at high tide.

In New Zealand, the endangered Shore Plover (*Charadrius novaeseelandiae*) is restricted to rocky shores where it nests in dense vegetation or occasionally in crevices among the boulders (Davis 1994). Several species of oystercatchers also use rocky shores for nesting, and some, such as the American Black Oystercatcher (Andres and Falxa 1995), rarely breed elsewhere. Oystercatchers come in two main color types: species that are predominantly black and species that have bold black-and-white plumage patterning. Where both types co-occur, the species that is predominantly black in color tends to be found in rocky shore habitats, whereas the pied species usually feeds in soft substrates (Hockey 1996). This pattern suggests that all-dark plumages, which are thought to be a derived characteristic (Hockey 1996), might confer an advantage in rocky shore habitats, and dark plumages certainly make oystercatchers difficult to see against basaltic, seaweed-covered rocks. Despite different phylogenetic affinities, many of the other rocky-shore specialists also have dark upperparts during the times of year when they use this habitat, supporting the idea that this coloration might provide benefits. There also are morphological similarities among species found on rocky shores.

18.3.2 INFLUENCE OF TIDES

For many coastal shorebirds, the greatest single influence on their local distribution and behavior is the state of the tide, since water above certain levels covers feeding habitat and alters prey availability (Burger 1984b). Since many shorebirds feed on mudflats that get covered during high tides, habitat use between high and low tides is frequently different. For example, Western Sandpipers at San Francisco Bay primarily use mudflats on low tides and move to seasonal wetlands and salt ponds on high tides (Warnock and Takekawa 1995). Similarly, Northern Lapwing (*Vanellus vanellus*) chicks in Sweden use mudflats at low tides and pastures next to the mudflats on high tides (Johansson and Blomqvist 1996). Many other studies from around the world have noted the same pattern: mudflats are used on low tides; pastures, marshes, mangrove, sand beaches, etc. on high tides (see Burger 1984b for list). In some areas with low tidal amplitude such as the southern Baltic Sea in Germany and Poland and coastal lagoons in southern Brazil, shorebirds rely on wind to expose mudflats (Piersma 1996b).

Foraging behavior is significantly influenced by tides. In Australia, calidridine sandpipers spend a greater proportion of their time feeding in months when mudflat exposure is higher

(December–March, Dann 1999). Prey choice can also change for shorebirds through the tide cycle. At the Dutch Wadden Sea, the bivalve *Macoma* predominated in the diet of Eurasian Oystercatchers during high-falling and high-rising tides, while *Nereis* worms predominated during low tides (De Vlas et al. 1996). Feeding rates of some species can also change during the tide cycle, undoubtedly due to changes in invertebrate behavior (Pienkowski 1981). For instance, feeding rates of Black-bellied Plovers (*Pluvialis squatarola*; also called Grey Plover) increased as the tide fell and then decreased abruptly about 2 h after low tide (Baker 1974). In contrast, no significant differences were detected in foraging rates of Common Redshanks on rising or falling tides (Goss-Custard 1977).

Depending on the timing of the tides, birds may be faced with little daylight during which they can feed on mudflats. This problem is especially acute at high latitudes where days are short. If there is insufficient time to find enough food to survive, these birds must either find alternative foraging sites during high tide, or feed at night when the tide is low. Nocturnal feeding is common among shorebirds and some species switch from visual foraging to tactile methods at night, while others appear capable of visual hunting in the dark (McNeil et al. 1992).

Because of their propensity to nest at the tidal interface, plovers and oystercatchers are apt to have nests destroyed by tides, especially during storms (Burger 1984b). Up to 10% of American Black Oystercatcher nests may be lost to storm surges (Andres and Falxa 1995) and high tides were responsible for 78% of 27 nest failures during a 3-year study of American Oystercatchers (*Haematopus palliatus*, Nol 1989). In Russia, almost half of 40 Common Ringed Plover nests at Kandalaksha Bay were destroyed by tides (Bianki 1967 in Burger 1984b), while over 3% of Snowy Plover nests (n = 901 nests) around Monterey Bay, CA were lost due to high tides between 1984–1989 (L. Stenzel and G. Page unpublished data).

18.3.3 INFLUENCE OF OCEANOGRAPHY AND CLIMATE

It is well known that seabird distributions are influenced by particular water masses and currents (see this volume), but how these factors influence shorebird distributions is less understood. Several studies have suggested that shorebirds are more abundant along coasts near marine upwelling than in adjacent areas without upwelling. All large concentrations of coastal shorebirds in the neotropics coincide with areas of upwelling (Duffy et al. 1981, Schneider 1981). Similar relationships between upwelling and bird abundance have been noted along the Atlantic coast of Africa (Tye 1987, Alerstam 1990) and in the Bay of Panama (Butler et al. 1997). The abundance of wintering Sanderlings is correlated with major coastal upwellings, presumably because the higher productivity translates into greater food supplies on beaches where they feed (Morrison 1984).

Interannual variation in the marine environment brought on by climatic conditions has a well-documented, profound effect on marine birds (Duffy et al. 1988, Ainley 1990; E. A. Schreiber, Chapter 7, this volume). El Niño–Southern Oscillation (ENSO) events are perhaps the best studied of these phenomena, yet little is known of how they affect shorebird populations. Since ENSO events cause shifts in ocean currents, upwelling, and weather patterns, all variables that impact the distribution and abundance of the marine prey of shorebirds, it is probable that these periodic events have profound impacts on shorebird populations. Briggs et al. (1987) noted that in the fall of 1982, phalarope densities off the coast of California were almost a quarter of their normal level, and attributed this decline to the ENSO event of 1982–1983. During this same ENSO event, Red-necked Phalaropes were absent from the waters around the Galapagos Islands (Duffy 1986), a site where they are common in normal years (Rubega et al. 2000). Displacement of shorebird species other than phalaropes by ENSO events is not uncommon. In 1998, an unprecedented number of Bristle-thighed Curlews (*Numenius tahitiensis*), as well as Grey-tailed Tattlers, Eurasian Whimbrel (*N. phaeopus variegatus* or *N. p. phaeopus*), and Bar-tailed Godwits (*Limosa lapponica*), occurred along the northwest coast of North America, far outside their normal range. These vagrants were apparently displaced while migrating by the climatic conditions associated with the 1997–1998

El Niño and the Western Pacific Oscillation (Mlodinow et al. 1999). To what degree these climatic events also result in significant mortality of shorebird populations is unknown.

18.4 SHOREBIRDS ON ISLANDS

18.4.1 ENDEMISM

Biologists often think of islands as important sites for evolutionary differentiation of terrestrial species. Seabirds as a group do not seem to be obvious candidates for isolation in an oceanic setting, but high rates of breeding site fidelity have resulted in considerable differentiation and speciation among Procellariiformes and Pelecaniformes on different island groups (e.g., Siegel-Causey 1988, Warham 1996). Shorebirds also have evolved a number of distinct species and subspecies that are confined to oceanic islands, many of which are among the most endangered shorebirds in the world. These include the St. Helena Plover (*Charadrius sanctaehelenae*), Chatham Snipe (*Coenocorypha pusilla*), and Tuamotu Sandpiper (*Prosobonia cancellata*), each restricted to the islands after which they are named and each with populations of only a few hundred birds (Piersma 1996a, b, Wiersma 1996). Even more threatened are the Black Stilt (*Himantopus novaezelandiae*) and the Chatham Oystercatcher (*Haematopus chathamensis*) of New Zealand, with populations of 100 birds or fewer (Hockey 1996, Pierce 1996). Three of the few shorebird species that have become extinct within recent centuries were oceanic island species: White-winged and Ellis's sandpipers (*Prosobonia leucoptera* and *P. ellisi*) of the Society Islands went extinct in the late 1800s (Piersma 1996b), and Canarian Black Oystercatcher (*Himantopus meadewaldoi*) of the Canary Islands disappeared early in the 20th century (Hockey 1996). The Shore Plover is largely restricted to rocky shores on one small island in the South Pacific, having been extirpated from the rest of New Zealand. Shore Plovers breed, feed, and roost along the shoreline. Foods include gastropods, bivalves, polychaetes, and various crustaceans, and much foraging occurs on large wave-cut platforms where they search among the seaweed and at the edges of tide pools (Marchant and Higgins 1993).

18.4.2 VISITORS

Oceanic islands provide vital nonbreeding habitat for a number of shorebird species. The insular Pacific hosts the world's population of Bristle-thighed Curlews, plus large numbers of Pacific Golden Plovers (*Pluvialis fulva*), Wandering Tattlers, and Ruddy Turnstones (Figure 18.2). The annual transoceanic migrations of these birds often involve nonstop flights of several thousand kilometers. Bristle-thighed Curlews travel overwater from their western Alaska breeding grounds

FIGURE 18.2 Ruddy Turnstone turns over a rock while looking for invertebrates to eat. (Drawing by J. Zickefoose.)

to spend the winter on atolls and small islands in the tropical Pacific Ocean (Marks and Redmond 1994), where they occasionally prey upon seabird eggs (Marks and Hall 1992). These migrations are even more remarkable in that birds like Bristle-thighed Curlews and Pacific Golden Plovers show strong interyear fidelity to wintering sites (Johnson and Connors 1996, Marks and Redmond 1996). On Laysan Island, of 16 marked adult Bristle-thighed Curlews, only one bird changed its nonbreeding home range area in 3 years of study, and that bird only moved 1 km after storm waves swept over its home range (Marks and Redmond 1996).

Lengthy overwater flights are unusual among birds in general, so why do some shorebirds make journeys to remote islands? One likely reason is that remote islands offer a safe environment. Wintering Bristle-thighed Curlews even undergo a simultaneous molt of their primary feathers, which leaves many birds flightless and vulnerable for short periods (Marks 1993), a trait that would be unimaginable for most shorebirds. Other advantages of wintering on distant oceanic islands may include reduced competition for space (Marks and Redmond 1996), and moderate temperatures which are energetically less expensive than colder climates (Kersten and Piersma 1987, Piersma et al. 1995). Conditions on islands are changing, however, with detrimental effects to shorebirds. These effects include increased urbanization spurred by human population growth and the introduction of mammalian predators. As this happens these sites become much less suitable for birds that have adapted to a predator- and human-free environment (see Conservation of Marine Shorebirds, below; J. C. Coulson, Chapter 4, this volume).

18.5 SHOREBIRDS AT SEA: PHALAROPES

While the majority of shorebirds interact with the marine environment during at least some part of their annual cycle, phalaropes are the only shorebirds that inhabit oceanic waters for much of their lives. Red and Red-necked phalaropes spend up to 9 months of the year swimming on the open ocean. Wilson's Phalarope (*Phalaropus tricolor*), the only other species in the genus, is not marine in life history or distribution. However, it is similarly aquatic and spends the nonbreeding season on saline lakes in the interior of North and South America (Colwell and Jehl 1994). Unless otherwise stated, "phalaropes" hereafter refers to Red and Red-necked phalaropes.

18.5.1 MORPHOLOGICAL ADAPTATIONS OF PHALAROPES TO LIFE AT SEA

As pelagic birds go, phalaropes are small (~20 cm long and weighing no more than 45 g; Cramp and Simmons 1983, Rubega et al. 2000) and brightly colored during the breeding season. While at sea, they wear the countershaded coloring so common to seabirds (Bretagnolle 1993), with white underparts, tail, neck, and face, gray mantle, and black eyepatches. They look like tiny gulls. Close examination of phalaropes reveals a combination of morphological adaptations for an aquatic life not seen in any other shorebirds, including modified legs, feet, and plumage.

Phalaropes are surface swimmers, which propel themselves by paddling, and their feet and legs reflect this. Their legs are laterally flattened to minimize drag and their toes are lobed, like those of loons and grebes, rather than webbed like most seabirds. The lobes fold behind the toe on the upstroke through the water, again reducing drag, and open on the backstroke to increase thrust (Obst et al. 1996). All shorebirds have plumage that is waterproof to some degree, but phalaropes are exceptionally waterproofed, with strikingly heavy belly and breast plumage (M. Rubega personal observation).

18.5.2 PELAGIC FEEDING BIOLOGY OF PHALAROPES

Confined to surface waters by their buoyancy, phalaropes will eat almost anything that floats and is small enough to ingest (including crustaceans, hydrozoans, molluscs, polychaetes, gastropods, insects, small fish, fish eggs, seeds, sand, and plastic particles), but they are first and foremost planktivores (see Cramp and Simmons 1983, Rubega et al. 2000, and references therein). While at sea they specialize on copepods, euphausiids, and amphipods, apparently rejecting plankters larger than about 6 mm by 3 mm (Baker 1977, Mercier and Gaskin 1985).

They locate prey by swimming along, looking into the water, and pecking at prey spotted on or near the surface. Where prey densities are high, peck rates as high as 180/min have been reported (Mercier and Gaskin 1985). A feeding phalarope uses the water clinging to prey to suspend a drop between its jaws; the prey ends up suspended in the drop. Spreading its jaws stretches the drop, and the surface tension drives the drop to the back of the bill where drop volume is minimized, and the prey can be swallowed; this whole process can take as little as 0.01 sec (Rubega and Obst 1993). To feed, phalaropes take advantage of the surface tension of water, a measure of the attraction of water molecules to one another, and the property which causes water drops to assume shapes with the smallest possible volume (Rubega and Obst 1993, Rubega 1997).

Phalaropes are well known for drawing prey to the surface using a feverish, toy-like spinning behavior. Historically, spinning has been explained as a way of "stirring up" prey from the bottom of pools or ponds (e.g., Tinbergen 1935). It is now known that spinning does generate water flow that lifts prey to the surface, although not in the manner originally thought. Instead, phalaropes create miniature upwellings by kicking surface water away from the center of the loop that their spin inscribes. This deflection of surface water causes water to flow up from beneath to replace the water at the surface (Obst et al. 1996). Spinning can draw water to the surface from as deep as 0.5 m in the water column. Per unit of water inspected for prey, this Herculean effort is nearly twice as expensive energetically as swimming in a straight line (B. Obst unpublished data).

18.5.3 DISTRIBUTION OF PHALAROPES AT SEA

Both marine phalaropes have circumpolar breeding distributions in the subarctic and Arctic, and essentially only come ashore to breed (Cramp and Simmons 1983, Rubega et al. 2000). In the nonbreeding season, like any good seabird they congregate in waters that are productive, and their at-sea distributions are largely tied to areas of upwelling where surface productivity is increased or food is brought to the surface. Thus, the California Current off western North America (Briggs et al. 1984, 1987), the Humboldt Current off western South America (Murphy 1936), and the Benguela Current off West Africa (Stanford 1953) are important foraging areas.

At smaller spatial scales, phalaropes feed at physical features in the ocean. They are a familiar component of the marine avifauna at convergences, drift lines, fronts, slicks, thermal gradients, and upwellings where food is concentrated and brought to the surface (Briggs et al. 1984, Brown and Gaskin 1988, Tyler et al. 1993, Wahl et al. 1993; see D. A. Shealer, Chapter 6, this volume). The size of phalarope flocks at predictable prey patches caused by these kinds of oceanographic features can reach staggering proportions; a single upwelling near Mount Desert Rock off the Maine coast in the northwest Atlantic is reported to have attracted an estimated two to three million migratory phalaropes (Finch et al. 1978, Vickery 1978, Mercier and Gaskin 1985).

Biotic factors that concentrate prey can also affect distributions. In the Gulf Stream, off the east coast of North America, small numbers of phalaropes have been found at floating mats of the marine alga *Sargassum*. These birds presumably feed on zooplankton associated with the mats, which are themselves often concentrated by various oceanographic features (Haney 1986). Red Phalaropes also feed in the muddy surface slicks created by Gray Whales (*Eschrichtius robustus*) feeding on the ocean floor (Harrison 1979, Obst and Hunt 1990, Elphick and Hunt 1993).

18.6 SHOREBIRD MIGRATION ACROSS THE MARINE ENVIRONMENT

One of the most spectacular yet poorly understood aspects of shorebirds is their biology during passage over large bodies of water. With the exception of phalaropes, shorebirds rarely touch the water during migration. Well over half of all shorebird species are migratory (Table 18.1), and many

FIGURE 18.3 Major marine migration routes of shorebirds and known wintering areas of Red-necked and Red phalaropes. Data sources include Thompson 1973, McNeil and Burton 1977, Summers et al. 1989, Alerstam 1990, Alerstam et al. 1990, Myers et al. 1990, Williams and Williams 1990, Summers 1994, Burger 1996, Wiersma 1996, Riegen 1999, Alerstam and Gudmundsson 1999, Underhill et al. 1999, and Williams and Williams in press.

species undertake significant passages over large marine water bodies (Figure 18.3). These migrations are generally associated with favorable wind patterns (Alerstam 1990) where overwater routes provide energetic savings compared to migrating along coastal paths (Williams and Williams 1990). Trans-oceanic migrants commonly embark on journeys of 3,000 to 5,000 km (McNeil and Burton 1977, Alerstam et al. 1990, Piersma and Davidson 1992); others appear to be capable of flights from 5,000 to 10,000 km (Thompson 1973, Tulp et al. 1994, Johnson and Connors 1996) or further (Williams and Williams in press). Notable long-distant transoceanic migrants include the Pacific Golden Plover, Ruddy Turnstone, Red Knot (*Calidris canutus*), the phalaropes, Bar-tailed Godwit, Hudsonian Godwit (*Limosa haemastica*), and the Far Eastern Curlew (*Numenius madagascariensis*). Transoceanic flights are not limited to the larger shorebirds, however, as evidenced by flights of Semipalmated Sandpipers and Least Sandpipers across large sections of the Atlantic Ocean from northern North America to South America (McNeil and Burton 1977), and flights of the Red-necked Stint (*Calidris ruficollis*) across large sections of the Pacific Ocean to and from Australia (Minton 1996).

18.6.1 COMMON OVERWATER MIGRATION ROUTES

18.6.1.1 Arctic Ocean

Shorebirds regularly migrate along the shores of the Arctic to access either breeding grounds or migration corridors heading south (Johnson and Herter 1990). Radar studies have revealed that some

FIGURE 18.4 Wandering Tattlers migrate from their breeding grounds in Alaska to along the American Pacific coast sites as far south as the coast of Peru, and to islands throughout the Central Pacific. (Photo by J. R. Jehl, Jr.)

shorebirds fly from central Siberia, east over the Arctic Ocean, to reach migration corridors in northern North America that take them to wintering grounds in the Americas (Alerstam and Gudmundsson 1999a, b). Red Knots breeding in northeast Ellesmere Island, Canada, may fly over parts of the Arctic Ocean north of Greenland, then down to Iceland before continuing on to sites in Europe (Davidson and Wilson 1992). Some shorebirds occasionally pass near the North Pole (Vuilleumier 1996).

18.6.1.2 Pacific Ocean

The longest transoceanic flights by shorebirds occur in the Pacific region (Williams and Williams in press). A fat female Bar-tailed Godwit appears to be capable of flying directly from southeastern Australia to South Korea (9200 km) on its way to breed in northern Russia or Alaska (Barter 1989), while Pacific Golden Plovers, Bristle-thighed Curlews, and other shorebirds commonly fly transoceanic routes from islands in the Central and South Pacific to tundra breeding grounds (Thompson 1973, Marks and Redmond 1994, Johnson et al. 1997; see Islands: Visitors, above). The major shorebird routes in the western Pacific go from New Zealand and Australia to the island regions of Malaysia, Indonesia, the Philippines, and Papua New Guinea to Korea, China, and Japan (especially the Yellow Sea region) and then northward (Figure 18.3), crossing stretches of ocean up to 6000 km long (McClure 1974, Tulp et al. 1994, Wilson and Barter 1998, Riegen 1999).

Along the eastern Pacific Ocean, shorebirds such as Dunlin (*C. alpina*) and Wandering Tattlers fly from the Alaska Peninsula and elsewhere in western Alaska across the Gulf of Alaska to wintering sites from British Columbia to Mexico (Warnock and Gill 1996; Figure 18.4). Red-necked and Wilson's phalaropes follow inland routes southward through central Canada and the western United States, heading west and southwest to reach the Pacific, where they join Red Phalaropes moving south offshore from British Columbia to South America (Cramp and Simmons 1983, Colwell and Jehl 1994, Rubega et al. 2000). Red Phalaropes are numerous in the California Current off the western coast of the United States from May to March. These birds are joined by migrating Red-necked Phalaropes from July to November (Briggs et al. 1984, Tyler et al. 1993, Wahl et al. 1993). Red and Red-necked phalaropes winter off the western coast of South America; most in this sector of the Pacific are found in or near the Humboldt Current. Red Phalaropes wintering in the Humboldt Current come from breeding populations in North America, and possibly the Siberian Arctic. Red and Red-necked phalaropes are also consistently found around the Galapagos Islands (R. Pittman, in Rubega et al. 2000) and have been reported as far south as the southern tip of Chile (Murphy 1936, Rubega et al. 2000).

18.6.1.3 Gulf of Mexico and the Caribbean Sea

Many shorebirds migrate across the Gulf of Mexico and the Caribbean Sea (Myers 1985, Harrington 1999). Migration over the Gulf of Mexico occurs mainly in the spring due to favorable wind conditions, with the major jumping off points being central Yucatán (Gauthreaux 1971, 1999, Byrkjedal and Thompson 1998) and northern Venezuela (McNeil and Burton 1977). Several species, including Semipalmated, Baird's (*C. bairdii*), Least, and White-rumped (*C. fuscicollis*) sandpipers, Short-billed Dowitchers (*Limnodromus griseus*), Semipalmated Plovers, and Red Knots, appear to take this route north from South America. Major landing spots in the southeastern United States are Texas and Louisiana with birds continuing north to exploit the invertebrate-rich wetlands through the Central Flyway (Harrington 1999, Skagen et al. 1999), or for birds that land farther east, along the Atlantic coast (McNeil and Burton 1977, Morrison 1984, Harrington 1999).

18.6.1.4 Atlantic Ocean

In the Atlantic Ocean, direct transoceanic flights by shorebirds may exceed 4000 km, especially along the southbound flight path from northeastern North America to the northeastern coast of South America. This fall route is taken by many species which in the spring travel north through the Central Flyway (McNeil and Burton 1977, Morrison 1984, Byrkjedal and Thompson 1998). Some species, like the Sanderling, may fly south across the Atlantic, veering over the Caribbean and down to the west side of South America (Myers et al. 1990).

Red-necked and Red phalaropes from Nearctic breeding populations formerly staged in huge flocks in the western Bay of Fundy (Finch 1977, Vickery 1978, Mercier and Gaskin 1985; see Conservation of Marine Shorebirds, below). Their movements south from there and their wintering destinations, however, are poorly known. Both species only occur in small numbers farther south along the western Atlantic seaboard (Bull 1974, Haney 1985, Lee 1986). American Red Phalaropes may cross the Atlantic to join European birds at wintering areas in the Benguela Current off the coast of west Africa (Stanford 1953, Cramp and Simmons 1983), but other wintering grounds may also exist. The wintering sites of Atlantic Red-necked Phalaropes are even more uncertain (Rubega et al. 2000). An overland crossing from the Atlantic to winter in the Humboldt Current of the Pacific is plausible for both species, but few phalaropes are seen in the Caribbean or in Central America (e.g., Cooke and Bush 1989); thus this alternative seems unlikely.

Other shorebirds from the northern Atlantic Ocean including Red Knots, Sanderlings, Purple Sandpipers, and Ruddy Turnstones commonly fly more than 2000 km, much over water, from North America, Greenland, and Iceland, to Britain and continental Europe (Dick et al. 1976, Gudmundsson et al. 1991, Summers 1994). Some of these birds continue down the eastern Atlantic Ocean coast to parts of Africa (Wilson 1981, Piersma et al. 1987, Summers et al. 1989), notably the Banc d'Arguin and Guinea-Bissau (Wymenga et al. 1990), or farther to Sierra Leone (Tye and Tye 1987). South of Sierra Leone, there are few significant nonbreeding sites for shorebirds (Tye and Tye 1987) until southern Africa. Some species, such as Red Knots and Ruddy Turnstones, may migrate across the Gulf of Guinea on their way to southern Africa (Summers et al. 1989, Piersma et al. 1992, Underhill et al. 1999), but the route is not well described.

18.6.1.5 Indian Ocean

The migration of shorebirds across the western section of the Indian Ocean is poorly described, but probably not substantial, except around the Arabian Sea. The European and western Siberian breeding populations of Red-necked Phalaropes migrate through the Caspian Sea, overland across Russia and Iran, through the Gulf of Oman, and thence to the Arabian Sea, where they winter

(Cramp and Simmons 1983). Williams and Williams (1990) note that some shorebirds breeding in the eastern Palearctic cross the Indian Ocean from the coasts of Iran and Pakistan on the way to southern Africa. Small numbers of shorebirds such as Ruddy Turnstones, Ruffs, and Curlew Sandpipers (*Calidris ferruginea*) may cross the western Indian Ocean, particularly over parts of the Arabian Sea (Bailey 1967), although the majority of these species in Africa appear to either migrate across the continent or up the west coast (Underhill et al. 1999). In general, only small numbers of shorebirds use the islands in the western Indian Ocean, Ruddy Turnstones being the most abundant species (Bailey 1967, Summers et al. 1987).

The eastern Indian Ocean coastline is frequently traversed by shorebirds, especially travelers between western Australia and Asia (Parish et al. 1987, Minton 1996, 1998). Large numbers of Black-bellied Plovers cross the North Australian Basin to Indonesia and some may continue across the Bay of Bengal as they head north to breeding grounds (Byrkjedal and Thompson 1998). Some southbound Curlew Sandpipers coming from breeding grounds may follow a similar route in reverse (Minton 1998). Other birds that may travel similar paths include Wood, Marsh (*Tringa stagnatilis*), and Broad-billed (*Limicola falcinellus*) sandpipers (McClure 1974, Lane 1987, Watkins 1993), although more information about migration through this region is needed (Parish et al. 1987).

18.6.2 BEHAVIOR WHILE MIGRATING

18.6.2.1 Orientation and Timing

When considering the incredible feats of migration that transoceanic shorebirds achieve, often as juvenile birds with no previous experience and unaccompanied by adults, the looming question is: how do birds "know" how to move from a breeding site to a nonbreeding site while crossing thousands of kilometers of ocean with no obvious landmarks? There have been few migration experiments with shorebirds. How much of this behavior is learned and influenced by the environment and how much is purely genetic are still unknown. However, innate control must be important for birds like juvenile Bristle-thighed Curlews and Pacific Golden Plovers, which travel thousands of kilometers across the Pacific Ocean, without the company of adults (Marks and Redmond 1994, Byrkjedal and Thompson 1998). Ruffs migrating from Siberia to northwest Africa and a variety of shorebirds migrating between Siberia and North America appear to follow the Great Circle route, using the sun for orientation (Alerstam 1990, Alerstam and Gudmundsson 1999a, b). On the other hand, Red Knots and Ruddy Turnstones migrating from Iceland toward northern Canada appear to follow a rhumbline route (Alerstam et al. 1990), as do Red Knots flying from Siberia to western Europe (Dick et al. 1987), and other shorebirds crossing large parts of the Atlantic and Pacific Oceans (Williams and Williams 1990).

18.6.2.2 Flock Size, Flight Speed, and Altitude

Flock sizes of migrating shorebirds range from fewer than ten to hundreds of individuals, with most flocks including 50 to 400 birds, usually all the same species (Alerstam et al. 1990, Tulp et al. 1994). Radar studies have revealed that migrating shorebirds seek out favorable wind currents and fly at heights from just above sea level to over 6000 m. Birds traveling south over the western Atlantic typically fly below 2000 m, climbing to 4000 to 6000 m over the Caribbean to avoid headwinds (Richardson 1976, Williams et al. 1977). Departures along this route are associated with the passage of a cold front with accompanying northwest winds (Williams et al. 1977). In the Gulf of Finland, Red Knots migrate at altitudes of up to 3000 m with most flying between 1000 and 1500 m (Dick et al. 1987).

Flight speeds of shorebirds can be impressive, ranging from 20 to 90 kph (Lane and Jessop 1985, Tulp et al. 1994), depending on accompanying wind speeds. Golden Plovers are some of the fastest migrants with estimated flight speeds of over 100 kph (Youngsworth 1936, Johnson and Connors 1996). A radio-tagged Western Sandpiper flew about 3000 km from San Francisco,

California to the Copper River Delta, Alaska within 42 h, or over 70 kph (Iverson et al. 1996), although flying this far in one apparent movement is not typical of the species (Iverson et al. 1996, Warnock and Bishop 1998). Far Eastern Curlews equipped with satellite platforms to track their movements from nonbreeding grounds in Australia to Siberian breeding grounds have been documented averaging 50 to 80 kph, and flying over 5500 km nonstop (Minton and Driscoll 1999).

18.7 CONSERVATION OF MARINE SHOREBIRDS

18.7.1 Problems at the Ocean–Continent Interface

In general terms, the litany of threats that shorebirds face is no different from that confronting other birds: habitat loss and degradation, direct and indirect persecution, and the many other problems a growing human population brings. Several species groups are especially threatened. Of greatest concern are those birds with populations restricted to a few oceanic islands. Such species have small geographic ranges and are inherently vulnerable to extinction, and island ecosystems are particularly susceptible to large-scale biotic changes following the introduction of exotic plants and animals. Many of these island species are already critically endangered (see Islands: Endemics, above), and continued human influences seem likely to exacerbate the situation.

Concerns also center around shorebird species that use beaches for nesting (e.g., Lambeck et al. 1996). Disturbance is inevitable given the fondness of humans for beach-associated recreation (Figure 18.5). Consequently, species such as the Piping Plover and Snowy Plover in North America or the Red-breasted Plover (*Charadrius obscurus*) require extensive management to protect breeding areas (Melvin et al. 1992, Lord et al. 1997, Paton and Bachman 1997). In addition to the direct effects of human trampling and disturbance, the rubbish that people leave on beaches may attract predators which then prey on shorebird eggs. This problem is not unique to shorebirds, and Piping and Snowy Plover management often goes hand in hand with protection efforts for endangered terns (Burger 1987, Koenen 1995; P. D. Boersma, Chapter 17, this volume).

FIGURE 18.5 Oystercatchers, like this Magellanic Oystercatcher, sometimes nest on beaches where they are often disturbed by humans. (Photo by J. R. Jehl, Jr.)

There are species that remain numerous, but which are nonetheless vulnerable because they gather in dense concentrations (Myers et al. 1987). Especially during the nonbreeding season, many shorebirds flock together to feed, migrate, and roost. Very often these congregations occur in areas where there is potential for conflict with human activities (Burger 1986, Pfister et al. 1992, Lambeck et al. 1996), increasing the frequency with which birds are flushed and temporarily preventing access to potential habitat (Burger 1981, Lord et al. 1997). Whether such disturbance is a conservation problem depends on the circumstances. If the birds can resort to alternative sites and the costs of movement are small, then these short-term behavioral modifications may not translate into a reduced population size (Gill et al. 1996). To date, few studies have evaluated whether human disturbance of feeding or roosting birds actually inflicts sufficient costs to reduce survival rates or to limit population sizes. Research on Black-tailed Godwits (*Limosa limosa*) and Eurasian Oystercatchers has shown that human-caused disturbances are unlikely to affect populations (Gill and Sutherland 2000, Goss-Custard et al. 2000). In the future, identifying cases where disturbance is truly a problem will require detailed analyses that link behavior to population dynamics.

18.7.1.1 Commercial Harvesting of Shorebird Prey

Coastal fisheries and shellfish harvesting create direct competition for food, alter abiotic conditions and community dynamics, and cause disturbance, all with unknown effects on shorebird population dynamics (Smit et al. 1987, Lambeck et al. 1996, Goss-Custard et al. 2000). The commercial harvest of Horseshoe Crabs (*Limulus polyphemus*) at Delaware Bay in the eastern United States threatens hundreds of thousands of Red Knots (Figure 18.6), Ruddy Turnstones, and other shorebirds that fatten on these eggs while en route to northern breeding grounds every spring (Kerlinger 1998, Tsipoura and Burger 1999, Weidensaul 1999). Similarly, management of the baitworm harvest is a conservation concern at the Bay of Fundy in Canada, where 50 to 90% of the world's Semipalmated Sandpipers stop to refuel for their transoceanic flight to wintering grounds in South America (Shepherd and Boates 1999), as well as being a conservation concern for shorebird populations in other parts of the world (e.g., Smit et al. 1987, Lambeck et al. 1996, Barter et al. 2000).

18.7.1.2 Hunting

Hunting is a human activity with obvious detrimental consequences for shorebirds. Game hunters in the United States and Canada devastated populations of shorebirds between 1870 and 1927 and, along with habitat alterations, caused permanent declines in species such as the Eskimo Curlew (*Numenius borealis*), which may be extinct (Gill et al. 1998). While hunting of shorebirds is now either illegal or strictly regulated in many places, it remains a problem in parts of the world. In Mexico and Central and South America, shorebirds are still hunted, but the significance of this take is unknown (Page and Gill 1994). In Europe, hunting pressure on shorebirds in coastal wetlands is still high (Smit et al. 1987). Worldwide, the highest threat to shorebird populations from hunting appears to come from Asia (Parish 1987, Bamford 1992, Johnson and Connors 1996). Along the East Asia Flyway, it has been estimated that the annual kill of shorebirds is between 250,000 and 1,500,000 birds, or 5 to 30% of the flyway population, with the most serious pressure coming from China (Parish 1987), where the demand for hunted shorebirds continues to climb (Ming et al. 1998). This hunting has significant impacts on individual species. Seven to 18% of the annual mortality of the Great Knot (*Calidris tenuirostris*), a heavily hunted species in Asia, may result from hunting (Bamford 1992).

18.7.1.3 Pollution

Pollution of marine habitat also has the potential to severely affect the distribution and abundance of shorebirds worldwide. Estuaries are often surrounded by urban development with associated rivers providing shipping access to major ports (e.g., Davidson et al. 1991). Oil and chemical

FIGURE 18.6 Red Knots feeding on Horseshoe Crab eggs in Delaware Bay, New Jersey. (Drawing by J. Zickefoose.)

refineries often are located in estuarine areas and may pollute the latter with a broad array of toxic chemicals (White et al. 1980, Prater 1981, J. Burger, Chapter 15, this volume). Another source of toxic chemicals is agricultural runoff (O'Connor and Shrubb 1986, Schick et al. 1987).

Biotic pollution is another problem, as ships flushing ballast water introduce exotic plant and animal species into estuarine waters. For example, more than 230 introduced species have become established in the San Francisco Bay ecosystem since 1850, with the rate of introductions increasing over time; on average, one new species has been added every 14 weeks since 1960 (Cohen and Carlton 1998). The effects of these introductions on shorebird populations are generally unknown. In Britain, the spread of the plant *Spartina anglica* (the hybrid progeny of an introduced species) has been implicated in the national decline of Dunlin. Dunlin declined primarily on estuaries where *Spartina* flourished (Goss-Custard and Moser 1988) and clearing the plant increased shorebird use locally (Evans 1986).

18.7.1.4   Coastal Development

With up to 70% of the world's human population living in the coastal zone (Cherfas 1990), it is not surprising that losses of coastal habitats to development have been acute. In California, over 90% of wetlands have been destroyed or altered (Dahl 1990). In Asia, where mangrove/mudflat habitat is the most important habitat for shorebirds, loss of that habitat has exceeded 70% in some countries (Parish 1987). Such high levels of habitat destruction along coasts are common throughout the world, as coastal habitats are reclaimed for other purposes.

Structures including seawalls, jetties, and piers can alter patterns of tidal flow, tidal range, and sedimentation, with subsequent effects on shorebird habitat quality, quantity, and availability. A type of habitat alteration that can have positive effects for shorebirds is the construction of salt ponds or salinas. These ponds are built for the commercial extraction of salt from saltwater by evaporation and are found in areas that experience warm temperatures year round and little rain in the summer. Salt ponds are used for nesting, foraging, and roosting by shorebirds and provide important shorebird habitat in southern Portugal, Spain, and France (Rufino et al. 1984, Grimmett and Jones 1989), the western United States (Anderson 1970, Warnock and Takekawa 1995, Terp 1998), southern Africa (Velasquez 1993), India (Sampath and Krishnamurthy 1989), and Australia (Lane 1987), among other places. In Australia, three of the ten most important areas for shorebirds encompass commercial salt ponds (Lane 1987). While the use of salt ponds by shorebirds varies widely depending on salinity and water depth (Velasquez 1993, Terp 1998), the major determinants of shorebird abundance in this habitat seem to be high concentrations of invertebrate prey along with shallow water allowing birds access to these prey.

18.7.2   PROBLEMS AT SEA: PHALAROPES

The population status of marine phalaropes is poorly known as their unusual life histories contribute to the lack of data. They do not nest colonially, which makes monitoring inefficient, and when they are not breeding, they are at sea, thus global population estimates are difficult to compile. In the few places where breeding populations have been monitored, reports about population trends are contradictory. Male Red-necked phalaropes declined 94% between 1980 and 1993 at La Pérouse Bay, Churchill, Manitoba (Reynolds 1987, Gratto-Trevor 1994 and unpublished data); in contrast, nesting densities increased over the same time period near Prudhoe Bay, Alaska (Troy 1996 and unpublished data).

What is known about populations at sea is not encouraging. Red and Red-necked phalaropes historically staged in the western Bay of Fundy during migration; estimates ran as high as two million birds. By the mid-1980s few birds could be found (Duncan 1995). Whether this crash indicates a true population decline is unclear. Other possible explanations are that birds followed shifting prey-bearing currents to elsewhere in the bay, or have been forced to move out into the western Atlantic by a crash in plankton stocks. A similar crash in the numbers of phalaropes detected migrating in Japanese waters in spring (N. Moores, in Rubega et al. 2000) and declines in numbers of phalaropes off the coast of the northwestern United States (Wahl and Tweit 2000) raises the specter of a more widespread problem. The disappearance of millions of birds suggests that they are as vulnerable as any marine seabird to anthropogenic alterations of marine environments.

18.7.3 INFLUENCE OF CLIMATE CHANGE AND SEA-LEVEL RISE

One of the biggest unknowns in modern conservation planning is the impact that global climate change will have on populations and ecosystems. Changing ocean conditions might already be responsible for massive declines or relocation of phalaropes (see above) and it is difficult to predict the long-term consequences for their food supply. In the California Current off western North America, an area where many phalaropes are found, temperatures have risen significantly between 1950 and 1992, resulting in a 70% decline in zooplankton abundance (Roemmich and McGowan 1995).

Rising sea levels clearly would impact shorebird habitat in coastal areas, at least in the short term (Lindström and Agrell 1999). Of course, sea levels have risen and fallen in the past, and shorebird populations have survived the changes. Today, however, many populations are smaller than they were in the past and much closer to the point where extinction becomes likely (Brown et al. 2000a). Human development also has greatly reduced the total amount of habitat and the possibility for habitats to "migrate" with changing conditions. In the past, as traditional staging or breeding sites disappeared, it is likely that birds had other options as new wetlands and other suitable habitat were created through natural processes.

Beaches are not the only habitat susceptible to sea-level rise. Loss of saltmarsh will remove nesting, foraging, and roosting areas, and the reduction in area of intertidal foraging habitats could be devastating for many shorebird populations (Moss 1998). Predicted reductions in the area of low-lying tundra and changes in Arctic plant communities could alter the abundance of breeding habitat for many shorebirds and the effects of climate change on invertebrate food supplies are essentially unknown (Lindström and Agrell 1999). For species that winter on tropical islands, where habitat is often at low elevation, important wintering grounds could disappear. Another effect of global warming is a likely change in the distribution, frequency, and intensity of storms (Michener et al. 1997). The impact of hurricane-force storms on coastal bird populations can be especially severe with birds killed and habitat destroyed (Michener et al. 1997).

18.7.4 FUTURE SHOREBIRD PROTECTION IN MARINE ENVIRONMENTS

Increasing awareness of the need for shorebird protection within countries through the completion of comprehensive shorebird conservation plans (e.g., Watkins 1993, Brown et al. 2000b) a positive sign that steps are being taken to understand and protect shorebird populations. A number of international conservation efforts that benefit shorebird populations have been initiated (Davidson et al. 1998), such as the African-Eurasian Waterbird Agreement under the Bonn Convention (Boere and Lenten 1998), the Odessa Protocol on International Cooperation on Migratory Flyway Research and Conservation (Hötker et al. 1998), the East Asian–Australasian Shorebird Reserve Network (Watkins 1997), and the Western Shorebird Reserve Network (Western Shorebird Reserve Network 1990).

However, the future of shorebirds living in the marine environment will be closely tied with the continued impact of humans on the environment. In North America, the majority of shorebird species show evidence of declines (R. I. G. Morrison personal communication; see also Brown et al. 2000b), yet little is known about the specific causes of these declines. Assessing the long-term consequences of these potential problems for shorebird population dynamics, and devising ways in which shorebirds can be protected that are compatible with human activities, are some of the big challenges for the future. Until these challenges are met, shorebird populations will continue to face largely adverse impacts, many of them human induced, throughout the marine environment.

ACKNOWLEDGMENTS

We thank T. Alerstam, R. Butler, M. Exo, J. Gill, R. Gill, C. Gratto-Trevor, N. Moores, R. I. G. Morrison, B. S. Obst, G. Page, R. Pittman, L. Stenzel, D. Troy, and T. Williams for providing us with additional information about their studies. Important and constructive comments in review of this manuscript were provided by E. A. Schreiber, J. Burger, and an anonymous reviewer. This is Contribution No. 895 of Point Reyes Bird Observatory.

LITERATURE CITED

AINLEY, D. G. 1990. Seasonal and annual patterns in the marine environment near the Farallones. Pp. 23–50 *in* Seabirds of the Farallon Islands (D. G. Ainley and R. J. Boekelheide, Eds.). Stanford University Press, California.

ALBERICO, J. A. R., J. M. REED, AND L. W. ORING. 1991. Nesting near a Common Tern colony increases and decreases Spotted Sandpiper nest predation. Auk 108: 904–910.

ALCORN, M. AND R. ALCORN. 2000. Seasonal migration of Banded Stilt *Cladorhynchus leucocephalus* to the Natimuk-Douglas salt pans in western Victoria, Australia. Stilt 36: 7–10.

ALERSTAM, T. 1990. Bird Migration. Cambridge University Press,

ALERSTAM, T. AND G. A. GUDMUNDSSON. 1999a. Bird orientation at high latitudes: flight routes between Siberia and North America across the Arctic Ocean. Proceeding of the Royal Society of London B 266: 2499–2505.

ALERSTAM, T. AND G. A. GUDMUNDSSON. 1999b. Migration patterns of tundra birds: tracking radar observations along the Northeast Passage. Arctic 52: 346–371.

ALERSTAM, T., G. A. GUDMUNDSSON, P. E. JÖNSSON, J. KARLSSON, AND Å. LINDSTRÖM. 1990. Orientation, migration routes and flight behaviour of Knots, Turnstones and Brent Geese departing from Iceland in spring. Arctic 43: 201–214.

ALLENG, G. P., AND C. A. WHYTE-ALLENG. 1993. Survey of Least Tern nesting sites on the south coast of Jamaica. Colonial Waterbirds 16: 190–193.

AMERICAN ORNITHOLOGISTS' UNION. 1998. Check-List of North American Birds. 7th ed. Allen Press, Inc., Lawrence, KA.

ANDERSON, W. 1970. A preliminary study of the relationship of saltponds and wildlife — south San Francisco Bay. California Fish and Game 56: 240–252.

ANDRES, B. A., AND G. A. FALXA. 1995. Black Oystercatcher (*Haematopus bachmani*) *in* The Birds of North America No. 155 (A. Poole and F. Gill, Eds.). Academy of Natural Sciences, Philadelphia; American Ornithologists' Union, Washington, D.C.

BAILEY, R. S. 1967. Migrant waders in the Indian Ocean. Ibis 109: 437–439.

BAKER, M. C. 1974. Foraging behavior of Black-bellied Plovers (*Pluvialis squatarola*). Ecology 55: 162–167.

BAKER, M. C. 1977. Shorebird food habits in the eastern Canadian Arctic. Condor 79: 50–62.

BAKER-GABB. 1996. Family Pedionomidae (Plains-Wanderer). Pp. 534–537 *in* Handbook of the Birds of the World. Vol. 3. Hoatzin to Auks (J. del Hoyo, A. Elliott, and J. Sargatal, Eds.). Lynx Edicions, Barcelona.

BAMFORD, M. 1992. The impact of predation by humans upon waders in the Asian/Australasian Flyway: evidence from the recovery of bands. Stilt 20: 38–40.

BARTER, M. 1989. Bar-tailed Godwit *Limosa lapponica* in Australia Part 1. Races, breeding areas and migration routes. Stilt 14: 43–48.

BARTER, M. A., J. R. WILSON, Z. W. LI, Z. G. DONG, Y. G. CAO, AND L. S. JIANG. 2000. Yalu Jiang National Nature Reserve, north-eastern China — a newly discovered internationally important Yellow Sea site for northward migrating shorebirds. Stilt 37: 13–20.

BIANKI, V. V. 1967. Gulls, shorebirds, and alcids of Kandalaksha Bay. Proceedings of Kandalaksha State Research. 6: 1–247.

BISHOP, M. A., P. M. MEYERS, AND P. F. MCNELEY. 2000. A method to estimate migrant shorebird numbers on the Copper River Delta, Alaska. Journal of Field Ornithology.

BOERE, G. C., AND B. LENTEN. 1998. The African-Eurasian Waterbird Agreement: a technical agreement under the Bonn Convention. International Wader Studies 10: 45–50.

BREAREY, D. M., AND O. HILDÉN. 1985. Nesting and egg-predation by Turnstones *Arenaria interpres* in larid colonies. Ornis Scandinavica 16: 283–292.

BREIEHAGEN, T. 1989. Nesting biology and mating system in an alpine population of Temminck's Stint *Calidris temminckii*. Ibis 131: 389–402.

BRETAGNOLLE, V. 1993. Adaptive significance of seabird coloration: the case of procellariiforms. American Naturalist 142: 141–173.

BRIGGS, K. T., K. F. DETTMAN, D. B. LEWIS, AND W. B. TYLER. 1984. Phalarope feeding in relation to autumn upwelling off California. Pp. 51–62 *in* Marine Birds: Their Feeding Ecology and Commercial Fisheries Relationships (D. N. Nettleship, G. A. Sanger, and P. F. Springer, Eds.). Canadian Wildlife Service, Ottawa.

BRIGGS, K. T., W. B. TYLER, D. B. LEWIS, AND D. R. CARLSON. 1987. Bird communities at sea off California: 1975 to 1983. Studies in Avian Biology 11: 1–74.

BROWN, R. G. B., AND D. E. GASKIN. 1988. The pelagic ecology of the Grey and Red-necked Phalaropes *Phalaropus fulicarius* and *P. lobatus* in the Bay of Fundy, eastern Canada. Ibis 130: 234–250.

BROWN, S., C. HICKEY, R. GILL, L. GORMAN, C. GRATTO-TREVOR, S. HAIG, B. HARRINGTON, C. HUNTER, G. MORRISON, G. PAGE, P. SANZENBACHER, S. SKAGEN, AND N. WARNOCK. 2000a. National Shorebird Conservation Assessment: Shorebird Conservation Status, Conservation Units, Population Estimates, Population Targets, and Species Prioritization. Manomet Center for Conservation Sciences. http: //www.Manomet.org/USSCP/files.htm

BROWN, S., C. HICKEY, AND B. HARRINGTON. 2000b. The U.S. Shorebird Conservation Plan. Manomet Center for Conservation Sciences, Manomet, MA.

BULL, J. 1974. The Birds of New York State. Doubleday Natural History Press, Garden City, New York.

BURGER, A. E. 1996. Family Chionidae (Sheathbills). Pp. 546–555 *in* Handbook of the Birds of the World. Vol. 3. Hoatzin to Auks (J. del Hoyo, A. Elliott, and J. Sargatal, Eds.). Lynx Edicions, Barcelona.

BURGER, J. 1981. The effect of human activity on birds at a coastal bay. Biological Conservation 21: 231–241.

BURGER, J. 1984a. Shorebirds as marine animals. Pp. 17–81 *in* Behavior of Marine Animals. Vol. 5. Shorebirds: Breeding Behavior and Populations (J. Burger and B. L. Olla, Eds.). Plenum Press, New York.

BURGER, J. 1984b. Abiotic factors affecting migrant shorebirds. Pp. 1–72 *in* Behavior of Marine Animals. Vol. 6. Shorebirds: Migration and Foraging Behavior (J. Burger and B. L. Olla, Eds.). Plenum Press, New York.

BURGER, J. 1986. The effect of human activity on shorebirds in two coastal bays in Northeastern United States. Environmental Conservation 13: 123–130.

BURGER, J. 1987. Physical and social determinants of nest-site selection in Piping Plover in New Jersey. Condor 89: 811–818.

BURGER, J., AND M. GOCHFELD. 1991. Human activity influence and diurnal and nocturnal foraging of Sanderlings (*Calidris alba*). Condor 93: 259–265.

BURTON, P. J. K. 1974. Feeding and the Feeding Apparatus in Waders: a Study of Anatomy and Adaptations in the Charadrii. British Natural History Museum, London.

BUTLER, R. W., R. I. G. MORRISON, F. S. DELGADO, R. K. ROSS, AND G. E. J. SMITH. 1997. Habitat associations of coastal birds in Panama. Colonial Waterbirds 20: 518–524.

BYRKJEDAL, I., AND D. THOMPSON. 1998. Tundra Plovers: The Eurasian, Pacific and American Golden Plovers and Grey Plover. T. & A. D. Poyser Ltd., London.

CHERFAS, J. 1990. The fringe of the ocean — under siege from the land. Science 248: 163–165.

CLARK, N. A. 1997. Estuarine wader impact assessments: possibilities and pitfalls. Pp. 17–23 *in* Effect of Habitat Loss and Change on Waterbirds (J. D. Goss-Custard, R. Rufino, and A. Luis, Eds.). Institute of Terrestrial Ecology Symposium No. 30, The Stationery Office, London.

COHEN, A. N., AND J. T. CARLTON. 1998. Accelerating invasion rate in a highly invaded estuary. Science 279: 555–557.

COLWELL, M. A., AND J. R. JEHL, JR. 1994. Wilson's Phalarope (*Phalaropus tricolor*). No. 83 *in* The Birds of North America (A. Poole and F. Gill, Eds.). Academy of Natural Sciences, Philadelphia; American Ornithologists' Union, Washington, D.C.

COOKE, R. G., AND M. BUSH. 1989. Out-of-season Red–necked Phalaropes in central Panama. Journal of Field Ornithology 60: 39–42.

COOPER, J. M. 1994. Least Sandpiper (*Calidris minutilla*). No. 115 *in* The Birds of North America (A. Poole and F. Gill, Eds.). Academy of Natural Sciences, Philadelphia; American Ornithologists' Union, Washington, D.C.

CRAMP, S., AND K. E. L. SIMMONS. (Eds.). 1983. Handbook of the Birds of Europe, the Middle East and North Africa: The Birds of the Western Palearctic. Vol. 3. Oxford University Press, Oxford.

CRESSWELL, W. 1996. Surprise as a winter hunting strategy in Sparrowhawks *Accipiter nisus*, Peregrines *Falco peregrinus*, and Merlins *F. columbarius*. Ibis 138: 684–692.

CROSSIN, R. S., AND L. N. HUBER. 1970. Sooty Tern egg predation by Ruddy Turnstones. Condor 72: 372–373.

DAHL. T. E. 1990. Wetlands of the United States 1780's to 1980's. Unpublished report of the U. S. Fish and Wildlife Service, Washington, D.C.

DANN, P. 1999. Feeding periods and supratidal feeding of Red-necked Stints and Curlew Sandpipers in Western Port, Australia. Emu 99: 218–222.

DAVIDSON, N. C., AND J. R. WILSON. 1992. The migration system of the European-wintering Knot *Calidris canutus islandica*. Wader Study Group Bulletin 64 Supplement: 39–51.

DAVIDSON, N. C., D. D'A. LAFFOLEY, J. P. DOODY, L. S. WAY, J. GORDON, R. KEY, M. W. PIENKOWSKI, R. MITCHELL, AND K. L. DUFF. 1991. Nature Conservation and Estuaries in Great Britain. Nature Conservancy Council, Peterborough.

DAVIDSON, N. C., D. A. STROUD, P. I. ROTHWELL, AND M. W. PIENKOWSKI. 1998. Towards a flyway conservation strategy for waders. International Wader Studies 10: 24–38.

DAVIS, A. 1994. Breeding biology of the New Zealand Shore Plover *Thinornis novaeseelandiae*. Notornis (Supplement) 41: 195–208.

DEL HOYO, J., A. ELLIOTT, AND J. SARGATAL. (EDS.). 1996. Handbook of the Birds of the World. Vol. 3. Hoatzin to Auks. Lynx Edicions, Barcelona.

DE VLAS, S. E. J., J. BUNSKOEKE, B. J. ENS, AND J. B. HULSCHER. 1996. Tidal changes in the choice of *Nereis diversicolor* or *Macoma balthica* as main prey species in the diet of the Oystercatcher *Haematopus ostralegus*. Ardea 84A: 105–116.

DICK, W. J. A., M. W. PIENKOWSKI, M. WALTNER, AND C. D. T. MINTON. 1976. Distribution and geographical origins of Knot *Calidris canutus* wintering in Europe and Africa. Ardea 64: 22–47.

DICK, W. J. A., T. PIERSMA, AND P. PROKOSCH. 1987. Spring migration of the Siberian Knots *Calidris canutus canutus*: results of a co-operative Wader Study Group project. Ornis Scandinavica 18: 5–16.

DUFFY, D. C. 1986. Seabird densities and aggregations during the 1983 El Niño in the Galapagos Islands. Wilson Bulletin 98: 588–591.

DUFFY, D. C., N. ATKINS, AND D. C. SCHNEIDER. 1981. Do shorebirds compete on their wintering grounds? Auk 98: 215–229.

DUFFY, D. C., W. E. ARNTZ, H. T. SERPA, P. D. BOERSMA, AND R. L. NORTON. 1988. A comparison of the effects of El Niño and the Southern Oscillation on birds in Peru and the Atlantic Ocean. Proceedings of the International Ornithological Congress 19: 1740–1746.

DUNCAN, C. D. 1995. The migration of Red-necked Phalaropes: ecological mysteries and conservation concerns. Birding 28: 482–488.

ELPHICK, C. S., AND G. L. HUNT, JR. 1993. Variations in the distribution of marine birds with water mass in the northern Bering Sea. Condor 95: 33–44.

ELPHICK, C. S., AND T. L. TIBBITTS. 1998. Greater Yellowlegs (*Tringa melanoleuca*). No. 355 *in* The Birds of North America (A. Poole and F. Gill, Eds.). Academy of Natural Sciences, Philadelphia; American Ornithologists' Union, Washington, D.C.

EMLEN, S. T., AND L. W. ORING. 1977. Ecology, sexual selection, and the evolution of mating systems. Science 197: 215–223.

ENS, B. J., K. B. BRIGGS, U. N. SAFRIEL, AND C. J. SMIT. 1996. Life history decisions during the breeding season. Pp. 186–218 *in* The Oystercatcher: From Individuals to Populations (J. D. Goss-Custard, Ed.). Oxford University Press, Oxford.

EVANS, P. R. 1986. Use of the herbicide 'Dalapon' for control of *Spartina* encroaching on intertidal mudflats: beneficial effects on shorebirds. Colonial Waterbirds 9: 171–175.

EVANS, P. R. 1991. Seasonal and annual patterns of mortality in migratory shorebirds: some conservation implications. Pp. 346–359 *in* Bird Population Studies (C. M. Perrin, J.-D. Lebreton, and G. J. M. Hirons, Eds.). Oxford University Press, Oxford.

EVANS, P. R., AND M. W. PIENKOWSKI. 1984. Population dynamics in shorebirds. Pp. 83–123 *in* Shore-birds: Breeding Behavior and Populations (J. Burger and B. L. Olla, Eds.). Plenum Press, New York.

EXO, K.-M. 1993. Höchsalter eines beringten Austernfischers (*Haematopus ostralegus*): 44 Jahre. Vogelwarte 37: 144.

FINCH, D. W. 1977. The autumn migration, New England Regional Report. American Birds 31: 225–232.

FINCH, D. W., W. C. RUSSELL, AND E. V. THOMPSON. 1978. Pelagic birds in the Gulf of Maine. American Birds 32: 281–294.

FJELDSÅ, J. 1996. Family Thinocoridae (Seedsnipes). Pp. 538–545 *in* Handbook of the Birds of the World. Vol. 3. Hoatzin to Auks (J. del Hoyo, A. Elliott, and J. Sargatal, Eds.). Lynx Edicions, Barcelona.

GAINES, E. P., AND M. R. RYAN. 1988. Piping Plover habitat use and reproductive success in North Dakota. Journal of Wildlife Management 52: 266–273.

GAUTHREAUX, S. A., JR. 1971. A radar and direct visual study of passerine spring migration in southern Louisiana. Auk 88: 343–365.

GAUTHREAUX, S. A., JR. 1999. Neotropical migrants and the Gulf of Mexico: the view from aloft. Pp. 27–49 *in* Gathering of Angels: Migrating Birds and Their Ecology (K. P. Able, Ed.). Cornell University Press, Ithaca, NY.

GILL, J. A., W. J. SUTHERLAND, AND A. R. WATKINSON. 1996. A method to quantify the effects of human disturbance on animal populations. Journal of Applied Ecology 33: 786–792.

GILL, J. A., AND W. J. SUTHERLAND. 2000. Predicting the consequences of human disturbance from behavioural decisions. Pp. 51–64 *in* Behaviour and Conservation (L. M. Gosling and W. J. Sutherland, Eds.). Cambridge University Press, Cambridge.

GILL, R. E., JR., P. CANEVARI, AND E. H. IVERSEN. 1998. Eskimo Curlew (*Numenius borealis*). No. 347 *in* The Birds of North America (A. Poole and F. Gill, Eds.). Academy of Natural Sciences, Philadelphia; American Ornithologists' Union, Washington, D.C.

GOSS-CUSTARD, J. D. 1977. The ecology of the Wash. III. Density related behaviour and the possible effects of a loss of feeding grounds on wading birds (Charadrii). Journal of Applied Ecology 14: 721–739.

GOSS-CUSTARD, J. D., AND M. E. MOSER. 1988. Rates of change in the numbers of Dunlin, *Calidris alpina*, wintering in British estuaries in relation to the spread of *Spartina anglica*. Journal of Applied Ecology 25: 95–109.

GOSS-CUSTARD, J. D., R. M. WARWICK, R. KIRBY, S. MCGRORTY, R. T. CLARKE, B. PEARSON, W. E. RISPIN, S. E. A. LE V. DIT DURELL, AND R. J. ROSE. 1991. Towards predicting wading bird densities from predicted prey densities in a post-Barrage Severn estuary. Journal of Animal Ecology 28: 1004–1026.

GOSS-CUSTARD, J. D., R. A. STILLMAN, A. D. WEST, S. MCGRORTY, S. E. A. LE V. DIT DURELL, AND R. W. G. CALDOW. 2000. The role of behavioural models in predicting the ecological impact of harvesting. Pp. 65–82 *in* Behaviour and Conservation (L. M. Gosling and W. J. Sutherland, Eds.). Cambridge University Press, Cambridge.

GRATTO-TREVOR, C. L. 1994. Monitoring shorebird populations in the Arctic. Bird Trends 3: 10–12.

GRIMMETT, R. F. A., AND T. A. JONES. 1989. Important Bird Areas in Europe. International Council for Bird Preservation Technical Publication No. 9, Cambridge.

GUDMUNDSSON, G., Å. LINDSTRÖM, AND T. ALERSTAM. 1991. Optimal fat loads and long-distance flights by migrating Knots *Calidris canutus*, Sanderlings *C. alba* and Turnstones *Arenaria interpres*. Ibis 133: 140–152.

HAIG, S. M. 1990. Piping Plover (*Charadrius melodus*). No. 2 *in* The Birds of North America (A. Poole and F. Gill, Eds.). Academy of Natural Sciences, Philadelphia; American Ornithologists' Union, Washington, D.C.

HANEY, J. C. 1985. Wintering phalaropes off the southeastern United States: application of remote sensing imagery to seabird habitat analysis at oceanic fronts. Journal of Field Ornithology 56: 321–333.

HANEY, J. C. 1986. Seabird patchiness in tropical oceanic waters: the influence of *Sargassum* reefs. Auk 103: 141–151.

HARRINGTON, B. A. 1999. The hemispheric globetrotting of the White-rumped Sandpiper. Pp. 119–133 *in* Gathering of Angels: Migrating Birds and Their Ecology (K. P. Able, Ed.). Cornell University Press, Ithaca, NY.

HARRISON, C. 1978. A Field Guide to the Nests, Eggs and Nestlings of North American Birds. Collins, New York.

HARRISON, C. S. 1979. The association of marine birds and feeding Gray Whales. Condor 81: 93–95.

HEEZIK, Y. M. VAN, A. F. C. GERRITSEN, AND C. SWENNAN. 1983. The influence of chemoreception on the foraging behaviour of two species of sandpiper, *Calidris alba* and *Calidris alpina*. Netherlands Journal of Sea Research 17: 47–56.

HEG, D., B. J. ENS, T. BURKE, L. JENKINS, AND J. P. KRUIJT. 1993. Why does the typically monogamous Oystercatcher *Haematopus ostralegus* engage in extra-pair copulations? Behaviour 126: 247–289.

HEPBURN, I. R. 1987. Conservation of wader habitats in coastal West Africa. Wader Study Group Bulletin 49, Supplement/IWRB Special Publication 7: 125–127.

HILDÉN, O. 1975. Breeding system of Temminck's Stint *Calidris temminckii* (Leisl.) Ornis Fennica 52: 117–146.

HOCKEY, P. A. R. 1996. Family Haematopodidae (Oystercatchers). Pp. 308–319 *in* Handbook of the Birds of the World. Vol. 3. Hoatzin to Auks (J. del Hoyo, A. Elliott, and J. Sargatal, Eds.). Lynx Edicions, Barcelona.

HOGLÜND, J., AND R. V. ALATALO. 1995. Leks. Princeton University Press, Princeton, NJ.

HÖTKER, H., E. LEBEDEVA, P. S. TOMKOVICH, J. GROMADZKA, N. C. DAVIDSON, J. EVANS, D. A. STROUD, AND R. B. WEST (EDS.). 1998. Migration and International Conservation of Waders: Research and Conservation on North Asian, African, and European Flyways. International Wader Studies 10.

HUME, R. A. 1996. Family Burhinidae (Thick-knees). Pp. 348–363 *in* Handbook of the Birds of the World. Vol. 3. Hoatzin to Auks (J. del Hoyo, A. Elliott, and J. Sargatal, Eds.). Lynx Edicions, Barcelona.

ISLEIB, M. E. "PETE." 1979. Migratory shorebird populations on the Copper River Delta and eastern Prince Williams Sound, Alaska. Studies in Avian Biology No. 2: 125–129.

IVERSON, G. C., S. E. WARNOCK, R. W. BUTLER, M. A. BISHOP, AND N. WARNOCK. 1996. Spring migration of Western Sandpipers (*Calidris mauri*) along the Pacific coast of North America: a telemetry study. Condor 98: 10–21.

JACKSON, D. B. 1994. Breeding dispersal and site-fidelity in three monogamous wader species in the Western Isles, U.K. Ibis 136: 463–473.

JEHL, J. R., JR. 1975. *Pluvianellus socialis*: biology, ecology, and relationships of an enigmatic Patagonian shorebird. Transactions of the San Diego Society of Natural History 18: 25–74.

JENNI, D. A. 1996. Family Jacanidae (Jacanas). Pp. 276–291 *in* Handbook of the Birds of the World. Vol. 3. Hoatzin to Auks (J. del Hoyo, A. Elliott, and J. Sargatal, Eds.). Lynx Edicions, Barcelona.

JOHANSSON, O. C., AND D. BLOMQVIST. 1996. Habitat selection and diet of lapwing *Vanellus vanellus* chicks on coastal farmland in S. W. Sweden. Journal of Animal Ecology 33: 1030–1040.

JOHNSON, O. W., AND P. G. CONNORS. 1996. American Golden Plover (*Pluvialis dominica*) and Pacific Golden Plover (*Pluvialis fulva*). No. 201–202 *in* The Birds of North America (A. Poole and F. Gill, Eds.). Academy of Natural Sciences, Philadelphia; American Ornithologists' Union, Washington, D.C.

JOHNSON, O. W., N. WARNOCK, M. A. BISHOP, A. J. BENNETT, P. JOHNSON, AND R. J. KLEINHOLZ. 1997. Migration by radio-tagged Pacific Golden Plovers from Hawaii to Alaska, and their subsequent survival. Auk 114: 521–524.

JOHNSON, S. R., AND D. R. HERTER. 1990. Bird migration in the Arctic: a review. Pp. 22–43 *in* Bird Migration: Physiology and Ecophysiology (E. Gwinner, Ed.). Springer-Verlag, Berlin.

KERLINGER, P. 1998. Showdown at Delaware Bay. Natural History Magazine 107: 56–58.

KERSTEN, M., AND T. PIERSMA. 1987. High levels of energy expenditure in shorebirds: metabolic adaptations to an energetically expensive way of life. Ardea 75: 175–187.

KIRWAN, G. M. 1996. Family Rostratulidae (Painted-Snipes). Pp. 292–301 *in* Handbook of the Birds of the World. Vol. 3. Hoatzin to Auks (J. del Hoyo, A. Elliott, and J. Sargatal, Eds.). Lynx Edicions, Barcelona.

KNYSTAUTAS, A. J. 1996. Family Ibidorhynchidae (Ibisbill). Pp. 326–331 *in* Handbook of the Birds of the World. Vol. 3. Hoatzin to Auks (J. del Hoyo, A. Elliott, and J. Sargatal, Eds.). Lynx Edicions, Barcelona.

KOENEN, M. T. 1995. Breeding Ecology and Management of Least Terns, Snowy Plovers, and American Avocets. M.S. thesis, Oklahoma State University, Stillwater.

LAMBECK, R. H. D., J. D. GOSS-CUSTARD, AND P. TRIPLET. 1996. Oystercatchers and man in the coastal zone. Pp. 289–351 *in* The Oystercatcher: From Individuals to Populations (J. D. Goss-Custard, Ed.). Oxford University Press, Oxford.

LANCTOT, R. B., AND C. D. LAREDO. 1994. Buff-breasted Sandpiper (*Tryngites subruficollis*). No. 91 *in* The Birds of North America (A. Poole and F. Gill, Eds.). Academy of Natural Sciences, Philadelphia; American Ornithologists' Union, Washington, D.C.

LANE, B. 1987. Shorebirds in Australia. Nelson Publishers, Melbourne.

LANE, B., AND A. JESSOP. 1985. Tracking of migrating waders in north-western Australia using meteorological radar. Stilt 6: 17–29.

LEE, D. S. 1986. Seasonal distribution of marine birds in North Carolina waters, 1975–1986. American Birds 40: 409–412.

LINDSTRÖM, Å., AND J. AGRELL. 1999. Global change and possible effects on the migration and reproduction of arctic-breeding waders. Ecological Bulletins 47: 145–159.

LOFTIN, R. W., AND S. SUTTON. 1979. Ruddy Turnstones destroy Royal Tern colony. Wilson Bulletin 91: 133–135.

LORD, A., J. R. WAAS, AND J. INNES. 1997. Effects of human activity on the behaviour of Northern New Zealand Dotterel *Charadrius obscurus aquilonius* chicks. Biological Conservation 82: 15–20.

MACIVOR, L. H., S. M. MELVIN, AND C. R. GRIFFIN. 1990. Effects of research activity on piping plover nest predation. Journal of Wildlife Management 54: 443–447.

MACLEAN, G. L. 1996. Family Glareolidae (Coursers and pratincoles). Pp. 364–383 *in* Handbook of the Birds of the World. Vol. 3. Hoatzin to Auks (J. del Hoyo, A. Elliott, and J. Sargatal, Eds.). Lynx Edicions, Barcelona.

MARCHANT, S., AND P. J. HIGGINS. (EDS.). 1993. Handbook of Australian, New Zealand and Antarctic birds. Vol. 2. Raptors to Lapwings. Oxford University Press, Melbourne.

MARKS, J. S. 1993. Molt of Bristle-thighed Curlews in the northwestern Hawaiian islands. Auk 110: 573–587.

MARKS, J. S., AND C. S. HALL. 1992. Tool use by Bristle-thighed Curlews feeding on albatross eggs. Condor 94: 1032–1034.

MARKS, J. S., AND R. L. REDMOND. 1994. Migration of Bristle-thighed Curlews on Laysan Island: timing, behavior and estimated flight range. Condor 96: 316–330.

MARKS, J. S., AND R. L. REDMOND. 1996. Demography of Bristle-thighed Curlews *Numenius tahitiensis* wintering on Laysan Island. Ibis 138: 438–447.

MCCLURE, H. E. 1974. Migration and Survival of the Birds of Asia. U. S. Army Med. Comp., SEATO Med. Project, Bangkok, Thailand.

MCNEIL, R., AND J. BURTON. 1977. Southbound migration of shorebirds from the Gulf of St. Lawrence. Wilson Bulletin 89: 167–171.

MCNEIL, R., P. DRAPEAU, AND J. D. GOSS-CUSTARD. 1992. The occurrence and adaptive significance of nocturnal habits in waterfowl. Biological Review 67: 381–419.

MELVIN, S. M., L. H. MacIVOR, AND C. R. GRIFFIN. 1992. Predator exclosures: a technique to reduce predation on Piping Plover nests. Wildlife Society Bulletin 20: 143–148.

MERCIER, F., AND D. E. GASKIN. 1985. Feeding ecology of migrating Red-necked Phalaropes (*Phalaropus lobatus*) in the Quoddy region, New Brunswick, Canada. Canadian Journal of Zoology 63: 1062–1067.

MICHENER, W. K., E. R. BLOOD, K. L. BILDSTEIN, M. M. BRINSON, AND L. R. GARDNER. 1997. Climate change, hurricanes and tropical storms, and rising sea level in coastal wetlands. Ecological Applications 7: 770–801.

MILLER, E. H., AND R. MCNEIL. 1988. Longevity record for the Least Sandpiper: a revision. Journal of Field Ornithology 59: 403–404.

MING, M., L. JIANJIAN, T. CHENGJIA, S. PINGYUE, AND H. WEI. 1998. The contribution of shorebirds to the catches of hunters in the Shanghai area, China during 1997–1998. Stilt 33: 32–36.

MINTON, C. D. T. 1996. The migration of the Red-necked Stint *Calidris ruficollis*. Stilt 29: 24–35.

MINTON, C. D. T. 1998. Migratory movements of Curlew Sandpipers *Calidris ferruginea* that spend the non-breeding season in Australia. Stilt 32: 28–40.

MINTON, C., AND P. DRISCOLL. 1999. Eastern Curlews on track again. Tattler 19: 2.

MLODINOW, S. G., S. FELDSTEIN, AND B. TWEIT. 1999. The Bristle-thighed Curlew landfall of 1998: climatic factors and notes on identification. Western Birds 30: 133–155.

MORRISON, R. I. G. 1984. Migrations systems of some New World shorebirds. Pp. 125–202 *in* Behavior of Marine Animals. Vol. 6. Shorebirds: Migration and Foraging Behavior (J. Burger and B. L. Olla, Eds.). Plenum Press, New York.

MORRISON, R. I. G., R. W. BUTLER, F. S. DELGADO, AND R. K. ROSS. 1998. Atlas of Nearctic Shorebirds and Other Waterbirds on the Coast of Panama. Special Publication, Canadian Wildlife Service, Ottawa.

MOSS, S. 1998. Predictions of the effects of global climate change on Britain's birds. British Birds 91: 307–325.

MURPHY, R. C. 1936. Oceanic Birds of South America, Vol. II. American Museum of Natural History, New York.

MYERS, J. P. 1985. Sanderlings do not drive across the Gulf of Mexico. Wader Study Group Bulletin 44: 42.

MYERS, J. P., P. G. CONNORS, AND F. A. PITELKA. 1979. Territoriality in non-breeding shorebirds. Studies in Avian Biology No. 2: 231–246.

MYERS, J. P., R. I. G. MORRISON, P. Z. ANTAS, B. A. HARRINGTON, T. E. LOVEJOY, M. SALLA-BERRY, S. E. SENNER, AND A. TARAK. 1987. Conservation strategy for migratory species. American Scientist 75: 19–26.

MYERS, J. P., M. SALLABERRY, E. ORTIZ, G. CASTRO, L. M. GORDON, J. L. MARON, C. T. SCHICK, E. TABILO, P. ANTAS, AND T. BELOW. 1990. Migration routes of New World Sanderlings (*Calidris alba*). Auk 107: 172–180.

NOL, E. 1989. Food supply and reproductive performance of the American Oystercatcher in Virginia. Condor 91: 429–435.

NOL, E., AND M. S. BLANKEN. 1999. Semipalmated Plover (*Charadrius semipalmatus*). No. 444 *in* The Birds of North America (A. Poole and F. Gill, Eds.). Academy of Natural Sciences, Philadelphia; American Ornithologists' Union, Washington, D.C.

OBST, B. A., AND G. L. HUNT, JR. 1990. Marine birds at gray whale mud plumes in the Bering Sea. *Auk* 107: 678–688.

OBST, B. S., W. HAMNER, E. WOLANSKI, M. HAMNER, M. RUBEGA, AND B. LITTLEHALES. 1996. Kinematics of phalarope spinning. Nature 384: 121.

O'CONNOR, R. J., AND M. SHRUBB. 1986. Farming and Birds. Cambridge University Press, Cambridge.

ORING, L. W., AND D. B. LANK. 1984. Breeding area fidelity, natal philopatry, and the social systems of sandpipers. Pp. 125–147 *in* Behavior of Marine Animals. Vol. 5. Shorebirds: Breeding Behavior and Populations (J. Burger and B. L. Olla, Eds.). Plenum Press, New York.

PAGE, G. W., AND R. E. GILL JR. 1994. Shorebirds in western North America: late 1800s to late 1900s. Studies in Avian Biology 15: 147–160.

PAGE, G. W., AND D. F. WHITACRE. 1975. Raptor predation on wintering shorebirds. Condor 77: 73–83.

PAGE, G. W., J. S. WARRINER, J. C. WARRINER, AND P. W. C. PATON. 1995. Snowy Plover (*Charadrius alexandrinus*). No. 154 *in* The Birds of North America (A. Poole and F. Gill, Eds.). Academy of Natural Sciences, Philadelphia; American Ornithologists' Union, Washington, D.C.

PARISH, D. 1987. Conservation of wader habitats in East Asia. Wader Study Group Bulletin 49, Supplement/IWRB Special Publication 7: 132–134.

PARISH, D., B. LANE, P. SAGER, AND P. TOMKOVITCH. 1987. Wader migration systems in East Asia and Australasia. Wader Study Group Bulletin 49, Supplement/IWRB Special Publication 7: 4–14.

PATON, P. W. C., AND V. C. BACHMAN. 1997. Impoundment drawdown and artificial nest structures as management strategies for Snowy Plovers. International Wader Studies 9: 64–70.

PAULSON, D. R. 1993. Shorebirds of the Pacific Northwest. University of Washington Press, Seattle.

PFISTER, C., B. A. HARRINGTON, AND M. LAVINE. 1992. The impact of human disturbance on shorebirds at a migration staging area. Biological Conservation 60: 115–126.

PIENKOWSKI, M. W. 1981. How foraging plovers cope with environmental effects on invertebrate behaviour and availability. Pp. 179–192 *in* Feeding and Survival Strategies of Estuarine Organisms (N. V. Jones and W. J. Wolff, Eds.). Plenum Press, New York.

PIERCE, R. J. 1996. Family Recurvirostridae (Stilts and avocets). Pp. 332–347 *in* Handbook of the Birds of the World. Vol. 3. Hoatzin to Auks (J. del Hoyo, A. Elliott, and J. Sargatal, Eds.). Lynx Edicions, Barcelona.

PIERSMA, T. 1996a. Family Scolopacidae (Sandpipers, snipes and phalaropes). Pp. 444–487 *in* Handbook of the Birds of the World. Vol. 3. Hoatzin to Auks (J. del Hoyo, A. Elliott, and J. Sargatal, Eds.). Lynx Edicions, Barcelona.

PIERSMA, T. 1996b. Family Charadriidae (Plovers). Pp. 384–409 in Handbook of the Birds of the World. Vol. 3. Hoatzin to Auks (J. del Hoyo, A. Elliott, and J. Sargatal, Eds.). Lynx Edicions, Barcelona.

PIERSMA, T., AND N. DAVIDSON. 1992. The migrations and annual cycles of five subspecies of knots. Wader Study Group Bulletin 64, Supplement: 187–197.

PIERSMA, T., A. J. BEINTEMA, N. C. DAVIDSON, O. A. G. MUNSTER, AND M. W. PIENKOWSKI. 1987. Wader migration systems in the East Atlantic. Wader Study Group Bulletin 64, Supplement/IWRB Special Publication 7: 35–56.

PIERSMA, T., P. PROKOSCH, AND D. BREDIN. 1992. The migration system of Afro-Siberian Knots Calidris canutus canutus. Wader Study Group Bulletin 64, Supplement: 52–63.

PIERSMA, T., N. CADÉE, AND S. DAAN. 1995. Seasonality in basal metabolic rate and thermal conductance in a long-distance migrant shorebird, the knot (Calidris canutus). Journal of Comparative Physiology B 165: 37–45.

PIERSMA, T., R. VAN AELST, K. KURK, H. BERKOUDT, AND L. R. M. MAAS. 1998. A new pressure sensory mechanism for prey detection in birds: the use of principles of seabed dynamics? Proceedings of the Royal Society of London B 265: 1377–1383.

PRATER, A. J. 1981. Estuary Birds of Britain and Ireland. T. & A. D. Poyser Ltd., Carlton.

RANDS, M. R. W. 1996. Family Dromadididae (Crab Plover). Pp. 302–307 in Handbook of the Birds of the World. Vol. 3. Hoatzin to Auks (J. del Hoyo, A. Elliott, and J. Sargatal, Eds.). Lynx Edicions, Barcelona.

REED, J. M., C. S. ELPHICK, AND L. W. ORING. 1998. Life-history and viability analysis of the endangered Hawaiian stilt. Biological Conservation 84: 35–45.

REYNOLDS, J. D. 1987. Mating system and nesting biology of the Red-necked Phalarope Phalaropus lobatus: what constrains polyandry? Ibis 129: 225–242.

RICHARDSON, W. J. 1976. Autumn migration over Puerto Rico and the western Atlantic: a radar study. Ibis 118: 309–332.

RIEGEN, A. C. 1999. Movements of banded Arctic waders to and from New Zealand. Notornis 46: 123–142.

ROEMMICH, D., AND J. A. MCGOWAN. 1995. Climatic warming and the decline of zooplankton in the California Current. Science 267: 1324–1326.

RUBEGA, M. A. 1997. Surface tension prey transport in shorebirds: how widespread is it? Ibis 139: 488–493.

RUBEGA, M. A., AND B. S. OBST. 1993. Surface-tension feeding in phalaropes: discovery of a novel feeding mechanism. Auk 110: 169–178 + frontispiece.

RUBEGA, M. A., D. SCHAMEL, AND D. TRACEY. 2000. Red-necked Phalarope (Phalaropus lobatus). No. 538 in The Birds of North America (A. Poole and F. Gill, Eds.). Academy of Natural Sciences, Philadelphia; American Ornithologists' Union, Washington, D.C.

RUFINO, R., A. ARAUJO, J. P. PINA, AND P. MIRANDA. 1984. The use of salinas by waders in the Algarve, south Portugal. Wader Studies Group Bulletin 42: 41–42.

SAFRIEL, U. N., B. J. ENS, AND A. KAISER. 1996. Rearing to independence. Pp. 219–250 in The Oyster-catcher: From Individuals to Populations (J. D. Goss-Custard, Ed.). Oxford University Press, Oxford.

SAMPATH, K., AND K. KRISHNAMURTHY. 1989. Shorebirds of the salt ponds at the Great Vedaranyam Salt Swamp, Tamil Nadu, India. Stilt 15: 2–23.

SANDERCOCK, B. K. 1997. Factors Affecting the Breeding Demography of Western Sandpipers (Calidris mauri) and Semipalmated Sandpipers (C. pusilla) Breeding at Nome, Alaska. Ph.D. dissertation, Simon Fraser University, Burnaby.

SANDERCOCK, B. K., AND C. L. GRATTO-TREVOR. 1997. Local survival in Semipalmated Sandpipers Calidris pusilla breeding in La Pérouse Bay, Canada. Ibis 139: 305–312.

SCHICK, C. T., L. A. BRENNAN, J. B. BUCHANAN, M. A. FINGER, T. M. JOHNSON, AND S. G. HERMAN. 1987. Organochlorine contamination in shorebirds from Washington state and the significance for their falcon predators. Environmental Monitoring and Assessment 9: 115–131.

SCHNEIDER, D. C. 1981. Food supplies and the phenology of migratory shorebirds: a hypothesis. Wader Study Group Bulletin 33: 43–45.

SIEGEL-CAUSEY, D. 1988. Phylogeny of the Phalacrocoracidae. Condor 90: 885–905.

SHEPHERD, P. C., AND J. S. BOATES. 1999. Effects of a commercial baitworm harvest on Semipalmated Sandpipers and their prey in the Bay of Fundy Hemispheric Shorebird Reserve. Conservation Biology 13: 347–356.

SIBLEY, C. G., AND B. L. MONROE, JR. 1990. Distribution and Taxonomy of Birds of the World. Yale University Press, New Haven, CT.

SKAGEN, S. K., P. B. SHARPE, R. G. WALTERMIRE, AND M. B. DILLON. 1999. Biogeographical Profiles of Shorebird Migration in Midcontinental North America. Biological Science Report USGS/BRD/BSR — 2000–0003. U.S. Government Printing Office, Denver, CO.

SKAGEN, S. K., AND H. D. OMAN. 1996. Dietary flexibility of shorebirds in the Western Hemisphere. Canadian Field-Naturalist 110: 419–444.

SMIT, C. J., R. H. D. LAMBECK, AND W. J. WOLFF. 1987. Threats to coastal wintering and staging areas of waders. Wader Study Group Bulletin 64, Supplement/IWRB Special Publication 7: 105–113.

STANFORD, W. P. 1953. Winter distribution of the Grey Phalarope *Phalaropus fulicarius*. Ibis 95: 483–491.

SUMMERS, R. W. 1994. The migration patterns of the Purple Sandpiper *Calidris maritima*. Ostrich 65: 167–173.

SUMMERS, R. W., L. G. UNDERHILL, D. J. PEARSON, AND D. A. SCOTT. 1987. Wader migration systems in southern and eastern Africa and western Asia. Wader Study Group Bulletin 49, Supplement/IWRB Special Publication 7: 15–34.

SUMMERS, R. W., L. G. UNDERHILL, C. F. CLINNING, AND M. NICOLL. 1989. Populations, migrations, biometrics and moult of the Turnstone *Arenaria i. interpres* on the East Atlantic coastline, with special reference to the Siberian population. Ardea 77: 145–168.

SUTHERLAND, W. J., B. J. ENS, J. D. GOSS-CUSTARD, AND J. B. HULSCHER. 1996. Specialization. Pp. 56–76 *in* The Oystercatcher: From Individuals to Populations (J. D. Goss-Custard, Ed.). Oxford University Press, Oxford.

TERP, J. M. 1998. Habitat Use Patterns of Wintering Shorebirds: The Role of Salt Evaporation Ponds in South San Diego Bay. M.S. thesis, San Diego State University, San Diego, CA.

THOMPSON, M. 1973. Migratory patterns of Ruddy Turnstones in the central Pacific region. Living Bird 12: 5–23.

TINBERGEN, N. 1935. Field observations of East Greenland Birds. I. The behavior of the Red-necked Phalarope (*Phalaropus lobatus* L.) in spring. Ardea 26: 1–42.

TROY, D. 1996. Population dynamics of breeding shorebirds in Arctic Alaska. International Wader Studies 8: 15–27.

TSIPOURA, N., AND J. BURGER. 1999. Shorebird diet during spring migration stopover on Delaware Bay. Condor 101: 635–644.

TULP, I., S. McCHESNEY, AND P. De GOEIJ. 1994. Migratory departures of waders from northwestern Australia: behaviour, timing and possible migration routes. Ardea 82: 201–221.

TYE, A. 1987. Identifying the major wintering grounds of Palearctic waders along the Atlantic coast of Africa from marine charts. Wader Study Group Bulletin 49: 20–27.

TYE, A., AND H. TYE. 1987. The importance of Sierra Leone for wintering waders. Wader Study Group Bulletin 49, Supplement/IWRB Special Publication 7: 71–75.

TYLER, W. B., K. T. BRIGGS, D. B. LEWIS, AND R. G. FORD. 1993. Seabird distribution and abundance in relation to oceanographic processes in the California Current System. Pp. 48–60 *in* The Status, Ecology and Conservation of Marine Birds of the North Pacific (K. Vermeer, K. T. Briggs, K. H. Morgan, and D. Siegal-Causey, Eds.). Canadian Wildlife Service Special Publication, Ottawa.

UNDERHILL, L. G., A. J. TREE, H. D. OSCHADLEUS, AND V. PARKER. 1999. Review of Ring Recoveries of Waterbirds in Southern Africa. Avian Demography Unit, University of Cape Town, Cape Town, South Africa.

VAN GILS, J., AND T. PIERSMA. 1999. Day- and night-time movements of radiomarked Red Knots staging in the western Wadden Sea in July–August 1995. Wader Study Group Bulletin 89: 36–44.

VAN GILS, J., AND P. WIERSMA. 1996. Family Scolopacidae (sandpipers, snipes and phalaropes), species accounts. Pp. 488–533 *in* Handbook of the Birds of the World. Vol. 3. Hoatzin to Auks (J. del Hoyo, A. Elliott, and J. Sargatal, Eds.). Lynx Edicions, Barcelona.

VAN RHIJN, J. G. 1991. The Ruff. T. & A. D. Poyser, Ltd., London.

VELASQUEZ, C. 1993. The Ecology and Management of Waterbirds at Commercial Saltpans in South Africa. Ph.D. dissertation, University of Cape Town, Cape Town, South Africa.

VICKERY, P. D. 1978. The fall migration; Northeastern Maritime region. American Birds 32: 174–180.

VON BOLZE, G. 1968. Anordnung und Bau der herbstschen Körperchen in Limicolenschnäbeln im Zusammenhang mit der Nahrungsfindung. Zoologischer Anzeiger 181: 313–343.

VUILLEUMIER, F. 1996. Birds observed in the Arctic Ocean to the North Pole. Arctic and Alpine Research 28: 118–122.

WAHL, T. R., K. H. MORGAN, AND K. VERMEER. 1993. Seabird distribution off British Columbia and Washington. Pp. 39–47 *in* The Status, Ecology and Conservation of Marine Birds of the North Pacific (K. Vermeer, K. T. Briggs, K. H. Morgan, and D. Siegal-Causey, Eds.). Canadian Wildlife Service Special Publication, Ottawa.

WAHL, T. R., AND B. TWEIT. 2000. Seabird abundances off Washington, 1972–1998. Western Birds 31: 69–88.

WARHAM, J. 1996. The Behaviour, Population Biology and Physiology of the Petrels. Academic Press, London.

WARNOCK, N., AND R. E. GILL, JR. 1996. Dunlin (*Calidris alpina*). No. 203 *in* The Birds of North America (A. Poole and F. Gill, Eds.). Academy of Natural Sciences, Philadelphia; American Ornithologists' Union, Washington, D.C.

WARNOCK, N., G. W. PAGE, AND B. SANDERCOCK. 1997. Local survival of Dunlin *Calidris alpina* wintering in California. Condor 99: 906–915.

WARNOCK, N., AND M. A. BISHOP. 1998. Spring stopover ecology of migrant Western Sandpipers. Condor 100: 456–467.

WARNOCK, S. E., AND J. Y. TAKEKAWA. 1995. Habitat preferences of wintering shorebirds in a temporally changing environment: Western Sandpipers in the San Francisco Bay estuary. Auk 112: 920–930.

WATKINS, D. 1993. A National Plan for Shorebird Conservation in Australia. Australasian Wader Studies Group, Royal Australasian Ornithologists Union and World Wildlife Fund for Nature, RAOU Report No. 90.

WATKINS, D. 1997. East Asian–Australasian Shorebird Reserve Network. Pp. 132–137 *in* Shorebird Conservation in the Asia-Pacific Region (P. Straw, Ed.). Australasian Wader Studies Group, Victoria.

WEIDENSAUL, S. 1999. Living on the Wind: Across the Hemisphere with Migratory Birds. North Point Press, New York.

WESTERN HEMISPHERE SHOREBIRD RESERVE NETWORK. 1990. Going global. Conservation and biology in the Western Hemisphere. WHSRN Network News 3: 1–11.

WHITE, D. H., K. A. KING, AND R. M. PROUTY. 1980. Significance of organochlorine and heavy metal residues in wintering shorebirds at Corpus Christi, Texas, 1976–1977. Pesticide Monitoring Journal 14: 58–63.

WIERSMA, P. 1996. Family Charadriidae (Plovers), species accounts. Pp. 411–442 *in* Handbook of the Birds of the World. Vol. 3. Hoatzin to Auks (J. del Hoyo, A. Elliott, and J. Sargatal, Eds.). Lynx Edicions, Barcelona.

WILLIAMS, T. C., AND J. M. WILLIAMS. 1990. The orientation of transoceanic migrants. Pp. 7–21 *in* Bird Migration: Physiology and Ecophysiology (E. Gwinner, Ed.). Springer-Verlag, Berlin.

WILLIAMS, T. C., AND J. M. WILLIAMS. In press. The migration of land birds over the Pacific Ocean. Proceedings of the International Ornithological Congress 22.

WILLIAMS, T. C., J. M. WILLIAMS, L. C. IRELAND, AND J. M. TEAL. 1977. Autumnal bird migration over the western North Atlantic Ocean. American Birds 31: 251–267.

WILSON, J. R. 1981. The migration of High Arctic shorebirds through Iceland. Bird Study 28: 21–32.

WILSON, J. R., AND M. A. BARTER. 1998. Identification of potentially important staging areas for "long jump" migrant waders in the east Asian-Australasian flyway during northward migration. Stilt 32: 16–27.

WOLFF, W. J., AND C. J. SMIT. 1990. The Banc D'Arguin, Mauritania, as an environment for coastal birds. Ardea 78: 17–38.

WYMENGA, E., M. ENGELMOER, C. J. SMIT, AND T. M. VAN SPANJE. 1990. Geographical breeding origin and migration of waders wintering in West Africa. Ardea 78: 83–112.

YOUNGSWORTH, W. 1936. The cruising speed of the golden plover. Wilson Bulletin 48: 53.

Mixed-Species Feeding Group of Wading Birds and Other Species

19 Wading Birds in the Marine Environment

Peter C. Frederick

CONTENTS

0-8493-9882-7/02/$0.00+$1.50
© 2002 by CRC Press LLC

19.1 INTRODUCTION

Many kinds of birds walk in water, or wade. This chapter is about the long-legged wading birds, which are here defined as the herons, egrets, ibises, storks, and spoonbills, all of which are in the order Ciconiiformes. Although shorebirds are referred to as "waders" in Europe and other parts of the globe, ciconiiform birds are quite distinct from shorebirds. Cranes (family Gruidae) and flamingos (family Phoenicopteridae) are also long-legged birds that wade, but not in the marine environment and they are not covered in this chapter. Long-legged wading birds are long in most dimensions, having long legs, toes, bills, and necks. With few exceptions, wading birds are strongly associated with shallowly flooded wetlands, in which they generally breed and feed.

Long-legged wading birds are one of the largest and most diverse groups of large birds, comprised of members of three main families (Figures 19.1 and 19.2). The Ardeidae, or herons, egrets and bitterns, are the most diverse, with approximately 60 species (Hancock and Kushlan 1984). These birds have straight, harpoon-like bills, generally narrow heads, a comb-like (pectinate) middle toe, and a modified 6th cervical vertebrae that allows the long neck to be held in an S-shape in flight. These species range from the diminutive Least Bittern (*Ixobrychus exilis*, 28 cm length) to the large and stately Goliath Heron (*Ardea goliath*, 140 cm). The Threskiornithidae (ibises and spoonbills, approximately 30 species) generally are shorter-legged, with distinctive down-curved or spatulate bills, grooved bill surfaces for cleaning feathers, a lack of powder down, a cupped middle toenail, and a slit-like cranial morphology (schizorhinal). Representatives include the brilliant Scarlet Ibis (*Eudocimus ruber*, 58 cm long), the Giant Ibis (*Thaumatibis gigantea*, to 103 cm), and the Roseate Spoonbill *(Ajaia ajaja*, 80 cm tall). The Ciconiidae, or storks (20 species), have massive straight or slightly decurved bills, and typically defecate on their legs for evaporative cooling. These are the giants of the order, including Wood Storks *(Mycteria americana*, 100 cm tall), the massive Marabou Stork of the African plains (*Leptopilos crumeniferus*, 120 cm tall), and the immense Jabiru Stork of Central and South American wetlands (*Jabiru mycteria*, 145 cm tall).

Although it is clear that the three main families of wading birds should be grouped together taxonomically within Ciconiiformes, there is considerable debate about other groups within Ciconiiformes. DNA evidence suggests that wading birds, flamingos, and pelicans are descended from a common ancestor (Sibley and Ahlquist 1990), and possibly that new-world vultures should be included within the order. Some taxa of wading birds are known to be quite old: ibises and herons date at least to the Miocene, about 25 million years ago. Some extinct island ibises on Jamaica and the Hawaiian Islands were flightless (Hancock et al. 1992).

19.2 REPRODUCTIVE BIOLOGY

Many species of long-legged wading birds are gregarious and may breed colonially in large, conspicuous, mixed-species aggregations, which can include up to 500,000 birds (Robertson and Kushlan 1974, Ogden 1994). Like many of the adaptations and life-history features of wading birds, coloniality is thought to be, in part, the result of needing to find and exploit patches of food that are unpredictable in space and time (Krebs 1974).

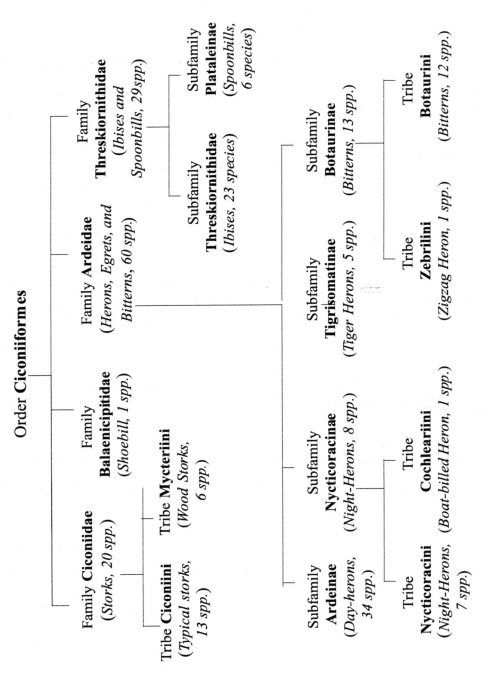

FIGURE 19.1 Schematic classification of ciconiiform birds, following Peters (1931), Hancock and Kushlan (1984), and Hancock et al. (1992).

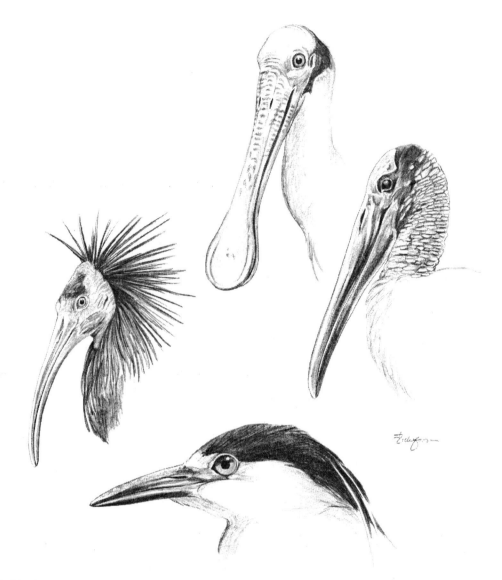

FIGURE 19.2 Illustrations of heads and bills of representatives of the major groups of long-legged wading birds: Roseate Spoonbill (*Ajaia ajaja*, Threskiornithidae, top), Wood Stork (*Mycteria americana*, Ciconidae, right), Black-crowned Night Heron (*Nycticorax nycticorax*, Ardeidae, bottom), and Waldrapp Ibis (*Geronticus eremita*, Threskiornithidae, left). (Drawing by J. Zickefoose.)

19.2.1 Pair Bonds and Parental Care

Wading birds are socially monogamous, with pair bonds that last at least one breeding attempt. Most wading birds probably acquire new mates every season (Simpson et al. 1987), though some species of storks may remain with the same mate for many years. Pair-formation displays often are elaborate (Meyerriecks 1960, McCrimmon 1974, Wiese 1976, Mock 1980, Hancock et al. 1992) and usually are performed from small territories defended by the male near eventual nest sites. Both members of the pair typically help build the nest, incubate, and care for young. As in many socially monogamous, colonial-nesting birds (Birkhead and Moller 1992), copulations between members of different pairs can occur (Fujioka and Yamagishi 1981, Frederick 1987b), though the extent of this behavior remains poorly studied.

19.2.2 NESTS, INCUBATION, AND YOUNG

Breeding colonies and roosts usually are formed on islands, either surrounded by water or by some vegetative buffer, or are in tall trees. These features may serve as a form of protection from terrestrial predators (Rodgers 1987). Nest substrate requirements are generally broad and well researched in this group of birds (McCrimmon 1978, Bjork 1986, Hafner 1997). Nesting wading birds are not very picky about the vegetation type in which they nest, though they may be more specific about nest height. Burger (1978) found that nest height within a colony reflected interspecies dominance hierarchies, with the most submissive species nesting closest to the ground.

Nests are built of sticks and other vegetation and may or may not be re-used between years (Hancock et al. 1992). Large aggregations of nesting wading birds can have direct effects on the vegetation in and around colonies. For example, Siegfried (1971) estimated that over 1.5 million sticks weighing over 2000 kg were needed to support a Cattle Egret (*Bubulcus ibis*) colony of 5000 pairs. As nest densities increase and the availability of nest material decreases, the size of individual nests decreases (Arendt and Arendt 1988), making nests less sturdy and more vulnerable to adverse weather. In addition, the deposition of excreta in colonies can kill shrubs and trees through excess nutrients (Wiese 1978).

Incubation begins with the laying of the first or second egg, resulting in hatching asynchrony and a size disparity between first- and last-hatched young. This pattern leads to unequal division of food resources, and often to high mortality of the smaller young (see also "Life History" below). Incubation of eggs ranges from 19 days in the smallest herons to 30 days in the largest storks. Young are semialtricial, usually hatched with some down but are unable to move much around the nest for the first couple of days. Feeding is by regurgitation of food from parents, either onto the surface of the nest or (usually later) directly into the chicks' bills. In herons, the young "scissor" the adult's bill by grasping on the outside of the parent's mandibles; the parent then regurgitates through partially open bill into the gape of the chick. In ibises and spoonbills, young place their bill directly into the gape of the parents.

Growth of young is rapid; legs and feet grow disproportionately faster than other body parts (McVaugh 1975), an adaptation interpreted as the need to rapidly gain locomotor abilities in order to climb away from predators (Werschkul 1979). Unlike many birds, young ciconiiform birds leave the nest some weeks in advance of the development of flight abilities, and up to half the period between hatching and leaving the colony may be spent in treetops and the vicinity of the nest (tens to >100 m from the nest site, Frederick et al. 1992). Thus in wading birds "fledging" refers to the time at which young actually fly away from the colony, rather than the departure of young from the nest. Parents also encourage young to follow them at feeding time, starting from hops between branches, to short, and then long flights in pursuit of the parent. The period from hatching to independence from the colony may take from 40 to 100 days.

19.2.3 REPRODUCTIVE SUCCESS

As with most birds, success of nesting attempts varies, depending on ecological and environmental conditions. Although nesting is rarely affected directly by weather (nests blown down or nest contents scattered, but see Quay 1963 in Parnell et al. 1988, Bouton 1999), indirect effects on foraging are more widespread (see below). Wading birds do not display much in the way of individual or group nest defense, and nesting success may be strongly affected by predatory reptiles, mammals, and birds (Shields and Parnell 1986, Rodgers 1987, Burger and Hahn 1989). Although some avian and reptilian scavengers may be considered normal associates of wading bird nesting aggregations (Shields and Parnell 1986, Burger and Hahn 1989, Frederick and Collopy 1989b, Bouton 1999, see "Management" below), large mammalian predators, particularly nocturnal ones, can cause widespread abandonment of colonies (Rodgers 1987, Post 1990). Measuring the effect of predation, however, has been a challenge, since the presence of researchers in colonies can result in opportunities for scavengers to rob nests. Several approaches have managed to get around this difficulty. One is to observe nests remotely (Pratt and Winkler 1985, Bouton 1999).

Productivity of nests may increase with age of nesting pairs in some species. For example, Fernandez-Cruz and Campos (1993) reported that in Grey Herons (*Ardea cinerea*), brood size increased from 1.8 to 2.8 in nests where parents were 2 and >4 years of age, respectively. There is evidence that clutch size increases at inland compared with coastal sites, and with increasing latitude (Rudegeair 1975, Kushlan 1977, Frederick et al. 1992). Explanations for the former pattern include energetic costs of salt excretion in coastal zones and increased availability of food resources at inland sites (Rudegeair 1975).

19.2.4 PREY AVAILABILITY AND NESTING SUCCESS

Access to rich food resources is probably the single most often cited factor affecting reproductive success. Annual fluctuations in availability of prey have been linked with date of nest initiation in Wood Storks (Ogden 1994) and number of nesting birds in White Ibises (*Eudocimus albus*, Frederick and Collopy 1989a, Bildstein et al. 1990) and Wood Storks (Ogden 1994). Similarly, events which interrupt the supply of food seem to lead to the abandonment of nesting events. These can include sharp increases in the surface water depth (Kahl 1964, Kushlan et al. 1975, Frederick and Collopy 1989a), droughts (Bancroft et al. 1994), and sudden onset of cold temperatures (Frederick and Loftus 1993). Availability of food therefore seems to be a powerful cue in the sequence leading to the instigation of nesting, as well as a direct cause of the cessation of nesting.

Food availability also affects nesting productivity. Powell (1983) compared Great Blue Herons (*Ardea herodias*) in Florida Bay that received food supplementation via handouts from local residents, with birds foraging in the estuary. "Panhandler" birds laid larger clutches and produced more young than did unsupplemented birds, indicating a strong effect of food availability. Similarly, Hafner et al. (1993) found that productivity of Little Egrets (*Egretta garzetta*) in the Camargue Delta of France was linked to access to high densities of prey in particular habitats.

Rainfall in the weeks or months preceding breeding has been correlated with reproductive effort and success by wading birds (Ogden et al. 1980, Maddock 1986, Bildstein et al. 1990, Hafner et al. 1994, Kingsford and Johnson 1998). This relationship appears to be related directly to the size of flooded wetland areas, and consequently to the productivity of aquatic fish and macroinvertebrates. The dynamics of aquatic prey communities may also be affected by fluctuations in populations of large, predatory fishes. Secondary productivity (production of fish and invertebrates that are primary grazers) may be strongly adapted to, and affected by, cycles of drought and flood. Droughts tend to result in direct mortality of wetland vegetation, either directly through desiccation or through the action of fires. These processes may lead to the release of nutrients stored in vegetation or in the surface layers of the soil and detritus. Nutrient release during re-flooding may fuel pulses of both primary and secondary productivity. An understanding of prey animal ecology remains crucial to understanding the linkage between wading birds and their wetland environments.

Given the importance of food availability to wading bird reproduction, it is not surprising that colony site choice is linked with the location, quality, and size of foraging habitat (Fasola and Barbieri 1978, Moser 1984, Gibbs et al. 1987, Gibbs 1991). In Illinois, the availability of lacustrine and emergent wetland was the primary determinant for location and size of Great Blue Heron colonies, with degree of isolation from human disturbance being of secondary importance (Gibbs and Kinkel 1987; see also Grull and Ranner 1998). Ogden (1994) demonstrated that the estuarine zone of the Everglades was abandoned by wading birds in favor of inland areas between 1975 and 1992, as a result of the loss of freshwater flows due to upstream water management (Walters et al. 1992, McIvor et al. 1994).

Adult wading birds often fly considerable distances from breeding colonies to foraging sites (Figure 19.3). Ogden et al. (1988) recorded Wood Storks flying up to 130 km from breeding

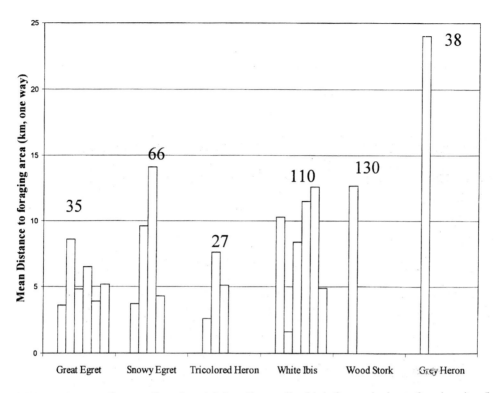

FIGURE 19.3 Average distances flown by adult breeding wading birds from colonies to foraging sites (km, one way); maximum distances are indicated above the bars. These data are from a mix of studies that variously used radio telemetry, marked birds, or light aircraft to document foraging distances of individual birds. Note that maximum distances for most species are much larger than means — up to 110 km for White Ibises and 130 km for Wood Storks.

colonies to feed, and Bildstein (1993) reported cases of White Ibises regularly traveling 110 km one way. These large distances traveled may be accomplished by direct, solitary flight (Smith 1995a), or may involve energetic savings through the use of formation flights or the use of thermals (Kahl 1964).

Wading birds are usually very flexible in choice of foraging sites, and foraging locations used while breeding may change frequently, both within a breeding season and between years (Custer and Osborn 1978, Hafner and Britton 1983, Bancroft et al. 1994, Frederick and Ogden 1997). It is not clear at what point distance to food has an effect on reproductive success. Certainly the large distances recorded by Ogden and Bateman (1970, above) were associated with successful breeding. Little is known of the ecology of young wading birds following departure from the colony. Many young wading birds disperse long distances shortly following fledging, and may be found hundreds of kilometers from their natal sites (Coffey 1943, 1948, Byrd 1978, P. Frederick unpublished), possibly allowing young to identify sources of food that are unpredictable in space and time (van Vessem and Draulans 1986).

19.3 FORAGING ECOLOGY

The foraging ecology of wading birds has been particularly well studied. The resulting body of literature offers a fascinating variety of scientific approaches involving the fields of sensory physiology, social behavior, cost–benefit analysis, predator–prey relationships, energy flow, niche partitioning, and nutrient ecology.

19.3.1 FORAGING BEHAVIOR

When feeding, wading birds use a variety of foraging techniques. Herons and egrets employ a range of behaviors that include slow stalking, sit-and-wait, active pursuit, and, more rarely, aerial foraging at the surface, or aerial plunges (Meyerriecks 1962, Kushlan 1976b; see Figure 19.4.). The Green-backed Heron (*Butorides striatus*) uses bait of various kinds to attract fish to within striking range (Higuchi 1986, 1988). Ibises and storks stalk or pursue prey, but are more likely to probe with partly open bills into soft substrate, using tactile means and sensory pits in the bill to detect prey. Both Snowy Egrets and Wood Storks frequently use their feet to stir up prey hidden in sediments or vegetation. Spoonbills swing their bills in a horizontal arc through the water, often in unison, a technique that when coupled with the unique configuration of the shape of bill, acts to pull small particles into the bill by creating an area of lower pressure in the bill opening into which small food items may be swept (Weihs and Katzir 1994).

Sight-foraging birds must contend with the dual problems of surface glare and the need to correct for the refraction caused by items being underwater (Figure 19.5). Both Little Egrets and Reef Herons (*Ardea gularis*) are able to correct for differences in actual position of prey due to refraction (Katzir and Intrator 1987, Lotem et al. 1991). Glare may be reduced by extending one or both wings during foraging (Frederick and Bildstein 1992), or by tilting the head (Krebs and Partridge 1973). One of the most extreme foraging behaviors is "canopy feeding," described primarily for Black Herons (*Ardea ardesiaca*), in which the wings are spread in a circle with the head and neck beneath the canopy, creating an area of darker water into which the egret looks for prey.

Although many wading birds are diurnal feeders, some, like the night-herons and Boat-billed Herons (*Cochlearius cochlearius*), are most frequently nocturnal, foraging in the daylight only when the energetic demands of nesting require it. Many species choose to forage during crepuscular hours at both ends of the day, in some cases despite weather and tidal conditions (Draulans and Hannon 1988).

Many wading birds forage early in the morning and are more likely to forage in flocks at that time. Although early-morning feeding is explained in part by the preceding nightlong fast, early feeding may also be the result of a predictable and temporary increased availability of prey. Hafner et al. (1993) found that timing of flock feeding and temporal variation in foraging success of Little Egrets in the Camargue of France were explained by low dissolved oxygen levels in water during the morning (nocturnal respiration by macrophytes depleted the water of oxygen, forcing fish to breathe in the more oxygenated layers at the surface). Soon after sunrise, dissolved oxygen increased as a result of the diurnal portion of plant respiration, and capture rates decreased rapidly.

19.3.2 FLOCK-FORAGING DYNAMICS

Wading birds often feed in dense mixed-species flocks with other waterbirds (Figure 19.6). Some species, like Snowy and Little Egrets, are rarely found foraging solitarily (Hafner et al. 1982, Master et al. 1993), while others, such as Tricolored Herons (*Egretta tricolor*) and Goliath Herons, are typically solitary when foraging (Mock and Mock 1980, Hancock and Kushlan 1984). Many species forage solitarily and breed colonially (Marion 1989). Individuals may switch from solitary to social foraging depending on the richness, predictability, and defensibility of the food source, as well as stage of nesting (Simpson et al. 1987, Draulans and Hannon 1988, Marion 1989). In South Florida, White Ibises and Snowy Egrets tended to travel in flocks and land together or near other birds, but Great Egrets (*Ardea albus*) and Tricolored Herons tended to forage solitarily whether they departed the colony in a flock or not (Smith 1995b; see also Strong et al. 1997). Master et al. (1993) suggested that Snowy Egrets were obligate in their use of dense foraging aggregations because their active foraging behaviors were, for a variety of reasons, most efficient in those situations.

Foraging flocks of up to several hundred individuals often are formed of several species of waterbirds. For example, Frederick and Bildstein (1992) observed foraging flocks in Venezuela containing up to seven species of ibises, five of herons, two storks, one spoonbill, two species of ducks, and three raptors. These large aggregations are a mix of conflicting pressures for individuals

FIGURE 19.4 Foraging behaviors displayed by Reddish Egret (*Egretta rufescens*), showing running (top), double-wing feeding (right), and peering into water (left). (Drawing by J. Zickefoose.)

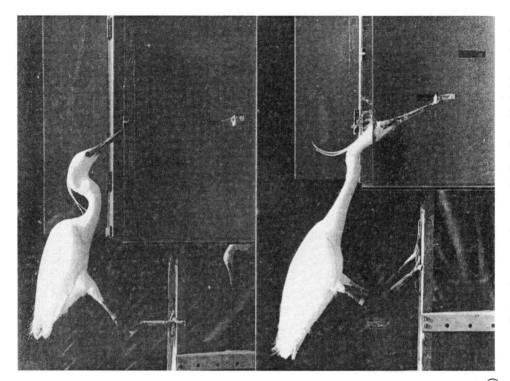

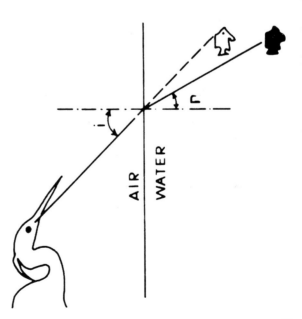

(b)

(a)

FIGURE 19.5 (a) Disparity between the actual and apparent position of prey in water due to light refraction at the water/air interface. (b) Striking of underwater prey by a Reef Heron (*Egretta garzetta gularis*), showing approach and aiming (above) and prey capture (below). (From Katzir and Martin [1994], reprinted with permission.)

FIGURE 19.6 Illustration of a dense multispecific feeding flock, showing standing (Great Egret, top right), foot dragging (Snowy Egret, bottom right), head swinging (Roseate Spoonbills, center), and foot stirring and groping (Wood Stork, bottom left). High densities of birds in such groups often lead to confusion of prey, as well as interference, competition, food piracy, and interspecific aggression. (Drawing by J. Zickefoose.)

trying to reap the benefits (dense prey, increased foraging success, decreased patch search time, increased overall vigilance for and safety from predators) and avoid the costs (increased attraction of predators, competition for prey, dominance interactions, interruption of foraging bouts, theft of prey items) of social foraging. A central problem in documenting the costs and benefits of flocking in waterbirds has been to separate the effects of quality of foraging site from the effects generated by the fact of many birds foraging together (competition or social facilitation). Wading birds have evolved a variety of bill structures, sizes, behavioral patterns, and prey preferences, a fact that suggests that interspecific competition may have resulted in partitioning of the feeding niche space through adaptation. Master et al. (1993) described dramatic interspecific differences in the degree of advantage in flock foraging of herons, suggesting that some species stand more to gain from these aggregations than do others. However, other studies suggest that avoidance of interspecific competition may occur even within mixed-species aggregations. Frederick and Bildstein (1992) found little evidence of overlap in various measures of foraging niche (behavior, depth, microhabitat, prey species) among seven species of ibises that were forced into foraging flocks by drying water in the Llanos of Venezuela. Caldwell (1981) found almost complete overlap of prey species amongst four socially foraging heron species studied in Panama.

Petit and Bildstein (1987) found that White Ibises, on the periphery of foraging flocks and solitary birds, stepped faster and looked up more often for predators than did individuals in the center. By comparison with solitary foragers, Master et al. (1993) found that species foraging actively in the center of a group showed the greatest improvements over solitary foraging; those using more sedentary behaviors and those on the periphery of the flock showed the least improvement.

Several authors have described species and individuals that seem to specialize on the theft of prey items procured by others (kleptoparasitism, Ens et al. 1990). In the Venezuelan Llanos, Scarlet Ibises frequently stole large aquatic water beetles from Glossy Ibises (*Plegadis falcinella*), and individual Scarlets even defended groups of Glossy Ibises from other potentially parasitic Scarlets (Frederick and Bildstein 1992). Primary thefts were often followed by secondary theft of the same items from conspecific Scarlet Ibises or by Yellow-headed Caracaras (*Milvago chimachima*). Similarly, Gonzalez (1996) found that 7 of the 15 species of wading birds studied in the llanos attempted either inter- or intraspecific food piracy, and that over 20% of the food consumed by Jabiru Storks came from food piracy behavior, with over 77% of piracy attempts successful.

19.3.3 SOLITARY FORAGING

Although multispecific feeding flocks are a conspicuous and frequent feature of wading bird foraging behavior, solitary and territorial feeding also is typical for many species (Powell 1983, Butler 1997). Not surprisingly, territorial feeders tend to forage by stalking, a strategy that is hindered by the activity of other individuals nearby. Hafner et al. (1982) noted that foraging success of the sedentary foraging Squacco Heron (*Ardeola ralloides*) decreased with flock size, suggesting that foraging in flocks is not generally advantageous for this species. Wiggins (1991) found that there were significant energetic costs to Great Egrets defending individual feeding territories, but that solitary birds tended to catch larger fish than did flock-foraging egrets.

19.3.4 FEEDING FROM HUMAN SOURCES

Wading birds may forage on food left by humans. In Africa, Marabou Storks frequently eat offal from slaughterhouses (Hancock et al. 1992), an easy extension of their natural habit of eating carcasses of large wild animals. Powell and Powell (1986) described routine consumption of bait fish from local human residents ("panhandling") among Great Blue Herons in Florida Bay, and showed that some birds specialize in begging bait fish from residents. Reliance on human food sources may become particularly important when other foraging choices become restricted. For example, Smith (1995b) found that 5 to 9% of Great Egrets on Lake Okeechobee, FL foraged by panhandling in nondrought years but 24% did so during a severe drought.

19.3.5 Conditions Affecting Foraging Success

Foraging success of wading birds is, in most situations, constrained by water depth (Powell 1987) and density of prey (Draulans 1987). Renfrow (1993) showed experimentally that foraging success of egrets in Texas impoundments was explained by both water depth and prey density. Surdick (1998) compared the characteristics of foraging sites in the Everglades with choice of foraging site and with foraging success of individual birds. He concluded that prey density, water depth, and vegetation density explained the vast majority of variation in foraging success and choice of foraging site. In vegetation-free impoundments, Gawlik (in press) showed experimentally that both water depth and prey density strongly affected choice of foraging site of ibises, storks, and herons with a clear increase in "giving-up density" with increasing depth. Gawlik also showed that some species were sensitive to depth, some sensitive to density, and others to both parameters.

19.3.6 Prey Animals

Most wading birds are opportunistic feeders and tend to specialize on whatever is locally abundant. Diets include a wide range of aquatic taxa, including fish, amphibians, crustaceans, aquatic insects, and other invertebrates. Even so, small mammals, lizards, and the occasional bird can be taken by some of the larger species when foraging on land (Butler 1997). In rice fields, the Glossy Ibis forages on rice for up to 58% of its diet (Acosta et al. 1996). Many species, such as the Tricolored Heron and Great Egret, are almost entirely piscivorous, while others, such as Yellow-crowned Night Herons (*Nycticorax violacea*), specialize on crustaceans. Sizes of prey taken are quite variable with Little Egrets specializing on tiny tadpole shrimp in some seasons (Hafner et al. 1982), while Goliath Herons take fishes of up to 50 cm length. The Shoebill (*Balaeniceps rex*) takes particularly large prey (Hancock et al. 1992). Overall, little is known about food habits or energetics during the nonbreeding season.

19.4 LIFE-HISTORY CHARACTERISTICS

In general, herons and ibises tend to be somewhat smaller and quicker to reach maturity than seabirds, and are probably more fecund and shorter-lived than the larger storks (Table 19.1).

19.4.1 Longevity and Fecundity

Wading birds tend to be short-lived in comparison with seabirds (Figure 19.7) and also tend to be more fecund, laying two to six eggs, rather than the one to three eggs common in most seabirds (Table 19.1).

19.4.2 Asynchronous Hatching

In most species, eggs hatch asynchronously, with older chicks 1 to 6 days older than younger chicks in the brood. Broods of wading birds often are reduced during the nestling period, either as a result of starvation of the younger chicks that beg less effectively, or through older chicks killing younger ones or forcing them from the nest (Mock et al. 1987a, b). The mechanism and degree of brood reduction may be species specific, or may be mediated by whether the prey animals fed to the young are of a size and shape that can be easily swallowed by older birds (many small items) or are indefensible (fish too large to swallow whole, Mock et al. 1987b). The most common explanation for the evolution of this pattern of brood reduction is that it allows adults a mechanism for adjusting brood size to the availability of prey, which is difficult to predict at the time of clutch formation (O'Connor 1978, Stenning 1996).

19.4.3 Breeding-Site Fidelity

Wading birds tend to have variable breeding-site fidelity, and annual turnover rates in colony occupancy can be high (Bancroft et al. 1988). Storks and solitary nesting species can be quite site

TABLE 19.1
Life-History and Reproductive Parameters of Selected Long-Legged Waders

Species	Clutch Size (range)	Incubation Period (d)	Nestling Period (d)[a]	Maximum Age (year)	First Breed (year)	Adult Survival (%/year)	First Year Survival (%/year)	Source
Black-crowned Night Heron	2–5	24-26	42-49	21	1–2	69	39	1
Yellow-crowned Night Heron	2–6	24-28	57		2			2
White-faced Ibis	3–4	20-26	56	14	2			3
Little Egret	2–5	21-25	45-50		1–2	75	6–55	4
Snowy Egret							<35	1
Green Heron	3–5	19-21	25	7	2			5
Tricolored Heron	3–5	22-25	51-59	18	2	68.4	21	6
White Ibis	2–5	20-21	47-56	16	2			7
Great Egret	2–4	26-27	75-85	23	2	74	24-66	8
Grey Heron	3–5	21-26	64	16	2	72–74	33–22	9
Great Blue Heron	3–6	28	60	23		71	29	10
Roseate Spoonbill	2–4	23-24	50-56	28	3			11
Wood Stork	3–4	28-32	75		3–4	80	60	11
Reddish Egret	3–4	26	56-70	12	2–3			11
African White Ibis	2–5	28-29	35-48		3			12
Glossy Ibis	2–6	21	25-28					12
Yellowbilled Stork	2–4	30	50-55		>3			12
Milky Stork	3–4	27-30	>60					12

[a] Includes postfledging period of dependence upon adult feedings at the breeding colony.

References: 1, Erwin et al. 1996; 2, Watts 1995; 3, Ryder and Manry 1994; 4, Hafner et al. 1998; 5, Davis and Kushlan 1994; 6, Frederick 1997; 7, Kushlan and Bildstein 1992; Palmer 1962; Kahl 1963; 8, Kahl 1963, Hancock and Kushlan 1984, Sepulveda et al. 1999; 9, Lack 1949, North 1979; 10, Owen 1959, Butler 1997, Hancock and Kushlan 1983; 11, Palmer 1962; 12, Hancock et al. 1992.

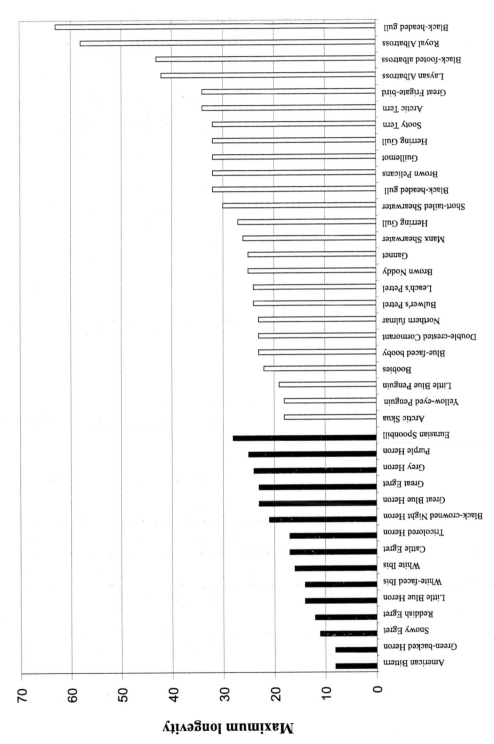

FIGURE 19.7 Comparison of maximum longevity records for free-ranging wading birds (15 species, in black) and seabirds (20 species, in white). Although there have been fewer attempts to band wading birds, ciconiiform birds appear to be shorter-lived in general than are seabirds.

faithful, while some ibises are nearly obligate nomads (Hancock et al. 1992, Frederick et al. 1996a, Frederick and Ogden 1997). Philopatry may be related to the predictability of food resources. Most storks are much more site faithful than the smaller ibises, probably because these large birds may compensate for spatial unpredictability of prey by flying long distances from more permanent colonies (Kushlan 1986, Frederick and Ogden 1997). It is also true that coastal colonies tend to be more stable in occupancy than are inland colonies, almost certainly because coastal habitats offer more predictable access to food (Kushlan 1977, Ogden et al. 1980).

19.4.4 SURVIVAL

As a result of logistical problems with banding and mark-recapture studies in wading birds (see "Management" below), there is much less information on movements and survival of wading birds than there is for seabirds. Important exceptions include Butler (1997), who used the relative proportions of adults and juveniles in the northwestern population of the Great Blue Heron to estimate annual survival in a nonmigratory population. In the northwestern Mediterranean, relatively high site fidelity and low number of potential breeding areas for Little Egrets allowed researchers to measure survival and life-history characteristics of this species (Hafner et al. 1998). In general, these studies demonstrate relatively high mortality in the first year of life with stabilization of survival rates thereafter (Table 19.1), and have shown that annual survival rates may be strongly affected by over-winter conditions (Kamyanibwa et al. 1990, Hafner et al. 1994, Cezilly et al. 1996). However, these studies were performed in temperate Europe, and survival of populations experiencing different climatic conditions or which are nonmigratory remains to be investigated (Cezilly 1997).

Long-term banding efforts are extremely valuable and productive programs. As the Little Egret banding program in France has demonstrated, high quality data can result even in the face of low site fidelity and high dispersal rates. In that program, only 9% of 3000 birds banded were re-sighted as breeders, yet this information was sufficient for the estimation of survival, and has led to strong insights into the importance of environmental constraints and management for population trends (Hafner et al. 1998, Thomas et al. 1999).

19.4.5 POPULATION REGULATION

While the preceding information emphasized the effects of food on reproduction and survival of wading birds, there is abundant evidence that predation on eggs and young also plays an important role (Rodgers 1987, Simpson et al. 1987), and that the evolution of adult foraging and flocking behavior was molded by this selective force (Caldwell 1986, Petit and Bildstein 1987). Despite these obvious adaptations to reduce predation risk, there is little evidence that predation on adults or on nest contents currently has any large or even measurable effect on wading bird population dynamics. Similarly, although there are relatively few studies of the effects of disease on wading birds (Forrester and Spalding in press), the available evidence suggests that disease is rarely a driving force in wading bird demography. Although hunting by humans has certainly been responsible for the decimation of some species and populations (Ogden 1978a, Hancock et al. 1992), and harvesting for food may be an important cause of disturbance and mortality in some third world countries (Gonzalez 1999), hunting is probably not widespread enough to function as a general limit on wading bird populations.

The mechanisms by which food limits wading bird populations is not obvious and is not necessarily the same in all species. In Great Blue Herons, Butler (1988) did not find strong evidence of competition for food or foraging sites or of density-dependent effects on food supply within the breeding season (Figure 19.8). Van Vessem and Draulans (1986) found no differences in reproductive success of Grey Herons that were related to colony size. Butler suggested, instead, that population regulation was probably achieved during the nonbreeding season through differential survival of first-year birds who had limited access to food due to adult aggression.

FIGURE 19.8 Little Egrets (Dimorphic Heron; a dark and a light phase bird), Crab Plovers, and other wading bird and shorebird species all forage in the shallow waters around Aldabra Island, Indian Ocean. Most researchers have not found strong evidence of competition for food or foraging sites during the breeding season (see text). (Drawing by J. Busby.)

Several studies demonstrated that weather and hydrological conditions during the nonbreeding season have large effects on survival of young birds (den Held 1981, Hafner et al. 1994, North 1979, Cezilly et al. 1996). Very little information is available on the subject of population regulation of tropical and subtropical species.

19.5 WADING BIRDS AS MARINE ANIMALS

Very few wading birds are found exclusively in marine habitats. These birds are typically found close to the immediate coastline except during migration, they rarely or never swim, and they show no morphological adaptations for open-water plunge or surface diving. Some species are capable of excreting salt through a salt gland (Shoemaker 1966, Johnston and Bildstein 1990), though the extent of this ability is not well known in this group.

Long-legged wading birds are a key component of the avifauna of many shallow coastal marine habitats, such as mudflats, tidal marshes, river deltas, salt pannes, and mangrove forests, in both tropical and temperate zones. In some of these areas, wading birds are the dominant shallow-water avian predator on small fishes and invertebrates (Bildstein et al. 1982, Berruti 1983, Howard and Lowe 1984, Butler 1997), to the extent that as a group, wading birds can be important determinants of energy flow in wetland ecosystems (Berruti 1983, Bildstein et al. 1982, Bildstein et al. 1991).

19.5.1 EFFECTS OF WADING BIRDS ON MARINE AND ESTUARINE ECOSYSTEMS

In large numbers, wading birds may exert strong effects on coastal ecosystems through direct predation, such as the alteration of abundance and community composition of fish communities (Kushlan 1976a), or alteration of size and sex ratio of prey populations by selective predation (Britton and Moser 1982, Trexler et al. 1994). Howard and Lowe (1984) found that Royal Spoonbills

(*Platalea regia*) consumed approximately 13% of the biomass production of shrimps (*Macrobrach-ium intermedium*) in an Australian seagrass bed. In addition, the birds were highly selective of adult female shrimps, resulting in up to 25% predation of that age class. Master (1992) found that mixed-species foraging aggregations of wading birds reduced populations of fishes in salt marsh pannes by up to 80%. However, Erwin (1985) found evidence of only very short-term resource depression in Great and Snowy Egrets, followed by rapid redistribution of prey.

Nutrient spikes from excreta resulting from colonial nesting may affect local animal and vegetative community composition and density, including changes in density and species compo-sition of aquatic plant and animal communities surrounding the colony (Powell et al. 1991, Frederick and Powell 1994), and increases in vegetative growth and attractiveness of nesting vegetation to herbivores (Onuf et al. 1977). Since shallow inshore habitats are important for the >90% of commercially important fisheries in the U.S., wading birds show the potential for affecting community structure and nutrient dynamics of marine communities and should therefore be considered an ecologically important part of coastal and nearshore ecosystems. Wading birds also redistribute contaminants through their feces. For example, Klekowski et al. (1999) noted that the mangroves in a Scarlet Ibis roost had significantly higher mutation rates than in the surrounding area, probably due to concentration of mercury in bird feces at the coast.

19.5.2 DEPENDENCE OF WADING BIRDS ON COASTAL ZONE HABITATS

Coastal areas are important feeding and breeding habitat for wading birds. Comprehensive statewide surveys in the U.S. demonstrated that 38% of all breeding aggregations of wading birds were found within 2 km of the coast in South Carolina (Dodd and Murphy 1996), 61 to 69% in Florida (Ogden et al. 1980), and 73% in Texas (Texas Colonial Waterbird Society 1982). Similarly, within the historical Everglades complex of fresh and estuarine habitats, the majority of breeding was located in the coastal zone (Ogden 1994), as was true for the ecologically similar Usamacinta delta in Mexico (Ogden et al. 1988). In Honduras and Nicaragua, coastal wetlands host the majority of breeding Jabiru Storks in the region (Frederick et al. 1996b).

There are probably several reasons for the apparent attraction of wading birds to coastal areas. First, coastal areas often show high primary and secondary productivity. The productivity of estuarine areas may result from: (1) the availability of nutrients when fresh- and saltwater mix and the energetic subsidy of tidal action; (2) the abundance of early life stages of marine creatures attracted by the refugia created by multihaline conditions; (3) the variety of estuarine submerged and aquatic vegetation; or (4) the influx of nutrients from freshwater rivers and streams. These conditions may operate together to create zones of high secondary productivity in shallow waters.

Second, wading birds depend on the availability of extensive shallow-water habitats, created as a result of inlets from the ocean (salt marshes) and outlets to the sea (deltas). High-energy beaches and rocky intertidal zones offer poor foraging conditions for birds that wade in shallow water. Wading birds may need a variety of shallow-water habitats to allow foraging under highly variable hydrologic conditions that change on scales of days (tidal conditions), seasons (tidal and sea surface elevation fluctuations, Powell 1987, Butler 1997), and years (Bancroft et al. 1994).

Third, coastal areas are usually tidal, resulting in a predictable daily exposure of shallow pools, flats, and riffles where prey may be concentrated, trapped, or otherwise made available by receding water. This is particularly striking when one compares coastal areas with inland marshes, in which drying conditions are seasonal rather than daily and flooding is extremely unpredictable from year to year (Kushlan 1976a, Ogden et al. 1980, Lowe 1981, Kingsford and Johnson 1998).

Fourth, coastal areas usually offer island nesting sites that are predictably surrounded by water, offering protection from mammalian predators. Predator protection may be augmented in some locations around the world by the presence of crocodilians in and near colonies (Frederick and Collopy 1989b). Islands in freshwater ponds and marshes may dry during droughts, and coastal marine islands are one of the few habitats that can offer wading birds dependably inundated colonies throughout the season. Kushlan (1977) compared White Ibises in coastal and freshwater areas of the Everglades, and concluded that even though some very large colonies occurred in freshwater areas, reproduction in the coastal zone was probably demographically more important to the population because annual reproduction and recruitment were far more predictable than in inland areas.

19.5.3 MARINE SPECIES

Despite frequent association with coastal habitats, wading birds are rarely strictly marine. Over 57% of the world's 109 wading bird species are often found in marine or estuarine habitats; 19% show marked preference for coastal habitat; only 9% live almost exclusively in marine habitat (Hancock and Kushlan 1984, Hancock et al. 1992). Within some species, races or subspecies are known to be almost exclusively marine, like the white color morph of the Great Blue Heron (*Ardea herodias occidentalis*), which occurs in coastal areas of southern Florida and eastern Mexico, and the *fannini* subspecies of the coastal Pacific Northwest (Butler 1997). In the Green Heron (*Butorides striatus*), most races are typically freshwater or more rarely estuarine, but some island races exist in completely marine habitats (Hancock and Kushlan 1984).

19.5.4 PHYSIOLOGY AND ECOLOGY IN THE COASTAL ZONE

19.5.4.1 Salt Balance

Other than the obviously saline conditions and daily fluctuations in water level, the rigors of coastal and marine life for wading birds are probably not very much different from those in freshwater habitats. Salt balance is maintained in most species through a combination of occasional freshwater availability, choice of nonsalty prey, and some ability to excrete salt through a nasal gland (Shoemaker 1966). The extent to which ciconiiform birds as a group are able to excrete salt is not well understood, though long-legged waders do have functional nasal salt glands (Figure 19.9). Both Grey Herons (Lange and Stalled 1966) and White Ibises (Johnston and Bildstein 1990) have been shown to excrete concentrated saline fluid through their nasal salt glands, though the ability to excrete salt loads in both species can be overwhelmed by drinking only seawater. Some wading birds do live apparently without fresh water indicating that they cope with ionic imbalance somehow. Yellow-crowned Night Herons and many seabird species occur on oceanic islands where there is virtually no access to fresh water for months or years, thus, it seems very likely that they are able to excrete salt as well as seabirds do.

FIGURE 19.9 Hypertrophied salt gland, shown above and to the right of the eye, in a 6-week-old White Ibis nestling. The salt gland is greatly enlarged due to experimental feeding on a high-salt diet of Fiddler Crabs (*Uca* spp). (Drawing by M. Davis, from Bildstein 1993.)

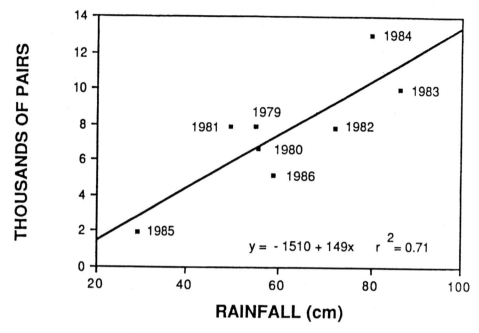

FIGURE 19.10 Relationship between the numbers of White Ibis pairs breeding on Pumpkinseed Island, South Carolina, and the amount of rainfall during the preceding winter-to-spring wet season, 1979–1989. The relationship is significant, at $p < 0.05$. (From Bildstein et al. 1990, reprinted with permission from the Wilson Ornithological Society.)

Coastal-nesting White Ibises feed their young largely on freshwater crustaceans from inland areas (Kushlan and Kushlan 1975, Kushlan and Bildstein 1992). In low-rainfall years, however, inland swamps dry up prior to the end of the nesting season, and adults are forced to feed their young on fiddler crabs (*Uca* spp.), which are salty. During those years, large numbers of nestlings die. This physiological constraint results in fewer nesting attempts in years when inland marshes are shallow or dry (Bildstein et al. 1990, see Figure 19.10). In an extreme example of this dependence, Bildstein (1990) found that Scarlet Ibises in Trinidad ceased to nest following the diversion of freshwater flow away from the formerly estuarine Caroni mangrove swamp. Management of freshwater outflows into estuaries is therefore critical for the conservation of some wading bird species. It is of note that this constraint is unlikely to be as severe for species that eat primarily fishes, since fishes are osmoregulators and their flesh is considerably less salty than surrounding marine waters.

19.5.4.2 Tidal Entrainment

Wading birds are highly dependent on shallow water for foraging, and in many cases benefit from rapidly receding surface waters for the entrapment or availability of prey (Kushlan 1986, Frederick and Collopy 1989a). The presence of a tidal influence in coastal areas assures coastal birds of a relatively predictable daily drying trend. One of the most obvious behavioral differences between populations of inland and coastal wading birds is the entrainment of feeding cycles to the tidal pattern (Powell 1987, Butler 1997, Ntiamoa-Baidu et al. 1998, Draulans and Hannon 1988). In the Pacific Northwest, Butler (1993) found that seasonal differences in the timing of tides (more low tide during the day in summer than in winter) allowed Great Blue Herons to catch more fish in summer than winter, and that this process was important in determining the timing of breeding.

In an area with little tidal influence, Powell (1987) demonstrated that wading birds in the subtropical Florida Bay estuary were sensitive to seasonal fluctuations in sea-surface level, rather

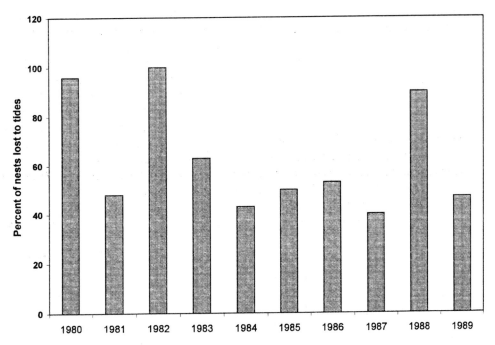

FIGURE 19.11 Percent of nests lost due to tidal inundation over a period of 10 years, at the Pumpkinseed Island colony in South Carolina. Extreme tides occur at this site during the conjunction of spring tides and strong northeast winds. (From data in Frederick 1987a and Bildstein 1993.)

than timing of weak daily tides. This seasonal effect is generated by the expansion and contraction of the ocean volume in response to seasonal fluctuations in temperature, leading to a consequent reduction in the use of the estuary by wading birds during months with deeper water. The effects of daily tides, seasonal fluctuations, and wind-driven tides on wading bird foraging habitat depend on many factors including geographic location, shape of bays and inlets, volume of freshwater flows, and topography (e.g., Berruti 1983).

19.5.4.3 Effects of Storms

The use of coastal habitat also puts wading birds at some risk from the effects of severe storm systems, such as hurricanes in the tropics and subtropics. In a coastal South Carolina colony over a 10-year period, an annual average of 63% (range 40 to 100%) of White Ibis nests on a low marsh island were destroyed by flooding during astronomical high tides with strong wind events (Figure 19.11; Frederick 1987a, Bildstein 1993). Despite this very high loss rate, ibises showed no sign of abandoning the colony site, which this species typically does in response to nest predation and human nest disturbance (Rodgers 1987, Post 1990).

In Florida Bay (southern Florida), mortality of a Great Blue Heron population during two separate hurricanes was measured at 30 to 40% (Powell et al. 1989). The indirect effects of hurricanes on nesting and foraging habitat may be much more important than direct mortality (Michener et al. 1997). Following the passage of Hurricane Hugo, Bildstein (1993) and Shepherd et al. (1991) described widespread salinization of formerly freshwater coastal feeding areas near a large colony of White Ibises in South Carolina. This degradation of foraging habitat was thought to cause a sharp decline in the local nesting population of White Ibises in the years following the hurricane.

Hurricanes and strong storm events can also have positive effects by creating new mud and grass flats necessary for foraging, and by opening new inlets (Paul 1991, Arengo and Baldassare 1999), or by keeping vegetation on nesting colonies in an early successional state that is preferred by the birds. Both the intensity and frequency of tropical storms and hurricanes are projected to increase with sea-surface warming resulting from global climate change (Michener et al. 1997), though it is unclear whether these changes will result in net positive or net negative effects on coastal habitats and wading bird populations.

19.6 MANAGEMENT OF WADING BIRDS

As wading birds and humans are drawn into ever closer contact through increasing human demands placed on diminishing land, water, and coastal resources, informed management of these species will become more critical, both for preservation of the birds and of the wetland ecosystems they depend upon (Parnell et al. 1988, Erwin 1996).

19.6.1 MANAGEMENT OF BREEDING SITES

Breeding-site protection and creation are thought to have been a main factor in the rapid recovery of wading birds in the U.S. from the devastating plume trade at the turn of the last century (Ogden 1978b, Parnell et al. 1988). One of the most direct actions that managers can take is in the protection and maintenance of nesting habitat (Erwin 1996). The vegetation in wading bird colonies often degrades naturally with time, both as a result of natural vegetative succession, and degradation of vegetative cover (Weseloh and Brown 1971, Parnell et al. 1988). Control measures may include either suppressing or replanting vegetation as necessary, or creating new breeding sites. A rapid northward range expansion of several species of wading birds in eastern North America during the 20th century has been attributed in part to the construction of hundreds of small dredge-spoil islands in estuarine and coastal areas as a result of the construction of the intracoastal waterway system (Parnell et al. 1986, Ogden 1978b).

Predation, especially by mammals, can result in destruction of entire areas of nesting within colonies, or even lead to the abandonment of the colonies themselves (see also "Dependence of Wading Birds on Coastal Zone Habitats" above). Although nesting at predation-prone sites (those close to or with access to dry land) should not be encouraged, mammalian predation is often due to one or a few individuals (Allen 1942, Rodgers 1987). Active trapping or fencing to protect colonies in these situations may have large payoffs. Crocodilians are frequent residents in and around wading bird colonies in some parts of the world, and although not direct predators of wading bird nests, they may play an important role in dissuading mammalian predators from entering colonies (Attwell 1966, Hopkins 1968). Frederick and Collopy (1989b) found no statistical association between wading bird use of colonies and presence of alligators (*Alligator mississippiensis*). In contrast, there are now several examples in Florida of persistent wading bird colonies having formed in tourist parks where large numbers of crocodilians are displayed. In each case, the wading birds are apparently choosing to nest over crocodilians, despite extremely close proximity (1 to 2 m) to heavy human foot traffic on boardwalks. Thus it seems likely that wading birds are attracted to areas with high crocodilian densities.

It also should be recognized that predation at low levels is probably a natural phenomenon in wading bird colonies. Snakes may commonly visit colonies and have a relatively small impact on nesting (Frederick and Collopy 1989b). Similarly, there is often a suite of opportunistic birds, snakes, and crocodilians that scavenge abandoned or temporarily unguarded nest contents (Frederick and Collopy 1989b, Wharton 1969). These opportunists may wreak havoc when wading birds are forced from their nests by close human approach, leading to the impression that they are predators rather than scavengers (Bouton 1999). Indeed, widespread predation in wading bird colonies may signal that some other form of stress is affecting the colony. For example, Turkey Vultures (*Cathartes aura*) are known to kill and eat wading bird nestlings, but their depredations within colonies in southern Florida are nearly always associated with widespread abandonment due to interruptions in the food supply of wading birds (Allen 1942, P. C. Frederick unpublished).

Although solitary avian scavengers are not usually able to displace adult wading birds from their nests, large numbers of scavengers can overwhelm the mild nest defense behavior of adults. Post (1990) described the role of large flocks of Fish Crows (*Corvus ossifragus*) during the demise of a large wading bird colony in Charleston, South Carolina. The crows were drawn in large numbers to the vicinity of the colony by a large municipal garbage dump nearby. Conversely, Frederick and Collopy (1989b) attributed the lack of avian scavengers in freshwater Everglades colonies to the absence of nearby human sources of food. Thus the management of unnatural food resources may be a key component of managing unnatural predation at wading bird colonies.

The maintenance of many unused colony sites within any nesting management area is probably wise, since colonies often shift locations as a result of predation and changes in colony vegetation (Erwin et al. 1995). As breeding colonies become unsuitable for various reasons, the need to create new habitat or to induce birds to move to new islands may occur. This has been accomplished in various ways, through the use of decoys and playbacks (Dusi 1985), by keeping caged adults at new colony sites, or by raising young in semicaptive conditions at new sites (McIlhenny 1939).

19.6.2 HUMAN DISTURBANCE ISSUES

Breeding and nonbreeding wading birds are sensitive to human disturbance in various forms (Gotmark 1992, Carney and Sydeman 1999). Wading bird colonies are very often the target of research activities, and direct human entry of colonies may result in loss of nest contents, reduced nesting success, reduced settlement of breeders in the colony, retarded growth of nestlings, and changes in nesting behavior (Allen 1942, Portraj 1978, Tremblay and Ellison 1979, Erwin 1980, Burger et al. 1995, Carlson and McLean 1996, Bouton 1999). However, these effects seem most severe during the early part of the nest cycle (Frederick and Collopy 1989c).

Many wading bird chicks leave the nest at the close approach of humans, and the effect of scattering chicks prematurely in this way may be devastating (Figure 19.12). Alternatively,

FIGURE 19.12 Young Wood Storks in their nest in Florida. Chicks may scramble out of nests if disturbed and may starve to death. (Photo by R. W. and E. A. Schreiber.)

Black-crowned Night Heron chicks become conditioned to the approach of researchers and chicks are actually less disturbed if approaches are regular and start at an early age (Parsons and Burger 1982). However, this response appears to be species specific (Davis and Parsons 1991).

In temperate-zone colonies, nesting often is relatively synchronous within and among species, and stress due to human intrusions can be minimized by confining visits to the later part of the breeding cycle. However, in tropical and subtropical locations, nesting may be spread out over many months. In some situations, colonies can be profitably studied by remote observation using blinds (Cairns et al. 1987, Fernandez-Cruz and Campos 1993, Bouton 1999) or vantage points (Pratt and Winkler 1985). Human disturbance can also be reduced by visiting colonies only during early morning and evening, when thermal stress on eggs and chicks is likely to be decreased.

Wading birds and their colonies are increasingly the subject of ecotourism enterprises throughout the world (Giannechinni 1993), and the potential for widespread human disturbance through these activities is tremendous. Burger et al. (1995) reported 15 to 28% mortality of heron nests in colonies that were entered by tourists. Using well-separated groups within the same Wood Stork colony as treatment plots, Bouton (1999) demonstrated that disturbance due to ecotourism reduced reproductive success in a colony of Wood Storks in the Brazilian Pantanal, even though the tourists in that study were carefully managed. Buffer zones of >75 m were recommended in the Brazilian study, and 50 m in the study by Burger et al. (1995). Ecotourism also affects wading birds at their foraging grounds. In a Florida drive-through wildlife refuge, Klein et al. (1995) found strong interspecific differences in the responses of various species of waterbirds, with threshold distances from roadways of 0 to 80 m, and threshold disturbance levels of 150 to 300 cars per day, depending on species.

Erwin (1989) and Rodgers and Smith (1995) derived minimum approach distances for boat traffic for waterbirds in several situations by noting the distances at which birds showed stress and avoidance behaviors in response to approaches by boats. Both studies recommended an approach distance of 100 m for wading birds. Even with reliable approach distances, the regulation and enforcement of watercraft approaches to waterbirds remains a thorny management issue, particularly in multiple-use areas. It is also of note that wading birds may be more sensitive to human approaches by land than by water (Vos et al. 1985).

It is tempting (and frequently correct) to assume that wading birds are usually affected by disturbances of many kinds, and that the wise management decision is to simply disallow ecotours and research. This reaction must be balanced by the value of the research results and public education, both of which can be of immeasurable benefit to managers. Further, it is not always true that wading birds are incompatible with disturbance. Several studies have shown that wading birds prefer to nest in sites well away from human activities (Erwin 1980, Gibbs et al. 1987, Watts and Bradshaw 1994), and that productivity of young is related to proximity to and buffer protection from disturbance (Carlson and Mclean 1996). However, there also are numerous examples of colonial wading birds nesting successfully in close proximity to humans. For example, wading birds now nest within 2 m of the edges of heavily traveled boardwalks in at least three large tourist attractions in Florida. Similarly, a large Great Blue Heron colony has persisted at the Stanley Park Zoo site in Vancouver, British Columbia for over 78 years (Butler 1997). Large mixed-species colonies have persisted for many years in noisy, heavily used industrial shipping channels in Tampa Bay and the harbors of Charleston, SC and Baltimore, MD. Thus, there is some hope that if the necessary conditions for nesting (good prey base, adequate nesting habitat, lack of predators, lack of direct disturbance) exist, that nesting wading birds can be conditioned to breed in close proximity to some kinds of intense human activity. The willingness to nest in proximity to human activity is almost certainly species specific, however, and protected refugia will probably always be necessary for some taxa.

19.6.3 FORAGING HABITAT

The link between wading bird nesting and the availability of prey animals at feeding sites is well established (see "Reproductive Biology"). Often, foraging areas may be well outside of areas protected for nesting, and managers may consequently feel that the management of foraging sites is essentially out of their hands. Managers need to identify where the birds they are protecting are feeding, to make those lands priorities for conservation. For example, the successful conservation of flamingos in the northern Mediterranean basin began with a partnership between a biological station trying to conserve a breeding colony, and a local salt works that managed the foraging and breeding sites (Johnson 1997). Identifying foraging sites may be done in a cost-effective manner by using radio telemetry, or by following adults from colonies to foraging areas using light aircraft (Smith 1995a).

It is difficult to recommend specific actions for managing foraging areas for wading birds, since studies of foraging ecology come from such a diversity of sites. One common factor seems to be that wading birds are attracted to dense aggregations of prey — these may be formed by the combination of both high standing stocks of fish or invertebrates and conditions which make those prey available. Water must be shallow enough for foraging (5 to 25 cm depth, depending on species). Many wading birds prefer to forage in open areas with relatively little emergent vegetation (Chavez-Ramirez and Slack 1995, Surdick 1998), since plants serve to obstruct the bird's view of prey and to offer hiding places for fish and invertebrates. Vegetation management may therefore be essential to keeping foraging areas productive for long-legged wading birds. Making prey "available" may be fairly easily accomplished if water levels can be decreased. Foraging opportunities should be optimal at two different times during the period of nesting. Nesting is often apparently cued by good food availability — perhaps as much as 2 months, and as little as a week, prior to initiation of courtship (Allen 1942, Babbitt 2000, P. C. Frederick unpublished). The second period during which food must be abundant is late chick rearing, during and through the time when young are leaving the colony. In subtropical wetlands, it has been demonstrated that interruptions in the food supply, particularly during the early part of the nesting cycle, result in nest abandonment (Kushlan et al. 1975, Frederick and Collopy 1989a). There may also be a trade-off between drying and prey standing stocks that operate on multiyear scales. Repeated annual drying of the freshwater marsh surface can result in depauperate prey animal populations (Loftus and Eklund 1994), leading to a declining carrying capacity. The ability to manage hydrology for prey availability is relatively easy by comparison with managing for high standing stocks of the prey animals themselves. The best course for wetland managers is to initiate monitoring or research which will better elucidate the local and site-specific drivers of prey animal populations.

The management of foraging habitat must be at a scale appropriate for the movements of the birds. In the Yucatan of Mexico, Arengo and Baldassarre (1999) reported large differences in the density and communities of aquatic prey of Greater Flamingos (*Phoenicopterus ruber ruber*) within an 8000-km² wetland complex. They concluded that much of this geographic variability was the result of hurricanes and hydrological variability, and that the long-term survival of flamingos in the area depended on a geographically widespread complex of habitats to provide appropriate feeding opportunities at any given time.

19.6.4 MONITORING WADING BIRD POPULATIONS

Monitoring reproductive and population responses to management of foraging and breeding sites is an essential part of managing nesting areas, as well as a key part of adaptive management. The scale of the survey attempted is of critical importance to the response measured, and entirely different answers may result depending on how much area is surveyed (Sadoul 1997, Bennetts and Kitchens 1997). In most cases in which population size or dispersion are the target of measurement, the mobility and lack of breeding-site fidelity of wading birds call for surveys that include entire ecosystems or regions. Survey techniques must be tailored to the size and goals of the survey program. Common techniques include systematic aerial survey, ground counts, roost counts, and mark–recapture studies (Dodd and Murphy 1995, Rodgers et al. 1995, Gibbs et al. 1988, Frederick et al. 1996).

Measurement of survival in wading birds is difficult, because it is hard to re-sight or recapture marked birds if breeding colonies commonly move, and especially difficult if there is an almost infinite number of potential nesting sites available to the birds. This situation is in direct contrast to the relatively high site fidelity and low number of potential nesting sites characteristic of many seabirds, where it is possible to estimate survival through the use of re-sightings and intercolony movements (Spendelow et al. 1995). Initially, estimates of survival in wading birds were derived from returns of birds banded as young (Lack 1949, Henny 1972). However, this method relies on the untenable assumption that recovery rates do not vary with age at banding (Brownie et al. 1985, Pollock et al. 1995), leading to a tendency toward underestimating survival (Clobert and Lebreton 1991). More recently, the development of more robust capture–mark–recapture models (Lebreton et al. 1992, 1993, Nichols 1992) has enabled the separate estimation of re-sighting and survival probabilities for individuals.

19.7 CONSERVATION OF WADING BIRDS IN THE COASTAL ZONE

The preceding descriptions illustrate that wetland habitats within the coastal zone serve as primary and often critical habitat for breeding, feeding, and migratory wading birds. Coastal zones may support the majority of nesting for many wading bird species, and may provide the most stable and productive habitats for reproduction. Thus the conservation of coastal and nearshore habitats and the maintenance of normal ecological processes in the coastal zone seem critical to the conservation of most wading bird species. For a discussion of methods and examples of conservation of coastal avian foraging habitats, see Bildstein et al. (1991) and Kushlan and Hafner (2000).

There are a number of threats that are likely to affect wading birds particularly. The most obvious of these threats may be placed into six categories discussed below. It is obvious that the causes and effects of these artificial categories may be strongly interwoven.

19.7.1 FRESHWATER FLOW AND DEGRADATION OF WETLAND PRODUCTIVITY

The productivity of estuarine areas derives from several sources, and the mixing of fresh- and saltwater is often a central and critical process. The availability of fresh water has been defined as perhaps the single most critical resource for the growth of human populations in the coming century. In many parts of the globe, the control, diversion, and reduction in flows of freshwater already are a threat to the productivity and ecological functions of estuaries (Alleem 1972, Nichols et al. 1986, Stanley and White 1993, McIvor et al. 1994, Jay and Simenstad 1996), and it is likely that wading birds will be affected by this process on a global scale. The Everglades of Florida has seen a very large reduction in freshwater flows to the coastal zone (Fennema et al. 1994) and a reduction in wading bird nesting attempts of over 90% (Bancroft 1989, Ogden 1994).

19.7.2 RISING SEA LEVEL

The effects of rising sea level on wading birds are likely to be several. Island nesting habitats are likely to become severely eroded (Erwin et al. 1995), and coastal habitat may be altered by human protection of coastline. This may be offset by the creation of nesting islands through the use of dredge-spoil material in waterways (Parnell et al. 1986, Erwin et al. 1995). In many cases, the renewal and maintenance of dredge-spoil islands are likely to be in increasing competition for materials with beach nourishment projects as sea level rises (Titus 1996). Of greater concern is the effect of sea level rise on shallow-water wetland foraging habitats that wading birds are so dependent upon (Michener et al. 1997). Global sea-level increases may be rapid enough that estuarine vegetation, such as mangroves, may not be able to move inland fast enough to keep pace, and large areas of estuarine vegetation may be lost (Snedaker 1995, Ellison and Farnsworth 1996). It therefore seems likely that shallow-water estuarine foraging areas for wading birds will decrease in area as sea levels rise. In the United States alone, the loss of coastal wetland areas is predicted to be over

30,000 km^2 (Titus 1996). Well-documented examples within the U.S. include the Everglades, the low country of South Carolina, coastal Louisiana, and the Chesapeake Bay.

19.7.3 LOSS OF COASTAL FORAGING HABITAT

Wading bird populations also may suffer from the loss of either inland or coastal foraging habitat because there is a relationship between size of feeding area and size and diversity of breeding colonies (see above). In British Columbia, loss of habitat is considered the main threat to the long-term viability of the population of Great Blue Herons in the Strait of Georgia ecosystem (Butler 1997). Inland freshwater wetlands are under considerable threat from agricultural and residential development, and given the importance of these wetlands to coastal nesting wading birds, coastal managers should be concerned about this linkage. The loss of coastal nesting and feeding habitat is also of concern as coastal wetlands are converted to aquaculture impoundments, salt works, and urban and industrial development (Je 1995).

Aquaculture holds both potential threats and benefits for wading birds. Production aquaculture is in direct conflict with any avian species that is likely to prey upon the product, and fish-eating birds have been at the center of conservation crises at aquaculture centers throughout the past two decades (Fleury 1994). In the United States, it has been estimated that several thousand wading birds are killed legally at aquaculture facilities, and that many more are killed illegally (Kushlan 1997).

On the other hand, aquaculture facilities and impoundments present rich sources of food that may fuel productive breeding aggregations and population growth, and enhance survival. Similar potential for positive interactions may be occurring with production agriculture and aquaculture in South America (Blanco and Rodriquez-Estrella 1998), Europe, and southern Asia. In Louisiana, wading birds often feed on undesirable fishes within crayfish culture ponds, and during periods of drying take "overstocked" crayfish which would normally die as ponds dry. However, there are some cases of real economic losses to farmers, and these have overwhelmingly dominated the industry attitude toward wading birds (Fleury 1994).

Wading birds are migratory and during the course of their lifetimes they rely on a mosaic of wetland habitats that may be hundreds of kilometers apart (Frederick et al. 1996b). These areas are only partly identified in most parts of the world, yet the conservation of wading birds demands that wintering and breeding areas be managed as a cohesive whole. For species breeding in the temperate zone, hydrological conditions at wintering sites in the tropics have been shown to be important to the annual survival of herons (den Held 1981, Hafner et al. 1994). In the United States, Mikuska et al. (1998) identified important wintering areas for herons by using clusters of band recoveries (see Figure 19.13).

19.7.4 DISEASE AND CONTAMINATION

Increases in disease outbreaks in wading birds derive from several sources. First, as a result of wetland loss and degradation of coastal habitat through various kinds of development, aquatic birds are found in increasingly dense concentrations in coastal refuges (Butler 1997). This crowding creates conditions favorable for the transmission of communicable diseases such as botulism. Disease outbreaks may also be related to pollution of various types. Toxic algal and dinoflagellate blooms are becoming increasingly common in coastal waters worldwide, and have been linked to increases in nutrient pollution (Cloern 1996, Burkholder 1998). Such blooms have the potential to grossly alter the abundance and community composition of fishes and crustaceans upon which wading birds normally prey (Burkholder 1998), and the blooms may be toxic to wading birds themselves, as well as to other predatory animals (Epstein et al. 1998).

As human demand for freshwater increases, it is likely that fresh and estuarine surface waters will contain more contaminants as a result of multiple use and re-use. These contaminants may include pesticides, herbicides, heavy metals, PCB, dioxins, silt, and nutrients. The effects of these

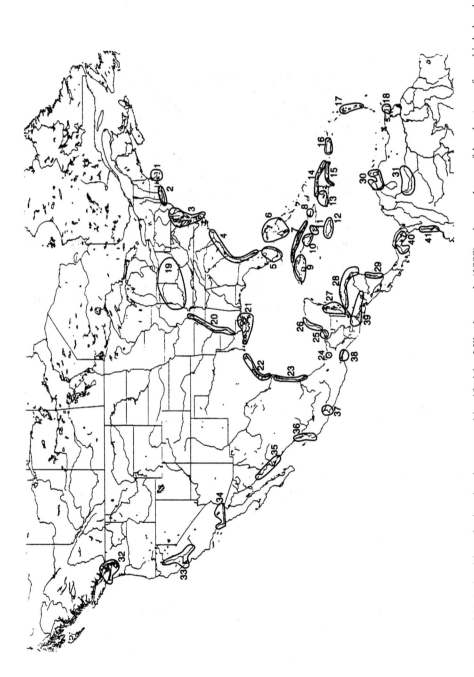

FIGURE 19.13 Locations of high densities of wintering herons, as identified by Mikuska et al. (1998) using analyses of band returns from birds banded in North America. Numbers refer to geographic units designated in Mikuska et al. (Reprinted with permission from the Waterbird Society.)

pollutants may be either directly toxic (including sublethal effects such as reproductive impairment and endocrine disruption; Spalding et al. 2000a, b) or indirect (alterations of aquatic animal and plant communities), and may well act in concert with other stresses (Erwin and Custer 2000, Custer 2000, Bennet et al. 1995, Chapter 15, this volume). Wading bird tissues and population fluctuations are often proposed as biomonitors of contaminants and of the effects of contaminants (Custer and Mulhern 1983, Kushlan 1993, Erwin and Custer 2000, Burger and Gochfeld 1993).

19.7.5 HUMAN DISTURBANCE

As a result of rapidly growing densities of humans in coastal zones, wading birds are being affected by human disturbance at increasing rates, having a significant impact on choice of breeding sites, reproductive success, and use of foraging sites (Tremblay and Ellison 1979, Erwin 1980, Powell and Powell 1986, Butler 1997, Bouton 1999). Although some species and individuals seem to adapt to, and even profit from, human presence (Powell 1983), the effects of disturbance can result in dramatic changes in distribution of birds and in population declines. There are several strategies for counteracting disturbance effects, including restriction of access to feeding and breeding areas (Erwin 1980), enforcement of approach distances (Rodgers and Smith 1995), and restrictions on various types of activity.

19.8 FUTURE RESEARCH PRIORITIES

The degree to which most wading bird species can tolerate saline conditions and saline diets is poorly understood. However, the example of White Ibises indicates that some coastal wading birds are intolerant of saline conditions, and that this condition imposes a limitation on suitable breeding and feeding habitat. Since this factor has such important implications for surface-water management in the coastal zone, it seems of immediate importance to understand the extent of this limitation among aquatic birds. As in the White Ibis example, the tolerances of young prefledging birds may tell much more about the impacts of salt on the reproductive ecology of the species than the tolerances of adults.

The net effects of both contaminants and toxic marine blooms on wading birds are likely to be extreme enough that this area merits considerable research attention (Figure 19.14). For example, there is little or no scientific basis on which to establish safe exposure levels of wading birds to most contaminants, and sublethal effects are those that have garnered the least attention. Wading birds in coastal areas are clearly under multiple stresses, and there is an immediate need for field studies which can measure the cumulative effects of these multiple stresses, such as increasing salinity, human disturbance, food limitation, and exposure to disease, and as increased aggression as a result of crowding.

It is still not clear whether, and at what spatial scale, wading birds are good geographic indicators of food supply. Simply stated, do aggregated foraging flocks indicate a healthy food supply, and how often are dense food supplies missed by wading birds? In addition, although the physical processes by which food becomes available to wading birds seem well described (hydrology, vegetation, temperature, oxygenation), the mechanisms by which prey abundance fluctuates in wetlands are very poorly researched. Research on the relative importance of nutrients, competitive predators, vegetative density, and community structure on prey abundance is needed to achieve any predictive understanding of wading bird ecology.

Finally, it seems extremely important to continue and expand monitoring of wading bird populations, since these animals can be used as low-cost biomonitors of stresses in estuarine and coastal ecosystems. Of particular interest is the establishment of more long-term studies of survival, since there are so few data on this critical parameter of life history, and because survival parameters can be used both as long-term monitoring tools and as methods to evaluate the effects of specific management actions. Studies of survival also may be the only way to demonstrate the demographic importance of rare ecological events.

FIGURE 19.14 Great Blue Heron (white phase) in southern Florida. Wading birds are dependent on healthy, uncontaminated coastal wetlands and estuaries for feeding. (Photo by R. W and E. A. Schreiber.)

ACKNOWLEDGMENTS

Many people have contributed unwittingly to this chapter both through their publications and through discussions with me over the years. In particular I wish to thank Keith Bildstein, John Ogden, Rich Paul, Bill Robertson, Marilyn Spalding, Tom Atkeson, Don Axelrad, Rob Bennetts, and Heinz Hafner for sharing central ideas and discussions. I would also like to recognize Keith Bildstein for providing constructive critique of an earlier draft of this chapter. The Florida Department of Environmental Protection and the U.S. Army Corps of Engineers have provided support of my research over the past 10 years. This is publication R-08145 of the Journal Series, Florida Agricultural Experiment Station.

LITERATURE CITED

ACOSTA, A., L. MUGICA, AND X. RUIZ. 1996. Resource partitioning between Glossy and American White Ibises in a ricefield system in southcentral Cuba. Colonial Waterbirds 19: 65–72.

ALLEEM, A. A. 1972. Effect of river outflow management on marine life. Marine Biology 15: 200–208.

ALLEN, R. P. 1942. The Roseate Spoonbill, National Audubon Society Research Report No. 2. National Audubon Society, New York.

ARENDT, W. J., AND A. I. ARENDT. 1988. Aspects of the breeding biology of the Cattle Egret *Bubulcus ibis* in Montserrat, West Indies, and its impact on nest vegetation. Colonial Waterbirds 11: 72–84.

ARENGO, F., AND G. A. BALDASSARRE. 1999. Resource variability and conservation of American Flamingos in coastal wetlands of Yucatan, Mexico. Journal of Wildlife Management 63: 1201 — 1212.

ATTWELL, R.I.G. 1966. Possible bird-crocodile commensalism. Ostrich 37: 54–55.

BABBITT, G. 2000. Morphologic, energetic, and behavioral aspects of reproduction in a capture group of Scarlet Ibises (*Eudocimus ruber*). Unpublished M.S. thesis, University of Florida, Gainesville.

BANCROFT, G. T. 1989. Status and conservation of wading birds in the Everglades. American Birds 43: 1258–1265.

BANCROFT, G. T., J. C. OGDEN, AND B. W. PATTY. 1988. Wading bird colony formation and turnover relative to rainfall in the Corkscrew Swamp area of Florida during 1982 through 1985. Wilson Bulletin 100: 50–59.

BANCROFT, G. T., A. M. STRONG, R. J. SAWICKI, W. HOFFMAN, AND S. D. JEWELL. 1994. Relationships among wading bird foraging patterns, colony locations, and hydrology in the Everglades. Pp. 615–87 *in* Everglades: The Ecosystem and Its Restoration (S. Davis and J. C. Ogden, Eds.). St. Lucie Press, Delray Beach, FL.

BATEMAN, D. L. 1970. Movement-Behaviour in Three Species of Colonial-Nesting Wading Birds: A Radio-Telemetric Study. Ph.D. dissertation, Auburn University, Auburn, AL.

BAYER, R. D. 1985. Bill length of herons and egrets as an estimator of prey size. Colonial Waterbirds 8: 104–109.

BENNET, W. A., D. J. OSTRACH, AND D. E. HINTON. 1995. Larval striped bass condition in a drought-stricken estuary — evaluating pelagic food-web limitation. Ecological Applications 5: 680–692.

BENNETTS, R. E., AND W. M. KITCHENS. 1997. Population dynamics and conservation of Snail Kites in Florida: the importance of spatial and temporal scale. Colonial Waterbirds 20: 324–329.

BERRUTI, A. 1983. The biomass energy consumption and breeding of waterbirds relative to hydrological conditions at Lake St. Lucia. Ostritch 54: 65–82.

BILDSTEIN, K. L. 1990. Status, conservation and management of the Scarlet Ibis (*Eudocimus ruber*) in the Caroni Swamp, Trinidad, West Indies. *Biological Conservation* 54: 61–78.

BILDSTEIN, K. L. 1993. White Ibis, Wetland Wanderer. Smithsonian Institution Press, Washington, D.C.

BILDSTEIN, K. L., G. T. BANCROFT, P. J. DUGAN, D. H. GORDON, R. M. ERWIN, E. NOL, L. X. PAYNE, AND S. E. SENNER. 1991. Approaches to the conservation of coastal wetlands in the western hemisphere. Wilson Bulletin 103: 218–254.

BILDSTEIN, K. L., R. CHRISTY, AND P. DECOURSEY. 1982. Size and structure of a South Carolina salt marsh avian community. Wetlands 2: 118–137.

BILDSTEIN, K. L., W. POST, J. JOHNSTON, AND P. FREDERICK. 1990. Freshwater wetlands, rainfall, and the breeding ecology of White Ibises in coastal South Carolina. Wilson Bulletin 102: 84–98.

BIRKHEAD, T. R., AND A. P. MOLLER. 1992. Sperm Competition in Birds. Academic Press, London.

BJORK, R. B. 1986. Reproductive ecology of selected ciconiiformes nesting at Battery Island, North Carolina. Unpublished M.S. thesis, University of North Carolina.

BLANCO, G., AND R. RODRIGUEZ-ESTRELLA. 1998. Human activity may benefit White-faced Ibises overwintering in Baja California Sur, Mexico. Colonial Waterbirds 21: 274–276.

BOUTON, S. N. 1999. Ecotourism in wading bird colonies in the Brazilian Pantanal: biological and socio-economic implications. Unpublished M.S. thesis, University of Florida, Gainesville.

BRITTON, R. H., AND M. E. MOSER. 1982. Size specific predation by herons and its effect on the sex-ratio of natural populations of the Mosquito-fish *Gambusia affinis* Baird and Girard. Oecologia 53: 146–151.

BROWNIE, C., D. R. ANDERSON, K. P. BURNHAM, AND D. S. ROBSON. 1985. Statistical Inference from Band Recovery Data Handbook, 2nd ed. U.S. Fish and Wildlife Service Resource Publication No. 156, Washington, D.C.

BURGER, J. 1978. The pattern and mechanism of nesting in mixed species heronries. Pp. 45–58 *in* Wading Birds. Research Report No. 7 (A. Sprunt, J.C. Ogden, and S. Winckler. Eds.). National Audubon Society, New York.

BURGER, J., AND M. GOCHFELD. 1993. Heavy metal and selenium levels in young egrets and herons from Hong Kong and Szechuen, China. *Archives of Environmental Contamination and Toxicology* 25: 322–327.

BURGER, J., M. GOCHFELD, AND L. J. NILES. 1995. Ecotourism and birds in coastal New Jersey: contrasting responses of birds, tourists, and managers. Environmental Conservation 22: 56–65.

BURGER, J., AND C. HAHN. 1989. Crow predation on Black-crowned Night Heron eggs. Wilson Bulletin 101: 350–351.

BURKHOLDER, J. M. 1998. Implications of harmful microalgae and heterotrophic dinoflagellates in management of sustainable marine fisheries. Ecological Applications, Supplement 8: s36–s62.

BUTLER, R. W. 1988. Population regulation of wading ciconiiform birds. Colonial Waterbirds 17: 189–199.

BUTLER, R. W. 1993. Time of breeding in relation to food availability of female Great Blue Herons (Ardea herodias). Auk 110(4): 693–701.

BUTLER, R. W. 1997. The Great Blue Heron. University of British Columbia Press, Vancouver, British Columbia.

BYRD, M. A. 1978. Dispersal and movements of six North American ciconiiforms. Wading Birds (A. Sprunt, J. C. Ogden, and S. Winckler, Eds.). National Audubon Society Research Report No. 7. National Audubon Society, New York.

CAIRNS, D. K., K. A. BREDIN, AND V. L. BIRT. 1987. A tunnel for hidden access to blinds at high latitude seabird colonies. Journal of Field Ornithology 58(1): 69–72.

CALDWELL, G. S. 1981. Attraction to tropical mixed-species heron flocks: proximate mechanism and consequences. Behavioral Ecology and Sociobiology 8: 99–103.

CALDWELL, G. S. 1986. Predation as a selective force on foraging herons: effects of plummage color and flocking. Auk 103: 494–505.

CARLSON, B. A., AND E. B. McLEAN. 1996. Buffer zones and disturbance types as predictors of fledging success in Great Blue Herons (Ardea herodias). Colonial Waterbirds 19: 124–127.

CARNEY, K. M., AND W. J. SYDEMAN. 1999. A review of human disturbance effects on nesting colonial waterbirds. Waterbirds 22: 68–79.

CEZILLY, F. 1997. Demographic studies of wading birds: an overview. Colonial Waterbirds 20: 121–128.

CEZILLY, F., A. VIALLEFONT, V. BOY, AND A. R. JOHNSON. 1996. Annual variation in survival and breeding probability in greater flamingos. Ecology 77: 1143–1150.

CHAVEZ-RAMIREZ, F., AND R. D. SLACK. 1995. Differential use of coastal marsh habitats by nonbreeding wading birds. Colonial Waterbirds 18: 166–171.

CLOBERT, J., AND J. D. LEBRETON. 1991. Estimation of bird demographic parameters in bird populations. Pp. 75–104 in Bird Population Studies (C. M. Perrins, J. D. Lebreton, and G. J. M. Hirons, Eds.). Oxford University Press, Oxford, U.K.

CLOERN, J. E. 1996. Phytoplankton bloom dynamics in coastal ecosystems: a review with some general lessons from sustained investigation of San Francisco Bay, California. Reviews of Geophysics 34: 127–168.

COFFEY, B. B., JR. 1943. Post-juvenile migration of herons. Birdbanding 14: 34–39.

COFFEY, B. B., JR. 1948. Southward migration of herons. Birdbanding 19: 1–15.

CUSTER, T. W. 2000. Environmental Contaminants. Pp. 251–268 in Heron Conservation (J. A. Kushlan and H. Hatner, Eds.). Academic Press, London.

CUSTER, T. W., AND B. L. MULHERN. 1983. Heavy metal residues in prefledgling Black-crowned Night Herons from three Atlantic coast colonies. Bulletin of Environmental Contamination and Toxicology 30: 178–185.

CUSTER, T. W., AND R. G. OSBORN. 1978. Feeding habitat use by colonially-breeding herons, egrets and ibises in North Carolina. Auk 95: 733–743.

DAVIS, W. E., JR., AND K. C. PARSONS. 1991. Effects of investigator disturbance on the survival of Snowy Egret nestlings. Journal of Field Ornithology 62: 432–435.

DAVIS, W. E., AND J. A. KUSHLAN. 1994. Green Heron (Butorides virescens). No. 129 in The Birds of North America (A. Poole and F. Gill, Eds.). Academy of Natural Sciences, Philadelphia; American Ornithologists Union, Washington, D.C.

DEN HELD, J. J. 1981. Population changes in the Purple Heron in relation to drought in the wintering area. Ardea 69: 185–191.

DODD, M. G., AND MURPHY, T. 1995. Accuracy and precision of techniques for counting Great Blue Heron nests. Journal of Wildlife Management 59: 667–673.

DODD, M. G., AND T. M. MURPHY. 1996. The Status and Distribution of Wading Birds in South Carolina, 1988–1996. Report SG9610-A, South Carolina Marine Resources, Columbia, SC, 66 pp.

DRAULANS, D. 1987. The effect of prey density on foraging behaviors of adult and first-year grey herons (Ardea cinerea). Journal of Animal Ecology 56: 479–493.

DRAULANS, D., AND J. HANNON. 1988. Distibution and foraging behaviour of Grey Herons (Ardea cinerea) in adjacent tidal and non-tidal areas. Ornis Scandinavica 19: 297–304.

DUSI, J. L. 1985. Use of sounds and decoys to attract herons to a colony site. Colonial Waterbirds 8: 178–180.

ELLISON, A. M., AND E. J. FARNSWORTH. 1996. Anthropogenic disturbance of Caribbean mangrove ecosystems: past impacts, present trends, and future predictions. Biotropica 28: 549–565.

ENS, B. J., P. ESSELINK, AND L. ZWARTS. 1990. Kleptoparasitism as a problem of prey choice: a study on mudflat-feeding curlews, Numenious arquata. Animal Behaviour 39: 219–320.

EPSTEIN, P., B. SHERMAN, E. SPANGER-SIEGFRIED, A. LANGSTON, S. PRASAD, AND B. MCKAY. 1998. Marine Ecosystems — Emerging Diseases as Indicators of Change. Center for Health and the Global Environment, Harvard University, Cambridge, MA, Grant Report #NA56GP 0623, 85 pp.

ERWIN, R. M. 1980. Breeding habitat use by colonially nesting waterbirds in two mid-Atlantic United States regions under different regimes of human disturbance. Biological Conservation 18: 39–51.

ERWIN, R. M. 1985. Foraging decisions, patch use, and seasonality in egrets (Aves: Ciconiiformes). Ecology 66: 837–844.

ERWIN, R. M. 1989. Responses to human intruders by birds nesting in colonies: Experimental results and management guidelines. Colonial Waterbirds 12: 104–108.

ERWIN, R. M. 1996. Dependence of waterbirds and shorebirds on shallow-water habitats in the mid-Atlantic coastal region: an ecological profile. Estuaries 19: 213–219.

ERWIN, R. M., AND T. W. CUSTER. 2000. Herons as indicators. Pp. 311–330 in Heron Conservation (J. A. Kushlan and H. Hafner, Eds.). Academic Press, London.

ERWIN, R. M., F. S. HATFIELD, AND T. WILMERS. 1995. The value and vulnerability of small estuarine islands for conserving metapopulations of breeding waterbirds. Biological Conservation 71: 187–191.

ERWIN, R. M., J. G. HAIG, D. B. STOTTS, AND J. S. HATFIELD. 1996. Reproductive success, growth and survival of Black-crowned Night-heron (Nycticorax nycticorax) and Snowy Egret (Egretta thula) chicks in coastal Virginia. Auk 113: 119–130.

FASOLA, M., AND F. BARBIERI. 1978. Factors affecting the distribution of heronries in northern Italy. Ibis 120: 537–540.

FENNEMA, R. J., C. J. NEIDRAUER, R. A. JOHNSON, T. K. MacVICAR, AND W. A PERKINS. 1994. A computer model to simulate natural Everglades hydrology. Pp. 249–289 in Everglades: The Ecosystem and Its Conservation (S. M. Davis and J. C. Ogden, Eds.). St. Lucie Press, Delray Beach, Florida.

FERNANDEZ-CRUZ, M., AND M. CAMPOS. 1993. The breeding of Grey Herons (Ardea cinerea) in western Spain: the influence of age. Colonial Waterbirds 16: 53–58.

FLEURY, B. E. 1994. Crisis in the crawfish ponds. Living Bird 1994: 28–34.

FORRESTER, D. W., AND M. G. SPALDING. (in press). Diseases and Parasites of Wild Birds in Florida. University Presses of Florida, Gainesville, FL.

FREDERICK, P. C. 1987a. Chronic tidally-induced nest failure in a colony of White Ibis. Condor 89: 413–419.

FREDERICK, P. C. 1987b. Extrapair copulations in the mating system of White Ibis (Eudocimus albus). Behaviour 100: 170–201.

FREDERICK, P. C. 1997. Tricolored Heron (Egretta tricolor). No. 306 in The Birds of North America (A. Poole and F. Gill, Eds.). Academy of Natural Sciences, Philadelphia; American Ornithologists' Union, Washington, D.C.

FREDERICK, P. C., AND M. W. COLLOPY. 1989a. Nesting success of five ciconiiform species in relation to water conditions in the Florida Everglades. Auk 106: 625–634.

FREDERICK, P. C., AND M. W. COLLOPY. 1989b. The role of predation in determining reproductive success of colonially nesting wading birds in the Florida Everglades. Condor 91: 860–867.

FREDERICK, P. C., AND M. W. COLLOPY. 1989c. Researcher disturbance in colonies of wading birds: effects of frequency of visit and egg-marking on reproductive parameters. Colonial Waterbirds 12: 152–157.

FREDERICK, P. C., AND K. L. BILDSTEIN. 1992. Foraging ecology of seven species of neotropical ibises (Threskiornithidae) during the dry season in the central llanos of Venezuela. Wilson Bulletin 104: 1–21.

FREDERICK, P. C., AND W. F. LOFTUS. 1993. Responses of marsh fishes and breeding wading birds to low temperatures: a possible behavioral link between predator and prey. Estuaries 16: 216–222.

FREDERICK, P. C., AND J. C. OGDEN. 1997. Philopatry and nomadism: contrasting long-term movement behavior and population dynamics of White Ibises and Wood Storks. Colonial Waterbirds 20: 316–323.

FREDERICK, P. C., AND G. V. N. POWELL. 1994. Nutrient transport by wading birds in the Everglades. Pp. 571–584 in Everglades: The Ecosystem and Its Restoration (S. Davis and J. C. Ogden, Eds.). St. Lucie Press, Delray Beach, FL.

FREDERICK P. C., R. BJORK, G. T. BANCROFT, AND G. V. N. POWELL. 1992. Reproductive success of three species of herons relative to habitat in southern Florida. Colonial Waterbirds 15: 192–201.

FREDERICK, P. C., M. G. SPALDING, AND G. V. N. POWELL. 1993. Evaluating methods to measure nestling survival in Tricolored Herons. Journal of Wildlife Management 57: 34–41.

FREDERICK, P. C., K. L. BILDSTEIN, B. FLEURY, AND J. C. OGDEN. 1996a. Conservation of large, nomadic populations of White Ibises (Eudocimus albus) in the United States. Conservation Biology 10: 203–216.

FREDERICK, P. C., J. C. CORREA-SANDOVAL, C. LUTHIN, AND M. G. SPALDING. 1996b. The importance of the Caribbean coastal wetlands of Nicaragua and Honduras to Central American populations of waterbirds and Jabiru Storks (*Jabiru mycteria*). Journal of Field Ornithology 68: 287–295.

FREDERICK, P. C., T. TOWLES, R. J. SAWICKI, AND G. T. BANCROFT. 1996c. Comparison of aerial and ground techniques for discovery and census of wading bird (Ciconiiformes) nesting colonies. Condor 98: 837–840.

FUJIOKA, M., AND S. YAMAGISHI. 1981. Extramarital and pair copulations in the Cattle Egret. Auk 98: 134–144.

GAWLIK, D. E. (In press). The effects of prey availability on the feeding tactics of wading birds. Ecological Monographs.

GIANNECCHINI, J. 1993. Ecotourism: new partners, new relationships. Conservation Biology 7: 429–432.

GIBBS, J. P. 1991. Spatial relationships between nesting colonies and foraging areas of Great Blue Herons. Auk 108(4): 764–770.

GIBBS, J. P., AND L. K. KINKEL. 1997. Determinants of the size and location of Great Blue Heron colonies. Colonial Waterbirds 20: 1–7.

GIBBS, J. P., S. WOODWARD, M. L. HUNTER, AND A. E. HUTCHINSON. 1987. Determinants of Great Blue Heron colony distribution in coastal Maine. Auk 104(1): 38–47.

GIBBS, J. P., S. WOODWARD, M. L. HUNTER, AND A. E. HUTCHINSON. 1988. Comparison of techniques for censusing Great Blue Heron nests. Journal of Field Ornithology 59: 130–134.

GONZALEZ, J. A. 1996. Kleptoparasitism in mixed-species foraging flocks of wading birds during the late dry season in the Llanos of Venezuela. Colonial Waterbirds 19: 226–231.

GONZALEZ, J. A. 1999. Effects of harvesting of waterbirds and their eggs by native people in the northeastern Peruvian Amazon. Waterbirds 22: 217–224.

GOTMARK, F. 1992. The effect of investigator disturbance on nesting birds. Pp. 63–104 *in* Current Ornithology, Vol. 9 (D. M. Power, Ed.). Plenum Press, New York.

GRULL, A., AND A. RANNER. 1998. Population parameters of the Great Egret and Purple Heron in relation to ecological factors in the Reed Belt of the Neusielder See. Colonial Waterbirds 21: 328–334.

HAFNER, H. 1977. Contribution a l'etude ecologique de quatre especes de herons (Egretta g. garzetta L,. Ardeola r. ralloides Scop., Ardeola i. ibis L., Nycticorax nycticorax L.). Ph.D. dissertation, University of Toulouse, France.

HAFNER, H., O. PINEAU, AND Y. KAYSER. 1994. Ecological determinants of annual fluctuations in numbers of breeding Little Egrets (Egretta garzetta) in the Camargue, France. Revue D'Ecologie — La Terre et La Vie. 49: 53–62.

HAFNER, H., Y KAYSER, V. BOY, M. FASOLA, A. C. JULLIARD, R. PRADEL, AND F. CEZILLY. 1998. Local survival, natal dispersal, and recruitment in Little Egrets (Egretta garzetta). Journal of Avian Biology 29: 216–227.

HAFNER, H., V. BOY, AND G. GORY. 1982. Feeding methods, flock size and feeding success in the Little Egret *Egretta garzetta* and the Squacco Heron *Ardeola ralloides* in Camargue, Southern France. Ardea 70: 45–54.

HAFNER, H., AND R. H. BRITTON. 1983. Changes of foraging site by nesting Little Egrets (Egretta garzetta) in relation to food supply. Colonial Waterbirds 6: 24–30.

HAFNER, H., P. J. DUGAN, M. KERSTEN, O. PINEAU, AND J. P. WALLACE. 1993. Flock feeding and food intake in Little Egrets *Egretta garzetta* and their effects on food provisioning and reproductive success. Ibis 135: 25–32.

HANCOCK, J. A., J. A. KUSHLAN, AND M. P. KAHL. 1992. Storks, Ibises and Spoonbills of the World. Academic Press, New York.

HANCOCK, J., AND J. A. KUSHLAN. 1984. The Herons Handbook. Harper & Row, New York.

HENNY, C. J. 1972. An analysis of the population dynamics of selected avian species. U. S. Fish and Wildlife Service Special Publication #1, Patuxent, MD.

HIGUCHI, H. 1986. Bait-fishing by the Green-backed Heron *Ardeola striata* in Japan. Ibis 128: 285–290.

HIGUCHI, H. 1988. Individual differences in bait-fishing by the Green-backed Heron *Ardeola striata* associated with territory quality. Ibis 130: 39–44.

HOPKINS, M., JR. 1968. Alligators in heron rookeries: *Alligator mississippiensis* is cohabitant. Oriole 33(2): 28–29.

HOWARD, R. K., AND K. W. LOWE. 1984. The impact of avian predation on the caridean shrimp *Macrobrachium intermedium*. Journal of Experimental Marine Biology and Ecology 74: 35–52.

JAY, D. A., AND C. A. SIMENSTAD. 1996. Downstream effects of water withdrawal in a small, high-gradient basin: erosion and deposition on the Skokomish River delta. Estuaries 19: 501–517.

JE, O. 1995. The ecology of mangrove conservation and management. Hydrobiologia 295: 343–351.

JOHNSON, A. R. 1997. Long-term studies and conservation of Greater Flamingos in the Camargue and Mediterranean. Colonial Waterbirds 20: 306–315.

JOHNSTON, J. W., AND K. L. BILDSTEIN. 1990. Dietary salt as a physiological constraint in White Ibis breeding in an estruary. Physiological Zoology 63: 190–207.

KAHL, M. P. 1963. Mortality of Common Egrets and other herons. Auk 80: 295–300.

KAHL, M. P., JR. 1964. Food ecology of the Wood Stork *Mycteria americana* in Florida. Ecological Monographs 34(2): 97–117.

KAMYANIBWA, S., A. SCHIERER, R. PRADEL, AND J. D. LEBRETON. 1990. Changes in adult survival in a western European population of White Stork (Ciconia ciconia). Ibis 132: 27–35.

KATZIR, G., AND N. INTRATOR. 1987. Striking of underwater prey by reef herons (Egretta gularis schistacea). Journal of Comparative Physiology A 160: 517–523.

KATZIR, G., AND G. R. MARTIN. 1994. Visual-fields in herons (Ardeidae) — panoramic vision beneath the bill. Naturwissenshaften 81(4): 182–184.

KINGSFORD, R. T., AND W. JOHNSON. 1998. Impact of water diversions on colonially-nesting waterbirds in the Macquarie marshes of arid Australia. Colonial Waterbirds 21: 159–170.

KLEIN, M. L., S. R. HUMPHREY, AND H. F. PERCIVAL. 1995. Effects of ecotourism on distribution of waterbirds in a wildlife refuge. Conservation Biology, 9: 1454–1465.

KLEKOWSKI, E. J., S. A. TEMPLE, A. M. SIUNG-CHANG, AND K. KUMARSINGH. 1999. An association of mangrove mutation, Scarlet Ibis, and mercury contamination in Trinidad, West Indies. Environmental Pollution 105: 185–189.

KREBS, J. R. 1974. Colonial nesting and social feeding as strategies for exploiting food resources in the Great Blue Heron. Behavior 51: 99–131.

KREBS, J. R., AND B. PARTIDGE. 1973. The significance of head tilting in the great blue heron. Nature, London 245: 533–535.

KUSHLAN, J. A. 1976a. Environmental stability and fish community diversity. Ecology 57: 821–825.

KUSHLAN, J. A. 1976b. Feeding behavior of North American Herons. Auk 93: 86–94.

KUSHLAN, J. A. 1977. Population energetics of the American White Ibis. Auk 94: 114–122.

KUSHLAN, J. A. 1986. Responses of wading birds to seasonally fluctuating water levels: strategies and their limits. Colonial Waterbirds 9: 155–162.

KUSHLAN, J. A. 1993. Colonial waterbirds as bioindicators of environmental change. Colonial Waterbirds 16: 223–251.

KUSHLAN, J. A. 1997. The conservation of wading birds. Colonial Waterbirds 20: 129–137.

KUSHLAN, J. A., AND K. L. BILDSTEIN. 1992. White Ibis. *In* The Birds of North America (A. Poole, P. Stettenheim, and F. Gill, Eds.). Academy of Natural Sciences, Philadelphia.

KUSHLAN, J. A., AND H. HAFNER. 2000. Heron Conservation. Academic Press, London.

KUSHLAN, J. A., AND M. S. KUSHLAN. 1975. Food of the White Ibis in southern Florida. Fla. Field Nat. 3: 31–38.

KUSHLAN, J. A., J. C. OGDEN, AND A. L. HIGER. 1975. Relation of Water Level and Fish Availability to Wood Stork Reproduction in the Southern Everglades. U.S. Geological Survey Open-File Rept., U.S. Geological Survey, Tallahassee, FL.

LACK, D. 1949. The apparent survival-rate of ringed herons. British Birds 42: 74–79.

LANGE, R., AND H. STAALAND. 1966. Anatomy and physiology of the salt gland in the Grey Heron, Ardea cinerea. Nytt Magasin for Zoologi 13: 5–9.

LEBRETON, J. D., K. P. BURNHAM, J. CLOBERT, AND D. R. ANDERSON. 1992. Modeling survival and testing biological hypotheses using marked animals: a unified approach with case studies. Ecological Monographs 62: 67–118.

LEBRETON, J. D., R. PRADEL, AND J. CLOBERT. 1993. The statistical analysis of survival in animal populations. Trends in Ecology and Evolution 8: 91–95.

LOFTUS, W. F., AND A. M. EKLUND. 1994. Long-term dynamics of an Everglades small-fish assemblage. Pp. 461–484 in Everglades, the Ecosystem and Its Conservation (S. M. Davis and J. C. Ogden, Eds.). St. Lucie Press, Delray Beach, FL.

LOTEM, A., E. SCHECHTMAN, AND G. KATZIR. 1991. Capture of submerged prey by little egrets (Egretta garzetta): strike depth, strike angle, and the problem of light refraction. Animal Behaviour 42: 341–346.

LOWE, K. W. 1981. Feeding behaviour and diet of Royal Spoonbills Platalea regia in Westernport Bay, Victoria. Emu 82: 163–168.

MADDOCK, M. 1986. Fledging success of egrets in dry and wet seasons. Corella 10: 101–107.

MARION, L. 1989. Territorial feeding and colonial breeding are not mutually exclusive: the case of the Grey Heron (Ardea cinerea). Journal of Animal Ecology 58: 693–710.

MASTER, T., M. FRANKEL, AND M. RUSSELL. 1993. Benefits of foraging in mixed-species wader aggregations in a southern New Jersey saltmarsh. Colonial Waterbirds 16: 149–157.

MASTER, T. M. 1992. Composition, structure and dynamics of mixed-species foraging aggregations in a southern New Jersey salt marsh. Colonial Waterbirds 15: 66–74.

McCRIMMON, D. A., JR. 1978. Nest-site characteristics among five species of herons on the North Carolina coast. Auk 95: 267–280.

McCRIMMON, D. A., JR. 1974. Stretch and snap displays in the Great Egret. Wilson Bulletin 86: 165–167.

McILHENNY, E. A. 1939. The Autobiography of an Egret. Hastings House, New York.

McIVOR, C. C., J. A. LEY, AND R. D. BJORK. 1994. Changes in freshwater inflow from the Everglades to Florida Bay including effects on biota and biotic processes: a review. Pp. 117–148 in Everglades: The Ecosystem and Its Restoration (S. Davis and J. C. Ogden, Eds.). St. Lucie Press, Delray Beach, FL.

McVAUGH, W., JR. 1975. The development of four North American herons. Living Bird Quarterly 14: 163–183.

MEYERRIECKS, A. J. 1960. Comparative breeding behaviour of four species of North American herons. Nuttall Ornithological Club # 2. Cambridge, MA.

MEYERRIECKS, A. J. 1962. Diversity typifies heron feeding. Natural History 72: 48–59.

MICHENER, W. K., E. R. BLOOD, K. L. BILDSTEIN, M. M. BRINSON AND L. G. GARDNER. 1997. Climate change, hurricanes and tropical storms, and rising sea level in coastal wetlands. Ecological Applications 7: 770–801.

MIKUSKA, T., J. A. KUSHLAN, AND S. HARTLEY. 1998. Key areas for wintering North American Herons. Colonial Waterbirds 21: 125–134.

MIRANDA, L., AND J. A. COLLAZO. 1997. Food habits of four species of wading birds (Ardeidae) in a tropical mangrove swamp. Colonial Waterbirds 20: 413–418.

MOCK, D. W. 1980. Communication strategies of Great Blue Herons and Great egrets. Behaviour 72: 156–170.

MOCK, D. W., T. C. LAMEY, AND B. J. PLOGER. 1987a. Proximate and ultimate roles of food amount in regulating egret sibling aggression. Ecology 68: 1760–1772.

MOCK, D. W., T. C. LAMEY, C. F. WILLIAMS, AND A. PELLETIER. 1987b. Flexibility in the development of heron sibling aggression: an intraspecific test of the prey-size hypothesis. Animal Behavior 35: 1386–1393.

MOCK, D. W., AND K. C. MOCK. 1980. Feeding behaviour and ecology of the Goliath Heron. Auk 97: 443–448.

MOSER, M. 1984. Resource partioning in colonial herons with particular reference to the Grey Heron Ardea cinerea L. and the Purple Heron Ardea purpurea L., in the Camargue, S. France. Ph.D. dissertation, University of Durham, U.K.

NICHOLS, F. H., J. E. CLOERN, S. N. LOUMA, AND D. H. PETERSON. 1986. The modification of an estuary. Science 231: 567–573.

NICHOLS, J. D. 1992. Capture-recapture models. Bioscience 42: 94–102.

NORTH, P. M. 1979. Relating Grey Heron survival rates to winter weather conditions. Bird Study 26: 23–28.

NTIAMOA-BAIDU, Y., T. PIERSMA, P. WIERSMA, M. POOT, P. BATLEY, AND C. GORDON. 1998. Habitat selection, daily foraging routines and diet of waterbirds in coastal lagoons in Ghana. Ibis 140: 89–103.

O'CONNOR, R. J. 1978. Brood reduction in birds: selection for fratricide, infanticide, and suicide? Animal Behaviour 26: 79–96.

OGDEN, J. C. 1994. A comparison of wading bird nesting dynamics, 1931–1946 and 1974–1989 as an indication of changes in ecosystem conditions in the southern Everglades. Pp. 533–570 in Everglades: The Ecosystem and Its Restoration (S. Davis and J. C. Ogden, Eds.). St. Lucie Press, Delray Beach, FL.

OGDEN, J. C. 1978a. Endangered Wood Stork. Pp. 3–4 in Rare and Endangered Biota of Florida, Vol. 2, Birds (H. W. Kale, Ed.). University Presses of Florida, Gainesville, FL.

OGDEN, J. C. 1978b. Recent population trends of colonial wading birds on the Atlantic and Gulf coastal plains. Wading Birds (A. Sprunt, J. C. Ogden, and S. Winckler, Eds.). Research Report No. 7. National Audubon Society, New York.

OGDEN, J. C., H. W. KALE, II, AND S. A. NESBITT. 1980. The influence of annual variation in rainfall and water levels on nesting by Florida populations of wading birds. Transactions of the Linnaean Society of New York 9: 115–126.

OGDEN, J. C., C. E. KNODER, AND A. SPRUNT, IV. 1988. Colonial wading bird populations in the Usamacinta Delta, Mexico. Pp. 595–605 in Ecologia de los rios Usamacinta y Grijalva. Tabasco, Mexico: Instituto Nacional de Investigaciones sobre Recursos Bioticos.

ONUF, C. P., J. M. TEAL, AND I. M. VALIELA. 1977. Interaction of nutrients, plant growth and herbivory in a mangrove ecosystem. Ecology 58: 514–526.

OWEN, D. F. 1959. Mortality of the Great Blue Heron as shown by banding recoveries. Auk 76: 464–70.

PALMER, R. S. 1962. Handbook of North American Birds, Vol. I. Yale University Press, New Haven, CT.

PARNELL, J. F., D. G. AINLEY, H. BLOKPOEL, B. CAIN, T. W. CUSTER, J. L. DUSI, S. KRESS, J. A. KUSHLAN, W. E. SOUTHERN, L. E. STENZEL, AND B. C. THOMPSON. 1988. Colonial waterbird management in North America. Colonial Waterbirds 11: 129–169.

PARNELL, J. F., R. N. NEEDHAM, R. F. SOOTS, JR., J. O. FUSSEL, III, D. M. DUMOND, D. A. McCRIMMON, JR., R. D. BJORK, AND M. A. SHIELDS. 1986. Use of dredged-material deposition sites by birds in coastal North Carolina, USA. Colonial Waterbirds 9: 210–217.

PARSONS, K. C., AND J. BURGER. 1982. Human disturbance and nestling behavior in Black-crowned Night-Herons. Condor 84: 184–187.

PAUL, R. T. 1991. Status Report — Egretta rufescens (Gmelin) Reddish Egret. U.S. Fish and Wildlife Service Report, Houston, TX.

PETERS, J. L. 1931. Check-List of Birds of the World. Harvard University Press, Cambridge, MA.

PETIT, D. R., AND K. L. BILDSTEIN. 1987. Effect of group size and location within the group on the foraging behavior of White Ibises. Condor 89: 602–609.

POLLACK, K. H., M. J. CONROY, AND W. S. HEARN. 1995. Separation of hunting and natural mortality using ring return models: an overview. Journal of Applied Statistics 22: 557–566.

PORTRAJ, J. J. 1978. The effects of human interference on heron growth and development. Unpublished M.S. thesis, Towson State University, Baltimore, MD.

POST, W. 1990. Nest survival in a large ibis-heron colony during a three-year decline to extinction. Colonial Waterbirds 13: 50–61.

POWELL, G. V. N. 1983. Food availability and reproduction by Great White Herons (Ardea herodias): a food addition study. Colonial Waterbirds 6: 139–147.

POWELL, G. V. N. 1987. Habitat use by wading birds in a subtropical estuary: implications of hydrography. Auk 104: 740–749.

POWELL, G. V. N., AND A. H. POWELL. 1986. Reproduction by Great White Herons Ardea herodias in Florida Bay as an indicator of habitat quality. Biological Conservation 36: 101–113.

POWELL, G. V. N., J. W. FOURQUREAN, W. J. KENWORTHY, AND J. C. ZIEMAN. 1991. Bird colonies cause seagrass enrichment in a subtropical estuary. Estuarine, Coastal and Shelf Science 32: 567–579.

PRATT, H. W., AND D. W. WINKLER. 1985. Clutch size, timing of laying and reproductive success in a colony of Great Blue Herons and Great Egrets. Auk 102: 49–63.

RENFROW, D. H. 1993. The effects of fish density on wading bird use of sediment ponds on an east Texas coal mine. Unpublished M.S. thesis, Texas A&M University, College Station, TX.

ROBERTSON, W. B., AND J. A. KUSHLAN. 1974. The southern Florida avifauna. Miami Geological Society Memoirs 2: 414–452.

RODGERS, J. A. 1987. On the antipredator advantages of coloniality: a word of caution. Wilson Bulletin 99: 269–270.

RODGERS, J. A., AND H. T. SMITH. 1995. Set-back distances to protect nesting bird colonies from human disturbance. Conservation Biology 9: 89–99.

RODGERS, J. A., S. B. LINDA, AND S. A. NESBITT. 1995. Comparing aerial estimates with ground counts of nests in Wood Stork colonies. Journal of Wildlife Management 59: 656–666.

RUDEGEAIR, T. J., JR. 1975. The reproductive behavior and ecology of the White Ibis (Eudocimus albus). Ph.D. dissertation, University of Florida, Gainesville.

RYDER, R. R., AND D. E. MANRY. 1994. White-faced Ibis. No. 130 *in* The Birds of North America (A. Poole and F. Gill, Eds.). Academy of Natural Sciences, Philadelphia; American Ornithologist's Union, Washington, D.C.

SADOUL, N. 1997. The importance of spatial scales in long-term monitoring of colonial Charadriiforms in southern France. Colonial Waterbirds 20: 330–338.

SEPULVEDA, M. S., G. E. WILLIAMS, P. C. FREDERICK, AND M. S. SPALDING. 1999. Effects of mercury on health and first year survival of free-ranging Great Egrets (Ardea albus) from southern Florida. Archives of Environmental Contamination and Toxicology 37: 369–376

SHEPHERD, P., T. CROCKETT, T. L. DE SANTO, AND K. L. BILDSTEIN. 1991. The impact of Hurricane Hugo on the breeding ecology of wading birds at Pumpkinseed Island, Hobcaw Barony, South Carolina. Colonial Waterbirds 14: 150–157.

SHIELDS, M. A., AND J. F. PARNELL. 1986. Fish Crow predation on eggs of the White Ibis at Battery Island, North Carolina. Auk 103: 531–539.

SHOEMAKER, V. H. 1966. Osmotic regulation and excretion in birds. Avian Biology 2: 527–574.

SIBLEY, C. G., AND J. E. AHLQUIST. 1990. Phylogeny and Classification of Birds. Yale University Press, New Haven, CT, 1008 pp.

SIEGFRIED, W. R. 1971. The nest of the Cattle Egret. Ostrich 42: 193–197.

SIMPSON, K., J. N. N. SMITH, AND J. P. KELSALL. 1987. Correlates and consequences of coloniality in Great Blue Herons. Canadian Journal of Zoology 65: 572–577.

SMITH, J. P. 1995a. Foraging flights and habitat use of nesting wading birds (Ciconiiformes) at Lake Okeechobee, Florida. Colonial Waterbirds 18: 139–158.

SMITH, J. P. 1995b. Foraging sociability of nesting wading birds (Ciconiiformes) at Lake Okeechobee, Florida. Wilson Bulletin 107: 437–451.

SNEDAKER, S. C. 1995. Mangroves and climate-change in the Florida and Caribbean region — scenarios and hypotheses. Hydrobiolgia 295: 43–49.

SPALDING, M. G., P. C. FREDERICK, H. C. McGILL, S. N. BOUTON, AND L. R. McDOWELL. 2000a. Methylmercury accumulation in tissues and effects on growth and appetite in captive Great Egrets. Journal of Wildlife Diseases 36: 369–376.

SPALDING, M. G., P. C. FREDERICK, H. C. McGILL, S. N. BOUTON, I. SCHUMACHER, C. G. M. BLACKMORE, L. RICHEY, AND J. HARRISON. 2000b. Histologic, neurologic, and immunologic effects of methylmercury in captive Great Egrets. Journal of Wildlife Diseases 36: 423–435.

SPENDELOW, J. A., J. D. NICHOLS, I. C. T. NISBET, H. HAYS, G. D. CORMONS, J. BURGER, C. SAFINA, J. E. HINES, AND M. GOCHFELD. 1995. Estimating annual survival and movement rates of adults within a metapopulation of Roseate Terns. Ecology 76: 2415–2428.

STANLEY, D. J., AND A. G. WHITE. 1993. Nile Delta: recent geological evolution and human impact. Science 260: 628–634.

STENNING, M. J. 1996. Hatching asynchrony, brood reduction and other rapidly reproducing hypotheses. Trends in Ecology and Evolution 6: 243–246.

STRONG, A. H., G. T. BANCROFT, AND S. D. JEWELL. 1997. Hydrological constraints on tricolored heron and snowy egret resource use. Condor 99: 894–905.

SURDICK, J. A. 1998. Biotic and abiotic indicators of foraging site selection and foraging success of four cicioniiform species in the freshwater Everglades of Florida. Unpublished M.S. thesis, University of Florida, Gainesville.

TEXAS COLONIAL WATERBIRD SOCIETY. 1982. An Atlas and Census of Texas Waterbird Colonies, 1973–1980. Caesar Kleberg Wildlife Research Institute, Kingsville, TX.

THOMAS, F., Y. KAYSER, AND H. HAFNER. 1999. Nestling size rank in the little egret (Egretta garzetta) influences subsequent breeding success of offspring. Behavioral Ecology and Sociobiology 45: 466–470.

TITUS, J. G. 1996. The risk of sea level rise. Climatic Change 33: 151–212.

TREMBLAY, J., AND L. N. ELLISON. 1979. Effects of human disturbance on breeding of Black-crowned Night Herons. Auk 96: 364–369.

TREXLER, J. C., R. C. TEMPE, AND J. TRAVIS. 1994. Size-selective predation of sailfin mollies by two species of heron. Oikos 69: 250–258.

VAN VESSEM, J., AND D. DRAULANS. 1986. The adaptive significance of colonial breeding in the Grey Heron (Ardea cinerea): inter- and intra-colony variability in breeding success. Ornis Scandinavica 17: 356–362.

VOS, D. K., R. A. RYDER, AND W. D. GRAUL. 1985. Response of breeding Great Blue Herons to human disturbance in Northcentral Colorado. Colonial Waterbirds 8: 13–22.

WALTERS, C. J., L. GUNDERSON, AND C. S. HOLLING. 1992. Experimental policies for water management in the Everglades. Ecological Applications 2: 189–202.

WATTS, B. D. 1995. Yellow-crowned Night-heron (Nyctanassa violacea). No. 161 *in* The Birds of North America (A. Poole and F. Gill, Eds.). Academy of Natural Sciences, Philadelphia; American Ornithologists' Union, Washington, D.C.

WATTS, B. D., AND D. S. BRADSHAW. 1994. The influence of human disturbance on the location of Great Blue Heron colonies in the lower Chesapeake Bay. Colonial Waterbirds 17: 184–86.

WEIHS, D., AND G. KATZGIR. 1994. Bill-sweeping in the spoonbill Platalea leucordia: evidence for a hydrodynamic function. Animal Behavior 47: 649–654.

WERSCHKUL, D. F. 1979. Nestling mortality and the adaptive significance of early locomotion in the Little Blue Heron. Auk 96: 116–130.

WESELOH, D. V., AND R. T. BROWN. 1971. Plant distribution within a heron rookery. American Midland Naturalist 86: 57–64.

WHARTON, C. H. 1969. The Cottonmouth Moccasin on Sea Horse Key, Florida. Bulletin of the Florida State Museum 14: 226–272.

WIESE, J. H. 1976. Courtship and pair formation in the Great Egret. Auk 93: 709–724.

WIESE, J. H. 1978. Heron nest-site selection and its ecological effects. Pp. 27–34 *in* Wading Birds (A. Sprunt, J. C. Ogden, and S. Winckler, Eds.). National Audubon Society Research Report No. 7.

WIGGINS, D. A. 1991. Foraging success and aggression in solitary and group-feeding Great Egrets (Casmerodius albus). Colonial Waterbirds 14: 176–179.

Appendix 1

List of Seabird Species with the Current International Union for the Conservation of Nature Red List Status of Those Species Which Are Considered Threatened

Order Sphenisciformes

Family Spheniscidae: Penguins

Pygoscelis papua	Gentoo Penguin	
P. antarctica	Chinstrap Penguin	
P. adeliae	Adelie Penguin	
Aptenodytes patagonicus	King Penguin	
A. forsteri	Emperor Penguin	
Eudyptes chrysocome	Rockhopper Penguin	VU
E. chrysolophus	Macaroni Penguin	VU
E. schlegeli	Royal Penguin	VU
E. pachyrhynchus	Fiordland Penguin	VU
E. robustus	Snares Penguin	
E. sclateri	Erect-crested Penguin	EN
Megadyptes antipodes	Yellow-eyed Penguin	EN
Spheniscus demersus	Jackass Penguin [Black-footed, African]	VU
S. humboldti	Humboldt Penguin	VU
S. magellanicus	Magellanic Penguin	
S. mendiculus	Galapagos Penguin	EN
Eudyptula minor	Blue Penguin [Fairy]	

Order Procellariiformes

Family Diomedeidae: Albatrosses

Diomedea exulans	Wandering Albatross	VU
D. antipodensis	Antipodean Albatross	VU
D. amsterdamensis	Amsterdam Albatross	CR
D. dabbenena	Tristan Albatross	EN
D. sanfordi	Northern Royal Albatross	EN
D. epomophora	Southern Royal Albatross	VU
Phoebastria irrorata	Waved Albatross	VU
P. albatrus	Short-tailed Albatross	VU
P. nigripes[1]	Black-footed Albatross	VU
P. immutabilis[1]	Laysan Albatross	
Thalassarche melanophris	Black-browed Albatross	
T. impavida	Campbell Albatross	VU
T. cauta	Shy Albatross	
T. eremita	Chatham Albatross	CR
T. salvini	Salvin's Albatross	VU
T. chrysostoma	Gray-headed Albatross	VU
T. chlororhynchos	Atlantic Yellow-nosed Albatross	
T. carteri	Indian Yellow-nosed Albatross	VU
T. bulleri	Buller's Albatross	VU
Phoebetria fusca	Sooty Albatross	VU
P. palpebrata	Light-mantled Sooty Albatross	

Family Procellariidae: Gadfly Petrels, Shearwaters, Fulmars, and Allies

Macronectes giganteus	Southern Giant Petrel	VU
M. halli	Northern Giant Petrel	
Fulmarus glacialoides	Southern Fulmar	
F. glacialis	Northern Fulmar	
Thalassoica antarctica	Antarctic Petrel	
Daption capense	Cape Pigeon [Cape Petrel]	
Pagodroma nivea	Snow Petrel	
Lugensa brevirostris	Kerguelen Petrel	
Pterodroma macroptera	Great-winged Petrel	
P. lessonii	White-headed Petrel	
P. incerta	Atlantic Petrel	VU
P. solandri	Providence Petrel	VU
P. magentae	Magenta Petrel [Taiko]	CR
P. ultima	Murphy's Petrel	
P. mollis	Soft-plumaged Petrel	
P. feae	Cape Verde Petrel [Fea's]	
P. madeira	Madeira Petrel	CR
P. cahow	Cahow or Bermuda Petrel	EN
P. hasitata	Black-capped Petrel	EN
P. caribbaea	Jamaica Petrel	CR
P. externa	Juan Fernandez Petrel	VU
P. cervicalis	White-necked Petrel	VU
P. baraui	Barau's Petrel	EN
P. phaeopygia	Galapagos Petrel	CR
P. sandwichensis	Hawaiian Dark-rumped Petrel [Hawaiian]	VU
P. neglecta	Kermadec Petrel	
P. arminjoniana	Trindade Petrel	VU
P. heraldica	Herald Petrel	
P. atrata	Henderson Petrel	EN
P. alba	Phoenix Petrel	VU
P. inexpectata	Mottled Petrel	
P. hypoleuca	Bonin Petrel	
P. nigripennis	Black-winged Petrel	
P. axillaris	Chatham Island Petrel	CR
P. cookii	Cook's Petrel	EN
P. defilippiana	Mas A Tierra Petrel [De Filippi's]	VU
P. longirostris	Stejneger's Petrel	VU
P. pycrofti	Pycroft's Petrel	VU
P. leucoptera	Gould's Petrel	VU
P. brevipes	Collared Petrel	
Pseudobulweria macgillivrayi	Macgillivray's Petrel [Fiji]	CR
P. rostrata	Tahiti Petrel	
P. becki	Beck's Petrel	CR
P. aterrima	Mascarene Petrel	
Halobaena caerulea	Blue Petrel	CR
Pachyptila vittata	Broad-billed Prion	
P. salvini	Salvin's Prion	
P. desolata	Antarctic Prion	
P. belcheri	Narrow-billed Prion	
P. turtur	Fairy Prion	
P. crassirostris	Fulmar Prion	
Bulweria bulwerii	Bulwer's Petrel	
B. fallax	Jouanin's Petrel	
Procellaria aequinoctialis	White-chinned Petrel	VU
P. conspicillata	Spectacled Petrel	CR

P. westlandica	Westland Petrel	VU
P. parkinsoni	Parkinson's Petrel [Black]	VU
P. cinerea	Grey Petrel	
Calonectris diomedea	Cory's Shearwater	
C. leucomelas	Streaked Shearwater	
Puffinus pacificus	Wedge-tailed Shearwater	
P. bulleri	Buller's Shearwater	VU
P. carneipes	Flesh-footed Shearwater	
P. creatopus	Pink-footed Shearwater	VU
P. gravis	Greater Shearwater	
P. griseus	Sooty Shearwater	
P. tenuirostris	Short-tailed Shearwater	
P. nativitatis	Christmas Shearwater	
P. puffinus	Manx Shearwater	
P. yelkouan	Levantine Shearwater	
P. mauretanicus	Balearic Shearwater	
P. auricularis	Townsend's Shearwater	CR
P. newelli	Newell's Shearwater	VU
P. opisthomelas	Black-vented Shearwater	VU
P. gavia	Fluttering Shearwater	
P. huttoni	Hutton's Shearwater	EN
P. lherminieri	Audubon's Shearwater	
P. heinrothi	Heinroth's Shearwater	VU
P. assimilis	Little Shearwater	

Family *Pelecanoididae*: Diving Petrels

Pelecanoides garnotii	Peruvian Diving-petrel	EN
P. magellani	Magellanic Diving-petrel	
P. georgicus	Georgian Diving-petrel	
P. urinatrix	Common Diving-petrel	

Family *Hydrobatidae*: Storm Petrels

Oceanites oceanicus	Wilson's Storm-petrel	
O. gracilis	Elliot's or White-vented Storm-petrel	DD
Garrodia nereis	Gray-backed Storm-petrel	
Pelagodroma marina	White-faced Storm-petrel	
Fregetta tropica	Black-bellied Storm-petrel	
F. grallaria	White-bellied Storm-petrel	
Nesofregetta fuliginosa	White-throated Storm-petrel [Polynesian]	VU
Hydrobates pelagicus	European or British Storm-petrel	
Halocyptena microsoma	Least Storm-petrel	
Oceanodroma tethys	Wedge-rumped Storm-petrel	
O. castro	Madeiran or Band-rumped Storm-petrel	
O. monorhis	Swinhoe's Storm-petrel	
O. leucorhoa	Leach's Storm-petrel	
O. macrodactyla	Guadalupe Storm-petrel [extinct]	CR
O. markhami	Markham's Storm-petrel	DD
O. tristrami	Tristram's Storm-petrel	
O. melania	Black Storm-petrel	
O. matsudairae	Matsudaira's Storm-petrel	DD
O. homochroa	Ashy Storm-petrel	
O. hornbyi	Hornby's or Ringed Storm-petrel	DD
O. furcata	Fork-tailed Storm-petrel	

Order Pelecaniformes
Sub-order *Phaethontes*, Family *Phaethontidae*: Tropicbirds

Phaethon lepturus	White-tailed Tropicbird

P. aethereus	Red-billed Tropicbird	
P. rubricauda	Red-tailed Tropicbird	

Sub-order Pelecani, Family Pelecanidae: Pelicans

Pelecanus onocrotalus	Eurasian Pelican [Great White]	
P. crispus	Dalmatian Pelican	LR
P. conspicillatus	Australian Pelican	
P. erythrorhynchos	American White Pelican	
P. rufescens	Pink-backed Pelican	
P. philippensis	Spot-billed Pelican	VU
P. occidentalis	Brown Pelican [includes Peruvian]	

Family Fregatidae: Frigatebirds

Fregata aquila	Ascension Frigatebird	VU
F. ariel	Lesser Frigatebird	
F. minor	Great Frigatebird	
F. andrewsi	Christmas Island Frigatebird	CR
F. magnificens	Magnificent Frigatebird	

Family Sulidae: Gannets and Boobies

Morus serrator	Australian Gannet	
M. capensis	Cape Gannet	VU
M. bassanus	Northern Gannet	
Papasula abbotti	Abbott's Booby	CR
Sula variegata	Peruvian Booby	
S. nebouxii	Blue-footed Booby	
S. dactylatra	Masked Booby	
S. granti	Nazca Booby	
S. leucogaster	Brown Booby	
S. sula	Red-footed Booby	

Family Phalacrocoracidae, Sub-family Phalacrocoracinae: Cormorants

Microcarbo africanus	Long-tailed Cormorant	
M. coronatus	Crowned Cormorant	
M. pygmaeus	Pygmy Cormorant	
M. niger	Little Cormorant	
M. melanoleucos	Little Pied Cormorant	
Compsohalieus penicillatus	Brandt's Cormorant	
C. harrisi[2]	Flightless Cormorant [Galapagos]	EN
C. neglectus	Bank Cormorant	VU
C. fuscescens	Black-faced Cormorant	
Hypoleucos brasiliensis[3]	Neotropic Cormorant	
H. auritus	Double-crested Cormorant	
H. fuscicollis	Indian Cormorant	
H. varius	Pied Cormorant	
H. sulcirostris	Little Black Cormorant	
Phalacrocorax carbo	Great Cormorant	
P. capillatus	Japanese Cormorant	
Leucocarbo nigrogularis	Socotra Cormorant	VU
L. capensis	Cape Cormorant	
L. bougainvillii	Guanay Cormorant	
Notocarbo verrucosus	Kerguelen Shag	
N. atriceps	Imperial Shag [Incl. Blue-eyed and King]	
N. bransfieldensis	Antarctic Shag	
N. georgianus	South Georgia Shag	
Nesocarbo campbelli	Campbell Island Shag	VU
Euleucocarbo carunculatus	New Zealand King Shag	VU

E. chalconotus	Stewart Island Shag	VU
E. onslowi	Chatham Island Shag	EN
E. colensoi	Auckland Island Shag	VU
E. ranfurlyi	Bounty Island Shag	VU
Stictocarbo magellanicus	Rock Cormorant	
S. pelagicus	Pelagic Cormorant	
S. urile	Red-faced Cormorant	
S. aristotelis	European Shag	
S. gaimardi	Red-legged Cormorant	
S. punctatus	Spotted Shag	
S. featherstoni	Pitt Island Shag	VU

Sub-family Anhinginae: Anhingas or Darters

Anhinga anhinga	Anhinga
A. rufa	African Darter
A. melanogaster	Indian Darter
A. novaeholandiae	Australian Darter

Order Charadriiformes
Sub-order Lari, Family Stercorariidae: Skuas and Jaegers

Stercorarius parasiticus	Parasitic Jaeger [Arctic Skua]
S. longicaudus	Long-tailed Jaeger [Skua]
S. pomarinus	Pomarine Jaeger [Skua]
Catharacta skua	Great Skua
C. chilensis	Chilean Skua
C. antarctica	Brown Skua [incl.Lonnberg's & Antarctic]
C. maccormicki	South Polar Skua

Family Laridae, Sub-family Larinae: Gulls

Larus scoresbii[4]	Dolphin Gull	
L. pacificus[5]	Pacific Gull	
L. belcheri	Band-tailed Gull	
L. atlanticus	Orlog's Gull	VU
L. crassirostris	Black-tailed Gull	
L. modestus	Gray Gull	
L. heermanni	Heerman's Gull	
L. leucophthalmus	White-eyed Gull	
L. hemprichi	Sooty Gull	
L. canus	Common Gull [Mew]	
L. audouinii	Audouin's Gull	
L. delawarensis	Ring-billed Gull	
L. californicus	California Gull	
L. marinus	Great Black-backed Gull	
L. dominicanus	Kelp Gull [Southern Black-backed]	
L. glaucescens	Glaucous-winged Gull	
L. occidentalis	Western Gull	
L. livens	Yellow-footed Gull	
L. hyperboreus	Glaucous Gull	
L. glaucoides	Iceland Gull [Including Thayer's Gull]	
L. argentatus	Herring Gull	
L. cachinnans	Yellow-legged Gull	
L. armenicus	Armenian Gull	
L. schistisagus	Slaty-backed Gull	
L. fuscus	Lesser Black-backed Gull	
L. ichthyaetus	Great Black-headed Gull	
L. brunnicephalus	Brown-headed Gull	
L. cirrocephalus	Gray-headed Gull	

L. hartlaubii	Hartlaub's Gull	
L. novaehollandiae	Silver Gull	
L. scopulinus	Red-billed Gull	
L. bulleri	Black-billed Gull	VU
L. maculipennis	Brown-hooded Gull	
L. ridibundus	Common Black-headed Gull	
L. genei	Slender-billed Gull	
L. philadelphia	Bonaparte's Gull	
L. saundersi	Saunder's Gull	VU
L. serranus	Andean Gull	
L. melanocephalus	Mediterranean Gull	
L. relictus	Relict Gull	VU
L. fuliginosus	Lava Gull	VU
L. atricilla	Laughing Gull	
L. pipixcan	Franklin's Gull	
L. minutus	Little Gull	
Creagrus furcatus	Swallow-tailed Gull	
Rissa tridactyla	Black-legged Kittiwake	
Rissa brevirostris	Red-legged Kittiwake	VU
Xema sabini	Sabine's Gull	
Pagophila eburnea	Ivory Gull	
Rhodostethia rosea	Ross's Gull	

Sub-family Sterninae: Terns

Sterna nilotica[6]	Gull-billed Tern	
S. aurantia	River Tern	
S. caspia[7]	Caspian Tern	
S. maxima[8]	Royal Tern	
S. elegans[8]	Elegant Tern	
S. bengalensis[8]	Lesser Crested Tern	
S. bergii[8]	Crested Tern [Swift]	
S. bernsteini[8,9]	Chinese Crested Tern	CR
S. sandvicensis[8]	Sandwich Tern [including Cayenne Tern]	
S. dougallii	Roseate Tern	
S. striata	White-fronted Tern	
S. sumatrana	Black-naped Tern	
S. hirundinacea	South American Tern	
S. hirundo	Common Tern	
S. paradisaea	Arctic Tern	
S. vittata	Antarctic Tern	
S. virgata	Kerguelen Tern	
S. forsteri	Forster's Tern	
S. trudeaui	Trudeau's Tern [Snowy-crowned]	
S. albifrons	Little Tern	
S. lorata	Peruvian Tern	
S. saundersi	Saunders' Tern	
S. antillarum	Least Tern	
S. superciliaris	Yellow-billed Tern	
S. nereis	Fairy Tern	
S. balaenarum	Damara Tern	
S. repressa	White-cheeked Tern	
S. acuticauda	Black-bellied Tern	
S. aleutica	Aleutian Tern	
S. lunata	Gray-backed Tern	
S. anaethetus	Bridled Tern	
S. fuscata	Sooty Tern	
Chlidonias niger	Black Tern	

C. hybridus	Whiskered Tern	
C. leucopterus	White-winged Tern	
C. albostriata[10]	Black-fronted Tern	EN
Phaetusa simplex	Large-billed Tern	
Larosterna inca	Inca Tern	
Anous stolidus	Brown Noddy	
A. minutus	Black Noddy	
A. tenuirostris	Lesser Noddy	
Gygis alba	White Tern	
G. microrhyncha	Lesser White Tern	
Procelsterna cerulea	Blue Noddy	
P. albivitta	Gray Noddy	

Family Rhynchopidae: Skimmers

Rhynchops niger	Black Skimmer
R. flavirostris	African Skimmer
R. albicollis	Indian Skimmer

Sub-order Alcae, Family Alcidae: Auks

Pinguinus impennis	Great Auk (extinct)	
Alle alle	Dovekie	
Alca torda	Razor-billed Auk	
Uria aalge	Common Murre or Guillemot	
Uria lomvia	Thick-billed Murre or Guillemot	
Cepphus grylle	Black Guillemot	
C. columba	Pigeon Guillemot	
C. carbo	Spectacled Guillemot	
Endomychura hypoleuca	Xantus' Murrelet	VU
E. craveri	Craveri's Murrelet	VU
Synthliboramphus antiquum	Ancient Murrelet	
S. wumizusume	Japanese Murrelet	VU
Ptychoramphus aleuticus	Cassin's Auklet	
Cyclorrhynchus psittacula	Parakeet Auklet	
Aethia cristatella	Crested Auklet	
A. pusilla	Least Auklet	
A. pygmaea	Whiskered Auklet	
Brachyramphus perdix	Long-billed Murrelet	
B. marmoratus	Marbled Murrelet	VU
B. brevirostris	Kittlitz's Murrelet	
Cerorhinca monocerata	Rhinoceros Auklet	
Fratercula cirrhata[11]	Tufted Puffin	
F. arctica	Atlantic Puffin [Common]	
F. corniculata	Horned Puffin	

[1] Historically placed in genus *Diomedea*.

[2] Sometimes placed in genus *Nannopterum*.

[3] Often placed in species *olivaceus*.

[4] Often placed in the genus *Leucophaeus*.

[5] Often placed in the genus *Gabianus*.

[6] Often placed in the genus *Gelochelidon*.

[7] Often placed in the genus *Hydroprogne*.

[8] Often placed in the genus *Thalasseus*.

[9] Usually listed with the species name *zimmermanni*.

[10] Often listed in the genus *Sterna*.

[11] Historically placed in Genus *Lunda*.

Note: CE = critically endangered, EN = endangered; VU = vulnerable; NT = near threatened; LP = low risk; DD = data deficient.

Appendix 2

Data on Life-History Characteristics, Breeding
Range, Size, and Survival for Seabird Species

Common Name	Scientific Name	Adult Mass (kg) Mean:range or Mean:female(male) Female(Male)	Breeding Distribution	Nest Location	Clutch Size (range)	Hatchling Type	Incubation Period (d)	Fledging Period (d)	Maximum Growth Rate (g/d)	Post-Fledging Care (d)	Age of First Breeding (yrs)	Foraging Distance	Wing Span (cm)	Maximum Age (yrs)	Annual Survival %	Index to Literature Cited
Order	**SPHENISCIFORMES**															
1 Gentoo Penguin	Pygoscelis papua	5.15 (5.58)	P,SP	O	2	SA	35–36	74–100	85	5–50	2	NS			80	43, 44, 129, 219
2 Chinstrap Penguin	Pygoscelis antarctica	3.8	P,SP	O,SI	2	SA	33–37	49–59	86	none	3–5	NS+				43, 129
3 Adelie Penguin	Pygoscelis adeliae	4.7 (5)	P	O	1.9(1–2)	SA	31–38	41–60	116	none	3–6	NS+			89	129, 152, 174, 209, 232
4 King Penguin	Aptenodytes patagonicus	14.3 (16)	P–Tm	None	1	SA	52–54	313–350	130	38	5 (4–5)	OS			82–93	11, 42, 129, 130, 152, 204, 226
5 Emperor Penguin	Aptenodytes forsteri	30 (38)	P	None	1	SA	62–67	147–190	113		5 (3–9)	OS			92–95	42, 114, 120, 129
6 Rockhopper Penguin	Eudyptes chrysocome	2.3–2.7 (2.8–3.4)	Tm,SP	O,Cr	2	SA	35–39	60–70	49	none		OS				129, 204, 232
7 Macaroni Penguin	Eudyptes chrysolophus	5 (6)	P,SP	SI,O	2	SA	35–37	62	69	none	6(5–8)	NS+,OS			77–86	43, 130, 232
8 Royal Penguin	Eudyptes schlegeli	5–6	SP	O,SI	2	SA	32–37	65			5	OS			86	129, 232
9 Fiordland Penguin	Eudyptes pachyrhynchus	3.4 (3.7)	Tm	O,Cr	2	SA	31–36	75	52	none	3–4	NS+				129, 130, 232
10 Snares Penguin	Eudyptes robustus	2.8 (3.4)	SP,Tm	O,Cl	2	SA	31–37	75			5	OS				129, 232
11 Erect-crested Penguin	Eudyptes sclateri	2.8–5.4 (3.4–6.3)	Tm,SP		2	SA	35	75				OS				129, 204
12 Yellow-eyed Penguin	Megadyptes antipodes	5.2	Tm	O	1–2	SA	39–51	106	67	none	2(2–3)	NS			87	129, 176, 232
13 Jackass Penguin	Spheniscus demersus	3.10:2.96 (3.30)	Tm	Bu	2	SA	38	70–80	35	none	3	NS			61.7	53, 130, 182, 232
14 Humboldt Penguin	Spheniscus humboldti	4.5	Tm,STr	Cr	2	SA	40					NS				130
15 Magellanic Penguin	Spheniscus magellanicus	2.7–4.1 (2.9–4.8)	Tm	Bu	2	SA	40	60–100		none	2	NS			85	129, 130, 232
16 Galapagos Penguin	Spheniscus mendiculus	2.03:1.87 (2.18)	Tr	Bu	2	SA	38	60–80	23	none	4–5	NS			85	130, 232
17 Blue Penguin	Eudyptula minor	0.79–2.1 (0.75–2.1)	Tm	Bu	1–2 D	SA	33–50	54–63		none	2–3	NS			74.5	175, 204, 232
Order	**PROCELLARIIFORMES**															
18 Wandering Albatross	Diomedea exulans	6–11	Tm, SP	O	1 B	SP	75–83	278	56	none	9 (7–11)	OS	311	35	96	14, 43, 129, 194, 213
19 Antipodean Albatross	Diomedea antipodensis		Tm	O	1	SP										101
20 Amsterdam Albatross	Diomedea amsterdamensis	4.8–8.0	Tm	O	1 B	SP	79	235					300	c40	94–95	32, 101
21 Tristan Albatross	Diomedea dabbenena		Tm	O	1	SP										
22 N. Royal Albatross	Diomedea sanfordi		TM	O	1 B	SP	79	240								101, 222
23 S. Royal Albatross	Diomedea epomophora	8–10	Tm, SP	O,SI	1 B	SP	78–80	241	83	none	7–11	OS		58+	97	129, 213
24 Waved Albatross	Phoebastria irrorata	3.5	Tr	O, SI	1 B	SP	60	167		none–11	8.3		230–240	c40	95	32, 95
25 Short-tailed Albatross	Phoebastria albatrus		Tm	O,SI	1 B?	SP	49	c180					213–229			32
26 Black-footed Albatross	Phoebastria nigripes	3.0 (3.4)	STr	O	1	SP	64–65	145.6		40	4	OS	193–227	41	95	14, 227
27 Laysan Albatross	Phoebastria immutabilis	2.85 (3.23)	STr	O	1	SP	65	165		weeks	8(8–9)	OS	195–202	c53	95	14, 67, 228
28 Black-browed Albatross	Thalassarche melanophris	3.22 (3.92)	SP	O,SI	1	SP	65–72	116–125	70		7–11	NS,OS	216–240		92	41, 43
29 Campbell Albatross	Thalassarche impavida	2.90:2.70 D61(3.10)	Tm	O,SI,Cl	1	SP	68–75	c120		none	10	OS	212–256		94.5	129, 225
30 Shy Albatross	Thalassarche cauta	4.00:3.4–3.8 (4–4.4)	Tm	O,SI,Cl	1	SP				none		NS,OS				32, 129
31 Chatham Albatross	Thalassarche eremita		Tm		1	SP										101
32 Salvin's Albatross	Thalassarche salvini		Tm		1	SP										101

No.	Common name	Scientific name														
33	Gray-head Albatross	Thalassarche chrysostoma	3-3.6 (3.4-3.7)	Tm,SP	O	1 B	SP 69-78	c141	63		7-13	OS	180-218	30+	95	41, 43, 129
34	Atlantic Yellow-nosed Alb.	Thalassarche chlororhynchos	2.5;2.4-2.9	Tm,SP	O,Cl,Sl	1	SP 71-78	c120	60		c9	OS	200-256	37+		129
35	Indian Yellow-nosed Alb.	Thalassarche carteri		Tm,SP	O,Sl	1	SP 7-78	c130			c9	OS				32
36	Buller's Albatross	Thalassarche bulleri	2.70-2.78 (3.12)	Tm,SP	O,Sl	1	SP 69-72	140-167	37	none	5+	NS	205-213	30+	91.3	32, 129, 182, 183
37	Sooty Albatross	Phoebetria fusca	2.4 (2.7)	Tm	Cl,Sl	1 B	SP 65-75	c160	45		5-15	OS	203			129, 222
38	Light-mantled Sooty Alb.	Phoebetria palpebrata	2.8-3.1	Tm-P	O,Cl,Sl	1	SP 66-69	140-157	45		7	OS	218		97.3	129, 222
39	Southern Giant Petrel	Macronectes giganteus	3.94 (5.19)	P-Tm	O	1	SP 59-66	112-123	79	none	4-10	OS	150-210		90-96	41, 111, 129, 152
40	Northern Giant Petrel	Macronectes halli	3.8 (5)	SP	O,Sl	1	SP 57-62	106-120	66		4-11	NS	199			129, 222
41	Southern Fulmar	Fulmarus glacialoides	0.7-1	P,SP	Sl,Cl	1	SP 43-50	48-56	32		9-11	OS	106		90-95	41, 129, 139, 222
42	Northern Fulmar	Fulmarus glacialis	0.58 (0.65) up to 0.8	P-Tm	O,Cl	1	SP 48.4(47-49)	53(49-58)	31	none	5(5-12)	OS	102-112	c31	94-97	14, 59, 60, 102, 103, 139, 165
43	Antarctic Petrel	Thalassoica antarctica	0.69 (0.81)	P,SP	Cr,Sl	1	SP 40-45	42-62	16	none	6	NS,OS	104		90	129, 222
44	Cape Pigeon, Cape Petrel	Daption capense	0.36-0.51 (0.38-0.55)	P,SP	Cl,O	1	SP 44-47	47-68	16	none	6	OS	87	25	94	14, 110, 129, 222
45	Snow Petrel	Pagodroma nivea	0.24-0.46	P	Cl,Cr	1	SP 41-49	41-55	17	none	3-7	NS,OS	83		93	19, 32, 41, 110, 129
46	Kerguelen Petrel	Lugensa brevirostris	0.29-0.36	Tm,SP	Bu	1	SP 46,51	59-62		none		OS	80-88			17, 32, 129, 222
47	Great-winged Petrel	Pterodroma macroptera	0.46-0.75	Tm	Bu,Cr,Un	1	SP 55-57	125		none	6-7	OS	97-102			129, 221, 222
48	White-headed Petrel	Pterodroma lessonii	0.58-0.81	Tm, SP	Bu	1	SP 60 +/-	112	13		5.5	OS	109			17, 129, 222
49	Atlantic Petrel	Pterodroma incerta		Tm	Bu	1	SP					OS?	100			32, 86, 222
50	Providence Petrel	Pterodroma solandri	0.41-0.43 (0.44-0.6)	Tm,STr	Bu	1	SP 56					OS	95-105			32, 129, 222
51	Magenta Petrel	Pterodroma magentae	0.42-0.51	Tm,STr	Bu,Un	1	SP 52	90?					102			32, 129, 222
52	Murphy's Petrel	Pterodroma ultima		STr,Tr		1	SP					OS?	97			32, 86, 222
53	Soft-plumaged Petrel	Pterodroma mollis	0.28-0.31	SP, Tm	Bu	1	SP 50	90-115		none		OS	83-95			32, 129, 222
54	Cape Verde Petrel	Pterodroma feae		STr	Bu	1	SP						83-95			32, 222
55	Madeira Petrel	Pterodroma madeira		STr		1	SP									32, 222
56	Bermuda Petrel	Pterodroma cahow		STr	Bu,Cr	1	SP 52	92				OS	89			32, 222
57	Black-capped Petrel	Pterodroma hasitata		STr		1	SP 51-54	90-100				OS	95			32, 222
58	Jamaica Petrel	Pterodroma caribbaea		Tr	Bu,Cr	1	SP						c95			32
59	Juan Fernandez Petrel	Pterodroma externa	0.41-0.54	STr,Tr	Bu	1	SP					OS	95-97			32, 129
60	White-necked Petrel	Pterodroma cervicalis	0.38-0.54	Tm	Bu,Sl	1	SP	100-115				NS	100			86, 129
61	Barau's Petrel	Pterodroma baraui	0.4	STr, Tr		1	SP					OS	96			129
62	Galapagos Petrel	Pterodroma phaeopygia	0.43	Tr	Bu,Cr	1	SP 55.3(54-58)	100-119	10	none	5(5-6)	OS	91		93	32, 200, 222, 223
63	Hawaiian Dark-rumped Petrel	Pterodroma sandwichensis	0.45 (0.33-.63)	Tr	Bu,Cr	1	SP 55.3(54-58)	100-119		none	5-6	OS			93	198, 199, 200
64	Kermadec Petrel	Pterodroma neglecta	0.37-0.59	STr,Tr	O	1	SP 50-52	110-130				NS	92			129, 222
65	Trindade Petrel	Pterodroma arminjoniana	0.29-0.46	STr,Tr	Bu,Sl	1	SP					OS	88-102			32, 129
66	Herald Petrel	Pterodroma heraldica	0.32	STr,Tr	Bu,Sl	1	SP					NS	88-102			32, 129
67	Henderson Petrel	Pterodroma atrata		STr		1	SP					OS				32
68	Phoenix Petrel	Pterodroma alba	0.27	STr,Tr	Un,Cr	1	SP					NS	83			32, 129
69	Motled Petrel	Pterodroma inexpectata	0.24-0.44	SP,Tm	Bu,Sl	1	SP 49-53	90-105				OS	78-84			32, 129, 222, 223
70	Bonin Petrel	Pterodroma hypoleuca	0.24 (0.152-0.308)	STr	Bu	1	SP 49	82(77-89)				OS	63-71	21		14, 85, 193, 222
71	Black-winged Petrel	Pterodroma nigripennis	0.17-0.2	STr,Tr	Bu	1	SP					OS	63-71			129
72	Chatham Island Petrel	Pterodroma axillaris	0.2	Tm	Sl	1	SP 10-15						63-71			129
73	Cook's Petrel	Pterodroma cookii	0.19	STr	Bu	1	SP 47-51	88				OS	65			129
74	Mas A Tierra Petrel (De Filippi's)	Pterodroma defilippiana		STr		1	SP						66?			32
75	Stejneger's Petrel	Pterodroma longirostris		SP-STr		1	SP					OS	53			32, 129

#	Common Name	Scientific Name	Adult Mass (kg) Mean:range or Mean:female(male) Female(Male)	Breeding Distribution	Nest Location	Clutch Size (range)	Hatching Type	Incubation Period (d)	Fledging Period (d)	Maximum Growth Rate (g/d)	Post-Fledging Care (d)	Age of First Breeding (yrs)	Foraging Distance	Wing Span (cm)	Maximum Age (yrs)	Annual Survival %	Index to Literature Cited
76	Pycroft's Petrel	Pterodroma pycrofti	0.13–0.2	Tm	Bu	1	SP	45	80					53–66			72	32, 129
77	Gould's Petrel	Pterodroma leucoptera	0.17–0.22	Tm–Tr	Cr	1	SP	42–49	77–84				OS	70				129
78	Collared Petrel	Pterodroma brevipes		Tr	Sl,Cr,U	1	SP											32
79	Macgillivray's Petrel (Fiji)	Pseudobulweria macgillivrayi		STr		1	SP											32, 222
80	Tahiti Petrel	Pseudobulweria rostrata		Tm,STr	Bu	1	SP							84				129, 222
81	Beck's Petrel	Pseudobulweria becki		Tr		1	SP											101, 222
82	Mascarene Petrel	Pseudobulweria aterrima		STr	Bu,Cr?	1	SP						OS?					32, 222
83	Blue Petrel	Halobaena caerulea	0.13–0.23	SP–Tm	Bu,Sl	1	SP	45–49	48–60	7.2			NS	62–71				129, 222, 223
84	Broad-billed Prion	Pachyptila vittata	0.17–0.24	Tm	Bu	1	SP	56	50–53				OS	61				129, 222, 223
85	Salvin's Prion	Pachyptila salvini	0.12–0.21	SP,Tm	Bu,Cr	1	SP	44–55	52–63	6.9			OS	59		84		41, 129, 222, 223
86	Antarctic Prion	Pachyptila desolata	0.10–0.19	P,SP	Bu,Cr	1	SP	44–46	51	6.9		5(3–6)	OS	63				41, 129
87	Narrow-billed Prion	Pachyptila belcheri	0.12–0.18	SP,Tm	Bu	1	SP	43–47	43–54, 69	6.5		6.7	NS+	56–59				17, 129
88	Fairy Prion	Pachyptila turtur	0.11–0.16 (0.12–0.18)	SP–STr	Bu,Cr	1	SP	44–54	50	6.6		4–5	NS	56				129, 222, 223
89	Fulmar Prion	Pachyptila crassirostris	0.10–0.19	Tm	Bu,Cr	1	SP						NS	60				129
90	Bulwer's Petrel	Bulweria bulwerii	0.99	STr,Tr	Cr	1	SP	44.2(38–46)	62(57–67)		none	4–6	OS	65–73		94.7		5, 129, 132, 140, 222
91	Jouanin's Petrel	Bulweria fallax		Tr		1	SP							76–83				32
92	White-chinned Petrel	Procellaria aequinoctialis	1.28 (1.39)	SP,Tm	Bu	1	SP	57–62	87–106	24		6.5	OS	134–147		79		17, 129, 222, 223
93	Spectacled Petrel	Procellaria conspicillata		Tm	Bu	1	SP						OS					32, 222, 223
94	Westland Petrel	Procellaria westlandica	1.1–1.2 (1.2–1.3)	Tm	Bu	1	SP	51–68	130		none	12	OS	135–140				32, 129, 222
95	Parkinson's Petrel	Procellaria parkinsoni	0.70:0.68 (0.72)	Tm	Bu	1	SP	56	96–122	18	none	6–8	OS	115		94		32, 113, 129, 140, 141
96	Grey Petrel	Procellaria cinerea	1.13	SP,Tm	Bu	1	SP	52–61	93, 110–120		none	7	OS	115–130				17, 129, 222, 223
97	Cory's Shearwater (Crete)	Calonectris d. diomedea	0.55:0.51(0.59)	STr	Bu,Cr	1	SP	53	92		none	7.4	OS			89		32, 142, 143, 144, 145, 222, 223
98	(Corsica)	Calonectris d. diomedea	0.66:0.60(0.71)	Tm,STr	Bu,Cr	1	SP	51	91		none	7.7				93		142, 143, 144, 145, 211, 222, 223
99		Calonectris d. borealis	0.89:0.82(0.96)	STr,Tr	Bu,Cr	1	SP	54	97	14	none	9	OS			95.6		86, 88, 129
100	Streaked Shearwater	Calonectris leucomelas	0.44–0.55	Tm,STr		1	SP	64	66–80					122				32, 129, 222
101	Wedge-tailed Shearwater	Puffinus pacificus	0.30–0.57	STr,Tr	Bu	1	SP	48–56	98–115	10	none	4	OS	97–105	26			14, 69, 129, 222, 229
102	Buller's Shearwater	Puffinus bulleri	0.41	STr	Bu	1	SP	51	100					97–99				32, 129, 222
103	Flesh-footed Shearwater	Puffinus carneipes	0.58–0.75	STr	Bu, Sl	1	SP	60	92				NS,OS	99–107				32, 129, 222
104	Pink-footed Shearwater	Puffinus creatopus		STr	Bu	1	SP							109				32, 129, 222
105	Greater Shearwater	Puffinus gravis	0.72–0.95	Tm,STr	Bu,Cr	1	SP	53–57	105					100–118				32, 129, 222
106	Sooty Shearwater	Puffinus griseus	0.63–0.95	Tm	Bu	1	SP	52.7	97	17	none	5–7	OS	103			93	32, 129, 177
107	Short-tailed Shearwater	Puffinus tenuirostris	0.47–0.64 (0.51–0.73)	STr	Bu	1	SP	52–55	94	13	none	6(5–7)	OS	93		93–94		41, 89, 129, 192
108	Christmas Shearwater	Puffinus nativitatis	0.32–0.34	STr,Tr	Cr,Un	1	SP	50–54	96					71–81				32, 129, 222

109	Manx Shearwater	Puffinus puffinus	0.35–0.58 (0.30–0.45)	SP;Tm	Bu	1	SP	51.3(47–66)	69(62–76)	12	none	6	OS	79	5	93	14, 18, 86, 93, 94, 125, 129
110	Levantine Shearwater	Puffinus yelkouan	0.34–0.42	Tm	Cr,Un	1	SP	52	72				NS	76–93			32, 222
111	Balearic Shearwater	Puffinus mauretanicus		Tm	Cr,Un	1	SP	c50	c72								32
112	Townsend's Shearwater	Puffinus auricularis	0.34 (0.32–0.36)	Tr	Bu,Cr	1	SP	62(47–66)	92.4(88–100)		none	4–7	OS				4, 115
113	Newell's Shearwater	Puffinus newelli	0.39 (0.34–0.43)	Tr	Bu,Cr	1	SP	47–66			none	4–7	OS			90	4, 122
114	Black-vented Shearwater	Puffinus opisthomelas	0.41	STr	Bu	1	SP	50	69				NS	76–89			32, 222
115	Fluttering Shearwater	Puffinus gavia	0.23–0.42	Tm,STr	Bu	1	SP						NS,OS	76			32, 129
116	Hutton's Shearwater	Puffinus huttoni	0.37	STr	Sl	1	SP	50–60	80			4–6	NS,OS	72–78			129
117	Audobon's Shearwater	Puffinus lherminieri	0.15–0.23	STr,Tr		1	SP	49–51	62–75			8	NS+	64–74			32, 129
118	Heinroth's Shearwater	Puffinus heinrothi		Tr		1	SP										32, 222
119	Little Shearwater	Puffinus assimilis	0.18–0.26	Tm,STr	Bu,Sl	1	SP	52–58	70–75				OS	55–67			32, 129
120	Peruvian Diving-petrel	Pelecanoides garnotii		Tm,STr	Bu	1	SP						NS				32, 222
121	Magellanic Diving-petrel	Pelecanoides magellani		Tm		1	SP						NS				32
122	Georgian Diving-petrel	Pelecanoides georgicus	0.09–0.15	Tm	Bu	1	SP	44–52	42–60	4.8		2	NS	30–38		75–87	32, 17, 129, 222, 223
123	Common Diving-petrel	Pelecanoides urinatrix	0.11–0.15	Tm	Bu,Sl	1	SP	53.5	49–55	4.3	none	2–3	NS	43			41, 129, 222, 223
124	Wilson's Storm-petrel	Oceanites oceanicus	0.03–0.04	P,SP	Bu,Cr	1	SP	33–53	51–97	2.8	none	4(4–5)	OS	39		91	41, 129, 222, 223
125	Elliot's (White-vented) S-p.	Oceanites gracilis		STr,Tr		1	SP										32
126	Grey-backed Storm-petrel	Garrodia nereis	0.03–0.04	Tm	Un	1	SP	40–45	75				OS	39			129, 222, 223
127	White-faced Storm-petrel	Pelagodroma marina	0.04–0.07	STr	Bu	1	SP	45–59	52–67, 80	2.2	none		OS	43			129, 222, 223
128	Black-bellied Storm-petrel	Fregetta tropica	0.05–0.06	SP,Tm	Cr	1	SP	35–44	55, 65–71		none		OS	46			129, 222, 223
129	White-bellied Storm-petrel	Fregetta grallaria	0.04–0.06	STr	Cr,Bu	1	SP	40	68				OS	45			129, 222, 223
130	White-throated Storm-petrel	Nesofregatta fuliginosa	0.055–0.115	STr,Tr	Un,Cr	1	SP						OS				32, 185
131	European (British) Storm-petrel	Hydrobates pelagicus	0.03	Tm, STr	Cr	1	SP	41	66	0.8	none		OS				222, 223
132	Least Storm-petrel	Halocyptena microsoma	0.02	STr	Cr	1	SP							32			32, 88
133	Wedge-rumped Storm-petrel	Oceanodroma tethys		Tr	Cr,Un	1	SP		76				OS				32
134	Madeiran Storm-petrel	Oceanodroma castro	0.038–0.040	Tm–Tr	Cr,Bu	1	SP	42	64–78		none			44–46			32, 222
135	Swinhoe's Storm-petrel	Oceanodroma monorhis		Tm–Tr	Bu	1	SP				none			44–46			32
136	Leach's Storm-petrel	Oceanodroma leucorhoa	0.04–0.05	Tm,STr	Bu,Cr	1	SP	43.3(37–50)	56–79	1.6	none	3–6	OS	45–48	36	79–93	3, 14, 39, 112, 129, 178, 222, 223
137	Markham's Storm-petrel	Oceanodroma markhami		Tr	Un,Cr	1	SP						OS				32, 222
138	Tristram's Storm-petrel	Oceanodroma tristrami		STr	Bu,Cr	1	SP						OS	56			32, 222
139	Black Storm-petrel	Oceanodroma melania		STr	Bu, Cr	1	SP							46–51			32, 222
140	Matsudaira's Storm-petrel	Oceanodroma matsudairae		STr	Bu	1	SP							56			32, 129
141	Ashy Storm-petrel	Oceanodroma homochroa	0.039	STr	Cr	1	SP	44.8(42–59)	84.4(72–119)		none		NS,OS		8		1, 2, 14, 222, 223
142	Hornby's Storm-petrel	Oceanodroma hornbyi		STr		1	SP						OS				32
143	Forked-tailed Storm-petrel	Oceanodroma furcata	0.06	Tm	Bu,Cr	1	SP	40	58	1.5	15–20	4–6	OS		17		14, 222, 223
	Order PELECANIFORMES																
144	White-tailed Tropicbird	Phaethon lepturus	0.32 (0.22–0.36)	STr,Tr	Un,Cr	1	SA	42(40–43)	70–85		none	4(est.)	OS	90–95			126, 129, 203
145	Red-billed Tropicbird	Phaethon aethereus	0.70	STr,Tr	Un,Cr	1	SA	42–44	80–90		none		OS	99–106			159, 203
146	Red-tailed Tropicbird	Phaethon rubricauda	0.50–0.85	STr,Tr	Un,Cr	1	SA	45(39–51)	85(72–91)	15	none	2–5	OS	113	29	90	14, 68, 129, 187
147	White Pelican	Pelecanus onocrotalus	6–9 (9–15)	Tm–Tr	O	2–4	A	29–31	65–70				NS,NS+				32, 106, 116, 172
148	Dalmatian Pelican	Pelecanus crispus	7.25–10 (9.5–12)	Tm,STr	O	1–3	A	31–34	85		15–20		NS				40, 116
149	Australian Pelican	Pelecanus conspicillatus	4–6.8	Tm,STr	O	2(1–4)	A	32–35	90			4–6	NS,OS	230–260			116, 129

#	Common Name	Scientific Name	Adult Mass (kg) Mean:range or Mean:female(male) Female(Male)	Breeding Distribution	Nest Location	Clutch Size (range)	Hatchling Type	Incubation Period (d)	Fledging Period (d)	Maximum Growth Rate (g/d)	Post-Fledging Care (d)	Age of First Breeding (yrs)	Foraging Distance	Wing Span (cm)	Maximum Age (yrs)	Annual Survival %	Index to Literature Cited
150	American White Pelican	Pelecanus erythrorhyncous	4.0–5.9 (5.0–7.1)	Tm	O	2(1–4)		30?	60		20+	3			26		64, 116
151	Pink-backed Pelican	Pelecanus rufescens	3.9–6.2 (4.5–7)	STr,Tr	O,T	1–3		30–35	70–75		10–21					c80	20, 50, 116
152	Spot-billed Pelican	Pelecanus philippensis	4.65 (5.1–5.7)	STr,Tr	T	3		29–31	120?								116, 146
153	Brown Pelican	Pelecanus occidentalis	3.2 (3.7)	Tm,STr	O,T	2.6(2–3)	A	30	79.5(71–88)	115	30–84	3(2–4)	NS,OS	210	28		116, 163, 190
154	Ascension Frigatebird	Fregata aquila	1.25	STr	O	1 B	A	43–51	180?		90–120		OS	196–201			159, 205
155	Lesser Frigatebird	Fregata ariel	0.78–1.11	Tr	T	1 B	A	50 +/–	145–179		140–180		OS	175–195	28		14, 48, 129, 185
156	Great Frigatebird	Fregata minor	1.10–1.89 (0.90–1.35)	Tr	T	1 B	A	56(51–60)	160–169		158–428	3–7	OS	198	35		14, 48, 129
157	Christmas Island Frigatebird	Fregata andrewsi	1.55 (1.4)	Tr	T	1 B	A	46–54	177		120–200		NS,OS	205–230			129, 159
158	Magnificent Frigatebird	Fregata magnificens	1.67 (1.28)	Tr	T	1 B	A	59(58–60)	165	15	150	5–7	OS	229			34, 49, 163
159	Australian Gannet	Morus serrator	2.3	Tm	O,Cl	1	A	43.6(37–50)	102(95–109)		none	3–7	OS	171–185	30+		88, 148
160	Cape Gannet	Morus capensis	2.7	Tm,STr	O	1	A	44	97.2		none	3–4	OS	185	30+		129, 148
161	Northern Gannet	Morus bassanus	3.07 (2.93)	Tm	Cl,O	1	A	43.6	91	80	180–230	4(3–6)	OS		30	94	97, 137, 148
162	Abbott's Booby	Sula abbotti	1.46	Tr	T	1 B	A	56–57	151–168		180–230		OS		30+		129, 148
163	Peruvian Booby	Sula variegata	1.3	Tm,STr	Cl,O	1.5(1–2)	A	42	78		62	2–3	NS,OS	156	20+		148
164	Blue-footed Booby	Sula nebouxii	1.80 (1.28)	Tr	Cl,O	2(1–3)	A	41	102		56	!2–6	NS,OS	156	20+		148
165	Masked Booby	Sula dactylatra	2.10 (1.88)	STr,Tr	Cl,O	2(1–2)	A	43(38–49)	130(109–151)	35	50–150	2–4	OS	156	30+	92.5	6, 24, 51, 129, 148, 235
166	Nazca Booby	Sula granti	1.75:1.88 (1.63)	Tr	O	1.6	A	43	120		30–62		NS,OS			83.2	97, 148
167	Brown Booby	Sula leucogaster	0.9–1.8	Tr	O,Cl	2(1–3)	A	42.8	95(85–105)	27	118–259	2–4	NS,OS	148	27	92–96	14, 51, 129, 148, 185
168	Red-footed Booby	Sula sula	0.8–1.5	Tr	T	1	A	44.5(43–49)	101.5(91–139)	20	190	2–4	OS	151	30+	90	129, 148, 189
169	Long-tailed cormorant	Microcarbo africanus	0.43–0.60 (0.50–0.88)	STr,Tr	O,T	2, 3, 8	A	23–25	35				NS+	80–90			116, 159
170	Crowned Cormorant	Microcarbo coronatus	0.67–0.78 (0.48–0.88)	Tm	O,T	1–5	A	23	30–35				NS				116, 159
171	Pygmy Cormorant	Microcarbo pygmaeus	0.56–0.64 (0.65–0.87)	Tm,STr	T	4–6	A	27–30	70				NS	80–90			116, 159
172	Little Cormorant	Microcarbo niger	0.36–0.53	STr,Tr	T	3–5	A						NS				116, 159
173	Little Pied Cormorant	Microcarbo melanoleucos	0.4–0.9 (0.7–0.9)	Tm,STr	T,(O)	3–5	A						NS				129, 159
174	Brandt's Cormorant	Compsohalieus penicillatus	1.4–2.3 (2.4–2.7)	Tm,STr	O	2.6(2–5)	A	29.2(28–32)			yes	3(2–9)	NS+	108	18	74	14, 15, 16, 62, 116, 220
175	Flightless Cormorant	Compsohalieus harrisi	2.5–2.9 (3.8–4.1)	Tr	O	2–4	A	35	60		30–90		NS			87.6	96, 116, 159
176	Bank Cormorant	Compsohalieus neglectus	1.80	Tm,STr	O,Cl	2(1–3)	A	29–30					NS	132			159
177	Black-faced Cormorant	Compsohalieus fuscescens		Tm	O,Cl	2–3	A						NS				129, 159
178	Neotropic Cormorant	Hypoleucos brasiliensis	1.07 (1.26)	STr,Tr	T,(O)	3–4(2–6)	A	24.6(23–26)	63		yes	1(1–3)	NS	93–102			129, 159
179	Double-crested Cormorant	Hypoleucos auritus	1.8–2.6 (2.0–3.0)	Tm,STr	O,Cl,T	4(1–7)	A	25–28	42–56		28	2.7(1–4)	NS		13		160, 210
180	Indian Cormorant	Hypoleucos fuscicollis	0.60–0.79	STr,Tr	T	3–6	A								18		35, 56, 62,104, 197
181	Pied Cormorant	Hypoleucos varius	1.1–1.9 (1.5–2.9)	Tm	O	2–4	A	25–33	47–60					110–130			159
182	Little Black Cormorant	Hypoleucos sulcirostris	0.5–1.2	Tm,STr	T	4–6	A						OS	95–105			129, 159
183	Great Cormorant	Phalacrocorax carbo	2–2.5	Tm,STr	O,T,Cl	3–6	A	30 (28–31)	53(49–56)		50	3(2–3)	NS,OS	130–150		88	35, 70, 116, 129

No.	Common Name	Scientific Name															
184	Japanese Cormorant	Phalacrocorax capillatus	3.1	STr	Cl	3	A	34	40				NS	102–110			116
185	Socotra Cormorant	Leucocarbo nigrogularis		STr		2–4	A	22–28	49				NS	109			116, 159
186	Cape Cormorant	Leucocarbo capensis	1.16–1.31	STr	Cl,O,T	2.4(2–3)	A					3					116, 159
187	Guanay Cormorant	Leucocarbo bougainvilli		Tm,Str	Sl,O	3(2–3)	A						NS,OS	110			159
188	Kerguelen Shag	Notocarbo verrucosus	1.7–2.0	P,SP	Cl	2–4	A										129
189	Imperial Shag	Notocarbo atriceps	1.5–2.4 (1.7–3.5)	SP	O,Cl	2.7(1–3)	A	28.8	77.5(75–80)		yes	2–4	NS	110–125			116, 129
190	Antarctic Shag	Notocarbo bransfieldensis	2.5–3.0	P,SP	O,Sl,Cl	2.5(2–3)	A		40–45			4+	NS			85–91	129, 159
191	South Georgia Shag	Notocarbo georgianus	2.47 (2.88)	SP	O,Sl,Cl	2.3, 2.8	A	28–31	65		yes	3+	NS,OS	105			129, 159
192	Campbell Island King Shag	Nesocarbo campbelli	2.0 (1.6–1.9)	Tm	O,Cl,Cr		A										129, 159
193	New Zealand King Shag	Euleucocarbo carunculatus	2.5	Tm	O,Sl	1–3	A						NS				129, 159
194	Stewart Island Shag	Euleucocarbo chalconotus	1.79–3.88	Tm	O,Sl	2–3	A						NS				159
195	Chatham Island Shag	Euleucocarbo onslowi	1.79 (2.4)	Tm	O,Sl,Cl	3	A						NS				129, 159
196	Auckland Island Shag	Euleucocarbo colensoi		Tm	O,Cl	3	A	26–32					NS,OS	105			129, 159
197	Bounty Island Shag	Euleucocarbo ranfurlyi	2.3–2.9	Tm	Cl	2–3	A						NS,OS				129, 159
198	Rock Cormorant	Strictocarbo magellanicus		Tm	Cl	3(2–4)	A						NS	92			159
199	Pelagic Cormorant	Strictocarbo pelagicus	1.53–1.70 (1.75–2.03)	Tm	O,Cr	3(1–6)	A	30 (29–31)	45(40–50)		weeks	2	NS,OS	95	18		14, 35, 109
200	Red-faced Cormorant	Strictocarbo urile	1.55–2.05 (1.9–2.27)	Tm	Cl,Sl	2–4	A	33(32–34)	59(54–64)		62	3	NS,OS	122		80–88	35, 116, 159
201	European Shag	Strictocarbo aristotelis	1.60 (1.94)	Tm	Cl,O	3(1–6)	A	34(33–35)	53(48–58)	52		2–3	NS,OS				35, 116, 171
202	Red-legged Cormorant	Strictocarbo gaimardi	1.3	STr,Tr	Cl	3(3–4)	A							91			116, 159
203	Spotted Shag	Strictocarbo punctatus	0.7–1.6 (.9–1.7)	Tm	Cl,Sl	2.7(2–4)	A	32(28–35)	62(57–71)			2	NS,OS	91–99			116, 129, 159
204	Pitt Island Shag	Strictocarbo featherstoni	0.645–1.33	Tm	Cl	3?	A						NS,OS				159
	Order	**CHARADRIIFORMES**															
205	Parasitic Jaeger	Stercorarius parasiticus	0.4,0.3–0.6(0.5: 0.3–0.6)	P–Tm	O	2 (1–2)	SP	26(24–29)	31(26–38)		14–28	4(3–7)	NS	104	18	89	71, 72, 157, 158, 230
206	Long-tailed Jaeger	Stercorarius longicaudus	0.28–0.44 (0.21–0.35)	PSP	O	2	SP	24.2	25(24–28)		24	3	NS	96	8		14, 158
207	Pomarine Jaeger	Catharacta pomarinus	0.68–1.0 (0.52–0.85)	PSP	O	2(1–3)	SP	25(23–27)			14+	2–3	NS+	150			71, 158, 231
208	Great Skua	Catharacta skua	1.418:1.48 (1.356)	Tm	O	2	SP	29 (26–32)	46 (40–51)	40		6.8	NS	130–138		93	71, 87, 158
209	Chilean Skua	Catharacta chilensis	1.1–1.7	Tm	O	2	SP	28–32	45–50				NS				27, 158
210	Brown Skua	Catharacta antarctica	1.2–2.1	P–Tm	O	2	SP	28–31	36–53			6+	NS+	126–160		90–96	27, 158
211	South Polar Skua	Catharacta maccormicki	0.9–1.6	P	O	2	SP	24–27			6+	6+	NS+	130–140		90–95	27, 158
212	Dolphin Gull	Larus scoresbii	0.52	Tm	O	2(2–3)	SP	24–27					NS	c89			236
213	Pacific Gull	Larus pacificus	0.9–1.18	Tm,STr	O	3	SP	23–26						137–157			27
214	Band-tailed Gull	Larus belcheri		STr,Tr	O	3	SP							120			27
215	Ortog's Gull	Larus atlanticus	0.9–0.96	STr,Tr	O	2–3	SP				yes			130–140			27
216	Black-tailed Gull	Larus crassirostris	0.43–0.64	Tm,STr	O	2–3	SP	24–25	35–40					126–128			27
217	Gray Gull	Larus modestus	0.36–0.40	STr	O	1–3	SP	29–31	40								27
218	Heermann's Gull	Laus heermanni	0.37–0.64	STr	O	1–2	SP		45			4		117–124			27
219	White-eyed Gull	Larus leucophthalmus	0.27–0.41	Tr	O	2–3	SP							110–115			27
220	Sooty Gull	Larus hemprichi	0.40–0.51	Tr	O	1–3	SP	25									27
221	Common (Mew) Gull	Larus canus	0.29–0.55	SP,Tm	O	3(1–3)	SP	24–26	30–35		Brief	3(2)		394–586	24		27
222	Audouin's Gull	Larus audouini	0.58–0.77	STr,Tr	O	3(1–4)	SP	26–33	35–40		75				32		27
223	Ring-billed Gull	Larus delawarensis	0.471 (0.566)	Tm	O	3(2–4)	SP	25.5(23–29)	35–40	24	no?	3(3–4)	NS	121–127			27, 63, 181, 217
224	California Gull	Larus californicus	0.556 (0.657)	Tm	O	2(2–4)	SP	25(23–27)	48(42–60)		yes	3–4	NS	121–140	25	79–92	14, 27, 233, 234
225	Great Black-backed Gull	Larus marinus	1.488 (1.829)	SP,Tm	O	2.8(2–3)	SP	26–32	45–55		20–40	4–5	NS,OS	169	27		14, 27, 39, 84

#	Common Name	Scientific Name	Adult Mass (kg) Mean:range or Mean:female(male) Female(Male)	Breeding Distribution	Nest Location	Clutch Size (range)	Hatchling Type	Incubation Period (d)	Fledging Period (d)	Maximum Growth Rate (g/d)	Post-Fledging Care (d)	Age of First Breeding (yrs)	Foraging Distance	Wing Span (cm)	Maximum Age (yrs)	Annual Survival %	Index to Literature Cited
226	Kelp Gull	Larus dominicanus	1.0:0.83 (1.05)	P-Tm	O	2.4	SP	28	61		41-54	4	NS	152-169		81	27, 108, 133, 134, 135
227	Glaucous-winged Gull	Larus glaucescens	0.946 (1.180)	SP;Tm	O	2(1-4)	SP	25-28	40-53		21	5.4(4-7)	NS	132-146	24	83-87	14, 27, 215, 216
228	Western Gull	Larus occidentalis	1.01:0.8-1.2	Tm,STr	O,SI	3(1-3)	SP	30-32	47.5(45-50)		30-90	4-6	NS	132-144	28	90	12, 14, 27, 33, 170
229	Yellow-footed Gull	Larus livens	1.32:1.1-1.4	STr	O	2.7(1-4)	SP	28(24-30)	37(30-45)		30-90	4-5	NS	142-155			27, 33, 162
230	Glaucous Gull	Larus hyperboreus	1.07-1.82		O	2-3	SP	26-30	49		26+			142-152			27
231	Iceland Gull	Larus glaucoides	0.56-0.86	SP;Tm	Cl,O	2-3	SP							125-130			27
232	Thayer's gull	Larus glaucoides thayeri	0.85-1.15		Cl	2-3?								130-140			27
233	Herring Gull	Larus argentatus	1.044 (1.226)	SP;Tm	Cl,O	3(1-3)	SP	30-32	47.5(45-50)	33	45+	4-5	NS	135-149	32	91	14, 27, 57, 121, 167, 168, 169
234	Yellow-legged Gull	Larus cachinnans	0.8-1.5			2.7(1-3)	SP	26-29	42-49					140-155			27
235	Armenian Gull	Larus armenicus	0.95-1.05			3	SP	28-30	40-45					140			27
236	Slaty-backed Gull	Larus schistisagus	1.05-1.69	Tm		3(1-4)	SP	28-30	40-45					132-148			27
237	Lesser Black-backed Gull	Larus fuscus	0.83:0.55-1.2	SP;Tm	O,Cl	2.7(1-3)	SP	24-28	30-40		yes			124-127	26		27, 92
238	Great Black-headed Gull	Larus icthyaetus	0.9-2.0	Tm	O	2.9(1-3)	SP	25			40	4-5		155-170			27, 92
239	Brown-headed Gull	Larus brunnicephalus	0.45-0.71	Tm		3(1-4)	SP	c24									27
240	Gray-headed Gull	Larus cirrocephalus	0.25-0.33	STr;Tr	O,F	2-3	SP	c25	c40					100-105			27
241	Hartlaub's Gull	Larus hartlaubii	0.23-0.34	STr		1-3	SP	c25		8		3		89-92			27
242	Silver Gull	Larus novaehollandiae	0.26-0.35	Tm,STr		D	SP	24	35-42				NS	91-96			27, 101, 153
243	Red-billed Gull	Larus scopulinus	0.28:0.259 (0.299)	Tm	O	1.87	SP	24				3.7				85.6	108, 134, 135
244	Black-billed Gull	Larus bulleri	0.19-0.27	Tm			SP										27
245	Brown-hooded Gull	Larus maculipennis	0.29-0.36	SP;Tm			SP										27
246	Common Black-headed Gull	Larus ridibundus	0.19-0.32	Tm-Tr	O,F	2.6(1-3)	SP	22-26	35		14-30	2-3	NS	94-110	32		27
247	Slender-billed Gull	Larus genei	0.22-0.35	Tm-Tr	O	2-3	SP	22	30-37		7+	2-3		102-110			27
248	Bonaparte's Gull	Larus philadelphia	0.17-0.23	SP;Tm	T	2	SP	22-25			14+	2-3	NS	78-84			27
249	Saunder's Gull	Larus saundersi		Tr	O	3	SP						OS				27
250	Andean Gull	Larus serranus	0.48	STr,Tr	O	2-3	SP										27
251	Mediterranean Gull	Larus melanocephalus	0.21-0.39	Tm,STr	O	2-3	SP	23-26	35-40		30+		NS	98-105	15+		27
252	Relict Gull	Larus relictus	0.42-0.66	Tm	O	2.6(2-3)	SP	24-26				3					27
253	Lava Gull	Larus fuliginosus		Tr	O	2.8(2-3)	SP	33	60		28						27
254	Laughing Gull	Larus atricilla	0.289 (0.327)	Tm-Tr	O,F	2.8(2-3)	SP	24.5(22-25)	42.5(35-50)		14-30	3-4	NS	102-107	19		25, 186, 188
255	Franklin's Gull	Larus pipixcan	0.26-0.28	Tm	F	3(2-4)	SP	24.5(23-26)	35-40		7+	2-3	NS	85-106	10		14, 22, 23, 26
256	Little Gull	Larus minutus	0.10-0.156	SP;Tm	O,F	2.7(1-3)	SP	21-22	28?		14+	2-3	NS				66, 214
257	Swallow-tailed Gull	Creagrus furcatus	0.55-0.74 (0.63-0.78)	Tr	O,Cl	1	SP	32(29-38)	60-70	46	30+	5(2-6)	OS			97	27
258	Black-legged Kittiwake	Rissa tridactyla	0.4:0.3-0.5(0.4:0.3-0.5)	SP;Tm	Cl	2(1-3)	SP	26.2(25-28)	41.5(34-58)	20	Yes	3-5	NS	104	18	88-93	10, 14, 37, 38, 105
259	Red-legged Kittiwake	Rissa brevirostris	0.38 (0.40)	SP;Tm	Cl	1(1-2)	SP	25-32 est	37-45 est		Yes		OS	90	19		14, 27, 31, 105

No.	Common name	Scientific name																
260	Sabine's Gull	Xema sabini	0.13–0.22		O,Cl	2(1–3)	SP	23–25	58–65		Yes		NS	81–87	19			27
261	Ivory Gull	Pagophila eburnea	0.448–0.687	P,SP	O	1(1–2)	SP	24–26	30–35		Yes	2–3	NS	108–120	19			39, 90
262	Ross's Gull	Rhodostethia rosea	0.12–0.25		O	3(1–5)	SP	19–21	21					82–92				27
263	Gull-billed Tern	Sterna nilotica	0.15–0.29	Tm–Tr	O	3(3–4)	P	22.5(22–23)	31.5(28–35)		60	5	NS	85–103	15.8			13, 14, 161, 191
264	River Tern	Sterna aurantia				3(3–4)	SP				21	2		80–85				27
265	Caspian Tern	Sterna caspia	0.662–(0.649)	Tm,STr	O	2(1–3)	SP	27(25–28)	40(35–45)		90+	3(2–4)	NS	127–146	30	87–91		14, 46, 81, 173
266	Royal Tern	Sterna maxima	0.32–0.50	Tm,STr		1	SP	25–31	30		200+	3–4	NS	100–135	24			14, 81
267	Elegant Tern	Sterna elegans	0.26(0.20–0.33)	STr	O	1	SP	26(23–33)	35		120	3	NS,OS	76–81				29, 184
268	Lesser Crested Tern	Sterna bengalensis	0.18–0.24	STr,Tr		1	SP	21–26	30–35		90+			88–105				81
269	Crested Tern	Sterna bergii	0.32–0.40	ST,Tr		1–2	SP	25–30	35–41		80+		NS	100–130				81, 124
270	Chinese Crested Tern	Sterna bernsteini					SP											81
271	Sandwich Tern	S. sandvicensis acuflavida	0.14–0.300	STr,Tm	O	1(1–2)	SP	24(23–29)	28(27–29)	9.8	30–120	4(3–4)	NS	86–105	23.5	75		14, 123, 196
272	Cayenne Tern	S. s. eurygnatha	0.176–0.30	STr	O	1(1–2)	SP	29	28		7+		NS		24			196
273	Roseate Tern	Sterna dougallii	0.11(0.088–0.139)	Tm–Tr	O	2(1–4)	SP	23.3(21–24)	26(22–30)	7	7–14	3(2–4)	NS	72–80	25	74–91		14, 82, 155, 202
274	White-fronted Tern	Sterna striata		STr	O	1–2	SP	29–35	29–35		90–180	2	NS		20			81
275	Black-naped Tern	Sterna sumatrana	0.098–0.11	Tr	O	1.8(1–3)	SP	21–23	24					84–86				81, 124
276	South American Tern	Sterna hirundinacea	0.17–0.19	Tm–Tr	O	2–3	SP	21–23	27									81
277	Common Tern	Sterna hirundo	0.13:0.097–0.146	SP–Tm	O	2.4(1–4)	SP	24(22–28)	26	8	21+	3.2(2–4)	NS	72–83	25	81–89		8, 14, 83, 154
278	Arctic Tern	Sterna paradisaea	0.11(0.08–0.127)	P,SP	O	1.8(1–2)	SP	23(22–27)	20–24	7	7+	3+	NS	76–85	34	90		14, 45, 124, 179
279	Antarctic Tern	Sterna vittata	0.15	STr,P	O	1(1–2)	SP	23–25	27–32					74–79				81, 108
280	Kerguelen Tern	Sterna virgata	0.13	Tm	O	1(1–2)	SP	24	35(31–39)		up to 20			c71				81, 108
281	Forster's Tern	Sterna forsteri	0.13–0.19	Tm,STr	O,F	2.8(2–3	SP	23–26						73–82				81
282	Trudeau's Tern	Sterna trudeaui	0.15–0.16	Tm,STr	F	3(2–4)	SP							76–70				81
283	Little Tern	Sterna albifrons	0.05–0.06	Tm–Tr	O	2–3	SP	21–24	20–24			3(2)		47–55	21			81
284	Peruvian Tern	Sterna lorata	0.04–0.05	Tr	O	1–2	SP	22–23										81
285	Saunder's Tern	Sterna saundersi	0.04–0.05			2	SP				26–56	3(2)	NS	50–55	24			81
286	Least Tern	Sterna antillarum	0.036–0.057	Tm–Tr	O	2(2–3)	SP	22(19–25)	20(17–21)		26–56	3(2)	NS	51	24	80–93		14, 36, 107, 131, 212
287	Yellow-billed Tern	Sterna superciliarus	0.04–0.057	STr,Tr	O	1.9(1–3)	SP	20–25						45–51				81
288	Fairy Tern	Sterna nereis	0.057	STr,Tr	F	1.7(1–2)	SP	18–22	20		75			51				81
289	Damara Tern	Sterna balaenarum	0.046	STr,Tr	F	1	SP	18–22	20		75			51				81
290	White-cheeked Tern	Sterna repressa	0.113–0.142	Tr	O	2–3	SP	21–23			Yes			73–83				81
291	Black-bellied Tern	Sterna acuticauda		STr,Tr	O	3(2–3)	SP											81
292	Aleutian Tern	Sterna aleutica	0.12:0.08–0.14	SP	O	2(1–3)	SP	22.5(20–27)	28(25–31)		14+	3+	NS+	75–80	26			9, 156
293	Gray-backed Tern	Sterna lunata	0.09–0.14 (0.099–0.165)	STr,Tr	O	1	SP	30(29–32)	38–49		Yes	4–5	NS+	73–76				5, 138
294	Bridled Tern	Sterna anaethetus	0.131:0.093–0.167	STr,Tr	Cr,Un,(O)	1	SP	29(26–33)	60(55–65)		20–40	4(2–5)	NS,OS	76–81	18	78–83		54, 55, 91, 108
295	Sooty Tern	Sterna fuscata	0.18:0.14–0.24	Tr	O	1	SP	29(26–33)	c60		17+	3(3–7)	OS	84(82–94)	36			14, 81
296	Black Tern	Chlidonias niger	0.065:0.063(0.064)	Tm	F	2.6(1–4)	SP	20.5(19–23)	22(20–25)		40	2(2–3)	NS	57–65	17			14, 58, 81, 195
297	Whiskered Tern	Chlidonias hybridus	0.06–0.101	Tm–Tr	F	2–3	SP	18–20	23		Yes			64–70				81
298	White-winged Tern	Chlidonias leucopterus	0.066:0.042–0.079	Tm	O	2.8(2–3)	SP	18–22	20–25			2		58–67				81
299	Black-fronted Tern	Chlidonias albostriata		Tm	O	.088–.096	SP	21–23										81
300	Large-billed Tern	Phaetusa simplex	0.208–0.247	STr,Tr		2.3(2–3)	SP				Yes							81
301	Inca Tern	Larosterna inca	0.189:0.18–0.21	Tr	Cl,Cr	2	SP		c28		28+		NS+					81
302	Brown Noddy	Anous stolidus	0.167 (0.178)	Tr	T,O	1 D	SP	34.5(33–36)	45(40–56)	6	30–100	3(3–7)	NS+	75–86	27			14, 21

	Common Name	Scientific Name	Adult Mass (kg) Mean:range or Mean:female(male) Female(Male)	Breeding Distribution	Nest Location	Clutch Size (range)	Hatchling Type	Incubation Period (d)	Fledging Period (d)	Maximum Growth Rate (g/d)	Post-Fledging Care (d)	Age of First Breeding (yrs)	Foraging Distance	Wing Span (cm)	Maximum Age (yrs)	Annual Survival %	Index to Literature Cited
303	Black Noddy	Anous minutus	0.098–0.144	STr,Tr	O,Cr,T	1 D	SP	34(31–39)	48–60		30–60 est	3(2–4)	NS+	66–72	25	75	80, 208
304	Lesser Noddy	A. m. tenuirostris	0.097–0.12	STr,Tr	T	1	SP	c35	c55					58–63			81
306	White Tern	Gygis alba	0.08–0.16	STr,Tr	T	1 D	SP	35.6(33–41)	48.6(31–67)		35–60	5, 3+	NS,OS	73	37		14, 147, 185
305	Lesser White Tern	Gygis microrhyncha	0.045:0.041–0.069		Cl,S,Cr	1	SP	c35				1?					81, 108
307	Blue Noddy	Procelsterna cerulea				1	SP		32–45					46–60			81
308	Gray Noddy	Procelsterna albivitta		STr,Tr		1	SP							46–60			81
309	Black Skimmer	Rynchops niger	0.25:0.23–0.29 (0.35:0.3–0.37)	Tr	O	4(1–6)	SP	22.9(21–26)	29(28–30)		14+	3	NS	107–127	20		14, 81, 173
310	African Skimmer	Rynchops flavirostris	0.111–0.204	STr,Tr		2.6(2–3)	SP	c21	c28					c106			81
311	Indian Skimmer	Rynchops albicollis		Tr		3(2–4)		c21						102–114			81
312	Great Auk	Pinguinus impennis		Tm	O	1	SP					7					136
313	Dovekie	Alle alle	0.16	P	Cl,Cr	1	SP	29	27.5(26–29)	6.5	none	5	NS,OS				76
314	Razor-billed Auk	Alca torda	0.72	Tm	Cr,Cl	1	SP	35	22.5(15–30)	12.3		4(4–5)	NS	66		88–90	76, 98
315	Common Murre	Uria aalge	1.1:0.87–1.3 (1.0:0.83–1.3)	Tm	Cl	1	SP	33	21.5(18–25)	11.4	yes	4–5	NS,OS	73	26	87–95	14, 99, 100, 201
316	Thick-billed Murre	Uria lomvia	1.0:0.76–1.2 (0.99:0.84–1.2)	P–Tm	Cl,Cr	1	SP	33(31–36)	22.5(15–30)	11.5	30–60	4(4–5)	NS,OS	45	29	89–90	14, 74, 77, 79
317	Black Guillemot	Cepphus grylle	0.43:0.30–0.53 (0.43:0.37–0.50)	P–Tm	Bu,O	2	SP	29	36.5(34–39)	13.8	none	3(3–4)	NS	58	13	87	14, 164
318	Pigeon Guillemot	Cepphus columba	0.45–0.55	SP,Tm	Cr,Bu	2(1–2)	SP	29(26–32)	40(35–54)	10–20	none	3	NS		14	80	14, 65, 218
319	Spectacled Guillemot	Cepphus carbo		Tm		1–2	SP	30			none						101, 151
320	Xantus' Murrelet	Endomychura hypoleuca	0.17	STr	Cr	2(1–2)	P	34(27–44)	2(1–4)		yes	3(2–3)	OS	42	15		14, 52, 101
321	Craveri's Murrelet	Endomychura craveri	0.15	STr	Cr	2	P		2		yes			41			151
322	Ancient Murrelet	Synthliboramphus antiquum	0.206 (0.206)	Tm	Bu	2(1–2)	P	32.7(29–37)	2(1–4)		30	3(2–4)	NS,OS	45			73, 78
323	Japanese Murrelet	S. wumizusume	0.183	Tm	Cr,Bu	2(1–2)	P	32–33	2		c30		NS+				101, 151
324	Cassin's Auklet	Ptychoramphus aleuticus	0.19	SP,Tm	Bu,Cr	1 D	SP	39(37–57)	46(41–50)		none	3(1–4)	NS,OS	44	16	75	14, 127, 128, 207
325	Parakeet Auklet	Cyclorrhynchus psittacula	0.26	Tm	O	1	SP	35	35	10.6	none			50			151
326	Crested Auklet	Aethia cristatella	0.26	SP,Tm	Cr	1	SP	33.8(29–40)	33.2(27–36)	11–13	none	3(est.)	NS,OS	48		89	76, 118, 166
327	Least Auklet	Aethia pusilla	0.084	SP,Tm	Cr	1	SP	30.1(25–39)	28.6	6	none	3	NS,OS	33		80.8	117, 119, 180
328	Whiskered Auklet	Aethia pygmaea	0.12	Tm	Cr	1	SP	35.5	40.5(39–42)			3(est.)	NS				31, 151
329	Long-billed Murrelet	Brachyramphus perdix			T	1	SP		27–40		none					77	151
330	Marbled Murrelet	Brachyramphus marmoratus	0.22	Tm	T	1	SP	29	33.5(27–40)			3	NS,OS	44			149, 150
331	Kittlitz's Murrelet	Brachyramphus brevirostris	0.224–0.243	Tm	O	1	SP	30?	24?		?		NS				47, 151
332	Rhinoceros Auklet	Cerorhinca monocerata	0.456 (0.510)	SP–STr	Bu	1	SP	44.9(39–52)	50–60	10	none	3(est.)	NS,OS	62	10		14, 75, 76, 224

		Mass	Breeding dist.	Nest Loc.	Clutch Size			Incub. Period									Index
333	Tufted Puffin	*Fratercula cirrhata*	0.78	Tm	Bu	1			47–55	16.1			OS	66			101, 151
334	Atlantic Puffin	*Fratercula arctica*	0.51:0.31–0.57 (0.48:0.31–0.57)	Tm	Bu	1	SP	42	39	11.2	none	6	NS,OS	55	32	94.2	7, 14, 151
335	Horned Puffin	*Fratercula corniculata*	0.62	Tm	Bu	1	SP		38	11.1	none		OS	61			151

Notes: Numbers in the Index to Literature Cited can be found below. Blank cells are due to a lack of information. The first editor welcomes updates of information and appropriate references for future revisions at SchreiberE@aol.com. Taxonomy follows that used in Appendix 1 (see Chapter 3).

Mass is given in kilograms (kg), as a mean. If not available by sex, a mean, or mean and range, are given. If available by sex, females are listed first and males in parentheses — mean:range (mean:range). Values are for birds during the breeding season. Dunning (1993) was used for body mass for many species.

Breeding distribution: Polar (P) — for Antarctica, Antarctic continent; for the Arctic, pole to 80°N. Subpolar (SP) — south, just off edge of Antarctic continent to about 50°S; north, 80°N to Arctic Circle. Temperate (Tm) — south, about 50°S to 30°S; north, below Arctic circle to 25–30°N (to freeze line in Florida for instance). Subtropical (STr) — south, about 30°S-15°S; north = about 25-30°N to 15°N. Tropical (Tr) — between 15°N and 15°S.

Nest Location: Bu, burrow; Cl, cliff or cliff ledge; Cr, crevice or natural hole (these may be on flat ground [some shearwaters], slopes [most puffins] or on cliff faces); F, floating in marsh; O, open ground; T, on a tree or bush; Sl, steep slope (not as steep as a cliff and generally not stone); Un, under tree or bush.

Clutch Size is given as a mean, a range, or both. B, biennial breeder; D, known to raise two clutches in a year.

Hatching Type: P, precocial; SP, semiprecocial; A, altricial; SA, semialtricial (see Chapter 13, Table 13.1 for explanation).

Incubation Period is given in days (d), begins with the laying of the first egg.

Fledging Period applies to the time (in days) it takes for chicks to first have sustained flight (or for penguins and Flightless Cormorants, when they go to sea). For seven species of alcids that leave the nest before they can fly and go to sea with a parent, it applies to the time in the nest before going to sea.

Maximum growth rate (in grams per day [g/d]) is the maximum amount gained by a chick in a day.

Postfledging Care: The actual length of post-fledging care is often difficult to determine. Most research is carried out in colonies and feeding of young may still occur after birds leave the colony.

Foraging Distance during the breeding season: OS, offshore; NS, nearshore; NS+, between about 0.5 km offshore and offshore (intermediate between the two).

Wing Span indicated may encompass both sexes.

Maximum Age is the maximum age recorded for the species. This figure is higher than adult life expectancy.

Annual Survival data are for adults.

c = *circa* or about.

APPENDIX 2, LITERATURE CITED

1. AINLEY, D. G. 1995. Ashy Storm-Petrel (*Oceanodroma homochroa*). No. 185 *in* The Birds of North America (A. Poole and F. Gill, Eds.). Academy of Natural Sciences, Philadelphia; American Ornithologists' Union, Washington, D.C.

2. AINLEY, D. G., S. H. MORRELL, AND T. J. LEWIS. 1974. Patterns in the life histories of Storm-Petrels on the Farallon Islands. Living Bird 13: 295–312.

3. AINLEY, D. G., R. P. HENDERSON, AND C. S. STRONG. 1990. Leach's Storm-Petrel and Ashy Storm-Petrel. Pp. 128–162 *in* Seabirds of the Farallon Islands: Ecology, Dynamics, and Structure of an Upwelling-System Community (D. G. Ainley and R. J. Boekelheide, Eds.). Stanford University Press, Stanford, CA.

4. AINLEY, D. G., T. C. TELFER, AND M. H. REYNOLDS. 1997. Newell's and Townsend's Shearwater (*Puffinus auricularis*). No. 297 *in* The Birds of North America (A. Poole and F. Gill, Eds.). Academy of Natural Sciences, Philadelphia; American Ornithologists' Union, Washington, D.C.

5. AMERSON, A. B., JR., AND P. C. SHELTON. 1976. The natural history of Johnston Atoll, central Pacific Ocean. Atoll Research Bulletin 192: 1–479.

6. ANDERSON, D. J. 1993. Masked Booby (*Sula dactylatra*). No. 73 *in* The Birds of North America, (A. Poole and F. Gill, Eds.). Academy of Natural Sciences, Philadelphia; American Ornithologists' Union, Washington, D.C.

7. ASHCROFT, R. E. 1979. Survival rates and breeding biology of Puffins on Skomer Island, Wales. Ornis Scandinavica 10: 100–110.

8. AUSTIN, O. L., AND O. L. AUSTIN, JR. 1956. Some demographic aspects of the Cape Cod population of Common Terns (*Sterna hirundo*). Bird-Banding 27: 55–66.

9. BAIRD, P. H. 1986. Arctic and Aleutian Terns. Pp. 349–380 *in* The Breeding Biology and Feeding Ecology of Marine Birds in the Gulf of Alaska (P. H. Baird and P. J. Gould, Eds.). Environmental Assessment of the Alaska Continental Shelf, Final Report, Vol. 45, Juneau, AK.

10. BAIRD, P. H. 1994. Black-legged Kittiwake (*Rissa tridactyla*). No. 92 *in* The Birds of North America (A. Poole and F. Gill, Eds.). Academy of Natural Sciences, Philadelphia; American Ornithologists' Union, Washington, D.C.

11. BARRAT, A. 1976. Quelques aspects de la biologie et de l'écologie du Manchot Royal (Aptenodytes patagonica) des Lisle Crozet. Com. Nat. Français Rech. Antarct. 40: 9–52.

12. BELLOSE, C. A. 1983. The breeding biology and ecology of a small mainland colony of Western Gulls. Master's thesis, California State University, San Jose.

13. BENT, A. C. 1921. Life Histories of North American Gulls and Terns. U.S. National Museum Bulletin 113.

14. BIRD BANDING LABORATORY. Patuxent Wildlife Research Center, U.S. Geological Survey.

15. BOEKELHEIDE, R. J., AND D. G. AINLEY. 1989. Age, resource availability, and breeding effort in Brandt's Cormorant. Auk 106: 389–401.

16. BOEKELHEIDE, R. J., D. G. AINLEY, S. H. MORRELL, AND T. J. LEWIS. 1990. Brandt's cormorant. Pp. 163–194 *in* Seabirds of the Farallon Islands (D. G. Ainley and R. J. Boekelheide, Eds.). Stanford University Press, Stanford, CA.

17. BRIED, J. Unpublished.

18. BROOKE, M. de L. 1978. Weights and measurements of the Manx Shearwater *Puffinus puffinus*. Journal Zoology London. 186: 359–374.

19. BROWN, D. A. 1966. Breeding biology of the Snow Petrel Pagodroma nivea. Australian National Antarctic Research Expedition Scientific Report B (1) Zoology 89: 1–63.

20. BROWN, L. H., AND V. E. M. BURKE. 1970. Observations on the breeding of the Pink-backed Pelican *Pelecanus rufescens*. Ibis 112: 409–412.

21. BROWN, W. Y. 1976. Growth and fledging age of the Brown Noddy in Hawaii. Condor 78: 263–264.

22. BURGER, J. 1972. Dispersal and post-fledging survival of Franklin's Gulls. Bird-Banding 43: 267–275.

23. BURGER, J. 1974. Breeding adaptations of Franklin's Gull (*Larus pipixcan*) to a marsh habitat. Animal Behavior 22: 521–567.

24. BURGER, J. 1980. The transition to independence and post-fledging parental care in seabirds. Pp. 624–666 *in* Behavior of Marine Animals: Current Perspectives in Research, Vol. 4. Marine Birds (J. Burger, D. L. Olla, and H. E. Winn, Eds.). Plenum Press, New York.

25. BURGER, J. 1996. Laughing Gull *(Larus atricilla)*. No. 225 *in* The Birds of North America (A. Poole and F. Gill, Eds.). Academy of Natural Sciences, Philadelphia; American Ornithologists' Union, Washington, D.C.

26. BURGER, J., AND M. GOCHFELD. 1994. Franklin's Gull *(Larus pipixcan)*. No. 116 *in* The Birds of North America (A. Poole and F. Gill, Eds.). Academy of Natural Sciences, Philadelphia; American Ornithologists' Union, Washington, D.C.

27. BURGER, J., AND M. GOCHFELD. 1996. Family Laridae (gulls). Pp. 572–622 *in* Handbook of Birds of the World (J. del Hoyo, A. Elliott, and J. Sargatal, Eds.). Lynx Editions, Barcelona.

28. BURKE, V. E. M., AND L. J. BROWN. 1970. Observation on the breeding of the Pink-backed Pelican *Pelecanus refescens.* Ibis 112: 499–512.

29. BURNESS, G. P., K. LEFEVRE, AND C. T. COLLINS. 1999. Elegant Tern *(Sterna elegans)*. No. 404 *in* The Birds of North America (A. Poole and F. Gill, Eds.). The Birds of North America, Inc., Philadelphia.

30. BYRD, G. V., AND J. C. WILLIAMS. 1993a. Red-legged Kittiwake *(Rissa brevirostris)*. No. 60 *in* The Birds of North America (A. Poole and F. Gill, Eds.). Academy of Natural Sciences, Philadelphia; American Ornithologists' Union, Washington, D.C.

31. BYRD, G. V., AND J. C. WILLIAMS. 1993b. Whiskered Auklet *(Aethia pygmaea)*. No. 76 *in* The Birds of North America (A. Poole and F. Gill, Eds.). Academy of Natural Sciences, Philadelphia; American Ornithologists' Union, Washington, D.C.

32. CARBONERAS, C. 1992. Order Procellariiformes. Pp. 198–270 *in* Handbook of Birds of the World (J. del Hoyo, A. Elliott, and J. Sargatal, Eds.). Lynx Editions, Barcelona.

33. CARMONA, R. 1993. Reproduccion y crecimiento de dos especies de gaviota (Larus livens y L. occidentalis) anidantes en Baja California Sur. M.Sc. thesis, Centro Interdisciplinario de Ciencias Mainas, Inst. Politech. Nacl., La Paz, Baja California, Sur.

34. CARMONA, R., J. GUZMAN, AND J. F. ELORDUY. 1995. Hatching growth and mortality of Magnificent Frigatebirds in southern Baja California. Wilson Bulletin 107: 328–337.

35. CODY, M. L. 1973. Coexistence, Coevolution and Convergent Evolution in seabird communities. Ecology 54: 1.

36. COLLINS, C. T., B. W. MASSEY, L. M. DARES, M. WIMER, AND K. GAZZINGA. 1996. Banding of adult California Least Tern at Marine Corps Base, Camp Pendleton between 1987 and 1995. Report to Department of Navy, Natural Resources Management Branch, Southwestern and Western Division, Naval Facilities Engineering Command, San Diego and San Bruno, CA.

37. COULSON, J. C., AND E. WHITE. 1956. A study of colonies of the Kittiwake *Rissa tridactyla.* Ibis 98: 63–79.

38. COULSON, J. C., AND E. WHITE. 1959. The post-fledging mortality of the Kittiwake. Bird Study 6: 97–102.

39. CRAMP, S., AND K. E. L. SIMMONS (Eds.). 1983. Handbook of the Birds of Europe, the Middle East, and North Africa, the birds of the Western Palearctic. Vol. 3. Waders to Gulls. Oxford University Press, Oxford.

40. CRIVELLI, A. J. 1987. The ecology and behaviour of the Dalmation Pelican *Pelecanus crispus*, a world-endangered species. Internal Report, Station Biologique de la Tour du Valat.

41. CROXALL, J. P. 1981. Aspects of the population demography of Antarctic and Subantarctic seabirds. Colloque sur les ecosystèmes subantarctiques, Paimpont, Com. Nat. Français Rech. Antarct.

42. CROXALL, J. P. (Ed.). 1987. Seabirds, feeding ecology and role in marine ecosystems. Cambridge University Press, Cambridge, England.

43. CROXALL, J. P., AND P. A. PRINCE. 1983. Antarctic penguins and albatrosses. Oceanus 26: 18–27.

44. CROXALL, J. P., AND P. ROTHERY. 1995. Population changes in the Gentoo Penguin Pygocelis papua at bird Island, South Georgia: potential roles of adult survival, recruitment and deferred breeding. Pp. 26–38 *in* Penguins: Advances in Research and Management (P. Dann, I. Norman, and P. Reilly, Eds.). Surrey, Beatty and Sons, Chipping Norton, Australia.

45. CULLEN, J. M. 1957. Plumage, age and mortality in the Arctic Tern. Bird Study 4: 197–207.

46. CUTHBERT, F. J., AND L. R. WIRES. 1999. Caspian Tern (*Sterna caspia*). No. 403 *in* The Birds of North America (A. Poole and F. Gill, Eds.). The Birds of North America, Inc., Philadelphia.

47. DAY, R. H., D. J. KULETZ, AND D. A. NIGRO. 1999. Kittlitz's Murrelet (Brachyramphus brevirostris). No. 435 *in* The Birds of North America (A. Poole and F. Gill, Eds.). The Birds of North America, Inc., Philadelphia.

48. DIAMOND, A. W. 1975. The biology and behaviour of frigatebirds *Fregata* spp. on Aldabra Atoll. Ibis 117: 302–323.

49. DIAMOND, A. W., AND E. A. SCHREIBER. In press. The Magnificent Frigatebird *Fregata magnificens*. *In* The Birds of North America (A. Poole and F. Gill, Eds.). The Birds of North America, Inc., Philadelphia.

50. DIN, N. A., AND K. ELTRINGHAM. 1974. Breeding of the pink-backed pelican *Pelecanus refescens* in Rwenzori National Park, Uganda. Ibis 116: 477–493.

51. DORWARD, D. F. 1962. Comparative biology of the white booby and the brown booby *Sula* spp. at Ascension. Ibis 103b: 174–220.

52. DROST, C. A., AND D. B. LEWIS. 1995. Xantus' Murrelet *(Synthliboramphus hypoleucus)*. No. 164 *in* The Birds of North America (A. Poole and F. Gill, Eds.). Academy of Natural Sciences, Philadelphia; American Ornithologists' Union, Washington, D.C.

53. DUFFY, D. C. 1987. Ecological implications of intercolony size variation in Jackass Penguins. Ostrich 58: 54–57.

54. DUNLOP, J. N., AND J. JENKINS. 1992. Known-age birds at a subtropical breeding colony of the Bridled Tern (*Sterna anaethetus*): a comparison with the Sooty Tern. Colonial Waterbirds 15: 75–82.

55. DUNLOP, J. N., AND J. JENKINS. 1994. Population dynamics of the Bridled Tern *Sterna anaethetus* colony on Penguin Island, southwestern Australia. Corella 18: 33–36.

56. DUNN, E. H. 1975. Growth, body components, and energy content of nestling double-creasted cormorants. Condor 77: 431–438.

57. DUNN, E. H., AND I. L. BRISBIN. 1980. Age-specific changes in major body components and caloric values of Herring Gull chicks. Condor 82: 398–401.

58. DUNN, E. H., AND D. J. ARGO. 1995. Black Tern (*Chlidonias niger*). No. 147 *in* The Birds of North America (A. Poole and F. Gill, Eds.). Academy of Natural Sciences, Philadelphia; American Ornithologists' Union, Washington, D.C.

59. DUNNET, G. M., A. ANDERSON, AND R. M. CORMACK. 1963. A study of the survival of adult Fulmars with observations on the pre-laying exodus. British Birds 56: 2–18.

60. DUNNET, G. M., J. C. OLLASON, AND A. ANDERSON. 1979. A 28-year study of breeding fulmars *Fulmars glacialis* in Orkney. Ibis 21: 293–300.

61. DUNNING, J. B., JR. 1993. CRC Handbook of Avian Body Masses. CRC Press, Boca Raton, FL.

62. ERWIN, R. M. 1995. The ecology of cormorants: some research needs and recommendations. Waterbirds 18 (Spec. Publ. 1): 240–246.

63. EVANS, R. M. 1990. Effects of low incubation temperatures during the pipped egg stage on hatchability and hatching times in domestic chickens and Ring-billed Gulls. Canadian Journal Zoology. 68: 836–840.

64. EVANS, R. M., AND F. L. KNOPF. 1993. American White Pelican (*Pelecanus erythrorhynchos*). No. 57 *in* The Birds of North America (A. Poole and F. Gill, Eds.). Academy of Natural Sciences, Philadelphia; American Ornithologists' Union, Washington, D.C.

65. EWINS, P. J. 1993. Pigeon Guillemot (*Cepphus columbia*). No. 49 *in* The Birds of North America (A. Poole and F. Gill, Eds.). Academy of Natural Sciences, Philadelphia; American Ornithologists' Union, Washington, D.C.

66. EWINS, P. J., AND D. V. WESELOH. 1999. Little Gull (*Larus minutus*). No. 428 *in* The Birds of North America (A. Poole and F. Gill, Eds.). The Birds of North America, Inc., Philadelphia.

67. FISHER, H. I. 1975. Mortality and survival in the Laysan Albatross Diomedea immutabilis. Pacific Science 29: 279–300.

68. FLEET, R. R. 1974. The Red-tailed Tropicbird on Kure Atoll. A.O.U. Monographs 16: 1–64.

69. FLOYD, R. B., AND N. M. SWANSON. 1983. Wedge-tailed Shearwaters on Muttonbird Island: an estimate of the breeding success and the breeding population. Emu 82: 244–250.

70. FREDERIKSEN, M., AND T. BREGNBALLE. 2000. Evidence for density dependent survival in adult cormorants from a combined analysis of recoveries and resightings. Journal of Animal Ecology 69: 737–752.

71. FURNESS, R. W. 1987. The Skuas. Poyser, Calton, U.K.

72. FURNESS, R. W. 1993. Arctic Skua. Pp. 196–197 in The New Atlas of Breeding Birds in Britain and Ireland: 1988–1991 (D. W. Gibbons, J. B. Reid, and R. A. Chapman, Eds.). Poyser, Berhamstead.

73. GASTON, A. J. 1994. Ancient Murrelet (*Synthliboramphus antiquus*). No. 132 in The Birds of North America (A. Poole and F. Gill, Eds.). Academy of Natural Sciences, Philadelphia; American Ornithologists' Union, Washington, D.C.

74. GASTON, A. J., AND D. N. NETTLESHIP. 1981. The Thick-billed Murres of Prince Leopold Island. Canadian Wildlife Service Monographs. No.6. Canadian Wildlife Service, Ottawa.

75. GASTON, A. J., AND S. B. C. DECHESNE. 1996. Rhinoceros Auklet (*Cerorhinca monocerata*). No. 212 in The Birds of North America (A. Poole and F. Gill, Eds.). Academy of Natural Sciences, Philadelphia; American Ornithologists' Union, Washington, D.C.

76. GASTON, A. J., AND I. L. JONES. 1998. The Auks. Oxford University Press, Oxford.

77. GASTON, A. J., AND J. M. HIPFNER. 2000. Thick-billed Murre (*Uria lomvia*). No. 497 in The Birds of North America (A. Poole and F. Gill, Eds.). The Birds of North America, Inc., Philadelphia.

78. GASTON, A. J., I. L. JONES, AND D. G. NOBLE. 1988. Monitoring Ancient Murrelet breeding populations. Colonial Waterbirds 11: 58–66.

79. GASTON, A. J., L. N. deFOREST, G. DONALDSON, AND D. G. NOBLE. 1994. Population parameters of Thick-billed Murres at Coats Island, Northwest Territories, Canada. Condor 96: 945–948.

80. GAUGER, V. H. 1999. Black Noddy (*Anous minutus*). No. 412 in The Birds of North America (A. Poole and F. Gill, Eds.). The Birds of North America, Inc., Philadelphia.

81. GOCHFELD, M., AND J. BURGER. 1996. Family Sternidae (terns). Pp. 624–666 in Handbook of Birds of the World (J. del Hoyo, A. Elliott, and J. Sargatal, Eds.). Lynx Editions, Barcelona.

82. GOCHFELD, M., J. BURGER, AND I. C. T. NISBET. 1998. Roseate Tern (*Sterna dougallii*). No. 370 in The Birds of North America (A. Poole and F. Gill, Eds.). The Birds of North America, Inc., Philadelphia.

83. GONZÁLEZ-SOLÍS, J., P. H. BECKER, AND H. WENDELN. 1999. Divorce and asynchronous arrival in Common Terns, *Sterna hirundo*. Animal Behaviour 58: 1123–1129.

84. GOOD, T. P. 1998. Great Black-backed Gull (*Larus marinus*). No. 330 in The Birds of North America (A. Poole and F. Gill, Eds.). The Birds of North America, Inc., Philadelphia.

85. GRANT, G. S., J. WARHAM, T. N. PETTIT, AND G. C. WHITTOW. 1983. Reproductive behavior and vocalizations of the Bonin petrel. Wilson Bulletin. 95: 522–539.

86. HAMER, K. C. In press. a. Birds: Procellariiformes. In Encyclopedia of Ocean Sciences (J. Steele, S. Thorpe, and K. Turekian, Eds.). Academic Press, London.

87. HAMER, K. C. In press. b. Great Skua *Catharacta skua*. In Birds of the Western Palearctic (D. Parkin, Ed.). Oxford University Press, Oxford.

88. HAMER, K. C. Unpublished.

89. HAMER, K. C., L. W. NICHOLSON, J. K. HILL, R. D. WOOLER, AND J. S. BRADLEY. 1997. Nestling obesity in procellariiform seabirds: temporal and stochastic variation in provisioning and growth of short-tailed shearwaters Puffinus tenuirostris. Oecologia 112: 4–11.

90. HANEY, J. C., AND S. D. MACDONALD. 1995. Ivory Gull (*Pagophila eburnea*). No. 175 in The Birds of North America (A. Poole and F. Gill, Eds.). Academy of Natural Sciences, Philadelphia; American Ornithologists' Union, Washington, D.C.

91. HANEY, J. C., D. S. LEE, AND R. D. MOORIS. 1999. Bridled Tern *(Sterna anaehetus)*. No. 468 in The Birds of North America (A. Poole and F. Gill, Eds.). The Birds of North America, Inc., Philadelphia.

92. HARRIS, M. P. 1964. Aspects of the breeding biology of the gulls, *Larus argentatus, L. fuscus and L. marinus*. Ibis 106: 432–456.

93. HARRIS, M. P. 1966a. Age of return to the colony, age of breeding and adult survival of Manx Shearwaters. Bird Study 13: 84–95.

94. HARRIS, M. P. 1966b. Breeding biology of the Manx Shearwater. Ibis 108: 17–33.

95. HARRIS, M. P. 1973. The biology of the waved albatross *Diomedea irrorata* of Hood Island, Galapagos. Ibis 115: 483–510.

96. HARRIS, M. P. 1979a. Population dynamics of the Flightless Cormorant *Nannopterum harrisi*. Ibis 121: 135–146.

97. HARRIS, M. P. 1979b. Survival and ages of first breeding of Galapagos seabirds. Bird-banding 50: 56–61.

98. HARRIS, M. P., AND S. WANLESS. 1989. The breeding biology of Razorbills Alca torda on the Isle of May. Bird Study 36: 105–114.

99. HARRIS, M. P., AND S. WANLESS. 1995. Survival and non-breeding of adult Common guillemots *Uria aalge*. Ibis 137: 192–197.

100. HARRIS, M. P., AND S. WANLESS. 1996. Differential responses of Guillemots Uria aalge and Shag Phalacrocorax aristotelis to a late winter wreck. Bird Study 43: 220–230.

101. HARRISON, P. 1983. Seabirds. Houghton Mifflin, Boston.

102. HATCH, S. A., AND M. A. HATCH. 1990. Breeding season of oceanic birds in a sub-arctic colony. Canadian Journal of Zoology 68: 1664–1679.

103. HATCH, S. A., AND D. N. NETTLESHIP. 1998. Northern Fulmar, *Fulmarus glacialis*. No. 361 *in* The Birds of North America (A. Poole and F. Gill, Eds.). Academy of Natural Sciences, Philadelphia; American Ornithologists' Union, Washington, D.C.

104. HATCH, J. J., AND D. V. WESELOH. 1999. Double-crested Cormorant *(Phalacrocorax auritus)*. No. 441 *in* The Birds of North America (A. Poole and F. Gill, Eds.). The Birds of North America, Inc., Philadelphia.

105. HATCH, S. A., G. V. BYRD, D. B. IRONS, AND G. L. HUNT. 1993. Status and ecology of kittiwakes (Rissa tridactyla and R. brevirostris) in the North Pacific. Pp. 140–153 *in* The Status Ecology and Conservation of Marine Birds in the North Pacific (K. Vermeer, K. T. Briggs, K. H. Morgan, and D. Siegel-Causey, Eds.). Canadian Wildlife Service, Special Publication, Ottawa, Ontario.

106. HATZILACOU, D. 1992. The breeding biology and feeding ecology of the Great White Pelican *Pelecanus onocrotalus*, at Lake Mikri Prespa (Northwestern Greece). Ph.D. dissertation, University of Athens, 183 pp. (in Greek).

107. HAYS, M. B. 1980. Breeding biology of the Least Tern, *Sterna albifrons,* on the Gulf Coast of Mississippi. M.S. thesis, Mississippi State University, Mississippi State.

108. HIGGINS, P. J., AND S. J. J. F. DAVIS. 1996. Handbook of Australian, New Zealand and Antarctic birds. Vol. 3. Oxford University Press, Melbourne.

109. HOBSON, K. A. 1997. Pelagic Cormorant (*Phalacrocorax pelagicus*). No. 282 *in* The Birds of North America (A. Poole and F. Gill, Eds.). Academy of Natural Sciences, Philadelphia; American Ornithologists' Union, Washington, D.C.

110. HUDSON, R. 1966. Adult survival estimates for two Antarctic petrels. British Antarctic Survey Bulletin 8: 63–73.

111. HUNTER, S. 1984. Breeding biology and population dynamics of giant petrels Macronectes at South Georgia (Aves Procellariiformes). Journal of Zoology, London 203: 441–460.

112. HUNTINGTON, C. E., R. G. BUTLER, AND R. A. MAUCK. 1996. Leach's Storm-Petrel *(Oceanodroma leucorhoa)*. No. 233 *in* The Birds of North America (A. Poole and F. Gill, Eds.). Academy of Natural Sciences, Philadelphia; American Ornithologists' Union, Washington, D.C.

113. IMBER, M. J. 1987. Breeding ecology and conservation of the Black Petrel *Procellaria parkinsoni*. Notornis 34: 19–39.

114. ISENMANN, P. 1971. Contribution á l'éthologie et á l'écologie du Manchot empereur *(Aptenodytes forsteri* Gray) á la colonie de Pointe Géologie (Terre Adélie). L'Oiseau et la R.F.O. 41: 9–64.

115. JEHL, J. R., JR. 1982. The biology and taxonomy of Townsend's Shearwater. Le Gerfaut 72: 121–135.

116. JOHNSGARD, P. A. 1993. Cormorants, darters and pelicans of the world. Smithsonian Institution Press, Washington, D.C.

117. JONES, I. L. 1993a. Least Auklet (*Aethia pusilla*). No. 69 *in* The Birds of North America (A. Poole and F. Gill, Eds.). Academy of Natural Sciences, Philadelphia; American Ornithologists' Union, Washington, D.C.

118. JONES, I. L. 1993b. Crested Auklet *(Aethia cristella)*. No. 70 *in* The Birds of North America (A. Poole and F. Gill, Eds.). Academy of Natural Sciences, Philadelphia; American Ornithologists' Union, Washington, D.C.

119. JONES, I. L., AND R. MONTGOMERIE. 1991. Mating and remating of Least Auklets (Aethia pusilla) relative to ornamental traits. Behavioral Ecology 2: 249–257.

120. JOUVENTIN, P., AND J.-L. MOUGIN. 1981. Les stratégies adaptatives des oiseaux de mer. Revue d'Ecologie (Terre & Vie) 35: 217–272.

121. KADLEC, J. A., AND DRURY, W. H. 1968. Structure of the New England Herring Gull population. Ecology 49: 644–676.

122. KING, W. B., AND P. J. GOULD. 1967. The status of Newell's race of the Manx Shearwater. Living Bird 6: 163–186.

123. LANGHAM, N. P. 1974. Comparative breeding biology of the Sandwich Tern. Auk 91: 255–277.

124. LANGHAM, N. P. 1983. Growth strategies in marine terns. Studies in Avian Biology 8: 73–83.

125. LEE, D. S., AND J. C. HANEY. 1996. Manx Shearwater (*Puffinus puffinus*). No. 257 *in* The Birds of North America (A. Poole and F. Gill, Eds.). Academy of Natural Sciences, Philadelphia; American Ornithologists' Union, Washington, D.C.

126. LEE, D. S., AND M. WALSH-McGEHEE. 1998. White-tailed Tropicbird (*Phaethon lepturus*). No. 353 *in* The Birds of North America (A. Poole and F. Gill, Eds.). The Birds of North America, Inc., Philadelphia.

127. MANUWAL, D. A. 1974. The natural history of Cassin's Auklet (*Ptychoramphus aleuticus*). Condor 76: 421–431.

128. MANUWAL, D. A., AND A. C. THORESEN. 1993. Cassin's Auklet (*Ptychoramphus aleuticus*). No. 50 *in* The Birds of North America (A. Poole and F. Gill, Eds.). Academy of Natural Sciences, Philadelphia; American Ornithologists' Union, Washington, D.C.

129. MARCHANT, S., AND P. J. HIGGINS. 1990. Handbook of Australian, New Zealand and Antarctic Birds, Vol. 1. Oxford University Press, Australia.

130. MARTINEZ, I. 1992. Order Sphenisciformes. Pp. 140–160 *in* Handbook of Birds of the World (J. A. del Hoyo, A. Elliott, and J. Sargatal, Eds.). Lynx Editions, Barcelona.

131. MASSEY, B. W., D. W. BRADLEY, AND J. L. ATWOOD. 1992. Demography of a California Least Tern colony including effects of the 1982–1983 El Niño. Condor 94: 976–983.

132. MEGYESI, J. L., AND D. L. O'DANIEL. 1997. Bulwer's Petrel (*Bulweria bulwerii*). No. 281 *in* The Birds of North America (A. Poole and F. Gill, Eds.). Academy of Natural Sciences, Philadelphia; American Ornithologists' Union, Washington, D.C.

133. MILLS, J. A. 1973. The influence of age and pair-bond on the breeding biology of the Red-billed Gull *Larus novaehollandiae scopulinus*. Journal of Animal Ecology 42: 147.

134. MILLS, J. A. 1989. Red-billed Gull. Pp. 387–404 *in* Lifetime Reproduction in Birds (I. Newton, Ed.). Academic Press, London.

135. MILLS, J. A., J. W. YARRALL, AND D. A. MILLS. 1996. Causes and consequences of mate fidelity in red-billed gulls. Pp. 286–304 *in* Partnerships in Birds. The Study of Monogamy (J. M. Black, Ed.). Oxford University Press, Oxford.

136. MONTEVECCHI, W. A., AND D. A. KIRK 1996. Great Auk (*Pinguinus impennis*). No. 260 *in* The Birds of North America (A. Poole and F. Gill, Eds.). The Birds of North America, Inc., Philadelphia.

137. MONTEVECCHI, W. A., R. E. RICKLEFS, I. R. KIRKHAM, AND D. GABALDON. 1984. Growth energetics of nestling Northern Gannets (*Sula bassanus*). Auk 101: 334–341.

138. MOSTELLO, C. S., N. A. PALASIA, AND R. B. CLAPP. 2000. Gray-backed Tern (*Sterna lunata*). No. 525 *in* The Birds of North America (A. Poole and F. Gill, Eds.). The Birds of North America, Inc., Philadelphia.

139. MOUGIN, J.-L. 1967. Etude ecologique des deux especes de fulmars le Fulmar Atlantique (*F. glacialis*) et le Fulmar Antarctique (*F. glacialoides*). Oiseau Review Fr. Ornithologie 37: 57–103.

140. MOUGIN, J.-L. 1989. Données préliminaires sur la structure et la dynamique de la population de Pétrels de Bulwer *Bulweria bulwerii* de l'île Selvagem Grande (30°09'N, 15°52'W). Compte-rendus de l'Académie des Sciences de Paris 308: 103–106.

141. MOUGIN, J.-L. 1990. La fidélité au partenaire et au nid chez le Pétrel de Bulwer *Bulweria bulwerii* de l'île Selvagem Grande (30°09'N, 15°52'W). L'Oiseau et la R. F. O. 60: 224–232.

142. MOUGIN, J.-L., B. DESPIN, C. JOUANIN, AND F. ROUX. 1987a. La fidélité au partenaire et au nid chez le Puffin cendré, *Calonectris diomedea borealis*, de l'île Selvagem Grande. Gerfaut 77: 353–369.

143. MOUGIN, J.-L., C. JOUANIN, AND F. ROUX. 1987b. Structure et dynamique de la population de Puffins cendrés *Calonectris diomedea borealis* de l'île Selvagem Grande (30°09'N, 15°52'W). L'Oiseau et la R. F. O. 57: 201–225.

144. MOUGIN, J.-L., C. JOUANIN, AND F. ROUX. 1988a. Les différences d'âge et d'expérience entre partenaires chez le Puffin cendré Calonectris diomedea borealis de l'île Selvagem Grande (30°09'N, 15°52'W). L'Oiseau et la R. F. O. 58: 113–119.

145. MOUGIN, J.-L., C. JOUANIN, AND F. ROUX. 1988b. L'influence des voisins dans la nidification du Puffin cendré Calonectris diomedea. Comptes-rendus de l'Académie des Sciences de Paris (III) 307: 195–198.

146. NAGULU, V. 1984. Biology of Spot-Billed Pelican, Pelecanus phillippensis, at Nelapattu, India. Ph.D. thesis.

147. NEITHAMMER, K. R., AND L. B. PATRICK-CASTILAW. 1998. White Tern (*Gygis alba*). No. 371 *in* The Birds of North America (A. Poole and F. Gill, Eds.). The Birds of North America, Inc. Philadelphia.

148. NELSON, J. B. 1978. The Sulidae: gannets and boobies. Oxford University Press. Oxford, UK.

149. NELSON, S. K. 1997. Marbled Murrelet (*Brachyramphus marmortus*). No. 276 *in* The Birds of North America (A. Poole and F. Gill, Eds.). Academy of Natural Sciences, Philadelphia; American Ornithologists' Union, Washington, D.C.

150. NELSON, S. K., AND T. E. HAMER 1995. Nesting biology and behavior of the Marbled Murrelet. Pp. 57–68 *in* Ecology and Conservation of the Marbled Murrelet (C. J. Ralph, G. L. Hunt, M. G. Raphael, and J. F. Piatt, Eds.). U.S.D.A. Forest Service Technical Report, PSW-152, Albany, CA.

151. NETTLESHIP, D. N. 1996. Family *Alcidae* (auks). Pp. 678–722 *in* Handbook of the Birds of the World (J. del Hoyo, A. Elliott, and J. Sargatal, eds.). Vol. 3. Lynx Edicions, Barcelona, Spain.

152. NETTLESHIP, D. N., AND G. L. HUNT, JR. 1988. Reproductive biology of seabirds at high latitudes in the northern and southern hemispheres. Proceedings of the International Ornithological Congress 19: 1141–1219.

153. NICHOLS, C. 1974. Double-brooding in a Western Australian population of the Silver Gull, *Larus novae-hollandiae*. Australian Journal of Zoology 22: 63–70.

154. NISBET, I. C. T. Unpublished.

155. NISBET, I. C. T. 1989. Status and biology of the northeastern population of the Roseate Terns (*Sterna dougllii*): a literature survey and update: 1981–1989. U.S. Fish Wildlife Service, Contract Report 50181-88-8105, Newton Corner, MA.

156. NORTH, M. R. 1997. Aleutian Tern (*Sterna aleutica*). No. 291 *in* The Birds of North America (A. Poole and F. Gill, Eds.). Academy of Natural Sciences, Philadelphia; American Ornithologists' Union, Washington, D.C.

157. O'DONALD, P. 1983. The Arctic Skua: A Study of the Ecology and Evolution of a Seabird. Cambridge University Press, Cambridge.

158. OLSEN, K. M., AND H. LARRSEN. 1997. Skuas and Jaegers. Yale University Press, New Haven, CT.

159. ORTA, J. 1992. Family Phaethontidae (Tropicbirds), Pp. 280–289; Family Phalacrocoracidae (Cormorants), Pp. 326–353; Family Fregatidae (Frigatebirds), Pp. 362–374 *in* Handbook of Birds of the World, Vol. 1. (J. del Hoyo, A. Elliott, and J. Sargatal, Eds.). Lynx Edicions, Barcelona.

160. PALMER, R. S. (Ed) 1962. Handbook of North American Birds. Vol. 1. Loons through Flamingos. Yale University Press, New Haven, CT.

161. PARNELL, J. F., R. M. ERWIN, AND K. C. MOLINA. 1995. Gull-billed Tern (*Sterna nilotica*). No. 140 *in* The Birds of North America (A. Poole and F. Gill, Eds.). The Academy of Natural Sciences, Philadelphia; and The American Ornithologists' Union, Washington, D.C.

162. PATTEN, M. A. 1996. Yellow-footed Gull (Larus livens). No. 243 *in* The Birds of North America (A. Poole and F. Gill, Eds.). Academy of Natural Sciences, Philadelphia; American Ornithologists' Union, Washington, D.C.

163. PENNYCUICK, C. J. 1983. Thermal soaring compared in three dissimilar tropical bird species, *Fregata magnificens, Pelecanus occidentalis* and *Coryagyps atratus*. Journal of Experimental Biology 102: 307–325.

164. PETERSEN, A. E. 1981. Breeding biology and feeding ecology of Black Guillemots. Unpublished Ph.D. thesis, Oxford University, Oxford.

165. PHILLIPS, R. A., AND K. C. HAMER. 2000. Growth and provisioning strategies of Northern Fulmars *Fulmaris glacialis*. Ibis 142: 435–445.

166. PIATT, J., B. ROBERTS, AND S. HATCH. 1990. Colony attendance and population monitoring of Least and Crested Auklets at St. Lawrence Island, Alaska. Condor 92: 97–106.

167. PIEROTTI, R. J. 1979. The reproductive behaviour and ecology of the Herring Gull in Newfoundland. Ph.D. dissertation, Dalhousie University, Halifax, Nova Scotia.

168. PIEROTTI, R. J. 1982. Habitat selection and its effect on reproductive output in the Herring Gull in Newfoundland. Ecology 63: 854–868.

169. PIEROTTI, R. J., AND T. P. GOOD. 1994. Herring Gull (*Larus argentatus*). No. 124 *in* The Birds of North America (A. Poole and F. Gill, Eds.). Academy of Natural Sciences, Philadelphia; American Ornithologists' Union, Washington, D.C.

170. PIEROTTI, R. J., AND C. A. ANNETT. 1995. Western Gull *(Larus occidentalis)*. No. 174 *in* The Birds of North America (A. Poole and F. Gill, Eds.). Academy of Natural Sciences, Philadelphia; American Ornithologists' Union, Washington, D.C.

171. POTTS, G. R. 1969. The influence of eruptive movements, age, population size and other factors on the survival of the shag (*Phalacrocorax aristotelis*). Journal of Animal Ecology 38: 53–102.

172. PYROVETSI, M. 1989. Foraging trips of White Pelicans (*Pelecanus onocrotalus*) breeding on Lake Mikri Prespa, Greece. Colonial Waterbirds 12: 43–50.

173. QUINN, J. S. 1990. Sexual size dimorphism and parental care patterns in a monomorphic and a dimorphic larid. Auk 107: 260–274.

174. REID, B. 1968. An interpretation of the age structure and breeding status of an Adelie Penguin population. Notornis 15: 193–197.

175. REILLY, P. N., AND P. BALMFORD. 1974. A breeding study of the Little Penguin (*Eudyptula minor*) in Australia. Pp. 161–188 *in* The Biology of Penguins (B. Stonehouse, Ed.). MacMillan and Co., London.

176. RICHDALE, L. E. 1957. A Population Study of Penguins. Oxford University Press, Cambridge, England.

177. RICHDALE, L. E. 1963. Biology of the Sooty Shearwater. Proceedings of the London Zoological Society 141: 1–117.

178. RICKLEFS, R. E., D. D. ROBY, AND J. B. WILLIAMS. 1986. Daily energy expenditure by adult Leach's Storm-Petrels during the nesting cycle. Physiological Zoology 59: 649–660.

179. ROBINSON, J. A., AND K. C. HAMER. 2000. Brood size and food provisioning in Common Terns Sterna hirundo and Arctic Terns Sterna paradisaea: consequences for chick growth. Ardea 88: 51–60.

180. ROBY, D. D., AND K. L. BRINK 1986. Decline of breeding Least Auklets on St. George Island, Alaska. Journal Field Ornithology 57: 57–59.

181. RYDER, J. P. 1993. Ring-billed Gull, *Larus delawaiensis*. No. 33 *in* The Birds of North America (A. Poole and F. Gill, Eds.). Academy of Natural Sciences, Philadelphia; American Ornithologists' Union, Washington, D.C.

182. SAGAR, P. M., AND J. WARHAM. 1993. A long-lived Southern Buller's Mollymawk (*Diomedea bulleri bulleri*) with a small egg. Notornis 40: 303–304.

183. SAGAR, P. M., AND J. WARHAM. 1997. Breeding biology of Southern Buller's Albatrosses at The Snares, New Zealand. Pp. 92–98 *in* The Albatross Biology and Conservation (G. Robertson and R. Gales, Eds.). Surrey Beatty and Sons, Chipping Norton.

184. SCHAFFNER, F. C. 1982. Aspects of the reproductive ecology of the Elegant Tern (Sterna elegans) at San Diego Bay. Master's thesis, San Diego State University, San Diego, CA.

185. SCHREIBER, E. A. Unpublished.

186. SCHREIBER, E. A., AND R. W. SCHREIBER. 1980. Breeding biology of Laughing Gulls in Florida. Part II: nesting parameters. Journal of Field Ornithology 51: 340–355.

187. SCHREIBER, E. A., AND R. W. SCHREIBER. 1993. The Red-tailed Tropicbird (*Phaethon rubricauda*). No 43 *in* The Birds of North America (A. Poole and F. Gill, Eds.). Academy of Natural Sciences, Philadelphia; American Ornithologists' Union, Washington, D.C.

188. SCHREIBER, E. A., R. W. SCHREIBER, AND J. J. DINSMORE. 1979. Breeding biology of Laughing Gulls in Florida. Part I: nesting, egg, and incubation parameters. Bird-Banding 50: 304–355.

189. SCHREIBER, E. A., R. W. SCHREIBER, AND G. A. SCHENK. 1996. The Red-footed Booby (*Sula sula*). No. 241 *in* The Birds of North America (A. Poole and F. Gill, Eds.). Academy of Natural Sciences, Philadelphia; American Ornithologists' Union, Washington, D.C.

190. SCHREIBER, R. W. 1976. Growth and development of nestling Brown Pelicans. Bird-Banding 47: 19–39.

191. SEARS, H. F. 1978. Breeding Behavior of the Gull-billed Tern. Ph.D. dissertation, University of North Carolina, Chapel Hill.

192. SERVENTY, D. L. 1967. Aspects of the population ecology of the Short-tailed Shearwater, *Puffinus tenuirostris*. Pp. 165–190 *in* Proceedings of the XIV International Ornithological Congress.

193. SETO, N. W. H., AND D. O. O'DANIEL. 1999. Bonin Petrel (*Pterodroma hypoleuca*). No. 385 *in* The Birds of North America (A. Poole and F. Gill, Eds.). The Birds of North America, Inc., Philadelphia.

194. SHAFFER, S. Unpublished.

195. SHEALER, D. Unpublished.

196. SHEALER, D. 1999. Sandwich Tern (*Sterna sandvicensis*). No. 405 *in* The Birds of North America, (A. Poole and F. Gill, Eds.). The Birds of North America, Inc., Philadelphia.

197. SIEGEL-CAUSEY, D. C., AND G. L. HUNT. 1986. Breeding-site selection and colony formation in Double-crested and Pelagic cormorants. Auk 103: 230–234.

198. SIMONS, T. R. 1984. A population model of the endangered Hawaiian Dark-rumped Petrel. Journal Wildlife Management. 48(4): 1065–1076.

199. SIMONS, T. R. 1985. Biology and behavior of the endangered Hawaiian Dark-rumped Petrel. Condor 87: 229–245.

200. SIMONS, T. R., AND C. N. HODGES. 1998. Dark-rumped Petrel (*Pteordroma phaeopygia*). No. 345 *in* The Birds of North America (A. Poole and F. Gill, Eds.). The Birds of North America, Inc. Philadelphia.

201. SOUTHERN, H. N., R. CARRICK, AND W. G. POTTER. 1965. The natural history of a population of guillemots (*Uria aalge Pont.*). Journal Animal Ecology 34: 649–665.

202. SPENDELOW, J. A., J. D. NICHOLS, I. C. T. NISBET, H. HAYS, AND G. D. CORMONS. 1995. Estimating annual survival and movement rates of adults within a metapopulation of Roseate Terns. Ecology 76: 2415–2428.

203. STONEHOUSE, B. 1962. The Tropic Birds (Genus *Phaethon*) of Ascension Island. Ibis 103b: 124–482.

204. STONEHOUSE, B. (Ed.). 1975. Biology of Penguins. Macmillan, London.

205. STONEHOUSE, B., AND S. STONEHOUSE. 1963. The frigatebird *Fregata aquila* of Ascension Island. Ibis 103b: 409–422.

206. SWATSCHEK, I., D. RISTOW, AND M. WINK. 1994. Mate fidelity and parentage in Cory's Shearwater *Calonectris diomedea*-field studies and DNA fingerprinting. Molecular Ecology 3: 259–262.

207. SYDEMAN, W. J., P. PYLE, S. D. EMSLIE, AND E. B. MCLAREN. 1996. Causes and consequences of long-term partnerships in Cassin's Auklets. Pp. 211–222 *in* Partnerships in Birds — The Study of Monogamy (J. M. Black, Ed.). Oxford University Press, Oxford.

208. TARBURTON, M. K. 1987. Some recent observations on seabirds breeding in Fiji. Notornis 25: 303–316.

209. TAYLOR, R. H. 1962. The Adelie Penguin, *Pygoscelis adeliae*, at Cape Royds. Ibis 104: 176–204.

210. TELFAIR, R. C., II, AND M. L. MORRISON. 1995. Neotropic Cormorant (*Phalacrocorax brasilianus*). No. 137 *in* The Birds of North America (A. Poole and F. Gill, Eds.). Academy of Natural Sciences, Philadelphia; American Ornithologists' Union, Washington, D.C.

211. THIBAULT, J.-C. 1994. Nest-site tenacity and mate fidelity in relation to breeding success in Cory's Shearwater *Calonectris diomedea*. Bird Study 41: 25–28.

212. THOMPSON, B. C., J. A. JACKSON, J. BURGER, L. A. HILL, E. M. KIRSCH, AND J. L. ATWOOD. 1997. Least Tern (*Sterna antillarum*). No. 290 *in* The Birds of North America (A. Poole and F. Gill, Eds.). Academy of Natural Sciences, Philadelphia; American Ornithologists' Union, Washington, D.C.

213. TICKELL, W. L. N., 1968. The biology of the great albatrosses *Diomedea exulans* and *Diomedea epomophora*. Antarctic Research Series 12: 1–55.

214. VEEN, J. 1980. Breeding behaviour and breeding success of a colony of the Little Gulls *Larus minutus* in the Netherlands. Limosa 53: 73–83.

215. VERBEEK, N. A. M. 1993. Glaucous-winged Gull (*Larus glaucescens*). No. 59 *in* The Birds of North America (A. Poole and F. Gill, Eds.). Academy of Natural Sciences, Philadelphia; American Ornithologists' Union, Washington, D.C.

216. VERMEER, K. 1963. The breeding ecology of the Glaucous-winged Gull (*Larus glaucescens*) on Mandarte Island. B.C. Occasional Papers, British Columbia Provincial Museum 13: 1–104.

217. VERMEER, K. 1970. Breeding biology of California and Ring-billed Gulls: a study of ecological adaptation to the inland habitat. Canadian Wildlife Service Report Series 12: 1–52.

218. VERMEER, K., K. H. MORGAN, AND G. E. J. SMITH. 1993. Nesting biology and mortality of Pigeon Guillemots in the Queen Charlotte Islands, British Columbia. Waterbirds 16: 119–127.

219. VOLKMAN, N. J., AND S. G. TRIVELPIECE. 1980. Growth in pygoscelid penguin chicks. Journal of Zoology 191: 521–530.

220. WALLACE, E. A. H., AND G. E. WALLACE. 1998. Brandt's Cormorant (*Phalacrocorax penicillatus*). No. 362 *in* The Birds of North America (A. Poole and F. Gill, Eds.). The Birds of North America, Inc., Philadelphia.

221. WARHAM, J. 1956. The breeding biology of the Great-winged Petrel *Pterodroma macroptera*. Ibis 98: 171–185.

222. WARHAM, J. 1990. The Petrels. Their Ecology and Breeding Systems. Academic Press, London.

223. WARHAM, J. 1996. The Behaviour, Population Biology and Physiology of the Petrels. Academic Press, London.

224. WATANUKI, Y. 1987. Breeding biology and foods of Rhinoceros Auklets on Teuri Island, Japan. Proc. NIPR Symp. Polar Biology 1: 175–183.

225. WAUGH, S. M., H. WEIMERSKIRCH, P. J. MOORE, AND P. M. SAGAR. 1999. Population dynamics of Black-browed and Grey-headed Albatrosses *Diomedea melanophrys* and *D. chrysostoma* at Campbell Island, New Zealand, 1942–96. Ibis 141: 216–225.

226. WIMERSKIRCH, H., J. C. STAHL, AND P. JOUVENTIN. 1992. The breeding biology and population dynamics of the King Penguin *Aptenodytes patagonica* on the Crozet Islands. Ibis 134: 107–117.

227. WHITTOW, G. C. 1993a. Black-footed Albatross (*Diomedea nigripes*). No. 65 *in* The Birds of North America (A. Poole and F. Gill, Eds.). Academy of Natural Sciences, Philadelphia; American Ornithologists' Union, Washington, D.C.

228. WHITTOW, G. C. 1993b. Laysan Albatross (*Diomedea immutabilis*). No. 66 *in* The Birds of North America (A. Poole and F. Gill, Eds.). Academy of Natural Sciences, Philadelphia; American Ornithologists' Union, Washington, D.C.

229. WHITTOW, G. C. 1997. Wedge-tailed Shearwater (*Puffinus pacificus*). No. 305 *in* The Birds of North America (A. Poole and F. Gill, Eds.). Academy of Natural Sciences, Philadelphia; American Ornithologists' Union, Washington, D.C.

230. WILEY, R. H., AND D. S. LEE. 1999. Parasitic Jaeger (*Stercorarus parasiticus*). No. 445 *in* The Birds of North America (A. Poole and F. Gill, Eds.). The Birds of North America, Inc., Philadelphia.

231. WILEY, R. H., AND D. S. LEE. 2000. Pomarine Jaeger (*Stercorarius pomarinus*). No. 483 *in* The Birds of North America (A. Poole and F. Gill, Eds.). The Birds of North America, Inc., Philadelphia.

232. WILLIAMS, T. D. 1995. The Penguins. Oxford University Press, Oxford.

233. WINKLER, D. W. 1983. Ecological and Behavioral Determinants of Clutch Size: The California Gull (*Larus californicus*) in the Great Basin. Ph.D. dissertation, University of California, Berkeley.

234. WINKLER, D. W. 1996. California Gull (*Larus californicus*). No. 259 *in* The Birds of North America (A. Poole and F. Gill, Eds.). Academy of Natural Sciences, Philadelphia; American Ornithologists' Union, Washington, D.C.

235. WOODWARD, P. W. 1972. The Natural History of Kure Atoll, Northwestern Hawaiian Islands. Atoll Research Bulletin No. 164.

236. YARIO, P., P. D. BOERSMA, AND S. SWANN. 1996. Breeding biology of the Dolphin Gull at Punta Tombo, Argentina. Condor 98: 208–215.

Index

A

Abbott's Booby, *Papasula abbotti*
  display, 308, 311, 335, 339
  distribution, 70
  feeding chicks, 338
  greeting ceremony, 334
  lack of aggression, 314
  mate fidelity, 332
  migration, 340
  nesting site, 279, 416
  postfledging care, 272
  taxonomy, 70
  vulnerability to hunting, 70
"Accidental loss" hypothesis, 290
Accipiter cooperi, Cooper's Hawk, 281
Accipiter nisus, Eurasian Sparrowhawk, 590
Acrocephalus aequinoctialis, Christmas Island Warbler, 202
Activity recorders, 140, 188, 189
Adaptations
  foraging, 140-143
  overview, 10-11
Adelie Penguin, *Pygoscelis adeliae*
  albumen proteins, 411
  breeding age, 220
  communication/display, 315, 317, 319, 321
  competition, 230
  divorce, 291
  egg water loss, 422
  foraging range, 155
  mate choice, 282, 284, 287
  metabolic rate, 362, 379
  nest site, 186, 269
  nest site and mate quality, 269
  photo, 105, 221
  quality segregation, 106
  removing water from prey, 469
  research effects, 565
  survival, 245
  swimming costs, 382
  synchronization between mates, 269
  thermal conductance, 374
  tobogganing, 383
  walking costs, 383
  water/salt balance, 473, 476
  weather effects, 204, 221

Adult experience
  care of eggs/young, 187, 188
  mate choice, 282, 283
  reproductive success, 220, 287, 290, 444, 622
Adult life expectancy (ALE), *see* Life expectancy
Adult mass
  mate choice and, 281-282
  measure of fitness, 182
Aerobic dive limit (ADL), 382-383
Aethia cristatella, *see* Crested Auklet
Aethia pusilla, *see* Least Auklet
Aethia pygmaea, Whiskered Auklet, 341
African-Eurasian Waterbird Agreement, 605
African Penguin, *see* Jackass Penguin
Age at first breeding, *see* Breeding age
Agelaius phoeniceus, Red-winged Blackbird, 283
Age-related vulnerabilities, 489-490
Age-specific fecundity, 248-249
Age-specific survival, 248-249
Age structure, 103
Agricultural runoff, 604
Ajaia ajaja, Roseate Spoonbill, 618
Albatross, *see also common names*
  adaptations, 10
  aggression, 324
  biennial breeding, 220, 222
  characteristics, 62
  clutch size, 118
  contaminant levels, 490
  deferred breeding, 96
  demographic tactics, 128
  demography and reproductive strategy, 271
  display, 310, 323, 324, 325, 326, 327
  distribution, 64, 79
  foraging, 137, 139
  latitude and, 276
  mate choice, 282
  mate/site fidelity, 103, 293
  migration, 250
  nest site/habitat, 224
  nest site defense, 105
  nest site prospecting, 323
  obtaining a mate, 232
  philopatry, 103, 293
  progesterone, 410
  species boundaries, 77
  squid, 152-153